2012 25th International Conference on VLSI Design

Hyderabad, India
7-11 January 2012

IEEE Catalog Number: CFP12041-PRT
ISBN: 978-1-4673-0438-2

Copyright © 2012 by the Institute of Electrical and Electronic Engineers, Inc
All Rights Reserved

Copyright and Reprint Permissions: Abstracting is permitted with credit to the source. Libraries are permitted to photocopy beyond the limit of U.S. copyright law for private use of patrons those articles in this volume that carry a code at the bottom of the first page, provided the per-copy fee indicated in the code is paid through Copyright Clearance Center, 222 Rosewood Drive, Danvers, MA 01923.

For other copying, reprint or republication permission, write to IEEE Copyrights Manager, IEEE Service Center, 445 Hoes Lane, Piscataway, NJ 08854. All rights reserved.

***This publication is a representation of what appears in the IEEE Digital Libraries. Some format issues inherent in the e-media version may also appear in this print version.**

IEEE Catalog Number: CFP12041-PRT
ISBN 13:   978-1-4673-0438-2
ISSN: 1063-9667

Additional Copies of This Publication Are Available From:

Curran Associates, Inc
57 Morehouse Lane
Red Hook, NY 12571 USA
Phone: (845) 758-0400
Fax: (845) 758-2633
E-mail: curran@proceedings.com
Web: www.proceedings.com

Proceedings

25th International Conference
on VLSI Design

Held jointly with

11th International Conference on Embedded Systems

Hyderabad, India   *7-11 January 2012*

Technical Co-Sponsorship
IEEE Circuits and Systems Society

Sponsored by
VLSI Society of India

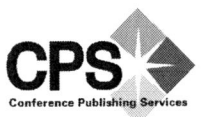

Los Alamitos, California

Washington • Tokyo

Proceedings

25th International Conference

on VLSI Design

VLSI Design 2012

2012 25th International Conference on VLSI Design

VLSID 2012

Table of Contents

Message from the Steering Committee Chair xiv

Message from the General Chair xv

Silver Jubilee Message from the Conference Founding Member xvi

Message from the Silver Jubilee Special Chair xvii

Message from the Silver Jubilee Conference Convener xviii

Message from the Program Chairs xix

Message from the Organizing Chair xxi

Message from the Tutorial Chairs xxii

Message from the Sponsorship and Exhibits Team xxiii

Message from the President, VLSI Society of India xxiv

VLSI Design Conference Steering Committee (2011) xxvi

VLSI Design 2012 Conference Committee xxvii

VLSI 2012 Technical Program Committee and Reviewers xxxi

VLSI Design 2011 Best Paper Awards xxxvi

VLSI Design Conference History xxxvii

Embedded Systems Design Conference History xxxviii

About the Cover xxxix

Keynote Speakers xl

Invited Keynote Talks

Keynote Talk: A History of the VLSI Design Conference 1
   Vishwani D. Agrawal

Keynote Talk: Semiconductor Industry: Best of Times, Worst of Times, and Nowhere
Else I Would Rather Be! 3
   Jaswinder S. Ahuja

Keynote Talk: A Wireless Sensor a Day Keeps the Doctor Away 5
   Bert Gyselinckx

Keynote Talk: The Variability Expeditions: Exploring the Software Stack
for Underdesigned Computing Machines 7
   Rajesh Gupta

Keynote Talk: Challenges in Automotive Cyber-physical Systems Design 9
   Samarjit Chakraborty

v

Full Day Tutorials

Tutorial T1: Design of Mixed-Signal Systems using SystemC AMS Extensions 11
Sumit Adhikari, Markus Damm, Christoph Grimm, and François Pecheux

Tutorial T2: Reversible Logic: Fundamentals and Applications in Ultra-Low Power,
Fault Testing and Emerging Nanotechnologies, and Challenges in Future 13
Himanshu Thapliyal and Nagarajan Ranganathan

Tutorial T3: DFM, DFT, Silicon Debug and Diagnosis – The Loop to Ensure Product
Yield 16
Srikanth Venkataraman and Nagesh Tamarapalli

Tutorial T4: Intellectual Property Protection and Security in System-on-Chip Design 18
Susmita Sur-Kolay and Swarup Bhunia

Tutorial T5: Advanced Analog-Mixed Signal System and Circuit Techniques 20
Pavan Hanumolu, Un-Ku Moon, and Terri Fiez

Tutorial T6: Variability-resistant Software and Hardware for Nano-Scale Computing 22
Nikil Dutt, Mani Srivastava, Rajesh Gupta, and Subhashish Mitra

Tutorial T7A: New Modeling Methodologies for Thermal Analysis of 3D ICs
and Advanced Cooling Technologies of the Future 25
David Atienza and Arvind Shridhar

Tutorial T7B: Optimally Addressing Verification Constraint Complexity for Effective
Functional Convergence 27
Shankar Hemmady

Tutorial T8A: Designing Silicon-Photonic Communication Networks for Manycore
Systems 28
Ajay Joshi

Tutorial T8B: Wireless System Design and Systems Engineering Challenges 29
Kameswara Rao B, Muralidhar Reddy B, and Ravi Kishore B

Embedded Tutorials

Embedded Tutorial ET1: Pole-Zero Analysis of Low-Dropout (LDO) Regulators: A
Tutorial Overview 31
Annajirao Garimella, Punith Surkanti, and Paul M. Furth

Embedded Tutorial ET2: Digital Subscriber Line 33
M Kalyana Kumar Rao, Shantha Kumari PV, and Boopalan Sellappan

Embedded Tutorial ET3: Packaging Trends, Die Package Co-Design Flow
and Challenges 35
Siva Kothamasu

Embedded Tutorial ET4: Advanced Techniques for Programming Networked
Embedded Systems 36
Vijay Raghunathan

Panel Discussion

Panel Discussion: SoC Realization – A Bridge to New Horizons or a Bridge
to Nowhere? 38
Sathyam K. Pattanam, P.P. Chakrabarti, Mahesh Mahendale, Srikanth Jadcherla, Seer Akademi,
Vikas Gautham, and Raju Bala Showry Pudota

Session A1: Application Driven Analog Design

Random Access Analog Memory (RA2M) for Video Signal Application 39
Nilanjan Chattaraj and Anindya Sundar Dhar

A 55-mW 300MS/s 8-bit CMOS Parallel Pipeline ADC 45
Manas Kumar Hati and Tarun K. Bhattacharyya

A 110-dB Dynamic Range, 76-dB Peak SNR Companding Continuous-Time $\Delta\Sigma$
Modulator for Audio Applications 51
Saravana Kumar and Shouri Chatterjee

Session B1: Application Specific Processing Architectures

Hardware Efficient Architecture for Generating Sine/Cosine Waves 57
Supriya Aggarwal and Kavita Khare

Power Aware Hardware Prototyping of Multiclass SVM Classifier Through
Reconfiguration 62
Rajesh A. Patil, Gauri Gupta, Vineet Sahula, and A.S. Mandal

A High Speed FIR Filter Architecture Based on Novel Higher Radix Algorithm 68
S.K. Sahoo and K. Srinivasa Reddy

Session C1: Low Power Analog-Mixed Signal Design

Impact of Body Bias Based Leakage Power Reduction on Soft Error Rate 74
Warin Sootkaneung and Kewal K. Saluja

An Area Efficient Diode and On Transistor Interchangeable Power Gating Scheme
with Trim Options for Low Power SRAMs 80
Ankur Goel, Donald Evans, Richard Stephani, Venkateswara Reddy, Dharmendra Rai,
Veerabadra Chary, and N. Sathisha

An Energy Efficient Oscillator Frequency Calibration Methodology Using Fraction
Phase Computation 85
Amitava Ghosh, Isha Das, and Achintya Halder

Session A2: High Speed Mixed Signal RF Design

Self-Induced Supply Noise Reduction Technique in GBPS Rate Transmitters 92
Nitin Gupta, Tapas Nandy, and Phalguni Bala

Buffer Design and Eye-Diagram Based Characterization of a 20 GS/s CMOS DAC 96
Mohit Singh and Shalabh Gupta

Analog Processing Based Equalizer for 40 Gbps Coherent Optical Links in 90 nm CMOS 101
Pawan Kumar Moyade, Nandakumar Nambath, Allmin Ansari, and Shalabh Gupta

Session B2: Designing Real Time Embedded Systems

HD Resolution Intra Prediction Architecture for H.264 Decoder 107
Jimit Shah, K.S. Raghunandan, and Kuruvilla Varghese

Design for Security of Block Cipher S-Boxes to Resist Differential Power Attacks 113
Bodhisatwa Mazumdar, Debdeep Mukhopadhyay, and Indranil Sengupta

Real-time Melodic Accompaniment System for Indian Music Using TMS320C6713 119
Prateek Verma and Preeti Rao

Session C2: Design Techniques for Power Management

Bidirectional Single-Supply Level Shifter with Wide Voltage Range for Efficient Power Management 125
Sujan K. Manohar, Vinod K. Somasundar, Ramakrishnan Venkatasubramanian, and Poras T. Balsara

Pole-Zero Analysis of Low-Dropout (LDO) Regulators: A Tutorial Overview 131
Annajirao Garimella, Punith R. Surkanti, and Paul M. Furth

Session A3: Analog/RF Design Techniques

3-D Parasitic Modeling for Rotary Interconnects 137
Vinayak Honkote, Ankit More, and Baris Taskin

Power Aware Post-Manufacture Tuning of MIMO Receiver Systems 143
Debashis Banerjee, Shreyas Sen, Shyam Kumar Devarakond, and Abhijit Chatterjee

Session B3: Communication Applications

GPU Implementation of a Programmable Turbo Decoder for Software Defined Radio Applications 149
Dhiraj Reddy Nallapa Yoge and Nitin Chandrachoodan

Session C3: Thermal Analysis and Temperature Aware Design

Run-time Prediction of the Optimal Performance Point in DVS-based Dynamic Thermal Management 155
Junyoung Park, H. Mert Ustun, and Jacob A. Abraham

Temperature-aware Task Partitioning for Real-Time Scheduling in Embedded Systems 161
Zhe Wang, Sanjay Ranka, and Prabhat Mishra

Towards Thermal Profiling in CMOS/Memristor Hybrid RRAM Architectures 167
Cory E. Merkel and Dhireesha Kudithipudi

Session A4: CMOS Sensors and MEMS

CMOS Gas Sensor Array Platform with Fourier Transform Based Impedance
Spectroscopy 173
  M. Pramod, Navakanta Bhat, Gaurab Banerjee, Bharadwaj Amrutur, K.N. Bhat,
  and Praveen C. Ramamurthy

A Compact Temperature Sensor at 1.8µA per Hz Conversion Rate and 1.1 °C
Accuracy for SOCs 179
  Subhajit Sen, Dan Babitch, and Noshir Dubash

Analysis of the Pull-In Phenomenon in Microelectromechanical Varactors 185
  Anindya Lal Roy, Anirban Bhattacharya, Ritesh Ray Chaudhuri, and Tarun Kanti Bhattacharyya

Session B4: Architecture and Logic Synthesis

Low-Latency No-Handshake GALS Interfaces for Fast-Receiver Links 191
  Jean-Michel Chabloz and Ahmed Hemani

Set-Cover Heuristics for Two-Level Logic Minimization 197
  Ankit Kagliwal and Shankar Balachandran

A Rapid Methodology for Multi-mode Communication Circuit Generation 203
  Liang Tang, Jorgen Peddersen, and Sri Parameswaran

Session C4: Energy Harvesting and Power Management

An Integrated CMOS RF Energy Harvester with Differential Microstrip Antenna
and On-Chip Charger 209
  Mahima Arrawatia, Varish Diddi, Harsha Kochar, Maryam Shojaei Baghini, and Girish Kumar

Low-Overhead Maximum Power Point Tracking for Micro-Scale Solar Energy
Harvesting Systems 215
  Chao Lu, Sang Phill Park, Vijay Raghunathan, and Kaushik Roy

Hybrid NEMS-CMOS DC-DC Converter for Improved Area and Power Efficiency 221
  Sujan K. Manohar, Ramakrishnan Venkatasubramanian, and Poras T. Balsara

Session A5: Physical Design and TCAD

A Heuristic Method for Co-optimization of Pin Assignment and Droplet Routing
in Digital Microfluidic Biochip 227
  Ritwik Mukherjee, Hafizur Rahaman, Indrajit Banerjee, Tuhina Samanta,
  and Parthasarathi Dasgupta

Clock Tree Skew Minimization with Structured Routing 233
  Pinaki Chakrabarti

Accurate Leakage Estimation for FinFET Standard Cells Using the Response Surface
Methodology 238
  Sourindra Chaudhuri, Prateek Mishra, and Niraj K. Jha

Session B5: System Level Design

Real-Time, Content Aware Camera–Algorithm–Hardware Co-Adaptation for Minimal
Power Video Encoding 245
    Joshua W. Wells, Jayaram Natarajan, Abhijit Chatterjee, and Irtaza Barlas

Session C5: Low Power Design Techniques

Way Sharing Set Associative Cache Architecture 251
    C.J. Janraj, T. Venkata Kalyan, Tripti Warrier, and Madhu Mutyam

A Novel Encoding Scheme for Low Power in Network on Chip Links 257
    Deepa N. Sarma, G. Lakshminarayanan, and K.V.R. Suryakiran Chavali

A Power Delivery Network Aware Framework for Synthesis of 3D Networks-on-Chip
with Multiple Voltage Islands 262
    Nishit Kapadia and Sudeep Pasricha

Session A6: Packaging and 3D Circuits

A Framework for TSV Serialization-aware Synthesis of Application Specific 3D
Networks-on-Chip 268
    Sudeep Pasricha

Session B6: Low Power IC Design I

An Ultra-low Power Symbol Detection Methodology and Its Circuit Implementation
for a Wake-up Receiver in Wireless Sensor Nodes 274
    Deepak Kumar Meher, Arunkumar Salimath, and Achintya Halder

Low-Power Self Reconfigurable Multiplexer Based Decoder for Adaptive Resolution
Flash ADCs 280
    Chetan Vudadha, Goutham Makkena, M. Venkata Swamy Nayudu, P. Sai Phaneendra,
    Syed Ershad Ahmed, Sreehari Veeramachaneni, N. Moorthy Muthukrishnan, and M.B. Srinivas

A 1.25GHz 0.8W C66x DSP Core in 40nm CMOS 286
    Raguram Damodaran, Timothy Anderson, Sanjive Agarwala, Rama Venkatasubramanian,
    Michael Gill, Dhileep Gopalakrishnan, Anthony Hill, Abhijeet Chachad,
    Dheera Balasubramanian, Naveen Bhoria, Jonathan Tran, Duc Bui, Mujibur Rahman,
    Shriram Moharil, Matthew Pierson, Steve Mullinnix, Hung Ong, David Thompson,
    Krishna Gurram, Oluleye Olorode, Nuruddin Mahmood, Jose Flores, Arjun Rajagopal,
    Soujanya Narnur, Daniel Wu, Alan Hales, Kyle Peavy, and Robert Sussman

Session C6: Diagnosis and Debug Techniques

A Reconfigurable On-die Traffic Generator in 45nm CMOS for a 48 iA-32 Core
Network-on-Chip 292
    Praveen Salihundam, Mohammed Asadullah Khan, Shailendra Jain, Yatin Hoskote, Satish Yada,
    Shasi Kumar, Vasantha Erraguntla, Sriram Vangal, and Nitin Borkar

Efficient Online RTL Debugging Methodology for Logic Emulation Systems 298
    Somnath Banerjee and Tushar Gupta

SCARE: Side-Channel Analysis Based Reverse Engineering for Post-Silicon
Validation 304
    Xinmu Wang, Seetharam Narasimhan, Aswin Krishna, and Swarup Bhunia

Session A7: Fast Algorithms for Nano CMOS AMS Optimization

Kriging-Assisted Ultra-Fast Simulated-Annealing Optimization of a Clamped Bitline
Sense Amplifier 310
    Oghenekarho Okobiah, Saraju P. Mohanty, Elias Kougianos, and Oleg Garitselov

Fast-Accurate Non-Polynomial Metamodeling for Nano-CMOS PLL Design
Optimization 316
    Oleg Garitselov, Saraju P. Mohanty, and Elias Kougianos

Circuit Optimization at 22nm Technology Node 322
    Angada B. Sachid, P. Paliwal, S. Joshi, M. Shojaei, D. Sharma, and V. Rao

Session B7: Low Power IC Design II

Synthesis of Reversible Circuits Using Heuristic Search Method 328
    Kamalika Datta, Gaurav Rathi, Indranil Sengupta, and Hafizur Rahaman

Minimum Cost Fault Tolerant Adder Circuits in Reversible Logic Synthesis 334
    Sajib Kumar Mitra and Ahsan Raja Chowdhury

Width-Aware Fine-Grained Dynamic Supply Gating: A Design Methodology
for Low-Power Datapath and Memory 340
    Lei Wang, Somnath Paul, and Swarup Bhunia

Session C7: Timing Issues in Test

Eliminating Performance Penalty of Scan 346
    Ozgur Sinanoglu

A Silicon Testing Strategy for Pulse-Width Failures 352
    Srinivas Vooka, Khushboo Agarwal, Abhijeet Shrivastava, Pranav Murthy,
    and Venkatraman Ramakrishnan

At-speed Testing of Asynchronous Reset De-assertion Faults...358
    Arvind Jain, Maheedhar Jalasutram, Srinivas Vooka, Prasun Nair, and Neeraj Pradhan

Session A8: Efficient Methods for AMS Design Optimization

A Library for Passive Online Verification of Analog and Mixed-Signal Circuits 364
Debjit Pal, Pallab Dasgupta, and Siddhartha Mukhopadhyay

A Fast Equation Free Iterative Approach to Analog Circuit Sizing 370
Supriyo Maji and Pradip Mandal

Iterative Performance Model Upgradation in Geometric Programming Based Analog
Circuit Sizing for Improved Design Accuracy 376
Samiran Dam and Pradip Mandal

Session C8: Formal Methods in Test and Verification

Analysis of Reachable Sensitisable Paths in Sequential Circuits with SAT and Craig
Interpolation 382
Matthias Sauer, Stefan Kupferschmid, Alexander Czutro, Sudhakar Reddy, and Bernd Becker

Formal Verification of Galois Field Multipliers Using Computer Algebra Techniques 388
Jinpeng Lv and Priyank Kalla

A Novel SMT-Based Technique for LFSR Reseeding 394
Sarvesh Prabhu, Michael S. Hsiao, Loganathan Lingappan, and Vijay Gangaram

Session A9: Circuit Simulation

Two Graph Based Circuit Simulator for PDE-Electrical Analogy 400
Yogesh Dilip Save, H. Narayanan, and Sachin B. Patkar

Modeling of Partially Depleted SOI DEMOSFETs with a Sub-circuit Utilizing
the HiSIM-HV Compact Model 406
Tarun Kumar Agarwal and M. Jagadesh Kumar

Implications of Halo Implant Shadowing and Backscattering from Mask Layer Edges
on Device Leakage Current in 65nm SRAM 412
H.C. Srinivasaiah

Session B9: Reconfigurable Architectures

Customizing Instruction Set Extensible Reconfigurable Processors Using GPUs 418
*Unmesh D. Bordoloi, Bharath Suri, Swaroop Nunna, Samarjit Chakraborty, Petru Eles,
and Zebo Peng*

Energy-Efficient Application Mapping in FPGA through Computation in Embedded
Memory Blocks 424
Anandaroop Ghosh, Somnath Paul, and Swarup Bhunia

Intra-Task Dynamic Cache Reconfiguration 430
Hadi Hajimiri and Prabhat Mishra

Session C9: Test Optimization

A Diagnosability Metric for Test Set Selection Targeting Better Fault Detection 436
    Subhadip Kundu, Santanu Chattopadhyay, Indranil Sengupta, and Rohit Kapur

Test Planning for Core-based 3D Stacked ICs with Through-Silicon Vias 442
    Breeta Sen Gupta, Urban Ingelsson, and Erik Larsson

Externally Tested Scan Circuit with Built-In Activity Monitor and Adaptive Test
Clock 448
    Priyadharshini Shanmugasundaram and Vishwani D. Agrawal

Author Index 454

Message from the Steering Committee Chair and Conference Founding Member

This is the Twenty-fifth International Conference on VLSI Design. Every year in this journey has been a milestone. Let us pause, reminisce, and enjoy the event. The conference began in Chennai in 1985. Since that time India's software and hardware industries have advanced. Indian companies have become global players. Academic landscape changed as well. The conference cannot take credit for any of these because these developments would have occurred independently of the conference. They occurred because the time was right. However, the conference played its part. We served the VLSI community in the best possible way. For the industry, we provided an annual meeting place, a place for exchanging ideas, a place to learn, and a place to make new acquaintances. For the academia we provided a window into the world for a two-way exchange. Indian academics could interact with other academics and their industry counterparts from within India as well as from abroad. Through an innovative fellowship program, initiated in 1994, the conference facilitated academic participation at a scale that would have been impossible otherwise. This conference, through the VLSI Society of India, has been a catalyst for initiating a large number of other technical meetings and working groups all across India. I think the strongest feature of this conference is its regular presence through those years. It has been a regular feature annually just as the sun is daily. VLSI professionals look forward to the next conference. We all know how many more things get done when plans include target dates.

The conference has created a brotherhood of dedicated enthusiasts who have devoted their time and effort year after year to organize it. Our organizers and participants are not different. If you are a participant today you become an organizer tomorrow. It is this group of organizers and participants that this conference is dedicated to.

Some of you may have heard this: Yesterday became history; tomorrow remains mystery; but today is a gift, that's why it is called present. So, let us enjoy this Twenty-fifth International Conference on VLSI design in its silver jubilee year.

Welcome again to Hyderabad. We can always find something new every time we come here. And once we leave, that will be the time to move on to new milestones, to golden, diamond and platinum jubilees.

With warm regards,

Vishwani D. Agrawal

Message from the General Chair

Conferences are like wine; some turn to vinegar, but the best improve with age. For 25 years now, the VLSI Design conference has gathered the best and the brightest in the semiconductor industry. Since the humble beginnings in 1986, the conference has been ably steered by the founder and the Steering Committee Chair, Prof. Vishwani Agrawal. Over the years, he has exemplified a unique balance of future vision and current needs, a balance that is also at the very core of the conference.

A lot has happened in the past 25 years. The dawn of the PC and the networked enterprise in the 80's, the Internet era in the 90's, and a decade of pervasive and cloud computing. In spite of the seismic shifts in technologies, the conference continues to be relevant. Why, you may ask.

First, the conference recognizes the power of simple things. A smile, a kind word, an honest compliment, an innocent question, or a little technology bet. All of them have the potential to turn our lives around. By showcasing little technology bets and their outcomes, by sharing the learning from lots of little failures, and by celebrating small wins year after year, we arrive at extraordinary outcomes! Second, the conference feeds innovation. Every child is an artist, an unwitting innovator. The problem is how to remain an artist once we grow up. It takes a community like the conference to keep the innovative spirit alive. And every year, the conference succeeds in bringing out the artist in all of us.

On behalf of the entire 2012 conference team, I am pleased to invite you to the 25th anniversary.

With warm regards,

Srimat Chakradhar

Message from the Conference Founding Member

After moving to IIT, Madras in 1973, I got an opportunity to visit VLSI labs in US, in particular AT&T labs. I wanted to teach a course on VLSI Design and set up a Lab. Dr. Vishwani Agrawal and Dr. Prathima Agrawals urged me to organize the first VLSI Design conference at IIT, Madras. They persuaded 18 leading VLSI professionals from different Labs in US to participate. SCI, Chandigarh was just starting their foundry. Surprisingly, there was large participation from within India. Thanks to the first VLSI conference, DOE (Department of Electronics, GOI) initiated a program to fund VLSI projects in universities and IIT's. Dr. Phadke led the program and a VLSI cell was started in DOE. I was interested in starting a "MOSIS" like project to support VLSI courses in college with support of foundries at SCL and ITI, Bangalore for providing fabrication support. Fortunately, our friends in US universities allowed us to download VLSI software. VLSI design activity picked up and VLSI conferences became an annual feature. I am very happy that younger generation of professionals and students have carried on the tradition. Thanks must be given to Drs. Agrawal, who actively supported the annual conferences.

I wish the Silver Jubilee conference VLSI 2012 all success and may India achieve the status of the World Design Centre for VLSI.

Best wishes,

Prof. H N Mahabala

Message from the Silver Jubilee Special Chair

A Silver Jubilee is a good opportunity to look back as well as to look forward.

A warm welcome to all the delegates to Hyderabad which has a potential to become the design house of the world and takes pride in the success of designing the iPod chip and also AMD's Fusion chip.

From its modest beginning in 1985 at IIT Madras under the visionary guidance of Dr. Vishwani Agrawal of Auburn University and Prof. H.N. Mahabala of IIT Madras, the workshop on VLSI Design has grown into an international conference on VLSI Design, which draws attendees from all over the world every year jointly with the International Conference on Embedded Systems. It is a great honor for me to be the Silver Jubilee Special Chair of this prestigious 25th International Conference on VLSI Design & 11th International Conference on Embedded Systems.

Despite its tremendous success and its obvious record of transforming the lives of millions of people across the world, the industry faces different complex issues when looking at new opportunities to grow globally both in product design and development as each country poses different challenges. We expect this year's event will offer great networking and knowledge sharing opportunity not only for industry leaders but also for students, researchers and academia.

I would like to congratulate each person who contributed to the success of this annual event which creates a platform for technical exchanges by experts from all over the world on the advancements in semiconductor research and development.

J A Chowdary

Message from the Silver Jubilee Conference Convener

It is a joy and a privilege to convene once again in Hyderabad the 25th International Conference on VLSI Design and 11th International Conference on Embedded Systems.

These are special days in Hyderabad but Hyderabad has always been special for the semiconductor industry. We expect this silver jubilee conference to bring back the spotlight to Hyderabad's VLSI Design and Embedded System sector advancement. Our focus this year is on Embedded Solutions for Emerging Markets – Consumer, Energy and Automotive. This is a premier global level annual event that provides a platform for Students, Industry Leaders and Design Experts to discuss the growth strategies in the context of VLSI Design and Embedded Systems both in the region and globally. I welcome you to this third grand event in a short span of five years.

It is proof of the importance and significance of this event that it has the presence and active participation of the most important researchers, technologists and business entities in the VLSI sector. I hope that the platform that we have created for ourselves for learning from each other and sharing the excitement of the profession will also be a launching pad for the future collaborations and fascinating results.

The last 25 years have brought us from the 'microelectronics era' to the 'nanoelectronics era'. Looking ahead to the next 25 years and preparing the ground for it will be a challenge.

I hope the deliberations, the interactions and the exchange of the knowledge and the facilitating of collaboration amongst the world's leading players in the VLSI domain will provide a road map for the next 25 years of our industry.

With warm regards,

Dasaradha Gude

Message from the Program Chairs

Welcome to Hyderabad! This historic city is once again playing host to the 25th International Conference on VLSI Design and the 11th International Conference on Embedded Systems that will be held jointly during January, 7-11, 2012. The conference is being held at India's largest convention center, the Hyderabad International Convention Center (HICC). With a seating capacity of over 4000 (and inbuilt flexibility to expand to 6500), HICC is also billed as South Asia's first world-class convention center.

VLSI Design 2012 begins with a two-day tutorial program that showcases the latest trends in technology. This year, our tutorial chairs Preeti Ranjan Panda and Nagi Naganathan have assembled a world-class tutorial program that covers many hot and relevant topics.

After the tutorials, there is an exciting three-day technical program which includes 71 regular technical paper presentations in which various researchers will present their new results to their peers. These papers have been selected from a pool of 239 submissions. Out of these, only 218 submissions were considered for review after accounting for withdrawals and excluding the duplicate and invalid/incomplete submissions. Though this number does not represent the highest this conference has seen, the submissions made it up in terms of quality as evident from the outstanding reviews received by many papers. The Technical Program Committee had to leave out several good papers in order to be within the limits of 3 days with 3 parallel sessions. The papers were grouped into 9 tracks and the review process was managed by the program chairs together with 17 track chairs at two physical TPC meetings in Rutgers, NJ and New Delhi, India. A team of 173 people (including members of the Technical Program Committee) across the world was involved in reviewing the papers. They really worked hard and submitted a total of 831 reviews. This helped us in ensuring that all the papers had a good number of reviews for the Program Committee to make an informed judgment. Nearly two thirds of the papers had 4 or more reviews per paper.

These technical papers have been arranged into 27 technical sessions scheduled in 3-parallel-session format. Further value addition to these sessions is provided by the 4 embedded tutorials and an invited talk by experts who will share their insights into some of the trendy areas. Apart from these, there is a panel discussion session on a topic of current interest that is expected to provide a forum for lively exchange of thoughts.

The highlight of the technical program of the 25th anniversary is a set of keynote addresses by industry and academia leaders who will provide big pictures spanning across time and space in their respective fields of expertise. The keynote speakers are: Prof. Vishwani Agrawal, Auburn University, USA who will talk about 25 years VLSI Design Conference, Mr. Jaswinder Ahuja, Cadence, India, Dr. Bert Gyselinckx, IMEC, Belgium, Prof. Sri Parameswaran, University of New South Wales, Australia, Prof. Rajesh Gupta, University of California at San Diego, USA and Prof. Samarjit Chakraborty from Technical University of Munich, Germany.

Welcome again to the historic city of Hyderabad and enjoy the exciting program of the 25th International Conference on VLSI Design and the 11th International Conference on Embedded Systems!

With warm regards,

Anshul Kumar   *Jörg Henkel*

Message from the Organizing Chair

It is with pleasure and anticipation that I welcome you all to the 25th VLSI Conference. There is something for each of us in the program that has been put together. As the world's top VLSI specialists converge on Hyderabad, we look forward to meaningful interactions that would propel our industry and profession to greater heights. The highlights of the conference are the keynote speech by Dr. Vishwani Agrawal; two full day tutorials by academia & industry experts; three day technical paper sessions. A unique opportunity created for the students to mingle with the stars of the industry and to get inspired to join their ranks. That our annual gathering and celebration of our industry has the fullest support of the people and the administration in Andhra Pradesh, India is evidenced by the fact that we are congregating in Hyderabad for the third time in a short span of five years. The success of a conference such as ours will depend on the involvement of all the stakeholders including that of the youngsters aspiring to take our industry into a more exciting phase.

My sincere thanks and congratulations to all the student volunteers as well as the distinguished industry leaders who led the conference work by chairing the different committees.

May I thank you for your participation in and contribution to the Silver Jubilee Conference on VLSI Design.

Yours sincerely,

Manoher Bommena

Message from the Tutorial Chairs

It is our great pleasure to welcome you to the 25th International Conference on VLSI Design and 11th International Conference on Embedded Systems. We are excited to present this year's tutorial program on January 7&8 for two days. We have 6 full day tutorials and 4 half day tutorials from eminent academic and industry experts on various topics.

We have 2 full day tutorials in the area of mixed signal design. The full day tutorial on mixed signal design with AMS and SystemC modeling will be delivered by speakers from Vienna University of Technology and the full day tutorial on the advanced design mixed signal circuits and system will be delivered by speakers from Oregon State University. A full day tutorial on variability resistant software and hardware will be delivered by speakers from UC Irvine, UCLA and Stanford. A full day tutorial on a new topic of reversible logic with applications of low power will be delivered by speakers from University of South Florida. A full day tutorial on DFT, DFM and silicon debug will be delivered by industry experts from AMD and Intel. Another full tutorial on a timely topic of intellectual property protection and security in SoC will be delivered by speakers from Case Western Reserve University and Indian Statistical Institute. We have 4 half day tutorials on a variety of topics. A half day tutorial on modeling for thermal analysis will be delivered by speakers from EPFL, Switzerland. Another half day tutorial on designing silicon photonic communication for many core systems will be delivered a speaker from Boston University. Another tutorial on verification will be delivered by an industry expert from Synopsys. Another tutorial on wireless system design will be delivered by speakers from HCL.

Overall we have an exciting line up of tutorials for two days. We sincerely hope that you will enjoy the tutorial program and a unique overall experience at the 25th International Conference on VLSI Design Conference and 11th International Conference on Embedded Systems.

Preeti Panda   *Nagi Naganathan*

Message from the Sponsorship and Exhibits Team

Semiconductor is the oil of information age that enables the embedded solutions as the life line of modern world. Semiconductors are traversing the opposite of Moore's law curve – with increasing embedded-semiconductor role in all solutions - increasing consumerization, wide ranging gadgets, all aspects of our lives are increasingly embedded with embedded-semiconductor-electronics solutions. Maturing of many technologies, increasing importance to efficiencies, increasing relevance of end user experience – all are mapping to very many new solutions coming out. India is in a unique position – to be the creator, provider to the world and as consumer, thus is at the head of the growth curve in embedded-semiconductor-domain. All of which make VLSI2012 more relevant, more game changing than ever before.

This IEEE International Conference on VLSI Design and Embedded Systems has been the catalytic platform over the past 25 years who bring together all the stake holders to exchange ideas, share thoughts, debate the trends and influence the agenda of the future that makes such trends a reality. VLSI2012 is the silver jubilee celebration of this confluence.

The vision of VLSI2012 is made a reality by the exemplary, invaluable, participation, contributions and support through sponsorships & exhibitions – of all those who not only have a dream, but walk the talk to realize it, despite these challenging times

Enabled and emboldened by the supportive hand of all the sponsors, exhibitors, we invite you for what we believe will be among the most game changing of conferences whose impact will reverberate for the next 25 years.

Welcome to VLSI2012!!

S Uma Mahesh

Message from the President, VLSI Society of India

Welcome to the Silver Jubilee edition of the International Conference on VLSI Design!

Twenty five years ago two visionaries - Dr. Vishwani Agrawal, Prof. H. N. Mahabala, saw the potential of India becoming a powerhouse in VLSI Design. In their discussions, they identified several gaps such as specialized education, research, infrastructure etc that needed to be addressed to realize the potential. Of the identified gaps, they realized that education and research was the foundation on which this vision could be realized. They realized that there had to be some mechanism to make this growth visible to the outside world. This led them to start a small annual meet to bring together professionals, academicians and students from India and abroad. Over the years this event blossomed into a full blown international conference and added several new dimensions to broaden and deepen the program and its impact. I have had the pleasure and honor of being associated with this process for most of the past twenty five years and have seen how this conference has been a key catalyst that has enabled India to realize its potential and emerge as the go-to hub for VLSI design and related activities!

This is a very special milestone that we must all take pride in and celebrate. In the last 25 years, we have seen the growth of VLSI related education in our universities, generations of well-trained fresh engineers and entry of several big and small corporations into India, leveraging this talent pool for their R&D work. But our job is not done. As we look forward into the next twenty five years, there is so much more that needs to be done. At the VLSI Society of India, our vision is to develop the VLSI related work on two fronts. The first one is to work with all stakeholders to enable university students to take an idea from concept to silicon and if the idea has merit, to take it to market. The second one is to develop the "Created in India" brand such that we have an entire ecosystem that supports the development of electronic products, from idea to manufacturing to distribution. This is a vision that all my peers in industry, academia and the government share. The government is already committed to and is working on an enabling policy framework, but there are several other milestones we have to achieve to realize this vision. We need to leverage the university network in India to create a rich and comprehensive portfolio of IP that can be rapidly assembled into SoCs to target new applications and product ideas. This will require us to further strengthen the academic and research back-bone of our nation and encourage innovation and collaboration. We will have to link the innovators to sources of funding, manufacturing and go-to-market infrastructure. We will have to encourage entrepreneurship at various levels and support it through a network of incubators. We will have to develop the capability and mindset to approach this in a very deliberate, focused, open and collaborative manner. Researchers and analysts, such as Goldman Sachs, have predicted that emerging economies such as India and China will be the growth drivers for next 25 years. Given this scenario and the great foundation laid down by this body over the last 25 years, I am excited about the possibilities and I hope you share in my excitement.

In conclusion, I would like to express my gratitude to everyone who has been involved with this conference over the past twenty five years to take it to such great heights and enabling the growth of the VLSI design and related industry in India. I congratulate this year's committee for putting together another excellent conference in the finest traditions established by their predecessors.

Finally, I welcome you to the city of Hyderabad and this special Silver Jubilee edition of this conference!

Best wishes for wonderful conference to everyone.

Warm regards,

Jaswinder Ahuja

VLSI Design Conference Steering Committee (2011)

Vishwani D. Agrawal
(Chair) Auburn University, USA

Jaswinder S. Ahuja
(President, VSI), Cadence, India

M. Balakrishnan
(Embedded Systems), IIT Delhi, India

Srimat Chakradhar
NEC Labs, USA

Dasaradha R Gude
SoCtronics Technologies Pvt. Ltd., India

Anshul Kumar
IIT Delhi, India

A. Prabhakar
(Past President, VSI)
Datanet Corp., India

N. Ranganathan
(IEEE) University of South Florida, USA

Srivaths Ravi
Texas Instruments, India

Anurag Seth
Kawasaki Microelectronics, India

VLSI Design 2012 Conference Committee

Steering Committee Chair

Vishwani D. Agrawal
Auburn University, USA

General Chair

Srimat Chakradhar
NEC Labs, USA

Silver Jubilee Special Chair

J A Chowdary
Talent Sprint Pvt. Ltd., India

Silver Jubilee Conference Convener

Dasaradha R Gude
SoCtronics Technologies Pvt. Ltd., India

Program Chairs

Anshul Kumar
IIT Delhi, India

Jörg Henkel
Karlsruhe Institute of Technology, Germany

Organizing Chair

Manoher Bommena
AMD, India

Organizing Committee

Sahana Kanjula
SoCtronics

Archna Vyasam
SoCtronics

Annajirao Garimella

xxvii

Tutorial Chairs

Preeti Ranjan Panda
IIT Delhi, India

Nagi Naganathan
LSI Corporation, USA

Publication Chairs

Annajirao Garimella

PA Govindacharyulu

Finance Chairs

Prem Nivasa
Qualcomm, India

Raghavendra Puligadda
SoCtronics Technologies Pvt. Ltd., India

Registration Chair

Krishna Prasad Pinapala
Incube Solutions

Exhibit Chair

S. Uma Mahesh
Indrion Technologies India Pvt. Ltd., India

Publicity Chairs

Shashidhar Reddy
Qualcomm, India

Rachna Jain
Qualcomm, India

Sponsorship Chairs

S. Uma Mahesh
Indrion Technologies India Pvt. Ltd., India

K Murali Krishna
SoCtronics Technologies Pvt. Ltd., India

Fellowship Chairs

K. Lal Kishore
JNTU Hyderabad, India

M.B. Srinivas
BITS, Hyderabad, India

Industry Forum Chairs

Mohan Narasimhan
Cypress Semiconductor, India

Pidugu Lakshmi Narayana
Xilinx, India

Design Contest Chairs

Student Conference Chair

Madhavi Latha
JNTU Hyderabad, India

Balaji Kanigicherla
Ineda Systems Pvt.Ltd.

K. Subbarangaiah
JNTU Hyderabad, India

North America Liaison

Anantha Chandrakasan
MIT, USA

Europe Liaison

Dimitris Gizopoulos
University of Athens, Greece

Asia-Pacific Liaison

Kazumi Hatayama
Nara Institute of Sci. and Tech. (NAIST), Japan

IEEE Liaison

N. Ranganathan
University of South Florida, USA

ISA Liaison

Pradip K. Dutta
Synopsys, India

VSI Liaison

Jaswinder Ahuja
Cadence, India

SSCS Liaison

Sreedhar Natarjan
TSMC, Canada

VLSI 2012 Technical Program Committee and Reviewers

Program Chairs

Anshul Kumar, *Indian Institute of Technology Delhi, India*
Jörg Henkel, *Karlsruhe Institute of Technology, Germany*

Track Chairs

Analog and Mixed Signal Design

Abhijit Chatterjee, *Georgia Institute of Technology, United States*
G S Visweswaran, *Indian Institute of Technology Delhi, India*

Architectures and Embedded Applications

Mona Mathur, *ST Microelectronics Pvt Ltd, India*
Jiang Xu, *Hongkong University of Science & Technology, China*

Arithmetic circuits and EDA

Kaushik De, *Synopsys, India*
Anand Raghunathan, *Purdue University, United States*

CAD for Analog and Mixed Signal Circuits

Saraju Mohanty, *University of North Texas, United States*

Low-power Analog-mixed Signal and Physical Design

Amit Patra, *Indian Institute of Technology Kharagpur, India*
Ruchir Puri, *IBM Research, United States*

Low-power design and DVS

Ajit Pal, *Indian Institute of Technology Kharagpur, India*
Rajiv Joshi, *IBM Research, United States*

Parallel and Reconfigurable Systems

David Atienza, *Ecole Polytechnique Fédérale de Lausanne, Switzerland*
Kolin Paul, *Indian Institute of Technology Delhi, India*

Test and Verification

Masahiro Fujita, *University of Tokyo, Japan*
Virendra Singh, *Indian Institute of Science, India*

VLSI Technology and TCAD

Mohammad A. Al Faruque, *Siemens Corporate Research, United States*
Shankar Balachandran, *Indian Institute of Technology Madras, India*

Technical Program Committee Members

Prathima Agrawal, *Auburn University, United States*

Vishwani Agrawal, *Auburn University, United States*

Swapna Banerjee, *Indian Institute of Technology Kharagpur, India*

Shabbir Batterywala, *Synopsys India Pvt Ltd, India*

Bernd Becker, *University of Freiburg, Germany*

Sanjuktha Bhanja, *University of South Florida, United States*

Sambuddha Bhattacharya, *Synopsys, India*

Oliver Bringmann, *Forschungszentrum Informatik, Germany*

Srimat Chakradhar, *NEC Labs, United States*

Nitin Chandrachoodan, *Indian Institute of Technology Madras, India*

Naehyuck Chang, *Seoul National University, Korea*

Karam Chatha, *Arizona State University, United States*

Santanu Chattopadhyay, *Indian Institute of Technology Kharagpur, India*

Jian-Jia Chen, *Karlsruhe Institute of Technology, Germany*

Yiran Chen, *University of Pittsburgh, United States*

Pallab Dasgupta, *Indian Institute of Technology Kharagpur, India*

Vijay Degalahal, *Intel Corporation, India*

Basant Dwivedi, *VirtuQ Education, India*

Prakash Easwaran, *Cosmic Circuits, India*

Malay Ganai, *NEC Labs, United States*

Anup Gangwar, *VirtuQ Education, India*

Annajirao Garimella, *United States*

Manoj Gaur, *National Institute of Technology Jaipur, India*

Swaroop Ghosh, *Intel Corporation, United States*

Padmini Gopalakrishnan, *Xilinx Inc, India*

Tarun Goyal, *Mentor Graphics, India*

Andreas Herkersdorf, *Technical University of Munich, Germany*

Michiko Inoue, *Nara Institute of Science and Technology, Japan*

Alok Jain, *Cadence Design Systems, India*

Niraj Jha, *Princeton University, United States*

Ajay Joshi, *Boston University, United States*

Rouwaida Kanj, *IBM Research, United States*

Harish Krishnaswamy, *Columbia University, United States*

Chidamber Kulkarni, *Xilinx Inc, India*

Eren Kursun, *IBM Research, United States*

Hai(Helen) Li, *New York University, United States*

Loganathan Lingappan, *Intel Corporation, United States*

Rabi Mahapatra, *Texas A&M University, United States*

Dmitri Maslov, *University of Waterloo, Canada*

Vinod Menezes, *Texas Instruments India Ltd, India*

Prabhat Mishra, *University of Florida, United States*

Raj Mitra, *Texas Instruments India Ltd, India*

Subhasish Mitra, *Stanford University, United States*

Tulika Mitra, *National University of Singapore, Singapore*

Jayanta Mukherjee, *Indian Institute of Technology Bombay, India*

Saibal Mukhopadhyay, *Georgia Institute of Technology, United States*

Madhu Mutyam, *Indian Institute of Technology Madras, India*

Nagi Naganathan, *LSI Corp, United States*

NS Nagaraj, *Texas Instruments Inc, United States*

Vijaykrishnan Narayanan, *Pennsylvania State University, United States*

Amit Pande, *University of California at Davis, United States*

Jongsun Park, *Korea University, Republic of Korea*

Sudeep Pasricha, *Colorado State University, United States*

Shanthi Pavan, *Indian Institute of Technology Madras, India*

Nachiket Potlapally, *Intel Corporation, United States*

Vijay Raghunathan, Purdue University, United States

Kewal Saluja, *University of Wisconsin Madison, United States*

Li Shang, *University of Colorado at Boulder, United States*

Mohit Sharma, *Texas Instruments India Ltd, India*

Aviral Shrivastava, *Arizona State University, United States*

Adit Singh, *Auburn University, United States*

K. Sridharan, *Indian Institute of Technology Madras, India*

Susmita Sur-Kolay, *Indian Statistical Institute Kolkata, India*

Juergen Teich, *University of Erlangen-Nuremberg, Germany*

Manish Vachharajani, *LineRate Systems, United States*

Shobha Vasudevan, *University of Illinois at Urbana Champaign, United States*

Kamakoti Veezhinathan, *Indian Institute of Technology Madras, India*

Zhonglei Wang, *Karlsruhe Institute of Technology, Germany*

List of Reviewers

Abhilash Goyal
Abhisek Dey
Achintya Halder
Ahmad E. Islam
Akshay Visweswaran
Alex Doboli
Amal Kundu
Aman Kokrady
Amitava DasGupta
Amrutur Bharadwaj
Anil Prabhakar
Anindya Dhar
Anoop Nair
Aravind NV
Arquimedes Canedo
Arun Chandorkar
Arun Kumar
Ashis Maity
Ashok Srivastava
Asudeb Dutta
Bibhu Datta Sahoo
Byunghoo Jung
Changzhi Li
Chao Lu
Deabshis Mandal
Debasis Banerjee
Debdeep Mukhopadhyay
Dhruva Ghai
Dinesh Sharma
Elias Kougianos
Enakshi Bhattacharya
Erik Larsson
Gaurab Banerjee
Geng Zheng
Görschwin Fey
H. S. Jamadagni

Hafizur Rahaman
Ilia Polian
Jacob Abraham
Jagdish Rao
Jawar Singh
Jimson Mathew
Jitendra Agrawal
Jyotirmoy Ghosh
Kailash Ray
Kalyan Goswami
Kaushik Bhattacharyya
Lakshmikantha Holla
Lars Bauer
Luo Sun
Malathi Mohan
Manas Hati
Maryam Shojaei Baghini
Mehdi Tahoori
Minhong Mi
Mohammed G. Khatib
Muhammad Shafique
Nagendra Krishnapura
Nagesh Tamarapalli
Nandita DasGupta
Nilanjan Chattaraj
Oghenekarho Okobiah
Oleg Garitselov
Pradip Mandal
Pranav Ashar
Preeti Ranjan Panda
Qunzeng Liu
Raghavendra RG
Rainer Buchty
Rajan Konar
Rajarshee Bharadwaj
Rajat Chakraborty

Rajiv Pandey
Rama Venkatasubramanian
Rishad A Shafik
Ritochit Chakraborty
Rolf Drechsler
Rubin Parekhji
Saket Gupta
Santosh Biswas
Saurav Bandyopadhyay
Seetal Potluri
Shabbir Batterywala
Shouri Chatterjee
Shreyas Sen
Shyam Kumar D. B. V. M.
Siva Sudani
Somnath Paul
Soumya Pandit
Sreedhar Prathy
Subho Chatterjee
Subir Roy
Sudhanshu Jamuar
Sumanth Gururajarao
Supratik Chakraborty
Suraj Sindia
Swarup Bhunia
Tamal Das
Uday Kumar
Vijaya Sankara Rao P
Wei Jiang
Wei Zhang
Weichen Liu
William Eisenstadt
Wreeju Bhaumik
Xinying Wang

List of Secondary Reviewers

A. Czutro
AHM Razibul Islam
Ajoy Mandal
Alexander Finder
Amit Kumar
Amlan Chakrabarti
Anandaroop Ghosh
Anuj Grover
Arvind Singh
Aswin Raghav Krishna
Ayan Mandal
Baijayanta Ray
Baohu Li
Benjamin Vogel
Bernhard Schmidt
Bhaskar Karmakar
Boris Alexandrov
Bryce Holton
Daniel Grosse
Daniel Ziener
Denny Lie
Farhana Rashid
Finn Haedicke
Heinz Riener
Hongyan Zhang
Jan Malburg
Jared Pager
Jayita Das
Jian Cai
Jing Lu
K. R. Viveka
Kamel Benaissa
Kaushik Ghoshal
Khondkar Ahmed
Mahesh Prabhu
Matthias Sauer
Mehdi Dehbashi
Min-Woo Lee
Mingliang Wang
Muhammad Usman Karim Khan
Nabeel Iqbal
Nand Kishore R
Nitin Yogi

Prasenjit Basu
Praveen Venkataramani
Priyadharshini Shanmugasundaram
R. R. Manikandan
Rahul Rithe
Rajat Chauhan
Ralf Wimmer
Ranjan Bose
Robert Wille
S. Aniruddhan
Sajish Sajayan
Schuyler Eldridge
Seetharam Narasimhan
Shashank Reddy Kaareddy
Shida Zhong
Stefan Frehse
Stefan Hillebrecht
Stephan Eggersglüß
Suneil Mohan
Thomas Ebi
Thomas Wild
Tobias Ziermann
Tushar Rawat
Vijay Reddy
Waheed Ahmed
Walter Stechele
Wen Yueh
Xiaodong Yang
Xinmu Wang
Yu Zhang
Zhen Wang

VLSI Design 2011 Best Paper Awards

Arun Kumar Choudhury Best Paper Award

"Efficient Trace Signal Selection for Post Silicon Validation and Debug"
Kanad Basu and Prabhat Mishra, University of Florida, Gainesville

Nripendra Nath Biswas Best Student Paper

"An Approach to Tolerate Process Related Variations in Memristor-Based Applications"
Jeyavijayan Rajendran, Harika Manem, Ramesh Karri, and Garrett S. Rose
Polytechnic Institute of New York University

Best Design Contest Entry Award

"Fast and Power Efficient 16x16 Array of Array Multiplier using Vedic Multiplication"
M. S. Prahalad, UVCE Bangalore

Honorable Mention for Design Contest Entry

"Targeting Small Hold Violations in Design by Increasing Net RC"
Sachin Mathur, ST Microelectronics

"An Area Efficient Implementation of Vector by Vector Shifter"
Ankit Garg, Santhosh Kumar Puram, Sri Raghava Kiran Mukku and Vasantha Srirambhatla
IKOA Semiconductor

2011 Best Paper Awards Committee

Vishwani Agrawal, Auburn University
Niraj Jha, Princeton University
Mahesh Mehendale, Texas Instruments
Preeti Ranjan Panda, IIT Delhi
Sachin Sapatnekar, University of Minnesota

Design Contest Awards Committee

C. Srinivasan, Cosmic Circuits
Vivek Pawar, Sankalp Semiconductor
S. N. Padmanabhan

VLSI Design Conference History

No.	Year	Location and Dates	Papers / Tutorials	Proceedings Pages	Approx. Attendance
1st	1985	Chennai, Dec 26-28	29/1	193	75
2nd	1988	Bangalore, Dec 15-18	26/4	496	150
3rd	1990	Bangalore, Jan 6-9	30/4	390	150
4th	1991	New Delhi, Jan 4-8	45/9	315	250
5th	1992	Bangalore, Jan 4-7	57/4	378	300
6th	1993	Mumbai, Jan 3-6	70/6	371	300
7th	1994	Kolkata, Jan 5-8	87/6	448	400
8th	1995	New Delhi, Jan 4-7	77/6	456	450
9th	1996	Bangalore, Jan 3-6	75/6	480	550
10th	1997	Hyderabad, Jan 4-7	84/6	608	550
11th	1998	Chennai, Jan 4-7	98/6	624	600
12th	1999	Goa, Jan 7-10	103/6	682	600
13th	2000	Kolkata, Jan 3-7	93/6	590	700
14th	2001	Bangalore, Jan 3-7	77/9	592	750
15th *	2002	Bangalore, Jan 7-11	109/8	834	1000
16th	2003	New Delhi, Jan 4-8	84/6	622	800
17th	2004	Mumbai, Jan 5-9	120/8	1132	800
18th	2005	Kolkata, Jan 3-7	113/9	922	850
19th	2006	Hyderabad, Jan 3-7	136/11	880	
20th	2007	Bangalore, Jan 6-10	147/15	990	
21st	2008	Hyderabad, Jan 4-8	108/10	780	
22nd	2009	New Delhi, Jan 5-9	79/9	632	
23rd	2010	Bangalore, Jan 3-7	79/8	461	
24th	2011	Chennai, Jan 2-7	66/8	391	450

*Jointly held with ASP-DAC

Embedded Systems Design Conference History

No.	Year	Location and Dates	No. of Papers	Proceedings Pages
1st	2002	New Delhi, Jan 2-4	8	70
2nd	2003	New Delhi, Jan 4-8	84	622
3rd	2004	Mumbai, Jan 5-9	120	1132
4th	2005	Kolkata, Jan 3-7	113	922
5th	2006	Hyderabad, Jan 3-7	136	880
6th	2007	Bangalore, Jan 6-10	147	990
7th	2008	Hyderabad, Jan 4-8	108	780
8th	2009	New Delhi, Jan 5-9	79	632
9th	2010	Bangalore, Jan 3-7	79	461
10th	2011	Chennai, Jan 2-7	66	391

About the Cover

Monolithic Buddha statue, an important landmark of Hyderabad city, adorns the front cover of this year's proceedings. This statue is located in the center of the tranquil waters of Hussain Sagar Lake on the rock of Gibraltar. It is approximately 22 meters (72 feet) in height and 450 tonnes (450,000 kilograms) in weight. Counted amongst the world's largest monoliths, this statue was chiseled out of a white granite rock. Starting from 1985, the statue was carved by 200 sculptors for more than two years, under the skilled guidance of Sri Ganapati Stapathi. The sculpting, the transport of the statue to its position and its erection had been a massive effort and had witnessed quite a few difficulties, including fall of the statue in the lake. Today, the statue stands majestically on a red lotus pedestal in the waters of the lake gracing the Hyderabad city.

On the occasion of this Silver Jubilee VLSI Design Conference, we hope that the conference also continues to stand tall, personifying the efforts that have been put in to bring the conference to its present stature, and serve the semiconductor community further over the coming years.

Welcome to the Pearl City of India, Hyderabad!

Annaji Row Garimella, Govindacharyulu

VLSID'2012 Publication Co-chairs

Keynote Speakers

Vishwani D. Agrawal, Auburn University, USA

Jaswinder Ahuja, Cadence, India

Bert Gyselinckx, IMEC, Belgium

Sri Parameswaran, University of New South Wales, Australia

Rajesh Gupta, University of California, San Diego, USA

Samarjit Chakraborty, Technical University of Munich, Germany

Keynote Talk

A History of the VLSI Design Conference

Vishwani D. Agrawal

James J. Danaher Professor of Electrical and Computer Engineering
Auburn University, USA

Abstract

In December 1985, a group of about eighty-five assembled on the campus of the Indian Institute of Technology in Chennai for the First International Workshop on VLSI Design. The sole purpose of that meeting was to sense the level of VLSI activities in India. We were looking for a focus for engineering education and research. We felt that VLSI technology will be an area of progress in the coming years. The workshop was concluded with great enthusiasm but with no plan for the future. Therefore, it was no surprise that no workshop was held in the following year or the year after. The Second and Third Workshops were held, with some difficulties, in 1988 and 1990, respectively. Fortunately, those initial workshops created enough synergy to get the ball rolling. They spun off their own sponsoring society, the VLSI Society of India. Today, the International Conference on VLSI Design has come a long way to become a conference with world-wide recognition. This talk traces the history of the conference, its beginning, the struggling years, successes, failures and some groundbreaking contributions that made it what it is today.

Speaker Biography

Vishwani D. Agrawal is the James J. Danaher Professor of Electrical and Computer Engineering at Auburn University, Alabama, USA. He has over forty years of industry and university experience, working at Bell Labs, Murray Hill, NJ; Rutgers University, New Brunswick, NJ; TRW, Redondo Beach, CA; IIT, Delhi, India; EG&G, Albuquerque, NM; and ATI, Champaign, IL. His areas of work include VLSI testing, low-power design, and microwave antennas. He obtained his BE degree from the University of Roorkee (renamed Indian Institute of Technology), Roorkee, India, in 1964; ME degree from the Indian Institute of Science, Bangalore, India, in 1966; and PhD degree in electrical engineering from the University of Illinois, Urbana-Champaign, in 1971. He has published over 300 papers, has coauthored five books and holds thirteen United States patents. His textbook, *Essentials of Electronic Testing for Digital, Memory and Mixed-Signal VLSI Circuits*, co-authored with M. L. Bushnell, was published in 2000. He is the founder and Editor-in-Chief (1990-) of the *Journal of Electronic Testing: Theory and Applications*, past Editor-in-Chief (1985-87) of the *IEEE Design & Test of Computers* magazine and a past Editorial Board Member (2003-08) of the *IEEE Transactions on VLSI Systems*. He is the Founder and Consulting Editor of the *Frontiers in Electronic Testing* Book Series of Springer. He is a co-founder of the *International Conference on VLSI Design*, and the *VLSI Design and Test Symposium*, held annually in India. He was the invited **Plenary Speaker** at the *1998 International Test Conference*, Washington D.C., and the **Keynote Speaker** at the *Ninth Asian Test Symposium*, held in Taiwan in December 2000. During 1989 and 1990, he served on the Board of Governors of the IEEE Computer Society, and in 1994, chaired the Fellow Selection Committee of that Society. He has received *eight* **Best Paper Awards** and *two* **Honorable Mention Paper Awards**. In 2006, he received the **Lifetime Achievement Award** of the VLSI Society of India, *in recognition of his contributions to the area of VLSI Test and for founding and steering the International Conference on VLSI Design in India.* In 1998, he received the **Harry H. Goode Memorial Award** of the IEEE Computer Society, *for innovative contributions to the field of electronic testing,* and in 1993, received the **Distinguished Alumnus Award** of the University of Illinois at Urbana-Champaign, *in recognition of his outstanding contributions in design and test of VLSI systems.* Dr. Agrawal is a **Fellow of the IETE-India** (elected in 1983), a **Fellow of the IEEE** (elected in 1986) and a **Fellow of the ACM** (elected in 2003). He has served on the advisory boards of the ECE Departments at University of Illinois, New Jersey Institute of Technology, and the City College of the City University of New York. See his website *http://www.eng.auburn.edu/~vagrawal*

Keynote Talk

Semiconductor Industry: Best of Times, Worst of Times, and Nowhere else I Would Rather Be!

Jaswinder S. Ahuja

Corporate VP & MD, Cadence Design Systems, India
President, VLSI Society of India

Abstract

Best of times: Semiconductors have been at the heart of most technological progress in the past few decades and have had a profoundly positive impact on the human condition. It is hard to argue when many say that we are living in the "golden age of electronics". Human ingenuity has enabled the creation of products that customers never asked for, but now cannot live without - everything from computers to digital cameras, mp3 players and mobile phones. The most successful products and applications are the creative use and convergence of technologies and application domains. Truly speaking, this is only the beginning. The consumerization of electronics, globalization of markets and the human aspiration for a better quality of life is creating unprecedented opportunities that only the semiconductor industry can help realize.

Worst of times: 2008 was an inflexion point in the global economic boom we witnessed since the dot-com bubble burst earlier in the millennium. Across the board, businesses are still struggling to grow and constantly looking at opportunities to improve operational efficiencies. Unemployment in the developed world is at a multi-decade high and GDP growth rates at a multi-year low. We are yet to feel the full impact of the economic down-turn, and the social impact that will follow has not even really started. One thing is certain, this is not a transitory phase; this is the new normal.

Nowhere else I would rather be: The next wave of economic growth will stem from integrating the people who belong to the "base-of-the-economic-pyramid" into the formal economy and dramatically improving their productivity. There are limitless possibilities and opportunities that exist for the semiconductor industry to continue to be a key driver to improve the human condition further over the next several decades. It is still the best place to be, and the "golden age of electronics" has only just started.

Speaker Biography

Jaswinder S. Ahuja, 48, is Corporate Vice President and Managing Director of Cadence Design Systems in India. Jaswinder leads Cadence's India Operating Region which includes R&D, Sales, Technical Field Operations, Marketing, Services, Global Support and Global IT.

Jaswinder is a founding member of VLSI Society of India and at present serves as its President. He was Chairman of the India Semiconductor Association in 2008-09 when he initiated the Electronic System Design and Manufacturing (ESDM) agenda for the industry. Jaswinder also serves on the board of advisors of FirstRain, Inc. and Zafesoft, Inc.

Jaswinder has a B.Tech in Electronics Engineering from IT-BHU, Varanasi and an MS in Computer Engineering from Northeastern University, Boston, USA. He also holds an Executive MBA from Stanford University, USA.

978-1-4673-0438-2/12 $31.00 © 2012 IEEE

Keynote Talk

A Wireless Sensor a Day Keeps the Doctor Away

Bert Gyselinckx
IMEC, Belgium

Abstract

Chronic diseases are predicted to be the leading cause of disability, and will become the most expensive problem affecting all countries (source: WHO). Many of these chronic diseases can be prevented, but this requires a paradigm shift to integrated and preventive healthcare (source: WHO). The focus of future healthcare systems should be on keeping people healthy, raising each individual's awareness on his own health and inducing efficient behavioral changes. The patient of the future is a healthy patient. Wearable sensors are instrumental in managing chronic conditions, providing real-time diagnostics and patient-centric therapies.

In this talk we will review recent technology breakthroughs in wireless sensors, and demonstrate their impact in a few pilot studies. New micro-electronic technologies lead to miniaturized low-power wireless patches, allowing 24/7 monitoring of ECG and other physiological signals for weeks or months. This is a game changing opportunity for epileptic patients, who are given the means to better manage their seizures. This is changing the life of Atrial Fibrillation patients, who can be diagnosed earlier and be given a more optimal treatment. Brain activity monitors are now integrated in headgear and headphones, allowing their use in the home environment without special skin preparation. These provide unprecedented opportunities to measure emotional valence and how one feels about his environment. Combined with other wearable physiological sensors, they provide feedback on one's emotional and stress level and may be use to interact with our surroundings.

Speaker Biography

Bert Gyselinckx combines his role as Human++ program director with that of general manager of imec at the Holst Centre. Human++ is pioneering a unique open-innovation ecosystem in the field of wireless sensors for healthcare and lifestyle. As a result, today over 200 scientific staff and a dozen of residents from industrial partner companies are creating groundbreaking future generation healthcare, lifestyle and life science solutions. In his early career, Bert made pioneering contributions to wireless OFDM communications that were the seed for a couple of startup companies leading to our current generation of WiFi modems.

Bert lives by the golden rule "working hard, playing hard". In 2001, he replaced his office chair for a bike saddle and went on a 12 month odyssey in the Asia Pacific region. 15000km later, he was inspired to create technologies that can have a true impact on society. Bert received the M.S. degree in Electrical Engineering from the Rijksuniversiteit Gent, Belgium, in 1992 and the M.S. degree in Air and Space Electronics from the Ecole Nationale Superieure de l'Aeronautique et de l'Espace, Toulouse, France, in 1993. At this time, he was also a trainee at the Research and Development group of Siemens in Munich, Germany.

978-1-4673-0438-2/12 $31.00 © 2012 IEEE

Keynote Talk

The Variability Expeditions: Exploring the Software Stack for Underdesigned Computing Machines

Rajesh Gupta
University of California San Diego, USA

Abstract

As microelectronic devices scale down to the level molecular assemblies the resulting circuits and systems do not behave like the precisely chiseled machines with tight tolerances. Modern computing is ignorant of the variability in the behavior of underlying components from device to device, chip to chip, its wear over time, or the environment in which the computing system is placed. The 'guardbands' used to guarantee component behavior (for power, performance) have gone to ridiculous margins accounting for as much as two-thirds of the chip area to meet performance 'specs' and is already undermining the gains from continued device scaling. Changing the way software interacts with hardware offers the best hope to recover the advantages from process scaling. In this talk I will describe our approach and progress in the Variability Expeditions project that fundamentally rethinks the rigid, deterministic hardware-software interface, to propose a new class of computing machines that rely on an opportunistic software stack to adapt to the conditions in an underdesigned hardware.

Speaker Biography

Rajesh K. Gupta is a professor and chair of Computer Science and Engineering at UC San Diego, and holds the QUALCOMM endowed chair. His research interests are in energy efficient systems that have taken turn towards large-scale energy use in recent years. His recent contributions include SystemC modeling and SPARK parallelizing high-level synthesis, both of which are publicly available and have been incorporated into industrial practice. Earlier Gupta lead or co-lead DARPA-sponsored efforts under the Data Intensive Systems (DIS) and Power Aware Computing and Communications (PACC) programs that demonstrated architectural adaptation and compiler optimizations in building high performance and energy efficient system architectures. His ongoing efforts include energy-efficient data-centers and large scale computing using memory-coherent algorithmic accelerators and non-volatile storage systems. In recent years, Gupta and his students have received a best paper award at IEEE/ACM DCOSS'08 and a best demonstration award at IEEE/ACM IPSN/SPOTS'05. Gupta received a BTech in EE from IIT Kanpur, MS in EECS from UC Berkeley and a PhD in Electrical Engineering from Stanford University. He currently serves as EIC of IEEE Embedded Systems Letters. Gupta is a Fellow of the IEEE.

Keynote Talk

Challenges in Automotive Cyber-physical Systems Design

Samarjit Chakraborty
Technical University of Munich, Germany

Abstract

Systems with tightly interacting computational (cyber) units and physical systems are generally referred to as cyber-physical systems. They involve an interplay between embedded systems, control theory, real-time systems and software engineering. A very good example of cyber-physical systems design arises in the context of automotive architectures and software. Modern high-end cars have 50-100 processors or electronic control units (ECUs) that communicate over a network of buses such as CAN and FlexRay. In such complex settings, traditional control-theoretic approaches -- where control engineers are only concerned with high-level plant and controller models -- start breaking down. Instead it becomes necessary to adopt a more holistic, cyber-physical systems design approach where the semantic gap between high-level control models and their actual implementations on multiprocessor platforms is quantified and consciously closed. We will give several examples on how this may be done and the current research challenges facing both academia and the industry.

Speaker Biography

Samarjit Chakraborty is a professor of Electrical Engineering at TU Munich in Germany, where he heads the Institute for Real-Time Computer Systems (RCS). Prior to joining TU Munich, from 2003 - 2008 he was an Assistant Professor of Computer Science at the National University of Singapore. He obtained his Ph.D. in Electrical and Computer Engineering from ETH Zurich in 2003. His research interests cover all aspects of system-level design of real-time embedded systems and software, including automotive electronics and software, advanced automotive driver assistance systems, e-mobility and electric vehicles. In addition to his Chair at TU Munich, Prof. Chakraborty leads a research program on embedded systems design for electric vehicles, at the TUM CREATE Centre for Electromobility in Singapore, where he also serves as a Scientific Advisor.

Prof. Chakraborty has published over 100 articles in various top-tier research forums in the real-time and embedded systems domain, including DAC, DATE, CODES+ISSS, EMSOFT, ASP-DAC, RTSS and RTAS, and regularly serves on the technical program committees of many of these conferences. He has served as the general chair of the Embedded Systems Week (ESWeek) 2011, was one of the general chairs of the 9th International Conference on Embedded and Ubiquitous Computing (EUC) 2011 and is a general chair of the IEEE International Symposium on Industrial Embedded Systems (SIES) 2012. He has also served as the technical program committee chair of the International Conference on Embedded Software (EMSOFT) 2009, and has been a Track/Topic Chair in several editions of DATE, ASP-DAC and RTSS.

For his Ph.D. thesis, he received the ETH Medal and the European Design and Automation Association's Outstanding Doctoral Dissertation Award in 2004. His work has also received Best Paper Awards at ASP-DAC 2011, EUC 2010, a HiPEAC Paper Award in 2009 and Best Paper Award nominations at EMSOFT 2010, CODES+ISSS 2008, ECRTS 2007, CODES+ISSS 2006, and DAC 2005.

Tutorial T1

Design of Mixed-Signal Systems using SystemC AMS Extensions

Sumit Adhikari, Vienna University of Technology, Austria
Markus Damm, Vienna University of Technology, Austria
Christoph Grimm, Vienna University of Technology, Austria
François Pecheux, UPMC, Paris, France

Abstract

SystemC has become an accepted standard for design of HW/SW Systems at system level. How-ever, nowadays systems include more and more analog/RF components such as transceivers, sensor interfaces, or PLL. To enable design of such mixed-signal systems, OSCI has standardized AMS extensions for SystemC in 2010. In addition to the capabilities of SystemC for modelling multi-processor HW/SW systems, the AMS extensions enable modelling the behaviour of analog/RF parts, physical environment and digital signal processing methods.

The tutorial gives an introduction into SystemC AMS extensions. The introduction includes the language itself, some simple examples, use cases and (top-down) methodology for refinement of AMS systems, starting from an executable specification.

Speaker Biographies

Sumit Adhikari completed Master of Technology from Indian Institute of Technology, Kharagpur. After working for several years with QualCore Logic Inc. and Austriamicrosystems AG, hc joined ICT, TU Vienna on April 2010 as a scientist. He is also working as AMSWG member at OSCI, contributing towards standardization of SystemC AMS. His interest lies on ASP designing, DSP designing and modelling, development of analogue-mixed signal HDL. His domain expertise lies on Automotive sensor actuator systems, RF-Wireless systems, Industrial electronic systems and Instrumentation applications. He has authored several publications in analogue designing, high level analogue system designing and modelling.

Markus Damm studied Mathematics and Computer Science at the Goethe University in Frankfurt am Main in Germany. After that, he joined the Computer Engineering group of Prof. Waldschmidt in Frankfurt. Since 2006, he works in the Embedded Systems Group of Prof. Grimm at the Institute of Computer Technology (ICT) at the TU Vienna. He is part of the SystemC AMS working group. His research interest includes the coupling of transaction level models with SystemC AMS models, and applying transaction level concepts for the simulation of wireless communication.

978-1-4673-0438-2/12 $31.00 © 2012 IEEE

Christoph Grimm works on the design and design methodology of embedded mixed-signal systems. He has authored more than 100 scientific publications. He is editor of the book "Languages for System Specification", Kluwer 2004 and co-author or contributor to many standards such as the "SystemC AMS" Language Reference Manual, or the standards IEEE 1076.1 (VHDL-AMS) and IEEE 1076.6 (VHDL-SIWG). In 2003, he was General Chair, and in 2005/6 Program Chair (AMS Topic) of the Forum on Specification and Design Languages, and vice chair of the OSCI SystemC-AMS WG and chairs the scientific advisory board of OVE. Since 2006 he is full professor for Embedded Systems at the Institute for Computer Technology, Vienna University of Technology.

François Pêcheux is an associate professor in the UPMC/LIP6 Laboratory in Paris, France. He is responsible for the SystemC training activities of UPMC/LIP6. He also coordinated several French and European projects like TSC, ADAM and played an active role in SocLib, a library of interoperable models for Multi Processor SystemC on Chip modelling. From 1995 to 2002 François Pêcheux worked in the Laboratoire de Physique et Applications des Semiconducteurs in Strasbourg, France. He also joined the Ecole Nationale Supérieure de Strasbourg (ENSPS). He developed several computer-aided design tools and dedicated software for the efficient modelling of systems at the CNRS and participated in the development of some mixed-signal models for sub-micron devices, taking into account physical interaction with the direct environment. In September 2002, he joined the LIP6 Laboratory Integrated Systems Department at the UPMC. He is the author or co-author of multiple articles and conference contributions on (SystemC-based) IC design methodology for homogeneous and heterogeneous systems. He has introduced a new course at UPMC dedicated to the modelling and simulation of heterogeneous systems.

978-1-4673-0438-2/12 $31.00 © 2012 IEEE

Tutorial T2

Reversible Logic: Fundamentals and Applications in Ultra-Low Power, Fault Testing and Emerging Nanotechnologies, and Challenges in Future

Himanshu Thapliyal, University of South Florida, USA
Nagarajan Ranganathan, University of South Florida, USA

Abstract

Reversible logic is emerging as a promising computing paradigm with applications in ultra- low power nanocomputing and emerging nanotechnologies such as quantum computing, quantum dot cellular automata (QCA), optical computing, etc. Reversible circuits are similar to conventional logic circuits except that they are built from reversible gates. In reversible gates, there is a unique, one-to-one mapping between the inputs and outputs, not the case with conventional logic. One of the primary motivations for adopting reversible logic lies in the fact that it can provide a logic design methodology for designing ultra-low power circuits beyond KTln2 limit for those emerging nanotechnologies in which the energy dissipated due to information destruction will be a significant factor of the overall heat dissipation. Further, logic circuits for quantum computers must be built from reversible logic components. Several important metrics need to be considered in the design of reversible circuits the importance of which needs to be discussed. Quantum computers of many qubits are extremely difficult to realize thus the number of qubits in the quantum circuits needs to be minimized. This sets the major objective of optimizing the number of ancilla inputs and the number of the garbage outputs in the reversible logic based quantum circuits. The constant input in the reversible quantum circuit is called the ancilla input, while the garbage output refers to the output which exists in the circuit just to maintain one-to-one mapping but is not a primary or a useful output. The reversible circuit has other important parameters of quantum cost and delay which need to be optimized.

In this tutorial, the speakers will introduce fundamentals of reversible logic and its promising applications in ultra-low power nanocomputing, fault testing and emerging nanotechnologies, as well as the current state of the art and challenges in future. First, a brief overview of reversible logic basics will be given. Next, the speakers will introduce basic reversible logic gates and the key metrics that need to be optimized in reversible logic design and synthesis. Reversible gates require constant ancilla inputs for reconfiguration of gate functions and garbage outputs that help in keeping reversibility. Thus, it is important to minimize parameters such as ancilla and garbage bits, quantum cost and delay in the design of reversible circuits. Next, the speakers will introduce a number of basic reversible gates used in design and a new reversible gate namely the TR gate (Thapliyal-Ranganathan) which has the unique structure that makes it ideal for the realization of arithmetic circuits such as adders, subtractors and comparators, efficient in terms of the parameters such as ancilla and garbage bits, quantum cost and delay. The general synthesis methods proposed in the existing literature target combinational and sequential logic synthesis in general and are not suitable for synthesis of reversible arithmetic units. This is because in arithmetic units such as adders, multipliers, shifters, etc, the choice of the hardware algorithm or the architecture has an impact on the performance and efficiency of the circuit. In this tutorial, the design methodologies and a framework to synthesize reversible data path functional units, such as binary and BCD adders, subtractors, barrel shifters, binary comparators and floating point units will be presented. The objective behind the proposed design methodologies is to synthesize arithmetic and logic functional units optimizing key metrics such as ancilla inputs, garbage outputs, quantum cost and delay.

Next, the speakers will present a set of methodologies for the design of reversible sequential circuits such as reversible latches, flip-flops and shift registers. Next, the application of reversible logic in online and offline testing of single as well as multiple faults in traditional and reversible nanoscale VLSI circuits is introduced. As reversible logic has applications in various emerging technologies such as quantum computing, quantum dot cellular automata, optical computing etc, thus this tutorial will also cover the introductory material on these emerging nanotechnologies. A brief overview will be presented of the design of low power circuits based on reversible computing paradigm such as split charge recovery logic (SCRL), reversible energy recovery logic (RERL), nmos reversible energy recovery logic (nRERL), r-MOS, etc. The tutorial will conclude with discussions on current state of the art, progress and future challenges, such as reversible programming languages, design of reversible CPU, and the open questions about the implementation technologies that could pave the way of ultra-low power nano-processor. In conclusion, this tutorial will touch on reversible logic having application in traditional CMOS for low power computing to non-classical computing domain such as quantum computing, QCA computing etc.

Speaker Biographies

Himanshu Thapliyal received the B.Tech. degree in Computer Engineering from G.B. Pant University, India, in 2004, and the M.S. degree in VLSI and embedded systems from IIIT Hyderabad, India, in 2006. He will obtain the Ph.D. degree in Computer Science and Engineering from University of South Florida, Tampa, in 2011. He has authored or coauthored more than 40 articles in refereed conferences and journals and is a co-owner of a U.S. patent issued recently. In 2009, he received the Distinguished Graduate Achievement Award from the Graduate and Professional Student Council at USF for outstanding research and academic achievement. In 2010, the IEEE Computer Society awarded him with the Richard E. Merwin Scholarship in recognition of his contributions to student chapter activities, academic achievement and for serving as a student ambassador. In 2011, he received 2010 UPE/CS Award for Academic Excellence from the Upsilon Pi Epsilon Honor Society for the Computing Sciences and the IEEE computer Society. His Ph.D. dissertation work on reversible logic has been featured in MIT Technology Review, ACM TechNews, New Scientist Magazine, insideHPC, Softpedia News, etc. He is currently serving as the IEEE Computer Society representative in IEEE GOLD.

Nagarajan "Ranga" Ranganathan (S'81-M'88-SM'92-F'02) received the B.E. (Honors) degree in Electrical and Electronics Engineering from Regional Engineering College (National Institute of Technology) Tiruchirapalli, University of Madras, India, 1983, and the Ph.D. degree in Computer Science from the University of Central Florida, Orlando in 1988. He is a Distinguished University Professor of Computer Science and Engineering at the University of South Florida, Tampa. During 1998-99, he was a Professor of Electrical and Computer Engineering at the University of Texas at El Paso. His research interests include VLSI circuit and system design, VLSI design automation, multi-metric optimization in hardware and software systems, computer architecture, and parallel computing. He has developed many special purpose VLSI circuits and systems for computer vision, image and video processing, pattern recognition, data compression and signal processing applications. He has co-authored over 275 papers in refereed journals and conferences, four book chapters and co-owns seven U.S. patents and one pending.

978-1-4673-0438-2/12 $31.00 © 2012 IEEE

He has edited three books titled *VLSI Algorithms and Architectures: Fundamentals, VLSI Algorithms and Architectures: Advanced Concepts*, IEEE CS Press, 1993, *VLSI for Pattern Recognition and Artificial Intelligence,* World Scientific Publishers, 1995 and co-authored a book titled, *Low Power High Level Synthesis for Nanoscale CMOS Circuits*, Springer, June 2008.

Dr. Ranganathan was elected as a Fellow of IEEE in 2002 for his contributions to algorithms and architectures for VLSI systems. He is a member of the IEEE, IEEE Computer Society, IEEE Circuits and Systems Society and the VLSI Society of India. He has served on the editorial boards for the journals: Pattern Recognition (1993-97), VLSI Design (1994-present), IEEE Transactions on VLSI Systems (1995-97), IEEE Transactions on Circuits and Systems (1997-99), IEEE Transactions on Circuits and Systems for Video Technology (1997-00), IEEE Transactions on Computers (2008-10), IEEE Transactions on CAD (2008-10) and ACM Transactions on Design Automation of Electronic Systems (2007-09). He was the chair of the IEEE Computer Society Technical Committee on VLSI during 1997-01. He is on the steering committee of the IEEE Transactions on VLSI Systems during 1999-01, 2007-present and the steering committee chair during 2002-03 and the Editor-in-Chief for two consecutive terms during 2003-06. He served as the program co-chair for ICVLSID'94, ISVLSI'96, ISVLSI'05, and ICVLSID'08 and as general co-chair for ICVLSID'95, IWVLSI'98, ICVLSID'98, ISVLSI'05 and ISVLSI'09. He has served on technical program committees of international conferences including ICVLSID, ICCD, ICPP, IPPS, SPDP, ICHPC, HPCA, GLSVLSI, ASYNC, ISQED, ISLPED, CAMP, ISCAS, MSE, DATE and ICCAD.

Dr. Ranganathan received the USF Outstanding Research Achievement Award in 2002, USF President's Faculty Excellence Award in 2003, USF Theodore-Venette Askounes Ashford Distinguished Scholar Award in 2003, the SIGMA XI Scientific Honor Society Tampa Bay Chapter Outstanding Faculty Researcher Award in 2004, and the Distinguished University Professor honorific title and the university gold medallion honor in 2007, and the USF Outstanding Undergraduate Teaching Award in 2008. He was a co-recipient of three Best Paper Awards at the Intl. Conf. on VLSI Design in 1995, 2004 and 2006 and the *IEEE Circuits and Systems Society* VLSI Transactions Best Paper Award in 2009. He has been appointed as the Faculty Liaison for the University of South Florida Board of Trustees for the term 2011-2014.

Tutorial T3

DFM, DFT, Silicon Debug and Diagnosis – The Loop to Ensure Product Yield

Srikanth Venkataraman, Intel, Oregon, USA
Nagesh Tamarapalli, AMD, India

Abstract

Semiconductor yield has traditionally been limited by random particle-defect based issues. However, as the feature sizes reduced to 90nm and below, systematic mechanism-limited yield loss began to appear as a substantial component in yield loss. In addition, it is becoming clear that ramping yield would take longer and final yields would not reach historical norms. A key factor for not reaching previously attained yield levels is the interaction between design and manufacturing. Yield losses in the newer processes include functional defects, parametric defects and issues with testing. Each of these sources of yield loss needs to be analyzed and understood by designers and tool developers. In addition, new techniques and methods must be devised to minimize the impact of these yield loss mechanisms.

After an introduction of the issues involved in the first section, the second section covers Design-for-Manufacturing (DFM) techniques to analyze the design content, flag areas of design that could limit yield, and make changes to improve yield. However, once the changes are made it is necessary to quantify their impact so that knowledge about yield contribution of different features can be fed back to design and DFM tools. Test presents an opportunity to close the loop by crafting test patterns to expose the defect prone features during automatic test pattern generation (ATPG) and by analyzing silicon failures through diagnosis to determine the features that are actually causing yield loss and their relative impact. The third section covers design techniques (DFX) to improve testability, debuggability and diagnosability, and DFM and defect-aware test generation to both meet product quality and expose yield issues at test. Section four covers the basic concepts and theoretical aspects of debug and diagnosis including algorithmic IC diagnosis, scan chain diagnosis, critical path based techniques and diagnosis of delay defects. The applications of the basic concepts and techniques for silicon debug are covered in section five. Section six covers the application of statistical diagnosis techniques to determine the features that are actually causing yield loss and their relative impact. Finally, in section seven, future trends, challenges and directions are covered.

The proposed tutorial includes the basics to be of interest to students, new engineers and managers but covers new and recent advances in hot topics in Design-for-Manufacturing, yield, test and diagnosis to be of interest to researchers and practicing engineers. The selection of topics cover a broad spectrum and will be of interest to a wide audience including design, test, product, validation, yield, debug and FA engineers.

Speaker Biographies

978-1-4673-0438-2/12 $31.00 © 2012 IEEE

Dr. Srikanth Venkataraman is a Principal Engineer at Intel Corporation in Hillsboro, OR. He manages an R&D group responsible for developing CAD tools for diagnosis, debug and test quality applications in the Design and Technology Solutions group. He has successfully developed and deployed several tools in test and diagnosis used all across Intel. His research interests include the areas of VLSI Test (product design for testability and test CAD), Fault diagnosis, Design Verification and Debug, CAD for VLSI, S/W Engineering and Development. He received his Ph.D. in Electrical and Computer Engineering from the University of Illinois at Urbana-Champaign. He has worked at Texas Instruments and ViewLogic Systems (Sunrise Test System). He has over 60 publications, 3 patents issued and 2 patents pending. He received the best paper award at IEEE VLSI Test Symposium 2000, top 10 papers at IEEE International Test Conference 2000 and the best panel at the IEEE VLSI Test Symposium 99. Intel awards include an Intel Achievement Award (2006), five Divisional Recognition Award (2000, 2002, and 2004), Technical Recognition Award (2002), Excellence Award (2001), Discover Award (2000), best papers at Intel Design and Test Technology Conference (2002, 2003). He has presented a tutorial on diagnosis and debug at the IEEE VLSI Test Symposium 2006, 2004 and 2003, IEEE International Test Conference 2004, Design Automation and Test in Europe 2004, European Test Symposium 2006, VLSI Design Conference 2006 and International Symposium on Testing and Failure Analysis 2003, 2004 and 2005. He is a member of IEEE, IEEE Computer Society and ACM.

Dr. Nagesh Tamarapalli is a Fellow with AMD India Design Center in Bangalore, India, where he is engaged in DFT and manufacturing test development for the next generation low-power Accelerated Processing Units (APUs). Prior to AMD, he was with Mentor Graphics DFT group where he worked on logic BIST, test compression and diagnosis tools. He has published in leading test conferences such as International Test Conference, Asian Test Symposium and journals such IEEE Transactions on CAD. A paper he co-authored at International Test Conference 1999 on logic BIST has been selected for Honorable Mention Award. This and another paper he co-authored at International Test Conference have been selected for "significant papers from the past 35 years". He is the co-inventor of 14 approved US patents in the area of testing. He has delivered DFT seminars in India and USA at several venues including VLSI Design conference 2006, ISQED 2007 and DAC 2008. He holds MS in Electrical Engineering from Indian Institute of Technology, Kharagpur, India and PhD in Electrical Engineering from McGill University, Montreal, Canada.

Tutorial T4

Intellectual Property Protection and Security in System-on-Chip Design

Susmita Sur-Kolay, Indian Statistical Institute, India
Swarup Bhunia, Case Western Reserve University, Cleveland, USA

Abstract

Gigascale integration in recent semiconductor technology mandates design reuse in order to meet the design specifications in time. Electronic description of VLSI design being an intellectual property (IP), may be infringed upon during design reuse. This calls for incorporating techniques for intellectual property protection in the VLSI design flow. The IP of VLSI design, which culminates in fabrication of the integrated circuit, differs from other sources of IPs because in addition to its physical and structural description, it has also a behavioral specification which should remain unaltered after application of IP protection techniques. Security in activation of chips, especially in embedded systems, is an equally grave issue and has led to the paradigm of design-for-security.

This tutorial aims at presenting the major concerns related to IP security that are significant to both the circuit designers and developers of CAD tools. The nature of threats are broadly categorized as (i) misappropriation by hacking during electronic commerce and intentional reselling mostly at design level, and (ii) unauthorized design retrieval. Various attack models and the mechanisms for effective counter measures such as encryption, obfuscation, watermarking and fingerprinting, and certain analytic methods derived from the behavioral aspect, specific to chip designs, will be discussed. First, the scenario of digital rights management, attack models and security goals will be described. Next, the existing approaches for protection of soft IPs such as HDL codes, firm IPs especially at the value-added layout level, and hard IPs including DFM-enhanced layout will be presented. This will include a number of published research results by the presenters. Finally, the recent advances in tackling security issues for design of smart cards and crypto processors will be surveyed.

Speaker Biographies

Susmita Sur-Kolay received the B.Tech degree from Indian Institute of Technology, Kharagpur and the Ph.D. from Jadavpur University, India. She was a research assistant at Massachusetts Institute of Technology, post-doctoral fellow at University of Nebraska-Lincoln, and visiting faculty at Intel Corp., USA. She is presently a Professor in the Advanced Computing and Microelectronics Unit of the Indian Statistical Institute, Kolkata, India. Her research publications in peer-reviewed journals and premier conferences, and a book chapter, are in the areas of algorithmic CAD for VLSI physical design, fault modeling and testing, and IP protection of VLSI design. She has served in several program committees and editorial boards. She is a Distinguished Visitor of IEEE Computer Society (India), Senior Member of IEEE, Member of IET and VLSI Society of India. Among many other awards, she was the recipient of the President of India Gold Medal (summa cum laude) at IIT Kharagpur, and IBM Faculty Award. Dr. Sur-

978-1-4673-0438-2/12 $31.00 © 2012 IEEE

Kolay has presented a number of tutorials on physical design, and intellectual property protection (CASCOM 2009, ICED 2008, VLSI Design 2004).

Swarup Bhunia received his B.E. (Hons.) from Jadavpur University, Kolkata, India, and the M.Tech. degree from the Indian Institute of Technology (IIT), Kharagpur. He received his Ph.D. from Purdue University, IN, USA, in 2005. Currently, Dr. Bhunia is an assistant professor of Electrical Eng. and Computer Sc. at Case Western Reserve University, USA. He has over ten years of research and development experience with over 120 publications in peer-reviewed journals and premier conferences. His research interests include low power and robust design, hardware security and protection, and novel test methodologies. Dr. Bhunia has given a number of tutorials on nanometer design issues in premier conferences (VTS 2010, ITC 2009, DATE 2009, ISLPED 2008). He is a senior member of IEEE.

Tutorial T5

Advanced Analog-Mixed Signal System and Circuit Techniques

Pavan Hanumolu, Oregon State University, USA
Un-Ku Moon, Oregon State University, USA
Terri Fiez, Oregon State University, USA

Abstract

This tutorial begins with a broad overview of challenges in emerging mixed signal systems. After describing the system-level requirements along with the architecture and circuit needs, specific circuit and system solutions will be discussed to highlight promising approaches. Design techniques for advanced analog- and mixed signal circuit blocks such as phase-locked loops and analog-to-digital converters will be covered in detail. Finally, the modeling and analysis of substrate noise coupling in mixed-signal integrated circuits is addressed.

This day long tutorial addresses both the system- and circuit-level aspects of emerging mixed-signal systems. Analysis and design techniques to implement analog to digital converters, phase-locked loops, and the impact of substrate noise on these circuits in large system-on-chips will be discussed. The tutorial is categorized into the following four categories.

1. Challenges in Emerging Mixed-Signal Systems and Applications: With the successful integration of systems-on-a-chip for a wide range of ubiquitous applications including cell phones and gaming devices, new applications for SOCs are arising that create unique challenges at the circuit and system level. In this talk several emerging applications for integrated mixed-signal systems will be highlighted including sensor networks, solar electronics, and tracking and monitoring devices. The system level requirements will be discussed along with the architecture and circuit needs. Specific circuit and system solutions will be discussed to highlight promising approaches to address both application and process challenges in the coming decade.

2. Phase-Locking Techniques for Frequency Synthesis: Phase-locked loops (PLLs) are essential building blocks in all digital, analog, and radiofrequency integrated circuits (ICs). The noise, power, and area of PLLs determine many of the key performance parameters in all such ICs. This tutorial describes the fundamental principles and concepts of PLL design. After reviewing the operation of a simple type-1 PLL and the characteristics of its building blocks, the operating and design principles of a charge-pump PLL will be discussed in detail. Phase noise analysis using a small-signal model will be described and noise-bandwidth-power tradeoffs will be presented. Existing and emerging techniques to alleviate these tradeoffs will be briefly discussed.

3. Digitally-Enhanced Phase-Locking Techniques: Implementing analog phase-locked loops (PLLs) in deep sub-micron processes pose many design challenges. In this talk we elucidate such challenges and address those using highly digital architectures. Implementation details of the building blocks in a digital

PLL such as time-to-digital converters, digital loop filters, and digitally-controlled oscillators will be described. Digital PLL specific design issues such as limit cycles and the dither jitter caused by them will be discussed.

4. Advanced and Emerging ADCs: Many analog IC designers and students are drawn to ADCs. While some ADC realizations have had a lasting impact, examples including pipelined ADCs with digital redundancy, flash ADCs with folding and interpolation, and multi-bit delta-sigma modulators with dynamic element matching, there are many more recent, advanced and emerging ADC design techniques that are receiving much attention and also gaining momentum in some areas. Many of these ideas are showered with doubts and honest criticism. However, we may also be entering a new era where some of these developments would help resolve the toughest submicron scaling challenges that analog designers face today. This tutorial will summarize and ponder the impact of a few selective as well as random slices of these advanced and emerging ADC designs.

Speaker Biographies

Pavan Kumar Hanumolu received the Ph.D. degree in electrical engineering from Oregon State University in 2006. Currently, he is an Assistant Professor in the School of Electrical Engineering and Computer Science at the same University. His research interests include high-speed I/O interfaces, digital techniques to compensate for analog circuit imperfections, time-based signal processing, and power-management circuits.

Un-Ku Moon has been with the Oregon State University since 1998. Prior to that, he was with Bell Labs (Reading & Allentown) 1988-1989 and 1994-1998. He received a bachelor's degree from the University of Washington, a master's degree from Cornell University, and a Ph.D. from the University of Illinois, Urbana-Champaign. His current research activities are found at
http://eecs.oregonstate.edu/~moon/research.

Terri S. Fiez is Professor and Head of Electrical Engineering and Computer Science at Oregon State University. From 2008 until mid 2009 she co-founded and served as CEO of Azuray Technologies, a startup developing micro-inverters for solar applications. Since returning to OSU in September 2009, she has taken on a leadership role for OSU's Sustainable Energy and Infrastructure (SENERGI) research thrust. After receiving her Ph.D. from OSU in 1990 she was a faculty member at Washington State University before returning to OSU to lead the department in 1999. She is an IEEE Fellow and her research interests are in analog and mixed-signal IC design and innovative engineering education approaches.

978-1-4673-0438-2/12 $31.00 © 2012 IEEE

Tutorial T6

Variability-resistant Software and Hardware for Nano-Scale Computing

Nikil Dutt, UC Irvine, USA
Mani Srivastava, University of California, Los Angeles, USA
Rajesh Gupta, University of California, San Diego, USA
Subhashish Mitra, Stanford University, USA

Abstract

As semiconductor manufacturers build ever smaller components, circuits and chips at the nano scale become less reliable and more expensive to produce – no longer behaving like precisely chiseled machines with tight tolerances. Modern computing tends to ignore the variability in behavior of underlying system components from device to device, their wear-out over time, or the environment in which the computing system is placed. This makes them expensive, fragile and vulnerable to even the smallest changes in the environment or component failures. This tutorial presents an approach to tame and exploit variability through a strategy where system components -- led by proactive software -- routinely monitor, predict and adapt to the variability of manufactured systems. Unlike conventional system design where variability is hidden behind the conservative specifications of an "over-designed" hardware, we describe strategies that expose spatiotemporal variations in hardware to the highest layers of software. After presenting the background and positioning the new approach, the tutorial will proceed in a bottom-up fashion. Causes of variability at the circuit and hardware levels are first presented, and classical approaches to hide such variability are presented. The tutorial then presents a number of strategies at successively higher levels of abstraction – covering the circuit, microarchitecture, compiler, operating systems and software applications – to monitor, detect, adapt to, and exploit the exposed variability. Adaptable software will use online statistical modeling to learn and predict actual hardware characteristics, opportunistically adjust to variability, and proactively conform to a deliberately underdesigned hardware with relaxed design and manufacturing constraints. The resulting class of UnO (Underdesigned and Opportunistic) computing machines are adaptive but highly energy efficient. They will continue working while using components that vary in performance or grow less reliable over time and across technology generations. A fluid software-hardware interface will mitigate the variability of manufactured systems and make machines robust, reliable and responsive to changing operating conditions - offering the best hope for perpetuating the fundamental gains in computing performance at lower cost of the past 40 years.

Speaker Biographies

Nikil Dutt is a Chancellor's Professor of CS and EECS at the University of California, Irvine. He received a B.E.(Hons) from BITS Pilani in 1980, MS from Penn State in 1983 and PhD from the University of Illinois at Urbana-Champaign in 1989. His research interests are in embedded systems design automation, computer architecture, optimizing compilers, system specification techniques, distributed embedded systems, and brain-inspired architectures and computing. He has received numerous best paper awards and is coauthor of 7 books. Professor Dutt served as Editor-in-Chief of ACM Transactions on Design Automation of Electronic Systems (TODAES) (2003-2008) and currently serves as Associate Editor of ACM Transactions on Embedded Computer Systems (TECS) and of IEEE Transactions on VLSI Systems (IEEE-TVLSI). He was an ACM SIGDA Distinguished Lecturer during 2001-2002, and an IEEE Computer Society Distinguished Visitor for 2003-2005. He has served on the steering, organizing, and program committees of several premier CAD and Embedded System Design conferences and workshops, and serves or has served on the advisory boards of ACM SIGBED and ACM SIGDA. Professor Dutt is a Fellow of the IEEE, an ACM Distinguished Scientist, and recipient of the IFIP Silver Core Award.

Rajesh Gupta is a professor and holder of the QUALCOMM endowed chair in Embedded Microsystems in the Department of Computer Science & Engineering at UC San Diego, California. He received his B. Tech. in Electrical Engineering from IIT Kanpur, India in 1984, MS in EECS from UC Berkeley in 1986 and a Ph. D. in Electrical Engineering from Stanford University in 1994. Earlier he worked as a circuit designer at Intel Corporation, Santa Clara, California as a member of three successful processor design teams; and on the Computer Science faculty at University of Illinois, Urbana-Champaign and UC Irvine. His current research is focused on energy efficient and mobile computing issues in embedded systems. He is author/co-author of over 150 articles on various aspects of embedded systems and design automation and four patents on PLL design, data-path synthesis and system-on-chip modeling. Professor Gupta serves as an advisor to Tallwood Venture Capital, RealIntent, Calypto and Packet Digital Corporation.

Prof. Subhasish Mitra directs the Robust Systems Group in the Department of Electrical Engineering and the Department of Computer Science of Stanford University. Before joining Stanford, he was a Principal Engineer at Intel Corporation. Prof. Mitra's research interests include robust system design, VLSI design, CAD, validation and test, and emerging nanotechnologies. His X-Compact technique for test compression has been used in more than 50 Intel products, and has influenced major CAD tools. The IFRA technology for post-silicon validation, created jointly with his student, was characterized as "a breakthrough" in the Communications of the ACM. His work on the first demonstration of imperfection-immune carbon nanotube VLSI circuits, jointly with his students and collaborators, was selected by NSF as a Research Highlight to the US Congress, and was highlighted as "a significant breakthrough" by the Semiconductor Research Corporation and the MIT Technology Review. Prof. Mitra's major honors include the Presidential Early Career Award for Scientists and Engineers from the White House, the highest US honor for early-career outstanding scientists and engineers, ACM SIGDA Outstanding New Faculty Award, IEEE CAS/CEDA Pederson Award for the IEEE Transactions on CAD Best Paper, IEEE/ACM Design Automation Conference Best Paper Award, Intel Design and Test Technology Conference Best Paper Award, IBM Faculty Awards,Terman Fellowship, and the Intel Achievement Award, Intel's highest corporate honor. At Stanford, he was honored multiple times by graduating seniors "for being important to them during their time at Stanford." Prof. Mitra also serves as an invited member on DARPA's Information Science and Technology Board.

Mani Srivastava is on the faculty at UCLA where he is a Professor in the Electrical Engineering Department, and is also affiliated with the Computer Science Department and the Center for Embedded Networked Sensing. He received both the M.S. and Ph.D. degrees from the University of California,

Berkeley, in 1987 and 1992, respectively. His graduate research was on silicon compilation, and hardware-software rapid prototyping and co-design of embedded VLSI systems for signal processing and control applications. Prior to joining the UCLA Electrical Engineering Department faculty in 1996, Srivastava worked on mobile and wireless multimedia networking at the Networked Computing Research Department at AT&T/Lucent Bell Labs at Murray Hill, NJ. At UCLA, Prof. Srivastava directs the Networked and Embedded Systems Laboratory (http://nesl.ee.ucla.edu), where his students work on diverse aspects of embedded and cyber-physical systems, distributed sensor networks, mobile computing, wireless networking, and pervasive communications. His research spans hardware, software, and algorithms, and emphasizes experimental systems and applications in domains such as mobile health, sustainability, participatory sensing, and defense. Srivastava has published extensively on his research with more than 240 papers many of which have been highly cited, holds five patents for his work on low-power and wireless networking, and has received many awards from top conferences. He is a Fellow of the IEEE, and has received the prestigious Okawa Foundation Grant, and the NSF CAREER Award. He has served as the EIC of the IEEE Trans. on Mobile Computing and the ACM Mobile Computing and Communications Review.

Tutorial T7A

New Modeling Methodologies for Thermal Analysis of 3D ICs and Advanced Cooling Technologies of the Future

David Atienza, EPFL, Switzerland
Arvind Shridhar, EPFL, Switzerland

Abstract

Increasing circuit densities, the proliferation of Multi-Processor Systems-on-Chips (MPSoCs) and high performance computing systems have resulted in an alarming rise in electronic heat dissipation levels, making the conventional thermal management strategies, including air cooled heat sinks, obsolete. The latest advancements in 3D Integration of IC dies have only aggravated this problem, creating a strong worldwide research interest in the development of advanced cooling technologies, such as interlayer microchannel liquid cooled heat sinks, to maintain ICs under safe operating temperatures. While this research has helped create a substantial amount of knowledge base pertaining to the heat transfer mechanism in advanced liquid cooling systems as applied to electronic circuits, this knowledge is yet to be transferred to the EDA community for it to be incorporated in the IC thermal simulators of the future. The existence of such tools becomes absolutely essential when IC designers are faced with the challenge of ascertaining the thermal reliability of their designs in the presence of liquid cooling systems. This tutorial aims to introduce the attendees to the key concepts that are needed to compute IC temperatures with and without microchannel liquid cooling and the principles behind compact modeling of forced convective heat transfer in advanced IC cooling technologies. A major part of this tutorial is based on the 3D-ICE thermal simulator, which has been built by the Embedded Systems Laboratory in EPFL, Switzerland (URL: http://esl.epfl.ch/3D-ICE). This simulator is based on the Compact Transient Thermal Modeling for forced convective cooling advanced by our research group. Since its release in 2010, more than 50 research groups across the world have downloaded it and are actively using it for their research.

Speaker Biographies

Prof. David Atienza received his MSc and PhD degrees in Computer Science and Engineering from Complutense University of Madrid (UCM), Spain, and Inter-University Micro-Electronics Center (IMEC), Belgium, in 2001 and 2005, respectively. Currently, he is Professor and Director of the Embedded Systems Laboratory (ESL) at EPFL, Switzerland, and Adjunct Professor at the Computer Architecture and Engineering Department of UCM. Additionally, he is Scientific Counselor of long-time research of IMEC Nederland (IMEC-NL), Holst Centre, Eindhoven, The Netherlands. His research interests focus on design methodologies for high-performance embedded systems and low-power Systems-on-Chip (SoC), including new thermal management techniques for 2D/3D Multi-Processor SoCs, wireless body sensor networks, dynamic memory management and memory hierarchy optimizations for embedded systems, novel architectures for logic and memories in forthcoming 3D nano-scale electronics, as well as Networks-on-Chip (NoC) design. In these fields, he is co-author of more than

150 publications in prestigious journals and international conferences. He has received a Best Paper Award at the IEEE/IFIP VLSI-SoC 2009 Conference and three Best Paper Award Nominations at the HPCS-WEHA 2010, ICCAD 2006 and DAC 2004 conferences. He is an Associate Editor of IEEE Transactions on CAD (in the area of System-Level Design), IEEE Letters on Embedded Systems and Elsevier Integration: The VLSI Journal. He is also an elected member of the Executive Committee of the IEEE Council of Electronic Design Automation (CEDA) since 2008 and an elected member of the Board of Governors of the IEEE Circuits and Systems Society (CASS) since 2010.

Arvind Sridhar graduated with Bachelors in Electronics Engineering from Anna University, India in 2006 and with a Masters in Applied Science from the Department of Electronics, Carleton University, Canada in 2009, majoring in EDA. Since then, he has been working at the Embedded Systems Laboratory in EPFL, Switzerland as a doctoral researcher under the supervision of Dr. David Atienza. At EPFL, he developed the compact modeling methodology for forced convection in microchannels, which forms the basis for 3D-ICE, an open source thermal simulator which is currently being used by researchers in more than 50 research labs across the world. His research interests include Simulation and Optimization CAD for VLSI, Thermal Modeling and Computational Fluid Dynamics. Arvind Sridhar has handled courses in Circuit Theory, Signals and Systems and Electromagnetic Theory as a teaching assistant in Carleton University. In June 2011, he was one of the course instructors for the Microscale Heat Transfer course for doctoral students at EPFL, introducing the concept of compact modeling for microchannel liquid cooling of ICs and 3D-ICE.

Tutorial T7B

Optimally Addressing Verification Constraint Complexity for Effective Functional Convergence

Shankar Hemmady, Synopsys, USA

Abstract

As SoC design becomes larger and more complex, verification engineers are expanding constrained-random testing to meet the validation demand. This expansion creates a new set of challenges: engineers now face exponentially growing verification performance and capacity issues. It is no longer enough to write constraints that simply function to validate a design. Today, engineers must also optimize the constraints they write for performance if they wish to have any hope of both successfully validating their design and meeting their deadlines. In this tutorial, we discuss a scalable methodology for writing constraints and optimizing performance is which will improve engineers' productivity to write and debug ever-increasing amounts of larger, more complex sets of constraints. Our goal is to reduce the time needed for functional convergence, and later for debug and volume manufacturing. We will also discuss the vital role of Verification IP providers and users in this scenario.

Speaker Biography

Shankar Hemmady is a Principal Engineer at Synopsys responsible for knowledge sharing and social networking. Over the past four years, Shankar was also responsible for power-aware verification, and verification planning and management solutions. He has managed the functional closure of over 25 commercial chips during the past 2 years as an engineering manager or consultant at 12 companies including AMD, Cirrus Logic, Fujitsu, Hewlett Packard, Intel, S3, Sun and Xerox. He has authored over 50 research papers and articles. In 2007, he authored the book, "Metric Driven Design Verification: An Engineer's and Executive's Guide to First Pass Success" which was named a best-seller at DAC 2007-08 by ESNUG and Springer Publications. Currently, Mr. Hemmady is currently serving as the Vice Chair of DVCon 2012 and as the Tutorials Chair of DesignCon 2012. He also serves on the Technical Program Committees of ISQED and IEDEC 2012.

Mr. Hemmady holds a B.S. in Electrical Engineering from the Indian Institute of Technology and an M.S. in Electrical & Computer Engineering from the University of Iowa. He completed Stanford's Advanced Management College executive program in 2002.

978-1-4673-0438-2/12 $31.00 © 2012 IEEE

Tutorial T8A

Designing Silicon-Photonic Communication Networks for Manycore Systems

Ajay Joshi, Boston University, USA

Abstract

The goal of this tutorial is to explain the limits and opportunities of using silicon-photonic link technology for inter-chip and intra-chip communication in manycore systems. Silicon-photonic links have larger bandwidth density and lower energy than equivalent electrical links. In this tutorial, I will first provide an overview of the silicon-photonic device technology and link transceiver/tuning circuits. Using three silicon-photonic network case studies { on-chip tile-to-tile network, process-to-DRAM network and DRAM memory channel, the various silicon-photonic network design issues at the physical level, micro-architecture level and architecture level will be explained in detail. An iterative design process, where we move between these three levels to meet the power-performance specications under the silicon-photonic technology constraints will also be presented. At the end of the tutorial, attendees will have a broad understanding of the capabilities of silicon-photonic technology, and they will be able to design and analyze silicon-photonic networks for Manycore systems.

Speaker Biography

Ajay Joshi received M.S. and Ph.D. in Electrical and Computer Engineering from Georgia Institute of Technology in 2003 and 2006, respectively, and B.Engg. in Computer Engineering from University of Mumbai in 2001. He is currently an Assistant Professor in the Electrical and Computer Engineering department at Boston University. Prior to joining Boston University, he worked as a postdoctoral researcher in the Electrical Engineering and Computer Science department at Massachusetts Institute of Technology. His research interests span across various aspects of VLSI design including circuits and systems for communication and computation, and emerging device technologies including silicon photonics and carbon nanotubes.

Tutorial T8B

Wireless System Design and Systems Engineering Challenges

Kameswara Rao B, HCL, India
Muralidhar Reddy B, HCL, India
Ravi Kishore B, HCL, India

Abstract

Pervasiveness of wireless technology is impacting every aspect of the society and is becoming a de facto feature of any electronic product. This tutorial provides a comprehensive overview of wireless system design and brings out the systems engineering issues through real-life design case studies. It begins with an overview of a typical wireless system and the underlying RF technology. Next, a detailed view of wireless system design with mathematical underpinning of concepts and design of sub-systems will be presented; it highlights systems engineering issues in wireless product development. Subsequently, the tutorial presents the relevant system design case studies, for various wireless applications viz; medical electronics, consumer electronics, defense, telecommunications etc., and explains the methodologies to address the systems engineering issues. Finally, the tutorial explains the trends in future wireless systems design and the associated challenges. The last session concludes the tutorial, followed by a discussion.

Speaker Biographies

Kameswara Rao B is Project Manager at HCL Technologies Limited, Chennai, India. He handles RF/wireless Projects and is responsible for design and delivery of RF and Microwave products / systems. He has had extensive experience of around 10 years in RF/Microwave Circuit & System Design across various domains like Wireless Infrastructure, Defense, Consumer Electronics and Industrial. He has a Bachelors Degree in Electronics and Communication Engineering from Institute of Engineers (India).

Muralidhar Bandi is Project Manager at HCL Technologies Limited, Chennai, India. He is responsible for developing complex wireless systems to address the Telecom and Defense market needs. He has worked on 2G/3G Base stations, multiband repeaters, optical wireless circuits to meet the Indoor & outdoor Distribution Antenna System requirements. Prior to the current assignment in HCL Technologies, he had worked with Bharat Electronics Limited on analog and digital television transmitters He has over 12 years of experience in RF and Microwave circuit designs and developing wireless transceivers for 2G/3G and 4G requirements. He had Bachelors in Technology Degree in Electronics and Communication Engineering from Nagarjuna University.

978-1-4673-0438-2/12 $31.00 © 2012 IEEE

Dr. B. Ravi Kishore is Associated General Manager at HCL Technologies Ltd., Chennai. He had completed his Ph.D. from Tokyo Institute of Technology, Tokyo in the area of fault-tolerant computing and design for testability for asynchronous circuits, in March, 1998. He had worked with Texas Instruments (Japan) from 1998 to 2002 in design automation group for analog and mixed-signal design. He worked in the area of power analysis, design for manufacturability of mixed-signal circuits. Since 2003, he has been working with HCL Technologies, Chennai, India. From 2003 to 2006, he was responsible for the defining and implementing the system level reliability process for various high-integrity applications in medical, avionics and networking. He successfully initiated several large customer engagements in the areas of wireless infrastructure, consumer electronics, defense & security and headed the product development teams to build end-to-end solutions. Currently, he heads the Centre of Excellence (CoE) team for RF Systems Engineering since 2009. As a part of CoE activities, the team is responsible for new technology initiatives and developing wireless solutions for RF Systems Engineering, in the application areas of wireless infrastructure, consumer electronics, medical electronics, defense & security, industrial electronics etc.

Embedded Tutorial

Pole-Zero Analysis of Low-Dropout (LDO) Regulators: A Tutorial Overview

Annajirao Garimella
Punith Surkanti
Paul M. Furth

New Mexico State University, Las Cruces, NM, USA

Abstract

We discuss the low-dropout (LDO) voltage regulator pole-zero analysis in this tutorial. A priori knowledge of poles and zeros assist in choosing the right topology and appropriate frequency compensation techniques before implementing the transistor level design, as the location of poles move with output load current. The objective of this tutorial is to provide a step-by-step procedure for analyzing poles and zeros in LDO regulators. To this end, two recent state-of-the-art LDO regulators from the literature are analyzed, explaining several intricacies involved. We explain step-by-step procedure in developing the small-signal model, breaking the voltage/current loop and techniques for quickly arriving at simple and approximate pole-zero equations. During the process, several frequency compensation techniques are elucidated. The derived analytic expressions for poles and zeros help in developing intuition of circuit behavior.

Speaker Biographies

Annajirao Garimella (S'99–M'10) received the B.E. degree in electrical and electronics engineering from the University of Madras, Chennai, India, in 2000, the M.S. degree in VLSI from Manipal University, Manipal, India, in 2002, and the M.S. and Ph.D. degrees in electrical engineering with a minor in management from New Mexico State University, Las Cruces, NM in 2009 and 2010, respectively. In 2006, he worked as an intern at Freescale Semiconductor, Inc., Austin, TX. In 2007, he held a co-op position at Texas Instruments, Dallas, TX. He has authored or coauthored more than 30 papers, published in conferences and journals. His research interests lie in the area of Circuits and Systems, with emphasis on analog, mixed-signal IC design, power management IC design, SoC design, and testing. He is recipient of HENAAC (Hispanic Engineer National Achievement Award Corporation) AMD Scholarship in 2005 and the HENAAC DaimlerChrysler Scholarship in 2006. Dr. Garimella received the Outstanding Ph.D. Graduate Alumni Association Award at New Mexico State University for the class of fall 2010. He is on the technical program committee and organizing committee of the 25th International Conference on VLSI Design, VLSID' 2012 and also the publication co-chair.

Punith R. Surkanti (S'07) is a Ph.D. candidate in electrical engineering at New Mexico State University, Las Cruces, NM. He obtained his B.Tech. from JNTU, Hyderabad in 2008 and M.S. degree in electrical engineering from New Mexico State University, Las Cruces, NM in 2011. He is the recipient of the Outstanding Graduate Teaching Assistantship Award from NMSU in 2011. He was selected for the Preparing Future Faculty Graduate Assistantship Award at NMSU in Fall 2009. His areas of interest include Analog, Mixed-signal and Power Management IC Design.

Paul M. Furth (S'90–M'96–SM'11) received a B.A. from Grinnell College, Ginnell, IA in 1984, the B.S.E.E. degree from the California Institute of Technology, Pasadena, CA in 1985 and the M.S. and Ph.D. degrees in electrical engineering from the Johns Hopkins University, Baltimore, MD, in 1992 and 1996, respectively. In 1995, he joined the Klipsch School of Electrical and Computer Engineering, New Mexico State University, Las Cruces, where he is currently an Associate Professor. He has work experience at Sandia National Labs, Micron, and Motorola. His areas of interest include analog and mixed-signal VLSI design, power management circuits, and CMOS image sensors. Dr. Furth received the Bromilow Teaching Excellence Award in 2008. He was the Technical Program Chair at the 42nd IEEE Midwest Symposium on Circuits and Systems Conference, MWSCAS 1999. He has chaired Special Sessions at MWSCAS 2010 and MWSCAS 2011 and currently serves as a Steering Committee Member for MWSCAS.

2012 25th International Conference on VLSI Design

Embedded Tutorial

Digital Subscriber Line

M Kalyana Kumar Rao
Shantha Kumari PV
Boopalan Sellappan

LSI R&D India Pvt Ltd, India

Abstract

Digital Subscriber Line (DSL) is a family of standards that allow existing twisted pair copper lines to carry modulated digital signals which uses telephone network and unused frequency Spectrum. In this tutorial we describe DSL (Digital Subscriber Line), comparing with POTS and ISDN. We explain the different DSL flavors available with the modulation techniques used and also discuss challenges in getting high performance and throughput, achieving xDSL rate and meeting the DSL standard.

We begin by introducing DSL having a pair of modems CO (Central office) and CPE (Customer Premises Equipment) and talk about frequency spectrums. We explain the limitations of POTS (Plain Old Telephone System) with Dial-up Connection and ISDN (Integrated Services Digital Network), and advantages of DSL over POTS. Next we describe various flavors of DSL with their frequency spectrums, power enhancements, standardization, profiles, and band plans. We discuss different modulation techniques with their advantages and disadvantages, including Single carrier Modulation like CAP and QAM and Multi Carrier Modulation like DMT. We explain the interferers/noise: Line Attenuation, The channel attenuation, Bridged taps, Impulse noise, White noise, NEXT, FEXT, RF Interference. We conclude the tutorial with a description of full activation and initialization phases.

The targeted audience for the tutorial is designers, developers, testers, people working in the area of VDSL, ADSL technology and People working on different platforms like DSLAM's, Central office, Customer premises equipment, Gateway products as well as products related to access network who are familiar with copper line and modems.

Speaker Biographies

M Kalyana Kumar Rao, M.Tech in Embedded Systems from DAVV, Indore, M.P with GATE 2003-96.44 percentile and having 7 years of experience in development and testing with 4+ years in DSL technology and around 2+ years in Network Processors. Paper on QOS AND LOW BANDWIDTH selected as one of the best paper in the event LSI India Annual Technical Conference. Worked more than a year for PHILIPS, Bangalore on porting Mobile Application framework for Philips Nexperia System Solution (Device Layer MSCA Implementation) middleware. Worked more than three and half years for

978-1-4673-0438-2/12 $31.00 © 2012 IEEE

IKANOS Communication Pvt Ltd, Bangalore on IKANOS VDSL2 Residential Gateway, CO and CPE with end-to-end networking. (The Vx180 processor family offers complete, high performance, highly-integrated, cost-effective solutions with VDSL2, ADSL2+, ADSL support, security engines and network processing capability.) Presently working for LSI India Research & Development Pvt. Ltd, Bangalore on The ACP3400 (Multicore PowerPC Communication Processor/Network Processor) also working on R&D with packet-processing tasks such as classification, scheduling, encryption, QoS and deep-packet inspection.

Shantha Kumari PV received the Bachelor of Engineering degree in Computer science and engineering from Visvesvaraya Technological University, India, in 2003. She is currently working for LSI India Research & Development Pvt. Ltd, Bangalore, India as Senior QA engineer.

Boopalan Sellappan is currently pursuing Ph.D. in Computer Science, Network Processor, in Anna University Trichirapalli. Received the Master Degree from Anna University, Chennai, in 2006. He is currently working for LSI India Research & Development Pvt. Ltd, Bangalore, India as QA engineer.

Embedded Tutorial

Packaging Trends, Die Package Co-Design Flow and Challenges

Siva Kothamasu

Texas Instruments, Bangalore

Abstract

Packaging has become one of the critical areas of the SoC design flow in recent years. Miniaturization of package with the ever increasing interface speeds and massive integration have opened up a lot of challenges for packaging engineers. This talk will cover different popular packages available in the industry and recent package trends. Traditionally, package design and analysis always followed SoC design, but it is becoming more and more imperative to design die and package in conjunction. Designing the die and package together in the loop to optimize die and package design and closure is popularly referred to as die package co-design. This talk will focus on the key aspects of co-design flow, package modeling/analysis aspects specifically for high performance designs. This talk will also cover some key issues and challenges from practical designs.

Speaker Biography

Siva Kothamasu is an elected Member Group Technical Staff at Texas Instruments (India), and leads the Package co-design, Electrical and Reliability analysis team in the TI India ARM-MPU business unit. He has more than 10 years industry experience working in various aspects of chip design covering Reliability CAD, Physical Design, Signal/Power Integrity Analysis and Package Co-Design. He has several papers in internal and external forums and is a member of IEEE. He holds a Bachelor of Engineering degree from Osmania University.

Embedded Tutorial

Advanced Techniques for Programming Networked Embedded Systems

Vijay Raghunathan

Purdue University, USA

Abstract

Introduction: Cyber-Physical Systems (CPSs) are poised to play a pivotal role in engineering new solutions to a variety of societal-scale problems, such as energy conservation, climate change, healthcare, transportation, etc. Their importance is reflected in the overall theme for VLSI Design 2012, which is "Embedded Solutions for Emerging Markets - Infrastructure, Energy, and Automotive." Networked embedded systems, such as wireless sensor networks, form a crucial building block for realizing large-scale CPSs and have, therefore, received considerable research attention over the past few years. While this has resulted in numerous technological advances (e.g., a plethora of tiny, cheap, and low-power sensor platforms is now available), the problem of programming a distributed wireless sensor network still remains a major challenge and a potential show stopper to widespread adoption. This challenge is best exemplified by the following quote from a recent EE Times article [1]:

"Programming the software that manages applications running on wireless sensor and control networks is currently so technically intricate, complex and laborious that it can take months of work by specialized programmers just to deploy the simplest application. That process takes even longer for more complex deployments..."

As wireless sensor networks continue to rapidly evolve from lab-scale prototypes towards real applications, it becomes even more crucial to address the programmability problem so that designers can realize large-scale deployments of these systems without scaling the human effort involved in programming them.

Overview of the Embedded Tutorial: This tutorial will present advanced, state-of-the-art techniques to program networked embedded systems such as wireless sensor networks. We will start with a quick introduction to distributed wireless sensor nets and the salient characteristics of today's sensor node hardware platforms. This will lead us to the key challenges involved in writing software for these systems and why they are so difficult to program correctly. We will then present various programming techniques that have been proposed to address these challenges. These techniques utilize new and novel programming abstractions, languages, and frameworks that address the unique requirements and characteristics of networked embedded systems. In particular, we will discuss an approach called macroprogramming, in which entire groups of embedded sensor nodes can be programmed as a collective, as opposed to programming individual nodes. The techniques that will be presented have all been validated using real sensor network deployments and example programs from these deployments

978-1-4673-0438-2/12 $31.00 © 2012 IEEE

will also be discussed. Finally, we will also highlight recent industrial activity in this area, such as the IBM Mote Runner programming framework and run-time environment for wireless sensor networks.

This embedded tutorial is targeted at students, researchers, and software engineers working in the area of embedded systems, who want to learn about the key issues and challenges involved in programming networked embedded systems and state-of-the-art solutions to these challenges.

References:
[1] T. Enwall, "Deploying Wireless Sensor Networks for Industrial Automation and Control," URL: http://www.eetimes.com/design/industrial-control/4013661/Deploying-Wireless-Sensor-Networks-for-Industrial-Automation-Control

Speaker Biography

Vijay Raghunathan is an Assistant Professor in the School of Electrical and Computer Engineering at Purdue University, where he leads the Embedded Systems Lab. His research interests include hardware and software architectures for embedded computing systems, wireless sensor networks, and system-on-chip design with an emphasis on reliable system design, low power design, and micro-scale energy harvesting. He has co-authored several book chapters, journal and conference papers, and has presented full-day and embedded tutorials on the above topics. He serves on the organizing and technical program committees of numerous leading ACM and IEEE conferences in the areas of embedded systems, VLSI, and sensor networks. In 2011, he served as a technical program co-chair for the ACM/IEEE International Conference on Information Processing in Sensor Networks (IPSN). Vijay is a recipient of several awards, including the best paper award at the ACM Conference on Embedded Networked Sensor Systems (SenSys) in 2011, the NSF CAREER award in 2010, the design contest award at the ACM International Symposium on Low Power Electronics and Design (ISLPED) in 2005, and the best student paper award at the IEEE International Conference on VLSI Design (VLSID) in 2000.

978-1-4673-0438-2/12 $31.00 © 2012 IEEE

Panel Discussion

SoC Realization – A Bridge to New Horizons or a Bridge to Nowhere?

Organizer

Sathyam K. Pattanam
Senior Director, Atrenta, India

Abstract

System on Chip (SoC) design promises to revolutionize a vast array of products and markets. Access to advanced semiconductor technology is opening up new markets and facilitating innovation at a rate which we have not seen before. All of this bodes well for a vibrant semiconductor industry its associated EDA industry.

Yet, there is a problem. The cost of design for SoC devices is growing at a rapid pace. Complexity is contributing to this cost rise. Because of shrinking market windows, it also becomes necessary to reuse semiconductor IP, either from third party sources or prior internal designs – there is just not enough time to design much of these chips from scratch. But the quality and completeness of this IP is often not completely known, further contributing to the cost and schedule uncertainty of SoC devices.

Structured design methodologies that focus on early identification and correction of design issues, rigorous methods to qualify semiconductor IP and automated approaches to assembling the components of an SoC promise to address many of the cost and schedule challenges. This early analysis methodology has been called "SoC Realization". In this panel discussion, we will explore the meaning of SoC Realization and discuss its impact on the cost and schedule for advanced SoC designs.

Panelists

1. P.P. Chakrabarti, *IIT Kharagpur, India* (Moderator)
2. Mahesh Mahendale, TI Fellow and Director, *Center of Excellence for VLSI Architectures, Texas Instruments, India*
3. Srikanth Jadcherla, Chairman and CEO, *Seer Akademi, USA*
4. Vikas Gautham, Director, Verification products, *Synopsys, India*
5. Raju Bala Showry Pudota, Group Director of IP Development, *Cadence, India*
6. Sathyam K. Pattanam, Senior Director, GenSys SoC and SpyGlass Low-Power Products, *Atrenta India*

978-1-4673-0438-2/12 $31.00 © 2012 IEEE

Random Access Analog Memory (RA^{2}M) for video signal application

Nilanjan Chattaraj
Electronics and Electrical Communication Engineering
Department
Indian Institute of Technology Kharagpur
Kharagpur, India
nilanjan.chattaraj@ieee.org

Anindya Sundar Dhar
Electronics and Electrical Communication Engineering
Department
Indian Institute of Technology Kharagpur
Kharagpur, India
asd@iitkgp.ac.in

Abstract— **This paper proposes a novel memory architecture, introducing Random Access Analog Memory (RA^{2}M), to store unquantized samples of video signal of maximum 5 MHz bandwidth for storing time duration in order of millisecond by implementing periodic memory refreshing mechanism in it. At 16.5 MHz sampling frequency with 25 frames/s frame rate, this implemented design can store voltage signal sample of up to 200 mV for 40 ms with 8 bit resolution. The proposed architecture contains unit RA^{2}M cell of 250 fF capacitance occupying 21 μm × 21 μm area with 4.1 mW average power dissipation per cell in 0.18 μm standard CMOS fabrication process. The improvement in signal storage time duration into analog memory by introducing periodic memory refreshing mechanism in voltage mode is implemented for the first time. The circuit implementation is based on switched capacitor technique and is compatible with conventional fabrication process. This architecture facilitates random location data accessibility and includes common mode noise rejection by its differential signal implementation.**

Keywords- Random Access Analog Memory; analog memory; sampled data architecture; semiconductor memory; memory for video application.

I. INTRODUCTION

Analog memory can be classified in two categories. One is for storing electrical signal for short time duration i.e., in order of nanosecond, using capacitor charging technique with conventional MOS transistor and the other one is for storing electrical signal for long time duration i.e., in order of years using floating gate transistor. Signal storage for short time duration using capacitor charging technique [1]-[7] is compatible with conventional fabrication process and is faster than those designed with floating gate transistor. But the memory leakage problem of capacitor charging technique using conventional MOS transistor refrains from storing electrical signal for long time duration. Additionally high susceptibility to noise also affects the resolution in this technique. Whereas the analog memory based on floating gate transistor [8]-[13] makes charge leakage so small, that this can store signal for several years without any noticeable loss in signal voltage. But requirement of special type of fabrication process for implementing floating gate transistor and its

inherently slower memory operations due to its dependence on quantum tunneling effect refrain from storing electrical signal for low time duration. The publication [14] based on capacitor charging technique in current mode and with memory refreshing mechanism does not contain any specifications of analog memory explicitly, and is organized in multiple-input - multiple-output fashion, instead of having all input cells multiplexed to a single input bus and all output cells multiplexed to a single output bus. The same publication also suffers from the limitation of using variable reference voltages corresponding to variable sample voltage for memory refreshing. So according to the literature, the existing analog memories can store electrical signal either for short time duration i.e., in order of nanosecond or for long time duration i.e., in order of year with the above mentioned limitations. But neither of the above types is suitable to store analog signal for medium time duration i.e., in order of milliseconds with conventional fabrication process compatibility and fast memory operations. This is a proposal of new architecture of analog memory to store analog signal for medium time duration, i.e., in order of millisecond by implementing periodic memory refreshing in voltage mode signaling. This is based on conventional capacitor charging mechanism and switched capacitor circuit. The circuit implementation is compatible with conventional fabrication process and memory operations are faster here as this technique uses conventional MOS transistor instead of floating gate MOS transistor.

This is a proposal of new architecture of analog memory, i.e., Random Access Analog Memory (RA^{2}M) to store video signal for time duration in the order of millisecond. This analog memory is based on conventional capacitor charging technique and the memory leakage problem due to capacitor discharging effect is compensated by implementing periodic memory refreshing mechanism manipulating the time constants of the memory cell. This architecture is based on voltage mode signaling with single-input - single-output data bus compatibility and the memory refreshing requires single reference voltage for all sample voltages unlike [14]. This RA^{2}M architecture is compatible with conventional fabrication process and provides faster memory operations than floating gate transistor based analog memory. Additionally this architecture provides random location data accessibility and common mode noise rejection by differential signaling. This is

978-1-4673-0438-2/12 $31.00 © 2012 IEEE

based on sampled data architecture and can be used as storage device in application specific video signal application. This will be able to store unquantized samples of video signal of maximum 5 MHz bandwidth. These are implemented using switched capacitor technique. In this architecture, 16.5 MHz sampling frequency and 25 frames/s frame rate give 8 bit resolution for up to 200 mV input signal voltage in standard 0.18 μm CMOS fabrication process. Resolution and input signal voltage handling capability can be increased by improving the performance of differential-input differential-output operational amplifier. This RA^2M can be used in several video applications where high speed, low cost, and reduced size of memory chip are preferred negotiating with resolution. For its less accuracy this is not preferred for numerical calculation.

II. RA^2M ARCHITECTURE

The video signal is sampled differentially around a common mode reference voltage and then is stored as differential voltage into the RA^2M cells. Each RA^2M cell contains single floating capacitor. A single block, designed with switched capacitor circuit and differential-input differential-output opamp, serves both the purposes of differential sampling and periodic memory refreshing. This architecture provides row and column addressing like digital memory. The complete view of RA^2M architecture is shown in Fig. 1.

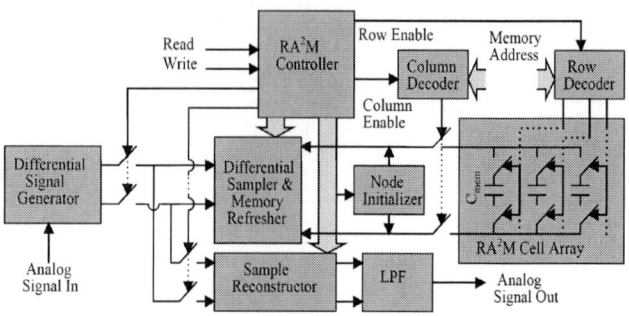

Fig. 2. Conceptual view of RA^2M architecture

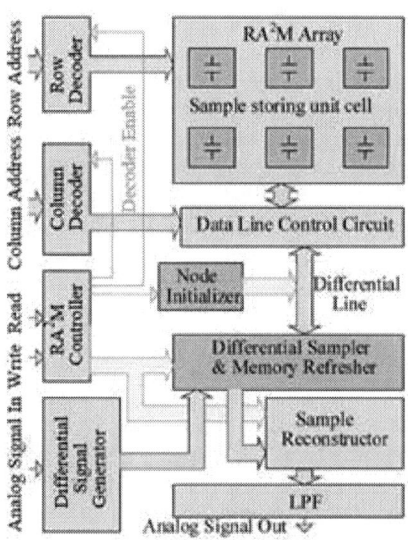

Fig. 1. The RA^2M architecture

The logical architecture of RA^2M (shown in Fig. 2) consists of six basic building blocks: differential signal generator, differential sampler & memory refresher, RA^2M memory cell, sample reconstructor, node initializer and RA^2M controller. A single block designed with switched capacitor circuit serves differential sampling and memory refreshing both. The video signal is fed to differential signal generator block for generating differential signal around a common mode reference voltage and then followed by differential sampling it is stored into the selected RA^2M cell. Node initializer initializes the nodes at reference voltage before each memory operation. The memory cells discharge with time, so those

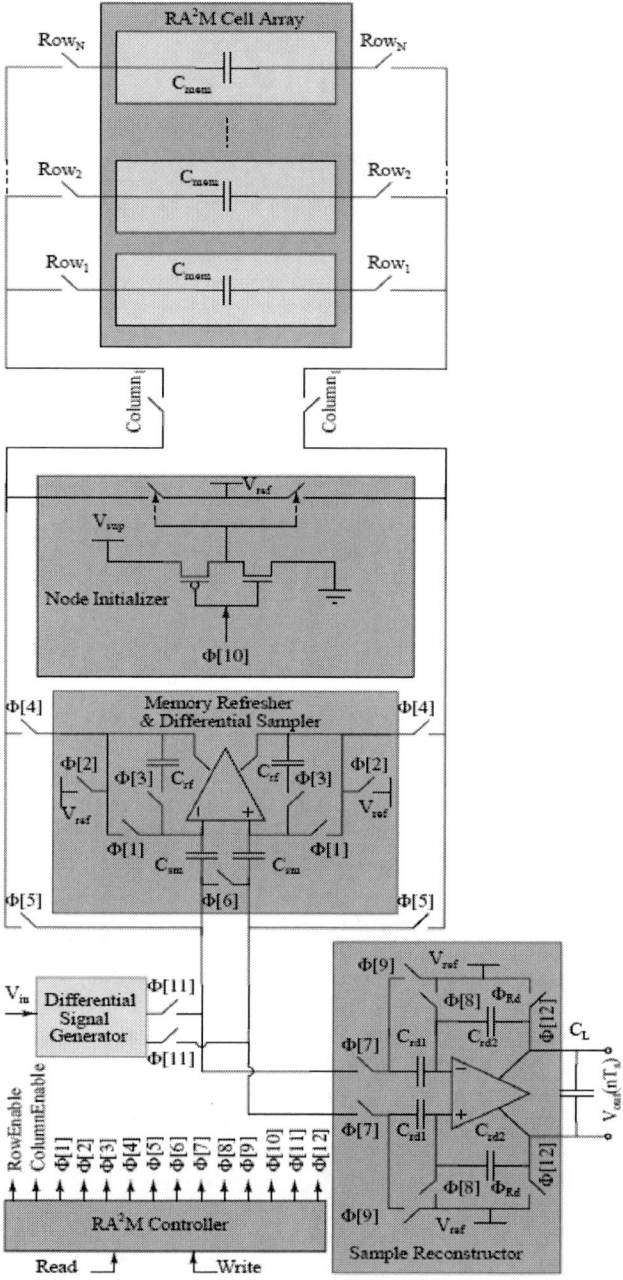

Fig. 3. Architecture of single column of RA^2M

978-1-4673-0438-2/12 $31.00 © 2012 IEEE   40

discharged samples are periodically fetched into the memory refresher block to reproduce those original samples and are restored again. While behaving as memory refresher, the capacitors inside the memory refresher block are calibrated to meet the gain requirement to reproduce the sample voltage. At the time of memory read, the sample reconstructor block fetches samples from memory cells and then reconstructs the original signal in differential signal mode to eliminate the effect of common mode noise.

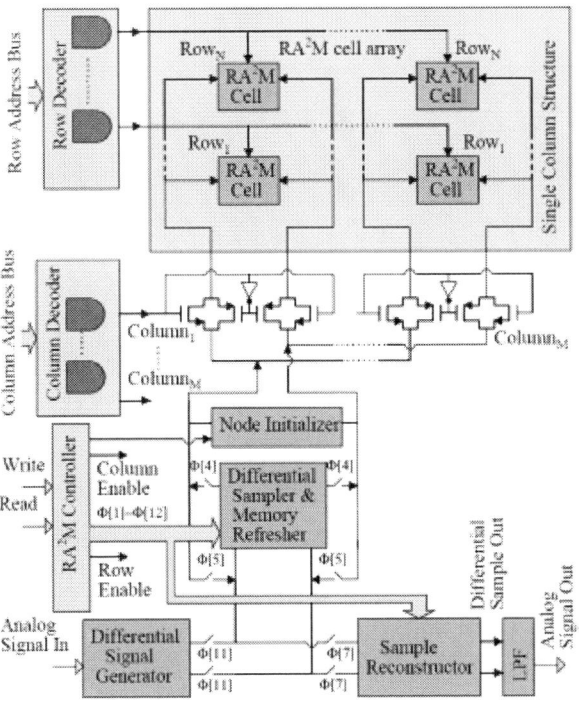

Fig. 4. Multiple column accommodation in RA^{2}M architecture

The single column structure of RA^{2}M architecture is shown in Fig. 3 and Fig. 4. The RA^{2}M controller is consisted of sequential and combinational logic circuits and provides all controlling signals to switches during memory operations. For its similarity with digital memory controller, it is not discussed here. The differential-input differential-output opamp used in this architecture is shown in Fig. 5. The open loop gain of that opamp is 50 dB with UGB 450 MHz.

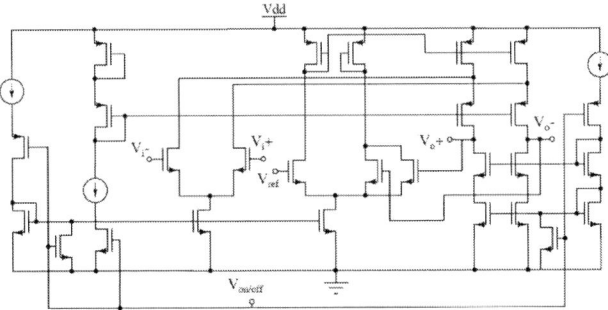

Fig. 5. Differential-input differential-output folded cascode opamp.

III. MEMORY OPERATION

Memory read/write along with periodic memory refresh operation in RA^{2}M can be easily understood studying a single column of this architecture. The RA^{2}M can store signal for long time by refreshing memory cells periodically. This has four states as shown in Fig. 6. In idle state, no operation is performed on RA^{2}M. After completing each memory write/refresh/read operation, it returns back to its idle state. Only one memory operation is done at a time. The maximum time, to write samples into RA^{2}M cells, is equal to the frame refreshing period T.

TABLE I
SIGNIFICANCE OF THE SYMBOLS USED IN MATHEMATICAL EXPRESSIONS

Symbol	Significance
n	instance of sample
T_s	sampling time period
$V_i(nT_s)$	n^{th} sample of input voltage signal
T_c	capacitor charging time during write/refresh/read
ΔT_c	inter pulse interval
R_s	average on resistance of CMOS transmission gate
C_{mem}	the capacitance of RA^{2}M cell
C_{sm}	sampling capacitor, equal to memory cell capacitor
C_{rf}	amplifying capacitor of memory refresher block
G	open loop gain of differential ended opamp
M	maximum number of memory cell refreshing
T	frame refreshing time period
τ_{off}	average switched off time constant
k	simulated gain of memory refresher block
k_0	theoretical gain of memory refresher block
Δk	deviation in gain of memory refresher block
b	simulated gain of sample reconstructor block
b_0	theoretical gain of sample reconstructor block
Δb	deviation in gain of sample reconstructor block
C_{rd1}	capacitor of sample reconstructor block
C_{rd2}	capacitor of sample reconstructor block
v_{mem}	differential sample voltage, stored across RA^{2}M cell
N	equivalent digital bit accuracy
C_{int}	capacitance of inter connection
$v_{out}(nT_s)$	n^{th} sample of input voltage signal at the output of memory reconstructor block during sample read
n	instance of sample
T_s	sampling time period
$V_i(nT_s)$	n^{th} sample of input voltage signal
T_c	capacitor charging time during write/refresh/read

A timer will count the time and once the value T is reached, the memory cell refreshing starts, and the timer count is set to zero. To read all samples, stored into the memory cells, it takes similarly maximum time of T. As shown in Fig. 7, each memory write/refresh/read operation takes 3 clock cycles i.e., $3T_c$ to complete each memory operation. Here, T_c is the switch-on time duration of each fundamental signal, as is shown in Fig. 7, Fig. 8, Fig. 9 and Fig. 10. To complete each memory operation $3T_c$ time is required. So if any write/refresh/read signals are enabled but if time is insufficient to perform that operation, in that case, the state of that memory will enter into idle state for that moment and then will go to the next state as per availability of time. This analog memory gives best performance for random memory access, based on the random algorithm such that the lifetime of the content is known. The performance can be improved here as the compensation for each discharged sample will be equal in this case by means of equally spaced memory cell refreshing in time domain. Fully random memory access is also possible but that deteriorates the performance. The controlling signals of memory write/refresh/read operations are shown in Fig. 8, Fig. 9 and Fig. 10. If $2v_i(nT_s)$ is the differential sample voltage of nth sample of a signal , assuming all capacitances and resistances as lump parameters, the amplification factor of memory refresher block is modeled as (1). The term Δk is the deviation in gain and is defined by (2). Similarly at the time of data read, after M memory refresh cycle, the amplification factor of sample reconstructor block is given by (3). The term Δb is the gain deviation of this block is defined by the equation (4). The final output voltage (5) contains the actual sample voltage along with a second term as error voltage, which is linearly proportional to the input signal voltage and increases accordingly.

$$k_0 = \frac{(G+1)}{\left[e^{-\frac{T}{\tau_{off}}}\frac{(2C_{mem}+C_{int})}{C_{sm}+(2C_{mem}+C_{int})}\left(1-\left(1-\frac{2}{5}\frac{C_{sm}+(2C_{mem}+C_{int})}{(2C_{mem}+C_{int})}\right)e^{-\frac{(C_{sm}+(2C_{mem}+C_{int}))T_c}{5R_sC_{sm}(2C_{mem}+C_{int})}}\right)\times\left(1-e^{-\frac{T_c}{3R_s(2C_{mem}+C_{int})}}\right)G\right]-1} \tag{1}$$

$$\Delta k = (k-k_0) = \left(\left.\frac{C_{sm}}{C_{rf}}\right|_{actual} - \left.\frac{C_{sm}}{C_{rf}}\right|_{theoritical}\right) \tag{2}$$

$$b_0 = \frac{(G+1)}{\left[2e^{-\frac{t}{\tau_{off}}}\frac{(2C_{mem}+C_{int})}{C_{rd1}+(2C_{mem}+C_{int})}\left(1-\left(1-\frac{1}{5}\frac{C_{rd1}+(2C_{mem}+C_{int})}{(2C_{mem}+C_{int})}\right)e^{-\frac{(C_{rd1}+(2C_{mem}+C_{int}))T_c}{5R_sC_{rd1}(C_{mem}+C_{int})}}\right)\times\left(1-\frac{1}{3}e^{-\frac{T_c}{3R_sC_{sm}}}\right)\left(1-e^{-\frac{T_c}{3R_s(2C_{mem}+C_{int})}}\right)\left(1+\frac{\Delta k}{k_0}\right)^M G\right]-1} \tag{3}$$

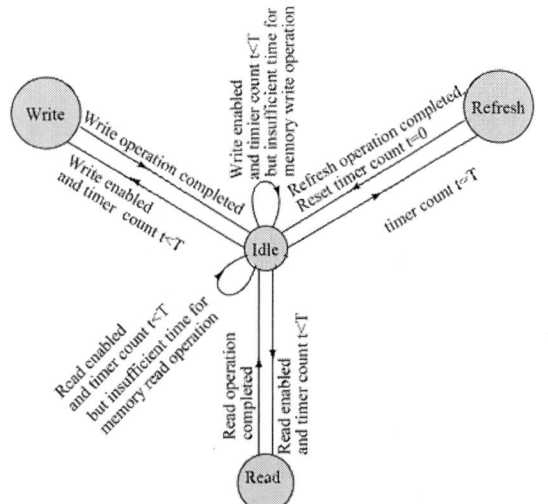

Fig. 6. State diagram of memory operations

		Switch1:Φ[1]	Switch2:Φ[2]	Switch3:Φ[3]	Switch4:Φ[4]	Switch5:Φ[5]	Switch6:Φ[6]	Switch7:Φ[7]	Switch8:Φ[8]	Switch9:Φ[9]	Switch10:Φ[10]	Switch11:Φ[11]	Switch12:Φ[12]	ColumnEnable	RowEnable
Controlling signals	**Write**														
	WriteTimeInitialize	√	√	√	√	√	√				√			√	
	WriteFirstPhase	√	√	√							√	√			
	WriteSecondPhase				√	√					√			√	√
	Refresh														
	RefreshTimeInitialize	√	√	√	√	√	√				√			√	
	SampleRead	√	√	√		√					√			√	√
	SampleRefresh				√	√		√			√			√	√
	Read														
	ReadTimeInitialize		√		√	√		√	√	√	√			√	
	ReadFirshPhase				√			√	√		√		√	√	√
	ReadSecondPhase									√					

Fig. 7. Combinational logic of RA²M controller

$$\Delta b = b-b_0 = \left(\left.\frac{C_{rd1}}{C_{rd2}}\right|_{actual} - \left.\frac{C_{rd1}}{C_{rd2}}\right|_{theoritical}\right) \tag{4}$$

$$v_{out}(nT_s) = v_i(nT_s) + \frac{\Delta b}{b_0} v_i(nT_s) \qquad (5)$$

Comparing the error voltage of (5) with quantization noise, we find the equivalent N bit resolution (6).

$$N = \frac{1}{\ln 2} \ln \left(\frac{V_{sup} b_0}{2\Delta b v_i(nT_s)} \right) \qquad (6)$$

Considering memory cell charging is up to 99.9 % of the sample voltage and interconnect capacitance $C_{int}(N)$ is a function of number of memory cells connected to a column, the maximum number of memory cells per column, which can be handled by the memory refresher block, is given by

$$C_{int}(N) \le \left(\frac{T_c}{21R_s} - 2C_{mem} \right). \qquad (7)$$

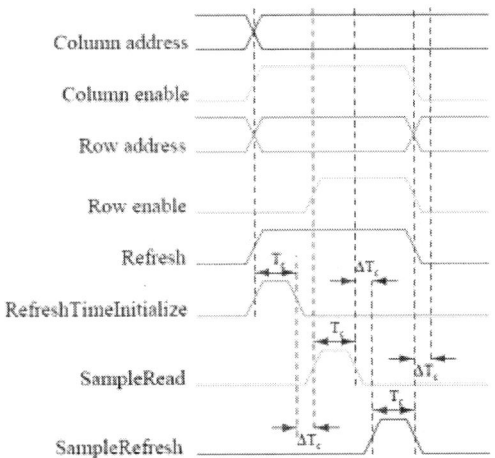

Fig. 8. Timing diagram of memory write operation

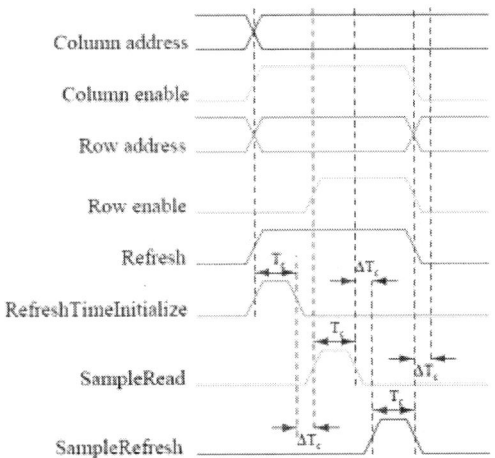

Fig. 9. Timing diagram of memory refresh operation

IV. SIMULATED RESULTS

The simulated result is tabulated in TABLE II based on 0.18 µm standard CMOS fabrication process. The Fig. 11

plots resolution of the output voltage of RA²M architecture in terms of equivalent digital bit with respect to sample voltage.

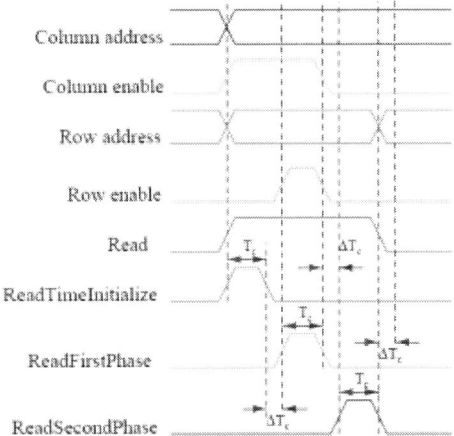

Fig. 10. Timing diagram of memory read operation

The performance falls at higher voltage level for several reasons, one is because of the limited output swing of that opamp. It maintains minimum resolution equivalent to 8 bit for up to 200 mV input signal voltage in 0.18 µm standard CMOS fabrication process. The effect of noise on the output sample voltage under process variation is shown from Fig. 12 to Fig. 19. The proposed architecture contains unit RA²M cell of 250 fF capacitance occupying 21 µm × 21 µm area with 4.1 mW average power dissipation per memory cell.

TABLE II
SIMULATED RESULT OF PROPOSED ARCHITECTURE

Parameter	Targeted Specifications	Simulation Results
Supply voltage	1.8 V	1.8 V
Common mode reference voltage	0.9V	0.9V
Input video signal voltage swing	0 V - 400 mV	0 V - 200 mV
Video signal band width	5 MHz	5 MHz
Sampling frequency	15 MHz	16.5 MHz
Frame rate	25 frames/s	25 frames/s
Sample hold time into RA²M cell	40 ms	40 ms
Memory refresh period	10 ms	10 ms
Read signal period	40 ms	40 ms
Resolution	8 bit	8bit equivalent

CONCLUSION

The proposed RA²M architecture is implemented with all the advantages of existing analog memory added with the advantages of signal storage capability of time duration in order of millisecond by implementing memory refreshing mechanism. This makes memory read/write operations faster compared to the analog memory, designed with floating gate transistor and is compatible with conventional fabrication process. As this is implemented by differential signaling, common mode noise is removed. This provides random location data accessibility also. This type of analog memory i.e., RA^{2}M can be used in several video applications where high speed, low cost, and reduced size of memory chip are preferred negotiating with resolution.

For its less accuracy than digital memory this is not preferred for numerical calculations. As a conclusion, in spite of having some limitations, several advantages of RA^{2}M motivate us to develop this architecture further for storing other natural signals also with high resolution as well as to produce more economic memory chip for several appliances in future.

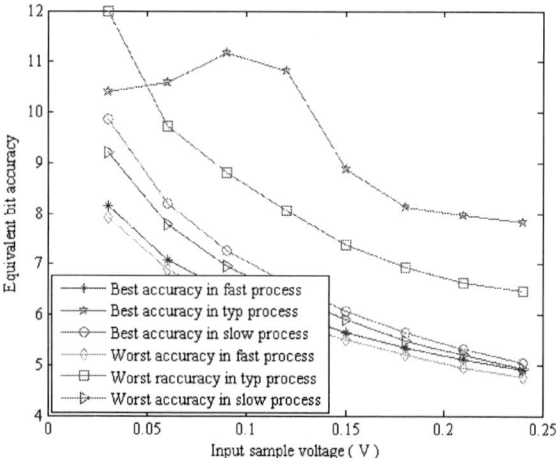

Fig. 11. Comparison of best and worst equivalent bit accuracy of RA^{2}M architecture after 40 ms, under typical fabrication process, based on memory read time instance for refreshing time period of 5 ms

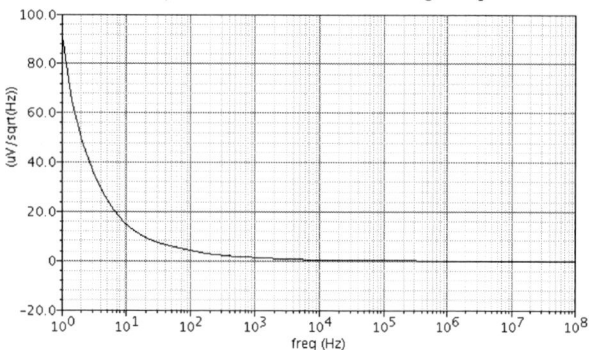

Fig. 12. Noise voltage variation with frequency of RA^{2}M architecture for memory refreshing time period of 5 ms in typical fabrication process

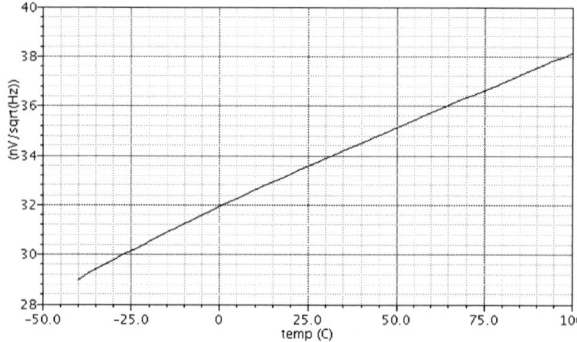

Fig. 13. Noise voltage variation with temperature of RA^{2}M architecture for memory refreshing time period of 5 ms in typical fabrication process

REFERENCES

[1] S. Hong, and W. E. Stark, "Performance effects of using analog memory in baseband signal processing systems design," IEEE International Symposium on Circuits and Systems, vol. 1, pp. 703 - 706 , May 2001

[2] K. L. Lee, A. A. Arthur, R. W. Jones, H. S. Matis, M. Nakamura, S. A. Kleinfelder, H. G. Ritter, and H. H. Wieman, "Analog-to-digital Conversion Using Custom Cmos Analog Memory For The EOS Time Projection Chamber," IEEE Nuclear Science Symposium, pp. 631 – 634, Oct. 1990

[3] S. A. Kleinfelder, "A 4096 cell switched capacitor analog waveform storage integrated circuit," IEEE Trans. Nucl. Sci., vol. 37, no. 3, pp. 1230 - 1236 , June 1990

[4] A. E. Stevens, R. P. Van Berg, J. Van der Spiegel, and H. H. Williams, "A time-to-voltage converter and analog memory for colliding beam detectors," IEEE J. Solid-State Circuits, vol. 24, no. 6, pp. 1748 – 1752, Dec. 1989

[5] G. Anelli, F. Anghinolfi, and A. Rivetti, "A large dynamic range radiation tolerant analog memory in a quarter micron CMOS technology," IEEE Nuclear Science Symposium, vol. 2, pp. 9-14 , Oct. 2000

[6] E. Delagnes, F. Feinstein, P. Goret, P. Nayman, J.P. Tavernet, F. Toussenel, and P. Vincent, "A Multigigahertz Analog Memory with Fast Read-out for the H.E.S.S.-II Front-End Electronics," IEEE Nuclear Science Symposium , vol. 1, pp. 332 – 336, Oct. 2006-Nov. 2006

[7] D. Gerna, M Brattoli, E. Chioffi, G. Colli, M. Pasotti and A. Tamasini, "An Analog memory For a QCIF format image frame storage" , IEEE International Symposium on Circuits and Systems, vol. 1, pp. 289 – 292, May 1996

[8] L.R. Carley, "Trimming analog circuits using floating-gate analog MOS memory," IEEE J. Solid-State Circuits, vol. 24, no. 6, pp. 1569 - 1575 , Dec. 1989

[9] T. Blyth, S. Khan, and R. Simko, "A Non-volatile Analog Storage Device Using EEPROM Technology," IEEE International Solid-State Circuits Conference, pp. 192 – 315, Feb. 1991

[10] E. Sackinger, and W. Guggenbuhl, "An analog trimming circuit based on a floating-gate device," IEEE J. Solid-State Circuits, vol. 23, no. 6, pp. 1437 – 1440, Dec. 1988

[11] Yong-Yoong Chai, and L.G. Johnson, "A 2×2 analog memory implemented with a special layout injector," IEEE J. Solid-State Circuits, vol. 31, no. 6, pp. 856 – 859, June 1996

[12] Kyu-Hyoun Kim, Kwyro Lee, Tae-Sung Jung, and Kang-Deog Suh, "An 8-bit-resolution, 360-µs write time nonvolatile analog memory based on differentially balanced constant-tunneling-current scheme (DBCS)," IEEE J. Solid-State Circuits, vol. 33, no. 11, pp. 1758 - 1762 , Nov. 1998

[13] R.R. Harrison, P. Hasler, and B.A. Minch, "Floating-gate CMOS analog memory cell array," IEEE International Symposium on Circuits and System, (ISCAS) vol.2, pp. 204 – 207, May 1998 –June 1998

[14] D. A. Panagiotopoulos, R. W. Newcomb, and S. K. Singh, "A current-mode analog memory for neurocomputing," Midwest Symposium on Circuits and Systems, vol. 2, pp. 787 – 790, Aug. 1999

2012 25th International Conference on VLSI Design

A 55-mW 300MS/s 8-bit CMOS Parallel Pipeline ADC

Manas Kumar Hati

Advanced Technology Development Centre
Indian Institute of Technology
Kharagpur, India-721302
E-mail: mkhati@ece.iitkgp.ernet.in

Tarun K. Bhattacharyya

Dept. of E&ECE
Indian Institute of Technology
Kharagpur, India-721302
E-mail: tkb@ece.iitkgp.ernet.in

Abstract—This paper describes 8-bit 300MS/s 7-stages parallel pipeline ADC with 1.5-bit per stage and power efficient architecture is designed by sharing an amplifier between two pipeline stages, introducing the proper clock timing between the two parallel stages. This architecture is realized by eight no. of amplifier and the sample hold architecture is designed by using double sampling sample hold (DSSH) technique. A wide swing and wide bandwidth regulated folded cascode and power efficient dynamic comparator has been developed to further reduce the power consumption of the pipeline ADC. The ADC is implemented in 0.18 µm CMOS process technology, achieves 57.40 dB spurious free dynamic range (SFDR), 49.078 dB signal to noise distortion ratio (SNDR), 7.86 effective no of bit (ENOB) and consumes 55 mW from 1.8 V supply. The resulting figure of merit (FOM) is 0.789 PJ/conversion step.

Keywords- Analog CMOS circuits, folded cascode OTA, DSSH, dynamic comparator, RSD block.

I. INTRODUCTION

High-speed medium resolution analog-to-digital converters (ADCs) are widely used in commercial application including data communication, instrumentation and consumer electronics. ADC is achieving a resolution of approximately 8-bits and sampling rates well above the hundreds of MHz find application in the Gigabit Ethernet and code conversion for flat-panels displays. In such applications, the reduction of power consumption associated with high-speed sampling is one key design issue in enhancing portability and battery operation.

The most promising topology for a high-resolution A/D converter implemented in a CMOS process is the pipeline architecture [1]-[3]. Unfortunately, the performance of the time interleaved ADCs are sensitive to offset and gain mismatch as well as aperture errors between the interleaved channels. Digital foreground calibration can remove the offset mismatch [4], [5] but remains interrupting the conversion of the input.

Employing double sampling and parallelism with the time interleaved pipeline ADCs, a very competitive power and area consumption can be obtained [7]. It is obvious that the parallelism enhances the conversion rate and lowers the slew rate. Several enhancement techniques have been used to enhance the sampling speed and accuracy [5], [6]. The performance of the ADC can be measured with three of its most important metrics –power dissipation, sampling speed

and effective number of bits. A figure of merit "FOM" can be obtained using these metrics as $FOM = \dfrac{P_{diss}}{2^{ENOB} \cdot f_{sample}}$.

The designed pipeline ADC's architecture produces most effective or impressive figure of merit among other architecture of ADCs. This paper discusses a parallel pipeline ADC with sampling speed 300 MS/s and resolution 8-bits. By appropriately sharing the OTA between the two pipeline stages power consumption and die area can be significantly reduced [8].

Section II described the architecture of the ADC. Section, III deals with the clock generation circuit. In section, IV discusses on generation of reference bias voltages. Section V described about the two main building block of ADC. Section VI presents the simulation results and conclusion is in the last section.

II. ADC ARCHITECTURE

As shown in Fig. 1 is parallel pipeline ADC architecture with two pipeline ADC's channel working in parallel.

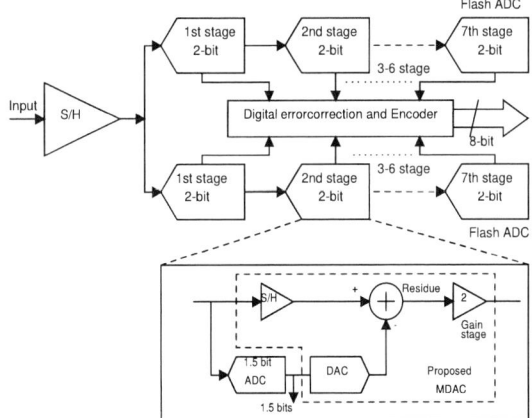

Figure 1: Parallel pipeline ADC Architecture.

Here in Fig.1 input signal is connected to a Double sampling sample hold circuit (DSSH) and the sampled signal transform into 1.5-bit code output. This output digital bits stream then transform into analog signal, subtracted from the input sampled signal and multiplied by a factor two to reach the full scale. There is no redundant sign digit (RSD) correction block after the last stage of the ADC so to get the actual value, the last stage is full flash type ADC. For reducing the common mode noise level here each stage is

978-1-4673-0438-2/12 $31.00 © 2012 IEEE   45

fully differential type also there is overlapped differential signal driver which transformed the single ended to double ended output signal as shown in Fig. 2 .

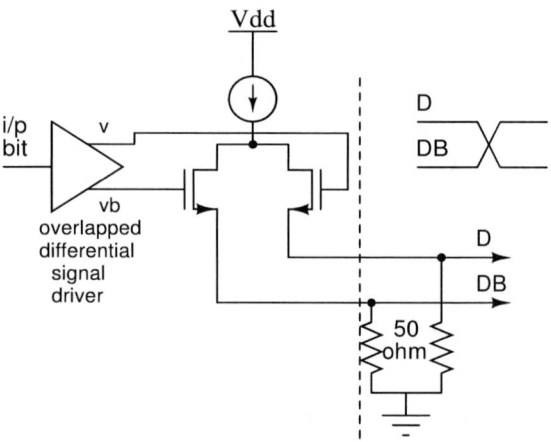

Figure 2: Output driver circuit used in the ADC.

The digital output from each channel is time interleaved and converted into overlapped differentials signals driving off-chip 50 Ω resistors. To get the data stream exactly what it is coming a redundant sign digit (RSD) correction has been implemented [8]. The single front rank S/H circuit with bootstrapped NMOS sampling switches minimizes the gain mismatch of the ADC and enhances the input signal bandwidth [9]. The magnitude of the sampling capacitors in S/H circuits is designed considering the KT/C noise [10].

Conventional transmission gates with any usable signal swing were not directly realizable because of the PMOS and NMOS transistors have different threshold voltage. Therefore, a bootstrap switch was required. The switch is single NMOS switch M10 as shown in Fig. 3. In the off state, the gate of M10 is grounded and the device is in cutoff. In the on state, a constant voltage of V_{dd} is applied across the gate to source terminals and a low on resistance is established in the M10 drain to source terminal that is independent of the input signal. Although the absolute voltage of the gate terminal may exceeds V_{dd} due to positive input signal but none of the terminal-to-terminal device voltage exceeds V_{dd} [12].

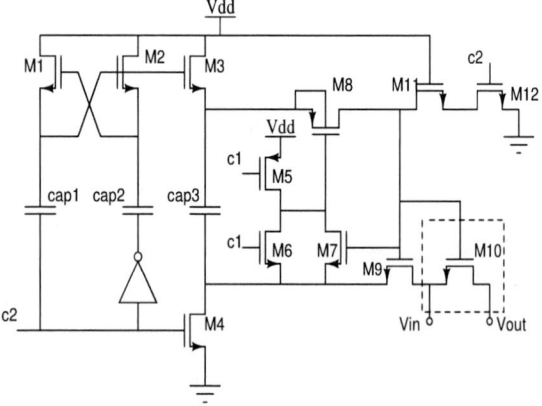

Figure 3: Bootstraps switch schematic.

A double sampling sample hold technique has been used to enhance the speed of the overall ADC and the DSSH has designed based on the skew insensitive technique as in Fig. 4.

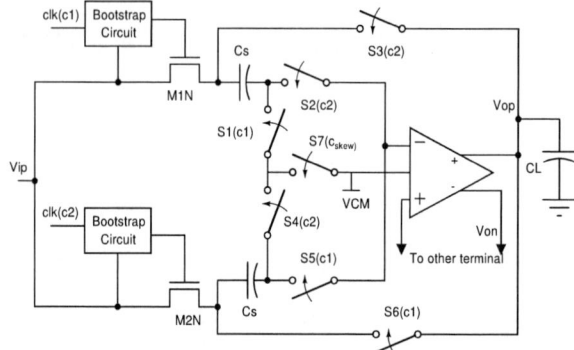

Figure 4: Skew insensitive double sampling S/H (DSSH) circuit.

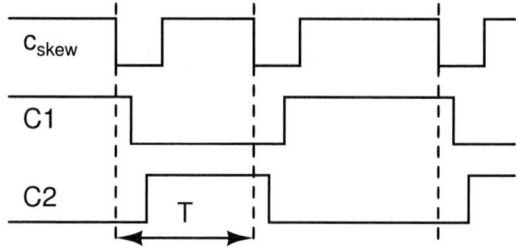

Figure 5: Clock timing diagram for DSSH circuit.

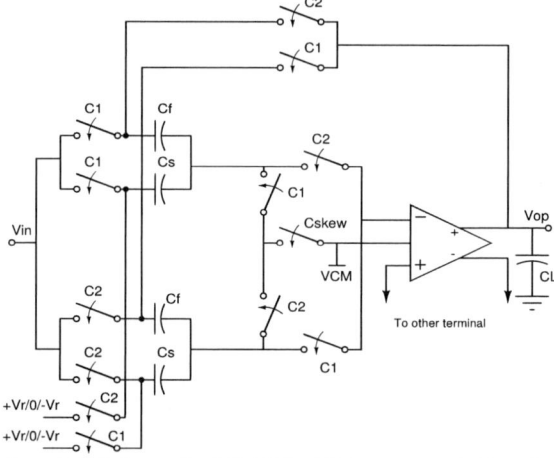

Figure 6: Switch capacitor MDAC for 1.5-bit per stage pipeline ADC.

The block diagram of the ADC is shown in Figure 1. It consists of seven stages, each of which produces 1-bit plus one redundant bit, which overlaps with the bits of the next stage. A flash ADC, providing 2-bits, follows the last pipeline stage. The incoming voltage signal is sampled by the S/H circuit and simultaneously digitized by the sub-ADC [13]. The result of the A/D conversion is immediately converted back to analog form and subtracted from the sampled hold signal. The resulting residue voltage is amplified by "two". In a switch capacitor realizing the S/H operation, the D/A conversion, the subtraction, and the

978-1-4673-0438-2/12 $31.00 © 2012 IEEE   46

amplification are all performed by a single circuit block called multiplying digital to analog converter (MDAC), actually which consists of OTA and set of switch capacitors. The consecutive stages operate in opposite clock phases as a result one-sample traverse two stages in one clock cycle.

The switch capacitor based MDAC implemented for 1.5-bit pipeline ADC is shown in Fig. 6. During the clock phase c1 input signal is connected to the almost equal capacitor named C_s and C_f. For the second phase c2 the C_f connected to the OTA output and the Cs connected to the reference voltage (QV$_{ref}$). The resulting output voltage can be written as

$$V_{out} = \frac{C_s + C_f}{C_f} V_{in} + Q \cdot \frac{C_s}{C_f} V_{ref} \qquad (1)$$

Where, $Q \in \{-1, 0, 1\}$

The second form of above equation can be written as

$$V_{out} = \begin{cases} 2V_{in} - V_{ref}; ..... Vin \geq \dfrac{V_{ref}}{4} \\[2mm] 2V_{in}; .... -\dfrac{V_{ref}}{4} \leq Vin \leq \dfrac{V_{ref}}{4} \\[2mm] 2V_{in} + V_{ref}; .... Vin \leq -\dfrac{V_{ref}}{4} \end{cases} \qquad (2)$$

The capacitor for MDAC circuit is calculated based on the noise model [10].

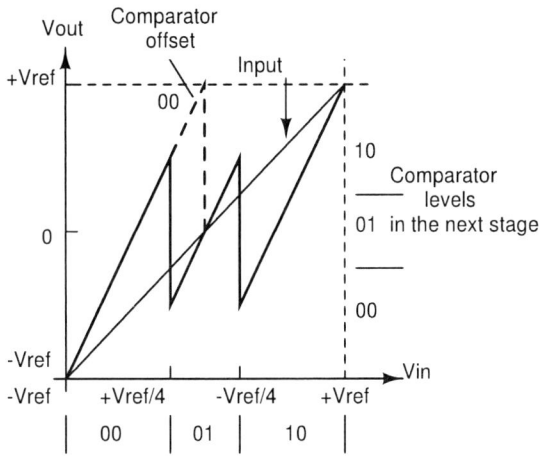

Figure 7: Transfer function of the 1.5-bit pipeline ADC.

The simplest pipeline stage is a 1-bit stage with one redundant quantization level, because of which it is often referred to as the 1.5-bit stage. Its transfer function is shown in Fig. 7. The signal range for both the input and output is form $-V_{ref}$ to $+V_{ref}$. Normally the comparator threshold level are set to $-V_{ref}/4$ to $+V_{ref}/4$ and the ADC output codes for the three regions are "00", "01" and "10". Due to comparator offset, the decision level shifted as shown in the Fig. 7. Consequently, the ADC output code remains "00" instead of

"01". However, residue remains in the input range of the next stage.

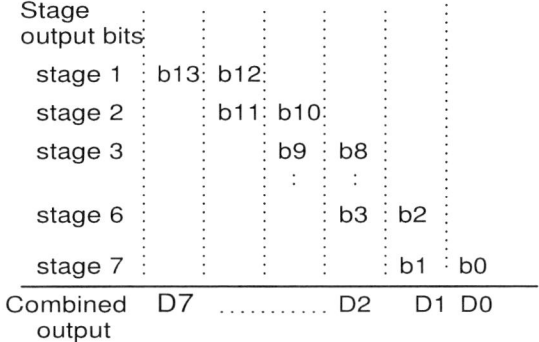

Figure 8: RSD correction in digital domain.

In the digital domain, the RSD correction [11] has been implemented as shown in Fig. 8. In addition, Fig. 9 shows the conceptual diagram of the digital error correction. The delay block is implemented with a p/n transmission gate and a static inverter; the adder is designed also by using p/n transmission gate. Delay blocks are properly clocked in such a way that it synchronizes the all data available at the output at the same time before get into full adder input. Each delay block has two-bit delay characteristics and clocked with half period of c1 or c2. Full adder addition technique is implemented properly to match the RSD algorithm for 2-bit per stage pipeline ADC is shown in Fig. 9.

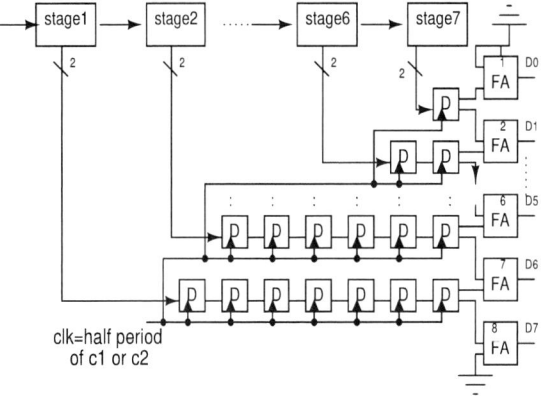

Figure 9: conceptual diagram for the digital error correction.

III. CLOCK GENERATOR

The new clock generator circuit is shown in Fig. 10. The circuit generating the non-overlapping signals is basically the same as used in the first S/H circuit. The short pulses for the common sampling switch are constructed with a circuit consisting of an inverter, a delay element, and a NAND gate. The D flip-flop generates the complementary half-speed clock signals. To reduce the jitter in the sampling clock (c1), the buffer chain can be made shorter by connecting the clock input of the D flip-flop directly to the incoming clock.

978-1-4673-0438-2/12 $31.00 © 2012 IEEE

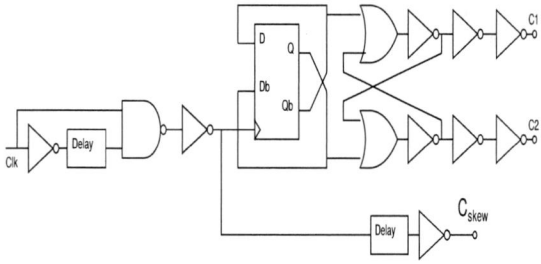

Figure 10: Clock generator for the skew insensitive circuit.

IV. RESISTOR STRING FOR VOLTAGE REFERENCE

A resistor string, dimensioned properly to provide proper tap voltages, is a simple solution for a voltage reference. To keep constant the VCM due to process and supply voltage variation a feedback amplifier and a PMOS switch is used as in Fig. 11. As a feedback amplifier, a single-stage differential pair gives sufficient gain, but, for stability reasons, an extra capacitor is typically needed at the amplifier output. This reference topology has been successfully implemented in single-chip direct conversion receivers [14], [15]. Error amplifier negative terminal voltage VCM can be achieved as the resistor string terminal voltage VCM by this error amplifier and PMOS switch.

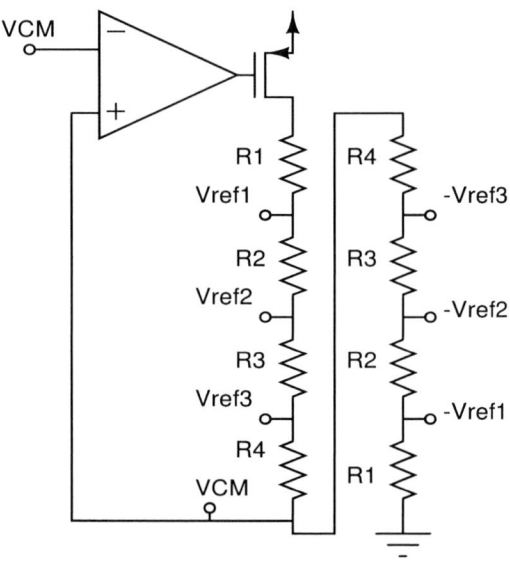

Figure 11: Reference voltage generating circuits.

V. CIRCUIT DESCRIPTION

In this section, we discussed on the two main building block of the pipeline ADC, OTA and Dynamic comparator.

A. Operational Transconductance Amplifier (OTA):

A high speed, low power, wide Gain bandwidth (GBW) regulated folded cascode with complementary input OTA has been designed for this ADC. The design criteria of this ADC are completely fulfilled by this OTA [10]. The schematic of the OTA is shown in Fig. 12. Bode plot of the

OTA shows that the dc gain is 90.39 dB and UGB frequency is 700.7 MHz with 500 fF load capacitor. Phase margin of the OTA is 63.85 deg. Table I shows the OTA performance metrics.

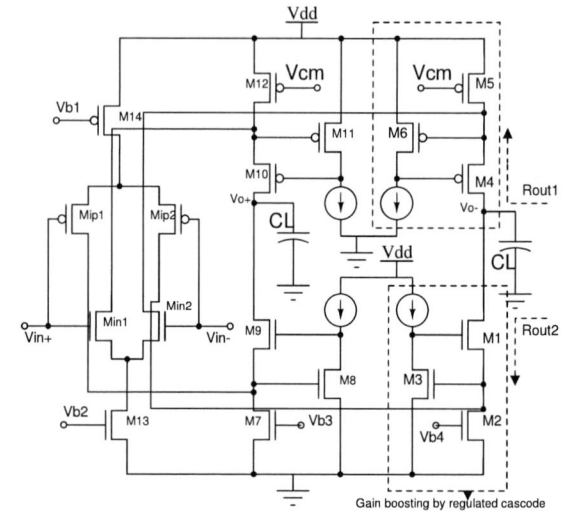

Figure 12: OTA Schematic view.

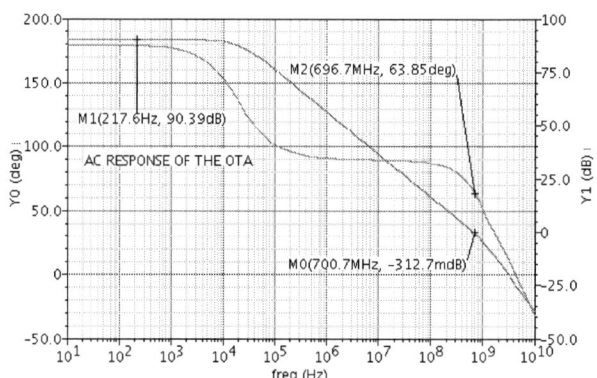

Figure 13: AC Response of the OTA.

TABLE I
OTA characteristics

Main characteristics	Achieved value
Power supply(V)	1.8
DC Gain(dB)	90.39
UGB(MHz)	700.7
Settling time(ns)	2
Power consumption(mW)	3.24

B. Dynamic Comparator

Dynamic comparator with differential input and output is suitable to reduce the area and power consumption of the ADC, due to large no of comparators are used in sub-ADC. Using 1.5-bits per stage one can tolerate a comparator offset of up to $\pm \dfrac{V_{ref}}{4}$, where V_{ref} is the reference voltage. In general, the comparators can tolerate an offset up to $\pm\dfrac{V_{ref}}{2^b}$ for a b bit stage. Notice that this comparator does not need multiple clocks or inverted clocks, one clock signal is sufficient to trigger transition from one phase to another and back again [8].

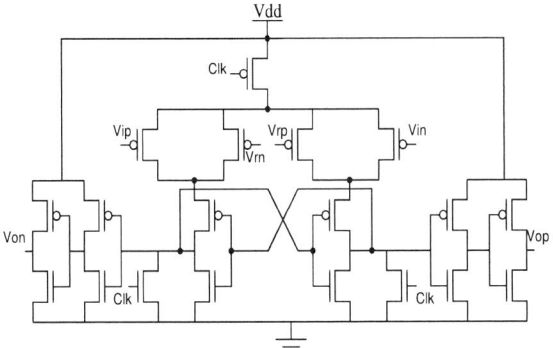

Figure 14: Dynamic Comparator schematic.

Some of the key parameters for this dynamic comparator are offset, delay and power dissipation. The offset needs to be within plus/minus one-fourth of the reference voltage, which in our case corresponds to ±75mV. We aimed for a speed of 300 MHz at 1.8V. This corresponds to a maximum delay of 400 ps from Clk goes low to output is valid. Here the offset of the comparator is very well below the required value. The drawback of the differential pair comparator is a large kickback noise. However, in a low-resolution pipeline stage this can usually be tolerated. The maximum power dissipation is 0.58 mW.

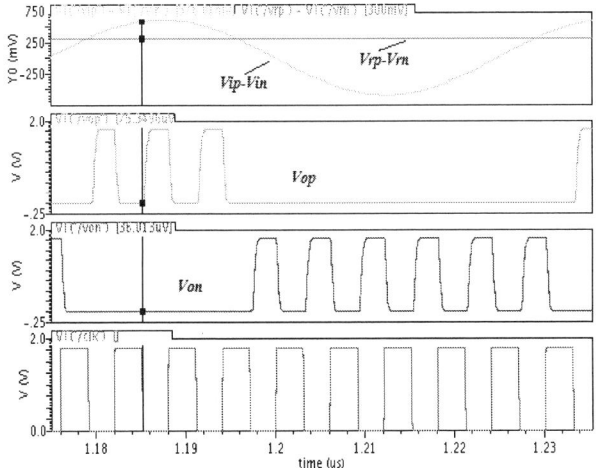

Figure 15: Transient response of the Dynamic comparator.

The transient response of the dynamic comparator is shown in Fig. 15. When Clk is "zero" its enable the output and remain in zero when Clk is in "on" state (Vdd). It shows that when $(V_{ip}-V_{in})$ is greater than $(V_{rp}-V_{rn})$ then the Output V_{op} is one (Vdd) and V_{on} is in zero state.

VI. SIMULATION RESULT OF THE ADC

The total power consumption of the ADC including on chip reference circuits is 55 mW from a 1.8 V supply. For static performance of the ADC, a ramp input with amplitude -300 mV to +300 mV was applied as the input signal to find out the differential nonlinearity (DNL) and integral non-linearity (INL) of the ADC. In addition, it includes a high accuracy DAC at the output of the designed ADC. It is found that -0.5LSB≤DNL≤+0.5LSB and -0.24LSB≤INL≤+0.4LSB. For finding the DNL it is very cleared from the ADC ramp input response as shown in Fig. 16. The maximum error is 0.5 LSB in magnitude. For the dynamic response of the ADC, a sinusoidal signal with frequency 1 MHz was fed at the input with 300 MS/s. Digital output of the ADC was converted to analog signal using an ideal DAC. In addition FFT plot is shown in Fig. 18. The proposed ADC achieved spurious free dynamic range (SFDR) of 57.40 dB and signal to noise and distortion ratio (SNDR) of 49.078dB. Fig. 19 shows the plot of SFDR and SNDR versus input frequency. The conversion rate is constant at 300 Msamples/s. The SFDR is down by 3 dB when the input frequency almost nears the 12MHz. Moreover the SNDR is down by 3 dB when the input frequency is 15MHz.

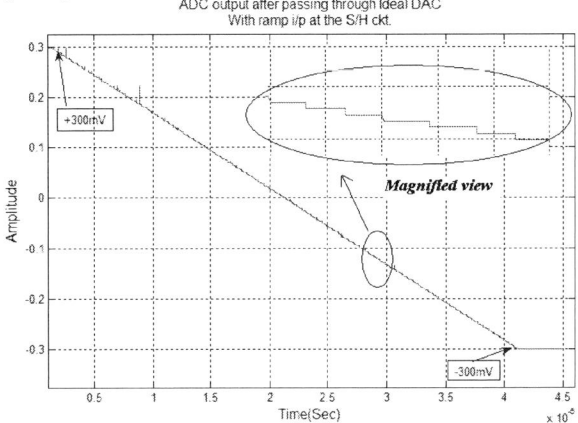

Figure 16: ADC output from an ideal DAC when the i/p signal is a ramp with amplitude +300 mV to -300 mV.

VII. CONCLUSION

This work presents the design techniques using double sampling with skew insensitive and bootstrapped switch used at the input of the S/H circuits. As we increase the input frequency the SFDR value gradually decreases and similarly with sampling frequency. The novel-clock generation scheme is expected to reduce the mismatch variation related to clock skew and compensate the gain mismatch of the parallel pipeline ADCs. The dynamic gain mismatch still restricts the ADC performance. The use of dynamic comparator and this sharing techniques also greatly reduces the overall power consumption of the ADC. The ADC consumes 55 mW at 300 Msamples/s from 1.8 V power supply.

978-1-4673-0438-2/12 $31.00 © 2012 IEEE

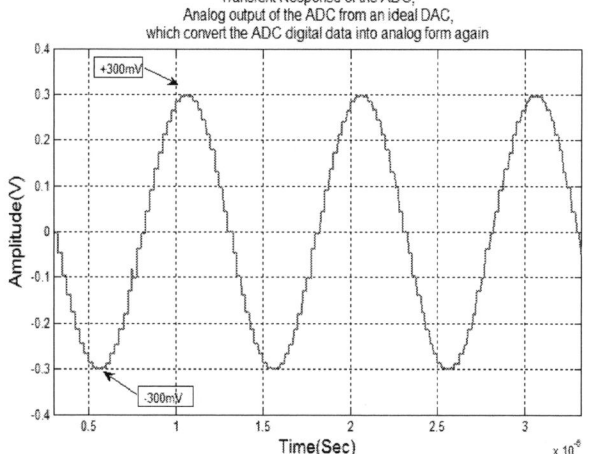

Figure 17: ADC output from a high accuracy DAC, when the input signal is sinusoidal with +300 mV to -300 mV.

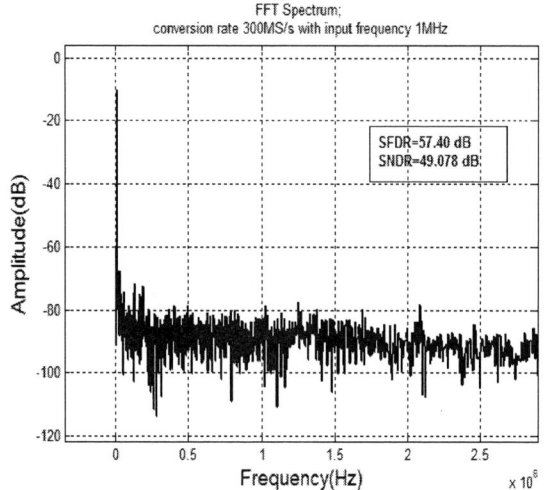

Figure 18: FFT plot of the ADC with input frequency 1 MHz and conversion rate 300 MS/s.

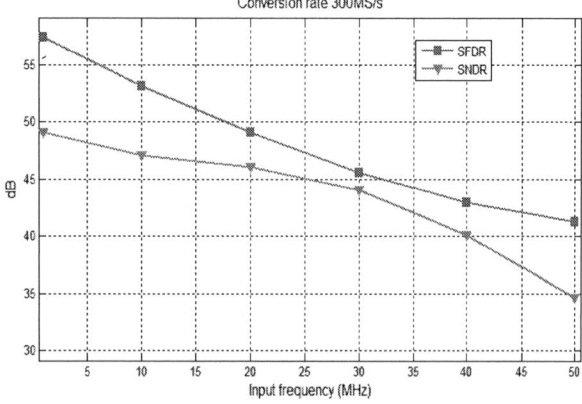

Figure 19: SFDR and SNDR versus input frequency.

ACKNOWLEDGMENT

The authors would like to thank the members of the Advanced VLSI Design Laboratory for their fruitful suggestions and National semiconductor U.S.A for giving the fabrication facility. Comments from the reviewers are also greatly appreciated.

REFERENCES

[1] L. Singer, S. Ho, M. Timko, and D. Kelly, "A 12-b 65-MS/s CMOS ADC with 82-dB SFDR at 120 MHz," in *ISSCC Dig. Tech.* Papers, pp. 38-39, Feb. 2000.

[2] J. Clerk Maxwell, Y. -T. Wang, and B. Razavi, "An 8-b 150-MHz CMOS A/D converter," *IEEE J. Solid-State Circuits*, vol. 35, pp. 308–317, Mar. 2000.

[3] K. C. Dyer, D. Fu, S. H. Lewis, and P. J. Hurst, "An analog background calibration technique for time-interleaved analog-to-digital converters," *IEEE J. Solid-State Circuits*, vol. 33, pp. 1912–1919, Dec. 1998.

[4] C. S. G. Conoroy, D.W. Cline, and P. R. Gray, "An 8-bit 85 MS/s parallel pipeline A/D converter in 1-um CMOS," *IEEE J. Solid-State Circuits*, vol. 28, pp. 447-454, Apr. 1993.

[5] L. Sumanen, M. Walteri, and K. A. I. Halonen, "A 10-bit 200 MS/s CMOS parallel pipeline A/D converter," *IEEE J. Solid-State Circuits*, pp. 1048-1055, July 2001.

[6] S. Limotyralis, S. D. Kulchycki, D. K. Su and B. A. Wooley, "A 150-MS/s 8-b 71-mW CMOS Time-Interleaved ADC," *IEEE J. Solid-State Circuits*, vol. 40, pp. 1057-1067, May 2005.

[7] W. BRIGHT, "8-B 75-MS/s 70 mW parallel pipelined ADC incorporating double sampling," in *ISSCC Dig. Tech* Papers, pp. 146-147, Feb. 1998.

[8] M. K. Hati and T. K. Bhattacharyya, "Design of Low power Parallel Pipeline ADC in 180nm standard CMOS Process," *IEEE International conference on communicatoion and signal processing (ICCSP)*, pp. 9-13, Feb. 2011.

[9] T. Cho and P. Gray, "A 10 b 20Msamples/s, 35mW Pipeline A/D converter," *IEEE J. Solid-State Circuits*, vol. 30, pp. 166-172, Mar. 1995.

[10] M. K. Hati and T. K. Bhattacharyya, "Design of a low power, high speed complementary input folded regulated cascode OTA for a parallel pipeline ADC," *IEEE Computer Society International symposium on VLSI (ISVLSI'11)*, pp. 114-119, July 2011.

[11] CMOS data converters for communication, by Mikael Gustavsson, J. Jacob Wikner and Nianxiong Nick Tan; Kluwer Academic Publishers, 2002.

[12] T. Cho and P. R. Gray, "A 10 b 20 Msamples/s, 35 mW pipeline A/D converter," *IEEE J. Solid-State Circuits,* vol.30, pp. 166–172, Mar. 1995; see also [Online]. Available: http://kabuki.eecs.berkeley.edu/~tcho/.

[13] Stephen H. Lewis, H. Scott Fetterman, G. F. Gross, R. Ramachandran, and T. R. Viswanathan,"A 10-b 20-Ms/s Analog-to-Digital Converter," *IEEE Journal of Solidstate Circuits*, Vol. 27, March 1992.

[14] A. Pärssinen, J. Jussila, J. Ryynänen, L. Sumanen, and K. Halonen, "AWide-Band Direct Conversion Receiver with On-Chip A/D Converters," in S*ymposium on VLSI Circuits Digest of Technical Papers,* pp. 32–33, June 2000.

[15] L. Sumanen and K. Halonen, "A Single-Amplifier 6-bit CMOS Pipeline A/D Converter for WCDMA Receivers," in Proceedings of the *IEEE Int. Symposium on Circuits and Systems (ISCAS'01)*, pp. I-584-587, May 2001.

A 110-dB Dynamic Range, 76-dB Peak SNR Companding Continuous-Time $\Delta\Sigma$ Modulator for Audio Applications

Saravana Kumar, Shouri Chatterjee

Department of Electrical Engineering, Indian Institute of Technology Delhi, New Delhi 110016, India

Abstract—This paper presents a companding continuous-time $\Delta\Sigma$ ADC for audio applications. The 3rd-order modulator uses a 3-bit companding quantizer and has an oversampling rate of 64. The companding quantizer is implemented by a log amplifier followed by a flash ADC. The modulator, in simulation, achieves a peak signal-to-noise ratio of 76 dB, a dynamic range of 110 dB in a 24 kHz bandwidth and dissipates 860 μW of power.

I. INTRODUCTION

Delta-Sigma modulators are widely used for achieving very high resolution analog-to-digital conversion, particularly when the signal bandwidth is small. Oversampling reduces the noise floor while noise-shaping removes the quantization noise from the signal band and shifts it to high frequencies. Continuous-time $\Delta\Sigma$ modulators are popular because of their inherent anti-aliasing properties. Compared to their switched-capacitor counterparts, continuous-time $\Delta\Sigma$ modulators require less power. For robust higher order $\Delta\Sigma$ modulator designs, multi-bit quantization is almost unavoidable. Further, multi-bit designs are insensitive to clock jitter. Several low-power multi-bit continuous-time $\Delta\Sigma$ modulators have been designed over the past decade, for example [1], [2].

An analog-to-digital converter (ADC) with a large dynamic range is particularly useful since it can be pushed closer to the transducer, thereby reducing the requirements of the signal conditioning circuits. Improvement in dynamic range can be achieved if the internal ADC adds quantization noise proportional to the input signal level [3]. Due to the non-linear time-varying quantization, the SNR can improve for smaller signal levels, which increases the dynamic range. This paper presents the design details of a companding 3rd-order continuous-time $\Delta\Sigma$ ADC. A log amplifier precedes the internal quantizer. The internal digital-analog converter (DAC) is designed to be closely matched to the log amplifier, to reverse the log operation. The companding $\Delta\Sigma$ ADC achieves a dynamic range of 110 dB over a 24 kHz signal bandwidth, for a power consumption of 860 μW, in simulation.

The paper is organized as follows. Section II deals with Simulink modeling of the companding $\Delta\Sigma$ modulator. In Section III we discuss the different circuit blocks and their implementation. Simulation results are presented in Section IV, and the paper is concluded in Section V.

II. SIMULINK MODELING OF COMPANDING $\Delta\Sigma$ MODULATOR

At the initial phase, the entire design was implemented on Simulink®. For the $\Delta\Sigma$ modulator a basic target of 100 dB dynamic range for an oversampling ratio of 64, was set. An initial design using the Delta-Sigma Toolbox [4] prompted a third-order noise-transfer-function (NTF). An out-of-band gain (OBG) of 2.0 was chosen as a compromise between stability and noise filtering. The simulated peak signal-to-noise ratio was 102 dB and an approximately equal amount of dynamic range was obtained.

A companding quantizer was further used to enhance the dynamic range. The quantizer levels for companding are set at -16 LSB, -4 LSB, -1 LSB, 0, 1 LSB, 4 LSB, 16 LSB, while the full scale value of the quantizer is from -64 LSB to 64 LSB.

Cascaded integrators with feed-forward (CIFF) summation was chosen to implement the loop filter. The CIFF architecture reduces the requirement of multiple feedback DACs and reduces area and power consumption. Simulink models were developed to analyze the behavior of the loop with the companding quantizer. Modulator non-idealities such as finite operational-amplifier gain-bandwidth, ADC and DAC mismatch, RC time-constant variation and clock jitter were also modeled as in [5] and their effects on peak SNR and dynamic range were observed.

The required unity-gain bandwidth for the integrators were evaluated from Simulink simulations. The design was simulated in the presence of 10 ppm of clock jitter, and 1% mismatch in the ADC. Fig. 1 shows a conceptual block diagram of the modulator, including the CIFF structure. A simulation of the companding $\Delta\Sigma$-modulator model (Fig. 2) resulted in a peak SNR of 91.2 dB and a dynamic range of 119.5 dB.

III. CIRCUIT DESIGN

A. Loop filter

As discussed in Section II, the loop filter of the $\Delta\Sigma$ modulator has a CIFF architecture. The third order loop filter requires three active-RC integrators and a summing amplifier. The resistor at the input of the first integrator dominates the thermal noise of the modulator and is chosen carefully for a thermal noise less than 110 dB below the full-scale of the modulator.

978-1-4673-0438-2/12 $31.00 © 2012 IEEE   51

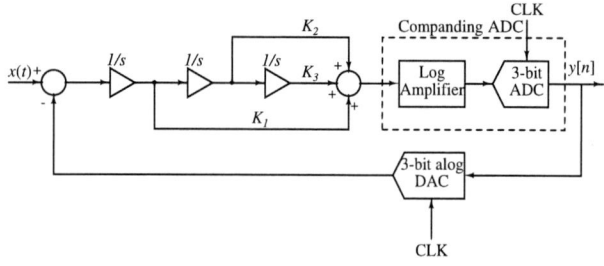

Fig. 1. Conceptual block diagram of the companding $\Delta\Sigma$ modulator.

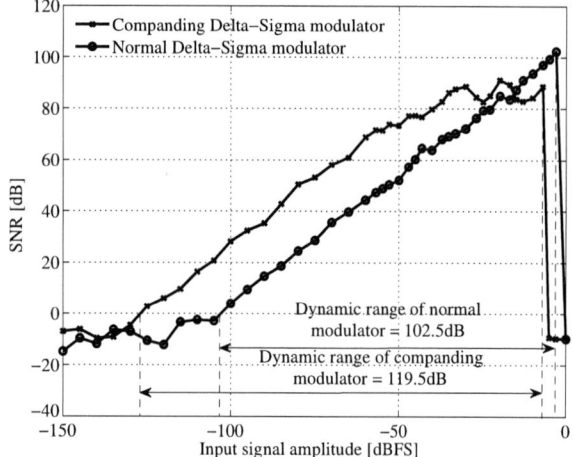

Fig. 2. Comparison of dynamic range of normal and companding $\Delta\Sigma$ modulator models.

Dynamic-range scaling is done by adjusting the resistor and capacitor values to prevent the saturation of the integrators. In particular, the resistor and capacitor values of the second and the third integrators are adjusted, with the overall transfer function remaining unchanged. Fig. 3 shows a detailed schematic of the loop filter, including the summing stage.

Excess loop delay [6] has adverse effects on the speed and on the stability of the $\Delta\Sigma$ modulator and is compensated for by using a weighted path directly from the DAC to the summing stage. The path is implemented indirectly from the DAC^+ and DAC^- signals through the first integrator, and then through C_d [1]. The gain of this path is $R_f C_d / R_1 C_1$. C_d is chosen to be 333 fF to compensate for the excess loop delay.

B. Operational transconductance amplifiers

The design requirements for the operational transconductance amplifiers (OTAs) were estimated using Simulink. All OTAs were required to have a dc gain of at least 60 dB and and the unity-gain bandwidths of the first, second and third integrators are required to be at least $3.5f_s$, $1.75f_s$ and $0.44f_s$ respectively.

The fully-differential OTAs used are two-stage designs using Miller compensation as shown in Fig. 4. The gain

requirements, estimated through Simulink simulations were not very high and so the first stage is a differential amplifier with the load as a current source. To minimize flicker noise, the first integrator uses a pMOS input (not shown) pair while the other integrators use an nMOS input pair (M_{1N}, M_{1P}).

M_{2P}, M_{2N} amplify the input differential signal across a pMOS load (M_{3P}, M_{3N}). Further amplification is provided in the second stage by M_{4P}, M_{4N} across M_{5P}, M_{5N}. M_1 draws a tail current of 15 μA for the first stage. Since the load of the first OTA is significant (the integrating resistors in the loop filter are small), the output stage has a comparatively large quiescent current of 60 μA. The second and third OTAs have lesser unity-gain bandwidth requirements than the first OTA. To optimize the power consumption in the system, all the OTAs were designed independently. Table I indicates the widths and lengths of all the transistors in the OTAs.

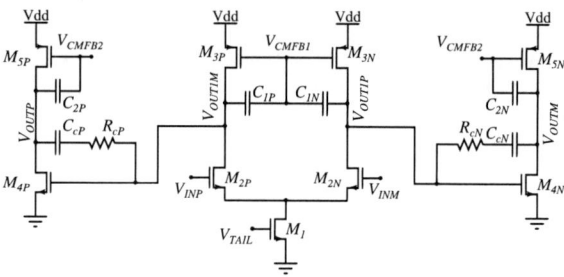

Fig. 4. Operational transconductance amplifier used in the integrators.

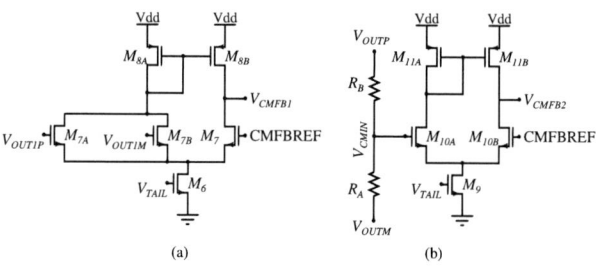

Fig. 5. CMFB circuits: (a) first stage CMFB (b) second stage CMFB.

Two distinct common-mode feedback (CMFB) loops are used for the first stage and for the second stage of the OTA, to provide lower common-mode gain, as shown in Fig. 5. The CMFB amplifier is a differential-input single-ended-output circuit. Fig. 5(a) shows the CMFB for the first stage of the OTA. The swing of the first stage of the OTA being less, the gate terminals of M_{7A}, M_{7B} are used to sense the common mode voltage. Fig. 5(b) shows the CMFB for the second stage of the OTA. The swing at the output of the second stage being high, large resistors R_A, R_B, are used to sense the output common mode voltage. Miller compensation through capacitors C_{1N}, C_{1P} and C_{2N}, C_{2P} in Fig. 4 maintain high phase-margin for the first stage and second stage common-mode loops.

978-1-4673-0438-2/12 $31.00 © 2012 IEEE

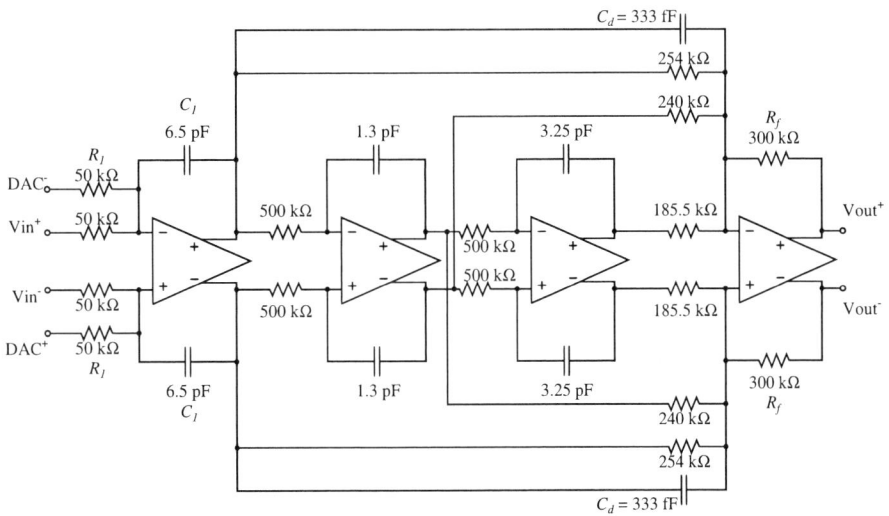

Fig. 3. Detailed schematic of the loop filter.

TABLE I
DEVICE SIZES OF THE OTAS

Device geometries in μm, in the first integrator (OTA-1), second integrator (OTA-2), third integrator (OTA-3), and summer (OTA-4) are given below.

Device	OTA-1 W/L	OTA-2 W/L	OTA-3 W/L	OTA-4 W/L
M_1	58/2	20/1	3.5/1	20/1
M_{2P}, M_{2N}	494/1	73/0.5	34/0.5	73/0.5
M_{3P}, M_{3N}	11/1	166/0.5	72/0.5	166/0.5
M_{4P}, M_{4N}	22/1	302/0.5	180/0.5	504/0.5
M_{5P}, M_{5N}	139/1	17/1	10/1	29/1

For OTA-1, M_1 is a pMOS device, M_{2P}, M_{2N} are pMOS devices, M_{3P}, M_{3N} are nMOS devices. The rest of the devices are as shown in Fig. 4.

C. Companding flash ADC

The companding quantizer is required to have characteristics as shown in Fig. 6. For an input full-scale voltage of ± 1.4 V, the least-significant-bit (LSB) size is 22 mV. For such a low LSB size, comparator offsets [7], [8] become dominant and therefore, power hungry pre-amplifiers are required to reduce them. Kick-back noise from the comparators also becomes significant, and sometimes a decision level can be missed.

To mitigate these problems, we use a logarithmic amplifier which amplifies small signals and attenuates large signals. The output of this log-amplifier is then given to a uniform-quantization flash ADC. By careful design of the log-amplifier, the decision levels indicated in Fig. 6 can be accurately obtained. Fig. 7 compares the performance of a $\Delta\Sigma$ modulator with a logarithmic amplifier and a uniform quantizer as opposed to a direct companding quantizer. The comparison demonstrates marginal increase in the noise floor because of inaccuracies in the logarithmic amplifier. The performance benefits far outweigh this increase in the noise floor.

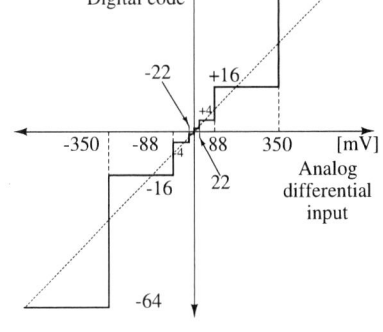

Fig. 6. Quantizer requirements. LSB is ± 22 mV for a full scale of ± 1.4 V differential.

1) Log-amplifier: The architecture of a true-logarithmic amplifier is a parallel-amplification parallel-summation topology as in Fig. 8. The G_m block is implemented by the differential amplifier topology given in Fig. 9. As the input voltage increases, at one point the tail current flows entirely into one arm of the amplifier. Beyond this input voltage the output current of the stage saturates. By carefully choosing bias currents of the G_m stages, a piecewise approximation of the log curve is obtained. The tail current for the G_m stage is fixed at 1.1 μA. A detailed design procedure is given in [9].

2) Flash ADC: The flash ADC specifications are relaxed due to the log amplifier and the decision levels are placed uniformly at a spacing of 200 mV. The schematic of the comparator [10] is shown in Fig. 10. The comparator is designed around a latch consisting of two inverters. When CLK is low, the latch is set and both outputs are at Vdd. When CLK is high, the sources of M_{3P} and M_{3N} are driven by a pseudo-differential stage, with the current in M_{3P} proportional

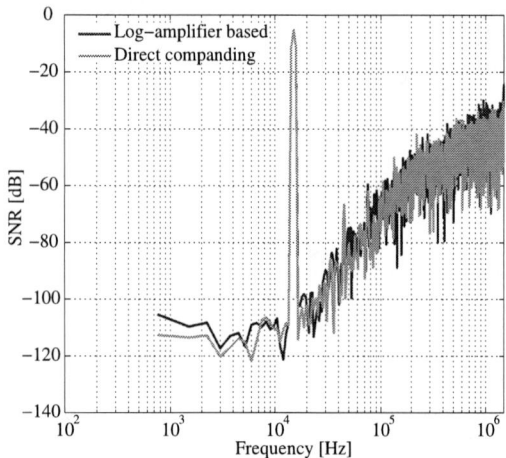

Fig. 7. A comparison of the $\Delta\Sigma$ modulator performance with a logarithmic amplifier and a uniform quantizer, as opposed to a direct companding quantizer.

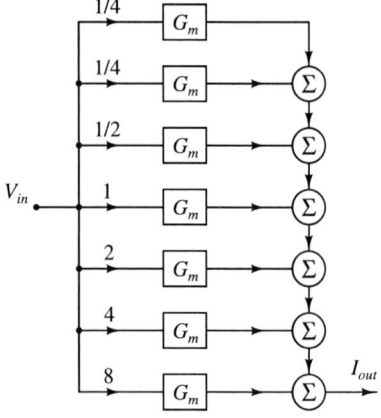

Fig. 8. A block diagram representation of a true-logarithmic amplifier.

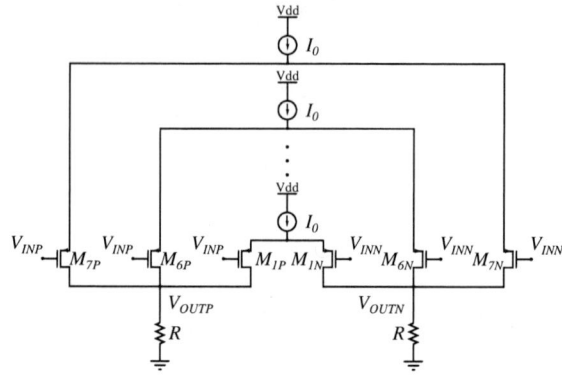

Fig. 9. G_m-current limiter stage.

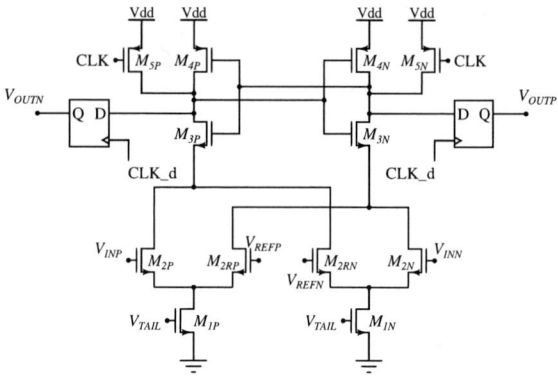

Fig. 10. High speed comparator.

to $V_{INP} + V_{REFN}$ and the current in M_{3N} proportional to $V_{INN} + V_{REFP}$. D-flip-flops, clocked by CLK_d, a delayed version of CLK, sample the correct value of the compared output.

D. DAC design

The feedback DAC uses a current steering architecture with a cascode current source as shown in Fig. 11. The current sources in the DAC are weighted according to the companding requirements. The current values range from 220 nA to 14.08 μA (differential). As a result, data weighted averaging cannot be applied directly to minimize the adverse effects of mismatches between current sources.

The cascode unit-current source provides high output resistance of the order of $g_m r_{ds}^2$, where g_m is the transconductance of a device and r_{ds} is the output resistance of the device. Specifically, in Fig. 11, the current source has an output

resistance of $g_{m_{5N}} r_{ds_{5N}} r_{ds_{6N}}$, the current sink has an output resistance of $g_{m_{2N}} r_{ds_{2N}} r_{ds_{1N}}$. The sizes of the devices in the unit-current source are made to be large, to reduce mismatches. The DAC is required to drive 50 kΩ resistors (Fig. 3). The device small signal parameters are chosen such that the combined output resistance of all the unit-current sources and the combined output resistance of all the unit-current sinks are much larger than the 50 kΩ load of the DAC.

The current mismatch equation between two equally sized MOS devices is given by [7], [8]:

$$\left(\frac{\sigma_I}{I}\right)^2 = \frac{4A_{V_T}^2}{V_{OV}^2 WL} + \frac{A_\beta^2}{WL}$$

where W, L are the width and the length of the MOS device respectively, V_{OV} is the gate overdrive voltage, and σ_I/I is the standard deviation of the mismatch in the current through the device. A_{V_T}, A_β are technology dependent proportionality constants that correspond to mismatches in the threshold voltage of the devices (V_T) and mismatches in the current factors of the devices (β), respectively.

The widths and lengths of the devices in the unit-current cell of the DAC were chosen such that the standard deviation of the current mismatch (σ_I/I) was less than 1%. The cascode devices, M_{2P}, M_{2N} and M_{5P}, M_{5N} are not critical to current

978-1-4673-0438-2/12 $31.00 © 2012 IEEE

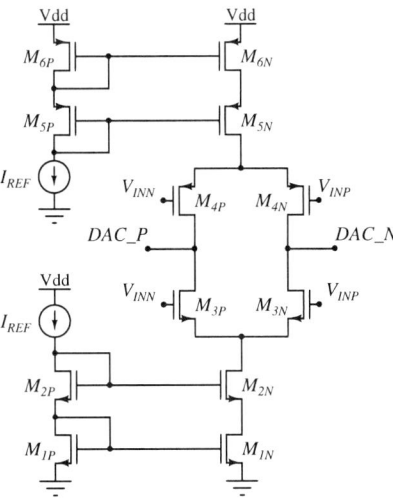

Fig. 11. DAC current cell.

matching. These devices were designed to be smaller in size.

In this design a large number of unit current cells, each of 220 nA, are used. To generate 14.08 μA current, for example, 64 unit current cells switch on together. The different DAC levels are indicated in Table II.

TABLE II
DESIGN DETAILS OF THE DAC.

Level	Code	DAC level	DAC current
-64	0000000	-700 mV	-14.08 μA
-16	0000001	-175 mV	-3.52 μA
-4	0000011	-44 mV	-0.88 μA
-1	0000111	-11 mV	-0.22 μA
+1	0001111	11 mV	0.22 μA
+4	0011111	44 mV	0.88 μA
+16	0111111	175 mV	3.52 μA
+64	1111111	700 mV	14.08 μA

E. Modified data-weighted averaging

The data-weighted averaging (DWA) algorithm [11] uses an accumulator and a barrel shifter. In this design the DWA algorithm is used with a small modification, as shown in Fig. 12.

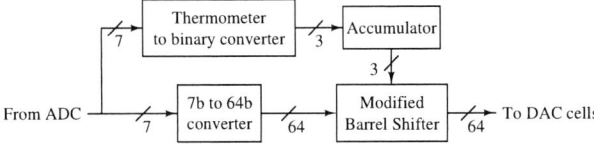

Fig. 12. Block diagram of modified data-weighted averaging circuit.

Since each current cell has a value of 220 nA, 64 current cells are required to produce the full range of required currents. The thermometer code from the ADC is first converted into a 64-bit code which is input to the barrel shifter. The barrel shifter is modified to produce shifts in such a way that the same set of current cells are not driven immediately again for the same input code. The barrel shifter is controlled by the accumulator outputs. In this manner, a first-order noise shaping of DAC mismatches is obtained.

IV. SIMULATION RESULTS

The design was implemented in a 0.18 μm CMOS technology. The true-logarithmic amplifier voltage-transfer characteristics is shown in Fig. 13. Fig. 14 shows the transfer characteristics of the logarithmic amplifier in comparison to an ideal scaled logarithm. The errors introduced are matched by the quantization levels of the flash ADC. With accurate design methods available for logarithmic amplifiers, this adjustment is easier with logarithmic amplifiers, as opposed to regular amplifiers. Further, since the logarithmic amplifier is within the noise-shaping loop just before the flash ADC, noise or errors introduced by the logarithmic amplifier is not a big concern.

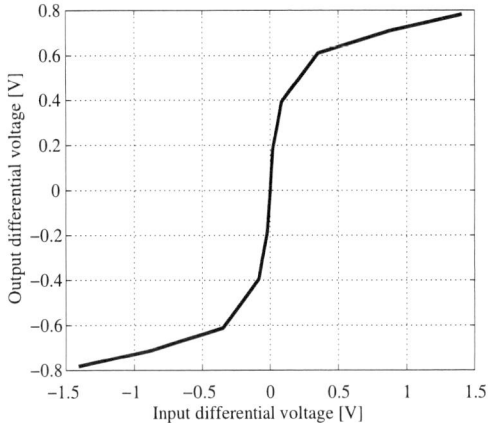

Fig. 13. True-logarithmic amplifier voltage-transfer characteristics.

The noise shaping property of the modulator is shown in Fig. 15. The simulations showed a peak SNR of 76.1 dB and a dynamic range of 110.9 dB (without incorporating thermal noise into the simulations) as shown in Fig. 16. Monte-Carlo analysis was not run for including random mismatches in the DAC levels. However, dynamic-element-matching in the DAC is expected to compensate for mismatches in the DAC. Dynamic-element-matching was simulated to compensate for systematic mismatches in the DAC. Analytical calculations predict 3 dB degradation in dynamic range due to thermal noise. The peak SNR does not increase significantly with increase in signal amplitudes.

Since the quantization noise added becomes large at large signal levels, it also reduces the maximum stable amplitude

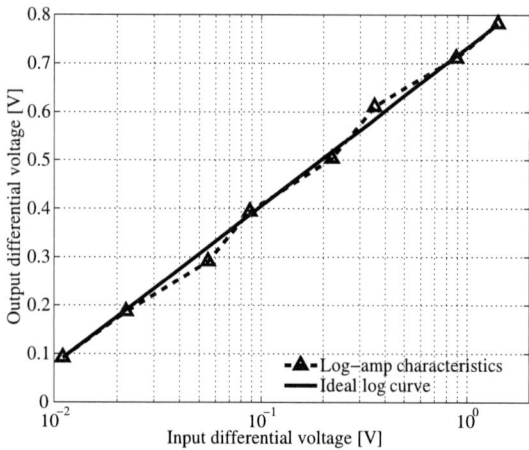

Fig. 14. Transfer characteristics of the logarithmic amplifier in comparison to an ideal logarithm.

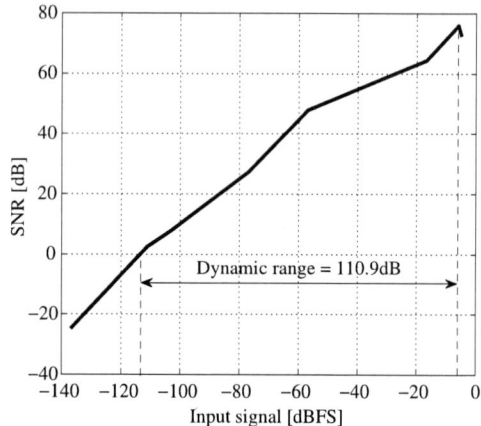

Fig. 16. Dynamic range of the companding $\Delta\Sigma$ modulator.

(MSA) of the modulator. The MSA of the companding $\Delta\Sigma$ modulator is -4.9 dBFS. By using the log-amplifier, the power-hungry preamplifier has been removed from the flash ADC. This has resulted in significant power reduction. The total power consumed by the $\Delta\Sigma$ modulator is 860 μW.

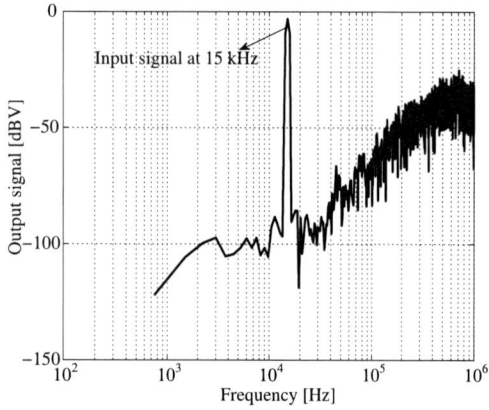

Fig. 15. Spectrum of output signal of the companding $\Delta\Sigma$ modulator, demonstrating noise shaping.

V. CONCLUSIONS

In this paper we have presented a companding continuous-time $\Delta\Sigma$ modulator with a simulated dynamic range of 110.9 dB, and peak SNR of 76.1 dB, for a signal bandwidth of 24 kHz and an oversampling ratio of 64.

As part of the $\Delta\Sigma$ modulator, the companding quantizer was implemented using a logarithmic-amplifier and a uniform quantization flash ADC. The use of the log-amplifier resulted in reduction of accuracy requirements of the quantizer. A matched feedback DAC with a modified DWA algorithm is incorporated in the design. Extensive simulations over a range of process corners, power-supply voltages, and ambient

temperatures were performed to validate the design. The $\Delta\Sigma$ modulator dissipates 860 μW of power.

REFERENCES

[1] S. Pavan, N. Krishnapura, R. Pandarinathan, and P. Sankar, "A power optimized continuous-time $\Delta\Sigma$ ADC for audio applications," *IEEE Journal of Solid-State Circuits*, vol. 43, no. 2, pp. 351–360, Feb. 2008.

[2] F. Gerfers, M. Ortmanns, and Y. Manoli, "A 12-bit power efficient continuous-time $\Delta\Sigma$ modulator with 250 μW power consumption," in *Proceedings of the 27th European Solid-State Circuits Conference, 2001*, Sep. 2001, pp. 538–541.

[3] Z. Zhang and G. Temes, "Multibit oversampled $\Sigma\Delta$ A/D convertor with nonuniform quantisation," *Electronics Letters*, vol. 27, no. 6, pp. 528–529, Mar. 1991.

[4] R. Schreier, "Delta sigma toolbox." [Online]. Available: http://www.mathworks.com/matlabcentral/fileexchange/19

[5] P. Chopp and A. Hamoui, "Analysis of clock-jitter effects in continuous-time $\Delta\Sigma$ modulators using discrete-time models," *IEEE Transactions on Circuits and Systems I*, vol. 56, no. 6, pp. 1134–1145, Jun. 2009.

[6] J. Cherry and W. Snelgrove, "Excess loop delay in continuous-time $\Delta\Sigma$ modulators," *IEEE Transactions on Circuits and Systems II: Analog and Digital Signal Processing*, vol. 46, no. 4, pp. 376–389, Apr. 1999.

[7] M. Pelgrom, A. Duinmaijer, and A. Welbers, "Matching properties of MOS transistors," *IEEE Journal of Solid-State Circuits*, vol. 24, no. 5, pp. 1433–1439, Oct. 1989.

[8] P. Kinget, "Device mismatch and tradeoffs in the design of analog circuits," *IEEE Journal of Solid-State Circuits*, vol. 40, no. 6, pp. 1212–1224, Jun. 2005.

[9] C. Holdenried, J. Haslett, J. McRory, R. Beards, and A. Bergsma, "A DC-4-GHz true logarithmic amplifier: theory and implementation," *IEEE Journal of Solid-State Circuits*, vol. 37, no. 10, pp. 1290–1299, Oct. 2002.

[10] J. Lin and B. Haroun, "An embedded 0.8 V/480 μW 6b/22 MHz flash ADC in 0.13μm digital CMOS process using a nonlinear double interpolation technique," *IEEE Journal of Solid-State Circuits*, vol. 37, no. 12, pp. 1610–1617, Dec. 2002.

[11] R. Baird and T. Fiez, "Linearity enhancement of multibit $\Delta\Sigma$ A/D and D/A converters using data weighted averaging," *IEEE Transactions on Circuits and Systems II: Analog and Digital Signal Processing*, vol. 42, no. 12, pp. 753–762, Dec. 1995.

978-1-4673-0438-2/12 $31.00 © 2012 IEEE

Hardware Efficient Architecture for Generating Sine/Cosine Waves

Supriya Aggarwal
Dept. of Electronics & Comm. Engg., NIT-Bhopal
Maulana Azad National Institute of Technology,
Bhopal, India
sups.aggarwal@gmail.com

Kavita Khare
Dept. of Electronics & Comm. Engg., NIT-Bhopal
Maulana Azad National Institute of Technology,
Bhopal, India
kavita_khare1@yahoo.co.in

Abstract—**This paper presents a hardware efficient architecture for generating sine and cosine waves based on the CORDIC (Coordinate Rotation Digital Computer) algorithm. In its original form the CORDIC suffers from major drawbacks like scale-factor calculation, latency and optimal selection of micro-rotations. The proposed algorithm overcomes all these drawbacks. We use leading-one bit detection technique to identify the micro-rotations. The scale-free design of the proposed algorithm is based on Taylor series expansion of the sine and cosine waves. The 16-bit iterative architecture achieves approximately 4.5% and 6.7% lower slice-delay product as compared to the other existing designs. The algorithm design and its VLSI implementation are detailed.**

Keywords-Cosine, CORDIC Algorithm, Leading-One Bit, Sine, Taylor Series

I. INTRODUCTION

The CORDIC algorithm [1], [2] is a generalized technique for computing various functions like sine/cosine [3], [4], transforms [5], [6], exponents/logarithms, square-roots, eigen values [7] etc. During the past 50 years [8], there have been major advances in the design of the algorithm to overcome its major drawbacks. In [9], [10], [11] authors suggest the use of greedy search algorithms for identifying the micro-rotations. But the efficiency of their approach depends on the probability of rotation angles being known prior to implementation; moreover, greedy search architectures are difficult to incorporate in hardware. Also, the associated variable scale-factor implementation cancels their advantage of reduced latency.

The low complexity technique for eliminating the scale-factor is the use of Taylor series expansion. The Scaling-Free CORDIC and modified scale-free CORDIC [12, 13] are techniques based on Taylor series approach. The former suffers from low range of convergence (RoC) which renders it unsuitable for practical applications, while the latter extends the RoC but introduces predictable but constant scale-factor of $1/\sqrt{2}$. The other hardware efficient architectures in [14, 15] also require scale-factor compensations to extend the range of convergence to the entire coordinate space.

Through this paper we present a novel completely scale-free CORDIC algorithm for generating sine/cosine waves. We use Taylor series expansion to obtain scale-free CORDIC equations. Using wave symmetry properties of sine and cosine functions, the RoC can be extended to the entire coordinate space. We use a leading-one bit detector to identify the micro-rotation sequence.

The rest of the paper is organized as: Section II gives a brief overview of the CORDIC algorithm. Section III presents the design of the scale-free CORDIC equations while Section IV discusses the micro-rotation sequence identification. Section V details the iterative architecture implementation. FPGA implementation and comparison of area-time complexities with existing techniques is presented in Section VI. Finally, Section VII concludes the paper.

II. CORDIC ALGORITHM

The underlying principle of the CORDIC algorithm is based on two-dimension geometry. The algorithm operates either in, rotation or vectoring mode, following linear, circular or hyperbolic trajectories. We focus on rotation mode of operation in circular trajectory.

A. Conventional CORDIC Algorithm

Let the vector $V_B[x_B, y_B]$ be derived by rotating the vector $V_A[x_A, y_A]$ through an angle 'θ', then:

$$\begin{bmatrix} x_B \\ y_B \end{bmatrix} = R_p \cdot \begin{bmatrix} x_A \\ y_A \end{bmatrix}, \quad R_p = \begin{bmatrix} \cos\theta & -\sin\theta \\ \sin\theta & \cos\theta \end{bmatrix} \quad (1)$$

Equation (1) forms the basic principle for iterative coordinate calculation in CORDIC algorithm [1]. The key concept in realizing rotations using CORDIC is to express the angle of rotation 'θ' as an aggregation of pre-defined elementary angles defined as:

$$\theta = \sum_{i=0}^{b} \mu_i \cdot \alpha_i \quad \text{where } \mu_i = \{-1, 1\}, \alpha_i = \tan^{-1} 2^{-i}$$

where b is the word-length in bits. $\quad (2)$

978-1-4673-0438-2/12 $31.00 © 2012 IEEE

The RoC of the conventional CORDIC is [-99.9°, 99.9°] and using extra iteration step can be extended to the entire coordinate space.

The rotation matrix R_p, in its original form is computation intensive; it requires computing sine and cosine functions with four multiplication and two addition operations. Factoring the cosine term simplifies the rotation matrix R_p (1) by converting the multiplication to shift operations, as tangent of the elementary angles is defined in negative powers of two (2). But the penalty paid is the introduction of the scale-factor which varies according to the cosine of the elementary-rotation.

$$R_p = K_i \cdot \begin{bmatrix} 1 & -\mu_i \cdot 2^{-i} \\ \mu_i \cdot 2^{-i} & 1 \end{bmatrix}, \quad K_i = 1/\sqrt{1 + 2^{-2i}} \quad (3)$$

As seen from (3), the scale factor K_i is independent of the direction of micro-rotation. With sequential execution of large number of iterations it tends to a constant, referred to as the gain of the CORDIC algorithm. The scale-factor is thus compensated either in the post or pre processing unit.

B. Review of Scale-Free CORDIC Algorithm

Scaling-free CORDIC [12] was the first attempt to design scale-free coordinate CORDIC equations using Taylor series expansion. Here, the micro-rotations are restricted to anti-clockwise direction only, such that, any angle of rotation is represented as the algebraic sum of elementary angles. In [12, 13], the sine and cosine functions are approximated to:

$$\sin \alpha_i = 2^{-i}, \qquad \cos \alpha_i = 1 - 2^{-(2i+1)} \quad (4)$$

However, the approximation imposes a restriction on the start iteration index $i = \left\lceil \frac{(b-2.585)}{3} \right\rceil$. Although, it avoids the costly scaling network, it suffers from a major drawback of very low RoC. For 16-bit data the start iteration index $i = 4$, which leads to very low RoC amounting to 7.16°.

A modification over this algorithm is presented in [12, 13], to extend the RoC over the entire coordinate space by using domain folding technique. The rotation angles are mapped in to the scope of scaling-free CORDIC using a pre-processing unit. But it requires a post-processing unit for implementation of adaptive scale factor $(1/\sqrt{2})$.

III. PROPOSED SCALE-FREE CORDIC ALGORITHM

The design of the proposed algorithm is based on the following ideas: (i) use Taylor series approximation for design of scale-free coordinate equations; (ii) eliminate redundant micro-rotations by restricting the iterations to uni-direction. A leading-one bit detector simplifies the identification of the micro-operations; and (iii) exploit symmetry properties of sine and cosine waves to extend the RoC to the entire coordinate space.

Design of CORDIC processor is divided in two parts, the coordinate calculation unit and the micro-rotation sequence identification. In this section we elaborate on the design of the scale-free CORDIC coordinate equations. While, the micro-rotation sequence identification is detailed in the next section.

A. Taylor Series Expansion of Trigonometric Terms

The Taylor series expansion of sine and cosine terms is given as:

$$\begin{aligned} \sin \alpha &= \alpha - (3!)^{-1} \cdot \alpha^3 + (5!)^{-1} \cdot \alpha^5 \cdots \\ \cos \alpha &= 1 - (2!)^{-1} \cdot \alpha^2 + (4!)^{-1} \cdot \alpha^4 \cdots \end{aligned} \quad (5)$$

For hardware implementation, the series in (5) needs to be approximated implying a compromise in accuracy. To determine the order of approximation for the design of coordinate equations, we design the CORDIC processor using the various orders of approximation in the coordinate rotation matrix R_p (1). The MSE error in sine and cosine values for angles in the range $[0, \pi/4]$ is tabulated in Table I. This range can be extended to the entire coordinate space using octant wave symmetry [16].

TABLE I. MSE ERROR IN SINE AND COSINE VALUES

Order of Approximation	Mean Square Error	
	Sine Values	Cosine Values
3	0.9279×10^{-6}	0.5206×10^{-6}
4	0.9639×10^{-6}	0.4863×10^{-6}
5	0.9380×10^{-6}	0.5447×10^{-6}
6	0.9381×10^{-6}	0.5446×10^{-6}
7	0.9381×10^{-6}	0.5446×10^{-6}

The errors for various Taylor series orders of approximation are small enough. In order to keep the hardware complexity to minimum, we choose third order of approximation.

B. Design of Coordinate Calculation Equations

The third order of approximation provides the desired accuracy (Table I). The rotation matrix R_p using third order of approximation is modified in (6).

$$R_p = \begin{bmatrix} 1 - \frac{\alpha_i^2}{2!} & -\left(\alpha_i - \frac{\alpha_i^3}{3!}\right) \\ \alpha_i - \frac{\alpha_i^3}{3!} & 1 - \frac{\alpha_i^2}{2!} \end{bmatrix} \quad (6a)$$

Assuming $\alpha_i = 2^{-i}$, the rotation matrix R_p (6a) can be simplified to (6b). To support shift-add implementation 3! is approximated to 2^3. With the approximated factorial value the mean square error in cosine value is 0.1135×10^{-6} and in sine value is 0.2369×10^{-6}. Thus, approximation of the factorial value does not affect the accuracy.

$$R_p = \begin{bmatrix} 1 - 2^{-(2i+1)} & -\left(2^{-i} - 2^{-(3i+3)}\right) \\ 2^{-i} - 2^{-(3i+3)} & 1 - 2^{-(2i+1)} \end{bmatrix} \quad (6b)$$

The block diagram for implementing the coordinate calculation unit is shown in Fig. 1.

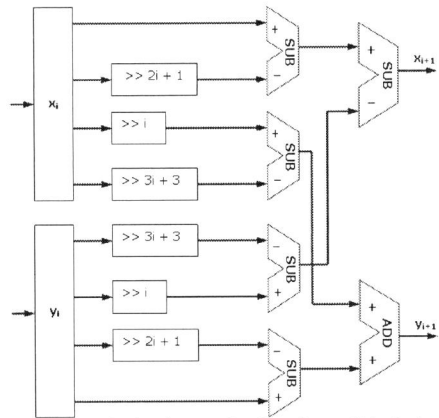

Figure 1 – Block Diagram for Coordinate Calculation

IV. MICRO-ROTATION SEQUENCE GENERATION

We propose realizing rotations using CORDIC by following certain rules: (i) uni-direction micro-rotations, as a result the angle of rotation is expressed as summation of selective elementary angles; (ii) allowing repetition of elementary angles for realizing rotations; and (iii) identifying the rotation based on the radix-2 representation of the rotation angle using a leading-one bit detector. As a result, we do not require any ROM for storing the elementary angles of rotation.

A. Redefining the Elementary Angles

In original CORDIC algorithm the elementary angles (α_i) are represented in (2). We redefine the elementary angles as:

$$\alpha_i = 2^{-i} \qquad (7)$$

The start iteration index (i) will determine the highest elementary angle that can be used for realizing rotations. The start iteration index depends on the Taylor series approximation used [9]. For third order of approximation the highest term neglected is ($\alpha^4/4!$). To prevent this term from affecting the accuracy, it should be zero during all coordinate calculations. This implies ($\alpha^4/4!$) should get a shift greater than the word-length. Hence, the lowest value of iteration index, such that this fourth order term is zero, is:

$$i = \left\lfloor \frac{\{b - \log_2(4!)\}}{4} \right\rfloor, \text{ where } b \text{ is the wordlength} \qquad (8)$$

For 16-bit, data the start iteration index is '$i=2$'. Therefore the highest elementary angle is $7\pi/88$ radians.

B. Identifying the Micro-rotation Sequence

The fixed point binary representation of the elementary angles has one and only one bit set, as they are represented in negative powers of two (7). The micro-rotation sequence identification exploits this property of the elementary angles. We generate the micro-rotation sequence using the algorithm in Table II.

The LOP (lead one position) refers to the leading one bit number of any input string counting from the Most Significant Bit (MSB). A step-wise execution for angle of 35° (0.6108 radians), represented in radix-2 representation as 1001_1100_0110_0001, is shown in Table III.

TABLE II
PSEUDO CODE FOR GENERATING THE MICRO-ROTATION SEQUENCE
LOP = Lead One Position

Input: angle to be rotated (θ_i)
Begin
LOP = Leading-One Bit of θ_i
   if (LOP == 15) then
      α = 0.25 radians
      Shift, i = 2
      $\theta_{i+1} = \theta_i - \alpha$
   Else
      Shift, i = 16 – LOP
      $\theta_{i+1} = \theta_i$ with $\theta_i[\text{LOP}]$ = '0'
End

TABLE III
DETERMINING THE MICRO-ROTATION SEQUENCE FOR ANGLE 35°
*Values are in Hexadecimal, (0,16) Fixed Point Format

Stg	θ_i *	LOP	Shift (i)	θ_{i+1} *
1	9C61	15	2	9C61 – 4000 = 5C61;
2	5C61	14	16 – 14 = 2	0C61; Set bit no. 14 to '0'
3	0C61	11	16 – 11 = 5	0461; Set bit no. 11 to '0'
4	0461	10	16 – 10 = 6	0061; Set bit no. 10 to '0'
5	0061	6	16 – 6 = 10	0021; Set bit no. 6 to '0'

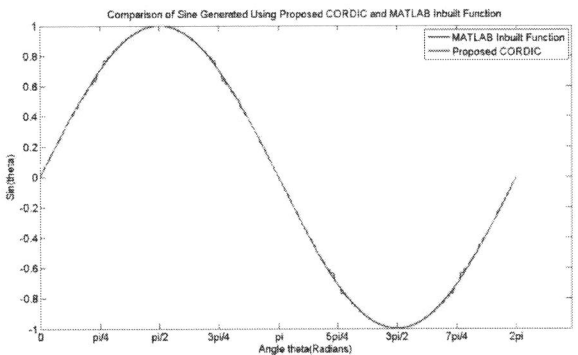

Figure 2 – Sine Wave Generated Using the Proposed CORDIC

C. Defining the Number of Iterations for Attaining the Desired RoC

The maximum angle that can be handled by the micro-rotation sequence generation unit is $\pi/4$ radians. Thus the RoC of the proposed algorithm is [0, $\pi/4$]. This can be extended to the entire coordinate space using the octant mapping technique in [16].

978-1-4673-0438-2/12 $31.00 © 2012 IEEE

For realizing the rotations in the range [0, $\pi/4$] maximum of three iterations corresponding to the highest elementary angle $7\pi/88$ are required. For convergence, we require additional four iterations, i.e., with a total of seven iterations any angle in the range [0, $\pi/4$] can be realized. The Fig. 2 shows the sine wave plot generated using the proposed CORDIC processor. The error percentage in the sine wave generated using the proposed CORDIC with respect to MATLAB is less than 0.04%.

V. Iterative Architecture of Proposed CORDIC Processor

Iterative Architecture uses the same stage for all the CORDIC iterations. The block diagram of the iterative architecture is shown in Fig. 3. The stage is a combination of the coordinate calculation and the micro-rotation sequence generation. The hardware implementation of coordinate calculation is show in Fig. 1 while the algorithmic implementation of micro-rotation sequence generator is shown in Table II.

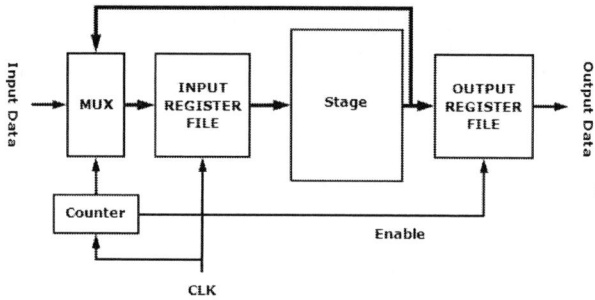

Figure 3: Iterative Architecture of the Proposed CORDIC Processor

The number of iterations required in a CORDIC processor decides the rollover count of the counter. For the proposed algorithm, the rollover count is seven. The expiry of the counter signals the completion of a CORDIC operation; depending on this signal, the multiplexer either loads a new data-set (rotation angle, initial value of 'x' and 'y') to start a fresh CORDIC operation, or recycles the output of the stage to begin a new iteration for the current CORDIC operation. The input and output register files act as latches for synchronization.

The feedback path comprises of the intermediate x, y, z and *shift* values. The number of feedback bits for 16-bit word-length are, $16 + 16 + 16 + 5 = 53$ bits.

During the first cycle of operation for any rotation angle only the micro-rotation sequence generation is active in the stage. In the subsequent cycles, the coordinate calculation uses the iteration index (i) from the feedback path for computing the coordinates, while the sequence generator computes the iteration index (i) for the next feedback cycle.

VI. FPGA Implementation and Complexity Issues

A. Area Complexity

The coordinate computation unit of the iterative architecture requires six N-b adders and six N-b shifters. Along with it to generate the required number of shifts it requires ($\log_2 N + M$)-b adder and a ($\log_2 N + 1$)-b shifter; M are extra bits required to store the sum. The micro-rotation sequence generator requires a $\log_2 N$-b adder and a N-b subtractor.

B. FPGA Implementation Results

The proposed CORDIC processor is coded in Verilog and synthesized and simulated using Xilinx ISE 9.2i. The processor is mapped onto the device Spartan 2E (XC2S200E-PQ208-6). It consumes 191 slices, 359 LUTs and attains a maximum operation frequency of 59.766 MHz. A comparative study with other existing architectures is tabulated in Table VII.

The worst cases iterations in case of approaches mentioned in [14] is ten, while for [13] is fifteen. The proposed approach takes a fixed (seven) clock cycles to converge for all rotation angles.

TABLE VII
Slice Delay Product Comparison on Spartan 2E XC2S200E Device For Different Approaches
Word Length = 16 bit (* Values taken from [])

Approach	LUTs	Slices	Max. Freq. MHz	Slice Delay Product
Algorithm – I [14]*	350	186	54.35	3.42
Algorithm – II [14]*	378	203	60.80	3.34
Scaling-Free CORDIC [13]*	1658	945	52.54	17.99
Xilinx Core*	1165	702	58.28	12.04
Proposed	359	191	59.766	3.19

VII. Conclusion

The proposed algorithm provides an area-time efficient processor for generating sine and cosine values in hardware. The FPGA implementation of the proposed processor has the least slice-delay product. Apart from hardware efficiency, the proposed CORDIC algorithm is a completely scale-free in nature. The elimination of redundant iterations using a leading-one bit technique improves the throughput of the conventional CORDIC. The order of approximation of Taylor series used not only meets the accuracy requirements but also attains adequate RoC. By using leading-bit detection we eliminate the need for complex search algorithms for identifying the micro-rotations.

VIII. References

[1] J. E. Volder, "The CORDIC Trigonometric Computing Technique", IRE Trans. Electronic Computing, Vol. EC-8, Sept 1959, pp. 330-334.

[2] J. S. Walther, "A unified algorithm for elementary functions", Spring Joint Computer Conf., 1971, Proc., pp. 379-385.

978-1-4673-0438-2/12 $31.00 © 2012 IEEE

[3] S. F. Hsiao, Y. H. Hu and T. B. Juang, "A Memory Efficient and High Speed Sine/Cosine Generator Based on Parallel CORDIC Rotations," IEEE Signal Processing Letters, Vol. 11, No.2, Feb-2004, pp. 152-155.

[4] N. Takagi, T. Asada, and S. Yajima, "Redundant CORDIC methods with a constant scale factor for sine and cosine computation," *IEEE Trans. Computers*, vol. 40, pp. 989–995, Sept. 1991.

[5] K. Maharatna and S. Banerjee, "A VLSI array architecture for hough transform," *Pattern Recognit.*, vol. 34, pp. 1503–1512, 2001.

[6] K. Maharatna, A. S. Dhar, and S. Banerjee, "A VLSI array architecture for realization of DFT, DHT, DCT and DST," *Signal Process.*, vol. 81, pp. 1813–1822, 2001.

[7] Y. H. Hu and H. M. Chern, "VLSI CORDIC array structure implementation of Toeplitz eigensystem solvers," in *Proc. IEEE Int. Conf. Acoust. Speech, Signal Processing*, NM, 1990, pp. 1575–1578.

[8] Pramod K. Meher, Javier Walls, Tso-Bing Juang, K. Sridharan, Koushik Maharatna, "50 Years of CORDIC: Algorithms, Architectures and Applications," IEEE Transactions on Circuits and Systems – I: Regular Papers, Vol. 56, No. 9, Sept. 2009, pp. 1893-1907.

[9] C. S. Wu and A. Y. Wu, "Modified vector rotational CORDIC (MVR-CORDIC) algorithm and architecture," *IEEE Trans. Circuits Syst. II*, vol. 48, pp. 548–561, June, 2001.

[10] Cheng-Shing Wu, An-Yeu Wu and Chih-Hsiu Lin, "A High-Performance/Low-Latency Vector Rotational CORDIC Architecture Based on Extended Elementary Angle Set and Trellis-Based Searching Schemes," *IEEE Transcations on Circuits and Systems–II : Analog and Digital Signal Processing,* Vol. 50, pg. 589–601, No. 9, Sept. 2003.

[11] Y. H. Hu and S. Naganathan, "An angle recoding method for CORDIC algorithm implementation," *IEEE Trans. Computers*, vol. 42, pp. 99–102, Jan. 1993.

[12] K. Maharatna, S. Banerjee, E. Grass, M. Krstic, and A. Troya, "Modified virtually scaling-free adaptive CORDIC rotator algorithm and architecture," *IEEE Trans. Circuits Syst. Video Technol.*, vol. 11, no. 11, pp. 1463–1474, Nov. 2005.

[13] K. Maharatna, A. Troya, S. Banerjee, and E. Grass, "Virtually scaling free adaptive CORDIC rotator," *IEE Proc.-Comp. Dig. Tech.*, vol. 151, no. 6, pp. 448–456, Nov. 2004.

[14] Leena Vachhani, K. Sridharan and Pramod K. Meher, "Efficient CORDIC Algorithms and Architectures for Low Area and High Throughput Implementation," IEEE Transactions on Circuit and Systems–II: Express Briefs, Vol. 56, No. 1, pg. 61-65., January 2009.

[15] F.J. Jaime, M. A. Sanchez, J. Hormigo, J. Villalba and E. L. Zapata, "Enhanced scaling-free CORDIC," IEEE Trans. Circuits and Systems I: Regular Papers, Vol. 57 No.7, pp. 1654-1662. July 2010.

[16] J. Vankka, Digital Synthesizers and Transmitters for Software Radio, Dordrecht, Netherlands. Springer: 2005.

Power Aware Hardware Prototyping of Multiclass SVM Classifier Through Reconfiguration

Rajesh A Patil
MNIT Jaipur
Email:rapatil_rtg@yahoo.co.in

Gauri Gupta
MNIT Jaipur
Email:gauri777@gmail.com

Vineet Sahula
MNIT Jaipur
Email:sahula@ieee.org

A. S. Mandal
CEERI Pilani
Email:atanu@ceeri.ernet.in

Abstract—**This paper presents power aware hardware implementation of multiclass Support Vector Machine on FPGA using systolic array architecture. It uses Partial reconfiguration schemes of XILINX for power optimal implementation of the design. Systolic array architecture provides efficient memory management, reduced complexity, and efficient data transfer mechanisms. Multiclass support vector machine is used as classifier for facial expression recognition system, which identifies one of six basic facial expressions such as smile, surprise, sad, anger, disgust, and fear. The extracted parameters from training phase of the SVM are used to implement testing phase of the SVM on the hardware. In the architecture, vector multiplication operation and classification of pairwise classifiers is designed. A data set of Cohn Kanade database in six different classes is used for training and testing of proposed SVM. This architecture is then partially reconfigured using difference based approach with the help of XILINX EDA tools. For feature classification power reduction is achieved using reconfiguration.**

Index Terms—**SVM, dynamic partial reconfiguration, systolic array.**

I. Introduction

Support Vector Machine (SVM) is a powerful machine-learning tool, introduced by Cortes and Vapnik [1]. SVMs are based on the concept of decision planes that define decision boundaries. A decision plane is one that separates between a set of objects having different class memberships. Due to their high classification accuracy rates, in many cases it outperforms well established classification algorithms such as neural networks. Software implementations of SVMs yield high accuracy rates but, they cannot efficiently meet real-time requirements of embedded applications. Hence, SVM hardware architectures emerged as a potential solution to achieve real-time performance. The majority of these emerging architectures is directed towards specific applications and cannot be easily modified to adapt in other scenarios. A significant advantage of SVMs is that whilst neural networks can suffer from multiple local minima, the solution to an SVM is global and unique. Two more advantages of SVMs are that that have a simple geometric interpretation and give a sparse solution. Unlike neural network, the computational complexity of SVMs does not depend on the dimensionality of the input space. Neural network use empirical risk minimization, whilst SVMs use structural risk minimization.

Hardware implementation of any design has its own significant advantages, especially for Image Processing applications, where size of image matrices is quite large. For software implementation, the total simulation and synthesis time are very large, hence hardware implementation is required. Literature suggests that existing general-purpose SVM architectures do not scale well in terms of required hardware resources, complexity, data transfer (wiring) and memory management, primarily because of two important constraints; the number of support vectors (SVs) and their dimensionality. An SVM with a small number of SVs, requires only a few computational modules, however, several applications may require a large number of high dimensional SVs. As such, a general-purpose SVM architecture with a large number of modules faces design challenges such as fan-out and routing, increased complexity, and other performance limitations. An efficient SVM architecture needs to be scalable, provide real-time performance, and be easy to design and implement.

Normally, SVM classifies data into two classes. But we have extended it for six classes using One versus All approach of SVM in which six classifiers are constructed during training. While testing for unknown sample, the class for which the classifier gives highest value of decision function, will be declared as the class of unknown. We classify facial expression into one of the six basic facial expression such as smile, surprise, sad, anger, disgust, fear using six class SVM. Simple hardware architecture for implementation of multiclass SVM classifiers on FPGA is presented. In MATLAB, SVM is trained for six classes, and the extracted parameters are used to implement SVM on the hardware. In the architecture, kernel operation and classification of pairwise classifiers is designed. Cohn Kanade database which consists of image sequences of different facial expressions is used for training and testing of SVM. Difference based approach of partial reconfiguration is applied to this design with the help of EDA tools by XILINX.

The technique presented in this paper depict that after reconfiguration, power (static as well as dynamic) is reduced by a significant amount. Though the work has been done targeting Xilinx Virtex-6 FPGA, the method is general enough to be applied on any FPGA featuring Partial Run Time Reconfiguration (PRTR).

This paper briefly explains feature classification using SVM in section II. In the same section we explain basic principles of SVM and how we are using SVM as classifier for facial expression recognition system. Section III provides a brief overview of other hardware implementations of SVM. Our architecture is presented in section IV. Results are presented

978-1-4673-0438-2/12 $31.00 © 2012 IEEE

in section V. Finally section VI provide conclusions and future work.

II. FEATURE CLASSIFICATION USING SVM

A. Support Vector Machine

Basically SVM is a binary classifier that classifies data into two classes. During the training phase it constructs separating hyperplanes between training samples of two classes, labeled as +1 and -1. The separating hyperplane that best separates the two classes is called the maximum-margin hyperplane and forms the decision boundary for classification. The data samples that lie at the boundary of each class are called support vectors (SVs). During the testing phase only support vectors are involved in the computation. When two data classes are not linearly separable, a kernel function is used to project data to a higher dimensional space (feature space), where linear classification is possible. This is known as the kernel trick and allows an SVM to solve nonlinear problems.

An n-dimensional pattern x has n coordinates, x = $(x_1, x_2, \ldots, x_n)$, where each $x_i \in R$ for $i = 1, 2, \ldots, n$. Each pattern x_j belongs to a class $y_i \in \{-1, +1\}$. Consider a training set T of m patterns together with their classes, $T = \{(x_1, y_1), (x_2, y_2), \ldots, (x_m, y_m)\}$. Consider a dot product space S in which the patterns x are embedded, $x_1, x_2, \ldots x_m \in S$. Any hyperplane in the space S can be written as

$$\{x \in S \mid w \cdot x + b = 0\}, \quad w \in S, \ b \in R \quad (1)$$

The dot product w·x is defined by

$$w \cdot x = \sum_{i=1}^{n} w_i x_i \quad (2)$$

A hyperplane $w \cdot x + b = 0$ can be denoted as a pair (w, b). A training set of patterns is linearly separable if at least one linear classifier exists defined by the pair (w, b), which correctly classifies all training patterns. All patterns from class +1 are located in the space region defined by $w \cdot x + b > 0$, and all patterns from class -1 are located in the space region defined by $w \cdot x + b < 0$. Using the linear classifier defined by the pair (w, b), the class of a pattern x_k is determined with

$$\text{class}(x_k) = \left\{ \begin{array}{l} +1 \ \text{if} \ w \cdot x + b > 0 \\ -1 \ \text{if} \ w \cdot x + b < 0 \end{array} \right\} \quad (3)$$

The distance from a point x to the hyperplane defined by (w, b) is

$$d(x \ ; \ w, b) = \frac{|w \cdot x + b|}{\|w\|} \quad (4)$$

where $\| w \|$ is the norm of vector w. Of all the points on the hyperplane, one has the minimum distance d_{min} to the origin

$$d_{min} = \frac{|b|}{\|w\|} \quad (5)$$

We consider the patterns from the class -1 that satisfy the equality $w \cdot x + b = -1$ and that determine the hyperplane H_1, the distance between the origin and the hyperplane H_1 is equal to $|-1 - b| / \|w\|$. Similarly, the patterns from the class +1 satisfy the equality $w \cdot x + b = -1$ and that determine the hyperplane H_2. the distance between the origin and the hyperplane H_2 is $|+1 - b| / \|w\|$. Hyperplanes H, H_1, and

H_2 are parallel and no training patterns are located between hyperplanes H_1 and H_2. So the margin of linear classifier H is $2/\|w\|$. The optimum separation hyperplane conditions can be formulated into the following expression that represents a linear SVM, minimize $f(x) = \frac{\|w\|^2}{2}$ with the constraints $g_i(x) = y_i(w \cdot x_i + b) - 1 \geq 0, i = 1, \ldots, m$. The optimization problem represents the minimization of a quadratic function under linear constraints (quadratic programming). A convenient way to solve constrained minimization problems is by using a Lagrangian function. When the Lagrange function is introduced, a Lagrange multiplier λ_i is assigned to each training pattern via the constraints $g_i(x)$. The training patterns from the SVM solution that have $\lambda_i > 0$ represent the support vectors. The training patterns that have $\lambda_i = 0$ are not important in obtaining the SVM model, and they can be removed from training without any effect on the SVM solution. After training, the classifier is ready to predict the class membership for new patterns, different from those used in training. The class of a pattern x_k is determined with

$$\text{class } (x_k) = \left\{ \begin{array}{l} +1 \ if \ w \cdot x_k + b > 0 \\ -1 \ if \ w \cdot x_k + b < 0 \end{array} \right\} \quad (6)$$

Therefore, the classification of new patterns depends only on the sign of the expression $w \cdot x + b$. To predict new patterns without computing the vector w explicitly we will use the support vectors from the training set and the corresponding values of the Lagrange multipliers λ_i.

$$\text{class } (x_k) = \text{sign} \left(\sum_{i=1}^{m} \lambda_i y_i x_i \cdot x_k + b \right) \quad (7)$$

Patterns that are not support vectors do not influence the classification of new patterns. To classify a new pattern x_k, it is only necessary to compute the dot product between x_k and every support vector.

Some common kernels include [2]:
1) Linear : $K(\vec{x}, \vec{z}) = (\vec{x}.\vec{z})$
2) Polynomial : $K(\vec{x}, \vec{z}) = (1 + (\vec{x}.\vec{z}))^d$
3) Sigmoid : $K(\vec{x}, \vec{z}) = \tanh((\vec{x}, \vec{z}) + \theta)$
4) Radial Basis Function : $K(\vec{x}, \vec{z}) = \exp(||(\vec{x} - \vec{z})||^2/(2\sigma^2))$

Binary SVM can be modified for Multiclass classification using following approaches:
1) One Vs One : This method constructs $K(K-1)/2$ classifiers. Classifier ij, named f_{ij}, is trained using all the patterns from class i as positive instances, all the patterns from class j as negative instances, and disregarding the rest. When classifying a new instance each one of the base classifiers casts a vote for one of the two classes used in its training. The class which get maximum votes, will be declared as a class of new instance.
2) One Vs All : OVA method constructs K binary classifiers. Classifier i is trained using all the patterns of class i as positive instances and the patterns of the other classes as negative instances. Test sample is classified in the class whose corresponding classifier has the highest output.

978-1-4673-0438-2/12 $31.00 © 2012 IEEE

B. Facial expression recognition System

In this system we are using Viola Jones algorithm for face detection and Candide wire frame model, and active appearance model for tracking facial expressions in image sequences and SVM for classification. The first frame of the image sequence is assumed to be of neutral facial expression and last frame corresponds to full facial expression. After face detection, Candide wire frame model is fitted on face image of first frame. Candide model has 113 vertices and 184 triangles. In the subsequent frames as the facial expressions changes, animation parameters of the model will change and model deforms. When we reach to last frame of image sequence, which corresponds to full facial expression, model will be fully deformed. Difference between the model vertices coordinates of first frame and last frame is given as input to SVM [3]. Only geometrical information is given as input to SVM, no texture information is required. Out of 113 vertices only 60 vertices are contributing to facial expressions. So we consider coordinates of only these 60 vertices, in order to reduce the dimensions of the vectors to $60 \times 2 = 120$.

The training phase of SVM produced the following data.

Table I
NUMBER OF SUPPORT VECTORS FOR EACH CLASS

S. No.	Facial Expression	No of Training Samples	No of Support Vectors	Bias Values
1	Smile	25	24	-2.6214
2	Surprise	25	7	-9.3837
3	Sad	21	12	-0.7933
4	Anger	20	26	-3.1861
5	Disgust	23	36	-1.8676
6	Fear	21	87	-0.9999

Class of any test image can be found out by (8), where $K(x_i, x_k)$ represent the kernel function. We use polynomial Kernel given by (9)

$$Decision\ Function\ D(\vec{x}) = \text{sign}\left(\sum_{i=1}^{m} \alpha_i y_i K(\vec{x}, \vec{s}_i) + b\right) \quad (8)$$

$$K(\vec{x}, \vec{s}_i) = (1 + \vec{x}.\vec{s}_i)^d \quad (9)$$

Here $\vec{s}$ represents the support vector (which is independently a column vector) of dimension $120 \times$ (no of support vectors of any particular class), and $\vec{x}$ is the test vector which is a row vector of dimension 1×120. α_i is the weight of support vector obtained after training. Here d is considered to be 1.

III. RELATED WORK ON HARDWARE IMPLEMENTATION OF SVM

Hardware implementations of SVMs have gained noticeable interest in recent years, primarily because of the potential real-time performance benefits they offer. Different efforts have been made towards the design of efficient architectures for SVM training and evaluation. Probably the first account of a hardware design for an SVM system is found in [4], an efficient implementation of a kernel based perception is shown in fixed-point digital hardware. The performance of the proposed hardware implementation with the results of a simulated SVM with both fixed and floating point underlying math is compared. Similar performance metrics is obtained among the different models, and considered the hardware design a better alternative due to the small amount of bits utilized for coding and computation. In [5] the same authors have proposed a digital architecture for nonlinear SVM learning. Its implementation on a field programmable gate array (FPGA) was discussed, along with the evaluation of the quantization effects on the training and testing errors. The authors established that a minimum of 20 bits for individual coefficient coding was required to maintain adequate performance. A different approach has been proposed by Khan [6], who presented an implementation of a digital SVM linear classifier using logarithmic number systems (LNS). They used LNS in order to transform the multiplication operations involved in evaluating the decision function into addition computations, which is a less consuming task. Their design was implemented into a Xilinx FPGA Spartan3 XL3S50pq208-5 device. Their analysis showed that the LNS hardware implementation had a classification accuracy equivalent to a LNS software simulation and to a floating point software implementation. Its performance was equivalent to a 20-bit fixed-point implementation, as was reported by Anguita, but the LNS version required 25% less slices of the Xilinx FPGA. Recent studies with different data sets have allowed these authors to conclude that even a 10-bit LNS architecture was guaranteed to match the performance of a double precision floating point alternative **[7]**.

Most implementations have targeted either specific applications, or require a large amount of computational resources, and thus are not suitable for general-purpose embedded environments. An attempt to design a massively parallel architecture to accelerate learning algorithms was also presented, which utilized arrays of vector processing elements controlled through a host CPU in SIMD fashion. The centralized nature of the CPU, however, creates a bottleneck, emphasizing the need for distributed control. An FPGA implementation utilizing a Microblaze processor as a control unit and custom hardware coprocessors to perform the SVM classification was presented in **[8]**. The design was restricted to 8-dimensional vectors, limiting the application space. An integrated vision system for object detection and localization based on SVM was presented in **[9]**, however, the system offered limited scalability. An analog custom processor was presented in **[10]**. FPGA-based coprocessor for Support Vector Machine training and classification in [11], their architecture uses custom low arithmetic precision and is based on clusters of vector processing elements that leverage the FPGA's DSP. In [12] they presented a new systematic approach based on the concept of scalable effort hardware, for the design of efficient hardware implementations for algorithms that demonstrate inherent error resilience. Scalable effort design is based on the identification of mechanisms at each level of design abstraction (circuit, architecture and algorithm) that can be used to vary the computational effort expended towards generation of the correct (exact) result. SVM implementations have also been explored using embedded processors in **[13]** and **[14]**.

IV. PROPOSED APPROACH

Decision function of SVM for classification is given by equation (8). The data flow of this computation is illustrated in Figure 1.

- Kernel Function Calculation: In this calculation two matrices (one is a test vector and another is support vector) are multiplied and a row vector is the output
- In the next step α_i and y_i are multiplied separately.
- Multiplication of α_i and y_i is further multiplied by the kernel function output, producing a scalar value.
- Now this scalar value is added with bias value to produce the final value of the class for which calculation is done.

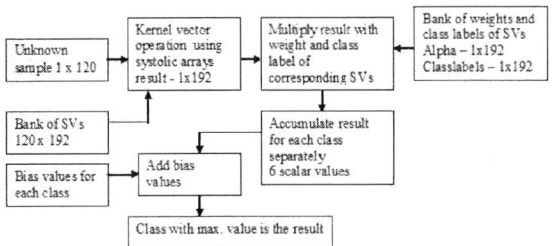

Figure 1.   Data flow of proposed approach

The above calculation is done for each and every class. Test Sample $\vec{x}$ belongs to the class for which value of Decision Function is maximum.

Table II
DIMENSIONS FOR DIFFERENT MATRICES

S No	Class	Test Vector	Support Vector	Kernel	$\alpha_i y_i$	Decision Function
1	Smile	1×120	120×24	1×24	1×1	1×1
2	Surprise	1×120	120×7	1×7	1×1	1×1
3	Sad	1×120	120×12	1×12	1×1	1×1
4	Anger	1×120	120×26	1×26	1×1	1×1
5	Disgust	1×120	120×36	1×36	1×1	1×1
6	Fear	1×120	120×87	1×87	1×1	1×1

A. Matrix Multiplication using Systolic Arrays

A systolic array is a computing network possessing [15] with the following features: Systolic Arrays are regular arrays of simple Processing Elements (PEs), where each processing element in the array is identical. Systolic algorithm relies on data from different directions arriving at PEs in the array at regular intervals and being combined (This combination may mean any computation like successive multiplication or addition etc.). Systolic Arrays are the architectures preferably used in the matrix multiplication operations. This chain of PEs operates in a pipelined manner and can be expanded vertically or horizontally with minor modifications to operate in a systolic manner. Typical systolic array structure is shown

in Figure 2. Block diagram of our proposed systolic array architecture is shown in Figure 3. Total number of support vectors is 192. So we need 192 Processing elements (PEs). In each PE we store one support vector, and its corresponding alpha and class label values. Each support vector has 120 elements. Internal architecture of PEs is shown in Figure 4. In each PE vector operation of SV and input vector is performed, generating a scalar value, which is then multiplied with alpha and class label, simultaneously input vector elements are passed to next PE.

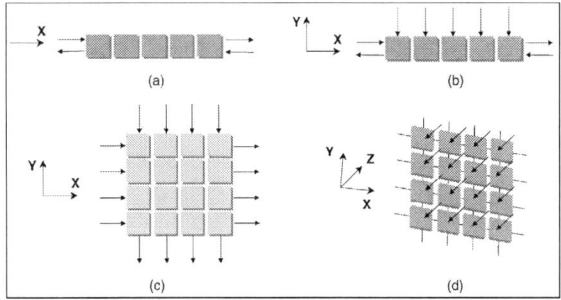

Figure 2.   Systolic Array Classification based on Data I/O. (a) Linear array with one-dimensional I/O. (b) Linear array with two-dimensional I/O. (c) Planar array with Perimeter I/O. (d) Planar array with three-dimensional I/O, offering planar data streams with area I/O. [16]

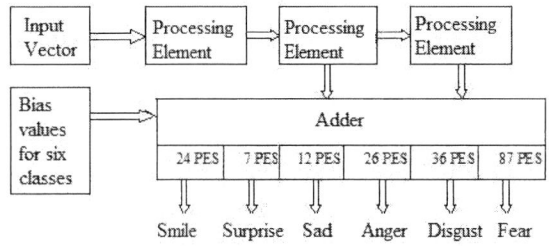

Figure 3.   Block diagram of Systolic array implementation

All PEs gives 192 scalar values which are stored in registers. Then class wise these values are added together along with bias values, e.g. first 24 values are added together along with first bias value. Then next 7 values are added together along with second bias values and so on till sixth class. The class which gives maximum value is the class of unknown input vector. Then we perform partial reconfiguration. Partial Reconfiguration is the ability to dynamically modify blocks of logic by downloading partial bit files while the remaining logic continues to operate without interruption. Partial Reconfiguration enables system flexibility, perform more functions while maintaining communication links. Other advantages are size and cost Reduction, time-multiplex the hardware to require a smaller FPGA and power reduction using shutting down power-hungry tasks, when not needed. High Performances, special purpose computer systems are typically used to meet specific application requirements or to off-load computations that are especially taxing to general-purpose computers.

978-1-4673-0438-2/12 $31.00 © 2012 IEEE   65

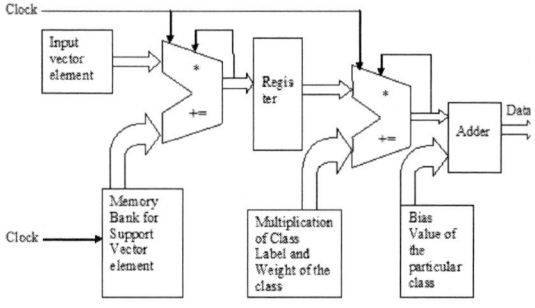

Figure 4. Internal architecture of PE

As hardware cost and size continue to drop and processing requirements become well-understood in areas such as signal processing and image processing, more special purpose systems are being constructed. Because the knowledge gained from individual experiences is neither accumulated nor properly organized, the same errors are repeated. I/O and computation imbalance is a notable example-often, the fact that I/O interfaces cannot keep up with device speed is discovered only after constructed a high speed, special-purpose device.

B. Low Power Reconfiguration Strategy

Partial reconfiguration (PR) is the ability for a portion of an FPGA to be reprogrammed while the remainder of the system remains unchanged. A partial bit stream loads only a portion of the design onto the FPGA rather than rewriting the entire design. Partial reconfiguration is especially useful for reprogramming a portion an FPGA during operation without affecting the rest of the system. This practice is called dynamic partial reconfiguration. Static partial reconfiguration refers to reprogramming a portion of the FPGA while the rest of the board is in a reset state [17]. Difference-based partial reconfiguration is used to only make small changes in the FPGA design. The generated bit stream only includes differences between designs. Difference-based partial reconfiguration allows for faster reprogramming of the device since only the changes must be rewritten, but it has limited applications as compared to the module-based solution. Systolic array implementation of SVM classifier is reconfigured as difference based approach. In a system implementing module-based partial reconfiguration, modules that are to be kept in continuous operation without the capability of being partially reprogrammed are referred to as static modules. One or more modules can be designated as the partially reconfigurable module(s), which require additional considerations in the design, synthesis, and implementation stages. Specifically, a modular design flow must be used which will synthesize and create separate bit streams for each of the reconfigurable modules as well as a total system bit stream including all of the static logic and one implementation of each of the reconfigurable modules. Partial reconfiguration can be implemented through a JTAG connection to a PC or internally through custom logic or an on-board processor, such as the embedded PowerPC in the Virtex II Pro FPGA. Partially

reprogramming the FPGA through internal circuitry, referred to as self-reconfiguration, is a much more useful method of partial reconfiguration since it eliminates the need for an external PC. The partial bit streams are stored in memory and are written to the Internal Configuration Access Port (ICAP) of the FPGA in order to reconfigure the specified region of the board with the new logic.

C. Reconfiguration Approach and Low Power Implication

Partial Reconfiguration of any design on FPGA leads to save the area and power (both static and dynamic in most of the cases). When design is implemented on FPGA, a particular area is occupied by the logics and different components used in the design. While performing Difference based approach of PR for Systolic Array implementation of SVM Classifier, routing of the nets is changed to reconfigure the particular design on FPGA. This further produces new native circuit description (ncd) file to further produce new bit streams. This partial bitstream is actually the difference of the previous ncd file and new ncd file. When new bit file is loaded to FPGA, using different EDA tools, it is seen very frequently that power (dynamic and static) is lowered than that of before.

V. IMPLEMENTATION AND SYNTHESIS RESULTS

All Implementations are done for target device XILINX 6vlx240tff1156-2. Systolic array matrix multiplication is a resource hungry approach because, in each step, a register is required to store the intermediate data. Slices LUTs used is 536 (in number), which can be justified because of large number of registers usage. Also, since register contents are toggling, hence number of LUTs with or without flip flops is significantly large.

Table III
PRIMITIVE AND BLACK BOX USAGE

GND	1	MUXF7	104
INV	4	VCC	1
LUT1	31	XORCY	224
LUT2	196	FD	45
LUT3	11	FDE	192
LUT4	10	FDE_1	224
LUT5	86	BUFGP	1
LUT6	198	OBUF	224
MUXCY	261	DSP48E1	7

Table III shows the primitive and black box usage of LUTs, DSP48s and clocks. Since output value travels to the end PE, hence by table VIth, it can be seen that minimum period is significantly large. Device utilization summary is given in Table IV. Timing summary is given in Table V. Power Optimization is done by using different EDA tools having input as native circuit description files. On chip power summary is given in Table VI. Without reconfiguration, ncd files are obtained using implementation and synthesis tool of XILINX (ISE). Some constraints have been added to produce the switching activity file. These constraints then further be the cause of the power reduction in the design after reconfiguration. Run Time Dynamic Partial Reconfiguration is performed on the design to produce comparison in the total power consumption, before and after the strategy. The total power consumption is the sum

978-1-4673-0438-2/12 $31.00 © 2012 IEEE

of static power and dynamic power. Power result for Systolic Array Implementation of SVM classifier is obtained by the use of difference based PR terminology. In this terminology, slice logic is changed i.e. routed differently than the original design to partially reconfigure it. Also, some nets are changed to produce new partial bit file, which is the difference of the previous one and the current one. Power supply summary is given in Table VII.

Table IV
DEVICE UTILIZATION SUMMARY

Device: XILINX 6vlx240tff1156-2	Utilized	Available	Percentage Utilization
Number of Slice Registers	461	301440	0.15%
Number of Slice LUTs	536	150720	0.35%
Number of LUT- Flip Flop Pairs with an unused Flip Flop	289	750	38%
Number of LUT- Flip Flop Pairs with an unused LUT	214	750	28%
Number of LUT- Flip Flop Pairs with fully used LUT-FF	247	750	32%
Number of IOBs	225	600	37%
Number of BUFG/BUFCTRLs	1	32	3%
Number of DSP48E1s	7	768	0.91%

Table V
TIMING SUMMARY

Period / Delay	Time (ns)
Minimum Period	6.844 ns
Maximum output required time after clock	0.654 ns

Table VI
ON-CHIP POWER SUMMARY

On-Chip	Power before Reconfiguration (mW)	Power after Reconfiguration (mW)
Clock	3.57	3.57
Logic	2.96	2.96
Signals	2.03	2.03
I/Os	0.18	0.18
DSPs	1.69	1.69
Leakage	1942.17	1921.73

Table VII
POWER COSUMPTION SUMMARY

Power (W)	Total (W)	Dynamic (W)	Quiescent (W)
Before Reconfiguration	2.042	0.010	2.031
After Reconfiguration	2.021	0.010	2.011

VI. CONCLUSIONS AND FUTURE WORK

SVMs exhibit high classification accuracy for different training sets and good generalization performance on data that is highly variable and difficult to separate, making them particularly suitable for expression recognition. Initially hardware description is written for implementation of SVM Classifier. Synthesis result for the design is obtained to have a look in the utilization percentage for different logic circuits and slices etc. on FPGA. Partial Reconfiguration of FPGA device (target board 6vlx240tff1156-2) is then performed. Difference based technique is used to perform partial reconfiguration for Systolic Array implementation of SVM Classifier. Power consumption (static and dynamic) is then evaluated before and after reconfiguration of the design. Power reduction up to 3 to 5 percent has been observed by using XILINX power analyzer EDA tool, due to Reconfiguration of the device. The goal of the technique has been to achieve power reduction due to PR. Hence, metric used for optimization was power. Area and Time also could be a metric for the designs to be performed. In the SVM classification domain, different kernel functions can be used to enhance the accuracy for test samples given for all the facial expressions. Also to perform matrix multiplication operation, faster and efficient methods could be explored in spite of Systolic Array approach. We are modifying the work using module based partial reconfiguration to achieve more power reduction.

REFERENCES

[1] C. Cortes and V. Vapnik, "Support-vector networks," *Machine Learning*, vol. 20, pp. 273–297, 1995.

[2] C. Kyrkou and T. Theocharides, "SCoPE: Towards a systolic array for SVM object detection," *Embedded Systems Letters, IEEE*, vol. 1, no. 2, pp. 46 –49, Aug. 2009.

[3] R. Patil, V. Sahula, and A. Mandal, "Features classification using support vector machine for a facial expression recognition system," in *Springer Machine Vision and Applications. (submitted)*.

[4] D. Anguita, A. Boni, and S. Ridella, "Digital kernel perceptron," *Electronics Letters*, vol. 38, no. 10, pp. 445 –446, May 2002.

[5] ——, "A digital architecture for support vector machines: theory, algorithm, and FPGA implementation," *IEEE Transactions on Neural Networks*, vol. 14, no. 5, pp. 993 – 1009, Sept. 2003.

[6] F. Khan, M. Arnold, and W. Pottenger, "Hardware-based support vector machine classification in logarithmic number systems," in *IEEE International Symposium on Circuits and Systems*, vol. 5, May 2005, pp. 5154 – 5157.

[7] ——, "Finite precision analysis of support vector machine classification in logarithmic number systems," in *Euromicro Symposium on Digital System Design*, Aug 2004, pp. 254 – 261.

[8] I. Biasi, A. Boni, and A. Zorat, "A reconfigurable parallel architecture for SVM classification," in *IEEE International Joint Conference on Neural Networks*, vol. 5, July 2005, pp. 2867 – 2872.

[9] R. Reyna, D. Esteve, D. Houzet, and M.-F. Albenge, "Implementation of the SVM neural network generalization function for image processing," in *Proceedings Fifth IEEE International Workshop on Computer Architectures for Machine Perception*, 2000, pp. 147 –151.

[10] R. Genov and G. Cauwenberghs, "Kerneltron: support vector "machine" in silicon," *IEEE Transactions on Neural Networks,*, vol. 14, no. 5, pp. 1426 – 1434, Sept. 2003.

[11] S. Cadambi, I. Durdanovic, V. Jakkula, M. Sankaradass, E. Cosatto, S. Chakradhar, and H. Graf, "A massively parallel FPGA-based coprocessor for support vector machines," in *17th IEEE Symposium on Field Programmable Custom Computing Machines*, april 2009, pp. 115 –122.

[12] V. Chippa, D. Mohapatra, A. Raghunathan, K. Roy, and S. Chakradhar, "Scalable effort hardware design: Exploiting algorithmic resilience for energy efficiency," in *Design Automation Conference (DAC), 2010 47th ACM/IEEE*, june 2010, pp. 555 –560.

[13] R. Pedersen and M. Schoeberl, "An embedded support vector machine," in *International Workshop o Intelligent Solutions in Embedded Systems*, June 2006, pp. 1 –11.

[14] S. Dey, M. Kedia, N. Agarwal, and A. Basu, "Embedded support vector machine : Architectural enhancements and evaluation," in *20th International Conference on VLSI Design*, Jan. 2007, pp. 685 –690.

[15] S. Y. Kung, *VLSI Array Processors*, Mar. 1988.

[16] S. Chai and D. Wills, "Systolic opportunities for multidimensional data streams," *IEEE Transactions on Parallel and Distributed Systems*, vol. 13, no. 4, pp. 388 –398, Apr. 2002.

[17] A. Zeineddini and K. Gaj, "Secure partial reconfiguration of FPGAs," in *IEEE International Conference on Field-Programmable Technology*, Dec. 2005, pp. 155 –162.

[18] S. Rao and T. Kailath, "Regular iterative algorithms and their implementation on processor arrays," *Proceedings of the IEEE*, vol. 76, no. 3, pp. 259 –269, Mar. 1988.

[19] T. Tuan and B. Lai, "Leakage power analysis of a 90nm FPGA," in *Proceedings of the IEEE Custom Integrated Circuits Conference*, Sept. 2003, pp. 57 – 60.

[20] J. Anderson and F. Najm, "Active leakage power optimization for FPGAs," *IEEE Transactions on Computer-Aided Design of Integrated Circuits and Systems*, vol. 25, no. 3, pp. 423 – 437, March 2006.

2012 25th International Conference on VLSI Design

A High Speed FIR filter Architecture based on Novel Higher Radix Algorithm

S.K.Sahoo, Srinivasa Reddy K

Birla Institute of Technology and Science,
Department of Electrical and Electronics Engineering, Pilani, India,
Email: subhendu_k@yahoo.com, srinivas.nrt@gmail.com

Abstract—**Redundant binary (RB) number systems are becoming popular because of its unique carry propagation free addition property. A finite impulse response (FIR) filter computes its output using multiply and accumulate operations. In the present work, a FIR filter based on novel higher radix-256 and RB arithmetic is implemented. The use of radix-256 booth encoding reduces the number of partial product rows in any multiplication by 8 fold. In the present work inputs and coefficients are considered of 16-bit. Hence, only two partial product rows are obtained in RB form for each input and coefficient multiplications. These two partial product rows are added using carry free RB addition. Finally the RB output is converted back to natural binary (NB) form using RB to NB converter. The performance of proposed multiplier architecture for FIR filter is compared with computation sharing multiplier (CSHM) implementation in 90nm technology. The proposed multiplication method for FIR filter is found to be faster approximately by 42% in comparison to CSHM implementation, however with 0.5% and 11% increase in area and power respectively.**

Keywords- Multiplier, Radix-256, Redundant binary addition (RBA).

I. INTRODUCTION

Rapid growth in mobile communication systems, wireless communication systems leads to high performance and low power VLSI digital signal processing systems. Filters and other signal processing units used in these systems need to have high speed and low power realization. Finite impulse response (FIR) filters play a crucial role in many signal processing applications. Various tasks such as spectral shaping, matched filtering, interference cancellation, channel equalization, etc. can be performed with these filters due to its absolute stability and linear-phase property. Hence, various architectures and implementation methods have been proposed to improve the performance of filters in terms of speed and complexity.

Since complexity reduction of FIR filter leads to high performance as well as low-power design, authors of [1]-[4] have focused on the implementation of FIR filters with discrete coefficient values selected from the powers-of-two coefficient space. Canonical sign digit codes, numbers can be represented as sums or differences of powers-of-two, are often used to represent the FIR filter coefficients. By using add and shift operations FIR filter structure can be simplified. Optimization algorithm is used by Mustafa et. all in [5] to find the coefficients with reduced number of signed-power- of-two. The Differential Coefficients Method is proposed by Sankarayya in [6] and by A. P. Vinod in [7]. This involves

using various orders of differences between coefficients, along with stored intermediate results, rather than using the coefficients themselves directly for computing the partial products in the FIR equation. In [8] authors have tried to implement the FIR filter in cascade form. However the numerical design and optimization of such structure are of much more difficult than the single-stage.

In the present work, simple and high performance FIR filter architecture is proposed. This is based on the Novel Partial product Generation method using Higher radix-256 Booth encoding (NPGHB). Use of radix-256 encoding reduces the number of partial product rows by 8-fold. Hence for a 16-bit coefficient multiplication with input, only two rows of partial products will be obtained. Again RB number system has unique characteristic of carry propagation free addition. The use of RB addition to add two partial product rows makes multiplication operation much faster.

As the proposed work is based on CSHM algorithm [9], same is discussed in brief in section II. The RB number system and the higher radix booth encoding is discussed in section III. The proposed architecture and its implementation with comparison are discussed in section IV and V. Finally the concluding remark is given in section VI.

II. COMPUTATION SHARING MULTIPLICATION IN FIR FILTER

A simple filter computation using convolution is given as:

$$Y(n) = \sum_{k=0}^{M-1} C_k X(n-k) \qquad (1)$$

Where C_k 's are filter coefficients and $X(n-k)$ are the input to the filter delayed by k units.

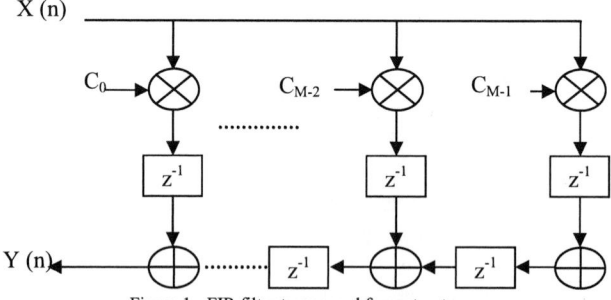

Figure 1. FIR filter transposed form structure

978-1-4673-0438-2/12 $31.00 © 2012 IEEE   68

This can be implemented using the direct or the using the transposed direct form realization as given in Figure 1. In this FIR filter the coefficient multiplication C_0, C_1,---,C_{M-1} with input X(n) using conventional multiplier is hardware expensive and time consuming. For high performance a computation sharing multiplication (CSHM) approach for FIR filter implementation as shown in Figure. 2 is proposed in [5]. In this the coefficient is divided in to group of small bit sequence named as *alphabet*. Multiplication of input with each alphabet is obtained by selecting from the outputs of a precomputer and then shifting as required. Finally a multiplication result of a coefficient with input is obtained by adding the alphabet multiplication results with required shifting.

For illustration of use of CSHM algorithm, a 16-bit coefficient C_0 = 1010 1101 1011 0110 can be divided into four alphabets α_3=1010, α_2=1101, α_1=1011 and α_0 = 0110. An alphabet of 4-bit, can have possible decimal values of 0, 1, ----, 14 or 15. All these alphabet multiplications can be obtained from few fundamental alphabet multiplications i.e. 1X, 3X, 5X, 7X, 9X, 11X, 13X and 15X, where X is the input. Hence these fundamental alphabet multiplications are obtained using pre-computer stage as shown in Figure 2. Then based on the alphabets of each coefficient, outputs of pre-computer are selected and finally added to get the coefficient multiplication using select and add units(S &A). For the present illustration, the multiplication of coefficient with X is:

$$C_0 * X = 2^{12}(\alpha_3 * X) + 2^8(\alpha_2 * X) + 2^4(\alpha_1 * X) + (\alpha_0 * X)$$

Here $\alpha_3 * X$ = $(1010)_2$ = $(10)_{10} * X$ can be obtained by selecting the fundamental alphabet multiplication 5X from precomputer output and shifting by one bit. Similarly all other alphabet multiplications can be obtained by selecting the proper fundamental alphabet multiplication and then shifting as required using the S & A unit. Then alphabet multiplications are aligned properly and then added to get coefficient multiplication of C_0 with X.

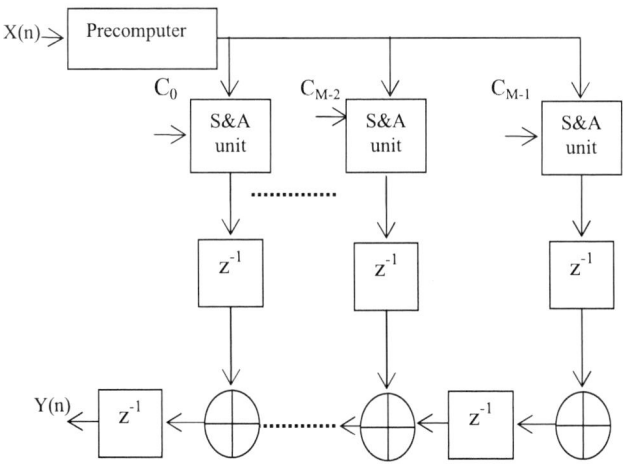

Figure 2. FIR filter using CSHM algorithm

III. REDUNDANT BINARY NUMBER SYSTEM AND ARITHMETIC BASICS

A. RB number system

RB number system is a part of signed digit representation proposed by Avizienis [10]. It has a fixed radix 2 and a digit set { $\bar{1}$, 0, 1} where $\bar{1}$ denotes –1. An n-digit RB integer Y = [y_{n-1} y_n . . . . y_0] SD2 where $y_i \in$ { $\bar{1}$, 0, 1}, and SD2 represent signed digit with fixed radix-2. This Y has the value $\sum_{i=0}^{n-1} y_i * 2^i$. It is similar to an unsigned binary integer except that y_i can be $\bar{1}$. To represent these three symbols two bits are required. The most commonly used coding scheme, which is also used in this work, is shown in Table I.

TABLE I. RB CODING SCHEME

X^-	X^+	RB digit
0	0	0
0	1	$\bar{1}$
1	0	1
1	1	0

B. Addition of two NB number to give RB number

The addition of A and B is expressed as [11]

$$A + B = A - (-B) = A - (\bar{B} + 1) = (A - \bar{B}) - 1 = (A, \bar{B}) - 1$$

Hence this expression indicates that two NB numbers can be added to give the result by just taking the combination of A and complement of B. However to find out the actual value of A+B, 1 is to be subtracted.

C. Carry propagation free addition of two RB umbers

The RB number representation allows representing an integer in several ways. For example, [0101] SD2, [011$\bar{1}$,] SD2, [1$\bar{1}$01] SD2, [1$\bar{1}$1$\bar{1}$] SD2, [10$\bar{1}$ $\bar{1}$] SD2 all represents decimal '5'. Because of this property addition of two RB numbers can be performed without carry propagation in two steps [12].

Carry propagation free addition is performed in two steps:

1) First step: In this step intermediate carry C_i {$\in$ ($\bar{1}$, 0,1)} and the intermediate sum digit S_i {$\in$ ($\bar{1}$, 0,1)} are determined to satisfy the equation $a_i + b_i = 2 C_i + S_i$ where a_i and b_i are augend and addend digits, respectively. However C_i and S_i are determined such that both S_i and C_{i-1} are neither 1's, nor $\bar{1}$'s. Hence the rules for computation for intermediate sum S_i and carry C_i looking to the augend and addend digit (X_i, Y_i) and the digits at the next lower –order position (X_{i-1}, Y_{i-1}) is given in Table II.

2) Second step: In the second step, the sum digit Z_i {$\in$ ($\bar{1}$, 0, 1)} is obtained at each position by adding the intermediate sum digit S_i and the intermediate carry C_{i-1} from lower-order position without generating a carry. An example of addition of two RB numbers is given in Example I. Here the two numbers 67 and 157 are represented in RB form. In the first

978-1-4673-0438-2/12 $31.00 © 2012 IEEE

step digits of same bit location of augend and addend are added to get the intermediate sum and intermediate carry following the rule given in Table II, just looking at the digits of next lower-order position. In the next step intermediate sum and intermediate carry are added without any carry generation to give the final result.

TABLE II.   COMPUTATION RULES FOR THE FIRST STEP IN CARRY PROPAGATION FREE ADDITION

Augend Digit (X_i)	Addend digit (Y_i)	Digits at the next-lower-order position (X_{i-1}, Y_{i-1})	Intermediate carry (C_i)	Intermediate Sum (S_i)
1	1	Any value	1	0
1 0	0 1	Both are nonnegative.	1	$\bar{1}$
		Otherwise	0	1
0	0			
1 $\bar{1}$	$\bar{1}$ 1	Any value	0	0
0 $\bar{1}$	$\bar{1}$ 0	Both are nonnegative.	0	$\bar{1}$
		Otherwise	$\bar{1}$	1
$\bar{1}$	$\bar{1}$	Any value	$\bar{1}$	0

.

Augend	$1\ \bar{1}\ 0\ 1\ \bar{1}\ \bar{1}\ 0\ \bar{1}$	(67)
Addend	$+\ 1\ 1\ \bar{1}\ 0\ 0\ 0\ \bar{1}\ \bar{1}$	(157)

Intermediate Sum Si	$0\ 0\ \bar{1}\ 1\ 1\ 1\ 1\ 0$	
Intermediate Carry Ci	$+\ 1\ 0\ 0\ 0\ \bar{1}\ \bar{1}\ \bar{1}\ \bar{1}$	
Sum Zi	$1\ 0\ 0\ \bar{1}\ 0\ 0\ 0\ 0\ 0$	(224)

Example I.   Carry propagation free addition.

Thus in redundant binary number system, parallel addition of two numbers by a combinational circuit is performed in a constant time, independent of word length of operands.

D. RB to NB conversion

RB numbers are advantageous because of its unique carry propagation free addition property. However the real world interface must be in NB form. Hence finally the results obtained in RB form must be converted to NB form [13-14]. The conversion of an n-digit redundant binary integer $Y = [y_{n-1} \ y_{n-2} \ldots\ldots\ y0]$ SD2 where $y_i \in \{\bar{1}, 0, 1\}$ into the equivalent (n+1) bit 2's complement binary integer $U = [\ u_n u_{n-1} \ldots u_0]^{\bar{2}}$ where $u_i \in \{0,1\}$ is performed by subtracting Y^- from Y^+ where Y^+ and Y^- are n bit unsigned binary integers formed from the positive digits and negative digits in Y, respectively. Such an example is shown in Example II.

$Y\ = 1\ \bar{1}\ \bar{1}\ 1 = 3$
$Y^+ = 1\ 0\ 0\ 1 = 9$ $Y^+ = 1\ 0\ 0\ 1$
$Y^- = 0\ 1\ 1\ 0 = 6$ $\overline{Y^-} = 1\ 0\ 0\ 1$
$U = Y^+ - Y^- = Y^+ + (\overline{Y^-} + 1) = 3$ 1

$\boxed{1} \downarrow \quad 0\ 0\ 1\ 1 = 3$

Carry is neglected

Example II.   RB to NB conversion

E. Booth encoding for high speed multiplication

In a non Booth conventional multiplication, the number of partial product rows is same as the number of bits in multiplier. An N-bit two's complement binary multiplier C can be represented as $(C_{N-1} \ C_{N-2} \ldots\ldots\ C_2\ C_1\ C_0)$. The decimal value of this number will be:

$$C = -C_{N-1} \cdot 2^{N-1} + \sum_{i=0}^{N-2} C_i \cdot 2^i \qquad (2)$$

Booth algorithm [15] divides C into $\left\lceil \frac{N}{K} \right\rceil$ groups, each with k-bits using radix-2^k encoding. Hence each group of k+1 overlapping multiplier bits are mapped to a singed digits D_i, where i = 0, 1, $\frac{N}{K} - 1$. The digit D_i can be obtained from k+1 bits, i.e. $D_i = C_{(i.k)+k-1} \ldots\ldots\ C_{(i.k)+1} C_{i.k} C_{(i.k)-1}$ where $C_{(i.k)-1}$ is overlapping bit. This $C_{(i.k)-1}$ is the most significant bit of the previous digit D_{i-1} and is zero for i =0. The decimal value of digit D_i can be given as:

$$D_i = -C_{(i.k)+k-1} \cdot 2^{K-1} + \sum_{j=0}^{K-2} C_j \cdot 2^j + C_{(i.k)-1} \qquad (3)$$

A digit D_i for a radix-2^k encoding can take one value from $(-2^{k-1}, -2^{k-1} +1, \ldots, 1, 0, 1, \ldots, 2^{k-1}-1, 2^{k-1})$. Hence a multiplication of multiplier/coefficient C with multiplicand X can be given as:

$$C \cdot X = \sum_{i=0}^{\left\lceil \frac{N}{k} \right\rceil - 1} 2^{k.i} \cdot D_i . X$$

In this $2^{k.i}$ can be obtained just by shifting. Hence only $D_i.X$ has to obtained. A table for different higher radix booth encoding is given below.

TABLE III.   DIFFERENT RADIX BOOTH ENCODING

K	Radix 2^k	No of PP rows for 16-bit operand	Possible $D_i.X$ values	FDMPP required to be computed
2	Radix-4	8	-2.X, -X,0,X,2.X	All can be obtained from X. No FDMPP computation required
3	Radix-8	6	-4.X, -3.X, ..., -X, 0, X, ..., 3.X, 4.X	3.X
4	Radix-16	4	-8.X, -7.X, ..., -X, 0, X, ..., 7.X, 8.X	3.X, 5.X and 7.X
5	Radix-32	4	-16.X,-15.X, ..., -X, 0, X, ...,15.X,16.X	3.X, 5.X, 7.X, 11.X, 13.X and 15.X
6	Radix-64	3	-32.X, -31.X, ..., -X, 0, X, ..., 31.X, 32.X	3.X, 5.X, 7.X, 13.X, 15.X, ..., 27.X, 29.X and 31.X
7	Radix-128	3	-64.X, -63.X, ..., -X, 0, X, ..., 63.X and 64.X	3.X, 5.X, 7.X, 13.X, 15.X, ..., 59.X, 61.X and 63.X
8	Radix-256	2	-128.X, -127.X, ..., -X, 0, X, ..., 127.X, 128.X	3.X, 5.X, 7.X, ..., 125.X and 127.X. Use of RB arithmetic will need 3.X, 5.X and 7.X.

For various values of k, we can get different higher radix booth encoding. For k = 3, the encoding is Radix-8 booth encoding. The number of partial product rows will be $\left\lceil \frac{16}{3} \right\rceil = 6$. Hence this need possible partial product values to be obtained are -4.X, -3.X, …, -X, 0, X, …., 3.X, 4.X. All these can be obtained by shifting from X and 3.X. Hence the fundamental digit multiplied partial product (FDMPP) row need to be computed is 3.X. Similarly for k = 2, 4, 5, 6, 7 and 8 all those values are obtained and noted in Table III. For Radix-256, the number of FDMPP is large. However the use of RB arithmetic using the proposed algorithm will require only 3.X, 5.X and 7.X as FDMPP as are required in Radix-16. This also will generate only two partial product rows. Hence, Radix-256 booth encoding is used for the 16-bit input operand.

F. Novel Partial product Generation using Higher Radix-256 Booth encoding

In the present work a Novel Partial product Generation method using Higher radix-256 Booth encoding (NPGHB) is used. Use of this new scheme reduces the number of partial product rows to only two, with small additional complexity. For generating the partial product rows, addition of S-groups and T-groups, is used. Hence the four partial product rows generated using radix-256 can be added using RB adders in three stages. This reduces both delay and hardware requirement.

In radix-256 encoding, the binary number in two's complement form is partitioned to overlapping group of nine bits to form digit D_i. This D_i can have one of the 257 values from –128, -127,……..,-1, 0, 1,……, 127, 128. The partial product generation for digit D_i in the form of 2^m (where m=1, 2, 3, 4, 5, 6, 7) is easy, as can be obtained just by shifting. Generation of other partial products for digits, like 3, 5, 7, 9, 11, ---, 127 will be difficult. In the present work all partial products corresponding to different digits D_i in RB form are generated by only adding a S-group digit multiplied partial product(SGDMPP) with T-group digit multiplied partial product(TGDMPP). The SGDMPP are ZERO, X, 2X, 3X, 4X, 5X, 6X, 7X, 8X and the TGDMPP are ZERO, 16X, 32X, 48X, 64X, 80X, 96X, 112X, 128X. All SGDMPP and TGDMPP coefficients are obtained from the nine bits of the digit D_i using the algorithm given below.

Algorithm:

The following is the algorithm to find the fundamental S-group and T-group digits to represent a digit D_i.
Assume the digit D_0 is made of lower eight bits of coefficient and an overlapping bit $C_7C_6C_5C_4C_3C_2C_1C_0C_{-1}$.
Here S_group and T_group are temporary variables to store intermediate value. The S_digit and T_digit are the final digit values.

$S.group = 4 \times C_2 + 2 \times C_1 + 1 \times C_0 + 1 \times C_{-1}$
$T.group = 16 \times (4 \times C_6 + 2 \times C_5 + 1 \times C_4 + 1 \times C_3)$

If $C_7 = 0$ and $C_3 = 0$
    Then S_digit = S_group and T_digit = T_group

else if $C_7 = 0$ and $C_3 = 1$
    Then S_digit = S_group-8 and T_digit = T_group
else if $C_7 = 1$ and $C_3 = 0$
    Then S_digit = S_group and T_digit = T_group-128
else if $C_7 = 1$ and $C_3 = 1$
    Then S_digit = S_group-8 and T_digit = T_group-128
End

An example of multiplication using radix-256 encoding is given below.
Here inputs are X and Y, where

Y = 206 =00000000000000000000000011001110.

This coefficient can be grouped into four digits with D_0 = -50, D_1= 1, D_2 = 0, D_3 = 0.

Hence $Y = 2^{32}.D_3 + 2^{16}.D_2 + 2^8. D_1 + D_0$ and

$Y.X = 2^{32}. (D_3.X) + 2^{16}. (D_2.X) + 2^8. (D_1.X) + (D_0.X)$

Here $D_0.X$ can be obtained by adding -2X (the SGDMPP) with - 48X (the TGDMPP).
This SGDMPP can be obtained by shifting -X, 4-bit position. Similarly the TGDMPP can be obtained by shifting -3X by 3-bit position.
Similarly $D_1.X$ also can be obtained by selecting proper SGDMPP and TGDMPP and grouping them for addition. Finally $D_0.X$, and $D_1.X$ shifted by 8-bit positions are added using RB adders.

IV. PROPOSED FIR FILTER ARCHITECTURE BASED ON NPGHB

The architecture of improved FIR filter using CSHM multiplier with RB arithmetic and Radix-256 encoding is given in Figure 3. The input to this filter is X and the coefficients at each tap point are C_0, C_1, ……,C_{M-1}. Both input and the coefficient are assumed to be of 16-bits. Each 16-bit coefficient is grouped in to two digits D0 and D1. To generate a DMPP, D0.X or D1.X, a SGDMPP need to be subtracted from a TGDMPP. The SGDMPP and TGDMPP are obtained from FDMPP. The FDMPP are generated using a precomputer stage. To select proper SGDMPP and TGDMPP from all FDMPPs, control signals are generated using control signal generator from the bits of digit Di. The select and add unit consists of a select unit and a RB adder. The selector selects proper SGDMPP and TGDMPP from FDMPPs, which forms the DMPP in RB form. The two DMPPs, D0X and D1X shifted by 8-bit positions are added using the RB adder in select and add unit to get coefficient multiplication with input. The coefficient multiplication results from each tap point are added using RB adders as shown. As external interface understands NB number, finally the FIR filter output in RB form can be converted into NB form using RB to NB converter. The detailed circuit and architecture of different units are discussed in this section.

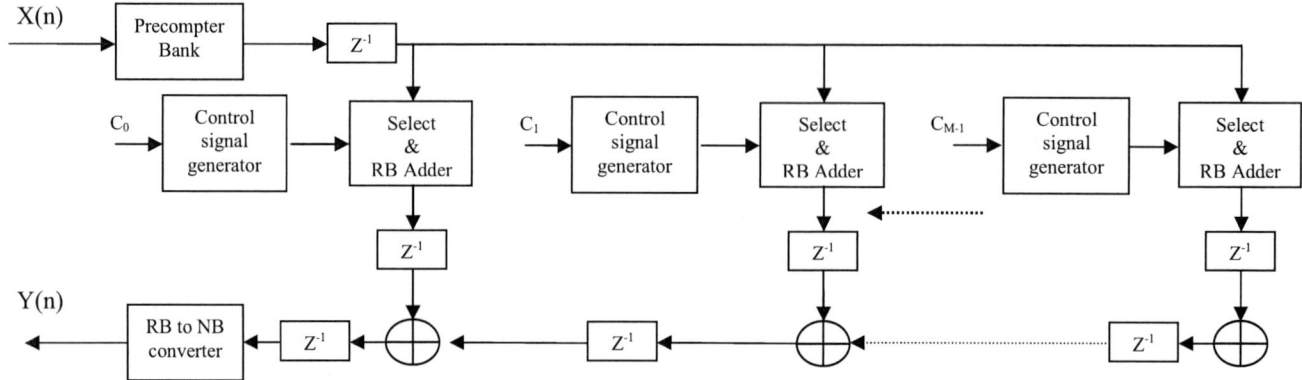

Figure 3. FIR filter architecture using radix-256 booth encoding and RB adder

A. Precomputer

The precomputer, computes FDMPPs i.e -1X, 3X, -3X, 5X, -5X, 7X and -7X. From these all other SGDMPP and TGDMPPs can be obtained by proper shifting. The computation method for these basic FDMPPs is given in TABLE IV. All of them are implemented using ripple carry adder except -5 X computations. The generation of -5X needs two stage additions. To have uniform propagation delay for all FDMPP computations in pre-computer -5X is implemented using square root select adder [16].

B. Control signal generator:

A DMPP corresponding to digit Di can be obtained by selecting proper SGDMPP and TGDMPP and grouping them for subtraction. For proper selection of SGDMPP and TGDMPP, specific control signals need to be given to selector unit of select and add unit. These control signals are obtained by control signal generator using the bits of digit. For a digit $D_i = c_7 c_6 c_5 c_4 c_3 c_2 c_1 c_0 c_{-1}$, the boolean expressions of the control signals which will select one each from TGDMPP(16X, 32X, 48X, 64X, 80X, 96X, 112X, 128X) and SGDMPP (-8X, -7X, ...,-1X, 0, 1X, ..., 7X, 8X) can be obtained using the algorithm discussed in section III.

TABLE IV. ALL FDMPP COMPUTATION METHOD USED IN PRE-COMPUTER

FDMPP	Computation method
-X	~X+1
3X	X<<1+1
-3X	X+~(X<<2)+1
5X	X<<2+X
-5X	~5X+1
7X	X<<3+~X+1
-7X	X+~8X+1

C. Selector:

Selector is a part of select and add unit. This selects one SGDMPP and other TGDMPP based on select signal for generation of a DMPP. The T-selector and S-selector circuits for getting single bit of a RB digit of DMPP is shown in Figure 4. Here P_{Ti} and P_{Si} are pair of bits corresponding to the ith RB digit of first row of partial product row corresponding to digit D0. In this work, the coefficient is considered of 16-bits, hence for two digits D0 and D1, two selector arrays are needed for generation of two RB partial product rows.

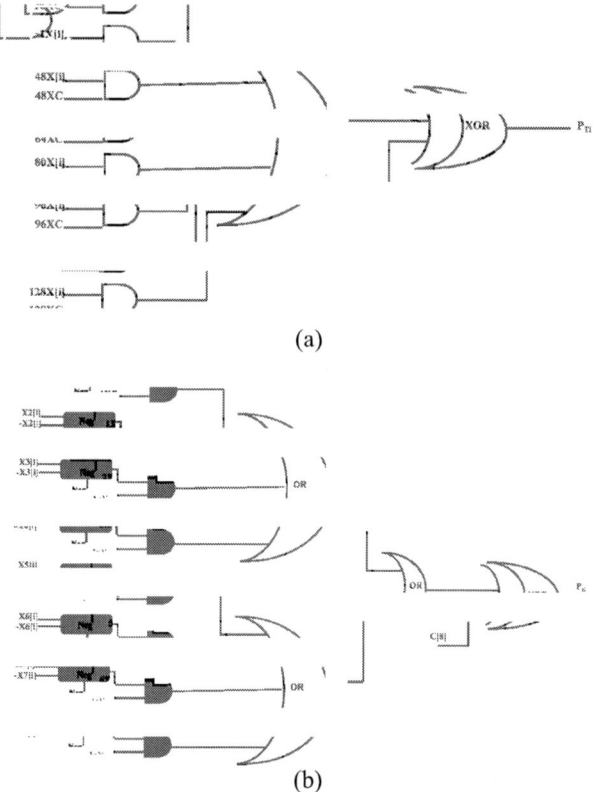

Figure. 4. The selector circuits for single digit generation of (a)T-group and (b)S-group digit

D. RB adder

A RB adder has unique characteristic of carry propagation free addition [17]. This is implemented using the algorithm given in Table II. RB adders are used in two places of FIR filter. First the two DMPPs are obtained in RB form. Then, from the selector circuits, these are added using RB adder and latched in registers. Then in the next clock cycle the multiplication result of one tap point and the previous tap point are added using RB adders.

E. RB to NB converter

To convert the final output of FIR filter into NB form, RB to NB converter is used. This can be implemented using a carry look ahead adder.

V. IMPLEMENTATION AND COMPARISON

In this section, an 11-tap FIR filter using the proposed architecture discussed in section IV was implemented. An 11-tap FIR filter based on CSHM architecture have also been implemented and compared in terms of delay, power and area. In the verification process identical inputs are used for both the architectures.

A. Implementation method used for FIR filter

The FIR filter using the NPGHRB and RB arithmetic as explained in Figure. 3 is implemented at gate level using Verilog coding. The functionality of the filter architecture was verified for large number of random input patterns using ModelSim simulator. Then the architecture is synthesized using Cadence. For synthesis purpose standard 90nm technology library is used [18]. Same methodology and technology is also used to synthesize FIR filter based on CSHM architecture. The post synthesis layout results are summarized in Table V.

TABLE V.   SUMMARY OF POST SYNTHESIS LAYOUT RESULTS

Filter based on	Delay(T) in ns	Area (A) in μm^2	Power(P) in mW	Power Delay Product
NPGHRB	7.138	0.141796	5.89	42.05
CSHM	12.486	0.141166	5.31	66.30

B. Post synthesis results and comparisons

After synthesizing the proposed architecture, the delay summary is obtained. Based on delay summery, the critical delay (T) is found to be 7.138ns in proposed architecture. The critical delay for filter based on CSHM is found to be the12.486ns. Hence the proposed filter architecture is 42% faster than the filter based on CSHM implementation. In the proposed architecture the critical delay path is, through select and RB adder to the latch where the coefficient multiplied input in RB form is stored. As control signals for selection purpose are already available and RB addition is carry free addition, the delay is small in the proposed architecture. However in terms of area (A) and power (P) the proposed architecture is slightly consuming more than CSHM implementation. This is because of the hardware expensive RB adders used in our architecture. For high speed applications in digital signal process (DSP), this architecture can be used.

VI. CONCLUSION

An FIR filter based on NPGHB and RB arithmetic is proposed. The use of radix-256 reduces the number of partial product rows in any multiplication by 8 fold. Thus a multiplication of 16-bit input results only two partial product rows in RB form. The obtained partial products in RB form are added using RB adder, which is much faster as no carry propagation occurs. The proposed multiplication method based on NPGHRB for FIR filter is found to be faster

approximately by 42% in comparison to CSHM implementation, however with 0.5% and 11% increase in area and power respectively. The idea presented in this paper can be used by designers to implement DSP algorithms for high speed applications.

REFERENCES

[1] Y. C. Lim, S. R. Parker, and A. G. Constantinides, "Finite word length FIR filter design using integer programming over a discrete coefficient space," IEEE Trans. Acoustics, Speech Signal Processing,vol. ASSP-30, pp. 661–664, Aug. 1982.

[2] Y. C. Lim and S. R. Parker, "FIR filter design over a discrete power-of-two coefficient space," IEEE Trans. Acoustics, Speech Signal Processing, vol. ASSP-31, pp. 583–591, June 1983

[3] H. Samueli, "An improved search algorithm for the design of multiplierless FIR filter with powers-of-two coefficients," IEEE Trans. Circuits Syst., vol. 36, pp. 1044–1047, July 1989

[4] Hai Huyen Dam, Cantoni A., Kok Lay Teo, Nordholm S.,"FIR Variable Digital Filter With Signed Power-of-Two Coefficients," IEEE Transactions on Circuits and Systems I, Volume: 54 , Issue: 6, pp. 1348 - 1357, June 2007.

[5] Mustafa Aktan, Arda Yurdakul, and Günhan Dündar,"An Algorithm for the Design of Low-Power Hardware-Efficient FIR Filters," IEEE Transactions on Circuits and Systems I, VOL. 55, NO. 6, pp. 1536-1545, JULY 2008

[6] N. Sankarayya, K. Roy, and D. Bhattacharya, "Algorithms for low power and high speed FIR filter realization using differential coefficients," IEEE Trans. Circuits Syst., vol. 44, pp. 488–497, June 1997.

[7] Vinod A.P., Singla A., Chang C.H., "Low-power differential coefficients-based FIR filters using hardware-optimised multipliers, " IET, Circuits, Devices & Systems, Volume: 1 Issue:1, pp.13-20, February 2007.

[8] Dong Shi, Ya Jun Yu, "Design of Discrete-Valued Linear Phase FIR Filters in Cascade Form," IEEE Transactions on Circuits and Systems I, Volume: 58 , No. 7, pp. 1627 - 1636, July 2011.

[9] Jongsun Park, Woopyo Jeong, Hamid Mahmoodi-Meimand, Yongtao Wang, Hunsoo Choo, Kaushik Roy., Computation sharing programmable fir filter for low-power and high-performance applications. IEEE Journal of Solid-State Circuits. 2004, 39: 348-357.

[10] Avizienis, "Signed-digit number representation for fast parallel arithmetic," IRE Trans. Electron. Computer., vol.EC-10, pp. 389-400, Sept.1961.

[11] Hiroshi Makino,Yasunobu Nakase, Hiroaki Suzuki, Hiroyuki Morinaka, Hirofumi Shinohara, and Koichiro Mashiko, "An 8.8-ns 54 x 54-Bit multiplier with high speed redundant binary architecture," IEEE J. Of Solid State Circuits, Vol.31, No.6, pp. 773-783, June 1996.

[12] N. Takagi, H. Yasura and S. Yajima., "High-speed VLSI multiplication algorithm with a redundant binary addition tree," IEEE Transactions on Computers, C-34(9): 217-220, Sept.1985.

[13] S. Kuninobu, T. Nishiyama, T. Edamatsu, T. Taniguchi, and N. Takagi, "Design of high speed MOS multiplier and divider using redundant binary representation," In Proc. 8th Symp. Computer Arithmetic, Italy, pp. 80-86, May 1987.

[14] Yum Kim, Bang-Sup Song, John Grosspietsch, and Steven F. Gilling , " A carry-free 54b x 54b multiplier using equivalent bit conversion algorithm, " IEEE J. Of Solid State Circuits, Vol. 36, No.10, pp. 1538-1544, Oct. 2001.

[15] A.D. Booth, "A signed binary multiplication technique," Quarterly J. Mech. Appl. Math, Vol. 4, Part 2, pp. 236-240, 1951.

[16] J. M. Rabaey, Digital Integrated Circuits: A Design Perspective. Englewood Cliffs, NJ: Prentice-Hall, 1996..

[17] Y. Harata, Y. Nakamura, H. Nagese, M. Takigawa, and N. Takagi, "A high-speed multiplier using a redundant binary adder tree," IEEE J. Solid-State Circuits, Vol. 22, pp. 28-34, Feb. 1987.

[18] 90-nano meter silicon process technology library, UMC

Impact of Body Bias Based Leakage Power Reduction on Soft Error Rate

Warin Sootkaneung and Kewal K. Saluja

Department of Electrical and Computer Engineering, University of Wisconsin-Madison
Madison, WI 53706, USA
sootkaneung@wisc.edu, saluja@ece.wisc.edu

Abstract— As device geometries shrink to nanometers, increasing leakage current coupled with particle induced soft errors is exasperating the circuit reliability issues. In this paper, we first establish that independent solutions to these two problems can not lead to a good final solution. A more thoughtful and integrated design methodology is required to reconcile these two challenging issues. Next, we investigate the dependency of soft error rate on the body bias based leakage reduction method and introduce a novel body bias-dependent soft error model. We propose an optimization based and a heuristic driven approach to reduce leakage while satisfying the soft error rate limit. Our methods provide appropriate body bias configurations that lead to near-optimal total mean time to failure improvement of a circuit.

Keywords-body bias; leakage power; soft error

I.   INTRODUCTION

Reliability requirement of modern processors is drawing considerable attention of the researchers working in the areas of power reduction methods and the soft error (SE) management. Yet, to the best of our knowledge, there is no study in literature that discusses these two problems together. The importance of addressing these two problems, namely leakage power reduction and soft error rate (SER) control, together stems from the fact that some of the methods that reduce the leakage current increase circuit vulnerability to SEs as discussed in this paper.

Three dominant components of leakage are subthreshold leakage, gate leakage, and reverse-biased, drain- and source-substrate junction band-to-band-tunneling (BTBT) leakage. In submicron regime, the contribution of each leakage portion plays an important role in total leakage power [1]. Technology scaling from generation to generation has lowered the device supply voltage in order to limit dynamic power consumption. On the other hand, device threshold voltage (V_{th}) has also been depressed to maintain the performance [2] and as a result, the reduced threshold voltage causes transistor leakage current to rise exponentially. Short term consequences of this are an increase in cooling costs and reduction in battery life. But, the long term effect is that an increase in device temperature due to elevated leakage can degrade the reliability and lifetime of a circuit. Hence, to reconcile leakage power issue in current nanoscale processors, novel low power design techniques, which employ acceptable performance penalty, are required.

Recently, several techniques to reduce the leakage by tuning the body bias (BB) during design time and runtime have been proposed [3], [4], [5], [6], [7], [8], [9], [10]. The key idea behind these techniques is based on the fact that the degree of BB profoundly affects device V_{th} and leakage power

dissipation. In particular, reverse BB (RBB) can be applied on high-speed devices to raise their V_{th} which reduces the leakage power; however, it causes device delay to increase. On the other hand, forward BB (FBB) can be applied on low-speed devices to lower their V_{th} and boost their performance, but this results in an increase in leakage power. Design time techniques predefine the delay of a circuit and selectively adjust BB of each or selected cluster of CMOS transistors to achieve desired leakage saving without increasing overall delay. The objective of these approaches is to assign high V_{th} to transistors on noncritical paths to reduce the leakage power, while the critical delay is maintained. Runtime techniques adaptively set the BB during operation in both idle (standby or sleep) and active states. These techniques employ the adaptive BB (ABB) which varies transistor BB on-the-fly from zero BB (ZBB) towards FBB or RBB depending on the performance demand and required leakage reduction.

A single event upset (SEU) from particle strikes can cause a transient error or SE that induces a flip at the output of logic circuit. In nanometer era, neutron-induced SE is becoming one of the most challenging reliability issues. Circuit designers are actively researching for good SER reduction techniques for their designs. To mitigate SEs in processor's storage and combinational units, various hardware, software, and coding based techniques have been developed. Examples of SER reduction techniques for storage parts include error correcting code (ECC) based techniques for conventional memories, hardware based techniques [11], [12] for high-performance memories, and pure software and compiler optimization based techniques [13], [14], [15] for register files. On the other hand, some of the techniques that have been studied for SE problems in combinational parts of a processor include inserting latch protection hardware [16], [17], [18] and device sizing [19], [20], [21].

In this paper, we investigate the impact of body bias based leakage power reduction on SER in combinational circuits. We first show through simulation that even though RBB of each transistor reduces leakage, it may cause a decrease in the amount of critical charge (Q_{crit}) of the device, which in turn can lead to an increase in transistor SER. In order to determine the impact of both factors, we use a mean time to failure (MTTF) as a common metric, which helps us integrate the leakage power and SER into reliability term and we can afterward maximize the overall circuit reliability. In this paper, we first conduct two sets of experiments based on optimization and heuristic approaches to reduce the leakage under a desired SER bound. Next, the leakage and SER are expressed in terms of MTTFs and hence, we can select a proper BB configuration such that it can provide the maximum total MTTF for a circuit.

978-1-4673-0438-2/12 $31.00 © 2012 IEEE

The remainder of this paper is organized as follows. Section II gives the motivation behind this study. Section III reviews some related theories. Sections IV and V discuss our reliability-aware leakage reduction by optimization and heuristic based approaches, respectively. The experimental results and discussion are reported in section VI. We finally conclude the study in section VII.

II. MOTIVATION

Leakage power dissipation increases device temperature which momentously influences lifetime of a circuit. In [22], [23], [24], and [25], permanent failure rates due to power and temperature were modeled in terms of MTTFs. Failure mechanisms discussed in those studies included electromigration, negative bias temperature instability, thermal cycling, and time-dependent dielectric breakdown. Although those studies dealt with modeling hard errors or defects due to elevated power and temperature, none of them considered the consequence of using their techniques on SE immunity. On the other hand, in [26], the dependency of SE immunity on the BB of a 10T SRAM cell was investigated. That study also took into account the variation of the supply voltage and indicated that SER is less sensitive to the BB at low supply voltages. However, leakage-related long term degradation issue was not addressed.

Our attempt is to adapt the previously proposed hard error model due to temperature and integrate it into our SE model. Unlike the studies above, we meticulously analyze the benefit from diminished leakage power and the detriment to SE immunity from the use of BB based leakage reduction technique. Thus, in this paper, we merge two different natures of circuit reliability, namely unceasingly escalated leakage and SE, two major design issues in nanometer era. We believe that this integration can improve the circuit reliability better than handling each problem separately.

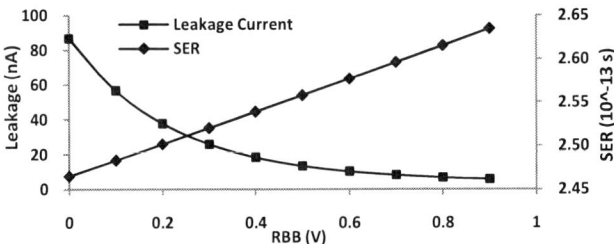

Figure 1. Leakage current and SER vs. RBB voltage plots for circuit c17

We performed a simple experiment on the circuit c17 in which the RBB for all gates in the circuit was varied, and the SER of the circuit was determined. Figure 1 shows the leakage current and SER as a function of RBB for circuit c17 mapped with 32-nm technology nodes. The leakage value was directly captured by SPICE simulation, whereas the average SER was calculated using SPICE (to obtain Q_{crit} of each device under RBB variation), neutron flux information [27], and logic simulation [20], [21] (to obtain the logical masking probability). It is evident from this figure that as RBB voltage increases, the leakage current decreases exponentially, whereas the SER increases steadily. In nanometer processors, a transient glitch from a particle strike, causing a circuit to fail, naturally occurs at higher rate than a permanent failure due to temperature from raised leakage power. Thus, in order to efficiently trade these two opposite reliability outcomes, we need a careful formulation based on the cost and benefit of the two effects.

In this paper, we use a common metric, MTTF to capture the effect of power-related hard errors and particle strike-induced SEs. We calculate the circuit MTTF of various fine-grained BB configurations in which different BB voltage levels are selectively assigned to each gate. We believe that we can obtain proper configurations that provide the highest reliability gain or largest MTTF to the circuit.

III. RELATED THEORIES

A transient glitch from neutron strikes can be modeled as a single exponential current injecting into the drain of a victim transistor as given in (1) [28].

$$I(t) = \frac{Q}{\Gamma}\sqrt{\frac{t}{\Gamma}}e^{-t/\Gamma} \qquad (1)$$

In this equation, Q is the amount of charge deposition and Γ is the technology dependent charge-collection time constant. The amount of charge Q that can bring the output voltage to VDD/2 and thus cause a circuit to fail is denoted as Q_{crit}. Using an energy transfer model proposed in [29], we can trace Q_{crit} back to the amount of terrestrial neutron energy needed to induce the SE. Further, we can obtain neutron flux as a function of neutron energy from the JEDEC89A standard [27].

In additional to the dependency of Q_{crit} or SER on BB, the SER is input and area dependent. From the study in [21], [30], input to a circuit/gate affects both electrical masking (the amount of Q_{crit}) and logical masking probabilities. Moreover, because neutron flux defines neutron strike rate per unit area, the SER is also proportional to the *active area*, the region near reverse-biased junction of each CMOS transistor. Equation (2) below gives the SER of a transistor t of a gate i when an input vector j is applied [21].

$$TrSER_{i(j,t)} = \frac{1}{k} * \emptyset_{i(j,t)}(Q_{crit}) * Ad_{i(t)} * w_{i(t)} * E_{i(j)} \quad (2)$$

In (2), k is the total number of simulated input vectors, $\emptyset_{i(j,t)}(Q_{crit})$ is the total neutron flux (as a function of Q_{crit}) which can produce the charge deposition greater than Q_{crit} at transistor t of gate i and gate input vector j, $Ad_{i(t)}$ is the *active area* which is approximately the drain area of transistor t of gate i, $w_{i(t)}$ is the *weighting factor* which is the ratio of the *active area* to the *circuit area*, and $E_{i(j)}$ is the *error count* which reflects the number of times a transient glitch (error) can propagate from gate i to the primary output(s). Since we assume that an upset occurs on one transistor at a time, we can sum SER of each transistor in a gate in (2) to obtain SER of a gate i, $Gate\ SER_i$ as given in (3)

$$Gate\ SER_i = \sum_{i(j)}\sum_{i(t)} Tr\ SER_{i(j,t)} \qquad (3)$$

The total SER of a circuit, $Circuit\ SER$, can be written as the summation of SER of each gate i as shown in (4).

$$Circuit\ SER = \sum_i Gate\ SER_i \qquad (4)$$

Finally, the *Circuit SER* in (4) can be used to obtain the MTTF due to SE (*MTTF_SE*) as given below:

$$MTTF_{SE} = \frac{1}{Circuit\ SER} \qquad (5)$$

As a result of the leakage, the equilibrium temperature due to the change in leakage of a circuit block can be obtained, using a cell level model [24]. In this model, when the leakage power, P_{leak0}, consumed by a cell at the operating temperature, T_0 increases to P_{leak1}, a new temperature, T_1 can be calculated by the following equation.

$$T_1 = T_0 + R_{\theta s}(P_{leak1} - P_{leak0}) \qquad (6)$$

where $R_{\theta s}$ is the thermal resistance of the substrate. The thermal resistance is dependent on material and dimension of the circuit block.

An increase in temperature of a device predominantly causes reduction in the MTTF of the circuit. Since in this study, we compare the MTTF due to leakage (*MTTF_leak*) of a circuit at initial and new operating temperature, a large thermal cycle model for MTTF [22] shown in (7) is used.

$$MTTF_{leak} \propto \left(\frac{1}{T_1 - T_0}\right)^{2.35} \qquad (7)$$

Since the effects of SE and leakage are assumed to be independent, the total MTTF for the entire circuit due these two effects is the reciprocal of the summation of both failure rates as given in (8).

$$MTTF_{circuit} = \frac{1}{\frac{1}{MTTF_{SE}} + \frac{1}{MTTF_{leak}}} \qquad (8)$$

IV. OPTIMAL SOLUTION FOR RBB BASED SE-AWARE LEAKAGE REDUCTION

In this section, we first introduce an optimization based approach for leakage power reduction which determines an optimal BB configuration without incurring a delay penalty. The formulation consists of RBB voltage and output arrival delay of each gate as variables, and the minimization of the leakage as an objective function. The critical delay, gate propagation delay, and SER are constraints of the problem. All experimental circuits, investigated in this study, are mapped using 32-nm technology from the predictive model [31].

The leakage current of each gate i, $I_{leak,i}$, is obtained from SPICE simulation. All gates in the cell library are separately tested for the worst case leakage power consumption. As a function of RBB voltage of the gate i, $V_{RBB,i}$, $I_{leak,i}$ is fitted into an exponential function as expressed in (9)

$$I_{leak,i} = a + be^{-cV_{RBB,i}} \qquad (9)$$

where a, b, and c are curve fitting constants.

We also perform SPICE simulation to capture the worst case propagation delay of each gate in the cell library. The results from SPICE reveal that the relationship between worst case propagation delay and BB of all the mapping gates is linear. Note that for the rest of this paper, the term "propagation delay" is referred to the "worst case propagation delay".

The critical delay, T_{max}, of the combinational circuit is used as the timing constrain in the optimization process. The value of T_{max} is precalculated using gate propagation delays and an

algorithm for finding the critical paths in the circuit graph. Thus, T_{max} for each circuit is conserved.

The optimization problem is formulated as follows. For a gate i with propagation delay d_i, a variable a_i representing the arrival time at the output of the gate i is defined. Given a gate f which is connected to an input of the gate i, we define an objective and specify gate/circuit delay constraints of an optimization problem [32] as given in (10), (11), (12), and (13).

$$min \sum_{i=1}^{total\ \#\ of\ gates+PIs} I_{leak,i} \qquad (10)$$

$$a_f + d_i \le a_i, \forall f \in fanin\ of\ i \qquad (11)$$

$$\begin{cases} a_i = 0 \\ d_i = 0 \\ I_{leak,i} = 0 \end{cases}, \forall i \in PI \qquad (12)$$

$$a_i \le T_{max}, \forall i \in set\ of\ gates\ connected\ to\ PO \qquad (13)$$

In the above problem statement, each primary input is considered as a dummy gate which satisfies the condition given in (12). At the primary output, the arrival delay must not exceed T_{max} as stated in (13). In addition to the timing constraints we have already defined, this study also limits the *Circuit SER* in (4) to the maximum SER, *Circuit SER_MAX* as given in (14) below.

$$Circuit\ SER \le Circuit\ SER_{MAX} \qquad (14)$$

The upper and lower limits of the RBB voltage variable are also required for the problem. Hence, equation (15) is provided.

$$0 \le V_{RBB,i} \le V_{RBB,MAX} \qquad (15)$$

Although the RBB voltage is a discrete variable, we allow the optimizer to generate a solution with continuous $V_{RBB,i}$ values. This relaxation significantly improves the solvability of the problem. However, the optimal results are then required to be discretized to desired voltage levels. We will discuss our approach for obtaining discrete values in the next section.

We use MATLAB optimization toolbox to solve this nonlinear optimization problem. The results for some circuits selected from the ISCAS'85, ISCAS'89, and ITC benchmark suits are provided in section VI.

V. HEURISTIC SOLUTION FOR RBB BASED SE-AWARE LEAKAGE REDUCTION

The optimal solution proposed in the previous section was very inefficient in computation time and could not solve large circuits in a reasonable time (see section VI). To improve computation time efficiency, we develop a heuristic based leakage reduction method. Basically, the method consists of two steps. The first step primarily focuses on maintaining the critical delay of a circuit, while the amount of leakage current is lowered as much as possible. Without SER consideration, large leakage reduction provided by the first step may cause the circuit SER to rise. As a result, second step is used to bring the increased SER down to a desired level.

Algorithm 1, as given in Figure 2, matches the RBB voltage to each of the gates and confines the circuit delay. Initially, the original critical delay of a circuit under ZBB condition is computed. Since we would like to conserve the critical delay, all the gates in the original critical path(s) must receive the ZBB. On the other hand, all noncritical gates are then assigned with the largest RBB level to have maximum leakage reduction. Since the change in RBB affects the circuit delay, the following steps bound the excessive delay to the required level. We start the first iteration by extracting new critical path(s) and critical delay. If the new critical delay is not greater than the original critical delay, we can obtain the maximum leakage reduction. However, if the critical delay is violated, we need to set RBB of some gates back towards ZBB. The criterion for selecting those gates and procedure for decreasing the RBB are as follows.

For each round of iterations, we search for a gate in new critical path(s) for which the increase in the amount of leakage at its currently assigned RBB value compared to the next lower level is the smallest. Once such a gate is found, its RBB is then reduced to the new lower level. We iteratively search for new critical path(s), new critical delay, and readjust the RBB of the candidate gates until the critical delay requirement is satisfied. This method indeed reduces the leakage current substantially, but it is insensitive to SER.

```
Algorithm 1

  1. Initialization: extract original critical
     path(s)/delay at ZBB.
  2. Assign the largest RBB to all noncritical gates.
  3. Extract new critical path(s) and gate delays.
  4. Check whether the new critical delay is greater than
     the original delay. If not, end the algorithm, else
     go to step 5.
  5. Determine the increase in leakage current for each
     gate in the critical path by changing the RBB to
     the next lower value.
  6. Identify a gate in the critical path that causes the
     smallest increase in leakage current and reduce its
     RBB to the next lower level.
  7. Iterate-go back to step 3.
```

Figure 2.   Algorithm 1: Leakage Reduction

```
Algorithm 2

  1. Initialization: obtain a BB configuration of a
     circuit from Algorithm 1
  2. Calculate the SER of gate/circuit using equations
     (3) and (4) for corresponding BB.
  3. Identify the most sensitive gate (having the largest
     SER) that does not receive the ZBB
  4. Check whether the current SER of the circuit is
     greater than the maximum bound. If not, end the
     algorithm, else go to step 5.
  5. Reduce the RBB of the most sensitive gate to the
     next lower level.
  6. Iterate-go back to step 2
```

Figure 3.   Algorithm 2: SER Reduction

To limit the increased SER which results from Algorithm 1, Algorithm 2 is used next to set the RBB of some of the gates back towards ZBB. This algorithm estimates the SER of each gate using the SE model discussed in section III, and selects most sensitive gates for further treatment as shown in Figure 3. The input to the Algorithm 2 is the BB configuration of a circuit provided by Algorithm 1. Afterwards, we calculate the SER of each gate and SER of the circuit using (3) and (4)

corresponding to the assigned RBB voltage, and identify the most sensitive gate (having the largest SER). The SER of the circuit is then compared to the maximum bound defined by the user. If the SER is within the limit, we terminate the simulator and output the BB configuration. If, on the other hand, the SER of the circuit increases beyond the maximum SER, we reduce the current RBB of the most sensitive gate to the next lower level. For each configuration, we repeatedly calculate and check the circuit SER until the criterion is met. Note that the delay constraint does not need to be checked in Algorithm 2 because timing performance cannot degrade when the RBB is moved towards ZBB.

Discretization of the RBB values obtained from the optimization based approach discussed in the previous section is done as follows. We make some small modifications to the Algorithm 1. Instead of assigning the maximum RBB to all noncritical gates, as in step 2 of Algorithm 1, we set the RBB values to the nearest available discrete values. Other steps of the algorithm are left intact. After this, Algorithm 2 is used to obtain the final solution.

VI.   EXPERIMENTAL RESULTS AND DISCUSSION

We evaluated the methodologies proposed in this paper using various benchmark circuits mapped to the cell library containing 2-, 3-and 4-input NAND and NOR gates and Inverters (NOT gates) in 32-nm technology node [31]. The optimization based approach was applied to the small circuits containing up to 700 gates, since the runtime was prohibitive for large circuits. However, we extended the investigation domain to large circuits (contain up to 12000s gates for the circuit S15850) for the heuristic based approach. In addition, the ratio of $MTTF_{leak}$ and $MTTF_{SE}$ at ZBB was assumed to be 30:1 [21], [25].

A.   Experimental Results

The normalized results of SER, leakage current, and MTTF from the optimization and heuristic based approaches are reported in Table I and Table II, respectively. In this experiment, we set the maximum RBB voltage to 0.4 V because beyond this point, leakage reduction gains of most circuits saturates (e.g., see Figure 1). The SER bound constraint is varied in steps of 0.25% until it reaches 1%; i.e., SER never exceeds 1.01 X (SER at ZBB). We also allow the levels of RBB voltages to be available at 0, 0.1, 0.2, 0.3, and 0.4 V. It can be noticed from the above tables that the circuits achieve the maximum MTTF at different SER bounds. However, for some circuits such as C1355 and i3, the minimum leakage occurs at a very low SER bound (less than 0.25%). Therefore, no change in leakage and MTTF can be seen at higher SER bounds.

We also investigate the impact of various RBB levels on leakage and MTTF. Figure 4 plots the decrease in leakage and increase in MTTF for the circuit C499. This circuit reaches the maximum MTTF at 0.50% SER bound. It can be seen in Figure 4 that the largest leakage reduction and MTTF gain can always be obtained by the configuration that consists of the RBB voltage levels of {0, 0.1, 0.2, 0.3, 0.4} V. This is intuitively evident. Yet, we also notice that solution consisting of {0, 0.3, 0.4} V levels also provides the best solution.

978-1-4673-0438-2/12 $31.00 © 2012 IEEE

TABLE I.   Normalized SER, Leakage, and MTTF Values from The Optimization Based Approach with Maximum RBB = 0.4 V

Circuit	SER Bound 0.25%			SER Bound 0.50%			SER Bound 0.75%			SER Bound 1.00%		
	SER	Leakage	MTTF	SER	Leakage	MTTF	SER	Leakage	MTTF	SER	Leakage	MTTF
C499	1.0023	0.7899	1.0116	1.0050	0.7240	1.0125	1.0055	0.7189	1.0122	1.0055	0.7189	1.0122
C880	1.0021	0.7803	1.0124	1.0048	0.6444	1.0163	1.0075	0.5546	1.0172	1.0097	0.4997	1.0168
C1355	1.0024	0.8281	1.0093	1.0024	0.8281	1.0093	1.0024	0.8281	1.0093	1.0024	0.8281	1.0093
S208	1.0025	0.7128	1.0155	1.0044	0.6321	1.0173	1.0074	0.5400	1.0179	1.0098	0.4865	1.0171
S420	1.0024	0.7094	1.0158	1.0046	0.6136	1.0178	1.0074	0.5338	1.0180	1.0099	0.4813	1.0172
S838	1.0023	0.7111	1.0158	1.0045	0.6169	1.0178	1.0075	0.5333	1.0180	1.0094	0.4891	1.0174
i1	1.0022	0.7708	1.0128	1.0047	0.6453	1.0164	1.0072	0.5678	1.0171	1.0095	0.5210	1.0163
i2	1.0025	0.7640	1.0129	1.0049	0.7298	1.0123	1.0059	0.7263	1.0114	1.0059	0.7263	1.0114
i3	1.0014	0.5028	1.0252	1.0014	0.5028	1.0252	1.0014	0.5028	1.0252	1.0014	0.5028	1.0252

TABLE II.   Normalized SER, Leakage, and MTTF Values from The Heuristic Based Approach with Maximum RBB = 0.4 V

Circuit	SER Bound 0.25%			SER Bound 0.50%			SER Bound 0.75%			SER Bound 1.00%		
	SER	Leakage	MTTF	SER	Leakage	MTTF	SER	Leakage	MTTF	SER	Leakage	MTTF
C499	1.0025	0.8027	1.0107	1.0050	0.7252	1.0124	1.0055	0.7189	1.0122	1.0055	0.7189	1.0122
C880	1.0025	0.8128	1.0101	1.0050	0.7134	1.0130	1.0075	0.6220	1.0146	1.0100	0.5502	1.0149
C1355	1.0024	0.8281	1.0093	1.0024	0.8281	1.0093	1.0024	0.8281	1.0093	1.0024	0.8281	1.0093
C5315	1.0025	0.6840	1.0169	1.0050	0.5689	1.0192	1.0075	0.5196	1.0184	1.0100	0.4725	1.0173
C6288	1.0025	0.8186	1.0098	1.0050	0.7493	1.0112	1.0075	0.6878	1.0118	1.0100	0.6318	1.0118
S208	1.0025	0.7463	1.0138	1.0050	0.6468	1.0161	1.0074	0.5843	1.0162	1.0100	0.5302	1.0156
S420	1.0025	0.7325	1.0145	1.0050	0.6328	1.0167	1.0075	0.5714	1.0167	1.0100	0.5287	1.0156
S838	1.0025	0.7348	1.0144	1.0050	0.6387	1.0165	1.0074	0.5709	1.0167	1.0100	0.5270	1.0157
S13207	1.0025	0.6889	1.0167	1.0050	0.6177	1.0173	1.0075	0.5597	1.0170	1.0100	0.5077	1.0163
S15850	1.0025	0.6896	1.0166	1.0050	0.6161	1.0174	1.0075	0.5583	1.0171	1.0100	0.5062	1.0163
i1	1.0024	0.8154	1.0100	1.0050	0.6706	1.0150	1.0068	0.6056	1.0160	1.0098	0.5255	1.0159
i2	1.0023	0.7854	1.0118	1.0047	0.7347	1.0122	1.0047	0.7347	1.0122	1.0047	0.7347	1.0122
i3	1.0014	0.5028	1.0252	1.0014	0.5028	1.0252	1.0014	0.5028	1.0252	1.0014	0.5028	1.0252
i7	1.0025	0.8041	1.0106	1.0050	0.6785	1.0147	1.0075	0.5879	1.0160	1.0100	0.5413	1.0152
i8	1.0025	0.6987	1.0162	1.0050	0.6185	1.0173	1.0075	0.5737	1.0165	1.0100	0.5161	1.0160

Furthermore, we compare the optimization based solution with the heuristic based solution for the circuit C499 in Figure 4. We note that the heuristic based solution is very close to the optimization based solution, though the configuration {0, 0.2, 0.4} V provides the best solution.

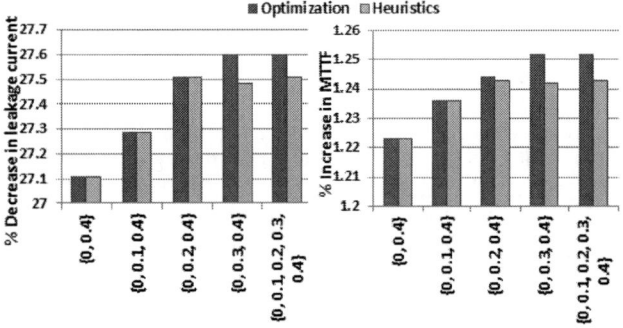

Figure 4.   Percentage decrease in leakage current and increase in MTTF of the circuit C499 for different sets of RBB voltages with Maximum RBB = 0.4 V and SER bound = 0.50%

B. Discussion

In our study, since we used the MTTF to evaluate the solution for few points of predefined SER bound (0.25%, 0.50%, 0.75%, and 1%), the optimal configuration for a design, which returned the largest MTTF, might not be very exact. For this reason, more integrated techniques, which can directly maximize the total MTTF due to leakage and SER, need to be developed. Furthermore, we assumed that the $MTTF_{leak}$ at ZBB is 30 times longer than the $MTTF_{SER}$ and used this assumption to estimate the absolute value of the $MTTF_{leak}$. However, if more accurate $MTTF_{leak}$ can be modeled, we can simply adopt it. Our proposed techniques can accommodate any new model for $MTTF_{leak}$ without any changes in the algorithms.

It is evident from Tables I and II that for the small circuits, the results of our heuristic driven leakage reduction are comparable with the optimal results. However, the runtime of the heuristic based approach is relatively very small. For the large circuits, the heuristic based approach took only a few seconds compared to the optimization based approach which was unable to find solution even after hours of computing time. Therefore, we believe our heuristic based solution is scalable for circuits with hundreds of thousands of gates. In addition, use of many BB voltage levels can increase the cost of the routing and placement substantially. Fortunately, the results of our investigation disclose that reduction in the number of available BB voltage levels does not significantly decrease leakage reduction; hence, MTTF improvement is quite insignificant even at fewer RBB levels. For instance, the circuit C499 designed with the dual-BB, {0, 0.4} V decreases the leakage gain and MTTF about 2% of the design allowing five available BB voltage levels, {0, 0.1, 0.2, 0.3, 0.4} V.

VII.   CONCLUSION

This study is one of the first efforts which establish that SE degradation happens when leakage reduction through tuning device RBB is performed. We found that raising the RBB voltage to lessen the leakage power can cause SER of a circuit to increase. In order to trade the benefit of one for the other for nanometer circuit designs, we developed optimization and heuristic driven BB based approaches for leakage reduction in which the SER is used as a constraint of the problem. The experimental results show that the specified maximum SER bound profoundly influences the change in the total MTTF of the circuit. The best solution is achieved when the total MTTF is maximized.

VIII. REFERENCES

[1] A. Agarwal, S. Mukhopadhyay, A. Raychowdhury, K. Roy, and C. H. Kim, "Leakage Power Analysis and Reduction for Nanoscale Circuits," *IEEE Micro*, vol. 26, no. 2, pp. 68-80, March-April 2006.

[2] J. W. Tschanz et al., "Dynamic Sleep Transistor and Body Bias for Active Leakage Power Control of Microprocessors," *IEEE Journal of Solid-State Circuits*, vol. 38, no. 11, pp. 1838-1845, November 2003.

[3] V. Khandelwal and A. Srivastava, "Active Mode Leakage Reduction Using Fine-Grained Forward Body Biasing Strategy," *Integration, the VLSI Journal*, vol. 40, no. 4, pp. 561-570, July 2007.

[4] Y. Nakamura et al., "1/5 Power Reduction by Global Optimization Based on Fine-Grained Body Biasing," in *the IEEE Custom Integrated Circuits Conference (CICC)*, San Jose, CA, 2008, pp. 547-550.

[5] H. Xu, R. Vemuri, and W. B. Jone, "Run-time Active Leakage Reduction by Power Gating and Reverse Body Biasing: An Energy View," in *the IEEE International Conference on Computer Design (ICCD)*, Lake Tahoe, CA, 2008, pp. 618-625.

[6] S. H. Kulkarni, D. M. Sylvester, and D. T. Blaauw, "Design-Time Optimization of Post-Silicon Tuned Circuits Using Adaptive Body Bias," *IEEE Transactions on Computer-Aided Design of Integrated Circuits and Systems*, vol. 27, no. 3, pp. 481-494, March 2008.

[7] A. Sathanur, A. Pullini, L. Benini, G. De Micheli, and E. Macii, "Physically Clustered Forward Body Biasing for Variability Compensation in Nanometer CMOS Design," in *the Design, Automation & Test in Europe Conference & Exhibition (DATE)*, Nice, France , 2009, pp. 154-159.

[8] P. Y. Chen, C. C. Fang, T. T. Hwang, and H. P. Ma, "Leakage Reduction, Variation Compensation Using Partition-Based Tunable Body-Biasing Techniques," in *the International Symposium on VLSI Design, Automation, and Test (VLSI-DAT)*, Hsinchu, Taiwan, 2009, pp. 170-173.

[9] A. Ghosh, R. M. Rao, and R. B. Brown, "A Centralized Supply Voltage and Local Body Bias-Based Compensation Approach to Mitigate Within-Die Process Variation," in *the 14th ACM/IEEE international symposium on Low power electronics and design (ISLPED)*, New York, NY, 2009, pp. 45-50.

[10] H. Mostafa, M. Anis, and M. Elmasry, "A Novel Low Area Overhead Direct Adaptive Body Bias (D-ABB) Circuit for Die-to-Die and Within-Die Variations Compensation," *IEEE Transactions on Very Large Scale Integration (VLSI) Systems*, vol. 19, no. 10, pp. 1848-1860, October 2011.

[11] H. R. Zarandi and S. G. Miremadi, "Soft Error Mitigation in Cache Memories of Embedded Systems by Means of a Protected Scheme," *Lecture Notes in Computer Science, Springer Berlin / Heidelberg*, vol. 3747, pp. 121-130, 2005.

[12] V. Gherman, S. Evain, M. Cartron, N. Seymour, and Y. Bonhomme, "System-Level Hardware-Based Protection of Memories against Soft-Errors," in *the Design, Automation and Test in Europe (DATE)*, Nice, France, 2009, pp. 1222-1225.

[13] N. Oh, P. P. Shirvani, and E. J. McCluskey, "Error Detection by Duplicated Instructions in Super-Scalar Processors," *IEEE Transactions on Reliability*, vol. 51, no. 1, pp. 63-75, March 2002.

[14] G. A. Reis, J. Chang, N. Vachharajani, R. Rangan, and D. August, "SWIFT: Software Implemented Fault Tolerance," in *the International Symposium on Code Generation and Optimization (CGO)*, Washington, DC, 2005, pp. 243-254.

[15] J. Lee and A. Shrivastava, "A Compiler Optimization to Reduce Soft Errors in Register Files," in *the Conference on Languages, Compilers, and Tools for Embedded Systems (LCTES)*, Dublin, Ireland, 2009, pp. 41-49.

[16] R. R. Rao, D. Blaauw, and D. Sylvester, "Soft Error Reduction in Combinational Logic Using Gate Resizing and Flipflop Selection," in *the IEEE/ACM International Conference on Computer-Aided Design (ICCAD)*, San Jose, CA, 2006, pp. 502-509.

[17] E. L. Hill, M. H. Lipasti, and K. K. Saluja, "An Accurate Flip-Flop Selection Technique for Reducing Logic SER," in *the International Conference on Dependable Systems and Networks (DSN)*, Anchorage, AK, 2008, pp. 128-136.

[18] S. Mitra, "Robust System Design," in *the 23rd International Conference on VLSI Design*, Bangalore, India, 2010, pp. 434-439.

[19] W. Sheng, L. Xiao, and Z. Mao, "Soft Error Optimization of Standard Cell Circuits Based on Gate Sizing and Multi-Objective Genetic Algorithm," in *the 46th Annual Design Automation Conference (DAC)*, San Francisco, CA, 2009, pp. 502-507.

[20] W. Sootkaneung and K. K. Saluja, "On Techniques for Handling Soft Errors in Digital Circuits," in *the International Test Conference (ITC)*, Austin, TX, 2010, pp. 1-9: Paper 25.2.

[21] W. Sootkaneung and K. K. Saluja, "Soft Error Reduction through Gate Input Dependent Weighted Sizing in Combinational Circuits," in *the 12th International Symposium on Quality Electronic Design (ISQED)*, Santa Clara, CA, 2011, pp. 603-610.

[22] J. Srinivasan, S. V. Adve, P. Bose, and J. A. Rivers, "Lifetime Reliability: Toward An Architectural Solution," *IEEE Micro*, vol. 25, no. 3, pp. 70-80, May-June 2005.

[23] D. Brooks, R. P. Dick, R. Joseph, and L. Shang, "Power, Thermal, and Reliability Modeling in Nanometer-Scale Microprocessors," *IEEE Micro*, vol. 27, no. 3, pp. 49-62, May-June 2007.

[24] B. Greskamp, S. R. Sarangi, and J. Torrellas, "Threshold Voltage Variation Effects on Aging-Related Hard Failure Rates," in *the IEEE International Symposium on Circuits and Systems (ISCAS)*, New Orleans, LA, 2007, pp. 1261-1264.

[25] P. Mangalagiri, S. Bae, R. Krishnan, Y. Xie, and V. Narayanan, "Thermal-Aware Reliability Analysis for Platform FPGAs," in *the IEEE/ACM International Conference on Computer-Aided Design (ICCAD)*, San Jose, CA, 2008, pp. 722-727.

[26] H. Fuketa, M. Hashimoto, Y. Mitsuyama, and T. Onoye, "Alpha-Particle-Induced Soft Errors and Multiple Cell Upsets in 65-nm 10T Subthreshold SRAM," in *the IEEE International Reliability Physics Symposium (IRPS)*, Garden Grove (Anaheim), CA, 2010, pp. 213-217.

[27] JEDEC89A Standard, "Measurement and Reporting of Alpha Particles and Terrestrial Cosmic Ray-Induced Soft Errors in Semiconductor Devices," Joint Electron Device Engineering Council, Solid State Technology Association, 2006.

[28] P. Hazucha and C. Svensson, "Impact of CMOS Technology Scaling on the Atmospheric Neutron Soft Error Rate," *IEEE Transactions on Nuclear Science*, vol. 47, no. 6, pp. 2586-2594, December 2000.

[29] D. G. Mavis and P. H. Eaton, "Soft Error Rate Mitigation Techniques for Modern Microcircuits," in *the 40th International Reliability Physics Symposium*, Dallas, TX, 2002, pp. 216-225.

[30] W. Sootkaneung and K. K. Saluja, "Gate Input Reconfiguration for Combating Soft Errors in Combinational Circuits," in *the International Conference on Dependable Systems and Networks Workshops (DSN-W)*, Chicago, IL, 2010, pp. 107-112.

[31] HSPICE PTM website. [Online]. Available: http://www.eas.asu.edu/~ptm.

[32] T. H. Wu, L. Xie, and A. Davoodi, "A Parallel and Randomized Algorithm for Large-Scale Discrete Dual-Vt Assignment and Continuous Gate Sizing," in *the ACM/IEEE International Symposium on Low Power Electronics and Design (ISLPED)*, Bangalore, India, 2008, pp. 45-50.

978-1-4673-0438-2/12 $31.00 © 2012 IEEE

2012 25th International Conference on VLSI Design

An Area Efficient Diode and On Transistor Interchangeable Power Gating Scheme with Trim Options for Low Power SRAMs

Ankur Goel[1], Donald Evans2, Richard Stephani[2], Venkateswara Reddy[1], Dharmendra Rai[1], Veerabadra Chary[1], Sathisha N[1]

[1]{ankur.goel, venkateswara.reddy, dharmendra.rai, Veerabadra.Chary, sathisha.n}@lsi.com

LSI India R&D Pvt. LtD, Bangalore, India

[2]{donald.evans, richard. Stephani} @lsi.com
LSI Corporation, USA.

Abstract— Reducing the leakage power in embedded SRAM memories is critical for low-power applications. Raising the source voltage of SRAM cells through diode transistor in standby mode reduces the leakage currents effectively. However, in order to preserve the state of the cell in standby mode, the source voltage cannot be raised beyond a certain level. To achieve that, the size of the required diode transistor becomes larger, as the supply voltage shrinks in the nano-CMOS technologies. In this work, an area efficient power gating technique with capability of post-silicon trimming of the voltage across SRAM cell is presented. Proposed interchangeable on transistor and diode scheme reduces the area overhead by 40% compared to conventional schemes, when applied to a 16Kb SRAM macro at 28nm CMOS technology at 0.85V supply voltage. Trimmable power gating scheme provides many options to trim the SRAM source voltage (ranging from 50mV to 150 mV in steps of approx. 25mV) with approx. 3% area overhead and more flexibility over conventional schemes.

Keywords-cmos;SRAM; 6T Cell, leakage; snm; power gating;

I. INTRODUCTION

The data retention current of SRAM cells has been dramatically increasing as the device sizes are shrinking and thus resulting in high power consumption. The dominant leakage mechanisms in a 6T SRAM cell in CMOS nano-meter technologies are sub-threshold leakage and gate leakage. Since last few years, gate leakage has been well controlled at the process level through the use of HIGH-K metal gates [7]. Sub-threshold leakage is still a challenge for low power SRAMs and the transistors which dissipate this leakage in a 6T SRAM cell are shown in Fig.1. Scaling the cell bias voltage by raising the source voltage of SRAM cells reduces the sub-threshold leakages. Raising the source voltage can be achieved by ground-gating the cells using a sleep transistor [2].

In general, the source voltage should be raised as much as possible so that maximum leakage reduction

can be achieved. However, in order to preserve the state of the cells in the standby mode, this voltage must not exceed a certain level. As the virtual ground (VSSC) node is raised, the SRAM cell SNM degrades as shown in Fig.2. Also, with the technology scaling down to smaller geometries, the exacerbated variation of device parameters causes significant inter-die and intra-die variations which lead to instability in the SRAM cells.

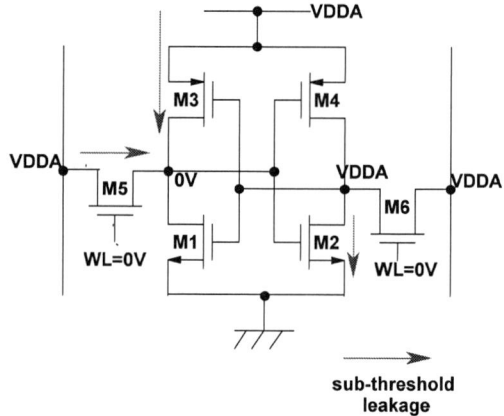

Fig.1. Leakage currents in 6T CMOS SRAM cell

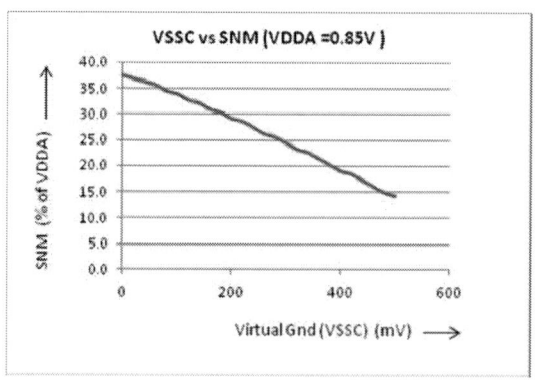

Fig.2. SRAM Cell Static noise margin vs. virtual gnd.

978-1-4673-0438-2/12 $31.00 © 2012 IEEE

The within-die variation results in cells which have different hold stabilities even on a single chip. Generally, the cells with larger noise margin are able to tolerate higher source voltages than those with smaller noise margin. Thus, SRAM array needs to be set to the voltage level determined by the weakest cell ensuring both stability as well as maximum leakage savings.

There are several schemes proposed in literature on power gating. Using a diode-connected PMOS bias transistor to control the virtual gnd has been proposed as shown in Fig. 3(a) [1]. Similar biasing schemes using nmos diode were also proposed [2,3].

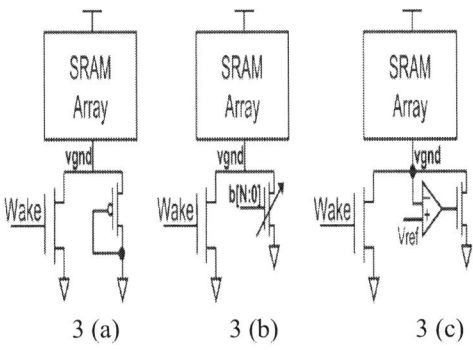

3 (a)   3 (b)   3 (c)

Fig.3. Conventional power gating schemes (a) Diode-connected PMOS bias transistor. (b) Programmable bias transistors (c) Active feedback with Op-Amp-based control.

There are three major issues associated with these schemes. Firstly, the area overhead is more; secondly these techniques do not provide a mechanism to fine-tune the source voltage after silicon results. The virtual gnd raise limits become even more stringent when NBTI/PBTI factors are taken into account [6]. Thirdly, the schemes don't provide an option to cut off the memory leakage in the shut down mode, where SRAM cell contents needn't be retained. Programmable bias transistors shown in Fig. 3(b) are effective in addressing the post silicon trimming problems [4]. There are two main benefits of using programmable bias transistors. First, the bias setting or VGND voltage level can be optimized based on the actual silicon results to achieve maximum leakage reduction. Second, different bias settings can be dynamically chosen at different supply voltages to allow a good design window under different voltage conditions. Although the programmable biasing transistors provide a good flexibility in controlling the VGND, they are still subject to within-die-, die-to-die and temperature-induced variations. Diode is a more adaptive solution than the fixed voltage bias transistors. A control scheme with an active feedback based on Op Amp is also proposed to overcome the

variations [5], as shown in Fig. 3(c). The reference voltage (*V*ref.) can be either generated internally or supplied externally. The major drawback of this scheme is the dc current consumed by the Op Amp that has to be replicated along each data bank to provide the needed granularity. Also, the area overhead of the scheme is larger, which makes it unattractive for use in SRAM memory compilers.

In this paper, a technique called 'Diode and on transistor interchangeable, trim enabled power gating scheme' is proposed. The scheme enables power gating at a lesser area compared to conventional schemes and trim option enables post-silicon trimming of the SRAM source voltage. The rest of the paper is organized as follows: the proposed scheme is described in section II. Section III illustrates application of the proposed power gating scheme to a 16Kb SRAM macro at 28nm CMOS technology. Simulation results are presented in Section IV. Finally, Section V concludes the paper.

II. Diode and On transistor inter-changeable Power gating scheme

The proposed 'Area efficient diode and on transistor inter-changeable power gating scheme' is shown in Fig.4. The scheme is applicable to SRAM memories in both the modes whether data retention is important or not. These two modes are explained in the next sections.

II.A. Light Sleep mode

Light sleep mode is defined as the leakage saving mode where SRAM data needs to be retained, so voltage across SRAM cell should be more than the minimum data retention voltage.

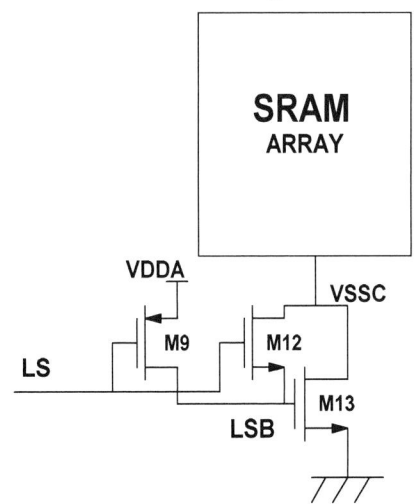

Fig. 4. Diode and On transistor Interchangable Power Gating Scheme

As shown in Fig.4., same transistor (M13) is used as 'on transistor' as well as diode. For illustration, with light sleep mode entry, LS will be logic '1' which makes M9 'off' and M12 'on'. When M12 is 'on', drain and gate of M13 are connected through M12 and form a diode structure. When SRAM comes out of light sleep, LS goes to logic '0', which makes M9 'on' and M12 'off'. LSB node goes to logic '1', M13 is 'on' and VSSC node goes to 0 and thus SRAM goes to active mode.

II.B.   Shut down mode

Shut down mode is defined as the mode where SRAM data need not be retained. So, in this mode, the VSSC level can be raised further to achieve better leakage savings. The logic which controls gate of 'on' transistor (M13), is modified to incorporate shut down mode. The circuit is as shown in Fig.5. This circuit supports both light sleep mode as well as shut down mode. The truth table for both the modes is shown in table1. When SD='1', irrespective of LS signal, memory goes to shut down mode and hence leakage savings are more compared to light sleep case. LSD signal is generated from LS and SD in the control block.

II.C Post silicon trimming

In the initial phase of a new technology offering, there are chances that silicon behavior deviates much from models. Leakage current of the SRAM cell and/or transistor threshold voltage can change, which can lead to diode voltage variation. In this case, if the diode voltage rises above the data retention limit, then SNM of SRAM cell degrades and SRAM data may get lost. To address this problem, a trimming circuit is added to circuit shown in Fig.5. The trimmable power gating scheme is as shown in Fig.6. In the default case, A=B=C=0, so that the diodes define the voltage at VSSC Node. If there are SNM issues in the standby mode, this VSSC voltage can be trimmed by changing the logic values of A, B and C

inputs so as to address the SNM issues. Also, if the wake-up latency (time taken by VSSC node to go from standby voltage to 0V), is more and memory is not able to meet the desired frequency of operation, then delay can be reduced by limiting the maximum value of voltage through programming of A, B, C nodes. The area overhead because of the trimming circuit is very small. The settings of trim options A, B and C can be obtained through fuses or scan registers. The virtual ground voltage (VSSC) level can be fine-tuned by programming the overall strength of the biasing transistors to achieve a desired value based on the silicon results. All the trim options are mixed with the shut down signal (SD) in the control block, so that when memory goes to shut down mode, A=B=C=0. For trimming of VSSC voltage, weak transistors were used (higher length), as they need to pass very low current (leakage current of SRAM array).

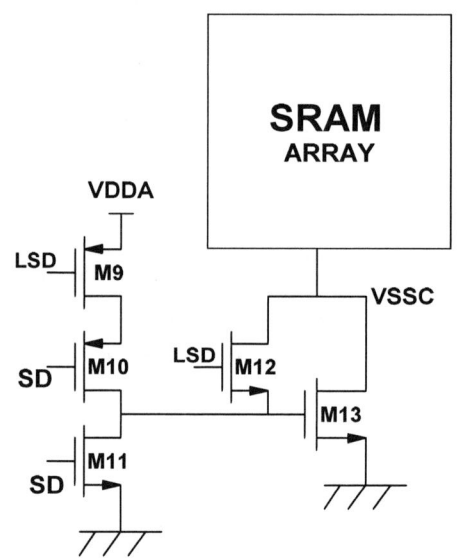

Fig. 5. SRAM Light Sleep and Shut down mode with the proposed scheme

LS	SD	LSD	Mode	Transistor State Description
0	0	0	Active	**On** :M9,M10,M13 ,**Off**:M11,M12 ,**Diode**: nil
0	1	0	Shutdown	**On**:M9,M11 ,**Off**=M10,M12,M13 ,**Diode**: nil
1	0	1	Light Sleep	**On**:M10,M12 ,**Off**=M9,M11 , **Diode**:M13
1	1	0	Shutdown	**On**:M9,M11 ,**Off**=M10,M12,M13 , **Diode**: nil

Table 1. Light sleep and Shut Down Mode truth table with transistor state description.

978-1-4673-0438-2/12 $31.00 © 2012 IEEE

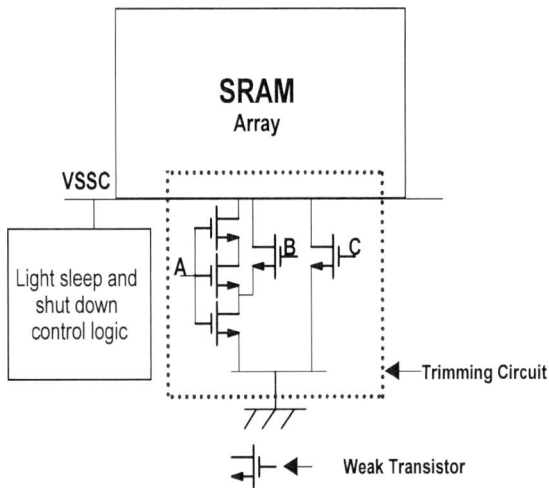

III. 16Kb SRAM macro with proposed power gating scheme

The proposed power gating scheme for leakage reduction was applied to 16 Kb (128RowsX128Cols, column-mux 4), SRAM macro in 28nm CMOS technology. The complete power gating circuit with SRAM array is shown in Fig.7. Light sleep and shut down logic is shared between 2 IOs' to save area. Trimming logic is also shared between 2 IOs' to save area, while the pull down transistors (M13R and M13L in Fig.6) were sized individual for each IO.

Fig. 6. Trimming Circuit for Post silicon voltage trimming

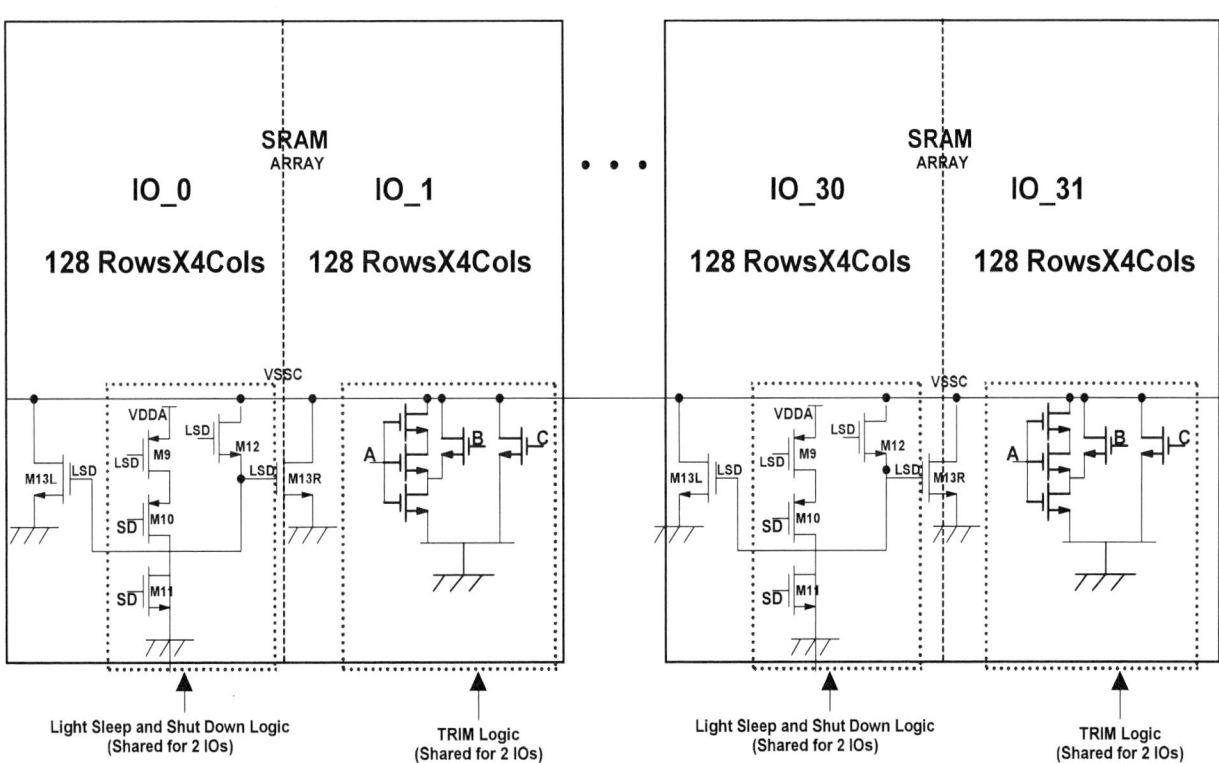

Fig.7. 16Kb SRAM Macro With proposed power gating and trimming logic

978-1-4673-0438-2/12 $31.00 © 2012 IEEE   83

IV.   RESULTS

Simulation results for the 16Kb SRAM macro show that leakage reduces by 52% in the 'light sleep' mode and 92% in the 'shut down' mode. The VSSC level (virtual gnd) is set to 150mV in the default case where all trim options are 0, with voltage supply being at 0.85V. For further tuning, trim option can be altered and VSSC level in the range of 50mV to 150mV can be achieved. Table 2. represents the VSSC level corresponding to each trim option. The step size of VSSC trim is kept around 25mV.

Trim Option			VSSC (mV) VDDA= 0.850V
C	B	A	mV
0	0	0	150
0	0	1	123
0	1	0	150
0	1	1	91
1	0	0	66
1	0	1	57
1	1	0	66
1	1	1	50

Table 2. VSSC level with different trim options

Area overhead of the proposed scheme is only 2.5% without trim circuit, which is approx. 40% lesser than the conventional schemes [1-3]. The total area overhead of the power gating circuit with trim is 3%. The on transistor is sized for achieving a bounce of less than 15mV at VSSC (virtual gnd) node, when word line is activated during read or write operation. IR drop is ensured to be less than 7mV.

V.   Conclusion

An area efficient 'Post Silicon Voltage-Trimmable Power Gating Scheme' for optimum leakage reduction and Yield Enhancement in SRAMs is presented. Light sleep and shut down modes of leakage savings are discussed.

References

[1] A. Bhavnagarwala, S. V. Kosonocky, M. Immediato, D. Knebel, and A.-M. Haen, "A pico-joule class, 1 GHz, 32 kB × 64 b DSP SRAM with self reverse bias," in *Proc. Symp. VLSI Circuits Dig. Tech. Papers,* Jun. 2003, pp. 251–252.

[2] T.Enomoto, Y.Oka and H.Shikano, "Self-controllable voltage level (SVL) circuit and its low power high speed CMOS circuit applications," *IEEE Journal of Solid State Circuits,* 38(7):1220- 1226, July 2003.

[3] Ankur Goel and Baquer Mazhari. Gate leakage and its reduction in deep submicron SRAM. *In proceedings of International Conference on VLSI Design,* Jan. 2005.

[4] K. Zhang *et al.,* "SRAM design on 65-nm CMOS technology with dynamic sleep transistor for leakage reduction," *IEEE Journal Solid-State Circuits,* vol. 40, no. 4, pp. 895–901, Apr. 2005.

[5] M. Khellah *et al.,* "A 256-Kb dual-VCC SRAM building block in 65-nm CMOS process with actively clamped sleep transistor," *IEEE J. Solid- State Circuits,* vol. 42, no. 1, pp. 233–242, Jan. 2007.

[6] Hao-I Yang, Wei Hwang, "Impacts of NBTI/PBTI and Contact Resistance on Power-Gated SRAM With High-Metal-Gate Devices," *IEEE Transactions on VLSI systems,* VOL. 19, NO. 7, July,2011.

[7] Pramod Kolar, Eric Karl, Uddalak Bhattacharya, Fatih Hamzaoglu, Henry Nho, Yong-Gee Ng, Yih Wang, Kevin Zhang, "A 32 nm High-k Metal Gate SRAM With Adaptive Dynamic Stability Enhancement for Low-Voltage Operation," *IEEE Journal of Solid State Circuits,* VOL. 46, NO. 1, Jan. 2011.

2012 25th International Conference on VLSI Design

An Energy Efficient Oscillator Frequency Calibration Methodology using Fraction Phase Computation

Amitava Ghosh, Isha Das, Achintya Halder

Department of Electronics & Electrical Communication Engineering
Indian Institute of Technology, Kharagpur
Kharagpur, India
09ec9403@iitkgp.ac.in, isha.das10@gmail.com, achintya@ece.iitkgp.ernet.in

Abstract—Wireless Sensor Nodes (WSNs) require a frequency calibration unit for correcting the spectral mask errors in its Radio Frequency (RF) front-end, which is caused by process, voltage and temperature (PVT) variations of the RF oscillator. In WSN applications, the conditions dictate such a frequency calibration unit to be energy efficient (i.e. both low power and having a fast settling time). In this paper, we propose an on-chip, fully embedded frequency calibration methodology, the corresponding algorithm and the hardware architecture, which are suitable for ultra-low power WSNs that use Frequency Shift Keying (FSK) as its modulation scheme. The architecture is based on computation of the fraction-phase of an RF oscillator frequency, when the former is divided by a reference frequency value. The proposed technique, though analogous to a fraction-N frequency synthesizer, requires no phase-locking mechanism and, therefore, requires no Phase Locked Loop (PLL). Simulation results have been presented showing the improved performance metrics over the state-of-the-art frequency calibration techniques used in WSNs.

Keywords-RF front-end, Voltage Controlled Oscillator (VCO); Digitally Controlled Oscillator (DCO); Frequency Synthesizer; Frequency Calibration; Fraction Phase; Energy Efficient; Low-power Wireless Communication; Wireless Sensor Nodes; Unconstrained Phase-Frequency Correction

I. INTRODUCTION

Wireless sensor nodes (WSNs) are a special class of communication systems, where energy efficiency while maintaining a certain Quality of Service (QoS) is the primary requirement. Majority of such systems use Frequency Shift Keying (FSK) as its modulation scheme [1-3]. In an FSK system, a frequency calibration unit is required at the transmitter to correct the spectral mask errors originating from frequency variations of the Radio Frequency (RF) oscillator under various environmental, process, temperature and supply voltage conditions. Ideally, the frequency calibration circuitry (i) must consume low power, (ii) must have a fast settling time, (iii) must have a low residual frequency-error, and (iv) should occupy a small chip area. In this paper, a new frequency calibration methodology is proposed, which has the following salient features:

- Both fast settling time and low power
- Low hardware cost – simple digital baseband circuitry and a few delay cells at RF frequency is needed

- Final frequency-error (in ppm) is programmable. Here an exact frequency match is not obtained as in a phase locked loop, however a frequency value within the targeted error specification is guaranteed (true for WSN scenarios)
- The methodology is usable for calibrating an oscillator frequency of any RF band
- There exists no requirement of the oscillator frequency tuning curve vs. its control voltage or the corresponding frequency control word (FCW) except that it only needs to be monotonic.

The paper is organized as follows: in Section II, a review of various RF oscillator frequency calibration techniques, is provided. In Section III, the theoretical background of the proposed fast and low-power frequency calibration scheme is presented. The design of the calibration algorithm is presented in Section IV, followed by the hardware architecture in Section V. The simulation results are presented in Section VI, where the (i) residual frequency error, (ii) energy consumption, (iii) calibration time and (iv) the resulting hardware cost have been presented and compared with prevalent frequency calibration techniques. This is followed by the conclusion in Section VII.

II. A COMPARATIVE REVIEW OF VARIOUS FREQUENCY CALIBRATION TECHNIQUES

The proposed frequency calibration methodology and the corresponding algorithm are generic and can be used to correct a particular target frequency error in any frequency band. However, the low hardware complexity and the fast calibration time make the proposed methodology particularly attractive for low-energy wireless communication applications.

Table I summarizes the state-of-the-art frequency calibration architectures, including the low-power ones that are used in WSN applications.

978-1-4673-0438-2/12 $31.00 © 2012 IEEE

TABLE I. FREQUENCY CALIBRATION STRUCTURES – A REVIEW

METHOD	FREQUENCY RESOLUTION	POWER CONSUMPTION (FOR THE GIVEN RESOLUTION)	HARDWARE COMPLEXITY INVOLVED	FREQUENCY CALIBRATION TIME TAKEN	REMARK
Classical PLL [4]	No average frequency error	14.4 mW (2.4 GHz)	VCO (voltage controlled oscillator), phase-frequency detector, charge pump and RF divider	Worst-case settling time (frequency jump from RF channel-1 to channel-10 of MICS band) 72.51 µs [5]	Power and settling time are high – unsuitable for WSNs
All Digital PLL (ADPLL) [6-7]	No average frequency error	42 mW (2.4 GHz)	DCO (Digitally controlled oscillator); complex digital sub-blocks, for e.g. IIR filters, phase accumulator, etc.	< 50 µs	Fast calibration, but high power consumption
Time to Digital Based Calibration [8]	0.75 ps	121 mW (3.3 GHz)	DCO; Time to digital converter, sigma delta modulator, counter, low pass filter	Not reported	Good frequency resolution, but much high power
Out-of-body calibration [9]	1-50 KHz (100 ppm @ 400 MHz)	Low Power. total Tx power 350 µW	Complexity of frequency-error computation is shifted to base-station / network controller node	Within 4 baseband cycles	Good energy efficiency. Calibration loop through external medium – chances of self-sustained, indefinite link failure.
Duty-Cycled PLL [10]	< 0.25 % from 300 MHz to 1.2 GHz	260 µW	DCO along with associated digital circuitry	2 µs at 900 MHz	Low power but poor frequency resolution
Integer Counting Scheme [11]	Not reported	220 µW	DCO and simple baseband circuitry	116 µs	Simple hardware. Unknown frequency-error bound.

Various other implementations of frequency calibration units have been reported in [12-15]. However, these architectures either consume power in tens of milliwatt range or have significantly slow frequency settling time values. Overall, the need for a low power and a fast settling frequency calibration methodology using low-cost hardware remains an open problem in the context of WSN applications.

III. THEORETICAL BACKGROUND

Similar to the All Digital PLL (ADPLL) concept, the proposed method is a time domain technique for frequency calibration of an RF oscillator, though much less hardware and computation intensive than the ADPLL technique.

In terms of working principle, if the technique in [11] has been akin to an integer-N synthesizer, then the proposed technique is analogous to a fraction-N synthesizer. The proposed technique measures the fraction phase of the RF carrier, where the output frequency of the RF oscillator is divided by a reference (whose frequency is much lower compared to that of the RF oscillator). While maintaining a low power during calibration, the proposed work further lowers the frequency calibration time compared to [11].

A. Principle of Operation:

The basic principle of operation relies on measuring the edge-separation (the rising edges or the falling edges) time between two RF carriers and comparing the edge-separation time with a set of ideal values. If the edge-separation time is smaller (higher) than that in the LUT, the frequency control word (FCW) has to be altered such that the digitally controlled oscillator (DCO) output frequency increases (decreases). However, the basic principle may not be applied directly as stated above, since in radio frequency the edge-separation is in sub-picosecond range and the power budget is a fraction of a mW. The proposed approach utilizes the fact that the edge separation between two frequency tones increase in time domain (time amplification). For this purpose, two reference time-boundaries (derived from a stable baseband clock reference) is used, which has a much longer time-period (lower frequency) compared to that of the RF oscillator output. The measured edge separation (equivalent to the fraction phase difference) in RF carrier then becomes wide enough to be resolved using low power techniques. The fractional phase-error accumulated over the known time-boundary is indicative of the oscillation frequency. The FCW driving the DCO is then altered in such a direction so as to bring it close to the targeted value. The principle is illustrated in Fig. 1. However, in practice, only one RF and reference frequency combination can be used at a particular time instant. The reference frequency (usually in the range of tens of MHz) is readily available from a system clock, which is anyway necessary for transmitting the digital bits and running the baseband digital circuits of a WSN. In order to ensure that the fraction-phase error computed using the above technique corresponds to one and unique frequency of the RF oscillator output, the maximum time-period of phase-accumulation i.e. the minimum frequency of the reference clock ('r') has to be higher than the frequency-span over which the oscillator can be calibrated.

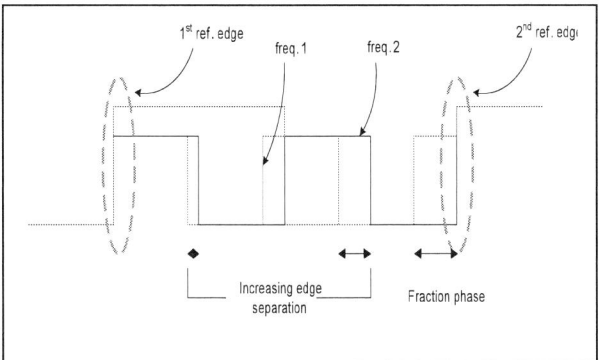

Fig. 1. Fraction phase generation principle

B. Coarse Calibration:

For a given reference frequency 'r', there will be 'r' groups of frequencies, each having a span of '1' freq. unit (for e.g. *1 MHz*) and a unique fraction phase value. An initial binary search would require $\lceil log_2 r \rceil$ iterations at the reference frequency in order to hit the correct group that contains the target frequency. In this process, the obtained RF oscillator fraction phase is the compared against the fraction phase values of the desired frequency. Since the latter values are constant, they are stored in a Look Up Table (LUT). The maximum frequency error after this 'coarse' calibration step is '1' frequency unit.

C. Fine Calibration:

In order to achieve the required target specification (say the error should be within 'x' frequency units), the reference frequency is next divided by two (using a counter). This magnifies the frequency groups (obtained after coarse calibration step based on fraction phase values) by a factor of '2'. Now the maximum error uncertainty becomes '0.5' frequency units. Hence the number of levels 'd' that need to be gone down (number of times, the reference frequency is to be divided) is given by:

$$d = \left\lceil log_2 \left(\frac{1}{x} \right) \right\rceil \qquad (1)$$

For reference frequency 'r', the normalized fraction values range from '0 to $r-1$(r-ary system)'. They are given by:

$$y = \frac{f_{tar}}{r} 2^l \qquad (2)$$

$$LUT(l) = fix[\{y - fix(y)\} * r] \qquad (3)$$

In (2) and (3), 'l' is the level of operation and starts from '0' to '$d-1$' . '$ftar$' is the target frequency, 'r' is the reference frequency and also the system radix. The values obtained from (3) are stored in an LUT and is used to tune to a specified frequency and to any given error specification. The fraction phase is sensed with a chain of inverters like in ADPLL which produce a thermometric code. The number of inverters required depends on the duty-cycle of the incoming RF waveform and is given by:

$$NoI = round(DC \cdot \frac{T_{nom}}{T_{inv}}) \qquad (4)$$

In (4), 'NoI' is the number of inverters required; 'DC' is the duty-cycle of the RF waveform as a fraction, T_{nom} is the average period of the RF waveform and 'T_{inv}' is the delay of each inverter. The inverters' output obtained at two reference edges are subtracted to obtain the fraction phase. This is required because the incoming RF frequency can have any phase relation w.r.t. to the reference.

IV. ALGORITHM DESIGN

The algorithm has been implemented in both MATLAB and C. An event driven model of DCO has been made which is called on whenever the fraction phase computation is carried out. In the event driven DCO model, a sampled time vector is created for a duration of T_{nom}. A random phase is generated from (0 - 2π). The random phase decides the point where transition from high to low value and vice-versa is made. A random generator generates jitter from the jitter specification that is added at each transition points. A second time vector is created in the same way which contains the phase shifted value at the second reference edge. The two vectors are sampled at the inverter sampling points where again, an uncertainty is added corresponding to a certain inverter delay variation value. The two obtained code-words are subtracted to give the phase in thermometer coded form. The DCO control bits are then calibrated by the proposed method.

```
Parameters:
  • Depth, 'd'
Inputs:
  • Target frequency, f_tar
  • LUT containing the fraction-phase values
    corresponding to the given f_tar
  • Reference clock frequency value, r
  • Radix, m which is equal to 'r'


Initialize:
FCW='0';
FCW(1)='1';
LUT fraction phase entries stored for 'f_tar'


// Frequency coarse tuning:
for i: 1 to [log_2 m]
do
   1.  Call event driven DCO to compute fraction
       phase for present FCW setting and reference
       frequency 'r'
   2.  Compare fraction phase value with LUT(i)
   if fraction phase value = LUT(i) then
       continue
   elseif fraction phase greater than LUT
   (considering cyclic phase shift case) then
       Add '1' to i+1^th bit position of FCW
   else
       Subtract '1' from i+1^th bit position of FCW
   end if
end for
```

```
Algorithm Contd….

//Frequency fine tuning:
for i = ⌈log₂m⌉ + 1    to   depth'd'
do
   1.  Divide 'r' by 2
   2.  Call event driven DCO to compute fraction
       phase for present FCW setting and reference
       frequency 'r'
   3.  Compare returned value with LUT(i)
       if  values are same then
                continue
       else if fraction phase greater than LUT
(considering cyclic phase shift case) then
           Add '1' to iᵗʰ bit position of FCW
       else
           Subtract '1' from iᵗʰ bit position of FCW
           end if
end for
return FCW
end
```

The cyclic phase shift in the algorithm means that a fraction phase of '2' signifies that the first phase vector is '0' and the second '2' in an *8-ary* system. Alternatively, the first and second vectors can also be '7' and '1'. Although the algorithm uses a modified form of the binary search, the main novelty lies in the concept of fraction phase generation and to tune a desired frequency by comparing it with a set of known values. For a non-monotonic tuning curve, some other form of search (linear search) may be used.

A pictorial depiction of the algorithm (for a radix-10 system) which calibrates to an error of less than 40 KHz is depicted in Fig. 2.

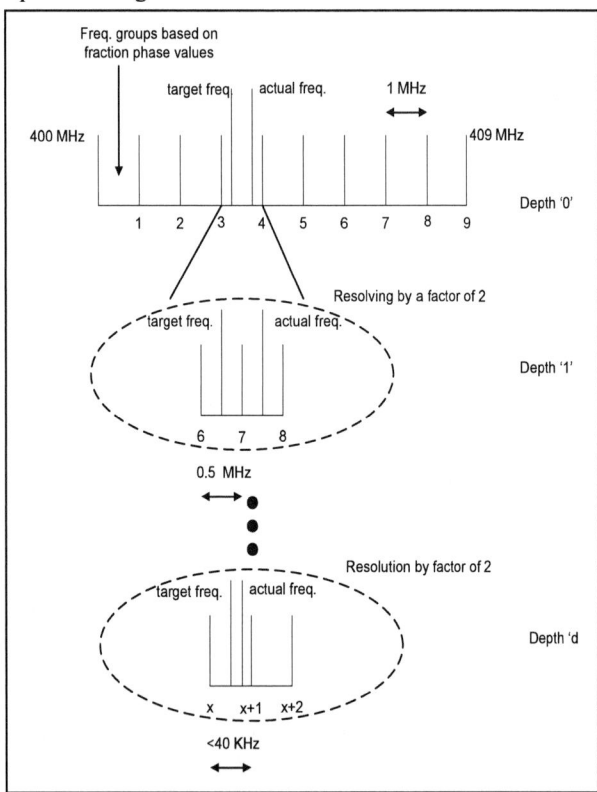

Fig. 2. Indexing structure of FCW and operation of the algorithm for a radix-10 system (MICS band specifications)

V. ARCHITECTURE OF THE FREQUENCY CALIBRATION TECHNIQUE

The architecture (Fig. 3) of the calibration unit is a mixed-signal system consisting of the following parts:

Analog/RF signal path: This section consists of a delay chain, output of which when sampled using the reference clock produces the fraction phase in thermometer coded form.

Digital computation block: This section consists of the logic circuitry that takes the thermometer coded data from the delay line, performs the necessary arithmetic and logical computation for updating the DCO frequency control word. In this section, the thermometer coded data is stored in two flip-flop arrays corresponding to two different reference-clock edges. The difference between these two values (i.e. the measured fractional phase value of the DCO output) is compared with the (targeted) fractional phase values stored in an LUT, corresponding to the particular reference clock frequency value (see Fig. 2). The comparator output acts as a control input to the add-subtract block, which either increments the FCW (if DCO frequency has to be increased) or decrements the FCW (if DCO frequency has to be lowered) or does nothing (if DCO frequency is already within the required frequency-search range corresponding to the particular reference clock frequency value).

Controller and reference clock generator: The final part consists of a controller. This section is very much essential for it keeps track of the depth (*d*) information in frequency search algorithm and uses a clock divider logic to generate the necessary reference clock of $r, \frac{r}{2}, \frac{r}{4}, \ldots \frac{r}{2^d}$ frequency values . The *start* input signal and the *stop* output signal are also used by this block to start and stop the entire algorithm, which may be interfaced with any general purpose microcontroller in hardware.

VI. SIMULATION RESULTS

The proposed frequency calibration methodology and the corresponding algorithm are generic and can be used to correct a target frequency error in any frequency band. However, the low hardware complexity and the fast calibration time make the proposed methodology particularly attractive for low-energy communication applications. In this application, ultra-low power MICS band (*402-405* MHz) wireless sensor network application has been selected and the proposed methodology has been applied to correct the MICS band spectral mask errors. In MICS band the transmit frequency must be within an error value of 100 ppm or approximately 40 KHz. For MICS band, the calibration depth that needs to be gone down to achieve the frequency error specification is $\left\lceil log_2 \left(\frac{1MHz}{40KHz} \right) \right\rceil =$ 5. More details of MICS band specification can be found in [16].

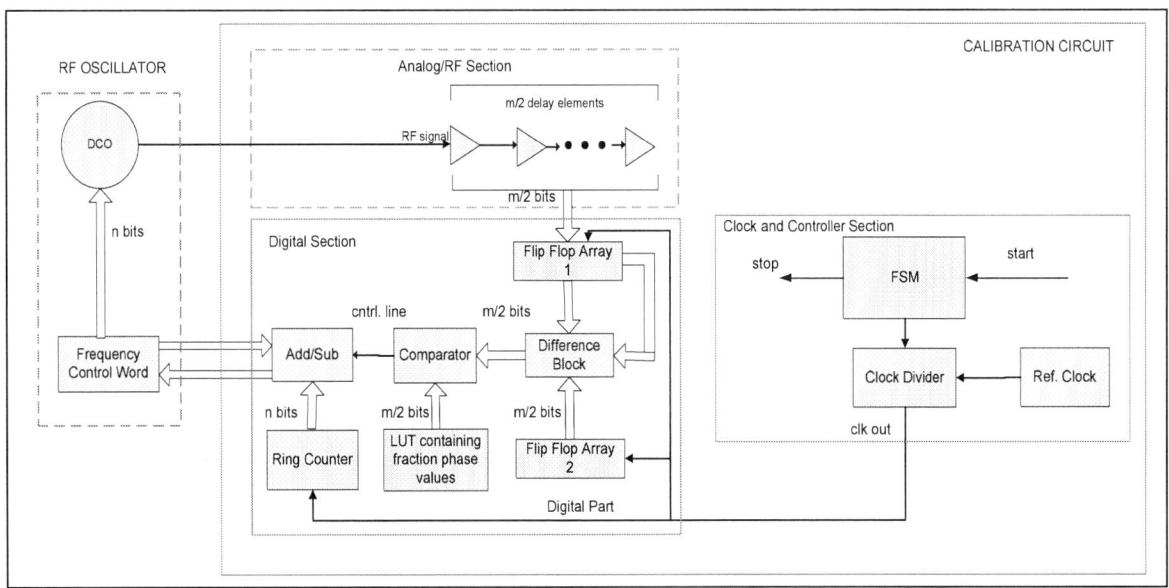

Fig. 3. Block Level implementation of the calibration circuit showing the interface with oscillator (main components only)

An MICS band LC DCO's frequency was calibrated using the proposed method. The DCO's frequency of oscillation using varactor based tuning is given in [17]:

$$ f = \frac{1}{2\pi\sqrt{LC_0}}\left(1 + \frac{V_R}{\phi_B}\right)^{m/2} \qquad (5) $$

where, L = the inductance of the LC-tank, C_0 = the zero bias capacitance of the varactor, V_R = the reverse voltage applied across the varactor, ϕ_B = the built-in potential of the p-n junction of the varactor, m = the exponent, which typically varies from 0.3 to 0.4. The primary parameters chosen were: $L = 4$ nH, $C_0 = 39.58$ pF, $\phi_B = 0.7$ V, $m = 0.35$ so that $f_0 \approx 400$ MHz. The span of the DCO was taken to be 8 MHz (400 to 408 MHz, free-running frequency being 400 MHz). Although (5) is non-linear yet monotonic, the narrow DCO span, makes it piecewise linear. The performance of the proposed frequency calibration technique has been evaluated from the point of view of i) Final frequency error, ii) Calibration time iii) Hardware cost and power consumption

A. Frequency Error Plots

The reference frequency chosen was 8 MHz. The system was tested with an 8 bit FCW (3 coarse tuning and 5 fine tuning bits). The T_{nom} value was considered for a '400' MHz frequency such that its value is 2.5 ns. The delay of each inverter is taken to be 312.5 ps; the number of inverters then comes out to be '4' assuming a 50% duty cycle of the RF waveform. The thermometer code is of '4' bit length which is converted to a '3' bit binary word by a simple code-converter. A snapshot of the thermometric code for the present example is shown in Table II.

TABLE II. THERMOMETRIC CODE AND CORRESPONDING PHASE

Fraction Phase	Thermometer code	Binary word
0	0000	000
1	1000	001
2	1100	010
3	1110	011
4	1111	100
5	0111	101
6	0011	110
7	0001	111

Initially, a perfectly linear and binary weighted tuning curve having a span of 8 MHz (400-408 MHz) has been used to test the convergence of the algorithm (for a particular frequency 402.749 MHz in Table III) and the absolute ppm error plot across the MICS band (Fig. 4). In Table III, 'fe' is the intermediate DCO frequency and 'fp' is the corresponding fraction phase value.

TABLE III. EXAMPLE OF A PARTICULAR FREQUENCY CONVERGENCE USING THE PROPOSED ALGORITHM

bit	codeword	fe(MHz)	fp	lut	decision	fref
1	10000000	404.03	4	3	>:-1	8
2	01000000	402.03	2	3	<:+1	8
3	01100000	403.03	3	3	=	8
3	01100000	403.03	6	5	>:-1	4
4	01010000	402.53	2	3	<:+1	2
5	01011000	402.78	6	6	=	1
6	01011000	402.78	4	4	=	.5
7	01011000	402.78	1	0	>:-1	.25
8	01010111	402.75	-	-	-	-

In Table III, the small deviation in 'fe' from the ideal value (10000000 should correspond to 404 MHz instead of 404.03MHz) is due to quantization effects in tuning curve simulation of the DCO.

Result:
Estimated frequency is 402.75 MHz
ppm error   2.48
Codeword   0 1 0 1 0 1 1 1

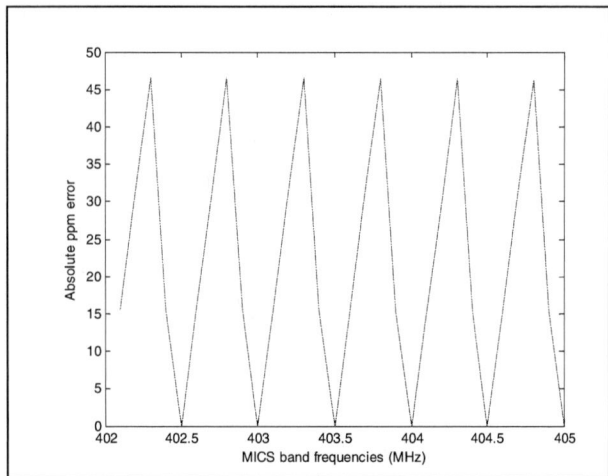

Fig. 4. Absolute ppm error across the MICS band

In Fig. 4, the error (similar to the quantization error curve of an analog-to- digital converter) is within the bound for a calibration depth of '5' as discussed in the beginning of the section. The system non-idealities (tuning curve non-linearity (5), RF waveform jitter, Inverter delay variation, systematic inverter delay-time period error and initial random phase) were next modeled. The algorithm was run with an extra FCW bit which produced the error plot of Fig. 5.

The jitter specification used was '15 ps' and each inverter had a delay variation of '20 ps'. Process variation was incorporated as a shift in center frequency from the nominal value, both negative (left) and positive (right) respectively, without loss in generality. The codeword length was '9' (1 bit extra to accommodate the non-idealities) and each bit of FCW was calibrated twice (for better resolution). In this scenario, the conversion period turns out to be '6.4' bit-times (for 100 kbps data rate) for calibrating a single tone and the final ppm error was within MICS band specifications under the given constraints. The convergence time is relative to the bit-rate and as the data rate is increased, the frequency error

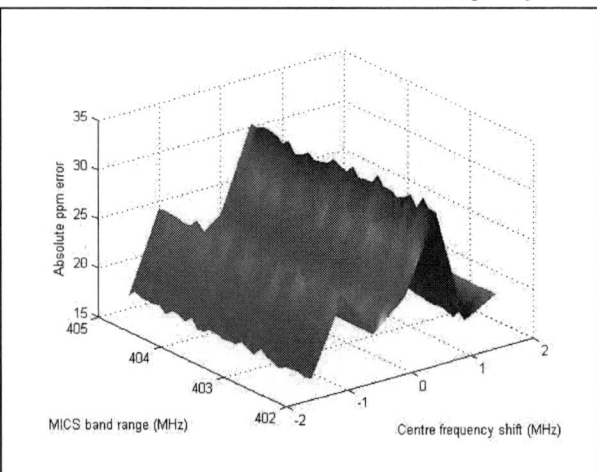

Fig. 5. Absolute ppm error plot incorporating circuit non-idealities

criterion is also relaxed and hence the relative conversion time value remains the same. The work in [11] requires 116 bit-times for calibration at 1 Mbps data rate, which is significantly slower than our proposed scheme.

B. Frequency calibration time:

During frequency calibration (without non-idealities), the time spent at any depth of frequency search is, two times of the time spent in the previous step of frequency search. For 'n_1' coarse tuning bits, and 'n_2' fine tuning bits, the formula turns out to be:

$$t_{calib} = 2 * (\frac{n_1}{f_{ref}} + \frac{2}{f_{ref}} + \frac{4}{f_{ref}} + \cdots n_2 \, terms) \quad (6)$$

which simplifies to

$$t_{calib} = \frac{2}{f_{ref}} * (n_1 + 2^{n_2+1} - 2) \quad (7)$$

An initial factor of '2' comes before the calibration time expression because, as the DCO frequency is changed, one reference period is reserved so that the DCO frequency settles and in the next cycle, fraction phase is computed.

C. Power Consumption Calculation:

The main power consumption contributor are the elements in the RF path apart from the DCO. The present event-driven model uses '4' delay cells having an average delay of '$T_{nom}/8$'. On the other hand, a classical PLL requires an divider before any comparison (phase or frequency) with the reference can be carried out. Taking into consideration only the switching power consumption of the digital blocks (true for CMOS logic) and the fact that the divider is composed of edge-triggered flip-flops and few combinatorial gates, the power consumption is bounded by: $2CV_{DD}^2f$ + *power consumption of few combinatorial gates* (C is the effective load capacitance; V_{DD} is the power supply and f is the RF frequency). It is assumed in the above calculation; each flip-flop is composed of two latches with each latch having a pair of complementary outputs. For a binary counter, for every half clock cycle (asynchronous counter considered), one pair will undergo a power consuming transition. The baseband section in a PLL consists of the charge pump and low pass filter. In the case of delay-cell based approach, a common technique is to design a current-starved inverter (as in [18]) followed by a static inverter (in order to make the wave-form squarish). The overall power (in RF path) would be: $4C'V_{DD}'^2f + I_D \cdot V_{DD}'$ (I_D is the charging/discharging current and the " ' " indicates modified load capacitance and supply values). The baseband section requires approximately '600' 2-input NAND gates [19] (the flip-flops have been decomposed into NAND gates [20]). If the RF path power consumption of the two approaches are made equal through circuit design (neglecting the impact of process variations which is beyond the scope of paper); present literature [4,13] suggest analog baseband section consumes substantially more power than the digital counterpart. The ADPLL technique [6,7] uses '24' delay cells. Even if the load capacitances are sized so that the overall dynamic power in RF path remains same as in the fraction phase based method, the base-band section would consume more power because the bus-width has increased to '24' in this case. The Vernier delay line concept [15] would require approximately '70' delay cells to achieve <40 KHz resolution in MICS band at '200' MHz clock frequency. The integer counting scheme's RF path power consumption would be similar to that of the PLL because it also uses a counter, however, the baseband frequency need to start '10' times or even '20' lower than our proposed scheme in order to achieve the same resolution. This would increase the conversion time by the same factor.

Another modification from the architecture proposed in Fig. 3 that would even reduce power consumption is the movement of the delay cells from the RF path to the clock reference path. The operation of the architecture would not be hampered by this movement because the delays are relative to the reference clock period; however the switching frequency (as there are now no more delay elements in the RF path) and hence the overall power consumption of the frequency calibration unit gets lowered as there are now no more elements in the RF path.

D. Hardware Complexity

The architecture requires approximately *600* digital two input NAND gates and few transistors in the analog section for proper functioning. This is easily realizable in monolithic form through the latest state-of-the-art VLSI fabrication process.

The performance of the fraction based scheme has been summarized in Table IV

TABLE IV. PERFORMANCE METRICS OF OUR PROPOSED SCHEME

Method	Fraction phase computation
Frequency Resolution	Any resolution can be achieved, depends on the final depth
Power Consumption	Comparable to [11]
Hardware Complexity	DCO, delay cells and baseband blocks
Calibration time taken	2* reference frequency at final depth
Remark	Proves to be the most energy efficient from a theoretical standpoint

VII. CONCLUSION

In this paper, the design of an energy efficient frequency calibration methodology, the corresponding algorithm and the hardware architectures have been presented. For WSNs, where low energy operation is of paramount importance, the proposed technique can be used for fast frequency calibration of the RF front-end for correcting spectral mask errors. Comparisons with the contemporary WSN frequency correction techniques show that the proposed technique is more energy efficient as well as has the flexibility of achieving any arbitrary frequency resolution, which is limited primarily by the phase jitter of the RF oscillator itself.

ACKNOWLEDGMENT

The authors would like to thank Advanced VLSI Design Lab, Indian Institute of Technology, Kharagpur for supporting the research.

RFERENCES

[1] C. Shunguang, A. J. Goldsmith, and A. Bahai, 'Energy-Constrained Modulation Optimization', *IEEE Trans., Wireless Commun.*, vol. 4, no.5, pp. 2349-2360, Sep. 2005

[2] P. D. Bradley, 'An Ultra Low Power, High Performance Medical Implant Communication System (MICS) Transceiver for Implantable Devices', in *IEEE Biomed. Circuits and Systs. Conf., BioCAS, London*, pp. 158-161, Nov./Dec. 2006

[3] A. Y. Wang, S. H. Cho, C. G. Sodini, and A. P. Chandrakasan, 'Energy Efficient Modulation and MAC for Assymetric RF Microsensor Systems', in *Intl. Symp. Low Power Electronics and Design, ISLPED, California*, pp.106-111, Aug. 2001

[4] C. T. Liu, H. H. Hsieh, and L. H. Hu, 'A Low-Power Quadrature VCO and It's Application to a 0.6V 2.4 GHz PLL', *IEEE Trans. Circuits Syst.-1,Reg. Papers*, vol. 57, no. 4, pp. 793-802, Apr. 2010

[5] A. Ghosh, A. Halder and A. S. Dhar, 'A Variable RF Carrier Modulation Scheme for Ultra Low Power Wireless Body-Area Network', accepted for publication in *IEEE Systems Jnl.*

[6] R. B. Staszewski, et.al., 'All digital Tx- frequency synthesizer and discrete time receiver for Bluetooth radio in 130 nm CMOS', *IEEE J. Solid-State Circuits*, vol. 39, no. 12, pp. 2278-2291, Dec. 2004

[7] R. B. Staszewski and T. Balsara, 'All-Digital PLL with ultra fast settling', *IEEE Trans. Circuits Syst.–II,Expr. Briefs*, vol. 54, no. 2, pp. 181-185, Feb. 2007

[8] M. Lee, M. E. Heidari and A. A. Abidi, 'A Low-Noise Wideband Digital Phase-Locked Loop Based on a Coarse-Fine Time-to-Digital Converter with Subpicosecond resolution', *IEEE J. Solid-State Circuits*, vol. 44, no. 10, pp. 2808-2816, Oct. 2009

[9] J. L. Bohorquez, A. P. Chandrakasan, and J. L. Dawson, 'A 350 uW CMOS MSK Transmitter and 400uW OOK Super-regenerative Receiver for Medical Implant Communications', *IEEE J. Solid-State Circuits*, vol. 44, no. 4, pp. 1248-1259, Apr. 2009

[10] S. Dragao, D. Leenaerts, B. Nauta, F. Sebastiano, K. Makinwa and L. J. Breems, 'A 200μW Duty-Cycled PLL for wireless sensor nodes in 65nm CMOS', *IEEE J. Solid-State Circuits*, vol. 45, no. 7, pp.1305-1315, Jul. 2010

[11] J. Ayers, K. Mayaram and T. S. Fiez, 'An Ultralow-Power Receiver for Wireless Sensor Networks', *IEEE J. Solid-State Circuits*, vol. 45, no. 9, pp. 1759-1769, Sep. 2010

[12] A. Aktas and M. Ismail, "CMOS PLL calibration techniques," *IEEE Circuits Devices Mag.*, vol. 20, no.5, pp. 6-11, Sep./Oct. 2004

[13] W. Rahajandraibe, L. Zaid, V. Cheynet de Beaupre and G. Bas, 'Frequency Synthesizer and FSK modulator for IEEE 802.15.4 based Applications', in *IEEE RFIC Symp , Honolulu*, pp. 229-232, Jun. 2007

[14] T. H. Lin and Y. J. Lai, 'An Agile VCO Frequency Calibration Technique for a 10-GHz CMOS PLL', *IEEE J. Solid-State Circuits*, vol. 42, no. 2, pp. 340-349, Feb. 2007

[15] G. H. Li and H. P. Chou, 'A High Resolution Time-to-Digital Converter Using Two Level Vernier Delay Line Technique', in *Nucl. Sci. Symp. Conf. Record NSS'07, Honolulu*, pp. 276-280, Oct./Nov. 2007

[16] FCC Rules and Regulations, Table of Frequency Allocations, Part 2.106, Nov. 2002

[17] B. Razavi, 'Design of Analog CMOS Integrated Circuits', 4th Ed : New Delhi, Tata McGraw Hill, 2002, 11th Reprint 2006 pp. 522

[18] M. Jahan and J Holleman, 'An Ultra-Low Power 400 MHz VCO for MICS Band Application', in *Intl. Conf. on Electrical & Computer Engineering, (ICECE,) Dhaka*, Dec. 2010, pp. 318-321

[19] M.M. Mano, 'Digital Logic and Computer Design', New Delhi Prentice Hall of India Pvt. Ltd., 22nd Reprint 2002, pp. 124-125,, 166, 375,

[20] S.M. Kang and Y. Leblebici 'CMOS Digital Integrated Circuits Analysis and Design', 3rd Ed. New Delhi, Tata Mcgraw Hill, 2003, 14th Reprint 2006, pp. 349

978-1-4673-0438-2/12 $31.00 © 2012 IEEE

2012 25th International Conference on VLSI Design

Self-Induced Supply Noise Reduction Technique in GBPS rate Transmitters

Nitin Gupta, Tapas Nandy and Phalguni Bala
STMicroelectronics Pvt. Ltd.
Greater Noida, India
e-mail: nitin.gupta@st.com, tapas.nandy@st.com and falguni.bala@st.com

Abstract- **In high speed link transmitters, one major contributor of jitter is the data-dependant switching of the transmitters. Such switching leads to oscillations in the supply R-L-C network. This paper presents an area-efficient way to reduce this supply noise by shifting the switching beyond the resonance frequency of the supply network, irrespective of the data-pattern. This scheme is implemented in HDMI transmitter in 65nm technology.**

Keywords-HDMI; Transmitter; High Speed Serial Link; Jitter

I. INTRODUCTION

Jitter is a key performance criterion in high-speed serial links (HSL). One of the major contributions of jitter comes from the switching noise in the supply, which is often induced by switching of the circuit itself. It is a common practice to use large on-chip decoupling capacitor to reduce switching noise. But using on chip decoupling capacitor costs lot of silicon area. Also the optimum value of decoupling capacitor is often linked to a particular package parasitic value, hence the reusability of the same design in multiple System-on-Chip (SoC) is compromised.

Here, a methodology has been proposed to reduce the self-induced supply noise related jitter. It is area efficient and the HSL can be reused in various SoCs, across a range of packages.

II. DATA DEPENDANT JITTER

HSL's transmitters have big output drivers preceded by the predriver horns, (which are sized-up long chain of inverters (Fig. 1)). The predrivers are large, hence cause significant switching noise on supply-ground pair; this effect is more pronounced in the case of large packages. This noise in turn, depends on the data pattern fed to the predriver chain, and increases jitter in the transmitter data as it propagates through this chain. The resultant jitter contribution by each component of the predriver has been shown in the inset of (Fig. 1).

The package parasitic R-L-C model, as lumped components on supply and ground, is shown in Fig. 2 (a). The s-parameters from a 764 pin PBGA package (representative of big SoC's package) have been used for supply and ground, The plots of impedance (magnitude and phase) vs frequency is shown in Fig. 2(b) & Fig. 2(c) respectively. It shows R-L-C tank behavior too. A network with practical values of package R-L-C and decoupling capacitor has a resonance frequency, ω_R, lower than the clock rate used in the high speed link circuits. In a random data, all frequency components falling in the proximity of ω_R trigger oscillations which last longer due to lower damping factor, ζ. The next switching current trigger

comes before the first oscillations die and so on [1]; This gives rise to complex oscillations with peak to peak amplitude large enough to cause significant jitter on the output.

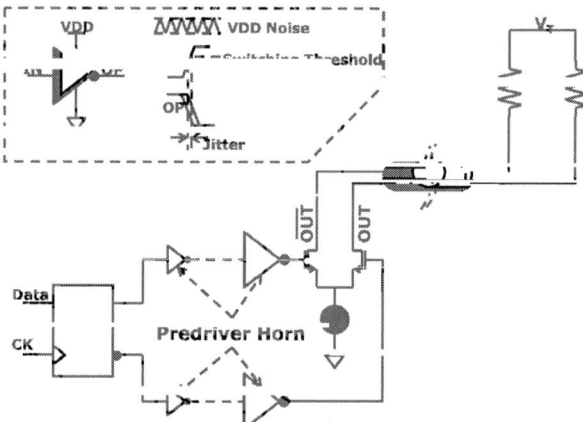

Figure 1. Standard Transmitter and supply noise induced jitter.

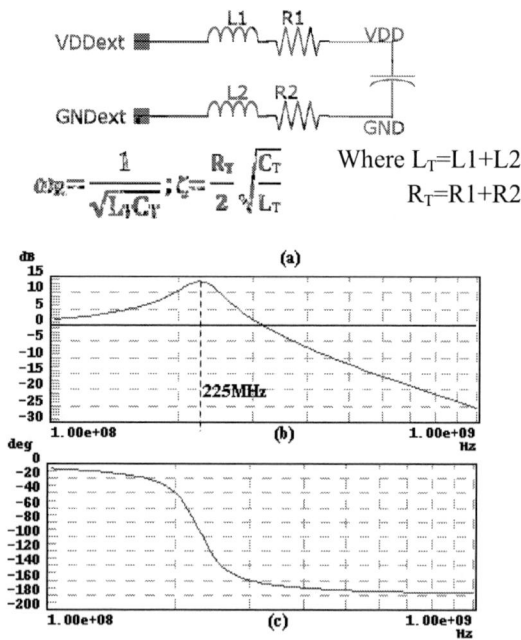

$$\omega_R = \frac{1}{\sqrt{L_T C_T}}; \zeta = \frac{R_T}{2}\sqrt{\frac{C_T}{L_T}} \quad \text{Where } L_T = L1 + L2 \quad R_T = R1 + R2$$

Figure 2. Package and decoupling lumped model & Supply-Ground impedance from S-param.

978-1-4673-0438-2/12 $31.00 © 2012 IEEE

Data Dependant Jitter (DDJ) can also be due to the bandwidth limitation of the communication channel. To overcome this particular problem pre-emphasis is used [4]. But the usage of pre-emphasis adds extra predrivers and drivers, which further contribute to the supply noise. So, the full advantage of pre-emphasis cannot be seen.

Other switching noise contributors are clock tree buffers and the remaining logic in the serial link, sharing the same supply-ground pair. Clock tree buffers always create a high frequency clock related switching noise which is usually higher than the ω_R, hence, it does not trigger oscillations. The remaining logic's switching noise is lower as compared to the big pre-drivers. In cases where it is large, it is a common practice to feed the logic by a dedicated supply-ground pair isolated from the one used for transmitter section. Hence, in this work the focus for the solution is on the predrivers.

III. EXISTING SOLUTIONS

The standard method suggested is to use optimum amount of on-chip decoupling capacitors and insert a small resistance on supply path or to have a DC consumption to attain faster damping [1],[2]. There are modern tools also available to estimate and place optimum amount of decoupling capacitors at the proper place. In all these methods, the optimum amount of decoupling is package dependant, which is against the reusability of the design. The big SoCs have large package parasitic and require large decoupling capacitors, so it costs significant amount of silicon area. Inserting small resistance in supply path introduces drop in supply voltage which is not desirable in high speed applications. Having a DC-consumption is also not desirable for any power optimized design.

The proposal of reducing supply noise by using digitally switched decoupling capacitor circuit [3] also depends on package and requires extra digital circuitry for controlling.

IV. PSEUDO SHADOW SWITCHING

If the frequency components of the switching current are shifted beyond the resonance frequency of the package R-L-C

network, by some method which is independent of the data pattern, then oscillations do not start. This can be achieved by implementing a similar shadow circuit and switching it when the real data does not switch. In this case, shadow data must be generated at speed, and it's data-path must be balanced with the real data-path. Although it does not contribute to the real data transmission, but it consumes area and power same as the real data-path. A smaller amount of decoupling capacitance can still be used to limit the L(di/dt) related noise on the supply.

The proposed circuit is shown in Fig. 3(a). It creates the same switching current on every clock edge, independent of the data pattern. The entire circuitry contributes to the real data transmission.

The serial data/datab is broken up into a pair of even and odd data/datab, launched on positive edges of the half-rate clocks, CkHalfBar and CkHalf, respectively. The even and odd data register outputs are gated by the opposite phases of the same half-rate clocks, i.e., CkHalf and CkHalfBar, respectively. At positive edge of CkHalf, one of the predriver sections, EA or EB, switches from '0' to '1' and one of the predriver sections, OA or OB, switches from '1' to '0' and vice versa for the rising edge of CkHalfBar. It ensures one pair of the predriver switching at every edge of the half rate clock, i.e., every rising edge of the data-rate clock. This way, the switching current frequency moves to data-rate frequency and beyond. Here, 2-to-1 data serialization happens at the two pairs of output multiplexing switch.

The proposed circuit for one tap pre-emphasis [shown in Fig. 3(b)] works in a similar manner with half rate clocks and with odd and even pre-emphasis data paths (namely POA, POB, PEA, PEB) and pre-emphasis bypass paths (namely BYPOA and BYPEA) to have the same switching spike on every clock edge, independent of the pre-emphasis data pattern. This strategy has the same effect as that of the data path described earlier by moving the switching frequencies to data rate frequency and beyond.

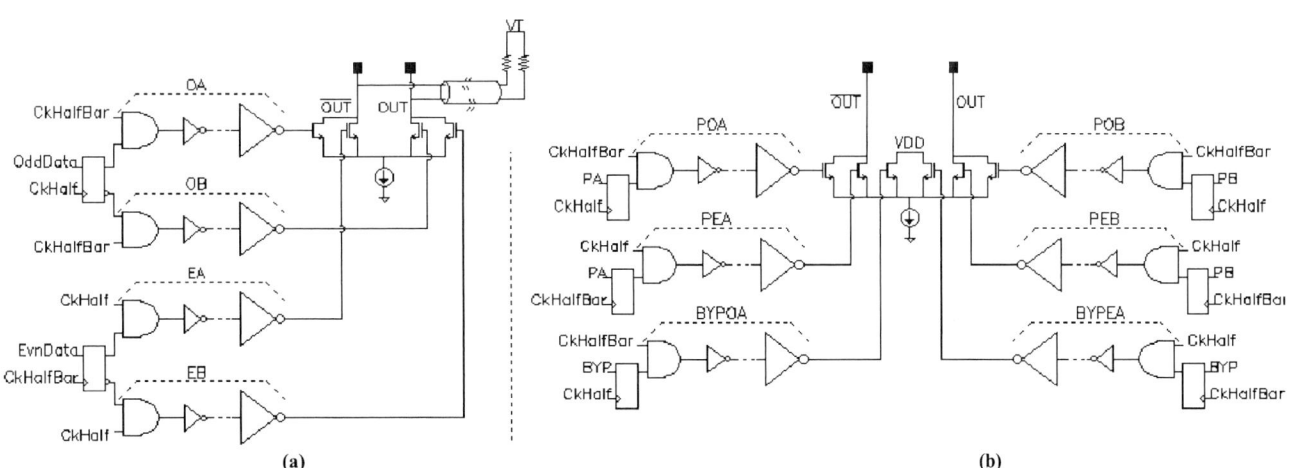

Figure 3. Proposed circuit to reduce data-dependent switching noise

The waveforms of Fig. 4 depict the switching current at every clock, independent of data. The data dependent switching current for the conventional circuit of Fig. 1 is also shown for comparison in the indicated section of Fig. 4. The current spikes at the falling edge of Ck are ignored as the predrivers, which are the major current spike contributors do not switch at these edges.

Using more parallel data with lower rate multi-phase clock and serializing at the final output multiplexer, is another way of achieving similar current profile at every clock-edge. Here more pairs of output multiplexing switches would be required, but this will lead to higher output parasitic capacitance, which in turn is detrimental to the high speed operation.

V. SIMULATION & SILICON RESULTS

The FFT of the simulated power supply current for a random data (2.5Gbps) with the conventional scheme of Fig. 1 is shown in Fig. 5(a). The same is shown in Fig. 5(b) with the proposed scheme of Fig. 3. There are clear peaks at the data rate (2.5GHz) and its harmonics with both the schemes. But at the lower frequencies (less than data rate), the power is well distributed for the conventional scheme and is enough to trigger oscillations at the resonance frequency of the supply network. On the other hand, in the same frequency range, the power in the proposed scheme is ~20dB less as compared to the conventional scheme. Higher frequency noise in the supply is better filtered and reduced by small amount of decoupling capacitors. This small amount of decoupling capacitor further reduces the resonance frequency to take advantage of the lower noise power at lower frequencies. The increase in area due to extra predrivers is offset by great reduction in required decoupling capacitor.

As the resonance frequency for big packages are usually in the range of few hundred MHz and the switching noise power of the proposed scheme moves to GHz range (data rate), supply network oscillation problem never occurs irrespective of the package selection. Hence, the decoupling capacitor strategy need not be reconsidered for every package, making the design reusable.

The simulated power supply waveforms for random data with the proposed scheme, in 764 PBGA package is shown in Fig. 6(a). The same is shown in Fig. 6(b) for the conventional scheme; the supply noise improvement in peak to peak amplitude is ~2.5 times. The corresponding simulated eye diagrams at 2.5GBPS are also shown in Fig. 7 (a) and (b) respectively with remarkable improvement in the jitter [Similar sized circuits for predriver and driver are used for these simulation-comparisons].

The scheme has been implemented in HDMI [5] transmitter (36bit 1080p 60Hz) in 65nM technology. The target specification for this development was 2.5Gbps/Channel. The eye diagram holds good even beyond the target data rate of 2.5GBPS. The measured eye diagram at 2.8GBPS is shown in Fig. 8(a). To achieve reasonable supply noise and acceptable jitter, larger on-chip decoupling capacitor had used in an earlier silicon implementation, with conventional method of Fig. 1, in the same technology, with 35% more area. The eye diagram has just passed the mask at the target data-rate of 2.5GBPS.This eye diagram from the earlier silicon implementation, is shown in Fig. 8(b).

The die photo of the prototype with the proposed scheme is shown in Fig. 9.

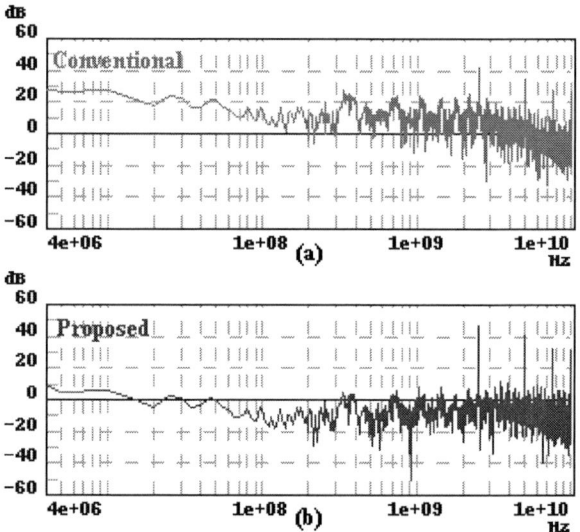

Figure 5. Simulated FFT of Supply Current of Fig. 1 and Fig. 3.

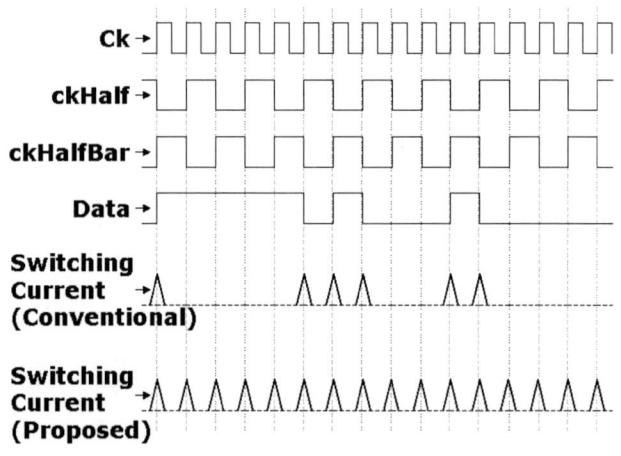

Figure 4. Switching current comparison in standard & proposed scheme

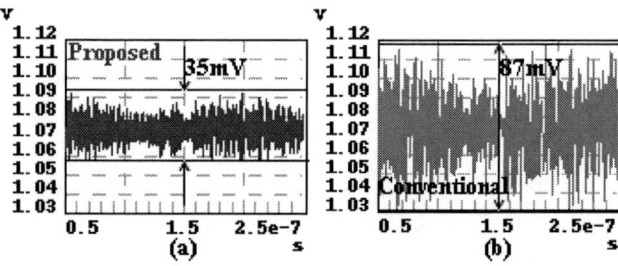

Figure 6. Simulated supply noise of proposed and conventional scheme.

978-1-4673-0438-2/12 $31.00 © 2012 IEEE

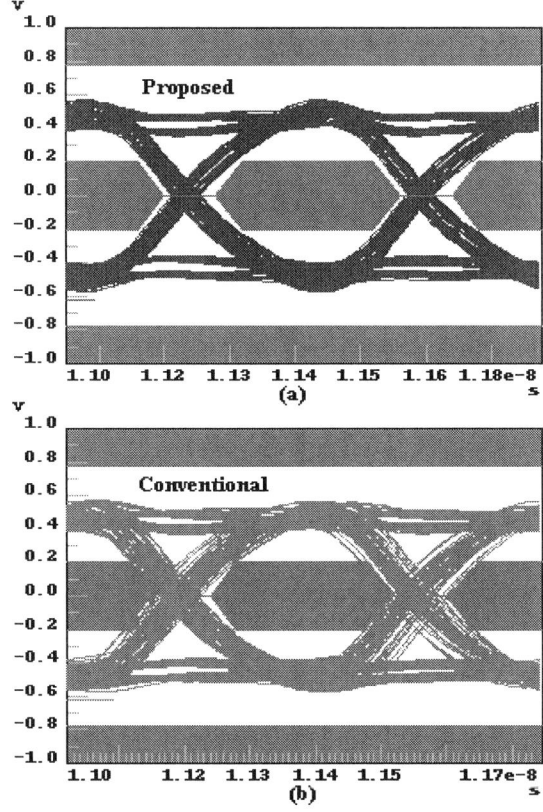

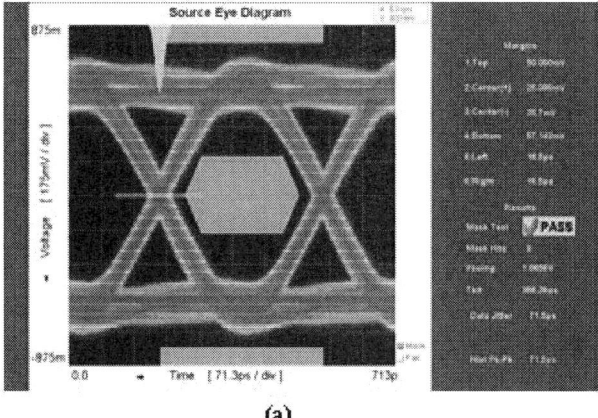

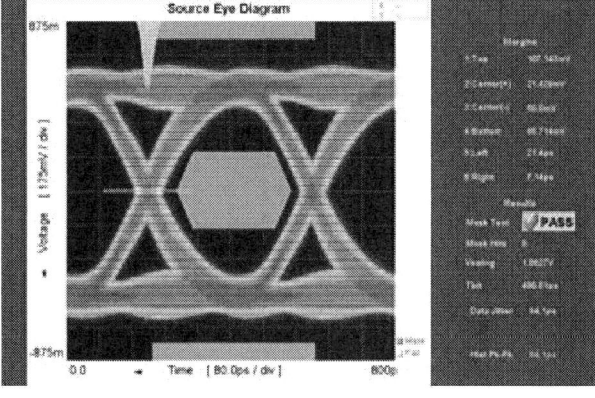

Figure 7. Simulated eye-diagram of proposed and conventional scheme.

VI. CONCLUSION

We introduced this scheme in the design of a 36bit, 1080P, 60Hz capable HDMI Transmitter in 65nM technology. Decoupling capacitor usage is very low as compared to the previous HDMI transmitter implementation. There is significant improvement in Jitter performance seen on the silicon, in addition to significant gain in area.

Figure 8. Measured Eye diagram at (a) 2.8GBPS with the proposed scheme and (b) 2.5GBPS with conventional scheme.

REFERENCES

[1] Patrik Larsson, "Resonance and Damping in CMOS Circuits with On-Chip Decoupling Capacitance", IEEE transactions on circuits & systems—I, vol. 45, no. 8, August 1998.

[2] Charles Hough et al,, "New Approaches for On-Chip Power Switching Noise Reduction", Custom Integrated Circuits Conference, 1995.

[3] Jie Gu, Hanyong Eom, and Chris H. Kim "On-Chip Supply Noise Regulation Using a Low-Power Digital Switched Decoupling Capacitor Circuit", IEEE JOURNAL OF SOLID-STATE CIRCUITS, VOL. 44, NO. 6, JUNE 2009.

[4] Chih-Hsien Lin, Chang-Hsiao Tsai, Chih-Ning Chen and Shyh-Jye Jou, "4/2 PAM Serial link Transmitter with Tunable Pre-Emphasis", ISCAS 2004.

[5] "High-Definition Multimedia Interface Specification Version 1.4a", http://www.hdmi.org.

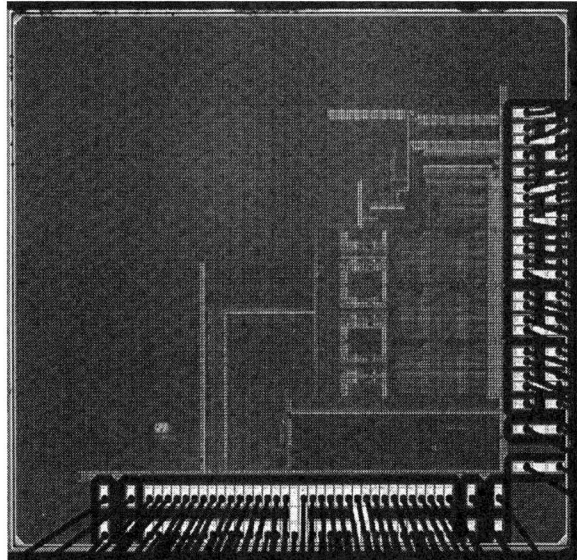

Figure 9. The die-photo of the proposed implementation.

978-1-4673-0438-2/12 $31.00 © 2012 IEEE

Buffer Design and Eye-Diagram Based Characterization of a 20 GS/s CMOS DAC

Mohit Singh, and Shalabh Gupta*
Department of Electrical Engineering, IIT Bombay
Powai, Mumbai-400076, India.
*Email: shalabh@ee.iitb.ac.in

Abstract—**High-Speed Digital-to-Analog Converters (DACs) are inevitable due to the advent of multi level modulation formats to meet the increasing demand of high data rates in communication systems. In this paper, a 4-bit 20GS/s DAC has been designed in 90nm CMOS technology. CMOS based DACs provide a low cost single IC solution as compared to compound semiconductor counterparts by fully integrating digital and RF blocks. In this paper, an on-chip Linear Feedback Shift Register (LFSR) is used to generate the required high-speed broadband data and eye-diagram of the DAC output is used for characterization. In order to drive the high capacitive loads along with routing, Electro Static Discharge (ESD) and pad capacitance ($\approx$ 800 fF) at a speed of 20 GS/s (13.1 GHz bandwidth), a new buffer architecture has also been implemented.**

Index Terms—**Characterizing High-Speed DACs, Eye-Diagram, High-speed buffer**

I. INTRODUCTION

High-speed Digital-to-Analog Converters (DACs) are expected to play a crucial role in emerging multi-gigabit/second digital communication systems that target higher channel capacity by improving spectral efficiency (measured in bits/sec/Hz). In particular, very high-speed (low resolution) DACs will be required for next generation optical communication systems [1] (for example to achieve transmission rates of more than 100 Gb/s using dual-polarization 64-QAM [2], [3]). With improving speed of CMOS circuits due to technology scaling, lower costs, and the option to integrate them as part of large VLSI circuits, implementation of such DACs in deep-submicron CMOS technologies becomes attractive. However, a number of challenges are associated with the design of high-speed DACs in general, and CMOS technology in particular. Until recently, to our knowledge, in literature DACs operating only up to 12 GS/s had been reported in CMOS technology [4]. Recently, a 56 GS/s DAC was published [5], however, it has not been fully characterized to evaluate its operation for communication systems employing multi-level signaling.

One of the major design bottlenecks is high capacitive loading due to internal routing, ESD protection diodes and bond pads. To counter the effect of capacitance due to ESD protection, broadband ESD protection networks are required. T-coil network based ESD structure can be used for broadband operation [6], however it cannot provide a frequency response that is flat enough to achieve high quality signal from a DAC. Moreover, it consumes significant chip area. Increasing driving

transistor sizes to drive these capacitive loads leads to self loading, further degrading the bandwidth. To overcome these issues, we propose a high-speed buffer that can drive highly capacitive loads at the DAC output.

Evaluation of such high-speed DACs is yet another challenge faced by the designers. Ramp and sinusoids are commonly used for DAC characterization [7]–[10]. Ramp signals don't capture the dynamic response of the DACs while sinusoidal test pattern generation is not easy at high speeds. More importantly, these test schemes do not emulate the random nature of the data used in broadband communication links and fail to characterize the dynamic behavior of these DACs. They also don't account for signal distortion due to group delay dispersion and/or dynamic non-linearity [11]. Because of these problems, though DACs in [12], [13] have been designed for high speed operation, they could only be tested for lower frequencies.

To facilitate evaluation of ultra-high speed DACs easily and in a more realistic situation, a pseudo-random test-pattern generation technique using LFSR, followed by eye-diagram measurement, has been suggested in [14]. In this paper, we propose a method to interpret the output eye-diagram produced by the input pseudo random test vectors and use it for characterizing high-speed DACs.

The organization of paper is as follows: Section 2 explains the design of 20 GS/s DAC in which specifically in subsection B the design of our proposed buffer has been discussed. In section 3, the new DAC characterization method using output eye-diagram has been discussed. Section 4 discusses the post-layout simulation results while section 5 concludes the paper.

II. 20 GS/s DAC DESIGN

Though DACs up to 43 GS/s have been reported using SiGe-BiCMOS-LSI technology [15], conversion rates of only up to 12 GS/s [4] have been reported for CMOS technology. A 56 GS/s CMOS DAC has been published [5], however, it has not been fully characterized for its operation employing multi-level signaling for communication links. Due to unavailability of DACs in standard CMOS technology at high conversion rates, electrical-optical-electrical (E-O-E) conversion technique had to be used to generate high-speed 8-level signals for a 64-QAM 112-Gb/s optical transmission experiment [16]. Recently, generation of 21.4 Gbaud 8-level signals was reported using an IC designed in InP technology

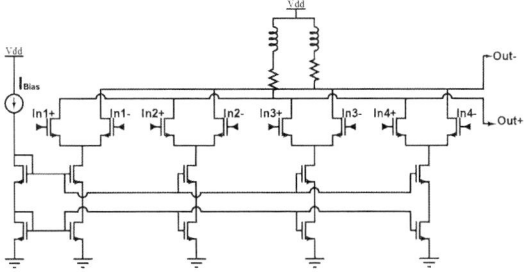

Fig. 1. Schematic of the designed 4-bit 20-GS/s current-steering DAC stage.

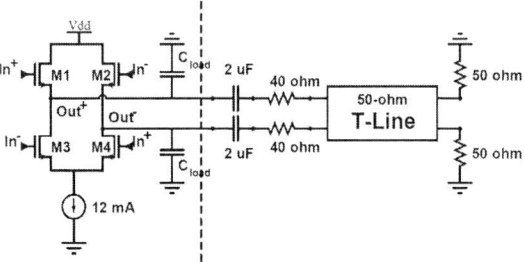

Fig. 2. Designed output buffer for 20 GS/s DAC connected to off-chip matching circuit.

[13]. This is because the f_T (transit frequency) of a typical CMOS process is much less than that of transistors in compound semiconductors. One method for increasing speed in CMOS technology could be to reduce load resistance and increase drive current. But with lower load capacitance, the layout resistance becomes significant and so is the impact of resistance variation on signal. This results in dynamic non-linearity (DNL) and limits the reduction of load resistance beyond a limit. To overcome these issues, use of inductors is a popular choice, which increases the DAC sampling rate, though it may distort the eye-diagram.

A. The DAC Stage

The prime application aimed for the designed DAC is in optical links, which requires low resolution and high-speed. The switched-current DAC is a popular choice for high-speed applications as they offer high linearity over broad bandwidths. It can be either binary-weighted or thermometer coded or a hybrid of both (segmented architecture). Thermometer coded DACs have lesser matching requirements for same DNL as compared to binary coded DACs, but because of the required low resolution and to avoid the extra decoding circuitry, binary weighted architecture has been chosen (Fig. 1). This extra decoding circuitry if chosen would also lead to additional area and power consumption. Integral non-linearity (INL) and Spurious-free dynamic range (SFDR) respectively have a direct and an inverse dependence on the output resistance of current source [17], [18]. Hence, to have a high output resistance, a configuration with cascoded current source has been used (Fig. 1). Maintaining the current sources in saturation is a difficult task in 90 nm with 1 Volt of supply. To overcome this limitation low V_T NMOS transistors have been used in switches and cascode, and regular V_T NMOS transistors for current source. To avoid the effect of process variations and mismatch between current sources, the current source has been designed with transistor length of 500 nm using common centroid topology. The low V_T switch transistors have been designed with a minimum allowed length of 80nm, so that it offers minimum capacitance and helps faster switching. For a load resistance of 25 Ω and a desired peak-to-peak differential output swing of 440 mV, current sources of 650 μA, 1.3 mA, 2.6 mA and 5.2 mA have been used. In order to assist the high-speed input data, it has been pre-emphasized with inductive

peaking using two 150 pH inductor in series with the load resistance.

B. The Output Buffer

The output produced by the designed DAC is weak and is capable of driving a load capacitance of only up to 150 fF properly. In order to drive a load capacitance of the order of 1pF at a speed of 20 GS/s, the load resistance has to be as low as 10 Ω. This increases the size of the driving transistor, which further adds on to the load capacitance. In order to ensure that the DAC output is capable of driving this output load capacitance (and the pad & ESD protection circuit), the output needs to be buffered. To ensure the above requirements are met, new buffer architecture is designed, in which the current from common drain configuration is bled into the common source configuration as shown in figure 2.

The buffer uses a push-pull configuration in which the bottom transistors (M3 & M4) assist the top ones (M1 & M2). The NMOS transistors (M1 & M2) on the top (w=150 um, l=90 nm) do not load the DAC even after its large transistor size due to there use in common drain configuration (like a source follower). The input common mode of DAC is pretty high (800 mV approx.), which reduces the size of the bottom NMOS transistors (M3 & M4) used in common source configuration (w=40um, l=90 nm) and avoids loading the previous stage. Though one of the two the bottom common source NMOS transistors (M3 and M4) may not be always in saturation, the performance as the output impedance offered by the buffer mainly depends upon the sizes of the common drain transistors (M1 and M2) at the top. One more added advantage of this configuration is that based on the load capacitance, the required output impedance of the buffer can be varied by tuning the tail current source. The output swing obtained at the buffer output was approximately 360 mV peak-to-peak differential. Use of low V_T transistors (M1, M2, M3 & M4) instead of regular V_T would have been slightly beneficial in increasing the output swing and reducing the buffer current.

After the buffer stage, the output impedance of the circuit falls down to approximately 10 Ω. In order to match this to the 50 Ω transmission line, series termination has been used on the source side by placing a tunable off-chip resistor (40 Ω approx.) in series to the transmission line (Fig. 2). Along with the resistor, a series off-chip capacitor of 2 uF is also used to

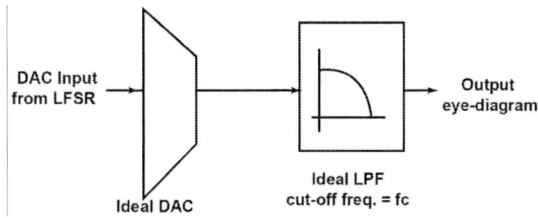

Fig. 3. Setup for calculation of DAC bandwidth based on eye-diagram.

block the DC of the output buffer. The drawback of using this series matching is reduction in the peak-to-peak differential output swing from 360mV to 200mV. As the DAC is designed for wide-band applications, no other matching method was preferable.

III. EYE-DIAGRAM CHARACTERIZATION METHOD

For high speed DACs, dynamic response is as important as the static characteristics such as DNL and INL, as the output signals do not settle down to the desired levels because of transients. Here we propose a formal method which can be used to characterize a DAC based on its output eye-diagram. Efficient generation of PRSS (Pseudo Random Symbol Sequence) required to produce this multi-level eye diagram can be implemented using alternate bits tapped from a maximal length sequence generated by an LFSR (Linear Feedback Shift Register) [14].

To estimate the bandwidth of the broadband DAC under test, a setup comprising of an ideal DAC with an ideal low-pass filter (LPF) is taken (Fig. 3) and simulated by applying broadband input similar to that applied to the DAC under test at the same sampling rate. The cut-off frequency (f_C) of the ideal LPF is tuned so that we get similar eye-openings in the output eye diagram (Fig. 4). Now, the cut-off frequency of the ideal LPF roughly equals the bandwidth of the DAC under test.

Figure 4(a) shows the output eye-diagram obtained for a designed 4-bit DAC tested using 8-bit LFSR at 20 GS/s while Figure 4(b) shows the output of an ideal DAC using the test setup shown in fig. 3 tested under similar test conditions. Both the figures show similar eye-openings. In order to test the effectiveness of this method, bandwidth of the DAC was calculated using sinusoudal input produced using an ideal sinusoid generator and ideal ADC. The bandwidth of the designed DAC is calculated to be 13.1 GHz, whereas the cut-off frequency of the ideal 1st order LPF comes out to be 11.5GHz. Thus we can see that the LPF cut-off frequency gives a good pessimistic approximation of the bandwidth of the broadband DAC under test.

A. C. Eye-based DNL and INL

Once the output eye has been plotted, the DAC resolution is calculated based on the output swing by the formula,

$$Resolution = \frac{Full\ scale\ range}{2^N - 1} \quad (1)$$

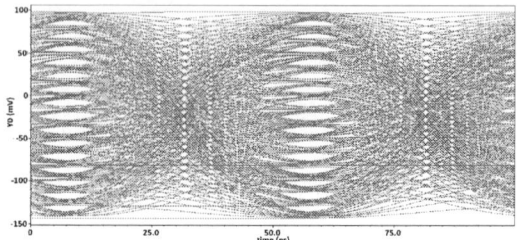

(a) Output eye-diagram of designed 4-bit DAC at 20 GS/s

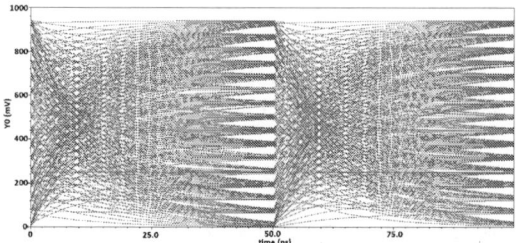

(b) Output eye-diagram of an ideal 4-bit DAC followed by an ideal LPF at 20 GS/s

Fig. 4. Output eye-diagram of (a) a 4-bit designed DAC (b) an ideal 4-bit DAC followed by and ideal LPF, under similar test conditions using 8-bit PRBS generator

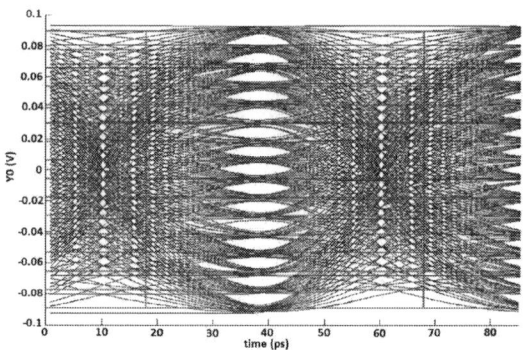

Fig. 5. Creating an ideal Eye-Grid based on the calculated resolution.

This calculated resolution is used to make the ideal eye-grid as shown in figure 5 for a 4-bit DAC.

Based on the mean (L_i) and the standard Deviation (σ_i) values at each level, the Q-factor for the ith level is calculated as:

$$Q - factor_i = \left(\frac{L_{i+1} - L_i}{\sigma_{i+1} + \sigma_i} \right) \quad (2)$$

$$where, L_i = Mean\ value\ of\ i^{th}\ level$$
$$\sigma_i = Standard\ deviation\ for\ i^{th}\ level$$

A range of $L_i \pm 3\sigma$ is referred as the error-band for that level. This error-band is then superimposed on the output eye-diagram for further calculations (Fig. 6). Now, referring to the standard DNL and INL definitions, we say that Eye-based

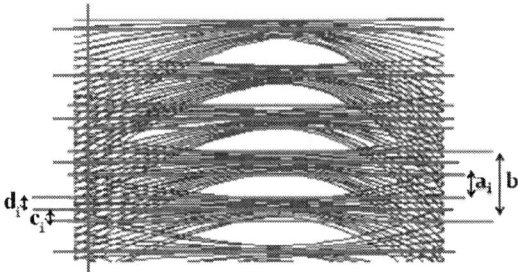

Fig. 6. Error-band of $\pm3\sigma$ superimposed on output eye-diagram

DNL is the deviation of minimum and maximum distance between two error-bands from 1 LSB (Least significant bit). Worst case deviation defines the overall DNL.

$$DNL_i = -(b_i - 1LSB) \; to \; (1LSB - a_i) \qquad (3)$$

$$DNL = Max(Max(|b_i - 1LSB|), Max(|1LSB - a_i|)) \quad (4)$$

where, a_i & b_i are the minimum and maximum distance between two error bands for i^{th} level, as shown in Fig. 6.

In a similar fashion, we define Eye-based INL as the maximum deviation of values in error band from that on the ideal eye-grid for a level. So, we take the worst case deviation to define the overall INL of the DAC.

$$INL_i = -c_i \; to \; d_i \qquad (5)$$

$$INL = Max.(max.|c_i|, max.|d_i|) \qquad (6)$$

This Eye-Diagram analysis method works only if the values lying within two error-bands do not overlap. Also, the DAC output must be monotonic and should not contain any missing codes. This method assumes that noise level is not dominant and is not of much significance as compared to DAC non-linearities. Hence, this technique will not work if noise level significantly degrades the eye quality. This method is also not a good option for high resolution DACs.

IV. RESULTS

Post-layout simulations were carried out for the designed DAC. The layout of the DAC designed in 90 nm CMOS technology is shown in figure 7.

The post-layout extracted DAC was run using a 20 GS/s 9-bit LFSR schematic in cadence. The LFSR output waveform and eye-diagram of all its outputs (DAC inputs) is shown in figure 8. The LFSR output has a swing of 550 mV peak-to-peak differential.

The 4-bit DAC was designed with LSB enable/disable option, so the output eye is plotted for 3-bit and 4-bit configurations of DAC as shown in figure 9. Proposed DAC characteristics were calculated for the designed DAC and have been tabulated in table I. The DNL values mentioned here have been calculated without any mismatch analysis. Though, for this low resolution DAC large transistor lengths (500nm) and common centroid topology was chosen to avoid mismatches,

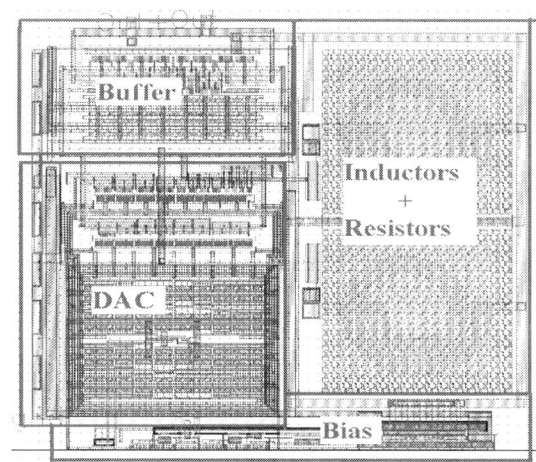

Fig. 7. Layout of the design comprising of the DAC block, load resistors & inductors

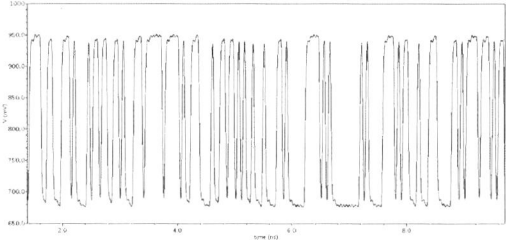

(a) One of the eight PRBS generator output waveforms used as input for the testing of designed DAC

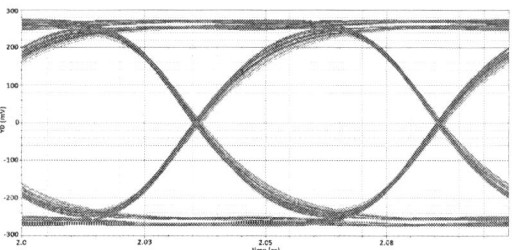

(b) Eye-diagram plot of all PRBS generator outputs fed to the DAC

Fig. 8. PRBS generator outputs (a) Output waveform (b) Eye-diagram of output

but this cannot compensate for random mismatches. The Eye-based DNL and INL were found to be 0.3995 and 0.2591 LSB respectively for 3-bit configuration of the DAC.

V. CONCLUSION

Output buffer for a DAC running on PRBS at 20 GS/s driving 800fF of output load has been designed. To characterize the output of such high-speed DACs, a method using eye-diagram has also been proposed in this paper. This method

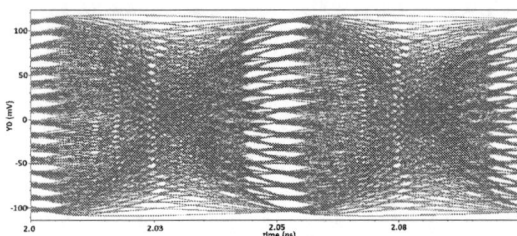

(a) Output eye-diagram of designed 4-bit DAC

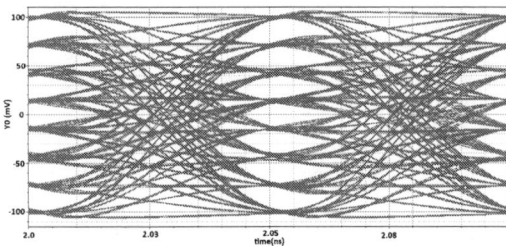

(b) Output eye-diagram of designed 4-bit DAC, run for 3-bits (LSB disabled)

Fig. 9. Output eye-diagrams plots of post-layout simulations

TABLE I
SUMMARY OF THE REALIZED 4-BIT DAC CHARACTERISTICS WITH POST LAYOUT SIMULATIONS

Resolution	4-bit
Sampling Rate	20 GS/s
Voltage supply	1 V
DNL (without mismatch)	0.04 LSB
INL	0.1 LSB
Eye-based bandwidth	10.5 GHz
Eye-based DNL (without mismatch)	1 LSB
Eye-based INL	0.6 LSB
Output Load	800 fF
Power Consumption	22.4mW
Technology	90 nm CMOS

has been used to estimate the DAC bandwidth and calculate characteristics like Eye-based DNL & Eye-based INL (analogous to traditional INL and DNL). Post layout extractions were carried out for the designed 4-bit 20 GS/s DAC and it was simulated for testing using designed 9-bit LFSR (schematic without parasitic extraction) in 90 nm CMOS techonolgy. A good eye opening was observed, promising a good DAC performance for broadband data.

REFERENCES

[1] S. Yamanaka, T. Kobayashi, A. Sano, H. Masuda, E. Yoshida, Y. Miyamoto, T. Nakagawa, M. Nagatani, and H. Nosaka, "11 x 171 gb/s pdm 16-qam transmission over 1440 km with a spectral efficiency of 6.4 b/s/hz using high-speed dac," in *European Conference and Exhibition on Optical Communication (ECOC)*, sept. 2010, pp. 1 –3.

[2] T. Kobayashi, A. Sano, A. Matsuura, E. Yamazaki, E. Yoshida, Y. Miyamoto, T. Nakagawa, Y. Sakamaki, and T. Mizuno, "120-Gb/s PDM 64-QAM transmission over 1,280 km using multi-staged nonlinear compensation in digital coherent receiver," in *OFC/NFOEC*, march 2011, pp. 1 –3.

[3] A. Sano, T. Kobayashi, K. Ishihara, H. Masuda, S. Yamamoto, K. Mori, E. Yamazaki, E. Yoshida, Y. Miyamoto, T. Yamada, and H. Yamazaki, "240-Gb/s polarization-multiplexed 64-QAM modulation and blind detection using PLC-LN hybrid integrated modulator and digital coherent receiver," in *ECOC*, sept. 2009, vol. 2009-Supplement, pp. 1 –2.

[4] J. Savoj, A. Abbasfar, A. Amirkhany, M. Jeeradit, and B.W. Garlepp, "A 12-GS/s Phase-Calibrated CMOS Digital-to-Analog Converter for Backplane Communications," *IEEE Journal of Solid-State Circuits*, vol. 43, pp. 1207 –1216, 2008.

[5] Y. M. Greshishchev, D. Pollex, S. C. Wang, M. Besson, P. Flemeke, S. Szilagyi, J. Aguirre, C. Falt, N. Ben-Hamida, R. Gibbins, and P. Schvan, "A 56GS/S 6b DAC in 65nm CMOS with 256x6b memory," in *ISSCC*, feb. 2011, pp. 194 –196.

[6] S. Galal and B. Razavi, "Broadband ESD protection circuits in CMOS technology," *ISSCC*, vol. 1, pp. 182 – 486, 2003.

[7] P. Schvan, D. Pollex, and T. Bellingrath, "A 22GS/s 6b DAC with integrated digital ramp generator," *Digest of Technical Papers. ISSCC*, pp. 122 –588 Vol. 1, feb. 2005.

[8] W. Jiang and V. D. Agrawal, "Built-in Self-Calibration of On-chip DAC and ADC," *in IEEE International Test Conference*, pp. 1 –10, Oct. 2008.

[9] A. Baccigalupi, M. D'Arco, A. Liccardo, and M. Vadursi, "Test Equipment for DAC's Performance Assessment: Design and Characterization," *in IEEE Transactions on Instrumentation and Measurement*, vol. 59, no. 5, pp. 1027 –1034, May. 2010.

[10] H. S. Yu, J. A. Abraham, S. Hwang, and J. Roh, "Efficient loop-back testing of on-chip ADCs and DACs," *In Proc. IEEE Asia and South pacific Design Automation Conference*, pp. 651 – 656, Jan. 2003.

[11] Y. Cong and R. L. Geiger, "A 1.5-V 14-bit 100-MS/s self-calibrated DAC," *IEEE Journal of Solid-State Circuits*, vol. 38, no. 12, pp. 2051, 2003.

[12] D. Baranauskas and D. Zelenin, "A 0.36W 6b up to 20GS/s DAC for UWB Wave Formation," *Digest of Technical Papers. ISSCC*, pp. 2380 –2389, feb. 2006.

[13] M. Nagatani, H. Nosaka, S. Yamanaka, K. Sano, and K. Murata, "A 32-GS/s 6-Bit Double-Sampling DAC in InP HBT Technology," in *Compound Semiconductor Integrated Circuit Symposium, 2009*, oct. 2009, pp. 1 –4.

[14] M. Singh, M. Sakare, and S. Gupta, "Testing of high-speed DACs using PRBS generation with "Alternate-Bit-Tapping"," in *Design, Automation Test in Europe Conference Exhibition (DATE)*, march 2011, pp. 1 –6.

[15] T. Sugihara, T. Kobayashi, Y. Konishi, S. Hirano, K. Tsutsumi, K. Yamagishi, T. Ichikawa, S. Inoue, K. Kubo, Y. Takahashi, K. Goto, T. Fujimori, K. Uto, T. Yoshida, K. Sawada, S. Kametani, H. Bessho, T. Inoue, K. Koguchi, K. Shimizu, and T. Mizuochi, "43 Gb/s DQPSK pre-equalization employing 6-bit, 43GS/s DAC integrated LSI for cascaded ROADM filtering," in *OFC/NFOEC*, march 2010, pp. 1 –3.

[16] J. Yu, X. Zhou, Y. K. Huan, S. Gupta, M. F. Huang, T. Wang, and P. Magill, "112.8-Gb/s PM-RZ-64QAM Optical Signal Generation and Transmission on a 12.5GHz WDM Grid," in *Optical Fiber Communication Conference*. 2010, p. OThM1, Optical Society of America.

[17] B. Razavi, *Principles of Data Conversion System Design*, NJ Press, 1995.

[18] J. J. Wikner and N. Tan, "Modeling of CMOS digital-to-analog converters for telecommunication ," *IEEE Transactions on Circuits and Systems II: Analog and Digital Signal Processing*, vol. 46, no. 5, pp. 489 –499, may. 1999.

Analog Processing Based Equalizer for 40 Gbps Coherent Optical Links in 90 nm CMOS

Pawan Kumar Moyade, Nandakumar Nambath, Allmin Ansari, Shalabh Gupta
Department of Electrical Engineering, IIT Bombay, Mumbai – 400076, India
p.k.moyade@gmail.com, npnandakumar@iitb.ac.in, ansari.allmin@gmail.com, shalabh@ee.iitb.ac.in

Abstract—Inter symbol interference introduced by fiber non-idealities such as polarization mode dispersion and chromatic dispersion would be one of the major limiting factors in achieving higher data rates in the existing Gigabit fiber-optic links. Receivers based on high speed ADCs followed by DSPs will be limited by the need for massive parallelization and interconnects. We propose analog signal processing based coherent optical link receiver to drastically reduce its power consumption, size and cost. A 40 Gbps analog processing adaptive DP-QPSK (dual polarization quadrature phase shift keying) equalizer in 90 nm CMOS technology is demonstrated using simulations, which dissipates 450 mW of power. A complete analog processing receiver is expected to consume less than one-tenth of the power consumed by chip using ADCs followed by signal processing in DSP.

Index Terms—Adaptive equalizers, CMOS integrated circuits, analog signal processing, coherent optical communications.

I. INTRODUCTION

Coherent modulation/detection and polarization multiplexing will play a crucial role in future optical links to satisfy the rapidly growing speed requirements. Electronics, which rely on ADCs (analog-to-digital converters) followed by DSP (digital signal processor) to remove channel impairments and recover transmitted data, have become the bottleneck in practical receivers for such links. While each ADC generates data at a very high speed (typically at ~56 GBytes/s for a 100 Gbps coherent DP-QPSK link [1]), the DSP speed is limited to ~500 MHz clock rates. Therefore, digital data from each ADC has to be massively parallelized, resulting in thousands of parallel interconnects and concurrent logic operations. Hence, there is an excessive amount of power consumption in high-speed ADCs, parallel interconnects, logic gates, parallel-to-serial/serial-to-parallel conversion blocks. In addition to power and chip area, a significant power/heat management effort is required that can finally make the solution very bulky.

As an example, a 40 Gbps DP-QPSK receiver employing this approach is implemented in 90 nm CMOS technology and dissipates 21 Watts of power [2]. A coherent receiver in the more advanced (65 nm CMOS) technology is expected to consume 50 Watts or more for a 100 Gbps DP-QPSK link [1]. A single chip 100 Gbps coherent receiver has not been demonstrated till date because of the problems mentioned above, even though 56 GS/s CMOS ADC by Fujitsu was first reported in early 2009.

To drastically reduce the power and chip area requirement by overcoming the ADC and DSP limitations, we propose

processing these signals in analog domain itself. In coherent reception, the most important channel impairments (that are due to dispersion effects) are linear in nature, and hence can easily be compensated using tapped delay line equalizers (that act as linear transversal filters). Analog circuits are, in general, also preferred over digital for carrying out clock recovery and carrier phase compensation operations in high speed transceivers. In this paper, we focus on the design of adaptive linear equalizer for 40 Gbps DP-QPSK coherent optical link.

The paper is organized as follows. A brief description of DP-QPSK system and channel impairments is given in section II. Section III gives the system level description of the Analog Signal Processing Based Equalizer. In section IV block level implementation and the schematics of major building blocks are detailed. Section V describes the simulations carried out and the results. Conclusions are drawn in Section VI.

II. DUAL POLARIZATION QUADRATURE PHASE SHIFT KEYING SYSTEM

DP-QPSK modulation is the basic format which incorporates diversity in both polarization and phase. The block diagram of a DP-QPSK system is given in Fig. 1. At the transmitter end, the optical carrier from the laser is split into two mutually orthogonal polarizations (X and Y), using a polarization beam splitter (PBS) and sent to two nested Mach-Zehnder modulators. This allows independent modulation of the in-phase, I, and the quadrature, Q, of the optical electric fields for both X and Y polarizations. The 40 Gbps data stream is de-interleaved into four 10 Gbps streams and fed to the modulators. Again, the QPSK modulated X and Y polarization optical signals are combined using a polarization beam combiner (PBC). This signal is then transmitted through the optical channel.

At the receiver side, the incoming signal is split into two orthogonal polarizations using a PBS and each polarization is fed to a separate 90° optical hybrid where it is mixed with the local oscillator (LO) laser. The signals on the I and Q phases of the orthogonally polarized received optical fields are delivered to four photo detectors (PD), where those are converted to electrical signals. These four signals are fed to an electrical signal processing unit, which essentially consists of a multidimensional equalizer, carrier-phase recovery and clock-data recovery modules.

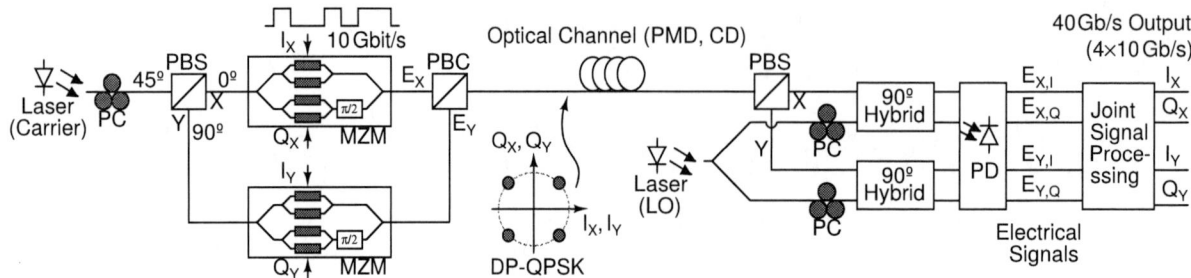

Fig. 1. A 40-Gbit/s transmission system using a DP-QPSK coherent optical link. PC - Polarization controller, PBS - Polarization beam splitter, MZM - Mach-Zehnder modulator, PBC - Polarization beam combiner, PMD - Polarization mode dispersion, CD - Chromatic dispersion, LO - Local oscillator and PD - Photo detector. E_x and E_y are the electric fields in X and Y polarizations respectively.

Chromatic dispersion (CD) and polarization mode dispersion (PMD) are the primary linear channel impairments which create ISI in the fiber-optic channel. CD is caused by the wavelength dependent refractive index profile of the fiber. Different wavelengths of the optical carrier travel with different group velocities, thereby producing a delay spread at the receiver situated at a distance. In the frequency domain CD can be represented as a scalar multiplication [3]:

$$\mathbf{H}_{CD}(\omega) = e^{-j(\frac{1}{2}\beta_2 L_{fiber}(\omega-\omega_s)^2 + \frac{1}{6}\beta_3 L_{fiber}(\omega-\omega_s)^3)}\mathbf{I} \quad (1)$$

where, L_{fiber} is the length of the fiber, β_2 is the dispersion parameter, β_3 is the dispersion slope, and ω_s is the optical carrier frequency. PMD is produced by the random fluctuations in the refractive index of the fiber along its mutually orthogonal sections, which result in different group velocities for the signals in horizontal and vertical polarizations. The differential group delay produced by PMD is given by $\delta\tau = \tau_{DGD} \times \sqrt{L_{fiber}}$, where, τ_{DGD} is the PMD parameter.

Since in coherent reception, dispersion is linear in nature, we can compensate for the same in the receiver section using linear filters. The proposed receiver structure based on analog signal processing is shown in Fig. 2. As opposed to the digital implementation the proposed receiver does not require the expensive high speed, high precision ADCs. The need for retiming is also not necessary since, the data is not parallelized.

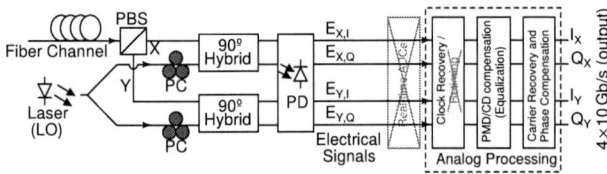

Fig. 2. Proposed receiver structure is shown. The electrical signals from the photo-detectors are processed in the analog domain itself, obviating the need for power hungry ADCs and DSP. Clock-data recovery and carrier phase compensation operations are also easier to implement in analog.

III. ANALOG SIGNAL PROCESSING BASED EQUALIZER

The adaptive linear equalizer is realized as a transversal filter which has primarily three unit operations – delay, multiplication and addition. In analog domain, high-speed multiplication can easily be performed using a variable gain

amplifier or a Gilbert cell with less than 10 transistors. Similarly summation of signals can be performed by simply adding currents onto a resistor. In our implementation, the adaptation is performed using the Least Mean Square Algorithm (LMS) in analog domain. However, part of LMS training block can be implemented in digital domain also (for example, see ref. [4]), since the update of weight coefficients can be performed at a much lower speed.

The transversal filter used to equalize the linear dispersive effects can be described by the equations [5]:

$$x' = \mathbf{h}_{xx}^T\mathbf{x} + \mathbf{h}_{xy}^T\mathbf{y} \quad (2)$$

$$y' = \mathbf{h}_{yx}^T\mathbf{x} + \mathbf{h}_{yy}^T\mathbf{y} \quad (3)$$

where, $\mathbf{x}$ and $\mathbf{y}$ are the column vectors of the delayed complex input electrical signals corresponding to the two polarization channels and x' and y' are the complex equalized outputs. Here $\mathbf{h}_{xx}, \mathbf{h}_{xy}, \mathbf{h}_{yx}, \mathbf{h}_{yy}$ are the column vectors of the complex LMS coefficients and can be computed as given in [5].

The in-phase and quadrature components of x' can be represented as,

$$x_0' = \mathbf{h}_{xx0}^T\mathbf{x}_0 - \mathbf{h}_{xx90}^T\mathbf{x}_{90} + \mathbf{h}_{xy0}^T\mathbf{y}_0 - \mathbf{h}_{xy90}^T\mathbf{y}_{90} \quad (4)$$

$$x_{90}' = \mathbf{h}_{xx0}^T\mathbf{x}_{90} + \mathbf{h}_{xx90}^T\mathbf{x}_0 + \mathbf{h}_{xy0}^T\mathbf{y}_{90} + \mathbf{h}_{xy90}^T\mathbf{y}_0 \quad (5)$$

where, the subscripts 0 and 90 denote the in-phase and quadrature(or real and imaginary) components of the complex quantity respectively. A similar set of equations can be obtained for the y channel also.

The LMS update equations for the filter coefficients are given by,

$$\mathbf{h}_{xx} = \mathbf{h}_{xx} + \mu e_x \mathbf{x^*} \quad (6)$$

$$\mathbf{h}_{xy} = \mathbf{h}_{xy} + \mu e_x \mathbf{y^*} \quad (7)$$

$$\mathbf{h}_{yx} = \mathbf{h}_{yx} + \mu e_y \mathbf{x^*} \quad (8)$$

$$\mathbf{h}_{yy} = \mathbf{h}_{yy} + \mu e_y \mathbf{y^*} \quad (9)$$

where, $e_x = \mathbf{d}_x - \mathbf{x}$ is the error signal for the x channel and $e_y = \mathbf{d}_y - \mathbf{y}$ is the error signal for the y channel. Here $\mathbf{d}_x$ and $\mathbf{d}_y$ are the reference signals in the x and y channels respectively.

978-1-4673-0438-2/12 $31.00 © 2012 IEEE

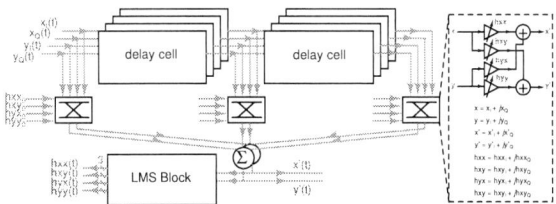

Fig. 3. Equalizer Architecture: Outputs x' and y' are complex signals. Inset shows block diagram of butterfly multiplier structure.

The update equations of the real and imaginary parts of the weight h_{xx0} (which is the first element in the weight vector $\mathbf{h}_{xx}$) are given by,

$$h_{xx0} = h_{xx0} + 2 * \mu \cdot \{e_0 \cdot x_0 + e_{90} \cdot x_{90}\} \qquad (10)$$

$$h_{xx0} = h_{xx0} + 2 * \mu \cdot \{e_{90} \cdot x_0 + e_0 \cdot x_{90}\} \qquad (11)$$

For continuous time implementation [6] the accumulation operation in the discrete update equation is replaced by the integration operation. A generic LMS equation such as $w(n+1) = w(n) + \mu \cdot e(n) \cdot x(n)$ is modified as

$$w(t) = \beta \int^t x(\tau) \cdot e(\tau) \cdot d\tau \qquad (12)$$

where, $x(t)$ is the continuous time input and $e(t)$ is the continuous time error signal given by $e(t) = d(t) - x(t)$.

IV. IMPLEMENTATION DETAILS

The block diagram of the Equalizer is shown in Fig. 3. It consists of the delay cells, butterfly multipliers, and the LMS circuit. The tapped delay line using active elements adds constant group delay to the incoming signals and feeds the butterfly multipliers. The use of an active delay cell reduces area considerably as opposed to a passive LC delay line. The butterfly multiplier block multiplies the complex filter weights with the delayed input signal values. It uses multipliers based on the folded version of the Gilbert cell topology with the weights being low speed signals. Each tap uses 16 such multipliers. The training signals along with the estimated outputs are fed to the LMS block. The weight update is carried out in this block. The LMS block is implemented using high frequency Gilbert cell multipliers and an integrator with each tap again using a total of 16 multipliers. A common mode feedback circuit is used to maintain the dc level at the output of the integrator stage. A digital offset correction circuit is used to correct the offset at the input of the differential integrator.

A detailed description of the major building blocks is given in the following subsections.

A. Delay Cells

The delay cell must provide constant group delay over the frequency range of interest. The differential delay unit schematic is shown in the Fig. 4. It is a variant of the circuit found in [7] .

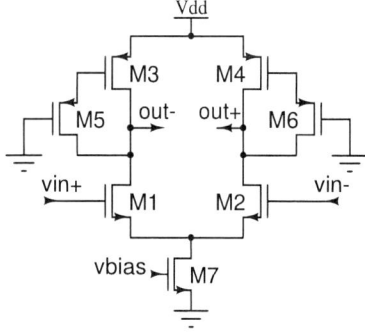

Fig. 4. Schematic of the delay cell. M3(M4) and M5(M6) act as an active inductor. The pole is primarily due to the capacitance at the drain of M1(M2).

An active load consisting only of PMOS transistors is used for common mode output level considerations. The common mode output level of the delay cell must be the same as the input common mode level so that the delay stages can be cascaded without using coupling capacitors. This necessitates the use of PMOS active load, since for an NMOS active load the sizes of the NMOS transistors become very large, causing capacitive loading.

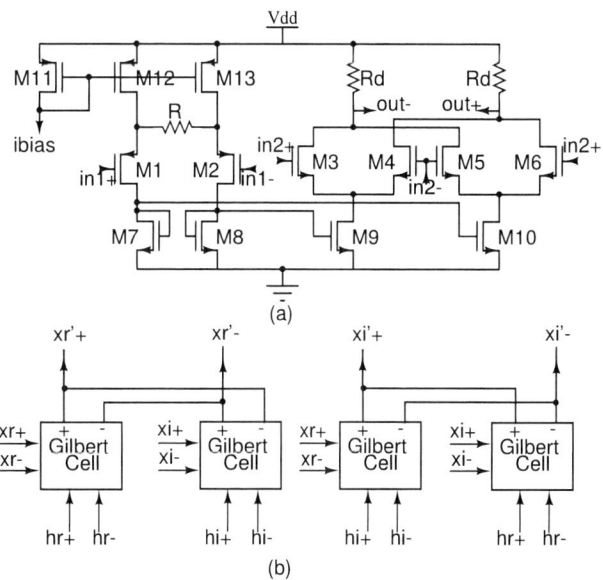

Fig. 5. Schematics of (a) the Gilbert cell multiplier (The differential inputs are denoted as in1+, in1-, in2+ and in2- and the output as out+ and out-) and (b) Complex multiplier (which uses two multipliers each to compute the real and imaginary parts of the complex multiplication $h \cdot x$)

The Cgs of M3(M4) in series with the channel resistance of M5(M6) adds a zero to the transfer function of the circuit. Thus M3(M4) and M5(M6) act as an active inductor. The pole is primarily due to the capacitance at the drain of M1(M2). By placing the zero close to the pole the bandwidth of the circuit can be optimized. The use of an active inductor instead of a passive inductor helps in reducing the area of the circuit.

The transfer function of the delay cell given in Fig. 4 can

978-1-4673-0438-2/12 $31.00 © 2012 IEEE   103

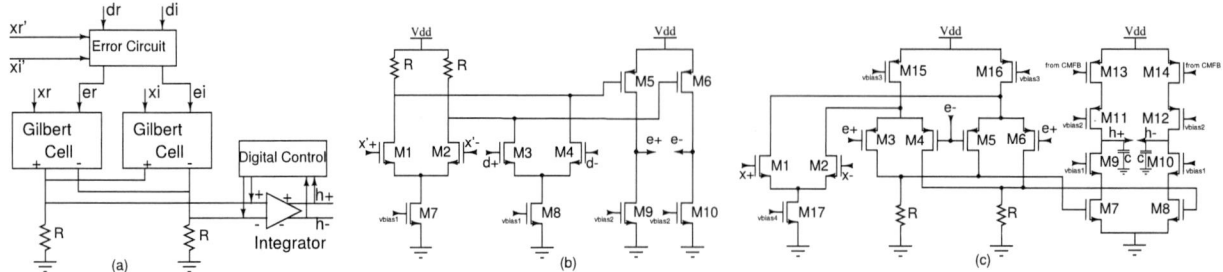

Fig. 6. LMS block (a) Block diagram (b) Schematic of the error circuit, which calculates $e(t) = d(t) - x'(t)$ and (c) Schematic of the multiplier-cascode integrator block, which updates the filter weights as $h(t) = \beta \int^t x(\tau) \cdot e(\tau) \cdot d\tau$

be written as [8]:

$$H(s) = \frac{\frac{gm_1}{C_l}\left[s + \frac{1}{RCgs}\right]}{s^2 + s\frac{1}{RCgs} + \frac{gm_2}{C_l}\frac{1}{RCgs}} \qquad (13)$$

Each delay unit gives a 13 ps delay with 5.7% variation over the signal bandwidth (due to inductive peaking) and a bandwidth of 12.6 GHz. Two of these delay units are cascaded to get an average group delay of 17.5 ps with a variation of ±1 ps. Each delay cell was drawing a 22 mA current from a 1 V supply.

B. Butterfly Multipliers

Multiplier stages are arranged in a butterfly structure to carry out the multiplication of the filter coefficients with the delayed input signal. The multiplier is based on the folded version of the Gilbert cell topology described in [9]. The schematic of the multiplier is shown in Fig. 5 (a). For high speed, passive loads are chosen over active loads as current summers, since those are having the advantage of higher bandwidth.

A single multiplier stage draws 6 mA of current from a 1 V supply.

1) Complex Multiplier: The complex multiplication is accomplished by adding the currents in a resistor with appropriate polarities. The schematic of the multiplier is shown in Fig. 5 (b).

C. LMS Block

The LMS block consists of an error circuit for calculating the difference between the reference signal and the estimated output signal, a multiplier-cascode integrator block and a digital control block for offset correction as shown in Fig. 6 (a).

1) Error Circuit: The error circuit calculates the difference $e(t) = d(t) - x'(t)$ where $d(t)$ is the reference input and $x'(t)$ is the equalizer output for each channel. The schematic of a single error circuit is shown in Fig. 6 (b). The error circuit is implemented as a two differential stages feeding a common resistor. The subtraction occurs in the current domain in the resistor. The output stage is a source follower circuit. The source follower stage is used to adjust the dc levels and to drive the capacitance of the input stage of the LMS circuit.

Two of these circuits are combined to calculate the combined error for both the real and imaginary parts of a channel.

The error circuit had a bandwidth of 12.5 GHz and a gain of ~1.5.

2) Multiplier-Cascode Integrator Block: The LMS weight update is carried out using a Gilbert cell multiplier followed by a cascode circuit as shown in Fig. 6 (c). The multiplier is a high frequency multiplier that performs the operation $e \cdot x$. To integrate the expression $e \cdot x$ the cascode stage is used that works as a first order low pass filter. The dc level at the capacitors is maintained through a common mode feedback circuit.

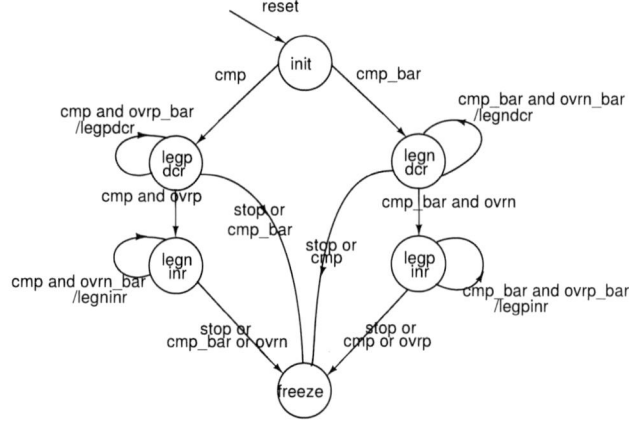

Fig. 7. State diagram of the control logic module. *cmp* is the output of the comparator. *legpinr* and *legpdcr* are the outputs of the control module that cause the thermometer counters to count up or down. *ovrp* and *ovrn* are the signals from the thermometer counter that indicate an overflow condition

The LMS block generates the complex weights h_{xx}, h_{xy}, h_{yx} and h_{yy}. Three of such LMS blocks are used in the entire circuit for each of the three taps. For finding the complex weights, outputs of the two Gilbert cells are added through a resistor to get $x_r e_r + x_i e_i$ and $x_r e_i - x_i e_r$ which are then integrated to get the differential real and imaginary weights respectively.

3) Digital Offset Correction Block: Under dc conditions, without offsets, the positive and negative outputs of the integrator must be equal. Due to mismatches and transistor offsets there will be a differential offset at the integrator output. A

978-1-4673-0438-2/12 $31.00 © 2012 IEEE

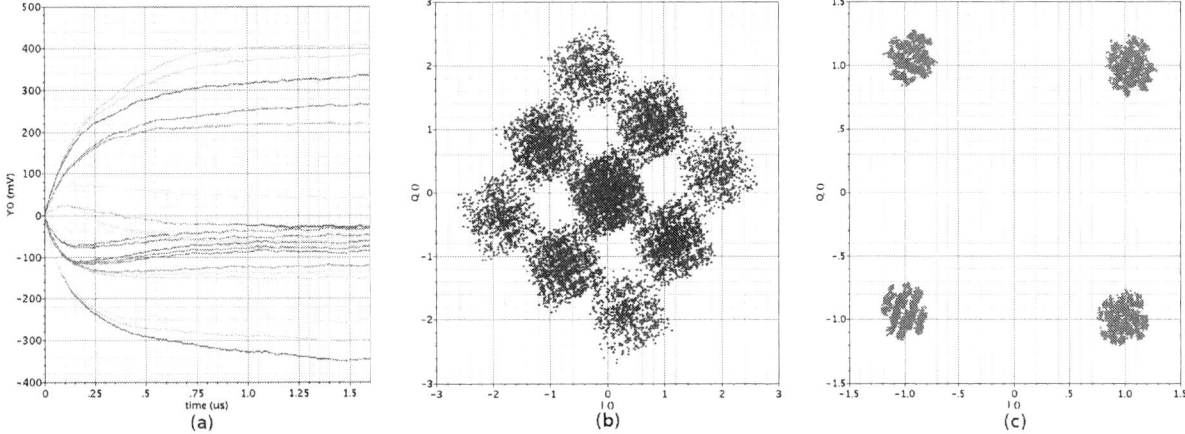

Fig. 8. Circuit simulation results: (a) The adaptive equalizer weight coefficients converging in $\sim 1\,\mu sec$, (b) Representative unequalized input constellation (horizontal polarization), and (c) Representative equalized output constellation (horizontal polarization).

digital module is used to correct for the offset error at the input of the integrator block. This is done by adjusting the current through the resistive load of the Gilbert cells in the LMS block using thermometer coded current sources, which is carried out during the startup of the circuit in dc conditions.

The current sources are controlled by the use of a error control module which consists of a comparator and a digital control logic. A comparator senses which of the two weight outputs is larger under dc conditions. The digital control logic increments or decrements the current in the load resistors based on the output of the comparator.

The offsets at the input of the integrators are corrected by changing the dc current through the resistors at the input of the integrators. Ten current sources are used at both the positive and negative input resistors. At reset the state of the current sources is '1111100000' ie. five of them are on and five are off. This allows us to increase or decrease the current through the resistors.

Through simulations it was found that the maximum input offset error that the integrator can tolerate is 2 mV. The value of the load resistor is 380Ω and the current through them is $700\mu A$. Hence 10 current sources each having a value of $2.6\mu A$ are used to correct for the input error at each resistor. Thus they correct for a $\pm 8 mV$ of offset error with a residual error of 0.8 mV.

The state diagram for the control logic is shown in Fig. 7. *Legp* refers to the resistor at the positive input of the integrator and *Legn* to the resistor at the negative input. Consider the output of the comparator to be positive. This implies that the voltage at the positive input of the integrator is more than the voltage at the negative input. So the control logic begins by reducing the current through the resistor at the positive input of the integrator (thus reducing the voltage) and then increasing the current through the resistor at the negative input. If at any point the comparator switches, the control block stops its operation and considers the offset to be well within the tolerance limit. A similar logic holds when the comparator

output is negative. An external *stop* signal forces the control logic to finish all its operations and go to the freeze state.

A thermometer counter controls the current sources for each leg. The control logic communicates with the counter through the signals *ovr* which signifies an overflow in the counter and *inr* and *dcr* which cause the counter to increment or decrement its count.

V. SIMULATIONS AND RESULTS

The optical system simulations were carried out using VPItransmissionMaker^{TM} followed by circuit level simulations in Cadence Design Environment. The circuit level simulations are carried out in TT corner at 80°C. At the transmitter side, the optical carrier from a DFB laser source with $\lambda = 1550$ nm and 100 kHz line-width is modulated using a nested Mach-Zehnder modulator. The combined optical signal from the modulator is transmitted through a single mode fiber with 30 km length. The fiber had an effective CD of 480 ps/nm and a PMD with a mean DGD of 0.55 ps. The receiver uses a PBS and PC followed by 90° hybrids and PDs as is shown in Fig.2.

In the absence of a carrier frequency recovery circuit, the LO optical signal was derived from the transmitter laser in the simulations. The electrical signals from the photodetectors are fed to the analog equalizer circuit. Fig. 8 shows the coefficient convergence and the input and output constellations for the horizontal polarization signals. A similar output constellation diagram is obtained for vertical polarization also. The power dissipation of the equalizer was found to be 450 mW and it consumes an area of 1.8 mm×1.1 mm.

VI. CONCLUSIONS

Analog signal processing approach for coherent optical receivers is an attractive solution in terms of power, area and cost. The first circuit level prototype of an analog signal processing equalizer has been implemented using 90 nm CMOS technology for 40 Gbps DP-QPSK link. It consumes 450 mW of power and has an area of 1.8 mm ×1.1 mm. Power

978-1-4673-0438-2/12 $31.00 © 2012 IEEE

consumption of a complete analog processing based receiver, which includes other modules, i.e. carrier phase correction and clock and data recovery blocks, is estimated to be about 1 Watt. This power number is much less in comparison to a chip based on ADC+DSP approach, which consumes 21 Watts of power in 90 nm CMOS process, as reported in [2]. If implemented in more advanced processes, such as SiGe technology or sub-40 nm CMOS, the analog implementation should easily work for 100 Gbps coherent links.

Though the analog equalizer may ultimately be limited in the number of taps it can use (due to distortion added to the signal per delay stage), for short haul or CD compensated optical links, it is shown to be a very efficient implementation.

ACKNOWLEDGEMENT

The authors would like to thank IIT Bombay for student support and DST for funding the project.

REFERENCES

[1] I. Dedic, "56Gs/s ADC: Enabling 100GbE," in *2010 Conference on OFC/NFOEC*. IEEE, 2010, pp. 1–3.

[2] H. Sun, K. Wu, and K. Roberts, "Real-time measurements of a 40 Gb/s coherent system," *Optics Express*, vol. 16, no. 2, pp. 873–879, 2008.

[3] E. Ip, A. P. T. Lau, D. J. F. Barros, and J. M. Kahn, "Coherent detection in optical fiber systems," *Optics Express*, vol. 16, no. 2, pp. 753–791, 2008.

[4] A. Momtaz, D. Chung, N. Kocaman *et al.*, "A fully integrated 10 Gbps receiver with adaptive optical dispersion equalizer in 0.13 μm CMOS," in *IEEE Symposium on VLSI Circuits Dig Tech Paper*, 2006.

[5] S. Savory, G. Gavioli, R. Killey, and P. Bayvel, "Transmission of 42.8 Gbit/s polarization multiplexed NRZ-QPSK over 6400km of standard fiber with no optical dispersion compensation," in *OFC/NFOEC 2007*. IEEE, 2007, pp. 1–3.

[6] S. Karni and G. Zeng, "The analysis of the continuous-time LMS algorithm," *Acoustics, Speech and Signal Processing, IEEE Transactions on*, vol. 37, no. 4, pp. 595–597, 1989.

[7] C. Lin and C. Chiu, "A 2.24 GHz wide range low jitter DLL-based frequency multiplier using pMOS active load for communication applications," in *ISCAS 2007*. IEEE, 2007, pp. 3888–3891.

[8] X. Lin, H. Lee, and J. Liu, "A continuous-time adaptive fir equalizer with lnv-ail delay line for 2.5gb/s data communication," in *Custom Integrated Circuits Conference, 2005. Proceedings of the IEEE 2005*, sept. 2005, pp. 413 –416.

[9] J. Babanezhad and G. Temes, "A 20-v four-quadrant cmos analog multiplier," *Solid-State Circuits, IEEE Journal of*, vol. 20, no. 6, pp. 1158 – 1168, dec 1985.

HD Resolution Intra Prediction Architecture for H.264 Decoder

Jimit Shah
CEDT, Indian Institute of Science,
jimitj@gmail.com

K.S. Raghunandan,
CEDT, Indian Institute of Science,
raghunandan85@gmail.com

Kuruvilla Varghese,
CEDT, Indian Institute of Science,
edkuru@cedt.iisc.ernet.in

Abstract— High performance video standards use prediction techniques to achieve high picture quality at low bit rates. The type of prediction decides the bit rates and the image quality. Intra Prediction achieves high video quality with significant reduction in bitrate. This paper presents novel area optimized architecture for Intra prediction of H.264 decoding at HDTV resolution. The architecture has been validated on a Xilinx Virtex-5 FPGA based platform and achieved a frame rate of 64 fps. The architecture is based on multi-level memory hierarchy to reduce latency and ensure optimum resources utilization. It removes redundancy by reusing same functional blocks across different modes. The proposed architecture uses only 13% of the total LUTs available on the Xilinx FPGA XC5VLX50T.

Keywords - Intra prediction; H.264 Decoder; 1080p HD; FPGA; Virtex-5; Video Processing

I. INTRODUCTION

The H.264 [1] offers excellent compression and video quality at the cost of increased decoding complexity. The paper describes the design of an area optimised intra prediction block for H.264 decoder that supports HDTV (1920x1088) resolution. Real-time decoding is achieved by implementing the intra prediction block in hardware. Our architecture achieves the high throughput required for HD resolution, while saving area by aggressive resource sharing across prediction modes.

Intra prediction is highly compute-intensive. The prediction is done based on the pixels along the periphery of the macroblock under prediction [3]. Intra prediction exploits the redundancy within a frame, like backgrounds of single colour with varying gradients or certain patterns like skin, roads etc. These commonly occurring pixel patterns are modelled as various modes of intra prediction. Intra prediction uses 4x4 and 16x16 block sizes, for representing fine and coarse details respectively. The standard and various features related to video compression are given in [3]. H.264 provides 13 modes of intra prediction, 9 modes for 4x4 and 4 for 16x16 block sizes. For 1920x1080 frame size with a rate of 60 fps, an intra-predicted macroblock needs to be decoded within 2.04 μs.

Related Work

Currently, work done in architecture of Intra prediction is towards reducing local memory usage for storing neighbouring pixels, scaling the frequency of operation for different resolutions and reusing functional blocks for lesser resource utilisation. The architecture in [4] uses independent data paths for chroma prediction, 16x16 luma and 4x4 luma predictions. The throughput is sufficient only to decode VGA (640x480) resolution H.264 video at 27 fps. Architecture presented in [5] utilizes separate local memory

for storage of 4x4 and 16x16 blocks, causing inefficient usage of on-chip memory. Also, their architecture does not reuse functional blocks to exploit the similar computations across different block sizes. The architecture in [6] is similar to our architecture. However, they have targeted decoding of QCIF (176x144) resolution video at 30 fps on 180 nm CMOS ASIC process.

Our work is targeted towards making a resource optimised H.264 intra prediction block for 1080p HD resolution on a Xilinx Virtex-5 FPGA based platform. Our architecture makes efficient use of FPGA Block RAM (BRAM) as cache memory, for storing temporary pixel data, reducing the access to the external DDR memory to fetch pixels. We also introduce a novel way of predicting intra prediction modes as explained in the following sections. We have achieved low Look Up Tables (LUT) utilization within FPGA, while providing a high throughput of 64 fps at HDTV resolution.

Rest of the paper is organised as follows. Section II gives a description of our intra prediction architecture and section III discusses the validation of our implementation. The result and comparison is given in section IV and section V concludes the paper.

II. ARCHITECTURE OF INTRA PREDICTION BLOCK

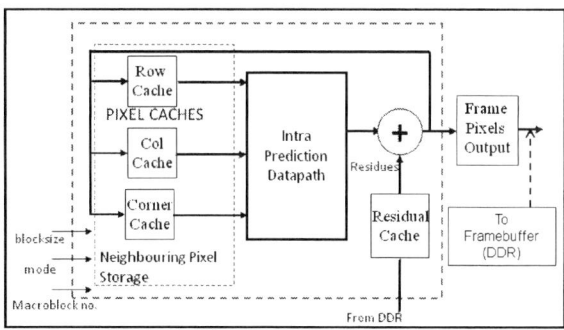

Fig 2.1: Block Diagram of the Intra Prediction module

In intra prediction mode, a macroblock is processed as a single 16x16 block or as 16, 4x4 blocks. This macroblock partitioning information is embedded into the H.264 bitstream by the encoder. The intra prediction block has an initial cost of parsing of bitstream and accessing DDR memory before and after processing. Parsing separates various modes and residual data. We assume that bitstream parsing is done concurrently with intra prediction in software and parsed data is stored in the DDR memory. The DDR latency also limits throughput of the intra prediction block. This latency is considered for the performance estimation of our hardware. Our intra prediction block is divided into two main units, Neighbouring Pixels Storage and Intra Prediction Datapath. This is shown in Fig 2.1.

A. Neighbouring Pixels Storage (NPS) Block

This block is a cache to reduce access to high latency on-board memory. The major bottleneck for any multimedia processing is the access to DDR memory, as all functional blocks read or store the data from and to DDR RAM. The throughput is limited by the access latency of DDR RAM. To overcome this limitation, on-chip caches are used for storing the neighbouring pixels of the current macroblock. Our caching scheme eliminates the need to fetch neighbouring pixels from DDR memory during intra prediction.

The intra prediction is done in three steps; read the neighbouring pixels from the cache, perform the macroblock prediction based on the syntax elements, and the contents of the cache are overwritten with the pixels from the current macroblock. Three caches are used for this purpose viz. *row*, *column* and *corner* caches, to store the Top Neighbouring Pixels (TNPs), Left Neighbouring Pixels (LNPs) and top corner pixel respectively. The *row* and *column* caches are implemented using BRAMs of FPGA. BRAMs are dedicated, configurable memory with address, data and control ports. The BRAMs are configured as Dual Port RAMs; one port for storing the pixels and other port for reading the pixels. The *row* BRAM stores one row of HD resolution i.e. 1088 pixels each of 8 bits. The size of the *row* memory is 8.5 KB. The *row* memory is configured for reading and writing 4 pixels/clock. The *column* BRAM is used to store one column of pixels of a macroblock. Thus, the size of column BRAM is 128 bits (16 pixels). The *column* memory has a read data width of 4 pixels/clock and a write data width of 1 pixel/clock. The read and write data widths are determined by the throughput requirement. The *corner* cache contains a set of three pixel registers to supply *intersection pixel*, the pixel at the intersection of the top row and left column neighbouring pixels. To achieve a throughput of 60 fps, four row pixels are predicted simultaneously. The cache replacement is simple; it replaces the current pixels with the predicted pixels. Replacement policy is same across all block sizes achieving lesser area. The pixel storage for both block sizes uses the three caches.

The Fig 2.2(a) shows the set of 33 pixels used for predicting current macroblock when the block size is 16x16. The 16 TNPs, marked **a**, are stored in *row* cache while 16 LNPs, marked **b**, are stored in *column* cache. All the 32 neighbouring pixels are fetched concurrently from the *row* and *column* BRAM in 5 clock cycles, which include one clock cycle for BRAM access latency. The *intersection pixel* (**m**) is stored in *top-corner* register within the *corner* cache.

Since, our architecture processes pixels row wise, the *column* cache is updated with one pixel during the processing of the last four pixels of each row. All the pixels predicted for the last row of a macroblock is stored in the

row cache, four pixels at a time. The *top-corner* register is updated, at the end of the processing of the current macroblock. This pixel is the right most pixel of the TNPs, as shown in the Fig 2.2(b). The pixels shown with diagonally hashed shading will replace the pixels indicated by **a**, and pixels shaded with dots will replace pixels indicated by **b**.

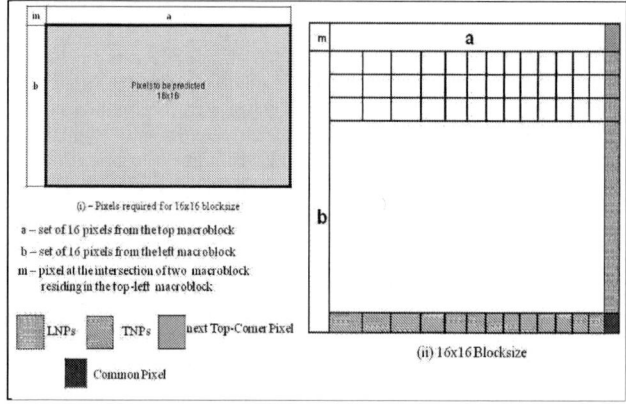

Fig 2.2: Neighbouring Pixels to be used for next macroblock

In H.264, number of prediction modes for 4x4 blocks are more to accommodate a greater variety of patterns. The H.264 specifies the use of 13 neighbouring pixels for predicting 4x4 blocks, as shown in Fig 2.3(a). There are four TNPs, labelled as **a**, four LNPs labelled as **b**. The rest of the pixels are the four pixels adjacent to the TNPs labelled as **c** and an *intersection pixel* **m**. The order of predicting sub-macroblocks is shown by the arrows in Fig 2.3(b) and the blocks are numbered in this order. The blocks 0, 1, 2, 4, 5, 8, 10 require data from the previous macroblock while the other blocks require data from the previously decoded sub-macroblocks within the current macroblock. For example, when processing block number 6, the TNPs are the bottom boundary pixels of block numbers 4 and 5. The LNPs are the right boundary pixels of block number 3. Hence, the pixels in the locations **c**, **d** and **f** will be replaced during the processing of blocks 4, 5 and 3 respectively for prediction of block 6. The dotted pixels indicate column pixels that will replace **e**, **f**, **g** and **h** while diagonal hashed pixels will replace **a**, **b**, **c** and **d**.

The storing of *intersection pixel* for the blocks is not as simple as in the case of 16x16 block size. For instance, the *intersection pixel* stored during block 1 prediction is used for predicting block 6, which does not immediately follow block 1 in the processing order. The *intersection pixel* would get replaced during the processing of block 3 in the normal case. Due to the zig-zag order of processing, two more registers, AUX1 and AUX2 are needed to store *intersection pixel* in the corner cache. The AUX1 register is updated, while processing the blocks 0, 2, and 8 and is used as *intersection pixel* for predicting blocks 2, 8, and 10

978-1-4673-0438-2/12 $31.00 © 2012 IEEE   108

respectively. Similar to AUX1 case, AUX2 is updated for the blocks 1, 4, 6 and 12 and is used for predicting blocks 4, 6, 12 and 14. For the rest of the blocks, *top-corner* register is used. This scheme uses only two extra registers to implement the caching scheme for 4x4 block size. This helps in saving logic and memory resources within FPGA. The boundary pixels of blocks 5, 7, 10, 11, 13, 14, and 15 are required, for predicting the next macroblock. These pixels are continually updated in *row* and *column* caches.

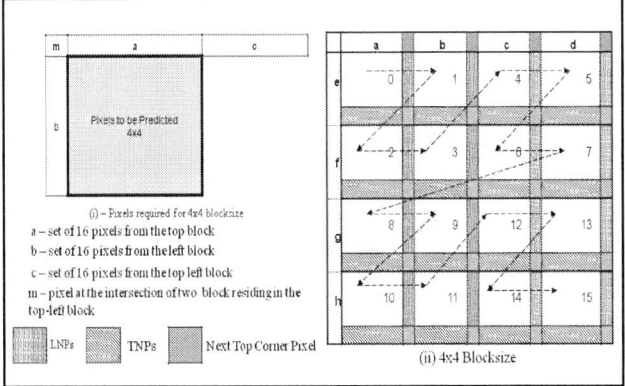

Fig 2.3: Boundary intra prediction for 4x4 macroblock

An equivalent scheme is used for chroma processing. The chroma prediction uses only one block size, viz. 8x8, processing of which is similar to 16x16 luma block processing. Three caches each are used for luma and chroma processing. Hence, six caches are used to store neighbouring pixels for subsequent macroblock processing and this reduces DDR latency. The sequencing of luma and chroma processing is controlled by one state machine, which also controls the block that fetches data from BRAM for all block sizes (luma and chroma) and controls the data path for intra prediction.

B. Intra Prediction Data Processing

This unit docs the computation on the pixels received from the Neighbouring Pixel Storage (NPS) block. Intra prediction supports 13 luma modes, 9 for 4x4 blocks and 4 for 16x16 blocks. It also supports 4 chroma modes, which are similar to 16x16 luma modes. These are classified into 4 categories for implementation, based on the hardware needed to perform each of them, viz. Average (4x4 luma and 16x16 luma, and 8x8 chroma), Horizontal and Vertical (4x4 luma, 16x16 luma, and 8x8 chroma), Plane (16x16 luma, and 8x8 chroma) and Diagonal (4x4 luma).

The architecture for the intra prediction is shown in Fig 2.4. Based on the intra prediction mode, the input data fetched from the pixel caches, is fed to the respective hardware units. The number of clock cycles for computation depends on the prediction mode. The five blocks in this unit are:

Register Array: This module is used to store all the pixel inputs required by the computation unit.

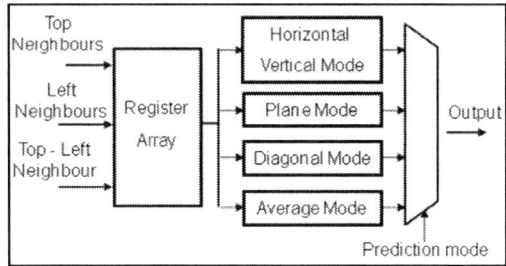

Fig 2.4: General Architecture of the Intra Prediction block

Horizontal and Vertical Mode block: Horizontal and Vertical modes are the two simplest modes and are common to all modes of prediction. This block has a multiplexer which passes the appropriate input pixels to the output. The select lines of the multiplexer are based on the mode and a counter. This counter is controlled by the block size. The counter size is flexible to support 4x4, 16x16 luma and 8x8 chroma blocks, making it reusable and area optimal.

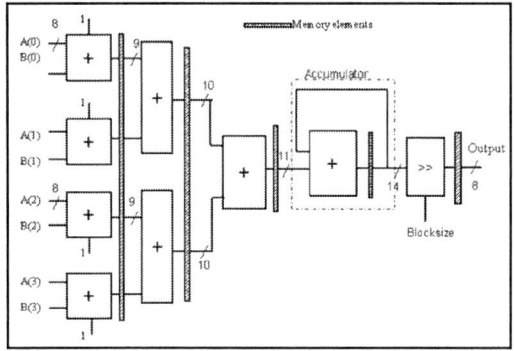

Fig 2.5: Architecture of Average mode block for 16x16, 8x8 and 4x4

Average Mode block: This block consists of tree of adders and an accumulator unit. The maximum data processed is 4 pixels per clock, as the maximum width of the on-chip BRAM is 4 pixels. The formula for the averaging mode is shown in (1)

$$P = \frac{\left(\sum_{i=0}^{N-1} \left(A(i) + B(i) \right) + N \right)}{2N}, \qquad (1)$$

where A(i)'s are TNPs, B(i)'s are LNPs, N is Block size and P is predicted pixel

The division is performed by shifting. To eliminate the rounding errors, a constant (N/2) is added. This constant is integrated into the first stage of adders as carry input. In every cycle of addition, the divisor is incremented by 8, since 8 pixels are added per clock. The final shifting will then be equivalent to the rounding. A pipelined binary tree adder structure performs the addition operation as shown in

978-1-4673-0438-2/12 $31.00 © 2012 IEEE

the Fig 2.5. The average block supports multiple block sizes, allowing reuse across modes and saving area. It is also possible to support the new 8x8 Luma block size, introduced in FRExt [9], without any modifications.

Diagonal Mode Block: This block performs six intra prediction modes for 4x4 blocks. The set of prediction equations are given in [1] and [3]. We notice these equations fall into three basic patterns.

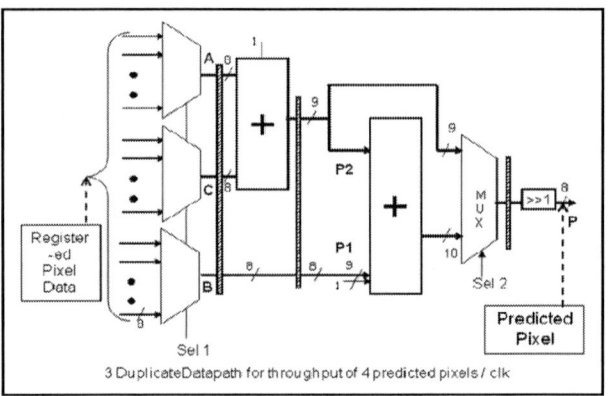

Fig 2.6: Architecture for 4x4 others mode

The first pattern of prediction equations are of the form

$$P = (A + 2B + C) / 4 \qquad (2)$$
$$= (A + 2B + C + 2) >> 2 \quad = (P_1 + P_2) >> 1 \qquad (3)$$

where $P1 = A + C + 1$ and $P_2 = 2B + 1$

P is the predicted pixel; A, B, C are the neighbouring pixels based on the mode of prediction and co-ordinate of the pixel to be predicted; P_1, P_2 are the partial sums used for implementation. The rounding operation is efficiently done by adding 2 to the pixel data and doing the right shift by one bit. In order to save area, this is done by splitting the expression into two as shown above.

The second pattern of equations is similarly reduced

$$P = \text{Average } (A, C) \text{ with equal weights}$$
$$= (A + C + 1) >> 1 = P_1 >> 1 \qquad (4)$$

The third pattern is

$$P = A$$
$$= \text{Average } (A, A) = (A + A + 1) / 2 = P_1 >> 1 \qquad (5)$$

The architecture given in Fig 2.6 shows the computation of P_1 using a two input adder with the carry input tied to 1. The implementation of P_2 is evident from the figure. P is obtained by adding P_1 and P_2 and by discarding the LSB of addition. The control signal for the input multiplexers (*Sel1*) is generated using mode of prediction and pixel position as inputs. The control signal (*Sel2*) is used to bypass the second stage adder, to satisfy the second and third expression. The architecture is pipelined to achieve 60 fps throughput and to reduce power dissipation. The described data path predicts one pixel/clock with a latency of one clock. We have replicated the block to achieve a throughput

of four pixels per clock. Our architecture supports the new prediction modes introduced in FRExt, as the new prediction equations can also be decomposed into the three patterns as above. FRExt specifies an additional block size of 8x8 luma pixels.

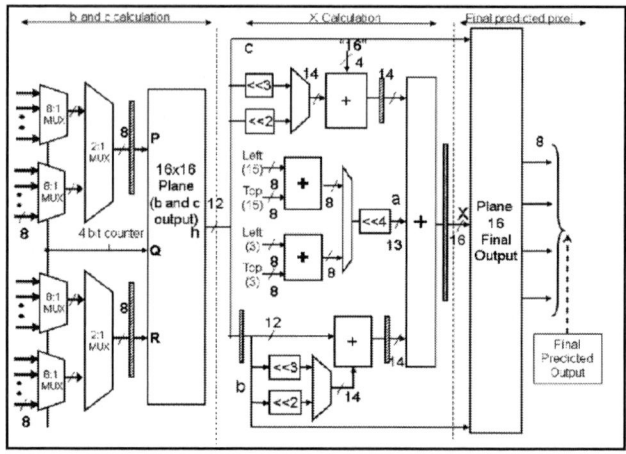

Fig 2.7: Architecture for Plane Mode

Plane Mode (Luma and Chroma modes) Block: This block performs the plane mode prediction for 16x16 blocks. It is the most complex of the intra prediction modes. The equation in section 8.3.3.4 in [1] for luma pixels is reduced to

$$\text{pred}_L[x, y] = \text{Clip1}((X + (b * x) + (c * y)) >> 5),$$
$$\text{with } x = 0,1, ...,15; \ y = 1,2, ...,16; \qquad (6)$$

where $X = a - 7b - 8c + 16$ (7)

$\text{pred}_L[x,y]$ is the predicted luma value at co-ordinates x,y. x and y are relative co-ordinates with respect to the top-left most pixel of the macroblock to be predicted.

Chroma pixels are also computed in similar fashion but are scaled appropriately to 8 pixels. The equations (8) and (9) are reduced from the equation 8.3.4.4 in [1]

$$\text{pred}_C[x, y] = \text{Clip1 }((X + (b * x) + (c * y)) >> 5),$$
$$\text{with } x = 0,1, ...,7; \ y = 1,2, ...,8; \qquad (8)$$

where $X = a - 3b - 4c + 16$ (9)

$\text{pred}_C[x,y]$ is the predicted chroma value at co-ordinates x,y. x and y are relative co-ordinates with respect to the top-left most pixel of the macroblock to be predicted.

The variables *a*, *b* and *c* are computed from the neighbouring pixels. The variables *b* and *c* are summation of the weighted differences of LNPs and TNPs respectively. The weights are dependent on the distance between the two pixels. For optimised resource utilisation, the same block is reused to calculate b and c. There is a two-level hierarchy of multiplexers; one is used for selecting the pixels for difference calculation according to the weight that is obtained from a counter, and the second for calculating either *b* or *c*. The architecture for the calculation of *b* and *c* is shown in the Fig 2.8.

978-1-4673-0438-2/12 $31.00 © 2012 IEEE   110

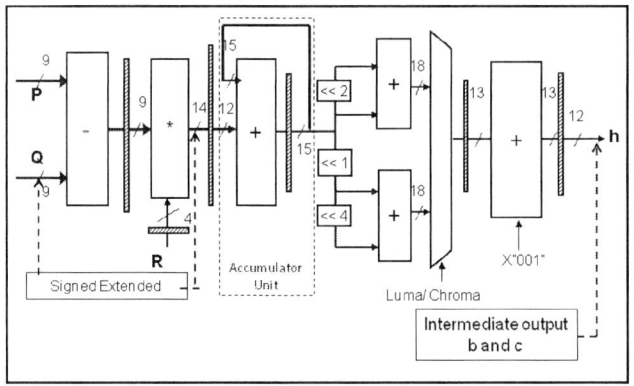

Fig 2.8: Architecture for calculating the intermediate results b and c

The output of input multiplexer structure shown in Fig 2.7 is given to subtract-multiply unit to calculate the weighted difference. The 15-bit accumulator unit performs the summation. The multiplication is done by suitable addition or subtraction of powers of two to save area. For e.g. the multiplication by five is performed as addition of *4x* and *x*. The division of powers of two are done by shifting e.g. division of *64* is done by right shifting by 6 bits. In case of chroma, the multiplying factor is *34, (32x + 2x)* while the rest of the equation remains the same. To make the block suitable for both luma and chroma mode, the bit width of all the signals in the datapath were increased by 1 so as to accommodate the complete dynamic range of both luma and chroma pixels.

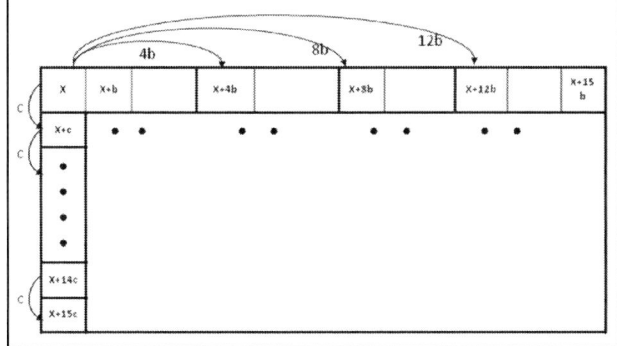

Fig 2.9: Implementation of 16x16 block

Once *b* is calculated, it is stored and calculation of *c* is done in the same hardware block (16x16 plane mode). *X* is calculated as per (7) or (9). It uses a pipelined binary tree adder structure for implementation, which is self-explanatory from the Fig 2.7.

The predicted pixels are obtained using (6) or (8) and its calculation is shown in Fig 2.9. The first four pixels values are calculated as *X, X + b, X + 2b* and *X + 3b*. These four pixels act as the seed for the row calculation. Each

successive row differs from the previous row by the value of *c*. To obtain the seed for the next row, *c* is added to the previous seed. Once the seed is obtained for that particular row, pixels are predicted by adding *4b, 8b* and *12b*, till all the rows are predicted.

The architecture in Fig 2.10 implements the above described flow. The 4-to-1 multiplexer selects *4b, 8b, 12b* for current row or *c* for new row. The first seed for the predicted block is calculated as *X, X+b, X+2b* and *X+3b* and adding *c* from 4-to-1 multiplexer in the output adder. This value is fed back (u, v, w, y) and stored for the rest of the row calculation as explained above. At the end of processing of each line of pixels, the register is loaded with the first four pixels of the next line by adding *c* to the previous line seeds. For chroma mode, the operation remains same except that instead of adding till *12b* the operation is completed by addition of *4b* for each row.

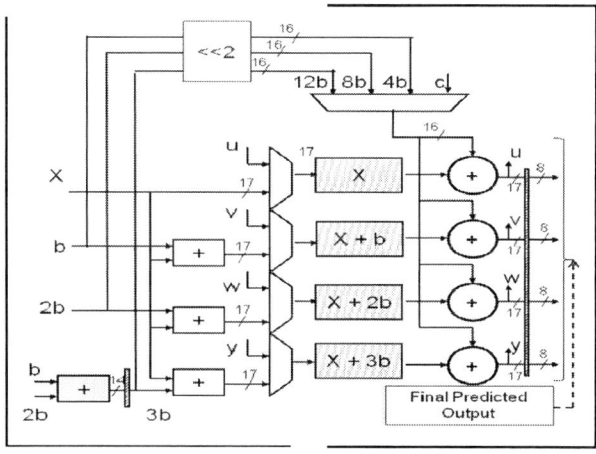

Fig 2.10: Architecture for pixel prediction from X, b and c

III. IMPLEMENTATION

Our architecture was implemented on an ML-505 board with Xilinx Virtex-5 FPGA (XC5VLX50T-FF1136). We have verified the functionality by comparison with the JM12.2 reference software decoder [10].

Since, we tested only the intra prediction block, supplying the input data to the block and verifying the output was done as follows. We extracted the parameters required for intra prediction, viz. block size, intra prediction modes and residual data, from a sample H.264 video using the JM12.2 reference code on a PC. Reference code parses H.264 bit stream and stores data in a format that can be passed directly to our intra prediction block. This bit stream is stored in a Compact Flash memory module. The JM12.2 reference code is also used to generate the final video frames to validate the output of our intra prediction block.

The FPGA test hardware consists of a soft core MicroBlaze processor, which copies the test data from the Compact Flash to DDR memory of the FPGA Board. Our intra prediction hardware reads the data from DDR memory, does intra prediction and the output frame data is written back to DDR memory. The MicroBlaze processor then copies these frames back to Compact Flash. The frame pixel data from our intra prediction block was verified to be bit accurate by matching it with output data from the JM12.2 reference code.

IV. RESULTS

A. Throughput Computation

We have estimated the clock cycle for computing one macroblock based on a worst case scenario for the intra prediction. The DDR transfer clock (200 MHz) cycles are not deterministic since the response of the controller will depend on the memory transfers due to other blocks. We have considered the typical values for this computation. The calculation is as shown below.

Intra Prediction Operations	Clocks
Input DDR read cycles per macroblock	58
Computation of output luma pixels (worst case - 4x4 average mode)	192
Computation of output chroma pixels (worst case - chroma plane mode)	72
Output DDR write cycles	58
Total clock cycles	**380**

Table 1: Worst Case clock cycle analysis of the intra predictor block

$$\text{Frame Rate at 1080p} = \frac{\text{No. of macroblocks/sec}}{\text{No. of macroblocks in one frame}}$$

$$= (380*5ns)^{-1}/ ((1920 \times 1088) / 256)$$
$$= 64.5 \text{ frames per second}$$

If we remove DDR access latencies from the computation, our block does the computation in 264 clock cycles. This translates to a throughput of 92.8 fps, following the similar calculations as shown above.

B. Area Computation

The area comparison of Intra Prediction block with [5] is listed in Table 2 along with the results obtained on the Virtex-5 XC5VLX50T-FF1136 FPGA.

FPGA	Xilinx Virtex-II Pro		Virtex-5
Blocks	Reference Paper[5]	Intra Prediction	Intra Prediction
LUT	5516	2627	2083
BRAM	3	12	4Kbytes
Clock (MHz)	40	200	200
Throughput (M samples/s)	160	194	194

Table 2: Comparison of Resource Utilisation

The current state-of-art picture quality in multimedia application is 1080p HD at 30 fps. The next stage is to progress to 60 fps. Our proposed intra prediction architecture achieves a throughput of 64 fps for 1080p HD resolution video, including memory access latency. From Table 2, the proposed design noted 21% increase in throughput while using only 48% of LUT resources as reported by [5]. Their clock frequency of 40 MHz implies that the critical path delay is high to work at higher frequencies. Our pipelined architecture works for HD resolution and can be effectively scaled down to lower resolution and frame rate. Their architecture achieved 30 fps while our block support frame rate above 60. The design in [7] supports 1080p HD at 30 fps using 7K LUT resources, which is three times more than our design. To decode 1080p HD video at 30 fps, our intra prediction block frequency can be scaled to 100 MHz, achieving 378K macroblocks/sec decoding rate. The architecture in [8] supports 260K macroblocks/sec, which is 30% less than our architecture.

V. CONCLUSION

In this paper, we have presented an area optimized, high throughput architecture for intra prediction block of H.264 for HD resolution. The architecture can work at different frequencies, based on the resolution and the frame rate. The area optimisation is done by using same functional block for the similar modes in 4x4 luma, 16x16 luma and chroma predictions. The design is validated on a Virtex-5 FPGA based platform. The architecture also has the flexibility to accommodate the changes introduced as part of Fidelity Range Extension (FRExt), an extension to H.264 standard.

References

[1] International Telecommunication Union ITU-T Series H - Advanced Video coding for generic audiovisual services 2005

[2] MPEG – 4, ISO/IEC 14496, http://www.itu.int/ITU-D/tech/digital-broadcasting/kiev/References/mpeg-4.html

[3] I. G. Richardson, H.264 and MPEG-4 Video Compression, Wiley, 2003

[4] E. Sahin and I. Hamzaoglu, "An Efficient Hardware Architecture for H.264 Intra Prediction Algorithm", *Design, Automation and Test in Europe (DATE) Conference*, April 2007.

[5] W. T. Staehler, E. A. Berriel, A. A. Susin, and S. Bampi, "Architecture of an HDTV Intraframe Predictor for a H.264 Decoder," *2006 IFIP International Conference*, Oct. 2006, Page(s):229 –233.

[6] Ke Xu, Chiu-Sing Choy, "A Power-Efficient and Self-Adaptive Prediction Engine for H.264/AVC Decoding", *IEEE Transactions on VLSI Systems March 2008, vol. 16, issue 3, pp. 302-313*

[7] S-H Wang, T. Lin, Z-H Lin, "Macroblock-Level decoding and deblocking method and its pipeline implementation in H.264 decoder and SOC design", *J. Zhejiang Univ. Sci. A*, vol. 8, no. 1, pp. 36–41, Jan.2007

[8] T.A. Lin, S.Z. Wang, T.M. Liu and C.Y. Lee, "An H.264/AVC Decoder With 4x4-Block Level Pipeline", *Proc. IEEE ISCAS, 2005, pp 1810-1813*

[9] D. Marpe, T. Wiegand, and S. Gordon, "H.264/MPEG–4 AVC fidelity range extensions: Tools, profiles, performance, and application areas,", *Proc. IEEE Int. Conf. on Image Proc. (ICIP)*, Sept. 2005, pp. 593–596

[10] Joint Model (JM) - H.264/AVC Reference Software, http://iphome.hhi.de/suehring/tml/download

Design for Security of Block Cipher S-Boxes to Resist Differential Power Attacks

Bodhisatwa Mazumdar, Debdeep Mukhopadhyay and Indranil Sengupta

{bodhisatwa,debdeep,isg}@cse.iitkgp.ernet.in
Department of Computer Science and Engineering
Indian Institute of Technology Kharagpur
India

Abstract—This paper proposes an S-box construction of AES-128 block cipher which is more robust to differential power analysis (DPA) attacks than that of AES-128 implemented with Rijndael S-box while having similar cryptographic properties. The proposed S-box avoids use of countermeasures for thwarting DPA attacks thus consuming lesser area and power in the embedded hardware and still being more DPA resistive compared to Rijndael S-box. The design has been prototyped on Xilinx FPGA Spartan device XC3S400-4PQ208 and the power traces of the two different running AES-128 algorithms with the proposed and Rijndael S-boxes have been analyzed separately. The experimental results of the FPGA implementations show a lesser gate count consumption and increased throughput for the AES-128 with proposed S-box as that when implemented with Rijndael S-box on the same FPGA device. The requirement of higher number of power traces to perform DPA analysis on AES-128 with *RAIN* S-box as compared to that implemented with Rijndael S-box is an experimental validation of the theoretical claim of lower transparency order computed for *RAIN* S-box as being more DPA resistant than that of Rijndael S-box.

Index Terms—S-box; Side-Channel Attacks; Differential Power Analysis; Transparency Order; Linear Cryptanalysis; Differential Cryptanalysis

I. INTRODUCTION

As communication networks are spreading by leaps and bounds, the need for secured communication and hence fast but secured cryptographic systems is growing bigger. This necessitates the call for *Design for Security* which entails design, implementation and security of cryptographic hardware and embedded systems. Block cipher cryptosystems embedded in cryptographic devices are susceptible to attacks which show that security cannot be an afterthought. Like as performance, testability are important issues which the designer takes care in the design cycle, security also has to be taken into consideration early in the design cycle. As a motivating example, differential power analysis (DPA) [1] attacks and vulnerability modeling of cryptographic devices [2] have shaken the strength of cryptographic implementations and have baffled the designers for the past fifteen years, that a very strong mathematical algorithm can be compromised using these powerful techniques. The Advanced Encryption Standard (AES) was found to be compromised, inspite of being a mathematically strong cipher. Research shows that SBoxes,

the non-linear components, in the cipher are responsible for the weakness against DPA. Literature shows that the asymmetry in the power consumption of transitions from 0 to 1, and 1 to 0 in the CMOS library are the root cause for such attacks. Counter-measures [3], [4], [5] have been discovered but they were found to be costly. Most prominent amongst the countermeasures use gate level masking to avoid glitches occurring in unmasked data [6]. Also algorithmic countermeasures involving computations exist which actually do not assist in encryption but mask the intermediate values of the block cipher with random values. But these countermeasures are found to be vulnerable to first-order differential side-channel attacks and are much expensive to implement [7] in terms of hardware footprint and the presence of glitches in datapath of circuits does not help masking of AES like block ciphers [8], [9]. Some countermeasures based on leakage-resistant logic styles [10], [11] exist but they lead to increased area and hence power consumption which is damaging in area of embedded cryptography. Also these countermeasures did not scale and were also extremely cumbersome. Most importantly, they make the AES algorithm so poor in performance that they rob the algorithm from its mathematical elegance and efficient performance, some of the prime reasons why the Rijndael algorithm emerged as the AES.

With this motivation, the present paper investigates whether the Boolean functions involved in the AES-Sbox can be *designed with security* against DPA attacks as an objective from the beginning. In this context, we use a specially designed Boolean function, namely *RAIN* [12] for an S-Box construction and design which is more DPA resistant than the Rijndael S-box while having similar classical cryptographic properties like *strict avalanche criteria (SAC)*, *propagation characteristic (PC)* and *correlation immunity (CI)* [13]. Our work shows that the design helps to develop a cryptographic system which is more strong against DPA without the requirement of any counter-measures, while maintaining similar cryptographic properties, footprint on FPGA resources, and critical path delay. This confirms that the nature of the Boolean functions have a very strong role on not only the mathematical robustness, but also on the attacks which exploit side channels like power. Hence, this leaves us with the open problem of developing design techniques for the SBoxes, which provide more robustness against DPA while preserving their security against conventional cryptanalysis as linear and differential cryptanalysis [14], [15]. Such a *Design for Security* will provide us with more scalable and efficient solutions against attacks.

This paper is organised as follows. Section II focuses on the desired S-box characteristics for the block cipher to be resistant against linear and differential cryptanalysis as well as algebraic attacks. It also throws light on the basic aspects of DPA attacks and the construction of AES-128 Rijndael S-box. Section III proposes the *RAIN* Sbox and its construction. Section IV mentions the FPGA implementation AES-128 cipher with both Rijndael and *RAIN* S-boxes and compares the implementation results in terms of hardware resources consumed and throughput in terms of maximum operating frequency. A comparison of the cryptographic properties of the Rijndael S-box and *RAIN* S-box is also stated in this section. Section V has the experimental results of DPA attacks on the on-board power traces captured from the FPGA implementation on the AES-128 using both the S-boxes. The final section draws up the conclusion of the paper.

II. PRELIMINARIES

An (n,n)-S-box is a map $f : \{0,1\}^n \rightarrow \{0,1\}^n$. It comprises of n-variable component Boolean functions $(f_1(x_1,\ldots,x_n), f_2(x_1,\ldots,x_n),\ldots,f_n(x_1,\ldots,x_n))$ each of which need to satisfy following cryptographic properties:

Balancedness: An n-variable Boolean function f is said to be balanced if $ham\_wt(f) = 2^{n-1}$ where ham_wt is the Hamming weight and f is a binary string of length 2^n.

Nonlinearity: The nonlinearity of n-variable Boolean function f is $nl(f) = min_{l \in A(n)} d(f,l)$ where $A(n)$ is the set of affine functions of n variables and $d(f,l)$ is the Hamming distance between functions f and l. Nonlinearity is the minimum hamming distance of f and the set of all n-variable affine functions.

Algebraic Degree: An n-variable Boolean function f represented as a multivariate polynomial over *GF(2)* called Algebraic Normal Form (ANF) is expressed as

$$f(x) = a_0 + \sum_{1 \leq i \leq n} a_i x_i + \sum_{1 \leq i \leq j \leq n} a_{i,j} x_i x_j + \ldots + a_{i,\ldots,n} x_i \ldots x_j$$

where the coefficients $a_0, a_1, \ldots, a_n, a_1 a_2, \ldots, a_{i,\ldots,n} \in \mathbb{F}_2$. The degree of this polynomial i.e. the number of variables in the largest monomial is called as algebraic degree or simply $deg(f)$. For an S-box with component functions, its algebraic degree is the maximum algebraic degree of all its component functions.

Strict Avalanche Criteria(SAC): For an n-variable Boolean function $f(\bar{X})$ where $\bar{X}$ is n-tuple $x_1,\ldots,x_n$ and $\bar{\alpha} \epsilon \{0,1\}^n$, f satisfies SAC if $f(\bar{X}) \oplus f(\bar{X} \oplus \bar{\alpha})$ is balanced for $\bar{\alpha}$ such that $ham\_wt(\bar{\alpha}) = 1$.

Propagation Characteristic (PC(k)): An n-variable Boolean function $f(\bar{X})$ satisfies PC of degree k if $f(\bar{X})$ changes with probability of half when i$(1 \leq i \leq k)$ bits of $\bar{X}$ are complemented i.e. $f(\bar{X}) \oplus f(\bar{X} \oplus \bar{\alpha})$ is balanced for any $\bar{\alpha}$ such that $1 \leq ham\_wt(\bar{\alpha}) \leq k$. The function f satisfies $PC(k)$ of order l if f satisfies $PC(k)$ when any l input bits of f are kept constant.

Correlation Immunity (CI): An n-variable Boolean function $f(\bar{X})$ is said to be correlation immune (CI) of order m if its Walsh Transform $W_f(\bar{\omega}) = 0$ for $1 \leq ham\_wt(\omega) \leq m$. Also if f is balanced then $W_f(\bar{0}) = 0$. Balanced m-th order correlation immune functions are called m-resilient functions.

Robustness to Differential Cryptanalysis: Let $F = (f_1, f_2, \ldots, f_n)$ be an $n \times n$ S-box, where f_i is a component function of S-box mapping $f_i : \{0,1\}^n \rightarrow \{0,1\}$. In the XOR distribution table [15] of F, let L be the largest value and N be the number of nonzero entries in the first column of the table. In both cases, the value 2^n is not counted. F is said to be *R-robust* against differential cryptanalysis where

$$R = (1 - \frac{N}{2^n})(1 - \frac{L}{2^n})$$

Bias for linear approximation of component functions: For an $n \times n$ S-box, $1 \leq \alpha \leq 2^n - 1$, $1 \leq \beta \leq 2^n - 1$ and component function of S-box f_i, $NS_i(\alpha, \beta)$ as the number of times out of 2^n patterns of f_i, such that the XOR-ed value of input bits masked by $\alpha \in \{0,1\}^n$ coincides with the the XOR-ed value of the output bits masked by $\beta \in \{0,1\}^n$ [14]. The table of values of NS_i corresponding to $2^n \times 2^n$ possible combinations of (α, β) form a linear approximation table [16].

A. Differential Power Analysis

Differential Power Analysis (DPA) is a power analysis attack which makes use of the fact that the instantaneous power consumption of CMOS based cryptographic devices is correlated to the intermediate values of the cryptographic algorithm. A power trace is a series of power consumption values. In this paper, each power trace coresponds to a plaintext while the secret key is kept hidden in an FPGA register. In DPA, the attacker measures the power consumption with a lot of plaintexts and the secret key embedded within the cryptographic hardware. The attacker then divides the power traces in two (single-bit DPA) or multiple (multi-bit DPA) sets on the basis of a hypothesis which is a hypothetical intermediate value (for e.g. an Sbox output bit value of the last round of AES-128) for every possible guessed value of round key of 10th round corresponding to that S-box. Then statistical methods like difference of means (DOM)[17] is applied to verify the correctness of the hypothesis. There exist distinguishable peaks whose values are called the DPA bias values in the processed differential traces if and only if the hypothesis is correct. The guessed key value for which the

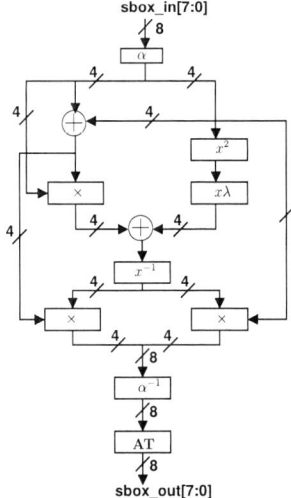

Fig. 1. Architecture of AES-128 Rijndael S-box

DPA bias value is the highest corresponds to the correct round key. The DPA attack in this paper is a single-bit DPA attack, based on the difference of means test.

B. AES-128 Rijndael S-Box

The AES-128 S-Box also called the Rijndael S-box exists in forward substitute byte transformation called SubBytes which contains a permutation of all possible 256 8-bit values. The S-box is generated by determining the multiplicative inverse in $GF(2^8)$ (block x^{-1} in Fig. 1) for the 8-bit input to the S-box. The multiplicative inverse is then transformed using the following affine transformation (block AT in Fig. 1) for each bit in the byte:

$$b_i' = b_i \oplus b_{(i+4)mod8} \oplus b_{(i+5)mod8} \oplus b_{(i+6)mod8} \oplus b_{(i+7)mod8} \oplus c_i$$

where c_i is the ith bit of byte c with value 0x63. In Fig. 1, blocks α and α^{-1} represent the isomorphic and inverse isomorphic mapping to $GF(2^8)$ whereas block $x\lambda$ is multiplication with a constant in $GF(2^4)$. The block $\times$ performs multiplication operation in $GF(2^4)$.

III. RAIN S-Box Construction

The proposed S-Box is based on vectorial function called Reversible Addition with Increased Nonlinearity (RAIN)[12] embedded within the Feistel structure of four rounds. The function RAIN is a mapping:

$$RAIN : \{0,1\}^n \times \{0,1\}^n \to \{0,1\}^n.$$

Assume $X = (x_{n-1}, x_{n-2}, .., x_0)$ and $K = (k_{n-1}, k_{n-2}, .., k_0)$ are two n-bit inputs and $Y = (y_{n-1}, y_{n-2}, .., y_0)$ is the n-bit output of the RAIN function where (x_{n-1}, x_0), (k_{n-1}, k_0), (y_{n-1}, y_0) are the respective (MSB, LSB) tuples. The function RAIN is denoted by the operator $\dagger$ and is defined as $Y = (X \dagger K)$ where

$$y_i = x_i \oplus k_i \oplus c_{i-1}; c_i = \bigoplus_{j=0}^{i} x_j.k_{i-j}$$

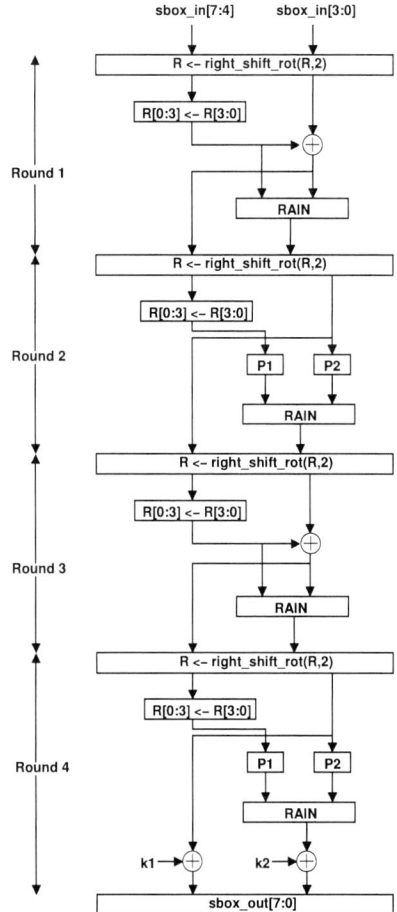

Fig. 2. Architecture of proposed *RAIN* S-box

where $\oplus$ is the modulo-2 sum, . represents AND operation, $0 \le i \le n-1, c_{-1} = 0$ and c_i is the carry term propagating from $i-th$ bit position to $(i+1)th$ bit position. The end carry c_{n-1} is neglected. For RAIN S-box proposed here we consider $n = 4$.

The RAIN S-box as shown in Fig. 2 is a mapping

$$RAIN\_SBOX : \{0,1\}^8 \to \{0,1\}^8$$

and comprises of four rounds of Feistel structure. All signals in Fig. 2 are of 4 bits unless otherwise mentioned. At the start of each round, a right-shift rotation of two bits $R \leftarrow right\_shift\_rot(R,2)$ operation is done. In rounds 1 and 3, the most significant nibble of $R \leftarrow right\_shift\_rot(R,2)$ output is first flipped $R[0:3] \leftarrow R[3:0]$ and then *XOR-ed* with the least significant nibble of $R \leftarrow right\_shift\_rot(R,2)$ output to form the input of permutation box $P2$. The input to the permutation box $P1$ is the output of flipping of most significant nibble of the right-shift rotation operation. The outputs of these permutation boxes are fed as input to the *RAIN* function block. For permutation box $P1$, if bitstring $\bar{C} = (c_0, c_1, c_2, c_3)$ is the input, the permuted bitstring $P1(\bar{C}) = (c_2, c_3, c_0, c_1)$. For permutation box $P2$, if bitstring $\bar{C} = (c_0, c_1, c_2, c_3)$ is

978-1-4673-0438-2/12 $31.00 © 2012 IEEE   115

the input, the permuted bitstring $P2(\bar{C}) = (c_3, c_2, c_1, c_0)$. The lower-nibble of output of these rounds is the output of *RAIN* function block as shown in Fig. 2. Whereas the most significant nibble of the round output is the output of XOR operation of flipped most significant nibble with the least significant nibble of the right-shift operation as mentioned above. In rounds 2 and 4, the most significant nibble of the right-shift rotation operation output is first flipped and then fed as input to the permutation box $P1$ while the least significant nibble of the rotation operation forms the input to the permutation box $P2$. The *RAIN* function block output forms the least significant nibble of the output of thesse rounds while the least significant nibble of the right-shift rotated output is taken as the most significant nibble output of these rounds. At the end of round 4, constants k1 = 0x06 and k2 = 0x03 are *XOR-ed* to prevent the fixed point representation corresponding to sbox input zero.

IV. FPGA IMPLEMENTATION OF AES-128

The AES-128 block cipher was implemented in sequential mode of operation on Xilinx FPGA Spartan device XC3S400-4PQ208 with both Rijndael S-box and *RAIN* S-box as shown in Fig. 3. The 128-bit *initial_key* and *plaintext* gets loaded when *load_plaintext* and *load_key* signals are asserted. The *key_scheduler* module generates *round_key* for each round and stores it in *key_memory*. After *round_key* for all the 10 rounds, *k_done* signal is asserted. This round key generation occurs whenever a new *initial_key* is loaded. The *round_function* module performs the *AddRoundKey*, *SubBytes*, *ShiftRows* operations of AES-128 for all the 10 rounds and *MixColumns* for the first 9 rounds of AES-128 and generates *round_output* after each round execution. The *round_output* is fed back as input to the next round for round number 2 to 10. The architectures for the two S-boxes differ only in the *SubBytes* operation where the Rijndael S-box is used in one implementation and the *RAIN* S-box is used in the other. The controller has a *counter* module which keeps a track of the current round number of AES-128 and then selects the corresponding *round_key* from the *key_memory* to the *round_function*. After the 10th round, the *counter* module asserts *output_valid* signal for one clock period indicating the corresponding *round_output* signal is the valid 128-bit *ciphertext*. The placement of the entire design on the Xilinx FPGA device by the floorplanner is as shown in Fig. 4. The colored regions in green correspond to the *round_function* block while those in pink and yellow belong to *key_scheduler* and *controller* blocks respectively. Fig. 5 and Fig. 6 show the power traces for FPGA implementations of AES-128 with Rijndael S-box and *RAIN* S-box respectively for a single plaintext. As each power trace pertains to a single plaintext, a power trace is captured whenever a new plaintext gets loaded into FPGA. The spikes in the traces occur when either a round output or key loading of AES-128 gets stored onto a register at a clock edge. The first trace spike corresponds to key loading phase in *key_memory* while the remaining trace spikes are for the 10 rounds of AES-128. So an entire encryption cycle comprises of 11 trace spikes.

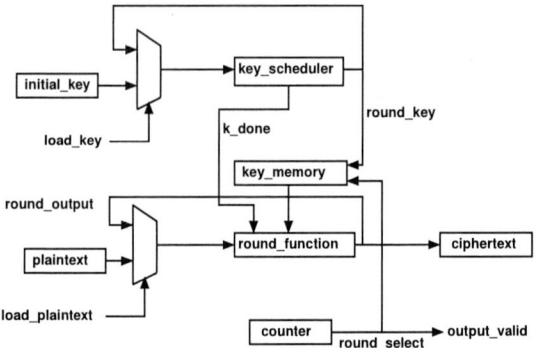

Fig. 3. FPGA implementation of AES-128

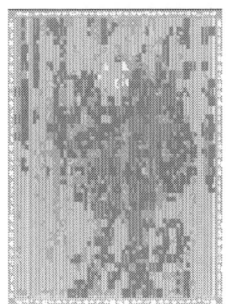

Fig. 4. Placement of the design using Xilinx floorplanner

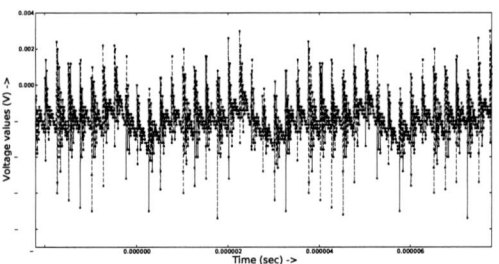

Fig. 5. Power trace sample of AES-128 with Rijndael S-box

An observation of both sample power traces reveal that the power trace profile between two consecutive rounds are almost similar for all rounds in AES-128 implemented with *RAIN* S-box. The corresponding power trace profile for AES-128 with Rijndael S-box varies from round to round. This means that power consumption of AES-128 with Rijndael S-box has a greater dependency on intermediate data operands than the data dependency of AES-128 implemented with *RAIN* S-box. As DPA attacks mainly exploit this feature, this exposes the increased vulnerability of AES-128 with the Rijndael S-box to DPA attacks compared to that implmented with *RAIN* S-box. In the experimental setup, the clock is being driven from an on-board crystal oscillator of 4 MHz. The power traces captured from the setup onto the digital sampling oscilloscope, are in terms of voltage drops across 1Ω resistance between V_{ccint} pin and ground pin on the FPGA board.

978-1-4673-0438-2/12 $31.00 © 2012 IEEE

TABLE I

FPGA IMPLEMENTATION RESULTS FOR AES-128 ON XILINX SPARTAN DEVICE XC3S400-4PQ208

Comparison Parameter	AES-128 with RAIN S-box	AES-128 with Rijndael S-box
Maximum Operating Frequency	71.5 MHz	64.5 MHz
Number of Slices	3026	3406
Number of Slice Flip Flops	2114	2246
Number of 4 input LUTs	5190	5672
Total equivalent gate count for the design	54552	60831

TABLE II

PERFORMANCE COMPARISON OF CRYPTOGRAPHIC PROPERTIES

Cryptographic properties	AES-128 with RAIN S-box	AES-128 with Rijndael S-box
Balancedness	Yes	Yes
Nonlinearity (MSB,..,LSB component functions)	(112,112,112,112,104,108,112,112)	(112,112,112,112,112,112,112,112)
Algebraic Degree	4	7
SAC satisfiability(MSB,..,LSB component functions)	(Yes,Yes,Yes,Yes,No,No,No,Yes)	(No,No,No,No,No,No,No,No)
PC(order satisfiability) (MSB,..,LSB component functions)	(1,3,5,7),(1,3,5,7),(1,3,5,7),(1,3,5,7), None,None,(7),(1,3,5,7)	None,(7),(3),None, None,None,(5),None
Correlation Immunity(order) (MSB,..,LSB component functions)	(1),None,None,None, None,None,None,(1)	None,None,None,None, None,None,None,None
Bias for Linear Approximation of component functions	0.0625, 0.0625, 0.0625, 0.0625, 0.0781, 0.0934, 0.0625, 0.0625	0.0625, 0.0625, 0.0625, 0.0625, 0.0625, 0.0625, 0.0625, 0.0625
Robustness to Differential Cryptanalysis	0.9453	0.9844
Transparency Order	7.79	7.86

TABLE III

NUMBER OF POWER TRACES REQUIRED TO EXTRACT THE 10-TH ROUND KEY OF AES-128

Initial key	Number of power traces for AES-128 with RAIN S-box	Number of power traces for AES-128 with Rijndael S-box
0x2b7e151628aed2a6abf7158809cf4f3c	14700	7200
0x739ed21a0fb689c54b2de13a6d187ccb	15600	7500
0x5d72ae9c18d43061b75f8b370dc325e9	14400	6900

A. Implementation Results and Performance Comparison

The FPGA implementation results for AES-128 using Rijndael S-box and *RAIN* S-box are summed up in Table I. As results show, AES-128 block cipher using *RAIN* S-box has smaller slice, 4-input look-up table(LUT) consumption and gate count compared to when it contains Rijndael S-box. Also as indicated in Table I, AES-128 block cipher has a higher maximum operating frequency for *RAIN* S-box compared to Rijndael S-box. Also the comparison of cryptographic properties satisfied by the both the S-boxes are mentioned

in Table II. The cryptographic properties listed in Table II correspond to the individual output bits of the respective S-boxes i.e. the component functions of the S-box. In the past, it was proposed that a robust S-box should have high algebraic degree, must be 0/1 balanced and complete, and satisfy strict avalanche criteria [18]. While none of the component functions of Rijndael S-box satisfy *SAC*, five component functions of *RAIN* S-box are found to satisfy *SAC* as shown in Table II. Also compared to Rijndael S-box, more number of component functions of *RAIN* S-box satisfy *PC* of higher order. The

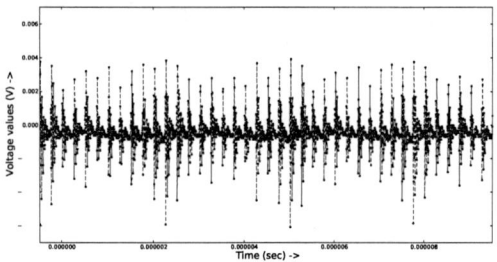

Fig. 6. Power trace sample of AES-128 with *RAIN* S-box

component functions corresponding to the most significant output bit and least significant output bit of *RAIN* S-box is found to have *correlation immunity* of order 1 while this property is absent in component functions of Rijndael S-box.

V. COMPARISON OF DPA ATTACK RESISTIVITY

In terms of robustness to DPA attacks, [19] proposed that the S-boxes with lower *transparency order* has greater resistance to DPA attacks. The *transparency order* of the *RAIN* S-box is found to be smaller than the AES-128 Rijndael S-box as mentioned in Table II. This quantification of the resistance to DPA attacks indicates that the *RAIN* S-box has higher resistance to DPA attacks than the Rijndael S-box. This has been experimentally validated by the requirement of larger number of power traces (for two different initial key values) to obtain the 10*th* round key of AES-128 for the FPGA implmentation of AES-128 block cipher with the *RAIN* S-box compared to the Rijndael S-box as shown in Table III. The AES-128 block cipher was implemented with separately using *RAIN* as well as Rijndael S-box on Xilinx Spartan device XC3S400-4PQ208 and the power traces were captured using the Tektronix current probe TCP0030.

VI. CONCLUSION

This paper focuses on *Design for Security* aspect which is recently becoming a point of paramount importance. This guarantees the level of security of cryptographic systems implemented on reconfigurable hardware like FPGA against side-channel attacks like DPA analysis while restoring cryptographic properties of block cipher primitives. This paper can be seen as a case study of S-box design for block ciphers where Boolean component functions of S-boxes has a bearance on DPA attack resistivity. In present literature, proposing and validating better Boolean functions for S-box construction for resisting side-channel attacks like DPA is still an open problem where this paper is a first attempt with practical implementation results. In this respect, this paper proposes an S-box construction based on *RAIN* function for an AES-128 like block ciphers. As results show, the component functions of proposed S-box have important cryptographic properties of *SAC, PC of higher orders, correlation immunity* which have lesser presence in the Rijndael S-box. The FPGA implementation of AES-128 with *RAIN* S-box also has lower

hardware resource consumption in terms of slices and look up tables (LUTs) and the maximum operating frequency is also found to be higher in comparison to the block cipher implemented with Rijndael S-box. The proposed S-box does not use any countermeasures developed on Rijndael S-box but still being more DPA resistant while avoiding higher area and power consumption. Also, the *transparency order* of the proposed S-box is found to be smaller than Rijndael S-box which quantifies the larger robustness of the *RAIN* S-box compared to the Rijndael S-box to single bit or multibit DPA attacks. This is evidently shown by the requirement of larger number of power traces to extract the 10th round key of AES-128 when it is implemented with *RAIN* S-box compared to Rijndael S-box.

REFERENCES

[1] Kocher P.C., Jaffe J., and Jun B., "Differential Power Analysis," in *CRYPTO*, ser. Lecture Notes in Computer Science, Wiener M. J., Ed., vol. 1666. Springer, 1999, pp. 388–397.
[2] A. Moradi, M. Salmasizadeh, M. T. M. Shalmani, and T. Eisenbarth, "Vulnerability modeling of cryptographic hardware to power analysis attacks," *Integration*, vol. 42, no. 4, pp. 468–478, 2009.
[3] J. D. Golic and C. Tymen, "Multiplicative Masking and Power Analysis of AES," in *CHES*, ser. Lecture Notes in Computer Science, B. S. K. Jr., Çetin Kaya Koç, and C. Paar, Eds., vol. 2523. Springer, 2002, pp. 198–212.
[4] E. Trichina, D. D. Seta, and L. Germani, "Simplified Adaptive Multiplicative Masking for AES," in *CHES*, 2002, pp. 187–197.
[5] T. Popp, S. Mangard, and E. Oswald, "Power Analysis Attacks and Countermeasures," *IEEE Design & Test of Computers*, vol. 24, no. 6, pp. 535–543, 2007.
[6] K. Kumar, D. Mukhopadhyay, and D. R. Chowdhury, "Design of a Differential Power Analysis Resistant Masked AES S-Box," in *IN-DOCRYPT*, ser. Lecture Notes in Computer Science, K. Srinathan, C. P. Rangan, and M. Yung, Eds., vol. 4859. Springer, 2007, pp. 373–383.
[7] E. Oswald, S. Mangard, N. Pramstaller, and V. Rijmen, "A Side-Channel Analysis Resistant Description of the AES S-Box," in *FSE*, ser. Lecture Notes in Computer Science, H. Gilbert and H. Handschuh, Eds., vol. 3557. Springer, 2005, pp. 413–423.
[8] S. Mangard, N. Pramstaller, and E. Oswald, "Successfully Attacking Masked AES Hardware Implementations," in *CHES*, ser. Lecture Notes in Computer Science, J. R. Rao and B. Sunar, Eds., vol. 3659. Springer, 2005, pp. 157–171.
[9] S. Mangard, T. Popp, and B. M. Gammel, "Side-Channel Leakage of Masked CMOS Gates," in *CT-RSA*, ser. Lecture Notes in Computer Science, A. Menezes, Ed., vol. 3376. Springer, 2005, pp. 351–365.
[10] K. Tiri and I. Verbauwhede, "Securing Encryption Algorithms against DPA at the Logic Level: Next Generation Smart Card Technology," in *CHES*, 2003, pp. 125–136.
[11] T. Popp, M. Kirschbaum, and S. Mangard, "Practical Attacks on Masked Hardware," in *CT-RSA*, 2009, pp. 211–225.
[12] J. Bhaumik, D. Mukhopadhyay, and D. R. Chowdhury, "Rain: Reversible Addition with Increased Nonlinearity," *To appear in IJNS*, 2011.
[13] P. Sarkar and S. Maitra, "Construction of Nonlinear Boolean Functions with Important Cryptographic Properties," in *EUROCRYPT*, ser. Lecture Notes in Computer Science, B. Preneel, Ed., vol. 1807. Springer, 2000, pp. 485–506.
[14] M. Matsui, "Linear cryptoanalysis method for des cipher," in *EURO-CRYPT*, 1993, pp. 386–397.
[15] E. Biham and A. Shamir, "Differential cryptanalysis of des-like cryptosystems," *J. Cryptology*, vol. 4, no. 1, pp. 3–72, 1991.
[16] D. R. Stinson, *Cryptography - theory and practice*, ser. Discrete mathematics and its applications series. CRC Press, 1995.
[17] S. Mangard, E. Oswald, and T. Popp, *Power Analysis attacks - Revealing the Secrets of Smart Cards*. Springer, 2007.
[18] B. Preneel, W. V. Leekwijck, L. V. Linden, R. Govaerts, and J. Vandewalle, "Propagation Characteristics of Boolean Functions," in *EURO-CRYPT*, 1990, pp. 161–173.
[19] E. Prouff, "DPA Attacks and S-boxes," in *FSE*, 2005, pp. 424–441.

978-1-4673-0438-2/12 $31.00 © 2012 IEEE

Real-time Melodic Accompaniment System for Indian Music Using TMS320C6713

Prateek Verma
Department of Electrical Engineering
IIT Bombay
Mumbai , India
prateekv@ee.iitb.ac.in

Preeti Rao
Department of Electrical Engineering
IIT Bombay
Mumbai , India
prao@ee.iitb.ac.in

Abstract— An instrumental accompaniment system for Indian classical vocal music is designed and implemented on a Texas Instruments Digital Signal Processor TMS320C6713. This will act as a virtual accompanist following the main artist, possibly a vocalist. The melodic pitch information drives an instrument synthesis system, which allows us to play any pitched musical instrument virtually following the singing voice in real time with small delay. Additive synthesis is used to generate the desired tones of the instrument with the needed instrument constraints incorporated. The performance of the system is optimized with respect to the computational complexity and memory space requirements of the algorithm. The system performance is studied for different combinations of singers and songs. The proposed system complements the already available automatic accompaniment for Indian classical music, namely the sruti and taala boxes.

Keywords - Pitch detection; DSP; Automatic accompaniment; Additive Synthesis .

I. INTRODUCTION

The purpose of this project is to build an automatic system that provides real-time melodic accompaniment in a music performance. Such a device can be useful in the Indian classical music performance context, in which there is a principal performer, rhythmic and melodic accompaniment and a drone instrument. Usually a *tanpura* is used as a drone instrument, which gives the tonal context to the melody. Melodic accompaniment is usually provided by a pitched instrument, such as a violin, *sarangi* or *harmonium*. The rhythmic accompaniment is usually provided by a percussive instrument, such as a *tabla* or *mridangam*. Typically the melodic accompaniment follows the singer's musical composition with some delay. For instance Carnatic vocalists are usually accompanied by the violin which shadows the singer's melody while also filling in the silent intervals. Likewise the harmonium tracks the Hindustani vocalist's melody. Pitch-continuous instruments like the violin and sarangi follow the vocal melodic contour closely including the expressive ornamentation (known as *gamaka*). On the other hand, the harmonium is operated via keys providing a set of discrete, pre-tuned pitch values only. Thus the harmonium can be constrained in following some continuous and rapid pitch modulations of the voice. However it is still favored by Hindustani musicians due to its rich tonal quality.

These days tanpuras are increasingly being replaced by electronic tanpuras or *sruti* boxes. Also virtual software for taals (rhythmic accompaniment) is available since it is pre-decided before a particular vocal performance. Some accompaniment systems such as MySong® [1] and LaDida® [2] can create accompanied music but they do not operate in real time but rather add only chords and beats after recording the music. This paper proposes an automatic accompaniment system based on a digital signal processor (DSP) which can synthesize the sound of a selected instrument such as violin or harmonium, and provide real-time melodic accompaniment. This can be useful for musicians to practice by providing a virtual concert like scenario. It can also be used in concerts as a virtual accompanying instrument. Some of the challenges faced in developing such an embedded system application are the response time, memory space, throughput and accuracy. All these factors will play a major role in the design of the algorithm and its implementation on the DSP.

Unlike composition-driven western classical music, an Indian classical vocal performance is heavy in improvisation. Since the Indian classical vocalist creates the melody during the performance within a decided Raga and Tala framework, there is no musical score already available to the accompanist. Rather the accompanist listens to the main performer and reproduces the melody on the fly with at most a short delay. This makes the design of a real-time automatic melody accompanier in Indian classical music difficult and challenging. Further, unlike Western classical music, which uses a fixed tuning system, in Indian classical music the singer is free to choose the actual pitch of the tonic note at the start of every performance. Consequently, the system must be able to automatically tune itself to the reference key. In our design the system imitates the actual pitch of the singer/lead artist. Therefore it works similarly for both improvisation and fixed compositions.

The organization of the paper is as follows: Section II gives a brief review of the algorithm involved, followed by the description of each sub-system. Section III describes the DSP implementation of the system. Section IV describes the evaluation of the accompaniment system followed by the conclusion and future work.

II. Algorithm Description

Fig 1. shows a block diagram of the melodic accompaniment system which uses the detected input singing voice signal parameters together with tuning parameters extracted from a short training audio sample to generate the accompaniment audio of a selected melodic instrument. In this section, we provide an overview of the main modules in Fig. 1.divided into A. Tuning, B. Pre-Processing, C. Pitch Extraction and D. Instrument Synthesis. More detailed descriptions as well as implementation aspects appear in Sec. III.

A. Tuning

A knowledge of the singer's volume (amplitude level) and level of background noise, if any, impinging on the microphone are required to reliably distinguish the signal from any background noise that may be present . The mean energy of the singer is calculated and used for thresholding of silence/non silence region. Further, knowledge of the singer's tuning i.e. the locations of notes in pitch space is important information if the selected accompaniment is a keyed instrument. To obtain this information a singer is first asked to sing a brief excerpt, e.g. a few musical notes, as training data. Next the assumption of musical equal-temperament scale is used to describe the entire pitch space. The computation of the pitches of the singer in terms of the note locations is done for a musical piece. In general singers do not exceed a three octave pitch range.

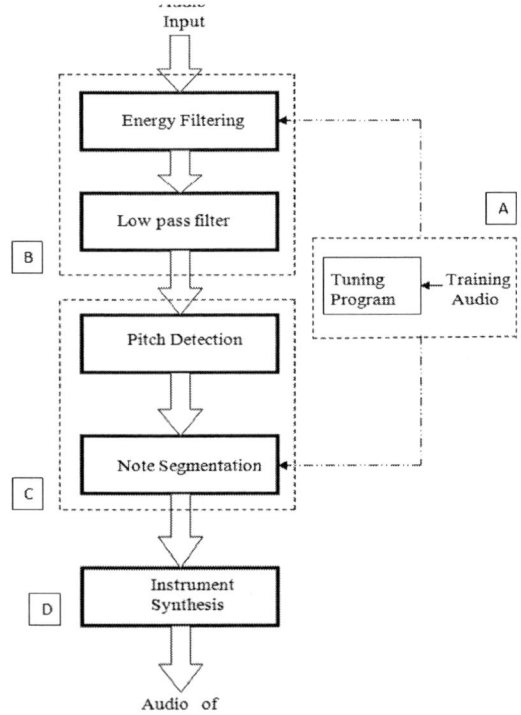

Fig 1. Block diagram of the proposed accompaniment system

B. Pre-Processing

It is very important to separate the silence and the non-silence regions of the signal in order to avoid the problem of spurious pitch values. Hence energy filtering is done wherein energy of each frame is calculated and then silence/voice determination is done. The average energy content of the musical piece is calculated from the tuning data with 0.55 of this value is taken as threshold on short-time energy to separate between the silence and non silence regions.

Voice signals have a rich harmonic structure spanning the frequency range 80 Hz to 5000 Hz. A pitched vocal utterance may contain 30-40 harmonic components. Moreover, the fundamental may not be the strongest one i.e. the first 2-8 harmonic components are usually stronger than fundamental component. The harmonic structure make pitch tracking complex with the possibility of octave errors. In order to improve the reliability and reduce the influence of strong higher order harmonics, low pass filtering is carried out.

C. Pitch Extraction and Note Segmentation

Pitch is detected every 40 ms by the short-time analysis of a windowed segment of the input signal. There are various pitch detection algorithms available both in the time and frequency domain based on the temporal periodicity and spectrum harmonicity respectively [4]. Table 1 depicts the comparison of well-known pitch detection methods, viz. the Average Magnitude Difference Function (AMDF), Auto Correlation Function (ACF) and Harmonic Product Spectrum (HPS). N denotes the number of samples in a frame and R is the number of harmonics considered in the HPS algorithm [5]. Generally the time domain approaches are computationally simpler. The AMDF algorithm only needs addition, subtraction and absolute value operations. Here we choose AMDF due to its low complexity and therefore suitability for real-time applications. A disadvantage of the AMDF method is that it is highly susceptible to background noise and intensity variation and therefore certain modifications are incorporated as described next.

TABLE I. COMPARISON OF VARIOUS PITCH DETECTION ALGORITHMS IN TERMS OF NUMBER OF COMPUTATIONS INVOLVED.

Algorithm	No. of Multiplications	No. of Additions
AMDF	-	N
ACF	N	N
HPS	2N(log N + R)	2N log N

The short time average magnitude difference function of a frame of the filtered input signal is defined as

$$x_w(m) = \frac{1}{N-m-1} \sum_{n=0}^{N-m-1} |s_w(n+m) - s_w(n)|$$

(1)

where $s_w(n)$ is the filtered voice signal, N is the analysis window duration, and the lag m varies between 0 and N-1. Here we see that AMDF calculations require no multiplications, a desirable property for real time applications. We see that the function $x_w(m)$ is minimized at the integer multiples of the pitch period (i.e. whenever m=0, ±T, ±2T….). Ideally $x_w(n)$ should approach zero at multiples of the period T but since the signal is quasi-periodic the difference only attains local minimum.

Further, the position of the global minimum may not match the pitch period because of the effect of the amplitude variations, noise and formants of the voice signal [3]. These can cause octave errors in the obtained pitch. There are several methods available to reduce these errors. YIN [3] has employed a cumulative mean difference function (CMNDF) in order to reduce the occurrence of sub-harmonic errors. It de-emphasizes higher period dips in the difference function. The CMNDF is given below.

$$x'_{nw}(m) = \begin{cases} 1 & if\ m = 0\ , \\ x_w(m) \Big/ \left[(1/m) \sum_{j=1}^{m} x_w(j) \right] & otherwise. \end{cases}$$

(2)

Theoretically $x'_{nw}(n)$ should tend to zero for integral multiples of the period but actually a local minima exists at these points. The position of the minimum value is found. Knowing the sampling frequency and this value, the pitch value of the frame is found. The minimum value will not always correspond to the pitch period. If we search for the global minimum then a lot of octave and sub harmonic errors occur. To avoid this and improve the accuracy of the algorithm, only local minima below a given absolute threshold are considered [3].

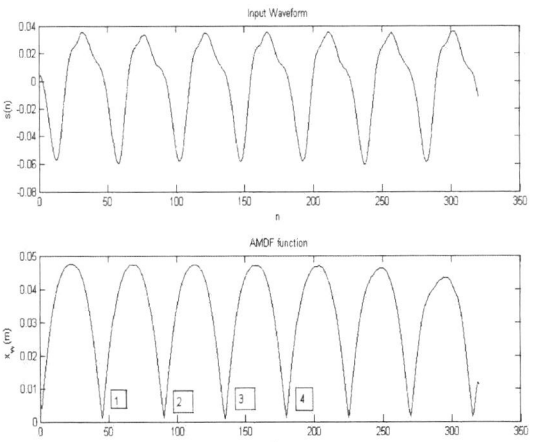

Fig 2. Example of a sub harmonic error.

For example, consider the AMDF function for a particular signal segment with actual pitch period = 45 samples in Fig 2. Clearly the global minimum, i.e. point 3 is not the pitch

period. Hence first we store the local minima occurring in a frame in increasing order. Then a value for the threshold and the number of minima points to look for is set from experimental observation. Now out of the first n number of local minima after arranging in the increasing order (7 in the above case), the lag corresponding to the local minimum closest to m=0, and lying below a threshold, is taken to be the pitch period.

For error correction post processing of the pitch values is done. Here it is assumed that there cannot be a pitch variation of more than two octaves between two successive frames as they are only 40 ms apart . Also the ground truth pitch value of the voice is assumed not to exceed 800 Hz. Hence in such cases the previous valid pitch value is assigned as pitch value of the current frame.

Now after this stage some rules are incorporated in making the rendition of the "virtual instrument" as close to real instrument as possible. For a pitch-continuous instrument such as the violin, no further processing is needed. However for a keyed instrument such as the harmonium, we need a step of note segmentation and labeling. Fig 3. shows a comparison between time aligned pitch contour of the vocalist to that of a harmonium player. The continuous pitch contour corresponds to the vocals whereas discrete valued graph correspond to the same track being played by a harmonium player with time alignment. Based upon such data, some simple rules are incorporated in the algorithm for making our system as close to the instrument as possible.

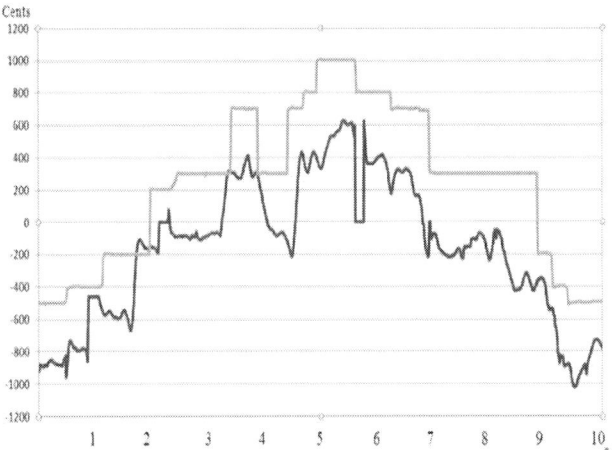

Fig.3 Comparison between the pitch contour of Harmonium player vs. Vocalist. A vertical offset has been applied to improve clarity.

If the raga is known beforehand (which is usually the case in case of classical music concerts) then all the swars which are not in a particular raga can be omitted at the time of note segmentation. We also see that there is a minimum duration for which a note is played in the actual instrument. This minimum note duration depends on the musical piece. Hence in our system also there is a provision of incorporating this, i.e. a note is played for a particular time duration which is adjustable. Also there is always some delay present between the accompanist and the vocalist. Each note has some link with its preceding and succeeding note. These linking notes are called grace notes or *Kan-swars*. It was observed that these Kan-swars are ignored most of the times by the accompanist. We thus use rules on allowed note locations and duration to convert the vocalist's tracked pitch contour into a note sequence suitable for harmonium. We thus use rules on allowed note locations and duration to convert the vocalist's tracked pitch contour into a note sequence suitable for harmonium.

D. Instrument Synthesis

In human ear, pitch perceived is logarithmically related to the physical fundamental frequency. The cent is a logarithmic unit of measure used for musical intervals. Twelve-tone equal temperament divides the octave into 12 semitones of 100 cents each. Typically, cents are used to measure extremely small finite intervals, or to compare the sizes of comparable intervals in different tuning systems. In a piano the adjacent keys are 100 cents apart. An octave comprises of 1200 cents.

If the selected accompaniment is a keyed instrument such as the harmonium, the next step involves gridding, i.e. snapping the instrumental pitch values to a pre-determined note-grid. This note-grid is determined as 100 cent intervals as computed in the Tuning block. At every analysis time-instant (spaced 40 ms apart) the singer's pitch value is snapped to the nearest grid location.

The system uses additive synthesis to generate the instrumental sound. Additive synthesis is defined as the addition of one or more pure tones to generate the required musical timbre. If we analyze the sound of different musical instruments we see that the same note (pitch) differs in sound quality (timbre) due to following:

1. Non-harmonic components present.
2. Amplitude of each harmonic component according to the spectral envelope corresponding to the instrument timbre .
3. The temporal volume envelope of the played note .

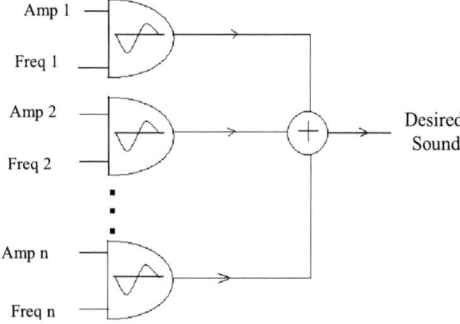

Figure 4. Additive synthesis

We simulate the desired sound accurately based on the following equation.

$$x(n) = \sum_{k=1}^{N} \left(a_k \sin \frac{2\pi f_0 k n}{F_s} + \varphi_k \right) \qquad (3)$$

where a_k denotes the amplitude of the kth harmonic (sinusoid) component, f_0 is the desired fundamental frequency, Fs is the chosen sampling frequency, φ_k is the phase of the kth harmonic, and N is the desired number of harmonics. The sound is synthesized using the implementation shown in Fig. 4 given the pitch, level (volume) and stored spectral envelope corresponding to the timbre of the sound. The harmonic phases are usually set to arbitrary initial values. Additive synthesis is a computationally complex but flexible sound modeling method. Inharmonic components are not considered here. Instrument temporal envelope modeling has not been considered here. But rather we assume that the harmonic amplitudes are constant over the note duration.

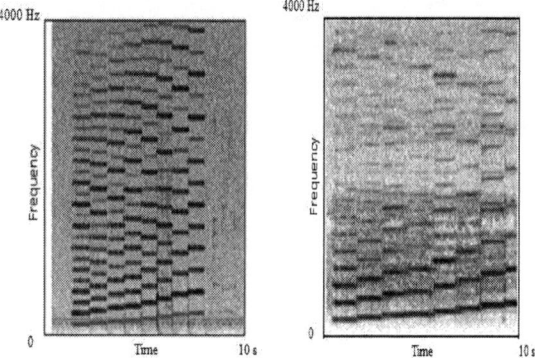

Fig 5. Spectrogram of an octave played on (a) harmonium and (b) flute.

From Fig 5. we see that harmonium is a spectrally rich instrument having dominant higher order harmonics. Thus synthesizing its sound takes up a lot of computation. On the other hand flute has at most five significant harmonics. Hence amount of computation time required for flute synthesis is less.

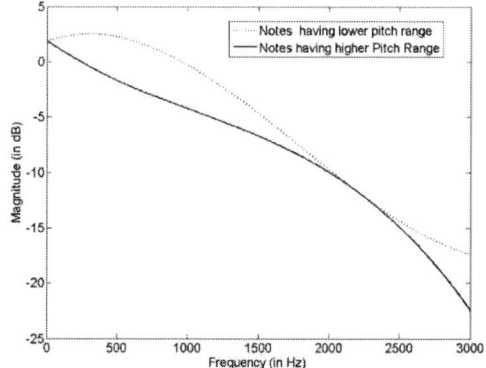

Fig.6 Comparison of spectral envelope of harmonium notes.

Fig.6. shows the smooth spectral envelope of a harmonium tone computed by averaging over a number of different notes in each of two frequency ranges:low (50-160 Hz) and high (170-512 Hz). Depending on the pitch of the desired note, the harmonic amplitudes are read off one of these stored envelope for additive synthesis.

The human ear is very sensitive to abrupt changes in phase. Hence phase of each final value of the frame is tracked in order to make the synthesized waveform shape continuous. Otherwise the synthesis will result in artifacts manifested as clicking sounds after each frame, resulting in an unpleasant and poor audio quality.

III. DSP IMPLEMENTATION OF THE ACCOMPANIMENT SYSTEM

In this section we discuss implementation of the proposed accompaniment system on TMS320C6713 floating point Digital Signal Processor from Texas Instruments [6]. It has an optimized architecture for fast operation capable of a large number of mathematical operations on large chunks of data with minimum latency in order to facilitate real time operations. TMS320C6713 has a clock frequency of 225 MHz but an architecture giving higher MIPS (Million Instruction Per Second) compared to other processors of similar speed [7]. The advantage of this processor over a fixed point processor is that there are reduced errors due to overflow, truncation and rounding of operands. Additionally this processor has the ability to handle a large dynamic range and also algorithms are easier to implement in floating point DSP rather than fixed point. The CCS (Code Composer Studio v3. 1) is used for compiling, and creating an assembly language code and linking.

The present algorithm involves block-based processing i.e. computations are carried out every 40ms throughout the non-silent region of the input audio signal to generate estimated pitch corresponding to the centre of the 40ms frame.

A. Programming aspects of DSP Implementation

For low pass filtering in the preprocessing stage, an elliptical filter is chosen as no other filter of equal order can have a faster transition in gain between the pass band and stop band. Lesser order means less complexity and memory consumption. A 10th order low pass elliptical filter with pass band as 600Hz and stop band attenuation of -80dB is designed. For implementing the 10th order filter, its transfer function is broken into second order sections and then cascaded. This is done so as to reduce the quantization effects. Each of the second order section is written in terms of its difference equation. This takes lesser number of CPU cycles as it involves only addition and storage elements. Linear buffers are used to store the delayed by one sample values. After each iteration the new value is stored in the first element of the storage buffers whereas the last value is discarded. Every iteration results in the buffer elements being right shifted by one (first in, last out).
Division generally takes much more time as compared to other mathematical operations. Hence it is avoided as much as possible. Synthesis of required audio needs a lot of harmonic

components to be generated each corresponding to a sine function. Hence we define a look up table for the values of the sine function. Again the intervals in the table must be optimized by experiments done without the loss of accuracy and quality. In order to save memory the look up table for only half the sinusoidal cycle is considered [8]. Thus 256 samples of a half cycle of sinusoid are stored. Sampling frequency cannot be too high since it means lesser time available to fill up the individual buffers and thus lesser processing time for the algorithm. Frame length too dictates the real time performance as well as pitch accuracy. Therefore repeated profiling was carried out till our system worked efficiently. It executes the code rapidly and uses fewer available resources on the chip.

Finally 8 kHz sampling was chosen as the best trade-off between sound quality and computation time. The size of the buffer for a 40ms analysis frame was thus equal to 320. Block-based processing with DMA was done instead of sample by sample processing due to computational complexity of the algorithm. Ping and Pong buffers alternately receive the samples or process and extract pitch. Instead of sorting the minima after AMDF computation which may take up a lot of computational cycles, a different approach is taken. First the global minimum of the function $x_w(m)$ is calculated for a particular frame. Then using a predetermined threshold only those minima that lie within the threshold are stored. Now out of these values the minima that has the least time index "m" is chosen as the desired value of the period. The values for the amplitudes of the Fourier coefficients of a particular instrument were stored in look up table to save the computational time.

This accompaniment system has many features which makes its similar to electronic tanpura in some ways. Its output pitch can be tuned to our desired pitch range providing flexibility. Since it was observed that the delay between the accompanying instrument and the lead vocalist varies depending on composition, mood and the need delay can be varied with . As of now there is a provision of playing one of the three accompanying instruments namely harmonium, violin and flute. But virtually any non percussion instrument can be synthesized by studying its spectrum for the amplitudes to be used in additive synthesis.

IV. EVALUATION OF THE ACCOMPANIMENT SYSTEM

For evaluation a set of five musical pieces is chosen. In order to cover a large number of variations songs having a variety of musical ornaments are selected. Five different singers were chosen. In the first stage the pitch accuracy is calculated. Here the accuracy depicts the total number of frames lying within a range of 50 cents of the original pitch value. A total of 745 seconds database of audio files was thus selected and the pitch accuracy was found to be 93.45% based on ground-truth pitch values obtained via a semi-automatic melody detection interface for polyphonic music [9].

After assigning fixed pitch values based on the detected user tuning, note names are assigned to the output of our "virtual instrument" for a particular song. Thus for each song we get a

978-1-4673-0438-2/12 $31.00 © 2012 IEEE

musical score wherein each note duration is assigned according to the known tempo of the audio file. To validate the note segmentation stage, it was necessary to get the ground truth note sequence corresponding to the songs. This was achieved by getting an expert harmonium player to reproduce the song on harmonium while listening to the corresponding singing audio over headphones. The harmonium audio was transcribed into notes by automatic pitch detection. Word error rate is often used to evaluate the performance of a speech recognition system. Here after getting these two strings (i.e. the automatically segmented note sequence and the human harmonium experts note sequence) they can be compared for word error rate viz. number of matches, number of substitutions, deletions and insertions between two optimally aligned strings. Word error rate is computed as :

$$WER = \frac{S+D+I}{N} \qquad (4)$$

where S is the number of substitutions, D is the number of deletions, I is the number of insertions, N is the number of words in the reference which are the notes played by an actual instrument player.

TABLE II. COMPARISON OF ACTUAL INSTRUMENT AND PROPOSED VIRTUAL ACCOMPANIMENT SYSTEM IN TERMS OF WORD ERROR RATE OF NOTES.

Song Name	% Notes Matching	% Note Substitutions	% Note Insertions	% Note Deletions	% WER
Kesar	78.32	5.59	11.89	4.2	21.68
Kaise Paheli	77.78	13.07	8.17	0.98	22.22
Total	77.88	11.66	8.87	1.59	22.12

Table II shows the results for a subset of the songs. Here we note that even though the accuracy was very high, the number of matching notes given by DSP and that of actual player is around 77%. This happens because the duration for which a note is played depends on individual player and style. Also musical ornaments *Kan-Swar* and *vibrato* are followed by our system whereas real instrument players often tend to ignore such notes. Thus there are greater number of substitutions as compared to deletions as the current system follows whatever the main artist does whereas the accompanist may follow these subtle changes depending on the expertise.

After this we move to evaluation of the accompaniment system from the point of view of hardware resources utilized. The hardware resources utilized towards the DSP implementation of the accompaniment system is reported in Table III. Here since a 40 ms frame was chosen with a sampling frequency of 8 kHz available clock cycles are 9 x 10^6.

TABLE III. HARDWARE UTILIZATION OF THE ACCOMPANIMENT SYSTEM

Resource Utilization Details	
DSP Device	TMS320C6713
Clock Frequency	225MHz
Program Memory Used	187kbytes
Clock Cycles consumed per execution	8.9671 x 10^6
%idle processor time per execution	0.3661
% Memory consumed	83.1 %

V. CONCLUSION AND FUTURE WORK

In this paper implementation of a music accompaniment system, particularly for Indian music, is implemented on a TMS320C6713. It gives satisfactory performance both in terms of sound quality and accuracy. The C-code is written keeping in mind the real time considerations and the limited memory space available on the DSP chip. We have managed to synthesize the sound of harmonium by formulation of some simple rules but these need to be further refined. The temporal envelope of the synthesized tones needs to be shaped for more authentic quality. Also from the experiments carried out, we see that the virtual harmonium is tracking the *vibrato* effect by switching between two notes whereas a harmonium player does it seldomly. There remain such differences for which musicological knowledge is required.

AKNOWLEDGEMENT

This work was done while Prateek Verma was with the Bharti Centre for Communication at IIT Bombay. He would also like to thank Vishweshwara Rao for his kind suggestions.

REFERENCES

[1] http://khu.sh/

[2] http://research.microsoft.com/en-us/um/people/dan/mysong/

[3] A. D. Chveigne And H. Kawahara, Yin, A Fundamental Frequency Estimator For Speech And Music, J. Accoust. Soc. Am. 111 (4), 2002, pp 1917-1930.

[4] D. Gerhard, Pitch Extraction And Fundamental Frequency: History And Current Techniques, Technical Report, University Of Regina, Canada, 2003.

[5] Patricio De La Cuadra, Aaronmaster And Craig Sapp, Efficient Pitch Detection Techniques For Interactive Music , Center For Computer Research In Music And Acoustics, Stanford University.

[6] TMS320C6713 Floating Point Digital Signal Processor, SPRS186L, Nov 2005.

[7] http://www.eecg.toronto.edu/~moshovos/ACA05/004-pipelining.pdf

[8] Francis Kua,Generation Of A Sine Wave Using A TMS320C54xx Digital Signal Processor, Application Report Texas Instruments SPRA 819, July 2004

[9] S. Pant, V. Rao and P. Rao, A Melody Detection User Interface For Polyphonic Music, Proc. of the National Conference on Communications (NCC), 2010, Chennai, India.

2012 25th International Conference on VLSI Design

Bidirectional Single-Supply Level Shifter with Wide Voltage Range for Efficient Power Management

Sujan K. Manohar*, Vinod K. Somasundar*, Ramakrishnan Venkatasubramanian*[+], and Poras T. Balsara*

* VLSI Circuits and Systems Laboratory, University of Texas at Dallas, Richardson TX 75080

[+]Texas Instruments Inc, Dallas TX 75243

e-mail: {sujan.manohar, vinodkadur, ramav, poras}@utdallas.edu

Abstract—Level shifter circuits are used to interface multiple voltage islands in many modern ICs or Systems-on-Chip (SoCs). Single-supply level shifters are being used to reduce the power routing resources and minimize the routing congestion at the chip level. A single-supply bidirectional level shifter aimed at low voltage which offers a wide voltage range (SS-WVRLS) is designed using standard commercial 90nm CMOS process. The proposed level shifter uses analog and digital circuit techniques to provide full voltage shifting range for any combination of supply voltages (VDDIN = VDD,VDDIN < VDD or VDDIN > VDD) in any step size (paper shows 25mv step) with no requirement for special low-V_T or high-V_T devices, thus reducing the process cost. Post layout SPICE simulation comparison results show that proposed circuit is functional for full core supply voltage range (0.6V - 1.32V) compared to other published level shifters. The circuit was tested for robustness under process mismatch conditions by 1000 point global and local Monte Carlo simulations.

Index Terms— Single–Supply, level shifter, routing congestion, bidirectional

I. INTRODUCTION

CMOS technology is being aggressively scaled to meet the ever increasing demand for low power, low area and high performance. To facilitate this, modern SoC's usually have multiple voltage islands to operate different parts of a chip at different voltages to achieve optimum speed/power ratio [1]. This has become a standard strategy for efficient power management [1][2]. A level shifter circuit is necessary to interface the signals crossing different supply voltage domains [1] [3].

Conventional level shifter circuits [1][4] require the supply voltage of both the domains, i.e. the power supplies of input as well as target voltage domain. Fig. 1 depicts a multi-voltage system using conventional level shifter for the signal interface. It may be noted that both power routing congestion and pin count for this system in a multi-voltage SoC would increase significantly [7]. Single supply level shifters, on the other hand, require only the power supply of the target domain to which the signal is being shifted. This eases the chip power routing complexity, reduces the pin count and hence the cost of the system as shown in Fig. 2. Existing single-supply level shifters published [3,5,7,8] are either unidirectional (i.e. only shift signals from low voltage domain to high voltage domain), have high leakage currents or are functional for a limited range of input and output supply voltages or have higher cost with additional

fabrication mask requirements for low-V_T and high- V_T devices. In this paper, we discuss existing single supply level shifter circuits and their limitations and propose a new bidirectional single-supply level shifter circuit having wide supply voltage shifting range (0.6V-1.32V). The proposed level shifter is also functional at low supply voltages (0.6V) and has flexibility to operate at all combinations of supply voltages in the range, making it a viable alternative for low voltage systems.

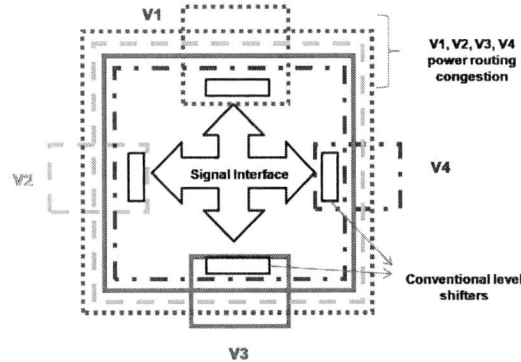

Fig. 1 Power routing congestion in multi supply system using conventional level shifters

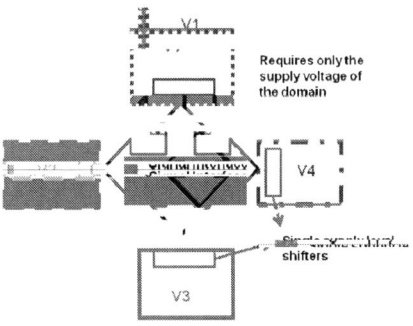

Fig. 2 Eased Power routing in multi supply system using single supply level shifters

II. BACKGROUND

A. Definition of Level shifter parameters/terms

VDDIN refers to the supply voltage of incoming signal and VDD refers to supply voltage of level shifter and its output signal.

978-1-4673-0438-2/12 $31.00 © 2012 IEEE

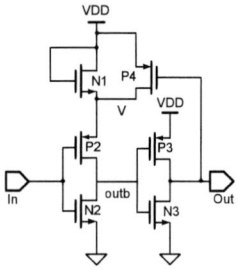

Fig. 3(a) Single supply level shifter [5]

B. Prior Art

Fig. 3(a) shows an implementation of single-supply shifter. This level shifter works simply by lowering output supply voltage by a threshold voltage (VDD-V_{Tn}) using a diode connected N-MOSFET N1 [5].This lower voltage at node *v* ensures lower gate to source voltage of P2 to turn it OFF when input is logic HIGH. When input is logic LOW, output gets discharged to logic LOW and feedback transistor P4 charges node *v* and hence node *out* to full VDD to turn OFF transistor P3.This circuit requires difference between input supply level and output supply level to be a maximum of threshold voltage of diode connected MOSFET N1 to avoid large leakage currents. This restricts the voltage translational range of the circuit and also the diode connected transistor limits the operating speed of the circuit as the first inverter operates initially with a reduced supply voltage of VDD- V_{Tn}.

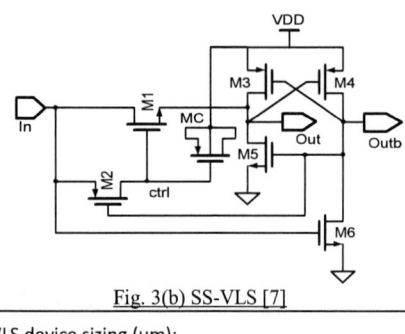

Fig. 3(b) SS-VLS [7]

SS-VLS device sizing (um):
M1=1.2/0.2;M2=0.24/0.1;MC=1.2/1.4;M3=0.3/0.1;
M4=1/0.1;M5=0.12/0.1;M6=3/0.1

Fig. 3(b) shows a unidirectional single supply level shifter SS-VLS, capable of shifting signals only from low input voltages to high output voltages [7]. The circuit works as follows: When input signal is logic HIGH at supply level VDDIN M6, M3 and M2 are turned ON and *out* is pulled to VDD. Node *ctrl* is charged to input supply level VDDIN through ON transistor M2. In the other case when input is at logic LOW, voltage at node *ctrl* held by capacitor MC turns ON M1, which turns ON M4 and M5 and discharges *out* to logic LOW.

SS-VLS alleviates the leakage current issue associated with shifting any input signal level to a higher target supply

voltage domain [7]. However it is unidirectional i.e. can shift input signals only from lower supply domain to higher supply domain. Circuit becomes non-functional in the other case as M1 will not turn OFF when input is at logic HIGH and results in a contention at node *out*. Also, transistor M2 should be sized weaker compared to M1 to prevent the node *ctrl* from getting discharged through M2 before the output node *out*, when the input signal is at logic LOW. Limitations on the relative sizing could aggravate the circuit robustness issues under process mismatch conditions.

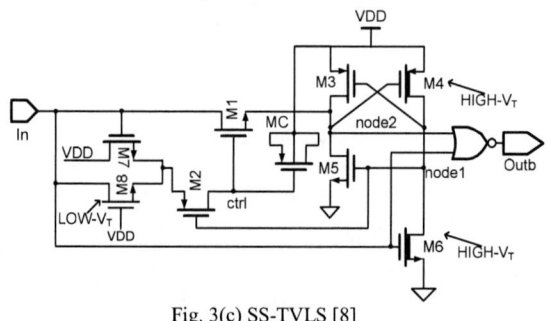

Fig. 3(c) SS-TVLS [8]

SS-TVLS device sizing (um):
M1=2/0.2;M2=0.24/0.1;MC=2/1.4;M3=0.12/0.1;
M4=0.32/0.1;M5=0.12/0.18;M6=M8=0.4/0.1;M7=0.3/0.1;

Single-supply true voltage level shifter (SS-TVLS) shown in Fig. 3(c) is an improved version of SS-VLS [8]. It is capable of shifting in both directions by using pass transistors M7 and M8. M4 and M6 are high-V_T devices to reduce short circuit leakage currents and M8 is low-V_T to reduce the V_T drop and enable node ctrl to charge to higher voltage.

Operation of this circuit is as follows:

a.   <u>VDDIN < VDD:</u> When input is logic HIGH, M6 and M3 turns ON and charges *node2* to VDD level. Output of the NOR gate *outb* goes logic LOW. Since VDDIN < VDD, series transistors M8 and M2 charge node *ctrl* to minimum of VDDIN and VDD-$V_{T.M8}$, where $V_{T.M8}$ is the threshold voltage of transistor M8. In the other case when input is logic LOW, voltage held at the node *ctrl* turns ON M1 and discharges *node2*. Positive feedback of M4 and M5 pulls down *node2* completely to ground (GND). *Outb* goes logic HIGH with both the inputs of nor gate at logic LOW, without any high leakage paths from VDD to GND.

b.   <u>VDDIN > VDD :</u> Functionality of SS-TVLS is same as described above in the first case VDDIN < VDD except that node *ctrl* charges through series transistors M7 and M2 to a minimum of VDD and VDDIN- $V_{T.M7}$, where $V_{T.M7}$ is the threshold voltage of transistor M7.

Voltage translational range of SS-TVLS is restricted by the threshold voltage V_T of the pass transistors M8/M7; as they restrict the maximum voltage to which node *ctrl* gets charged. This forces a requirement on transistor M8 to be low-V_T. However, circuit fails to function when VDDIN and VDD are low and close to each other as it gets harder to charge node *ctrl* above the V_T of M1.

For example:

Let $\frac{T}{2} \gg \tau$, where T is the time period of the input signal and τ is the time constant to charge node *outb* via switched current source M4. $V_{T.M8}$=360mV and $V_{T.M7}$=$V_{T.M1}$=420mV (V_T values are from commercial 90nm process used).

Case 1: VDDIN = 0.6V, VDD = 0.7V.
M8 is nearly off with only a V_{GS} of 100mV $< V_{T.M8}$
Max. Voltage on node *ctrl* = 0.7V-$V_{T.M8}$ = 0.34V $< V_{T.M1}$.
Case 2: VDDIN = 0.7V, VDD = 0.6V.
M7 is nearly off with only a V_{GS} of 100mV $< V_{T.M7}$
Max. Voltage on node *ctrl* = 0.7V-$V_{T.M7}$ = 0.28V $< V_{T.M1}$.

Circuit functionality at these points is further worsened by the lower gain of the latch operating at low VDD. Similar to SS-VLS, SS-TVLS also requires M2 to be sized weaker compared to M1 to prevent the node *ctrl* from getting discharged into input before *node2*.

III. Proposed SS-WVRLS Level Shifter

A. Level Shifter Architecture

This section presents the proposed single-supply wide voltage range level shifter (SS-WVRLS) design that addresses the limitations in earlier approaches is shown in Fig. 4. It comprises of the following key building blocks: input stage, output buffer, positive feedback and leakage sense. The operation of the circuit is as follows: When input signal IN goes logic HIGH, M5 is already ON with *fedz* at logic HIGH held by the internal loop, which enables switching current through M3, M1 and M5. This current is then mirrored from M3 to M4 and charges *outb* (capacitor CC) to full supply voltage VDD. OUT changes to logic HIGH at VDD level through output buffer. Internal signal *fedz* then goes to logic LOW after a certain delay to turn OFF M5 and M17 and turns ON M14 to enable the positive feedback. PMOS M2 with gate at VDDIN is not completely OFF and will charge the internal node *cm* to the same level as *outb* and stop conducting when drain and source reach same potentials.

The leakage sense circuit helps in restoring voltage level at node *outb* due to possible input signal glitches or leakage. In case of a glitch or perturbation at the input, node *outb* discharges slightly below VDD and turns ON M13 with just the required bias to mirror back the charging current to node *outb* via M11, M16, M15, M3 and M4. Here M16 is sourced by M3 via ON transistor M15. For glitch robustness, M16 is sized similarly as M12.

When IN goes to logic LOW, capacitor CC couples the node *outb* to input and discharges *outb*. Initial discharge of this node is sensed by transistor M13, which turns ON the positive feedback loop. M11 mirrors the same current back

to M12 and discharges *outb* to logic LOW and hence OUT through output buffer. IN at logic LOW turns OFF M15 and disconnects the positive feedback from the input stage. Internal signals *fedz* and *fed* goes logic HIGH and LOW respectively, so node X is pulled high to cut off any leakage current. Positive feedback loop gets turned off through M14 as *fedz* goes logic HIGH. M2 is fully ON to maintain logic LOW at *outb*.

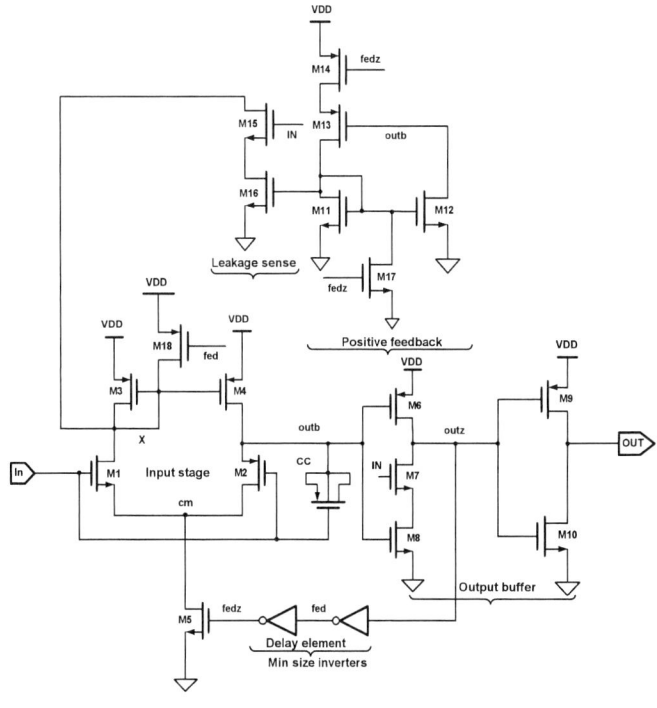

Fig. 4 Single-Supply Wide Voltage Range Level Shifter (SS-WVRLS).

SS-WVRLS device sizing (um):
M1=M3=M4=M5=0.3/0.1;M2=M7=M8=M9=M10=M11=M12=M13=
M14=M15=M16=0.2/0.1; CC=4(1/0.5); M6=0.5/0.2;
M17=M18=0.12/0.1; inverters (Pmos=Nmos=0.12/0.1)

B. Design Considerations

Important design considerations for SS-WVRLS include CC, node capacitance of *outb* (C_{par_outb}) and sizing of delay elements in the internal feedback. Voltage coupled from input to *outb* is given by (1):

$$V_{outb} = \frac{V_{in} \times CC}{CC + C_{par\_outb}} \qquad (1)$$

Hence to minimize the attenuation of V_{outb} due to C_{par_outb}, CC is chosen to be 3 to 5 times C_{par_outb} and also care is taken in the layout to minimize C_{par_outb}. Delay elements should be chosen such that their delay is less than half of the input time period.

C. Level Shifter operation

Fig. 5 shows transient simulation waveforms for 0.6V-1.32V up-conversion at 27C for an input frequency of 250MHz. It can be noted that TPLH=319.2ps,

TPHL=80.7ps and the circuit has no static current when input is not switching.

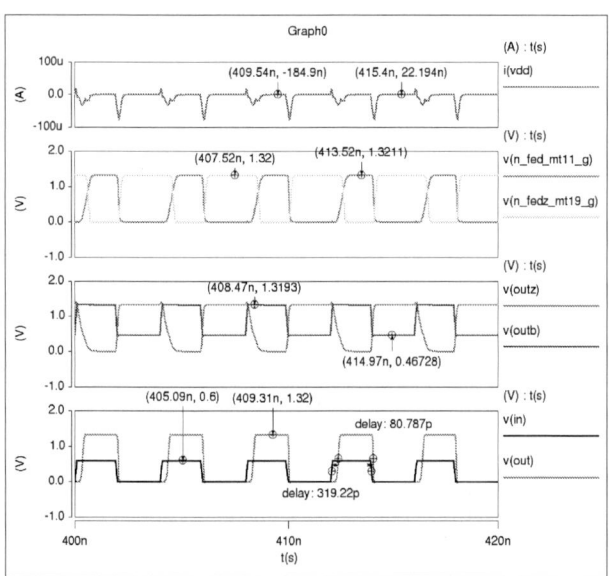

Fig. 5 SS-WVRLS: VDDIN=0.6V, VDD=1.32V up-shifting transient simulation.

IV. SIMULATION RESULTS AND COMPARISONS

SS-VLS, SS-TVLS and SS-WVRLS were designed in standard commercial 90nm CMOS process with following V_T parameters.

Nominal-V_T for L=0.1um	445mV
Nominal-V_T for L=0.2um	420mV
High- V_T	535mV
Low- V_T	360mV
Max core supply voltage	1.32V

SS-VLS and SS-TVLS were designed to meet the target delays as in [8] in the process used. Fig.10 shows the cadence layout for SS-WVRLS.

Tables 1 and 2 summarize the post layout performance parameters for voltage up-conversion from 0.8V-1.2V and down-conversion from 1.2V-0.8V for the three level shifter circuits, respectively. SS-VLS is unidirectional as stated before and hence not listed in tables 1, 2, 3 and 4 for voltage down conversion. It should be noted that, SS-TVLS has an inverted output compared to SS-VLS and SS-WVRLS. It can be observed that voltage up-conversion delays of SS-WVRLS are closer to that of SS-TVLS but have higher down conversion fall delay. This is because SS-TVLS fall transition path is faster with mainly a NOR gate delay and it improves for higher VDDIN, favoring down conversion. Tables 3 and 4 summarize 0.8V-1.32V up-conversion and 1.32V-0.8V down conversion performance metrics for three level shifter circuits subjected to 1000 point global plus local mismatch Monte Carlo simulation. It can be noted that SS-TVLS has slightly better standard deviation for the delays as compared to SS-WVRLS.

TABLE 1: 0.8V-1.2V UP-CONVERSION AT 27° C, 250MHZ INPUT.

Performance Parameter	SS-WVRLS	SS-TVLS	SSVLS
Worst case voltage translation range	0.6V-1.32V	0.972V	0.927V (unidirectional)
Special V_T devices	none	High -V_T and Low- V_T	none
Supply difference	Functional for all step size and for VDDIN=VDD	Non- functional for VDDIN=VDD until 0.927V	Functional for VDDIN=VDD but unidirectional
Output logic	Non-inverted	Inverted	Non-inverted
Delay Rise (ps)	48.40	30.57	133.25
Delay Fall (ps)	33.44	27.46	58.29
Energy Rise (fJ)	5.54	8.32	13.91
Energy Fall (fJ)	6.28	5.09	14.38
Leakage Current High (pA)	498.8	486.92	513.39
Leakage Current Low (pA)	328.6	210.79	355.9
Layout Area (µm²)	23.284 (5.94 X 3.92)	19.781 (6.475 X 3.055)	16.64 (6.54 X 2.545)

TABLE 2: 1.2V-0.8V DOWN-CONVERSION AT 27° C, 250MHZ INPUT.

Performance Parameter	SS-WVRLS	SS-TVLS
Delay Rise (ps)	39.80	48.50
Delay Fall (ps)	62.09	14.04
Energy Rise (fJ)	2.24	6.63
Energy Fall (fJ)	5.99	2.33
Leakage Current High (pA)	188.81	158.42
Leakage Current Low (pA)	98.35	75.38

TABLE 3: 0.8V-1.32V UP-CONVERSION WITH MISMATCH AT 27° C.

Performance Parameter	SS-WVRLS		SS-TVLS		SSVLS	
	μ	σ	μ	σ	μ	σ
Delay Rise (ps)	47.12	5.71	23.48	3.78	136.45	14.89
Delay Fall (ps)	23.33	4.19	29.74	4.82	62.24	5.23
Energy Rise (fJ)	7.22	0.93	9.83	1.25	21.83	2.56
Energy Fall (fJ)	7.26	0.46	6.81	1.88	18.31	1.37

TABLE 4: 1.32V-0.8V DOWN-CONVERSION WITH MISMATCH AT 27° C.

Performance Parameter	SS-WVRLS		SS-TVLS	
	μ	σ	μ	σ
Delay Rise (ps)	65.22	4.29	59.40	2.98
Delay Fall (ps)	59.91	4.03	9.56	1.92
Energy Rise (fJ)	3.39	0.63	12.08	0.73
Energy Fall (fJ)	4.63	0.46	2.55	0.40

Despite its slightly better performance, SS-TVLS has a limited level shifting range. Considering the case when VDDIN=VDD, the minimum supply voltage required to charge node *ctrl* above $V_{T.M1}$ (as explained in section II) is 0.927V and is depicted in Fig. 6. This is further worsened by the threshold voltage variations (In the commercial 90nm

978-1-4673-0438-2/12 $31.00 © 2012 IEEE

process used, 3σ base V_T variation not accounting for the variations due to device dimensions is 45mV. This increases the minimum supply voltage required to 0.972V). Thus the worst case level shifting range for SS-TVLS is 0.972V-1.32V, limiting its usage in low voltage applications and SoC's employing aggressive DVFS techniques where supply voltage is continuously scaled to an optimum value anywhere in full range to monitor the speed/power [2,9].

in full range supply variation and lower cost with no requirement of additional special VT devices.

Leakage sense circuit helps in restoring the output level when *outb* is discharged due to any high to low glitches in the input signal coupled through *CC* as explained in section III.

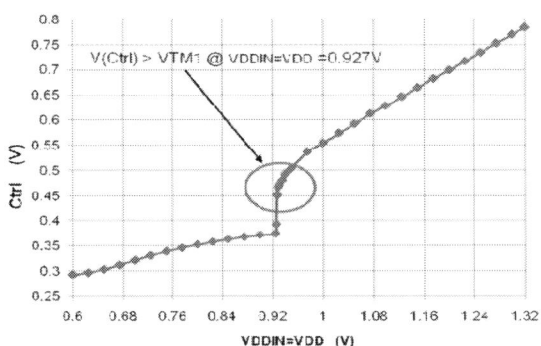

Fig. 6 SS-TVRLS: Voltage variation of node ctrl for VDDIN=VDD

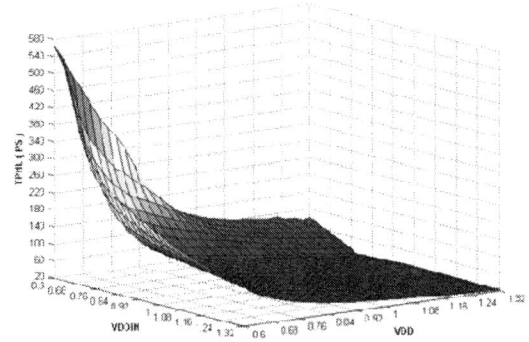

Fig.8 Fall delay of SS-WVRLS as VDDIN and VDD is varied from 0.6V-1.32V in 25mV step.

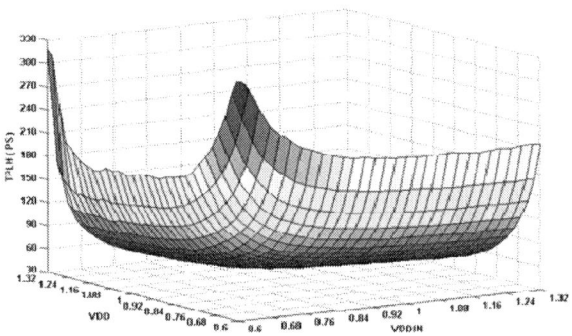

Fig. 7 Rise delay of SS-WVRLS as VDDIN and VDD is varied from 0.6V-1.32V in 25mV step.

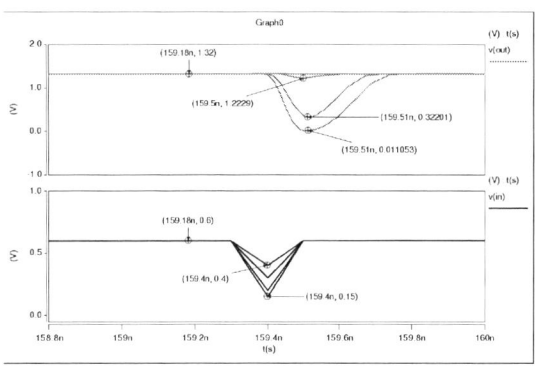

Fig. 9 SS-WVRLS: Glitch analysis with input glitch width = 200ps and glitch magnitude =0.2V, 0.3V, 0.4V, 0.45V for VDDIN=0.6V, VDD=1.32V up-conversion.

The proposed level shifter circuit has up and down voltage translation range of 0.6V-1.32V and is functional for any combinations of VDDIN and VDD in the range for any step size, imposing no supply restrictions for the system design. Fig.7 and Fig.8 show the plot of rise and fall delays for all combinations of VDDIN and VDD from 0.6V to 1.32V in steps of 25mV. Propagation delay is maximum when both supply voltages are minimum i.e. for VDD = VDDIN = 0.6V: TPLH = 345ps, TPHL = 564ps. Fall delay increases more sharply than rise delay at lower supply voltages due to lower coupled voltage and then improves for increasing up/down conversion. SS-WVRLS achieves performance close to SS-TVLS with ~17.7% increase in layout area, but offers additional low voltage (0.6V, below which the circuit enters sub-threshold) operation, flexibility

However input glitches of magnitude close to $V_{T.M13}$ or greater could turn ON positive feedback transistor M12 more than the leakage sense transistor M16 and further aggravate discharge of node *outb* and consequently flip the state of level shifter. Fig.9 shows the glitch sensitivity analysis with input glitch width of 200ps in steady state for 0.6V-1.32V up-conversion. Maximum glitch tolerance to recover the output state is observed to be 0.45V (~$V_{T.M13}$) and that which produces no transition at the output is 0.3V. However glitch tolerance will decrease for higher glitch widths. Glitch robustness could be further improved at cost of performance by increasing the strength of leakage sense circuit or choosing a lower value of *CC*.

Delay dependence on the input slew rate was checked for worst case condition (VDDIN= 0.6V, VDD=1.32V). As input slew rate varies from 50ps-500ps, fall delay increases monotonically from 311.4ps to 411.6ps (max. variation of 100.2ps) and rise delay increases monotonically from 78ps to 110ps (max. variation of 32ps).Variation of fall delay here is higher because VDDIN = 0.6V and fall delay relies on initial capacitive coupling from fall transition of the input signal. It improves for higher values of VDDIN.

Fig.11 shows power supply (VDD) ramp up simulation depicting the startup of SS-WVRLS for VDDIN=0.6V, VDD=1.32V. Input (IN) is kept logic low to simulate the worst case for startup. VDD is ramped from 0-1.32V in 10uS time, and then IN is toggled. It can be seen that OUT switches with input and all internal nodes settle having no steady state leakage.

V. CONCLUSION

This paper proposes a robust single-supply, wide voltage range level shifter circuit. SS-WVRLS is functional across full range of core supply voltages 0.6V-1.32V, for all combinations of VDDIN and VDD in any step size making it suitable for low voltage applications and has lower cost requiring no special V_T (low-V_T or high V_T which need extra process masks) transistors. Furthermore, in a multi-core SoC with DVS, the supply voltages of different voltage islands are not known a-priori. So there is a need for wide voltage range level shifters that can convert any voltage level to any other desired voltage level not imposing additional design constraints on the SoC for either power or performance. Proposed single supply level shifter circuit satisfies this requirement and offers complete range of supply voltage shifting unrestricted by threshold voltage limits owing to its current mode operation. Hence it is apt for DVS in the modern SoC's.

Typically, large Soc's with multiple cores and voltage domains have an area in the range of 100-350mm² [10][11]. For a reasonable level shifter usage (10,000 instances) the area overhead of SS-WVRLS over SS-TVLS would be a very small percentage (<0.01%) of the total SoC area. So, the small area increase over SS-TVLS is justified by the numerous other benefits of the architecture.

REFERENCES

[1] Kulkarni, S.H.; Sylvester, D.; "High performance level conversion for dual V/sub DD/ design," Very Large Scale Integration (VLSI) Systems, IEEE Transactions on, vol.12, no.9, pp.926-936, Sept. 2004.

[2] Ishihara, F.; Sheikh, F.; Nikolic, B.; "Level conversion for dual-supply systems," Very Large Scale Integration (VLSI) Systems, IEEE Transactions on, vol.12, no.2, pp.185-195, Feb. 2004.

[3] Tawfik, S.A.; Kursun, V.; "Multi-Vth Level Conversion Circuits for Multi-VDD Systems," Circuits and Systems, 2007. ISCAS 2007. IEEE International Symposium on, vol., no., pp.1397-1400, 27-30 May 2007.

[4] Tan, S.C.; Sun, X.W.; "Low power CMOS level shifters by bootstrapping technique," Electronics Letters, Aug 2002.

[5] Puri, R.; et. al, , "Pushing ASIC performance in a power envelope," Design Automation Conference, 2003. June 2003.

[6] Puri, R.; Pan, D.; Correale, A. Jr.; Kung, D.S; Joshi, Rajeev; "Single Supply Level Converter," IBM Research Disclosure YOR82003015.

[7] Khan, Q.A.; et. al.; "A single supply level shifter for multi-voltage systems," VLSI Design, 2006.

[8] Garg, R.; Mallarapu, G.; Khatri, S.P.; "A Single-supply True Voltage Level Shifter," Design, Automation and Test in Europe, DATE 2008.

[9] Tran, C.Q.; Kawaguchi, H.; Sakurai, T.; "Low-power high-speed level shifter design for block-level dynamic voltage scaling environment," Integrated Circuit Design and Technology, 2005. ICICDT 2005.

[10] Nawathe, U.M.; et. al., "An 8-Core 64-Thread 64b Power-Efficient SPARC SoC," Solid-State Circuits Conference, 2007. ISSCC 2007.

[11] Zongjian Chen; et. al., "A 25W SoC with Dual 2GHz Power Cores and Integrated Memory and I/O Subsystems," Solid-State Circuits Conference, 2007.

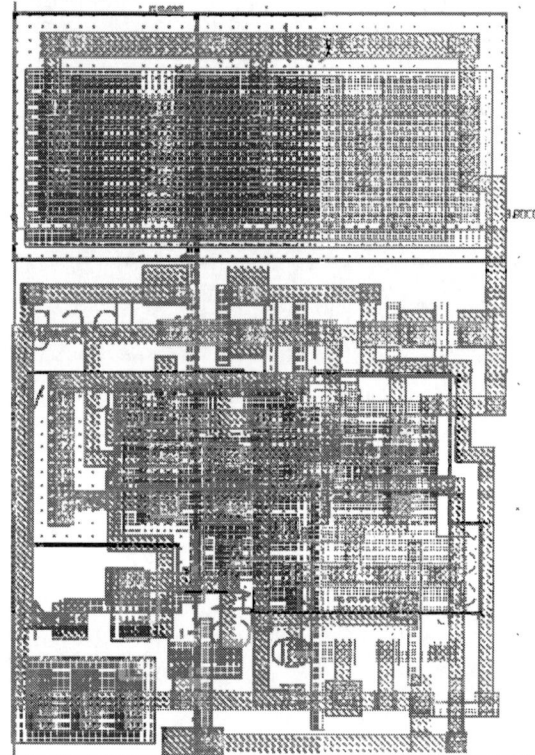

Fig. 10 Cadence layout of proposed SS-WVRLS.

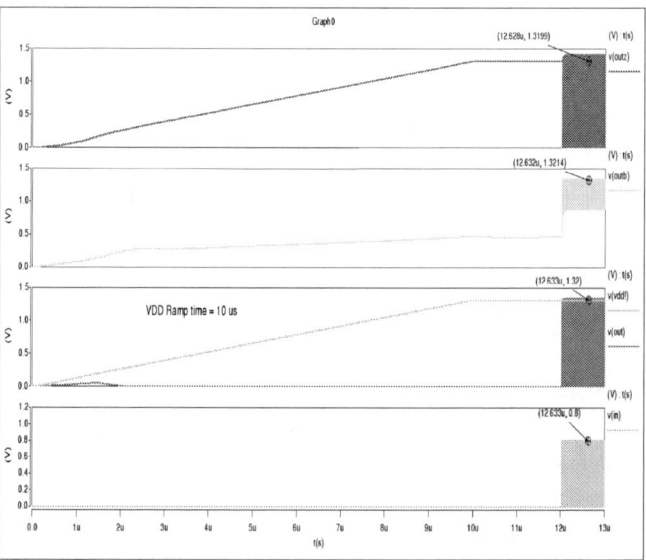

Fig. 11 Power Supply ramp-up simulation SS-WVRLS [VDDIN=0.6V, VDD=1.32, IN=0V initially, then switched: OUT follows IN.]

2012 25th International Conference on VLSI Design

Pole-Zero Analysis of Low-Dropout (LDO) Regulators: A Tutorial Overview

Annajirao Garimella, Punith R. Surkanti and Paul M. Furth
VLSI Laboratory, Klipsch School of Electrical and Computer Engineering
New Mexico State University, Las Cruces, NM 88003, USA
Email: garimella@ieee.org, punith@nmsu.edu and pfurth@nmsu.edu

Abstract—Analyzing poles and zeros of a circuit is often essential for (a) choose the appropriate topology for given specifications, (b) understanding the frequency response of the circuit and (c) stabilizing the circuit by choosing appropriate frequency compensation techniques. Analyzing poles and zeros of a low-dropout (LDO) voltage regulator is often intriguing as (a) the voltage/current control loop need to be broken for small-signal analysis and (b) the location of poles move with output load current. The objective of this tutorial is to provide a step-by-step procedure for analyzing poles and zeros in LDO regulators. To this end, two recent state-of-the-art LDO regulators from the literature are analyzed, explaining several intricacies involved. During the process, several frequency compensation techniques are elucidated.

Index Terms—Frequency compensation, pole-zero analysis, low-dropout (LDO) regulators, cascode compensation, current buffers, LHP zero, power-supply rejection (PSR).

I. INTRODUCTION

Stability at various loading conditions with improved line and load transient response is often the objective of low-dropout (LDO) voltage regulator design [1]–[9]. A priori knowledge of poles and zeros assists the designer in choosing the correct topology before implementing the transistor level design. Finding the transfer function, and thus poles and zeros, of the voltage and/or current loops of the LDO regulator often helps in developing intuition of the circuit and parameters affecting the loop gain, pole-zero locations, gain-bandwidth product, phase margin and other stability conditions [10], [11]. In this tutorial, we explain a step-by-step procedure for performing small-signal analysis and deriving the equations of poles and zeros in LDO regulators. Two recent state-of-the-art LDO regulators [9], [12] are analyzed. The outline of the paper is as follows: Section II outlines the steps for finding poles and zeros. Section III illustrates pole-zero analysis in a three-stage LDO with reverse nested Miller compensation using current buffers [9]. Sections IV and V laid a background for analyzing poles and zeros of a high PSR current-mode dual loop LDO [12] in Section VI. Conclusions are drawn in Section VII.

II. STEPS FOR FINDING POLES AND ZEROS

The steps for obtaining the DC gain, poles and zeros of a LDO regulator are
1) For the given topology, draw the small-signal model
2) Break the corresponding voltage or current loop

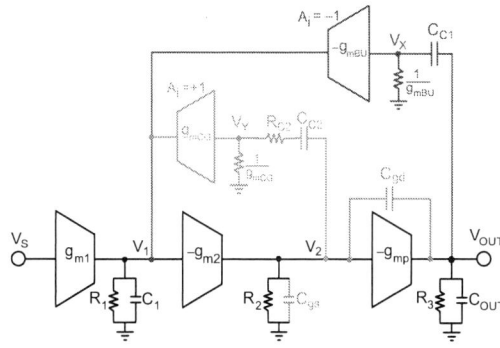

Fig. 1. Block-level architecture of LDO with reverse nested Miller Compensation using current buffers.

3) Apply Kirchhoff's Current Law (KCL) at each node
4) Solve the KCL equations using symbolic manipulation software to obtain the transfer function
5) Set $s (= j\omega) = 0$ to find the DC gain of the amplifier
6) Factor and solve the numerator to obtain zeros
7) Applying the assumption of widely-separated poles, solve the denominator to obtain the poles, OR applying the assumption of widely-separated poles with a complex pole pair, solve the denominator to obtain the poles

III. LDO WITH REVERSE NESTED MILLER COMPENSATION USING CURRENT BUFFERS

In this section, the three-stage LDO regulator of [9] is chosen as an example and detailed pole-zero analysis is performed. The block-level architecture of the LDO regulator is shown in Fig. 1 and the adapted transistor-level schematic is shown in Fig. 2. This regulator uses with reverse nested Miller compensation using current buffers (RNMCCB). RNMCCB allows cancellation of a non-dominant pole with left-half plane (LHP) zero, forming a stable LDO with adequate phase margin ($\geqslant 75^o$) with improved transient line and load response.

Referring to Fig. 2, the first stage is an NMOS folded-cascode differential amplifier (g_{m1}) formed by transistors M_1-M_9 and the second stage is an NMOS common-source amplifier (g_{m2}) formed by transistor M_{10} with diode-connected transistor load M_{11}. The third stage is the power stage (g_{mP}), formed by pass transistor M_{pass} and sampling resistors R_{f1} and R_{f2}. The output resistance of the first stage is R_1 and

978-1-4673-0438-2/12 $31.00 © 2012 IEEE   131

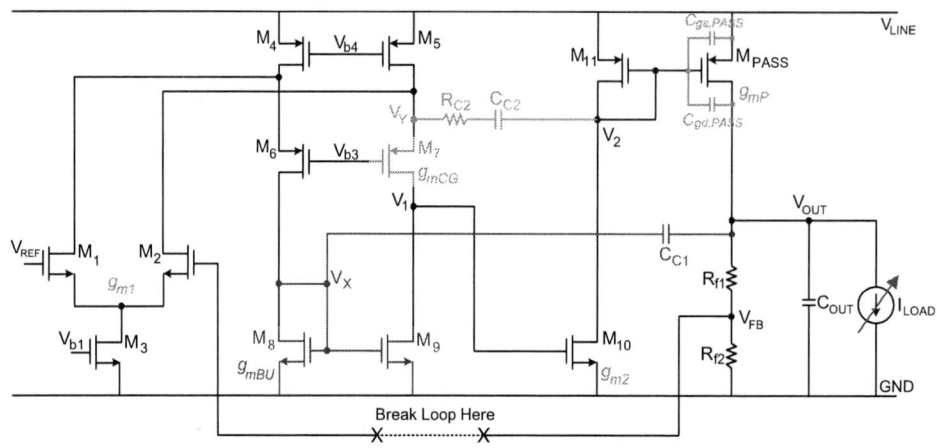

Fig. 2. Schematic of the three-stage LDO regulator with reverse nested Miller compensation using current buffers, adpated from [9]. Generation of two LHP zeros using 1) cascode compensation 2) current mirror (as inverting current buffer) compensation is highlighted.

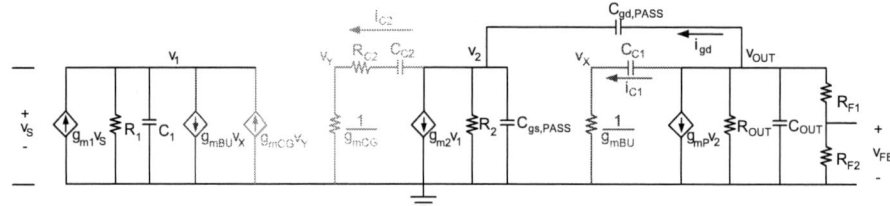

Fig. 3. Small-signal model of the LDO regulator of Fig. 2

the lumped parasitic capacitance is C_1. For the pass transistor M_{PASS}, significant parasitic capacitances $C_{gd,PASS}$ and $C_{gs,PASS}$ are considered for pole-zero analysis. The output resistance of the second stage at node V_2 is R_2 and the lumped parasitic capacitance can be approximated as $C_{gs,PASS}$. The output resistance of the power stage at node V_{OUT} is R_{OUT} and C_{OUT} is the output capacitor, either integrated or external capacitor.

A. Compensation Network

Reverse nested Miller compensation (RNMC) uses one inner loop and one outer loop [13]. Compensation capacitor C_{C1} is connected from node V_{OUT} to low-impedance input node V_X of the current mirror formed by transistors M_8-M_9 with transconductance g_{mBU}. The current mirror acts as an inverting current buffer [9], [14]–[17], denoted as g_{mBU} in Fig. 1. C_{C1} and g_{mBU} form the outer loop of the RNMC.

The inner loop is between nodes V_2 and V_1 formed by compensation capacitor C_{C2}, series resistor R_{C2} and common-gate cascode transistor M_7, whose transconductance is g_{mCG}. The compensation capacitor C_{C2} and resistor R_{C2} are connected between V_2 and V_Y, where V_Y is the source node of the common-gate transistor M_7, whose transconductance is g_{mCG} in Fig. 1. Transistor M_7 acts as a positive current buffer [9], [14], [18]–[22] and the compensation network is popularly known as cascode compensation or Ahuja compensation.

B. Small-signal Modeling

The feedback is broken between node v_{FB} and gate input of transistor M_2 as shown in Fig. 2. Let v_S be the differential

TABLE I
SMALL-SIGNAL PARAMETERS OF LDO USING RNMCCB(FIG. 3)

g_{m1}	g_m of M_1, M_2
R_1	$r_{o9}\|\|g_{mCG}r_{o7}\left(r_{o5}\|\|r_{o2}\right)$
C_1^*	$C_{gd7} + C_{gd9} + C_{gs10}$
g_{mBU}	g_m of M_8
g_{mCG}	g_m of M_7
g_{m2}, r_{o10}	g_m, r_o of M_{10}
R_2	$\frac{1}{g_{m11}}\|\|r_{o10}$
C_2^*	$C_{gd10} + C_{gs11} + C_{gs,PASS} \approx C_{gs,PASS}$
g_{mP}, r_{oP}	g_m, r_o of M_{PASS}
R_{OUT}	$r_{oP}\|\|\left(R_{f1} + R_{f2}\right)\|\|R_L$
C_3^*	$C_{gd,PASS} + C_{OUT} \approx C_{OUT}$

* Omitting bulk capacitances for simplicity

input voltage between the reference voltage v_{REF} and the feedback voltage v_{FB}. The open-loop small-signal diagram of Fig. 2 from v_S to v_{FB} is shown in Fig. 3. The first stage is non-inverting so as to ensure that the overall gain from v_{REF} to v_{OUT} is non-inverting. Equations for the small-signal parameters of the regulator of Fig. 3 are given in Table I.

The first-stage differential amplifier is represented with voltage-controlled current source (VCCS) $g_{M1}v_S$. The output of the first-stage is v_1. The second-stage is represented with VCCS $g_{M2}v_1$. The output of the third-stage is v_{OUT}. The compensation capacitor C_{C2} and resistor R_{C2} are connected between v_2 and v_Y, where v_Y is the source node of the common-gate transistor M_7, whose impedance to ground is

978-1-4673-0438-2/12 $31.00 © 2012 IEEE

$1/g_{mCG}$. The positive current buffer between v_Y and v_1 is represented with VCCS $g_{mCG}v_Y$. The compensation capacitor C_{C1} is connected between v_{OUT} and v_X, where v_X is the low impedance node with $1/g_{mBU}$ as the impedance to ground. The inverting current buffer between v_X and v_1 is represented with VCCS $g_{mBU}v_X$. $C_{gd,PASS}$ is connected between v_2 and v_{OUT}, which forms a feedforward path.

C. Applying Kirchhoff's current law (KCL)

We apply KCL at every node in Fig. 3. The set of KCL equations are

$$
\begin{aligned}
v_S &= v_{REF} - v_{FB} \\
v_1 &= \left(g_{m1}v_S - g_{mBU}v_X + g_{mCG}v_Y\right)Z_1 \\
v_2 &= \left(-g_{m2}v_1 - i_{C2} + i_{gd}\right)Z_2 \\
v_{OUT} &= \left(-g_{mP}v_2 - i_{C1} - i_{gd}\right)Z_{out} \\
v_{FB} &= \beta v_{OUT} \\
\beta &= \frac{R_{F2}}{R_{F1}+R_{F2}} \\
v_X &= \frac{i_{C1}}{g_{mBU}} \\
v_Y &= \frac{i_{C2}}{g_{mCG}} \\
Z_1 &= R_1 \| \frac{1}{sC_1} = \frac{R_1}{1+sC_1R_1} \\
Z_2 &= R_2 \| \frac{1}{sC_{gs}} = \frac{R_2}{1+sC_{gs}R_2} \\
Z_{out} &= R_{OUT} \| \frac{1}{sC_{OUT}} = \frac{R_{OUT}}{1+sC_{OUT}R_{OUT}} \\
i_{C1} &= sC_{C1}\left(v_{OUT}-v_X\right) \\
i_{C2} &= \frac{(v_2-v_Y)}{R_{C2}+\frac{1}{sC_{C2}}} = \frac{sC_{C2}(v_2-v_Y)}{1+sC_{C2}R_{C2}} \\
i_{gd} &= sC_{gd,PASS}\left(v_{OUT}-v_2\right)
\end{aligned}
\tag{1}
$$

D. Pole-Zero Equations

Solve the equations in (1) so as to eliminate v_1, v_2, v_{OUT}, v_X, v_Y, i_{C1}, i_{C2} and i_{gd} to obtain the transfer function $A_{LDO-RNMCB}(s) = v_{FB}(s)/v_S(s)$ using symbolic manipulation software. Assuming that $C_1 \ll (C_{C1}, C_{C2}$ and $C_{OUT})$, and the poles are widely separated, small-signal analysis yields the transfer function given by

$$
A_{LDO-RNMCCB}(s) \approx
$$

$$
A_{DC}\frac{\left(1+sR_{C2}C_{C2}\right)\left(1+s\frac{C_{C1}}{g_{mBU}}\right)}{\left(1+\frac{s}{\omega_{P1}}\right)\left(1+sR_{C2}C_{C2}\right)\left(1+s\frac{C_{OUT}}{g_{mBU}R_{C2}g_{mP}}\right)}
\tag{2}
$$

The open-loop dc loop gain of the regulator is given by

$$
A_{DC} = \beta g_{m1}R_1 g_{m2}R_2 g_{mP}R_{OUT},
\tag{3}
$$

where β is the feedback factor. The dominant pole ω_{P1} occurs at the output of the first stage and is given by

$$
\omega_{P1} \approx -1/R_1 C_{C1} g_{m2}R_2 g_{mP}R_{OUT}
\tag{4}
$$

The second pole ω_{P2} is due to the inner loop compensation network, given by

$$
\omega_{P2} \approx -1/R_{C2}C_{C2}
\tag{5}
$$

Two LHP zeros are formed, one due to the outer loop and one due to the inner loop, given by

$$
\omega_{Z1} = -\frac{g_{mBU}}{C_{C1}}
\tag{6}
$$

$$
\omega_{Z2} \approx -1/R_{C2}C_{C2}
\tag{7}
$$

From (5) and (7), we see that LHP zero ω_{Z2} effectively cancels the second pole ω_{P2}. The third pole ω_{P3} is a non-dominant pole situated much after ω_{GBW}. Higher order poles and right-half plane (RHP) zero due to $C_{gd,PASS}$ are neglected. The equations of poles and zeros of the LDO are given in Table II. The gain-bandwidth product is approximated as

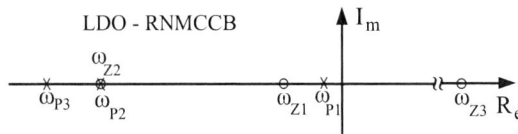

Fig. 4. Diagram illustrating pole-zero locations of Fig. 2 (not to scale).

$$
\omega_{GBW} \approx \frac{g_{m1}}{C_{C1}}
\tag{8}
$$

The Phase Margin of LDO can be expressed as

$$
\begin{aligned}
PM &= 180^o - \tan^{-1}\left(\frac{\omega_{GBW}}{\omega_{P1}}\right) + \tan^{-1}\left(\frac{\omega_{GBW}}{\omega_{Z1}}\right) \\
&\quad - \tan^{-1}\left(\frac{\omega_{GBW}}{\omega_{P3}}\right) \\
&\approx \tan^{-1}\left(\frac{g_{m1}^2 C_{OUT} + g_{mBU}^2 C_{C1} g_{mP}R_{C2}}{g_{m1}g_{mBU}C_{OUT}}\right)
\end{aligned}
\tag{9}
$$

As evident from Fig. 4, the LDO regulator's response is similar to that of a single-pole single-LHP-zero response with adequate phase margin ($\geqslant 75^o$), depending on the location ω_{Z1}.

TABLE II
POLE-ZERO EQUATIONS OF LDO WITH RNMCCB (FIG. 2)

Parameter	Equation
A_{DC}	$\beta g_{m1}R_1 g_{m2}R_2 g_{mP}R_{OUT}$
ω_{P1}	$-\frac{1}{R_1 C_{C1} g_{m2}R_2 g_{mP}R_{OUT}}$
ω_{Z1}	$-\frac{g_{mBU}}{C_{C1}}$
ω_{P2}, ω_{Z2}	$-\frac{1}{R_{C2}C_{C2}}$
ω_{P3}	$-\frac{g_{mBU}R_{C2}g_{mP}}{C_{OUT}}$
ω_{Z3}	$+\frac{g_{mP}}{C_{gd,PASS}}$

IV. SOURCE DEGENERATED DIFFERENTIAL AMPLIFIER

The schematic of a bipolar differential amplifier with source degeneration is shown in Fig. 5(a). The circuit contains degeneration resistor R_{DEG} and degeneration capacitor C_{DEG}. The source degenerated amplifier acts as a band-limited high pass filter whose high-pass corner frequency is defined by $1/(2\pi R_{DEG}C_{DEG})$.

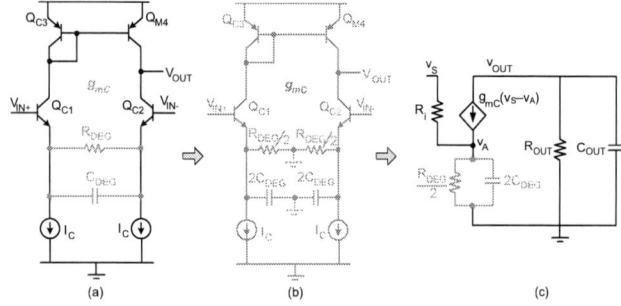

Fig. 5. (a) Schematic of the differential amplifier with source degeneration, (b) alternate representation, and (c) equivalent small-signal model.

Resistor R_{DEG} can be represented with two series resistors of value $R_{DEG}/2$ and similarly, capacitor C_{DEG} can be represented with two series capacitors of value $2C_{DEG}$ with the intermediate nodes as virtual ground as shown in Fig. 5(b). The equivalent small-signal model of the differential amplifier with source degeneration is shown in Fig. 5(c). The amplifier is modeled with a VCCS $g_{mC}(v_S - v_A)$ in series with $R_{DEG}/2$ and $2C_{DEG}$ in parallel to ground as shown in Fig. 5(c).

Let $v_S \equiv V_{IN+} - V_{IN-}$. Resistor R_{OUT} and capacitor C_{OUT} include any load attached to node V_{OUT}. Applying KCL at every node in Fig. 5(c), the set of equations are

$$0 = g_{mC}(v_S - vA) + \frac{v_{OUT}}{R_{OUT}} + sC_{OUT}v_{OUT}$$

$$0 = -g_{mC}(v_S - vA) + \frac{v_A}{0.5R_{DEG}} + 2sC_{DEG}v_A \quad (10)$$

Solving (10), we obtain the transfer function. The dc gain and the pole-zero equations of the source degenerated amplifier is given by

$$A_{DC} = \left(\frac{1}{1/g_{mC} + 0.5R_{DEG}}\right)R_{OUT} \quad (11)$$

$$\omega_{Z1} = \frac{1}{R_{DEG}C_{DEG}} \quad (12)$$

$$\omega_{P1} = \frac{1}{R_{OUT}C_{OUT}} \quad (13)$$

$$\omega_{P2} = \frac{g_{mC}}{2C_{DEG}} \quad (14)$$

Observing the poles and zeros in (12)-(14), depending on the values of R_{DEG} and C_{DEG}, zero ω_{Z1} is often designed appear first, followed by the first pole ω_{P1}, and the second pole ω_{P2}, as shown in Fig. 6, to obtain the characteristic of a band-limited high pass filter (Fig. 7).

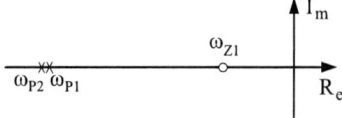

Fig. 6. Diagram illustrating pole-zero locations of Fig. 5 (not to scale).

V. PHASE-LEAD COMPENSATOR

Compensation techniques are used to achieve adequate phase margin for improving stability. Most of the compensation techniques introduce an LHP zero to increase phase margin (phase-lead). Often LDOs are compensated with external

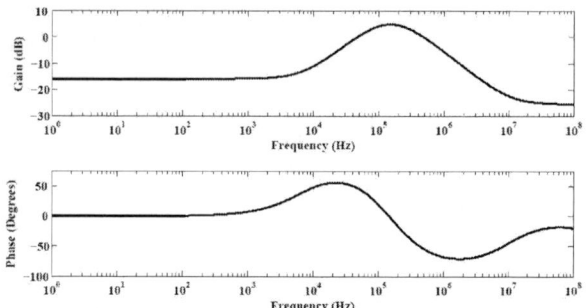

Fig. 7. Magnitude and phase plot of source degenerated differential amplifier, shown as an example.

output capacitor equivalent series resistance (ESR) LHP zero, achieving phase-lead by inserting an LHP zero just before a non-dominant pole. This technique can also be implemented at an internal node of LDO by adding a resistor R_Z in series with capacitor C_C to ground as shown in Fig. 8.

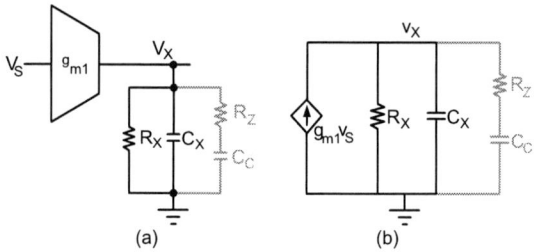

Fig. 8. (a) Phase-lead compensator, (b) equivalent small-signal model.

The small-signal model of the phase-lead compensator is shown in Fig. 8(b).

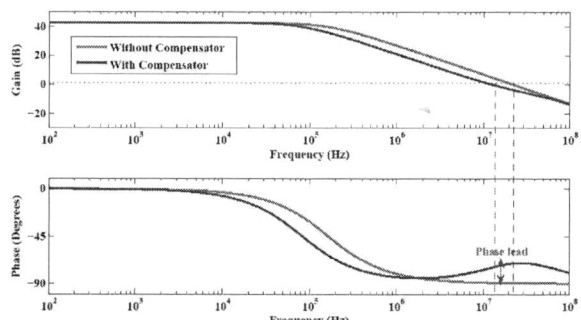

Fig. 9. Magnitude and phase response of the amplifier with and without compensator, shown as an example.

Applying KCL to the small-signal model and solving the equation, we obtain the transfer function. KCL equation is:

$$0 = -g_{m1}v_S + \frac{v_X}{R_X} + sC_Xv_X + \frac{v_X}{R_Z + 1/sC_C} \quad (15)$$

Assuming the magnitude of C_X and C_C are comparable and $R_1 \gg R_Z$, the DC gain and pole-zero equations of the

978-1-4673-0438-2/12 $31.00 © 2012 IEEE

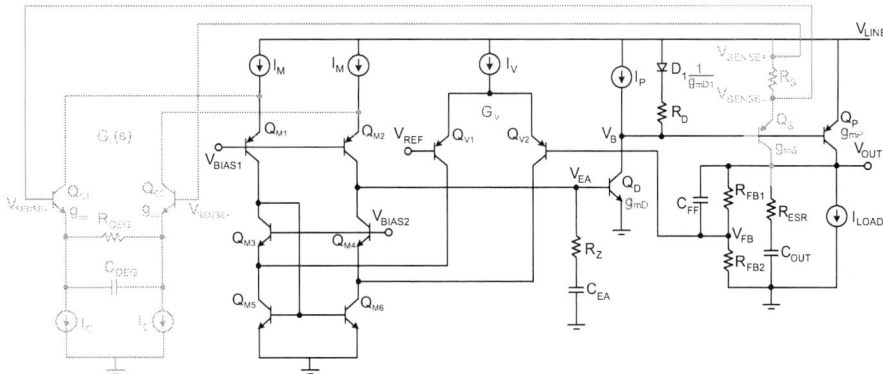

Fig. 10. Transistor-level schematic of high power-supply rejection (PSR) current-mode LDO Regulator, adapted from [12].

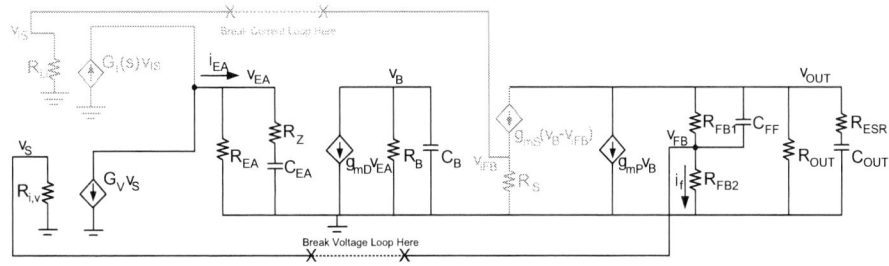

Fig. 11. Small-signal model of high PSR current-mode LDO Regulator in Fig. 10.

amplifier with phase-lead compensator are given by

$$A_{DC} = g_{m1}R_X \qquad (16)$$

$$\omega_{P1} = \frac{1}{R_X(C_X + C_C)} \qquad (17)$$

$$\omega_{P2} = \frac{1}{R_Z(C_X||C_C)} \qquad (18)$$

$$\omega_{Z1} = \frac{1}{R_Z C_C} \qquad (19)$$

where $C_X||C_C = C_X C_C/(C_X + C_C)$. Observing the expression for the non-dominant pole ω_{P2} and LHP zero ω_{Z1} in (18), we note that the zero appears approximately one octave before the non-dominant pole. This helps in achieving phase-lead as shown in Fig. 9.

VI. HIGH PSR CURRENT-MODE DUAL-LOOP LDO

Fig. 10 shows the transistor-level schematic of the bipolar current-mode dual-loop LDO of [12] with high power-supply rejection (PSR) performance. This LDO contains two feedback loops, a voltage loop and a current loop. The small-signal model and architecture of the LDO is shown in Figs. 11 and 12.

The voltage loop is formed by G_V with input pair Q_{V1} – Q_{V2}, common-emitter transistor Q_D ($-g_{mD}$), pass transistor Q_P ($-g_{mP}$) with sampling resistors R_{FB1}, R_{FB2}. The current loop is formed by sense transistor Q_S (g_{mS}) with degeneration resistor R_S, high-pass filtering transconductance $G_I(s)$ implemented with transistors $Q_{C1} - Q_{C2}$, degeneration

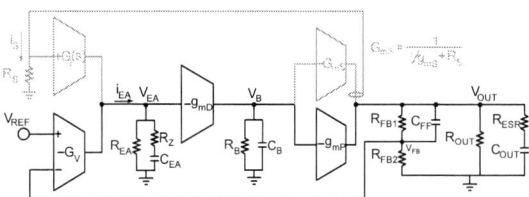

Fig. 12. Architecture of LDO with both voltage and current loops [12].

resistor R_{DEG} and capacitor C_{DEG}, and is fed into the voltage loop. $G_I(s)$ forms a high-pass filter characteristic with a band limited beyond the unity-gain-frequency of the voltage loop. The purpose of the amplifier Q_D is to decouple Q_P's large parasitic capacitance from node v_{EA}. Fig. 11 shows the small-signal model and the places to break the current and voltage loops. Equations of small-signal parameters of the LDO of Fig. 11 are given in Table III.

We apply KCL at every node in Fig. 11. The set of KCL equations obtained are

$$0 = i_{EA} + G_I(s) + G_V v_s$$

$$i_{EA} = \frac{v_{EA}}{R_{EA}} + \frac{v_{EA}}{R_Z + 1/sC_{EA}}$$

$$0 = g_{mD} v_{EA} + \frac{v_B}{R_B} + v_B s C_B \qquad (20)$$

$$0 = g_{mS}(v_B - v_{iFB}) + g_{mP} v_B + \frac{v_{OUT}}{R_{OUT}} +$$

$$\frac{v_{OUT}}{R_{ESR} + 1/sC_{OUT}} + i_f$$

$$g_{mS}(v_B - v_{iFB}) = \frac{v_{iFB}}{R_S}$$

$$i_f = \frac{v_{OUT} - v_{FB}}{R_{FB1}} + (v_{OUT} - v_{FB}) s C_{FF}$$

$$i_f = \frac{v_{FB}}{R_{FB2}}$$

978-1-4673-0438-2/12 $31.00 © 2012 IEEE   135

TABLE III
SMALL-SIGNAL PARAMETERS OF LDO IN FIG. 11

G_V	g_m of Q_{V1},Q_{V2}
$R_{i,V}$	β_{PNP}/G_V
g_{mC}	g_m of Q_{C1},Q_{C2}
$G_I(s)$	$\left(\dfrac{1}{\frac{1}{g_{mC}}+0.5R_{DEG}}\right)\dfrac{1+sR_{DEG}C_{DEG}}{1+s\frac{2C_{DEG}}{g_{mC}}}$
$R_{i,I}$	$\beta_{NPN}/G_I(s)$
g_{mD},r_{oD}	g_m, r_o of Q_D
g_{mS},r_{oS}	g_m, r_o of Q_S
G_{mS}	$\dfrac{1}{1/g_{mS}+R_S}$
g_{mP},r_{oP}	g_m, r_o of Q_P
g_{m4},r_{o4}	g_m, r_o of Q_{M4}
r_{o6}	r_o of Q_{M6}
R_{EA}	$\left(\frac{1}{2}r_{o6}g_{m4}r_{o4}\right)\|\beta_{NPN}/g_{mD}\approx\beta_{NPN}/g_{mD}$
R_B	$(1/g_{mD1}+R_D)\|r_{oD}\|\beta_{PNP}/G_{mS}\|\beta_{PNP}/g_{mP}$
C_B	$C_{beQP}+C_{beQS}+C_{bcQ_D}$
R_{OUT}	$r_{oP}\|r_{oS}\|R_L$
β_{FB}	$R_{FB2}/(R_{FB1}+R_{FB2})$

The pole-zero equations obtained are tabulated in Table IV. The zero ω_{Z1} from $G_I(s)$ precedes the pole at output ω_{P1}, as shown in Fig. 13. At higher frequencies, g_{mC} (g_m of Q_{C1}/Q_{C2}) increases and forms a pole ω_{P2} with degeneration capacitor C_{DEG}. Pole ω_{P3} is due to C_{EA} at node V_{EA} and zero ω_{Z2} is due to R_Z and C_{EA}. Zero ω_{Z4} is due to R_{ESR} and C_{OUT}.

TABLE IV
POLE-ZERO EQUATIONS OF LDO REGULATOR OF FIG. 10

Parameter	Location	Equation
A_{DC-LDO}		$\beta_{FB}G_V R_{EA}g_{mD}R_B g_{mP}R_{OUT}$
$A_{DC-G_I(s)}$		$\dfrac{g_{mC}}{1+0.5R_{DEG}g_{mC}}$
ω_{Z1}	$G_I(s)$	$-\dfrac{1}{R_{DEG}C_{DEG}}$
ω_{P1}	V_{OUT}	$-\dfrac{1}{R_{OUT}C_{OUT}}$
ω_{P2}	$G_I(s)$	$-\dfrac{g_{mC}}{2C_{DEG}}$
ω_{P3}	V_{EA}	$-\dfrac{1}{C_{EA}(R_{EA}+R_Z)}$
ω_{Z2}	V_{EA}	$-\dfrac{1}{R_Z C_{EA}}$
ω_{P4},ω_{Z3}	V_{FB}	$-\dfrac{1}{C_{FF}(R_{FB1}\|R_{FB2})}$
ω_{Z4}	V_{OUT}	$-\dfrac{1}{R_{ESR}C_{OUT}}$

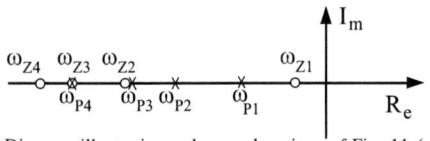

Fig. 13. Diagram illustrating pole-zero locations of Fig. 11 (not to scale).

VII. CONCLUSIONS

In this tutorial, we have analyzed poles and zeros of two state-of-the-art LDO regulators. We outlined the steps for finding poles and zeros and applied those steps to a three-stage LDO and a high PSR LDO. A wide range of compensation techniques were illustrated including RNMC using current buffers, phase-lead compensation and current-mode feedback with a high pass filtering characteristic. The derived analytic expressions for poles and zeros help in developing intuition of circuit behavior.

REFERENCES

[1] G. A. Rincon-Mora and P. E. Allen, "A low-voltage, low quiescent current, low drop-out regulator," *IEEE J. Solid-State Circuits*, vol. 33, no. 1, pp. 36–44, Jan. 1998.

[2] ——, "Optimized frequency-shaping circuit topologies for LDOs," *IEEE Trans. Circuits Syst. II: Analog and Digital Signal Processing*, vol. 45, no. 6, pp. 70–708, June 1998.

[3] K. N. Leung and P. K. T. Mok, "A capacitor-free CMOS low-dropout regulator with damping-factor-control frequency compensation," *IEEE J. Solid-State Circuits*, vol. 38, no. 10, pp. 1691–1702, Oct. 2003.

[4] P. Hazucha, T. Karnik, B. A. Bloechel, C. Parsons, D. Finan, and S. Borkar, "Area-efficient linear regulator with ultra-fast load regulation," *IEEE J. Solid-State Circuits*, vol. 40, no. 4, pp. 933–940, April 2005.

[5] V. Gupta and G. A. Rincon-Mora, "A 5mA 0.6μm CMOS Miller-compensated ldo regulator with -27db worst-case power-supply rejection using 60pF of on-chip capacitance," in *Proc. IEEE International Solid-State Circuits Conference, 2007. ISSCC 2007.*, Feb. 2007, pp. 520–521.

[6] S. K. Lau, P. K. T. Mok, and K. N. Leung, "A low-dropout regulator for SoC with Q-reduction," *IEEE J. Solid-State Circuits*, vol. 42, no. 3, pp. 658–664, March 2007.

[7] M. Al-Shyoukh, H. Lee, and R. Perez, "A transient-enhanced low-quiescent current low-dropout regulator with buffer impedance attenuation," *IEEE J. Solid-State Circuits*, vol. 42, no. 8, pp. 1732–1742, Aug. 2007.

[8] R. J. Milliken, J. Silva-Martinez, and E. Sanchez-Sinencio, "Full on-chip CMOS low-dropout voltage regulator," *IEEE Trans. Circuits Syst. I: Regular Papers*, vol. 54, no. 9, pp. 1879–1890, Sept. 2007.

[9] A. Garimella, M. W. Rashid, and P. M. Furth, "Reverse nested Miller compensation using current buffers in a three-stage LDO," *IEEE Trans. Circuits Syst. II*, vol. 57, pp. 250–254, April 2010.

[10] P. R. Surkanti, A. Garimella, and P. M. Furth, "Pole-zero analysis of multi-stage amplifiers: A tutorial overview," in *IEEE 54th International Midwest Symposium on Circuits and Systems (MWSCAS), 2011*, Aug. 2011, pp. 1–4.

[11] A. Garimella and P. M. Furth, "Frequency compensation techniques for op-amps and LDOs: A tutorial overview," in *IEEE 54th International Midwest Symposium on Circuits and Systems (MWSCAS), 2011*, Aug. 2011, pp. 1–4.

[12] A. P. Patel and G. A. Rincon-Mora, "High power-supply-rejection (PSR) current-mode low-dropout (LDO) regulator," *IEEE Trans. Circuits Syst. II: Express Briefs*, vol. 57, no. 11, pp. 868–873, Nov. 2010.

[13] A. D. Grasso, G. Palumbo, and S. Pennisi, "Advances in reversed nested Miller compensation," *IEEE Trans. Circuits Syst. I: Reg. Papers*, vol. 54, p. 1459, July 2007.

[14] P. J. Hurst, S. H. Lewis, J. P. Keane, F. Aram, and K. C. Dyer, "Miller compensation using current buffers in fully differential CMOS two-stage operational amplifiers," *IEEE Trans. Circuits Syst. I: Reg. Papers*, vol. 51, no. 2, pp. 275–285, Feb. 2004.

[15] G. A. Rincon-Mora, "Active capacitor multiplier in Miller-compensated circuits," *IEEE J. Solid-State Circuits*, vol. 35, no. 1, pp. 26–32, Jan. 2000.

[16] ——, *Analog IC Design with Low-Dropout Regulators*. McGraw-Hill Professional, 2009, United States., 2009.

[17] A. Garimella, M. W. Rashid, and P. M. Furth, "Single-Miller compensation using inverting current buffer for multi-stage amplifiers," in *Proc. IEEE International Symposium on Circuits and Systems, ISCAS 2010*, May 2010, pp. 1579–1582.

[18] P. R. Gray and R. G. Meyer, "MOS operational amplifier design-a tutorial overview," *IEEE J. Solid-State Circuits*, vol. 17, no. 6, pp. 969–982, Dec. 1982.

[19] B. K. Ahuja, "An improved frequency compensation technique for CMOS operational amplifiers," *IEEE J. Solid-State Circuits*, vol. 18, no. 6, pp. 629–633, Dec. 1983.

[20] R. J. Reay and G. T. A. Kovacs, "An unconditionally stable two-stage CMOS amplifier," *IEEE J. Solid-State Circuits*, vol. 30, no. 5, pp. 591–594, May 1995.

[21] D. B. Ribner and M. A. Copeland, "Design techniques for cascoded CMOS op amps with improved PSRR and common-mode input range," *IEEE J. Solid-State Circuits*, vol. 19, no. 6, pp. 919–925, Dec. 1984.

[22] G. Palumbo and S. Pennisi, *Feedback Amplifiers: Theory and Design*. Kluwer Academic Publishers: Boston, 2002.

978-1-4673-0438-2/12 $31.00 © 2012 IEEE

2012 25th International Conference on VLSI Design

3-D Parasitic Modeling for Rotary Interconnects

Vinayak Honkote, Ankit More and Baris Taskin

Department of ECE, Drexel University, Philadelphia, PA, USA, 19104

vh32,am434@drexel.edu, taskin@coe.drexel.edu

Abstract—Resonant rotary clocking is a high-frequency, low-power technology for high performance integrated circuits (IC). The implementation of the rotary clocking technology requires long interconnects with varying geometric shape segments on the chip, which are modeled by transmission lines. The parasitics exhibited by the transmission line interconnects play a major role in characterizing the high frequency operation. To this end, the impact of parasitics on the operating characteristics of the rotary rings due to the different interconnect segments are identified. The interconnect parasitics are analyzed using a 3D finite element method based full wave electromagnetic analysis. Simulations performed for the rotary ring with 3D full wave based parasitic analysis results in 23.68% reduced clock frequency when compared with a conventional 2D based parasitic analysis. The power dissipated on the rotary ring simulated using the 3D full wave based parasitic analysis is around 84% less than the clock tree and is within 5% of the power dissipated on the ring simulated using the 2D based parasitic analysis.

I. INTRODUCTION

Advances in deep-submicron (DSM) circuit design have led to increased performance gains in modern digital VLSI circuits. With the modern day multi-core and many-core processor architectures the design objective has shifted from high frequency implementations to low power cores partially at the expense of operating speed. Nevertheless, the operating frequency of each of these multi and many core processors is in GHz range. These GHz range frequency requirements and low power budgets of the DSM circuits make the task of clock signal distribution quite challenging. The prevailing methodology to generate high-frequency clock signals is to use on-chip frequency multiplication with phase-locked loop (PLL) components. The on-chip PLL components occupy chip area and lead to problems with capacitive loading and power dissipation [1]. To this end, adiabatic switching offers an appealing alternative by circulating the used energy back in the circuit [2]. Resonant clocking technologies of coupled LC [3, 4], two phase clocking [5], standing wave [6] and rotary clocking [7] have been presented in literature. Out of these technologies, resonant *rotary* clocking technology provides constant magnitude, varying phase clock signals. The rotary clocking oscillators, demonstrated in an array implementation of transmission line mobius rings in Fig. 1, rely on the wave traveling principle of transmission lines to generate high frequency clock signals. A rotary traveling wave at 18 *GHz* frequency is implemented in [8], and up to 70% power savings are reported in [9].

Rotary clocking is typically implemented with a grid of regular rings [7]. In [10], a grid of non-regular rotary rings

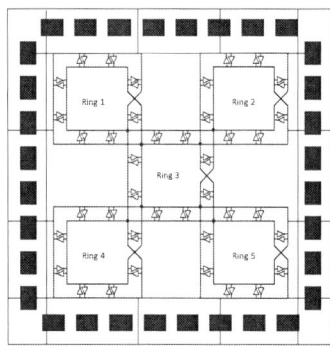

Fig. 1. A Rotary clock architecture with 5 rings.

is proposed for minimizing the total wirelength. The implementation of the rotary clocking technology with both regular rings [Fig. 1] and non-regular rings [10] requires long interconnects on the chip, which are modeled by transmission lines. These transmission lines constructing the regular or non-regular rings have varying geometric shape segments, such as, the corner segment or a cross-over segment. There is a very limited amount of work analyzing the transmission line parasitics (inductance and capacitance) associated with the segments of the high frequency rotary rings [7, 9, 11]. Previously in [9], the rotary ring structures are analyzed in detail with the objective of power minimization. Alternative SPICE models for rotary clocking have been proposed in literature in [7, 9, 12]. In [7], SPICE simulations are performed with lumped *RLC* models, which are inaccurate for transmission lines with a high frequency of operation. In [9], partial element equivalent circuits (PEEC) based closed form expressions are used for simplified simulation models. In [12], a U-element based SPICE model is proposed. A U-element [13] model uses a 2D solver to capture the lossy transmission line behavior. However, these methods are incomplete as they do not capture the parasitics due to the different geometries (corners and crossovers) in the rotary rings. Further, the accurate characterization of on-chip rotary interconnects requires 3D full wave electromagnetic analysis. Towards this end, a 3D finite element based full wave electromagnetic analysis is presented for the characterization of different transmission line segments constituting a rotary ring. The rotary ring is modeled in SPICE incorporating the parasitics extracted from the 3D electromagnetic analysis and is compared with the U-model and the PEEC models. Further, the power dissipated on the rotary ring is analyzed using the SPICE simulations.

978-1-4673-0438-2/12 $31.00 © 2012 IEEE

The rest of the paper is organized as follows. In Section II, the rotary clocking technology is reviewed. In Section III, interconnect modeling is presented. In Section IV, parasitic analysis for the rotary ring structures is presented. In Section V, the experimental results are presented. In Section VI, the work is summarized.

II. ROTARY CLOCKING TECHNOLOGY

Rotary clocking technology is implemented with a grid based structure of regular or custom rings [7, 10]. An oscillation on the rotary ring can start spontaneously upon any noise event or stimulated by a start up circuit for controlled operation [7]. When the oscillation is established, a square wave signal can travel along differential transmission lines of the ring without termination. Oscillations on the rings are locked in phase on the rotary oscillatory array, minimizing the effects of jitter. The anti-parallel inverter pairs between the interconnects [as shown in Fig. 1] serve to sustain the signal propagation on the transmission lines and aid in charge recovery process. Such rotary oscillator generated square waves present low jitter, controllable skew properties.

Since the traveling wave requires two rotations to complete a clock period, the rotary oscillation frequency is approximated as:

$$f = \frac{v_p}{2l}, \tag{1}$$

where v_p is the phase velocity of the wave and l is the length of the rotary ring [7]. v_p is calculated using the per-unit-length differential inductance L_l and capacitance C_l as:

$$v_p = \frac{1}{\sqrt{L_l C_l}}. \tag{2}$$

The per-unit-length differential inductance L_l is estimated as [7]:

$$L_l = \frac{\mu_0}{\pi} \log\left[\left(\frac{\pi s}{w+t}\right) + 1\right], \tag{3}$$

where s, w, t, and μ_0 are the separation between the wires, width of the wire, thickness of the wire, and permeability in vacuum, respectively. The per-unit-length capacitance C_l is given by [7]:

$$C_l = C_{inv_l} + C_{wire_l} + C_{tra_l}, \tag{4}$$

where C_{inv_l}, C_{wire_l}, and C_{tra_l} are the per-unit capacitance by the inverters, tapping wires and the transmission lines, respectively. The wires used in rotary clocking technology are wide enough such that the wire resistance is negligible.

III. MODELING INTERCONNECT PARASITICS

The frequency of each rotary ring is estimated using (1) and (2) as:

$$f_{osc} = \frac{1}{2l\sqrt{L_l C_l}}. \tag{5}$$

The oscillation frequency f_{osc} depends on the circuit parasitics in (3) and (4). Hence, to characterize the rotary operation, an accurate modeling of interconnect parasitics is necessary.

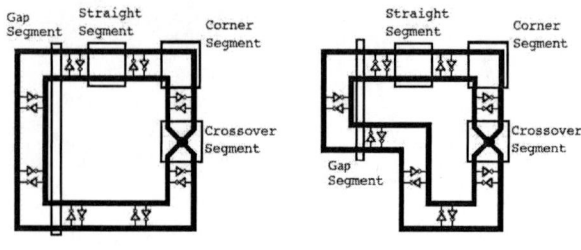

(a) Segments on a regular ring   (b) Segments on a custom ring

Fig. 2. Segments on the regular and custom rings.

A 3D full wave electromagnetic based analysis is the most accurate way of modeling the transmission line parasitics. However, they are computationally intensive and time consuming. Especially, for the array structure of rotary rings, full wave electromagnetic analysis is understandingly computationally expensive and time consuming. Hence, to speed up this computation, simple sub-structures of transmission line segments for the rotary oscillatory array are examined.

Consider the rotary rings shown in Fig. 2. In order to accurately analyze the parasitics, the interconnects forming the rotary rings are partitioned into straight, corner, crossover, and gap segments, which are categorized in Sections III-A, III-B, III-C, and III-D, respectively.

A. Straight Segments

Consider the straight segment on a rotary ring topology shown in Fig. 2. The magnified view of the straight segment is shown in Fig. 3(a). Each straight segment is composed of length l_{seg}, width w, and thickness t. The separation between the transmission lines is s. Straight segments are the most abundant geometric shapes in a regular or custom topology rotary clock network.

B. Corner Segments

Consider a corner segment on the rotary ring shown in Fig. 3(b). Due to the cross-connected arrangement (mobius topology of differential transmission lines) for rotary rings, each corner on the rotary ring has the outer transmission line and the inner transmission line. The length of the transmission line at the corner segment is composed of l_{seg} and l_{add}. Note that, l_{add} is the additional transmission line contributed by the corner segment as shown in Fig. 3(b). The number of corner segments is at minimum four and does depend on the custom topology in a non-regular ring.

C. Crossover Segments

The traveling wave in rotary clocking is not terminated due to the mobius crossing on the rotary ring. Consider a crossover segment on the rotary ring shown in Fig. 3(c). The length of the transmission line at the crossover segment is composed of l_{seg} and l_{add}. In an IC implementation, the crossover segment is fabricated on two metal layers to avoid a short circuit. Typically, each rotary ring has a unique crossover segment, although, higher number of crossover segments are possible.

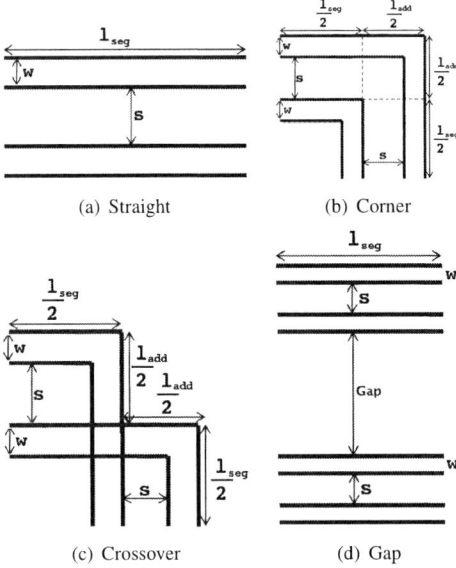

(a) Straight   (b) Corner

(c) Crossover   (d) Gap

Fig. 3. Different types of segments on a rotary ring.

D. Gap Segments

The distance between the opposite edges in a rotary ring, marked on Fig. 3(d), is termed a "gap". In particular, a custom ring can have multiple edges, and every opposite edge pair contributes towards the additional parasitics. In order to investigate the level of this contribution, the mutual inductance between the opposite pairs (i.e. gap) is analyzed. The mutual inductance computation for the gap segment is similar to the case of mutual inductance in a straight segment. The major difference is that the separation between the transmission lines is the length of an edge of the custom ring as opposed to the separation s in Fig. 3(a). It is projected that due to the distance between the transmission lines, the mutual inductance contributed by the gap is controllable. Based on this projection a minimum gap is devised for a rotary ring, either to eliminate the mutual inductance by keeping the gap long enough or by analyzing for the existing mutual inductance due to the gap. The gap segments exist in all rotary topologies. For the regular ring topology, the length of gap segment is $\frac{2l}{4}$. However, for the non-regular ring topology, the length of the gap segment depends on the custom topology.

IV. Parasitic Analysis

The equivalent circuit for the straight segment (Section III-A) can be modeled by the U-model in SPICE. However the corner segments (Section III-B) and the crossover segments (Section III-C) can not be modeled by the U-model in SPICE. The 90° bend at the corner segment causes reflection in the traveling waves due to which the wave velocity v_p is not uniform at the corner segments. Note that, the transmission line U-models in SPICE do not take into consideration the additional corner parasitics. Inductance estimation especially is a challenging problem as the wire (transmission line) inductances are defined over current loops, and the current

loops are dependent on the circuit context of the switching wires. The PEEC model [14] has been developed to solve this problem and this does not require the current return paths to be predetermined. The PEEC approach introduces the concept of partial inductance of a wire segment, corresponding to a return path at infinity. The partial self inductance is defined as the inductance of a wire segment that is in its own magnetic field, while the partial mutual inductance is defined between the two wire segments, each of which is in the magnetic field produced by the current through the other. For two wire segments k and m, the partial mutual inductance M_{km} is given by:

$$
\begin{aligned}
M_{km} &= \frac{1}{I_m a_k}\left(\int_{a_k}\int_{l_k}\vec{A}_{km}\bullet d\vec{l}_k da_k\right) \\
&= \frac{\mu_0}{4\pi a_k a_m}\int_{a_k}\int_{a_m}\int_{l_k}\int_{l_m}\frac{d\vec{l}_k\bullet d\vec{l}_m}{r_{km}}da_k da_m,
\end{aligned}
\tag{6}
$$

where l_i and $a_i(i=k\ or\ m)$ are the length and the cross section area of a wire segment i. r_{km} is the distance between any two points on the segments k and m. $\vec{A}_{km}$ is the magnetic vector potential along the segment k due to the current I_m in the segment m, given by:

$$
\vec{A}_{km} = \frac{\mu_0}{4\pi a_m}\left(\int_{a_m}\int_{l_m}\frac{I_m}{r_{km}}d\vec{l}_m da_m\right).
\tag{7}
$$

The closed form PEEC equations based on (6) and (7) are used to capture the corner parasitic effects at high frequencies [15, 16]. The corner mutual inductance between the two transmission lines is computed using [17]:

$$
\begin{aligned}
M = &\left[\frac{1}{2}(L_{s+t}+L_{s-t})-L_s\right]\cdot\left(\frac{s}{t}\right)^2 + (L_{s+t}-L_{s-t})\cdot\left(\frac{s}{t}\right) \\
&+\frac{1}{2}(L_{s+t}+L_{s-t}).
\end{aligned}
\tag{8}
$$

The self inductances L on the RHS of (8) is calculated by:

$$
\frac{L}{l_{seg}} = \frac{\mu}{2\pi}\left[\ln\frac{2l_{seg}}{0.2235\cdot(w+t)}-1\right],
\tag{9}
$$

where s, t, l_{seg}, w represent the separation, thickness, length and width of the transmission line segments, respectively, and $\mu = 4\pi$ nH/cm is the permeability in free space. Note that, the subscripts in (8) denote the thickness of the segment, whose width is w and length is l. The additional capacitance due to the corner is estimated using [18, 19]:

$$
C_{corner} = 0.5\times C_l\times w,
\tag{10}
$$

where C_l is the capacitance per unit length of the transmission line.

The crossover segment involves the interconnects crossing over multiple metal layers. The accurate modeling of the crossover requires the analysis of the electric and magnetic coupling due to the multiple metal layers and the substrate parasitics.

For the gap segment analysis (Section III-D), the opposite edges of the rotary ring are divided into regular segments. The mutual inductance between the two straight segments constituting the gap is computed using (8). The separation s is the gap in this case.

978-1-4673-0438-2/12 $31.00 © 2012 IEEE

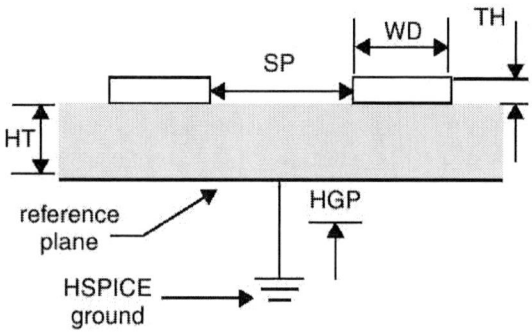

(a) 2D analysis topology (Spice U-model)

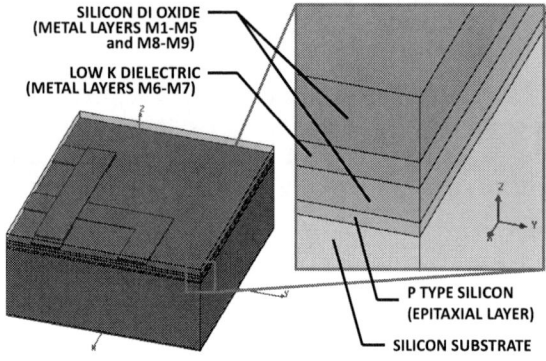

(b) Process based topology (90nm process)

Fig. 4.   Different types of interconnect modeling topologies.

TABLE I
DIFFERENT INTERCONNECT SEGMENT MODELING TOPOLOGIES AND
ANALYSIS METHODS.

		SPICE U-model (2D)	PEEC	HFSS 3D
2D analysis topology	Straight	✓	✓	✓
	Corner	×	✓	✓
	Crossover	×	×	×
Process based topology	Straight	×	×	✓
	Corner	×	×	✓
	Crossover	×	×	✓

the height of the dielectric. This topology lacks the multi-metal layers which are typically present in all IC modeling. Also, the 2D analysis topology does not model the lossy substrate and a high conductivity epitaxial layer present in most semiconductor processes. In order to model the environment of operation for an electromagnetic analysis, it is necessary to model the process topology. The electromagnetic analysis is performed on the 90nm low power process based topology. HFSS solver is used to perform the 3-D full wave electromagnetic analysis [20]. In Fig. 4(b), the basic structure used to compute the parasitics is illustrated.

In Table. I, the extraction methods for the straight segment, corner segment and the crossover segments are tabulated. The straight segments can be extracted using the 2D analysis topology [based on Fig. 4(a)], with a 2D solver (using U-model in SPICE), PEEC and HFSS 3D (3D full wave electromagnetic analysis tool). However, the corner segments can be extracted either using PEEC or using HFSS 3D modeling. SPICE U-model does not account for the corner parasitics. Also, the crossover segments cannot be characterized in the 2D analysis topology as the different metal layers and the substrate effects are absent in this topology.

SPICE circuit for the rotary ring is constructed with the parasitics extracted using the straight segments without incorporating the corner effects. The clock waveforms obtained are shown in Fig. 5(a). The oscillation frequency is 4.35GHz. Next, the corner parasitics are incorporated with the straight segment parasitics for the rotary ring circuit in SPICE. The clock waveforms obtained are shown in Fig. 5(b). The oscillation frequency in this case is 4.33GHz. Note that, there is a slight decrease in the frequency when the additional corner parasitics are included in simulation. For a ring with increased corners [10], the oscillation frequency further reduces due to the added parasitics of the corner segments.

Next, the process based topology is used to characterize the parasitics for different rotary segments. The HFSS 3D full wave electromagnetic analysis is used to model the straight, corner, crossover, and the segments. Note that, for the "gap" segment analysis in Section III-D, the opposite edges of the rotary ring are divided into regular segments. In Fig. 6, the plot shows the decrease in mutual inductance with increase in gap for the rotary ring methodology. With a segment size of 1000μm, if the minimum gap is approximately 70% of the segment size, the mutual inductance becomes negligible. For the custom ring implementations in [10], the gap has to be fixed as greater than 70% of the minimum length of the

V. EXPERIMENTAL RESULTS

The rotary ring is implemented on the 90nm CMOS IC process with the BSIMv4 transistor model. The perimeter of the rotary ring (3200μ) is fixed based on the desired oscillation frequency f_{osc} (4.5GHz). The different segments of the rotary ring corresponding to Sections III-A, III-B, and III-C are analyzed and included in the modified circuit models for rotary ring simulation. The simulation results are presented in V-A. The power analysis on the rotary ring is presented in Section V-C. Further, the effects of parasitics on the rotary oscillation frequency and the phase velocity are discussed in Section V-B.

A. Results for Interconnect Segment Modeling

Based on the interconnect segments, the corresponding parasitics can be extracted by using the 2D analysis topology used in the previous works (e.g. topology used by U-model in SPICE) and the multi-layered process based topology (e.g. 90nm process topology). The process based topology more accurately models a typical IC and the environment of operation for the rotary ring. In Fig. 4(a), the basic structure used to compute the parasitics using the 2D analysis topology used by the U-model in SPICE is illustrated. The parameters SP, WD, TH, correspond to the separation, width and thickness of the transmission lines, respectively. HT is

978-1-4673-0438-2/12 $31.00 © 2012 IEEE   140

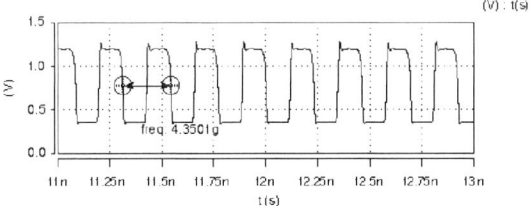

(a) Clock signal simulated for a rotary ring with no corner parasitics.

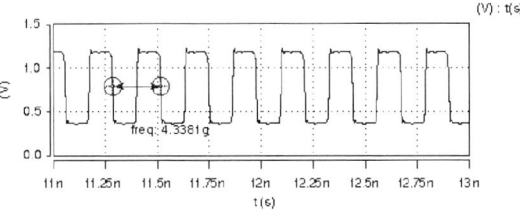

(b) Clock signal simulated for a rotary ring with additional corner parasitics.

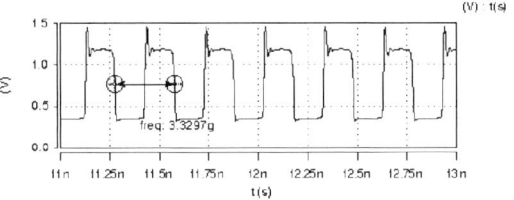

(c) Clock signal simulated for a rotary ring with additional corner parasitics and the cross over parasitics.

Fig. 5. Simulated clock waveforms.

ring edge to eliminate the "gap" effect. In a regular ring of the conventional ROA, the gap is $\frac{2l}{4}$ long. This gap is long enough so that the mutual inductance contributed by the gap can be safely neglected, which has been the norm. The rotary circuit is rebuilt in SPICE with the parasitic analysis using HFSS. The clock waveforms obtained are shown in Fig. 5(c). The observed oscillation frequency is $3.32GHz$.

B. Discussion on Oscillation Frequency and Phase Velocity

Let the simulated oscillation frequency using 2D modeling of parasitics and 3D modeling of the parasitics be f_{2D} and f_{3D}, respectively. From the simulation results shown in Fig. 5, for a design frequency of $f_{osc} = 4.5GHz$, the resulted frequencies f_{2D} and f_{3D} are $4.35GHz$ and $3.32GHz$, respectively. Note that, with the addition of corner and crossover parasitics the oscillation frequency is reduced by 23.68%. This drop in frequency can be attributed to the non-uniform velocity of the traveling wave due to the corner and crossover segments. The frequency in (1) is rewritten as:

$$f^{new} = \alpha \frac{v_p^{straight}}{2l}, \qquad (11)$$

where α is the compensation factor due to the corner and crossover parasitics and $v_p^{straight}$ is the phase velocity of the wave on a straight transmission line. In general, the compensation factor α is the slowdown of the propagation velocity

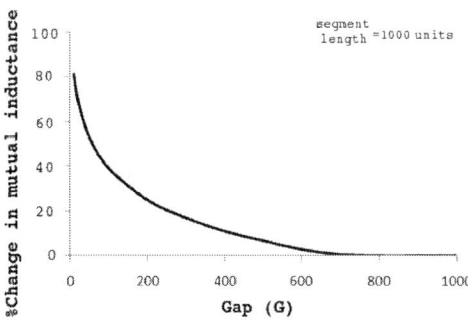

Fig. 6. Mutual inductance for varying "gap". 1unit=$1\mu m$.

due to parasitics and can be empirically estimated by:

$$\alpha = \frac{f_{3D}}{f_{2D}}. \qquad (12)$$

In the simulated rotary ring with 4 corners and a crossover, the compensation factor α is 0.76.

C. Power Analysis

One of the main characteristics of the rotary oscillators is the charge recovery property. The rotary oscillators store the energy in the inductors during the discharging stage so that the stored energy can be re-circulated during the charging stage– thus minimizing the dynamic power consumption. Hence, the power dissipation in the rotary oscillators is mainly the static power due to the resistance of the transmission line interconnects. The overall power dissipation with the rotary oscillators can be estimated:

$$P_{total} = P_{ring} + P_{wire}, \qquad (13)$$

where P_{ring} and P_{wire} are the power dissipated on the rotary ring and the power dissipation due to the capacitive loads exhibited by the tapping wires, respectively. P_{ring} is estimated as:

$$P_{ring} = P_{tra} + P_{inv}, \qquad (14)$$

where P_{tra} and P_{inv} are the power dissipated due to the transmission line parasitics and the inverter pairs, respectively. The static power (P_{tra}) dissipated due to the transmission line interconnects is further expressed as:

$$P_{tra} = \frac{V_{DD}^2}{Z_0^2} R_l, \qquad (15)$$

where V_{DD} is the power supply voltage, R_l is the total resistance of the rotary ring interconnects, and Z_0 is the transmission line impedance. Z_0 is approximated as:

$$Z_0 = \sqrt{\frac{L_l}{C_l}}. \qquad (16)$$

The power dissipated is measured using SPICE simulations. Rotary ring is simulated in SPICE using the U-models incorporating the parasitics of the segments characterized in Section III. The power dissipation is tabulated in Table II.

First, the SPICE circuit for rotary ring is constructed with the parasitics extracted using the straight segments without

TABLE II
POWER DISSIPATION ON THE RING WITH DIFFERENT SEGMENTS.

Type of segments in the ring	Power
Only straight	0.248w
Straight and Corner (4)	0.251w
Straight, Corner (4) and Cross-Over (1)	0.260w

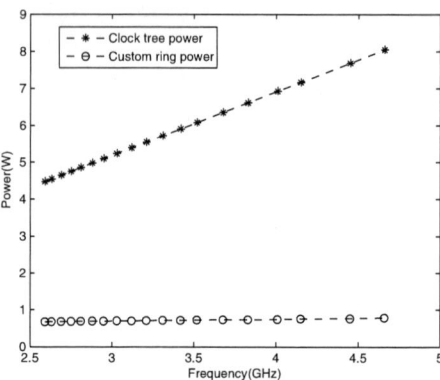

Fig. 7.   Comparison of power dissipated on rotary-rings with a clock tree.

incorporating the corner effects. This is the current state of research in [9, 12]. The power dissipation is 0.248w.

Second, the corner parasitics are incorporated with the straight segment parasitics for the rotary ring circuit in SPICE. The power dissipation is 0.251w. Note that, the additional 0.003w power dissipation in this case is due to the additional parasitics of the corner segments.

Third, the rotary ring is simulated using the straight (Section III-A), corner (Section III-B) and the crossover segments (Section III-C) characterized using the process based topology (90nm process). This is the most accurate characterization of the rotary ring, because, it includes the regular segments, corner segments (4 in the ring) and the crossover segment. The power dissipation observed is 0.260w. Note that, the additional 0.012w in this case compared to the case with straight segment and corner segments is due to the additional parasitics of the crossover segments. Thus, the proposed model with the increased accuracy—obtained through the proposed scalable application of 3D full wave electromagnetic simulations—leads to a 4.84% increase in the power dissipation projected for a 3200μm rotary ring operating at 3.32 GHz in a 90nm technology.

Finally, the power dissipation on the rotary rings is compared with the power dissipated on a clock tree network. A method similar to the one in [9] is adopted for power comparison. In Fig. 7, the power dissipated at varying frequencies on rotary rings and clock trees is shown. On an average, around 84% power saving is observed with the rotary rings when compared with the clock trees.

VI. CONCLUSION

In this paper, the different segments constituting the rotary rings are identified. A 3D finite element based electromagnetic analysis is adopted for accurately characterizing the additional parasitics contributed by the corners and crossovers in a rotary ring. The simulations demonstrate the accuracy of the 3D full wave based parasitic analysis which results in 23.68% reduction in the observed oscillation frequency when compared with the parasitics analyzed using the 2D based methodology. Thus, the proposed 3D based methodology is critical for timing. Further, the power dissipated on the rotary ring using the 3D full wave based parasitic modeling is analyzed. The power dissipated on the rotary rings is around 84% less than the clock tree power and is within 5% of the power dissipated in the 2D based methodology.

REFERENCES

[1] H. G. Chyun and J. Hung, "Phase-locked loop techniques. a survey," *IEEE Transactions on Industrial Electronics*, vol. 43, no. 6, pp. 609–615, Dec. 1996.

[2] V. L. Chi, "Salphasic distribution of clock signals for synchronous systems," *IEEE Transactions on Computers*, vol. 43, no. 5, pp. 597–602, May 1994.

[3] S. C. Chan, P. J. Restle, N. K. James, and R. L. Franch, "A 4.6 GHz resonant global clock distribution network," in *Proceedings of the IEEE International Solid State Circuits Conference*, Feb. 2004, pp. 341–343.

[4] A. Drake, K. Nowka, T. Nguyen, J. Burns, and R. Brown, "Resonant clocking using distributed parasitic capacitance," *IEEE Journal of Solid-State Circuits*, vol. 39, no. 9, pp. 1520–1528, Sept. 2004.

[5] J.-Y. Chueh, M. C. Papaefthymiou, and C. H. Ziesler, "Two-phase resonant clock distribution," in *Proceedings of the IEEE Computer Society Annual Symposium on VLSI*, May 2005, pp. 65–70.

[6] F. O'Mahony, C. Yue, M. Horowitz, and S. Wong, "A 10-GHz global clock distribution using coupled standing-wave oscillators," *IEEE Journal of Solid-State Circuits*, vol. 38, no. 11, pp. 1813–1820, Nov. 2003.

[7] J. Wood, T. Edwards, and S. Lipa, "Rotary traveling-wave oscillator arrays: A new clock technology," *IEEE Journal of Solid-State Circuits*, vol. 36, no. 11, pp. 1654–1665, Nov. 2001.

[8] G. D. Mercey, "A 18GHz rotary traveling wave VCO in CMOS with I/Q outputs," in *Proceedings of the European Solid-State Circuits Conference (ESSCIRC)*, Sept. 2003, pp. 489–492.

[9] Z. Yu and X. Liu, "Low-power rotary clock array design," *IEEE Transactions on Very Large Scale Integration (VLSI) Systems*, vol. 15, no. 1, pp. 5–12, Jan. 2007.

[10] V. Honkote and B. Taskin, "Custom rotary clock router," in *Proceedings of IEEE International Conference on Computer Design (ICCD)*, Oct. 2008, pp. 114–119.

[11] C. Zhuo, H. Zhang, R. Samanta, J. Hu, and K. Chen, "Modeling, optimization and control of rotary traveling-wave oscillator," in *Proceedings of the IEEE/ACM International Conference on Computer Aided Design (ICCAD)*, Nov. 2007, pp. 476–480.

[12] V. H. Cordero and S. P. Khatri, "Clock distribution scheme using coplanar transmission lines," in *Proceedings of the Design, Automation and Test in Europe (DATE)*, Mar. 2008, pp. 985–990.

[13] *HSPICE Signal Integrity User Guide*, Synopsys, 2009.

[14] A. E. Ruehli, "Inductance calculations in a complex integrated circuit environment," *IBM Journal of Research and Development*, pp. 470–481, Sept. 1972.

[15] ——, "Equivalent circuit models for three dimensional multiconductor systems," *IEEE Transactions on Microwave Theory and Techniques*, vol. 22, no. 3, pp. 216–221, Mar. 1974.

[16] F. W. Grover, *Inductance Calculations: Working Formulas and Tables*. Instrument Society of America, 1962.

[17] R. B. Wu, C. N. Kuo, and K. K. Chang, "Inductance and resistance computations for three-dimensional multiconductor interconnection structures," *IEEE Transactions on Microwave Theory and Techniques*, vol. 40, no. 2, pp. 262–271, Feb. 1992.

[18] E. Bogatin, *Signal Integrity - Simplified*. Prentice Hall, 2004.

[19] T. C. Edwards and M. B. Steer, *Foundations of Interconnect and Microstrip Design*. Wiley, 2004.

[20] *High Frequency Structure Simulator: User's Guide*, 10th ed., Ansoft Corporation, Jun. 2005.

978-1-4673-0438-2/12 $31.00 © 2012 IEEE

Power Aware Post-Manufacture Tuning of MIMO Receiver Systems

Debashis Banerjee[1], Shreyas Sen[2], Shyam Kumar Devarakond[1], Abhijit Chatterjee[3]

[1]Student Member, IEEE, [2]Member, IEEE, [3]Fellow, IEEE

[1]School of Electrical and Computer Engineering, Georgia Institute of Technology,

[2]Circuit Research Lab, Intel Corporation

debashis.banerjee@gatech.edu, shreyas.sen@ gatech.edu, shyamkumar@gatech.edu, chat@ece.gatech.edu

[1]*Abstract*— This paper presents a methodology for post-manufacture tuning of MIMO (Multiple-Input-Multiple-Output) wireless systems aimed at increasing device manufacturing yield under large process variations. The goal is to achieve specified system-level EVM (Error Vector Magnitude) targets for MIMO receiver systems by tuning the individual MIMO receiver subsystems whose combined gain, noise and nonlinearity parameters determine the overall system-level EVM metric. While there has been prior work on tuning of SISO systems, the current work is novel due to the fact that the performances of individual receiver subsystems can combine in different ways to result in the same system-level EVM value. As the systems are tuned to meet the system level end-to-end metric it is ensured that all devices that are tuned for EVM consume the *least amount of power* possible. Tuning infrastructure for MIMO receivers and underlying tuning algorithms are developed in this work. A 1x2 MIMO system for a 2.4 GHz OFDM WLAN is used to demonstrate the core ideas of this research. This work demonstrates a 23% increase in yield of the device with an average power increase of 9.18% for the system under consideration.

Keywords- MIMO, post-manufacture tuning, yield improvement

I. INTRODUCTION

With increasing demand for reliable, fast communication over adverse channel conditions, new technologies have come to the fore to ensure robust error-free data transmission and reception. A key concept that has had a large footprint in the area of reliable wireless communication that of multiple-in-multiple-out (MIMO) wireless systems. Within a relatively short time, MIMO has been adopted into modern communication standards such as WLAN, WiMax and LTE that are deployed across a variety of consumer products today with great cost and data rate benefits[1][2][3]. The key idea in a MIMO system is that of space-time signal processing. Digital circuitry is used to process time multiplexed data while spatially distributed antenna arrays enable spatial multiplexing of signals. A MIMO system can operate in several modes[4]. In spatial multiplexing mode, it effectively increases data throughput by transmitting different data streams across spatially distributed antennae. In diversity mode, it takes advantage of the space diversity of

the system (here, the same data is transmitted to multiple receiver subsystem), resulting in an improvement in the quality of the link without gaining any increase in data throughput. In addition, there are operational modes of MIMO systems that combine the above two modes in different ways. Often these schemes involve space-time coding (STC) of signal streams and involve joint encoding of the individual data streams across all antennae for enhanced wireless link performance[4].

Traditional wireless communication links suffer from multipath fading due to signals arriving at an antenna through different paths with random phase and amplitude. These signals add up to form a resulting received signal which can be severely degraded and "noisy". Depending on whether there is a significant line of sight component or not, the wireless the channel can be modeled as Rician or Rayleigh, respectively. The effect of such multipath propagation is known as channel fading. The main advantage of MIMO systems is that they not only overcome the negative impact of multipath fading but also utilize the presence of multiple receiver subsystems to boost the performance of MIMO transceivers across any channel[5]. This comes at the cost of more complex digital circuitry in the baseband processor. However, MIMO systems can achieve better wireless performance without the need for additional spectrum.

One of the principal problems in the manufacture of modern RF transceivers is production testing and tuning for production yield improvement. The performance of RF circuits fabricated in scaled CMOS technologies is highly susceptible to manufacturing process variations [6]. With increasingly aggressive RF specifications for high-speed communication, the specification margins for circuits today are already very small. Increasingly, a significant percentage of manufactured devices for new products fail to meet stringent performance specifications. Several methods have been used to solve this problem. In design centering [7], the device is designed to be at the centre of the specification space with the specification values of nominal devices lying in the center of the specification upper and lower bounds (for two-sided specifications). However, if the variability in the manufacturing process is large, then production yield can be severely compromised. The effect of process variation becomes even more critical in a MIMO system as the

---

[1] This work was supported by GSRC/FCRP 2009-DT-2049 & NSF grant number NSF CCR 0916270.

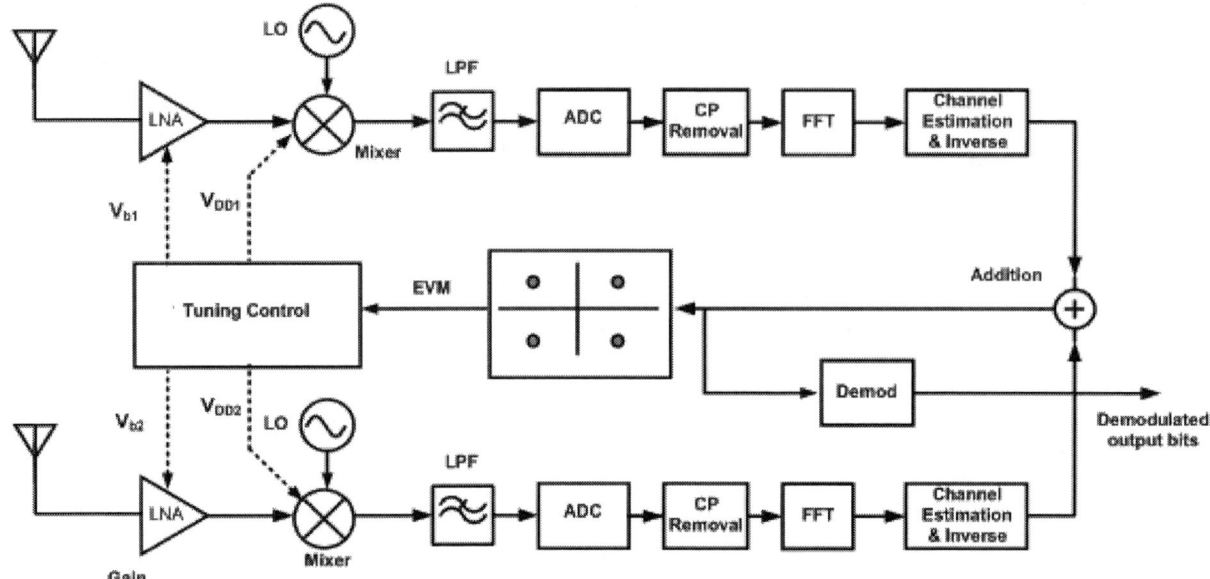

Figure 1: MIMO 1x2 receiver architecture

probability of multiple RF chains all meeting their individual specifications is even lower than for a single chain.

In a MIMO system since more than one chain contributes to the EVM in the diversity mode the imperfection in 1 chain can be compensated for by tuning the other making it a multi-dimensional optimization problem. In this work we have modeled a 1x2 system where the performance of the device for any given link is increased by having two receiver chains. The signals passing though the two chains are then combined together using Maximal Ratio Combining (MRC) to enhance the performance as given by the EVM metric. This system has a diversity order of 2 and strictly speaking is a single-input/multiple-output MRC system. However the concept is generic and can be easily generalized to MxN MIMO systems.

In the 2 chain receiver discussed in the paper it is shown with the help of Monte-Carlo simulations that due to process variation in the 2 LNAs and the 2 Mixers in the receiver a considerable fraction of the total device do not meet the performance specification of the device. The LNA and Mixer have in-built bias tuning knobs in them which can be used to change the receiver performance. Each such change in bias setting affects the performance of the aggregate receiver. Thus by changing the bias settings and constantly monitoring the health of the aggregate receiver one can attempt to tune the device to within acceptable performance threshold. In previous works, specification based tuning [8] and signature based tuning [9], [10] has been demonstrated for SISO systems. In our work, we propose a EVM based approach for MIMO systems. It has been established that Error Vector Magnitude (EVM) has a strong correlation with Bit Error Rate (BER). Hence EVM is taken as a metric for the health of the system. A suitable threshold is placed on the intrinsic EVM to determine whether the system performance is acceptable or not. However changing the bias settings lead to

a penalty in the form of a corresponding increase in power of the system. The paper presents an algorithm to determine the best way to turn these knobs so as to tune the device back to its performance specification bounds with a minimal increment in power. Thus over a batch of manufactured devices there is a considerable increase in yield with a very small average power increase.

The rest of the paper is arranged as follows: Section II discusses the MIMO 1x2 receiver architecture, Section III discusses the effect of process variation on the various specifications of the LNA and Mixer, Section IV discusses the post-manufacture tuning methodology, Section V presents the simulation results and Section VI concludes the paper.

II. MIMO 1x2 RECEIVER ARCHITECTURE

The receiver chains with signal combination and control logic for controlling the tuning knobs for post-production tuning based on measured EVM is shown in Figure 1. Each of the 2 chains is identical in terms of the blocks used. Each of the 2 chains consists of LNA, Mixer, Low-Pass filter followed by VGA and ADC. The Cyclic Prefix (CP) is then removed from the digitized output of the ADC and FFT is performed to get the symbols. Channel estimation is then performed using the Pilot tones sent from the transmitter followed by signal recombination using MRC. The composite signal is then demodulated to get the output bits. EVM is computed after performing MRC and is used as a metric for tuning of the control knobs so as to obtain the desired performance. Below we briefly describe some of the key blocks used in the tuning process.

A) LNA: An LNA designed in 0.18u is used with 2 tuning knobs bias(V_b) and supply(V_{DD}) as described in detail in [11]. The variation of the Gain, IIP3, NF and Power of the

978-1-4673-0438-2/12 $31.00 © 2012 IEEE   144

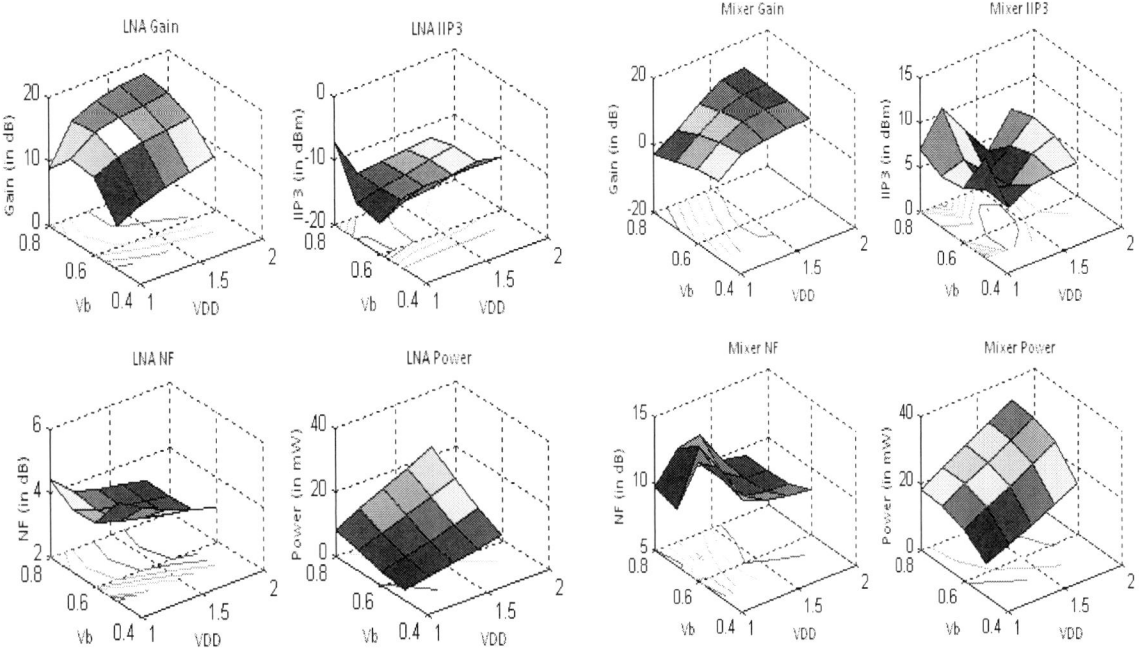

Figure 2: LNA and Mixer Specification surfaces over V_b and V_{DD} variation for a nominal device

LNA is shown in Figure 2. It is seen that for a particular V_{DD} the gain increases and NF decreases as power increases as V_b is increases from 0.5 V to 0.8 V. We choose V_b as a tuning knob keeping V_{DD} fixed to 1.8V to exploit this trade-off. A consequence of this choice is that IIP3 decreases as V_b is increased.

B) Mixer and Filter: A mixer with 2 tuning knobs V_b and V_{DD} is described in [11]. The variation of the Gain, IIP3, NF and Power of the mixer is shown in Figure 2. We can see that if we choose V_{DD} as the tuning knob we can exploit the gain and NF versus power trade-off in the mixer. So to demonstrate our concept we keep the V_b fixed at 0.5V and use V_{DD} as the tuning knob. The mixer output is fed to a filter. The filter extracts the down-converted baseband signal from the mixer output.

C) VGA: The output of the mixer is amplified by a voltage mode Variable Gain Amplifier (VGA) to the full scale of the ADC. The amount of gain provided by the VGA is controlled by the signal power at the ADC output. The power consumed by the VGA remains relatively constant over the entire tuning range of its gain.

D) ADC: The output of the VGA is digitized to an 8-bit word. The full scale of the ADC is 0.8V (i.e. from 0.4V to -0.4V).

E) Cyclic Prefix Removal: A cyclic prefix is a copy of the last part of the OFDM symbol that is attached to the transmitted OFDM symbol to overcome Inter-symbol interference in time-dispersive channels[12]. It is at least as long as the significant part of the impulse response experienced by the transmitted signal. This is removed before the FFT and demodulation.

F) FFT processor: One of the key features of a QPSK signal is that at the transmitter end several symbols each riding on an orthogonal subcarrier frequency is combined using IFFT to get a composite time domain signal. At the receiver end the inverse of this operation must be carried out. Hence a FFT processor is used to perform the Fourier transform of the signal after cyclic prefix removal. In this simulation framework we use a 64 point FFT signal.

G) Demodulation: The output of the FFT block is a train of complex symbols which can be demodulated to get the information bits. In this setup we have used a QPSK modulation scheme with 4 constellation points. Hence each complex symbol contains 2 bits. So from the output of a 64 point FFT we get a total of 128 information bits.

H) Channel estimation and inverse: In this setup we have performed Maximal ratio combining (MRC)[13] for the 2 branch receiver. The basic idea behind this technique the following. Let s_0 be the signal transmitted signal at any instant. Let the effect of channel between transmitter and the 2 receiver be denoted by h_1 and h_2 where h_1 and h_2 are complex quantities. Then the signal received at the 2 receivers are:

$$r_1 = h_1.s_0 + n_1 \qquad (1)$$

$$r_2 = h_2.s_0 + n_2 \qquad (2)$$

Here n_1 and n_2 represent complex noise and interference. At the receiver the signal is combined as follows:

$$s_0^* = h_1^*.r_1 + h_2^*.r_2 \qquad (3)$$

Where h_1^* and h_2^* are complex conjugates of the channel response and s_0^* is the estimated received symbol.

$$s_0^* = (\alpha_1^2 + \alpha_2^2).s_0 + h_1^*n_1 + h_2^*n_2 \qquad (4)$$

Here α_1 and α_2 are magnitudes of the complex quantities h_1 and h_2. The noise terms $h_1^*n_1$ and $h_2^*n_2$ being uncorrelated, do not add in power while the signal components add up and hence there is an increase in effective SNR. Thus the symbol can be reconstructed at the receiver. The main problem here is the estimation of h_1 and h_2. For this purpose the data in the OFDM signal is appended with a preamble. This preamble sequence is already known at the receiver end. Now by seeing the effect of the channel on the preamble sequence the receiver estimates the channel matrix and accordingly can reconstruct the transmitted signal even in the presence of noise and interference. Errors are caused due to imperfect channel estimation. For the purpose of post-manufacture tuning we shall assume that we can control the strength of the signal at the input of the LNA. For the case of symmetric paths h_1 and h_2 are equal since there is no channel. This implicitly assumes that we are tuning for the case when both channels to the 2 antennas are equally bad which is the worst case scenario is.

I) Performance metric estimation and Control block:
During post-manufacture tuning appropriate test signals are fed to the receiver and the corresponding EVM is calculated. The calculated EVM serves as metric of the health of the device. A bound is set on the maximum intrinsic EVM contribution of the receiver itself for an acceptable bit error rate during normal operation of the device. During post-manufacture tuning it is ascertained if the receiver meets this EVM threshold specification. If the device lies outside the bounds of acceptance the control block tries to tune the device in such a fashion such that it operates within the acceptable EVM threshold for the new tuning knob combination. The control law is designed in such a fashion that it tries to minimize the increase in power while bringing the device back within specifications.

III. EFFECT OF PROCESS VARIATION

Any manufacturing process has a certain amount of built in process variation associated with it. This can be intra-die or inter-die. Due to the variations in physical parameters such as oxide thickness, doping concentration etc circuit level parameters such as gain, IIP3, NF, power consumption are affected. Despite a lot of precautionary measures these variations can never be completely be eliminated.

In order to simulate the effect of process variation on the LNA and mixer specifications we perform Monte-Carlo simulations. In this fashion we create 50 process perturbed devices for each LNA and Mixer. As we expected the effect of physical parameters has an observable effect on the circuit specifications.

To obtain process perturbed receivers for a MIMO 1x2 system we now take 100 combinations of 4 devices, i.e. 2 LNAs and 2 mixers. The performance of the composite receiver is a function of the specifications of the individual building blocks like LNA, mixer, ADC etc. Thus clearly the performance of the 100 receivers would have a distribution due to process variations and some of these devices will not meet the performance specifications required for the particular device. The percentage of devices not meeting these specifications would depend upon the amount of variation in the processes and also to certain extent upon how sensitive the receiver is to each such variation.

IV. POST-MANUFACTURE TUNING

A. Channel Metric and Threshold

The basic premise for the post-manufacture tuning of a 1x2 MIMO receiver is that it must satisfy a certain intrinsic performance specification to be deemed suitable for real time operation. One such performance specification is the Bit Error Rate (BER) which indicates the probability of a bit being in error on reception using the receiver. However the BER testing for stringent specifications on the device requires us to send a very large number of bits. Hence we explore to alternative measure of BER. It has been shown in [11] that EVM has a strong correlation with BER . System level simulations in [14] have shown that EVM increases monotonically with BER and EVM thresholds can be determined for the maximum acceptable BER. For a target BER of 5e-4 it has been shown that the acceptable EVM for QPSK and 16-QAM are 35% and 12% respectively. It is shown in [15] that an EVM value of 5% gives acceptable performance upto 64-QAM. Hence for our work this figure can be deemed as the threshold of acceptance of intrinsic EVM of a receiver. A tighter bound ($< 5\%$) will result in higher average power among tuned devices and/or lower final yield, but yield increase might be higher.

B. Tuning Methodology

In our tuning methodology we have chosen EVM as the metric for the health of the device. The EVM of a receiver is dependent on the specifications like gain, IIP3 and NF of the blocks of each receiver chain like the LNA and the mixer. However gain-NF and IIP3 affect the EVM in different situations. When the signal is very weak, the EVM is dominated by the noise added by the devices and hence gain and NF have a major contribution towards the EVM. For this input since the signal never hits the non-linearities IIP3 has little influence on the EVM. Similarly when input signal is

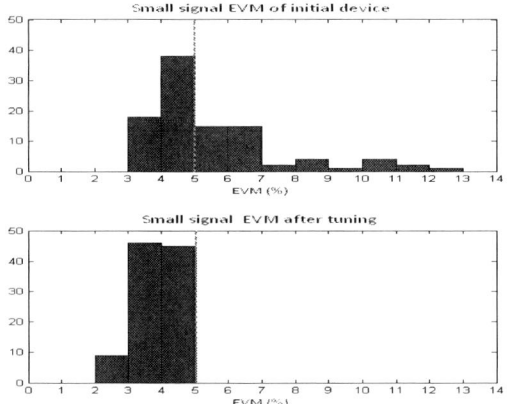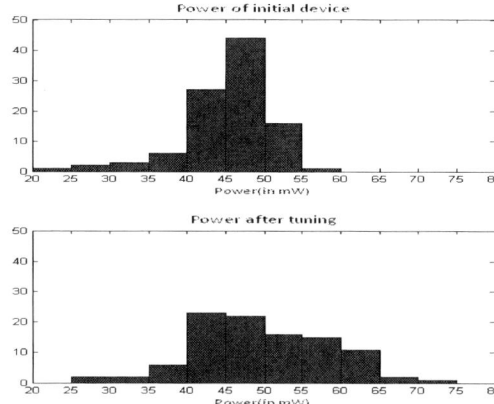

Figure 3: Small signal EVM and Power distributions before and after tuning

high the major contribution to EVM is due to the distortion and clipping in the signal due to non-linearity introduced by the LNA and IIP3. Since the signal is relatively large the noise added by the components are insignificant. Thus gain and NF of the LNA and mixer have little influence. We only need to test the receiver at the two extreme ends of signal strength. For any test signal of intermediate signal strength the EVM criterion would be automatically met if the device lies within EVM bounds at the two ends. This is because for any intermediate signal strength, the demand on Gain/NF and IIP3 is lesser than for extreme cases. We can observe the nature of variation of gain, NF and IIP3 of the LNA and mixer from the graphs Figure 2 and note that for the tuning knobs chosen IIP3 of LNA decreases with increasing power while the mixer IIP3 remains relatively constant and behaves without any observable trend. Thus there is no scope of trading off IIP3 with power. Hence for our case we suggest the following 2 step methodology:

(a) Do post-manufacture tuning for low power input signal.

(b) Using the tuning knob settings obtained in part (a) do testing for large amplitude of input signal and discard the devices which do not meet specifications for testing done in part (b).

For example in batch of 100 devices let the initial yield be 60%. For the remaining 40 devices we apply low power input signal as in part (a) and tune back the devices using the power optimal control algorithm. Let us say we could tune back 35 of the 40 devices back to specifications (within EVM threshold). Then we apply a stimulus (OFDM signal) with large amplitude and discard the devices for which the EVM>EVM$_{th}$ from the 100. If after this we find that 10 more devices fail our test then our improved yield is 60%+35%-10%=85%. The proposed method is suitable for a production test type environment and requires off-chip measurement components.

C. Test Stimulus

For the purpose of tuning we need to first calculate the EVM for each tuning knob setting. The test stimulus here is

OFDM-QPSK signal. As explained earlier in our methodology we need to choose two signal amplitudes with low and high power. We choose a signal amplitude for the low power input signal so that it is close to the noise floor such that the effect of gain and NF is reflected in the EVM of the device. After careful study we choose a signal level of -84dBm.

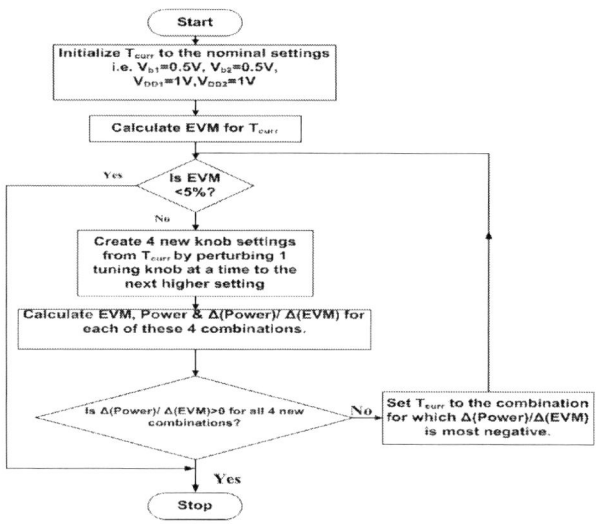

Figure 4: Tuning algorithm flowchart

This figure is consistent with the minimum sensitivity of WLAN MIMO receivers which exhibit the same order of sensitivity (-91dBm) as in [16]. For the signal with higher power we want the input signal to be close to P1dB of the system. For this we translate the minimum mixer IIP3 back by maximum LNA gain (across process and tuning knobs) and subtract 9.6dB to get P1dB. However to account for the high PAR of an OFDM symbol we back off by another 8.4 dB to get -44dBm as the signal level. Thus there is a difference of 40 dB between the 2 signal levels.

D. Tuning algorithm

The tuning algorithm is illustrated with the help of a flowchart in Figure 4. The basic idea of the tuning algorithm is that for each device we start at the nominal setting

978-1-4673-0438-2/12 $31.00 © 2012 IEEE   147

$V_{b,LNA1}$=0.5V, $V_{b,LNA2}$=0.5V, $V_{DD,Mixer1}$=1V, $V_{DD,Mixer2}$=1V and check whether our receiver meets the EVM specification for the current setting. If not we perturb each tuning knob to the next higher setting and create 4 new combinations of test settings. For each of these combinations we check which setting has the most negative gradient of EVM with Power and choose the next setting. We iterate this until we reach within the EVM threshold (~5%). T_{curr} is the optimum tuning knob combination .Thus the algorithm tunes the device back to within bounds of performance specification in a power optimal fashion.

V. RESULTS

We initially create 100 process instances of the MIMO 1x2 receivers and set the tuning knobs at the nominal point. Next we apply the small signal amplitude OFDM-QPSK stimulus to these devices and observe the power and EVM distributions. This is shown in Figure 3. We observe that for nominal setting we get a yield of 56% while the average power consumption is 45.63mW. Next we apply a small amplitude signal to this setup and tune the devices using the algorithm described earlier. This results in all of the outlying devices to tune back to within bounds. However the corresponding penalty is paid in terms of the average power consumption which goes upto 49.82mW, an increase of 9.18% as illustrated in Figure 3.

Next we apply a large amplitude input signal to the tuned devices and observe the EVM. We find that there are 21 devices which are lying outside the EVM threshold. A device is acceptable when it meets the stringent EVM specifications for both small and large input signals. So the devices lying outside the EVM threshold cannot be accepted .This is shown in Figure 5.

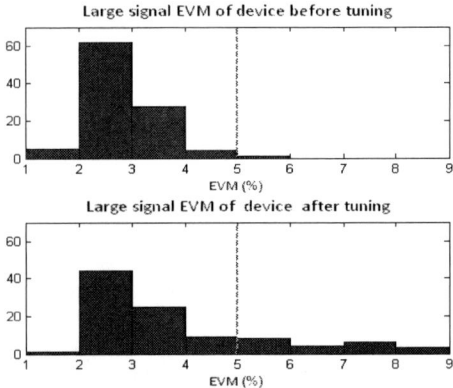

Figure 5: Large signal EVM before and after tuning

Thus our effective yield is 79% which is a considerable improvement from the original 56% yield that we had obtained. The percentage increase in yield is 23%.

VI. CONCLUSION

Our proposed power aware post-manufacture tuning technique shows a considerable improvement in yield of MIMO receivers from 56% to 79%. The exact yield improvement by this methodology would actually depend on

the tunability of the devices and the power profiles. But this work illustrates that significant improvement in yield can be obtained with minimal increase in power. Future work could include designing a combined stimulus which essentially reduces the two step process to a one step process. Thus this methodology can be employed in low-cost, low-power wireless communication devices employing MIMO.

REFERENCES

[1] "Performance Evaluation of WiMAX/IEEE 802.16 OFDM Physical Layer", Mohammad Azizul Hasan, MS thesis, Helsinki University of Technology.

[2] "An Analysis of the Benefits of Uplink MIMO in Mobile WiMAX Systems",June 2008,SEQUANS Communications, Bilel Bouraoui, Amélie Duchesne, Bertrand Muquet, Ambroise Popper(White Paper).

[3] "Overview of the 3GPP Long Term Evolution Physical Layer", Jim Zyren, Freescale. (White Paper)

[4] Gesbert, D.; Shafi, M.; Da-shan Shiu; Smith, P.J.; Naguib, A.; , "From theory to practice: an overview of MIMO space-time coded wireless systems," *Selected Areas in Communications, IEEE Journal on* , vol.21, no.3, pp. 281- 302, Apr 2003

[5] Catreux, S.; Greenstein, L.J.; Erceg, V.; , "Some results and insights on the performance gains of MIMO systems," *Selected Areas in Communications, IEEE Journal on* , vol.21, no.5, pp. 839- 847, June 2003

[6] "Analog-RF IP Integration challenges SoC Designers",April/May 2006 issue, Chip Design Magazine.

[7] R. J. Pratap, P. Sen, C. E. Davis, R. Mukhophdhyay, G. S. May, and J. Laskar, "Neurogenetic design centering," *IEEE Trans. Semiconductor Manufacturing*, vol. 19, pp. 173-182, 2006.

[8] Natarajan, V.; Sen, S.; Devarakond, S.K.; Chatterjee, A.; , "A holistic approach to accurate tuning of RF systems for large and small multiparameter perturbations," *VLSI Test Symposium (VTS), 2010 28th* , vol., no., pp.331-336, 19-22 April 2010

[9] S. Devarakond, S. Sen, V. Natarajan, A. Banerjee, H. Choi, G. Srinivasan, A. Chatterjee, "Digitally Assisted Concurrent Built-In Tuning of RF Systems Using Hamming Distance Proportional Signatures," Asian Test Symposium, pp. 283-288, 2010 19th IEEE Asian Test Symposium, 2010

[10] Natarajan, V.; Sen, S.; Banerjee, A.; Chatterjee, A.; Srinivasan, G.; Taenzler, F.; , "Analog Signature- Driven Postmanufacture Multidimensional Tuning of RF Systems," *Design & Test of Computers, IEEE* , vol.27, no.6, pp.6-17, Nov.-Dec. 2010

[11] Sen, S., Natarajan,V., Senguttuvan,R. and Chatterjee, A., "Pro-VIZOR: Process Tunable Virtually Zero Margin Low Power Adaptive RF for Wireless Systems", 45th ACM/IEEE Design Automation Conference, 2008. DAC 2008, 8-13 June 2008, pp. 492 – 497.

[12] "Wireless OFDM Systems: How to make them work?" ,Marc Engels(Ed.)

[13] Alamouti, S.M.; , "A simple transmit diversity technique for wireless communications ," *Selected Areas in Communications, IEEE Journal on* , vol.16, no.8, pp.1451-1458, Oct 1998.

[14] Senguttuvan, R.; Sen, S.; Chatterjee, A.; , "VIZOR: Virtually zero margin adaptive RF for ultra low power wireless communication," *Computer Design, 2007. ICCD 2007. 25th International Conference on* , vol., no., pp.580-586, 7-10 Oct. 2007

[15] Sen, S.; Senguttuvan, R.; Chatterjee, A.; , "Environment-Adaptive Concurrent Companding and Bias Control for Efficient Power-Amplifier Operation," *Circuits and Systems I: Regular Papers, IEEE Transactions on* , vol.58, no.3, pp.607-618, March 2011

[16] Nathawad, L et al. , "A Dual-Band CMOS MIMO Radio SoC for IEEE 802.11n Wireless LAN," *Solid-State Circuits Conference, 2008. ISSCC 2008. Digest of Technical Papers. IEEE International* , vol., no., pp.358-619, 3-7 Feb. 2008

2012 25th International Conference on VLSI Design

GPU Implementation of a Programmable Turbo Decoder for Software Defined Radio Applications

Dhiraj Reddy Nallapa Yoge and Nitin Chandrachoodan
Dept. of Electrical Engineering
IIT Madras
Chennai, India
ee06b066, nitin@ee.iitm.ac.in

Abstract—**This paper presents the implementation of a 3GPP standards compliant configurable turbo decoder on a GPU. The challenge in implementing a turbo decoder on a GPU is in suitably parallelizing the Log-MAP decoding algorithm and doing an architecture aware mapping of it on to the GPU. The approximations in parallelizing the Log-MAP algorithm come at the cost of reduced BER performance. To mitigate this reduction, different guarding mechanisms of varying computational complexity have been presented. The limited shared memory and registers available on GPUs are carefully allocated to obtain a high real-time decoding rate without requiring several independent data streams in parallel.**

Keywords-**Turbo Decoder; GPU implementation; CUDA; Parallel Log-MAP; Guarding Mechanisms;**

I. INTRODUCTION

Turbo codes are an important class of forward error correcting (FEC) codes because of their low bit-error rate (BER) and frame error rate (FER) performance, and are widely used in many 3G and 4G standards such as UMTS[1], 3GPP LTE[2] etc. The turbo decoder is one of the most computationally challenging and time consuming parts of the encoding-decoding process because of the complex iterative decoding algorithm. Hence, typically ASIC implementations of the decoder are preferred for achieving a high data throughput. Graphic Processing Units (GPUs) provide an alternative for achieving the same high data throughput as achieved on dedicated ASICs, but with the added benefit of programmability in software. Implementation of a high throughput configurable turbo decoder completely in software is an important step towards a pure software defined radio (SDR) realization for testing out various evolving 4G standards. Today, even many hand held devices contain GPUs and the decoder implementation may be extended to them as well.

GPUs, with their large number of processing cores, are very good at handling tasks that exhibit gratuitous amounts of data parallelism. In this paper, we utilize the different kinds of data parallelisms inherent to the computations in the Log-MAP algorithm and also additional data parallelisms brought about by splitting the code block into several small code blocks to achieve a high throughput. We distribute the workload of the parallel Log-MAP algorithm efficiently across all the processor cores and also effectively utilize the fast on chip shared memory. The splitting of the code block into several small code blocks comes at the cost of reduced BER and FER performance and hence different guarding mechanisms to mitigate this reduction are also presented. The relative merit of each of these mechanisms in mitigating

[1]Universal Mobile Telecommunications System
[2]Third Generation Partnership Project - Long Term Evolution

the degradation in performance and the associated impact on throughput are also discussed.

The turbo decoding algorithm is usually based on the BCJR algorithm and variants thereof [1], [2]. However, these algorithms are highly serial in nature, requiring a complete traversal of a block from end-to-end and then in reverse for each iteration. Windowing techniques [3] have been proposed to work around this problem to some extent, but the interleaver used in turbo codes forms yet another challenge for implementation in parallel. As a result, although there have been several VLSI implementations of decoders of different capabilities, very few parallel software approaches have been proposed, such as [4] and [5].

One important aspect that is typically used to obtain a high degree of parallelism in the decoding process is to consider the decoding of several code blocks in parallel. For example, [4] considers the decoding of 100 blocks in parallel to estimate the throughput of the decoder. While this is useful in terms of estimating resource usage and parallelism, it does not help in the case where we want a single stream of data to be decoded in software, as would be the case in a true SDR receiver.

In this paper, we have concentrated on the case of a single stream of data and tried to maximize the throughput, while keeping in consideration the time required for transfer of blocks of data into and out of device memory.

The rest of the paper is organized as follows: Section II overviews the Log-MAP algorithm and section III explores the possible parallelisms in it. Section IV describes an architecture aware mapping of the parallel Log-MAP algorithm onto the GPU. The BER performance results and the achieved throughput are presented in section V followed by the conclusion in section VI.

II. LOG-MAP ALGORITHM

For conciseness, we do not go into the full details of how a turbo encoder and decoder are implemented, and rather concentrate only on those aspects relevant to the discussion of the parallel implementation. For further details, including the structure of the encoder and decoder, use of the interleaver, etc., excellent tutorials are available in [1] and [2].

The turbo decoder consists of two half decoders, each of which exchange appropriately interleaved extrinsic log-likelihood ratios (LLRs) after each iteration. The extrinsic likelihood ratios in each half-decoder are estimated using the Maximum a posteriori probability (MAP) algorithm [1]. The MAP algorithm in its direct form involves the computation of exponentials, and hence a simplified version of it that uses log-likelihood ratios, the Log-Map algorithm, is often used for both hardware and software implementations [2].

978-1-4673-0438-2/12 $31.00 © 2012 IEEE   149

Let $\boldsymbol{u} = u_1, u_2, \ldots, u_N$ and $\boldsymbol{x^p} = x_1^p, x_2^p, \ldots, x_N^p$ denote the input bit-stream and the parity bit-stream generated by the constituent encoder respectively. For each half decoder, let $\boldsymbol{y^s} = y_1^s, y_2^s, \ldots, y_N^s$ and $\boldsymbol{y^p} = y_1^p, y_2^p, \ldots, y_N^p$ denote the noisy AWGN versions of $u_1, u_2, \ldots, u_N$ and $x_1^p, x_2^p, \ldots, x_N^p$ respectively. Let $L_a(k)$ denote the a priori LLR of bit u_k passed on from the other half-decoder and $L_e(k)$ denote the computed extrinsic LLR.

For a state transition from a state s_{k-1} at a trellis stage $k-1$ to a state s_k at the trellis stage k, the branch metric $\gamma_k(s_{k-1}, s_k)$ is defined as :

$$\gamma_k(s_{k-1}, s_k) = (L_c(y_k^s) + L_a(k))u_k + L_c(y_k^p)x_k^p \quad (1)$$

where L_c is the channel reliability value, and bit u_k causes the state transition $s_{k-1} \to s_k$ and generates the parity bit x_k^p. The possible state transitions for the 3GPP LTE standards compliant encoder are shown in Fig.1

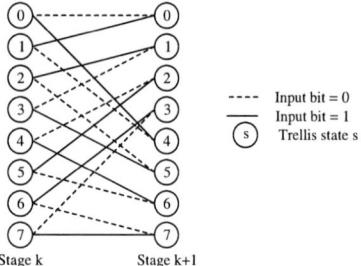

Figure 1: Trellis state transition diagram of 3GPP LTE encoder

The forward state metric for a state s_k at a stage k of the trellis $\alpha_k(s_k)$ is defined as:

$$\alpha_k(s_k) = max^*_{s_{k-1} \in S}(\alpha_{k-1}(s_{k-1}) + \gamma_k(s_{k-1}, s_k)) \quad (2)$$

where S is the set of all states from which a state transition is possible to the state s_k.

The backward state metric for a state s_k at a stage k of the trellis $\beta_k(s_k)$ is defined as:

$$\beta_k(s_k) = max^*_{s_{k+1} \in S}(\beta_{k+1}(s_{k+1}) + \gamma_{k+1}(s_k, s_{k+1})) \quad (3)$$

where S is the set of all states to which a state transition is possible from the state s_k.

After computation of α, β and γ, two LLRs per trellis state are computed. The state LLR $\Lambda(s_k | u_k = 0)$ and $\Lambda(s_k | u_k = 1)$, for a state s_k at a trellis stage k are defined as:

$$\Lambda(s_k | u_k = 0) = \alpha_{k-1}(s_{k-1}) + \gamma_k(s_{k-1}, s_k) + \beta_k(s_k) \quad (4)$$

$$\Lambda(s_k | u_k = 1) = \alpha_{k-1}(s_{k-1}) + \gamma_k(s_{k-1}, s_k) + \beta_k(s_k) \quad (5)$$

where $u_k = 0$ or 1 respectively causes the state transition $s_{k-1} \to s_k$ respectively.

The extrinsic LLR for u_k is computed as:

$$L_e(k) = max^*_{s_k \in S}(\Lambda(s_k | u_b = 1)) - max^*_{s_k \in S}(\Lambda(s_k | u_b = 0)) \\ - L_c(y_k^s) - L_a(k) \quad (6)$$

where S is the set of all possible states and max^* is defined as

$$max^*(S) = \ln(\sum_{s \in S} e^s). \quad (7)$$

A. max^* Function

The way in which the max* function is actually computed divides the MAP algorithm into two algorithms - the Full log-map algorithm and the Max log-map algorithm. In the full log-map algorithm, max^* is computed as:

$$max^*(a, b) = max(a, b) + \ln(1 + e^{-|b-a|}) \quad (8)$$

In the max log-map algorithm, max^* is computed as:

$$max^*(a, b) = max(a, b) \quad (9)$$

B. QPP Interleaver

The 3GPP LTE standard uses the Quadratic Permutation Polynomial (QPP) interleaver. The QPP interleaver is defined as:

$$\Pi(x) = f_1 x + f_2 x^2 \pmod{N} \quad (10)$$

where f_1 and f_2 satisfy several properties detailed in [6].
A computationally less expensive way of computing the QPP interleaver function defined in Eq. (10) is

$$\Pi(x) = (f_1 + f_2 x \pmod{N})x \pmod{N} \quad (11)$$

III. PARALLELISM IN LOG-MAP ALGORITHM

The Log-MAP turbo decoding algorithm, in its direct form, exhibits only a little amount of data parallelism. In this algorithm, there exists a very strong data dependency between adjacent trellis stages. The computation of both the forward and backward state metrics(SM) proceeds in a serial fashion along the trellis stages. The serial evaluation of the SMs forms the basis of the MAP algorithm and any attempts to parallelize this part would result in severe BER performance loss.

A. Trellis State Level Parallelism

The SMs α and β at each trellis stage can be computed concurrently for all the trellis states. For example, the $\alpha_k(s_k)$ or $\beta_k(s_k)$, for all the states at a stage k of the trellis can all be computed at once if, $\alpha_{k-1}(s_{k-1})$ or $\beta_{k+1}(s_{k+1})$ respectively, are known for all the states. The branch metric (BM) γ can be computed in parallel for all the possible state transitions at each stage of the trellis. In fact, the BM computation shows complete data parallelism i.e the BMs of all the trellis stages themselves can all be computed in parallel. In the 3GPP LTE standard encoder, since there are 8 trellis states and 16 possible state transitions, there is a 8-way parallelism in α and β and a 16-way parallelism in γ computations to exploit. The trellis state level parallelism is inherent to the Log-Map algorithm and hence does not cause any BER performance degradation.

B. Sub-block Level Parallelism

Though the trellis state level parallelism offers some amount of data parallelism, the degree of parallelism shown is small. Further data parallelism can be obtained by dividing the code block into several smaller sized code blocks called sub-blocks and performing the decoding on each of the sub-blocks independently in each iteration [7]. During each iteration, in each sub-block, the forward and backward recursions run only over the length of the sub-block. Hence, during each iteration, the forward and backward recursions of all the sub-blocks are completely data independent and can be done concurrently. As before, after each iteration, the extrinsic LLRs computed by each of the two half-decoders are exchanged with appropriate interleaving. Fig.2 contrasts the forward and backward recursions in the MAP algorithm with and without sub-block level parallelism.

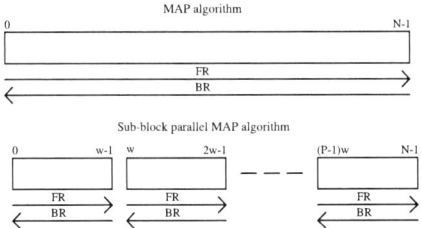

Figure 2: Recursions in parallel MAP algorithm

Let a code block of length N be split into P sub-blocks each of length $w = \frac{N}{P}$. The degree of data parallelism thus brought about is equal to the number of sub-blocks P. An increase in P increases the amount of data parallelism possible but it comes at the cost of reduced BER performance. The BER performance degradation occurs because the α and β metrics at the start and end of a sub-block respectively are no longer known. However, the decrease in BER performance can be reduced by adopting various guarding mechanisms to estimate the α and β metrics at the start and the end of a sub-block respectively.

C. Guarding Mechanisms

Simply initializing the α and β metrics of all states to equal values at the start and the end of the sub-blocks leads to severe performance degradation. To minimize the degradation in BER performance the following three guarding mechanisms as shown in Fig. (3) are used.

1) Previous Iteration Value Initialization (PIVI): Here, the α metrics at the beginning of a sub-block are initialized with the α metrics at end of the previous sub-block from the previous iteration. Similarly, the β metrics at the end of a sub-block are initialized with the β metrics at the beginning of the next sub-block from the previous iteration. This mechanism requires extra memory but involves no extra arithmetic computations.

2) Double Sided Training Window (DSTW): Here, training windows are run on either sides of the sub-block to allow the α and β metrics to develop into better estimates [3]. At the start of the training window, both α and β metrics for all the states are set to equal values. A larger training window results in better BER performance but would come at the cost of extra computations.

3) Previous Iteration Value Initialization with Double Sided Training Window (PIVIDSTW): Here, the features of both of the above mechanisms are combined. As in DSTW, this also has training windows running on either sides of the sub-block. But, at the beginning of the training window, instead of initializing the α and β metrics of all states to equal values, they are set equal to the α and β metrics from the trellis stages at the end of the training windows from the previous iteration. Naturally, since this

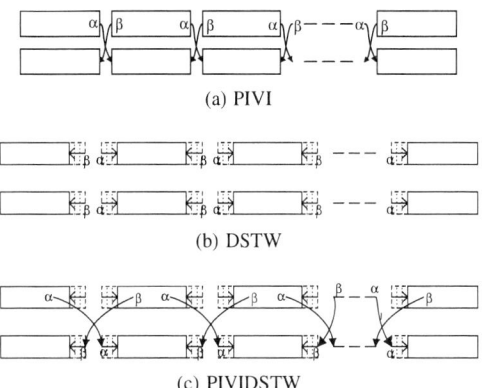

Figure 3: Types of Guarding Mechanisms

guarding mechanism combines both the above two mechanisms, it requires extra memory and involves extra computations.

IV. MAPPING THE LOG-MAP ALGORITHM ON TO GPU

To obtain a high throughput on the GPU, an architecture aware [8] mapping of the algorithm is paramount. The GPU architecture differs significantly from that of a CPU [9]. In this paper, we have made use of the Compute Unified Device Architecture (CUDA) processors from Nvidia, and some of the implementation decisions are influenced by the restrictions on registers, shared memory etc. imposed by the specific processor available to us. However, we have tried to explain the principles behind the parallelization and memory usage in general terms that make it easy to adapt the design to a different architecture with different constraints.

A CUDA compatible GPU has a large number of processing cores and is capable of running several hardware threads concurrently but has no hardware managed cache. Four different kinds of device memories are presented to the programmer [10]. The sizes and relative latencies of each of these memories directly impact the mapping of any algorithm on to the GPU. A fast on chip memory is available in the form of shared memory. The computations wherein all accesses are to the shared memory occur much faster than those wherein one or more accesses involve the global memory [11].

Occupancy, which is a measure of the number of threads being actually run concurrently on a streaming multi-processor (SMP) [12], is determined by the number of threads per thread block, the shared memory allocated per thread block and the number of registers per each thread. Higher the occupancy, better would GPU be in hiding the delays associated with accesses to high latency memories The principles followed here for mapping of the Log-MAP turbo decoding algorithm are general and apply to any generator function in the encoder, any code block length and any interleaver function. However, the mapping has been performed for the 3GPP standard specifications.

A. Half-Decoder Kernel

The decoding is done iteratively by the two half-decoders. In each half-decoder, the decoding proceeds with a forward traversal followed by a backward traversal through the length of the sub-block for all the sub-blocks. The same kernel definition is used for both the half-decoders since, both perform the same set of

Table I: Operands for α_k computation [4]

Thread id(i)	u=0		u=1	
	s_α^0	x_α^{p0}	s_α^1	x_α^{p1}
0	0	0	1	1
1	3	1	2	0
2	4	1	5	0
3	7	0	6	1
4	1	0	0	1
5	2	1	3	0
6	5	1	4	0
7	6	0	7	1

computations on the input data and are identical in all aspects other than in reading the input a priori LLRs $L_a(k)$ and the systematic channel values $\boldsymbol{y}^s$ in direct or interleaved order, and returning the computed extrinsic LLRs $L_e(k)$ in direct or deinterleaved order. The extrinsic LLRs returned by both the half-decoders are in direct order. A kernel is launched with 8 threads [5] and P thread blocks.

The computation of α, β and extrinsic LLRs all require the γ metrics. The gamma metrics can either be computed once, stored and fetched when required or be computed each time they are required. The latter approach has been chosen in the current implementation because of the shared memory size and global memory latency constraints [13].

B. Forward Traversal

In the forward traversal, $\alpha_k(s)$ for each trellis stage k is computed for the length of the sub-block w. The computation of $\alpha_k(s)$ is given by Eq. (2). For each state s, at a trellis state k, there are exactly 2 states at trellis stage $k-1$ from which a state transition to state s is possible, one each for $u_k = 0$ & 1. Each thread evaluates the $\alpha_k(s)$ for one state. To compute the $\alpha_k(i)$ for a state i, the thread needs to know the states s_α^0 and s_α^1 from which there is a state transition possible to the state i and the parity bits x_α^{p0} and x_α^{p1} associated with the state transitions $s_\alpha^0 \rightarrow i$ and $s_\alpha^1 \rightarrow i$. The Table. I summarizes the operands needed for α computation.

During the forward traversal the α metrics are computed and stored, since they are required for the computation of extrinsic LLRs computed during the backward traversal. The pseudo code for the α computation during the forward traversal is given by the Algorithm 1.

Algorithm 1 Forward Traversal - Thread i computes $\alpha_k(i)$

---

for $k = 1$ **to** w **do**
$\quad \gamma^0 \leftarrow 0.5 \times ((L_c y_k^s + L_k^e)(-1) + L_c y_k^p(x_\alpha^{p0} | u_k = 0))$
$\quad \gamma^1 \leftarrow 0.5 \times (L_c y_k^s + L_k^e + L_c y_k^p(x_\alpha^{p1} | u_k = 1))$
$\quad \alpha^0 \leftarrow \alpha_{k-1}(s_\alpha^0 | u_k = 0) + \gamma^0$
$\quad \alpha^1 \leftarrow \alpha_{k-1}(s_\alpha^1 | u_k = 1) + \gamma^1$
$\quad \alpha_k(i) \leftarrow max^*(\alpha^0, \alpha^1)$
$\quad$ SYNC
end for

C. Backward Traversal

In the backward traversal, the $\beta_k(s)$ and the extrinsic LLR $L^e(k)$ are computed for each trellis stage k along the length of

the sub-block. The extrinsic LLR is computed immediately after β computation, thereby, removing the need to store the β metrics for the entire length of the sub-block. The computation of $\beta_k(s)$ is given by Eq. (3). For each state s, at a trellis state k, there are exactly 2 states at trellis stage $k + 1$ to which a state transition from state s is possible, one each for $u_k = 0$ & 1. Each thread evaluates the $\beta_k(s)$ for one state. To compute the $\beta_k(i)$ for a state i, the thread needs to know the states s_β^0 and s_β^1 to which there is a state transition possible from the state i and the parity bits x_β^{p0} and x_β^{p1} associated with the state transitions $i \rightarrow s_\beta^0$ and $i \rightarrow s_\beta^1$. A table similar to the α computation applies for the β as well.

The state LLRs Λ_0 and Λ_1 for all the 8 states, as given by Eq. (4), are computed concurrently with each thread computing the state LLRs for one state. At each trellis stage, the computation of the extrinsic LLR from the state LLRs, as given by Eq. (6), can be done immediately after β metric computation. However, this would require the serial computation of the max^* function over all the the 8 states, thereby leaving all but one of the 8 threads in each block idle. To utilize all the threads, the state LLRs $\Lambda(s_k | u_k = 0)$ and $\Lambda(s_k | u_k = 1)$ for 8 consecutive trellis stages are stored and then the extrinsic LLRs for these 8 stages are computed concurrently using the 8 available threads. The pseudo code for β and extrinsic LLR computations is given in the Algorithm 2

Algorithm 2 Backward Traversal - Thread i computes $\beta_k(i)$ and $L^e(k)$

---

for $l = 1$ **to** $\frac{w}{8}$ **do**
$\quad$ **for** $j = 1$ **to** 8 **do**
$\quad\quad k \leftarrow 8l + j$
$\quad\quad \beta^0 \leftarrow \beta(s_\beta^0 | u_k = 0) + \gamma^0$
$\quad\quad \beta^1 \leftarrow \beta(s_\beta^1 | u_k = 1) + \gamma^1$
$\quad\quad \beta_{hold}(i) \leftarrow max^*(\beta^0, \beta^1)$
$\quad\quad$ SYNC
$\quad\quad \beta(i) \leftarrow \beta_{hold}(i)$
$\quad\quad$ SYNC
$\quad\quad \gamma^0 \leftarrow 0.5 \times ((L_c y_k^s + L_k^e)(-1) + L_c y_k^p(x_\beta^{p0} | u_k = 0))$
$\quad\quad \gamma^1 \leftarrow 0.5 \times (L_c y_k^s + L_k^e + L_c y_k^p(x_\beta^{p1} | u_k = 1))$
$\quad\quad \Lambda_0(8i + j) \leftarrow \gamma^0 + \beta(s_\beta^0 | u_k = 0)) + \alpha_k(i)$
$\quad\quad \Lambda_1(8i + j) \leftarrow \gamma^1 + \beta(s_\beta^1 | u_k = 1)) + \alpha_k(i)$
$\quad$ **end for**
$\quad$ SYNC
$\quad L_{e0}, L_{e1} \leftarrow 0$
$\quad$ **for** $j = 1$ **to** 8 **do**
$\quad\quad L_{e0} \leftarrow max^*(L_{e0}, \Lambda_0(8i + j))$
$\quad\quad L_{e1} \leftarrow max^*(L_{e1}, \Lambda_1(8i + j))$
$\quad$ **end for**
$\quad L_e(k) = (L_{e1} - L_{e0}) - L_c y_k^s - L_a(k)$
end for

D. Memory Allocation

1) Global Memory: Storing data in the global memory is the only way of exchanging data between different kernel launches. The extrinsic LLRs need to be exchanged between the different kernel launches corresponding to each of the half-decoders after every iteration and hence a copy of them is stored in the global memory. In addition, the PIVI and PIVIDSTW guarding mechanisms present the need to store α and β metrics at the ends of the sub-block in global memory as these need to be passed on to the next iteration.

2) Shared Memory: The shared memory is almost as fast as the registers and is the fastest available memory that the programmer can directly control. Shared memory is allocated for a thread block. The γ_k computation requires channel values and extrinsic LLRs. Hence, these are fetched from the global memory on to the shared memory. For a sub-block size of w, this would require storing $3w$ floats. The extrinsic LLR computation during the backward traversal requires the α_k values computed during the forward traversal. Hence, the α_k values for the entire size of the sub-block need to be stored in the shared memory. This requires storing another $8w$ floats. These two constitute the bulk of the shared memory allocated for each block. In addition, β_k computation requires β_{k-1} and hence 16 floats are required for β. In the extrinsic LLR computation given by Eq. (6), the $\Lambda(s_k \,|\, u_b = 0)$ and $\Lambda(s_k \,|\, u_b = 1)$ values for all states are stored for 8 trellis stages. This requires the storage of 64 floats each for $\Lambda(s_k \,|\, u_b = 0)$ and $\Lambda(s_k \,|\, u_b = 1)$. Extra shared memory is required for implementing the various guarding mechanisms. The PIVI requires 16 floats each for α and β initializations. The DSTW requires $6g$ floats and the PIVIDSTW requires $32 + 6g$ floats for a guard window of size g.

3) Constant Memory: The constant memory is ideal for storing all those values which remain constant during a kernel execution. It has been used for storing the operands mentioned in Table. I and the similar table required for the computation of α and β metrics respectively, as well as the indices required for deinterleaving.

V. BER PERFORMANCE AND THROUGHPUT

The BER performance and the throughput of the designed turbo decoder depends on the type of the decoding algorithm used - Full Log-MAP or Max Log-MAP, the number of parallel sub-blocks P and the type of guarding mechanism used - PIVI, DST or PIVIDSTW. The effect of each of these factors on the BER performance and the throughput has been presented below.

A. BER Performance

As is to be expected, the Full Log-MAP algorithm fares better than Max Log-MAP algorithm and this fact remains unchanged with the number of parallel sub-blocks. As shown in Fig.4, the BER performance degradation increases with a increase in the number of parallel sub-blocks. Among the three guarding mechanisms, the DSTW mechanism fares the worst and hence has been discarded. The PIVI mechanism can be used as a possible guarding mechanism. As shown in Fig.5, for the number of parallel sub-blocks P = 96, where the size of each sub-block w is only 64, the decoder with PIVI guarding mechanism provides a BER performance that is within 0.1dB and a FER performance that is within 0.2dB of the optimal case of a single code block. The PIVIDSTW guarding mechanism, which is a combination of both PIVI and DST, can be used to further improve the BER and FER performance. For $P = 96$, $w = 64$ and for a window size $g = 8$, the decoder with PIVIDSTW guarding mechanism provides a BER performance that is within 0.01dB and a FER performance that is within 0.02dB of the optimal case of a single code block.

B. Throughput

The BER performance and throughput of the designed turbo decoder have been evaluated by testing it on a 64-bit Linux

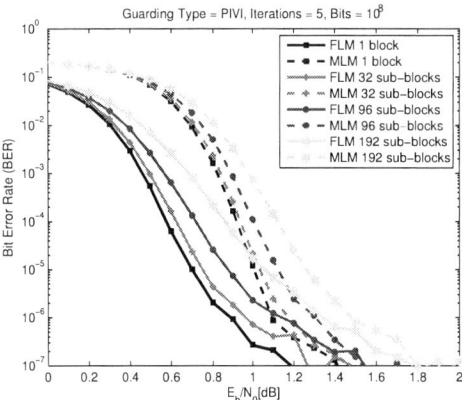

Figure 4: Effect of P on BER performance for Max Log-MAP and Full Log-MAP

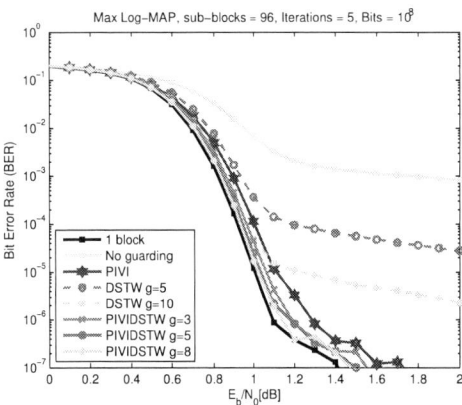

Figure 5: Effect of the type of guarding mechanism on BER performance

platform with 4GB DDR2 memory running at 800MHz and an AMD Phenom 9750 Quad-Core Processor running at 2.4GHz. The GPU used in the implementation is Nvidia GeForce 9800 GX2 graphics card of compute capability 1.1 running at 1.5GHz with 512MB of GDDR3 memory running at 1GHz. It has 128 streaming processors batched into 8SMPs; 8192 registers and 16KB of shared memory per SMP. The time measured for calculating throughput in our implementation is the cumulative sum of the times to transfer the input channel values from the host to the device, perform the decoding on the device and return the decoded bits back to the host. In addition, as would be the requirement for any real time usage of the turbo decoder, the data transfer is done separately for each code block. For a code block size of 6144 and for an Max Log-MAP implementation with sub-blocks P=96, the time for data transfer is 0.99ms and is quite significant compared to the time for one decoding iteration on the GPU, which is equal to 0.41ms.

The Table. II showcases the speed up achieved using the GPU over an implementation done purely on the CPU for both Max-Log-MAP and Full Log-MAP implementations. For a Max Log-MAP turbo decoder with 5 iterations, the GPU implementation with 96

Table II: Throughput on CPU vs GPU for $P = 96$

Iters	MLP/FLP Decoder Throughput (Kbps)		
	CPU	GPU	
		PIVI	PIVIDSTW g=5
1	991/116	4380/4050	4300/3990
2	524/56	3390/3010	3200/2840
3	351/37	2760/2400	2580/2240
4	264/28	2330/1990	2170/1850
5	211/21	2020/1700	1860/1580
6	176/19	1780/1490	1630/1370
7	151/17	1590/1320	1450/1210

Table III: Throughput vs P, with PIVI

Iters	MLP/FLP Decoder Throughput (Mbps)				
	P=32	P=64	P=96	P=128	P=192
1	3.0/2.6	3.9/3.5	4.4/4.1	4.6/4.3	4.7/4.4
2	1.9/1.6	2.9/2.5	3.4/3.0	3.7/3.3	3.8/3.4
3	1.4/1.2	2.3/1.9	2.8/2.4	3.1/2.7	3.1/2.7
4	1.1/0.9	1.9/1.6	2.3/2.0	2.7/2.3	2.6/2.3
5	1.0/0.8	1.6/1.3	2.0/1.7	2.3/2.0	2.4/2.0
6	0.8/0.7	1.4/1.1	1.8/1.5	2.1/1.8	2.1/1.8
7	0.7/0.6	1.2/1.0	1.6/1.3	1.9/1.6	1.9/1.6

parallel sub-blocks is more than an order of magnitude faster than the CPU implementation. The C code run on the CPU is compiled using gcc with -O3 optimization flag and is single threaded i.e it does not utilize any parallelism on multiple CPU cores.

As is to be expected, the throughput of the Full Log-MAP algorithm is lesser than that of the Max Log-MAP algorithm and the throughput of PIVIDSTW guarding mechanism is lesser than that of PIVI guarding mechanism. For 5 iterations, compared to the Max Log-MAP algorithm, the Full Log-MAP algorithm is slower by approximately 300Kbps. The throughput achieved by PIVIDSTW guarding mechanism with window size $g = 5$ is approximately 150Kpbs lesser than that achieved by PIVI guarding mechanism.

The throughput achieved by the designed turbo decoder for different number of sub-blocks P is shown in Table. III. The throughput increases initially with increase in P as the occupancy increases because of the decrease in the shared memory usage per each thread block. The throughput tends to saturate with a further increase in P as now the register usage and not the shared memory becomes the constraining factor for occupancy.

C. Comparison

As explained earlier, there are currently hardly any published results on implementation of turbo decoders on GPUs. [4] presents results of implementation on an Nvidia Tesla C1060 GPU, and they are able to obtain a speed of up to 6.77 Mbps for a Max Log-MAP implementation with $w = 64$ and 5 iterations on this machine. A direct comparison between the two architectures is complicated by the fact that they operate at different frequencies and the Tesla architecture has a greater number of registers, thus alleviating one of the main bottlenecks in the implementation.

Another important factor that makes comparisons difficult is the fact that the implementation in [4] uses 100 blocks loaded into GPU memory at a time. Even though they account for the memory transfer time, the presence of independent blocks makes it easier to find parallelism. In our implementation, we have considered only one data stream at a time, in order to focus on getting a practical SDR implementation suitable for single user terminals.

VI. CONCLUSION

A 3GPP standards compliant configurable turbo decoder has been implemented completely in software. More than an order of magnitude speed up over an implementation done purely on the CPU has been achieved. Different guarding mechanisms to mitigate the degradation in BER performance of the parallelized Log-MAP algorithm have been presented. The PIVIDSTW guarding mechanism as been shown to be capable of producing a BER performance that is within 0.01dB of the optimal case of a single code block. The principles used in the mapping of the Log-MAP algorithm can be readily extended to the newer architectures of Nvidia GPUs with larger number of cores and larger sized low latency memories to achieve a further increase in throughput.

REFERENCES

[1] W. E. Ryan, "A Turbo Code Tutorial," New Mexico State University, Tech. Rep., 1997.

[2] S. A. Abrantes, "From BCJR to turbo decoding: MAP algorithms made easier," University of Porto, Tech. Rep., 2004.

[3] M. Marandian, J. Fridman, Z. Zvonar, and M. Salehi, "Performance analysis of turbo decoder for 3GPP standard using the sliding window algorithm," in *Personal, Indoor and Mobile Radio Communications, 2001 12th IEEE International Symposium on*, vol. 2, sep/oct 2001, pp. E–127–E–131 vol.2.

[4] M. Wu, Y. Sun, and J. Cavallaro, "Implementation of a 3GPP LTE turbo decoder accelerator on GPU," in *Signal Processing Systems (SIPS), 2010 IEEE Workshop on*, oct. 2010, pp. 192–197.

[5] D. Lee, M. Wolf, and H. Kim, "Design space exploration of the turbo decoding algorithm on GPUs," in *Proceedings of the 2010 international conference on Compilers, architectures and synthesis for embedded systems*, ser. CASES '10. New York, NY, USA: ACM, 2010, pp. 217–226.

[6] J. Sun and O. Takeshita, "Interleavers for turbo codes using permutation polynomials over integer rings," *Information Theory, IEEE Transactions on*, vol. 51, no. 1, pp. 101–119, jan. 2005.

[7] O. Muller, A. Baghdadi, and M. Jezequel, "Exploring Parallel Processing Levels for Convolutional Turbo Decoding," in *Information and Communication Technologies, 2006. ICTTA '06. 2nd*, vol. 2, 0-0 2006, pp. 2353–2358.

[8] S. Hong and H. Kim, "An Analytical Model for a GPU Architecture with Memory-level and Thread-level Parallelism Awareness," Georgia Institute of Technology, Tech. Rep., 2009.

[9] D. Kirk, W. Hwu, and W. Hwu, *Programming massively parallel processors: a hands-on approach*, ser. Applications of GPU Computing Series. Morgan Kaufmann Publishers, 2010.

[10] J. Sanders and E. Kandrot, *CUDA by Example: An Introduction to General-Purpose GPU Programming*. Addison-Wesley, 2010.

[11] NVIDIA Corporation, *NVIDIA CUDA C Best Practices Guide Version 3.2*, 2010. [Online]. Available: http://developer.nvidia.com/cuda-downloads

[12] ——, *NVIDIA CUDA C Programming Guide Version 3.2*, 2010. [Online]. Available: http://developer.nvidia.com/cuda-downloads

[13] M. Harris, "Optimizing Parallel Reduction in CUDA," *NVIDIA Developer Technology*, 2007.

978-1-4673-0438-2/12 $31.00 © 2012 IEEE

2012 25th International Conference on VLSI Design

Run-time Prediction of the Optimal Performance Point in DVS-based Dynamic Thermal Management

Junyoung Park, H. Mert Ustun, and Jacob A. Abraham
Computer Engineering Research Center
The University of Texas at Austin, Austin, Texas, USA 78712
{jypark@cerc.utexas.edu, mustun@mail.utexas.edu, jaa@cerc.utexas.edu}

Abstract—Due to the increasing trend toward greater processor power density and computationally intensive applications, Dynamic Thermal Management (DTM) has become an essential technique in modern processors. Among many DTM techniques, Dynamic Voltage Scaling (DVS) is widely used because of its chief virtue – a cubic reduction in power at the relatively minor cost of a linear performance penalty. Because this reduction comes at a cost in execution speed, a key point of DVS-based DTM research is how accurately the processor predicts the optimal performance point where it can meet the thermal constraints while also minimizing the performance penalty. In this paper, we propose a new DVS-based DTM technique that makes the prediction of the optimal performance point more accurate. To achieve this, run-time prediction techniques are used and different power compositions due to process variations are considered from a VLSI perspective. The prediction process is performed by referring to one of the Look-Up Tables (LUTs) prepared during design time and also the average clock enable ratio that is dynamically calculated at run time. The simulation results show that we can achieve maximum processor performance while keeping the processor temperature from exceeding the threshold temperature.

I. INTRODUCTION

As process geometry shrinks and the total number of integrated transistors continues to increase, the power density of processors has been steadily rising. This growing power density has led to excessive heat in the processors, causing the die temperatures to rise. Furthermore, the increasing trend toward heavier workloads that run on these processors has been making them even hotter while also increasing the durations of such high temperatures during run time. Although normal operation often resumes once the operating conditions return to the temperature specifications, processors may fail to fulfill their intended functions when their operating conditions lie outside a threshold temperature. These occasional malfunctions may commonly arise from unacceptable propagation delay, large leakage current, or changes in noise margins. In addition, frequently exceeding the maximum operating temperature may cause irreversible changes in the operating characteristics, such as threshold voltage drift, electro-migration, gate oxide breakdown, and die fracture [1].

In order to prevent such reliability issues, numerous prior studies have been performed on Dynamic Thermal Management (DTM) techniques. Because temperature is proportional to power consumption, the basic concepts of DTM techniques are similar to those seen in existing low-power techniques [2]. One of the popular DTM approaches is the use of Dynamic Voltage Scaling (DVS), an open-loop technique where the processor is characterized for throughput at a given clock frequency and voltage [3]. This technique changes operating points when the die temperature goes over the fixed temperature threshold. Because the reduction in frequency combined with the quadratic effect of the supply voltage reduction results in cubic reduction of the dynamic power consumption, DVS is the most effective method of power reduction [4, 11]. This can be observed in the following equation:

$$P_{dynamic} \propto C_L V_{DD}{}^2 f \qquad (1)$$

where C_L is the total load capacitance on the chip, V_{DD} is the supply voltage and f is the switching frequency [5].

The challenge of a DVS-based DTM technique is that even though changing operating points effectively reduces power and, in turn, the die temperature, this reduction comes at a cost in execution speed [5]. Therefore, a key point in the design of DTM methods is how accurately the processor predicts the optimal performance point where it can meet the thermal constraints while also minimizing the performance penalty [2, 5, 6, 10]. Running at non-optimal operating points could result in a huge performance degradation or even lead to failure to lower the processor temperature.

In recent years, researchers have proposed a number of DVS-based DTM methods. However, most of these are impractical in real scenarios for two main reasons: (i) thermal simulators used for DTM design have limitations in regard to expressing the exact thermal behavior of processors, and (ii) the effect of process variation on DTM methods is not considered. These reduce the prediction accuracy and result in the misprediction of the optimal performance point when these DTM techniques are applied to real products.

In this paper, in order to alleviate the drawbacks stemming from previous DTM approaches, we suggest a new DVS-based DTM method that provides more accurate prediction. Our method considers different power compositions due to process variation during design time and monitors thermal stress of processors in run time. The prediction process is then performed by referring to one of the Look-Up Tables (LUTs) extracted from the Power-Frequency Tables and also the average clock enable ratio to find the voltage frequency operating point that would result in the best possible performance under current workload and temperature conditions.

The major contributions of our study can be summarized as follows: (i) we provide a novel run-time prediction scheme for a DVS-based DTM technique that has multiple operating points; (ii) this scheme does not require any complex post-measurement step to find a correlation between processor power and temperature; (iii) to our knowledge, this is the first paper that considers the process variation effect on DTM from a VLSI perspective.

The remainder of this paper is organized in the following manner. Section II reviews the previous related research. Section III shows the motivational studies. In Section IV, we present our DVS-based DTM method. In Section V, we implement our idea with a high-speed MIPS processor and show our simulation results. Finally, in the last two sections, we describe future work and present the conclusions of our work.

II. RELATED WORK

In order to maximize throughput while also meeting the thermal constraints, many researchers have considered finding the optimal performance points in DVS systems. Although many of these studies are meaningful in that they show the feasibility of the proposed techniques as DTM methods, most of them are inefficient or not optimal in real scenarios due to one or both of the two reasons specified earlier.

The first reason is that the thermal simulators used for DTM designs are limited in expressing the exact thermal behavior of processors. This is because thermal simulators are usually based on simplified forms of thermal models. Yeo et al. [7, 8] and Lee et al. [9] proposed the use of a DVS-based DTM scheme for a multimedia system in order to find the appropriate operating points. However, since they regard the total processor power as the dynamic power only, these methods are not suitable for modern processors that use nm-process technologies.

In addition, some factors are considered constant or omitted in thermal models due to their relatively small effect. For example, most DTM techniques have been built based on the thermal models which simply assume that heat transfer capability of active heat sinks is constant [13]. Such assumptions could lead to different results than those that we expected during design time if these DTM techniques are applied to real products.

Moreover, existing DTM techniques are designed with pre-defined thermal specifications during design time. These specifications are in fact determined by various environmental variables, such as the outside temperature, the package types and materials, the systems in which they are used, etc. Yet, since these variables can vary throughout the manufacturing process of processors and systems, it is difficult to design DTM methods with exact thermal specifications during design time.

In order to consider realistic thermal environments, some researchers [12, 15] have used post measurement steps to find a correlation between processor power and temperature. In their study, the decision to change operating points is based on a simple formula garnered via a regression test in the real processor. However, obtaining exact coefficients for this formula after chip fabrication is rather cumbersome, and post measurement steps introduce potential likelihoods of overestimation and underestimation, which in turn reduce accuracy. This is because the regression test is affected by the test environment and reproducing the exact die temperatures is impossible during test time. Moreover, since chips can have different coefficients according to the various process corners under which they were manufactured, this technique is not suitable for high-volume production of processors.

Skadron et al. [5] proposed a hybrid DTM technique that is based on combining fetch gating and binary DVS. The main idea here is to provide different DTM techniques according to the respective thermal stresses. Similar to our run-time prediction technique, they use run-time monitoring to capture the thermal stress of the processor in a given thermal environment. However, our study differs from theirs since we consider the selection of one among multiple DVS operating points while their study is limited to using either fetch gating or binary DVS. Our simulation results show that we can gain more benefit from a DVS with multiple operating points compared to binary DVS, especially for a set of periodic tasks.

The second reason that makes existing DTM methods impractical in real scenarios is that the process variation effect has not been taken into account in previous studies. In real cases, processors show quite different power compositions according to the process corners under which they were manufactured. If the DTM method acts in the same way for all processors made under various process corners, their intended DTM functions do not work for some processors. This means that DTM methods should be able to provide different appropriate responses even when the same method is implemented in

all processors, or in other words, the method should be able to adjust itself to the processor.

III. MOTIVATIONAL OBSERVATION

In this section, to better understand and further discuss our approach, we present the background of two topics; the thermal models and power compositions. In the first subsection, we scrutinize the equations that show the relationship between power and temperature. In the second subsection, we investigate power compositions under different process corners.

A. The Thermal Models

In order to obtain a formula for the relationship between processor temperature and power, we begin by reviewing two definitions in regard to energy. First, the total heat energy in a processor, E_{heat}, can be expressed by the integration of total power in the processor, $P(t)$, over run time, as in equation (2). Since $P(t)$ is the difference between the power of the heat source, $P_{source}(t)$, and the power transferred to the heat sink, $P_{sink}(t)$, we can rephrase equation (2) as in equation (3), as follows:

$$
\begin{aligned}
E_{heat} &= \int P(t)dt & (2) \\
&= \int [P_{source}(t) - P_{sink}(t)]dt. & (3)
\end{aligned}
$$

Second, the quantity of heat accumulated in the processor, Q_{heat}, can be written as in equation (4):

$$
Q_{heat} = cm[T(t) - T_0] \tag{4}
$$

where c is the specific heat, m is the mass of the processor die, $T(t)$ is the temperature of the processor at time t, and T_0 is the initial temperature of the processor. Since the total heat energy and the quantity of heat accumulated in the processor are the same, by using equations (3)∼(4) we can derive equation (5).

$$
T(t) = \frac{1}{cm} \int [P_{source}(t) - P_{sink}(t)]dt + T_0 \tag{5}
$$

The power consumption of the processor, $P_{source}(t)$, directly depends on the type of instructions being run by the processor and its operating frequency. On the other hand, the sink power, $P_{sink}(t)$, also known as the heat transfer, is determined by the package and the outside temperature.

Fig. 1 shows the simple heat transfer mechanism. Here, ΔX is the thickness of the package and ΔT is the temperature difference between the outside air and the processor. In a one-dimensional differential form, Fourier's Law of heat transfer can be expressed by equation (6), where k is the thermal conductivity of the package material and A is the surface area of the package. (To make this analysis simple, we assume here that k is constant.) Since ΔT and ΔX can be respectively rephrased as the difference between T_{out} and $T(t)$, and between X_{high} and X_{low}, we can replace these terms as in equation (7). This equation means that the greater the temperature

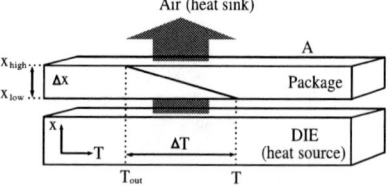

Fig. 1: A simple heat transfer mechanism between the heat source (processor) and heat sink (air)

difference between a processor and the outside, the larger the heat transfer becomes and hence the more effective the cool-down will be.

$$P_{sink}(t) = kA\frac{\Delta T}{\Delta X} \tag{6}$$

$$= kA\frac{[T(t) - T_{out}]}{[X_{high} - X_{low}]} \tag{7}$$

In order to keep the temperature constant, we need to satisfy the following condition:

$$\frac{dT(t)}{dt} = 0. \tag{8}$$

Applying the time derivative in equation (5), we can see that the temperature of a processor is constant when $P_{source}(t)$ is equal to $P_{sink}(t)$, as shown in the following equation.

$$P_{source}(t) = P_{sink}(t) = kA\frac{[T(t) - T_{out}]}{[X_{high} - X_{low}]} \tag{9}$$

Therefore, if we wish to keep the die temperature around the threshold temperature, T_{MAX}, and therefore have minimal performance degradation due to cool-down, the power consumption of a processor, $P_{source}(t)$, should be set close to $kA\frac{[T_{MAX} - T_{out}]}{[X_{high} - X_{low}]}$.

B. Power Compositions

DTM methods should pay attention to the leakage power as well as the dynamic power, especially when DVS controls the power consumption [14]. This is not only because the leakage power becomes greater as technology scales down, but also since the behavior of the leakage power exhibits different behavior than that of the dynamic power. Therefore, for an accurate prediction of the optimal performance point to be made, we should deal with the dynamic and leakage power separately.

Power composition can be affected by several thermal environmental factors. The voltage level affects both the dynamic and leakage power portion of the total power. While the dynamic power is affected by the operating frequency, the leakage power is influenced by the temperature. (The most significant leakage power source, the sub-threshold leakage, exponentially increases as the temperature rises [13].)

Process variations also affect the power composition of the processor by influencing the leakage power. Though all of the chips are topologically the same, the chips manufactured under the fast process corner (FF) have more leakage power than those manufactured under the slow corner (SS). Therefore, accurate DTM techniques should be adaptable to the process corners under which the processors were made. Fig. 2 shows the effect of process variations on the power composition of a MIPS processor which we designed using 45nm technology. All of the power values in this figure are normalized by

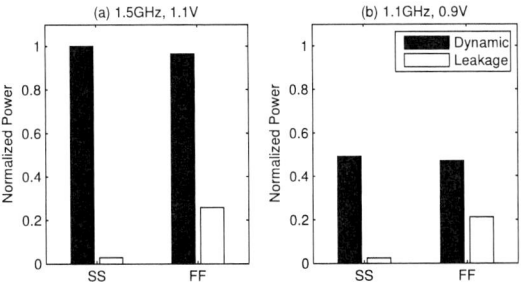

Fig. 2: Power compositions of our 45nm MIPS processor under different process corners

the dynamic power value of the operating point whose voltage is 1.1V and frequency is 1.5GHz at the SS corner.

IV. DESIGN CONCEPTS

In this section, the design process of our run-time prediction technique is presented.

A. Basic Components

The principal component of our technique, the Clock Controller, is made up of three main parts: the EN (enable) Generator, the Average Calculator and the LUTs. Fig. 3 offers an example of a configuration of this system.

Based on the temperature information coming from the sensors, the EN Generator produces the EN signal. This block implements the Stop-Go policy of clock gating [2]. As long as the temperature is lower than the threshold temperature, the EN value remains at 1. This keeps the system running at full-speed. However, if the temperature exceeds this threshold (a thermal emergency), the EN Generator changes the EN value to 0 in order to apply full clock gating (pausing the operation).

The Average Calculator works out the average EN value of the sampled EN values for the duration of the monitoring. The period of the monitoring time should be chosen accordingly in order to obtain the representative behavior of the thermal stress of processor applications in a given thermal environment. For example, for a set of periodic tasks that generate a high temperature, at least one iteration should be monitored.

The LUTs block in the Fig. 3 is composed of several Look-Up Tables which correspond to different process corners. Each LUT contains the information about the optimal performance points that correspond to the average EN values at a process corner. These LUTs are extracted from Power-Frequency Tables that are prepared during design time. The detailed process of creating LUTs is explained in the last subsection.

B. Operation

When the temperature exceeds the threshold temperature, the EN Generator and the Average Calculator start operating. In order to keep the temperature at around the threshold level, the EN signal will fluctuate between 1 and 0. Our Average Calculator counts the number of cases in which the EN signal is 1 and averages them over the total number of samples taken during a predefined period called monitoring time. Fig. 4 shows an example of an EN waveform during the monitoring period. After the monitoring time is completed, the processor can change its operating points. The destination operating point is determined by two factors; by referring to one of the LUTs that corresponds to the process corner under which the processor was manufactured, as well as to the calculated average EN value.

In our approach, DTM responses differ according to the severity of the thermal stress of the applications. If thermal stress is not severe, only clock gating is used; if not, the DVS system changes its operating point to the optimal performance point. By choosing appropriate DTM methods for given thermal stress, we can maximize the performance of the processor.

C. Average EN Value (η)

The average EN value, η, represents the thermal stress of the applications in a given thermal environment, since this is the ratio of the time at which the processor actually does work, i.e., when it is not clock gated. (The η has the range of 0~1.) Therefore, the actual performance of the processor is obtained by the effective frequency, which is equal to the product of the operating frequency and the value

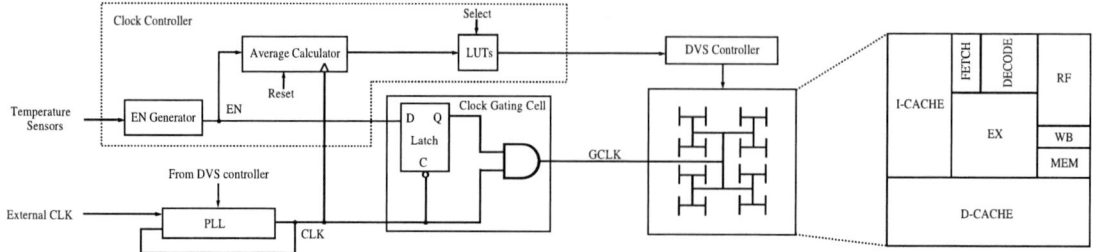

Fig. 3: (Left) Overall block diagram, and (Right) floorplan

of η. If the processor temperature is high, the average EN value will be low. On the other hand, if its temperature is low, the average EN value will be high.

Using the Stop-Go clock gating approach during the monitoring period fixes the processor temperature around the threshold temperature, T_{MAX}. According to equation (9), this constant processor temperature makes $P_{sink}(t)$ constant at the value of $kA\frac{T_{MAX}-T_{out}}{[X_{high}-X_{low}]}$. This means that during monitoring time, only $P_{source}(t)$ affects the EN signal generation that is composed of dynamic and leakage power.

Again, since both temperature and voltage are constant during the time of monitoring, we can assume that the leakage power is constant in this period. Therefore, the dynamic power is the only factor that affects EN signal generation. As a result, during monitoring time, we can say that $\eta \cdot f$ is the effective frequency that generates the same power level as that of a processor which has an average EN value of η and whose operating frequency is f. This is why we can use the value η as an index to find the optimal performance point.

D. Power-Frequency Tables

For different η values, there can be many child operating points that are derived from original operating points where the η values are 1.0. In order to evaluate each operating point from the perspective of power and performance, we prepare the Power-Frequency Table which is filled with measured and estimated Power-Frequency data pairs. Before further discussion of how to prepare the Power-Frequency Table, let us define the following terms that will be used throughout this paper.

- S_m^η: The derivative state (child state) derived from the original state (parent state) of S_m ($S_m^{1.0}$)
- D_m^η: The dynamic power of the state S_m^η
- L_m^η: The leakage power of the state S_m^η
- v_m^η: The operating voltage of the state S_m^η
- f_m^η: The effective frequency of the state S_m^η

TABLE I shows an example of a Power-Frequency Table that is prepared for a DVS processor with n-operating points. In this table, $S_0 \sim S_{n-1}$ are the original operating points of the processor. Here, S_0 is the default operating point in the DVS system which provides the highest frequency.

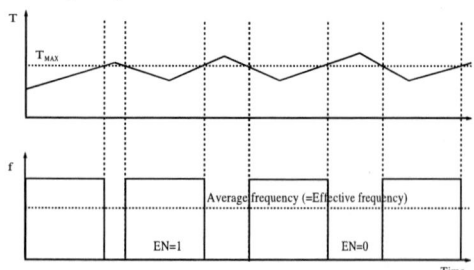

Fig. 4: Clock gating and average frequency

The power data for all the original operating points is measured by doing post-layout simulations with simple regression vectors. In all the simulations for different original operating points, the outside temperature should be kept at the threshold temperature, T_{MAX}. This is because in some cases, changing operating points would not reduce the power dissipation much, keeping the power almost the same as in the previous state, leading to a processor temperature that stays almost constant at T_{MAX}. The assembly codes that are used for our regression tests are based on previous research [15]. These codes contain instruction streams that heavily access the hot spots of the processor. In the case of our MIPS core simulations, the register file (RF) and the execution unit (EX) respectively dissipate around $45 \sim 50\%$ and $30 \sim 33\%$ of the total power for various instruction streams. Hence for our regression tests, we used R-type instructions which frequently activate RF and EX.

We can obtain the dynamic and leakage power values of the original states, D_m ($D_m^{1.0}$) and L_m ($L_m^{1.0}$), separately from post-layout simulations. (The symbol η for all notations can be omitted when its value is 1.0.) The Power-Frequency data pairs of the derivative operating points are estimated from those of the original operating points. Because the EN Generator determines effective frequency, we can rephrase D_m^η as the product of η and D_m ($D_m^{1.0}$) for the derivative states that have the same parent state. On the other hand, we can say the L_m^η values in the derivative states are similar to L_m ($L_m^{1.0}$) if they have the same parent state. This is because clock gating keeps the processor temperature at around the threshold temperature, T_{MAX}, during monitoring time. We can rephrase this information as seen in the following equations.

$$D_m^\eta \approx \eta \cdot D_m \qquad (10)$$

$$L_m^\eta \approx L_m \qquad (11)$$

In addition, we can write the effective frequency of the derivative operating points in terms of the original DVS operating point frequency as seen below.

$$f_m^\eta = \eta \cdot f_m \qquad (12)$$

Finally, in order to make the comparison of each value easier, we can simply normalize the power and the frequency values by the total power and the frequency of S_0 ($S_0^{1.0}$), respectively, as follows.

$$Normalized\ P_{source} = \frac{\eta \cdot D_m + L_m}{D_0 + L_0} \qquad (13)$$

$$Normalized\ f_{effective} = \eta \cdot \frac{f_m}{f_0} \qquad (14)$$

As previously mentioned, since process variation greatly affects power compositions, we need additional post-layout simulations in order to get the Power-Frequency Tables for the respective process corners. The whole process of obtaining Power-Frequency Tables is summarized in Algorithm 1. The Power-Frequency Table for our

TABLE I: The Power-Frequency Table for the DVS processor (n-operating points: $\{S_0, \cdots, S_m, \cdots, S_{n-1}\}$)

Operating points \ Average EN value (η)		1.0	0.9	...	0.1
$S_0(v_0, f_0)$	Power	1	$\frac{0.9D_0+L_0}{D_0+L_0}$	...	$\frac{0.1D_0+L_0}{D_0+L_0}$
	Frequency	1	0.9	...	0.1
$\vdots$	Power	$\cdot$	$\cdot$	...	$\cdot$
	Frequency	$\cdot$	$\cdot$	...	$\cdot$
$S_m(v_m, f_m)$	Power	$\frac{D_m+L_m}{D_0+L_0}$	$\frac{0.9D_m+L_m}{D_0+L_0}$	...	$\frac{0.1D_m+L_m}{D_0+L_0}$
	Frequency	$\frac{f_m}{f_0}$	$0.9\frac{f_m}{f_0}$	...	$0.1\frac{f_m}{f_0}$
$\vdots$	Power	$\cdot$	$\cdot$	...	$\cdot$
	Frequency	$\cdot$	$\cdot$	...	$\cdot$
$S_{n-1}(v_{n-1}, f_{n-1})$	Power	$\frac{D_{n-1}+L_{n-1}}{D_0+L_0}$	$\frac{0.9D_{n-1}+L_{n-1}}{D_0+L_0}$	...	$\frac{0.1D_{n-1}+L_{n-1}}{D_0+L_0}$
	Frequency	$\frac{f_{n-1}}{f_0}$	$0.9\frac{f_{n-1}}{f_0}$	...	$0.1\frac{f_{n-1}}{f_0}$

45nm MIPS processor at the SS corner was obtained as shown in TABLE II (Left).

E. Changing Operating Points and LUTs

In order to find the optimal performance point using the value of η, we should compare the normalized frequency and power values of the current derivative state, S_0^η, with values of all the derivative states belonging to $S_1 \sim S_{n-1}$. Among the derivative states that have equal or less power, the optimal performance point is the one that has the highest frequency. For example, if the average EN value, η, is 0.2, the candidate derivative states that correspond to each original state are $S_1^{0.5}$, $S_2^{0.8}$, $S_3^{1.0}$ and $S_4^{1.0}$ in TABLE II (Left). Among the candidates that have the same or less power, $S_3^{1.0}$ gives the best performance (the highest frequency). Therefore, S_3 is the optimal DVS operating point that the processor should move to.

In some situations, it is also possible that a better operating state cannot be found. This means that it is better to maintain the clock gating technique in the present operating point rather than switching to another. For example, if the state of the processor is $S_0^{0.9}$, the best choice for the DTM will be clock gating. Even though $S_1^{1.0}$ has a lower power than the existing state, its normalized frequency value (0.87) is lower than that of S_0 (0.90).

For a practical processor design, the LUTs are extracted from Power-Frequency Tables of each process corner. These LUTs contain information of destination operating points corresponding to different η values. TABLE II (Right) shows the LUT at the SS corner extracted from the Power-Frequency Table shown in TABLE II (Left). One of the LUTs is selected throughout the processor binning process which processor manufacturers typically perform for real products.

V. SIMULATION AND EVALUATION

To verify the proposed idea, we used the MIPS processor as the design base. The processor was implemented using a 45nm process

---

Algorithm 1 Obtaining the Power-Frequency Tables and LUTs for the DVS processor (n-operating points: $\{S_0, \cdots, S_m, \cdots, S_{n-1}\}$)

Input: $LIBs = \{SS, \cdots, FF\}$
Output: $LUTs = \{LUT_{SS}, \cdots, LUT_{FF}\}$

1: START:
2: **for all** $LIBs$ **do**
3:   **for all** n-operating points (original states) **do**
4:     Measure D_m ($D_m^{1.0}$) and L_m ($L_m^{1.0}$) through simulation
5:     Calculate D_m^η and L_m^η using the equations (13)~(14)
6:   **end for**
7:   Form the Power-Frequency Table and extract the LUT from it
8: **end for**
9: END:

---

and was highly optimized in order to meet the target frequency of 1.5GHz. To check the thermal behavior of this processor under various thermal stress and environments, we utilized an in-house thermal simulator that uses the same concepts as that of HotSpot [16]. Post-layout simulations were carried out using PrimeTime PX in order to obtain the power dissipation of the processor while running benchmark programs which show periodic operations. Then, such power values were fed into the thermal simulator. The simulation threshold temperature was selected as $70°C$ and the outside temperature was set between $15°C$ and $30°C$ with a step size of $5°C$. Finally, we assumed that an epoxy package was used and modeled the sink power accordingly.

We present the simulation results of a JPEG encoder program as an example to evaluate our proposed technique. Fig. 5 illustrates the EN signal waveforms and η values during the monitoring time. We can see that, as expected, the average EN value, η, decreases as the outside temperature increases. Based on this data, we can find the corresponding optimal point (the best performance point) from the LUT shown in TABLE II (Right). The η values of 0.44, 0.32, 0.26 and 0.15 respectively correspond to S_1, S_2, S_2 and S_3 in the table.

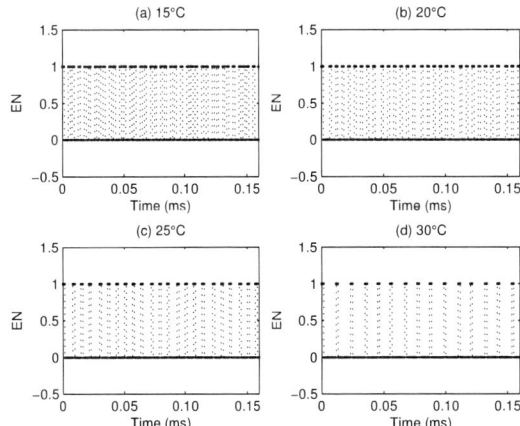

Fig. 5: The EN signal waveforms and the average EN values, η, obtained from the simulation of a JPEG encoder program: (a) $15°C$ (η=0.44); (b) $20°C$ (η=0.32); (c) $25°C$ (η=0.26); (d) $30°C$ (η=0.15)

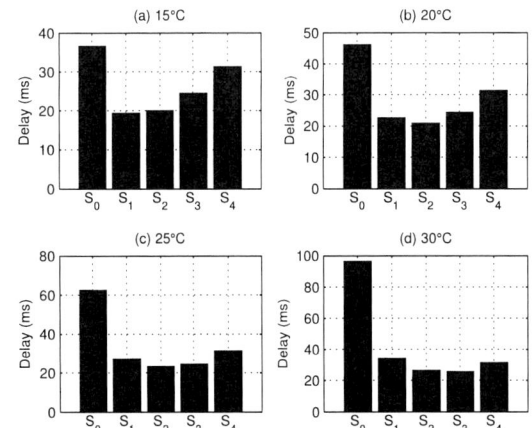

Fig. 6: The execution times of a JPEG encoder program at different operating points and temperatures: (a) $15°C$; (b) $20°C$; (c) $25°C$; (d) $30°C$

TABLE II: (Left) The Power-Frequency Table for our 45nm MIPS processor at the SS corner, and (Right) the corresponding LUT

Operating points / Average EN value (η)		1.00	0.90	0.80	0.70	0.60	0.50	0.40	0.30	0.20	0.10
$S_0(1.1V, 1.5GHz)$	Power	1.00	0.92	0.84	0.76	0.68	0.61	0.53	0.45	0.37	0.29
	Frequency	1.00	0.90	0.80	0.70	0.60	0.50	0.40	0.30	0.20	0.10
$S_1(1.0V, 1.3GHz)$	Power	0.58	0.53	0.49	0.44	0.40	0.36	0.31	0.27	0.22	0.18
	Frequency	0.87	0.78	0.69	0.61	0.52	0.43	0.35	0.26	0.17	0.09
$S_2(0.9V, 1.1GHz)$	Power	0.42	0.38	0.34	0.30	0.26	0.22	0.18	0.14	0.11	0.07
	Frequency	0.73	0.66	0.59	0.51	0.44	0.37	0.29	0.22	0.15	0.07
$S_3(0.8V, 0.9GHz)$	Power	0.33	0.30	0.27	0.23	0.20	0.17	0.14	0.11	0.08	0.04
	Frequency	0.60	0.54	0.48	0.42	0.36	0.30	0.24	0.18	0.12	0.06
$S_4(0.7V, 0.7GHz)$	Power	0.27	0.24	0.22	0.19	0.17	0.14	0.11	0.09	0.06	0.03
	Frequency	0.47	0.43	0.38	0.33	0.28	0.24	0.19	0.14	0.09	0.05

η	Optimal Performance Points
1.0	$S_0 \ (S_0^{1.0})$
0.9	$S_0 \ (S_0^{0.9})$
0.8	$S_1 \ (S_1^{1.0})$
0.7	$S_1 \ (S_1^{1.0})$
0.6	$S_1 \ (S_1^{1.0})$
0.5	$S_1 \ (S_1^{1.0})$
0.4	$S_1 \ (S_1^{0.9})$
0.3	$S_2 \ (S_2^{1.0})$
0.2	$S_3 \ (S_3^{1.0})$
0.1	$S_3 \ (S_3^{0.8})$
0.0	$S_4 \ (S_4^{1.0})$

TABLE III: The execution times of the three benchmark programs

Benchmarks / Outside Temp.		$15\,^{\circ}C$	$20\,^{\circ}C$	$25\,^{\circ}C$	$30\,^{\circ}C$
JPEG encoder	Worst	36.6 ms	46.2 ms	62.4 ms	96.3 ms
	Best (Predicted)	19.4 ms	20.8 ms	23.3 ms	25.5 ms
quick sort	Worst	12.6 us	15.9 us	21.5 us	33.2 us
	Best (Predicted)	6.5 us	7.3 us	8.0 us	8.9 us
matrix mult	Worst	48.1 us	60.6 us	81.9 us	126.4 us
	Best (Predicted)	25.8 us	28.1 us	31.4 us	34.5 us

Fig. 6 shows the total execution times of the JPEG encoder program corresponding to each operating point at various temperatures. We can see that the S_1, S_2, S_2 and S_3 provide the best performance points for $15\,^{\circ}C$, $20\,^{\circ}C$, $25\,^{\circ}C$ and $30\,^{\circ}C$ respectively, which are the same as the ones we had predicted during the monitoring time.

Although S_0 provides the fastest clock speed among the possible operating points, it causes a lot of heat dissipation in the processor, leading to the heavy use of the clock gating technique to reduce temperature, which in turn makes the total execution time longer than others. S_4 is the most effective operating point in decreasing the die temperature. However, its clock speed is too slow to reduce the total execution time. Since the optimal performance points provide the appropriate clock speed and voltage level, the power consumption of a processor can be made close to the heat transfer of the heat sink. As a result, a processor can keep its temperature constant around the threshold temperature while suffering the least performance penalty.

The execution times of the three benchmark programs are presented in TABLE III. The simulation results show that we can achieve maximum processor performance while keeping the processor temperature from exceeding the threshold temperature.

VI. FUTURE WORK

A strong enhancement to our proposed DTM method can be made by further improving the accuracy of LUTs which store the optimal operating points. This can be realized by making the LUT entries programmable and then programming them accordingly during the binning process based on the individual characterization of dies or wafers. Another extension can be made for multi-core microprocessors by having independent temperature monitors for each core and utilizing a temperature-aware workload distribution in the operating system.

VII. CONCLUSIONS

In this paper, we have designed a new DVS-based DTM technique from a VLSI perspective for DVS processors with multiple operating points. In order to increase the accuracy of the optimal performance point prediction, this technique monitors the thermal stress of a processor during run time and uses several LUTs for different process corners. The monitoring is performed while applying Stop-Go clock gating and the average EN value is recorded at the end of the monitoring time. Prediction of the optimal performance point is made using the average EN value and one of the LUTs that corresponds to the process corner under which the processor was manufactured. As a result, we maximize the performance of a processor while keeping the processor temperature from exceeding the threshold temperature.

REFERENCES

[1] P. Lall, M. G. Pecht and E. B. Hakim, "Influence of Temperature on Microelectronics and System Reliability," *CPC Press*, 1997.

[2] D. Brooks and M. Martonosi, "Dynamic Thermal Management for High-Performance Microprocessors," *International Symposium on High-Performance Computer Architecture*, pp. 171-182, Jan. 2001.

[3] M. Elgebaly and M. Sachdev, "Variation-Aware Adaptive Voltage Scaling System," *IEEE Transactions on VLSI Systems*, vol. 15, no. 5, pp. 560-571, May 2007.

[4] S. Zhang and K. S. Chatha, "Approximation Algorithm for the Temperature-aware Scheduling Problem," *International Conference on Computer-Aided Design*, pp. 281-288, May 2007.

[5] K. Skadron, "Hybrid Architectural Dynamic Thermal Management," *Design, Automation and Test in Europe Conference*, vol. 1, pp. 10-15, 2004.

[6] C. H. Lim, W. R. Daasch and G. Cai, "A Thermal-Aware Superscalar Microprocessor," *International Symposium on Quality Electronics Design*, pp. 517-522, 2002.

[7] I. Yeo, H. Lee, E. Kim and K. Yum, "Effective Dynamic Thermal Management for MPEG-4 Decoding," *International Conference on Computer Design*, pp. 623-628, 2007.

[8] I. Yeo and E. Kim, "Hybrid Dynamic Thermal Management Based on Statistical Characteristics of Multimedia Applications," *International Symposium on Low Power Electronics and Design*, pp. 321-326, 2008.

[9] W. Lee, K. Patel and M. Pedram, "GOP-Level Dynamic Thermal Management in MPEG-2 Decoding," *IEEE Transaction on VLSI Systems*, pp. 662-672, 2008.

[10] A. Cohen, L. Finkelstein, A. Mendelson, R. Ronen and D. Rudoy, "On Estimating Optimal Performance of CPU Dynamic Thermal Management," *Computer Architecture Letters*, vol. 2, 2003.

[11] P. Chaparro, G. Magklis, J. Gonzalez and A. Gonzalez, "Using MCD-DVS for Dynamic Thermal Management Performance Improvement," *The Tenth Intersociety Conference on Thermal and Thermomechanical Phenomena in Electronics Systems*, pp. 140-146, 2006.

[12] J. Lee, K. Skadron and S. Chung, "Predictive Temperature-Aware DVFS," *IEEE Transactions on Computers*, vol. 59, no. 1, pp. 127-133, 2010.

[13] D. Shin, S. Chung, E. Chung and N. Chang, "Energy-Optimal Dynamic Thermal Management: Computation and Cooling Power Co-Optimization," *IEEE Transactions on Industrial Informatics*, vol. 6, no. 3, pp. 340-351, 2010.

[14] R. Jejurikar, C. Pereira and R. Gupta, "Leakage Aware Dynamic Voltage Scaling for Real-Time Embedded Systems," *Proceedings of Design Automation Conference*, pp. 275-280, 2004.

[15] S. Chung and K. Skadron, "Using On-Chip Event Counters for High-Resolution, Real-Time Temperature Measurements," *The Tenth Intersociety Conference on Thermal and Thermomechanical Phenomena in Electronics Systems*, pp. 140-146, 2006.

[16] K. Skadron, M. R. Stan, K. Sankaranarayanan, W. Huang, S. Velusamy and D. Tarjan, "Temperature-Aware Microarchitecture: Modeling and Implementation," *ACM Transactions on Architecture and Code Optimization*, vol. 1, no. 1, pp. 94-125, Mar. 2004.

Temperature-aware Task Partitioning for Real-Time Scheduling in Embedded Systems*

Zhe Wang, Sanjay Ranka and Prabhat Mishra

Dept. of Computer and Information Science and Engineering

University of Florida, Gainesville, USA

Email: {zhwang, sanjay, prabhat}@cise.ufl.edu

Abstract—Both power and heat density of on-chip systems are increasing exponentially with Moore's Law. High temperature negatively affects reliability as well the costs of cooling and packaging. In this paper, we propose task partitioning as an effective way to reduce the peak temperature in embedded systems running either a set of periodic heterogeneous tasks with common period or periodic heterogeneous tasks with individual period. For task sets with common period, experimental results show that our task partitioning algorithms is able to reduce the peak temperature by as much as $5.8^\circ C$ as compared to algorithms that only use task sequencing. For task sets with individual period, EDF scheduling with task partitioning can also lower the peak temperature, as compared to simple EDF scheduling, by as much as $6^\circ C$. Our analysis indicates that the numbers of additional context switches (overhead) is less than 2 per task, which is tolerable in many practical scenarios.

I. INTRODUCTION

The power density of on-chip systems rises very rapidly and the rate will continue to grow [1]. This makes thermal management a significant design challenge for microprocessors. Rise in on-chip temperature directly impacts the performance and time-to-failure of switching devices. This is accentuated by the fact that the cooling and packaging cost rises super-linearly with the growth of power density [2]. Also, the cost of cooling and packaging is one of the significant contributors to computer system [3]. High temperature increases leakage power of the chip, and thereby potentially lead to thermal runaway. It has been shown that a $10\text{-}15^\circ C$ reduction in peak temperature can double the lifetime of the chip [4].

Existing power-aware techniques do not address the temperature issues in embedded systems. The main reason is that the power distribution of multiprocessors is not uniform. Localized heating rises much rapidly than the whole chip, leading to nonuniform temperature distribution on the chip with localized high-temperature hot spots and spatial gradients [5]. Traditional methods to control the on-chip temperature is to employ better packaging and cooling techniques (e.g., active fan cooling, water cooling and heat pipe). These active cooling systems may not always be suitable for embedded systems because of the space and battery limitations. Building thermal analysis ability into EDA flow allows the system to address the thermal impact on various on-chip parameters and incorporate effects of non-uniform thermal profiles during IC design process. However, such technologies are unable to deal with a variety of runtime situations.

In this paper, we focus on software approaches for thermal management. These approaches are flexible and do not have some of the limitations that are described above. The processor thermal behavior can be effectively modeled using RC model [5]. If the average power of a processor is P over a time period t, then the transient temperature $T(t)$ at the end of this period, using this model, is given by:

$$T(t) = P \times R + T_A - (P \times R + T_A - T_i)e^{-t/RC} \quad (1)$$

where R is the thermal resistance and C is the thermal capacitance, T_A is the ambient temperature and T_i is the initial temperature.

Given a particular chip and its outside environment, ambient temperature, thermal resistance and capacitance are fixed. Based on the parameters of Equation (1), there are three major factors affecting the on-chip transient temperature: average power of the processor, initial temperature and execution time. Dynamic Voltage and Frequency Scaling (DVFS) can be used to reduce the power consumption by lowering the supply voltage and operating frequency, thereby reduce the on-chip temperature [6]–[11]. However, DVFS faces a serious problem in time-constrained applications. Temperature aware task sequencing algorithm, which reduces the initial temperature, developed in [12] can reduce peak temperature compared to a random sequence. However, temperature aware task sequencing fails to reduce temperature in cases when one or more of the "hot" [1] tasks are long. The algorithm to defer execution of hot tasks [13] fails to reduce temperature in the same situation. This is because when the execution time of a "hot" task is too long, it can lead to a high steady-state temperature irrespective of the initial temperature.

We propose to partition the "hot" tasks into multiple subtasks and interleave these subtasks with "cool" tasks to reduce the overall maximum temperature (see Figure 2). The focus of this paper is to use this technique effectively to reduce the maximum temperature. To the best of our knowledge, our work is the first attempt to develop efficient task partitioning algorithms to demonstrate significant temperature reduction. In this paper, we propose a heuristic task partitioning algorithm using "cool" tasks to interleave "hot" tasks for a periodic set of tasks with common period. We also propose another heuristic task partitioning algorithm for a periodic set of tasks with individual period. Finally, we provide a thorough evaluation and comparison to show how task partitioning can assist in thermal-aware management problems. Experimental results show that our algorithm outperforms the task sequencing algorithm [12] by reducing the peak temperature by as much as $6^\circ C$.

The rest of the paper is organized as follows. Section II describes the background of thermal aware analysis. Section III introduces the problems of periodic tasks with common and individual period, and proposes heuristic task partitioning algorithms to solve them. Section IV compares these algorithms with task sequencing algorithm and EDF algorithm, respectively. Section V concludes the paper.

II. PRELIMINARIES

In this section, we briefly describe three related concepts. First, we discuss how to measure the thermal profile of a task. Next, we present how to analyze the thermal profile of a task sequence. Finally, we define the peak temperature of a task sequence.

---

*This work was partially supported by NSF grant CCF-0903430 and SRC grant 2009-HJ-1979.

[1] We define "hot" tasks as tasks with higher average power consumption, and "cool" tasks as tasks with lower average power consumption.

A. Thermal profile of individual tasks

The basic thermal equation has already been introduced in Equation (1). By letting $t \rightarrow \infty$ in Equation (1), we can get the steady-state temperature:

$$T_S = P \times R + T_A \qquad (2)$$

Based on our experiments, it takes less than 1 second to reach steady-state temperature for a $1.5GHz$ processor with product of thermal resistance and thermal capacitance to be $0.2053\ Joules/Watt$.

B. Thermal profile of task sequences

Consider a periodic set of heterogeneous tasks (i.e., tasks with different thermal profiles), with the execution time of these tasks given by $c_1, c_2, ..., c_N$. The periods of these tasks are given by $p_1, p_2, ..., p_N$. The average power consumption during execution time is given by $P_1, P_2, ..., P_N$. Suppose these tasks are ordered in a particular sequence $S = <\tau_1, ..., \tau_N>$. The hyper-period of these tasks is defined by $LCM_{i=1}^{N}\{p_i\}$, where LCM stands for the least common multiple. By executing these tasks in a hyper-period for a larger number of iterations (they are periodic tasks), the temperature of these tasks will rise from initial temperature and reach a final temperature, where the the hyper-period temperature profile repeated periodically [12]. We call this steady-state the hyper-period steady-state. Using the above arguments, we can analyze the thermal profile of one hyper-period sequence, other hyper-period sequences will have exactly the same thermal profile as this one.

C. Peak temperature of task sequences

Using the above simplifications, we have the final temperature of each task as follows:

$$T_1 = P_i \times R + T_A - (P_i \times R + T_A - T_N)e^{-c_1/RC}$$
$$\cdots$$
$$T_N = P_i \times R + T_A - (P_i \times R + T_A - T_{N-1})e^{-c_1/RC} \qquad (3)$$

As we can see, the temperature in Equation (1) is a monotonic function, when $P_i \times R + T_A > T_i$, the temperature of the processor increases during the task execution time and vice versa. Therefore, either T_i or $T(t)$ is the maximum temperature during the execution time of the task. Thus, we define the maximum final temperature of tasks in one hyper-period as the peak temperature of the sequence.

$$\text{peak temperature} = \max\{T_1, T_2, \cdots, T_N\} \qquad (4)$$

III. TASK PARTITIONING ALGORITHMS

There are two major challenges developing algorithms that use task partitioning to reduce peak temperature:

1) *Number of Partitions:* A task can be partitioned into a large number of very small pieces. However, this may result in significant overhead of preemption and restart. Choosing the right number of partitions that carefully tradeoffs the number of partitions and the resultant temperature reduction is important.

2) *Sequencing of Subtasks:* A reordering of "hot" and "cool" tasks has to ensure that the subtasks of a given task maintain the sequential order. For example if a task A is decomposed into subtasks A1, A2 and A3. A1 should always be executed before A2, and A2 should always be executed before A3.

Besides the obvious novelty of proposing a partitioning approach for addressing the thermal issues, the paper develops novel algorithms to address the following two broad scenarios:

1) A periodic set of tasks with common period. All the tasks have the same arrival time and deadline.
2) A set of periodic tasks with individual period. Each task may have different arrival time and deadline.

In this section, we first give an illustrative example showing that the peak temperature of task partitioning algorithm is less than that of task sequencing algorithm [12]. Assume that two tasks, τ_1 and τ_2, have average power consumption P_1 and P_2, respectively. Their execution times are t_1 and t_2, $t_1, t_2 > 0$. Without loss of generality, we assume $P_1 < P_2$. Therefore, τ_1 is a "cool" task and τ_2 is a "hot" task. Based on the task sequencing algorithm [12], as Figure 1 shows, the "hot" task is followed by a "cool" task. The temperature at time t is denoted as $T(t)$, $t \in [0, t_1 + t_2]$. Thus, the initial temperature is $T(0)$. The ambient temperature is T_A.

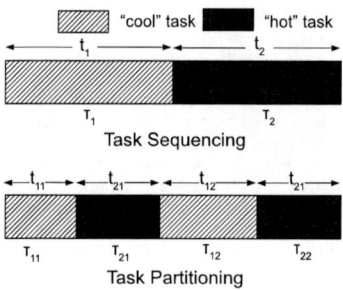

Fig. 1. An example of the task sequencing and task partitioning. There are two tasks in task sequencing: a "cool" task τ_1 followed by a "hot" task τ_2. The execution time of τ_1 and τ_2 are t_1 and t_2, respectively. These two tasks are partitioned into four subtasks in task partitioning. The (task, execution time) sequence is (τ_{11}, t_{11}), (τ_{21}, t_{21}), (τ_{12}, t_{12}), (τ_{22}, t_{22}), where $t_1 = t_{11} + t_{12}$, $t_2 = t_{21} + t_{22}$.

We also introduce the task partitioning. We assume that both task τ_1 and τ_2 are partitioned into two subtasks. By interleaving the subtasks, the (task, execution time) sequence is (τ_{11}, t_{11}), (τ_{21}, t_{21}), (τ_{12}, t_{12}), (τ_{22}, t_{22}) (see Figure 1). Figure 2 shows the transient temperature of task sequencing and task partitioning. We can see that task partitioning can achieve lower peak temperature because the "cool" task absorbs the heat generated by the "hot" task.

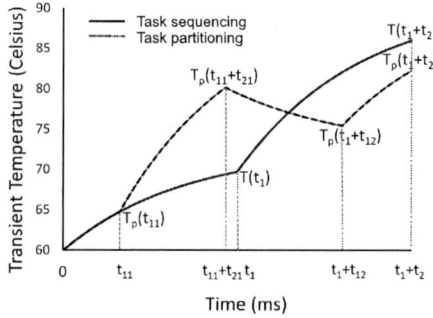

Fig. 2. Transient temperature comparison between the task sequencing and task partitioning. In task sequencing, the temperature after τ_1 finishes is $T(t_1)$. The temperature after τ_2 finishes is $T(t_1 + t_2)$. In task partitioning, the temperature after τ_{11} finishes is $T_p(t_{11})$. The temperature after τ_{21} finishes is $T_p(t_{11} + t_{21})$. The temperature after τ_{12} finishes is $T_p(t_1 + t_{12})$. The temperature after τ_{22} finishes is $T_p(t_1 + t_2)$.

A. Periodic tasks with common period

Consider a periodic set of N heterogenous tasks L, let P_i be the average power consumption during the execution time c_i of task τ_i. The goal is to find a sequence of these tasks using task partitioning to minimize the peak temperature. Due to the fact that all the tasks have the same common period, each task can be moved freely within the period. Furthermore, we need to analyze only one period, other periods will be the same as this one.

Methods that only reorder the sequence of tasks (such as TSA [12]) fail to further reduce the peak temperature when some "hot" tasks of long enough execution time can reach close to its steady-state temperature and this temperature is relatively independent of the initial temperature. We propose a task partitioning algorithm to reduce the temperature using preemption. The main idea of task partitioning algorithm is to partition the "hot" tasks into several subtasks and interleaving them with "cool" tasks to absorb the heat generated by "hot" subtasks. To partition "hot" tasks into more subtasks and generate enough "cool" subtasks to interleave with them, we divide the tasks into *categories* based on their power profile. The tasks in higher categories are partitioned into more subtasks. The subtasks in lower categories act as "cool" tasks and interleave with subtasks in higher categories. Details of the algorithm is shown in Algorithm 1.

Algorithm 1 Task Partitioning Algorithm (TPA)

1: Sort the tasks based on the power profile from coolest to hottest
2: Group the sorted tasks into k categories with equal number of tasks. These categories are numbered from 1 to k.
3: Partition tasks in category j, $2 \leq j \leq k$, into 2^{i-1} equal subtasks. Partition tasks in category 1 into 2 equal subtasks.
4: **for** $i = 1$ to $k - 2$ **do**
5:     Interleave tasks of ith category with tasks of $i+1^{th}$ category to form the new $i+1^{th}$ category
6: **end for**// *Now only two categories are left. The first one is category k and the second one is a new category $k-1$ derived by combining category 1 through $k-1$.*
7: Insert the subtasks in new category $k-1$ into the intervals of tasks in category k.

First, we sort the tasks based on their power consumption from coolest to hottest. In the second step, we group the sorted tasks into k *categories* from category 1 to category k. Category 1 is the category of coolest n/k tasks, and category k is the category of hottest n/k tasks. The reason that we divide the tasks into different categories is that we need to partition "hotter" tasks into more subtasks to reduce the peak temperature. We also need enough "cooler" subtasks to separate the "hot" subtasks (see Figure 3). By having different categories of tasks, we can achieve both targets simultaneously. In this paper, we assume that when a task is partitioned into several small subtasks, the subtasks have the same average power consumption as the original task. We used Wattch [14] to compute average power of the subtasks. As expected, these numbers are comparable with the average power of the original tasks in our benchmark set.

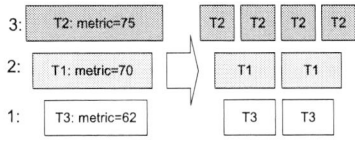

Fig. 3. Task Partitioning Algorithm (Step 1): Sort the tasks based on the power consumption, group the tasks into 3 categories. Task in category 3 is the hottest and task in category 1 is the coolest. Partition tasks into subtasks. Task in category 3 is partitioned into $2^{3-1} = 4$ equal subtasks. Task in category 2 is partitioned into $2^{2-1} = 2$ equal subtasks. Task in category 1 is partitioned into 2 equal subtasks.

In the next step, we partition tasks in category j, $2 \leq j \leq k$ into 2^{j-1} equal subtasks. The tasks in category 1 are partitioned into 2 equal subtasks. After recursively interleaving tasks in i^{th} category with tasks in $i+1^{th}$ category, there are only two categories left. The first one corresponds to category k and the second one is derived by combining category 1 through $k-1$. For the sake of convenience, we call this combined set as new category $k-1$. We now have $n/k \cdot 2^{k-1}$ tasks in category k, and $n/k \cdot 2^{k-1}$ intervals between these subtasks. We also have $n/k \cdot (2 + 2 + 4 + ... + 2^{k-2}) = n/k \cdot 2^{k-1}$ tasks in new category $k-1$. Therefore, we have enough tasks from category $k-1$ to interleave with the tasks in category k (see Figure 4). In the last step, we insert the tasks of new category $k-1$ into the intervals of category k (see Figure 5).

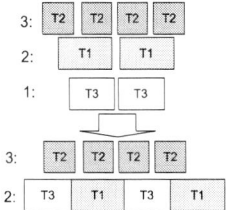

Fig. 4. Task Partitioning Algorithm (Step 2): Interleave subtasks of category 1 with subtasks of category 2 as the new category 2. Now we have enough subtasks from category 2 to interleave with the subtasks of category 3.

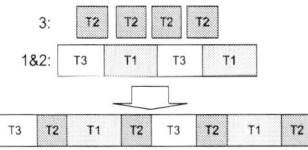

Fig. 5. Task Partitioning Algorithm (Step 3): We insert subtasks of new category 2 into intervals of subtasks of category 3 to get final sequence.

B. Periodic tasks with individual period

Consider a set of periodic N heterogeneous tasks in a set L where each task has its own period p_i. The arrival time a_i is equal to the start time of its period and the deadline d_i is equal to the end time of its period. Recall that the temperature profile of one hyper-period is identical to that of other hyper-periods. We only need to analyze the task instances within one hyper-period.

Theoretically, each periodic task corresponds to an infinite sequence of identical activities, called *instances*. The first instance of each periodic task arrives at time 0. Let P_i be the average power consumption during the execution time c_i of task τ_i. The goal is to find a sequence of these tasks using task partitioning to minimize the peak temperature.

The algorithm developed in the previous section does not apply to this scenario as the arrival and deadline constraints have to be carefully addressed. This additional constraint limits the task ordering that can be used. In this section, we develop a novel algorithm that integrates task partitioning technique into EDF scheduler[2]. We first use the EDF scheduler to schedule the tasks to get an initial sequence S_i. Based on the initial sequence, we can use Equation (3) to get the thermal profile of the task sequence. Using this thermal profile, let the task instance where peak temperature occurs be called "hot" task instance, denoted by τ_h.

---

[2]Earliest-Deadline-First (EDF) [15] is a dynamic scheduling algorithm that schedules the tasks according to their absolute deadlines. Tasks with earlier deadlines will be executed at higher priorities.

The peak temperature is reduced by partitioning τ_h into several subtask instances interleaved by other "cool" task instances. Because the "hot" task instance τ_h cannot move before its arrival time or after its deadline, we only need to analyze the interval between the arrival time and deadline of τ_h, represented as a_{τ_h} and d_{τ_h}, called *hot interval*. All other task instances except the "hot" task instance in the hot interval are called *"cool" task instances*, denoted by τ_c. It is worth noting that this definition of "hot" and "cool" tasks is substantially different as compared to the previous section that considered periodic tasks with common period. A high level description of the EDF scheduling with our task partitioning is given in Algorithm 2.

Algorithm 2 EDF with task partitioning

1: Use EDF scheduler to get the initial schedule of these tasks
2: **while** loop for M times **do**
3:     Calculate the thermal profile of task sequence, find the "hot" task instance τ_h where peak temperature occurs.
4:     Partition the task instances whose execution period overlap with the arrival time or deadline of the "hot" task instance.
5:     In the hot interval, remove all the subparts of τ_h and calculate the available slack for each "cool" task instance.
6:     **while** there are parts of τ_h unassigned and some "cool" task instance has available slack **do**
7:         **for** each "cool" task instance τ_{ci} in the hot interval **do**
8:             **if** $slack_i > 0$ **then**
9:                 Append one unit of τ_h into τ_{ci} and update the slack for all "cool" task instances
10:            **end if**
11:        **end for**
12:    **end while**
13:    If there is still some subparts of τ_h unassigned, scan the hot interval and assign them uniformly into the idle time.
14: **end while**

There are two major steps that are required for the Algorithm 2 and are as follows:

a) Partition the task instances whose execution period overlaps with the arrival time and/or deadline of the "hot" task instance: After finding the task instance τ_h, we can limit our analysis to the hot interval. It is likely that some task instances are across either a_{τ_h} or d_{τ_h}. We partition such task instances across these time lines using the following equation. If some task τ_k across the time line (either a_{τ_h} or d_{τ_h}) γ, that is, $s_{\tau_k} < \gamma < e_{\tau_k}$, where s_{τ_k} and e_{τ_k} are the start time and end time of τ_k, respectively. We have:

$$\tau_k|[s_{\tau_k}, e_{\tau_k}] \to \tau_k'|[s_{\tau_k}, \gamma], \ \tau_k''|[\gamma, e_{\tau_k}] \quad (5)$$

Where $\tau_k'|[a_{\tau_k}, \gamma]$ means task τ_k' starts executing at s_{τ_k} and will finish at γ.

The task instance τ_k is partitioned into at most 2 subtask instances, τ_k' and τ_k'' based on the time line. We limit our further analysis to the subtask instance that is in the hot interval.

b) Slack allocation: EDF, in general, can break up a task into many subtasks to ensure the arrival and deadline constraints. In particular, "hot" task instance may have been decomposed into multiple subtasks. Thus there may exist more than one part of "hot" task instance τ_h in the hot interval.

We first scan the hot interval and remove all the parts of the "hot" task instances from hot interval and combine them into one task (see Figure 6). Due to this step, other "cool" task instances in the hot interval will have more flexibility in time constraints. The

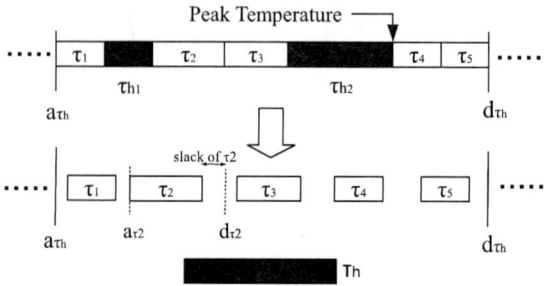

Fig. 6. First portion of the slack allocation step in the EDF based partitioning algorithm. Peak temperature occurs at task instance τ_h. a_{τ_h} and d_{τ_h} are the arrival time and deadline of τ_h, respectively. There are two parts of τ_h: τ_{h_1}, τ_{h_2}. Both τ_{h_1} and τ_{h_2} are removed from the sequence. Then other "cool" tasks instances (τ_{c1} - τ_{c5}) in the hot interval have more flexibility. Therefore, the slacks of these "cool" task instances can be calculated (For ease of presentation in this limited space, only τ_{c2}'s arrival time a_{τ_2}, deadline d_{τ_2} and slack are shown).

extent of flexibility available under time constraints is quantified by the *slack*. Slack is the difference between *Latest Start Time* (LST) and *Earliest Start Time* (EST). EST, LST and slack of "cool" task instance τ_i can be computed as follows:

$$EST_i = \max(a_{\tau_h}, a_{\tau_i}, EST_{pred_i} + c_{pred_i})$$
$$LST_i = \min(d_{\tau_h}, d_{\tau_i}, LST_{succ_i}) - c_i$$
$$slack_i = LST_i - EST_i \quad (6)$$

where $pred_i$ is the predecessor of τ_i defined by EDF schedule, $succ_i$ is the successor of τ_i defined by EDF schedule. Here, a_{τ_i} and d_{τ_i} are the arrival time and deadline of task instance τ_i, respectively, and c_i is the execution time of τ_i.

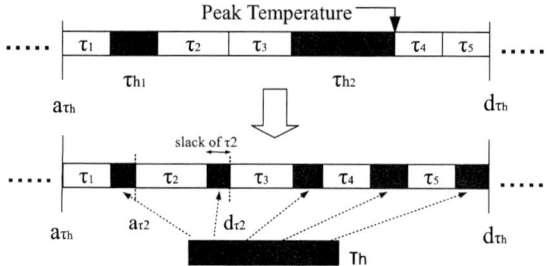

Fig. 7. Second portion of the slack allocation step in the EDF based partitioning algorithm. Calculating the slacks of all the tasks in the hot interval. For each task τ_{ci}, append one *unit* of τ_h into τ_{ci} at a time and update the slacks. When no slacks left and there are still some parts of τ_h unassigned, scan the hot interval and assign them uniformly into the idle time.

After the available slacks have been calculated for all the "cool" task instances, we insert the "hot" task instance back into these slacks as uniformly as possible. This is done by appending one *unit* slack of τ_h to each "cool" task instance at a time (one *unit* slack is a small constant representing a small period of time). If there is still some subparts of τ_h unassigned, there must exist some idle time that no "cool" task instance can have slack on it. The remaining part of the τ_h is uniformly decomposed into these idle times (see Figure 7).

The above process corresponds to a single iteration of the algorithm. This process is applied iteratively for several iterations. In each iteration, a potentially new "hot" task instance is chosen based on the thermal profile. The number of iterations or loops that should be iterated over can be fixed or chosen based on the level of improvement achieved. For our experiments, we found that 10-15 iterations are sufficient for deriving most of the benefits.

IV. EXPERIMENTAL RESULTS

We used a platform that is based on ARM Cortex A8 [16]: 2-width in-order issue, 32KB instruction and data caches for evaluating our algorithms. The clock speed was set to 1.5GHz. Using default thermal configurations in HotSpot [17] and the floorplan and silicon area of ARM Cortex A8, the thermal resistance and capacitance can be computed as $1.83^{o}C/Watt$ and $0.112J/^{o}C$, respectively [12]. The ambient temperature was set at $45.15^{o}C$. We used the architecture-level power simulator Wattch [14] to obtain the power consumption of tasks.

A. Tasks with Common Period

For tasks with common period, synthetic tasks are generated to find the profitable number of categories to achieve a lower peak temperature and real benchmarks are generated to compare the performance between task sequencing algorithm (denoted by TSA [12]) and our task partitioning algorithm (denoted by TPA). For tasks with individual period, real benchmarks are generated.

1) Synthetic Tasks: Tasks were generated to compare the thermal reduction achieved by our task partitioning algorithm with different number of categories. The numbers of clock cycles of tasks are uniformly distributed in $[1.5 \times 10^{8}, 1.45 \times 10^{9}]$, the power consumption of these jobs are uniformly distributed in $[5, 25]$ Watt. The numbers of jobs tested are $32, 64, 128, 192, 256$. The numbers of categories in task partitioning are $2, 3, 4, 5$.

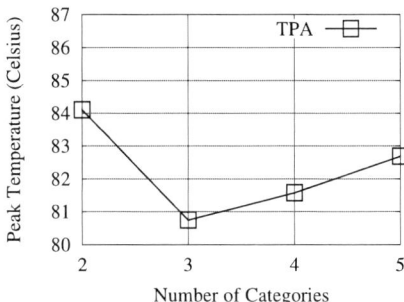

Fig. 8. Peak temperature comparison between different number of categories.

Figure 8 shows the peak temperature comparison using different number of categories for the task partitioning algorithm. The number of tasks used in this scenario is 64. Recall from the task partitioning algorithm, different number of categories will lead to different number of partitioning within each categories. It is necessary to find a profitable number of categories to achieve a lower peak temperature. The results show that 3 categories are enough to achieve most of the benefits for reducing the peak temperature. In fact using larger number of categories may result in slightly higher temperature. This is because when the number of categories is large, the number of tasks in the highest category is fewer. Some tasks originally in the highest category will be pushed into the second highest category. These tasks, who are already very hot, are treated as "cooler" tasks to interleave with tasks in the highest category. However, these tasks cannot absorb enough heat from the "hot" tasks. This can lead to the algorithm effectively not able to reduce the temperature significantly. Our experiments suggest that 3 categories should be ideal for most practical scenarios.

2) Real Benchmarks: We use Mibench [18] and Mediabench [19] to form four sets of the benchmark tasks. The characteristics of these benchmarks are shown in Table I.

TABLE I
THE FOUR SETS OF BENCHMARK TASKS WITH COMMON PERIOD

set1	patricia, adpcm, rijndael, susan, crc, FFT, dijkstra, epic
set2	patricia, djpeg, adpcm, sha, FFT, rijndael, susan, rijndael
set3	sha, djpeg, FFT, rijndael, dijkstra, epic, rijndael, susan
set4	rijndael, dijkstra, FFT, gsm, sha, patricia, pegwit, djpeg

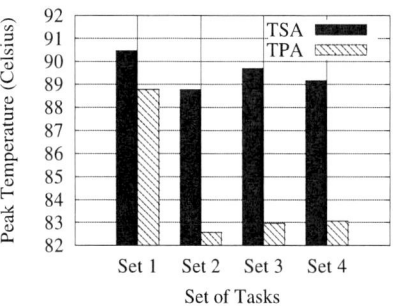

Fig. 9. Peak temperature comparison between task sequencing algorithm and task partitioning algorithm on real tasks. Number of categories is 3.

Figure 9 shows the peak temperature comparison between task sequencing algorithm and our task partitioning algorithm on real tasks for 3 categories. The task partitioning algorithm reduces the peak temperature by as much as $5.8^{o}C$ compared with task sequencing algorithm. Given the above results, a choice of 3 categories should generally provide a good tradeoff between low context switching overhead and high reduction in peak temperature as discussed in Section IV-C.

B. Tasks with Individual Period

For tasks with individual period, we also use Mibench [18] and Mediabench [19] to form four sets of the benchmark tasks. These benchmarks are shown in Table II. Each periodic task has individual periods. We set the deadline of all tasks to be the end of its own period. Also, we assume that arrival time of the first instance of all tasks is 0.

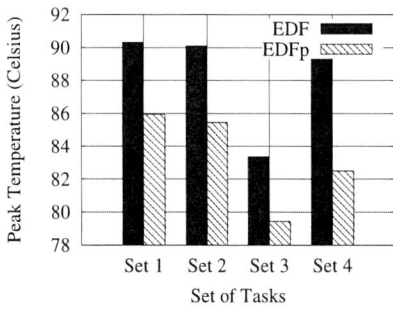

Fig. 10. Peak temperature comparison between EDF and EDFp ($M = 15$).

We compare our approach with EDF that does not directly address thermal issues. Figure 10 shows the peak temperature comparison between EDF and our approach, called EDFp. The experimental results show that the EDFp outperforms EDF by as much as $6^{o}C$.

TABLE II
THE FOUR SETS OF BENCHMARK TASKS WITH INDIVIDUAL PERIOD

Set Number	Benchmark	Execution Time (Clock Cycles)	Period (Clock Cycles)
set1	patricia	1.32×10^8	7.2×10^8
	adpcm	1.81×10^7	1.35×10^8
	susan	1.48×10^7	2.66×10^8
	crc	2.13×10^8	9.73×10^8
	dijkstra	2.9×10^7	1.74×10^8
set2	pegwit	2.12×10^7	2.72×10^7
	gsm	6.35×10^6	9.52×10^7
	epic	3.21×10^7	1.44×10^8
	crc	2.13×10^8	9.73×10^8
	patricia	1.32×10^8	7.2×10^8
set3	gsm	6.35×10^6	9.52×10^7
	patricia	1.32×10^8	7.2×10^8
	susan	1.48×10^7	2.66×10^8
	djpeg	1.57×10^7	9.42×10^7
	adpcm	1.81×10^7	1.35×10^8
set4	rijndael	4.5×10^7	3.6×10^8
	susan	1.48×10^7	2.66×10^8
	FFT	1.54×10^8	1.1×10^9
	sha	4.8×10^7	2.7×10^8
	adpcm	1.81×10^7	1.35×10^8

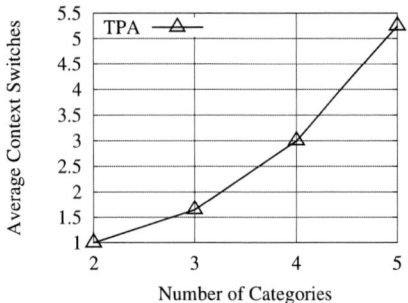

Fig. 11. Average number of context switches per task of task partitioning algorithm for various number of categories.

C. Context Switching Overhead

We first analyze the context switching overhead introduced by partitioning the tasks with common period. Figure 11 shows the number of context switch per task for task partitioning algorithm using variable number of categories. For task partitioning algorithm with 3 categories (which is shown most profitable in Figure 8), the number of context switches is about 1.8 per task, which is tolerable in many practical scenarios [3].

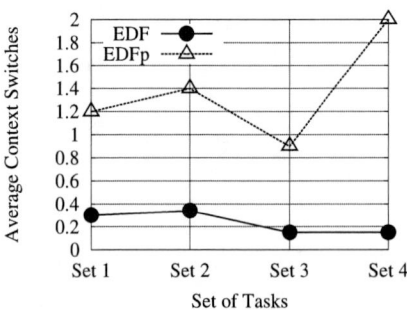

Fig. 12. Average number of context switches per task comparison between EDF and EDFp for various sets of tasks.

[3]Context switch time on ARM cpu can be less than $10us$ [20]

Second, we analyze the context switching overhead for tasks with individual periods. Figure 12 shows the average number of context switches per task between EDF and EDFp for various sets of tasks. The overhead for EDFp (less than 2 context switches per task) is tolerable for most practical scenarios.

V. CONCLUSION

Both power and heat density of on-chip systems are increasing exponentially with Moore's Law. High temperature negatively affects reliability as well the costs of cooling and packaging. In this paper, we propose task partitioning as an effective way to reduce the peak temperature in the embedded systems. We developed novel algorithms that address two broad scenarios: (1) a set of periodic tasks with common period and (2) a set of periodic tasks with individual period. Experimental results show that our first algorithm outperforms the task sequencing algorithm [12] by reducing the peak temperature by as much as $6°C$. For task sets with individual period, EDF scheduling with task partitioning can lower the peak temperature, as compared to simple EDF scheduling, by as much as $6°C$. Our analysis indicates that the number of additional context switches (overhead) is less than 2 per task, which is tolerable in many practical scenarios. These results are promising and clearly demonstrate that task partitioning is an effective way to reduce the peak temperature in embedded systems.

REFERENCES

[1] K. Skadron et al., "Temperature-aware computer systems: Opportunities and challenges," *IEEE Micro*, vol. 23, no. 6, pp. 52–61, 2003.
[2] S. Gunther et al., "Managing the impact of increasing microprocessor power consumption," *Intel Technology Journal*, 5(1), pp. 1–9, 2001.
[3] S. Gunther et al., "Managing the impact of increasing microprocessor power consumption," *Intel Technology Journal*, vol. 1, pp. 1–9, 2001.
[4] *Failure mechanisms and models for semiconductor devices*, jedec.org.
[5] K. Skadron et al., "Temperature-aware microarchitecture: Modeling and implementation," *TACO*, vol. 1, no. 1, pp. 94–125, 2004.
[6] D. Brooks and M. Martonosi, "Dynamic thermal management for high-performance microprocessors," in *HPCA*, 2001, p. 0171.
[7] R. Rao and S. Vrudhula, "Efficient online computation of core speeds to maximize the throughput of thermally constrained multi-core processors," in *ICCAD*, 2008, pp. 537–542.
[8] M. Kadin and S. Reda, "Frequency planning for multi-core processors under thermal constraints," *ISLPED*, 2008, pp. 213–216.
[9] S. Murali et al., "Temperature control of high-performance multi-core platforms using convex optimization," in *DATE*, 2008, pp. 110–115.
[10] T. Ebi et al., "Tape: thermal-aware agent-based power economy for multi/many-core architectures," *ICCAD*, 2009, pp. 302–309.
[11] R. Ayoub and T. Rosing, "Predict and act: dynamic thermal management for multi-core processors," in *ISLPED*, 2009, pp. 99–104.
[12] R. Jayaseelan and T. Mitra, "Temperature aware task sequencing and voltage scaling," in *ICCAD*, 2008, pp. 618–623.
[13] J. Choi et al., "Thermal-aware task scheduling at the system software level," in *ISLPED*, 2007, pp. 213–218.
[14] D. Brooks, V. Tiwari, and M. Martonosi, "Wattch: a framework for architectural-level power analysis and optimizations," *Computer Architecture News*, vol. 28, no. 2, p. 94, 2000.
[15] P. Pillai and K. Shin, "Real-time dynamic voltage scaling for low-power embedded operating systems," in *SOSP*, 2001, p. 102.
[16] ARM, www.arm.com/products/processors/cortex-a/cortex-a8.php.
[17] University of Virginia, http://lava.cs.virginia.edu/HotSpot/.
[18] M. Guthaus et al., "Mibench: A free, commercially representative embedded benchmark suite," in *WWC*, 2001, pp. 3–14.
[19] C. Lee, M. Potkonjak, and W. Mangione-Smith, "Mediabench: a tool for evaluating and synthesizing multimedia and communicatons systems," in *MICRO*, 1997, pp. 330–335.
[20] SEGGER, http://www.segger.com/cms/context-switching-time.html.

Towards Thermal Profiling in CMOS/Memristor Hybrid RRAM Architectures

Cory E. Merkel and Dhireesha Kudithipudi
Department of Computer Engineering
Rochester Institute of Technology, Rochester, USA
{cem1103, dxkeec}@rit.edu

Abstract—In this paper, we propose a hybrid temperature sensing resistive random access memory (TSRRAM) architecture composed of traditional CMOS components and emerging memristive switching devices. The architecture enables each RRAM switching element to be used both as a memory bit and a temperature sensor. The TSRRAM is integrated into an Alpha 21364 processor as an L2 cache. Its accuracy and performance were simulated using a customized simulation framework. SPEC2000 benchmarks were used to generate thermal profiles in the Alpha processor core. Active and passive sensing mechanisms are also introduced as means for DTM algorithms to determine the thermal profile of the RRAM switching layer. The proposed architecture yielded a 2.14 K mean absolute temperature error during passive sensing, which is well within the useful range of dynamic thermal management (DTM) algorithms. Furthermore, the proposed design is shown to have only an 8 cycle performance overhead.

I. INTRODUCTION

Static random access memory (SRAM) has traditionally been used for on-chip CPU memories such as caches and registers. However, reliability degradation caused by leakage currents, reduced static noise margins, and soft errors severely limits the scalability of SRAM past the 16 nm technology node [1]. Consequently, several new memory technologies, including phase-change memory (PCRAM), ferroelectric memory (FeRAM), magnetic memory (MRAM), molecular memory, carbon nanotube-based memory, and resistive random access memory (RRAM) have been proposed [2]. In particular, RRAM has received an explosive growth in research in the last few years. This has mainly been a consequence of various experimental demonstrations of thin-film memristors, which are promising devices for RRAM's physical storage mechanism. Furthermore, RRAM's high density, low power, speed, and non-volatility make it a viable technology for future on-chip memory applications.

Thin-film memristor-based RRAM can be integrated with CMOS processor cores as shown in Figure 1. The RRAM crossbar layer, which contains the storage media, is fabricated on top of CMOS layers. This configuration reduces the area overhead of the RRAM to the overhead of the CMOS-crossbar layer interface circuitry. However, because of their close proximity, the crossbar layer will be strongly affected by heat generation in the CMOS layers. Several failure mechanisms in CMOS devices have been shown to accelerate with high temperatures [3, 4]. It is also expected that interconnect wearout, access times, and retention [5] within the RRAM's crossbar

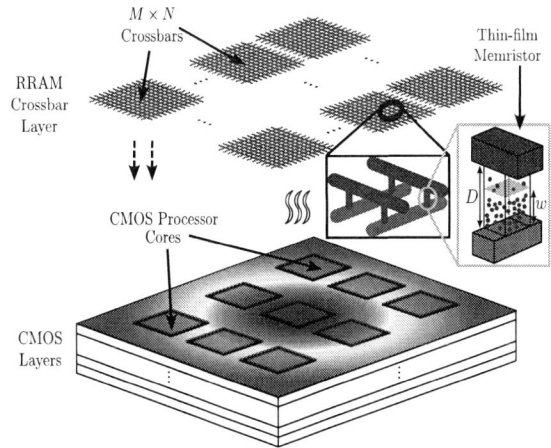

Fig. 1. 3D CMOS/memristor hybrid RRAM physical structure.

layer will be strongly correlated to temperature variations. As a result, accurate thermal profiles from both the CMOS and crossbar layers will be required for effective on-chip thermal management mechanisms.

Thermal profiling in 2D CMOS devices has typically been achieved using semiconductor thermal diodes [6]. However, their area and design overheads restrict the number and location of diodes in the CMOS layer [7]. In this work, we explore the integration of temperature sensors into an RRAM crossbar layer, which is fabricated on top of CMOS layers. Our specific contributions are:

- Design of a temperature sensing RRAM (TSRRAM) architecture in which each thin-film memristor (Figure 1) can store a memory bit and measure the temperature in the crossbar layer
- Introduce active and passive sensing techniques to utilize the TSRRAM for generating a thermal profile of the entire chip
- Design a custom simulation framework to assess the accuracy of the TSRRAM for several SPEC2000 CPU benchmarks

The rest of this paper proceeds as follows: Section II gives an overview of thin-film memristors, their temperature dependence, and their application in RRAM architectures. Section III discusses the TSRRAM that was designed in this work. It also introduces the active and passive sensing techniques that can be used for thermal profiling with the TSRRAM. The simulation setup and results are presented in Section IV, and Section V concludes this work.

II. RRAM Overview

This section gives a brief overview of thin-film memristors, their application in resistive random access memory (RRAM), and the temperature dependence of RRAM's write speed.

A. Thin-film Memristors

Hysteretic resistance switching in thin films has been observed for over 40 years [8, 9]. In 2008, Strukov *et al.* showed that the resistance switching behavior in vacancy-doped titanium dioxide thin films could be described by an electrical memristive system [10]. Thin film materials that exhibit resistance switching in an applied electric field have since become known as thin-film memristors. The exact physical processes that cause the switching phenomena depend on the materials and fabrication processes. In this work, we have considered devices that switch as a result of ion motion within the thin-film material.

As shown in Figure 1, these devices contain mobile dopant ions (red circles) in both high and low concentration regions, resulting in regions of low and high resistivity (respectively). A domain wall at location w separates these two regions. As shown in [10], the device can be modeled as a current-controlled memristive system [11]:

$$
\frac{dx}{dt} = \frac{\mu_I(T)R_{on}}{D^2}i_m(t)
$$
$$
v_m(t) = [R_{on}x + R_{off}(1-x)]\,i_m(t)
\tag{1}
$$

where D is the film thickness, $x = w/D$ is the state variable, μ_I is the ion mobility, R_{off} is the memristor resistance when $x = 0$, R_{on} is the memristor resistance when $x = 1$, $v_m(t)$ is the terminal voltage, and $i_m(t)$ is the current through the memristor. Most work that considers this model treats μ_I as a constant. However, μ_I is actually a temperature-dependent quantity, and is related to the ion diffusion coefficient D_I at low electric fields via the Nernst-Einstein relation [12]:

$$
\mu_I(T) = \frac{q_I D_I(T)}{k_B T},
\tag{2}
$$

where q_I is the ion charge, k_B is Boltzmann's constant, and D_I is given by

$$
D_I(T) = fa^2\exp\left(-\frac{E_A}{k_B T}\right)
\tag{3}
$$

In (3), f is the ion jump frequency, a is the ion jump distance, and E_A is the ion activation energy. In the rest of this section, we will show that the thin-film memristor's temperature dependence significantly affects RRAM write speeds. This relationship between temperature and write speed is leveraged in Section III to design a temperature sensing RRAM architecture.

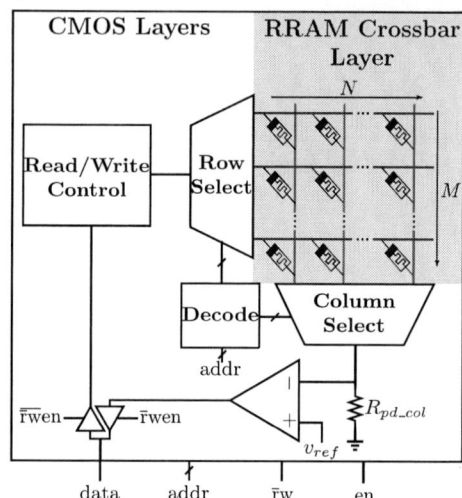

Fig. 2.   RRAM block architecture.

B. RRAM Architecture

RRAM data is stored in the form of resistance, making it immune to the soft error and charge leakage problems encountered in SRAM. Figure 2 shows a CMOS/memristor hybrid RRAM block architecture. An $M \times N$ crossbar contains MN thin-film memristors, each of which can store a high or low resistance state. As described in Figure 1, the crossbar circuit can be fabricated on top of the CMOS layers. Supporting circuitry for reading and writing to the memristors in the crossbar layer is implemented in standard CMOS technology. Multiplexers and demultiplexers select a specific row or column of the crossbar based on a decoded address, allowing a single memristor to be read or written.

1) Read Operation: During the read operation, $\bar{r}w$ is low, and a read voltage v_r is applied to the selected crossbar row. The read voltage is typically small in order to ensure that the read operation is non-destructive. The selected column is pulled to ground through R_{pd_col}. The voltage across the pull-down resistor is compared to a reference voltage v_{ref} to determine the state of the memristor. The reference voltage is given by

$$
v_{ref} = v_r\frac{R_{pd\_col}}{0.5R_{on} + 0.5R_{off} + R_{pd\_col}}
\tag{4}
$$

2) Write Operation: During the write operation, $\bar{r}w$ is high, and a positive or negative write voltage v_w is applied to the selected row to change the memristor resistance to either R_{on} or R_{off}. The write voltage is selected based on the value of the data (positive voltage for *data* = 1 and negative voltage for *data* = 0). Assuming a constant write voltage, the memristor write time can be derived from (1) as

$$
t_w = \frac{D^2}{\mu_I(T)v_w}\left(\frac{r_1-1}{2}\left(x_0^2 - x_f^2\right) + (r_1 + r_2)\left(x_f - x_0\right)\right)
\tag{5}
$$

where $r_1 = R_{off}/R_{on}$ and $r_2 = R_{pd\_col}/R_{on}$. The dependence of t_w on T allows the temperature of any location in

978-1-4673-0438-2/12 $31.00 © 2012 IEEE   168

the RRAM crossbar layer to be found by measuring the write time of the corresponding memristor.

III. TEMPERATURE SENSING RRAM DESIGN

In this section, we provide a design for a temperature sensing RRAM (TSRRAM), which leverages the RRAM's write time dependence on temperature, discussed in the previous section. The spatial distribution of the RRAM's crossbar layer over the CMOS processor cores (Figure 1) combined with the high density of the crossbar circuits will allow temperature measurements at almost any place on the chip. This flexibility enables several dynamic thermal management and transient fault detection routines that are not feasible using traditional temperature sensing methods. Figure 3 shows the top-level TSRRAM design and supporting components. The rest of this section discusses each component in more detail.

A. TSRRAM Block

The TSRRAM block is a bit-addressable RRAM block with additional circuitry for measuring each bit's write time. A specific bit is addressed by selecting a crossbar row and column. The upper $\log_2(M)$ bits of *addr* specify the row, and the remaining bits specify the column.

1) Idle and Read Operation: When the TSRRAM block is idle, the enable signal *en* should be low. This cuts off the TSRRAM block from the data bus and grounds both terminals of the currently-selected memristor. During the read operation, *en* is high and $\overline{rw}$ is low. A read voltage v_r is applied to the addressed row, and the voltage across R_{pd_col} is compared to a reference to determine the memristor's logic level. During the read operation, the reference voltage is given by the voltage divider

$$v_{ref} = v_{row} \frac{R_{pd\_ref}}{R_{ref\_r} + R_{pd\_ref}}, \qquad (6)$$

where $R_{ref\_r} = 0.5R_{on} + 0.5R_{off}$. The timer output *ren* is high during the read operation, which allows the read voltage (comparator output) to be placed on the data bus.

2) Write Operation and Temperature Sensing: In the write operation, *en* and $\overline{rw}$ are high. A write voltage will be selected depending on the data to be written. A positive write voltage is selected if the *data* signal is high, which will make the memristor domain wall move into a low resistance state. A negative write voltage is selected if the *data* signal is low, which will make the memristor domain wall move into a high resistance state.

The temperature is estimated by measuring the write time of a specific memristor. The existing read circuit is used to detect when a memory bit has been written. The finite state machine for the timer is shown in Figure 5. After a reset, the timer is in the IDLE state. Here, the timer's count value is zero. When the *ts* signal transitions to a logic high value, the state machine transitions to either the START0 or START1 state, depending on the current value of the memory bit being written. When the read comparator output switches from its original value, the timer is stopped, and the timer count value can be transmitted serially through the read select mux.

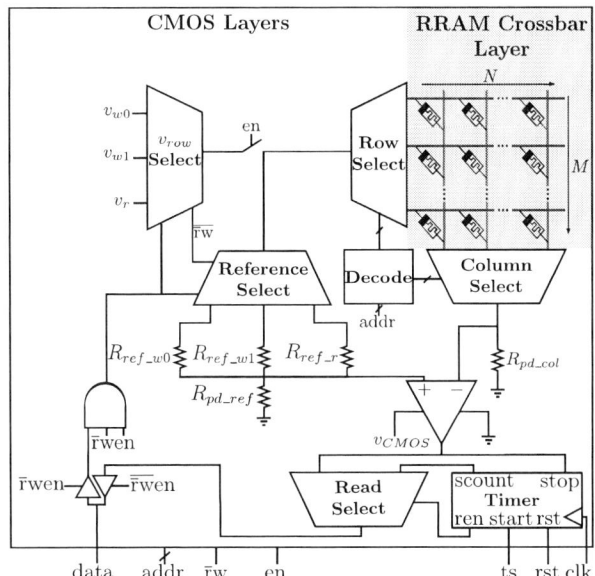

Fig. 4. TSRRAM block architecture.

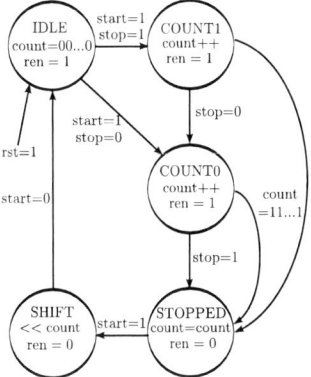

Fig. 5. Timer finite state machine.

B. TSRRAM Group

The TSRRAM block shown in Figure 4 is bit-addressable. Several of these blocks are combined to form a word-addressable TSRRAM group. Blocks within a group share identical *addr*, $\overline{rw}$, and *en* signals. This allows the bits within a word to be striped across the blocks. One or more TSRRAM groups makes up the TSRRAM. The number of groups is dictated by the desired memory size, as well as any restrictions on the size of the crossbars within a block.

C. Supporting Components

There are four supporting components required in order to use the TSRRAM for dynamic thermal management. When the processor core issues a write instruction, the TSRRAM controller translates the write address. The upper $G - 1$ bits are decoded to form the *en* signals for each TSRRAM group. The rest of the bits are used for the *addr* bus. The TSRRAM controller also pulses the *ts* signal to start the timers in each TSRRAM block. A state machine is used to enable the TSRRAM for the longest possible write time. The write

978-1-4673-0438-2/12 $31.00 © 2012 IEEE

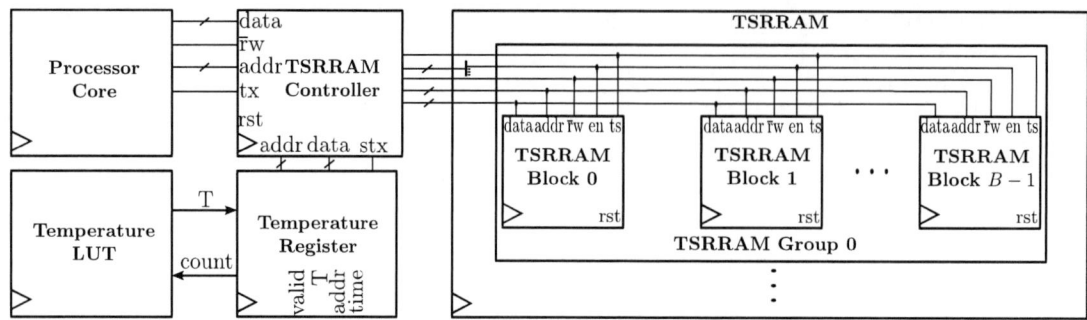

Fig. 3. Top-level TSRRAM architecture.

time is a monotonically decreasing function of temperature. Therefore, the longest write time corresponds to the bottom of the temperature range considered: 300 K. After the write operation is completed, the TSRRAM controller pulses the *stx*, and the timer data is transmitted serially from the TSRRAM to the temperature register using the data bus.

The timer information for each of the B TSRRAM blocks is then transformed into a temperature using a pre-calibrated temperature lookup table (LUT). The temperature register also captures the address and system time associated with the last write instruction. A *valid* bit is asserted after all of the timer data has been transformed into temperatures. At this point, the temperature register holds all of the necessary information for DTM procedures at both the hardware and system software levels.

D. Active and Passive Sensing

The TSRRAM architecture can be used to sense temperatures both actively and passively. In active temperature sensing, a software or hardware-level DTM algorithm would write data to certain memory locations in order to determine a global temperature profile, or to examine fine-grained temperature statistics for a specific region of the chip. Active sensing schemes should first read the data at a specific memory location, then write to it for a temperature measurement, and then write the original data back. During this entire process, context switching should be disabled to ensure that intermediate data modifications do not affect other tasks.

The write operations that exist in a set of processor tasks can also be used to gather thermal profile data using the TSRRAM. This idea is exploited for passive sensing. Assuming that several tasks are running on a single processor, the spatial distribution of memory write accesses should look approximately random, yielding temperature measurements that cover the entire chip. This method is preferred because there is no possibility of data corruption, and temperature measurements can be taken in parallel with tasks' write operations. However, this approach will not work when the processor is idle. In that case, active sensing can be used, or thermal profiling can be halted.

IV. RESULTS AND ANALYSIS

A. Simulation Setup

Crosstherm, a custom simulation framework, was developed to analyze the effects of temperature variation in

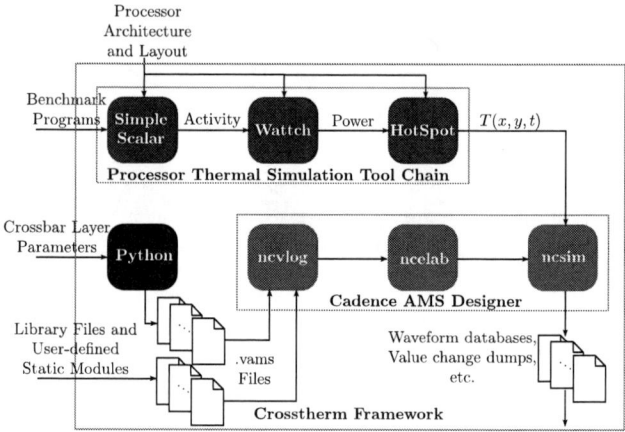

Fig. 6. Crosstherm simulation framework.

CMOS/memristor hybrid RRAM architectures. Figure 6 shows a high-level block diagram of the Crosstherm simulation framework. The framework is divided into three main modules. The first is the processor thermal simulation tool chain, which provides space and time-dependent thermal profiles for a given processor architecture and benchmark programs. SimpleScalar 2.0 [13], a well-known microarchitectural simulator, provides switching activities for a given benchmark program and processor type. This work uses the Alpha 21364 processor core [14] with its L2 cache removed for all simulations. Wattch [15] provides power estimation from the switching activities generated in SimpleScalar. For this work, all power estimations are based on a 45 nm technology node. Finally, HotSpot [16] estimates thermal profiles based on the power data from Wattch. HotSpot was modified to report grid-level (default is block-level) thermal data every 10000 CPU cycles. Cadence AMS Designer was used to simulate the TSRRAM and supporting components, which were implemented in Verilog AMS. A Python script was also written to automatically generate Verilog AMS source files based on different options and design parameters. This allowed designs to be highly generic and easy to customize.

B. Design Parameters

Each of the three modules in the Crosstherm framework is highly customizable. Table I shows the key design parameters used for simulation. The ion jump frequency, ion

$$x_{f0} = \frac{(r_1 + r_2) - \sqrt{(r_1 + r_2)^2 - (r_1 - 1)\left(\frac{r_1 - 1}{2}x_{f1}^2 - (r_1 + r_2)x_{f1} - \frac{r_1 - 1}{2} + r_1 + r_2\right)}}{r_1 - 1} \tag{7}$$

TABLE I
DESIGN PARAMETERS.

Parameter	Symbol	Value
Ion Jump Frequency	f	10 THz
Ion Jump Distance	a	0.15 nm
Ion Activation Energy	E_A	0.18 eV
Ion Charge	q_I	+2 e
Film Thickness	D	10 nm
ON Resistance	R_{on}	100 Ω
OFF Resistance	R_{off}	16 kΩ
Crossbar Rows	M	1
Crossbar Columns	N	1024
Column Pull-down Resistor	R_{pd_col}	1 kΩ
Reference Pull-down Resistor	R_{pd_ref}	1 kΩ
Read Reference Resistor	R_{ref_r}	8050 Ω
Write 0 Reference Resistor	R_{ref_w0}	15.2 kΩ
Write 1 Reference Resistor	R_{ref_w1}	4279 Ω
CMOS Voltage	v_{CMOS}	3 V
Read Voltage	v_r	0.5 V
Write 0 Voltage	v_{w0}	-10 V
Write 1 Voltage	v_{w1}	10 V
TSRRAM Group (Word) Size	B	8
Number of TSRRAM Groups	G	32
Clock Frequency	f_{clk}	1 GHz
Simulation Timestep	N/A	100 ps

jump distance, ion charge, film thickness, and on and off resistances were taken from [5]. The activation energy and write voltages were chosen to reflect write times on the order of those presented in experimental devices.

The parameters M, N, B, and G were chosen to achieve a 4 kB memory size with sneak path elimination. The read reference voltage is chosen to be the resistance at which the memristor domain wall is 0.5. R_{ref_w0} and R_{ref_w1} were chosen such that even with a 5% variation in R_{on} or R_{off} the timer circuit would still be able to detect the memristor switching states, and the measured write time from 0 to 1, $t_{w0\rightarrow1}$, would be the same as the write time from 1 to 0, $t_{w1\rightarrow0}$. The values of x_{f0} and x_{f1} for which $t_{w0\rightarrow1} = t_{w1\rightarrow0}$ are given by (7), where x_{f0} and x_{f1} are the final x values for which the timer circuit detects that data has been written to a memristor. The clock frequency was chosen to be 1 GHz. A 0.5 V read voltage was used for non-destructive reading. The simulation timestep was chosen to be 10% of the clock frequency.

C. Calibration and Theoretical Maximum Error

The TSRRAM's temperature LUT was calibrated by writing 1 and 0 to a memristor for several temperatures and capturing the associated write time. The results of the calibration are shown in Figure 7(a). The solid blue curve gives the theoretical write time vs. temperature from (5). The simulation data points (green circles) were fit to a 10th degree polynomial (dotted line). The final values for the temperature LUT were taken from the fit.

Since process variations were not considered in this work, the theoretical maximum error for any temperature $E_{max}(T)$

is completely due to time quantization:

$$E_{max}(T) = T(\lceil t_w(T) - 1/f_{clk}\rceil) - T(\lceil t_w(T)\rceil), \tag{8}$$

where $T(t_w)$ is the theoretical temperature corresponding to t_w, and $t_w(T)$ is the theoretical write time corresponding to T. Plotted on the inset is the theoretical maximum error versus temperature for the chosen simulation parameters. The results of (8) are plotted on the inset of Figure 7(a). Over the temperature range considered, the maximum theoretical error is expected to be ≈ 8 K. Note that this could be reduced by using a high-speed clock for the timer circuit.

D. Passive Sensing Results

The passive sensing capabilities of the TSRRAM architecture were tested by implementing a random mix of TSRRAM read and write instructions while running four benchmarks–GCC, Art, Gzip, and Parser–in the CMOS layer. The thermal profile in the crossbar layer was assumed to be identical to that in the CMOS layer. Data and addresses for each write instruction were chosen randomly. In general, the chosen set of instructions represents a memory-bound processor running several different tasks. This is the best-case scenario for passive sensing. In the case of CPU-bound loads, where few tasks are running, active sensing techniques will need to be employed.

Figure 7(b) shows example data from the GCC benchmark after 10,000 cycles. The left floorplan shows the raw thermal profile (CMOS layer), and the right floorplan shows the temperature measurements made by the TSRRAM (crossbar layer). A DTM routine should decide how much of this data to store in memory for thermal management. The error histogram for the GCC benchmark is shown in Figure 7(c). The distribution is bi-modal with the range of positive errors approximately three times larger than the range of negative errors. This is because time quantization has a positive error associated with it. Most of the negative errors are likely due to the polynomial fit that was used for calibration. Figure 7(d) shows the maximum absolute error versus temperature for the GCC benchmark. Ignoring the top three outliers, the error is increasing with temperature, as expected. The extra source of error associated with the polynomial fit also causes the maximum error to exceed the expected maximum error. Similar results were obtained for the other three simulated benchmarks. Table II summarizes the results. The mean error for each benchmark is well within the range of DTM procedures. Furthermore, the performance overhead of passive sensing is due only to the serial transfer of timer data from the TSRRAM to the temperature register. Since 8-bit timers were used, the additional write latency is only 8 cycles.

978-1-4673-0438-2/12 $31.00 © 2012 IEEE

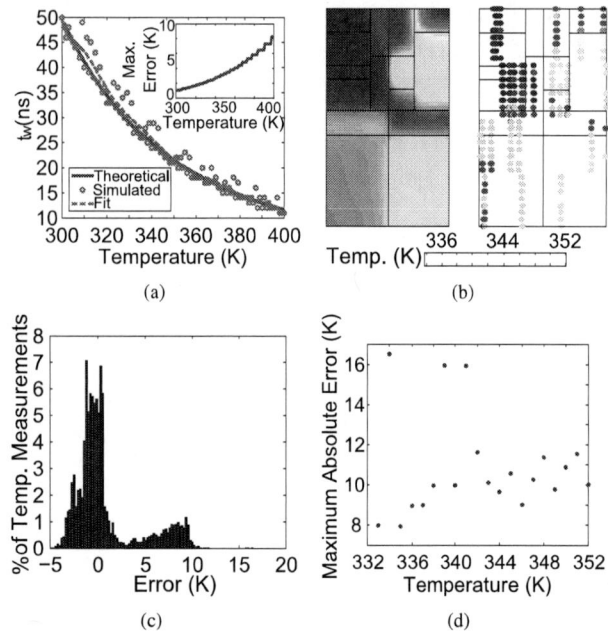

Fig. 7. (a) TSRRAM calibration and theoretical maximum error due to time quantization. (b) Example temperature data from the GCC benchmark after 10,000 cycles. On the left is the raw thermal profile from HotSpot, and on the right is the set of temperature measurements produced by the TSRRAM. (c) TSRRAM error histogram for the GCC benchmark. (d) Maximum TSRRAM error versus temperature for the GCC benchmark.

TABLE II
SPEC2000 CPU BENCHMARKS.

Benchmark	Max. Error (K)	Mean Error (K)	Std. Dev. (K)
GCC	16.49	2.20	2.65
Art	15.06	1.98	1.87
Gzip	15.11	2.23	2.16
Parser	17.88	2.16	2.55

V. CONCLUSIONS AND FUTURE WORK

This work explored thermal profiling in 3D CMOS/memristor hybrid RRAM architectures. We have designed an RRAM architecture with built-in temperature sensing capabilities. The architecture exploits the temperature dependence of the RRAM's write time, and allows each memory bit to also be used as a temperature sensor. A custom simulation framework was developed to test the architecture in the presence of realistic CMOS-layer temperature variations. Four benchmarks from the SPEC2000 CPU suite were used to assess the accuracy of the design. The average error across all benchmarks was 2.14 K, with a performance overhead of only 8 cycles. However, when process variations are considered, the magnitude of this error is likely to increase unless a more rigorous calibration process is employed.

Several avenues exist for extending this work. Here, we have assumed that the thin-film memristors in the TSRRAM follow a linear ionic drift model, where the ion drift velocity is linear in the applied electric field. Future work will explore the temperature sensing capabilities of the TSRRAM using more accurate models (such as those based on exponential ionic drift) or models that represent memristors composed of different materials, switching mechanisms, etc. In this work, only the effect of temperature on thin-film memristors was considered. Furthermore, only one of the temperature dependencies in thin-film memristors, ion mobility, was considered. An extension of this work could consider the effects of temperature in crossbar nanowires as well as CMOS components. The temperature dependence of ion diffusion and electron transport, as well as the effect of heat generation in the crossbar layer can also be considered in future models.

REFERENCES

[1] *International Roadmap for Semiconductors*, 2010 Update. [Online]. Available: http://www.itrs.net/reports.html

[2] A. Chung, J. Deen, J.-S. Lee, and M. Meyyappan, "Nanoscale memory devices," *Nanotechnology*, vol. 21, no. 41, p. 412001, 2010.

[3] A. H. Ajami, K. Banerjee, M. Pedram, and L. P. P. P. van Ginneken, "Analysis of non-uniform temperature-dependent interconnect performance in high performance ics," in *DAC '01: Proceedings of the 38th annual Design Automation Conference*. New York, NY, USA: ACM, 2001, pp. 567–572.

[4] J. Srinivasan, S. V. Adve, P. Bose, and J. A. Rivers, "The impact of technology scaling on lifetime reliability," in *Proc. Int. Conference on Dependable Systems and Networks*, Jun. 2004, pp. 1–10.

[5] D. Strukov and R. Williams, "Exponential ionic drift: fast switching and low volatility of thin-film memristors," *Appl. Phys. A: Materials Science & Processing*, vol. 94, pp. 515–519, 2009.

[6] E. Rotem, J. Hermerding, C. Aviad, and C. Harel, "Temperature measurement in the intel core duo processor," in *Proc. of 12th Int. Workshop on Thermal investigation of ICs*, Sep. 2006, pp. 1–5.

[7] J. Long, S. O. Memik, G. Memik, and R. Mukherjee, "Thermal monitoring mechanisms for chip multiprocessors," *ACM Trans. Archit. Code Optim.*, vol. 5, no. 2, pp. 1–33, 2008.

[8] F. Argall, "Switching phenomena in titanium oxide thin films," *Solid-State Electronics*, vol. 11, pp. 535–541, 1968.

[9] J. Blanc and D. L. Staebler, "Electrocoloration in SrTiO$_3$: Vacancy drift and oxidation-reduction of transition metals," *Phys. Rev. B*, vol. 4, no. 10, Nov. 1971.

[10] D. B. Strukov, G. S. Snider, D. R. Stewart, and S. R. Williams, "The missing memristor found," *Nature*, vol. 453, no. 7191, pp. 80–83, May 2008.

[11] L. Chua and S. M. Kang, "Memristive devices and systems," *Proc. IEEE*, vol. 64, no. 2, pp. 209–223, Feb. 1976.

[12] U. Weinert and E. A. Mason, "Generalized nernst-einstein relations for nonlinear transport coefficients," *Phys. Rev. A*, vol. 21, no. 2, 1980.

[13] D. Burger and T. M. Austin, "The simplescalar tool set, version 2.0," *SIGARCH Comput. Archit. News*, vol. 25, pp. 13–25, Jun. 1997.

[14] S. Mukherjee, P. Bannon, S. Lang, A. Spink, and D. Webb, "The alpha 21364 network architecture," in *Hot Interconnects 9, 2001.*, 2001, pp. 113–117.

[15] D. Brooks, V. Tiwari, and M. Martonosi, "Wattch: a framework for architectural-level power analysis and optimizations," *SIGARCH Comput. Archit. News*, vol. 28, pp. 83–94, May 2000.

[16] K. Skadron, M. R. Stan, K. Sankaranarayanan, W. Huang, S. Velusamy, and D. Tarjan, "Temperature-aware microarchitecture: Modeling and implementation," *ACM Trans. Archit. Code Optim.*, vol. 1, pp. 94–125, Mar. 2004.

978-1-4673-0438-2/12 $31.00 © 2012 IEEE

CMOS Gas Sensor Array Platform with Fourier Transform based Impedance Spectroscopy

Pramod M[*], Navakanta Bhat[*†], Gaurab Banerjee[*], Bharadwaj Amrutur[*], K N Bhat[*] and Praveen C Ramamurthy[‡]

[†]Department of Electrical Communication Engineering.
[*]Centre for Nano Science and Engineering.
[‡]Department of Materials Engineering.
Indian Institute of Science, Bangalore, 560012.
Email : pramodm@ece.iisc.ernet.in

Abstract—A digital readout CMOS interface platform for gas sensor array is demonstrated in 0.35 μm CMOS process. The chip contains 27 sensor pixels where each pixel can be functionalized with a distinct gas sensing material. A sensor pixel can amplify and digitize the sensor signal and perform impedance spectroscopy upto 10 KHz. Signal amplification is achieved by digitally programmable gain stages and digitization is done using continuous-time $\Delta\Sigma$ modulator. An on-chip reference pixel is provided to compensate for the phase shift introduced by the signal processing blocks. This chip is validated by functionalizing a sensor pixel with polycarbazole conducting polymer for sensing acetone. The chip consumes 57 mW of power at 3.3 V supply.

I. INTRODUCTION

In recent years, gas sensors have gained prominence since they can be used in a variety of applications such as monitoring green house gases, food processing, process control in industries etc. [1]. Integration of gas sensing materials such as conducting polymers (CPs) with CMOS chips significantly reduces the cost and size of hand held portable sensor systems. CPs are particularly attractive because of their ability to operate at room temperature, unlike inorganic metal oxides which operate at 300°C-500°C [2]. Also, the post processing is simpler when integrating a CP with CMOS chip. But most CPs show cross sensitivity to gasses, which can be mitigated by the use of an array of sensors along with techniques like principal component analysis to improve discrimination among different gasses.

Impedance spectroscopy (IS) is relatively new electrical approach to characterize interactions at the interface or within the bulk of different materials [3]. IS of the sensor provides information about the nature and concentration of analyte gas. Impedance spectrum also adds another dimension for classification of gas mixtures. Recently, an interface chip for polymer-carbon black composites based gas sensor array has been proposed in [4] assuming that the sensor response is purely resistive, which may not always be the case. IS has been implemented on CMOS chips using frequency response analyzer (FRA) approach [5]–[7]. In this approach, the sensor is excited with a tone of certain frequency and the sensor output is then correlated with in-phase and out-of-phase tones of the same frequency to extract the sensor impedance. Although this approach leads to efficient hardware

implementation, the impedance extraction has to be done for every frequency of interest and minimum duration of integration is not well defined. This makes FRA approach relatively slow and unsuitable for instantaneous impedance measurements of nonstationary sensor-analyte systems. On the other hand, a more efficient way of performing IS is to adopt fast Fourier transform (FFT) based approach [8], where the sensor is excited with a signal of wide frequency content and Fourier transform of the sensor output is computed to simultaneously extract the impedance for all frequencies of interest. This approach creates significantly larger amounts of data. But it is preferable for gas sensing applications because of moderate array density and availability of low cost off-chip DSPs for data processing. In this work, we present a gas sensor array in CMOS technology with FFT based IS.

This paper introduces a digital readout interface chip for CMOS gas sensor array fabricated in AMS 0.35 μm CMOS process. The chip contains 27 sensor pixels which can be functionalized with distinct conducting polymers by a **single post processing step**. The front end signal conditioning circuit at each pixel amplifies the sensor signal through digitally programmable gain stages and subsequently digitizes it using a continuous time $\Delta\Sigma$ modulator (CT-$\Delta\Sigma$M). Digital data is read out using memory addressing approach for ease of interfacing with an external DSP. An on-chip reference pixel to decouple the phase change of the sensor from phase shift introduced by the signal processing blocks is presented. The gas sensing ability of the chip is validated by functionalizing a sensor pixel with polycarbazole to sense acetone. The rest of the paper is organized as follow. Sec. II details the architecture of the chip. Sec. III presents the experimental results and Sec. IV presents conclusions.

II. CHIP ARCHITECTURE

The chip contains an array of 6×5 pixels of which there are 27 sensor pixels and a reference pixel which are introduced in subsequent sections. The chip also contains a pixel for ambient pressure measurement and a pixel for ambient temperature measurement which will be reported elsewhere.

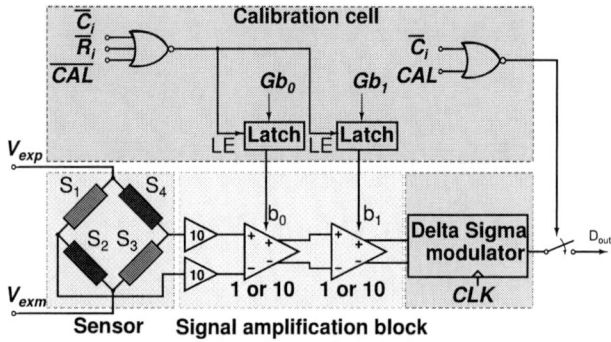

(a)

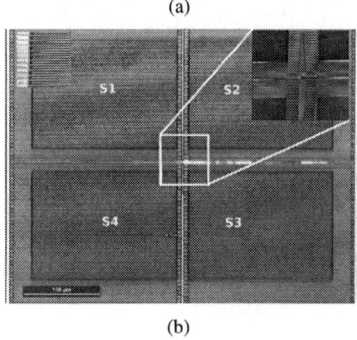

(b)

Fig. 1. (a) Architecture of sensor pixel. (b) Sensor structure configured as Wheatstone bridge, CP is drop coated on all the four sensing elements (S_1 to S_4). S_1 and S_3 are exposed to gas while S_2 and S_4 are passivated. Inset shows the SEM image of the wells defined by the Passivation layer. Horizontal lines are Metal 4 (top most metal) of the CMOS process.

A. Sensor Pixel

The architecture of the sensor pixel is shown in Fig. 1(a).

1) Sensor: The base impedance of CPs varies widely with the type of the polymer. It is sufficient to measure the relative change in sensor impedance rather than its absolute value. This can be conveniently done using Wheatstone bridge configuration as shown in Fig. 1(a). Each element of the bridge is realized using interdigitated capacitor (IDC) structures as shown in Fig. 1(b). The IDCs are designed with Metal 4 (top most metal) of the CMOS process. The width and gap between the digits is chosen as 0.6 μm (minimum dimensions allowed by the CMOS process). PAD layer is defined over these IDCs to etch away the Si_3N_4 and SiO_2 passivation layers. By this approach, we can avoid the post processing steps for creating wells for confining the polymers.

2) Signal Amplification Block: The impedance change of a given CP is different for different analyte gases. Hence, it is important for the signal processing circuits to account for different sensitivities of CPs. To achieve this, three gain levels are connected in cascade to cover wide range of sensor outputs as shown in Fig. 2. The first gain level uses single ended two stage miller compensated opamp connected in feedback which gives a fixed voltage gain of 10. The second and third gain levels are realized using fully differential two stage miller compensated opamps. Two common mode feedback circuits set the common mode voltage of each stage of these opamps.

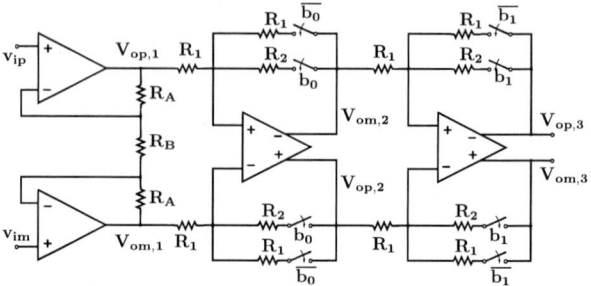

Fig. 2. Circuit schematic of digitally programmable gain stages for signal amplification from the sensor.

Circuit shown in Fig. 2 can provide overall gain of either 10, 100 or 1000.

The gain setting of second and third level is stored using two latches of calibration setup shown in Fig. 1(a). The contents of the latches are controlled by external pins $\overline{CAL}$, Gb_0 and Gb_1. R_i and C_i are row and column select lines from the on-chip address decoder.

3) Conversion to Digital Data: Multiplexing of amplified sensor signal for off-chip digitization leads to loss of signal integrity due to noise pick up and distortion due to multiplexing switches. Instead, the amplified sensor signal can be digitized at each pixel to avoid these sources of error. $\Delta\Sigma$ modulators are well suited for this because they provide higher conversion accuracy for lower precision requirements from passive components [9]. It is easier to readout data from an array of modulators. The decimation filtering required for modulator outputs can be performed efficiently using off-chip DSPs. In particular, having a continuous time (CT) loop filter in the $\Delta\Sigma$ modulator has the following advantages over switched capacitor loop filter. A CT-$\Delta\Sigma$ modulator consumes lesser power which reduces the effect of heating on CPs. The switching noise is significantly reduced. This is particularly important since the sensor element is placed directly above these circuits.

In this work a low pass second order CT-$\Delta\Sigma$ modulator with single bit quantizer is designed for 10 KHz signal bandwidth (BW), oversampling ratio (OSR) of 32 and clock frequency (f_{CLK}) of 640 KHz. The CT loop filter is realized using Cascade of Integrators with Feed Forward summation (CIFF) architecture. Fully differential opamps described in Sec. II-A2 are used for implementing the loop filter . The differential dynamic comparator circuit introduced in [9] is used for our design. The architecture of CT-$\Delta\Sigma$ modulator is shown in Fig. 3.

B. Reference Pixel

In this section we present a novel on-chip solution to decouple the phase change of the sensor impedance from the phase shift added by signal processing blocks.

Let Z be the base value of the impedance of each of the sensors S_1 to S_4 of the Wheatstone bridge (shown in Fig. 1(b)) and let ΔZ be the change in impedance of the sensors S_1 and

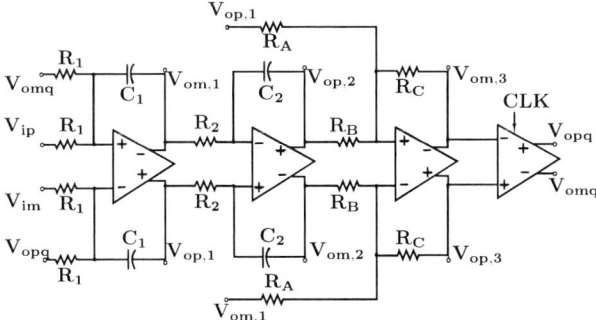

Fig. 3. Second order fully differential low pass CT-$\Delta\Sigma$ modulator employing CIFF loop filter realized using Active-RC integrators.

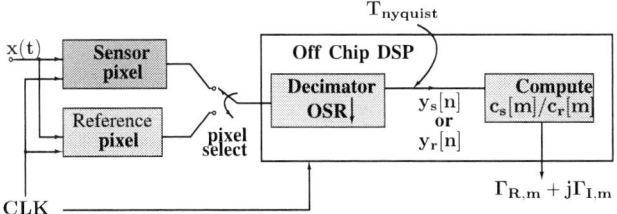

Fig. 4. Setup to determine the real and imaginary component of the sensor impedance.

S_3 also let $Z \gg \Delta Z$. Let $\Delta\phi = \angle(\Delta Z/2Z)$ denote the phase change of the sensor impedance on exposure to an analyte gas. The sensor is followed by gain stages, CT-$\Delta\Sigma$ modulator etc. which collectively add additional phase Ψ to $\Delta\phi$. Ψ is a function of frequency, time (drift) and temperature. Therefore it is essential to decouple the two. This is achieved by use of an on-chip reference pixel.

Consider that an input $x(t) = cos(2\pi f_0 t)$ is fed both to the reference pixel and a sensor pixel as shown in Fig. 4. The output of the sensor pixel and reference pixel after off-chip decimation filtering is given by (also refer to Fig. 1(a)).

$$y_s[n] = \left|\frac{\Delta Z}{2Z}\right| A_s cos(\omega_0 n + \Delta\phi + \Psi) \quad (1)$$

$$y_r[n] = A_r cos(\omega_0 n + \Psi) \quad (2)$$

Where $\omega_0 = 2\pi f_0 T_{nyquist}$, $T_{nyquist} = OSR/f_{CLK}$, A_s and A_r are the gain settings of the senor and reference pixel respectively (Sec. II-A2).

Let $f_0 T_{nyq} = m/N$ $m < N$ for relatively prime integers m and N so that $y_s[n]$ and $y_r[n]$ are periodic signals with period N.

The discrete time Fourier series coefficients of $y_s[n]$ and $y_r[n]$ will be,

$$c_s[k] = \frac{1}{N}\sum_{n=0}^{N-1} y_s[n] e^{-j2\pi n \frac{k}{N}} \quad (3)$$

$$c_r[k] = \frac{1}{N}\sum_{n=0}^{N-1} y_r[n] e^{-j2\pi n \frac{k}{N}} \quad (4)$$

$$k = 0, 1, 2, \ldots N-1.$$

It can be shown that the summation in Eqn. 3 and Eqn. 4 will be non zero only for $k = m$. Dividing Eqn. 3 by Eqn. 4 we get.

$$\frac{c_s[m]}{c_r[m]} = \left|\frac{\Delta Z}{2Z}\right| . \frac{A_s}{A_r} . e^{j\Delta\phi} = \Gamma_{R,m} + j\Gamma_{I,m} \quad (5)$$

From Eqn. 5, we can obtain the real ($\Gamma_{R,m}$) and imaginary ($\Gamma_{I,m}$) components of the relative change in sensor impedance for frequency f_0 independent of Ψ. This approach is validated through measurements by sweeping the excitation phase of five senor pixels relative to the reference pixel, it is found that the normalized RMS phase error is less than 0.12%.

FFT based IS using Reference Pixel: Suppose that the pixels are excited with a signal $x(t)$ with L frequency components f_0 to f_{L-1} such that,

$$x(t) = \sum_{i=0}^{L-1} b_i cos(2\pi f_i t) \quad (6)$$

where b_i is the amplitude of the sinusoid at f_i then we have,

$$y_{s,FFT}[n] = \sum_{i=0}^{L-1} \left|\frac{\Delta Z_i}{2Z_i}\right| . A_s . b_i . cos(\omega_i n + \Delta\phi_i + \Psi_i) \quad (7)$$

$$y_{r,FFT}[n] = \sum_{i=0}^{L-1} A_r . b_i . cos(\omega_i n + \Psi_i) \quad (8)$$

Here Z_i is the impedance of the sensor at f_i. By appropriate choice of f_i and f_{CLK}=640 KHz we can ensure that $f_i T_{nyq} = m_i/N$, $m_i < N$ for each m_i such that $f_i \leq 10$ KHz. With this choice, $y_{s,FFT}$ and $y_{r,FFT}$ are periodic with period N. The Fourier series coefficients can be expressed as,

$$c_{s,FFT}[k] = \frac{1}{N}\sum_{n=0}^{N-1} y_{s,FFT}[n] e^{-j2\pi n \frac{k}{N}} \quad (9)$$

$$c_{r,FFT}[k] = \frac{1}{N}\sum_{n=0}^{N-1} y_{r,FFT}[n] e^{-j2\pi n \frac{k}{N}} \quad (10)$$

$$k = 0, 1, 2, \ldots N-1.$$

Eqn. 9 and Eqn. 10 will be non zero for every $k = m_i$. Further, by computing $c_{s,FFT}[m_i]/c_{r,FFT}[m_i]$ as shown below, we can obtain the impedance change for every frequency f_i.

$$\frac{c_{s,FFT}[m_i]}{c_{r,FFT}[m_i]} = \left|\frac{\Delta Z_i}{2Z_i}\right| . \frac{A_s}{A_r} . e^{j\Delta\phi_i} = \Gamma_{R,m_i} + j\Gamma_{I,m_i} \quad (11)$$

In Sec. III we show results of a sinc waveform with 19 frequency components being used to excite a sensor pixel and the extracted impedance change for all these frequencies.

III. MEASUREMENTS

The chip is fabricated in AMS 0.35 μm 4 metal CMOS process through Europractice. The micrograph of the chip and the test board is shown in Fig. 5. The reference pixel and all the sensor pixels are individually characterized. The opposite arms of the Wheatstone bridge are shorted manually using a micromanipulator. A function generator (Agilent 81150A) is used to excite the pixels with a differential sinusoid of

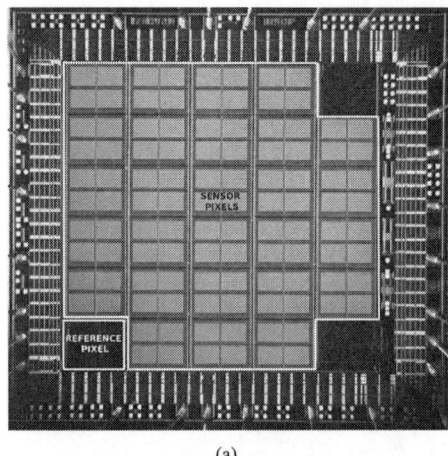

(a)

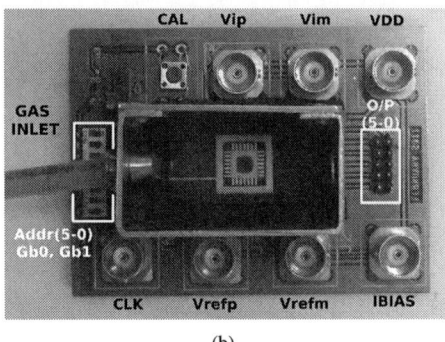

(b)

Fig. 5. (a) Micrograph of the chip. The die occupies an area of 3.3×3.3 mm^2. (b) Test board for testing the packaged chip.

3.125 KHz and clock of 640 KHz. The calibration bits (Gb_0, Gb_1 and $\overline{CAL}$) and address lines are controlled manually. The chip outputs are logged for 0.5 s using Lecroy MSP-500 Mixed Signal Oscilloscope and later analyzed on a PC.

Fig. 6 shows the mean power spectral density (PSD) of all the pixels for -24.3 dBFS and 3.125 KHz input. The PSD is obtained by averaging 32 sets of 8 K point Hann windowed FFT of the modulator output. The gain is set to 10 for all pixels. The PSD of an ideal second order single bit $\Delta\Sigma$ modulator obtained by Schreier Delta-Sigma toolbox [10] is also shown in the figure. It can be seen that the noise shaping closely resembles the ideal spectrum. While degradation of in-band singal to noise ratio (SNR) can be due to non ideal opamps and clock jitter. Distortion peaks present at the harmonics of the fundamental are also introduced by gain stages preceding the CT-$\Delta\Sigma$ modulator. Fig. 7 shows the variation of SNR and SNDR for gain setting of 10 and 100 across different pixels. Peak SNR of 55.2 dB is obtained for -16.4 dBFS input while peak signal to noise and distortion ratio (SNDR) of 49.6 dB occurs at -26.8 dBFS input for gain setting of 10. The measured dynamic range (DR) is 49.6 dB and 48.9 dB for gain setting of 10 and 100 respectively.

Polycarbazole in N-Methyl-2-pyrrolidone (NMP) is drop coated on all the four IDCs of a sensor pixel and air dried

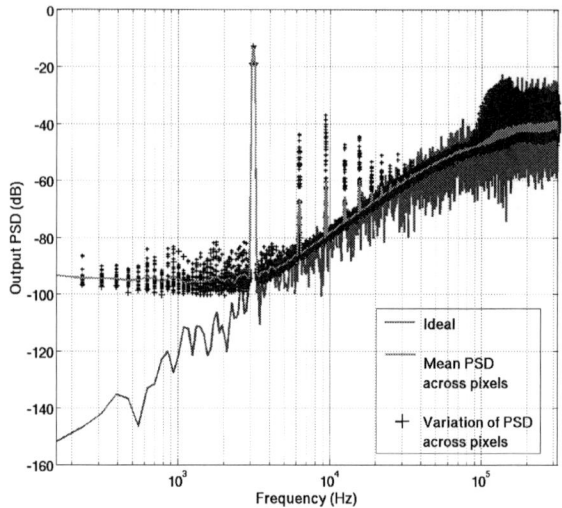

Fig. 6. Output power spectral density.

to remove residual solvent. Opposite arms of the Wheatstone bridge are passivated with polydimethylsiloxane (PDMS) using digital microinjector. Fig. 8 shows the time response for two different concentrations of acetone in ambient for 3.125 KHz sinusoid. Inset in Fig. 8 shows the impedance spectrum obtained by exciting the chip with a periodic sinc waveform of 19 tones from 468.75 Hz to 8906.25 Hz and exposing it to 10K ppm of acetone in ambient. Γ_R and Γ_I are obtained as described in Sec. II-B. Measurements shown in Fig. 8 indicate that impedance changes in polycarbazole, when exposed to acetone, are predominantly resistive.

Tab. I summarizes the experimental results. Tab. II compares this work with other CMOS gas sensor array architectures.

TABLE I
MEASURED PERFORMANCE SUMMARY

Technology	0.35 μm CMOS
Supply / Power	3.3 V / 57 mW
Number of sensor pixels	27
Die size	3.3×3.3 mm^2
Sensor signal bandwidth	10 KHz
f_{CLK} / CT-$\Delta\Sigma$M OSR	640 KHz / 32
Loop filter Order / Type	2 / Active-RC CIFF
Peak SNR / SNDR (gain 10)	55.2 / 49.6 dB
Peak SNR / SNDR (gain 100)	47.2 / 45.6 dB
Mean DR Gain 10 / Gain 100	49.6 / 48.9 dB

IV. CONCLUSION

In conclusion, a digital readout interface chip for gas sensor array applications has been fabricated in 0.35 μm CMOS process. The chip consumes 57 mW with 3.3 V supply and 640 KHz clock. It occupies 3.3×3.3 mm^2 and contains 27 sensor pixels. Sensor signal amplification and digitization is done at every pixel. Impedance spectroscopy of the sensor can be performed upto 10 KHz using FFT approach. SNDR greater than 49 dB (8 bit resolution) is achieved across all pixels for the gain setting of 10. A novel on-chip technique

978-1-4673-0438-2/12 $31.00 © 2012 IEEE

	[11], 2006	[12], 2006	[13], 2007	[14], 2007	[4], 2010	[15], 2011	**This work**
Impedance Spectroscopy	No	No	No	No	No	No	**Yes**
Digital readout	Yes	Yes	No	No	Yes	Yes	**Yes**
On-chip digital calibration	Capacitor array	Log converter	No	Prog. current	No	No	**Prog. gain**
Number of sensors	8	3	16	70	6	16	**27**
Read out channels	8	3	1	70	1	4	**27**
On-chip intergration of sensor	No	Yes	Yes	Yes	No	No	**Yes**
Sensing material	WO_3	SnO_2	SnO_2	CP	CP	SnO_2	**CP**
Operating temperature ($^\circ$C)	100-400	200-400	300	27	27	300	**27**
CMOS process (μm)	0.35	0.8	5	0.6	0.18	0.35	**0.35**

TABLE II

COMPARISON WITH OTHER CMOS GAS SENSOR ARRAY ARCHITECTURES.

for decoupling phase change of sensor impedance from phase shift introduced by signal processing blocks is demonstrated. Measurement results of a functionalized sensor pixel have been presented.

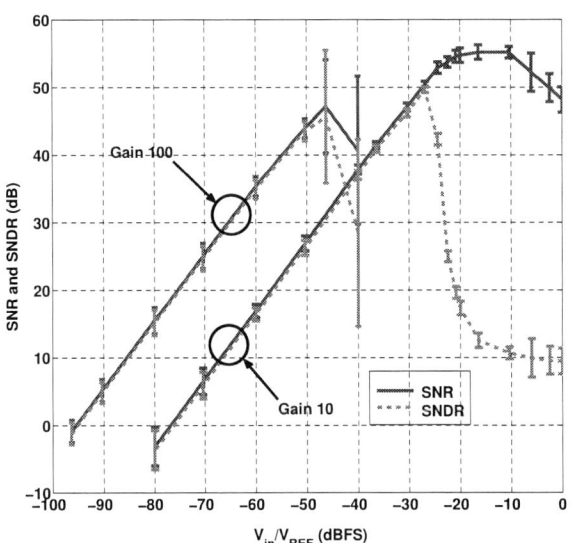

Fig. 7. Measured SNR and SNDR of 28 pixel outputs. Gain of each pixel is set to 10 or 100. The error bars show the variation of SNR and SNDR across different pixels.

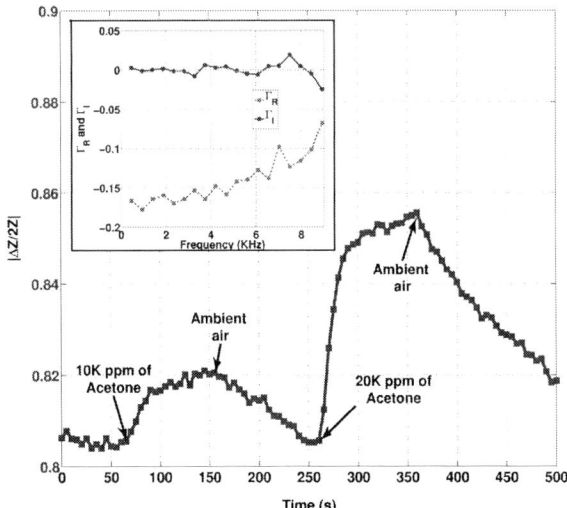

Fig. 8. Time response and impedance response of the sensor.

REFERENCES

[1] J. Gardner, P. Guha, F. Udrea, and J. Covington, "CMOS Interfacing for Integrated Gas Sensors: A Review," *Sensors Journal, IEEE*, vol. 10, no. 12, pp. 1833–1848, 2010.

[2] H. Bai and G. Shi, "Gas Sensors based on Conducting Polymers," *Sensors*, vol. 7, no. 3, pp. 267–307, 2007.

[3] E. Barsoukov and J. Macdonald, *Impedance spectroscopy: theory, experiment, and applications*. LibreDigital, 2005.

[4] C. Wu and K. Tang, "A Polymer-based Gas Sensor Array and its Adaptive Interface Circuit," in *VLSI Design Automation and Test (VLSI-DAT), 2010 International Symposium on*. IEEE, 2010, pp. 355–358.

[5] D. Rairigh, A. Mason, and C. Yang, "Analysis of On-Chip Impedance Spectroscopy Methodologies for Sensor Arrays," *Sensor Letters*, vol. 4, no. 4, pp. 398–402, 2006.

[6] X. Liu, D. Rairigh, C. Yang, and A. Mason, "Impedance-to-digital Converter for Sensor Array Microsystems," in *Circuits and Systems, 2009. ISCAS 2009. IEEE International Symposium on*. IEEE, 2009, pp. 353–356.

[7] X. Liu, D. Rairigh, and A. Mason, "A Fully Integrated Multi-channel Impedance Extraction Circuit for Biosensor Arrays," in *Circuits and Systems (ISCAS), Proceedings of 2010 IEEE International Symposium on*. IEEE, 2010, pp. 3140–3143.

[8] B. Chang and S. Park, "Electrochemical Impedance Spectroscopy," *Annual Review of Analytical Chemistry*, vol. 3, pp. 207–229, 2010.

[9] S. Pavan, N. Krishnapura, R. Pandarinathan, and P. Sankar, "A Power Optimized Continuous-Time $\Delta\Sigma$ ADC for Audio Applications," *Solid-State Circuits, IEEE Journal of*, vol. 43, no. 2, pp. 351–360, 2008.

[10] R. Schreier, "The Delta-Sigma Toolbox version 7.2," 2008.

[11] M. Malfatti, D. Stoppa, A. Simoni, L. Lorenzetti, A. Adami, and A. Baschirotto, "A CMOS Interface for a Gas Sensor Array with a 0.5%-Linearity over 500k-to-1G Range and$\pm$2.5 C Temperature Control Accuracy," in *IEEE ISSCC proceedings*, 2006, pp. 294–295.

[12] D. Barrettino, M. Graf, S. Taschini, S. Hafizovic, C. Hagleitner, and A. Hierlemann, "CMOS Monolithic Metal–Oxide Gas Sensor Microsystems," *Sensors Journal, IEEE*, vol. 6, no. 2, pp. 276–286, 2006.

[13] B. Guo, A. Bermak, P. Chan, and G. Yan, "A Monolithic Integrated 4×4 Tin Oxide Gas Sensor Array with On-chip Multiplexing and Differential Readout Circuits," *Solid State Electronics*, vol. 51, pp. 69–76, 2007.

[14] T. Koickal, A. Hamilton, S. Tan, J. Covington, J. Gardner, and T. Pearce, "Analog VLSI Circuit Implementation of an Adaptive Neuromorphic Olfaction Chip," *Circuits and Systems I: Regular Papers, IEEE Transactions on*, vol. 54, no. 1, pp. 60–73, 2007.

[15] K. Arshak, V. Velusamy, O. Korostynska, K. Oliwa-Stasiak, and C. Adley, "A CMOS Single-Chip Gas Recognition Circuit for Metal Oxide Gas Sensor Arrays," *Circuits and Systems, IEEE Transactions on*, vol. IEEE Early Access, 2011.

978-1-4673-0438-2/12 $31.00 © 2012 IEEE

2012 25th International Conference on VLSI Design

A Compact Temperature Sensor at 1.8µA per Hz Conversion rate and 1.1 °C accuracy for SOCs

Subhajit Sen (Senior Member, IEEE)
DAIICT,
Gandhinagar, Gujarat, India
sen@ieee.org

Dan Babitch,
CSR,
San Jose, CA, USA

Noshir Dubash
CSR,
Phoenix, AZ, USA

Abstract— **A compact (0.1 mm² area) temperature-recording system that is suitable for easy integration into an SOC is described. It includes a PTAT sensor, a pre-amplifier, and a first-order sigma-delta modulator based ADC all operating at 1.2V supply. The switched-capacitor pre-amplifier uses an auto-zeroing scheme based upon capacitive reset to avoid the need for shorting the op-amp outputs and inputs during reset. Errors due to transistor leakage are eliminated by selective use of thick-oxide transistors in the design. Another contribution of the paper is to illustrate a scheme that uses two reference voltages in the sigma-delta modulator ADC corresponding to the minimum and maximum temperatures measured to improve its effective resolution.**

Index Terms— system-on-chip (SOC), PTAT (proportional to absolute temperature), switched-capacitor, auto-zeroing, sub-threshold leakage current, MOSFET, sigma-delta modulator.

I. INTRODUCTION

The availability of accurate local temperature in the die of a large SOC is very useful for not only compensation for temperature drifts in the analog/RF sub-systems but also for thermal management/control of the entire SOC. A typical SOC typically has a complex power-management system with several power modes possibly including a low-power "keep-alive" (KA) mode in which some bare minimum functionality is permitted. In the KA mode of a GPS application, since the frequency of the crystal clock varies with temperature, a record of the temperature over time provides information how much the local time has drifted and when an energy-intensive synchronization process of local time with satellite time is necessary. Conversely, in high power modes it becomes important to limit die temperature so as to avoid hot-spots and catastrophic thermal-runaway. A temperature sensor for an SOC should therefore (a) dissipate very little power and (b) it should be compact enough to be easily integrated at any location of a complex SOC chip floor-plan. In order to avoid routing multiple power supplies in dense region of the floor-plan as well as for minimum power requirement it becomes

necessary to operate the sensor at the core supply voltage of the SOC (1.2V for 65nm CMOS).

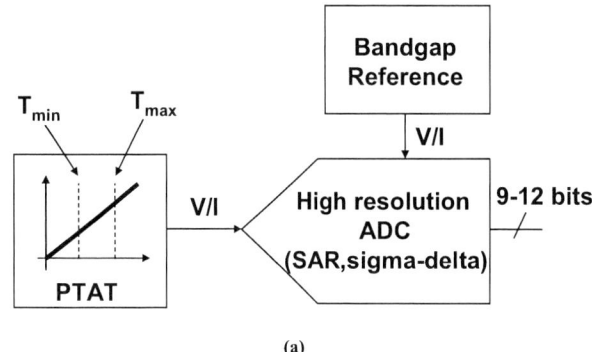

(a)

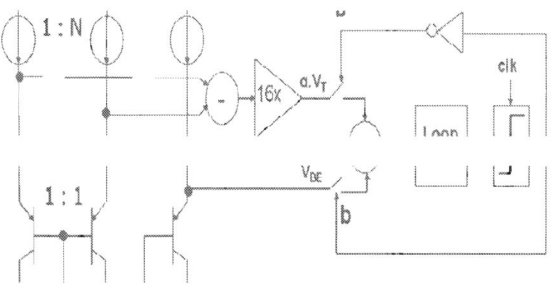

$$\text{Code density} = \mu = \frac{\alpha.V_T}{(V_{BE} + \alpha.V_T)}$$

(b)

Fig. 1: (a) A general temperature sensing and digitizing scheme (top) and (b) A temperature digitization scheme without explicit bandgap circuit(bottom)

Conventional temperature sensors and digitizers are built using two typical schemes shown in Figs. 1(a) and 1(b). In the scheme shown in Fig. 1(a) the PTAT sensor generates a voltage or current proportional to absolute temperature that

978-1-4673-0438-2/12 $31.00 © 2012 IEEE   179

lies between T_{min} and T_{max} and is converted into digital domain using an ADC using a temperature compensated band-gap reference. The scheme typically requires the subtraction of the offset temperature (T_{min}) either in the analog or digital domain[2]. Also it requires large area resistors for V-I conversion. In recent years a popular scheme shown in Fig. 1(b) has been used that avoids the need for generating an explicit band-gap voltage or current and therefore achieves better area and power efficiency[3,4]. In this scheme the PTAT ($\alpha.V_T$) and CTAT (V_{BE}) voltages are selected by bit and bit-complement and combined at the input summing node of a 1-bit sigma-delta loop in such a way that the resulting code density (μ) is equal to the ratio of the PTAT voltage expressed as a fraction of a temperature compensated band-gap voltage (Fig. 1(b)). Note that in both the schemes the PTAT voltage or current is proportional to absolute temperature and therefore some mechanism is required to subtract the offset temperature (T_{min}) either in the analog or digital domain. Since only 30% of the absolute temperature range is actually used [3], these schemes typically require a higher resolution sigma-delta modulator than necessary, possibly requiring an extra op-amp and/or larger biasing current for a given temperature conversion rate. Another feature in high-accuracy temperature sensors is that a significant design effort is made towards reducing the V_{BE} non-linearity of the BJT transistors and ratio-mismatch in current mirrors using analog techniques(such as dynamic-element-matching) to achieve better linearity in the PTAT and CTAT generators over temperature. These typically result in relatively complex circuits [3,4] that do not suit well for relatively lower accuracy applications in digitally dominated SOCs where good analog switches are difficult to implement due to low supply voltage as well as transistor leakage. Although [4] achieves good temperature resolution in a compact area using 65 nm CMOS it uses several op-amps in its circuits and does not explain how the switch-leakage issue is resolved.

This paper proposes a simple scheme for a temperature sensor and a digital interface for a temperature recorder suitable for SOC applications using a standard 65 nm CMOS technology. The digitizing scheme requires an explicit band-gap reference (as in Fig. 1(a)) that is shared with other analog sub-systems of the SOC. However, by deriving two references corresponding to the minimum(T_{min}) and maximum (T_{max}) temperatures it avoids the 70% loss of resolution in the schemes referred to earlier. Section-II introduces the architecture of the temperature recorder followed by details of the design of the low-voltage auto-zeroing switched-capacitor (SC) pre-amplifier and a one-bit first-order sigma-delta modulator. Section-III discusses the issues of gate and sub-threshold leakage when low threshold-voltage (V_{th}) core transistors are used and their mitigation. Section-IV discusses measurement results with and without calibration at the PTAT sensor and digital outputs.

II. TEMPERATURE RECORDER SUB-SYSTEM

Fig. 2 shows the temperature recorder scheme. It consists of a PTAT sensor that generates a voltage that is amplified by a switched-capacitor (SC) pre-amplifier. The amplified PTAT voltage is converted by a first-order sigma-delta modulator into a one-bit output that is accumulated using a counter to give a 9-bit digital representation of the temperature. At typical process corner the digital output covers 160 C temperature range between -50 °C to 110 °C. In order to make the temperature recorded more reliable and immune to rapid temperature fluctuations and to various noise sources in an SOC environment the 9-bit digital output of the sigma-delta ADC is applied at the input of a digital hysteresis circuit with 2 LSB hysteresis range. The hysteresis circuit simply ensures that the digital output code does not change unless the input code changes by at least 2 LSB's. This provides a +/- 1 LSB noise immunity and therefore helps to prevent the recorder memory from being filled up by frequent unreliable (noisy) readings. This 9-bit hysteresis circuit output is truncated and is recorded as a byte (8-bit) along with a time-stamp. Thus the complete 9-bit code range from 0 - 511 ($2^9 - 1$) is covered and maps on to the operable temperature range. This avoids the need for subtraction of the minimum absolute temperature offset ($T_{ref-min}$) in the analog or digital domains.

As shown in the timing diagram in Fig. 2, the temperature recorder begins a cycle of temperature conversion upon a system command (TS_ON). After the cycle is completed all the analog circuits of the sub-system are shut off. The cycle consists of a warm-up period during which analog circuits are allowed to establish bias and the 9-bit counter is disabled. The necessary warm-up period is conservatively assumed to be 20 cycles in the design but can be reduced down to a few (say 3) cycles by proper design. This is followed by a 512 (2^9) cycles of conversion at the end of which a 9-bit digital sample is obtained. Subsequently, the sub-system is shut off so that it dissipates a small (few nA) amount of leakage power.

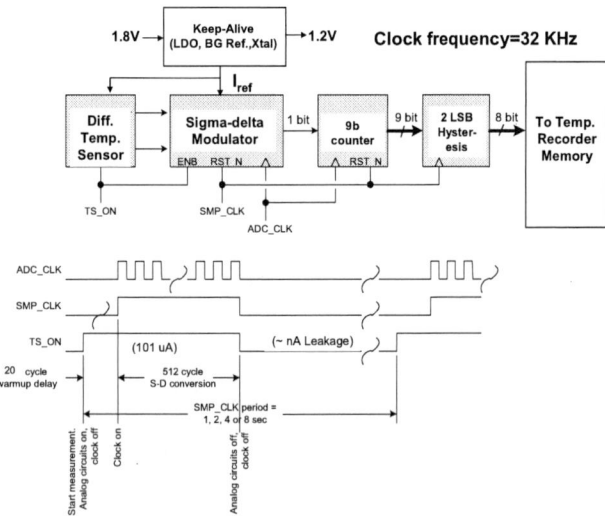

Fig. 2: Proposed Temperature Recorder scheme

A. Low Voltage PTAT Sensor and Pre-amplifier

The PTAT sensor and pre-amplifier is shown in Fig. 3. The PTAT voltage is obtained as the emitter-base voltage (V_{EB}) difference of two identical substrate PNP BJT transistors (available in standard digital CMOS processes) biased at 8 times emitter currents ratios obtained from ratio-ed PMOS cascoded current mirrors [5]. This gives a temperature sensitivity of 186 µV/ degree C. The output of the PTAT sensor is amplified by a SC amplifier [5] that uses auto-zeroing for op-amp offset cancellation and flicker noise reduction. A correlated-double-sampling (CDS) [1] arrangement that gives a 2X gain factor improvement was obtained by switching the polarity of the ΔV_{EB} PTAT signal in each half-cycle using the switching arrangement shown Fig. 3. Thus, with ($C_1/C_2 = 6$) the total sensitivity at the output of the pre-amplifier is improved to 2.2 mV/C.

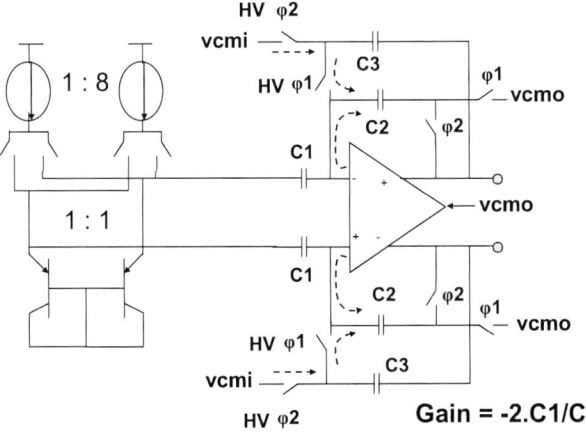

Fig. 3: PTAT sensor and capacitive-reset pre-amplifier circuit

One difficulty with conventional SC pre-amplifiers is that because the circuit operates at a low supply voltage (VDD=1.2V typical) and the differential output common-mode is required to be kept at mid-rail (VDD/2) for maximum output swing, the differential outputs cannot be shorted to inputs for offset storage during reset as in[5]. This is due to the high resistance of the CMOS switches when the sum of the PMOS and NMOS V_{th} exceeds VDD at high temperature and worst-case process corner. In addition since the output has to be reset every clock-cycle it imposes a slew-rate requirement on the op-amp even for slow moving inputs. Therefore, instead a capacitive-reset scheme shown in Fig. 3 is used. Here, during the reset phase φ_1 an offset voltage is developed across the op-amp input terminals that is stored in capacitors C_1, C_2 while C_3 provides capacitive feedback with the voltage sampled during the previous output sampling phase φ_2. The capacitive-reset scheme shown in Fig.3 is similar to a low-glitch SC amplifier proposed earlier [6,3] for reduced slewing requirement on the op-amp whereas the main requirement here is offset cancellation without using output-input shorting switches. Note also that the glitch-reduction capacitor used in [6,3] is not required here due to relaxed settling requirements(low clock frequency). The op-amp of the pre-amplifier is a fully differential PMOS input folded-cascode amplifier using low-voltage cascode biasing [7] with its input

common-mode voltage (vcmi) kept close to ground. These help to keep the op-amp input stage biased with high gain (75 dB), high CMRR and low flicker-noise. Lower vcmi also reduces the on resistance of the high-V_{th} NMOS transistor switches used for leakage current reduction described in section-III. At the op-amp output side, the inherent differential offset of the pre-amplified output at $T_{ref-min}$ helps to keep the switch resistance low during φ_2. The total integrated noise at the output of the pre-amplifier during the output sampling phase (φ_2) was 83 and 395 µV and at 30 and 100 °C respectively. The op-amp bias current is 45 µA and most of the transistors are biased in the sub-threshold region.

Because of the high output resistance of the pre-amplifier (due to low bias current) the sensor is susceptible to (high-frequency) impulsive transient power supply noise. This drawback has been eliminated by a suitable clock-generation scheme that ensures that any power transient impulse (generated by other blocks on the chip) occurs after the pre-amplifier output sampling phase.

B. SIGMA-DELTA MODULATOR

A modified first-order SC sigma-delta modulator [1] is used to convert the amplified PTAT voltage into a one-bit output and shown in Fig. 4. The two

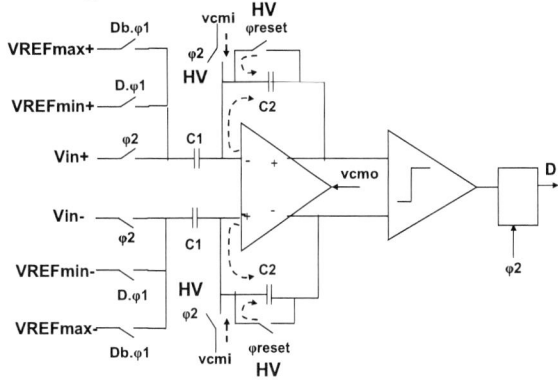

Fig. 4: Sigma-delta modulator with leakage paths (dashed arrows)

differential reference voltages $VREF_{min}$ and $VREF_{max}$ correspond to the pre-amplified sensor outputs at the minimum ($TREF_{min}$=-50 °C) and maximum ($TREF_{max}$=110 °C) temperatures respectively and add or subtract from the integrator output depending upon the logic level of the output code (D). Therefore at $TREF_{min}$ the output of the modulator is ideally an all-zeroes (0000..) code and at $TREF_{max}$ it is an all-ones (1111…) code with a toggle (1010..) pattern at the average or mid-point between the two at 30 °C. Since the ADC reference range covers the temperature range completely, this scheme results in a better power and area efficiency than [3] by lowering the requirement on ADC resolution, saving an extra op-amp and avoiding the need for an elaborate decimation-filter. A shorting switch is required to reset the integrating capacitor (C_2) at the beginning of the conversion cycle during warm-up. Even though the switch-resistance is likely to be large here, this is not an issue since several clock cycles (> 20) are available during the warm-up period. The op-amp used is the same as the one for the pre-

978-1-4673-0438-2/12 $31.00 © 2012 IEEE

amplifier except that it is biased at half the current. The reference and common-mode input/output voltages of the pre-amplifier and sigma-delta modulator are derived from a band-gap reference current $I_{ref}=5$ µA available from the KA section of the SOC (see Fig. 2) and dropped across a common resistor string.

The topmost plot of Fig. 5(a) shows the preamplifier op-amp output waveform as it settles from the power-up condition. The lowermost plot shows the waveform at the two op-amp inputs. The topmost plot of Fig. 5(b) shows the sigma-delta integrator output waveform at -40 °C and the waveform below it shows the sigma-delta bit output (D).

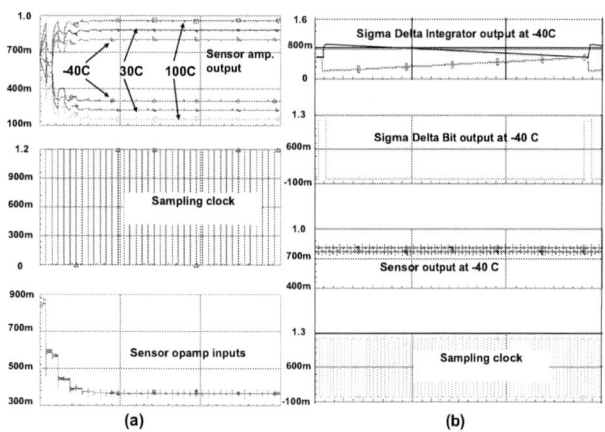

Fig. 5: PTAT Sensor, pre-amplifier and sigma-delta modulator waveforms

III. LEAKAGE CURRENT ERROR REDUCTION

As explained in section-I, the temperature recorder system is required to operate at core (65 nm) transistor supply voltage of 1.2V supply. Due to their small V_{th} and oxide-thickness these transistors suffer from significant sub-threshold and gate leakage currents. For example as shown in Fig. 3, sub-threshold leakage current in the switch connecting C2 and C3 when φ_1 is LOW results in error charge building up in the integrating capacitor C2(shown as dashed arrows). While this is a common-mode error if the circuit is completely matched, it results in significant temperature dependent error that degrades the accuracy of temperature measurement. Similarly, the mismatched gate leakage current flows continuously into the C2 resulting in significant errors in both the pre-amplifier (Fig. 3) and sigma-delta modulator op-amps (Fig. 4).

The above leakage problem has been resolved replacing the core 65nm transistors in the above mentioned switches with higher V_{th} low-leakage NMOS transistors (indicated by label HV in Figs. 3 and 4) and choosing the op-amp input common-mode voltage to be close to ground. The PMOS input pair transistors in both op-amps are replaced by low-leakage high V_{th} transistors whose large oxide-thickness ensures smaller gate leakage (of the order of a few pA). Low vcmi ensures sufficiently small on resistance of the NMOS switches while enabling the op-amp input stage to operate with only a marginal loss of headroom in its PMOS tail current source. Note however that leakages in the switches connected to the

left terminal of C1 and right terminal of C2 in the pre-amplifier (Fig. 3) and the left terminal of C1 in the sigma-delta modulator (Fig. 4) are not important since those terminals are always driven by a low-impedance input.

Table-1: Temperature Recorder Features

Supply Voltage	1.2V ± 10%
Temperature Range Required	-40 to 100 °C
Temp. Error (Absolute)	< 9 °C
Temp. Error (1 temp. calibration at 25 °C)	-1.0/+1.4 °C
Temp. Error (3 temp. calibration)	-1.1/+1.0 °C
Error due to KA VDD (1.8V) variation	+/- 2 LSB
Average Power consumtion at 1 sample/sec	1.77 uA @ 1.2V supply
Crystal clock frequency	32 KHz
Chip Area	0.1 mm^2

IV. MEASUREMENT RESULTS

The temperature recorder system is designed such that for the nominal process corner a temperature range of -50 °C to 110 °C is mapped to a stored memory code range of 0 to 255 (2^8-1). It was implemented in a 65 nm CMOS process with options for thick-oxide, low-leakage, high-voltage I/O transistors and MIM-capacitors. Packaged die measurements were done in a temperature-controlled oven. Fig. 6 plots the error (mV) in the measured PTAT sensor output voltage with respect to its ideal value ($\Delta V_{BE}=(kT/q).\log_e(8)$) across several packaged dies obtained from typical and two extreme corner lots. The corresponding simulated error at typical process corner is also shown. Fig. 7 shows the absolute error in the measured temperature when the 8-bit temperature representation of the stored code is compared against the expected temperature assuming a linear variation of code with temperature.

Fig. 8 shows the measured temperature after calibration at a single temperature (25 °C) and also assuming linear slope. The worst-case error is -1.0/+1.4 °C for 8 packaged dies across nominal and corner lots. When the temperature vs. code is modeled as a second order (quadratic) equation and calibrated (LMS fitted) at three temperatures (at 25 °C and near the two extreme ends of the range) a worst-case error of -1.1/+1.0 °C is obtained as shown in Fig. 9. The effect of ADC quantization noise (about 0.6 C) due to stored 8-bit resolution is clearly visible in Figs. 8 and 9 suggesting the possibility of improved accuracy if the raw 9-bit ADC output is directly observed. A second-order curve-fitting polynomial is optimal since most band-gap circuits exhibit quadratic curvature in their output voltage vs. temperature and only 3 measurements are required to determine the 3 polynomial coefficients.

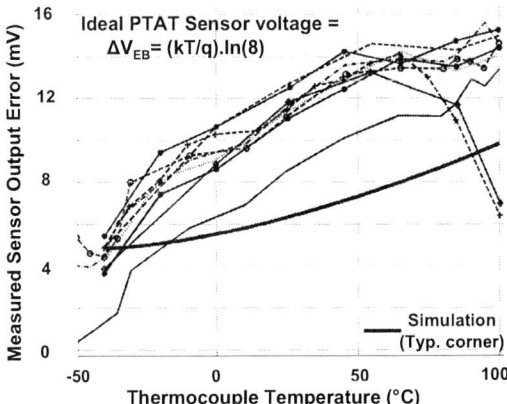

Fig. 6: Measured PTAT Sensor output error

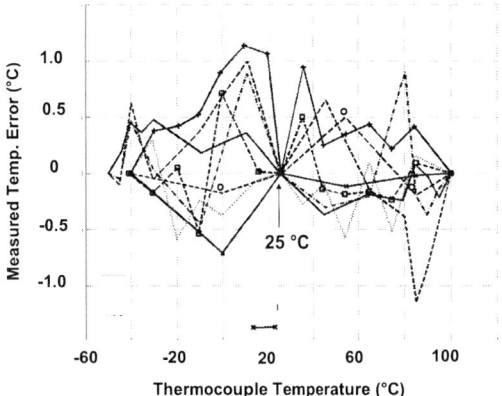

Fig. 9: Measured temperature error after three-temperature calibration

V. CONCLUSIONS

We have illustrated the design of a low-voltage and low-power temperature-sensor and recorder that achieves 8-bit temperature resolution and is compact enough to be easily integrated into an SOC floor-plan. A key strategy is to use a scheme whereby the entire operating temperature range (from T_{min} to T_{max}) is completely mapped to the ADC reference range as well as limit the number of op-amps in the design to two and re-use or share the band-gap reference with other sub-systems within the SOC.

ACKNOWLEDGEMENT

The authors would like to thank ex-SiRF (CSR) colleagues Peter Naji, Suhas Kulhalli, Narayan Prasad Ramachandran, Satheesh Kumar A.S. and Hemanth for useful discussions and layout support.

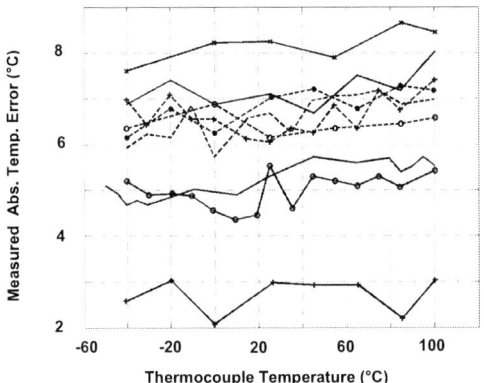

Fig. 7: Measured temperature error without calibration

REFERENCES

[1] D. Johns, K. Martin, "Analog Integrated Circuit Design", John Wiley.

[2] A. Bakker, J.H. Huijsing, "Micropower CMOS temperature sensor with digital output", *IEEE J. Solid-State Circuits*, vol. SC-31, No. 7, pp. 733-737, July 1996.

[3] M.A.P. Pertijs, A. Niederkorn, X. Ma, B. McKillop, A. Bakker, J.H. Huijsing, "A CMOS Smart Temperature Sensor with a 3-σ Inaccuracy of ±0.5 °C From -50 °C to 120 °C", *IEEE J. Solid-State Circuits*, vol. SC-40, No. 2, pp. 454-461, Feb. 2005.

[4] Fabio Sebastiano, Lucien J. Breems, Kofi A. A. Makinwa, Drago Salvatore, Domine M.W. Leenaerts, Bram Nauta, "A 1.2-V 10- μW NPN-Based Temperature Sensor in 65-nm CMOS With an Inaccuracy of 0.2 °C (3σ) From 70 °C to 125 °C", *IEEE J. Solid-State Circuits*, vol. 45, No. 2, pp. 2591 -2601, Dec. 2010.

[5] M. Tuthill, "A switched-current, switched-capacitor temperature sensor in 0.6-μm CMOS", *IEEE J. Solid-State Circuits*, vol. 33, No. 7, pp. 1117-1122, July 1998.

[6] H. Matsumoto, K. Watanabe, "Spike-Free Switched-Capacitor Circuits", *Electronic Letters*, Vol. 23, no. 8, pp. 428-429, April 1987.

[7] B.A. Minch, "A Low-Voltage MOS Cascode Bias Circuit for All Current Levels", *Proc. IEEE Int. Symp. Circuits and Systems*, May 2002, pp. 619-622.

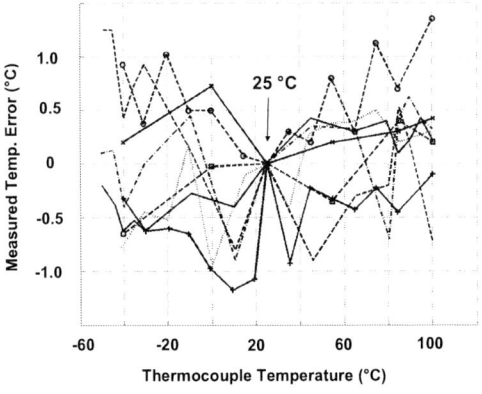

Fig. 8: Measured temperature error after calibration at 25 °C

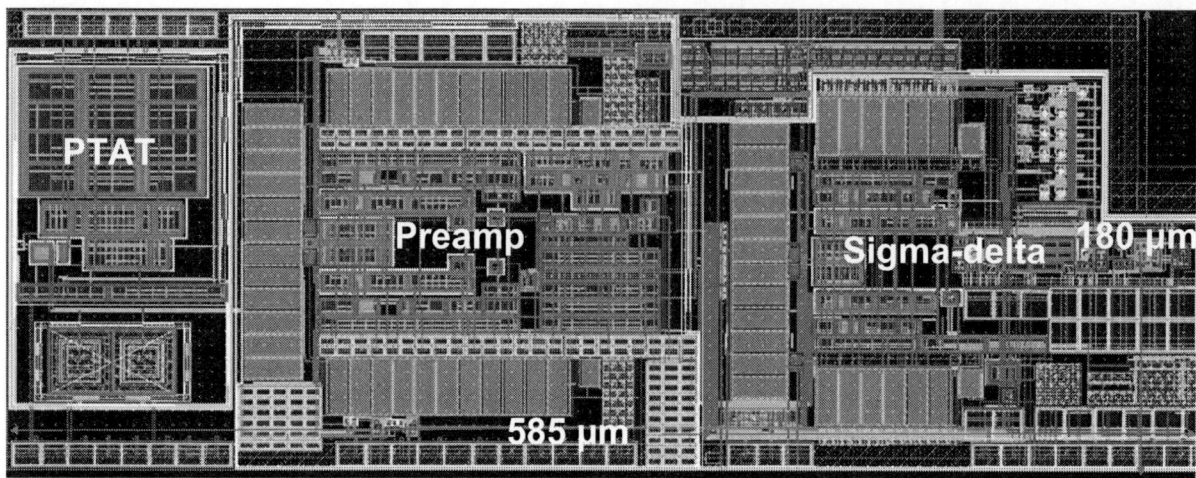

Fig. 10: Layout of Temperature Recorder (without recorder memory). Area= 585 μm X 180 μm = 0.1 mm^2

Analysis of the Pull-In phenomenon in Microelectromechanical Varactors

[1]Anindya Lal Roy, [1]Anirban Bhattacharya, [1]Ritesh Ray Chaudhuri
[1]Advanced Technology Development Centre
Indian Institute of Technology, Kharagpur
West Bengal, India

[2]Tarun Kanti Bhattacharyya**
[2]Department of Electronics & Electrical Communication Engineering
Indian Institute of Technology, Kharagpur
West Bengal, India

Abstract—High-Q factor voltage controlled oscillators (VCO) need a wide tuning range and low phase noise over gigahertz ranges of frequency which depends on the tunability of the capacitors in the LC tank circuit. The reasons behind the development of a microelectromechanical (MEM) varactor were the difficulties encountered in the realization of on-chip variable capacitors having low phase noise and high quality factors with a wide tuning range in the span of frequencies over process and temperature variations. This paper presents an efficient closed-form model for determination of the pull-in voltage in a surface micromachined MEM varactor which is a factor directly affecting the tunability of the device. The nonlinear spring hardening effects associated with proper load-deflection characteristics of clamped plates and the electrostatic spring softening effects due to the parallel-plate and fringing field capacitances have been taken into account with the dimensions of the device optimized through finite element analysis (FEA).

Keywords-MEM; varactor; pull-in; nanoindentation; electrostatic actuation

I. INTRODUCTION

Electrostatically actuated MEMS-based sensors or actuators are rapidly gaining acceptance over a wide range of applications due to their good scaling properties at smaller dimensions, higher sensitivities, high energy densities along with the possibility of realizing new designs and low-power consumption [1]. The MEM varactor presented in this work is an electrostatically actuated diaphragm-type device with the periphery clamped by flexures [2].

The operating principle of the varactor is the voltage-driven movement of the top plate under the force of electrostatic attraction due to a constant actuation voltage applied at the bottom electrode. It is this nonlinear electrostatic force of attraction associated with the constant voltage drive mode that introduces the phenomenon of "pull-in" which occurs at a certain geometry-dependent voltage limit [3], causing the top varactor plate to collapse on the bottom electrode and effectively rendering it non-functional as a tunable capacitor. The determination of the pull-in voltage of MEM varactors is of prime importance as it provides valuable insight into device material properties like residual stresses and elastic modulus as well as device parametric properties

like sensitivity, frequency response, dynamic range and distortion-induced instabilities [4]. Parallel-plate modelling of the pull-in phenomenon assumes a piston-like motion of the suspended structure with a linear spring constant while excluding fringing capacitance effects and approximately predicts the collapse of the suspended plate when the highest deformation exceeds 0.33% of the air-gap.

In this work, the surface micromachined MEM varactor has been subjected to nano-indentation and scratch tests for obtaining accurate load-deflection characteristics and also takes into account the fringing field capacitances wherever necessary. An empirical capacitance model of VLSI on-chip interconnects [5] has been mathematically extended to derive an expression for the nonlinear electrostatic pressure acting on the electrostatically actuated varactor plate and has been utilized along with the load-deflection characteristics to determine the pull-in voltage. The model developed here maintains good conformity with FEA results as well as simulated lumped parameter models.

II. MEM VARACTOR

A. Conceptual geometry of the mechanical model

The varactor studied in this work is of quad-beam geometry with the centrally placed rectangular polysilicon plate acting as a seismic mass suspended by four symmetrically aligned thin polysilicon flexures. The two flexures attached on either side of the central plate and rigidly clamped at the other ends to fixed walls form the suspended top plate of the varactor. The beams are of the identical dimensions and are positioned equidistantly from the plate edges. The bottom plate of the varactor is a polysilicon electrode fixed to passivated silicon substrate and this electrode also acts as the actuating electrode for device operation. Additionally, the central plate has a 10×10 array of perforations distributed symmetrically over the entire surface area in order to reduce the effects of squeezed-film damping, effectively reducing the voltage drive needed for electrostatic actuation and hence, lowering power consumption. The simulated and fabricated quad-beam structures are shown in Fig. 1 (a) and (b) and the dimensions are tabulated in Table 1.

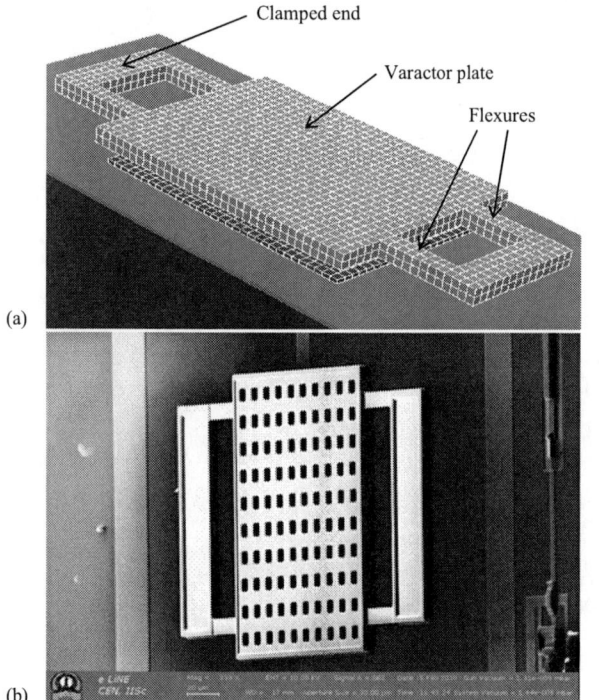

(a)

(b)

Fig. 1 (a) Solid model of the microelectromechanical varactor developed for finite element method-based CoSolveEM® simulation (b) Scanning electron microscope image of the microfabricated MEM varactor

B. Basis of the electrical model

The configuration of an electrostatically actuated top varactor plate separated by a dielectric spacer from a fixed bottom plate (electrode) can be thought of as the integration of infinitesimally thin VLSI on-chip interconnects (top plate) separated from the ground plane (bottom plate) by a dielectric layer (air). The fringing field capacitances owing to the clamped flexures on either side can be neglected by considering their fixed ends to be at infinity while the varactor plate attached to their moving ends and the perforations on the top plate contribute to the total capacitance.

TABLE I. GEOMETRY OF THE MEM VARACTOR

MEM Varactor Elements	Dimensions		
	Length	Breadth	Thickness
Varactor Plate	340 μm	240 μm	2 μm
Actuation Electrode	320 μm	220 μm	0.5 μm
Flexures	100 μm	20 μm	2 μm
Perforations	10 μm	10 μm	2 μm
Air-gap	2 μm		

III. ELECTROMECHANICAL VARACTOR MODEL

From the basic assumptions made in the previous section, the total capacitance of the varactor can be expressed as a composition of the parallel-plate capacitance, the fringing field capacitances of the plate breadth and thickness and those of the perforations of an approximately square geometry. By accounting for the average fringing field capacitances introduced by the edges and corners of the perforations as well

as the varactor plate itself, we formulate the total capacitance of the device as

$$C_{total} = \varepsilon_0\varepsilon_r L\left[\left[\int_0^W\left(\frac{dW}{h_0}\right) - \frac{mnab}{Lh_0}\right] + 1.06\left[\int_0^W\left(\frac{dW}{h_0}\right) - \frac{mnab}{Lh_0}\right]^{\frac{1}{4}}\right]$$
$$+ 1.06\varepsilon_0\varepsilon_r\left[L + 2mn(a+b)\right]\left[\frac{H}{h_0}\right]^{\frac{1}{2}} + 0.77 \quad (1)$$

where m and n are the total number of rows and columns of perforations on the plate, ε_0 and ε_r are the permittivities of free-space and the dielectric layer (air-gap: $\varepsilon_r = 1$); L, W, H are the length, breath and thickness of the varactor plate, a and b are the length and breadth of each perforation with h_0 being the air-gap between the top and bottom plates. The first term in brackets represents the parallel-plate consideration while the others represent the fringing field capacitances due to the periphery of the varactor plate, with corrections for the perforations as shown in Fig. 2 (a), (b) and (c).

The equation of motion of the top plate can be represented as the differential equation for a single degree of freedom spring-mass system with negligible squeezed-film damping effects due to the presence of a perforation matrix on the plate and is given by:

$$M\frac{d^2z}{dt^2} + (K_{spring} - K_{soft})z = F_{electrostatic} \quad (2)$$

where M is the mass of the plate, K_{spring} is the effective spring constant of the flexures, K_{soft} represents the softening effect of flexure springs under electrostatic actuation and F_e is the electrostatic force of attraction driving the varactor which is given by:

$$F_{electrostatic} = -\frac{d}{dz}\left(\frac{1}{2}C_{total}V^2\right) \quad (3)$$

which originates from V, the actuating voltage drive. It is noteworthy to mention here that the electrostatic force created due to the actuation voltage attempts to collapse the top plate onto the bottom plate, which in turn is resisted by the mechanical pressure generated by the restoring spring force supplied by the flexures.

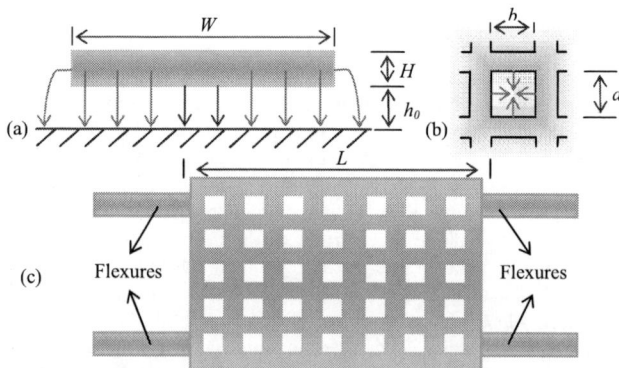

Fig. 2 (a) Breadth-edge view of the varactor plate showing the colour-coded parallel-plate and fringing fields (b) Magnified view of a perforation and the fringing field due to its edges (c) Top view of the varactor (not to scale), attached to the clamping flexures

Prior to delving further into the electrostatically actuated spring-mass system, let us take a closer look at the load-deflection characteristics of the suspended varactor plate as obtained through nano-indentation tests performed on the device.

A. Spring hardening effects in clamped plates

It has been reported for clamped-clamped flexures subjected to ramp loading that the rate of increase in deflection decreases as the corresponding applied load is increased [6]. By a logical extension, we can conclude that our observations might be similar for rigidly clamped plates under ramp loading conditions. This is a stiffness nonlinearity commonly referred to as hardening of the mechanical structure because the boundary conditions restrict the axial straining of the varactor plate as the deflection increases. The restoring spring force provided by the flexures with an attached plate mass in this case can be defined by the generalized polynomial expression:

$$F_{spring} = K_1 z + K_3 z^3 \approx K_{spring} z \qquad (4)$$

In the above equation, K_1 is the linear component which includes the Hookean assumption as well as the process-dependent constant arising from residual stresses and is represented by

$$K_1 = S_{res}\left(\frac{\sigma' H}{l^2}\right) + S_{flex}\left(\frac{E'H^3}{l^4}\right) \qquad (5)$$

where σ' is the residual stress in the varactor plate expressed as $\sigma_0(1-\upsilon)$, l is the length of the flexures and E' is the effective Young's modulus for plates given by $E' = E/(1-\upsilon^2)$ with υ being the Poisson's ratio for the given material while K_3 is the nonlinear component of the spring constant given by

$$K_3 = S_{nl}\left(\frac{E'H}{l^4}\right) \qquad (6)$$

S_{res}, S_{flex} and S_{nl} are constants pertaining to the residual stress, the flexure elasticity and the nonlinear spring hardening effect giving rise to the composite constant K_{spring}.

From the above considerations, the spring constant of the mechanical flexure denoted by K_{spring} can be defined as the gradient of the spring force F_{spring} as a function of the displacement z:

$$K_{spring} = \frac{dF_{spring}}{dz} = K_1 + 3K_3 z^2 \qquad (7)$$

The constants K_1 and K_3 can be experimentally determined from load-deflection characteristics for varactors subjected to ramp loading conditions. In general, small-deflection regimes are governed by the linear part of K_{spring} and the behaviour can be roughly modeled as that of a linear spring-mass system. However, as the magnitudes of the deflection approach the flexure/plate thickness dimensions, which for MEMS devices are arguably the same; the nonlinear part initiates its role in contributing to the hardening effect resulting in a Duffing spring-model of the flexures supporting the varactor plate.

B. Nano-indenter validation

The fabricated polysilicon MEM varactor has been subjected to progressive linear scratch test over the entire plate area and static nano-indentation tests applying a central load on the plate with a ramp loading rate of 4 mN/min over a range of 0.1 mN to 2 mN with a 10 μm diamond-tipped SD-A28 Rockwell® indenter at an acquisition rate of 10 Hz. The mapped load-deflection characteristics were obtained as shown in Fig. 3 (a) and can be compared with the theoretical nonlinearity of a generic cubic spring shown in Fig. 3 (b).

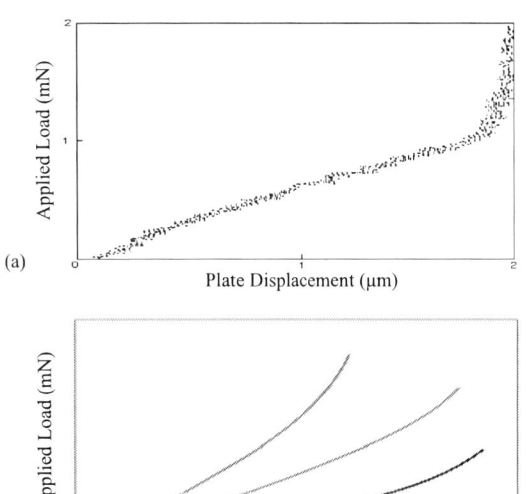

Fig. 3 (a) Load-deflection characteristics obtained from nano-indentation tests (b) Theoretical load-deflection characteristics of Duffing springs with nonlinear (cubic) stiffness K_1(red) > K_1(blue) > K_1(black) and K_3(red) > K_3(blue) > K_3(black)

From the experimental as well as the theoretical load-deflection characteristics plotted above, it is fairly reasonable to argue that the varactor presented here follows an approximate Duffing equation with cubic stiffness and that the nonlinearities introduced by the restoring spring force plays an important role in the determination of the mechanical and electrostatic equilibrium of the device. This forms the basis of the argument that the above mentioned factors are pivotal in the calculation of the pull-in voltage of a MEM varactor.

Returning to the electrostatically actuated spring-mass system, the force of actuation on the can be decomposed into a Taylor series expansion about an initial displacement of z_0 under no-load conditions and (2) reduces to

$$M\frac{d^2z}{dt^2} + (K_{spring} - K_{soft})z$$

$$= \frac{\varepsilon_0 \varepsilon_r L\left(W - \frac{mnab}{L}\right)V^2}{2(h_0 - z_0)^2}\left[1 - \frac{2z_0}{(h_0 - z_0)} + \ldots\right]$$

$$+ \frac{0.1325\varepsilon_0\varepsilon_r L\left(W - \dfrac{mnab}{L}\right)^{\!\!\frac{1}{4}} V^2}{\left(h_0 - z_0\right)^{\!\frac{5}{4}}}\left[1 - \frac{1.25z_0}{\left(h_0 - z_0\right)} + \ldots\right]$$

$$+ \frac{0.265\varepsilon_0\varepsilon_r \{L + 2mn(a+b)\}H^{\frac{1}{2}} V^2}{\left(h_0 - z_0\right)^{\!\frac{3}{2}}}\left[1 - \frac{1.5z_0}{\left(h_0 - z_0\right)} + \ldots\right] \quad (8)$$

where

$$K_{soft} = \frac{\varepsilon_0\varepsilon_r L\left(W - \dfrac{mnab}{L}\right) V^2}{\left(h_0 - z_0\right)^3}$$

$$+ \frac{0.17\varepsilon_0\varepsilon_r L\left(W - \dfrac{mnab}{L}\right)^{\!\!\frac{1}{4}} V^2}{\left(h_0 - z_0\right)^{\!\frac{9}{4}}}$$

$$+ \frac{0.4\varepsilon_0\varepsilon_r \{L + 2mn(a+b)\}H^{\frac{1}{2}} V^2}{\left(h_0 - z_0\right)^{\!\frac{5}{2}}} \quad (9)$$

Here, the first term represents the electrostatic softening due to the parallel-plate capacitance while the remaining terms denote the softening due to the fringing field capacitances.

IV. MODELLING THE PULL-IN VOLTAGE

A. The Nonlinear Restoring Force

The variational method [7], which provides a relatively accurate solution of the spring-mass differential equation of the varactor, leads us to an expression of the restoring spring force F_{spring} acting on the moving plate and is given by

$$F_{spring} = K_1 z + K_3 z^3$$

$$= \left\{S_{res}\left(\frac{\sigma' H}{l^2}\right) + S_{flex}\left(\frac{E' H^3}{l^4}\right)\right\}z$$

$$+ \left\{S_{nl}\left(\frac{E' H}{l^4}\right)\right\}z^3 \quad (10)$$

with K_1 and K_3 having been defined in (5) and (6). The constants S_{res}, S_{flex} and S_{nl} can be determined by experimental analysis as well as data fitting via finite element simulations over an extensive range of flexure dimensions while maintaining the minor approximation that the varactor plate remains parallel to the ground plane.

At pull-in, we replace $z = 0.33h_0$ in (10) to obtain the restoring spring force acting on the plate which is given by

$$F_{spring-PI} = \left\{S_{res}\left(\frac{\sigma' H h_0}{3l^2}\right) + S_{flex}\left(\frac{E' H^3 h_0}{3l^4}\right)\right\}$$

$$+ \left\{S_{nl}\left(\frac{E' H h_0^3}{27l^4}\right)\right\} \quad (11)$$

B. The Nonlinear Electrostatic Force

Electrostatic actuation by a constant drive voltage V creates a non-uniform force of attraction which pulls the varactor plate down towards the bottom electrode. This happens due to a higher charge concentration at the plate compared to the clamped flexures which contributes to an increased force of attraction on the top plate in comparison to the flexures. This causes a position-dependent deflection along the length of the flexures with the varactor plate, while parallel to the actuation electrode, being deflected the most. This non-uniformity of the electrostatic force increases as the flexures are further deformed due to the inverse-square law of Coulomb attraction.

Thus, in order to determine the plate deflection, it is necessary to derive an approximately uniform and linearized expression for the electrostatic force schematically depicted in Fig. 4. This can be achieved from (8) and (9) by linearizing the electrostatic force about the point of zero-deflection when the varactor plate was free of all forces ($z_0 = 0$) which also allows us to apply the parallel-plate approximation by ignoring the

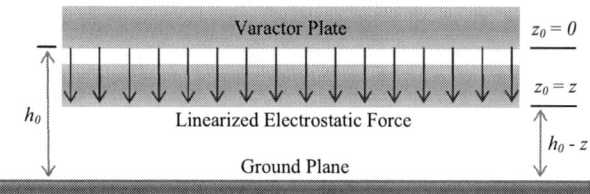

Fig. 4 Schematic of the assumed parallel-plate configuration from the initial no-load, no-deflection condition to an arbitrary deflection z under the approximated uniformly linearized electrostatic force acting on the varactor plate from the ground plane

flexure curvature as shown. Applying the above thought and neglecting the nonlinear terms of the Taylor expansion, we find that for $z_0 = 0$

$$M\frac{d^2 z}{dt^2} + (K_{spring} - K_{soft})z$$

$$= \frac{\varepsilon_0\varepsilon_r L\left(W - \dfrac{mnab}{L}\right) V^2}{2h_0^2} + \frac{0.1325\varepsilon_0\varepsilon_r L\left(W - \dfrac{mnab}{L}\right)^{\!\!\frac{1}{4}} V^2}{h_0^{\frac{5}{4}}}$$

$$+ \frac{0.265\varepsilon_0\varepsilon_r \{L + 2mn(a+b)\}H^{\frac{1}{2}} V^2}{h_0^{\frac{3}{2}}} \quad (12)$$

with the electrostatic softening constant K_{soft} reducing to

$$K_{soft} = \frac{\varepsilon_0\varepsilon_r L\left(W - \dfrac{mnab}{L}\right) V^2}{h_0^3} + \frac{0.17\varepsilon_0\varepsilon_r L\left(W - \dfrac{mnab}{L}\right)^{\!\!\frac{1}{4}} V^2}{h_0^{\frac{9}{4}}}$$

$$+ \frac{0.4\varepsilon_0\varepsilon_r \{L + 2mn(a+b)\}H^{\frac{1}{2}} V^2}{h_0^{\frac{5}{2}}} \quad (13)$$

Now, if we assume the motion of the varactor plate to be quasi-static, the time-dependent second derivative in (12) can be ignored, allowing us to solve for the approximately linear electrostatic force in the static case which turns out to be

$$F_{electrostatic} = Kz = \frac{\varepsilon_0\varepsilon_r L\left(W - \dfrac{mnab}{L}\right)V^2}{2h_0^2}$$

$$+ \frac{0.1325\varepsilon_0\varepsilon_r L\left(W - \dfrac{mnab}{L}\right)^{\frac{1}{4}}V^2}{h_0^{\frac{5}{4}}}$$

$$+ \frac{0.265\varepsilon_0\varepsilon_r\{L + 2mn(a+b)\}H^{\frac{1}{2}}V^2}{h_0^{\frac{3}{2}}} + K_{soft}z \qquad (14)$$

where K_{soft} is given by (13).

The linearized electrostatic force acting on the varactor plate at pull-in can be found from (14) by putting $z = 0.33h_0$. Substituting K_{soft} with the expression in (13), the electrostatic force of attraction which acts on the top plate at pull-in condition becomes

$$F_{electrostatic-PI} = \frac{5\varepsilon_0\varepsilon_r L\left(W - \dfrac{mnab}{L}\right)V_{PI}^{\,2}}{6h_0^2}$$

$$+ \frac{0.1892\varepsilon_0\varepsilon_r L\left(W - \dfrac{mnab}{L}\right)V_{PI}^{\,2}}{h_0^{\frac{5}{4}}}$$

$$+ \frac{0.3983\varepsilon_0\varepsilon_r\{L + 2mn(a+b)\}H^{\frac{1}{2}}V_{PI}^{\,2}}{h_0^{\frac{3}{2}}} \qquad (15)$$

C. The Pull-In Voltage

Having obtained the expressions for both the pull-in electrostatic as well as the pull-in restoring spring force, we now equate (11) and (15) to obtain a closed-form model of the pull-in voltage for the MEM varactor presented here since, at pull-in equilibrium, the electrostatic and the spring forces counter-balance each other. The pull-in voltage of the device as obtained is given by

$$V_{PI} = \sqrt{\frac{S_{res}\left(\dfrac{\sigma' H h_0}{3l^2}\right) + S_{flex}\left(\dfrac{E'H^3 h_0}{3l^4}\right) + S_{nl}\left(\dfrac{E'Hh_0^3}{27l^4}\right)}{0.8333\alpha h_0^{-2} + 0.1892\alpha h_0^{-\frac{5}{4}} + 0.3983\beta h_0^{-\frac{3}{2}}}}$$

(16)

where

$$\alpha = \varepsilon_0\varepsilon_r L\left(W - \frac{mnab}{L}\right) \text{ and } \beta = \varepsilon_0\varepsilon_r\{L + 2mn(a+b)\}H^{\frac{1}{2}}$$

V. VALIDATING THE MODEL

The pull-in voltage of the MEM varactor derived in the above sections was compared with the results of electromechanical finite element simulations in the MemMech and CoSolveEM module of Coventorware® [8] as well as simulations of the lumped parameter model in the Saber

Architect® platform. Lumped parameter simulation of the pull-in voltage was performed using the *Arc Length Continuation Voltage with Position Input* module which overcomes convergence discontinuities by moving along the force-balance curve. The schematic of the lumped parameter model is shown in Fig. 5 (a) and the deflection-volt characteristic from Architect simulations is shown in Fig. 5 (b) with the results compared with the FEA results as well as our developed model in Table 2.

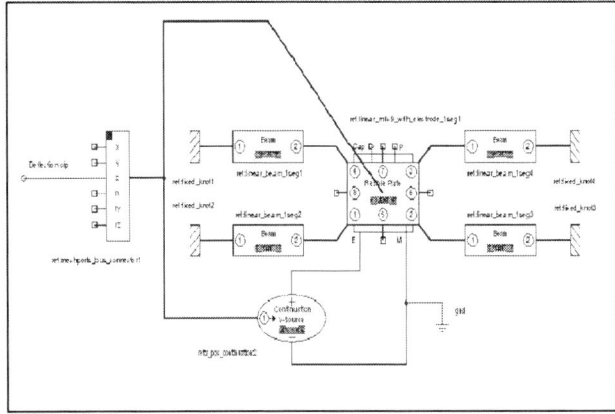

(a)

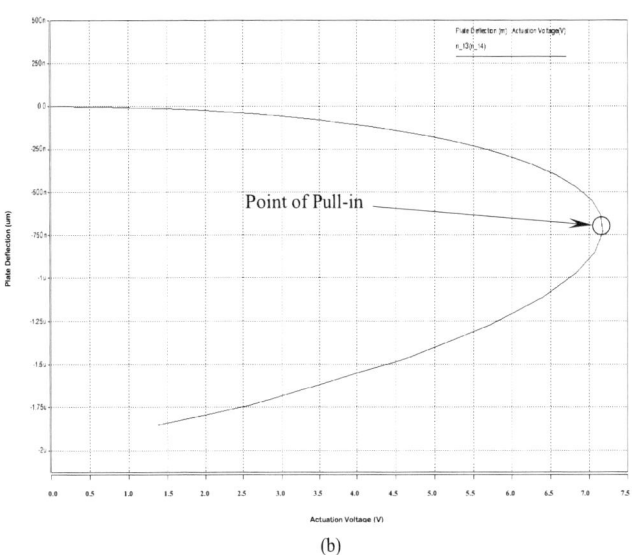

(b)

Fig. 5 (a) Saber Architect® lumped parameter schematic of the MEM varactor (b) The Saber Architect® plot of the varactor plate deflection against the applied actuation voltage is shown. The minor deviations are due to the chaotic behaviour observed beyond pull-in and prior to touchdown.

TABLE II.   COMPARISON OF THE ANALYTICAL MODEL WITH FINITE ELEMENT SIMULATIONS AND LUMPED PARAMETER MODELS

Model Used	Results	
	Pull-In Voltage	% Error
Quasi-analytical Model	6.93 V	−1.88%
CoSolveEM®	6.8 V	
Quasi-analytical Model	6.93 V	+3.9%
Saber Architect®	7.2 V	

It is clear from the data tabulated above that the analytical model developed here on the basis of experimental validation of the nonlinearities in the restoring spring force from the load-deflection characteristics and the empirical approximation of the nonlinear electrostatic force acting on the varactor plate through pertinent assumptions are in good agreement with the results generated by standard MEMS simulators.

VI. CONCLUSION

The work presented here comprises a closed-form quasi-analytical model of the pull-in voltage of a mechanically characterized, surface micromachined MEM clamped plate-type varactor and its comparison with FEA results and lumped parameter simulation. The small percentage of deviations can be accounted for by the fact that several empirical assumptions were made during the mathematical extension of the expression for VLSI on-chip interconnect capacitances as well as the approximations made due to the difficulty in exhaustive and accurate modelling of the fringing field capacitances of the perforations. The minor fringing effect contribution from the flexures was neglected in the derivation of our model, as were the parasitic capacitances due to the passivation layer separating the bottom electrode from the single-crystalline silicon substrate. However, in spite of all the above drawbacks, the model is seen to maintain a good consistency with the results obtained from standard MEMS simulators because parasitic effects are mostly ignored during finite element (FE) simulations as well as lumped parameter models. Further improvements shall be made to the current model after extensive electrical characterization to understand the effects of flexure fringing field capacitances while incorporating the effects of van der Waal forces, proper coefficient of damping and other erroneous factors affecting the device operation, which are absent in idealized simulators and models used today with the aim of reaching an accurate pull-in model for MEM varactors.

ACKNOWLEDGEMENT

The authors would like to thank the Defense Metallurgical Research Laboratory (DMRL), Hyderabad for the usage of their nanoindentation facility as well as the National Programme on Micro And Smart Systems (NPMASS) for the provision of computational facilities for the work presented here.

REFERENCES

[1] L. L. Mercado, S. Kuo, T.T. Lee, and L. Liu, "A mechanical approach to overcome RF MEMS switch stiction problems," in Proc. 2003 Electronic Components and Technology Conference, 2003, pp 377–384.

[2] M. –H. Bao, Micromechanical Transducers, Handbook of Sensors and Actuators 8, Elsevier, 2005, pp. 140–162.

[3] S. Pamidighantam, R. Puers, K. Baert, and H. A. C. Tilmans, "Pull-in voltage analysis of electrostatically actuated beam structures with fixed-fixed and fixed-free end conditions," Jornal of Micromechanics and Microengineering, vol. 12, no. 4, pp.458–464, 2002.

[4] R. Puers, and D. Lapadatu, "Electrostatic forces and their effects on capacitive mechanical sensors," Sensors and Actuators A, vol. 56, no. 3, 1996, pp. 203–210.

[5] N. V. D. Meijs, and J. T. Fokkema, "VLSI circuit reconstruction from mask topology," Integration, vol. 2, no. 2, pp. 85–119, 1984.

[6] K. Worden, and G. R. Tomlinson, Nonlinearity in Structural Dynamics–Detection, Identification and Modelling, Institute of Physics Publishing, 2001, pp 68–71.

[7] S. D. Senturia, Microsystems Design, Boston, MA: Kluwer Academic, pp. 249–259, 2000.

[8] Coventor Inc., "Pull-in voltage analysis of electrostatically-actuated beams verifying the accuracy of Coventor behavioral models," Available online,
http://www.coventor.com/media/fem_comparisons/pullin_voltage.pdf

Low-Latency No-Handshake GALS Interfaces for Fast-Receiver Links

Jean Michel Chabloz
KTH - Royal Institute of Technology
Stockholm, Sweden
chabloz@kth.se

Ahmed Hemani
KTH - Royal Institute of Technology
Stockholm, Sweden
hemani@kth.se

Abstract—In this paper we introduce a novel interface for Globally-Asynchronous, Locally-Synchronous systems which does not use any form of handshake to cross the gap between the clock domains. In particular, links in which the Receiver runs faster than the Transmitter are targeted. The interface works by finding an approximate ratio between the clock frequencies. Then, ratiochronous synchronizers that can tolerate clock drifts are employed to transmit data from the Transmitter to the Receiver clock domain. Thanks to the periodic properties of rationally-related systems, no handshake is employed and the average latency of the interface is decreased $\sim 75\%$ compared to state-of-the-art GALS interfaces. Additionally, the interface uses only standard cells and, save for a delay line, can be designed at Register Transfer Level.

I. INTRODUCTION

In the latest technology nodes, it has become very difficult to employ a Fully-Synchronous design style for all but the most simple electronic systems [1]. The main problem is that Fully-Synchronous systems require the presence of a global balanced clock tree with a number of leaves equal to the number of sequential cells in the whole system [2]. As the number of leaf cells roughly doubles with every new technology node [2], clock trees are rapidly getting unmaneageable [3]. The Fully-Asynchronous alternative design style [4] totally eliminates the concept of clock but requires the redesign of pre-existing IP libraries and, despite recent progress, still suffers from a lack of reliable and well-established design tools [5].

An alternative solution is given by the Globally-Asynchronous, Locally-Synchronous (GALS) design style [6], which has recently enjoyed success [7]. The design style breaks a system into modules, all of them synchronous but all running at independent clock frequencies. At chip-level, no global clock tree is present, which eliminates the need for global chip-wide timing closure. The modules are synchronous, thus easy to design and maintain. Clock-domain crossing between unrelated frequencies is a complex problem and GALS interfaces introduce a high latency penalty, as will be discussed in Section III. This drawback has hampered the deployment of GALS in important latency-sensitive applications such as Network-on-Chip (NoC) platforms [8], where often Mesochronous clocking (all modules run at the same frequency but there is no global phase alignment between the clocks) is deployed as a less flexible but more efficient alternative [9].

The Globally-Ratiochronous, Locally-Synchronous (GRLS) design style was recently introduced to build high-performance multiclock systems [10], [11]. A GRLS system is divided into synchronous independent modules, all running at rationally-related frequencies (i.e. all frequencies are a fractional multipe of each other such as $\frac{3}{4}, \frac{2}{5}$, etc.). In a GRLS system the alignment between the clocks is periodic; thus, the GRLS communication problem is much simpler than the GALS communication problem, and can be solved using interfaces that do not rely on handshake and are much more performant [10]. Despite the performance advantages over GALS, GRLS remains less flexible than GALS. Although it was shown that the flexibility drop is reasonable [11], designing an interface for GALS systems that does not rely on handshake and has the same performance advantages compared to GRLS is highly desirable.

The contribution of this paper is the design of a scheme to adapt the existing GRLS interface to a GALS clocking scenario in which the Receiver runs faster than the Transmitter. The scheme exploits the tolerance to clock drifts of the GRLS interface. The new interface is called GNH-FR interface (For GALS - No Handshake - Fast-Receiver interface). The GNH-FR interface allows to obtain the same performance benefits of GRLS in a GALS system, i.e. it allows to keep the same performance benefits of GRLS without requiring the Transmitter and Receiver frequencies to be rationally-related. A generalization of the GNH-FR interface to a GNH-FT interface, meant for scenarios in which the Transmitter runs faster than the Receiver is left to future works. The interface is validated through formal analysis considering also the presence of non-idealities such as clock jitters, propagation delay misalignments between the different lines, etc. The latency of the GNH-FT interface is then compared to that of state-of-the-art GALS interfaces showing $\sim 75\%$ average latency improvements.

The Remainder of the paper is organized as follows: in Section II the unidirectional Fast-Receiver GALS communication problem is formally defined; in Section III state-of-the-art GALS interfaces are described and discussed; in Section IV the main idea upon which the GNH-FR interface is based is presented; in Sections V the GNH-FR interface is introduced and analyzed; in Section VI the impact of non-idealities such as jitters and propagation delay misalignments is assessed; in

Section VII a realistic scenario is considered and the conditions necessary for the interface to work correctly are given; in Section VIII latency comparisons with a GALS Asynchronous FIFO interface are conducted; Section IX concludes the paper.

II. PROBLEM DEFINITION

We target the classical unidirectional Fast-Receiver GALS communication problem, i.e. the problem of interfacing a Transmitter module and a Receiver module, clocked respectively by clk_T and clk_R, running at two stable, unrelated and unknown frequencies

$$f_T = \frac{1}{T_T}; \ f_R = \frac{1}{T_R} \ with \ f_R > f_T$$

The GALS communication problem does not make any assumption on the phase alignment between any two edges of the two clocks. The GALS communication problem is a special case of the Generalized GALS communication problem. Unlike the GALS communication problem, the Generalized GALS communication problem does not assume the frequencies of the Transmitter and the Receiver to be constant, i.e. every cycle of both clocks has potentially an unpredictable length, and the time instants at which clock edges occur are unpredictable.

III. RELATED WORK

The GALS concept was first introduced in [6]. One of the first GALS interfaces to be introduced was the Pausible-Clock technique [6], [12], [13]. Pausible-Clock interfaces guarantee infinite Mean Time Between Failures (MTBF), i.e. it can be mathematically guaranteed that metastability failures never arise with such interfaces. When communication takes place, Pausible-Clock interfaces gate the clock of the whole module using a MUTEX until metastability is resolved. MUTEXes make performances unbounded and unpredictable (it is impossible to know for how long the clocks will be gated) and are non-standard components which must be designed at Transistor Level. These limitations have prevented pausible-clock techniques from finding widespread adoption in industry [14] despite some success in niche applications [13].

The Clock-Gating communication scheme [15], [16] for GALS systems is an evolution of Pausible-Clock interfaces. Clock-Gating interfaces use a clock-gating mechanism to stop an external clock source when communication takes place. Because there are no stoppable ring oscillators, the non-determinism and absence of frequency stability that are intrinsic to Pausible-Clock interfaces are not present. However, as for Pausible-Clock systems, Clock-Gating interfaces also introduce an efficiency penalty when communication takes place, because the clocks can be stopped for one or more cycles to allow metastability-free communication. Unlike Pausible-Clock interfaces, Clock-Gating interfaces are not mathematically guaranteed to be metastability-free; however, for realistic scenarios very long Mean Time Between Failures (MTBF) of thousands of years or more are typical [17]. Clock-Gating interfaces must be designed at Gate Level.

Asynchronous FIFOs [18] are efficient solutions for crossing clock domains, widely used in industry practice especially for streaming applications. A double-clock FIFO is introduced between the Transmitter and the Receiver modules. The Transmitter writes to the FIFO using its own clock, while the Receiver uses its own clock to read from it. Internally, the Tail and Head pointers of the FIFO are Grey-coded and cross the clock domain boundary using high-latency multistage synchronizers [19]. As for Clock-Gating interfaces, very high but finite MTBF are typical. Data can be read by the Receiver only after the information about the new incoming data item (encoded in the Head pointer) is synchronized to the Receiver clock domain. Asynchronous FIFOs interfaces can be designed at Register Transfer Level.

In literature, the work closer to ours is [20]: the STARI Mesochronous interface [21] was generalized to support GALS links using an approximation of the clock frequency ratio. However, this work gives only a general methodology for the estimation of the ratio between the clocks and does not establish how good the approximation should be for the interface to work correctly. Moreover, the Ratiochronous interface used in [20] is a complex Transistor-Level circuit that presents multiple design challenges which make it unpractical [10].

IV. THE KEY INSIGHT

With the exception of [20], all GALS interfaces that have been proposed in literature are based on handshake, i.e. when the Transmitter has something to send to the Receiver it first informs the Receiver about the upcoming data transmission; when the Receiver is ready to accept data, then data is transmitted. In Pausible-Clock interfaces, the mechanism is implemented using explicit handshake signals. In Asynchronous FIFOs interfaces, Grey-coded pointers are synchronized from the Transmitter to the Receiver clock domain using built-in multi-stage synchronizers and data is read by the Receiver only after the pointer has crossed the synchronizers (implicit handshake). In both cases, the explicit or implicit handshake carries an inherent latency penalty.

[10] argues that the GRLS communication problem is inherently simpler than the GALS communication problem, because the periodic properties of rationally-related systems make the relationship between the clock edges periodic, i.e. predictable. This allows to build a carefully-designed handshake-free GRLS interface which uses a learning phase to determine when the Receiver should sample data. The learning phase takes place continuously during operation using a strobe signal which is added to the data lines. Analysis of a strobe sample obtained at a given time instant t_0 takes several cycles to complete and determines a future time t_1 on which a new data item will be available for sampling. The Transmitter module outputs data as soon as it has data to output and the Receiver samples it on the first safe clock edge, drastically reducing latency.

We now extend this concept to the GALS communication problem. For a GALS communication problem, unlike a generalized GALS communication problem, the relationship between the clock edges can be predicted and an interface that does not rely on handshake can be designed. The solution we propose consists of approximating the ratio between the frequencies of the Transmitter and Receiver and use a variation of the GRLS interface (which can tolerate a non-perfect ratio between the clocks) to synchronize data from the Transmitter to the Receiver.

V. NO-HANDSHAKE FAST-RECEIVER INTERFACE

To interface a Transmitter module clocked by clk_T with a faster Receiver module clocked by clk_R, the GNH-FR interface employs a GNH-FR Transmitter inserted in the Transmitter clock domain and a GNH-FR Receiver inserted in the Receiver clock domain, as shown in Figure 1.

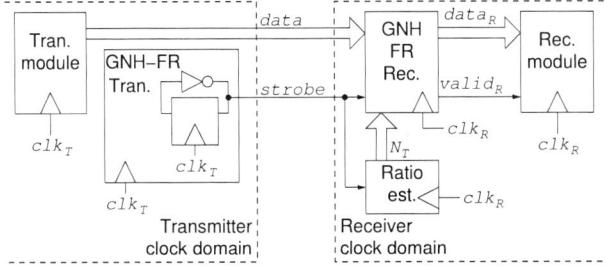

Fig. 1.   GNH-FR Interface schematic

The Transmitter module outputs data in every cycle, without backpressure or handshake mechanisms. The GNH-FR Transmitter generates the *strobe* signal, which toggles every time a data item is output by the Transmitter module (i.e. once per cycle). The GNH-FR Receiver uses the *strobe* signal during a learning phase, which takes place continuously during operation. The learning phase is used to determine on which clock edges (positive or negative) data will be available for sampling. When a data item is sampled, it is fed to the Receiver module. Because the Receiver clock is faster than the Transmitter clock, data will not generally be available in every cycle. To indicate valid data items, the GNH-FR Receiver uses an additional $valid_R$ signal to indicate that the data on the $data_R$ lines is a valid data item. To operate correctly, the GNH-FR Receiver needs to have an estimation of the ratio between the frequencies. The ratio between the frequencies is estimated upon reset by the Ratio Estimation block. Ideally, the GNH-FR Receiver is integrated with the Receiver module, i.e. the input registers of the Receiver module are removed and are substituted with the GNH-FR Receiver.

A. Operation

The structure of the GNH-FR Receiver is shown in Figure 3.

We first suppose that an estimated ratio between the clocks is available:

$$\frac{f_R}{f_T} = \frac{T_T}{T_R} \simeq \frac{N_T}{N_R}$$

How to obtain a sufficiently accurate estimate is a problem that will addressed in Subsection V-B.

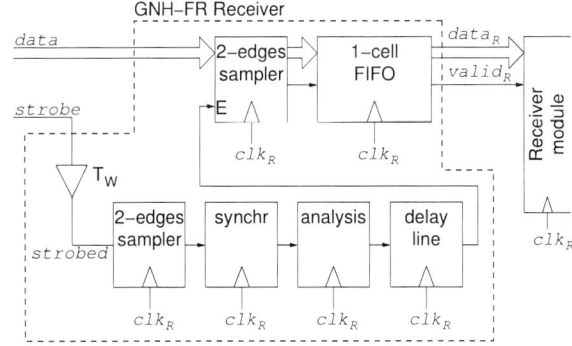

Fig. 2.   GNH-FR Receiver

We define as Periodicity Cycles PC_T and PC_R the time intervals:

$$PC_T = N_R T_T; \ PC_R = N_T T_R$$

PC_T corresponds to N_R Transmitter cycles while PC_R corresponds to N_T Receiver cycles.

If the estimated ratio $\frac{N_T}{N_R}$ is sufficiently close to the ratio between the frequencies $\frac{f_R}{f_T}$, then $PC_T \simeq PC_R$. We define as Periodicity Cycle Error ΔPC the time interval:

$$\Delta PC = PC_T - PC_R$$

With a good estimation of the ratio between the frequencies, the alignment between the clocks is almost the same (with a ΔPC error) every N_R Transmitter cycles and every N_T Receiver cycles.

The Transmitter module outputs one data item per clock cycle. The GNH-FR Transmitter generates a *strobe* signal that toggles once in every cycle. The strobe is routed through the channel bundled together with the data lines. At the Receiver end of the channel, the strobe is delayed by a delay line with a propagation delay $T_W \simeq \frac{T_R}{4}$, to generate the delayed strobe *strobed*. The *strobed* signal is continuously sampled on both positive and negative edges of the Receiver clock clk_R. Before the *strobed* samples are analyzed, they are synchronized to the Receiver clock domain using high-latency synchronizers made up by cascades of flipflops. The resulting sequence of *strobed* samples is denoted as:

$$s_0, s_1, ..., s_i, ...$$

Where sample s_i indicates the *strobed* sample that was obtained at time t_i. Because the Receiver samples the *strobed* signal on both positive and negative clock edges, the relation:

$$t_i - t_{i-1} = \frac{T_R}{2}$$

holds. If the setup/hold constraints of the *strobed* sampler at time t_i were violated, i.e. if *strobed* transitioned between $t_i - t_s$ and $t_i + t_h$ (t_s and t_h indicate respectively the setup and hold times of the sampler), given a sufficiently high number of synchronization stages (two are normally enough to ensure a very high MTBF [17]) *strobed* sample s_i is a stable but random value (referred to as a corrupted sample). Two consecutive corrupted samples can never occur, i.e. if s_{i-1} is corrupted, then s_i is not, and viceversa.

If $s_i \neq s_{i-1}$ is observed by the GNH-FR Receiver, then the unit can conclude that the *strobed* signal transitioned during the time interval $(t_{i-1} - t_s, t_i + t_h)$. In fact, if *strobed* was stable during the time interval, both samples s_{i-1} and s_i would not be corrupted and would necessarily be identical. Therefore, if $s_i \neq s_{i-1}$ is observed by the GNH-FR Receiver, the unit can conclude that the *strobe* signal toggled during the interval

$$(t_{i-1} - T_W - t_s, t_i - T_W + t_h)$$

and that a data item arrived at the same time.

Because the Transmitter module outputs data in every cycle, another data item will arrive N_R Transmitter cycles later, during the interval:

$$(t_{i-1} + PC_T - T_W - t_s, t_i + PC_T - T_W + t_h)$$

which is equivalent to:

$$(t_{i-1} + PC_R + \Delta PC - T_W - t_s,$$
$$t_i + PC_R + \Delta PC - T_W + t_h) =$$
$$(t_{i+2N_T-1} + \Delta PC - T_W - t_s,$$
$$t_{i+2N_T} + \Delta PC - T_W + t_h)$$

Because the Transmitter module outputs one data item per clock cycle, the data item will be stable on the channel until at least time instant:

$$t_{i+2N_T-1} + T_T + \Delta PC - T_W - t_s \geq$$
$$t_{i+2N_T} + \frac{T_R}{2} + \Delta PC - T_W - t_s$$

In other words, when the GNH-FR Receiver observes $s_i \neq s_{i-1}$, it is guaranteed that a data item will be stable on the channel during the interval

$$(t_{i+2N_T} + \Delta PC - T_W + t_h,$$
$$t_{i+2N_T} + \frac{T_R}{2} + \Delta PC - T_W - t_s) \qquad (1)$$

The GNH-FR interface employs the following scheme for data sampling at the Receiver end of the channel: if the Receiver detects $s_i \neq s_{i-1}$, then data is sampled N_T cycles after s_i was sampled, at time t_{i+2N_T}. Because of the high-latency multi-stage synchronizers, the analysis of the *strobed* samples takes several clock cycles to complete. However, as long as the analysis is completed in N_T cycles, the GNH-FR Receiver is able to complete the analysis by the time data should be sampled. Because two consecutive samples will never be both corrupted, every transition of the strobe signal will be detected; therefore, if the scheme is applied continuously, all data items are sampled.

Operation is shown in Figure 3. At time instant 2, the GNH-FR Receiver completes the analysis of the *strobed* samples s_0 and s_1. Because $s_1 \neq s_0$, strobe transition A is detected. At time instant 3, data item B is sampled by the GNH-FR Receiver. The alignment between *strobe* toggle A and clock edge 1 is close (with a ΔPC error) to the alignment between

the time at which data item B reaches the GNH-FR Receiver and clock edge 3. As long as certain conditions are satisfied, it is guaranteed that the data samplers will not encounter metastability when sampling data item B on clock edge 3.

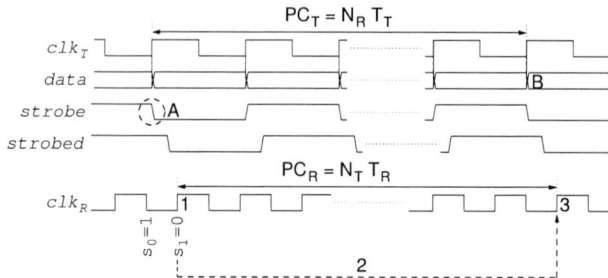

Fig. 3. Sampling in the GNH-FR Receiver

It is guaranteed that a data item is safely sampled as long as Interval 1 contains the metastability interval of the sampler at t_{i+2N_T}:

$$(t_{i+2N_T} - t_s, t_{i+2N_T} + t_h)$$

This translates into the constraint:

$$\Delta PC + (t_s + t_h) < T_W < \frac{T_R}{2} + \Delta PC - (t_s + t_h) \qquad (2)$$

If the constraint is satisfied, then the interface works correctly and every data item is sampled safely by the GNH-FR Receiver. The remaining problem is how to estimate the ratio between the frequencies sufficiently well to guarantee that the value of ΔPC satisfies Relation 2.

A 1-cell FIFO buffer is introduced in the GNH-FR Receiver, because the GNH-FR Receiver may sample two data items in a single cycle (one on the positive edge of the clock and one on the negative), while the Receiver module can consume a single data item per cycle. However, because the Transmitter module runs slower than the Receiver and outputs only one data item per clock cycle, a single-cell buffer is guaranteed never to overflow.

The whole interface can be designed at RTL save for the delay line, which requires Gate-Level design.

B. Ratio Estimation

Upon reset, the Ratio Estimation block at the Receiver samples the *strobe* signal on both the positive and the negative edges of clock clk_R for a sufficiently high number c of cycles. The samples are synchronized to the Receiver clock domain using high-latency multi-stage synchronizers and the Ratio Estimation block counts how many transitions of signal *strobe* are detected in the $2c$ samples that are obtained. Every transition is detected because the *strobe* signal toggles once every Transmitter cycle and the Transmitter runs slower than the Receiver.

The Ratio Estimation block examines the $2c$ consecutive samples st_i of the *strobe* signal, indicated as $st_1, ..., st_{2c}$, detecting t transitions of the *strobe* signal, i.e. t samples st_i

which are different compared to the sample received half a cycle earlier st_{i-1}. The values of c and t can be used to approximate the ratio between the clock frequencies of the Transmitter and the Receiver.

If the Receiver detects t transitions of the strobe in $2c$ samples, the two following relations must hold:

$$tT_T \geq cT_R - (t_s + t_h)$$
$$(t-1)T_T \leq (2c+1)\frac{T_R}{2} + (t_s + t_h) \quad (3)$$

The relations are shown graphically in Figure 4. Two limit cases are shown: for both examples, $c = 16$ and $t = 13$. The scenario at the top (bottom) of the Figure corresponds to the minimal (maximal) value f_{Tmin} (f_{Tmax}) that can result in 13 detected transitions. Values of f_T higher than f_{Tmax} or lower than f_{Tmin} cannot result in $t = 13$. Parenthesis indicate corrupted samples, which randomly stabilized to the value in the parenthesis.

To estimate the ratio between the frequencies from the values of c and t, Relations 3 can be transformed into

$$cT_R - (t_s + t_h) \leq tT_T \leq cT_R + T_T + \frac{T_R}{2} + (t_s + t_h)$$

To obtain accurate estimations of the ratio between the clocks, the Ratio Estimation block follows Algorithm 1. The algorithm

---

Algorithm 1 Ratio Estimation Algorithm

1: **while** t not divisible by J **do**
2:     Wait J clock cycles
3:     Calculate the number of transitions t detected between cycle 1 and the current cycle c.
4: **end while**
5: $N_T = \frac{c}{J}$;
6: $N_R = \frac{t}{J}$;

---

returns values of t and c that are both multiple of a given integer J. Therefore, the following relations hold:

$$\begin{cases} N_R T_T \geq N_T T_R - \frac{t_s + t_h}{J} \\ N_R T_T \leq N_T T_R + \frac{T_T}{J} + \frac{T_R}{2J} + \frac{t_s + t_h}{J} \end{cases} \Rightarrow$$

$$\begin{cases} PC_T \geq PC_R - \frac{t_s + t_h}{J} \\ PC_T \leq PC_R + \frac{T_T}{J} + \frac{T_R}{2J} + \frac{t_s + t_h}{J} \end{cases} \Rightarrow$$

$$-\frac{t_s + t_h}{J} \leq \Delta PC \leq \frac{T_T}{J} + \frac{T_R}{2J} + \frac{t_s + t_h}{J} \quad (4)$$

Relation 4 establishes bounds for ΔPC. The bounds can be arbitrarily reduced by making the learning phase longer by choosing a higher value for J. Given bounds for f_T and f_R, a value of J that guarantees to satisfy equation 2 (thus guaranteeing that the interface operates correctly) can easily be determined (see Section VII).

VI. NON-IDEALITIES

Constraint 2, derived in Section V and necessary for the GNH-FR interface to work correctly is meant for an ideal scenario with no clock jitter, no data jitter and no propagation delay misalignments between the strobe and the different data lines. In a real scenario, however, these issues must be considered. Noise in the system introduces jitter, and it is impossible to realize a bundle of data lines with perfectly-matched propagation delays.

We consider now a more realistic scenario:

- The arrival time of each data and strobe line to the Receiver end of the channel is subject to a maximal item-to-item jitter J_T compared to the ideal case. The jitter takes into account the jitter of the Transmitter clock and the jitter added during the propagation through the channel.
- The time at which the Receiver clock edges occur is subject to a maximal half-cycle-to-half-cycle jitter J_R.
- The maximal misalignment between the propagation delays of a data line and the strobe line is MIS.
- The Duty Cycle of clk_T and clk_R is not exactly 50% but deviates ΔDC from the ideal 50% value.

The GNH-FR interface builds on the assumption that, when a strobe transition is detected at time t_i, then the data lines are guaranteed to be stable during the interval:

$$(t_{i+2N_T} + \Delta PC - T_W + t_h, t_{i+2N_T+1} + \Delta PC - T_W - t_s)$$

However, conducting a worst-case analysis in presence of non-idealities, it can only be guaranteed that the data lines are stable in the interval:

$$(t_{i+2N_T} + \Delta PC - T_W + J_R + J_T + MIS + t_h;$$
$$t_{i+2K} + \frac{T_R}{2} + \Delta PC - T_W +$$
$$-T_R\,\Delta DC - J_R - J_T - MIS - t_s)$$

To guarantee that the interface operates correctly, it is necessary that this interval does not overlap with the metastability window of the data sampler at t_{i+2N_T}, i.e. with the interval $(t_{i+2N_T} - t_s; +t_{i+2N_T} + t_h)$.

To guarantee that this happens, the following two relations must hold:

$$T_W > \Delta PC + J_R + J_T + MIS + t_s + t_h$$
$$T_W < \Delta PC + \frac{T_R}{2} - T_R\,\Delta DC +$$
$$-J_R - J_T - MIS - t_s - t_h$$

The two relations, coupled with the impossibility to define perfect delay lines, determine the constraints on ΔPC and subsequently the minimal value of J that must be used during the Ratio Estimation phase.

VII. CASE STUDY

As an example, let us consider a 90 nm implementation of a multi-processor GALS system. The data jitter J_T is given by $J_T = J_C + J_P = 30ps$, where $J_C = 10ps$ is the clock jitter and $J_P = 20ps$ is the jitter of the propagation delay through the channel caused by crosstalk. The Receiver clock jitter J_R is given by $J_R = J_C = 10ps$. Let us consider also a misalignment between the data and strobe lines $MIS = 10ps$, and let us consider $t_s + t_h = 20ps$. Finally, we consider that

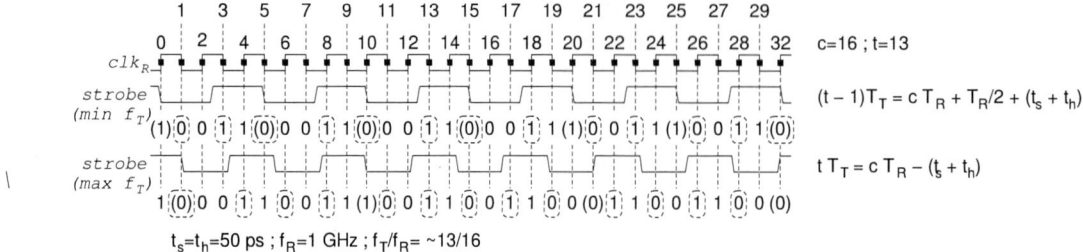

Fig. 4. Estimation of the ratio between the clocks

the Duty Cycle of the clocks has a maximal $\Delta DC = 5\%$ deviation from the nominal 50% value. With $f_R = 1 GHz$, the value of T_W that is optimal for the constraints on ΔPC is $T_W = 225 ps$. To keep into account non-idealities on the delay line, we consider $200 ps < T_W < 250 ps$. Under these conditions, the constraint $-130 ps < \Delta PC < 130 ps$ can be derived. If a bound $T_T < 1.25 ns$ is given, then equation 4 gives the bound:

$$-\frac{20 ps}{J} < \Delta PC < \frac{1.770 ps}{J}$$

A value $J = 20$ will satisfy equation 2 and guarantee that the interface operates correctly.

VIII. LATENCY ANALYSIS

Compared to traditional GALS interfaces, the absence of handshake in the GNH-FR interface unlocks latency benefits due to the fact that the GNH-FR Receiver is ready to receive data as soon as the Transmitter module outputs it. The benefits are similar to those of the GRLS interface introduced in [10].

In a GALS system, latency depends on the alignment between the clocks. Best-Case (BC), Average-Case (AC) and Worst-Case (WC) latencies can thus be defined. To make a general analysis, latencies are calculated considering a null propagation delay through the channel and no contention (all buffers are empty and a single data item travels from the Transmitter to the Receiver). Latencies are normally expressed in terms of Receiver clock cycles T_R.

In a GNH-FR interface, data items are output by the Transmitter module immediately on the channel. Considering null the setup and hold time constraints of the input strobe samplers, the Best-Case, Average-Case and Worst-Case Reception latencies of the GNH-FR interface are given by:

$$L_{GNH-FR,BC} = T_W$$
$$L_{GNH-FR,AC} = 0.5T_R + T_W$$
$$L_{GNH-FR,WC} = T_R + T_W$$

Typical GALS Asynchronous FIFO interfaces as implemented in [19] have Best-Case, Average-Case and Worst-Case latencies respectively equal to $L_{GALS,BC} = 2T_R$, $L_{GALS,AC} = 2.5T_R$ and $L_{GALS,WC} = 3T_R$. Considering the optimal-robustness value $T_W = \frac{T_R}{4}$, GNH-FR interfaces have latencies respectively equal to $L_{GNH-FR,BC} = 0.25T_R$, $L_{GNH-FR,AC} = 0.75T_R$ and $L_{GNH-FR,WC} = 1.25T_R$. The

benefits over GALS interfaces are thus 75% for Average-Case latency and 50% for Worst-Case latency.

IX. CONCLUSION AND FUTURE WORK

In conclusion, the standard-cell GNH-FR interface has latency figures which are much lower compared to GALS alternatives.

This paper aimed mostly at presenting a rigorous theoretical analysis of the method; a complete deployment of the interface in a computation platform is under way. Also, the interface will be generalized to cover scenarios in which the Transmitter runs faster than the Receiver.

REFERENCES

[1] International Technology Roadmap for Semiconductors Report, 2009
[2] J. M. Rabaey, "Digital Integrated Circuits: A Design Perspective," Prentice Hall, 1995
[3] P. Teehan et al., "A Survey and Taxonomy of GALS Design Styles," IEEE journal of Design & Test of Computers, Sept-Oct. 2007
[4] I. E. Sutherland and J. Ebergen, "Computers Without Clocks," Scientific American, Aug. 2002
[5] S. Borkar, "Does asynchronous logic design really have a future?," EE Times, 2003
[6] D. M. Chapiro, "Globally Asynchronous Locally-Synchronous Systems," PhD thesis, Stanford University, Oct. 1984
[7] N. Wingen, "What If You Could Design Tomorrow's System Today?," DATE 2007
[8] J. Oberg, "Clocking Strategies for Network-on-Chips," In "Networks-on-Chip", Kluwer Academic Publishers, 2003
[9] S. R. Vangal et al., "An 80-Tile Sub-100-W TeraFLOPS Processor in 65-nm CMOS," IEEE Journal of Solid-State Circuits, Jan. 2008
[10] J. M. Chabloz and A. Hemani, "A Flexible Interface for Rationally-Related Frequencies," ICCD 2009
[11] J. M. Chabloz, "Distributed DVFS with Rationally-Related Frequencies and Quantized Voltage Levels," ISLPED 2010
[12] K. Y. Yun and R. P. Donohue, "Pausible clocking: a first step toward heterogeneous systems," ICCD 1996
[13] J. Muttersbach et al., "Practical design of globally-asynchronous locally-synchronous systems," ASYNC 2000
[14] F. K. Gürkaynak et al., "GALS at ETH Zurich: Success or Failure?", ASYNC 2006
[15] E. Amini et al., "Globally asynchronous locally synchronous wrapper circuit based on clock gating," Symposium on Emerging VLSI Technologies and Architectures, 2006
[16] J. Carlsson et al., "A Clock Gating Circuit for Globally Asynchronous Locally Synchronous Systems," NORCHIP 2006
[17] R. Ginosar, "Fourteen ways to fool your synchronizer," ASYNC 2003
[18] D. Kim et al., "Asynchronous FIFO Interfaces for GALS On-Chip Switched Networks," SoC 2005
[19] C. E. Cummings and P. Alfke, "Simulation and Synthesis Techniques for Asynchronous FIFO Design with Asynchronous Pointer Comparisons," Synopsys Users Group Conference, 2002
[20] A. Chakraborty and M. R. Greenstreet, "Efficient self-timed interfaces for crossing clock domains," ASYNC 2003
[21] M. R. Greenstreet, "Implementing a STARI chip," ICCD 1995

978-1-4673-0438-2/12 $31.00 © 2012 IEEE

Set-Cover Heuristics for Two-Level Logic Minimization

Ankit Kagliwal
Computer Science and Engineering Department
Indian Institute of Technology Madras
Chennai - 600036, India
kagliwal@cse.iitm.ac.in

Shankar Balachandran
Computer Science and Engineering Department
Indian Institute of Technology Madras
Chennai - 600036, India
shankar@cse.iitm.ac.in

Abstract—Given a Boolean function, the Unate-Covering Problem (UCP) is NP-hard. This problem can be modeled as a set-cover problem where minterms are the elements and implicants form the sets. Traditional solutions in logic synthesis use set-cover algorithms that are oblivious to the special semantic of the elements and sets. We propose three new heuristics for the set-cover problem which are aware of the relationship between implicants and minterms. We show that the proposed heuristics are effective for breaking ties when a cyclic core is obtained. We evaluate the heuristics on a set of hard instances from BHOSLIB benchmark suite. We also replace ESPRESSO's set cover algorithm using these heuristics and compare the logic synthesis results. We further map the minimized Boolean equations using ABC's technology mapping tool using 2-input NAND gates and 4-input Lookup Tables (LUTs).

Index Terms—set-cover, heuristic, two-level logic minimization, ESPRESSO-II

I. INTRODUCTION

TWO-LEVEL logic minimization forms an essential part of logic synthesis. One of the steps in logic minimization is computing an equivalent minimum representation of the given Boolean function. Well known techniques designed for computing the minimum representation of a Boolean function include Karnaugh Map and Quine-McCluskey Tabulation method [1]. Quine-McCluskey method first generates all the prime implicants and then minimizes the generated implicants to obtain a simplified Boolean function. A detailed description of different algorithms for obtaining a minimum representation of a Boolean function is given in [2]–[4].

Given the set of implicants, finding a minimum set of implicants which represent the same Boolean function is referred to as Unate Covering Problem (UCP). The unate covering problem can be modeled as a set-cover problem. Many heuristics have been proposed to solve the set-cover problem, but all of them ignore the fact that the rows represent minterms and columns represent implicants.

This paper presents three new logic-aware heuristics for the unate covering problem. One heuristic approach uses the number of literals in the implicants to minimize the number of implicants. The other two heuristics optimize the number of implicants by measuring the sharing between various implicants, and between different outputs of the function.

ESPRESSO-II is used to evaluate the proposed heuristics against the existing ones. The minimized Boolean functions

are technology mapped to 2-input NAND gates and 4-input LUTs using ABC logic minimization tool [5]. It is found that the proposed heuristics help in optimizing the area better than the other heuristics.

II. PROBLEM DEFINITION

A. Background of logic synthesis

Let $B = \{0, 1\}$ and $Y = \{0, 1, -\}$. A **Boolean** *(logic)* **function** f defined on n variables $x_1, x_2, ..., x_n$ and having m outputs $y_1, y_2, ..., y_m$ is a mapping from $f : B^n \rightarrow Y^m$, $n \geq 1$, $m \geq 1$. Each element $x \in B^n$ is called a **cube** or **minterm**. The symbol '$-$' refers to the **don't care condition**, i.e. the output of the function for minterm does not matter. The **ON-set**, **OFF-set** and **DC-set** of a function refer to the sets of cubes for which the function value is **1**, **0** or **-** respectively . The ON-set, OFF-set and the DC-set are denoted by f^{ON}, f^{OFF} and f^{DC} respectively. For a multi-output function, each output has its own ON-set, OFF-set and DC-set. A variable x_i can appear in two forms in a minterm, called **literals**. The two categories are the original form x_i and the complemented form $\bar{x}_i$.

An **implicant** is a disjunction of minterms and the implicant is said to cover the minterms. An implicant may not necessarily contain all the variables (either in original or complemented form), whereas a minterm always contains all the variables. A **prime implicant** of a function is an implicant which makes the function true but if any one literal is removed from the prime implicant, the function cannot be true anymore because of that implicant. In other words, a prime implicant is a maximal implicant of the given function. For instance, for the function $f(x, y, z) = y\bar{z} + xyz + xy$, xyz is an implicant of f, but it is not a prime implicant because it is contained in the implicant xy.

A Boolean function in SOP form can be viewed as a collection of prime implicants which set the value of function as true. The challenge here is to minimize the number of prime implicants defining the underlying Boolean function. This problem is called the **Unate Covering Problem (UCP)**. Formally, UCP is defined as follows: Given a Boolean function f on a set of minterms $M = \{m_1, m_2, ..., m_s\}$, $M \subseteq B^n$ and a set of prime-implicants $P = \{P_1, P_2, ..., P_t\}$ defined as disjunctions on $m_i \in M$, then UCP is the problem of finding a minimum cardinality subset $C \quad P$, such that,

978-1-4673-0438-2/12 $31.00 © 2012 IEEE

$\forall m_i$, $m_i \in M$, $\exists C_j \in C$ and C_j covers minterm m_i. The set C is also called as **minimum prime cover** for the unate covering problem.

B. Set-cover Formulation

It is easy to see that any implicant can be modeled as an element of the power-set of the minterms. Hence a straight forward question is: *given the set of minterms and the set of prime implicants, can the number of prime implicants be minimized without affecting the truth-value of underlying Boolean formula?* The unate-covering problem can be abstracted as a set-cover problem as follows: Given a universe $M = \{m_1, m_2, ..., m_s\}$ of minterms and a collection $P = \{P_1, P_2, ..., P_t\}$ of implicants defined on M, find the minimum number of implicants from the collection P such that all the minterms in M are covered. The problem can be represented as a Boolean matrix (cover-matrix) where the rows correspond to the minterms and the columns correspond to the implicants. The cover-matrix $A_{s \times t}$ is defined as follows:

$a_{ij} = 0$: minterm i is not covered by implicant j

$a_{ij} = 1$: minterm i is covered by implicant j

Hence given a cover-matrix, the aim is to identify the minimum number of columns which cover all the rows, that is, for each row i there exists at least one column j with $a_{ij} = 1$ and the number of columns is the minimum such possible.

One of the approximation algorithms used for the set-cover problem is the greedy algorithm which approximates the set-cover with an approximation ratio of $O(\ln n)$ [6]. This was improved to $O(\ln n_o)$ in [7], where n_o is the cardinality of the largest set. Later, in [8], it was showed that the greedy set-cover algorithm has an approximation ratio of $(1 - o(1)) \ln n$.

Some of the reduction rules, [3], [9], which help in obtaining a smaller instance of the cover-matrix are:

- *Essential Implicant*: If any minterm m_i is contained in one and only one implicant, then the implicant has to be present in the minimum prime-cover as there is no other implicant containing m_i.
- *Implicant dominance*: If an implicant P_i is fully contained in another implicant P_j, then P_j can be removed as minterms covered by implicant P_j can still be covered using implicant P_i.
- *Minterm dominance*: If two minterms m_i and m_j are such that the set of implicants containing m_i form a super-set of the set of implicants containing m_j, then the minterm m_i can be removed.

These reduction rules can be applied repeatedly and in any order until the cover-matrix cannot be reduced further. If the resulting matrix is empty then the optimum solution is obtained which contains all the essential implicants computed so far. If the resulting matrix is not empty, then it is called a cyclic core. Figure 1 represents a cyclic core. The cyclic core exposes the NP-hard property of the unate covering problem and hence the question: *since no reduction is possible on a cyclic core, which implicant should be a part of the cover?* We propose three different heuristics in the paper which answer this question.

Figure 1: A Cyclic Core

Several techniques exist to resolve the cyclic core. One of the techniques which gives exact cover is branch-and-bound, where first a bound on the solution size is obtained, then an implicant is selected to branch upon. The branch has two sub-problems: one containing the selected implicant and one excluding it. As soon as an intermediate solution obtained by a branch contains more implicants than the initial bound, it is pruned. There are sophisticated algorithms based on branch and bound techniques which achieve exact cover [3], [9], [10], but they can take exponential time to obtain the solution for some instances. One way to improve the run-time is to use heuristics to select a subset to be a part of the cover. Some of the well-known heuristics are:

- *Greedy (*GR*)*: Pick the subset which covers most of the uncovered elements.
- *Least-Covered-Most-Covering (*LCMC*)*: Select the element which is present in least number of subsets. Then among the subsets containing this element select one which covers maximum uncovered elements.
- *Contributive-Selection (*CS*)*: Assign weights to elements in the inverse proportion of the number of subsets they are contained in. Then assign weights to columns such that each column gets a weight which is the sum of weights of elements covered by it. Then select the subset which has the maximum weight.

LCMC and CS were proposed in BOOM [11], [12].

All these heuristics ignore the fact that the set-cover instance is actually derived from a logic minimization problem. Note that the reductions performed on the cover-matrix are independent of any heuristic used. But, the applied heuristic influences the dominance-relationships in the remainder cover-matrix. Note that, once a cyclic core is broken, the remainder cover-matrix could either be a cyclic core of smaller size or it can result in a matrix which can be reduced using standard reduction rules.

III. SET-COVER HEURISTICS FOR LOGIC SYNTHESIS

We propose three new heuristics to solve the set-cover problem. These heuristics explicitly make use of the relationships between minterms and implicants that form the rows and columns of the cover matrix. We first define weights for literals, implicants and outputs. We then present algorithms that use different combinations of weights and the order in which they are considered. We also discuss the intuition behind the ordering for each heuristic.

A. Weights

The weights of literals are based on the number of implicants that contain the literals. The weight assigned to implicants use the weights on literals as well as other factors. Weights on outputs are designed based on implicants.

Weights of Literals (LW_x): Each variable occurs in two forms: original and complemented form. For each literal x, the weight LW_x assigned is the count of the implicants which contain the literal. If there are two literals x_1 and x_2 such that $LW_{x_1} > LW_{x_2}$, the implicants that contain x_1 could be taken up for cover earlier than those that contain x_2.

Weights of Outputs: Each output y_j is assigned a weight IC_j that is defined as the number of implicants in the DC or $ON-$set of y_j.

$$IC_j = |y_j^{ON} \cup y_j^{DC}| \qquad (1)$$

Weights of Implicants : Let n be the number of variables in a $m-$output Boolean function f. With each implicant P_i of f, we associate several weights based on the properties desired in the circuits.

- **Literal-Count** (LC_i): The number of literals present in the implicant.
- **Set-Cardinality** (SC_i): The number of minterms in the cover-matrix that are covered by the implicant i in the original set-cover instance. Note that this is the count of the number of 1's in column i and in general $SC_i \leq 2^{n-LC_i}$. As set-cover reductions proceed, some minterms may get covered but according to the definition given here, SC_i remains unaltered.
- **Weighted-Literal-Count** (WL_i): This is the weighted sum of the literals that are present in the implicant

$$WL_i = \sum_{x \in X_i} LW_x \qquad (2)$$

where X_i is the set of literals in implicant i.

- **Output-Count** (OC_i): The number of outputs containing the implicant i.
- **Weighted-Output-Count** (WO_i): The summation of weights of outputs that contain implicant i

$$WO_i = \sum_{y \in Y_i} IC_y \qquad (3)$$

where Y_i is the set of outputs that contain implicant i.

B. General Framework for Logic Minimization using Set-Cover

We present a general framework for set-cover for a Boolean function f in Algorithm 1.

Steps 1, 2 and 3 assign weights to the implicants and reduce the cover-matrix to a cyclic core. Step 4 breaks the cyclic core by picking a column based on rules 4(a) to 4(d). Step 4(a) picks the column with the most number of uncovered elements thereby covering most minterms in the core. Step 4(b) picks the implicant with the least number of literals as there is usually a direct correlation between less literal count and circuit area.

Note that the greedy method GR does not have steps 4(b) through 4(d).

Algorithm 1 $UC(A, H)$

Input : a Boolean matrix A, a heuristic approach H
Output : a minimal cover (all the essential columns)

1) Compute weights of each column according to heuristic H
2) Apply all the reduction rules on A in any order till no further reductions are possible. Obtain a matrix A'
3) If A' is empty, solution is obtained and is exactly equal to all the essential columns computed
4) If A' is non-empty, a cyclic core is obtained
    a) Select the column containing most number of uncovered elements
    b) If there is a tie, select the column that has the least literal count LC
    c) If there is a tie on the literal count, break using heuristic H
    d) If there are further ties, break them randomly
    e) Mark the selected column as an essential column
5) $A = A'$
6) Repeat steps 2-5 till A is empty

C. Heuristics to Break a Cyclic Core

We now discuss three new heuristics that we propose for H in Algorithm 1.

*1) Maximum Set Cardinality (*MXSC*):* The first heuristic uses the set-cardinality SC as a tie-breaker. If two implicants i and j cover the same number of uncovered minterms and have the same literal count and if $SC_i < SC_j$, heuristic MXSC picks j as the next essential column. This is done to differentiate implicants based on the actual number of minterms present in the set-cover matrix. For instance, if $SC_i = SC_j - k$, such that $k < 2^{n-LC_i}$, the tie-breaker in Step 4(b) will not distinguish i and j. MXSC selects an implicant covering more minterms.

*2) Literal Weights and Output Count (*WLOC*):* Heuristic MXSC does not differentiate between implicants that are present in multiple outputs. Covering an implicant which is present in more than one output is desirable because it induces sharing of gates after synthesis. This may result in reduction of area. A metric to measure this sharing is using the output count OC. For each implicant, we assign the weight

$$WLOC_i = WL_i \times OC_i \qquad (4)$$

We pick the implicant with the highest value of $WLOC$.

*3) Literal Weights and Output Weights (*WLWO*):* WLOC does not account for outputs that have many implicants. It is quite likely that an output y_j which contains the largest number of implicants shares many implicants with other outputs. Hence we assign a weight to the implicants that account for this sharing.

$$WLWO_i = WL_i \times WO_i \qquad (5)$$

We pick the implicant with the highest value of $WLWO$.

D. Example

We explain the relevance of the weights and the three heuristics using the example shown in Figure 2. f is a 6-

input-3-output function with input variables $\{a, b, c, d, e, f\}$ and output functions $\{y_1, y_2, y_3\}$. The cover matrix for the above function is shown in Figure 2(a), where the rows form the minterms and the columns are the implicants. In this example, there are no reductions possible and we start with a cyclic-core. Figure 2(b) shows which implicants are contained in the outputs. $LW[a, b, c, d, e, f, \bar{a}, \bar{b}, \bar{c}, \bar{d}, \bar{e}, \bar{f}] = [2, 0, 2, 0, 1, 1, 0, 2, 0, 2, 1, 1]$. The implicant set is $\{a\bar{d}e, c\bar{d}\bar{f}, \bar{b}c\bar{e}, a\bar{b}f\}$. For this set, $LC = [3, 3, 3, 3]$, $SC = [2, 2, 2, 2]$, $WL = [5, 5, 5, 5]$, $OC = [2, 1, 2, 3]$ and $WO = [6, 3, 5, 8]$. Weight IC for the outputs is $IC[y_1, y_2, y_3] = [3, 3, 2]$.

The weights of the implicants used in the three heuristics are displayed in Figure 2(c).

$$
\begin{array}{c}
\begin{array}{ccccc}
 & a\bar{d}e & c\bar{d}\bar{f} & \bar{b}c\bar{e} & a\bar{b}f \\
a\bar{b}\bar{c}\bar{d}ef & \begin{bmatrix} 1 & 0 & 0 & 1 \\ abcd\bar{e}\bar{f} & 1 & 1 & 0 & 0 \\ \bar{a}\bar{b}cd\bar{e}\bar{f} & 0 & 1 & 1 & 0 \\ a\bar{b}cd\bar{e}f & 0 & 0 & 1 & 1 \end{bmatrix}
\end{array}
\end{array}
$$

(a) A Cover Matrix

	y_1	y_2	y_3
$a\bar{d}e$	1	1	0
$c\bar{d}\bar{f}$	1	0	0
$\bar{b}c\bar{e}$	0	1	1
$a\bar{b}f$	1	1	1

(b) Output Details

	$a\bar{d}e$	$c\bar{d}\bar{f}$	$\bar{b}c\bar{e}$	$a\bar{b}f$
GR	2	2	2	2
LCMC	–	–	–	–
CS	1	1	1	1
MXSC	2	2	2	2
WLOC	10	5	10	15
WLWO	30	15	25	40

(c) Weights of Implicants based on Different Heuristics

Figure 2: Difference Between Proposed Heuristics

In the greedy approach (GR), the weights on the implicants are equal to the number of uncovered minterms for the implicant. The cover could be $\{a\bar{d}e, \bar{b}c\bar{e}\}$ or $\{c\bar{d}\bar{f}, a\bar{b}f\}$. For LCMC, since no weights are assigned on the implicants directly, there is no specific preference of one term over the other. The CS heuristic assigns weights which are equal for all the implicants and hence ends up picking an implicant chosen at random. The heuristics LCMC and CS can pick any one of the covers, $\{a\bar{d}e, \bar{b}c\bar{e}\}$ or $\{c\bar{d}\bar{f}, a\bar{b}f\}$. The MXSC approach depends on the original set-cover instance and selects implicant based on higher set-cardinality, SC. Since we assume the matrix in Figure 2(a) as the original instance, the weights of all implicants are equal. Hence, no clear winner in the choice of implicants can be obtained. However, if the matrix in Figure 2(a) is derived from a larger instance, the MXSC approach will select an implicant which covered most number of minterms. Notice that the literal count is same for any valid cover of the example.

All the heuristics discussed till now can potentially result in two different covers and can select any one randomly. The next two approaches help in identifying a potentially better cover. Given the output constraints in Figure 2(b), WLOC and WLWO

assign different weights to all the implicants and thereby break ties among the implicants. The choice is made based on the amount of sharing between literals, implicants and outputs. Both WLOC and WLWO pick the cover $\{c\bar{d}\bar{f}, a\bar{b}f\}$. The difference between WLOC and WLWO is that the latter considers weighted output count rather than simple output count and provides scope for finding a better cover in terms of implicant-sharing.

IV. EXPERIMENTS AND RESULTS

We perform two sets of experiments. First, we evaluate the heuristics on a set of hard-instances of benchmarks for the set-cover algorithms called BHOSLIB. These benchmarks are not derived from circuits and hence do not have any semantics like implicants, minterms or outputs attached to them. In the second set of experiments we evaluate the heuristics on MCNC benchmarks.

A. Evaluation of set-cover Heuristics on BHOSLIB

The BHOSLIB benchmarks [13] are hard set-cover instances where the optimum set-cover size is known before. While generating a BHOSLIB instance, first an instance with the known optimal set-cover size is selected and then the instance is grown in size without affecting the optimal set-cover size. Each initial optimal instance can be expanded to give different instances with the same optimal set-cover size. The cover-matrix size in these benchmarks varies from 17000 rows to 126000 rows and 450 columns to 1500 columns.

The three heuristics we proposed cannot be directly evaluated on these benchmarks because these benchmarks do not have any interpretation as circuits. However, the notion of literal count (LC) can be attached to the set-cover problem as follows: For a $s \times t$ set-cover matrix, the weights of the columns can be assigned as

$$LC_i = \lceil \log_2 s \rceil - \lceil \log_2 s_i \rceil + 1$$

where s_i is the number of rows covered by column i. When $s \to 2^n$ the formula captures the number of literals in a Boolean formula with n-variables. We can use this definition of LC_i in the MXSC heuristic for set-cover. Note that WLOC and WLWO cannot be evaluated as they depend on number of outputs of a Boolean function.

We implemented MXSC and three other heuristics (GR, LCMC and CS) and compare their performance on BHOSLIB benchmarks. The results are presented in Table I. Comparison with exact cover was not possible as running exact cover algorithm on the BHOSLIB benchmarks did not terminate even after 24 hours.

BHOSLIB benchmarks consists of 8 families, each containing 5 instances built on the optimum instance. Hence, we report the average of all the instances for a particular family. The different metrics used to evaluate the heuristics are set-cover size, summation of LC_i in the final cover obtained, number of cyclic cores obtained and run-time. Number of cyclic cores is an interesting metric as it gives a measure of how many times a column was selected heuristically or how many times step 4 was executed in Algorithm 1.

Table I: Set-Cover Results on BHOSLIB Benchmark Suite

Family	Avg. #Rows	#Cols	Set-cover Size (averaged over the family)					$\sum LC_i$ of the final cover (averaged over the family)				No. of cores in the set-cover instance (averaged over the family)			
			EXACT	GR	LCMC	CS	MXSC	GR	LCMC	CS	MXSC	GR	LCMC	CS	MXSC
frb30-15	17827	450	420	428.6	443.2	428.6	**428.2**	3916.8	4049.8	3916.8	**3912**	**396.4**	431	**396.4**	396.8
frb35-17	27923.8	595	560	570.6	587.8	570.6	**569.6**	5150.4	5302.4	5150.4	**5140.6**	534.2	575	534.2	**532.6**
frb40-19	41379.2	760	720	732	755.6	732	**731.6**	7203	7438.6	7203	**7199.4**	687.4	748.2	687.4	688.4
frb45-21	58636.6	945	900	914.4	940.6	**914.3**	915	**8759.6**	9016.2	**8759.6**	8764	**866.4**	933	**866.4**	867.2
frb50-23	80456.8	1150	1100	1117	1144.4	1117	**1116.2**	11539.2	11827.4	11539.2	**11530**	1066.2	1134.4	1066.2	**1063.6**
frb53-24	94235.4	1272	1219	**1237.6**	1264	**1237.6**	**1237.6**	12642.8	12914.4	12642.8	**12640.6**	1185	1249.6	1185	**1184.8**
frb56-25	109619	1400	1344	1363.8	1392.2	1363.8	**1363.4**	13855.6	14145.4	13855.6	**13850.4**	1306.4	1378	1306.4	**1305.6**
frb59-26	126358.6	1534	1475	**1494.6**	1524.4	**1494.6**	1495.2	**15092.8**	15391.4	**15092.8**	15097.4	**1431.6**	1506.8	**1431.6**	1435.8

Table I shows that MXSC produces better average set-cover size for 6 families. Also, it can be seen that the reduction of literal count has a direct correlation with set cover size thus justifying the use of the measure LC_i. MXSC has the least core size on 4 benchmark families. GR and CS performances are similar to each other while LCMC performs worse on all the parameters.

Table II: Runtimes on BHOSLIB Benchmarks (secs)

GR	LCMC	CS	MXSC
1931.87	2772.97	2013.82	1961.01

The runtimes for the different heuristics when applied on the complete BHOSLIB benchmark suite are shown in Table II. It is seen that MXSC runs 1.51% slower than GR heuristic, while LCMC and CS run 41.41% and 2.69% slower than MXSC heuristic.

From these results, we conclude that MXSC is a good candidate for set-cover heuristics for logic minimization. LCMC takes longer runtimes and produces worse results. Therefore, we exclude LCMC from further consideration in the rest of the comparisons. The results from MXSC are encouraging enough to try the heuristic and the other weighted variants in a logic minimization framework.

B. Two-Level Logic Minimization with ESPRESSO-II

To validate the performance of the proposed heuristic on the logic minimization problem, experiments were conducted on MCNC Benchmark suite [14] using ESPRESSO-II [14], [15] as a tool. ESPRESSO-II performs both exact and heuristic logic minimization. We use the option -Dexact which creates set-cover instances internally. The tool has a native heuristic which is a variation of the CS heuristic which we call EH. We then performed both exact and heuristic minimization of the set-cover problem. Since LCMC performed poorly on the set-cover problem, we implemented all the heuristics in ESPRESSO-II except LCMC.

The MCNC benchmarks were given to ESPRESSO-II in the PLA format. Each heuristic was evaluated on three parameters: the set-cover size, total literal count in the set-cover solution and the number of cyclic cores. Of the 148 benchmarks, ESPRESSO-II failed to generate all the prime implicants for 17 benchmarks and produced an empty cover-matrix for 11 other benchmarks. Hence, we evaluate the proposed heuristic on the remaining 120 benchmarks. Table III summarizes

results on all the 120 MCNC benchmarks. Average values are reported for each parameter and each heuristic. Results in **boldface** indicate that the method is the best among the class w.r.to the parameter under comparison.

From Table III, the weighted heuristic methods (MXSC, WLOC, WLWO) have reduced set-cover sizes and reduced number of cores compared to the greedy method. CS method seems to be performing well overall except for a small increase in runtime. It is also interesting to note that the ESPRESSO-II's heuristic method (which is also a variation of CS) does not fare as good as the other heuristics. CS is also able to produce the final implicant set with the least literal cost and number of cubes in spite of increase in the number of cores compared to our proposals. To investigate the effect of our methods, we took these implicant sets generated by ESPRESSO-II and performed technology mapping using ABC.

C. Technology Mapping using ABC

The optimized implicant sets from ESPRESSO-II were mapped to two different technologies: one using only 2-input NAND gates and another to LUT based FPGAs. ABC [5] is a logic synthesis tool based on And-Inverter Graphs (AIGs). We used the standard cell and LUT mapping utilities to map the implicant sets.

Technology Mapping using 2-input NAND Gates: To synthesize the logic circuit using 2-input NAND gates, a library consisting of inverter, buffer and 2-input NAND gate was specified and synthesis was performed using the command *map*. Table IV presents the results of technology mapping using 2-input NAND gates. Since all the library cells are equivalent, the delay reported by ABC is equivalent to the number of levels of logic in the resulting implementation.

Technology Mapping using FPGA Synthesis: The basic logic structure in a FPGA is a Look-up table or LUT. More precisely, a FPGA is composed of k-input-1-output look-up tables where each LUT can implement any k-input Boolean function. We used ABC to synthesize the PLA circuits on a 4-input LUT framework. ABC performs a optimal-delay based LUT mapping. A simple script comprising ABC commands *resyn2* and *fpga* was used to perform FPGA synthesis. The results of LUT synthesis are also presented in Table IV.

For both technologies, MXSC is able to produce an implementation with lesser gate count. Even though ESPRESSO-II's implicant sets were worse than the other heuristics, they get synthesized with the least delays when mapped to 2-input NAND gates. This advantage is lost when the circuits

Table III: Summary of Results on MCNC Benchmarks

Parameter	EH	GR	CS	MXSC	WLOC	WLWO
Set-Cover Size	70.62	71.13	**70.54**	71.16	71.09	71.12
No. of Cores	6.68	5.02	5.76	5.04	4.98	**4.93**
Run-time (sec)	0.40	**0.19**	0.21	0.20	0.20	**0.19**
Literal Count of the Set-Cover solution	656.03	659.13	**655.15**	658.65	658.54	658.58
Total Literal Count after Post-processing	1103.93	1106.54	**1103.39**	1106.32	1106.08	1106.01
No. of Cubes after Post-processing	121.50	121.94	**121.43**	122.00	121.93	121.94

(Average values are reported across all the 120 benchmarks.
Post-processing refers to cube-algebra done on set-cover solution and initial essential cubes.)

Table IV: Summary of Technology Mapping Results on MCNC Benchmarks using ABC

Parameter	EH	GR	CS	MXSC	WLOC	WLWO
No. of NAND gates	826.52	828.90	827.36	**824.51**	826.24	824.96
NAND-Synthesis Delay	**16.63**	16.69	16.67	16.72	16.73	16.71
No. of LUTs	242.32	243.40	242.83	241.28	241.23	**240.94**
LUT-Synthesis Delay	4.73	4.72	4.71	**4.70**	4.71	4.71
NAND-Synthesis Area-Delay Product	**19607.46**	19718.11	19705.39	19615.50	19716.64	19659.16
LUT-Synthesis Area-Delay Product	1695.22	1710.69	1695.54	**1690.64**	1694.21	1693.03

(Average values are reported across all the 120 benchmarks)

reduce in depth. This is evident from the rows marked NAND-Synthesis Delay and LUT-Synthesis Delay. Also note that all the heuristics have performances very close to each other with $< 1\%$ variation across the methods.

V. CONCLUSIONS

We presented three heuristics for the set-cover problem that can be used in two-level logic minimization algorithms. These heuristics understand the special semantics attached to rows and columns of the set-cover matrices which arise in logic synthesis. We showed how these heuristics perform on hard instances in the BHOSLIB benchmark suite. We also showed how these methods fare in a logic synthesis setup.

From our results, it is evident that the methods do offer modest improvements over other set-cover heuristics which are oblivious to the special relationship between the rows and columns.

In the future, we are planning to use these heuristics in a multi-level framework like SIS [16] which uses two-level optimization repeatedly. We also plan to study whether the minor improvements coming from the two-level logic minimization using our heuristics can result in significant reduction in area in the multi-level framework.

REFERENCES

[1] W. V. Quine, "The problem of simplifying truth functions," *The American Mathematical Monthly*, vol. 59, no. 8, pp. pp. 521–531, 1952.

[2] G. D. Micheli, *Synthesis and Optimization of Digital Circuits*, 1st ed. McGraw-Hill Higher Education, 1994.

[3] O. Coudert, "Two-level logic minimization: an overview," *Integration, the VLSI Journal*, vol. 17, no. 2, pp. 97–140, 1994.

[4] G. D. Hachtel and F. Somenzi, *Logic Synthesis and Verification Algorithms*, 1st ed. Norwell, MA, USA: Kluwer Academic Publishers, 2000.

[5] Abc: A system for sequential synthesis and verification (last accessed on 14/11/2011). [Online]. Available: http://www.eecs.berkeley.edu/ alanmi/abc/

[6] D. S. Johnson, "Approximation algorithms for combinatorial problems," in *Proceedings of the fifth annual ACM symposium on Theory of computing*, ser. STOC '73. New York, NY, USA: ACM, 1973, pp. 38–49.

[7] V. Chvatal, "A greedy heuristic for the set-covering problem," *Mathematics of Operations Research*, vol. 4, no. 3, pp. pp. 233–235, 1979.

[8] U. Feige, "A threshold of ln n for approximating set cover," *J. ACM*, vol. 45, pp. 634–652, July 1998.

[9] O. Coudert and T. Sasao, *Two-level logic minimization*. Norwell, MA, USA: Kluwer Academic Publishers, 2002, pp. 1–27.

[10] X. Y. Li, M. F. M. Stallmann, and F. Brglez, "Effective bounding techniques for solving unate and binate covering problems," in *Design Automation Conference*, 2005, pp. 385–390.

[11] J. Hlavička and P. Fišer, "Boom: a heuristic boolean minimizer," in *Proceedings of the 2001 IEEE/ACM international conference on Computer-aided design*, ser. ICCAD '01. Piscataway, NJ, USA: IEEE Press, 2001, pp. 439–442.

[12] P. Fišer. (2006) Boom-ii the boolean minimizer (last accessed on 14/11/2011). [Online]. Available: http://service.felk.cvut.cz/vlsi/prj/BOOM/

[13] K. Xu. (2005, Dec.) Benchmarks with hidden optimum solutions for set covering, set packing and winner determination (last accessed on 14/11/2011). [Online]. Available: http://www.nlsde.buaa.edu.cn/ kexu/benchmarks/set-benchmarks.htm

[14] Espresso logic minimization tool (last accessed on 14/11/2011). [Online]. Available: http://embedded.eecs.berkeley.edu/pubs/downloads/espresso/index.htm

[15] R. Rudell and A. Sangiovanni-Vincentelli, "Multiple-valued minimization for pla optimization," *Computer-Aided Design of Integrated Circuits and Systems, IEEE Transactions on*, vol. 6, no. 5, pp. 727 – 750, september 1987.

[16] Sis logic synthesis tool (last accessed on 14/11/2011). [Online]. Available: http://embedded.eecs.berkeley.edu/pubs/downloads/sis/index.htm

978-1-4673-0438-2/12 $31.00 © 2012 IEEE

2012 25th International Conference on VLSI Design

A Rapid Methodology for Multi-mode Communication Circuit Generation

Liang Tang,   Jorgen Peddersen,   Sri Parameswaran

School of Computer Science and Engineering, University of New South Wales, Sydney, Australia

{liangt, jorgenp, sridevan}@cse.unsw.edu.au

ABSTRACT

The need to integrate multiple wireless communication protocols into a single low-cost, low power hardware platform is prompted by the increasing number of emerging communication protocols and applications. This paper presents an efficient methodology for integrating multiple wireless protocols in an ASIC which minimizes resource occupation. A hierarchical datapath merging algorithm is developed to find common shareable components in two different communication circuits. The datapath merging approach will build a combined generic circuit with inserted multiplexers (MUXes) which can provide the same functionality of each individual circuit. The proposed method is orders of magnitude faster (well over 1000 times faster for realistic circuits) than the existing datapath merging algorithm (with an overhead of 3% additional area) and can switch communication protocols on the fly (i.e. it can switch between protocols in a single clock cycle), which is a desirable feature for seemingly simultaneous multi-mode wireless communication. Wireless LAN (WLAN) 802.11a, WLAN 802.11b and Ultra Wide Band (UWB) transmission circuits are merged to prove the efficacy of our proposal.

Keywords

Datapath merging, multi-mode, wireless baseband

1. INTRODUCTION

Numerous wireless communication protocols have appeared recently, each targeting a different application domain. Protocols such as WCD-MA and 3GPP Long Term Evolution (LTE) are used for wide area communication, WLAN for local area mobile internet, and UWB, Bluetooth for short distance data transfer. Modern mobile devices (such as mobile phones, tablets etc.) have to include a number of these communication protocols to meet stringent market requirements. These have to be small in area, consume little power and are often disposable in just a few years. Thus it is desirable to integrate multiple communication protocols into a single ASIC chip, which has small foot print and low power consumption. Current state-of-the-art communication solutions for mobile devices are able to integrate several protocols on a single chip. For example, TI's latest WiLink 7.0 chip includes WLAN, GPS, Bluetooth and FM communication protocols [1]. Although all these protocols are included in a single chip, each of them is still implemented as individual blocks, some components are manufactured unnecessarily as only some of the protocols are active at any one time. One example is the multiply accumulate (MAC) circuits from both WLAN and UWB are similar and can be shared. The baseband circuits of communication protocols have many such blocks which can be shared in order to reduce size and power. Note that only the digital baseband components are considered for sharing, since these are the circuits which can significantly benefit from sharing [2]. Currently, Software Defined Radios (SDRs) have been proposed by researchers to support switchable baseband protocols in a single hardware platform. SDRs have used FPGA, DSP or processor array architectures. All of these platforms are flexible, but consume significant amount of power, have large footprints, and not cost effective for mobile terminals [3, 4, 5, 6, 7]. Typically such solutions are useful for base stations. Cost effective solutions with such flexible circuits for mobile terminals usually result in inadequate performance. For example, it is difficult to achieve 480 Mbps data rate [8] required for UWB using FPGAs, DSPs

or processor arrays in mobile terminals. Some of the newer protocols even reaching 1Gbps data rate, certainly require the use of ASICs. One example is WLAN 802.11ac [9].

A purely ASIC implementation can perform better than FPGA, DSP or processor based approaches. Although the Non-recurring engineering (NRE) cost is high for an ASIC, the unit price of an ASIC is extremely low when mass produced. Since ASICs are not flexible, thus far they have not been used to create complete baseband circuits which are switchable and are area efficient. However, with datapath merging methodology [10], baseband circuits of different communication protocols can be merged and switched to make certain that each individual communication protocol is maintained. One simple example of datapath merging is shown in Figure 1 which merges two convolutional encoders, G1 and G2, into a generic circuit. The merged circuit has identical functionality of G1 and G2 by the selection of the inserted MUX. Besides the advantages of low area, reduced power consumption and high performance of an ASIC implementation, datapath merging also provides on the fly switching which is suitable for fast switching, such as convolutional encoder/decoder and LDPC encoder/decoder switching in 802.11n [11]. This on the fly switching feature also allows fast and frequent protocol switching of multiple communication protocols using a single antenna. Figure 2(a) represents the concept of simultaneous two communication modes baseband transmission using the methodology described in this paper. One example of two modes is WLAN and UWB. Although there are many MUXes to select mode one or mode two dedicated circuits as illustrated in Figure 2(b), all of these MUXes are controlled by a single bit mode selection signal which allows the mode switching to be completed in a single cycle theoretically (if different clocks are used in two merging circuits, a few more clock cycles are needed to eliminate clock glitches during mode switching). The merged circuit can transmit WLAN and UWB signal interlaced baseband packet frame by frame. From the user's perspective, both communications protocols are seemingly supported simultaneously.

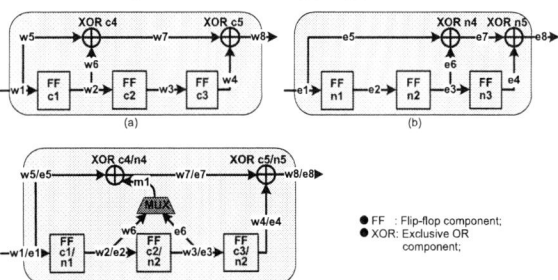

Figure 1: Example of data-flow graph merging. (a) Convolution encoder G1 circuit. (b) Convolution encoder G2 circuit. (c) G1/G2 merged circuit.

Datapath merging methodology has already been used in other areas, such as instruction set synthesis in [12, 13, 14] and reconfigurable systems [10]. Compared to circuits in an instruction set, communication circuits are enormously complicated. Therefore, the algorithms used in [12, 13, 14] will be extremely slow to converge or not converge at all when used for such complex circuits. The method proposed here is a

978-1-4673-0438-2/12 $31.00 © 2012 IEEE   203

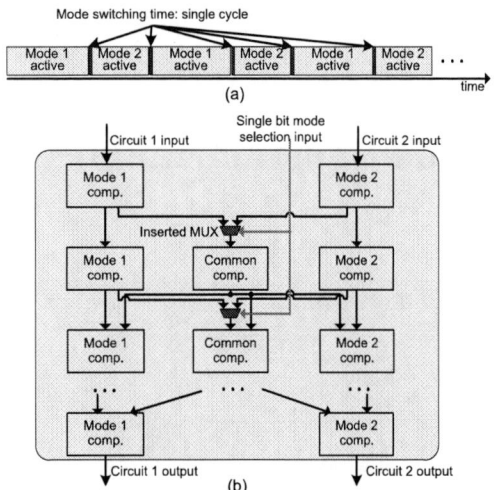

Figure 2: Concept of mode switching. (a) Mode switching in time domain. (b) Block diagram of switching circuit.

hierarchical datapath merging algorithm, which significantly improves the algorithm in [10] to converge quickly by slightly sacrificing area for faster runtimes. Note that the traditional sharing of components in high level synthesis shares components amongst a single circuit and schedule the components accordingly. In datapath merging, two separate circuits find the most suitable components to merge.

Motivation

We are motivated by the need for a methodology to design a multi-mode oriented baseband circuit which is low cost, small and low timing delay for use in a mobile terminal. Previous methods which are FPGA based, DSP based, or processor array based are not suitable for mobile terminals.

We are further motivated by the availability of datapath merging algorithms. However, existing datapath merging algorithms are used for merging of circuits with only a small number of components. Communication circuits are large, with hundreds of components and thousands of wires. Additionally, communication circuits contain loopback circuit, and have widely differing bit widths. These constraints make the naïve application of existing algorithms impractical.

Problem Statement

Given a pair of communication baseband circuits (with loopback and containing several hundred components), described at a hierarchical register transfer level (RTL), find a new merged circuit which can be switched between the functionality of either circuit and has minimal area, within a feasible time frame (in the order of hours).

Outline

The rest of this paper is organized as follows. In section 2 the current research on datapath merging is reviewed. In section 3 the hierarchical merging methodology is proposed. Section 4 and Section 5 show the experimental setup and results. The conclusion is given in Section 6.

2. RELATED WORK

Datapath merging algorithms can efficiently merge circuits represented by Data Flow Graphs (DFGs). The most optimal shareable component pairs between different circuits are shared, and MUXes are inserted to select other non-shareable circuits. The merged circuit can support each of the original circuits by simply switching a single bit selection signal which connects the various inserted MUXes. This topic has been intensively studied previously [12, 10, 13, 15]. Brisk et al. [12] modeled instruction sets as directed acyclic graphs (DAGs) instead of general graphs. By searching the longest common subsequence from a pair of input circuits, a merged circuit can be derived with reduced area. However, the loopback circuits are not supported which are very common in communication circuits. The Figure 3(a) gives a loopback circuit example: differential binary phase-shift keying (DBPSK) modulation circuit, which is one of modulation circuit in various communication systems. This circuit can be merged with

the XOR circuit in Figure 3(b) and the merged circuit is illustrated in Figure 3(c). However, this merging is not supported in [12] as only acyclic circuit is considered there. Moreano et al. [10] proposed a maximum weight clique algorithm based solution, to automatically insert MUXes for datapath merging. Compatibility graphs are first built to represent all possible sharing of components and wires between two input circuits. Then, each component or a wire pair which are shareable would be represented as a node in the compatibility graph. Each node would be assigned a weight, which represents the cost saved by sharing the two components. Any two nodes would be connected by an edge if there isn't a conflict between the two nodes (i.e., that these two nodes could be shared at the same time). By finding maximum weight cliques in this graph, the most optimal merging datapath can be found. This methodology has been extended to share components of different bit widths in [15]. As the maximum weight clique algorithm is a NP-Complete problem, large communication circuits rarely converge by proposals in [10, 15]. A tradeoff between area saving and timing delay constraints has been proposed by Zuluaga et al. [13]. With inserted MUXes, the timing delay of the merged circuit may be worse than the circuits before merging if the inserted MUXes are placed on the critical timing path. Sub-optimal area saving can be achieved by taking timing delay into account. However, in [13], only acyclic circuits are supported and loopback circuits, which are present in communication baseband cannot be optimized.

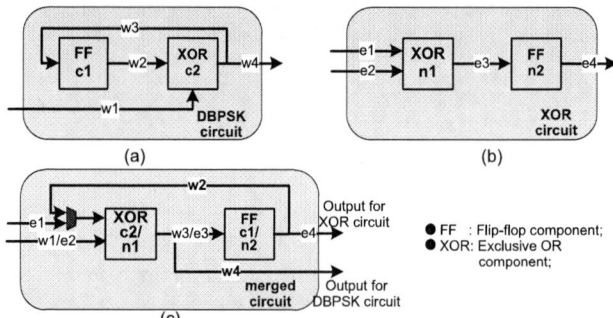

Figure 3: Loopback circuit merging example. (a) DBPSK circuit. (b) XOR circuit. (c) Merged circuit.

Signal processing circuit oriented circuits merging has been proposed by Andriamisaina et al. in [16]. With their GAUT tool, the C program based mutually exclusive algorithms will be converted to individual DFGs representing input circuits, then DFGs are merged to a single circuit with resource sharing. Although certain low-level circuit implementations have been considered, the DFG generated by the GAUT tool may not meet the strict requirements set by the hardware designer. Additionally, only certain digital signal processing module merging is discussed, lacking a complete solution which merges two communication systems.

Our work mainly differentiates itself from [12] and [13] by the ability to support loopback circuits. We take advantage of the nature of communication chain circuits where similar modularized circuits are implemented in one submodule. Examples of such similar sub-circuits are the convolution encoder circuit for 802.11a and UWB, and the modulation circuit for 802.11a, 802.11b and UWB. Our work differentiates from [10] and [15] in recursively partitioning the whole input communication chain circuits into submodule circuits, which significantly reduces the complexity of the maximum weight clique algorithm, as the size of clique graphs are reduced dramatically. Our method is also different from [16] as we focus on the whole communication chain instead of individual signal processing modules. Further, since the input and output of our methodology are well designed RTL descriptions, our solution can be easily verified, and can use existing ASIC design flows in industry. Please note that power analysis is beyond the scope of this paper.

Our Contribution

- Novel methodology is presented by which two communication circuits can be merged together to provide fast switching between those circuits. These merged circuits are area and timing efficient.

978-1-4673-0438-2/12 $31.00 © 2012 IEEE

- A hierarchical datapath merging algorithm based on [10] is presented, which enables the merging of large communication circuits with hundreds of components in reasonable time.

- Seemingly simultaneous communication is supported on single antenna platform through on the fly mode switching.

3. METHODOLOGY

The aim of the described methodology is to merge two complex base band communication circuits described by RTL. The starting point is two RTL descriptions, which are to be merged to a single switchable circuit. RTL was selected as the entry point for our methodology as it balances accuracy with ease of understanding. Since RTL based methods are already widely used in industry, we believe that the merged description can use the existing design flow within the industrial setting effectively. The proposed hierarchical datapath merging methodology is outlined in Figure 4 and it can be classified into three stages.

The first stage is the DFG extraction stage, which extracts DFGs from the input RTL description of each circuit. Each submodule of the input circuits will be extracted individually, and components (such as adder, multiplier, etc.) are replaced from a predefined component library (as shown in Figure 4 - expanded later). Each submodule construct (in VHDL submodules are given with a "submodule" construct) is replaced with a black box representation with its connections to other components. Each of the black boxes is then iteratively examined and DFGs are created. The result of this stage is multiple DFGs (called flat DFGs) for each submodule.

The second stage is hierarchy insertion which builds the hierarchical DFGs. The flat DFGs are analyzed to replace the black boxes with connections to the DFGs of their respective submodules. During this process the hierarchy information is embedded into the DFG (called the hierarchical DFG).

The last stage is hierarchical datapath merging which is the core of our proposal and takes these two circuits represented by hierarchical DFGs and merges them together. The hierarchical datapath merging algorithm is used to determine an efficient merged RTL circuit which converges algorithm execution in reasonable time. More detailed explanations of these stages are discussed in the following subsections.

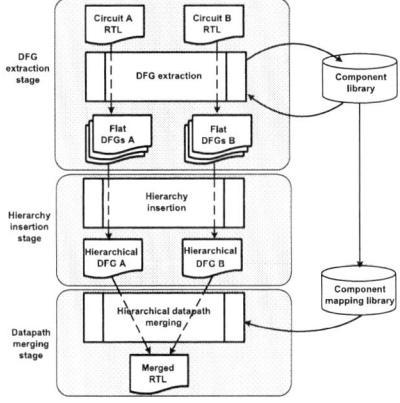

Figure 4: Methodology

3.1 DFG extraction

The first stage of Figure 4, the DFG extraction, is performed as follows. To represent the circuit at a suitable granularity, a component library is utilized. This library includes basic computing circuits which are implemented as RTL modules. Examples include adder, flip-flop, and comparator. Components can be large scale, such as the butterfly operator in FFT, Add-Compare-Select (ACS) operator in Viterbi decoder, etc. These larger coarse grained units accelerate the design merging process. The DFG will only consist of these components from component library and black boxes which represent submodules. The input of DFG extraction is the two communication circuits implemented in hierarchical RTL (i.e., containing submodules). The RTL description is modified to include only components from the component library and submodules. Then, the updated RTL descriptions based on

the component library are fed to a commercial tool (in our case Synopsys Design Compiler [17]) which synthesizes the descriptions into netlists. An in-house tool is developed to extract the top module's and each submodule's DFG from each input netlist. For any processing module, our tool only gives the components under this module and its child submodules are represented as black boxes without internal structure. As the hierarchical information is not explicitly disclosed here, these DFGs are called flat DFGs. An example of an RTL description is shown in the Figure 5(a). There is only one submodule, "mult_add", in the top module. Two flat DFGs - one for the top module and another one for the mult_add submodule - will be generated and can be seen in the Figure 5(b) and (c). The logic components under the top module are drawn in its DFG. However, the top module's DFG does not have any information or links to the contents of mult_add. The contents of mult_add are only present in the separate mult_add's DFG at this stage.

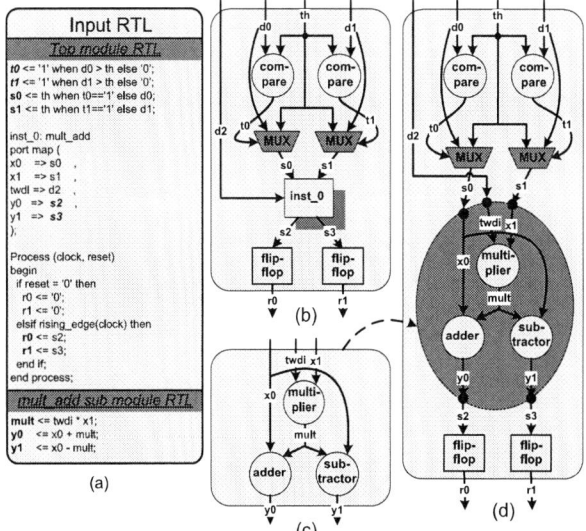

Figure 5: RTL updating by component library. (a) Input RTL. (b) Root module flat DFG. (c) submodule flat DFG. (d) Final hierarchical DFG

3.2 Hierarchy insertion

The second stage of Figure 4 inserts the hierarchy into the DFGs. Both flat DFGs and hierarchical DFGs are represented by a list of Components and Wires (or CW) in our system. Each submodule in the hierarchical DFG is represented as a module directive followed by a list of components within the submodule.

Throughout the hierarchy insertion stage, the flat DFGs are compiled into a single hierarchical DFG by another in-house tool which starts from the root module and concatenates all child modules. Figure 5(d) shows the result of such concatenation. The two flat DFGs from Figure 5(b) and 5(c) have been joined.

3.3 Hierarchical datapath merging

There are two motivations for hierarchical datapath merging. The first motivation is to reduce merging complexity. The most time consuming part of the datapath merging algorithm is to find the maximum weight clique from the compatibility graph which is NP-Complete [10]. Given N shareable CWs, the number of nodes of compatibility graph will be N, and number of edges will be in order of (N^2). By partitioning the circuit into multiple modules in hierarchical datapath merging, the size of compatibility graph will be reduced to M nodes and order of (M^2) edges, where M is the maximum shareable CWs in all submodule pairs. As $M << N$, the problem scope of maximum weight clique is reduce and our experiment shows the hierarchical datapath merging run time is more than 1000 times faster (for realistic circuits) with only 3% additional overhead in area saving compared to the method in [15]. The second motivation is to keep input circuit module hierarchy even after merging, and ensure whole submodules remains as one on the layout (and not scattered over the whole chip) allowing the designer to easily satisfy timing constraints. Sharing every component might result in scattered layout of a module which can significantly increase delay overhead. Note that designers use hierarchical methods to describe circuits in RTL. We exploit this use of hierarchy to simplify merging of circuits. The designer can choose to make certain submodules to be part of the parent module (using a flag called the *Flatten_Flag*). Such merging can occasionally result in improved sharing of components (not shown in this paper due to lack of space).

3.3.1 Hierarchical recursion algorithm

To merge the hierarchical datapaths of the two circuits, a recursive algorithm to selectively merge pairs of submodules from the two circuits is used. This algorithm is shown in Algorithm 1. The input of this algorithm is a pair of modules from two circuits. As the recursion starts from root modules of two circuits, the pair of root modules is selected and fed into algorithm firstly. Each module's DFG is represented as a list of Direct - Components and Wires or D-CW list. Submodules that have their *Flatten_Flag* set have their components included within the D-CW of its parent module, while those with the *Flatten_Flag* cleared are excluded. For example, submodule A1 has only one component a1_1 and a further submodule A11 which has a11_1 component. If submodule A11's *Flatten_Flag* is TRUE, then both components a1_1, a11_1 and their interconnection wires belong to D-CW of A1. However, if a11's *Flatten_Flag* is False, a11_1 and interconnection wires associated to it are excluded from D-CW of A1 .

This algorithm firstly checks if the input DFG is a module without any implementation, so called NULL module. When either of the inputs set to a NULL module, the size of the other module is returned as the new size (as it remains the same) as described in Algorithm 1 line 1 to line 6. D-CW lists of DFGs (i.e. d_cw_a and d_cw_b in the line 7 of Algorithm 1) are generated according to *Flatten_Flag*. These D-CW lists will be fed into DATAPATH-MERGE algorithm (described in the next section) to determine the merged size and merged circuit that could be achieved by merging d_cw_a and d_cw_b. If either circuit contains unflattened submodules (i.e. *Flatten_Flag* cleared), those submodules are then collected for each circuit. NULL modules are added to ensure that both circuits have same number of remaining unflattened submodules. Those submodules that are paired with NULL submodules represent unpaired submodules. This recursive algorithm then recursively calls itself with each combination of submodule pair from the two circuits to determine the merged size of each merging process.

With the results, the submodule combinations are searched to find the best combination of submodule merges which saves the most area. The result is a one-to-one mapping of selected submodule pairs (i.e. *pair_list* in Algorithm 1 line 18) that has the lowest area as described. This analysis is performed by the Hungarian algorithm [18], which searches all possible one-to-one mappings of submodules between the two circuits to find optimal area result. With the result of *pair_list*, the merged size and merged circuit from the submodule pairs can be achieved. Finally, the merged size and merged circuit from d_cw_a / d_cw_b pair and submodule pairs are combined and returned.

3.3.2 Datapath merging algorithm

The datapath merging algorithm will be invoked by the hierarchical recursion algorithm to perform submodule pair datapath merging. Datapath merging of the DFGs from a submodule pair is performed in a similar method to the datapath merging of [10] and [15], with additional constraints of interface wire mapping between a submodule and its parent module. To guarantee no loss of timing performance, MUXes will not be inserted on the critical path as proposed in [19]. Some minor, additional pruning techniques are used to accelerate the merging process. However, these are not described here due to lack of space.

This algorithm is represented in Algorithm 2. The inputs of this algorithm are the flat DFGs represented by D-CW lists. Operating upon the DFGs for the two circuits, a shareable CW pair list will be built with the help of the component mapping library. The criteria for a shareable component pair are: 1) both components in the different circuits belong to the same type of operation regardless bit width is same or not; 2) or both components in the different circuits belongs to different type of operation but them can be mapped to the same definition within a component mapping library; e.g., adders and subtractors are different components, however subtractors can be implemented by an adder with an additional inverter. So the adder and subtracter have the same definition in the mapping library and they are shareable components.

The determination of shareable wires is based on the result of shareable components. The criteria for a pair of shareable wires are: 1) both

---

Algorithm 1 Hierarchical recursion

HIERARCHY-RECUR (hierarchical DFG A and B)
1: **if** *circuit_A is NULL module* **then**
2:   *Return circuit_B's size as merging result;*
3: **end if**
4: **if** *if circuit_B is NULL module* **then**
5:   *Return circuit_A's size as merging result;*
6: **end if**
7: d_cw_a = D-CW of DFG A; d_cw_b = D-CW of DFG B;
8: $(sum\_1, merged\_1, circuit\_1)$ = DATAPATH-MERGE (d_cw_a, d_cw_b);
9: *num_sub_a = number of unflatten submodules in DFG A;*
10: *num_sub_b = number of unflatten submodules in DFG B;*
11: *num_sub = MAX(num_sub_a, num_sub_b);*
12: *Create NULL modules for DFG A/B if its number of submodule is less than num_sub;*
13: **for** $i = 1$ to num_sub **do**
14:   **for** $j = 1$ to num_sub **do**
15:     $(sum\_a[i,j], merged\_a[i,j], circuit\_a[i,j])$ = HIERARCHY-RECUR(DFG of submodule i in DFG A, DFG of submodule j in DFG B);
16:   **end for**
17: **end for**
18: *pair_list = Hungarian algorithm (sum_a , merged_a);*
19: $(sum\_2, merged\_2, circuit\_2)$ = MERGE-PAIR-BUILD(sum_a , $merged\_a$, $circuit\_a$, pair_list);
20: $(sum\_size, merged\_size)$ = $(sum\_1, merged\_1) + (sum\_2, merged\_2)$;
21: *merged_circuit = concatenation circuit_1 and circuit_2;*
22: *Return sum_size, merged_size and merged_circuit;*

Algorithm 2 Datapath merging

DATAPATH-MERGE(flat DFG A and B)
1: *find all shareable pairs between DFG A, B;*
2: *calculate possible area saving of each shareable pair;*
3: **for** all shareable pairs **do**
4:   **if** *MUX is required* **then**
5:     *area saving = area saving - area size of MUX;*
6:   **end if**
7:   **if** *area saving < threshold* **then**
8:     *remove this shareable pair;*
9:   **end if**
10: **end for**
11: *Construct compatibility graph from shareable pairs;*
12: *clique_circle = MAX-WEIGHT-CLIQUE(compatibility graph);*
13: *circuit_m = CIRCUIT_GEN(DFG A and B, clique_circle);*
14: *merged_m = size of circuit_m;*
15: *sum_m = DFG A size + DFG B size;*
16: *Return sum_m, merged_m and circuit_m;*

---

wires in the two circuits are sourced from the same sharable component pair; 2) and both wires in the two circuits output to the same sharable component pair; 3) and there is not a bit-range conflict between the two wires;

Each sharable CW pair is given a weight according to the area saving gained by sharing the pair of components (with due consideration to the inserted MUXes). CW pairs whose weight is smaller than a predefined threshold (the area of an inserted MUX in our system) are removed from the list as the line 7-9 of Algorithm 2. An interconnected compatibility graph is constructed using the remaining sharable CW pairs. The nodes of the compatibility graph should be shareable CW pairs, and those nodes are interconnected if there is no resource sharing conflict between the nodes [10]. Finding the maximum weight clique in this compatibility graph is an NP-complete problem and is solved using a heuristic polynomial-time algorithm MAX-WEIGHT-CLIQUE with approximation which is from [20]. The nodes (i.e. *clique_circle* in line 12 of Algorithm 2) in the resultant max weight clique are the near optimal result for datapath merging of the two input DFGs. Merged circuit is generated by CIRCUIT_GEN function with the input of DFGs and the calculated *clique_circle* which indicates the real shareable CWs.

4. EXPERIMENTAL SETUP

To evaluate the proposed methodology for the integration of multiple communication protocols, the whole baseband packet payload transmission chain from WLAN [11] (including 802.11a and 802.11b) and UWB [8] were built.

Modulations in the 802.11 family use Orthogonal Frequency Division Multiplexing (OFDM) in 802.11a and Direct Sequence Spread Spectrum (DSSS) in 802.11b. Thus both 802.11a and 802.11b protocols were chosen. UWB, a high speed wireless communication protocol, which can support 480Mbps data rates, was also selected.

High level block diagrams of 802.11a, 802.11b and UWB are shown in Figure 6. As common in practice, each block is implemented as a submodule. There are similar functional blocks in the transmission chain of different protocols, such as scrambler, convolution encoder and modulation. All DREG submodules which store parameters from and to the MAC layer, such as data rate, modulation scheme, payload length, etc. are also similar. However, the parameters for DREG submodules are different and thus the implementation circuits of these blocks are different. We implemented 8-bit parallel processing for these three communication protocols and strictly followed the specifications in [11] [8].

There are some special submodules excluded in the analysis presented in the results and analysis section. Since the CRC submodule in 802.11b and UWB are identical, they are excluded. iFFT submodules in 802.11a and UWB are also excluded as their size is much larger than other submodules, and our merging results for other submodules will be distorted. Note that 820.11a iFFT occupies about 90% of the area of whole UWB TX chain and is more than 20 times larger than the whole 802.11b TX chain in our developed circuits.

We developed the baseband packet payload data transmission circuits of these three protocols in RTL, then their datapaths were merged using the methodology proposed in this paper. Note that only BPSK and QPSK modulation schemes (the most common) are supported to simplify the development of 802.11a, 802.11b and UWB circuits.

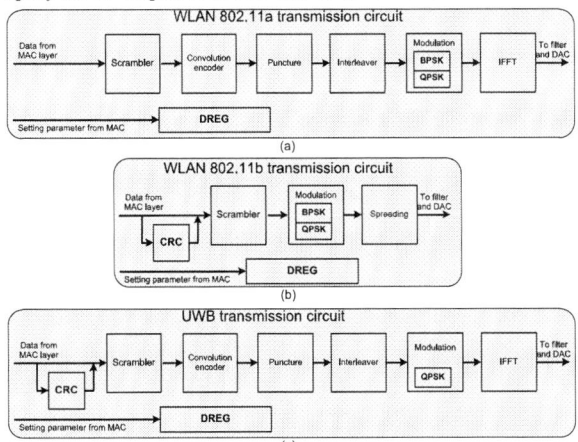

Figure 6: 802.11a/802.11b/UWB transmission chain block diagram.

The test vectors for the system were generated using Matlab [21]. Then, by feeding these test vectors to the individual circuits and the merged circuits in RTL simulation tool (ModelSim [22] in our system), we verified the functionality of merged generic circuits. Area and timing results are generated by Synopsys Design Compiler with TSMC 65nm technology. All experiments were conducted on an Intel Quad Core CPU running at 2.66GHz, with 4GB RAM.

5. RESULTS AND ANALYSIS

5.1 Analysis of merged circuit saving area

We compared our result with [15] because we also support different bit width merging of communication circuits. The merging results prove the hierarchical datapath merging algorithm can approach the area saving results of the method in [15], but is significantly faster.

The merging results of 802.11a / 802.11b, UWB / 802.11a and UWB / 802.11b baseband transmission chain are summarized in Table 1, 2 and 3 and (all three tables have the same format).

The first two columns of Table 1, 2 and 3 show the submodule pairs from two circuits which are selected to be merged. For each submodule pair, the *Sum* column gives the total area result of both submodule circuits together before being merged, while the *Merged* column give the merged circuit size. The saved circuit size is listed in *Saved* column. All these columns provide numbers in units of thousands of gates. The area saving percentage is given in the *Sav(%)* column in percentage representation. Digital register, scrambler, convolution encoder, puncture, interleaver, modulation and spreading submodules are represented

Table 1: 802.11a and 802.11b Merging Results

11a	11b	Sum (k gates)	Merged (k gates)	Saved (k gates)	Sav.(%)
DREG	DREG	6.45	3.23	3.22	49.9%
Scram	Scram	0.39	0.20	0.20	49.8%
Conv	N.A.	0.44	0.44	0.00	0.0%
Punc	N.A.	0.25	0.25	0.00	0.0%
Interl	Spread	1.35	1.31	0.04	2.8%
Mod	Mod	0.69	0.35	0.35	49.9%
Total		9.57	5.77	3.80	39.7%

Table 2: UWB and 802.11a Merging Results

uwb	11a	Sum (k gates)	Merged (k gates)	Saved (k gates)	Sav.(%)
DREG	DREG	6.45	3.23	3.22	49.9%
Scram	Scram	0.49	0.29	0.20	40.0%
Conv	Conv	1.04	0.60	0.44	42.0%
Punc	Punc	0.48	0.32	0.16	33.7%
Interl	Interl	8.60	7.38	1.22	14.2%
Mod	Mod	0.64	0.35	0.29	45.5%
Total		17.70	12.17	5.53	31.2%

Table 3: UWB and 802.11b Merging Results

uwb	11b	Sum (k gates)	Merged (k gates)	Saved (k gates)	Sav.(%)
DREG	DREG	6.45	3.23	3.22	49.9%
Scram	Scram	0.49	0.29	0.20	40.0%
Conv	N.A.	0.60	0.60	0.00	0.0%
Punc	N.A.	0.23	0.23	0.00	0.0%
Interl	Spread	7.51	7.47	0.04	0.5%
Mod	Mod	0.64	0.35	0.29	45.5%
Total		15.93	12.18	3.75	23.5%

as *DREG*, *Scram*, *Conv*, *Punc*, *Interl*, *Mod* and *Spread* respectively. For example, in Table 1, 802.11a and 802.11b submodules are being merged. The submodules being merged are shown to be the 11a DREG with 11b DREG, 11a scrambler with 11b scrambler, 11a interleaver to 11b spreading and so on. The convolution encoder and puncture submodule from 11a are not shared with the circuit in 11b because there are more submodules in 11a than in 11b and all of submodules in 11b have been shared with other submodules in 11a.

The circuit saving percentage is calculated according to Equations 1 and Equations 2:

$$Saved\_size = Sum\_size - Merged\_size \quad (1)$$

$$Saved\_percentage = (Saved\_size/Sum\_size) * 100\% \quad (2)$$

The *Saved_size* is the difference between sum of individual circuits size *Sum_size* and the total merged circuit size *Merged_size*. The *Saved_percentage* is calculated based on the *Saved_size* and *Sum_size*. Thus, the best *Saved_percentage* will be 50% when two input circuits are identical. The merged module pairs will be analyzed based on their *Merged_size* and the pairs will be selected which can contribute to the minimum total merged size.

From Table 1, it can be seen that the submodules with similar functionality in 802.11a and 802.11b are selected for final datapath merging, as these submodule pairs can result in maximal area savings. It is reasonable to expect that the DREG in 802.11a and DREG in 802.11b can achieve nearly 50% saved area which is near ideal score based on our saved_percentage definition, since the circuits of these two submodules are nearly identical as they have the same Media Access Control layer/Physical layer (MAC/PHY) interface and register banks of the same size. The only difference between them is the initial value of the register, which is controlled by the Pull Up and Pull Down cell that connect to the register via internal MUXes.

Due to the insertion of MUXes, the maximum timing delay of merged circuit may be longer than individual circuits before merging. However, when MUXes are not inserted in the critical path, the maximum timing delay will not be worse after merging. The maximum timing delay of UWB/802.11a, UWB/802.11b and 802.11a/802.11b baseband transmission circuit merging are 2.4ns, 2.4ns and 1.9ns respectively after merging. These numbers were the same as before merging. As mentioned previously, the algorithm proposed in [19] has been used here to avoid MUX insertion on critical path.

5.2 Analysis of merging algorithm execution time

To evaluate the algorithm execution speed, the datapath merging algorithm in [15] and hierarchical datapath merging described in this paper are compared.

Table 4: Execution time comparison and area overhead

Case	Non-hierarchy [15]			Our proposal				Speedup
	Sav. (%)	Exe. time	Graph size	Sav. (%)	Exe. time	Graph size	area over-head (%)	
syn1	15.1	12s	197/17k	13.2	4s	64/1k	1.9	3
syn2	15.7	26m	241/26k	14.5	4s	54/1k	1.3	392
syn3	37.2	50m	267/33k	37.1	4s	99/5k	0.1	761
syn4	49.8	61m	307/45k	47.4	4s	136/9k	2.5	915
11a/11b	40.8	83m	366/63k	39.7	5s	136/9k	1.1	996
uwb/11a	N/A	>4d	539/134k	31.2	7s	136/9k	N/A	49k+
uwb/11b	25.1	278m	331/52k	23.5	5s	136/9k	1.6	3336

There are seven test cases for algorithm execution time evaluation. Test cases syn1 - syn4 are synthetic examples based on 802.11b DFG. Each test case has two input circuits, named circuit one and circuit two. The circuits of circuit one from these four test cases are identical and they are merely copies of the 802.11b DFG, while the circuits of circuit two are altered version of the 802.11b DFG with modifications in several components to create differences. Test case syn1 has the most modifications while test case syn4 has the fewest modifications. The reason for these synthetic test cases is to evaluate the datapath merging described in [15] and our methodology with different size of compatibility graph which indicates the complexity of the maximum weight clique algorithm. Test cases 11a/11b, UWB/11a and UWB/11b are the original 802.11a/802.11b, UWB/802.11a and UWB/802.11b circuit pairs respectively.

The algorithm execution time of these test cases are shown in Table 4. The first column gives the test case name. The second, third and fourth columns show the area saving in percentage, algorithm execution time and size of compatibility graph by non-hierarchical datapath merging algorithm in [15]. The fifth column to the seventh column gives the area saving percentage, algorithm execution time and size of the largest compatibility graph for all submodule pairs combination by our hierarchical algorithm. The compatibility graph size is expressed by number of graph nodes/number of graph edges. This number of nodes/edges indicates the complexity of the problem. The area overhead is given in the eighth column and the last column shows speedup in the unit of how many times faster. The results in Table 4 indicate that although the hierarchical datapath merging maximally results in 3% poorer area saving, its execution time is 1000 times faster than datapath merging algorithm from [15]. For UWB/802.11a combination, we are unable to obtain a result in reasonable time with [15] as there are more shareable CWs, which increase the size of compatibility graph, whereas with the hierarchical merging algorithm, we are able to achieve such a result in seven seconds. Note that the graph sizes are same in our proposal for test case syn4, 11a/11b. uwb/11a and uwb/11b. DREG to DREG submodule pairs are selected to figure the graph sizes as their graph sizes are larger than all other submodule pairs in these cases. And the DREG submodules are very similar, thus the same graph sizes are calculated.

5.3 Discussion

The hierarchical datapath merging has drawback which only support a limited number of communication protocols. When the number of communication protocols increases, the overhead of area, power and delay due to inserted MUXes will be too great and will counteract the benefits of the ASIC implementation. However, this is of little concern, as in a real multi-mode product, only a small number of communication protocols will be supported, e.g. only four communication protocols are supported in [1]. Despite this limitation, it is our belief that hierarchical datapath merging is a viable solution for multi-mode baseband integration. With the rapid execution time for the proposed hierarchical datapath merging, a multi-mode baseband circuit designer is able to quickly evaluate the cost of hardware implementation allowing rapid hardware/software partitioning of the whole system. Note that at time to preserve the state (for rapid switching of protocols), not all registers can be shared.

We also believe our methodology is suitable for SDR baseband design. Our proposal is better than FPGA, DSP and processor approaches of SDR in terms of area saving, timing performance as what we proposed is a purely ASIC solution. However, only known communication protocols can be supported due to the number of components and wires that are fixed in ASIC. The inserted MUXes just select the datapath of predefined circuits. As a result, software downloads [23] cannot be supported. However, in reality, for SDR baseband implementations of

mobile devices, most circuits of individual communication protocols will not be changed, as SDR baseband provides part of the physical layer functionality, which is less likely to be altered after the release of protocol specification.

This paper only analyzes the area and delay of hierarchical datapath merging, the power consumption is not explored in this work (though it should be significantly less than FPGA, DSP or processor array methods [24]). In addition, only two circuits can be merged at this stage, though this limitation can be fairly easily overcome with the methodology described in Algorithm 2 in [15]. At this stage the work only merges submodules at the same level. It was assumed that these types of communication circuits are implemented in similar ways (due to similar design methodologies used for communication design generation). Our future work will aim to overcome these limitations.

6. CONCLUSION

In this paper, a hierarchical datapath merging algorithm is presented to build a generic circuit with inserted MUXes which has identical functionality to the individual circuits. We state that this is high efficient solution for multi-mode communication baseband integration. This methodology can save area significantly without worsening the circuit timing performance, while allowing rapid mode switching.

7. REFERENCES

[1] "Ti wilink single-chip wlan, gps, bluetooth and fm solution." Available at: www.ti.com/wilink7-pb.

[2] B. Mennenga, J. Guo, and G. Fettweis, "A component based reconfigurable baseband architecture," in *Mobile and Wireless Communications Summit*, 2007.

[3] L. S. Nagurney, "Software defined radio in the electrical and computer engineering curriculum," in *Proceedings of the 39th IEEE international conference on Frontiers in education conference*, FIE'09, (Piscataway, NJ, USA), pp. 1489–1494, IEEE Press, 2009.

[4] N. Himanshu Shekhar, C.B.Mahto, "Fpga implementation of tunable fft for sdr receiver," in *International Journal of Computer Science and Network Security*, pp. 186–190, 2009.

[5] Y. Lin, H. Lee, M. Woh, Y. Harel, S. Mahlke, T. Mudge, C. Chakrabarti, and K. Flautner, "Soda: A high-performance dsp architecture for software-defined radio," *IEEE Micro*, vol. 27, pp. 114–123, 2007.

[6] A. Duller, D. Towner, G. Panesar, A. Gray, and W. Robbins, "picoarray technology: The tool's story," in *DATE '05: Proceedings of the conference on Design, Automation and Test in Europe*, (Washington, DC, USA), pp. 106–111, IEEE Computer Society, 2005.

[7] M.I.Taj, O.Hammami, and M.Akil, "Sdr waveform components implementation on single fpga multiprocessor platform," in *ICECS*, pp. 790 – 793, Dec. 2010.

[8] "Multiband ofdm physical layer specification," 2005. Available at: http://www.wimedia.org/.

[9] "Official ieee 802.11 working group project timelines," 2011. Available at: http://www.ieee802.org/11/Reports/802.11_Timelines.htm.

[10] N. Moreano, E. Borin, C. D. Souza, and G. Araujo, "Efficient datapath merging for partially reconfigurable architectures," in *IEEE Transactions on Computer Aided Design of Integrated Circuits and Systems*, pp. 969–980, 2005.

[11] "Wireless lan medium access control (mac) and physical layer (phy) specifications," 2011. Available at: http://standards.ieee.org/.

[12] P. Brisk, A. Kaplan, and M. Sarrafzadeh, "Area-efficient instruction set synthesis for reconfigurable system-on-chip designs," in *Proceedings of the 41st annual Design Automation Conference*, DAC '04, (New York, NY, USA), pp. 395–400, ACM, 2004.

[13] M. Zuluaga and N. Topham, "Design-space exploration of resource-sharing solutions for custom instruction set extensions," *Trans. Comp.-Aided Des. Integ. Cir. Sys.*, vol. 28, pp. 1788–1801, December 2009.

[14] N. Pothineni, P. Brisk, P. Ienne, A. Kumar, and K. Paul, "A high-level synthesis flow for custom instruction set extensions for application-specific processors," in *Proceedings of the 2010 Asia and South Pacific Design Automation Conference*, ASPDAC '10, (Piscataway, NJ, USA), pp. 707–712, IEEE Press, 2010.

[15] Y. J. Chong and S. Parameswaran, "Custom floating-point unit generation for embedded systems," *Trans. Comp.-Aided Des. Integ. Cir. Sys.*, vol. 28, pp. 638–650, May 2009.

[16] C. Andriamisaina, P. Coussy, E. Casseau, and C. Chavet, "High-level synthesis for designing multimode architectures," *Trans. Comp.-Aided Des. Integ. Cir. Sys.*, vol. 29, pp. 1736–1749, November 2010.

[17] "Synopsys design compiler." Available at: http://www.synopsys.com/.

[18] "Munkres' assignment algorithm." Available at: http://csclab.murraystate.edu/bob.pilgrim/445/munkres.html.

[19] M. Fazlali, A. Zakerolhosseini, A. Shahbahrami, and G. Gaydadjiev, "High speed merged-datapath design for run-time reconfigurable systems," in *FPT '09: International Conference on Field-Programmable Technology*, pp. 339–343, 2009.

[20] "Cliquer." Available at: http://users.tkk.fi/pat/cliquer.html.

[21] "Mathworks matlab." Available at: http://www.mathworks.com/.

[22] "Mentor graphics modelsim." Available at: http://model.com/.

[23] M. Cummings and S. Heath, "Mode switching and software download for software defined radio: the sdr forum approach," *IEEE Communications Magazine*, vol. 37, pp. 104–106, 1999.

[24] I. Kuon and J. Rose, "Measuring the gap between fpgas and asics," in *FPGA '06: Proceedings of the 2006 ACM/SIGDA 14th international symposium on Field programmable gate arrays*, (New York, NY, USA), pp. 21–30, ACM, 2006.

An Integrated CMOS RF Energy Harvester with Differential Microstrip Antenna and On-Chip Charger

Mahima Arrawatia, Varish Diddi, Harsha Kochar, Maryam Shojaei Baghini and Girish Kumar

Department of Electrical Engineering, IIT Bombay
Powai, Mumbai-400076, India.

Email: mahima87@ee.iitb.ac.in, varish.bvb@gmail.com, harsha@ee.iitb.ac.in, mshojaei@ee.iitb.ac.in, gkumar@ee.iitb.ac.in

Abstract—This paper presents an energy harvesting system which extracts energy from radio frequency radiation for battery charging applications. It comprises of a new differential center tapped Microstrip antenna, off-chip matching circuit, on-chip novel CMOS rectifier and control circuitry in 180nm CMOS technology. For on-chip modules, thick oxide devices have been used so as to meet the charging requirements of the target batteries. The designed battery charging module charges either of 250μAh and 10μAh batteries according to the selected battery and available input power. The 250μAh battery is charged at input power of 2dBm and 10μAh battery is charged at input power of -6.5dBm. The fabricated antenna has a gain of 8.3dB and a VSWR less than 2 in the bandwidth from 844 to 970MHz. An efficient Periodic Steady State(PSS)-based power matching technique is also presented which improves the system efficiency. For 0dBm input power at 950 MHz to the antenna the proposed technique leads to 55.2% efficiency of RF to DC converter system. In addition to the integrated design a discrete rectifier using high-frequency Schottky diodes and differential microstrip antenna at the input are fabricated. The discrete system uses the proposed matching technique and exhibits efficiency of 40% for -11dBm received power by the antenna.

Index Terms—RF Energy Harvesting, Battery charging, Impedance Matching, CMOS, Rectifier, Antenna.

I. INTRODUCTION

Battery operated electronic systems have established themselves in various applications including wireless mobile phones and variety of hand-held devices for different applications. However extending durability of the battery is a matter of interest. Specifically these systems can also be employed in scenarios where frequent human intervention is not always possible. Hence, energy harvesting from ambient to charge the battery or even to empower the system without any battery has gained momentum recently [1]–[3].

Energy harvesting can be targeted towards applications which actually consume considerable power but are employed intermittently. In such applications, system is composed of battery which provides power when needed and harvested energy charges this battery [4]. Design of antennas with high gain and wide bandwidth is crucial to maximize the received power. Various antenna topologies have been reported in the literature for RF energy harvesting [5]–[7]. Maximum gain of 11.98dB is reported in [6] but for a bandwidth of only about 20MHz. In the context of integrated rectifiers various CMOS rectifier

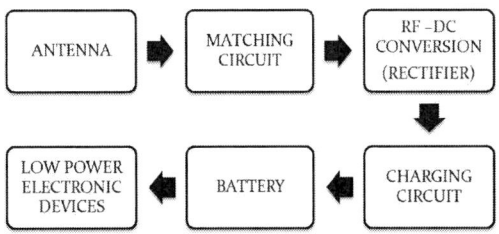

Fig. 1. Basic Block Diagram of RF Energy Harvesting System

topologies have been proposed in [8]–[10]. Efficiency of the overall system greatly depends upon the matching between the rectifier and antenna. Variable input impedance of the CMOS rectifier with frequency and input power further limits total system efficiency. Differential CMOS rectifiers provide better efficiency, compared to single-ended configurations. However, a fully differential configuration requires the use of a dipole or loop antenna which do not provide enough gain. Basic block diagram of RF energy harvesting system is shown in fig.1.

The proposed new CMOS rectifier is presented in Section II. Section III addresses the problem of variation of the input impedance in rectifiers with input power and frequency and provides a power-based input matching technique. Section IV presents a new fully differential high gain and bandwidth antenna topology. The complete energy harvesting system is given in section V. Section VI shows post layout simulations of the RF energy harvesting chip and measurement result of fully differential rectenna (antenna and discrete rectifier).

II. NOVEL CMOS RECTIFIER TOPOLOGY FOR BATTERY CHARGING REQUIREMENT

In this section a novel CMOS rectifier topology is proposed. To make a self sustained battery charging system, following issues should be taken care:

1. Available input power to the rectifier is low and there is no external bias voltage available.
2. The frequency of operation is high (RF frequencies); therefore parasitic capacitance of MOSFETs is major source of loss in the circuit.

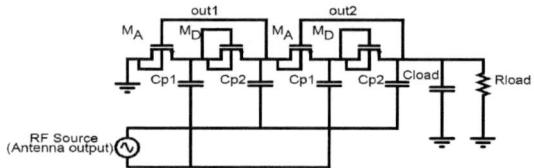

Fig. 2. Proposed Novel Differential Rectifier Topology (two-stage configuration is shown here)

3. Non-linear dependence of the rectifier input impedance on the frequency and input power.
4. The rectifier should have minimum power loss to increase the overall system efficiency.

Different techniques for realization of CMOS rectifier have been proposed. In [9] floating gate transistors are used for rectifier implementation which require external programming and also suffer from leakage due to very low threshold voltages. In [8] matching network is not considered while calculating the efficiency of the rectifier. In [8] rectifier is implemented in 0.18u technology using 1.8V devices hence are not suitable for battery charging application. In [10] an external dc bias is used to switch ON MOSFETS, which is not available for a self sustaining battery charging system.

The proposed novel rectifier topology is shown in Fig. 2. Transistor M_A has its gate connected to the alternate node and M_D has its drain connected to the gate. Each transistor has its drain-connected substrate for dynamic threshold voltage modulation. In the negative half cycle, M_A is switched ON and it gets biased by the next node voltage. In the positive half cycle, M_D is ON. In the proposed rectifier the external bias circuit is not needed. Every alternate transistor is biased using the node voltage from the next transistor. In the proposed topology diode connected transistors prevent reverse leakage.

A. DC Analysis of the Proposed Novel Rectifier

Referring to Fig.2 M_A is switched ON in the positive cycle as its gate observes a positive potential and source terminal is linked to the negative terminal. The effective voltage V_{gs} is equal to peak of the RF input signal applied, V_a, and the voltage on the capacitor is V_{cap}. The maximum voltage to which the capacitor charges (V_{cmax}) is given in (1)

$$V_a - V_{cap} > V_{tn} => V_{cmax} = V_a - V_{tn} \qquad (1)$$

In the second half cycle M_A turns OFF and M_D turns ON. The effective gate source voltage of M_D is

$$V_{in}/2 + V_a - V_{tn}(-V_{in}/2) = 2V_{in} - V_{tn} \qquad (2)$$

The second capacitor gets charged to $2(V_a - V_{tn})$. In the third half cycle source and drain of M_A interchange, hence it will not charge the capacitor further. If the capacitor is not charged completely then source and drain do not interchange and M_A will again charge the first capacitor to $V_a/2$. The parasitic capacitances and parasitic diodes are the main sources of reverse leakage which decreases the efficiency of CMOS rectifier considerably. Fig. 3 shows parasitic capacitances

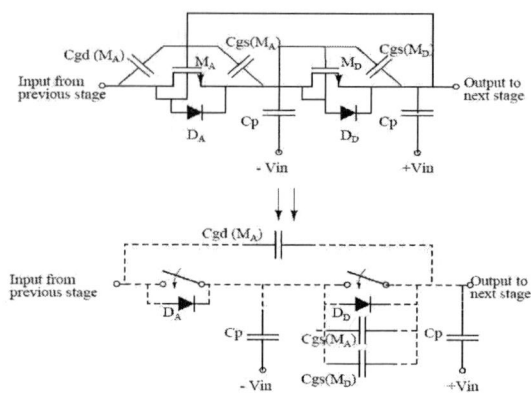

Fig. 3. Parasitic Capacitances and Diodes in the Proposed CMOS Rectifier

which play an important role in deciding amount of drop in the output voltage. The parasitic capacitance Cgd (M_A), Cgs (M_A) and Cgs(M_D) form a parallel combination while one of the transistors is in the ON state. Fig. 3 also shows the parasitic diode in parallel with the transistor but it switches ON only when transistor is in the ON state, therefore it does not contribute to the reverse leakage current. A single ended version of the topology is also proposed here. It uses an alternate combination of NMOS and PMOS transistors. NMOS transistor is used as a switch in the first cycle and PMOS bulk diode is utilized to rectify the input signal at the second half cycle. The efficiency of the system is also dependent on the matching of the rectifier to the antenna which is explained in the next section.

III. IMPEDANCE MATCHING

Many approaches have been reported in the literature for matching antenna impedance to the rectifier input impedance [11], [13], [15], [16]. Rectifier input impedance has to be determined and then matched to 50Ω resistance of antenna.

A tunable inductor can be used at the interface of antenna and rectifier [13], [14]. This will resonate out the input capacitance of rectifier. However the resistive parts also need to be matched. The transistors can be replaced by equivalent small-signal resistance averaged over all gate overdrive voltages [15]. However rectifier is a large-signal switching circuit where its transistors act as switches. They traverse through all possible regions of operation. Therefore small-signal model leads to errors in getting effective input impedance.

An approach to calculate the input impedance of the rectifier has been developed recently [16]. It uses Periodic Steady State (PSS) analysis to obtain the steady state input resistance and reactance, separately. PSS is a large-signal analysis. Therefore it can be appropriately used for simulation of non-linear circuits like rectifiers. The same methodology is used here for determining input impedance of the rectifier. Also a critical insight has been developed on power-dependent input impedance of the rectifier. Rectifier input impedance is a strong non-linear function of the input power. In applications where system has to work for a wide range of input power

978-1-4673-0438-2/12 $31.00 © 2012 IEEE

TABLE I
RESULTS OF THE PROPOSED TECHNIQUE FOR MATCHING A SINGLE
STAGE RECTIFIER TO THE ANTENNA AT $0dBm$ ANTENNA POWER AT
950MHz

Input Power P(dBm) at which impedance is obtained	Z(P) (Ω) input impedance of rectifier	V_{out} (V) output voltage of rectifier	Conversion Efficiency of rectifier with 0dBm Antenna Power matched to corresponding Z(P) (%)
0	$290-j5580$	0.42	1.8
5	$1460-j1130$	2.3	52.9
10	$890-j510$	2.14	45.8
15	$650-j360$	2.35	55.2
20	$620-j330$	2.3	52.9

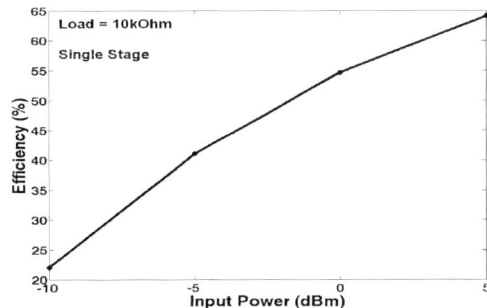

Fig. 4. Efficiency of the Proposed Novel Single Stage Rectifier with 10Kohm Load at 950MHz

levels it is advisable to use off-chip matching elements. This also provides flexibility to control the bandwidth of matching network. Off-chip matching network has high-Q elements (SMD devices) hence ensures efficient matching.

Transistors used in the rectifier traverse through all regions of operation cutoff, linear and saturation.Rectifier input impedance decreases rapidly with increasing input power. When PSS analysis is run for a set up with an antenna of given input power connected as a port to the rectifier, input impedance of the rectifier under unmatched conditions is obtained. In order to achieve matching while taking effect of input power on the impedance level into consideration, an algorithm is developed and presented here. To match the antenna to the rectifier input impedance at input power P the following algorithmic procedure is used.

1. Find unmatched input impedances $Z(P_i)$ for various antenna power levels $P_i >$ P.
2. Obtain matching network N_i, which matches antenna resistance to $Z(P_i)$.
3. Use various networks N_i between antenna and rectifier at input power level P to get various Power Conversion Efficiencies (PCE), η_i.
4. Choose optimum matching network, N_{opt}, at which PCE, η_i, is maximum.

Table I shows the simulation results for a single stage rectifier using thick-oxide, triple-well MOSFETs at 950 MHz. The target is to match this rectifier with load $10k\Omega$ at antenna power level of 0dBm at 950MHz with output resistance of 50Ω. Interestingly if matching network is designed according to [16] then the system would be matched will yield an efficiency of 1.8%. The proposed procedure here leads to maximum efficiency of 52.9% for input power of 0dBm if non-linear input impedance of the rectifier is matched according to the input impedance obtained at 15dBm. For 0dBm conjugate LC matching L = 109nH , 24nH and C = 500fF.

IV. RECTIFIER OPTIMIZATION FOR BATTERY CHARGING SYSTEM

For Li-ion battery charging a voltage of 3.3V is required hence thick oxide triple well devices are used to design the proposed rectifier. These devices have a higher threshold voltage hence dynamic threshold modulation is used in the

TABLE II
EFFICIENCY OF THE PROPOSED NOVEL RECTIFIER WITH NUMBER OF
STAGES AT 300KOHM LOAD $-10dBm$ POWER AT 950MHz

Number of stages of the proposed rectifier	Power Conversion Efficiency (%)
1	22.5
2	30
3	30
4	29
5	22.5

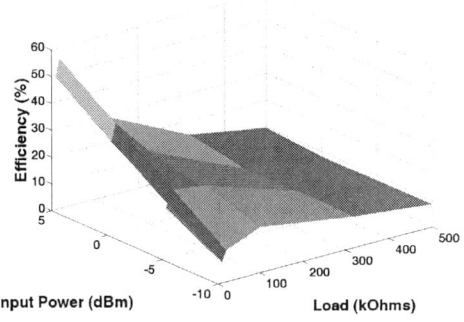

Fig. 5. Efficiency Contour of the Proposed Novel 2 stage Rectifier with Input Power and Load (Post Layout) at 950 MHz

proposed rectifier. Threshold voltage is reduced but remains high enough to avoid reverse leakage. Therefore rectifier efficiency is improved. Performance of the rectifier would reduce with increase in size of the MOSFET, hence the transistor size needs to be optimized for a particular load current taking into consideration loss due to parasitics. Transistor with W/L ratio of $1\mu/8\mu$ is used for triple-well thick oxide devices, obtained for maximum efficiency at 950 MHz.

Fig. 4 shows the efficiency of the single stage proposed rectifier with various power levels and 10Kohm load.A maximum efficiency of 64% is achieved at 5dbm input power.

Table II shows the efficiency of the proposed novel rectifier with number of stages at $300k\Omega$ load and -10dBm power.A maximum efficiency of 30% is achieved for 2 stages. Approximate load of the $10\mu A$ battery which is used to demonstrate

978-1-4673-0438-2/12 $31.00 © 2012 IEEE

one of the battery charging system is $320k\Omega$ for which 2-stage give maximum efficiency. Single stage rectifier gives maximum efficiency for $10k\Omega$ load which is the approximate load of $250\mu A$ battery. But it does not provide voltage boosting which would be required to maintain 3.3V for battery charging. Hence we have chosen 2 stages of the proposed rectifier for powering both battery charging system. Fig 5 shows the post layout efficiency contour of 2 stage proposed rectifier with load and input power. A maximum efficiency of 55.8% is achieved at an input power of 5dBm and $20k\Omega$ load.The efficiency of a single ended version of the proposed rectifier is only 10% for $10k\Omega$ load and 0dBm input power.

V. ANTENNA

Gap coupled Microstrip antenna [12] satisfies gain, efficiency and bandwidth criteria but it is single ended configuration. This limits the design of rectifier topologies as single ended configuration has an efficiency of only 10% in comparison to its differential counterpart which has an efficiency of 54.7% for 2 stage configuration. Generally, antennas having a fully differential configuration is limited to dipole or loop antennas. These antennas have a gain of about 2dB. To overcome these limitations,design of a fully differential center tapped electromagnetically coupled Microstrip antenna with high gain, efficiency and bandwidth is proposed. Figure

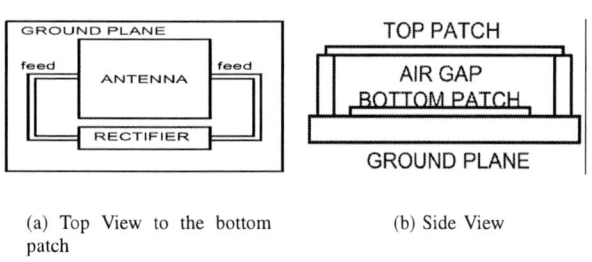

(a) Top View to the bottom patch

(b) Side View

Fig. 6.   Proposed Topology of Novel Fully Differential Microstrip Antenna

6 shows the proposed fully differential microstrip antenna. It consists of a two layered electromagnetically coupled Square microstrip antenna. The top patch is fabricated in inverting configuration so that the substrate provides a dielectric cover for the antenna. Both the patches are fabricated on glass epoxy substrate having ϵ_r = 4.4,thickness 1.6mm and tan δ= 0.02. Bottom patch has a dimension of 82mm×82mm which is chosen according to center frequency of the desired bandwidth. The dimensions of the top patch are 127mm×127mm. An air gap of 18mm is kept between top and bottom patch. The center point of the bottom patch is tapped to ground. RF signal output is taken at 41mm on both sides to provide a fully differential center tapped configuration. The antenna has gain of 8.3 dB and VSWR less than 2 bandwidth from 844 to 970MHz. The efficiency of the antenna is 80%.

VI. RF ENERGY HARVESTING SYSTEM

The designed energy harvesting system extracts the RF energy from ambient. In our target application the energy

TABLE III
CHARGER SYSTEM SELECTION ACCORDING TO CONTROL ANALOG INPUT

Analog Input Voltage	SW1a	SW2a	Remarks
0 - 650mV	OFF	OFF	Both Batteries are Isolated
650mV - 1.4V	ON	OFF	3.3V, $250\mu A$ battery charger system is enabled
1.4 - 3.3V	OFF	ON	3.3V, $10\mu A$ battery charger system is enabled

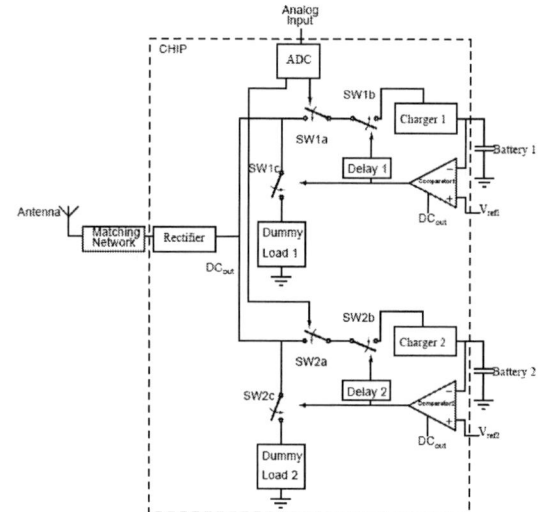

Fig. 7.   Block Diagram of RF Energy Harvesting System

charges one of the two batteries in constant current mode. Block diagram of the energy harvesting system, is shown in fig.7.

A. Novel Skewed Inverter based Flash ADC for Battery Charger System Selection

Architecture of the designed flash ADC is shown in Fig.8. It has a very low static-power consumption of around 445nW for a supply voltage of 3.3V. The reference inputs to ADC are taken from diode-based voltage divider, which produces enough accuracy for target charger application. It should be noted that the control analog input of the ADC is static in nature. The comparators are implemented using skewed-inverters of which skewed switching thresholds are determined by the supply voltage of ADC. A feedback mechanism is used so that these switching thresholds are immune to process variations as shown in Fig.8 [17]. The output of comparators is thermometer coded and is given to an encoder to obtain two-bit binary equivalent output. These bits control two switches SW1a and SW2a. The design is such that only three states of output are possible which leads to either both switches being OFF or only one of the two being ON. TableIII shows possible states of switches SW1a and SW2a with analog input range.

B. Control Circuitry and Charging System

The control circuitry consists of comparators, delay blocks, switches and dummy loads. The charging path from a rectifier is closed if two series switches SW1a and SW1b (or SW2a

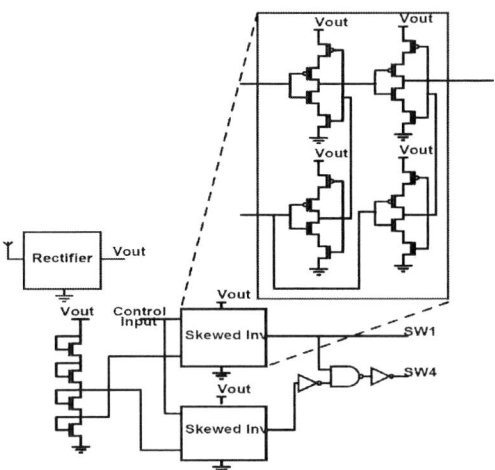

Fig. 8. Proposed Novel Skewed Inverter based Process Invariant Flash ADC for Energy Harvesting System

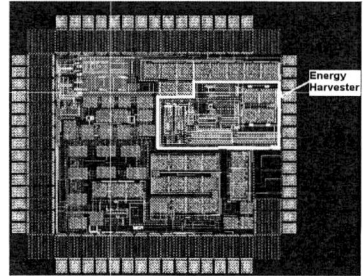

Fig. 9. Layout of RF Energy Harvesting System

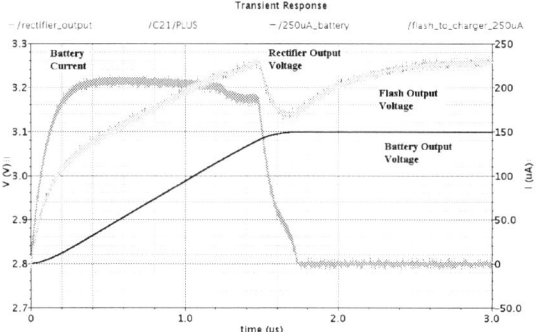

Fig. 10. Charging profile of 3.2V, 250μA battery

and SW2b) are closed. These switches are implemented using transmission gates of large size so as to reduce their ON-resistance and hence the voltage drop across them. Switch SW1a (or SW1b) is controlled by ADC and SW1b (or SW2b) is controlled by a comparator followed by a delay module. Switches SW1c (or SW2c) connect the rectifier to dummy loads which are designed according to the input power level. These switches are also driven by comparators. To save area SW1c and SW2c are of smaller size, compared to other switches. This is because voltage drop across these switches is not of as importance as other switches. Both chargers are designed for constant current charging which is the main part of charging profile. Due to limitations in the breakdown voltage of CMOS transistors in 180nm technology (3.3V) we have limited maximum voltage to which the battery is charged (3.2V here). Hence, constant voltage charging of the profile is not included. Every charger is a constant current generator and is supplied by the rectifier output. A cascode current charger has been used to avoid back flow of current from the battery. Both chargers can work under reduced supply of 3.1V with only 5% reduction in the charging current.

Differential input stage of the comparators is followed by a decision making latch circuit with a high gain. Third stage is another differential amplifier to further boost the gain of the circuit. Third stage is followed by two cascaded inverters which takes output to perfect logic level. The comparator consumes only 10.9μW static power at supply of 3.3V. The comparator functions correctly from supply of 1.7 to 3.3V. Each comparator compares the corresponding battery voltage level with a reference voltage at which the charging should stop according to battery requirements.

Before disconnecting the charger from comparator care should be taken to avoid rising output voltage of the rectifier to a very high value since it is going to become unloaded. For this purpose two dummy loads with appropriate switching arrangements are used as shown in Fig. 7. Comparator output

drives two switches SW1c and SW2c connecting rectifiers output to dummy loads. Two delay modules which drive SW1b and SW2b are also driven by the comparator output. The delay circuit is composed of cascaded inverters sized to get around 200ns delay. During battery charging switches SW1c and SW2c are OFF whereas SW1b and SW2b are ON. When the battery charges to the expected voltage comparator output toggles to disconnect charger. However first switch SW1c (or SW2c) is turned ON and then after a delay of 200ns switch SW1b (or SW2b) is turned OFF. This arrangement is important to achieve make-before-break action. Otherwise rise at the output voltage of the rectifier may lead to breakdown of devices in control circuitry. Dummy loads are designed according to the input power available at the input of the rectifiers and are switched ON to keep the output voltage of rectifiers below 3.3V.

VII. POST LAYOUT SIMULATION RESULTS AND MEASURED RESULTS

The RF energy harvesting system for battery charging application is designed and laid out in 180nm CMOS MM-RF technology using thick oxide devices. Fig.9 shows the layout of the energy harvesting system on a multi-project chip. The rectifier stages are optimized according to the load and battery requirements. The input to the antenna is RF signal of both polarities. So the ESD (Electrostatic Discharge) protection circuits are removed from I/O pads so as to allow such high-voltage signals to be given as input to the rectifier. Total area consumed for layout of energy harvesting system is 0.23mm^2.

978-1-4673-0438-2/12 $31.00 © 2012 IEEE   213

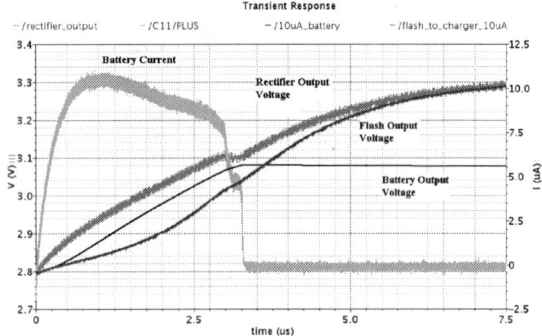

Fig. 11.   Charging profile of 3.2V, 10μA battery

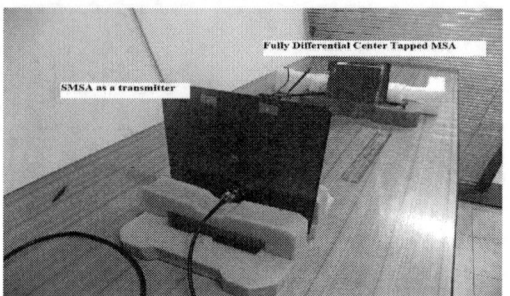

Fig. 12.   Experimental Set-up to Validate Working of Antenna

TABLE IV
MEASURED RESULTS OF FABRICATED RECTENNA WITH MATCHING
NETWORK AT 950MHz

Distance between Transmitting and Receiving Antenna (cm)	Received Power at Antenna Input (dBm)	Output Voltage of Rectifier (V_{out}) at 22$k\Omega$	Power Conversion Efficiency η (%)
19	3.5	2.89	17.3
50	-1.1	1.6	14.9
90	-11	0.84	40
208	-17	0.21	10

The limited energy at the antenna requires the circuits to consume low power. The capacitors used were MIM so as to minimize the leakage and output ripple in rectifier. These are the main factors affecting the area of layout. A single section differential off-chip matching network is employed. Passive components are used for the matching network in simulation. This is consistent with practical high-Q SMD inductors and capacitors which are used for such matching circuits.

Post-layout simulations of RF energy harvesting system charging external batteries are shown in Fig.10 & Fig.11. The charging profile, shown in Fig.10, is related to the on-chip charger for batteries with 3.2V and constant charging current of 250μA. We had to limit the maximum output voltage of the battery to 3.2V due to limited breakdown voltage of thick oxide devices in 180nm CMOS technology. The input power in this case is 2dBm. The reference voltage is kept to be 2.81V. As simulation results show the delay of comparator is 1.3μs and battery is charged from 2.8V to 3.06V in less than 2μs. Accordingly system efficiency is calculated to be 32.9%.

The charging profile for the on-chip charger with constant charging current of 10μA is shown in Fig.11. The input power is -6.5dBm. The battery is charged from 2.8V to 3.01V in 4.3μs. This leads to the system efficiency of 11%. Lower efficiency, compared to the previous case, is attributed to less required power. Otherwise output voltage of the rectifier increases beyond 3.3V which is a limit for reliable operation of thick oxide devices in 180nm CMOS technology. Fig.12 shows the experimental setup to validate the principle of fabricated rectenna using Schottky diode based rectifier. Table IV shows the measured results. The peak efficiency of 40% is observed at received power of -11 dBm at 950 MHz.

VIII. CONCLUSION

In this paper an energy harvesting system with a novel center tapped electromagnetically coupled microstrip antenna and a high efficiency CMOS rectifier have been proposed. A power based impedance matching network has been developed which improves the overall system efficiency as the input impedance of the rectifier changes with power.

REFERENCES

[1] Chao Lu; et al.; , "Micro-scale energy harvesting: A system design perspective," *ASP-DAC* , Jan. 2010
[2] Lu Chao; et al.; , "Vibration energy scavenging and management for ultra low power applications," *ISLPED*, Aug. 2007
[3] Jun Yi;et al.,"Analysis and Design Strategy of UHF Micro-Power CMOS Rectifiers for Micro-Sensor and RFID Applications," *Circuits and Systems I: Regular Papers, IEEE Transactions on* , Jan. 2007
[4] Torres, E.O.; et al.; "A 0.7-μ m BiCMOS Electrostatic Energy-Harvesting System IC," *IEEE JSSC* , Feb. 2010
[5] Z. W. Sim; et al.; "Investigation of PCB Microstrip patch receiving antenna for outdoor RF energy harvesting in wireless sensor networks ", *IEEE Conf. on Ant. and Prop.,*, Nov. 2009
[6] A.C. Patel; et a.l; " Power Harvesting for Low Power Wireless Sensor Networks. " *IEEE Conf. on Ant. and Prop.,*, Nov. 2009
[7] V.Rizzoli; et al.; "CAD Multi-Resonator rectenna for micro power generation. " *Proc. of 4th Euro. Microw. Conf.*, Sep. 2009.
[8] Kotani, K.et al.; "High-Efficiency Differential-Drive CMOS Rectifier for UHF RFIDs," *IEEE JSSC* , Nov. 2009
[9] Triet Le; et al.; "Efficient far-field radio frequency energy harvesting for passively powered sensor networks", *IEEE JSSC*, May 2008.
[10] T. Umeda; et al.; "A 950MHz Rectifier Circuit for Sensor Networks with 10m-Distance ", *Proc. of IEEE ISSCC, USA*, 2005
[11] Arrawatia. M; et al.; "RF Energy Harvesting System at 2.67 and 5.8GHz," *APMC, 2010* , Dec. 2010
[12] G. Kumar and K. P. Ray Broadband Microstrip Antenna, Artech House, USA 2003.
[13] Barnett, R.; et al.; "Design of multistage rectifiers with low-cost impedance matching for passive RFID tags," *(RFIC Symp., 2006 IEEE* , June 2006
[14] Barnett, R.E.;et al.;"A RF to DC Voltage Conversion Model for Multi-Stage Rectifiers in UHF RFID Transponders," *Solid-State Circuits, IEEE Journal of* , Feb. 2009
[15] Wilas, J.; et al.; , "Power harvester design for semi-passive UHF RFID Tag using a tunable impedance transformation," *Comm. and Info. Tech., 2009. ISCIT 2009. 9th Int. Symp. on* , Sept. 2009
[16] Mazzilli, F.; et al., "Design methodology and comparison of rectifiers for UHF-band RFIDs,"*RFIC Symp., 2010 IEEE* , May 2010
[17] Meng-Tong Tan; et al., "A process-independent threshold voltage inverter-comparator for pulse width modulation applications," *Electronics, Circuits and Systems. The 6th IEEE Int. Conf. on*, 1999

Low-Overhead Maximum Power Point Tracking for Micro-Scale Solar Energy Harvesting Systems

Chao Lu, Sang Phill Park, Vijay Raghunathan, and Kaushik Roy
School of Electrical and Computer Engineering, Purdue University
West Lafayette, IN 47907, USA
E-mail: {lu43, park143, vr, kaushik}@purdue.edu

Abstract—Environmental energy harvesting is a promising approach to achieving extremely long operational lifetimes in a variety of micro-scale electronic systems. Maximum power point tracking (MPPT) is a technique used in energy harvesting systems to maximize the amount of harvested power. Existing MPPT methods, originally intended for large-scale systems, incur high power overheads when used in micro-scale energy harvesting, where the output voltage of the transducers is very low (less than 500mV) and the harvested power is miniscule (only hundreds of µW). This paper presents a low-overhead MPPT algorithm for micro-scale solar energy harvesting systems. The proposed algorithm is based on the use of a negative feedback control loop and is particularly amenable to hardware-efficient implementation. We have used the proposed algorithm to design a micro-scale solar energy harvesting system, which has been implemented using IBM 45nm technology. Post-layout simulation results demonstrate that the proposed MPPT scheme successfully tracks the optimal operating point with a tracking error of less than 1% and incurs minimal power overheads.

I. INTRODUCTION

Rapid advances in nanoscale integration have resulted in a new class of highly miniaturized electronic systems (*e.g.,* wireless sensor networks, biomedical implants). Despite the extreme constraints on size (and hence battery capacity), these systems often need to operate for long periods of time without the need for battery replacement [1]. A promising approach to overcoming the problem of limited energy availability in these micro systems is to scavenge energy from ambient sources (*e.g.,* solar radiation, heat flow) [2-3]. Solar energy harvesting through photovoltaic (PV) conversion is particularly attractive because of the ubiquitous availability of light sources and its relatively high power density. The feasibility of using solar energy to power micro-scale systems has already been demonstrated (*e.g.,* retinal implants that are powered using energy harvested from the light entering the human eye [4]). This paper focuses on the efficient design and optimization of such micro-scale solar energy harvesting systems.

Micro-scale solar energy harvesting involves several design challenges. First, the form-factor constraint mandates the use of miniature PV modules (at most a few cm^2), which output very low voltage (less than 0.6V), especially under indoor lighting conditions. As a result, they cannot be used to directly power electronic systems and a voltage booster is required to step up the PV module's output voltage to a higher value. Voltage boosting from a micro-scale PV module requires innovations in circuit design [5]. Second, the amount of electrical power produced by a PV module depends on the effective load

impedance at its output. The optimal loading point of a PV module is called the maximal power point (MPP) and changes with varying light intensity. Ensuring that the PV module always operates at its MPP is done using a technique called MPP tracking. MPP tracking becomes even more important in micro-scale energy harvesting because the power output of a miniature PV module is miniscule (only hundreds of µW) to begin with. Therefore, it is crucial to squeeze every last ounce of power from the module. At the same time, the power overhead introduced by the MPPT scheme should be very small, in order to transfer as much of the extracted power as possible to the electronic system. The power overheads of previously proposed MPPT schemes (detailed in Section II) degrade overall power efficiency.

This paper makes the following contributions: (a) We propose a new, low-overhead MPP tracking algorithm that is well suited for micro-scale energy harvesting systems. The proposed algorithm is based on the use of a negative-feedback control loop and does not require additional hardware (such as voltage/current sensors, microcontrollers, or additional pilot PV cells, which are used by existing approaches). By eliminating hardware components, our proposed algorithm significantly decreases the power overhead of MPP tracking, (b) We present the design and implementation of a micro-scale solar energy harvesting system that uses the proposed algorithm to perform automatic MPP tracking, and (c) We present post-layout circuit simulation results that demonstrate that the proposed MPP tracking scheme successfully tracks the system's MPP voltage (tracking error of less than 1%) with minimal power overhead.

The remainder of this paper is organized as follows. Section II reviews related work and existing MPP tracking schemes. Section III describes the proposed MPP tracking algorithm and its circuit implementation in a micro-scale solar energy harvesting system. Section IV presents post-layout simulation results to evaluate the performance of the proposed MPP tracking scheme and Section V concludes the paper.

II. RELATED WORK

MPP tracking has been extensively investigated for macro-scale solar harvesting systems that use large PV panels [13]. Due to the abundant amount of power generated by the PV panels, the power overhead introduced by the MPP tracking scheme is usually not a primary design consideration. As described above, that is not the case in micro-scale systems. Techniques for MPP tracking in micro-scale solar energy harvesting systems are summarized in [6]. Among these techniques, the fractional open-circuit (FOC) method and the hill-climbing method are the most common.

The FOC method is based on the empirical observation that the MPP voltage of a PV module is an almost-constant fraction of its open circuit voltage. Techniques based on this method compute the MPP voltage at runtime by momentarily disconnecting the PV module from the load and sensing its open circuit voltage. This method does not require any intensive computation. However, the associated hardware cost and power overhead for time multiplexing between normal energy harvesting and open circuit voltage sensing is a concern. To address this, the authors of [7] propose using a second PV module as a pilot cell. The open-circuit voltage of the pilot cell is monitored instead of the main PV module, thus avoiding frequent interruption to the normal energy harvesting process. However, in addition to increasing system size and cost, the pilot cell should be chosen carefully to ensure that its MPP voltage is identical to that of the main PV module.

The second commonly used MPPT method, namely hill-climbing, is an iterative search-based approach. It involves a continuous perturbation of the duty cycle or switching frequency of a power converter (thereby perturbing the load impedance of the PV module) till the MPP is reached. The instantaneous output power is usually computed in software using a microcontroller as a product of the instantaneous output voltage and current [8]. In order to reduce hardware cost and power consumption, the authors of [9] proposed the use of dedicated hardware to replace the microcontroller. However, their system still requires the use of voltage/current sensors.

From the above discussion, it is clear that existing MPPT methods do not fully address the requirements of micro-scale energy harvesting systems. In the next section, we propose an automatic MPP tracking algorithm that is based on a joint analysis of the PV module and the power converter at its output, and present a hardware-efficient implementation of it.

III. MICRO-SCALE SOLAR ENERGY HARVESTING SYSTEMS

A. System Overview

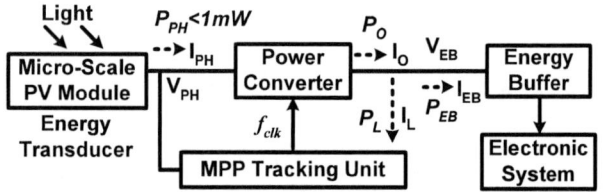

Figure 1. Block diagram of proposed micro-scale energy harvesting system

Fig. 1 shows the block diagram of the proposed micro-scale solar energy harvesting system. The PV module converts light energy into electrical energy. The power converter extracts the PV module's output current (I_{PH}) and transfers the harvested power to an energy storage buffer (e.g., battery). To avoid the use of bulky off-chip inductors [10-12] and achieve compact on-chip integration [9], a capacitive charge pump is used as the power converter. The proposed MPP tracking algorithm is implemented by the tracking unit, which adapts to variations in ambient light intensity by adjusting its output (f_{clk}) to track the

MPP. The power converter's output current (I_O) consists of two components: I_L that powers the operation of the MPP tracking unit, and I_{EB} that represents the net harvested current into the energy buffer. If I_L is less than I_O, then there is sufficient harvested power to charge the energy buffer. It is obvious that maintaining the entire system at its maximum power point is equivalent to maximizing P_{EB}. In order to maximize P_{EB}, P_{PH} should be maximized and the power loss in the power converter and MPP tracking unit should be minimized.

B. PV Module

The electrical characteristics of a PV module can be expressed as [6]:

$$I_{PH} = I_{PH,SC} - I_{SAT}[e^{\frac{q}{AKT}(V_{PH}+I_{PH}R_S)} - 1] - \frac{V_{PH}+I_{PH}R_S}{R_P} \quad (1)$$

When the PV module does not have a load (i.e., it operates under an open circuit), I_{PH} is zero and V_{PH} is replaced by the open circuit voltage (V_{OC}). Hence, we obtain:

$$I_{PH,SC} = I_{SAT}\left\{e^{\frac{q}{AKT}(V_{OC})} - 1\right\} + \frac{V_{OC}}{R_P} \quad (2)$$

It is well-known that a stronger light irradiance results in a larger $I_{PH,SC}$ as well as a higher open circuit voltage, V_{OC}. V_{OC} can be used as an electrical parameter that indicates light intensity variation. Substituting (2) back to (1), since R_P is usually large and R_S is very small, we obtain:

$$I_{PH} \approx I_{SAT}\left\{e^{\frac{q}{AKT}(V_{OC})} - e^{\frac{q}{AKT}(V_{PH})}\right\} \quad (3)$$

As the proposed energy harvesting system is intended for indoor environments, the temperature of the PV module does not vary significantly. In the absence of significant temperature variation, I_{PH} is a function of V_{OC} and V_{PH}. For any given light irradiance (V_{OC}), V_{PH} varies with the output loading. When the output loading matches with the PV module, P_{PH} is maximized. The corresponding output voltage is denoted as V_{MPP}, at which the maximum power is extracted from the PV module.

C. Power Converter

We aim to model the electrical behavior of a charge pump in its steady state operation. A step-up charge pump operates by boosting the PV module's voltage and transferring the harvested charge/energy to a higher-voltage energy buffer. The output current for linear topology [9] or tree topology [5] charge pumps can be modeled as:

$$I_{OUT} = f_{clk}Q_{avg} = \frac{1}{N-1}f_{clk}C(NV_{IN} - V_{OUT}) \quad (4)$$

where N is the ideal voltage step-up ratio, C is the capacitance used in each stage, and f_{clk} is the switching frequency. V_{IN} and

978-1-4673-0438-2/12 $31.00 © 2012 IEEE

V_{OUT} are the charge pump's input and output voltages, respectively. The input current can be modeled as:

$$I_{IN} = NI_{OUT} + I_{CP,LOSS} = \frac{N}{N-1}f_{clk}C(NV_{IN}-V_{OUT})+f_{clk}\beta \quad (5)$$

where $I_{CP,LOSS}$ is the charge pump's internal current loss that can be modeled by $f_{clk} \times \beta$, and β is a constant parameter depending on the specific design [9]. Varying the charge pump switching frequency is equivalent to adjusting the input impedance of the power converter. Note that the input impedance of the power converter is the output loading of the PV module. Thus, for a given light irradiance (V_{OC}), there exists an optimal switching frequency, at which the input impedance of the power converter is best matched with the PV module. Hence, the PV module's MPP is realized at this switching frequency. In the hill-climbing algorithm, this optimal frequency is found by tuning the switching frequency continuously and measuring the instantaneous output power. Such a perturb-and-observe approach is effective, but not power efficient, to track the optimal frequency.

D. MPP Tracking Algorithm Derivations

Based on the electrical models of the PV module and power converter, we can derive the optimal switching frequency. As depicted in Fig. 1, when the PV module is connected with the power converter, the PV module's output current I_{PH} equals the power converter's input current I_{IN}. Further, the PV output voltage (V_{PH}) equals the power converter's input voltage (V_{IN}). Jointly considering equations (3) and (5), the charge pump switching frequency can be calculated as follows:

$$f_{clk} = \frac{I_{SAT}(e^{\frac{q}{AKT}(V_{OC})} - e^{\frac{q}{AKT}(V_{PH})})}{NC(NV_{PH}-V_{EB})/(N-1)+\beta} \quad (6)$$

This equation is also illustrated in Fig. 2 to show how V_{PH} changes with a varying switching frequency f_{clk}. When f_{clk} is zero, no charge is transferred and the input current of a power converter is zero, thus, the PV module voltage is its open circuit voltage (V_{OC1} for light irradiance 1 and V_{OC2} for light irradiance 2). When the switching frequency increases, the power converter's input impedance decreases and V_{PH} decreases, as shown in Fig. 2.

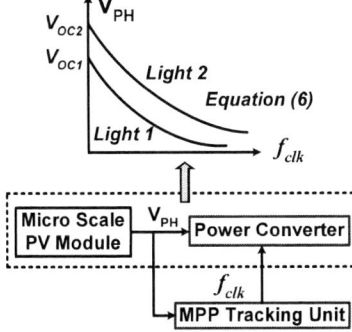

Figure 2. Joint modeling of the PV module and power converter

When V_{PH} equals V_{MPP}, the system is operating at the maximum power point. The corresponding optimal frequency ($f_{clk,optimal}$) is obtained by substituting V_{MPP} for V_{PH} in equation (6). On the other hand, prior work using the FOC approach has found that the PV module's MPP voltage is $\alpha \times V_{OC}$, where α is an almost constant parameter for a given PV module. Hence, replacing V_{OC} by V_{MPP}/α in (6) results in the equation (7), which models the desired optimal frequency for MPP operation. Note that this equation is a generic result, which is applicable for any given light irradiance (V_{OC}). If the MPP tracking unit in Fig. 1 is able to produce such a clock signal, whose switching frequency follows equation (7), the entire system will stabilize at $V_{PH} = V_{MPP}$. Fig. 3 illustrates the desired *V-f* relationship of the MPP tracking unit.

$$f_{clk,optimal} = \frac{I_{SAT}(e^{\frac{q}{\alpha AKT}V_{MPP}} - e^{\frac{q}{AKT}V_{MPP}})}{NC(NV_{MPP}-V_{EB})/(N-1)+\beta} \quad (7)$$

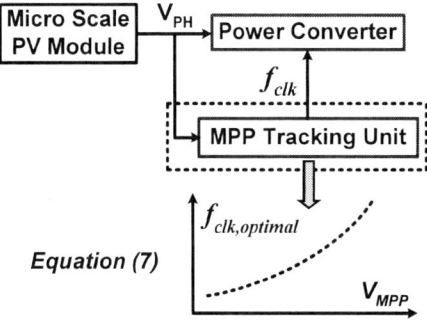

Figure 3. Behavior modeling of the desired MPP tracking unit

The closed loop tracking behavior is illustrated in Fig. 4. From the full-system perspective, negative feedback is the fundamental mechanism that enables the system to stabilize at V_{MPP}. For any given light irradiance, the relationship between V_{PH} and f_{clk} is modeled by equation (6). The MPP tracking unit exhibits the relationship between $f_{clk,\ optimal}$ and V_{MPP} according to equation (7). As a consequence, once closed loop operation starts, the only stable state that satisfies both equations (6) and (7) is the intersection point of the two curves (point A for light irradiance 1) in Fig. 4. If the light irradiance varies, the tracking algorithm spontaneously shifts its stable operating point and re-stabilizes at a new MPP point (*e.g.*, point A shifts to point B when the light irradiance varies from 1 to 2). This MPP tracking algorithm distinguishes itself from other existing tracking algorithms by its ability to thus perform automatic MPP tracking without any voltage/current sensing or additional computation.

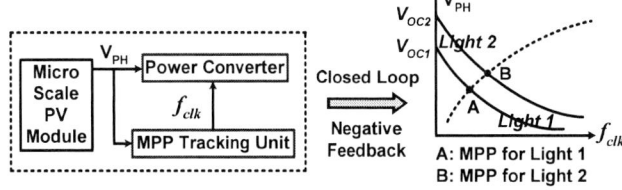

Figure 4. Proposed closed-loop MPP tracking process

978-1-4673-0438-2/12 $31.00 © 2012 IEEE

E. MPP Tracking Unit Implementation

In this subsection, we address the issue of designing a low overhead MPP tracking unit to implement the desired equation (7). From (7), it is clear that the optimal frequency varies exponentially with V_{MPP}. Note that the allowed power budget for MPP tracking in micro-scale energy harvesting systems is very limited (only several μW), hence, an exact hardware implementation of (7) is not practical, because it needs complicated circuits that will invariably involve unacceptable power loss. Therefore, it is critical to come up with a circuit structure that is simple while still being accurate. Since an exponential term can be approximately decomposed into the summation of a linear term and a quadratic term, equation (7) can be expressed as (8) for a given light intensity range. Such an approximation greatly facilitates efficient hardware implementation of equation (7).

$$f_{clk,optimal} = a \times f(V_{MPP}) + b \times f(V_{MPP}^2) \quad (8)$$

In [9], a voltage-controlled oscillator (VCO) is used to generate a switching frequency for the charge pump and its *V-f* relationship is linear. Here we add an extra transistor (M_1) to the conventional VCO and build a simple MPP tracking unit (*i.e.,* polynomial VCO) as shown in Fig. 5. The output switching frequency (f_{clk}) of this circuit is modeled in equation (9). For any given design scenario, various possible design parameters (*e.g.,* W/L of transistor M_1, R_{BIAS}) are substituted into the equation (9) and compared with the equation (7) for evaluating the approximation. Note it is also required to substitute V_{MPP} into V_{PH} in equation (9). Once the proper design parameters are found, the design process for the tracking unit is complete. The accuracy of this approximation will be verified in Section IV. In terms of hardware cost, the proposed approach only employs a polynomial VCO, while the design of [9] makes use of a linear VCO, an output current sensor, and a decision generation circuit. Hence, it is evident that the proposed tracking unit is simpler, leading to a lower power overhead. Note that the VCO's phase noise does not affect the MPP tracking in such energy harvesting systems [16].

$$f_{clk} = \frac{I_{CHARGE}}{2V_M C_S} = \frac{\frac{1}{2}\mu C_{OX}(\frac{W}{L})_{M1}(V_{PH}-V_{TH})^2 + \frac{V_{PH}}{R_{BIAS}}}{2V_M C_S} \quad (9)$$

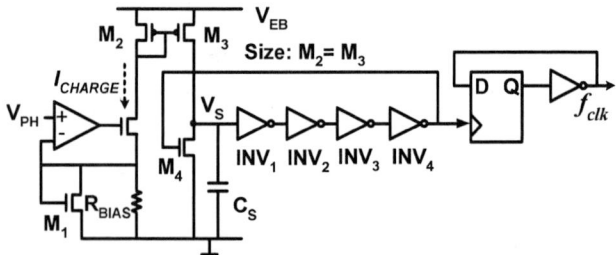

Figure 5.   The proposed MPP tracking unit implementation

IV.   SIMULATION RESULTS

In order to verify the proposed MPP tracking algorithm, we designed and implemented a micro-scale energy harvesting system using IBM 45nm technology. The layout of the entire design is shown in Fig. 6, occupying an area of 990μm$\times$620μm. This design is targeted at small form-factor wearable applications, which can only use a very tiny (sub cm^2) single-junction PV module and operate in indoor weak light conditions most of the time. Hence, a 0.64cm^2 single junction PV cell is used as the energy transducer. Its open circuit voltage is 0 - 0.45V under indoor weak irradiance. A tree topology charge pump [5] was implemented as the power converter and its ideal voltage conversion ratio is N=4. A 1μF capacitor with an initial voltage of 0.9V was used to mimic the energy buffer. HSPICE post-layout simulations were carried out using BSIM models.

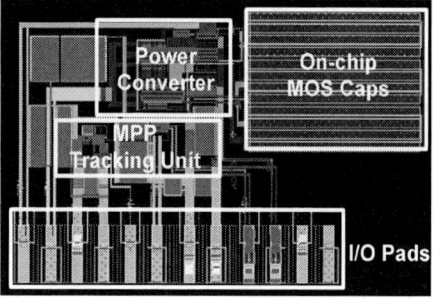

Figure 6.   Layout of the fabricated energy harvesting system

Our first simulation explored the relationship of the harvested power P_{EB} and the charge pump switching frequency f_{clk}. We disabled the MPP tracking unit and applied a varying switching frequency from an external source. Four different light intensities were simulated and the simulation results are plotted in Fig. 7. For a given light intensity, the harvested power P_{EB} changes with the applied charge pump frequency and there exists a peak power point at its optimal frequency $f_{clk,optimal}$. This figure also shows that $f_{clk,optimal}$ varies within a wide range (*e.g.,* 6.25MHz for 784LUX and 11.11MHz for 2152LUX) and validates the necessity for ultra-low power MPP tracking at runtime.

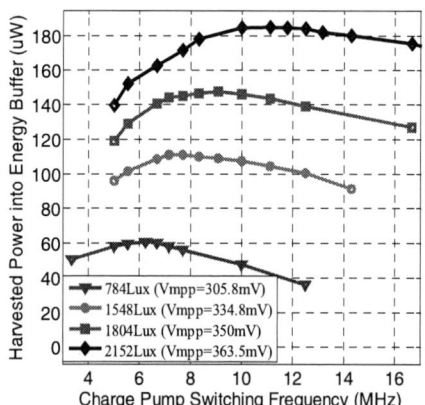

Figure 7.   Harvested output power vs. charge pump switching frequency

Our second simulation investigated the V-f relationship of our proposed tracking unit (*i.e.*, polynomial VCO). We disconnected the MPP tracking unit from the rest of the system and studied its behavior separately. By varying its input voltage, the output clock frequency of the tracking unit was obtained and plotted in Fig. 8 (shown by the dashed line). The results for V_{MPP} and $f_{clk,optimal}$ from the previous simulation are also plotted (solid line) for comparison. We can see that the proposed MPP tracking unit approximates the desired V-f curve well. The disparity between the two curves is only 2.45% on average.

Note that the proposed MPP tracking unit in Fig. 5 could be influenced by process variations, which makes circuit parameters such as transistor threshold voltage unpredictable. At different process corners, the V-f curve produced by the proposed MPP tracking unit drifts away from the expected values in a typical process corner. Although compensation techniques (*e.g.*, adaptive body biasing [14] [15]) may be applied to the proposed hardware structure to create a process variation tolerant MPP tracking unit, these methods increase design complexity and power overhead. To mitigate the effect of process variation without too much overhead, we applied adaptive transistor sizing to M1 (*i.e.*, M1 is implemented by a group of parallel transistors with different sizes and the actual size of M1 is determined by controlling the number of parallel conduction paths). Circuit simulation shows that the average disparity between the desired curve and the proposed approximation for different process corners is 4.57%. This validates our design decision to approximate the desired exponential equation (7) with a polynomial equation (9).

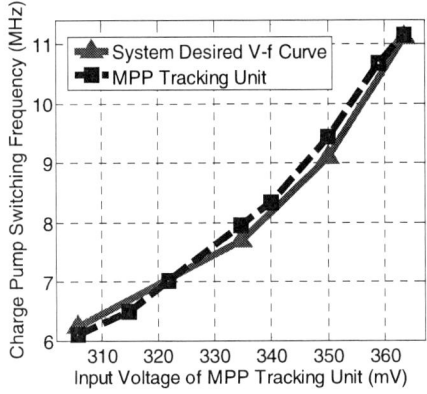

Figure 8. V-f relationship of the proposed MPP tracking unit

Our third simulation was to demonstrate the closed loop tracking performance of our approach. We enabled the MPP tracking unit and allowed the system to automatically track the MPP. The final stable voltages are recorded and summarized in Table I. The desired MPP voltages, obtained from Fig. 7, are also included in Table I. We can observe that the tracking error percentages are small (less than 1%) and the real harvested power from the PV module is at most only 0.2% less than the maximum available power.

TABLE I.   STEADY STATE MPP TRACKING PERFORMANCE

Lighting	System Desired Voltage	Our Tracked Voltage	Percentage Difference	Harvesting Efficiency ($P_{PH}/P_{PH,MAX}$)
784LUX	305.8mV	305.2mV	0.2%	99.9%
1548LUX	334.8mV	333.3mV	0.15%	99.9%
1804LUX	350mV	347.6mV	0.69%	99.8%
2152LUX	363.5mV	362.6mV	0.25%	99.9%

Simulation was also carried out to validate the dynamic tracking (*i.e.*, negative feedback) process. In this simulation, the light irradiance was fixed at 2152LUX. The previous simulations have shown that, for this light intensity, the desired MPP voltage is 363.5mV and the optimal frequency is 11.11MHz. V_{PH} is initially set at 390mV, which is higher than the optimal voltage (363.5mV). Fig. 9 shows the simulation waveforms for this tracking process. It is clear that V_{PH} gradually drops and moves towards the tracked stable voltage (362.6mV), demonstrating convergence of our negative feedback loop. The charge pump switching frequency f_{clk} also decreases towards the desired stable value. At T=4.5μs, V_{PH} (363.4mV) is already very close to its final stable value (362.6mV) and the tracking is almost complete. Fig. 10 demonstrates the simulated tracking process when the initial voltage V_{PH} (320mV) is below the desired value. Both figures show that the proposed tracking scheme is robust and that the tracking time is in the range of only a few microseconds.

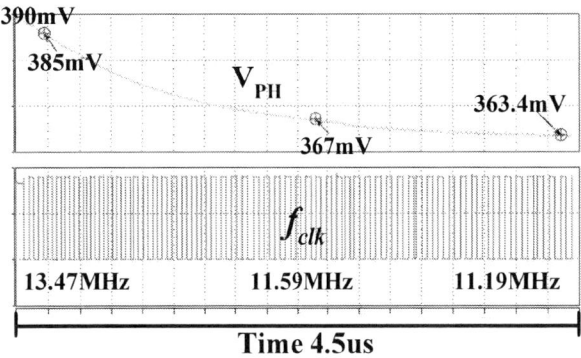

Figure 9. Dynamic tracking process (initial V_{PH} above V_{MPP})

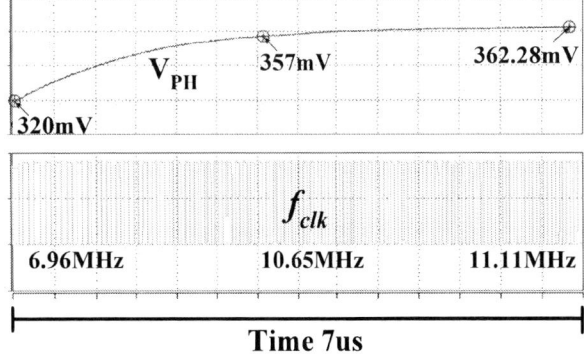

Figure 10. Dynamic tracking process (initial V_{PH} below V_{MPP})

Table II summarizes the simulation results of full-system power harvesting performance. Full-system power harvesting efficiency is defined as the ratio between the actual net-harvested power (P_{EB}, the power flowing into the energy buffer after all losses are considered) and the maximum available power ($P_{PH, MAX}$) for any given light irradiance. Here $P_{PH, MAX}$ was calculated based on equation (1). The actual harvested power P_{EB} is obtained from simulations. Table II shows the power harvesting efficiency is at least 30% for a wide range of weak light irradiance. Table II also reveals the average power loss of the entire energy harvesting system is only 180 μW.

TABLE II. SYSTEM POWER HARVESTING PERFORMANCE

Lighting	Maximum P_{PH} Available (μW)	P_{EB} (μW)	Power efficiency ($P_{EB} / P_{PH,MAX}$)
784LUX	140.72	46.65	33.15%
1548LUX	250.71	95.29	38%
1804LUX	333.8	131.87	39.51%
2152LUX	443.78	169.98	38.3%

Table III summarizes and compares recent related approaches to MPP tracking in micro-scale solar energy harvesting systems. As discussed before, it is crucial to perform MPP tracking while ensuring minimal hardware and power overhead. The design in [7] adopts the FOC tracking method and uses an additional PV module as a pilot cell to sense the variation of light intensity. In [8], an MCU is used to work with current/voltage sensors to implement hill climbing tracking algorithm. Then, in order to further reduce the hardware cost and power consumption, the authors of [9] implement a hill-climbing algorithm by replacing the MCU with a custom-designed current sensor and decision generation block. While this decreases the overhead, it still suffers the overhead of this current sensor and decision block. The work proposed in this paper provides a low-cost and energy-efficient solution for MPP tracking in micro-scale solar energy harvesting systems. The proposed tracking algorithm is based on system/circuit behavior analysis and derivations, and only uses a polynomial VCO without any additional sensing or computation. Thus, it results in lower hardware overheads compared to existing work.

TABLE III. QUALITATIVE OVERHEAD COMPARISON WITH EXISTING WORKS

	MPP Tracking Approach	Overheads
[7]	Fractional open circuit (FOC)	Additional PV module used as a pilot cell
[8]	Hill climbing algorithm	Current/Voltage sensor and MCU
[9]	Hill climbing algorithm	Current sensor, decision generation block, linear VCO
This Work	Negative-feedback automatic tracking	Polynomial VCO (without any sensor or decision block)

V. CONCLUSION

This paper presented a new maximum power point tracking algorithm for micro-scale solar energy harvesting systems. The proposed algorithm is based on a careful joint analysis and modeling of the behavior of the PV module and the power converter. By modeling the desired optimal frequency for MPP

and using a simple circuit structure for implementing it, the proposed approach avoids the need for additional hardware such as voltage or current sensors, microcontrollers, etc. Based on the proposed MPP tracking approach, a fully on-chip micro-scale solar energy harvesting system was designed. Simulation results verified the proposed MPP tracking scheme and demonstrated that the scheme is robust with a tracking time of only a few microseconds.

ACKNOWLEDGMENT

This research was funded in part by the National Science Foundation under grant CCF-1018358. The opinions expressed here represent those of the authors and not necessarily of NSF.

REFERENCES

[1] A. P. Chandrakasan, N. Verma, and D. C. Daly, "Ultra low power electronics for biomedical applications," *Annual Review of Biomedical Engineering*, vol. 10, pp. 247-274, August 2008.

[2] L. Mateu and F. Moll, "Review of energy harvesting techniques and applications for microelectronics," *Proc. SPIE Microtechnologies for the New Millennium*, pp. 359-373, 2005.

[3] V. Raghunathan and P. H. Chou, "Design and power management of energy harvesting embedded systems," *Proc. International Symposium on Low Power Electronics and Design*, pp. 369-374, 2006.

[4] Optobionics Corp. (http://optobionics.com/)

[5] C. Lu, S. P. Park, V. Raghunathan, and K. Roy, "Efficient power conversion for ultra low voltage micro scale energy transducers," *Proc. Design, Automation and Test in Europe*, pp. 1602-1607, 2010.

[6] C. Lu, V. Raghunathan, and K. Roy, "Maximum power point considerations for micro-scale solar energy harvesting systems," *Proc. International Symposium on Circuits and Systems*, pp. 273-276, 2010.

[7] D. Brunelli, C. Moser, and L. Thiele, "An efficient solar energy harvester for wireless sensor nodes," *Proc. Design, Automation, and Test in Europe*, pp. 104-109, 2008.

[8] E. Koutroulis, K. Kalaitzakis, and N. C. Voulgaris, "Development of a microcontroller-based, photovoltaic maximum power point tracking control system," *IEEE Trans. on Power Electronics*, vol. 16, no. 1, pp. 46-54, January 2001.

[9] H. Shao, C. Y. Tsui, and W. H. Ki, "A micro power management system and maximum output power control for solar energy harvesting applications," *Proc. International Symposium on Low Power Electronics and Design*, pp. 298-303, 2007.

[10] M. Chen and G. A. Rincon-Mora, "Single inductor multiple input multiple output (SIMIMO) power mixer-charger-supply system," *Proc. International Symposium on Low Power Electronics and Design*, pp. 310-315, 2007.

[11] E.O. Torres and G.A. Rincon-Mora, "Electrostatic energy-harvesting and battery-charging CMOS system prototype," *IEEE Trans. on Circuits and Systems I*, vol. 56, no. 9, pp. 1938-1948, September 2009.

[12] E. Carlson, K. Strunz, and B. Otis, "A 20mV input boost converter with efficient digital control for thermoelectric energy harvesting," *IEEE J. Solid-State Circuits*, vol. 45, no.4, pp. 741-750, April 2011.

[13] E. Esram and P. L. Chapman, "Comparison of photovoltaic array maximum power point tracking techniques," *IEEE Transactions on Energy Conversion*, vol. 22, no. 2, pp. 439-449, June 2007.

[14] J. W. Tschanz, S. Narendra, R. Nair and V. De, "Effectiveness of adaptive supply voltage and body bias for reducing impact of parameter variations in low power and high performance microprocessors," *IEEE J. Solid-State Circuits (JSSC)*, pp. 826-829, May 2003.

[15] K. Kang, S. P. Park, K. Kim, and K. Roy, "On-chip variability sensor using phase-locked-loop for detecting and correcting parametric timing failures, " *IEEE Transactions on Very Large Scale Integration (VLSI) Systems*, vol. 18, no. 2, pp.270-280, February 2010.

[16] H. Shao, C. Y. Tsui and W. H. Ki, "The design of micro power management system for applications using photovoltaic cells with the maximum output power control," *IEEE Trans. VLSI Systems*, vol. 17, pp. 1138-1142, August 2009.

2012 25th International Conference on VLSI Design

Hybrid NEMS-CMOS DC-DC converter for improved area and power efficiency

Sujan K. Manohar*, Ramakrishnan Venkatasubramanian*† and Poras T. Balsara*
* VLSI Circuits and Systems Laboratory, University of Texas at Dallas, Richardson TX 75080
†Texas Instruments Inc, Dallas TX 75243
Email: {sujan.manohar, ramav, poras}@utdallas.edu

Abstract—Nano-electromechanical (NEM) relays are a promising class of emerging devices that exhibit zero leakage operation. Numerous end applications of NEM relay logic circuits have been proposed recently [1][2]. This work explores the usage of NEM relays in on-chip DC-DC converters. As a feasibility study of using NEMS in integrated power electronics, discontinuous conduction mode (DCM) buck regulator with specifications suitable for portable applications has been implemented in a NEMS-CMOS hybrid design and the results are compared against a standard commercial $0.35\mu m$ CMOS implementation. R_{on} of the NEM relay switch is constant and is insensitive to the gate slew rate. This creates a paradigm shift in design of power switches. This coupled with infinite R_{off} offers significant area and power advantages over CMOS. Accurate Verilog-A models were developed based on published fabrication results of NEM relays [1] operating at 1V with a nominal air gap of $5 - 10nm$. This work shows that NEMS-CMOS hybrid DC-DC converter has an area savings of $60X$ over CMOS and achieves 95% efficiency at max load condition ($50mA$).

I. Introduction

Nanoelectromechanical (NEM) Relays are 4 terminal mechanical devices that are electrostatically actuated. When the mechanical switch is turned ON, a metal channel creates a conducting path between source and drain. When the switch is OFF, there is no drain-source current and hence the leakage through the device is zero. NEM relay behaves like an ideal switch. The electrostatic actuation of the mechanical switch results in a mechanical delay through the relay, which is orders of magnitude more than the electrical delay of the logic implemented using the relays. Numerous NEM relay implementations have been proposed recently that show significant energy efficiency improvements over CMOS circuits while operating at low frequencies [1][2].

A cantilever beam based NEM relay has been reported in [3], which has been further improved to a suspended gate NEM relay in [1]. A laterally actuated NEM relay device is reported in [2] in which a poly-silicon beam is laterally actuated to realize the mechanical switch. Carbon-nanotube based NEM relays are reported in [6]. Even though this work is based on suspended gate NEM relays, the concept can be extended to any relay technology that provides a resonably low R_{on}.

As a feasibility study on using of NEMS over CMOS in integrated power management, a DC-DC converter with identical specifications was designed in both CMOS and NEMS-CMOS hybrid technologies. The motivation to use

a DC-DC converter for this analysis is that there is an increasing demand to generate highly efficient on-chip supply voltages with good area-performance trade-off. Voltage regulation is required for supply desensitization of circuit-performance in almost all portable telecommunication and computer products that require higher current efficiency. As the result, the power efficiency is an important parameter that determines the longevity of a battery [4]. A buck converter is a step-down DC to DC converter. Switched mode buck regulators was chosen for this exploration since they are widely used to perform DC-DC conversions in battery-operated portable devices.

Another key motivation is that power MOSFET occupies a significant portion of the total area in any power management integrated circuit. Any potential area reduction arising from the hybrid design methodology would result in considerable cost savings. A comparison of NEMS vs CMOS charge pump shows that NEMS relays have significant area benefits over CMOS [5].

This work focuses on study of a step-down DC to DC converter using cantilever beam or suspended gate NEM relays. The device properties assumed and the design approach are generic enough that they can be extended to other relay technologies like the lateral actuation relay [2] and the carbon-nanotube based relay [6] as well.

II. Background

A. NEM Relay device

An electrostatically actuated cantilever beam that functions as a relay is described in [3]. This cantilever beam when actuated, touches the source and drain electrodes and creates a channel for conduction. An improved four-terminal NEM relay device has been reported by [1], which incorporates a movable poly-SiGe gate suspended by spring-like folded flexures above tungsten electrodes. A thin coating of titanium oxide is applied to the device to improve the reliability of the suspended gate. Relays operating at $10V$

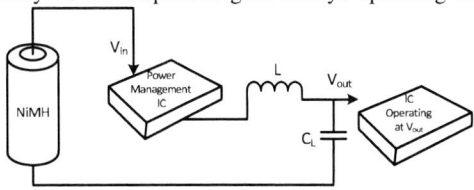

Figure 1. Typical Buck converter usage

978-1-4673-0438-2/12 $31.00 © 2012 IEEE

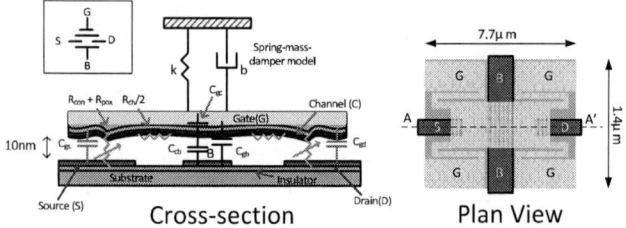

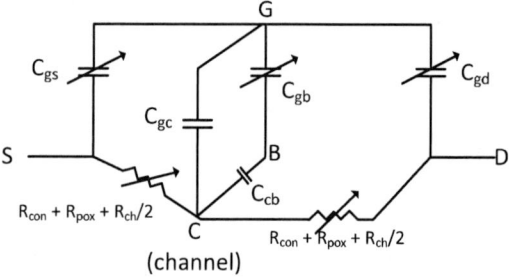

Figure 2. Suspended gate relay reported in [1]. NEM Relay symbol shown in the inset

Figure 3. Verilog-A electrical model of Suspended gate relay

with a $90nm$ air gap between the drain/source dimples and the suspended gate have been fabricated and reported. This work uses a scaled NEM relay device that has device dimensions scaled down to a 90 nm technology, operating at $1V$ with a nominal gap of $5\text{-}10nm$ between the gate and drain/source as reported by [1].

Fig 2 shows the device construction and basic electrical and mechanical elements in the relay circuit. The suspended gate is electrostatically actuated whenever there is a voltage between the gate and the base. When the electrostatic force is strong enough to overcome the spring-mass-damper system, the channel electrode touches the dimples on source and drain, thereby creating a metal channel between source and drain. When the voltage between gate and base is lowered, the electrostatic force reduces and will eventually open the relay once it crosses a release threshold voltage. Since there is no metal channel between source and drain when the relay is OFF, there is no leakage current.

B. CMOS-NEMS Hybrid design

In this NEMS based power electronics exploratory work, a hybrid CMOS-NEMS process is assumed. In the hybrid architecture, the power transistors and the associated power stage logic is architected using NEM relays and the feedback loop and gain in the system is architected using CMOS. In order to integrate NEM relays in a CMOS process, a method to post-fabricate the relays on top of the CMOS chip or integrate the relay into the backend metallization layers with no penalty in the overall die area is proposed in [7]. Since NEM relays can be fabricated using a low-temperature process,[8] proposes that they can be monolithically integrated on top of CMOS circuits using a back-end of line (BEOL) CMOS process (i.e., processing of all metal interconnects, vias and interlayer dielectric). Hence, NEM relays are proposed to be placed on top of CMOS transistors thereby resulting in substantial reduction in the footprint area of the chip.

Though promising, lots of challenges remain to be solved before NEM relays can be integrated with CMOS process. More research and experiments are needed to understand the manufacturability of NEM relays on top of CMOS, as well as the associated process costs, yield, and testing costs.

C. Definition of NEM Relay parameters

Pull-in voltage V_{pi} is the voltage applied between gate and base above which the electrostatic force overcomes the

spring-mass-damper system and the relay turns ON. Release voltage V_{rl} is defined as the voltage below which the relay opens and the switch is OFF. The NEM relay exhibits a hysteritic property for the pull-in and release threshold voltages. The pull-in voltage (V_{pi}) is larger than the release voltage (V_{rl}). The mechanical delay involved in switching ON the relay is denoted as t_{mon} and the mechanical delay involved in switching OFF the relay is denoted as t_{moff}. The mechanical delay is an order of magnitude larger than the electrical delay of the relay (t_e). As an example, the mechanical delay of the suspended gate relay used in this work is of the order of hundreds of nanoseconds, whereas the electrical delay is of the order of tens of picoseconds.

III. VERILOG-A MODEL OF SUSPENDED GATE RELAY

An accurate Verilog-A model of the $1V$ scaled device has been developed based on fabrication results reported in [1]. A spring-mass-damper system models the mass of the suspended gate and flexures. At any instant in time, the electrostatic force (F_{elec}) balances with the spring-mass-damper system. The force vector equation is expressed as follows:

$$m\ddot{x} = F_{elec}(x) - b\dot{x} - kx \qquad (1)$$

where, x is the displacement of the gate, m is the mass of the suspended gate, k is the spring constant of the gate structure, b is the damping coefficient of the motion of the gate.

The electrostatic force F_{elec} is expressed as:

$$F_{elec} = \frac{\epsilon_0 A_{ov} V_{gb}^2}{2(g_0 - x)^2} \qquad (2)$$

where ϵ_0 is the permittivity of free space, V_{gb} is the voltage between gate and body, A_{ov} is the area of overlap between gate and body electrodes and g_0 is the normal gap between electrodes when switch is OFF.

The voltage necessary to turn on the relay is the pull-in voltage and is derived to be

$$V_{pi} = \sqrt{\frac{8kg_0^3}{27\epsilon_0 A_{ov}}} \qquad (3)$$

The mechanical delay (t_m) is inversely proportional to the gate overdrive voltage ($|V_{gb}|/V_{pi}$) and the undamped angular

978-1-4673-0438-2/12 $31.00 © 2012 IEEE

frequency of the spring-mass-damper system $\sqrt{k/m}$. This is denoted by:

$$t_m \propto \sqrt{\frac{m}{k}} \frac{V_{pi}}{|V_{gb}|} \qquad (4)$$

The Verilog-A model of the NEM relay takes into account all the mechanical and electrical effects according to the above equations and incorporates the electrical model shown in Fig. 3. The model covers the self actuation effect but does not cover electro-thermal or mechanical-thermal effects. Also, Van der Waals' force on the suspended gate is not modeled.

C_{gb} is the gate to base capacitance. C_{gs} and C_{gd} are the gate-source and gate-drain capacitances respectively. Since the gate forms a true parallel plate capacitor with base and source/drain, these three capacitances contribute to bulk of the intrinsic capacitance in the device. Since the distance between the parallel plates is a variable, C_{gb}, C_{gs} and C_{gd} are variable capacitors in the model. C_{cb} is the channel to base capacitance and is insignificant. R_{con} is the contact resistance. R_{pox} is the resistance of the passivating oxide and R_{ch} is the channel resistance. The R_{on} of the device is about $1k\Omega$. The R_{off} of the device is infinity – due to absence of any current through the relay when it is OFF.

IV. CMOS BUCK REGULATOR DESIGN

A. Buck regulator specifications

Switched-mode buck regulators are widely used to perform DC-DC conversions in battery-operated portable devices. Maximizing the power efficiency, especially under light-load condition and minimizing the size of inductor are two major important requirements in buck regulators. Discontinuous-conduction mode (DCM) enables the buck regulator to use a small-value inductor, even at low switching frequency, thereby reducing the switching power loss and having the potential to achieve high light-load power efficiencies. The power efficiency of a DCM buck regulator depends on the controlled switching timing of power transistors. Both the CMOS and NEMS-CMOS hybrid buck regulators are designed for identical specifications shown in Table. I.

B. Architecture

The CMOS DCM buck regulator architecture described in [9] is used as the reference CMOS architecture for this exploratory work. Fig. 4 shows the synchronous-rectifier-based DCM buck regulator. The PMOS transistor M_p is the power transistor. The NMOS transistor M_n doubles up as the forward-biased diode that helps in minimizing the

Table I
BUCK REGULATOR SPECIFICATIONS

Output Voltage	1.5 V
Input Voltage range	1.8 V - 2.4 V (two-cell NiCd batteries)
Load current	50mA
Output Voltage Ripple (max)	40mV
Inductor L/Capacitor CL	1uH/6.8 uF

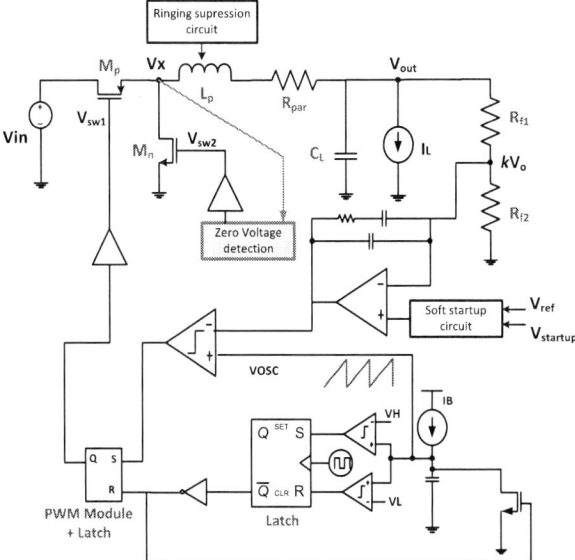

Figure 4. CMOS DCM Buck regulator architecture

Table II
CMOS SWITCH SIZES

Switches	W (m)	R_{on}(formula)
PMOS power switch (M_p)	40e-3	0.13 Ω
NMOS switch (M_n)	21.7e-3	0.08 Ω
PMOS Buffer	6.67e-3	0.26 Ω
NMOS Buffer	6.67e-3	0.26 Ω
Ring switch	5.00e-3	3.51 Ω

conduction power loss during the fraction of the switching period when it is conducting.

The operation of the CMOS buck regulator in discontinuous conduction mode is shown in Fig. 5. The buck converter goes through three distinct phases of operation as described in [10]. In Phase-1, the PMOS switch M_p is ON and the NMOS switch M_n is OFF. A current rise takes place across the inductor L_p. During the dead-time phase, the PWM modulator the generates V_{sw1} opens M_p. To fight against its collapsing magnetic field, the inductor reverses its voltage

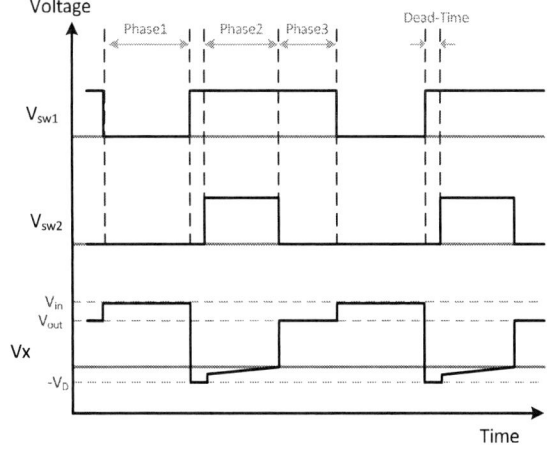

Figure 5. CMOS DCM Buck regulator operation

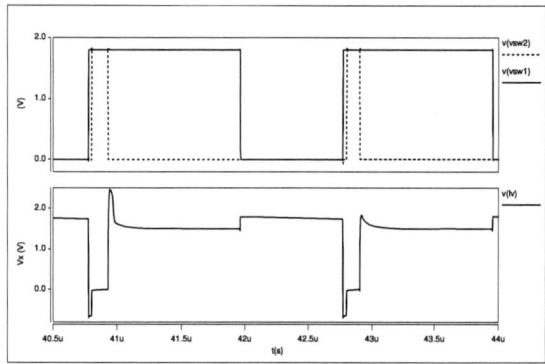

Figure 6. CMOS DCM Buck regulator waveform

and since the inductive current still needs to flow somewhere in the same direction, the body diode in M_n gets activated. The body diode helps in clamping the negative voltage at node V_X to $-V_D$. This duration when both M_p and M_n are OFF is denoted as dead-time in the buck converter. In phase-2, the NMOS M_n is turned ON. This pulls the node V_X up to ground. Minimizing the dead time is important to reduce the conduction power loss. In phase-3, both M_p and M_n are turned OFF. During this time, the node V_x will have ringing due to the parasitic capacitances. In order to suppress this ringing, a bypass path around the inductor L_p is activated in this phase. This ensures that the voltage at node V_X is same as the output voltage V_o. The cycle is then repeated starting with phase-1.

To ensure proper synchronous rectification, an adaptive dead-time control and a ringing suppression circuit are added [9]. These circuits help minimize the power losses due to inadequate switching timing of power transistors and electromagnetic interference (EMI) caused by the ringing at node V_X. The steady-state SPICE waveform showing the buck converter operation is shown in Fig.6.

V. NEMS-CMOS HYBRID BUCK REGULATOR

A. NEM-CMOS hybrid Buck regulator architecture

In a typical power management chip, the power transistors consume about 60%-70% of the total chip area. NEM relay based switches have low ON resistance ($\approx 1k\Omega$). From an area standpoint, the power transistors are ideal candidates for replacement with NEM relay devices. Numerous parallel relays can be added to support the total current that has to be drained through the power stage. DCM regulators typically have a low speed of operation($\approx 500kHz$). This is advantageous for the NEMS power stage since mechanical delay of the NEM relay devices is large and it cannot be used in high speed designs (1MHz+). Also, NEM relays can provide a more energy efficient implementation for low frequency of operations. For 500kHz operation of the DC-DC converter, the power stage can definitely be replaced with NEMS relay devices.

The proposed NEMS-CMOS hybrid buck regulator architecture is shown in Fig. 7. The power MOSFET M_p in

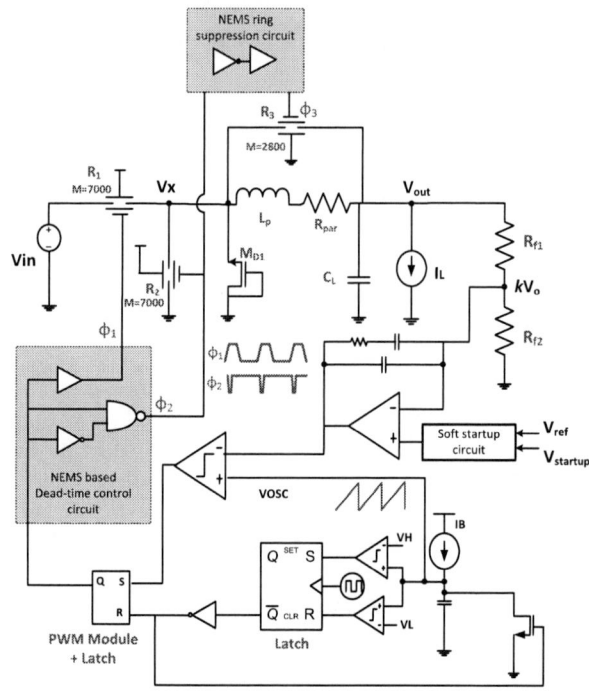

Figure 7. NEMS-CMOS Hybrid Buck regulator architecture

Fig. 4 is replaced with numerous parallel relays (collectively denoted as R_1) to support the same current density. Note that the base of the device is tied to V_{dd}. This configuration mimics the operation of a PMOS transistor.

The NMOS transistor M_n is replaced with a set of parallel relays R_2. This relay takes care of pulling node V_X to ground in phase-2. However, during the dead time, the relay does not double up as a body diode to arrest the negative swing on node V_X. So another NMOS device M_{D1} connected in diode configuration is added in parallel to relay R_2. Note that usage of relay R_2 is advantageous in area only

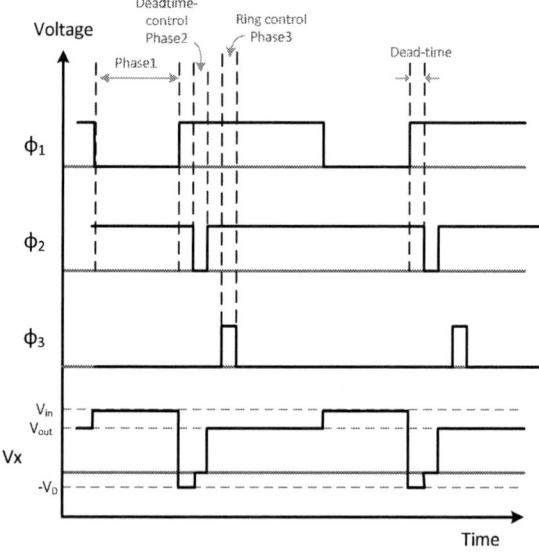

Figure 8. NEMS Buck regulator operation

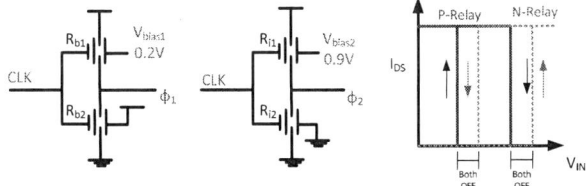

Figure 9. NEM Buffer inverter biasing to achieve non-overlapping clocks

if the combined area of the relays in R_2 and the NMOS transistor M_{D1} is less than the area of transistor M_n in the CMOS implementation. Since M_{D1} is configured as a diode, it need not present an ideal ground to node V_X and is sized to just sink the peak inductor current. Hence the area of transistor M_{D1} is much smaller as opposed to using an NMOS transistor as the low side switch. The operation of the hybrid NEMS-CMOS buck regulator is shown in Fig. 8.

B. NEMS based dead-time control circuit

The control for relays R_1 and R_2 is generated by a NEMS based pulse generation logic. The signals ϕ_1 and ϕ_2 are generated such that ϕ_2 is a falling pulse generated off of the rising edge transition of signal ϕ_1. So, as soon as the relay R_1 is OFF, the node V_X will start going negative. It will be clamped at $-V_D$ by the NMOS diode and the pulse generated (ϕ_2) will activate relay R_2, which in turn will pull node V_X back to zero. The idle time in this configuration is one mechanical delay – the delay incurred in the pulse generation logic. The idle time in the NEMS implementation is of the order of 40ns (comparable to the 20ns idle time in CMOS implementation [9]).

C. NEMS based ring suppression circuit

The ring suppression circuit, which essentially is a bypass of the inductor to get node V_X same as V_{out}, needs to be activated in phase-3. A NEMS based ring suppression logic generates signal ϕ_3, which is a delayed inverted version of signal ϕ_2. Note that if both relays R_2 and R_3 are ON at the same time, it will result a path from V_{out} to ground and should be avoided. So ϕ_2 and ϕ_3 are architected to be non-overlapping using relay biasing.

D. NEM Driver design

An important property of the NEM relay device is that R_{on} is constant (sum of channel and contact resistances) with comparatively low self-loading capacitance as opposed to CMOS switches, where, resistance varies with gate-source voltage (input gate slew rate). This fact simplifies the design of NEM relay drivers greatly. Unlike in CMOS technology, NEM relay drivers need not maintain the sharp slew rates across its internal nodes to curtail shoot-through current and guarantee constant R_{on} for the power switch. This property is utilized in the design of NEM power switches to achieve substantial area and power savings. One NEM relay buffer (two relays) can comfortably drive 1000+ NEM

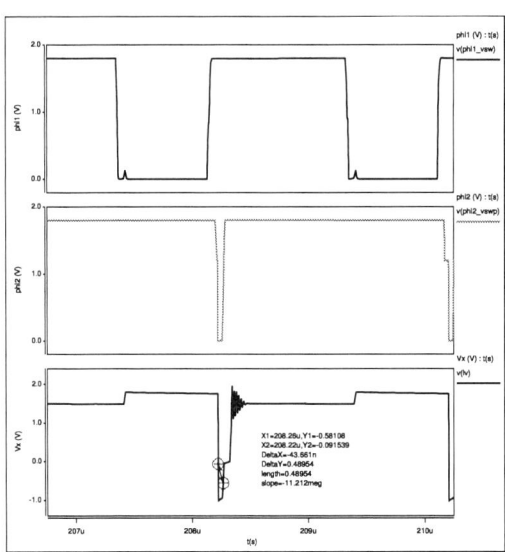

Figure 10. NEMS Buck regulator waveform

relay gate terminals. So the area of the buffer logic reduces significantly when implemented with NEM Relays.

E. Relay biasing

In order to generate non-overlapping clocks inside the deadtime control circuit, relay biasing technique [11][5] is used. When a relay is biased to a voltage above 0, equally more gate voltage is necessary to turn on the mechanical switch. So the relay will turn on slowly. This principle was used to create non-overlapping clocks with a separation of 30 ns (more than the dead time in the worst process corner). Fig. 9 shows the biasing of clock buffer and inverter to guarantee non-overlapping clocks.

F. Determination of number of parallel relays in R_1 and R_2

The width of each of the transistors in the power stage of the CMOS design is shown in Table. II. The ON resistance (R_{on}) of the each of the NEM relay devices is $1k\Omega$. The number of parallel relays required to support an R_{on} identical to Table. II determines the total number relays required in the hybrid NEMS-CMOS architecture. The number of relays required for the buffer is 16 and is negligible. Relay R_1 (PMOS power switch) requires 7000 relays. Relay R_2 (NMOS power switch) requires 7000 relays. Relay R_3 (Ring switch) requires 2800 relays. Total number of relays required for the proposed implementation is ≈ 16800.

VI. SIMULATION SETUP AND RESULTS

A. CMOS design simulations

The CMOS DCM buck regulator with the adaptive dead-time control and the dead-time-aware ringing suppression circuit has been implemented in a standard $0.35\mu m$ CMOS process. The buck regulator is soft-started to prevent large in-rush current. The buck regulator operates at a switching frequency of about 500 kHz and can provide a regulated output voltage of 1.5 V with a maximum load current of 50

Table III
CMOS VS HYBRID NEMS-CMOS BUCK CONVERTER COMPARISON

	$0.35\mu m$ CMOS	NEMS	Improvement
Area	$3.3mm^2$	$0.20mm^2$	60X
Efficiency (I_L=5mA)	83.3%	83.3%	
Efficiency (@ max load = 50mA)	91%	95%	

mA. The maximum output ripple voltage of the DCM buck regulator is about 40 mV.

B. NEMS-CMOS hybrid simulations

Accurate Verilog-A model of the NEM relay based on published fabrication results was used to model the NEM relay devices. The relay operates at 1V with $V_{pi} = 0.6V$, $V_{rl} = 0.55V$, $t_{mon} = 130ns$ and has an area of $12\mu m^2$ [1]. For an overdrive voltage of 1.8V, the mechanical delay is $t_{mon} = 40ns$. Larger the overdrive voltage, smaller the mechanical delay. However, this speedup is achieved at the expense of device reliability [12]. Ideally a NEMS relay optimized for 1.8V should be used in the DC-DC converter. The hybrid NEMS-CMOS simulation was run in HSPICE. The verilog-A model incorporates all the resistance and capacitances in the device. The efficiency of the buck converters is calculated from the power numbers reported by the Verilog-A simulations. The steady-state SPICE waveform showing the operation of the hybrid buck converter is shown in Fig.10.

C. Results

1) Area comparison of the power stage: The power stage in the CMOS buck converter requires an area of $\approx 3.3mm^2$. This is calculated using the length and overhang distances for $0.35\mu m$ technology and multiplying it by the total width of all the power transistors. Each NEM relay requires an area of $12\mu m^2$ and the total number of NEM relays required is 16800. Hence the total NEMS power stage area is $0.20mm^2$. This results in a 60X area reduction in the power stage when implemented using the hybrid approach.

2) Power efficiency comparison: The efficiency comparison between the CMOS and hybrid buck regulators is shown in Table. III. The no-load efficiency is almost the same between the two regulators since it is $\approx V_{out}/V_{in}$. In max load condition, the efficiency of the hybrid regulator is better than CMOS. Typically, the buffer efficiency is a significant contributor in determining the overall efficiency of the CMOS regulator. In NEM based design, the power switches do not need large number of buffers [5] and the buffers themselves do not have any leakage and shoot-through current due to inherent hysterisis, thereby having an efficiency of $\approx 100\%$. Hence the hybrid design has a higher efficiency of 95% at the max load condition as compared to the CMOS efficiency of 91%. The NEM power switch offers resistive impedence and hence does not modify the quality of current through the DC-DC converter.

VII. CONCLUSION

CMOS technology scaling has reached its limit on sub-threshold performance to the point that further reduction in the threshold voltage would actually increase the amount of energy consumed per operation. Threshold voltage of transistors has already been scaled to a value that balances leakage energy and dynamic energy optimally. Further, there is not much supply voltage scaling expected in smaller technology nodes due to the limits set by kT/q. So the need for new devices with steeper subthreshold slope like NEM relay is necessary to enable design of energy efficient circuits. This work explores usage of NEM relays in integrated power electronic circuits. It shows that NEM relay based hybrid DCM buck converter achieves $60X$ area reduction as compared to $0.35\mu m$ CMOS design with identical specifications. The efficiency of the NEMS based hybrid buck regulator is 95% at max load condition.

REFERENCES

[1] Spencer, M.; et.al.; , "Demonstration of Integrated Micro-Electro-Mechanical Relay Circuits for VLSI Applications," Solid-State Circuits, IEEE Journal of , vol.46, no.1, pp.308-320, Jan. 2011

[2] Soogine Chong, et.al.; , "Nanoelectromechanical (NEM) relays integrated with CMOS SRAM for improved stability and low leakage," In Proceedings of the 2009 International Conference on Computer-Aided Design (ICCAD '09)

[3] Chen, F.; et. al.; , "Integrated circuit design with NEM relays," Computer-Aided Design, 2008. ICCAD 2008.

[4] A. Stratakos et al., High-efficiency low voltage dc-dc conversion for portable applications, in Int. Workshop Low-Power Design, Apr. 1994.

[5] Venkatasubramanian, R.; et. al.,; , "Ultra low power high efficiency charge pump design using NEM relays," IEEE MWSCAS, 2011

[6] Dadgour, H.F.; Banerjee, K.; , "Hybrid NEMS-CMOS integrated circuits: A novel strategy for energy-efficient designs," Computers & Digital Techniques, IET 2009

[7] Fariborzi, H.; et. al.,; , "Analysis and demonstration of MEM-relay power gating," Custom Integrated Circuits Conference (CICC), 2010.
Syst., vol. 6, no. 1, pp. 39, Mar. 1997.

[8] Chen, C.; et. al., ; ,"Efficient FPGAs using nanoelectromechanical relays," Proceedings of the 18th annual ACM/SIGDA international symposium on Field programmable gate arrays (FPGA), 2010 ACM.

[9] Hoi Lee; Seong-Ryong Ryu; , "An Efficiency-Enhanced DCM Buck Regulator With Improved Switching Timing of Power Transistors," TCAS-II, March 2010

[10] C. Basso. Switch-Mode Power Supply Spice Cookbook, Mc-Graw Hill Education, 2001.

[11] Nathanael, R.; et. al.,; , "4-terminal relay technology for complementary logic," Electron Devices Meeting (IEDM), 2009

[12] Venkatasubramanian, R., et.al ; , "Improving performance of NEM relay logic circuits using integrated charge-boosting flip flop," Nanoscale Architectures (NANOARCH), 2011.

A Heuristic Method for Co-optimization of Pin Assignment and Droplet Routing in Digital Microfluidic Biochip

Ritwik Mukherjee[*], Hafizur Rahaman[*], Indrajit Banerjee[*], Tuhina Samanta[*], and Parthasarathi Dasgupta[†]

[*]Bengal Engineering & Science University, Shibpur, Howrah, India

Email: ritwik@rediffmail.com, rahaman_h@it.becs.ac.in, askforindra@yahoo.com, tuhina_samanta@yahoo.com,

[†]Indian Institute of Management Calcutta, India, Email: partha@iimcal.ac.in

Abstract—Design automation in Digital microfluidic biochip is of immense importance in todays clinical diagnosis process. In this paper, we try to build a heuristic algorithm to simultaneously perform droplet routing and electrode actuation. The proposed method is capable of performing (i)droplet routing with minimal electrode usages in optimized routing completion time, and (ii)minimal number of control pin assignment on the routing path for successful droplet transportation. The proposed method is a co-optimization technique that finds the possible shortest path between the source and the target pair for a droplet and assigns control pins in an optimal manner to actuate the routing path. Intersection regions for multiple droplets are also assigned with pins in an efficient manner to avoid unnecessary mixing between several droplets. The proposed method is tested on various benchmarks and random test sets, and experimental results are quite encouraging.

Keywords-: **Digital microfluidic biochip, Droplet routing, Pin assignment, Co-optimization heuristic**

I. INTRODUCTION

Digital microfluidic biochips, also referred to as a lab-on-chip (LOC) [4], provides freedom of performing biomedical operations and diagnosis in a miniaturized test sites providing reconfigurability and reusability. However, with ever increasing demand of resources and design complexity, a great attention is required for proper scheduling and resource binding for a digital microfluidic biochip [6]. Biochip operations are based on microfluidic technology, and works on very small amount of sample fluids (nanoliter or picoliter), called droplets, and are controlled as well as routed through proper electrode actuation by means of suitable control pins [2]. Generally, arrays of electrodes on biochip are controlled independently using individual control pin for each cell, and the mechanism is known as *direct addressing biochip* [8]. However, the complexity and cost of the design for *direct addressing mode* biochip increases linearly with the number of electrodes. Number of control pins reduces to a great extent for *cross-referencing* biochip [5], where electrodes are assigned common pins horizontally and vertically. Another variant is *pin-constrained* design [9] that tries to optimize the number of control pins as well as resource binding. *Pin-constrained digital microfluidic biochip* design alleviates the growth of control pins to a large extent.

In this paper, we try to deliver a pin-constrained droplet routing technique that performs the control pin design at the early stage of droplet routing and try to co-optimize both the

control pin allocation and resource utilization during droplet routing stage. The proposed method is capable of finding the shortest paths for multiple droplets with proper electrodes actuation along the routing paths. The rest of the paper is organized as follows. Section II gives a brief overview of the existing works. Section III formulates the co-optimization problem for pin-constrained droplet routing. Our proposed method is discussed in Section IV, followed by a formal description and complexity analysis of the algorithm in Section V. Experimental results are elaborated in Section VI. Finally, Section VII concludes the work with some possible future directives.

II. EXISTING WORKS

Pin-constrained biochip design was first proposed by Chakrabarty et al [10], [11], where broadcast technique by generating *"compatible sequences"* and clique partitioning, and array partitioning based on time span overlapping for the droplet routing path were proposed. Main objective of pin-constraint based design technique is to optimize number of control pin by proper modeling of assays and their operations. A repetitive control pin pattern can reduce the number of pins, hence the design complexity drastically. A complete design flow for pin-constrained microfluidic biochip was proposed by Lin et al [3], where they elaborated an elegant design flow based on a ILP formulation, where they completed the task in three stages of stage assignment, device assignment, and guided placement. As pin-constrained biochip design and droplet routing are interdependent, co-optimization of them improves the design substantially. In [12], Zhao et al proposed an ILP based array partitioning technique leveraging the concept proposed in [11] to optimize control pin assignment and resource allocation for droplet routing simultaneously. This brief survey motivates us to propose an algorithm for pin-constrained biochip design and co-optimization for the same. In the next section, mathematical formulation of the co-optimization problem for pin-constrained droplet routing is described.

III. PROBLEM FORMULATION

Pin-constrained droplet routing problem is treated as droplet transportation problem, where the main objective is to complete successful transportation and completion of routing of all the droplets towards their destinations or target locations by proper actuation of electrodes. Droplet routing problem can be formulated as follows.

---

[1]The work is supported by the grant from the Department of IT, Govt. of West Bengal, India, under Project: VLSI Design.

Optimal Pin-Constrained Droplet Routing Problem: Given a set of nets $N = \{N_1, N_2, \ldots, N_n\}$ specified by their source(S_i)-target(T_i) pair locations, $N_i = (S_i, T_i); i = 1, 2, \ldots, n$, construct a routing path for each of the nets, whose length is calculated as the total number of electrodes or unit cells traversed to complete the droplet routing, while minimizing the maximum number of electrodes used and the routing completion time. As the droplets are assumed to move in a concurrent fashion, routing completion time is the maximum of all the droplet routing completion time and is alternately termed as *Latest arrival time (La. time)*. Co-optimization of pin assignment and droplet routing is to simultaneously optimize routing resource utilization and number of pins assigned for the routing paths. Hence the objective function can be written as follows,

$$Minimize\ M := \{M_1(E), M_2(E), M_3(E)\} \qquad (1)$$

where, $M_1(E) = \sum_{i=1}^{n} U_{Ni}(E), U_{Ni}(E_i)$ is the total number of electrodes(E) on the ith routing path. $M_2(E) = max\{T_{la,N_i}(E)\}, T_{la,N_i}(E)$ is the La. time of the ith droplet. $M_3(E) = \sum_{i=1}^{n} P_j^i(E), P_j^i(E_i)$ is the number of pins assigned on E electrodes for the ith net, with $j = 4, 5, \ldots, k$. k is an integer accounting total number of pins on a routing path. Our aim is to minimize the three tuple M making a trade-off between multiple objective functions M_1, M_2, M_3 inside M.

Optimal pin assignment can be done by multiple use of the same control pin used for external biasing. It can be formulated as follows. Given a set of electrodes $U_{Ni}(E)$ assigned to a particular net N_i, and a set of pins $\{P_1, P_2, \ldots, P_k\}$, with $k = 1, 2, \ldots,$ and $k << U_{Ni}(E)$, use a set of $\{P_1, P_2, \ldots, P_k\}$ control pins repetitively to assign the entire routing path with the control pins obeying the actuation constraints discussed next such that,

$$U_{N1}(E) \leftarrow \{P_m\} \in \{P_1, P_2, \ldots, P_k\}; \qquad (2)$$

$$U_{N2}(E) \leftarrow \{P'_m\} \in \{P_1, P_2, \ldots, P_k\}; \qquad (3)$$

with $\{P_m\} \bigcap \{P'_m\} = \Phi$, when $U_{N1}(E)$ and $U_{N2}(E)$ are diagonally or non-diagonally adjacent, and $\{P_m\} \bigcap \{P'_m\} \neq \Phi$, when $U_{N1}(E)$ and $U_{N2}(E)$ are nonadjacent.

A. Source Requirement

Each net N_i consists of a pair of source and target locations for two-pin nets, and two sources and a target location for a three-pin net. For a net N_i, initially a droplet resides on the source electrode $E_i^S(x_i, y_i)$, at time $t = 0$, where (x_i, y_i) is the coordinate location of the source electrode on a $2D$ plane. The constraint is represented as, $S_i(E_i^S, t_0, x_i, y_i) = 1$.

B. Sink Requirement

After successful completion of droplet routing, each net ends at the target location $E_i^T(x'_i, y'_i)$, sink for the corresponding droplet, where (x'_i, y'_i) is the coordinate location of the target electrode on a $2D$ plane, within maximum routing completion time and stays there until all the droplets are routed to their corresponding target locations. The constraint is written as, $T_i(E_i^T, t_0, x'_i, y'_i) = 1$ for the ith droplet.

C. Fluidic Constraints

To optimize droplet movement and their reachability at the target cells, hence to optimize area, routability and throughput, droplets are moved concurrently in time-multiplexed manner [1]. To avoid any unwanted mixing during concurrent transportation of droplets, fluidic constraints need to be maintained as is stated next. Let d_i at (x_i^t, y_i^t) and d_j at (x_j^t, y_j^t) denote two independent droplets at time t. Then, the following constraints, generally called **Fluidic Constraint** [1], should be satisfied for any t during routing:

- Static Constraint: $|x_i^t - x_j^t| > 1$ or $|y_i^t - y_j^t| > 1$
- Dynamic Constraint: $|x_i^{t+1} - x_j^t| > 1$ or $|y_i^{t+1} - y_j^t| > 1$ or $|x_i^t - x_j^{t+1}| > 1$ or $|y_i^t - y_j^{t+1}| > 1$

which implies that no two droplets d_i and d_j can be on diagonal and non-diagonal adjacent electrodes at the same time instant t.

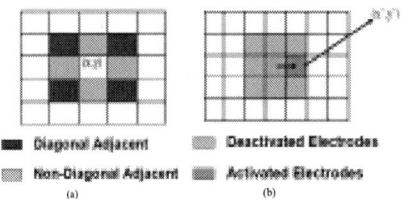

Fig. 1. Electrode Actuation Constraint

D. Electrode Actuation Constraint

When a droplet d_i resides on a particular electrode, say (x, y), the electrode needs to be activated by applying high voltage through an external pin. All the other diagonal and non-diagonal adjacent electrodes (shown in Figure 1(a)) need to be deactivated by connecting to a single pin supplied with a low voltage. No other droplet can reside on any of these deactivated electrodes at the same time instant. Successful fluid transportation through the chip area without unnecessary mixing necessitates this electrode activation sequence. To transport the droplet from its present location to the adjacent location, adjacent cell needs to be activated, while deactivating the previous position. The droplet then moves to the new location due to the surface tension gradient. The detail mechanism of droplet movement is discussed in [2]. When a new cell gets activated for a droplet to be moved there, corresponding adjacent electrodes are simultaneously deactivated as desired.

Routing architecture followed in droplet routing for microfluidic biochip is Manhattan Architecture [7], hence the net constructed always comprises of a rectilinear path. Manhattan architecture is followed because of restricted movements of the droplets over the chip area, where the droplets can move either in vertical or horizontal directions, and any diagonal movement in discarded to avoid mix-up or failure in routing. Hence, an optimal routing path between a source and a target locations for a droplet d_i is a half-perimeter of a rectangular bounding box bounding the source and the target locations for the droplet d_i. In the next section, co-optimization process of droplet routing path and pin-assignment is discussed in detail.

978-1-4673-0438-2/12 $31.00 © 2012 IEEE

IV. OPTIMAL PIN ASSIGNMENT

The following discussion focuses on the parallel problems of optimal droplet routing path formulation as well as optimization of electrode pin assignment on a microfluidic biochip array comprising of $m \times n$ cells. Let us assume two two-pin nets where d_1 and d_2 be two droplets with source locations $S_1(x_1, y_1)$, $S_2(x_2, y_2)$ and target locations $T_1(x_1', y_1')$, $T_2(x_2', y_2')$ respectively.

A. Co-optimization Through Routing Path

Any droplet is represented on a 3×3 matrix region on a biochip where a set of three control pin assignments (P_1, P_2, P_3) can be used repetitively for actuation of droplet control electrodes enabling droplet movements. Generally, a microcontroller I/O pins are dedicated for biasing these control pins. By applying proper bias on the target pins, bit sequences will be varied as, $100 \rightarrow 010 \rightarrow 001$, as is shown in Figure 2. A droplet stays on the electrode whose control pin is activated with bit '1'. Minimum three pins are required to activate the desired cell without any electrode actuation conflict. Instead of direct pin addressing, such sets of pin repetitions can remarkably optimize electrode pin assignment process.

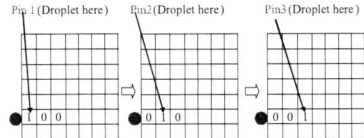

Fig. 2. Optimal Pin Assignment Process

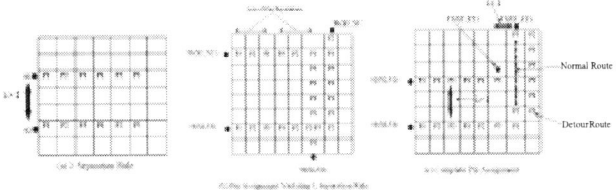

Fig. 3. Pin Repetition Pattern

For multiple droplets routing in a concurrent fashion, same repetition pattern can be applied to all the routing paths if they are separated by a specified λ distance with $\lambda \geq 2$, as is shown in Figure 3 (a). This λ distance is maintained to satisfy static and dynamic fluidic constraints stated in Section III. If there is any violation in λ distance through original routing path, detour path is chosen to satisfy the distance rule. As is shown in Figure 3(c), initial path from S_1 to T_1, marked by the dotted line violates the λ distance rule, hence another detour path is chosen. Detour path is selected by checking the availability of the nearest neighbor electrode. The new path abides by the fluidic constraints at the cost of increased electrode usages. Example pin assignment for two droplets routed concurrently is shown in Figure 3(b) and (c). λ separation is disregarded for the routing paths that do not overlap in same time span.

Also during routing, multiple droplets can traverse the same routing path at different time span to reuse electrodes, hence to minimize the resource consumption. However, during pin assignment electrode actuation constraint is always satisfied.

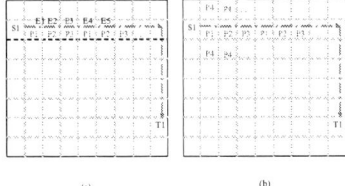

Fig. 4. Minimum Pin Requirement for Single Droplet Transportation (a)Set of Three Pins (b) Set of Four Pins

Theorem 1: Lower bound in number of control pins is four for successful transportation of a single droplet through a routing path.

Proof: Single control pin fails to actuate the array of electrodes in required fashion and two control pins connected to alternate electrodes along the route trace is not a feasible plan to satisfy electrode actuation constraint. Hence, we start with three control pins.

1) **Case1:** Let us consider three control pins P_1, P_2, and P_3, connected to the electrodes $\{E_1, E_2, E_3\}$, and the pattern is repeated for the next sets of electrodes, $\{E_4, E_5, E_6\}$ and so on (Figure 4(a)). Initially a droplet is on the electrode "1" by activating P_1, while P_2 and P_3 are deactivated. Then to move the droplet to the electrode "E_2", P_2 is activated deactivating P_1 and P_3. As both the adjacent electrodes of "E_2" are deactivated, the droplet attains stability at that electrode without splitting to other parts. Hence, three control pins are necessary for a single droplet transportation.

   However, in the present discussion we do not put emphasis on the other adjacent diagonal and non-diagonal electrodes of "E_2" not assigned with any pin. To satisfy electrode activation constraint, these electrodes need to be deactivated. Top three sets of electrodes and bottom three sets of electrodes of the set of electrodes $\{E_1, E_2, E_3\}$ are other adjacent and non-adjacent electrodes of "E_2" not assigned with any pins. Any of the three pins may be assigned to them for the purpose of activation, but will not provide with successful activation. Say, P_1 is assigned on the top and P_3 are assigned on the bottom electrodes to make them deactivated when P_2 (electrode "E_2") is activated. But in the next time instance, when P_3 becomes active to transfer the droplet from electrode "E_2" to electrode "E_3", both "E_3" and one of its adjacent electrodes becomes active causing unnecessary splitting. Hence, three control pins is not sufficient condition for successful droplet transportation.

2) **Case2:** If a set of four control pins (P_1, P_2, P_3, and P_4) are chosen for electrode actuation, the fourth pin P_4 can be assigned on both the sides of the electrodes that reside on routing paths without any activation conflict discussed in Case1.

The complete pin assignment is shown in Figure 4(b). Hence it is proved that minimum *four* pins are necessary and sufficient to activate electrodes for single droplet transportation. ∎

The droplet routing path can be formulated by obtaining a possible shortest path between S_i and T_i for a droplet d_i. Optimal routing path is the *Manhattan distance* between the source and the target. During each droplet movement, it has to be checked whether the distance between the present location of a droplet and its target location gets decreased gradually in terms of both x and y coordinates to avoid unnecessary drifting away from the right track. A droplet is assigned a new electrode in its path with lesser *Manhattan Distance*. Manhattan distance (M) between two electrodes $e_1(x_1, y_1)$ and $e_2(x_2, y_2)$ is calculated as, $M = |(x_1 - x_2)| + |(y_1 - y_2)|$. As the droplet moves, the pins are assigned accordingly, for electrode actuation. An example Manhattan path for a single droplet is shown in Figure 5. Possible shortest path from S_1 to T_1 is $A \rightarrow B \rightarrow D$. Any movement towards C or E is discarded by checking *Manhattan Distance* of the present location of the droplet. It is quite evident from the figure that *Manhattan Distance* increases towards both the directions (C and E). Original route deviates from the Manhattan distance due to routing conflict, or presence of any blockage. Stalling and detour is implemented in these cases to find the best possible solution.

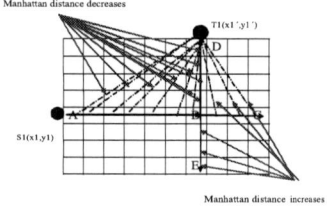

Fig. 5.   Manhattan Routing Path

B. Alternate Route:

Resource conflict arises when two droplets try to access the same electrode at the same instant of time. An example is shown in Figure 6(a). Here droplet d_1 from source S_1 is headed towards target T_1 and d_2 from S_2 towards T_2. The pin assignments for electrode actuation is done in $P_1, P_2, P_3, P_1, P_2, P_3, \ldots$ manner. At time instant t, there will be droplet d_1 on electrode A_1 (from source S_1) and d_2 on electrode A_2(from source S_2) respectively. The example in Figure 6 portrays that the electrode A between the electrodes A_1 and A_2 has to be actuated with pin P_3 as per the regular schedule of pin assignment, while both d_1 and d_2 try to access it at the same time. This pin assignment is erroneous since an active pin P_3 will produce an unnecessary mixing of the droplets d_1 and d_2. To avoid unnecessary mixing, electrode A is assigned with a pin P_A(say) different from the regular $P_1, P_2, P_3, P_1, P_2, P_3, \ldots$ sequence. Apart from separate pin assignment, d_1 and d_2 needs alteration in path to avoid any

overlap in their path at the same time span. Alteration in path keeps routing completion time same at the cost of use of some more resources. The alternate routes are shown by the solid arrowheads in Figure 6(b).

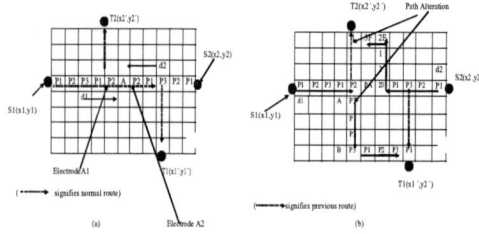

Fig. 6.   Alternate Routing Path (a)Normal Route (b)Alternate Path

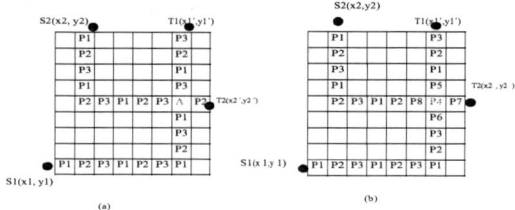

Fig. 7.   Intersecting Region of Two Droplet Routing Paths

C. Intersection Region

A major problem in pin constrained droplet routing phenomenon arises when two or more droplets travel in a path that intersect at a common point, at the same or different time instant. Because of a common intersection point, multiple control pins are assigned at the particular electrode, which hampers the rules of pin assignment. As is shown in Figure 7(a), pin P_2 is assigned at electrode A for the $S_1 - T_1$ pair, and P_1 is assigned for $S_2 - T_2$ pais. To avoid any such conflict in pin assignment, and to maintain electrode actuation constraint, all the intersecting regions are replaced with a $5 - color$ pin assignment as is represented in Figure 7(b). The common intersecting point is replaced with a pin, separate from the existing three pin sequences and four diagonally adjacent electrodes are assigned with four different electrodes to avoid unnecessary mixing. If stalling is required, a new pin (say P_9) needs to be assigned on the electrode where a droplet is stalled and is kept active till the droplet resides there. Hence, the following observation follows.

Observation 1: Atleast six pins are required for successful routing of multiple droplets through an intersection region of the routing paths over a $2 - D$ array of electrodes on a biochip.

V. PROPOSED ALGORITHM

A formal description of the proposed algorithm is shown in Figure 9. In a preprocessing stage, method $Generate\_AdjMatrix()$ constructs a grid graph and its adjacency matrix corresponding to the input of $2D$ array of electrodes. Each electrode is considered as a vertex. The edges of the graph are only horizontal and vertical satisfying droplet movement constraint, and any blockage over the chip area is represented as no edge in the grid graph. The graph is used as an input to the algorithm to find possible route between any source-target pair. At the onset of the co-optimization process, all the source locations are checked and assigned with pin 1 from the regular set of pins if separated by $\lambda \geq 2$, else assigned with a different pin. Step 14 in the algorithm checks for λ separation between routing paths discussed in Section IV and assigns pin in regular sequence of $\{P_1, P_2, P_3\}$. Alternate route is decided by checking the direction of movement of the ith and j^{th} droplet in step 17. Any intersecting region is detected in step 32 of the algorithm, and the pin assignment is done in step 33, and is performed in a post processing phase by inputing the common intersection region. Electrode count and pin count is updated each time the routing path is updated by accumulating unused electrodes and assigning pins to unused paths. Each target location is assigned with single separate pin to hold the droplet there until it is collected.

Lemma 1: Total time complexity of the proposed droplet routing algorithm is $O(K \times n \times M) + O(\frac{M}{2} \times M) + O(C \times n^2)$, where $M \times M$ is the size of the biochip array, and n is the total number of droplets.

Proof: Maximum time required to find the *Manhattan Distance* for a single droplet is $O(K \times M)$, where K is a constant. Each time to choose a new electrode q_i in step 13 of the proposed algorithm, checking is done three times for the adjacent left, top or bottom electrode of the q_{i-1} electrode, disregarding any choice of backward movement. This checking can be done maximum for a single row of M electrodes, and a single column of M electrodes to find the shortest possible path. Hence, total time complexity of checking is $O(3 \times (M + M))$. Total time required to compute Manhattan distance for n droplets is $O(3 \times n(M + M)) \cong O(K \times n \times M)$.

Alternate route (step 18) is required when two droplets face each other at the same time stamp. We do not consider the deadlock situation, when four droplets are headed towards a single electrode at the same time stamp and backtracking is required to avoid the deadlock situation. Hence, maximum number of alternate routes can be possible for $\frac{M}{2} + \frac{M}{2} = M$ droplets headed towards each other over the chip area of $M \times M$ electrodes, as minimum $\lambda = 2$ separation is maintained between the droplets at the same time instance to avoid unwanted mixing. Possible number of locations for checking alternate route is also $\frac{M}{2}$. Hence, time required to compute alternate route is $O(\frac{M}{2} \times M)$.

Diagonal and non-diagonal adjacent pin assignments can be done in constant time, say $O(C)$ for each present droplet location. For a complete routing path, the same checking requires $O(C \times K \times M)$ time, as each Manhattan routing path can be constructed in $O(K \times M)$ time. For all the droplet routing paths, adjacency checking can be done in $O(C \times n \times K \times M) \cong O(C' \times n \times M)$ time, where C' is a constant.

All the intersecting regions of a routing paths are replaced with an unique arrangement of $5 - pins$. Time required to compute intersection between n droplets of nets is $O(n \times n) =$

Droplet_Algo()

Input: Array of electrodes ($M \times M$) with prefix obstacle locations; Number of droplets (n) with source and target coordinates.
Output: Route all the two-pin nets from source to target having minimum possible paths and optimized pin assignment

(*Preprocessing*)
1. Generate_AdjMatrix($M \times M$);
(*Initialization*)
2. pin_count := 0; electrode_count := 0; Choose a regular sequence of pins {1, 2, 3}
(*Initial Source Pin Allocation*)
3. for i 1 to n,
4. if not already assigned with pin
5.   if dist(S_i) - dist(S_{i+1}) > 2
6.   assign pin 1 from regular set of three pins
7.   else
8.   assign separate pin to the source location
9. end for
(*Droplet Routing*)
10. for i 1 to n,
11.   Choose S_i = source and T_i = target
12.   While target T_i is not reached
13.   Choose next electrode q_i with shortest $Manhattan$ distance;
14.   if $\lambda \geq 2$
15.   assign pin from regular sequence;
16.   Update path $SP = SP \bigcup(q_i)$;
electrode_count = electrode_count + non visited electrodes;
17.   if q_i and q_j faces themselves for two different droplets (*q_j is any neighbor electrode for the jth droplet*)
18.   Use alternate path;
19.   Update path $SP_i = SP_i \bigcup(q_i)$; $SP_j = SP_j \bigcup(q_j)$;
electrode_count = electrode_count + non visited electrodes;
20.   assign pin from regular sequence;
21.   if $\lambda \leq 2$
22.   if no overlap of routing path in time span
23.   Update path $SP = SP \bigcup(q_i)$;
electrode_count = electrode_count + non visited electrodes;
24.   Continue with regular path and pin assignment if there is no common pin in diagonal and non-diagonal adjacent electrodes;
25.   Assign new pin if common pin in diagonal and non-diagonal adjacent electrodes;
26.   else
27.   Choose detour path to make $\lambda \geq 2$;
28.   Update path $SP = SP \bigcup(q_i)$;
electrode_count = electrode_count + non visited electrodes;
29.   Continue with regular path and pin assignment if there is no common pin in diagonal and non-diagonal adjacent electrodes;
30.   Assign new pin if common pin in diagonal and non-diagonal adjacent electrodes;
31.   Assign pin from regular sequence;
32.   else if q_i an intersection
33.   Intersect_Assign(q_i);
(*5 − color pin assignment*)
34.   end for
35. return;

Fig. 8. Co-optimization of Pin Assignment and Droplet Routing for $DMFB$

$O(n^2)$ considering intersection between each pair of nets. If a net intersects with another net multiple times, which is of lower probability of occurrence and comes into picture only when detour occurs, time required to calculate intersecting regions is $O(C \times n^2)$, where the constant term C having a small integer value, accounts for multiple intersections between a pair of nets. Stalling and detour requires nominal computation time and is omitted in the present discussion.

Worst case time complexity of our proposed algorithm is the sum of the complexities of all the above steps, which comes out as $O(K \times n \times M) + O(\frac{M}{2} \times M) + O(C' \times n \times M) + O(C \times n^2) \cong O(K \times n \times M) + O(\frac{M}{2} \times M) + O(C \times n^2)$. ∎

VI. EXPERIMENTAL RESULTS

Proposed heuristic is implemented in Linux platform on a Core i3 M370 Intel Processor, with 2.40GHz. processor speed and 3GB memory. Experimental studies are performed using multiplexed bioassay, in-vitro, protein benchmark suits and on

TABLE I
COMPARATIVE STUDY OF PIN ASSIGNMENT

Multiplexed Bioassay	15×15
Pins without TDPS [12]	38
Pins with TDPS [12]	20
Pins in our method	15

TABLE II
EXPERIMENTAL RESULTS FOR BENCHMARKS AND RANDOM TEST
INSTANCES

Name of benchmark (grid)	max. # of nets	Electrode used [1]	La. time[1]	Electrode used(Ours)	La. time (Ours)	Pin Used
invitro_1(16)	6	258	n/a	200	15	30
protein_1(21)	6	1688	n/a	372	24	19
test_1(12)	12	67	100	94	38	48
test_3(12)	12	n/a	n/a	102	60	43
test_6(16)	16	119	179	177	64	48
test_7(16)	16	113	183	174	63	69

some hard test sets from [1]. A comparative study shows that our method of pin-assignment minimizes total number of pins required for assay completion to a large extent compared to that in [12], and [1]. Results are elucidated in Table I for multiplexed bioassay. Comparative study of resource utilization is elaborated in Table II. Our algorithm gives optimized results in terms of routing completion time, called *La time*, compared to that in [1]. However, number of electrode usages in algorithm deteriorates because we prioritize detour in our approach to solve routing conflict. A difference of our work from that in [1] is that, our work is capable of optimizing number of pins required for droplet routing on a pin-constrained biochip. Total pin count for our work is also presented in the same table. A schematic diagram of pin assignment for a hard test set $test\_1$ in Table II is shown in Figure 10.

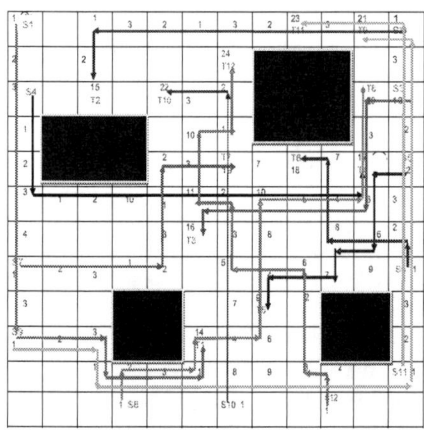

Fig. 9. Pin Assignment for a Hard Test Set Over 12×12 Microfluidic Array

VII. CONCLUSION

High performance droplet routing algorithm development is a major goal in today's DMFB design. We present an algorithm, which is computationally efficient, and co-optimizes resource utilization for droplet routing and pin-assignment in DMFB. In this paper we propose a heuristic method that is capable of providing optimized solution for different bioassays,

having two-pin and three-pin nets. Experimental results show improvement in results for our algorithm compared to some other existing works.

As in practice, a biochip may be used to route multiple number of droplets having several pins, one of the possible future scope is to modify our algorithm for n-pin nets. The usage of biochip is popular these days for its reconfigurability. An algorithm may be developed for a reconfigurable placement technique that may optimize the subsequent routing phase.

ACKNOWLEDGEMENT

The authors would like to thank Prof. Krishnendu Chakrabarty, Professor, Electrical and Computer Engineering department, and Dr. Yang Zhao, research scholar, Electrical and Computer Engineering department, Duke University, Durham, NC.

REFERENCES

[1] M. Cho and D. Pan. A high-performance droplet routing algorithm for digital microfluidic biochips. *IEEE Transaction on Computer-Aided Design of Integrated Circuits and Systems*, 27(10):1714 – 1724, 2008.
[2] Richard B. Fair Fei Su, Krishnendu Chakrabarty. Microfluidics-based biochips: Technology issues, implementation platforms, and design-automation challenges. *IEEE Transaction on Computer-Aided Design of Integrated Circuits and Systems*, 25(2):265 – 277, February 2006.
[3] Cliff Chiung-Yu Lin and Yao-Wen Chang. Ilp-based pin-count aware design methodology for microfluidic biochips. *IEEE Transaction on Computer-Aided Design of Integrated Circuits*, 29(9):1315 – 1327, September 2010.
[4] T. Mukherjee. Design automation issues for microfluidic biochips. In *Proceedings of Internationa Conference on Computer Aided Design*, pages 463 – 470, November 2005.
[5] Chia-Lin Yang Ping-Hung Yuh, Sachin Sapatnekar and Yao-Wen Chang. A progressive ilp based routing algorithm for cross-referencing biochips. In *Proceedings of Design Automation Conference*, pages 284 – 289, June 2008.
[6] H. Ren P. Paik V. Pamula R. B. Fair, V. Srinivasan and M. Pollack. Electrowetting-based on-chip sample processing for integrated microfluidics. In *Proceedings of IEDM*, page 32.5.132.5.4, 2003.
[7] M. Sarrafzadeh and C. K. Wong. *An Introduction To VLSI Physical Design*. McGraw-Hill Series in Computer Science, New York, 1996.
[8] F. Su T. Xu, W. Hwang and K. Chakrabarty. Automated design of pin-constrained digital microfluidic biochips under droplet-interference constraints. *ACM Journal on Emerging Technologies in Computin Systems*, 3(14), 2007.
[9] V. Pamula V. Srinivasan and R. Fair. An integrated digital microfluidic lab-on-a-chip for clinical diagnostics on human physiological fluids. *Journal of Lab Chip*, 4:310 – 315, 2004.
[10] Tao Xu and K. Chakrabarty. Broadcast electrode-addressing for pin-constrained multi-functional digital microfluidic biochips. In *Proceedings of Design Automation Conference*, pages 173 – 178. California, USA, June 2008.
[11] Tao Xu and Krishnendu Chakrabarty. Droplet-trace-based array partitioning and a pin assignment algorithm for the automated design of digital microfluidic biochips. In *Proceedings of International Conference on Hardware/Software Co-design and System Synthesis*, pages 112 – 117, October 2006.
[12] Yang Zhao and Krishnendu Chakrabarty. Co-optimization of droplet routing and pin assignment in disposable digital microfluidic biochips. In *Proceedings of ACM/IEEE International Symposium on Physical Design*, pages 69 – 76. California, March 2011.

Clock Tree Skew Minimization with Structured Routing

Pinaki Chakrabarti

Synopsys (India) Pvt. Ltd.

Bangalore, India, 560016

Email : pinaki@synopsys.com

Abstract— **One of the goals of clock tree synthesis in ASIC design flow is skew minimization. There are several approaches used in traditional clock tree synthesis tools to achieve this goal. However, many of the approaches create a large number of clock-buffer levels while others result in congested clock routing. Increase in buffer level and routing congestion essentially triggers the problem of increase in buffer area and total power. Also the performance of the circuit is degraded due to on-chip variation in such situations. For certain fan-out number restricted designs, a few proposals with H-tree routed clock nets have been proposed to reduce the skew, but those proposals can hardly be used across various designs used in industry. Here we propose a method where skew minimization is mainly achieved by structured routing of clock nets. Finally, we show that with this proposal, for a few real designs from industry, we could reduce the skew up to 6.5% with increase in total wire delay up to 1.89% compared to when simple H-tree routing was deployed.**

Keywords—**Clock tree, H-tree, Skew, Synthesis, Routing**

I. INTRODUCTION

Clock tree synthesis is one of the important steps in physical design flow for ASICs. It is challenging in many aspects and one of the goals of clock tree synthesis is minimizing the skew among the related flip-flops or sequential elements. Several industry standard tools perform clock tree synthesis by using different optimization and clustering techniques. In deep sub-micron designs where interconnect delay is very prominent, often the skew and insertion delay obtained using estimated interconnect routing, fail to correlate well with those values obtained after detailed routing. A considerable amount of re-work is then required to meet the original design goals.

This paper describes a method to consider structured routing of the clock paths which ensures that the clock travel time is almost same to all the end points (the flip-flops or the sinks) of the clock tree.

There has been a considerable amount of work done on clock routing. Jackson et al. [1] used a purely geometric approach in automatic clock synthesis. But this can result in large skew. Wire sizing was used to obtain variation immuned low skew for buffered clock tree by Tsai et al. [12]. Edahiro [3] tried to solve skew problem with better clustering and routing. But that essentially assumed uniform placement of flip-flops. Tsay [2] considered using good delay models to achieve minimal skew and Edahiro [4] proposed improved routing algorithm to reduce the skew but both the approaches did not consider much about the rise time constraints. Oh et al. [9] addressed power reduction techniques using good clock routing without putting much consideration to skew minimization and also it assumed that the clock cells can be placed uniformly. Vittal et al. [8] [5] proposed power optimal buffered clock tree design methodologies but those essentially compromised with the skew in the clock circuit. Among the structured routed clock trees, H-tree clock

distribution was a methodology well discussed by Narasimhan et al. [13]. Reaz et al. [14] proposed a zero skew clock routing method but it hardly considered realistic floor-plans with many routing and placement blockages.

In this paper we describe methods to create structured routing for the buffered clock tree in order to ensure very low skew even when the clock buffers are not placed uniformly, without violating any routing rules.

The rest of this paper is organized as follows. Section II formally defines the problem. In Section III, we describe the skew minimizing algorithm which creates structured routes on the unrouted clock paths of the clock tree built with just logical connections. The results are described in Section IV. In Section V, we conclude by giving a few directions for future improvement.

II. PROBLEM FORMULATION

In a clock tree the clock signal reaches to different end points through different paths. The skew in a clock tree is the time difference between the longest arrival time and shortest arrival time. Clearly, the skew results from unbalanced paths.

Let the total delay of any clock path P_i be δ_i. The clock skew ρ is defined as below:

$$\rho = \operatorname*{Max}_i(\delta_i) - \operatorname*{Min}_i(\delta_i); \forall P_i \qquad (1)$$

The delay δ_i, of a clock path P_i can be expressed as

$$\delta_i - \delta_{i_w} + \delta_{i_c} \qquad (2)$$

where δ_{i_w} is the total wire delay and δ_{i_c} is the total cell delay of the path P_i.

In deep sub-micron designs, around 80% of the delay on a path is contributed by the wire delay while remaining comes from the cell delay [10]. Thus a major portion of the skew ρ is contributed by the wire delay in the circuit. Hence, ρ can be reduced a lot if we can minimize the wire delay difference among all the paths. Let $\delta_{w_{max}}$ be the wire delay of the path having maximum wire delay and $\delta_{w_{min}}$ be the wire delay of the path having minimum wire delay. Let their difference be $\Delta_w = \delta_{w_{max}} - \delta_{w_{min}}$. Our aim is to minimize the skew by minimizing Δ_w. So, formally, the original problem is

$$Minimize(\rho) \qquad (3)$$

However, we will solve the following reduced problem :

$$Minimize(\Delta_w) \qquad (4)$$

III. PROPOSED METHOD

There have been several techniques proposed to solve the problem of minimizing the skew. Many of them depend on cell delay manipulation while a few depend on creating appropriate routing.

We propose a method of balancing wire delays so that the clock travels from the root to all the end points through paths having almost the same wire length. Hence, the delay due to interconnects for all paths remains the same and skew due to interconnect is reduced to a great extent. We achieve this by creating structured routing.

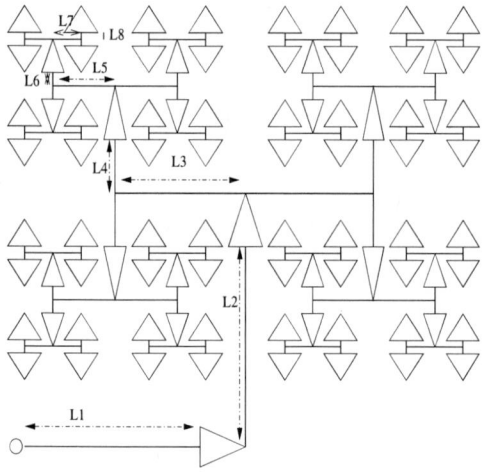

Fig. 1. Traditional H-tree routing

In Fig. 1, we present a traditional H-tree routing where the cell placement is very uniform and the interconnect routing between them results in regular H-shapes. Because of uniformity of the structure, from the root of the clock to each end point, the clock travels exactly the same wire length. In the same figure, the clock travels a total distance of $\sum_{i=1}^{8} L_i$ to reach to any end point from root. Hence the skew obtained by this clock structure is almost zero if we can ensure that the intermediate cell delays are same [11].

The basic problem of this kind of structure is ensuring the uniformity. In today's ASIC designs, a typical floor-plan contains various macros, memory blocks, IPs as well as routing and placement blockages. All these routing impediments factor into the problem as uniformity of cell placement and routing is hard to maintain in such designs. Thus the above approach does not adequately address to the skew optimization problem in many practical designs. Also, this approach requires that all the cells must have four loads which is a burdensome restriction for any clock tree synthesizer.

Now we discuss our approach using a more generic tree as shown in Fig. 2. In this tree, first few levels of a clock tree are shown. In such a clock tree, the output pin of a buffer at a level is connected to the load pins through a logical connection denoted as *clock net*. The load pins are the input pins of the buffers at the next level. In this figure, different number of loads are there for different clock nets. The number of loads are either one, two or four in different levels. Thus, for each net, H-shaped routing is not possible. It is to be noted that each clock net has single

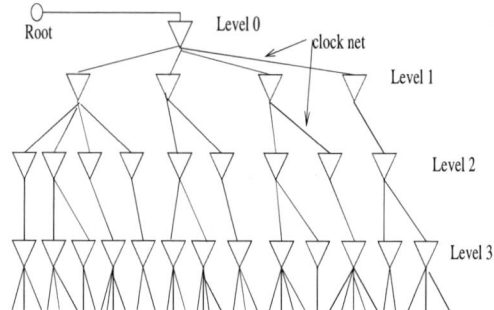

Fig. 2. Logical structure of a buffered clock tree

driver.

We denote a clock-net as H-net if it has four loads. Similarly, nets with one and two loads are referred to as SL- and I-nets, respectively. An H-net or an I-net is symmetric with respect to its driver in the sense that the driver drives an equal number of loads in either side. For example, in an H-net, a driver drives two loads in each side. An SL-net is not symmetric. It is to be noted that in case the cells are not placed uniformly, there can be skew in a single symmetric net due to unbalanced paths from the driver to loads.

To obtain a clock tree with only H, I, SL type logical nets, clock-cells/buffers must be placed appropriately. For that purpose we used a skew constrained clock cell placement algorithm which is a modified version of an algorithm proposed by Natesan et al. [6]. The proposed structured routing method uses already obtained logical nets with four, two or one load(s) by the above algorithm, to create appropriate routes on them.

The wire delay δ_{w_N} of a certain clock net N depends on resistance R_N and capacitance C_N of its wire layout path. Different metal layers, used for routing, have different resistance values per unit length (R_{u_l}) and different capacitance values per unit length (C_{u_l}). Typically, for clock routing, two neighbouring higher metal layers (e.g., metal5 and metal6) are reserved where C_{u_l} and R_{u_l} values don't differ too much between them. Without loss of generality, we consider here that C_{u_l} and R_{u_l} remain the same across the metal layers to be used for clock routing. Thus, the wire delay becomes the function of wire length, L_N. It can be shown [5] [7] that for a typical clock net in ASIC,

$$\delta_{w_N} \propto L_N{}^2 \qquad (5)$$

Thus when the length of a net increases, wire delay for that net also increases monotonically. Hence, the clock arrival time at each end point can be made almost the same if it traverses the paths with same lengths from the root. In order to ensure this, at a certain level Lev_i, we make sure that the length is the same for all the nets. This is done by inserting detours or intentional extra routes to all but the longest net in Lev_i to match the longest net in that same level. The detour amount for a particular net N at level Lev_i is determined by the following equation:

$$L_{dt_{i_N}} = \underset{\forall j}{Max}(L_{i_j}) - L_{i_N} \qquad (6)$$

Here, L_{i_N} denotes the estimated wire length from the driver to any of the loads of the net N in the level Lev_i and L_{i_j} is the estimated driver-to-a-load wire length of any net j in the same level. We estimate the detour amount before doing detailed routing for the nets at each level.

Once estimated, the extra amount of wire of length $L_{dt_{i_N}}$ is inserted as detour when doing the structured routing for the net N in H or I or SL shapes. This detour can typically be inserted in either side of the driver for symmetric nets and between the driver and the load for SL-net type.

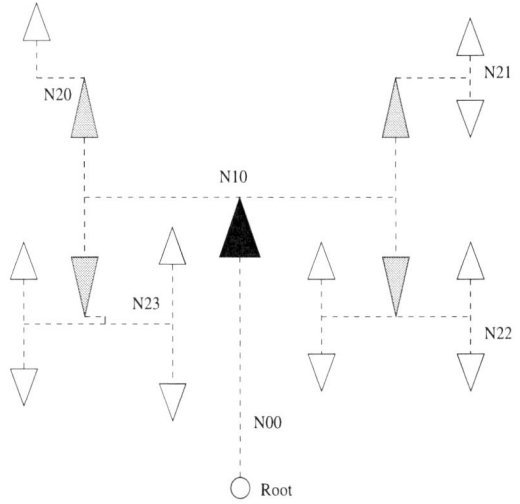

Fig. 3. Route estimation for detour calculation

In Fig. 3 the nets $N00$ and $N20$ are of type SL-net while $N10$, $N22$ and $N23$ are of type H-nets. The net $N21$ is of I-net type. The dotted lines show the possible routing structures of the nets. The net $N00$ which is driven by the root is the only net in that level of the tree. Hence, there is no need to balance this net's length against that of any other net. The same is true for $N10$. However, since it has four loads, the clock must traverse paths of the same length from the driver to each of the loads. Thus a well balanced H-shape is sufficient for the net $N10$. In Level-2, there are four nets and from the same figure it can be seen that each of these have different shape and unrouted length. In this case, the net $N23$ has the longest path for clock to travel from the driver to each of the loads. In detailed routing stage we insert detour appropriately as derived by the Eq.(6) on the shorter nets to ensure that the clock paths match for all the nets in a particular level.

We can also see that $N23$ can not have a perfect H-shape as the placement of its loads in left side does not form an exact mirror image with the loads in right side. So, to balance among its own loads we need to insert a small routing jog to tap the driver to a point other than the mid-point of the central wire of the H-net. Pre-detailed routing estimation considers this situation to avoid any future surprises.

Fig. 4 shows the detours that have been actually inserted on the nets $N20$, $N21$ and $N22$. As was estimated, a small routing jog has also been inserted to tap the driver to the central wire of the H-shape of $N23$.

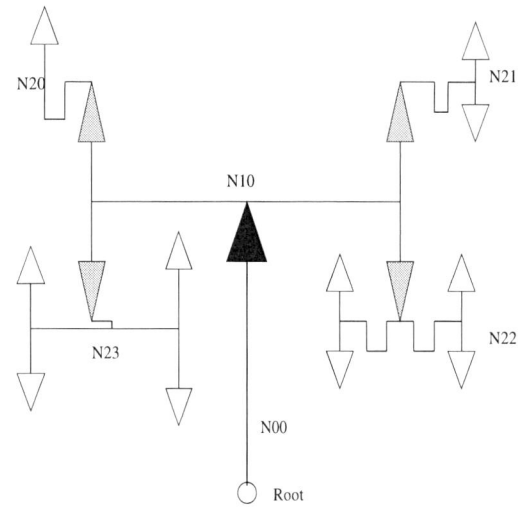

Fig. 4. Clock tree with detoured net routes

A. Detour insertion algorithm

Now we formally describe the algorithm of detour insertion. First, we describe the functions to create H-, I- and SL- routing shapes given the driver d, estimated driver tap point T_{dp_e} and the amount of detour L_{dt} of the net N.

We use a core function of syntax

$$\mathbf{M_{12} = Connect(p_1, p_2, p_3)}$$

Honouring blockages and routing rules, this function does the detailed routing between the pins/points p_1 and p_2 and creates a connection, represented by $S(p_1, p_2)$. This function uses p_3, which can be *NULL* as well, to return weighted mid point M_{12} of S.

do-H-route(N, d, T_{dp_e}, L_{dt})
- Let N has loads $p_{H_{11}}$, $p_{H_{12}}$ on one side of T_{dp_e}; $p_{H_{21}}$, $p_{H_{22}}$ are the loads on other side
- M_{ld_1} = Connect($p_{H_{11}}$, $p_{H_{12}}$, *NULL*)
- M_{ld_2} = Connect($p_{H_{21}}$, $p_{H_{22}}$, *NULL*)
- T_{pd_a} = Connect(M_{ld_1}, M_{ld_2}, T_{dp_e})
- If $L_{dt} \neq 0$, add detour L_{dt} on the connections $S1(T_{pd_a}, M_{ld_1})$ and $S2(T_{pd_a}, M_{ld_2})$
- M_{dummy} = Connect(d, T_{pd_a}, *NULL*).

do-I-route(N, d, T_{dp_e}, L_{dt})
- Let N has the loads p_{I_1}, p_{I_2}
- T_{pd_a} = Connect(p_{I_1}, p_{I_2}, T_{dp_e})
- M_{dummy} = Connect(d, T_{pd_a}, *NULL*).
- If $L_{dt} \neq 0$, insert detour either on the connection $S1(d, T_{pd_a})$ OR on the connections $S2(p_{I_1}, T_{pd_a})$ AND $S3(p_{I_2}, T_{pd_a})$

do-SL-route(N, d, L_{dt})
- Let N's load be p_{SL}
- M_{dummy} = Connect(d, p_{SL}, *NULL*)
- If $L_{dt} \neq 0$, insert detour on connection $S(d, p_{SL})$

Now we outline the main algorithm **do-clktree-route**():

```
1. Let R be the root of the tree.  Let
   N_R be the net driven by R
2. c = 0.  The level of N_R is Lev_c
3. Route N_R in H- or I- or SL-shape
   based on number of loads
4. If there is no more level to explore,
   Stop, else go to Step 5
5. For each load p_i ∈ Lev_c, traverse the
   corresponding cell arc to find the cell
   output pin which is the driver d_i of
   the net N_i in level L_{c+1}.  Insert each
   such N_is into the list q.  c = c + 1
6. For each net N_i ∈ q, estimate the route
   shape, length and find its detour L_{dt_{c_i}}
7. While all nets in q are not visited,
   get the next net N_i from q
 • Estimate the driver tap point T_{dp_e} if
     N_i is H- or I-net
 • N_i is H-net:do-H-route(N_i,d_i,T_{dp_e},L_{dt_{c_i}})
 • N_i is I-net:do-I-route(N_i,d_i,T_{dp_e},L_{dt_{c_i}})
 • N_i is SL-net:do-SL-route(N_i,d_i,L_{dt_{c_i}})
 • Mark N_i as visited
8. Go to Step 4
```

Above algorithm routes all the nets in the given tree level-by-level. Hence, its runtime depends on the number of nets to be routed in the tree. Since the core function **Connect()** considers routing congestion to do the blockage aware detailed routing, a congested design floor-plan has influence on the overall execution time of this algorithm.

IV. RESULTS

For our experiments, we have chosen a few designs at $65nm$ or below nodes, from industry. Each of these designs has complex floor-plans with multiple macros and blockages and large number of flops.

First, we have experimented on a design **A** with $65K$ flops driven by a clock. For the first ten levels of that clock-tree, starting from the root, we applied the detouring algorithm. In this portion of the tree, each buffer drives either one or two or four loads so that SL, I or H shaped routes are possible. At the 10-th level, there are 56 buffers. We used two metal layers, metal5 and metal6 for routing the nets in the tree. The design floor-plan contains a few macros and a few routing and placement blockages. We compare the skew up to 10-th level of the tree between without and with detouring.

In the above experiment we kept the buffers already inserted and placed by a previous synthesis step. Here, not all the buffers are same. The different buffer types introduce different cell delays and load capacitance at various levels. Thus we see non-monotonicity in the skew value changes in a few levels, e.g., level 8 and level 10. However, if we look into the Table I, we see that skew values are very different from level 5 onwards between the two experiments. The skew values in first 3 levels are 0 due to uniform cell placement. With detouring, we see that the skew in each level, from 5-th to 10-th, is less compared to the ones without detouring. Hence, the detoured interconnects

TABLE I

LEVEL-WISE SKEW(PS) COMPARISON BETWEEN WITHOUT AND WITH WIRE DETOUR FOR DESIGN A

	Without detour	With detour
ρ_{Lev_1}	0	0
ρ_{Lev_2}	0	0
ρ_{Lev_3}	0	0
ρ_{Lev_4}	2	2
ρ_{Lev_5}	20	15
ρ_{Lev_6}	47	25
ρ_{Lev_7}	48	39
ρ_{Lev_8}	41	39
ρ_{Lev_9}	86	52
$\rho_{Lev_{10}}$	54	51

TABLE II

TOTAL WIRE DELAY(PS) COMPARISON BETWEEN WITHOUT AND WITH WIRE DETOUR FOR DESIGN A

	Without detour	With detour
$\delta_{w_{010}}$	901	918

essentially reduced the interconnect skew for that portion of the circuit. At the end of 10-th level, with detouring, we improved the skew by 5.55% which is significant for such a design with complex floor-plan.

From the Table II we observe that with our proposed algorithm of detour insertion, the wire delay of the longest path from root to the level 10, $(\delta_{w_{010}})$ is 1.89% more compared to that without intentional detour insertion. This additional delay is directly contributed by the extra wire used in detouring. We found that the wire length has been increased by 2.2% in our approach.

In next experiment we chose another design **B** with 32K flops and we selected first five levels of the tree to do the experiment. There is only one macro but there are a few small blockages scattered amidst the floor-plan. Cell placement ensured that the wire lengths of each net in a level remained same (no detours) but the cells were placed in a certain pattern that the symmetricity is not neatly maintained (like H-net $N23$ shown in Fig. 3).

TABLE III

FINAL LEVEL SKEW(PS) AND TOTAL WIRE DELAY(PS) COMPARISON BETWEEN WITHOUT AND WITH WIRE DETOUR FOR DESIGN B

	Without detour	With detour
ρ_{Lev_5}	26	24
$\delta_{w_{05}}$	615	621

In this experiment, we ran traditional H-tree routing which assumes the symmetricity and insert no jog/detour and we also ran our detouring approach that can add required jogs while tapping the driver. We used metal5 and metal6 routing layers. We found that the overall skew for five-level tree has been improved by 6.5% with minimal route length increase of 1% in our method as shown in Table III.

V. CONCLUSION

In this paper, we proposed a structured way of routing to reduce the clock skew to a great extent in today's deep sub-micron designs. This detouring approach is very much applicable for shallow but spatially spread clock-trees. For example, for creating driver tree for clock-mesh, this type of tree structure is used extensively. We relaxed the condition of uniform cell placement. We devised the algorithm where H shaped routes can be balanced with other H as well as I and SL shapes. Finally, we showed in two designs from industry where skew has been reduced up to 6.5% with minimal route length and wire delay increases in the selected clock-sub-trees.

In future there is scope for enhancing the algorithm to handle other shapes as well. For example, we can extend our method for the nets with three loads. For further skew reduction, post-route optimizations can be deployed after such a structured routing.

Acknowledgement

The author would like to thank Ashok Vittal of Synopsys India Pvt. Ltd., Bangalore for valuable discussions and Sambuddha Bhattacharya & Kapa Venkateswarlu of Synopsys India Pvt. Ltd., Bangalore for their comments and constructive suggestions.

REFERENCES

[1] M. A. B. Jackson, A. Srinivasan and E. S. Kuh, *Clock Routing for High Performance IC's*, Proc. DAC, 1990, pp. 573-579

[2] R. -S. Tsay, *Exact Zero Skew*, Proc. International Conference on Computer-Aided Designs, 1991, pp. 336-339

[3] M. Edahiro, *A Clustering Based Optimization Algorithm in Zero Skew Routing*, Proc. Design Automation Conference, 1993, pp. 612-616

[4] M. Edahiro, *An Efficient Zero Skew Routing Algorithm*, Proc. Design Automation Conference, 1994, pp. 375-380

[5] Ashok Vittal and Malgorzata Marek-Sadowska, *Power Optimal Buffered Clock Tree Design*, DAC, June 1995, pp. 497-502

[6] V. Natesan, D. Bhatia, *Clock-Skew Constrained Cell Placement*, Proc. Internation Conference on VLSI Design, Jan 1996, pp. 146-149

[7] A. B. Kahng, K. Masuko, S. Muddu, *Analytical Delay Models for VLSI Interconnects Under Ramp Input*, Proc. ICCAD, 1996, pp. 30-36

[8] Ashok Vittal and Malgorzata Marek-Sadowska, *Low-Power Buffered Clock Tree Design*, IEEE Transaction on Computer-Aided Design of Integrated Circuits and Systems, Vol. 16, Issue 9, Sept. 1997, pp. 965-975

[9] Jaewon Oh and M. Pedram, *Power Reduction in Microprocessor Chips by Gated Clock Routing*, Proc. ASP-DAC, 1998, pp. 313-318

[10] J. Jeon, D. Kim, D. Shin and K. Choi *High-Level Synthesis Under Multi-Cycle Interconnect Delay*, Proc. ASP-DAC, 2001, pp. 662-667

[11] Sachin Sapatnekar, *Timing*, Kluwer Academic Publishers, 2004

[12] Jeng-Liang Tsai and C. Chung-Ping Chen, *Process-Variation Robust and Low-Power Zero-Skew Buffered Clock-Tree Synthesis Unsing Projected Sacn-Line Sampling*, Proc. ASP-DAC, 2005, pp. 1168-1171

[13] Ashok Narasimhan and Sridhar Ramalingam, *Impact of Variability on Clock Skew in H-tree Clock Networks*, International Symposium on Quality Electronic Design, 2007, pp. 458 - 466

[14] M. B. I. Reaz, N. Amin, M. I. Ibrahimy, F. Mohd-Yasin, A. Mohammad *Zero Skew Clock Routing For Fast Clock Tree Generation*, Proc. CCECE, 2008 pp. 23 - 28

2012 25th International Conference on VLSI Design

Accurate Leakage Estimation for FinFET Standard Cells Using the Response Surface Methodology

Sourindra Chaudhuri, Prateek Mishra and Niraj K. Jha
Department of Electrical Engineering
Princeton University
Princeton, NJ, 08544
{schaudhu, pmishra, jha}@princeton.edu

Abstract—Among different multi-gate transistors, FinFETs and Trigate FETs have set themselves apart as the most promising candidates for the upcoming 22nm technology node and beyond owing to their superior device performance, lower leakage power consumption and cost-effective fabrication process. Innovative circuit design and optimization techniques will be required to harness the power of multi-gate transistors, which in turn will depend on accurate leakage and timing characterization of these devices under spatial and environmental variations. Hence, in order to aid circuit designers, we present accurate analytical models using central composite rotatable design (CCRD) based on response surface methodology (RSM) to estimate the leakage current in FinFET standard cells under the effect of variations in gate length (L_G), fin thickness (T_{SI}), gate-oxide thickness (T_{OX}) and gate-workfunction (Φ_G). To the best of our knowledge, this is the first attempt to develop analytical models for leakage estimation of FinFET devices/logic gates based on TCAD simulations of adjusted 2D device cross-sections that have been shown to track TCAD simulations of 3D device behavior within a 1-3% error range. This drastically reduces the CPU time of our modeling technique (by several orders of magnitude) without much loss in accuracy. We present analytical leakage models for different logic styles, e.g., shorted-gate (SG) and independent-gate (IG) FinFETs, at the 22nm technology node. The leakage estimates derived from the analytical models are in close agreement with quasi-Monte Carlo (QMC) simulation results obtained for different adjusted-2D (3D) devices/logic gates with a maximum root mean square error (RMSE) of 5.28% (7.03%).

I. INTRODUCTION

Multi-gate transistors [1], e.g., FinFETs and Trigate FETs, are expected to start replacing conventional MOSFETs in the next few years owing to their smaller subthreshold leakage, superior gate-control over the channel and reduced sensitivity to process variations. Initially, the cost and complexity associated with fabrication were the prime challenges in making these devices the industry driver. These are no longer critical as the key players in the semiconductor industry have managed to fabricate these devices at roughly the same cost and with minor modifications to the conventional CMOS fabrication process. Superior performance, significantly lower leakage and low fabrication cost are the main reasons behind recent adoption of these devices by various companies.

In this paper, we focus on FinFETs, which are the most promising variant among different multi-gate transistors ex-

Acknowledgments: This work was supported by SRC under contract no. 2010-HJ-2079.

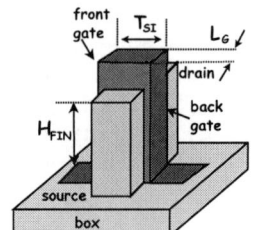

Fig. 1: A typical n/p-FinFET device

plored so far. FinFET is usually a silicon-on-insulator (SOI) structure with a thin vertical silicon fin placed on an insulating base and surrounded by metal gates (Fig. 1). The top gate is separated from the channel by placing a thick spacer material in between, which makes it a double-gate structure. By eliminating the spacer, the structure can be transformed into a Trigate device. The ratio of fin body thickness (T_{SI}) to gate length (L_G) is usually kept small to ensure tighter control over the channel by the gate. To address the problem of random dopant fluctuations, doping in the channel region is typically maintained at a very low level (close to undoped). FinFETs can be operated in two different modes: shorted-gate (SG) and independent-gate (IG). In the SG mode of operation, the front and back gates are electrically connected to deliver maximum gate drive, whereas in the IG mode, the top of the FinFET is etched away to make the gates independent so that they can be biased to different values. In the IG mode of operation, back-gate bias can be used to alter the threshold voltage (V_{th}) of the front gate. This is very useful in reducing the off current, i.e., leakage current (I_{LEAK}), of the transistor.

Though FinFETs can deliver superior performance at a lower leakage and similar cost relative to CMOS, they are still susceptible to process and environmental variations. As FinFETs become mainstream, characterizing their leakage and timing will become essential for circuit design and optimization. The key challenge in characterizing a FinFET is its complex structure, which requires much longer processing times as compared to traditional MOSFETs. However, we have shown that it is possible to find equivalent adjusted 2D cross-sections of FinFETs that closely (error within 1-3%) track 3D device behavior and remain valid even under process variations [15]. We still need to characterize leakage and timing behavior of these devices under the influence of process and environmental variations and adjusted 2D cross-sections

978-1-4673-0438-2/12 $31.00 © 2012 IEEE   238

can make this job significantly easier. The key contributions of this paper are as follows:

- Since, L_G, T_{SI}, T_{OX} and Φ_G have been shown to have a significant impact on I_{LEAK} [14], we consider variations in these parameters to develop leakage models for n/pFinFETs, inverters and NAND gates implemented in SG and IG modes of operation using CCRD based on RSM.
- Instead of simulating complete 3D devices, we deploy CCRD on adjusted 2D cross-sections of FinFET devices, which provide accuracy close to that obtained from 3D device simulation, yet in several orders of magnitude less time (the time reduces from CPU days to minutes).
- Finally, we show that the leakage estimates obtained from the RSM models presented in this paper are in close agreement [maximum RMSE of 5.28% (7.03%)] with the QMC [22] simulation results for adjusted-2D (3D) FinFET devices and logic gates.

The rest of the paper is organized as follows. In Section II, we review related work. In Section III, we give a detailed description of the simulation setup. In Section IV, we discuss design of experiments used for modeling leakage in FinFET devices/gates. In Section V, we discuss validation of our models. Finally, in Section VI, we conclude.

II. RELATED WORK

The promise of multi-gate transistors has motivated researchers to explore the impact of these devices at the circuit level. Research has already been done on obtaining different leakage-delay trade-offs based on the SG/IG modes of operation [2]–[5]. To optimize power of FinFET circuits, gate sizing and multiple supply/threshold voltages have been established as effective strategies in [5]–[8]. Use of Φ_G is recommended to control device V_{th} under process variations [9]. The dependence of I_{LEAK} on device temperature has also been investigated [10]–[12].

Researchers have developed a quasi-3D numerical model to simulate modern FinFET structures [17]. This model accounts for ballistic transport along the channel. It yields very accurate results for devices with minimum length below 10nm because it is developed based on non-equilibrium Green's function. L_G, T_{SI}, T_{OX} and Φ_G have been shown to be critical in determining device behavior under process variations [14]. The leakage-delay spectrum has been explored for various FinFET standard cells in [13], however, not under process variations. Simulations run on a fully self-consistent 3D quantum transport simulator show a clear discrepancy between 2D and 3D simulation of a FinFET device [18]. Hence, work has been done to bridge this accuracy gap between 3D and 2D simulations [15] through adjustment of 2D cross-sections.

III. SIMULATION SETUP

Our simulation platform integrates Sentaurus TCAD Structure Editor (SE), device simulator, Inspect [16], MATLAB and Python scripts. It automates mixed-mode device simulation to estimate leakage in FinFET standard logic cells under process variations. MATLAB and Python scripts are used to postprocess the output, develop analytical models using CCRD and validate them against QMC simulations.

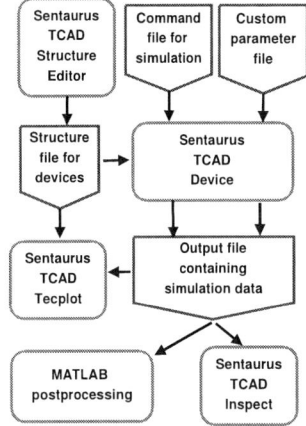

Fig. 2: Simulation flow

Fig. 2 shows the simulation flow, which includes generation of the desired 2D FinFET structure, simulation of the same to estimate leakage and postprocessing for the purpose of developing/validating the models. We target SG and IG FinFET devices and logic gates implemented in the 22nm technology node. Fig. 3 shows a 3D nFinFET and Fig. 4 its 2D cross-section. Heavily-doped extended source and drain regions are identified by $(H_{CON} \times L_{CON})$. They eventually terminate at the undoped channel region. The figure also makes it apparent that the doping concentration decreases gradually instead of changing in an abrupt fashion. This gives rise to an underlap (L_{UN}) region near both the source and drain. L_{UN} is defined as the distance from the physical gate edge to the point where the doping concentration starts decreasing from its peak value. For high-performance logic, a Φ_G of 4.4eV (4.8eV) is used for the nFinFET (pFinFET) and a back gate bias of -0.2V for IG-mode simulations [14].

Table I provides the values assumed for different physical parameters that describe a typical n/pFinFET at the 22nm technology node. Referring to Fig. 4, the process parameters are front physical gate length (L_{GF}), back physical gate length (L_{GB}), front gate oxide thickness (T_{OXF}), back gate oxide thickness (T_{OXB}), front gate thickness (H_{GF}), back gate thickness (H_{GB}), front gate spacer thickness (L_{SPF}), back gate spacer thickness (L_{SPB}) and L_{UN}. W_{NEFF} and W_{PEFF} denote the effective widths of an nFinFET and pFinFET, respectively, and are defined as $2nH_{FIN}$ (where n is the number of fins in the FinFET and H_{FIN} is the fin height). These effective widths take into account both the front and back inversion channels formed in a typical FinFET. Other device parameters are body doping (N_{BODY}), source/drain doping (N_{SD}), fin pitch (F_P) that denotes the minimum distance required between the midpoint of two adjacent fins and operating voltage (V_{DD}).

For all our simulations, we consider bandgap narrowing models, the Philips-unified model for carrier mobility, and Shockley-Read-Hall and band-to-band tunneling models to

TABLE I: FinFET device parameter values

Parameter name (Unit)	Value
L_{GF}, L_{GB} (nm)	20
T_{OXF}, T_{OXB} (nm)	1
T_{SI} (nm)	10
H_{FIN} (nm)	40
H_{GF}, H_{GB} (nm)	34
L_{SPF}, L_{SPB} (nm)	12.5
L_{UN} (nm)	10
F_P (nm)	60
N_{BODY} (cm^{-3})	10^{16}
Φ_G (eV)	nFET: 4.4, pFET: 4.8
N_{SD} (cm^{-3})	10^{20}
V_{DD} (V)	1.1

TABLE II: Adjusted parameter values for SG (IG) Fin-FETs [15]

Device	Parameter (Unit)	Unadjusted	Adjusted
nFinFET	L_{GF}, L_{GB} (nm)	20	19 (18)
	Φ_G (eV)	4.4	4.384 (4.385)
stacked nFinFET	L_{GF}, L_{GB} (nm)	20	19 (18)
	Φ_G (eV)	4.4	4.392 (4.393)
pFinFET	L_{GF}, L_{GB} (nm)	20	19.5 (18.5)
	Φ_G (eV)	4.8	4.818 (4.816)
pFinFET($2\times$)	L_{GF}, L_{GB} (nm)	20	21 (21)
	Φ_G (eV)	4.8	4.814 (4.817)

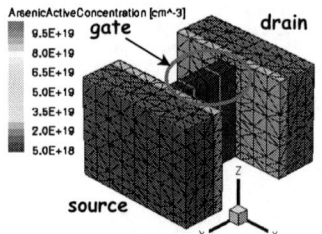

Fig. 3: 3D nFinFET device created by SE

accurately capture the effect of carrier recombination [16]. Table II identifies the adjusted parameter values for SG and IG devices ($2\times$ represents a device with two fins), which have been shown to effectively capture 3D device behavior [15]. We use CCRD on these adjusted cross-sections to model leakage under process variations.

IV. DESIGN OF EXPERIMENT (DOE)

To develop analytical leakage models for FinFET logic gates, a wide variety of DOEs are possible. In this work, we rely on CCRD based on RSM to develop such models for SG/IG n/pFinFET, SG/IG inverter (SG/IG-INV) and SG/IG NAND gate (SG/IG-NAND). DOE first determines the most influential input parameters through screening of all available input parameters, then analyzes the combined influence of those important input parameters on the output. DOEs can be divided into three main categories: screening, full factorial and response surface [19], with varying efficiencies in covering the design space.

The screening design eliminates less important input parameters to reduce the design space, thus compromising accuracy.

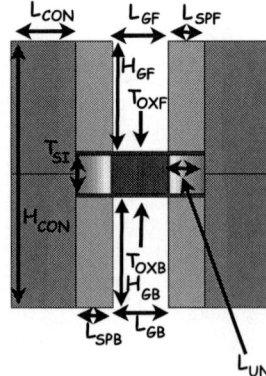

Fig. 4: 2D (X-Y) cross-section of the 3D nFinFET device

In the full factorial design, each input variable can have multiple levels, thus significantly increasing problem complexity. The response surface design usually combines different statistical and mathematical methods to model engineering problems. It optimizes a response surface characterized by various input parameters and determines its dependence on those parameters. The number of levels employed for each input variable is based on the dimension of the model under consideration. Though both full factorial and response surface designs yield highly accurate results because of their conservative nature, they become impractical for higher-order models that depend on a large number of input variables. In this work, we consider quadratic models. For modeling a quadratic response surface, each variable needs to have three levels. Thus, for a t-variable design, the total number of experiments required is 3^t, which makes this procedure impractical for large t.

CCRD addresses the problems discussed above and was first proposed by Box and Wilson and further improved in [20]. It is geometrically demonstrated in Fig. 5. It includes a center point, six axial points and eight factorial cube points. Hence, for a t-variable design, the total number of experiments required is $2^t + 2t + 1$, which can be reduced by replacing the factorial portion of the design with fractional factorial [21]. The inclusion of factorial and axial points helps generate the linear and quadratic terms, respectively. To maintain a constant variation of the model from the design center, axial points are chosen in such a fashion that they allow rotatability. After determining the range of variation in input parameters, typically assumed to be around $\pm 10\%$ of their mean values, we encode them as ± 1 for factorial points, $\pm \beta$ for axial points and 0 for the center point. To take into account process variations in Φ_G, a variation of $\pm 10\%$ in $|\Phi_G - \Phi_S|$ is actually considered, where Φ_S represents the workfunction of the intrinsic semiconductor material. Φ_S is typically 4.6eV. Table III summarizes all the formulas used to calculate the corresponding values of the encoded CCRD levels where the minimum (maximum) value is represented by x_{min} (x_{max}) [23]. The value of α is calculated using the formula $\alpha = 2^{t/4}$. To model leakage in a single n/pFinFET, we consider four input variables: adjusted gate length of n/pFinFET (L_{Gn}/L_{Gp}), T_{SI}, T_{OX} and adjusted gate workfunction of n/pFinFET (Φ_{Gn}/Φ_{Gp}). We consider full factorial CCRD in these cases. Hence, the total number of experiments required is 25 ($t = 4$) with the value of $\alpha = 2$. To model leakage in a FinFET logic gate, e.g., inverter/NAND, we consider six input variables: L_{Gn}, L_{Gp}, T_{SI}, T_{OX}, Φ_{Gn} and Φ_{Gp}, with full factorial CCRD. Hence, the total number

TABLE III: Relationship between coded and actual variable values

Code	Actual value of variable
$-\beta$	x_{min}
-1	$[(x_{max}+x_{min})/2] - [(x_{max}-x_{min})/2\alpha]$
0	$[(x_{max}+x_{min})/2]$
$+1$	$[(x_{max}+x_{min})/2] + [(x_{max}-x_{min})/2\alpha]$
$+\beta$	x_{max}

TABLE IV: Process parameters of SG FinFETs/logic gates along with their levels for CCRD

Device	Process parameter	Coded variable level				
		Lowest $-\beta$	Low -1	Center 0	High $+1$	Highest $+\beta$
nFinFET	L_{Gn} (nm)	17	18	19	20	21
	T_{SI} (nm)	9.0	9.5	10.0	10.5	11.0
	T_{OX} (nm)	0.9	0.95	1.0	1.05	1.1
	Φ_{Gn} (eV)	4.364	4.374	4.384	4.394	4.404
pFinFET	L_{Gp} (nm)	17.5	18.5	19.5	20.5	21.5
	T_{SI} (nm)	9.0	9.5	10.0	10.5	11.0
	T_{OX} (nm)	0.9	0.95	1.0	1.05	1.1
	Φ_{Gp} (eV)	4.798	4.808	4.818	4.828	4.838
Inverter	L_{Gn} (nm)	17	18.30	19	19.70	21
	L_{Gp} (nm)	19	20.30	21	21.70	23
	T_{SI} (nm)	9.0	9.64	10.0	10.36	11.0
	T_{OX} (nm)	0.9	0.96	1.0	1.04	1.1
	Φ_{Gn} (eV)	4.364	4.377	4.384	4.391	4.404
	Φ_{Gp} (eV)	4.794	4.807	4.814	4.821	4.834
NAND	L_{Gn} (nm)	17	18.28	19	19.72	21
	L_{Gp} (nm)	17.5	18.78	19.5	20.22	21.5
	T_{SI} (nm)	9.0	9.64	10.0	10.36	11.0
	T_{OX} (nm)	0.9	0.96	1.0	1.04	1.1
	Φ_{Gn} (eV)	4.364	4.377	4.384	4.391	4.404
	Φ_{Gn} (eV) (stacked)	4.372	4.385	4.392	4.399	4.412
	Φ_{Gp} (eV)	4.798	4.811	4.818	4.825	4.838

TABLE V: Process parameters of IG FinFETs/logic gates along with their levels for CCRD

Device	Process parameter	Coded variable level				
		Lowest $-\beta$	Low -1	Center 0	High $+1$	Highest $+\beta$
nFinFET	L_{Gn} (nm)	16	17	18	19	20
	T_{SI} (nm)	9.0	9.5	10.0	10.5	11.0
	T_{OX} (nm)	0.9	0.95	1.0	1.05	1.1
	Φ_{Gn} (eV)	4.365	4.375	4.385	4.395	4.405
pFinFET	L_{Gp} (nm)	16.5	17.5	18.5	19.5	20.5
	T_{SI} (nm)	9.0	9.5	10.0	10.5	11.0
	T_{OX} (nm)	0.9	0.95	1.0	1.05	1.1
	Φ_{Gp} (eV)	4.796	4.806	4.816	4.826	4.836
Inverter	L_{Gn} (nm)	16	17.30	18	18.70	20
	L_{Gp} (nm)	19	20.30	21	21.70	23
	T_{SI} (nm)	9.0	9.64	10.0	10.36	11.0
	T_{OX} (nm)	0.9	0.96	1.0	1.04	1.1
	Φ_{Gn} (eV)	4.365	4.378	4.385	4.392	4.405
	Φ_{Gp} (eV)	4.797	4.810	4.817	4.824	4.837
NAND	L_{Gn} (nm)	16	17.28	18	18.72	20
	L_{Gp} (nm)	16.5	17.78	18.5	19.22	20.5
	T_{SI} (nm)	9.0	9.64	10.0	10.36	11.0
	T_{OX} (nm)	0.9	0.96	1.0	1.04	1.1
	Φ_{Gn} (eV)	4.365	4.378	4.385	4.392	4.405
	Φ_{Gn} (eV) (stacked)	4.373	4.386	4.393	4.400	4.413
	Φ_{Gp} (eV)	4.796	4.809	4.816	4.823	4.836

TABLE VI: Coded process parameters along with their actual values for SG nFinFET

Code				L_{Gn}	T_{SI}	T_{OX}	Φ_{Gn}
-1	-1	-1	-1	18	9.5	0.95	4.374
-1	-1	-1	1	18	9.5	0.95	4.394
-1	-1	1	-1	18	9.5	1.05	4.374
-1	-1	1	1	18	9.5	1.05	4.394
-1	1	-1	-1	18	10.5	0.95	4.374
-1	1	-1	1	18	10.5	0.95	4.394
-1	1	1	-1	18	10.5	1.05	4.374
-1	1	1	1	18	10.5	1.05	4.394
1	-1	-1	-1	20	9.5	0.95	4.374
1	-1	-1	1	20	9.5	0.95	4.394
1	-1	1	-1	20	9.5	1.05	4.374
1	-1	1	1	20	9.5	1.05	4.394
1	1	-1	-1	20	10.5	0.95	4.374
1	1	-1	1	20	10.5	0.95	4.394
1	1	1	-1	20	10.5	1.05	4.374
1	1	1	1	20	10.5	1.05	4.394
$-\beta$	0	0	0	17	10.0	1.00	4.384
β	0	0	0	21	10.0	1.00	4.384
0	$-\beta$	0	0	19	9.0	1.00	4.384
0	β	0	0	19	11.0	1.00	4.384
0	0	$-\beta$	0	19	10.0	0.90	4.384
0	0	β	0	19	10.0	1.10	4.384
0	0	0	$-\beta$	19	10.0	1.00	4.364
0	0	0	β	19	10.0	1.00	4.404
0	0	0	0	19	10.0	1.00	4.384

of experiments required for logic gates is 77 ($t = 6$) with the value of $\alpha = 2.83$. The values for different levels of CCRD input variables are summarized in Tables IV and V for SG and IG modes of operation, respectively. In Table VI (VII), we summarize the values of all input variables at different levels for all the required experiments to model leakage in a single SG nFinFET (SG-INV). Tables can similarly be obtained for SG pFinFET, IG n/pFinFET, IG-INV and SG/IG-NAND.

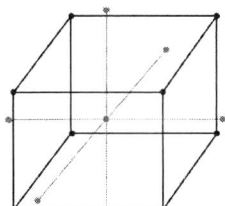

Fig. 5: CCRD for $t = 3$

We have now defined all the necessary components required for CCRD based on RSM. RSM based models are typically quadratic in predictor variables, which denote a valid relationship between the response variable and different predictor levels. A typical RSM polynomial can be realized as:

$$Y_i = \beta_0 + \beta_1 X_i + \beta_2 X_i X_j + \epsilon \tag{1}$$

where X_i's are the predictor variables that predict the response variable Y_i. This is achieved by finding the coefficients of regression, e.g., $\beta_0, \beta1$ and β_2, with ϵ being the predictor noise. Regression analysis generally finds these coefficients by minimizing ϵ over the whole sample space. We performed this analysis with the help of MATLAB 7.0. Since leakage

in a FinFET device or logic gate has been observed to be exponentially dependent on process parameters [14], Y_i represents $\ln(I_{LEAK})$ with I_{LEAK} measured in ampere. Table VIII shows the coefficients found for FinFETs, where $x1$, $x2$, $x3$ and $x4$, respectively, denote L_{Gn}/L_{Gp} (depending on whether it is an n/pFinFET), T_{SI}, T_{OX} and Φ_{Gn}/Φ_{Gp} (depending on whether it is an n/pFinFET). The coefficients are reported up to four decimal places and anything less significant than that is shown as 0. Tables IX, X and XI give the coefficients obtained for input vector dependent leakage models of both SG/IG-INV and SG/IG-NAND, where $x1$, $x2$, $x3$, $x4$, $x5$ and $x6$, respectively, represent L_{Gn}, L_{Gp}, T_{SI}, T_{OX}, Φ_{Gn} and Φ_{Gp}.

V. VALIDATION OF THE RSM MODEL

We next validate the leakage models presented in the previous section by comparing the results obtained from these models against QMC simulation results based on both adjusted-2D and 3D devices/gates.

TABLE VII: Coded process parameters along with their actual values for SG-INV

Code						L_{Gn}	L_{Gp}	T_{SI}	T_{OX}	Φ_{Gn}	Φ_{Gp}
−1	−1	−1	−1	−1	−1	18.30	20.30	9.64	0.96	4.377	4.807
−1	−1	−1	−1	−1	1	18.30	20.30	9.64	0.96	4.377	4.821
−1	−1	−1	−1	1	−1	18.30	20.30	9.64	0.96	4.391	4.807
−1	−1	−1	−1	1	1	18.30	20.30	9.64	0.96	4.391	4.821
−1	−1	−1	1	−1	−1	18.30	20.30	9.64	1.04	4.377	4.807
−1	−1	−1	1	−1	1	18.30	20.30	9.64	1.04	4.377	4.821
−1	−1	−1	1	1	−1	18.30	20.30	9.64	1.04	4.391	4.807
−1	−1	−1	1	1	1	18.30	20.30	9.64	1.04	4.391	4.821
−1	−1	1	−1	−1	−1	18.30	20.30	10.36	0.96	4.377	4.807
−1	−1	1	−1	−1	1	18.30	20.30	10.36	0.96	4.377	4.821
−1	−1	1	−1	1	−1	18.30	20.30	10.36	0.96	4.391	4.807
−1	−1	1	−1	1	1	18.30	20.30	10.36	0.96	4.391	4.821
−1	−1	1	1	−1	−1	18.30	20.30	10.36	1.04	4.377	4.807
−1	−1	1	1	−1	1	18.30	20.30	10.36	1.04	4.377	4.821
−1	−1	1	1	1	−1	18.30	20.30	10.36	1.04	4.391	4.807
−1	−1	1	1	1	1	18.30	20.30	10.36	1.04	4.391	4.821
−1	1	−1	−1	−1	−1	18.30	21.70	9.64	0.96	4.377	4.807
−1	1	−1	−1	−1	1	18.30	21.70	9.64	0.96	4.377	4.821
−1	1	−1	−1	1	−1	18.30	21.70	9.64	0.96	4.391	4.807
−1	1	−1	−1	1	1	18.30	21.70	9.64	0.96	4.391	4.821
−1	1	−1	1	−1	−1	18.30	21.70	9.64	1.04	4.377	4.807
−1	1	−1	1	−1	1	18.30	21.70	9.64	1.04	4.377	4.821
−1	1	−1	1	1	−1	18.30	21.70	9.64	1.04	4.391	4.807
−1	1	−1	1	1	1	18.30	21.70	9.64	1.04	4.391	4.821
−1	1	1	−1	−1	−1	18.30	21.70	10.36	0.96	4.377	4.807
−1	1	1	−1	−1	1	18.30	21.70	10.36	0.96	4.377	4.821
−1	1	1	−1	1	−1	18.30	21.70	10.36	0.96	4.391	4.807
−1	1	1	−1	1	1	18.30	21.70	10.36	0.96	4.391	4.821
−1	1	1	1	−1	−1	18.30	21.70	10.36	1.04	4.377	4.807
−1	1	1	1	−1	1	18.30	21.70	10.36	1.04	4.377	4.821
−1	1	1	1	1	−1	18.30	21.70	10.36	1.04	4.391	4.807
−1	1	1	1	1	1	18.30	21.70	10.36	1.04	4.391	4.821
1	−1	−1	−1	−1	−1	19.70	20.30	9.64	0.96	4.377	4.807
1	−1	−1	−1	−1	1	19.70	20.30	9.64	0.96	4.377	4.821
1	−1	−1	−1	1	−1	19.70	20.30	9.64	0.96	4.391	4.807
1	−1	−1	−1	1	1	19.70	20.30	9.64	0.96	4.391	4.821
1	−1	−1	1	−1	−1	19.70	20.30	9.64	1.04	4.377	4.807
1	−1	−1	1	−1	1	19.70	20.30	9.64	1.04	4.377	4.821
1	−1	−1	1	1	−1	19.70	20.30	9.64	1.04	4.391	4.807
1	−1	−1	1	1	1	19.70	20.30	9.64	1.04	4.391	4.821
1	−1	1	−1	−1	−1	19.70	20.30	10.36	0.96	4.377	4.807
1	−1	1	−1	−1	1	19.70	20.30	10.36	0.96	4.377	4.821
1	−1	1	−1	1	−1	19.70	20.30	10.36	0.96	4.391	4.807
1	−1	1	−1	1	1	19.70	20.30	10.36	0.96	4.391	4.821
1	−1	1	1	−1	−1	19.70	20.30	10.36	1.04	4.377	4.807
1	−1	1	1	−1	1	19.70	20.30	10.36	1.04	4.377	4.821
1	−1	1	1	1	−1	19.70	20.30	10.36	1.04	4.391	4.807
1	−1	1	1	1	1	19.70	20.30	10.36	1.04	4.391	4.821
1	1	−1	−1	−1	−1	19.70	21.70	9.64	0.96	4.377	4.807
1	1	−1	−1	−1	1	19.70	21.70	9.64	0.96	4.377	4.821
1	1	−1	−1	1	−1	19.70	21.70	9.64	0.96	4.391	4.807
1	1	−1	−1	1	1	19.70	21.70	9.64	0.96	4.391	4.821
1	1	−1	1	−1	−1	19.70	21.70	9.64	1.04	4.377	4.807
1	1	−1	1	−1	1	19.70	21.70	9.64	1.04	4.377	4.821
1	1	−1	1	1	−1	19.70	21.70	9.64	1.04	4.391	4.807
1	1	−1	1	1	1	19.70	21.70	9.64	1.04	4.391	4.821
1	1	1	−1	−1	−1	19.70	21.70	10.36	0.96	4.377	4.807
1	1	1	−1	−1	1	19.70	21.70	10.36	0.96	4.377	4.821
1	1	1	−1	1	−1	19.70	21.70	10.36	0.96	4.391	4.807
1	1	1	−1	1	1	19.70	21.70	10.36	0.96	4.391	4.821
1	1	1	1	−1	−1	19.70	21.70	10.36	1.04	4.377	4.807
1	1	1	1	−1	1	19.70	21.70	10.36	1.04	4.377	4.821
1	1	1	1	1	−1	19.70	21.70	10.36	1.04	4.391	4.807
1	1	1	1	1	1	19.70	21.70	10.36	1.04	4.391	4.821
−β	0	0	0	0	0	17.00	21.00	10.00	1.00	4.384	4.814
β	0	0	0	0	0	21.00	21.00	10.00	1.00	4.384	4.814
0	−β	0	0	0	0	19.00	19.00	10.00	1.00	4.384	4.814
0	β	0	0	0	0	19.00	23.00	10.00	1.00	4.384	4.814
0	0	−β	0	0	0	19.00	21.00	9.00	1.00	4.384	4.814
0	0	β	0	0	0	19.00	21.00	11.00	1.00	4.384	4.814
0	0	0	−β	0	0	19.00	21.00	10.00	0.90	4.384	4.814
0	0	0	β	0	0	19.00	21.00	10.00	1.10	4.384	4.814
0	0	0	0	−β	0	19.00	21.00	10.00	1.00	4.364	4.814
0	0	0	0	β	0	19.00	21.00	10.00	1.00	4.404	4.814
0	0	0	0	0	−β	19.00	21.00	10.00	1.00	4.384	4.794
0	0	0	0	0	β	19.00	21.00	10.00	1.00	4.384	4.834
0	0	0	0	0	0	19.00	21.00	10.00	1.00	4.384	4.814

TABLE VIII: RSM $\ln(I_{LEAK})$ model coefficients for SG/IG FinFETs

Variable	Coefficient values			
	SG nFinFET	SG pFinFET	IG nFinFET	IG pFinFET
const	-218.7566	-1446.5000	1111.9000	1436.7000
$x1$	0.2789	-2.5956	1.0390	-1.5115
$x2$	-3.0146	5.1270	-3.3721	2.4858
$x3$	-12.9068	22.6787	-22.2912	3.6547
$x4$	132.0084	552.2046	-475.8330	-641.6192
$x1.x2$	-0.0103	-0.0025	0.0022	0.0038
$x1.x3$	-0.0447	-0.0021	-0.0124	-0.0325
$x1.x4$	-0.1468	0.4707	-0.4318	0.1829
$x2.x3$	0.1645	0.1570	0.1112	-0.0464
$x2.x4$	0.7961	-0.9315	0.8209	-0.3558
$x3.x4$	3.3439	-3.8334	4.1933	-1.1053
$x1^2$	0.0103	0.0056	0.0190	0.0116
$x2^2$	0.0008	-0.0125	0.0091	-0.0077
$x3^2$	-0.4767	-1.9922	2.2887	2.2944
$x4^2$	-19.7836	-53.3875	49.9166	70.2763

TABLE IX: RSM $\ln(I_{LEAK})$ model coefficients for FinFET inverters

Variable	Coefficient values			
	SG-INV		IG-INV	
	0-$\ln(I_{LEAK})$	1-$\ln(I_{LEAK})$	0-$\ln(I_{LEAK})$	1-$\ln(I_{LEAK})$
const	630.1314	-313.9584	-5815.8000	202.8466
$x1$	0.4665	-0.0024	2.2287	-0.0357
$x2$	-0.0625	-1.6919	0.5783	-1.5950
$x3$	-3.8539	4.5179	-2.3163	4.7065
$x4$	-15.0982	19.2815	-4.1205	14.2519
$x5$	-95.1056	-5.4918	1212.7000	-86.9939
$x6$	-143.3855	84.6699	1326.5000	-56.4607
$x1.x2$	0.0000	0.0000	0.0000	0.0000
$x1.x3$	-0.0089	0.0000	0.0017	0.0000
$x1.x4$	-0.0407	0.0000	-0.0203	0.0000
$x1.x5$	-0.2164	0.0000	-0.5469	0.0000
$x1.x6$	0.0000	0.0000	0.0000	0.0000
$x2.x3$	0.0000	-0.0095	0.0000	-0.0096
$x2.x4$	0.0000	-0.0321	0.0000	-0.0338
$x2.x5$	0.0000	0.0000	0.0000	0.0000
$x2.x6$	0.0000	0.2817	0.0000	0.2009
$x3.x4$	0.1530	0.1600	0.1060	0.2962
$x3.x5$	0.9378	0.0000	0.9286	0.0000
$x3.x6$	0.0000	-0.9001	0.0000	-0.9078
$x4.x5$	3.4146	0.0000	3.3063	0.0000
$x4.x6$	0.0000	-3.8375	0.0000	-3.3176
$x5.x6$	-0.0003	0.0000	-0.0001	0.0000
$x1^2$	0.0129	0.0001	-0.0002	0.0010
$x2^2$	0.0015	0.0082	-0.0138	0.0146
$x3^2$	0.0110	0.0180	-0.0666	0.0117
$x4^2$	0.4850	0.0336	-4.7564	0.6672
$x5^2$	6.0085	0.6264	-142.3986	9.9195
$x6^2$	14.8927	-4.4157	-137.6870	10.3294

For QMC simulation of adjusted-2D devices/gates, we considered 1000 combinations of values for the input variables based on the concept of Sobol sequence, which ensures that points are evenly distributed in the testing space. This method, which is primarily based on low-discrepancy sequences, is known to converge quickly with much fewer samples relative to complete Monte Carlo simulation [22]. We assume that all input variables are normally distributed and varied simultaneously. To test the models developed for single FinFETs, we varied L_{Gn}/L_{Gp}, T_{SI}, T_{OX} and Φ_{Gn}/Φ_{Gp} simultaneously. Likewise, all six input variables were varied to verify input-vector dependent leakage models for inverters and NAND gates. The leakage distribution derived from such QMC simulations shows excellent agreement with the distribution obtained from the RSM model for SG n/pFinFETs, inverter and NAND gate, as shown in Figs. 6, 7 and 8, respectively. This is true for IG devices and logic gates as well. Table XII shows that the testing RMSE is at most 2.78% (5.28%) for devices/gates based on SG (IG) FinFETs. In Figs. 9,

TABLE X: RSM $\ln(I_{LEAK})$ model coefficients for FinFET SG-NAND

Variable	Coefficient values			
	00-$\ln(I_{LEAK})$	01-$\ln(I_{LEAK})$	10-$\ln(I_{LEAK})$	11-$\ln(I_{LEAK})$
const	73.2272	626.5669	517.1655	2957.3000
$x1$	0.2294	0.4279	-0.0131	-0.2870
$x2$	0.0000	-0.0571	-0.0468	-3.5143
$x3$	-2.9469	-3.7814	-2.7715	4.2243
$x4$	-12.9203	-14.8345	-13.8860	13.9818
$x5$	-1.2025	-92.3907	-72.0579	-678.8023
$x6$	0.0730	-145.0673	-118.7617	-649.3856
$x1.x2$	0.0000	0.0000	0.0000	0.0000
$x1.x3$	-0.0058	-0.0086	-0.0112	0.0000
$x1.x4$	-0.0266	-0.0392	-0.0490	0.0000
$x1.x5$	-0.0772	-0.2077	-0.0961	0.0000
$x1.x6$	0.0000	0.0000	0.0000	0.0000
$x2.x3$	0.0000	0.0000	0.0000	-0.0006
$x2.x4$	0.0000	0.0000	0.0000	0.0067
$x2.x5$	0.0000	0.0000	0.0000	0.0000
$x2.x6$	0.0000	0.0000	0.0000	0.5555
$x3.x4$	0.0663	0.1520	0.1671	0.1544
$x3.x5$	0.7598	0.9201	0.7042	0.0000
$x3.x6$	0.0000	0.0000	0.0000	-0.9551
$x4.x5$	3.1620	3.3508	3.1822	0.0000
$x4.x6$	0.0000	0.0001	0.0000	-3.9605
$x5.x6$	0.0000	-0.0005	-0.0002	-0.0001
$x1^2$	0.0030	0.0129	0.0126	0.0076
$x2^2$	0.0000	0.0015	0.0012	0.0178
$x3^2$	0.0000	0.0110	0.0091	0.0366
$x4^2$	-0.0095	0.4827	0.3784	2.5897
$x5^2$	-4.7155	5.7960	3.5023	77.4182
$x6^2$	-0.0076	15.0550	12.3249	71.1786

TABLE XI: RSM $\ln(I_{LEAK})$ model coefficients for FinFET IG-NAND

Variable	Coefficient values			
	00-$\ln(I_{LEAK})$	01-$\ln(I_{LEAK})$	10-$\ln(I_{LEAK})$	11-$\ln(I_{LEAK})$
const	-1667.9000	-5435.8000	-4163.2000	-5753.7000
$x1$	0.1320	2.1659	1.2948	0.4665
$x2$	0.1505	0.4649	0.3629	-0.9738
$x3$	-1.9735	-2.3841	-2.5473	4.2007
$x4$	-7.3814	-4.7218	-8.3780	22.6112
$x5$	346.8590	1135.3000	859.4257	1164.9000
$x6$	401.7406	1240.6000	968.3421	1272.7000
$x1.x2$	0.0000	0.0000	0.0000	0.0000
$x1.x3$	0.0068	0.0017	-0.0068	0.0000
$x1.x4$	0.0107	-0.0194	-0.0315	0.0000
$x1.x5$	-0.0876	-0.5424	-0.3345	0.0000
$x1.x6$	0.0000	0.0000	0.0000	0.0000
$x2.x3$	0.0000	0.0000	0.0000	0.0041
$x2.x4$	0.0000	0.0000	0.0000	-0.0880
$x2.x5$	0.0000	0.0000	0.0000	0.0000
$x2.x6$	0.0000	0.0000	0.0000	0.2320
$x3.x4$	0.0603	0.1047	0.1616	-0.0604
$x3.x5$	0.6403	0.9279	0.9008	0.0000
$x3.x6$	0.0000	0.0000	0.0000	-0.3843
$x4.x5$	2.3686	3.2724	3.5104	0.0000
$x4.x6$	0.0000	0.0000	0.0000	-1.7778
$x5.x6$	0.0000	-0.0003	0.0000	0.0000
$x1^2$	0.0016	0.0010	0.0029	-0.0130
$x2^2$	-0.0041	-0.0126	-0.0098	-0.0084
$x3^2$	-0.0282	-0.0630	-0.0442	-0.0859
$x4^2$	-1.3922	-4.3836	-3.3400	-4.9578
$x5^2$	-44.1707	-133.5775	-102.5974	-132.8314
$x6^2$	-41.7089	-128.8035	-100.5338	-128.4638

10 and 11, we compare the leakage distributions generated for SG n/pFinFETs, inverter and NAND gate, respectively, by the RSM model and QMC simulation of 3D device/gate with 100 combinations of values for the input variables. The RMSEs for these cases remain in the 3-7% range. This finding justifies the use of adjusted-2D devices/gates instead of 3D devices/gates in developing the RSM models. Thus, our macromodels accurately generate leakage distributions of SG/IG mode FinFET devices/gates within fraction of a second and make QMC simulation (adjusted-2D/3D), which may take CPU days/weeks, unnecessary.

VI. CONCLUSIONS

In this paper, we presented an efficient modeling technique for accurately estimating leakage in FinFET devices and standard cells under process variations. It uses CCRD based on RSM and employs adjusted 2D cross-sections of FinFETs that are good at tracking actual 3D device behavior. The leakage models were tested against QMC simulation results, showing very close agreement. Thus, this approach is both efficient and accurate.

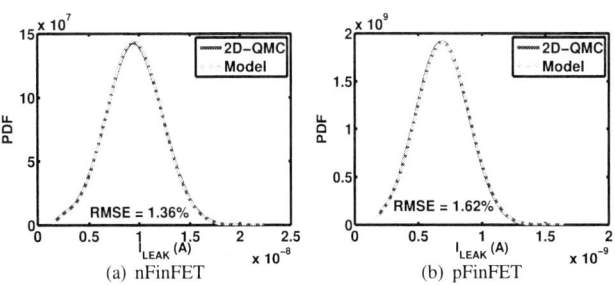

Fig. 6: 2D-QMC vs. RSM based leakage distributions for SG FinFETs

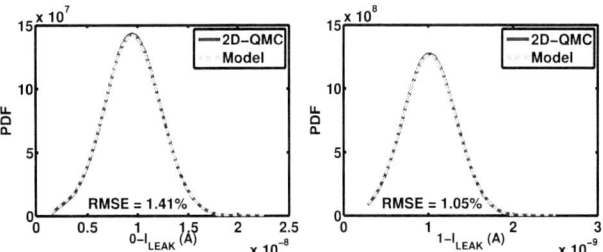

Fig. 7: 2D-QMC vs. RSM based leakage distributions for SG-INV

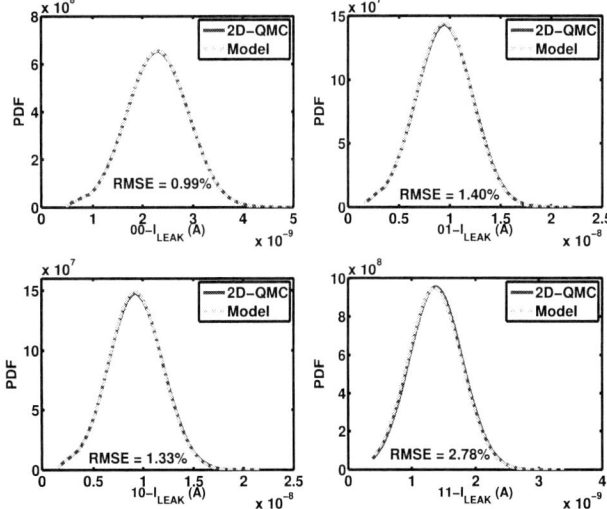

Fig. 8: 2D-QMC vs. RSM based leakage distributions for SG-NAND

TABLE XII: Testing I_{LEAK} RMSE for SG/IG FinFET devices/gates

Mode/Device	nFinFET	pFinFET	Inverter		NAND			
	I_{LEAK}	I_{LEAK}	$0\text{-}I_{LEAK}$	$1\text{-}I_{LEAK}$	$00\text{-}I_{LEAK}$	$01\text{-}I_{LEAK}$	$10\text{-}I_{LEAK}$	$11\text{-}I_{LEAK}$
SG	1.36%	1.62%	1.41%	1.05%	0.99%	1.40%	1.33%	2.78%
IG	3.97%	2.41%	3.76%	1.35%	1.16%	3.64%	2.78%	5.28%

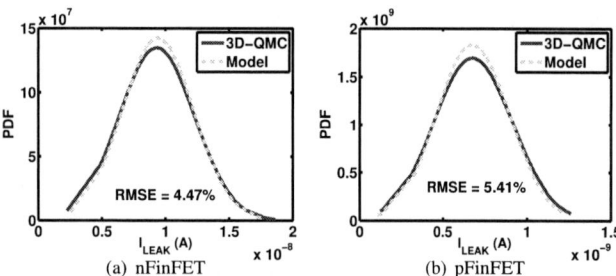

(a) nFinFET   (b) pFinFET

Fig. 9: 3D-QMC vs. RSM based leakage distributions for SG FinFETs

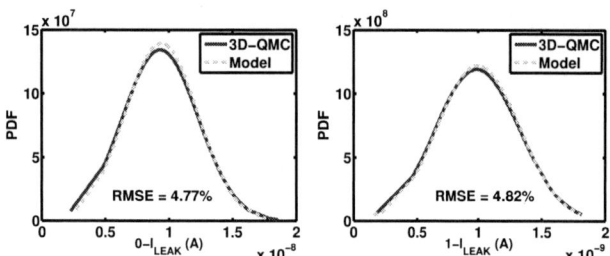

Fig. 10: 3D-QMC vs. RSM based leakage distributions for SG-INV

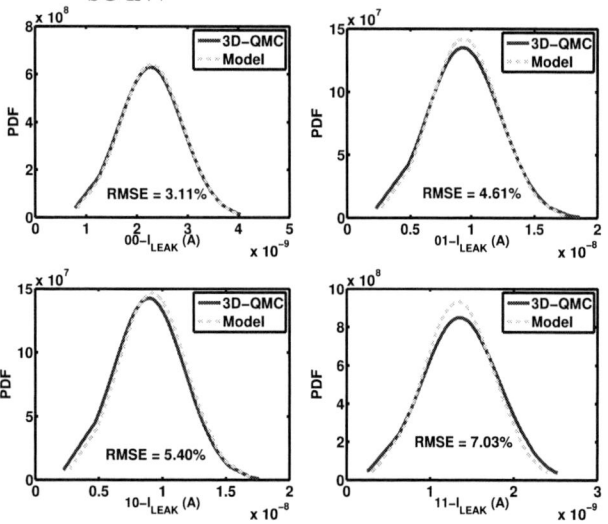

Fig. 11: 3D-QMC vs. RSM based leakage distributions for SG-NAND

REFERENCES

[1] E. J. Nowak, I Aller, T. Ludwig, K. Kim, R. V. Joshi, C. T. Chuang, K. Bernstein, and R. Puri, "Turning silicon on its edge," *IEEE Circuits and Devices Magazine*, vol. 20, no. 1, pp. 20-31, Jan.-Feb. 2004.

[2] M.-H. Chiang, K. Kim, C. Tretz, and C.-T. Chuang, "Novel high-density low-power logic circuit techniques using DG devices," *IEEE Electronic Device Lett.*, vol. 52, no. 10, pp. 2339-2342, Oct. 2005.

[3] A. Kumar, B. A. Minch, and S. Tiwari, "Low voltage and performance tunable CMOS circuit design using independently driven double gate MOSFETs," in *Proc. Int. SOI Conf.*, Oct. 2004.

[4] S. A. Tawfik and V. Kursun, "High speed FinFET domino logic circuits using independent gate-biased double-gate keepers providing dynamically adjusted immunity to noise," in *Proc. Int. Conf. Microelectronics*, Dec. 2007, pp. 175-178.

[5] P. Mishra, A. Muttreja, and N. K. Jha, "Low-power FinFET circuit synthesis using multiple supply and threshold voltages," *ACM Emerging Technologies in Computing Systems*, vol. 5, no. 2, pp. 1-23, July 2009.

[6] J. Ouyang and Y. Xie, "Power optimization for FinFET based circuits using genetic algorithms," in *Proc. IEEE Int. SOC Conf.*, Sept. 2008, pp. 211-214.

[7] B. Swahn and S. Hassoun, "Gate sizing: FinFETs vs. 32nm bulk MOSFETs," in *Proc. Design Automation Conf.*, July 2006, pp. 528-531.

[8] A. Muttreja, N. Agarwal, and N.K. Jha, "CMOS logic design with independent-gate FinFETs," in *Proc. IEEE Int. Conf. Computer Design*, Oct. 2007, pp. 560-567.

[9] S. Xiong and J. Bokor, "Sensitivity of double-gate and FinFET devices to process variations," *IEEE Trans. Electron Devices*, vol. 50, pp. 2255-2261, Nov. 2003.

[10] H. Ananthan and K. Roy, "A fully physical model for leakage distribution under process variations in nanoscale double-gate CMOS," in *Proc. Design Automation Conf.*, July 2006, pp. 413-419.

[11] J. Gu, J. Keane, S. Sapatnekar, and C. H. Kim, "Statistical leakage estimation of double gate FinFET devices considering the width quantization property," *IEEE Trans. VLSI Systems*, vol. 16, pp. 206-209, Feb. 2008.

[12] J. H. Choi, J. Murthy, and K. Roy, "The effect of process variation on device temperatures in FinFET circuits," in *Proc. Int. Conf. Computer-Aided Design*, Nov. 2007, pp. 747-751.

[13] A. N. Bhoj and N. K. Jha, "Design of ultra-low-leakage logic gates and flip-flops in high-performance FinFET technology," in *Proc. IEEE Int. Symp. Quality Electronic Design*, Mar. 2011, pp. 1-8.

[14] P. Mishra, A. N. Bhoj, and N. K. Jha, "Die-level leakage power analysis of FinFET circuits considering process variations," in *Proc. IEEE Int. Symp. Quality Electronic Design*, Mar. 2010.

[15] S. Chaudhuri and N. K. Jha, "3D vs. 2D analysis of FinFET logic gates under process variations" in *Proc. IEEE Int. Conf. Computer Design*, Oct. 2011, pp. 435-436.

[16] Sentaurus TCAD, HSPICE, Design Compiler manuals, http://www.synopsys.com.

[17] X. Shao and Z. Yu, "Nanoscale FinFET simulation: A quasi-3D quantum mechanical model using NEGF," *Solid-State Electronics*, vol. 49, no. 8, pp. 1435-1445, 2005.

[18] H. R. Khan, D. Mamaluy, and D. Vasileska, "3D NEGF quantum transport simulator for modeling ballistic transport in nano FinFETs," *Physics: Conf. Series*, vol. 107, no. 1, 2008.

[19] A. Mutlu and M. Rahman, "Statistical methods for the estimation of process variation effects on circuit operation," *IEEE Trans. Electronics Packaging Manufacturing*, vol. 28, no. 4, pp. 364-375, Oct. 2005.

[20] G. E. P. Box, W. G. Hunter, and J. S. Hunter, *Statistics for Experimenters: An Introduction to Design, Data Analysis and Model Building*, John Wiley and Sons, New York, 1978.

[21] *Engineering Statistics Handbook*, http://www.itl.nist.gov/div898/handbook.

[22] A. Singhee, and R. A. Rutenbar, "From finance to flip flops: A study of fast quasi-Monte Carlo methods from computational finance applied to statistical circuit analysis," in *Proc. Int. Symp. Quality of Electronic Design*, Mar. 2007, pp. 685-692.

[23] N. Aslan, "Application of response surface methodology and central composite rotatable design for modeling and optimization of a multigravity separator for chromite concentration," *Powder Technology*, vol. 185, no. 1, pp. 80-86, 2008.

Real-Time, Content Aware Camera–Algorithm–Hardware Co-Adaptation for Minimal Power Video Encoding

Joshua W. Wells, Jayaram Natarajan, and Abhijit Chatterjee
School of Electrical and Computer Engineering
Georgia Institute of Technology, Atlanta, GA
Email: {josh.wells, jayaram, chat}@gatech.edu

Irtaza Barlas
Impact Technologies, Atlanta, GA
Email: irtaza.barlas@impact-tek.com

Abstract—In this paper, a content aware, low power video encoder design is presented in which the algorithms and hardware are co-optimized to adapt concurrently to video content in real-time. Natural image statistical models are used to form spatiotemporal predictions about the content of future frames. A key innovation in this work is that the predictions are used as parameters in a feedback control loop to intelligently downsample (change the resolution of the frame image across different parts of the image) the video encoder input immediately at the camera, thus reducing the amount of work required by the encoder per frame. A multiresolution frame representation is used to produce regular data structures which allow for efficient hardware design. The hardware is co-optimized with the algorithm to reduce power based on the reduced input size resulting from the algorithm. The design also allows for selectable, graceful degradation of video quality while reducing power consumption.

Index Terms—adaptive video encoding, low-power DSP, self aware

I. INTRODUCTION

The goal of video encoding is to reduce the amount of redundant information in a video signal producing a more efficient, high-entropy signal. Modern video encoding standards such as H.264 and MPEG-4 are capable of producing excellent compression, but at the cost of high power consumption in the video encoder. Performing real-time video encoding with limited energy supply is a major issue, particularly for mobile camera systems. The presented work is capable of overcoming the problems associated with a limited energy system by using predictive video encoding (PVE). PVE uses previously encoded information to form a model and predict how future information in the video should be encoded. This reduces the cost associated with analyzing the signal to expose redundant information.

A. Prior Work

Prior work in low-power video encoding has concentrated on improving the energy efficiency of specific processing elements (PEs) within the video encoder system. The motion estimation (ME) kernel (described in Section II) typically accounts for 60-80% of the computational load of a video encoder [1]. Accordingly, prior works in low-power video compression have focused on the ME kernel or other individual PEs to conserve energy [2], [3], [4], [5], [6], [7], [8], [9], [10]. Notably, in [8], an error-tolerant, voltage over-scaled ME design is proposed. Past work has been effective at either reducing the workload of the ME kernel or reducing its voltage directly at the expense of video precision. Although meaningful work has been done to increase performance of the various PEs of a video encoder, little progress has been made to *reduce the power consumption of the encoder as a system*.

B. Novel Contributions

The proposed design *reduces the power consumed by the entire video compression system by extending algorithm control all the way to the camera sensor itself.* Rather than encoding all data captured by the camera, *raw data is pre-compressed at the sensor* without direct knowledge of the spatial redundancies that may be present in the frame, resulting in a smaller signal being presented to the core encoding system. There are three major contributions described in this paper:

- Pre-compression Camera Sensor
- Video Input Feedback Controller
- Dynamic Voltage and Frequency Scaling Control

Integrated into the typical video encoder design as shown in Fig. 1, each of these elements performs an important role in *pre-compressing a video signal and reducing system power consumption.*

Pre-compression of the signal is performed directly on the camera sensor by sensing and transforming the signal together [11]. Instead of the camera producing a typical full-resolution frame, it outputs a multi-resolution version of the frame referred to as a pre-compressed frame (PCF). Static regions of a video such as background are sensed with lower resolution while more dynamic regions are sensed with higher resolution *without loss of information or loss of detail in the decoded video sequence.*

This material is based upon work supported by DARPA project W31P4Q-09-C-0538 and partially supported by the National Science Foundation under Grant No. CCR-0834484.

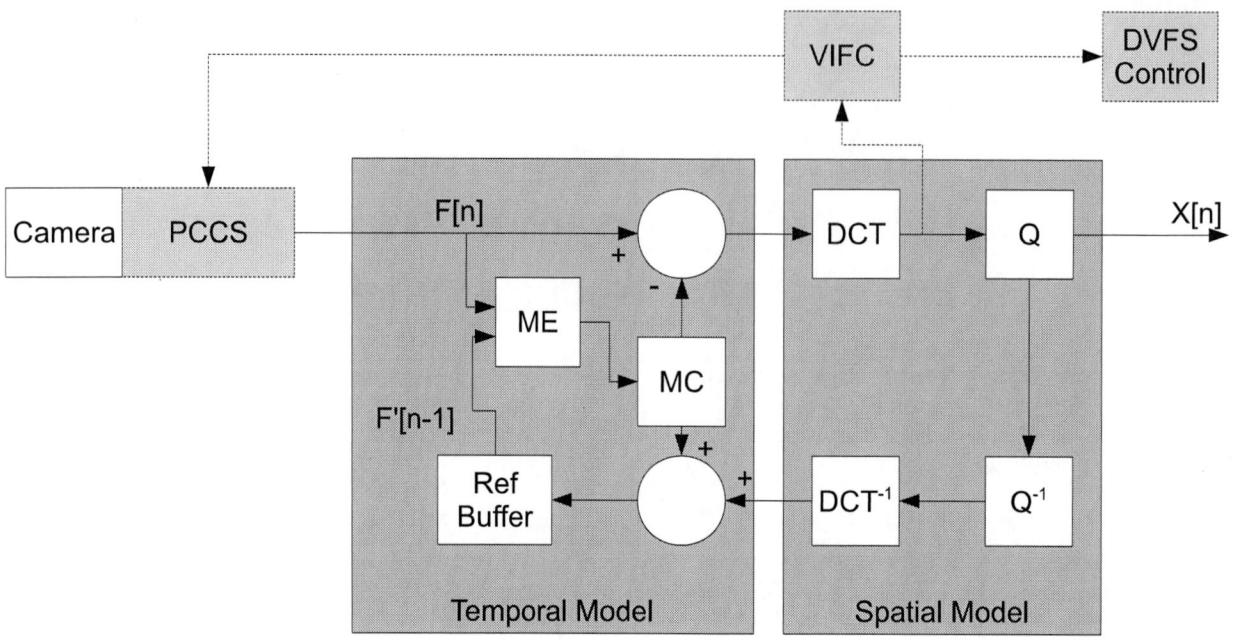

Fig. 1. Proposed video encoder design depicting pre-compression camera sensor (PCCS), motion estimation (ME), motion compensation (MC), discrete cosine transform (DCT), inverse DCT, quantizer (Q) and inverse quantizer

The video input feedback controller (VIFC) dynamically predicts the spatial redundancy of future frames based on previous, encoded frames. The control signal from the VIFC to the camera sensor indicates which regions of the next frame are likely to be redundant and should be sensed with lower resolution.

The dynamic voltage and frequency scaling control (DVFSC) is responsible for controlling the voltage and clock frequency of all core PEs which are shown in the blue regions of Fig. 1. The DVFSC changes the dynamic voltage and frequency scaling (DVFS) of each PE before each frame is encoded. This is possible and necessary because the timing requirements for each frame change based on its content. The more redundant the information in a frame is, the smaller the size of thePCF is. With less information to process by the core PEs of the encoder, the slower the information can processed and still meet the timing restrictions of the frame rate. Each PE is scaled so that the video encoder takes just under the frame period to encode each PCF. The result is a video encoder that is capable of large reductions in power based on the content of the video.

C. Paper Overview

In Section II, a basic video encoder design will be reviewed to help explain the details of the proposed design. Section III describes the method used to predictively decrease the information coming from the sensor. Section V explains how the system power consumption is reduced using DVFS. Sections VI and VII describe the methods used to evaluate the proposed

design and the conclusions drawn.

II. VIDEO ENCODER OVERVIEW

The general process of video encoding consists of first removing temporally redundant information, followed by removing spatially redundant information, using temporal and spatial models, respectively. Rather than processing frames as a whole, the frames are divided into equally sized blocks of adjacent pixels and the blocks are processed independently. As shown in Fig. 1, temporally redundant information is removed by performing subtractions on each block of data being encoded to reveal what has changed from the previous frame. The resulting residual blocks are passed to the spatial model. Spatial redundancy is reduced by performing a 2-D discrete cosine transform (DCT) based transform on the residual blocks. It is likely that a contiguous group of the highest frequency coefficients in the DCT domain will be near zero and can be neglected. Finally, a selected type of variable-length coding (VLC) is used for the DCT coefficients. VLC is an extensive topic in coding theory that will not be addressed in this paper.

The ME kernel shown in Fig. 1 searches areas of the previous frame for a group of pixels that is highly correlated to the current block being encoded. This search is necessary to account for moving objects in the video. Each block in the frame being encoded is processed independently by the ME kernel. Given the sheer volume of data, this is the most computationally intense operation of the entire encoder [1]. The location of the best matching area from the previous frame

978-1-4673-0438-2/12 $31.00 © 2012 IEEE

is described with a simple vector, is called the prediction, and is produced by the motion compensation (MC) kernel. A residual block is produced by subtracting the prediction from the raw data of the block being encoded. The residual block is passed on to the spatial model.

The spatial model transforms/quantizes each residual block of data into the frequency domain using a DCT based transformation. Typically, a portion of the highest frequency coefficients are very close to zero after transformation quantization and are considered insignificant. The insignificant coefficients are discarded. It is this insignificant high-frequency information that the proposed design eliminates at the video sensor itself. An inverse quantization and inverse DCT kernel are also required in the spatial model to provide accurate reference information for future frames to be encoded. The decoded information is stored in the ref. buffer of Fig. 1. The stored result is the exact data that is produced by the decoder and used as a reference by both encoder and decoder.

III. VIDEO SENSOR PRE-COMPRESSION

After a given block has been processed by the temporal model and the spatial model, the redundant data is revealed as insignificant high-frequency coefficients. The elimination of these coefficients is equivalent to applying a spatial low-pass filter to the residual block. The effect of typical video encoding can be described by

$$X[k,l] = \mathrm{DCT}\{b[m,n] - b'[m,n]\} \cdot H[k,l] \qquad (1)$$

where x, b, b', and h are the compressed block, raw input block, reference block, and low-pass filter respectively. The same result can be achieved by first decimating the raw input from the camera as represented by

$$X[k,l] = \mathrm{DCT}\{b[m,n] * h[m,n] - b'[m,n] * h[m,n]\} \quad (2)$$

To perform the pre-compression before the signal is passed through the temporal and spatial models, an estimate of the magnitude of the DCT coefficients is required which will be covered in Section IV.

Using CMOS separable transform image sensor (CSTIS), the pre-compression of blocks can be moved all the way to the input sensor itself [11]. Pre-compressed blocks that have been sensed in this way are already described in the DCT domain. This means the typical digital DCT function of the encoder can be bypassed. However, a consequence of sensing a pre-compressed block in the DCT domain is that normal ME and MC are no longer possible. This is of little consequence because as will be shown in Section IV, pre-compressed regions of frames are relatively static and can be assumed to have null motion vectors.

Non-contributing high-frequency content is removed before the video signal is passed to the temporal model by sensing the signal in the DCT domain for low frequency coefficients only. If a sensed block is allowed to have a completely variable number of low frequency coefficients included, blocks of varying sizes will be produced which requires an undesirable, complicated design for the PEs of the encoder. To mitigate

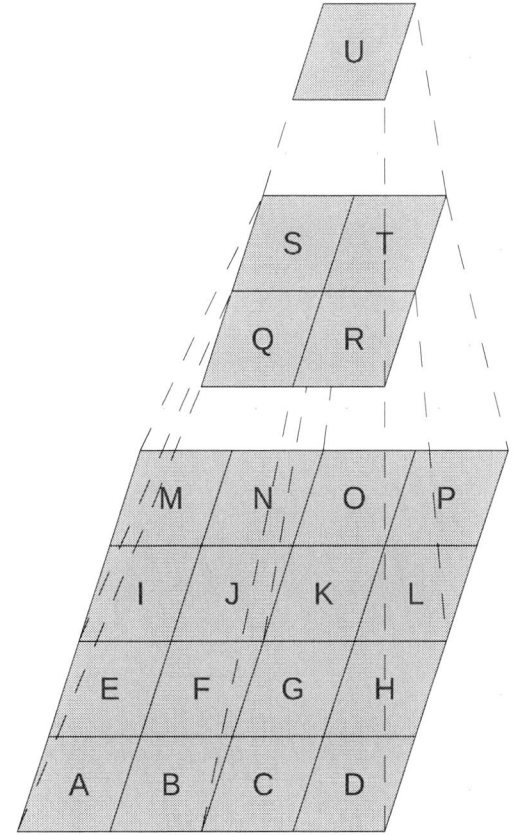

Fig. 2. Multiresolution quadtree structure

this problem, effective decimation of blocks is limited to a quadtree structure as shown in Fig. 2 allowing the PEs to remain unchanged. For a block to be effectively decimated by the smallest increment, it must be concatenated with three of its neighbors according the the quadtree structure and then decimated by a factor of 2. For example, in Fig. 2, blocks A, B, E, and F can be combined and decimated by a factor of 2 to yield block Q. Higher levels of decimation are allowed as long as they adhere to the quadtree structure meaning that all sibling blocks must be collapsed to a single parent. For example, in Fig. 2, blocks A-P can be combined and decimated by a factor of 4 to yield block U. The quadtree structure allows the input frame to be represented as a multi-resolution image where areas predicted to have more high-frequency residue can be represented with more resolution and other areas with less as illustrated by Fig. 3. Each block bounded by the black lines in the figure represents the same amount of information – a single block. To make viewing easier, the blocks have been interpolated to fill the whole area they represent. Blocks that occupy a larger area like the one in the top left quadrant contain less detailed information.

Fig. 3. Hypothetical multi-resolution image

IV. VIDEO INPUT FEEDBACK CONTROLLER

The pre-compression camera sensor (PCCS) described in Section III requires a control signal indicating the quadtree structure that should be applied to the next frame. The is accomplished by analyzing the spectrum of each encoded block in the current PCF. A model is fit to the spectrum of the block that predicts the amount of extra detail or missing detail in the residual block. Based on the model parameters, a decision is made to increase, decrease, or keep the resolution level associated with the corresponding area of the next frame.

The power spectrum of a natural image can be modeled using an exponential function of the form

$$|P_X(j\omega)| = \beta(\omega + 1)^\alpha \qquad (3)$$

where β and α are the parameters for the model and ω is the polar spatial frequency [12]. The expected magnitude of the spectrum decays exponentially with respect to increasing polar frequency. Once the two parameters for the model are determined, the spectrum can be extended beyond sampled frequencies to determine how much error is predicted to exist in each block. The error can be measured in peak signal-to-noise ratio (PSNR). The VIFC attempts to keep the PSNR of the signal at a specified level. By pre-computing a family of curves and storing it in a look-up table, power used to fit the model parameters can be made negligible. The goal of the image resolution control kernel is minimize overall resolution on the input image while maintaining a given PSNR.

V. SYSTEM DYNAMIC VOLTAGE AND FREQUENCY SCALING

The PCCS modulates the size of the signal for each frame presented to the video encoder. The video encoder also has a fixed frame rate. Since the size of the input signal is modulated, but the amount of time needed to process the signal is static, the video encoder may process smaller frames (less information) at a slower speed without error. Next, DVFS is used to scale back the power consumption of the encoder for all blocks (except for the control and decimation blocks). When the voltage and frequency are scaled down, PE delay time scales up. Based on the hardware implementation of the encoder, a table is computed with optimal voltage and frequency values for different frame sizes.

The DVFSC kernel is responsible for setting the proper voltage and clock frequency for each PE in the video encoder just prior to the encoding of a frame. Transistor-level simulations of each processing element reveal the relationship between scaled voltage, frequency, and delay on the critical path. In each PE, a static amount of work is required for each block processed. For a given critical path delay, the amount of time required to process a single block is known. The VIFC sends a control signal to the DVFSC indicating the number of blocks to be encoded in the next frame. The DVFSC uses a lookup table to translate the block count into a voltage and frequency level for each of the PEs and sets the voltage just prior to the encoding of the frame. Essentially, voltage and frequency are scaled down as the block count is scaled down from its largest count (full resolution.)

VI. DESIGN ANALYSIS

Three topics must be addressed regarding the validation of the proposed design. First, the VIFC prediction accuracy must be measured. If the VIFC is not capable of producing valid predictions, either power consumption or video quality will suffer. Second, the video quality must be analyzed. When a less than nominal resolution level is predicted, error will be introduced into the encoded signal which will affect the quality of the decoded sequence. Lastly, the power savings from the proposed technique must be evaluated. The savings will depend on the actual video content, so multiple test video sequences are used.

A. VIFC Performance

As discussed in Section IV, the VIFC predicts how much information in a transformed block is missing or alternatively, unnecessary. To evaluate the accuracy of the predictor, a sampling of frames is taken from the test video sequences and indiscriminately decimated by a factor of 2, meaning that groups of four blocks are combined into a single block. Prediction blocks are subtracted from the decimated blocks and the resulting residual blocks are transformed and passed to the VIFC. The VIFC predicts the amount of missing information in each residual test block. The directly computed (correct) amount of error in each block is compared to the predicted error, measured in PSNR as defined by

$$PSNR = 10 \cdot \log_{10} \left(\frac{MAX_I^2}{MSE} \right) \qquad (4)$$

978-1-4673-0438-2/12 $31.00 © 2012 IEEE

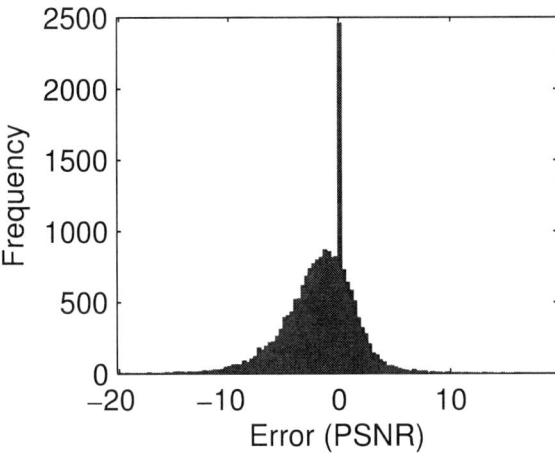

Fig. 4. Prediction accuracy for decimation by a factor of 2

$$MSE = \frac{1}{MN} \sum_{m=0}^{M-1} \sum_{n=0}^{N-1} |x_1[m,n] - x_2[m,n]|^2$$

where MAX is the maximum unsigned integer possible using 8 bits; $x_1[m,n]$ and $x_2[m,n]$ are the two signals being compared. The difference between the two values represents the prediction error. A histogram of the VIFC performance is shown in Fig. 4. The results indicate that the proposed VIFC design is valid.

B. Video Quality

Well known test video sequences in the signal processing community were used to evaluate the proposed design in terms of quality. Each raw test sequence was dynamically pre-compressed, simulating the method described in Section III. The reduced signal was processed as depicted in Fig. 1 with the exception of bypassing the quantizer. The quantizer is bypassed to avoid mistaking quantization error for pre-compression error. Each frame is decoded and evaluated against the original, raw test sequence. In each test signal, the VIFC is tasked with predicting the resolution for each region of the frame that will maintain a PSNR of at least 35dB. The resulting PSNR of the test sequences are shown in 5. The VIFC does an excellent job at maintaining the required signal quality. It should also be noted that modern video encoding methods are capable of producing an encoded sequence with a quality similar to the proposed encoder, but no power savings are achieved with modern methods.

C. Power Savings

Modern video compression standards do not specify how a signal should be encoded. Instead, they specify what criteria the encoded signal must meet. Therefore, many different hardware implementations have been developed, and as a result, it is very difficult to quantify power savings in terms of Watts. Instead, for the purposes of this research, estimated power savings are described in terms of the amount of power that can be reduced compared to nominal operation.

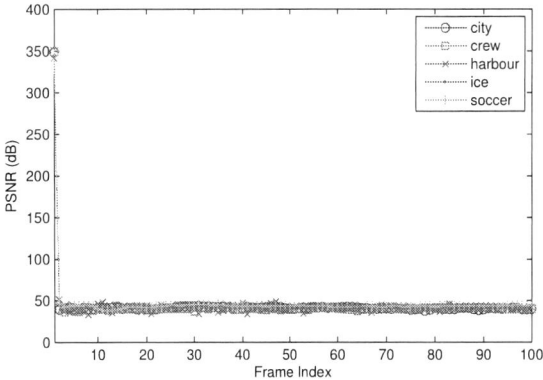

Fig. 5. Decoded video test sequence quality

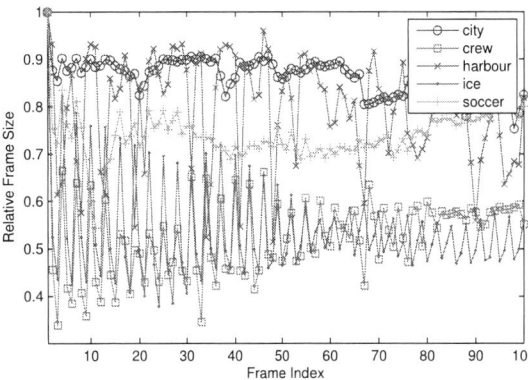

Fig. 6. Information content of each pre-processed frame relative to non-preprocessed information content

Power savings starts with pre-compression. Pre-compression alone is capable of saving dynamic power by reducing the digital signal size and therefore reducing the amount of switching in the circuit for each frame. The extent of the reduction in switching depends on the content of the video. If there is a large amount of temporal and spatial redundancy in the signal, the VIFC will perform well and pre-compression will also be effective, necessary switching in the digital circuit. For each frame in the test sequences, the size of the pre-compressed signal is shown relative to the size of the typical, non-pre-compressed signal. Fig. 6 shows relative frame sizes. Due to the difference in the content of each sequence, a different amount of work is required for different frames while the actual quality of each encoded signal remains constant as shown in Fig. 5.

Although the reduced signals result in reduced switching which decreases the dynamic power consumption of the system, power utilization is not maximized until DVFS is used to reduce the voltage for each PE. As described in Section V, fixed frame rates allow the proposed encoder to perform DVFS on each PE in the encoder. Since there is no standardized encoder design, nominal power savings using

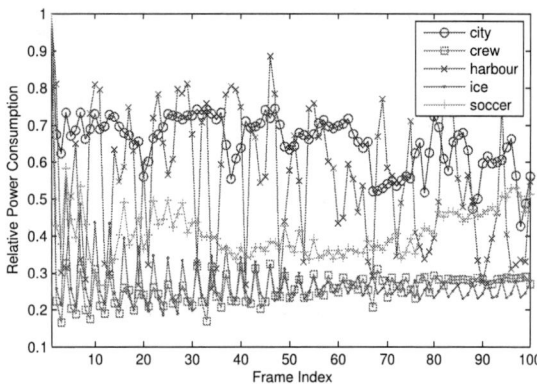

Fig. 7. Estimated dynamic power savings

DVFS for dynamic power are generalized to an estimated power relative to typical non adaptive encoding. The relative frame size is defined as

$$\gamma = \frac{S}{S_{max}} \qquad (5)$$

where S is the frame size of an adaptively acquired frame as shown in Fig. 6, and S_{max} is the size of the frame if it were acquired normally (full resolution.) The relative dynamic power of an adaptive encoder can now be computed as

$$P_{Rdyn} = \frac{(V\gamma)^2}{V^2}\frac{f\gamma}{f} = \gamma^3 \qquad (6)$$

where V is the nominal voltage of a non adaptive encoder and f is the average switching frequency. From (6), estimated power results for the sample video sequences are computed. The results are shown in Fig. 7. The presented estimates are ideal. They do not account for incremental steps in DVFS but as the number of allowed levels are increased, the savings should approach the computed estimates.

VII. CONCLUSION

The presented theory and results demonstrate the huge potential in power savings for real-time embedded video encoding systems. But to achieve such large savings, the system needs to be considered as a whole. Trying to improve the efficiency of each PE has reached its limits. Although the overall success of the video encoder depends on power efficient operations, the ultimate savings comes from reducing the number of operations needed through the entire system. For a real-time video encoder to truly use minimal power, it must adapt to the content of the video being processed. All parts of the video encoder system (sensors, algorithm, and hardware) must be able to adapt. The proposed design is capable of dynamically scaling the amount of work performed and power consumed throughout the system so that just enough energy is used to meet the required goals.

REFERENCES

[1] P. Kuhn, *Algorithms, complexity analysis, and VLSI architectures for MPEG-4 motion estimation.* Kluwer Academic Publishers, 1999.
[2] T. Chen, Y. Huang, and L. Chen, "Fully utilized and reusable architecture for fractional motion estimation of h. 264/avc," in *Proc. ICASSP*, vol. 5, 2004, pp. 9–12.
[3] M. Kim, I. Hwang, and S. Chae, "A fast vlsi architecture for full-search variable block size motion estimation in mpeg-4 avc/h. 264," in *Proceedings of the 2005 Asia and South Pacific Design Automation Conference.* ACM, 2005, p. 634.
[4] S. Yap and J. McCanny, "vlsi architecture for variable block size video motion estimation," *IEEE Transactions on Circuits and Systems Part 2: Express Briefs*, vol. 51, no. 7, pp. 384–389, 2004.
[5] H. Nakayama, T. Yoshitake, H. Komazaki, Y. Watanabe, H. Araki, K. Morioka, J. Li, L. Peilin, S. Lee, H. Kubosawa *et al.*, "An mpeg-4 video lsi with an error-resilient codec core based on afast motion estimation algorithm," in *2002 IEEE International Solid-State Circuits Conference, 2002. Digest of Technical Papers. ISSCC*, vol. 1, 2002.
[6] D. Mohapatra, G. Karakonstantis, and K. Roy, "Significance driven computation: a voltage-scalable, variation-aware, quality-tuning motion estimator," in *Proceedings of the 14th ACM/IEEE international symposium on Low power electronics and design.* ACM, 2009, pp. 195–200.
[7] H. Cheong, I. Chong, and A. Ortega, "Computation error tolerance in motion estimation algorithms," in *IEEE International Conference on Image Processing, ICIP'06*, 2006.
[8] G. Varatkar and N. Shanbhag, "Energy-efficient motion estimation using error-tolerance," in *Low Power Electronics and Design, 2006. ISLPED'06. Proceedings of the 2006 International Symposium on*, Oct. 2006, pp. 113–118.
[9] G. V. Varatkar and N. R. Shanbhag, "Variation-tolerant motion estimation architecture," in *Signal Processing Systems, 2007 IEEE Workshop on*, Oct. 2007, pp. 126–131.
[10] G. Varatkar and N. Shanbhag, "Error-resilient motion estimation architecture," *Very Large Scale Integration (VLSI) Systems, IEEE Transactions on*, vol. 16, no. 10, pp. 1399–1412, Oct. 2008.
[11] R. Robucci, J. Gray, L. Chiu, J. Romberg, and P. Hasler, "Compressive sensing on a cmos separable-transform image sensor," *Proceedings of the IEEE*, vol. 98, no. 6, pp. 1089–1101, 2010.
[12] A. Van der Schaaf and J. Van Hateren, "Modelling the power spectra of natural images: statistics and information," *Vision Research*, vol. 36, no. 17, pp. 2759–2770, 1996.

2012 25th International Conference on VLSI Design

Way Sharing Set Associative Cache Architecture

C J Janraj
janrajcj@gmail.com

T Venkata Kalyan
kalyantv@cse.iitm.ac.in

Tripti Warrier
tripti@cse.iitm.ac.in

Madhu Mutyam
madhu@cse.iitm.ac.in

Computer Architecture and Systems Laboratory
Department of Computer Science and Engineering
Indian Institute of Technology Madras
Chennai, India 600036

Abstract—**In order to minimize the conflict miss rate, cache memories can be organized in set-associative manner. The downside of increasing the associativity is increase in the per access energy consumption. In conventional n-way set-associative caches, irrespective of the set-wise demand, each set has n cache ways at its disposal, but cache sets may exhibit non-uniform demand for these cache ways. Exploiting this property, we propose a novel cache architecture, called *way sharing cache*, wherein by allowing sharing of cache ways among a pair of cache sets, we obtain dynamic energy savings as high as 41% in DL1 cache with negligible performance penalty.**

I. INTRODUCTION

Continuing advancements in semiconductor process technology help in achieving higher clock rates and very large scale integration of transistors. The large scale integration of transistors can be exploited for providing additional hardware support to tap instruction-level parallelism and memory locality available in applications. Thus, the process technology advancements can in turn help microprocessors to gain significant performance improvements. Switching capacitance increases with process technology scaling because more functionality is added. This in turn increases the dynamic power consumption which is proportional to both clock frequency and switching capacitance. As circuit level techniques alone may not keep power consumption to reasonable levels, there has been increasing interest in architectural techniques to reduce the switching capacitive power component. By reducing the number of signal transitions within a processor for a given workload, architectural techniques can reduce energy consumption. As an increasing percentage of die area is being used for cache memories and the on-chip caches alone can consume over 40% of the total chip power budget [9], several works have been proposed in literature to minimize energy consumption in on-chip cache hierarchy.

By exploiting the locality of reference property available in applications, cache memories help in reducing the speed gap between processor and main memory. As accessing main memory takes significant number of processor cycles and cache memories are effective in improving the system performance, modern computer systems employ multi-level cache organizations. Typically, set-associative cache organization is preferred as it effectively reduces conflict misses.

Figure 1 shows a conventional 128 set 4-way L1 data cache. For performance reasons, all the cache ways of a n-way set-associative cache are accessed in parallel to know whether or

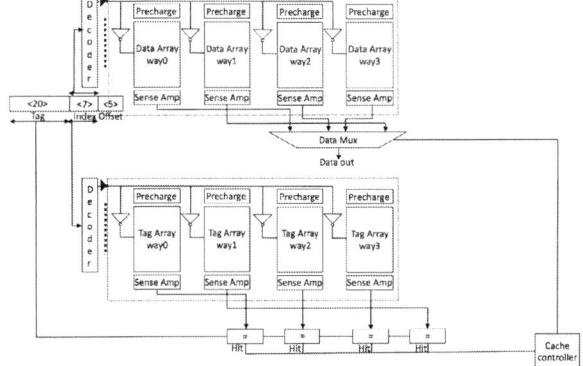

Fig. 1.   Conventional cache architecture for 128×4-way L1 data cache.

not a load instruction is going to hit in the data cache, but this process incurs significant dynamic power consumption. Several techniques have been proposed in literature to minimize power consumption. In *selective cache ways* [4] technique, cache ways are selectively activated based on applications need to minimize power consumption significantly, but profiling is required to know the number of cache ways to be selectively activated/deactivated. A *configurable unified set-associative cache* is proposed in [8], where each way can be individually shutdown to reduce dynamic power consumption and the cache can be configured as an instruction, data or unified cache.

A *cache with adaptive line sizes* is proposed in [13], where the cache line size is adjusted based on application behavior to reduce the memory traffic significantly. In *V-way cache* technique [10], by increasing the number of tag-store entries relative to the number of data lines, a mechanism is proposed to change the associativity on a per-set basis. In [14], a heuristic algorithm is proposed to select the best cache parameters such as cache size, associativity, line size, etc., for a given application and given an on-chip hardware mechanism to implement the heuristic. *Accounting cache* [5] is based on a re-sizable selective cache ways mechanism. The accounting cache first accesses part of the cache ways of a set-associative cache and if there is a miss, it accesses the remaining ways. A swap between the first and second accesses is needed whenever there is a miss in the first and a hit in the second access.

To exploit the underutilized sets for minimizing the conflict miss rate of direct-mapped caches, *balanced cache mechanism* [16] is proposed, wherein programmable decoders are used to

978-1-4673-0438-2/12 $31.00 © 2012 IEEE   251

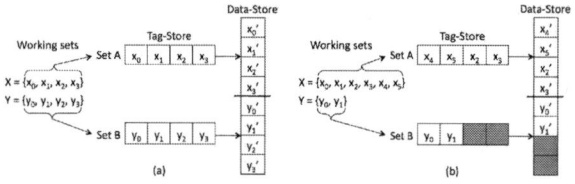

Fig. 2. Example showing a conventional 4-way set-associative cache.

redirect accesses to underutilized sets and hence improve the overall performance of cache. In [15], techniques such as way-concatenation, way-shutdown, and line-concatenation are used to configure the associativity, cache size, and cache line size, respectively. *Asymmetric set associative cache* [6] is a low power cache wherein different ways have different number of cache lines. In this mechanism, as long as the accesses hit in the smaller cache ways, which can terminate accesses to larger cache ways, power consumption can be minimized.

Way-level share cache [12] is designed in such a way that the number of ways of a conventional set-associative cache is reduced to half and the number of sets is doubled. The first half of the sets is implemented in the same way as that of the conventional set-associative cache, but the second half of the sets is used for sharing. The sharing mechanism is implemented by a pointer that points to the cache ways. The pointer-based redirect may increase the average access latency.

All the above mentioned approaches minimize power consumption at the cost of increased area and/or performance penalty. In this paper, we propose a novel cache architecture, called *way sharing cache*, wherein two sets share a subset of cache ways apart from having their own cache ways. The sharing of cache ways helps highly demanded sets to get more cache ways and the wastage due to underutilized sets is reduced. Unlike the above mentioned techniques, our technique does not require significant changes to the conventional cache architecture and can be easily applied to L1 as well as L2 caches. Experimental analysis show that our way sharing architecture indeed achieves 14% average and 41% maximum energy savings when implemented in DL1 cache, for benchmarks with high hit rates. For UL2 caches with fast type access, the savings are found to be around 6%.

II. MOTIVATION

When applications exhibit non-uniform memory access behavior [7] across different sets of a cache, some sets face heavy demand, which in turn cause significant conflict miss rate, and the other sets are underutilized, which in turn waste precious cache space. Lack of coordination between the heavy demand sets and the underutilized sets results in significant performance loss.

Let us take an example as given in Figure 2, which shows a traditional 4-way set-associative cache. Assume that all memory references of working sets X and Y are mapped to set A and set B, respectively. The data lines for addresses $x_0, x_1, \cdots$, are represented in the figure as $x_0', x_1', \cdots$.

Referring to Figure 2(a), as long as the cache access behavior is uniform, the demand on sets A and B is equal, and hence both halves of the data-store are equally utilized.

Actual applications may have variable demand on cache sets. As shown in Figure 2(b), assume that at a different phase in the program, working set X increases by two elements and working set Y decreases by two elements. Set A is unable to accommodate all the elements of the working set X, resulting in conflict misses. Set B, on the other hand, has two un-utilized ways. If set B shares its un-utilized ways with set A, the conflict misses can be avoided.

One way to share the cache ways among pairs of cache sets is to increase the associativity and reduce the number of sets. In other words, for a $m \times n$-way set-associative cache (where 'm' and 'n' represent the number of sets and associativity, respectively), one can consider $\frac{m}{2} \times 2n$-way set-associative cache so that a set in the later configuration refers to two sets in the former configuration. The problem with this approach is two-fold: 1) significant increase in the associativity can increase the access time and energy significantly; 2) fair sharing may not be possible between the shared sets as the heavy demand set can completely dominate the other set. These issues can in turn offset the benefits of sharing.

Solution to this problem is to share the cache ways without increasing the associativity significantly and provide fair sharing among the shared sets. We propose a way sharing technique, wherein we address these two issues. By allocating certain number of cache ways exclusively for each set of a shared set pair, we make sure that no set of a shared set pair dominates the other set. At the same time, providing certain number of cache ways for sharing between shared set pair, we make sure that heavy demand sets get more quota of cache ways as compared to the underutilized sets, which in turn minimizes the conflict miss rate.

III. (m, k)-WAY SHARING SET-ASSOCIATIVE CACHE

To exploit the set-wise nonuniform demand, we propose a novel cache mechanism, called *way sharing cache architecture*. The basic idea behind the way sharing technique is to organize the cache in such a way that, the cache sets can have variable associativity to satisfy the application's requirements during different phases of the execution. In a (m, k)-*way sharing set-associative cache*, k ways from the available m ways are shared among a pair of sets and the remaining $(m-k)$ ways are equally distributed to the sets in the shared pair. We call the sets which are sharing as *friend sets*. If a set s is sharing with another set p, we say that s and p are friends. A single set in our way sharing cache is actually representing two sets in a conventional cache, hence the total number of sets in our way sharing cache is exactly half of the total number of sets in the corresponding conventional set-associative cache. By considering cache set numbering in a conventional cache, each set s in the cache has a unique friend p such that $|p - s| = 2^l$, for some value of $l \geq 0$. In other words, friend sets are separated by a distance of 2^l. The reason for considering a distance of 2^l, $l \geq 0$, between friend sets is to make the implementation simple (our technique can work for any distance).

Unlike the conventional set-associative caches, where each

	Distance between friend sets						
Benchmark	1	2	4	8	16	32	64
bzip2	1.1762	1.1762	1.1762	1.1762	**1.1763**	**1.1763**	**1.1763**
crafty	0.8757	0.8762	0.8762	0.8763	0.8768	0.8768	**0.8769**
gap	1.1471	1.1468	1.1473	1.1477	**1.1478**	1.1474	**1.1478**
gcc	1.8025	1.8025	1.8025	1.8026	1.8025	1.8025	**1.8026**
gzip	1.2049	1.2049	1.2050	1.2050	1.2050	1.2050	**1.2051**
mcf	0.9002	0.8999	0.9007	0.9007	0.9007	**0.9008**	0.9007
parser	1.1826	1.1831	1.1833	1.1833	1.1834	1.1843	**1.1844**
perlbmk	**0.8071**	0.8066	0.8065	0.8067	0.8066	0.8066	0.8056
twolf	0.9692	0.9696	0.9696	0.9697	0.9701	0.9699	**0.9702**
vortex	1.5827	1.5834	1.5834	1.5832	1.5837	1.5835	**1.5845**
vpr	1.0087	1.0092	1.0094	**1.0095**	**1.0095**	1.0094	**1.0095**
applu	1.2049	**1.2072**	1.2030	1.2022	1.2022	1.2001	1.1973
apsi	**0.9726**	0.9725	0.9695	0.9693	0.9696	0.9689	0.9689
art	**0.9037**	0.9036	**0.9037**	**0.9037**	**0.9037**	0.9037	**0.9037**
galgel	1.4346	1.4383	1.4382	1.4380	1.4386	1.4384	**1.4391**
lucas	**1.9969**	**1.9969**	**1.9969**	**1.9969**	**1.9969**	**1.9969**	1.9969
mesa	1.5597	1.5598	1.5598	1.5598	1.5598	1.5598	**1.5599**
mgrid	1.8524	1.8510	1.8553	**1.8620**	**1.8620**	1.8445	1.8596
swim	**1.9834**	**1.9834**	**1.9834**	**1.9834**	**1.9834**	**1.9834**	**1.9834**
wupwise	1.4794	1.4795	1.4794	1.4794	**1.4796**	**1.4796**	**1.4796**

TABLE I

BENCHMARK-WISE IPC VALUES FOR DIFFERENT DISTANCES BETWEEN FRIEND SETS. HERE, WE CONSIDER $64 \times (6, 2)$-WAY SHARING DL1 CACHE.

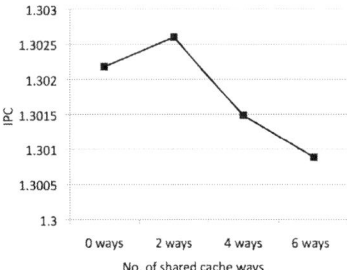

Fig. 3. Average IPC across all the benchmarks for different number of shared ways.

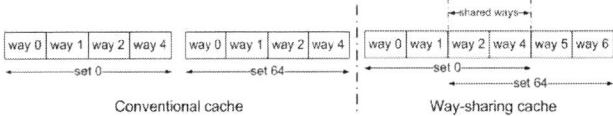

Fig. 4. Conventional 4-way set-associative cache Vs (6, 2)-way sharing cache.

set has a fixed number of ways, in (m, k)-*way sharing set-associative caches*, each set can possess variable number of ways based on the value of k chosen from the available m ways. Each set owns $\frac{(m-k)}{2}$ ways exclusively and the remaining k ways are shared by both the set and its friend set. Effectively, based on the application's requirement that is executing, at any given point of time a set can have the number of cache ways in the range of $\frac{(m-k)}{2}$ to $\frac{(m+k)}{2}$. The advantage with this organization is that during different phases of a program execution, if a set requires larger associativity, the shared ways can be utilized and this sharing of ways happens dynamically. To choose the best separation distance between friend sets and to determine the number of shared ways, we conduct simulation study for various values of l and k. Our simulation methodology is detailed in Section IV.

A. On the Distance between Friend Sets

As different applications exhibit different memory access behavior, a single value of l may not be suitable for all the applications. For instance, for applications which exhibit mostly sequential accesses, the distance between friend sets should be kept as large as possible so that the conflict miss rate can be minimized, whereas for applications whose memory access behavior is totally random, any value of l can be considered. By considering different l values, ranging from 0 to 6, and a $64 \times (6, 2)$-way DL1 cache having $32B$ blocks, we conduct the study to know the best l value. In order to reduce the total number of simulations to run, here we assume that the number of ways to be shared is 2 and once we obtain the best l value, we then validate whether the assumption of 2 ways to be shared is correct or wrong. Table I shows benchmark-wise *instructions per cycle* (IPC) values for different separation distance values. Benchmark-wise the maximum IPC values are shown in bold face. From the table, we can deduce that separation distance of 64 is better as 15 out of 20 benchmarks give the best IPC values for this distance. So, we choose $l = 6$ and consider sets which are separated by a distance of 64 as friend sets.

B. On the Number of Cache Ways to be Shared

As different applications may show different demand for the set-wise associativity, we conduct the study on the value of k to know the best value. By considering different k values from the set $\{0, 2, 4, 6\}$, and a $64 \times (6, k)$-way DL1 cache having $32B$ blocks, we obtain the average IPC values as shown in Figure 3. Results support that sharing of two ways out of available 6 ways gives the best performance.

Based on the above discussion, in this work, we consider the value of l as 6, so that the sets in any pair of friend sets are separated by a distance of 64. For example, set 0 and set 64 are friends, while set 4 and set 64 are not friends. Similarly, we consider the value of k as 2, so that two ways are shared among shared pair of sets. Please note that the work can be easily extended for any values of l and k.

C. Implementation Details

For the purpose of evaluation, with respect to the baseline cache (refer to Section IV), we consider (6, 2)-way sharing DL1 cache of size $12KB$ with $32B$ blocks. Our way-sharing cache has 64 pairs of shared sets with 6 ways in each pair. It is denoted as $64 \times (6, 2)$-way sharing DL1 cache. Sharing in our technique enables a flexibility of 2, 3, or 4-way associativity available to each set in the shared pair of sets. Each set s, $0 \leq s \leq 63$, of the conventional cache is sharing a set in our way sharing cache with its friend set p, $64 \leq p \leq 127$, such that s and p sets own two ways each and share two ways among them as illustrated in Figure 4.

Figure 5 shows the implementation details of a $64 \times (6, 2)$-way sharing set-associative cache. The LSB of the 21-bit tag is used as the *set_select_bit*, distinguishing a set s from its friend set p. This bit is used to enable four ways among the six ways in the accessed set for both tag and data arrays. The shared ways are not gated as they have to be enabled whenever either of the associated sets is accessed. For example, during a load access to set 0, the *set_select_bit* will be 0 and hence the first four ways are enabled. On the other hand, when there is a load access to set 64, the *set_select_bit* will be 1 and the

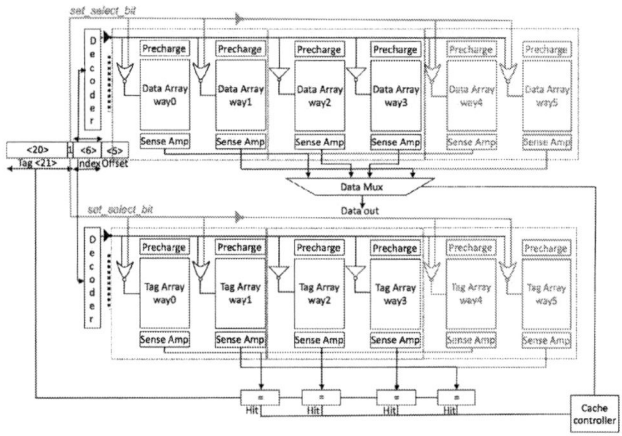

Fig. 5. Way sharing cache architecture for 64 × (6, 2)-way L1 data cache.

Parameter	Value
Instr. window	64-RUU, 32-LSQ
Issue width	4 instructions per cycle
Functional units	4 IntALU, 2 IntMult, 2 FPALU, 1 FPMult, 1 FPDiv
IL1 cache	16KB, 4-way, 32B block, 1-cycle latency, LRU
DL1 cache	16KB, 4-way, 32B block, 1-cycle latency, LRU
Way sharing DL1	12KB, 6-way, 32B block, 1-cycle latency, LRU
L2 cache	Unified, 512KB, 8-way, 64B block, 12-cycle latency, LRU
Way sharing L2	Unified, 384KB, 12-way, 64B block, 12-cycle latency, LRU
Memory latency	160 cycles
Instruction TLB	16-entry, 4KB block, 4-way, 30-cycle miss penalty
Data TLB	32-entry, 4KB block, 4-way, 30-cycle miss penalty
Branch predictor	Combined, Bimodal 2K table, 2-level, 1K table, 8 bit history, 1K choser
BTB	512-entry, 4-way, 18-cycle mis-predict penalty

TABLE II

PROCESSOR CONFIGURATION USED IN OUR SIMULATIONS.

INT	FFWD	FP	FFWD
bzip2	35.8B	applu	209.0B
crafty	36.5B	apsi	77.9B
gap	17.9B	art	13.6B
gcc	35.7B	galgel	51.5B
gzip	14.9B	lucas	48.0B
mcf	14.4B	mesa	182.7B
parser	24.3B	mgrid	297.3B
perlbmk	35.2B	swim	211.6B
twolf	179.4B	wupwise	181.0B
vortex	86.8B		
vpr	54.9B		

TABLE III

SPEC 2000 CPU BENCHMARKS USED IN OUR SIMULATIONS.

last four ways are accessed. This implementation ensures that at any point of time, an access to any set enables only four ways of the available six ways. The output from the selected four ways is sent to the data select multiplexor (Data Mux) and the output from the Data Mux is based on the result of tag matching (as is the case with the conventional cache). The 21-bit tag is chosen for a 64-set cache to ensure that the two friend sets do not have the same tag and hence one set will not access the data of the other set.

In the above implementation, each shared cache line is associated with two sets, so additional care is needed to identify the block that needs to be evicted during a cache miss. We consider the regular LRU replacement policy with a small change. Though two sets s and p are sharing a single set in our way sharing cache, we maintain a separate LRU status for each of these two sets. As a result of that, the status of shared ways is updated whenever an access is requested to either of the friend sets. We consider a 6-bit shift register for each way of the way-sharing cache. Consider a set pair (s, p) that is sharing a set in the (6, 2)-way sharing cache. Without loss of generality, we assume that set s is associated with way 0 to way 3 and set p is associated with way 2 to way 5. Whenever set s is accessed, the shift registers of all the ways (i.e., ways $0 - 3$) associated with that set (except the way in which the actual access is taken place) are shifted by one position to the right by inserting 0 and the shift register of the way in which the actual access is taken place is shifted by one position to the right by inserting 1 (similarly, access to set p is done accordingly). When there is a cache miss corresponding to set s (or p), we look at the shift registers of ways $0 - 3$ (or $2 - 5$) and choose a victim cache line corresponding to a way whose shift register with least count.

Maintaining the set-wise shift registers and updating in the above mentioned way makes sure that the cache sets which have high cache way utilization rate can get more number of cache ways whereas those sets which have low cache way utilization can get fewer cache ways. Though 6-bit shift registers are used per set in the way-sharing cache mechanism, as the number of sets in the way-sharing mechanism is half

of the number of the sets of the corresponding conventional cache, the overhead due to shift registers in the way-sharing cache mechanism is minimal as compared to that of the conventional cache.

IV. EXPERIMENTAL METHODOLOGY

We consider the base processor configuration as shown in Table II. We run 20 SPEC2000 CPU Benchmarks [3] (11 integer and 9 floating point) on modified SimpleScalar tool [2]. SPEC2000 CPU benchmarks are compiled for Alpha ISA. Fast forwarding count for the benchmarks is obtained from Simpoint [11] and is given in Table III. Each benchmark is fast-forwarded using the FFWD count and then simulated for the next 100 million instructions. We use Cacti 5.3 [1] to find the access energy values for both baseline cache and the way sharing cache considering $45nm$ technology node. We evaluate the effectiveness of our technique by considering L1 data cache and L2 unified cache. We consider the base L1 data cache configuration as $16KB$ DL1 cache with 4-way associativity and $32B$ blocks. Hence, the total number of sets in the base L1 data cache is 128. We denote the base DL1 cache as 128×4-way DL1 cache. Similarly, the base L2 unified cache configuration is referred as 1024×8-way cache with $64B$ block.

V. RESULTS AND ANALYSIS

We evaluate the effectiveness of our way-sharing mechanism by applying it to data L1 (DL1) and unified L2 (UL2) caches. The effectiveness of way-sharing can be very limited in instruction L1 (IL1) cache because of its high spatial locality.

Table IV shows the cache access time, energy, and area for different cache configurations. As DL1 cache access time is

	Values obtained using CACTI5.3						Used in our evaluation					
	16KB 4-way L1 (128× 4-way)	8KB 2-way L1 (64× 4-way)	512KB 8-way L2 (1024× 8-way)		256KB 8-way L2 (512× 8-way)		16KB 4-way L1 (128× 4-way)	64× (6,2)-way sharing DL1	512KB 8-way L2 (1024× 8-way)		512× (12,4)-way sharing UL2	
	Fast	Fast	Serial	Fast	Serial	Fast	Fast	Fast	Serial	Fast	Serial	Fast
Access time (ns)	0.5314	0.4955	1.9212	1.6627	1.5470	0.9608	0.5314	0.5451	1.9212	1.6627	1.7017	1.0569
Access energy (nj)	0.0307	0.0139	0.1613	0.5991	0.1116	0.2768	0.0307	0.0153	0.1613	0.5991	0.1228	0.3045
Area (mm^2)	0.2862	0.1293	3.3638	4.6705	1.7697	2.1237	0.2862	0.2106	3.3638	4.6705	2.9018	3.4757

TABLE IV

CACHE PARAMETERS OBTAINED FOR VARIOUS CACHE CONFIGURATIONS USING CACTI5.3.

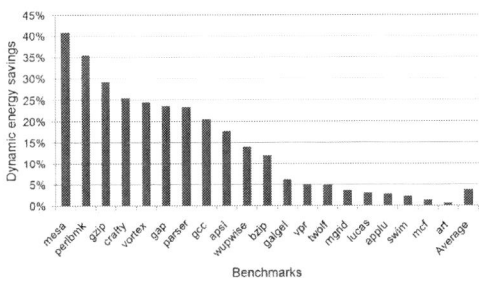

Fig. 6. Energy savings of way sharing DL1 cache over the conventional DL1 cache.

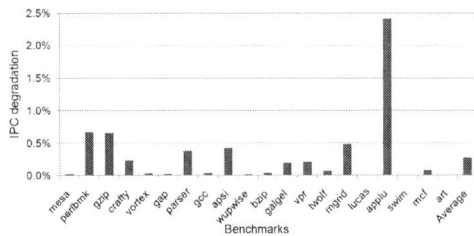

Fig. 7. IPC degradation incurred by way-sharing DL1 cache over the conventional 16KB 4-way DL1 cache.

critical for the system performance, we consider *fast-type* of cache for DL1 caches (refer to [1]). In fast-type caches, extra energy is traded for a potentially faster access. That is, both tag and data arrays are accessed simultaneously and then data is read out from the matched way. In the case of L2 cache, we consider both *fast-type* and *serial-type*. In serial-type caches, the cache tags are accessed first and if there is a match, data belonging to the hit way is read out.

We know that CACTI considers associativity in the order of 2^n, for some n. But the associativity of our way-sharing caches (both DL1 and L2) is not in the order of 2^n. Given that in the way sharing DL1 cache, for any given access, only four ways of a set are active, we consider its access time and dynamic energy consumption to be almost equal to those of a $8KB$ 4-way cache. Similarly, by observing the operation of a way sharing UL2 cache, we consider the access time and per access energy to be equal to those of a $256KB$ 8-way cache. From the area point of view, the $64 \times (6, 2)$-way sharing DL1 cache takes one-and-half times the area of $8KB$ 4-way DL1 cache and $512 \times (12, 4)$-way sharing UL2 cache takes one-and-half times that of $256KB$ 8-way L2 cache. Since these are more like approximations, we conservatively consider an additional 10% overhead each for access time, energy and area.

From Table IV, we can see that the way-sharing DL1 cache has better per-access energy, area values, and almost equal access time to those of conventional cache, respectively. In the case of UL2 cache with fast-type, the way-sharing mechanism gives better access time and energy with less area. The way-sharing mechanism may not be effective in UL2 caches with serial-type as serial-type caches are already optimized for energy consumption.

The dynamic energy savings (considering L1, L2 and main memory) achieved by $64 \times (6, 2)$-way sharing DL1 cache when compared with 128×4-way base DL1 cache are shown in Figure 6. The benchmarks are sorted in decreasing order of the savings obtained. We can see that energy savings range from

0.5% ("art") to 40.8% ("mesa") with an average almost equal to 3.7%. The huge variation in the obtained savings is mainly due to the variation in DL1 miss rates across the benchmarks. For example, the rightmost 6 benchmarks (with savings less than 5%) have DL1 miss rates above 9% ("art", "mcf" and "swim" have miss rates greater than 20%) both in base cache and way sharing cache. Due to high DL1 miss rates, the energy consumption contributed by DL1 is very less to the overall energy consumption, limiting the achievable savings. When we consider the remaining 14 benchmarks, the average savings is found to be 13.9%.

Figure 7 shows performance degradation for $64 \times (6, 2)$-way sharing cache over the base cache. Except for "applu" benchmark, our technique incurs negligible performance penalty for all other benchmarks as compared to the base cache. In the case of "applu" benchmark, our technique incurs a performance penalty of around 2.4% (we reason this behaviour shortly). From both the energy and performance results, we can say that our way-sharing technique is effective in reducing the dynamic energy of DL1 caches.

To understand the impact of way sharing on both performance and dynamic energy consumption, we compare our way sharing DL1 cache with a base DL1 cache having reduced number of ways (similar to the selective cache ways mechanism [4]). As the size of our $64 \times (6, 2)$-way sharing DL1 cache is $12KB$, we turn-off one cache way in the 128×4-way base DL1 cache so that we obtain a 128×3-way cache of size $12KB$. Figure 8 gives the dynamic energy savings (as primary Y-axis) and incurred performance penalty (as secondary Y-axis) of 128×3-way cache with respect to the 128×4-way cache. We can clearly see that way-sharing cache is better than 128×3-way cache, both in terms of energy savings and IPC degradation. To understand this, we look at the average memory access time (AMAT) of the base cache, way-sharing cache and base cache with a reduced way, and find that the way-sharing cache indeed has lower AMAT than the base cache with a reduced way (due to space constraints we do

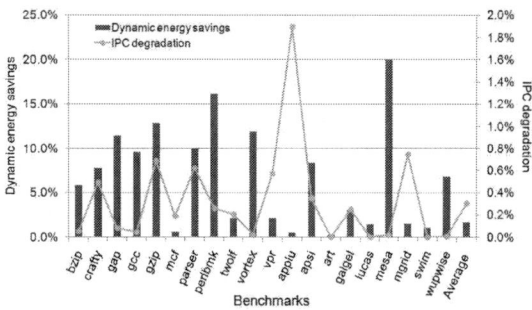
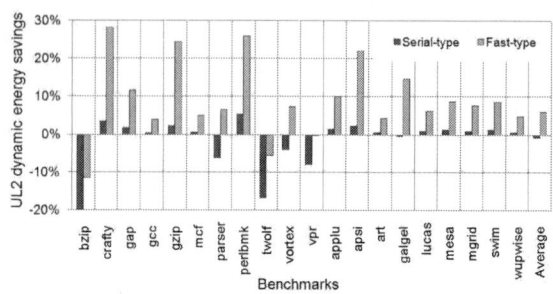

Fig. 8. Energy savings (primary Y-axis) and IPC degradation (secondary Y-axis) of 128 × 3-way DL1 cache over the conventional 128 × 4-way DL1 cache.

Fig. 9. Energy savings of way sharing mechanism over the conventional cache when it is applied to UL2 caches of serial-type and fast-type.

not present AMAT observations in more detail). Because of way sharing among sets, the number of misses incurred by the way-sharing cache is less than that of 128 × 3-way cache. This strengthens our motivation to go for sharing of ways.

Closer look at "applu" provides us with two reasons for the higher performance degradation it incurs (Figure 7). Firstly, Figure 8 shows that "applu" experiences performance loss (around 1.9%) when just one way per set is turned-off, indicating that it is a cache size sensitive application. Secondly, per-set statistics show that (due to space constraints we do not include these stats here), the accesses to all the sets are more or less uniform. This observation is also supported by our results in Table I wherein we can see that for any combination of friend sets the performance obtained is much less than the baseline (base IPC is 1.2269).

To know the effectiveness of our way-sharing mechanism when it is applied to unified L2 cache alone, we consider $512 \times (12, 4)$-way sharing UL2 cache and compare it with 1024×8-way base UL2 cache. Here we consider both serial-type and fast-type modes (as explained earlier in this section). Figure 9 shows energy savings for our $512 \times (12, 4)$-way sharing UL2 cache with respect to 1024×8-way base UL2 cache. As predicted earlier, the way-sharing mechanism is not effective when it is applied to UL2 caches with serial-type and it incurs an average energy overhead of 0.7% (for "bzip" overhead is as much as 19.8%). In the case of fast-type UL2 caches, the way-sharing mechanism reduces the overall memory energy by 6%. Fast-type L2 caches are generally used in systems when performance is more critical (example, Intel Core family processors). In such systems, we can apply the way-sharing mechanism in L2 caches also (we observe that it incurs negligible performance penalty).

Comparing Figures 6-9, we can say the way-sharing mechanism is effective in improving the energy savings with negligible performance loss by enabling the cache sets to utilize the available ways efficiently.

VI. CONCLUSION

We proposed a way-sharing cache architecture, wherein two cache sets share some cache ways so that highly demanded sets can get more cache ways than that of underutilized sets. Simulation results showed that our architecture achieves significant dynamic energy savings when applied in DL1 cache, across

different benchmarks, with negligible performance loss. We also found that this architecture can be implemented in UL2 caches targeted for performance. Existing low power cache mechanisms complement the proposed architecture and can easily be applied over it for further minimization of power consumption. We consider dynamic mapping of friend sets as future work.

ACKNOWLEDGMENTS

This work is supported in part by grant from Department of Science and Technology (DST), India, project no. SR/S3/EECE/0018/2009.

REFERENCES

[1] "Cacti", http://www.hpl.hp.com/research/cacti/.
[2] "Simplescalar", http://www.simplescalar.com.
[3] "Spec 2000 cpu benchmark suite", http://www.spec.org.
[4] D. H. Albonesi, "Selective Cache Ways: On-demand cache resource allocation", International Symposium on Microarchitecture, 1999, pp. 248-259.
[5] S. Dropsho and et. al, "Integrating adaptive on-chip storage structures for reduced dynamic power", International Conference on Parallel Architectures and Compilation Techniques, 2002, pp. 141-152.
[6] Z. Hu, M. Martonosi, and S. Kaxiras, "Improving cache power efficiency with an asymmetric set-associative cache", High Performance Memory Systems, Springer Verlag, 2003.
[7] J. kwon Peir, Y. Lee, and W. W. Hsu, "Capturing dynamic memory reference behavior with adaptive cache topology", International Conference on Architectural Support for Programming Languages and Operating Systems, 1998, pp. 240-250.
[8] A. Malik, B. Moyer, and D. Cermak, "A low power unified cache architecture providing power and performance flexibility", International Symposium on Low Power Electronics and Design, 2000, pp. 241-243.
[9] J. Montanaro, et. al, "A 160-mhz, 32-b, 0.5-w cmos risc microprocessor", IEEE Journal of Solid State Circuits, 2006, pp. 1703-1714.
[10] M. K. Qureshi, D. Thompson, and Y. N. Patt, "The V-Way cache : Demand-based associativity via global replacement", International Symposium on Computer Architecture, 2005, pp. 544-555.
[11] "Simpoint 3.0: Faster and more flexible program analysis", Journal of Instruction Level Parallelism, 2005, pp. 1-28.
[12] M. Tang and X. Lin, "A novel scheme to balance the cache sharing in high performance computing system", International Conference on High Performance Computing and Communications, 2008, pp. 695-701.
[13] A. V. Veidenbaum, et. al, "Adapting cache line size to application behavior", International conf. on Supercomputing, 1999, pp. 145-154.
[14] C. Zhang, F. Vahid and R. Lysecky, "A self-tuning cache architecture for embedded systems", ACM Trans. on Embedded Computing Systems, 2004, pp. 407-425.
[15] C. Zhang, F. Vahid and W. Najjar, "A highly configurable cache for low energy embedded systems", ACM Trans. on Embedded Computing Systems, pp. 363-387, 2005.
[16] C. Zhang, "Balanced cache: Reducing conflict misses of direct-mapped caches", International Symposium on Computer Architecture, 2006, pp. 155-166.

A Novel Encoding Scheme for Low Power in Network on Chip links

Deepa N.Sarma
Department of ECE
National Institute of technology
Tiruchirapalli,India

G. Lakshminarayanan
Department of ECE
National Institute of technology
Tiruchirapalli,India

Suryakiran Chavali K.V.R.
Department of ICE
National Institute of technology
Tiruchirapalli, India

Abstract— Dynamic power dissipation in interconnects is a major contributor to power consumption in Network on Chips (NoCs). This is mainly due to two factors, self switching activity of the particular link and coupling switching activity among adjacent links. Two novel techniques are proposed to reduce power consumption due to switching transition and crosstalk. First technique reorders the data in such a way that switching transition is brought down. In the second technique, it is ensured that power consumption due to cross coupling activity is reduced. An end to end encoding scheme facilitating two stage coding to reduce power consumption in wormhole routed network on chip is designed using the proposed power reduction techniques. Encoder and Decoder exhibiting the proposed scheme have been described in RTL level in Verilog HDL, synthesized and mapped into UMC180 nm technology library. It has been observed that the proposed technique (TSC) offers an average reduction in dynamic power consumption of 17.34%. Proposed scheme was compared with existing techniques and observations concluded that there was not much degradation in area, speed and static power dissipation. Power reduction when subjected to different kinds of data streams was analyzed and results indicate that proposed scheme offers uniform power reduction irrespective of the nature of data stream unlike the existing techniques.

Keywords- Network on Chip links; self switching; crosstalk; low power; two stage coding; analysis; uniform power reduction.

I. INTRODUCTION

Highly scalable modular structure of Network on Chips (NoCs) makes a fitting replacement for system on chip designs (SoCs) in designs incorporating large number of processing cores [1][2]. Amongst the communication resources, as technology shrinks, the power ratio between NoC links and routers increases making the links becoming more power hungry than routers [7].

Several encoding schemes have been proposed to reduce power in context of bus based architectures.Bus-invert method [4] can be applied to encode randomly distributed data patterns. [8], [9] and [11] deal with reduction of switching activity in serial links (caused due to serialisation of parallel data). A few encoding techniques have been

defined to take into consideration the contribution of cross-coupled capacitance [6], [12]. Jantsch *et al.* [5] analyzed the use of partial bus invert coding as link level low power encoding technique with the conclusion that it spends several times more power than no encoding at all. In [7], an end to end encoding technique has been suggested for reducing power dissipation.

Here, a two stage coding scheme (TSC) for power reduction has been devised, considering both self switching activity and cross coupling effect. The encoder and decoder structures are placed at the network interface level. The data packet passes through two stages of encoding before ensuring reduction in switching activity and crosstalk. First stage reorders and rearranges data in such a way that transition in each link is reduced. The second stage sends data as is or inverts the data entering each link, where the decision to invert the data is dependent on contribution of cross-coupled activity in the power dissipation of particular link.

The proposed scheme can only be applied to NoCs supporting wormhole routing as it does not support interleaving of flits. Encoder structure has been analysed with a set of representative data streams. A comparative study has been done with BI [4], CDBI [6], SC [7] techniques and it has been proved that the above scheme offers significant power reduction irrespective of nature of data stream unlike the above mentioned techniques. An average power reduction of 17.34% was obtained.

The rest of the paper is organized as follows. In Section II power reduction techniques incorporated in the proposed scheme are discussed. The architecture of the encoder and decoder is discussed in Section III. In Section IV technique is validated by analysis and experiments. Finally Section V cites the conclusion and possibility of future work. References are cited in Section V1.

978-1-4673-0438-2/12 $31.00 © 2012 IEEE

II. PROPOSED SCHEME

The dynamic power consumed by the interconnects and driver is given by:

$$P = [T_{0\to1}(C_s + C_l) + T_c C_c] Vdd^2 F_{clk} \qquad (1)$$

where Vdd is the supply voltage, F_{clk} is the clock frequency, C_s is the self capacitance (which includes the parallel plate capacitance and the fringe capacitance), C_l is the load capacitance and C_c is the coupling capacitance. $T_{0\to1}$ and T_c are the average number of effective transitions per cycle for C_s and Cc respectively. They are computed as follows. $T_{0\to1}$ counts the number of $0\to1$ transitions in the bus in two consecutive transmissions. T_c counts the correlated switching between physically adjacent lines [7]. The proposed technique uses two power reduction techniques to minimise correlation switching activity and self switching activity. These are stated as follows:-

A. Reducing self switching activity

From (1), it can be deduced that dynamic power dissipation in a link is directly proportional to switching activity in a link. As shown in [9], when an n bit parallel data is serialized, the switching transition in the links increases. Our encoding scheme rearranges the data prior to serialization so as to bring down the power consumption by reducing number of transitions. The encoding scheme is stated as follows:-

1. Consider an n bit data (n=2^m, m>1) B $_{(t)}$.

2. Before serialization n bit parallel data is checked for number of transitions.

3. If $N_0 > N_T$, $B_{(t)} = B_{s(t)}$ else $B_{(t)} = B_{(t)}$, where N_0 is number of transitions, N_T refers to threshold value, $B_{s(t)}$ is obtained by interchanging odd and even bits of input data .

4. The threshold was observed to vary as follows:-

$$N_T = n/2-1, \text{ where } n=2^m \text{ for } m=2$$
$$N_T = n/2, \text{ where } n = 2^m \text{ for } m=3$$
$$N_T = n/2+2(m-4)+1 \text{ where } n=2^m, \text{ for } m = 4, 5, 6 \qquad (2)$$

B. Eliminating crosstalk

Four types of coupling transitions are enumerated as follows. A Type I transition occurs when one of the lines switches while the other stays unchanged. In a Type II transition one line switches from low to high and the other from high to low. A Type III transition occurs when bothlines switch simultaneously. Finally, in a Type IV transition both lines do not switch. The coupling transition activity Tc is a weighted sum of the different type of

TABLE I
CHANGE IN TRANSITION WHEN ADJACENT LINK IS INVERTED

Time	Normal coding (b,b+1)					Inverted coding(inverting the bits on link b+1)				
T-1 T	00 01	00 10	11 01	11 10	*Type1*	01 00	01 11	10 00	10 11	*Type1*
T-1 T	01 10		10 01		*Type2*	00 11		11 00		*Type3*
T-1 T	00 11		11 00		*Type3*	01 10		10 01		*Type2*
T-1 T	00 00	01 01	10 10	11 11	*Type4*	01 01	10 10	00 00	10 10	*Type4*

coupling transition contributions.

$$Tc = k_1 T_1 + k_2 T_2 + k_3 T_3 + k_4 T_4 \qquad (3)$$

Here the Ti, $i = 1, 2, 3, 4$, are the average number of transition type i and ki are weights. According to [6] it is assumed $k1 =1$, $k2 =2$ and $k3 =k4 =0$. That is, $k1$ is assumed as reference for other types of transition. The effective capacitance in Type II transition is usually twice that of a Type I transition. In Type III transition, as both signal switch simultaneously. Finally, in Type IV transition there is no dynamic charge distribution over Cc. Based on this, Equation (1) can be expressed as follows:

$$P = [T_{0\to1}(C_s + C_l) + (T_1+2T_2) C_c] Vdd^2 F_{clk} \qquad (4)$$

To eliminate crosstalk, link b and adjacent link b+1(Refer to table 1) are considered. Each two bit word represents status of link b and b+1 respectively at specified time. It can be observed that if we invert the bits entering link b+1, the type 2 transition gets converted to type 3 and vice versa. After inversion equation (3) can be written as:-

$$Tc' = k_1 T_1 + k_2 T_2' + k_3 T_3' + k_4 T_4 \qquad (5)$$

Where $T_2' = T_3$, $T_3' = T_2$. Hence in (3) if $T_2 > T_3$, the data bits of the particular link is inverted. As k3=0, from (5) and (3) Tc'<Tc, as $T_2' < T_2$. This scheme can be applied along with the first scheme as inverting does not alter the effect of shuffling of bits.

III. ARCHITECTURE

Serial links for NoC data transport have been proposed to overcome the drawbacks of parallel links. They should not only allow savings in wire area and power dissipation and reduction of signal interference, noise and crosstalk, but also eliminate the need for multiple line drivers and buffers [13], but serialisation increases the latency of data. A compromise

---

Identify applicable sponsor/s here. If no sponsors, delete this text box. (sprs)

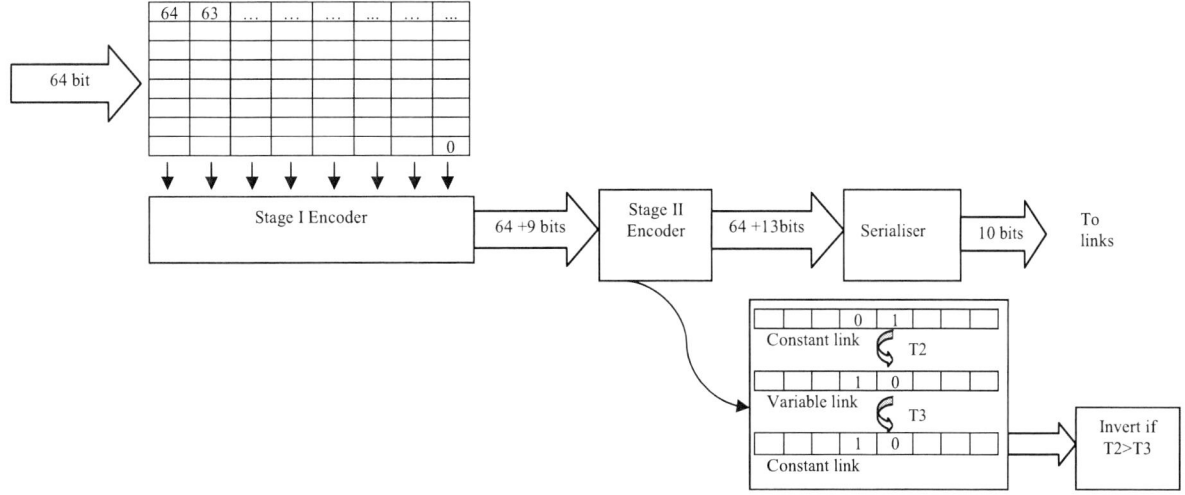

Figure 1. Encoder Block diagram with M=64, N=8

is chosen between two and partly serialised link similar to [10] is considered.

A. Design of Encoder and Decoder

Encoder is placed at network interface level. As shown in fig 1, an M (here M=64) bit packet passes through first encoding stage where data which has to pass through each link is rearranged (each column in the packet as in fig 1) or sent as it is in accordance with the encoding scheme described in section II A.

Additional M/N+1 (where N stands for number of links, in this design assumed to be 8) bits are added as control bits and data packet enters stage II. Here alternate links(four odd links) are inverted or send as they are after comparison with adjacent links as per encoding scheme in section II B. Note that power reduction acquired in the first stage remains intact as inverting ensures same number of switching transitions.

Data bits (M bits) and (M/N+1+M/2N) control bits are converted to (M/N+M/4N) bits by a serializer. Control bits are sent through M/4N extra links. First extra link consists of M/N control bits of first stage. These control bits are further encoded using encoding scheme in stage I encoder. This extra encoding information is carried by the extra bit from stage I. This extra bit, M/2N control bits generated in stage II constitute second extra link. Zeros are interleaved between the bits for synchronisation in second control link.

It is assumed that M, $N=2^k$ where k is any positive integer. As shown in fig 2.b, the first four bits of control word (C1 to C4) indicate whether particular link ha&s been inverted or not. Eight bit control words from stage I encoder are encoded further using scheme in section II A which constitute the last eight bits (D0 to D7) in fig 2. b. The extra bit (E) holds the encoding information of bits (D0 to D to D7) in fig 2. b. The extra bit (E) holds the encoding information of bits (D0 to D7) above.

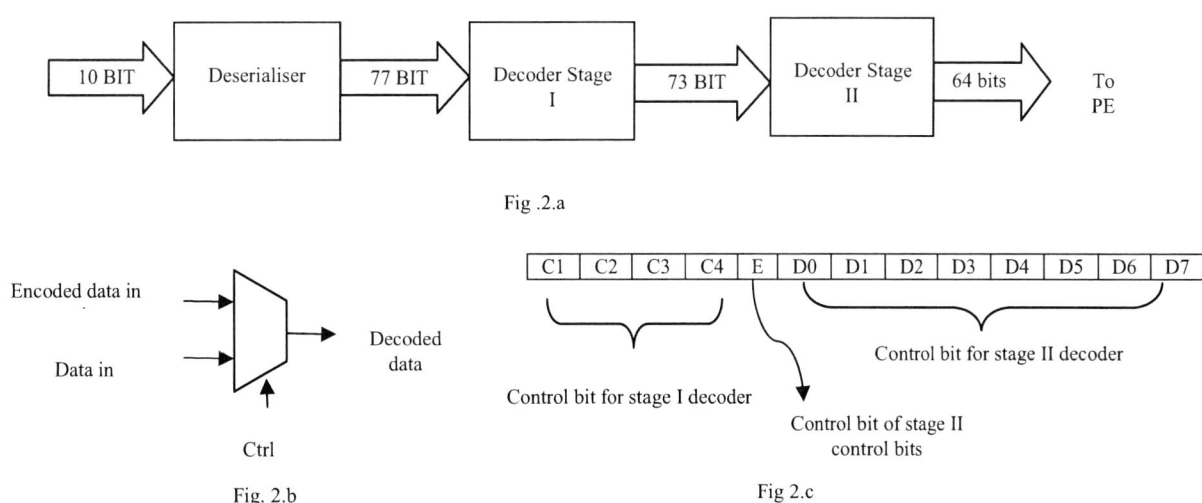

Fig .2.a

Fig. 2.b

Fig 2.c

Figure 2. a. Block diagram of the encoder b. Basic decoding structure to decode each link c. Control bits

978-1-4673-0438-2/12 $31.00 © 2012 IEEE   259

Decoder constitutes of a deserialiser which obtains the control word. As illustrated in fig 2.c, of the 13 bit control word, first four bits control the stage I decoder. Ninth bit decodes the eight control bits which decode the M bit data. Basic structure to decode each 8 bit data is illustrated in Fig 2.b where 'data in' refers to input data and 'encoded data in' stands for shuffled (stage II decoder) or inverted (stage I decoder) version of data in.is chosen between two and partly serialised link similar to [10] is considered.

B. Logic design

The encoder and the decoder for (M=64,N=8) have been designed in Verilog HDL described at the RTL level, synthesized with Synopsys Design Compiler and mapped onto a UMC 180 nm technology library. Both encoding and decoding structures are placed at the network interface before the serializer contrary to one on links in [9]. BI [4], CDBI [6] and SC [8] encoders operating on same data width has also been implemented and mapped into the same library. A comparison of area and power and timing is illustrated in Fig 3.It can be observed that proposed scheme (TSC) ensures minimum overhead in power and area (exception only in the case of BI encoder). It should also be noted that delay over head is much lesser compared to SC scheme. It can be deduced from the graphs that proposed scheme (TSC) ensures better performance parameters.

IV. EXPERIMENTS

Two stage encoder has been tested against different types of data streams and simulation results confirm that significant power reduction is obtained in all the cases. Also in order to compare the efficiency of proposed encoder, the encoder has been replaced with BI [4], CDBI [6] and SC [7] encoders and subjected to the same data streams. All the encoders have been synthesized with Synopsys Design Compiler (UMC 180 nm technology library) and operated at a frequency of 500MHz.It has been observed that distribution of power reduction in case of proposed scheme is uniform in all the cases.

A. Experiments

As indicated in Section III, encoder operating on 64 bit data was implemented in RTL level. This encoder was tested against different types of data streams. These include a PDF file of 774 KB, an image file of 6.77KB an audio file of 3.92 MB, and a video file of about 47.5MB. The contents of above mentioned files were applied to the proposed encoder and average power dissipation was calculated using (1) for each link. Also, average power dissipation in absence of encoder was calculated from which percentage of power reduction was obtained. This was replaced by BI, CDBI and SC encoders and same steps were repeated. Percentage of

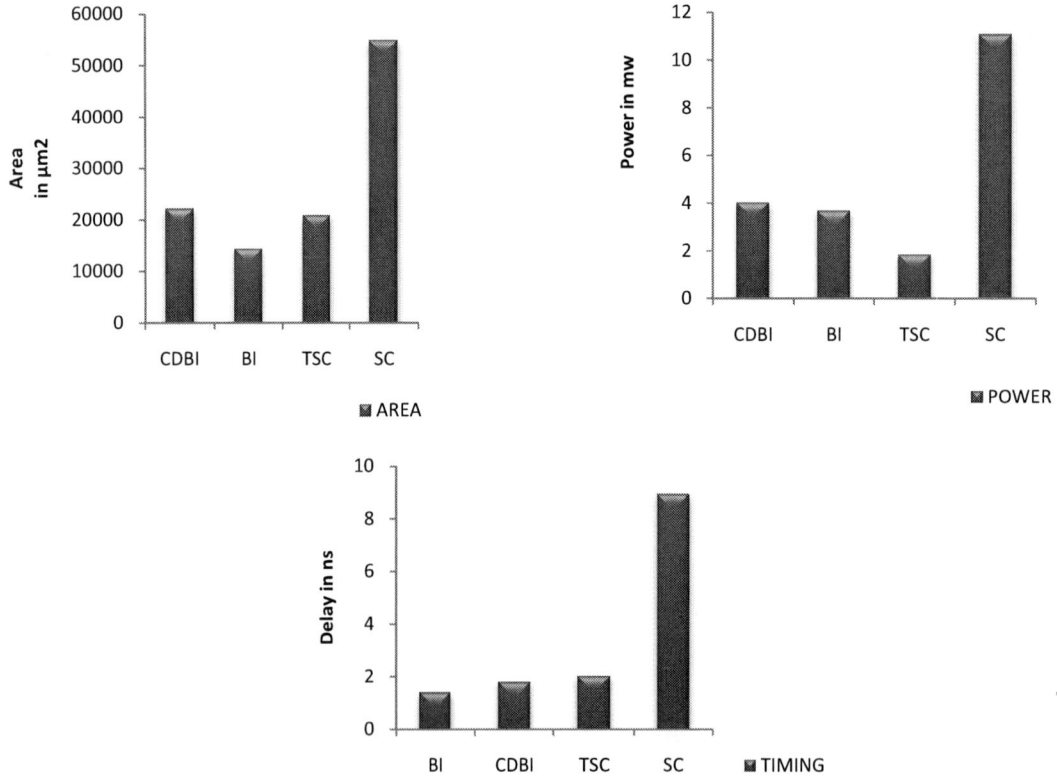

Fig 3.Comparison of area, power an delay with different encoders

power reduction in the case of each type of data stream was plotted as illustrated in fig 4.

B. Results

Proposed scheme (TSC) obtains a 19.6 % reduction in case of audio (MP3) file, 17.1 % for image (JPG), 19.5% for video (MPEG) file, 12.7 % for text and 17.8 % for PDF file.

As illustrated in figure 4, it is observed that proposed scheme (TSC) provides uniform power reduction irrespective of the nature of the data stream. For instance, in the case of SC, the results are relatively better for aggressively compressed data streams like MPEG, JPG and MP3 but relatively less for PDF and text file. Similarly for CDBI, there is no significant power reduction except in the case of audio file. Though BI gives good power reduction in case of PDF file, no satisfactory power reduction was obtained in the case of other data streams. Even though best power reduction is not guaranteed in all the cases TSC provides almost same power reduction with average of 17.34% among all cases. In case of BI, CDBI and SC average power dissipation comes down to 4.27 %, 11.27%, and 14.5 %.

V. CONCLUSION

In this paper, two encoding techniques are proposed to reduce power consumption due to self switching activity and crosstalk in NoC links. A two stage encoding scheme incorporating the proposed techniques has been devised. The proposed encoding structure has been implemented and its performance is compared with BI, CDBI and SC encoders. Our module has been tested against different types of data streams and it was observed that unlike other schemes uniform power reduction is guaranteed in all cases. An average power reduction of 17.34% has been obtained.

The proposed technique is not applicable in virtual channel routed networks on chip as the encoder is placed in network interface level. Future work should be towards making it flexible for different routing mechanisms and optimisation of power and area overhead caused by extra module.

VI. REFERENCES

[1] Vangal and Sriram et al., On An 80-Tile 1.28TFLOPS Network-on-Chip in 65 nmCMOS, Digest of Technical Papers, IEEE Intl. Solid State Circuits Conference, pp.98–589, 2007.

[2] Hoi Jun Hoo,Kangmin Lee and Jun Kyong Ki ,Low power network on chip for high perfomance SOC design,Volume 1 of system on chip design and technologies CRC press, 2008 edition.

[3] S. R. Vangal, J. Howard, G. Ruhl, S. Dighe, H. Wilson,J. Tschanz, D. Finan, A. Singh, T. Jacob, S. Jain, V. Erraguntla,C. Roberts, Y. Hoskote, N. Borkar, and S. Borkar, "An80-tile sub-100-W TeraFLOPS processor in 65-nm CMOS,"IEEE Journal of Solid-State Circuits, vol. 43, no. 1, pp. 29–41, Jan. 2008.

[4] M. R. Stan and W. P. Burleson, "Bus invert coding for low power I/O," *IEEE Transactions on Very Large Scale Integration Systems*, vol. 3, pp. 49–58, Mar. 1995.

[5] A. Jantsch, R. Lauter, and A. Vitkowski, "Power analysis of link level and end-to-end data protection in networks on chip,"in IEEE International Symposium on Circuits and Systems,vol. 2, May 2005, pp. 1770–1773.

[6] K. W. Kim, K. H. Baek, N. Shanbhag, C. L. Liu, and S. M.Kang, "Coupling-driven signal encoding scheme for low power interface design," in IEEE/ACM International Conference *on Computer-aided Design*, 2000, pp. 318–321.

[7] Maurizio Palesi, Fabrizio Fazzino Giuseppe Ascia, and Vincenzo Catania,"Data Encoding for Low-Power in Wormhole-Switched Networks-on-Chip"in 12th Euromicro Conference on Digital System Design / Architectures, Methods and Tools,pp. 119 -126,2009

[8] Jaesung Lee,"On chip serialisation method for low power communications " in ETRI journal,volume 32,Number 4,August 2010,pp. 540-547.

[9] K. Lee, S.-J. Lee, and H.-J. Yoo, "Low-power network-on-chip for high-performance soc design ," *IEEE Transactions on Very Large Scale Integration (VLSI) Systems*, vol. 14, pp. 148–160, 2006.

[10] S.-J. Lee *et al.*, "An 800 MHz star-connected on-chip network for application to systems on a chip," in IEEE Int. Solid-State Circuits Conf. *Dig. Tech. Papers*, Feb. 2003, pp. 468–469.

[11] K. Lee *et al.*, "SILENT: Serialized low-energy transmission coding for

on-chip interconnection networks," in *IEEE Int. Conf. Comput.-Aided Des. Dig. Tech. Papers*, Nov. 2004, pp. 448–451.

[12] J. Henkel, H. Lekatsas, and V. Jakkula, "Encoding schemes for address busses in energy efficient SOC design," in *VLSISOC*

2001 11th International Conference of Very Large Scale*Integration*, Montpellier, France, Dec 2.

[13] A. Morgenshtein, I. Cidon, "Comparative analysis of serial vs parallel links in NoC," 2004 International symposium on SoC, pp. 185-188,2004.

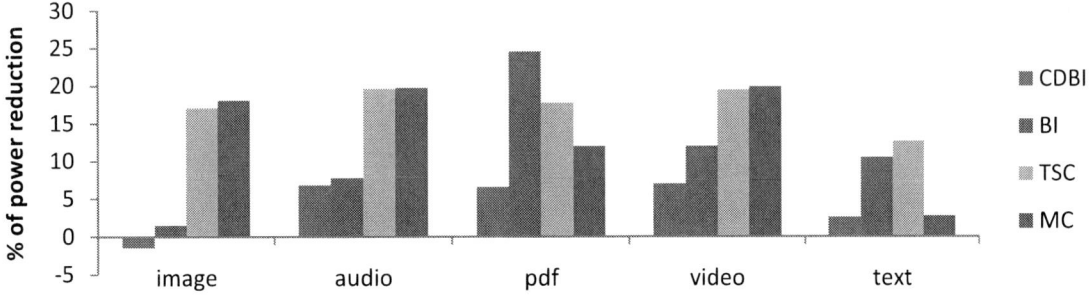

Fig 4.Comparison of performance with different encoders subject to various data streams

978-1-4673-0438-2/12 $31.00 © 2012 IEEE

2012 25th International Conference on VLSI Design

A Power Delivery Network Aware Framework for Synthesis of 3D Networks-on-Chip with Multiple Voltage Islands

Nishit Kapadia, Sudeep Pasricha
Department of Electrical and Computer Engineering
Colorado State University, Fort Collins, CO, U.S.A.
nkapadia@colostate.edu, sudeep@colostate.edu

Abstract - *IR drops in a Power Delivery Network (PDN) on chip multi-processors (CMPs) can worsen the quality of voltage supply and thereby affect overall performance. This problem is more severe in 3D CMPs with network-on-chip (NoC) fabrics where the current in the PDN increases proportionally to the number of device layers. Even though the PDN and NoC design goals are non-overlapping, both the optimizations are interdependent; for instance, each new core mapping on the 3D die will change traffic patterns and have a unique distribution of IR-drops in the PDN. Unfortunately, designers today seldom consider design of PDN while synthesizing NoCs. If NoC synthesis is carried out without considering the associated PDN design cost, it can easily result in an overall sub-optimal design. In this work, for the first time, we propose a novel PDN-aware 3D NoC synthesis framework that minimizes NoC power while meeting performance goals; and optimizes the corresponding PDN for total number of Voltage Regulator Modules (VRMs), current efficiency, and grid-wire width while satisfying IR-drop constraints. Our experimental results show that the proposed methodology provides more comprehensive results compared to a traditional approach where the NoC synthesis step does not consider the PDN costs.*

1. Introduction

Designing a robust Power Delivery Network (PDN) is critical to the overall performance of today's CMPs. The PDN is required to deliver a stable power supply across the chip, which is within a desired voltage range; and tolerate large variations in load currents [1]. Multiple voltage islands (*VIs*) are generally used in modern CMPs to minimize the total power dissipation while meeting performance constraints. The PDN is required to supply power at different voltage levels corresponding to the *VIs* while keeping power loss to a minimum. With increasing device density and supply voltage levels, the supply currents have risen; however the scaling of PDN impedance has not kept up with this trend [2]. IR drops can worsen the quality of voltage supply and thereby affect the ultimate performance of the CMP. This problem is more severe in 3D CMPs as the current in the PDN could be as many times more as the number of device layers compared to a 2D CMP. Besides, the number of I/O pins on an *n*-layered 3D CMP is about *n* times smaller than its 2D counter-part, thus exacerbating the problem of a degraded voltage supply in 3D designs [3].

Another critical component at the heart of emerging 3D CMPs is the network-on-chip (NoC) architecture that enables tens to hundreds of cores to communicate with each other at the intra- and inter-layer levels. As the power dissipated in the NoC has become a significant portion of the total on-chip power, optimizing the communication power in addition to computation power is critical [6]. Several recent works have proposed techniques to synthesize regular and custom 3D NoC topologies [9][28]-[32] to optimize communication power. However, these works do not consider the design of the PDN while mapping cores and designing the NoC fabric, and typically generate a single power-optimized configuration. Performing synthesis of the PDN for the best generated configuration in these cases puts stringent demands on the already strained PDN. This can either make it

extremely difficult to meet the PDN constraints such as maximum grid-width, maximum number of voltage regulator modules (VRMs), and minimum current efficiency; or lead to over-margining for the PDN, which can be wasteful. Thus, the traditional approach of synthesizing a NoC fabric without considering the PDN ends up severely constraining the PDN design space, often leading to sub-optimal or even completely infeasible designs.

In this work, for the first time, we propose an automated framework for PDN-aware synthesis of mesh-based NoC fabrics in 3D CMPs that optimizes communication power while meeting application performance constraints. We recognize the key insight that different instances of voltage partitioning and core-to-tile mapping (different configurations of the NoC synthesis process) can *significantly* alter the power/voltage distribution map seen by the PDN. Accordingly, our framework considers the interdependence between a synthesized NoC configuration and its corresponding VRM placement and power efficiency in the PDN. The novel contributions of our synthesis framework are as follows:

- We employ a novel branch and bound procedure that combines directed search and random search to produce multiple mapping solutions satisfying *VI* constraints while optimizing NoC power;
- We develop a linear programming formulation as well as a fast heuristic to synthesize a PDN comprised of a segmented power grid for cores running at multiple voltages; with a topological structure of VRMs that considers physical placement of VRMs on the 3D mesh to optimize current efficiency and VRM count;
- We generate a set of interesting design points (Pareto mappings) that allow a designer to weigh the PDN design cost against NoC design cost, and select a suitable solution that meets power, performance, and PDN design goals.

2. Related Work

Many researchers [7]-[11] have proposed custom topology synthesis techniques for NoC fabrics that improve overall performance at the cost of sacrificing the regularity of mesh-based structures. Although these custom architectures are expected to achieve better latency and area utilization, their design process is more complex and faces several challenges, such as greater crosstalk and uncertainty in link delays due to irregular interconnect structures. Thus, a conservative enough custom design may actually offset the advantages of better performance [12]; especially for medium to large sized (in terms of total number of cores) NoC architectures. The problem of NoC synthesis on regular structures with multiple supply *VIs* has been addressed in several works [6][13]-[19]. Given the promise of 3D technologies, 3D NoC synthesis in recent years has also attracted significant research efforts [9][28]-[32]. These works have proposed techniques to optimize the 3D NoC designs for power, temperature, and performance. However, none of the above approaches have considered the effects of NoC synthesis on the efficiency and overheads associated with the PDN design; in other words, these approaches are not PDN-aware.

Techniques for optimizing PDNs in 3D ICs have been studied in a few recent works [1]-[3][20][21]. Amelifard et al. [1] use dynamic

978-1-4673-0438-2/12 $31.00 © 2012 IEEE

programming to generate a multi-level tree topology of suitable Voltage Regulator Modules (VRMs) to improve the power efficiency in the PDN. Jain et al. [2] propose a multi-story power delivery technique which improves upon IR noise in the PDN by recycling current between different power supply domains. Falkenstern et al. [20] use simulated annealing to co-synthesize the floorplan and P/G network, optimizing wirelength, area, P/G routing area, and IR-drops. Chen et al. [21] propose an integrated 3D architecture of stacked-TSV, thermal and power distributed network (STDN); and use a simulated annealing floorplanner to minimize voltage drop, temperature, and other factors in STDN. None of these PDN optimization techniques considers the system level impact of the 3D NoC fabric and core mapping across the layers.

In this paper, we present novel techniques for PDN design as well as NoC synthesis; for mesh-based 3D CMPs. To the authors' knowledge, this is the first work which proposes a physically aware 3D NoC synthesis framework that also integrates PDN optimization to produce a more efficient overall CMP design.

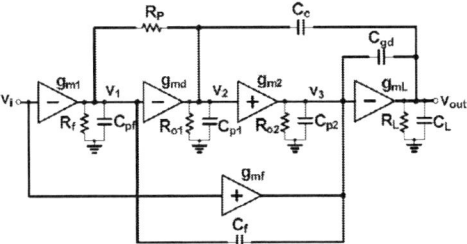

Figure 1: Schematic of an LDO-VRM [23]

3. PDN Design with Multiple Supply Voltages

High circuit density in smaller footprint 3D ICs presents a unique challenge for designers of PDNs, requiring the network to deliver significantly more current than in 2D ICs with fewer P/G bumps, while also circumventing increasingly daunting IR-drop issues. Voltage regulator modules (VRMs) are key components of any PDN, responsible for stepping down the high voltage of the power source. To cope with supply voltage variations in emerging 2D and 3D ICs, traditional off-chip voltage regulators require large decoupling (or bypass) capacitors and inductors that end up occupying excessive PCB-area. Moreover, the parasitic inductance and resistance between the regulator and the processor hinders the regulator from reacting quickly to load transients [4]. Bringing the voltage regulators on-chip (closer to the load) is one solution to the problem that would result in smaller decaps and inductors needed, as the parasitic elements fall. Additionally, an on-chip regulator can react quickly to the load transients, save on-board space, as well as reduce the number of external P/G pins needed. Low Drop-Out regulators (LDOs; Fig. 1) are particularly amenable for on-chip integration due to their small area overhead and low dropout voltage, as opposed to switching-type regulators which employ on-chip inductors that occupy valuable area [5]. The characteristics of LDO-VRMs assumed in our PDN design framework follow designs proposed in [23][24].

A PDN traditionally uses a single continuous power grid made of orthogonal interconnects (on the top wiring levels) running across the chip at the electrical potential of the external pin voltage. In systems with multiple voltage levels, VRMs can be inserted appropriately to step down from the single external voltage level to the different operating voltage levels of the cores/modules/VIs. As all voltages are stepped-down from a high external voltage level, the power conversion efficiency in this approach could be poor. Alternatively, as many parallel power grids as the number of supply voltages can be implemented [22]. But even with reasonable number of voltage levels, this approach could result in a prohibitively high PDN routing area overhead. Ultimately, the chosen design approach must cope with the IR-drop problem which is worse in 3D ICs by as much as

3.4× compared to 2D ICs [3], and becomes more severe as we move farther away (on the power grid) from the power source. This problem can either be rectified by inserting additional VRMs, which have an associated area overhead or by increasing the grid-wire width, which increases the PDN routing area on the chip.

In this work, we propose a PDN design structure with a segmented grid configuration that allows for more power efficient stepping down of the voltages derived from VRM outputs, while at the same time considering the overheads of VRM insertions and grid-wire width. To the best of our knowledge, the problem of PDN design that includes determining the locations of VRMs for 3D CMPs operating at multiple voltage levels has not been addressed before.

4. Problem Formulation

We are given the following inputs to our problem:
- A regular 3D mesh-based NoC with dimensions (dim_x, dim_y, dim_z) with the number of tiles $T = dim_x * dim_y * dim_z$ and each tile consisting of a compute core and a NoC router;
- A core graph $G(V,E)$; with a set of T vertices $\{V_1, V_2, ..., V_T\}$ representing homogenous cores on which tasks have already been mapped, and the set of M edges $\{e_1, e_2, ..., e_M\}$ that represent communication dependencies between cores;
- A set of triplets constituting operating voltages, operating frequencies and maximum supply currents for the T cores $\{(v_1, f_1, i_1), (v_2, f_2, i_2), (v_3, f_3, i_3), ..., (v_T, f_T, i_T)\}$;
- An external voltage supply EV to the PDN and a 3D segmented power grid;
- A set of r possible grid-wire resistance (PDN branch resistance) values: $\dot{R} = \{R_1, R_2, .., R_r\}$.

Given the above inputs, our goal is to obtain a core to die mapping and synthesize a regular 3D mesh NoC for a specific application, such that all application performance requirements as well as PDN IR-drop constraints are satisfied; while minimizing the total communication power in the NoC components (routers, links, voltage level converters or VLCs, mixed clock FIFOS or MCFIFOs), the external current (EI) drawn by the PDN, and the number of VRMs.

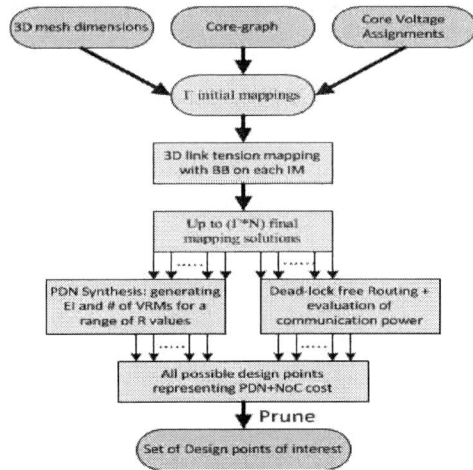

Figure 2: Flow of the PDN-aware NoC Synthesis Framework

5. PDN-aware NoC Synthesis Framework

The flow of our PDN-aware NoC synthesis framework is shown in Fig. 2. As a first step, Γ initial mappings are generated from the communication dependencies defined by the core-graph and the voltage assignments, where Γ is the total number of allowed voltage levels. A link tension based Branch & Bound (BB) procedure is then run on each Initial Mapping (IM) to generate multiple final mapping candidates. On each of these candidates, deadlock free YXZ routing is employed to compute the total power of all the NoC components: routers, links, as well as MCFIFOs and VLCs (which are needed when crossing *VIs*). Next, the PDN synthesis step generates a PDN design and evaluates corresponding PDN costs for a candidate, given the set $\acute{R}$ of grid-wire resistance values. The communication power represents the cost of the NoC whereas, the PDN cost includes the total number of VRMs needed, grid-wire width, and the external current (*EI*) drawn. This cost is computed for all the final mapping candidates to generate a set of final solution points. The points which have both the PDN costs and the NoC cost greater than some other point are pruned to finally produce a set of final design points, each optimized for the PDN and NoC design objectives by varying degrees. In the following subsections, we describe the steps in detail.

5.1 Initial Mapping (IM)

In this first step, we generate an initial core-to-tile mapping by traversing an Inter-Island Communication Graph $IICG(V_{isl}, E_{isl})$, where the vertices constitute the entire islands and the edges represent the aggregate communication bandwidth between the respective islands. A breadth-first search (BFS) starting with each of the Γ islands as the root node would produce Γ distinct sequences of islands, each of length Γ. The order of islands in each sequence is based on decreasing communication bandwidths with the island selected as the root node. Subsequently, the cores are mapped onto the tiles of the NoC in order of the island $sequence_j$ to generate IM_j.

We follow a pre-defined sequence of tile co-ordinates to generate the initial mapping as illustrated in Fig. 3 for a 64 core NoC. Notice in the figure that the mapping starts from the top-right corner of the topmost layer (layer 1) and ends at the top-left corner of the same layer. Such an ordering grows in x, y and z directions in a symmetrical way, thereby keeping the Manhattan distances between the currently placed core and recently placed cores short, at the same time, guaranteeing *VI* integrity (i.e., every core in a *VI* has at least one neighbor of the same voltage level as itself).

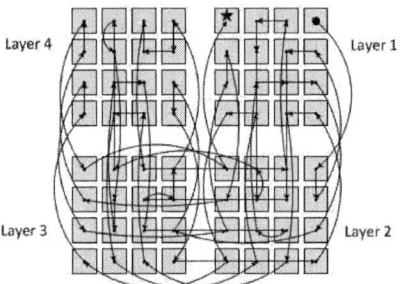

Figure 3: The order of placement of cores to generate an initial mapping (IM) for a 64-core 3D-mesh NoC

5.2 3D Link-Tension Mapping with BB

We propose a branch and bound (BB) technique that combines random search with directed search to generate multiple mapping candidates. Link-tension is the product of the communication bandwidth and post-mapping Manhattan distance for any edge in the core graph $G(V,E)$. The directed swaps are geared to reduce the highest tensions in the NoC in order to reduce communication power. The 'best swap' swaps the core under most tension (on the NoC link-tension map) in the direction of most tension. We combine the directed swaps with random swaps for effective exploration of the solution search space. Any swap which does not disintegrate *VIs* is considered valid; where we consider swaps between adjacent cores in horizontal and vertical planes, as well as horizontal-diagonal swaps.

Let n be the maximum branching degree and N, an upper bound on the total number of candidates which is a multiple of n; α be a positive fraction which governs the weight of the random component

in BB, and C be the number of current candidates. The pseudo code for the BB procedure is given below, which is run on each IM.

3D Link-tension Mapping with BB

input: core graph G(V, E) and an IM solution

1: **while** (($C<N$) && (at least one non-leaf node exists in C));**do**$\forall$non-leaf nodes on the current BB level {
2: Compute B: $B = (n + 1) - \lceil (n - 1) * (C + 1)/N \rceil$
3: Compute R and D: $R = \lfloor \alpha * B/2 \rfloor$; $D = B - R$
4: Find out the D best swaps (directed search) and check their validity
5: If one or more valid swaps found, proceed to step 8
6: Find the best valid swap while considering all cores
7: If no swap is valid, mark this candidate (node) as a leaf; else execute swap (branch out child) and delete current node; then, goto next iteration
8: Compute the R random valid swaps
9: Execute computed random and directed swaps, branching out a new child for each swap and delete current node; then, goto next iteration }
10: $\forall$ non-leaf nodes, perform only the best swaps until equilibrium is attained on each candidate

output : Up to N final mapping candidates

At any level of a B-way search tree (B is variable representing current degree of branching) of intermediate mapping candidates, D best swaps and R random swaps are considered for each node. With N as an upper bound on the total number of final candidates, the branching degree proportionally decreases with increasing number of intermediate candidates. An intermediate candidate node becomes a leaf node (signifying a final mapping candidate) when no more directed swaps are possible on it and is never again considered for further swaps. When the existing number of candidates reach the upper bound of N, only the best swaps are made on all the non-leaf solutions (B is reduced to 1) until they converge to equilibrium (a state where valid swaps which reduce total NoC tension are no longer available). Alternatively, if no non-leaf solutions remain, BB terminates as no random swaps are allowed on leaf-nodes. Finally, a set of up to N final mapping candidates are obtained from a single initial mapping.

5.3 PDN Synthesis

PDN synthesis is performed on each of the mapping candidates produced by the BB procedure. We propose a segmented power distribution grid for a 3D-mesh CMP with *VIs*. We address the PDN design problem for the global grid, where each grid-node supplies to a core in the 3D mesh and VRMs are integrated to scale down voltage to cores in *VIs*. Besides overheads of chip area and power dissipation of the VRM components, proper placement of these VRMs on a 3D-mesh is critical for better supply efficiency as well as for minimizing the grid-wire width needed to satisfy the IR-drop constraints. The performance of any core is highly dependent on the quality of voltage supply; besides, we do not evaluate IR-drops in the power grid at the sub-core level (one and only one core is supplied to by a grid-node); thus, a tolerance of just 1% in the voltage supply level is assumed. In this work, as we investigate the steady state effects of the PDN, time-varying network characteristics such as transient noise are not considered. We also assume that all PDN braches have uniform resistance R, which is the norm for PDNs. Fig. 4 shows an example of a 3D-mesh CMP with *VIs* and corresponding PDN with VRMs. Note that the orange branches run at the external voltage supply *EV*. The PDN should be able to restrict the IR drops at each core within the set tolerance limit of the rated core voltage.

We propose a novel topological structure of VRM placements for better current efficiency; where the stepped-down voltages from outputs of VRMs are used as either of the following: *(i)* as voltage supplies to cores of the same voltage, such that the IR-drop constraints are satisfied (e.g., from Core {001} to Core {002} in Fig. 4); or *(ii)* as inputs to the VRMs for cores of lower voltages (to

978-1-4673-0438-2/12 $31.00 © 2012 IEEE   264

further step-down the voltage level), such that the minimum drop-out voltage requirements of VRM are satisfied (e.g., from Core {011} to Core {012} in Fig. 4).

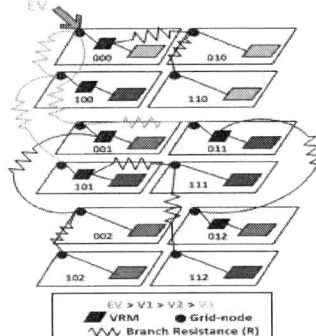

Figure 4: An Illustration of a Segmented PDN Structure

5.3.1 Linear Programming Formulation

We formulate the PDN synthesis problem as an exact Linear Programming problem with the following goal:

$$\textit{Minimize: } [\alpha.\sum B_{i,j,k} + \psi.(EI)]$$

where $\sum B_{i,j,k}$ is the total number of VRMs used and EI is the external current drawn from the power supply. We are given a set of T tile co-ordinates $T_{i,j,k}$, for $0 \leq i \leq dim_x\text{-}1$, $0 \leq j \leq dim_y\text{-}1$, $0 \leq k \leq dim_z\text{-}1$; on a 3D mesh with dimensions $\{dim_x, dim_y, dim_z\}$. The external voltage source (EV) is located at $T_{0,0,0}$ (upper left corner of the topmost layer in the 3D mesh). For a given core to tile mapping solution, $C_{i,j,k}$ and $CI_{i,j,k}$ are the operating voltage levels and the maximum current requirements of the cores at the respective co-ordinates. The design variables considered in our problem are as follows:

- VRM placements are represented with binary variables:
  - $B_{i,j,k} = 1$, if VRM is present at co-ordinates {i,j,k}
  - $B_{i,j,k} = 0$, if VRM is absent at co-ordinates {i,j,k}
- Branch currents emanating from the grid-nodes in d {x, y or z} direction: $I_{i,j,k\text{-}d}$ (Fig. 5(a))
- Branch currents emanating from VRMs in d {x, y or z} direction: $IV_{i,j,k\text{-}d}$ (Fig. 5(b))
- Grid-node voltages: $V_{i,j,k}$
- PDN branch resistance R, can take one of the r values from set $\acute{R}$ = $\{R_1, R_2,.., R_r\}$

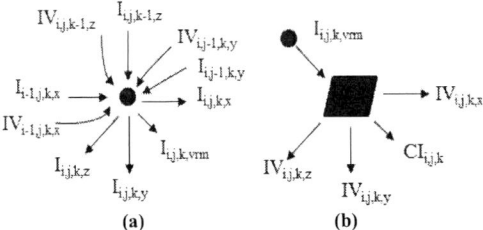

(a)   **(b)**

Figure 5: (a) Input/Output currents through a grid-node {i,j,k} (b) current across a VRM at co-ordinates {i,j,k}

Due to lack of space, we now present only the key program constraints.

Constraint 1: As voltage can only be stepped-down, voltage is derived from at least one neighboring grid-node with voltage no smaller than the current grid-node {i,j,k}. Thus,

$$[V_{i,j,k} \leq V_{i\text{-}1,j,k}] | [V_{i,j,k} \leq V_{i,j\text{-}1,k}] | [V_{i,j,k} \leq V_{i,j,k\text{-}1}]$$
$$\forall \left(\{i, j, k\} - \{0,0,0\}\right)$$
$$x_{i,j,k} = 1; \; if \; V_{i,j,k} \leq V_{i\text{-}1,j,k}$$

$$= 0; \quad otherwise$$

LP-Representation:

$$\left(V_{i\text{-}1,j,k} - V_{i,j,k}\right) - x_{i,j,k}.MAXVALUE < 0$$
$$\left(-V_{i\text{-}1,j,k} + V_{i,j,k}\right) - x'_{i,j,k}.MAXVALUE \leq 0$$

where MAXVALUE is a large positive value, and $x'_{i,j,k}$ is the inverse of $x_{i,j,k}$: $x'_{i,j,k} + x_{i,j,k} = 1$. Constraints in y and z directions are defined similarly. Thus,

$$x_{i,j,k} + y_{i,j,k} + z_{i,j,k} \geq 1.$$

Constraint 2: VRMs need to be placed wherever the core voltage is less than the corresponding grid-node voltage

$$B_{i,j,k} = 1; \; if \; C_{i,j,k} < V_{i,j,k}$$
$$= 0; \quad otherwise$$

$B'_{i,j,k}$ is the respective inverse: $B'_{i,j,k} + B_{i,j,k} = 1$
LP-Representation:

$$[V_{i,j,k} - C_{i,j,k}] - B_{i,j,k}.MAXVALUE \leq 0$$
$$[C_{i,j,k} - V_{i,j,k}] - B'_{i,j,k}.MAXVALUE < 0$$

Constraint 3: The VRM minimum drop-out voltage constraint is defined as:

$$V_{i,j,k} - C_{i,j,k} \geq 0.2; if \; B_{i,j,k} = 1$$

LP-Representation:

$$V_{i,j,k} - C_{i,j,k} + B'_{i,j,k}.MAXVALUE \geq 0.2$$

Constraint 4: The core voltages must be no greater than the respective grid-node voltages (with a tolerance of 1% of the rated core voltage):

$$(0.99) * C_{i,j,k} \leq V_{i,j,k}$$

Constraint 5: Current across a VRM is shown in Fig. 5(b). Power efficiency of a VRM is defined by the following equation [1]:

$$\eta_p = \frac{I_{out}}{I_{in}} * \frac{V_{out}}{V_{in}}$$

As current efficiency is basically the power efficiency at constant voltage values; to linearize, we consider current efficiency as:

$$I_{out} = \eta_c * I_{in} = IV_{i,j,k,x} + IV_{i,j,k,y} + IV_{i,j,k,z} + CI_{i,j,k}$$

as shown in Fig. 5(b). A minimum drop-out voltage of 0.2V is assumed for the VRMs based on [23][24].

$$I_{out} = 0.98 * I_{in}; \quad if, \; 0.4 > V_{out} - V_{in} \geq 0.2$$
$$I_{out} = 0.95 * I_{in}; \quad if, \; 0.6 > V_{out} - V_{in} \geq 0.4$$
$$I_{out} = 0.90 * I_{in}; \quad if, \; V_{out} - V_{in} \geq 0.6$$

Also, the input current of the VRM is defined as:

$$I_{i,j,k,vrm} = CI_{i,j,k}; \quad if, \; B_{i,j,k} = 0$$
$$I_{i,j,k,vrm} = I_{in}; \quad if, \; B_{i,j,k} = 0$$

The LP representations for the constraint are omitted for brevity.

5.3.2 PDN Synthesis Heuristic

We also propose a more scalable and near-optimal solution to the PDN synthesis problem which basically does a breadth first search (BFS) starting from the farthest node (tile) from the external power source (root node located at the lower right corner of the bottom-most layer) while assigning grid-node voltages and branch currents. The heuristic computes the placement of VRMs at grid co-ordinates while satisfying all the constraints discussed in the LP-formulation, including IR-drop constraints. In the BFS procedure, each level in the breath-first tree is termed as a *front*, thus, the node {0, 0, 0} becomes the final front. Also, any node in the current front derives the input current from its 'upstream neighbor(s)'. The heuristic attempts to minimize the number of VRM insertions at every front and at the same time, chooses as many incoming currents as possible at each node for a better IR-drop distribution. Let $CD_{i,j,k}$ be the current demand at the node {i,j,k} which is the sum of outgoing currents at the grid node. The basic flow of our heuristic is as follows:

PDN Synthesis Heuristic

input: $CI_{i,j,k}$ and $C_{i,j,k}$ values for the 3D mesh

1: Put root node in the front and assign: $V_{root}=C_{root}$ and $CD_{root}=CI_{root}$
2: **while** front is non-empty, **do**{
3: Assign upstream neighbors and processing priorities to all the nodes in the front by calling *Priority_assign()*
4: Sort all cores in the front in order of their processing priorities; for each core in the sorted-list, do{
5: Distribute $CD_{i,j,k}$ over all (one or more) of the incoming branches
6: Place VRMs at upstream or current node by calling *VRM_insert()*
7: Assign grid-node voltages at upstream nodes based on Ohms law
8: Advance the front, i.e. current front is deleted and the set of all upstream neighbors of the old front becomes the new front }

output : EI, $B_{i,j,k}$, $V_{i,j,k}$ and branch currents

The *VRM_insert()* function inserts a VRM at the upstream/current node when the upstream neighbor has a lower/higher voltage than the current node. In our PDN heuristic, we assume that the IR drop constraint will not be violated when at least one of any two consecutive nodes on the current path contains a VRM; therefore, an IR drop constraint violation is possible only when the voltages of upstream and current nodes are similar. In such a situation, an IR drop constraint violation (voltage required at the upstream node is higher than the 1% tolerance range) is rectified by the *VRM_insert()* function by inserting a VRM at the upstream node.

Given $B_{i,j,k}$ (VRM presence bit), $V_{i,j,k}$ and $CD_{i,j,k}$, the grid-node voltages and the current demands for all the nodes in the current front, the *Priority_assign()* function computes a set of upstream neighbors and processing priorities for each core in the current front. As VRM insertion is not needed (in absence of IR-drop violation) for upstream neighbor(s) of similar voltage as the current grid-node; whenever one or more upstream nodes with similar voltage are available, ($C_{upstream}=V_{i,j,k}$) they are used exclusively to supply the current node [Rule 1]. Also, all available incoming currents from upstream neighbors (with similar voltage levels) are utilized to reduce the overall effective resistance of the PDN. If no upstream nodes of similar voltage are available, and one or more prospective upstream neighbors have higher core voltages than the current grid-node, ($C_{upstream}>V_{i,j,k}$) the one with the lowest voltage amongst them is chosen to minimize current loss in the corresponding VRM [Rule 2]. Finally, if upstream nodes of only lower voltages are available, ($C_{upstream}<V_{i,j,k}$) the one with the highest voltage amongst them is chosen to minimize current loss in the corresponding VRM [Rule 3].

After a non-zero set of upstream neighbors are assigned to each node in the front, the relative order of processing of these nodes is determined by *Priority_assign()*. Any node in the current front which has already been assigned a VRM has a rigid voltage requirement because its grid-node voltage is assigned to supply to down-stream nodes. Therefore, nodes in the front which have VRMs inserted are given highest processing priority of 0 to be able to use the unassigned upstream neighbors. For the rest of the nodes; nodes assigned upstream neighbors through [Rule 3], [Rule 2] and [Rule 1] are assigned the processing priorities of 1, 2 and 3 respectively. The nodes in the front with processing priorities of 1 and 2 derive current from a single upstream neighbor and thus are given precedence over the ones with processing priority of 3 in the order of processing.

5.4 3D Routing

We employ YXZ routing in our 3D NoC fabric, which is not only deadlock-free but also uses minimal routes and has a low area footprint for implementation, thus enabling power efficient routing. The YXZ routing scheme is used on each of the mapping candidates produced by the BB procedure to compute NoC power. During the routing process, link-insertions are performed as needed to support application bandwidths, and router sizes are simultaneously updated. Also, VLCs and MCFIFOs are inserted for inter-*VI* links appropriately. After routing is done for the entire mesh, total power

dissipation in the NoC is computed taking into consideration the number of links inserted, link loads, router sizes, number of VLCs and MCFIFOs used, and the corresponding voltage/frequency values.

5.5 Solution Pruning

Once a set of solutions is output by the synthesis flow, a penultimate pruning is performed to remove solution points that are not relevant. For instance, if two solutions have the same number of VRMs and communication power, but different external current (EI) values, then the solution with the higher EI value is pruned.

6. Experiments

We used the ARM11MPCore multi-core processors [26] as the base compute cores in our experiments, which support three operating voltage levels. Our experiments were conducted on applications based on pseudo-random core graphs derived using TGFF [27] with edge weights annotated with bandwidths representing inter-core communication requirements. We conservatively assume that the square of the voltage scales linearly with the frequency, as in previous works [6]. The three core voltages and their corresponding frequencies and maximum supply current values we employed are: (0.8V, 195MHz, 0.52A), (1.0V, 304MHz, 0.50A) and (1.2V, 437MHz, 0.48A). The maximum supply current values are derived from the rated maximum compute core power values. Also, the value of external voltage source of the PDN is assumed to be 1.5V. The architecture and power values of routers and links as well as MCFIFOs and VLCs operating at different voltages and frequencies are taken from [6]. The branch resistance values used in the PDN design are: 43, 53 and 63 mΩ (based on [3]). In the BB procedure for 3D mapping, values of n=9 and N=800 are used so that the maximum values that variables B, D and R can take are 9, 5 and 4 respectively, with α=1. The PDN linear programming formulation is solved using the open-source tool lp_solve 5.5.2 [25].

6.1 Results

Our first set of experiments focus on the PDN synthesis problem and compare the fidelity of solutions obtained from the LP formulation and the heuristic. The PDN synthesis results (*EI* and # of VRMs needed for a range of *R* values) obtained using our heuristic are found to be within 15% of the corresponding LP-results for small benchmarks with less than 10 cores. Table 1 summarizes the results for 3D CMPs with 4 (2×2), 8 (2×2×2), and 9 (3×3) cores. The key advantage of using the heuristic is that it generates a solution in a matter of a few seconds, whereas the LP formulation takes in the order of several hours. For problem sizes with greater than 10 cores, the LP solver did not return a solution even after being left to run for an entire day. Thus the heuristic provides a more scalable and thus practical solution to the PDN synthesis problem.

Table 1: Comparison between results obtained from LP and the PDN Heuristic Implementations

#of cores	3D core config.	R (mΩ)	LP		Heuristic	
			# of VRMs	EI (Amp)	# of VRMs	EI (Amp)
4	2×2	43	2	0.203	2	0.208
	(1 layer)	63	2	0.203	2	0.208
8	2×2×2	43	3	0.409	4	0.421
	(2 layers)	63	4	0.411	4	0.422
9	3×3	43	6	0.464	6	0.480
	(1 layer)	63	6	0.469	7	0.483

Next we explore the results generated by our complete PDN-aware 3D NoC synthesis approach for a large 64 (4×4×4) core CMP. Fig. 6 shows the 3D solution space, with each candidate solution characterized by its communication power, number of VRMs, and external current (EI) drawn. The results are shown for the grid wire resistance values of 43 mΩ and 63 mΩ (results for the 53 mΩ case not shown). The most important insight from these results is that the solution with the minimum communication power (highlighted by red dots in each of the figures) does not necessarily have optimum PDN cost. For instance in Fig. 6(b), the solution (shown as red '*') with the lowest communication power (NoC cost) of 2374 mW, not only requires the highest number of VRMs (31) but also has a very high *EI* of 3.972 A. Traditional 3D NoC synthesis approaches output the lowest communication power solution. By the time the PDN is designed in the back end for this solution, it is too late to trade-off PDN complexity with communication power. For the lowest communication power solutions shown in Fig. 6 (a)(b), it would be quite a significant challenge to meet the area and power design-constraints during PDN design and designers may need to resort to over-margining which is wasteful and increases system cost.

In contrast, our proposed PDN aware NoC synthesis framework can produce a set of interesting design points that can enable trade-offs between NoC power dissipation, VRM count, and external current drawn. For instance, possible solutions which optimize all three cost metrics are the pentagrams highlighted in blue. Additionally, if either the minimal number of VRMs or the minimum EI is required as the final design solution, other design points (a square or a diamond in black, respectively) could be chosen. Ultimately, our framework allows designers to explore trade-offs in the PDN and NoC design space early on at the system level, and is invaluable to achieving better quality designs.

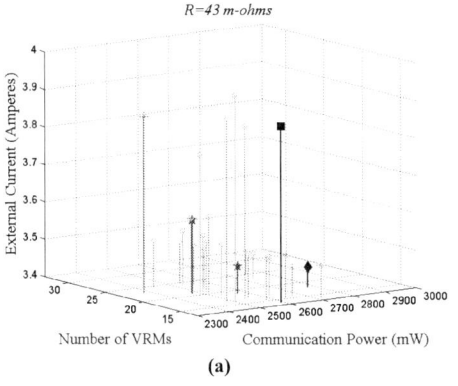

(a)

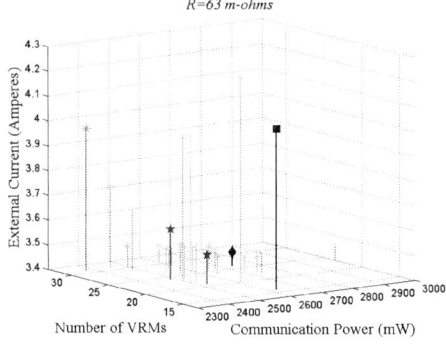

(b)

Figure 6: Results of the PDN-aware NoC synthesis framework for (a) R=43 mΩ (b) R=63 mΩ

7. Conclusion

In contrast to the traditional CMP design approach where PDN design is done on a mapping instance which is optimized exclusively for NoC costs (e.g. communication power) this work advocates an automated framework for PDN-aware synthesis of NoCs in 3D CMPs. Our framework enables the designer to weigh the PDN design costs against the NoC design costs and thereby obtain a more efficient overall solution. The experimental results show that the solution space uncovered by our framework can allow system level trade-off analysis of PDN and NoC design decisions which has never been attempted before. By accepting solutions with less than optimal NoC power dissipation characteristics, our framework reveals that it is possible to significantly reduce PDN design cost.

References

[1] B. Amelifard, et al., "Optimal Design of the Power-Delivery Network for Multiple Voltage-Island System-on-Chips," IEEE TCAD 28(6), May 2009.
[2] P. Jain et al. "A multi-story power delivery technique for 3D integrated circuits," Intl. Symp.on ISLPED pg. 57, Aug. 2008.
[3] N. Khan et al. "System-level comparison of power delivery design for 2D and 3D ICs," IEEE Intl. Conf. 3D-IC, Sept. 2009.
[4] W. Kim et al. "Enabling OnChip Switching Regulators for Multi-Core Processors using Current Staggering," Proc. ASGI, 2007.
[5] Z. Zenget al. "Tradeoff analysis and optimization of power delivery networks with on-chip voltage regulation," Proc. DAC pp. 831, June 2010.
[6] N. Kapadia, et al., "VISION: A Framework for Voltage Island Aware Synthesis of Interconnection Networks-on-Chip," Proc. GLSVLSI 2011.
[7] S. Murali et al., "Designing Application-Specific Networks on Chips with Floorplan Information," Proc. ICCAD. 2006.
[8] C. Seiculescu et al., "NoC Topology Synthesis for Supporting Shutdown of Voltage Islands in SoCs," Proc. DAC, July, 2009.
[9] P.Zhou et al., "Application-Specific 3D Network-on-Chip Design Using Simulated Allocation," Proc. ASPDAC. Jan., 2010.
[10] K. Srinivasan, K. Chatha, "A low complexity heuristic for design of custom network-on-chip architectures," Proc. DATE, 2006.
[11] K. Srinivasan et al., "Linear-Programming-Based Techniques for Synthesis of Network-on_Chip Architectures," IEEE TVLSI, 14(4) Apr 2006.
[12] U. Ogras, et al., "It's a Small World After All": NoC Performance Optimization Via Long-Range Link Insertion," IEEE TVLSI 14(7), Jul 2006.
[13] L.Leung, C. Tsui; "Energy-Aware Synthesis of Networks-on-Chip Implemented with Voltage Islands," Proc. DAC, 2007.
[14] P. Ghosh, et al. "Efficient Mapping and Voltage Islanding Technique for Energy Minimization in NoC under Design Constraints," Proc. SAC, 2010.
[15] U. Ogras et al, "Voltage-Frequency Island Partitioning for GALS-based Networks-on-Chip", Proc. DAC, pp: 110–115, 2007.
[16] J. Hu, R. Marculescu; "Communication and task scheduling of application-specific networks-on-chip," IEE CDT 152(5), Sep. 2005.
[17] C Chou et al., "Energy and Performance-Aware Incremental Mapping for Networks on Chip with Multiple Voltage Levels," IEEE Transactions on CAD (TCAD), 27(10): 1866–1879, Oct. 2008.
[18] J. Hu, R. Marculescu; "Energy and Performance-Aware Mapping for Regular NoC Architectures," IEEE TCAD 24(4), Apr. 2005.
[19] W. Jang, D. Ding, D. Pan, "A Voltage-Frequency Island Aware Energy Optimization Framework for Networks-on-Chip," Proc. ICCAD, pp: 264 – 269, Nov. 2008.
[20] P.Falkenstern et al., "Three-dimensional integrated circuits (3D IC) Floorplan and Power/Ground Network Co-synthesis," Proc. ASPDAC, 2010.
[21] H. Chen et al., "A New Architecture for Power Network in 3D IC," Proc. DATE, March 2011.
[22] M. Popovich et al., "On-Chip Power Distribution Grids with Multiple Supply Voltages for High-Performance Integrated Circuits," IEEE Transactions on VLSI Systems vol. 16(7), pp. 908-921, July 2008.
[23] Y. Lee, K. Chen; "A 65nm sub-1V multi-stage low-dropout (LDO) regulator design for SoC systems," IEEE Intl. MWSCAS pg. 584; Aug. 2010.
[24] J. Guo, K. Leung; "A 6-µW Chip-Area-Efficient Output-CapacitorlessLDO in 90-nm CMOS Technology," IEEE Journal of Solid State Circuits 45(9), pg. 1896, Sept. 2009.
[25] http://lpsolve.sourceforge.net/5.5/
[26] http://www.arm.com/products/processors/selector.php
[27] http://ziyang.eecs.umich.edu/~dickrp/tgff/
[28] X. Jiang, T. Watanabe, "An efficient 3D NoC synthesis by using genetic algorithms," Proc. TENCON Nov. 2010.
[29] C. Seiculescu, "SunFloor 3D: A tool for Networks On Chip topology synthesis for 3D systems on chips," DATE 2009.
[30] K. Siozios et al., "A High-Level Mapping Algorithm Targeting 3D NoC Architectures with Multiple Vdd," ISVLSI 2010.
[31] C. Seiculescu et al., "Comparative Analysis of NoCs for Two-Dimensional Versus Three-Dimensional SoCs Supporting Multiple Voltage and Frequency Islands," IEEE TCAS, May 2010.
[32] M. Arjomand et al., "Voltage-Frequency Planning for Thermal-Aware, Low-Power Design of Regular 3-D NoCs," VLSID 2010.

2012 25th International Conference on VLSI Design

A Framework for TSV Serialization-aware Synthesis of Application Specific 3D Networks-on-Chip

Sudeep Pasricha
Colorado State University, Fort Collins, CO, USA
sudeep@colostate.edu

Abstract – With increasing performance-per-watt implementation requirements for emerging applications and barriers in interconnect scaling for ultra-deep submicron (UDSM) technologies, traditional 2D integrated circuits (2D-ICs) are being pushed to their limit. Three dimensional integrated circuits (3D-ICs) have recently emerged as a promising solution that can overcome many of the performance, area, and power concerns in 2D-ICs. In this paper we propose a novel framework (MORPHEUS) for the synthesis of application-specific 3D networks on chip (NoCs). The goal is to generate 3D NoCs that meet application performance constraints while minimizing power dissipation. MORPHEUS incorporates thermal-aware core layout, 3D topology and route generation, and placement of network interfaces (NIs), routers, and serialized vertical through silicon vias (TSVs). Experimental studies on several chip multiprocessor (CMP) applications indicate that our generated solutions notably reduce power dissipation (up to 2.3×) and average latency (up to 1.2×) over 2D NoCs. Comparisons with a previous work on application-specific 3D NoC synthesis also show improvements in power dissipation (up to 1.9×) and average latency (up to 1.6×).

I. INTRODUCTION

The rise in application complexity in recent years together with the need for power efficiency in computing systems has led to more and more cores being integrated on a single chip. Such chip multiprocessors (CMPs) have demonstrated superior performance-per-watt than their uniprocessor counterparts. If the trend of integrating greater number of cores on a chip is to continue in the coming years, two major challenges need to be overcome. Firstly, the ongoing reduction in lithographic features is becoming increasingly more expensive. As a result, there is a practical limitation on the number of cores that can be integrated viably on the single active layer available to CMPs today. Secondly, with rising core counts, the amount of wires on chip has been steadily rising. Due to effects such as parasitic resistivity, crosstalk, and electromigration interference in UDSM nodes, long global interconnects have become a major delay and power bottleneck [1]. According to the ITRS [2], focus on new technologies and methodologies and not further reduction in feature size is the key to further performance-per-watt enhancements.

Of the several different disruptive technologies that are being investigated today, 3D integrated circuits (3D-ICs) with wafer-to-wafer bonding technology is one of the most promising candidates that can achieve power, performance, cost, and area demands of emerging applications in the coming years [3]-[6]. In wafer-to-wafer bonded 3D-ICs, active devices (processors, memories) are placed on multiple layers and vertical Through Silicon Vias (TSVs) are used to connect components across the stacked layers. Multiple active layers in 3D-ICs can enable increased integration of cores within the same area footprint as traditional single layer 2D-ICs. In addition, long global interconnects between cores can be replaced by much shorter inter-layer TSVs, improving performance and reducing on-chip power dissipation. Recent 3D-IC test chips from IBM [3][4] and Tezzaron [5] have confirmed the benefits of 3D integration technology.

With the advent of many-core CMPs in recent years, the on-chip communication fabric has been evolving to cope with increased bandwidth and reliability requirements. Traditionally used bus based architectures have given way to packet switched network on chips (NoCs) that offer higher bandwidth, reliability, and scalability in UDSM technologies. With the introduction of 3D-ICs, it is expected that NoC fabrics will be extended into the third dimension. Recent research has begun exploring various 3D NoC topologies [7]-[9] and shown significant performance improvements for these topologies over 2D NoCs. However, the design of such 3D NoC fabrics customized for specific applications has received very little attention to date. Even though a significant body of work exists for the synthesis of 2D bus-based [10]-[12] and 2D NoC [13]-[18] communication architectures, the techniques are not directly applicable for 3D NoC synthesis because of the peculiar challenges of 3D IC design. For instance, vertical TSVs have a larger pitch (5μm×5μm or more) that takes up space in the active layers and has at least an order of magnitude greater footprint than regular vias in the metal layers. These TSV interconnects are therefore expected to be limited in number. This heterogeneity and limited density of TSVs needs to be considered while synthesizing and optimizing 3D NoCs. Additionally, cores and routers can be placed on one of the many layers available, which dramatically increases design space complexity in 3D NoC based communication architectures.

In this paper, we propose a novel framework (MORPHEUS) for the application-specific synthesis of 3D NoCs, optimized for low power dissipation. MORPHEUS automates the process of thermal-aware core layout, 3D topology and route generation, and placement of network interfaces (NIs), routers, and serialized vertical TSVs. Experimental studies on several CMP applications indicate that our generated solutions notably reduce power dissipation (up to 2.3×) and average latency (up to 1.2×) over 2D NoCs. Compared to solutions generated by a previous work on application-specific 3D NoC synthesis, MORPHEUS generates solutions with lower power dissipation (up to 1.9×) and lower average latency (up to 1.6×).

II. RELATED WORK

Over the last several years, there has been a growing interest in 3D ICs as a means to alleviate the interconnect bottleneck problem currently facing 2D-ICs. A key challenge with 3D-ICs is its high thermal density due to multiple cores being stacked together, that can adversely impact chip performance and reliability. Therefore several researchers have proposed thermal-aware floorplanning techniques for 3D-ICs [19]-[21]. A few researchers have explored interconnect architectures for 3D-ICs such as 3D mesh and stacked mesh topologies [7] and a hybrid bus-NoC topology [8]. Some recent work has looked at decomposing cores (processors [24][25], NoC routers [26], and on-chip cache [27]) into the third dimension which allows reducing wire latency at the intra-core level, as opposed to the inter-core level. Circuit level models for TSVs were presented in [9].

The problem of custom interconnect architecture synthesis for 2D-ICs has received a lot of attention in the past, for point-to-point and bus based architectures [10]-[12] and NoC topologies [13]-[18]. Only recently have approaches been proposed for the synthesis of custom NoCs for 3D-ICs [28]-[32]. An ILP based synthesis technique for a 3D network with low-radix routers is proposed in [28]. However, the generated solution has many long links and the scalability of the approach is not clear. A methodology for application-specific topology synthesis and route computation for 3D-ICs that performs localized synthesis optimizations for every layer is proposed in [29], based on the author's previous work on 2D NoC synthesis. The 3D NoC synthesis approach in [30] extends [29] by additionally

978-1-4673-0438-2/12 $31.00 © 2012 IEEE

268

determining switches placement and iteratively adding TSV links during synthesis. A technique for application-specific 3D NoC synthesis based on a low-level greedy rip-up and reroute procedure for determining routes is proposed in [31]. Every core is initially allocated a router that is later merged using a greedy heuristic. However, the method used for router and TSV allocation in layers is not clearly specified. A multi-commodity flow formulation is used in [32] to solve a similar problem, but with analytical (e.g., queuing) models of the NoC and high level abstractions of applications, which may reduce the overall accuracy of the solution.

Unlike existing approaches, our application-specific 3D NoC synthesis framework (MORPHEUS) enables a more comprehensive exploration of the design space by additionally integrating TSV serialization, NI placement, and thermal-aware core placement, together with layout-aware partitioning and allocation of routers and TSVs, to optimize power while meeting performance constraints.

III. 3D INTEGRATED CIRCUITS

Before presenting details of the MORPHEUS synthesis flow for 3D NoCs, we briefly discuss two relevant issues that are important to consider when designing application-specific NoCs for 3D ICs.

A major concern in the adoption of 3D ICs is the increased power densities that can result from placing a core on top of another core. As high peak temperatures due to increasing power densities can cause catastrophic IC failure and this is already a major ckoncern in 2D architectures, the move to 3D will accentuate the thermal problem. The problem of thermal-aware 3D core layout is thus tightly coupled with the problem of 3D NoC synthesis, and cannot be ignored as in some previous 3D NoC synthesis approaches [28]-[31]. Another critical restriction that severely constrains the design space in 3D-ICs and must be considered during synthesis is the limitation on the number of TSVs between layers. This limitation is due to the high area overhead of TSV pads that are required to interface with TSVs in each active layer. It is clear today that TSV fabrication technology lacks maturity and has low yield due to unsuccessful wafer alignment prior to and during the wafer bonding process. This is expected to remain a major challenge in the years to come. A simple and effective way to improve yield is to add hardware redundancy by using larger (e.g., double area) TSV pads. As misalignments are caused by the unavoidable shift of bonding pads with respect to their nominal position, using larger square pads can improve misalignment tolerance by an order of magnitude [6]. However, the large pads can complicate routing in active layers. This motivates the need for some form of serialization of TSVs to reduce TSV pad footprint in active layers. In [33], it was shown that TSV serialization can reduce TSV pad area overhead in active layers by as much as 70% at a negligible performance and power overhead. This motivates utilizing TSV serialization in our application specific 3D NoC synthesis framework.

IV. MORPHEUS SYNTHESIS FRAMEWORK

A. Inputs and Problem Description

We assume that we are given a set of computational and memory cores onto which application tasks have already been mapped, after a hardware-software partitioning phase. The cores are arranged in a core dependency graph (CDG) which is one of the inputs to our framework. The CDG is an annotated directed graph $G(V, E)$ where each node $v_i \in V$ corresponds to a core, directed edge $e_{ij} \in E$ is a communication flow from v_i to v_j, and edge weight $w(e_{ij})$ is given by:

$$w(e_{ij}) = \sigma * \mu(e_{ij}) + (1 - \sigma) * \rho(e_{ij})$$

where $\mu(e_{ij})$ and $\rho(e_{ij})$ are the latency and bandwidth constraints respectively for e_{ij}, and σ is a designer-specified parameter based on application characteristics. Each core v_i is a rectangular shaped hard macro that has a fixed width (W_i) and height (H_i) associated with it, and the layer to which it is mapped in the 3D IC is specified by $layer_i$.

The number of layers (ζ) in the 3D IC on which the application is to be implemented and the maximum die dimensions ($W_{die} \times H_{die}$) are also specified as inputs to our framework. As the TSV density is limited due to practical implementation concerns, we assume a maximum TSV density threshold between adjacent layers (δ) as a designer-specified input that depends on the chosen 3D-IC implementation technology. The core to layer assignment $V \rightarrow \zeta$ is assumed as an input from the designer, and is usually based on temperature or power density concerns. For instance, a designer may choose to interleave high power density computational cores with cooler layers comprising of low power density memory cores. NoC architecture parameters (e.g., operating voltage, clock frequency, link width) can either be specified by the designer, or varied in steps in a user-defined range, with our framework being invoked at each step. Finally, the technology library node (e.g., 65nm, 45nm) is an input that enables accurate delay and power estimation in the framework.

Problem Definition: *Given an application CDG with bandwidth and latency constraints, a core to layer ($V \rightarrow \zeta$) mapping, number of 3D IC layers (ζ), TSV density threshold (δ), maximum die dimensions ($W_{die} \times H_{die}$), NoC architecture parameters, and a target technology node, the goal of the MORPHEUS framework is to synthesize an application-specific 3D NoC topology with a layout for all cores, NIs, TSVs, routers, and links that satisfies all performance (bandwidth, latency) constraints in the application while minimizing NoC power.*

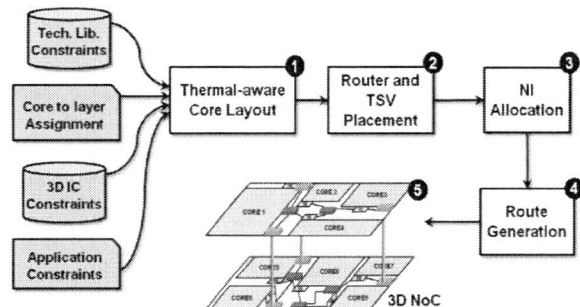

Figure 1. MORPHEUS 3D NoC synthesis framework

B. MORPHEUS Synthesis Flow Overview

Fig. 1 gives a high level overview of the MORPHEUS application-specific 3D NoC synthesis framework. In the first phase, a thermal-aware core layout using a 3D floorplanner is performed to place the cores assigned in each layer in a manner that minimizes peak temperature, wirelength, and chip area. The output of this step is a complete layout of the cores in every layer in the 3D-IC. In the next phase, the placement of NoC routers and TSV pads on the active layers is determined. A partitioning-based heuristic is used in this phase to explore the design space for a spectrum of router and TSV densities, while preserving the TSV density threshold (δ) constraint, in part by utilizing serialization. Next the NI placement for each core is determined to minimize critical path lengths. Finally, deadlock free routes are determined between cores. The output of the framework is a Pareto set of valid solutions (satisfying all performance constraints) that trade-off power with performance slack. The following subsections describe the various phases in the flow in more detail.

C. Thermal-aware Core Layout

Given a core to layer mapping, the MORPHEUS framework performs 3D floorplanning in the first phase to obtain a placement of cores in each layer. This core layout is often influenced by non-network-based interconnections (e.g., off-chip interface pin locations), and is an important step as it can have a significant impact on the quality of the synthesized NoC architecture. As stacked 3D-ICs can have significant reliability and performance issues due to higher

978-1-4673-0438-2/12 $31.00 © 2012 IEEE   269

power density and peak temperatures than their 2D-IC counterparts, it becomes essential to perform core placement with thermal-awareness. Communication bandwidth and latency constraints must also be satisfied, by minimizing wirelength. In addition, overall chip area should also be minimized.

To solve this multi-objective core layout problem, we make use of a *genetic algorithm* (*GA*) based approach. A genetic algorithm is an iterative exploration algorithm that is based on a computational analogy with biological adaptive systems. The algorithm starts with an initial random pool of solutions (each represented by a chromosome) that are evaluated at each iteration (generation) by a fitness score obtained from an objective cost function. A new generation is created by first increasing the population by generating new individual solutions, and then selecting a constant number of solutions based on their fitness criteria. Genetic operators such as *crossover* and *mutation* are used to create an evolution of the solutions at every iteration. The best solution is finally selected after a specified number of generations. A detailed discussion of genetic algorithms can be found in [34]. In our GA formulation, we encode the location and orientation of each core in the chromosome using a sequence-pair representation [35]. The GA objective function is aimed at a hybrid optimization of thermal, communication, and area costs. The thermal cost is represented by the peak temperature (T_{peak}) of the design, and the area cost (A_{cost}) is given by the chip area. The communication cost (C_{cost}) is a hybrid formulation that combines bandwidth and latency constraints as follows:

$$C_{cost} = \sigma * \sum_{\forall e_{ij}} \frac{\rho(e_{ij}) * l(e_{ij})}{max(\rho) * max(l)} + (1 - \sigma) * \sum_{\forall e_{ij}} \frac{min(\mu) * l(e_{ij})}{\mu(e_{ij}) * max(l)}$$

where $l(e_{ij})$ is the wirelength between cores v_i and v_j, $max(l)$ is the maximum wirelength between two cores with a communication flow, $max(\rho)$ and $min(\mu)$ are the maximum bandwidth, and minimum latency constraints among all flows, and σ is a weight parameter set by a designer based on application characteristics. Then the objective function for our GA formulation is given as:

$$F(T_{peak}, C_{cost}, A_{cost}) = \frac{a_1}{\log(T_{peak})} + \frac{a_2}{\log(C_{cost})} + \frac{a_3}{\log(A_{cost})}$$

where a_1, a_2, and a_3 are weighting parameters used to guide the optimization. The logarithmic values provide a more descriptive range and characterization for input variations. For any potential layout solution, C_{cost} can be calculated based on HPWL (half perimeter wire length) distances and μ/ρ constraints for each communication flow. The chip area A_{cost} can also be determined after a layout of every core is obtained. To obtain T_{peak} estimates, we make use of a 3D adaptation of Hotspot [36] which is a well-known tool for temperature estimation in 2D-ICs. Based on the average power dissipation of each core (known for every core based on technology library used and application characteristics), physical dimensions, and location of a core in the 3D-IC, Hotspot returns temperature estimates for the chip.

To accommodate placement of smaller components such as routers, NIs, and TSV pads in subsequent phases of the synthesis flow, we increase core dimensions by a small margin φ before core layout. After layout, a compaction step is performed to reduce core dimensions and create whitespace for inserting routers, NIs, and TSV pads later. φ is set to 5% in our framework, although higher values can also be used to ease wiring congestion. After the layout phase, each core v_i has a placement $P(v_i) = (x_i, y_i, z_i)$ with (x_i, y_i) referring to the bottom left coordinates of the core in a layer and z_i referring to the layer to which the core is mapped. Together with the core width and height, a unique placement is obtained for every core in the 3D-IC. Once such a layout is available, MORPHEUS can more accurately determine inter-core wiring delays and power dissipation.

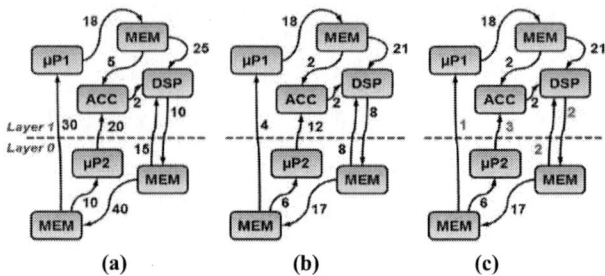

Figure 2. (a) CDG, (b) ECDG, (c) ESCDG

D. Router and TSV Pad Placement

In the next phase, we determine the number and location of routers and TSVs in the synthesized solutions. Our procedure for router synthesis extends the technique proposed in [30] for switch synthesis in 3D NoCs by adding support for serialization and using layout awareness to obtain lower average power and latency implementations. We make use of a min-cut partitioning approach to determine the optimal number of routers in the implementation.

Algorithm 1 describes the steps in this phase for router and TSV allocation and placement. In the initial steps we transform the core dependency graph $CDG(V,E)$ into an enhanced core dependency graph $ECDG(V,E')$ by updating the edge weights with a relative inter-core distance term to enable a more accurate estimation of flow criticality (Steps 2-3). Then we transform $ECDG(V,E')$ into an enhanced scaled core dependency graph $ESCDG(V,E'')$ by scaling the inter-layer links that need to traverse TSVs by a factor of $\omega * |layer_i - layer_j|$ (Step 4). This scaling is done so that the subsequent min-cut partitioning step does not lead to cores in different layers that do not have any flows between layers being assigned to the same partition, which can lead to an excessive number of inter-layer TSVs. Fig. 2 shows an example of how a CDG is transformed into an $ECDG$ and then into an $ESCDG$ (with $\omega=4$). Note how the criticality of the links changes when going from a CDG to an $ECDG$ due to the addition of more accurate inter-core wirelength information available after the core layout phase from Section IV.C.

Algorithm 1: Router and TSV Pad Allocation and Placement

1:   // Generate ESCDG(V,E") from CDG(V,E)
2:   **for** each $e_{ij} \in E$ **do**
3:      $w(e_{ij}) = w(e_{ij}) * l(e_{ij})/max(l)$
4:      **if** $layer_i \neq layer_j$ **then** $w(e_{ij}) = w(e_{ij})/(\omega * |layer_i - layer_j|)$
5:   **end for**
6:   // Create min-cut partitions
7:   **for** $t = 1$ to $|V|$ **do**
8:      using Kerninghan-Lin min-cut heuristic, create t partitions in ESCDG to create solution instance s_t, with δ' TSVs
9:      **while** $\delta' > \delta$ **do**
10:        select e_{ij} with $layer_i \neq layer_j$ and min $w(e_{ij})$
11:        serialize TSV(e_{ij}) by degree k
12:        $w(e_{ij}) = w(e_{ij}) * k$
13:        $\delta' = \delta' - (k-1)/k * link\_width$
14:      **end while**
15:      **if** $\delta' \leq \delta$ **then** FP_check_store()
16: **end for**

Next, a wide spectrum of router counts from 1 to the number of cores in the application ($|V|$) is swept (Steps 7-16). The routers are assumed to be wormhole switched with a predictive-forwarding enabled four-stage pipeline [22], and a parameterizable number of ports. In each iteration, t min-cut partitions are created using the Kerninghan-Lin algorithm (Step 8). The cores in each partition share the same router. If there exist multiple inter-partition edges between two partitions I and J, these are merged into a single edge e_{IJ} with weight $w(e_{IJ}) = \sum w(e_{ij})$, $\forall e_{ij}$, where i and j are cores such that $i \in I$ and $j \in J$. The idea behind the partitioning step is to ensure a minimal

978-1-4673-0438-2/12 $31.00 © 2012 IEEE   270

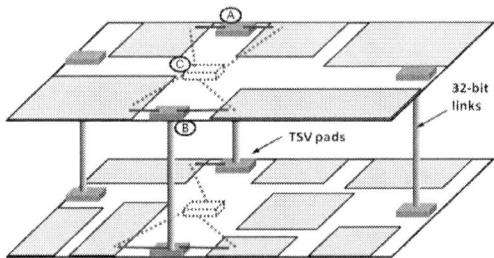

Figure 3. TSV serialization example

number of hops for communication flows that are more critical. Cores within a partition can reach each other within a single router hop.

If the total number of TSVs (δ') in the generated solution instance (s_i) is greater than the TSV density threshold (δ), we explore using serialization of TSVs to reduce the number of TSVs. Fig. 3 shows how TSV serialization can be beneficial during the synthesis process. Suppose the maximum number of TSVs is limited to 3*32=96 per layer (i.e., 3 links that are 32 bits wide each). Then the scenario shown in Fig. 3 is an invalid solution as it has 4*32=128 TSVs between layers. To get a valid solution, the number of TSVs must be reduced to 96. One way to do this is by merging and replacing the closest TSVs (A and B) with a single TSV (C). However, doing so can end up increasing wiring costs as shown in the figure, where solid links to TSV locations A and B from cores are now replaced by longer (dotted) links to TSV location C in both layers. This can dramatically increase the number of repeaters, pipeline buffers, and consequently increase power dissipation and delay. To avoid such a scenario, if serialization of degree 2 is employed at locations A and B, the TSV links are reduced to half at each location. The inter layer TSV threshold is thus satisfied without requiring TSVs at locations A and B to be replaced by a TSV at a location C. In this manner, serialization can prevent unnecessary routing congestion and reduce power dissipation in the 3D network.

The serialization process in our algorithm starts by selecting the TSV with the least communication cost (Step 10) and serializes it by a degree k (Step 11). We set $k=2$ in our approach, but higher degrees can also be considered. Next the communication cost of the edge with the serialized link is increased by the factor k (Step 12), and the number of TSV links in s_i is reduced by $(k-1)/k*link\_width$ which is the number of TSVs reduced due to serialization (Step 13). The serialization process continues (Steps 9-14) till the TSV threshold constraint is no longer violated. Finally, if $\delta' \le \delta$ either after serialization or as generated in Step 8, we invoke our GA floorplanner with *FP_check_store()* to perform router and TSV pad placement (Step 15). The floorplanner keeps the relative locations of cores fixed and performs a placement of the routers and TSV pads (with area overhead added for any serialization circuitry used) while optimizing the GA objective function as described in the previous section. The output layout is checked to ensure no latency constraint is violated by calculating the number of cycles to traverse pipelined wires and routers for each flow. If latency violations are detected, the floorplanning phase is repeated after increasing weights on violated edges. If a valid solution is obtained after the phase, it is stored, otherwise it is discarded. The final output of this phase is a set of m valid solutions $S = \{s_1, s_2, ..., s_m\}$ that meet all latency and bandwidth constraints of the application.

E. NI Allocation

The next phase is to determine the location of the network interface (NI) component for each core. The NI is the bridge between the core and the network, and is generally located at the core boundary. Depending on the core internals, pin layout, and core orientation, there can be some flexibility during NI allocation. For every core v_i, we define a set $P_i = \{p_1, p_2, ..., p_n\}$ that provides valid locations for NI

location at the core periphery. For instance, Fig. 4 shows three possible NI locations for *core9*. The location of a core's NI can have a significant impact on inter-core wire length and routing, and consequently communication delay and power dissipation. In Fig.4, suppose *core1* has a communication flow to *core3*. If NI location A is chosen for *core1*, the wirelength and cost will be much higher than if location B had been selected. The pitfalls of inefficient NI placement are exacerbated for 3D ICs. For instance, for a communication flow between *core4* and *core6*, and TSV location as shown in Fig. 4, if NI locations D and E are fixed in the two layers because of their proximity (i.e., small Manhattan distance), it may lead to an excessive wirelength than if locations F and G were chosen that have a higher Manhattan distance of separation. The anomaly exists due to the limited number and location of TSVs in 3D-ICs. This is the motivation for considering NI placement after the TSV and router allocation phase in the MORPHEUS framework.

To determine NI locations for all cores, we make use of a greedy shortest path heuristic for each solution in the valid solution set S. For a given valid solution, first a topological sort in non-ascending order is performed for all the cores based on aggregate incident communication costs $\sum w(e_{ij})$ on each core. This allows us to ascertain the relative criticality of the cores. Next, we select the core v_i at the top of the sorted list, select an NI location from P_i, calculate minimum cost paths to each core after floorplanning, and sum up the costs to obtain a single fitness cost for the NI location. The process is repeated for all possible NI locations in P_i. The NI location with the lowest fitness cost is selected and fixed for core v_i. The corresponding NIs at the destination cores that are part of the minimum cost paths from the selected source NI location are also fixed. The process is repeated by selecting the next core in the sorted list with an unselected NI location, or if its NI location has been previously fixed but at least one of its destination NIs remains unselected. The output after this process is a layout with fixed NI locations for every core.

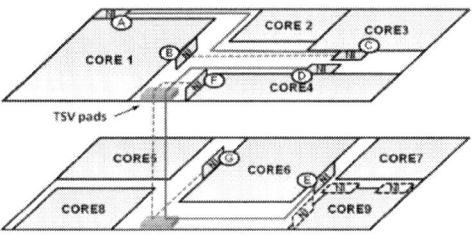

Figure 4. Network interface placement issues

F. Deadlock Free Route Generation

Finally, after the core, router, TSV, and NI locations have been fixed in the 3D-IC for each valid solution, it is important to check for possible deadlock conditions. To analyze deadlocks, we create a flow dependency graph and check for possible cycles. To avoid possible deadlock conditions, we make use of the rich body of literature in the area of deadlock avoidance in interconnection networks [16][18] and make use of escape virtual channels in the routers as a means to break deadlocks, wherever our analysis indicates a need.

V. EXPERIMENTS

To validate our proposed MORPHEUS synthesis framework, we used it to synthesize different CMP applications on a 3D IC. Six applications from the well-known SPLASH-2 benchmark suite (*Barnes, Water-NSq, FFT, Cholesky, Ocean, Raytrace*) [37] were selected, then parallelized, and mapped onto multiple irregular sized cores. These CMP applications were used as inputs to our synthesis approach. Table I summarizes the details of the CMP applications, such as number of cores (including processors and on-chip memories), and the number of layers on which the cores are to be mapped in the 3D IC implementation.

TABLE I. CMP Applications

CMP applications	Description	Cores	Layers
Barnes	Galaxy evolution	32	2
Water-NSq	Forces/potentials of H_2O molecules	38	2
FFT	FFT kernel	44	2
Cholesky	Cholesky factorization kernel	76	4
Ocean	Ocean movements	88	4
Raytrace	3-D ray tracing	112	4

The maximum TSV threshold (δ) between adjacent layers was fixed at 1024 (i.e., 32 32-bit links) and the maximum die area constraint was kept at 16mm×16mm. A pad size of 5µm×5µm was assumed for each TSV. We made use of an in-house SystemC-based 3D NoC simulator to simulate and generate performance and power results. Power estimation modules were integrated into the simulator from a modified version of Orion 2.0 [38] and CACTII [39]. Long links were pipelined to maintain high operating frequency operation. The latency and power impact of pipelining was considered in our final results. We targeted our results for the 45nm technology library, and the NoC was clocked at 1 GHz. For the GA floorplanner, we set the population size to 100, crossover and mutation probabilities to 0.9 and 0.01 respectively, and maximum generation to 100,000 based on our experience and guidelines from extensive simulations in [23].

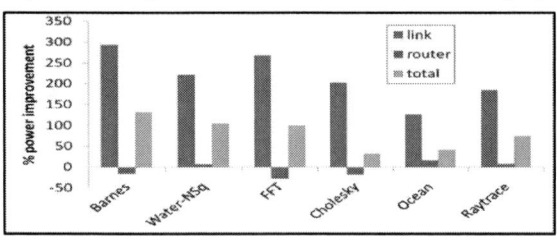

(a)

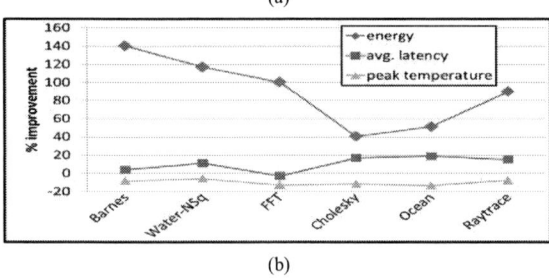

(b)

Figure 5. Application-specific 2D vs. 3D NoC comparison (a) average power dissipation, (b) energy and average latency

Our first experiment compares the results generated by the MORPHEUS framework for an application-specific 3D NoC, with an application-specific 2D NoC. The 2D NoC was synthesized using a subset of techniques in MORPHEUS that are relevant to 2D NoCs, to ensure a fair comparison. Fig. 5(a) shows the percentage improvement in power dissipation for the most power efficient application-specific 3D NoC solution, over the most power efficient application-specific 2D NoC solution, for each CMP application. In general, 3D-ICs replace long interconnects with much shorter TSVs, resulting in a significant reduction in wiring, and consequently link power dissipation. The routers however become more complicated, due to additional ports for vertical transfers and the overhead of serialization/de-serialization circuitry. Overall there is as much as a 2.3× reduction in average power dissipation for the generated application-specific 3D NoCs compared to their 2D counterparts. While a reduction in average power can improve reliability and reduce cooling costs (particularly relevant for 3D ICs that have high power densities due to active layer stacking), energy consumption is also an important metric relevant especially for battery-driven

devices. Fig. 5(b) shows the percentage improvement in total energy consumption and average latency for the synthesized application-specific 3D NoC solutions over the synthesized application-specific 2D NoC solutions. The average latency goes down slightly as high latency long interconnects are replaced by much shorter and low latency TSVs. The overall energy consumption also decreases by as much as 2.4× for the application-specific 3D NoC solutions, compared to the application-specific 2D NoC solutions. Finally, peak temperature estimates were found to be higher by 5.9%-13.3% for the 3D-IC implementations compared to the 2D-IC implementations, underscoring the need for additional thermal-aware design techniques such as throttling, adaptive voltage/frequency scaling in 3D-ICs.

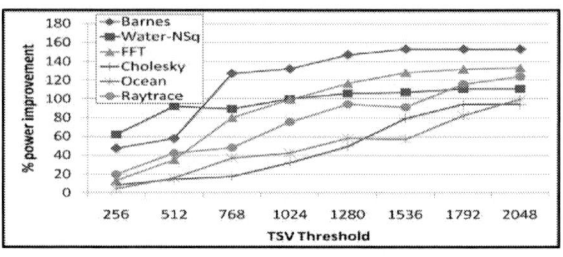

(a)

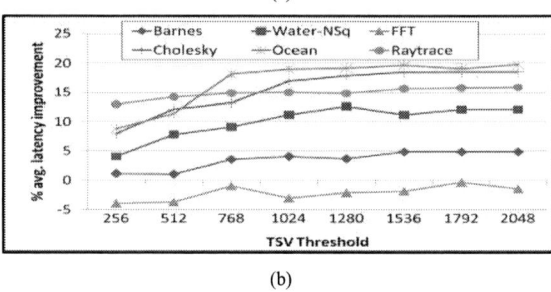

(b)

Figure 6. Impact of varying TSV threshold in MORPHEUS (a) power dissipation, (b) average latency

The next set of experiments show the impact of changing the TSV threshold (δ) on the power and average latency of the application-specific 3D NoC solutions generated by MORPHEUS. The solutions considered are the most power efficient ones from the set of synthesized solutions for each TSV threshold value. Fig. 6(a) shows the percentage power improvement for the synthesized application-specific 3D NoC over the synthesized 2D application-specific NoC baseline. It can be seen that for applications with fewer cores, the power dissipation reduces rapidly initially as the allowed number of TSVs is increased, but the improvements begin to saturate because of relatively few inter-layer communication flows that can take advantage of the increased number of allowed TSVs. For applications with larger numbers of cores and additional layers, the inter-layer communication demand is higher in general, leading to greater reduction in link power dissipation as the allowed number of TSVs is increased. However, the increased power dissipation in routers negates some of the benefits of lower link power dissipation. Fig 6(b) shows the percentage average latency improvement for the synthesized application-specific 3D NoC over the synthesized application-specific 2D NoC baseline. For almost all applications, the latency improvement saturates after a point. Except for *FFT*, the average latency is reduced for all the applications by as much as 19%. For FFT, the increase in complexity and greater traffic loading in the 3D-enabled routers translates into lesser opportunities for predictive forwarding in the router, and thus the latency increases slightly.

The final set of experiments compare the solutions generated by MORPHEUS with and without serialization, and with the solutions obtained from a previously proposed framework for synthesizing application-specific 3D NoCs [30]. Even though [30] does not address core layout, to ensure a fair comparison we use the same core layout for [30] as generated by MORPHEUS for our solutions. In this manner, we specifically compare the algorithmic effectiveness of the two approaches independent of the initial core layout step. For the comparison, we again select the most power efficient solution generated by the approach outlined in [30] and by MORPHEUS for the given applications. Fig 7(a) shows the percentage improvement in power dissipation, while Fig 7(b) shows the improvement in average latency for the solutions generated by MORPHEUS compared to the solutions generated by [30]. While applications with fewer cores (*Barnes, Water-NSq*) do not particularly benefit from serialization, other larger applications can be seen to clearly gain from using the TSV serialization technique. The improvements over [30] come from better NI allocation, serialization, and better allocation for routers and TSVs in the 3D IC. Overall our results indicate an up to 98% improvement in power dissipation and up to 62% improvement in average latency for the MORPHEUS framework, compared to [30]. These results highlight the effectiveness of the automated MORPHEUS application-specific 3D NoC synthesis framework for emerging CMP designs that utilize 3D-IC technology.

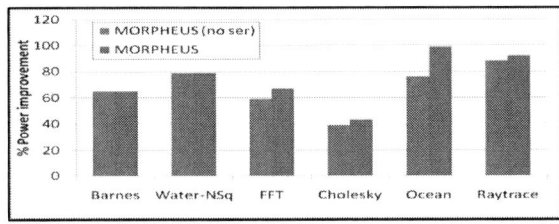

(a)

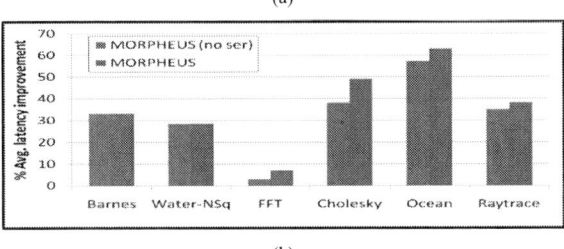

(b)

Figure 7. Comparing MORPHEUS vs. MORPHEUS without serialization vs. [30] (a) power dissipation, (b) average latency

VI. CONCLUSION

On-chip communication architectures are significant factors in determining performance and power dissipation for emerging CMP applications. However, designing an on-chip interconnect fabric especially for 3D ICs is a major challenge due its much larger design space compared to 2D ICs. In this paper, we proposed a framework (MORPHEUS) for the automated synthesis of application-specific 3D NoCs. MORPHEUS combines techniques for thermal-aware core layout, TSV serialization, allocation of routers, NIs, and TSVs, and deadlock-free path generation with the goal of generating low power solutions that satisfy all application performance constraints. The effectiveness of this framework can be seen from the notable power and latency improvements obtained over application-specific 2D NoCs, as well as solutions generated by heuristics from previously proposed work in the area of application-specific 3D NoC synthesis. Our future work will explore incorporating the reliability metric during 3D NoC synthesis within MORPHEUS.

REFERENCES

[1] S. Pasricha, and N. Dutt. "On-Chip Communication Architectures", Morgan Kauffman, ISBN 978-0-12-373892-9, Apr 2008
[2] International Technology Roadmap for Semiconductors (ITRS), 2007.
[3] A. W. Topol et al., "Three-dimensional integrated circuits," IBM J. Res. &

Dev. Vol. 50 No. 4/5 Jul/Sep 2006.
[4] K. Bernstein, et al., "Interconnects in the Third Dimension: Design Challenges for 3D ICs," Proc. DAC 2007, pp.562-567.
[5] R. S. Patti, "Three-Dimensional Integrated Circuits and the Future of System-on-Chip Designs", Proc IEEE, Vol 94, No. 6, Jun 2006.
[6] V. F. Pavlidis, E. G. Friedman, "Three-dimensional Integrated Circuit Design", Morgan Kaufmann, Sep 2008.
[7] B. Feero, P.P. Pande, "Performance Evaluation for Three-Dimensional Networks-On-Chip", Proc. ISVLSI 2007.
[8] F. Li et al., "Design and Management of 3D Chip Multiprocessors Using Network-in-Memory", Proc. ISCA 2006, pp. 130-141.
[9] I. Loi et al., "Supporting vertical links for 3D networks on chip: toward an automated design and analysis flow", Proc. NanoNet 2007.
[10] S. Pasricha, et al., "Floorplan-aware Automated Synthesis of Bus-based Communication Architectures", IEEE/ACM DAC 2005.
[11] J. Hu et al., "System-Level Point-to-Point Communication Synthesis Using Floorplanning Information", Proc. ASPDAC 2002.
[12] S. Pasricha, N. Dutt, M. Ben-Romdhane, "Constraint-Driven Bus Matrix Synthesis for MPSoC", Proc. ASPDAC 2006.
[13] S. Murali, et al. "Synthesis of Predictable Networks-on-Chip-Based Interconnect Architectures for Chip Multiprocessors", IEEE TVLSI 15:8, 2007
[14] A.Pinto et al., "Efficient Synthesis of Networks on Chip", ICCD 2003, pp. 146-150, Oct 2003.
[15] K. Srinivasan et al., "An Automated Technique for Topology and Route Generation of Application Specific On-Chip Interconnection Networks", Proc. ICCAD 2005.
[16] S. Kwon, S. Pasricha, "POSEIDON: A Framework for Application-Specific Network-on-Chip Synthesis for Heterogeneous Chip Multiprocessors", IEEE ISQED 2011
[17] J. Xu et al., "A design methodology for application-specific networks-on-chip", ACM TECS, 2006.
[18] S. Murali et al., "Designing Application-Specific Networks on Chips with Floorplan Information", pp. 355-362, ICCAD 2006.
[19] Z. Li, et al., "Efficient thermal-oriented 3D floorplanning and thermal via planning for two-stacked-die integration", ACM TODAES 11:2, Apr 2006.
[20] C. Addo-Quaye, "Thermal-aware mapping and placement for 3-D NoC designs," Proc. IEEE Int. Syst.-on-Chip Conf., 2005, pp. 25–28.
[21] E. Wong, et al. "3D Floorplanning with Thermal Vias" Proc. DATE 2006.
[22] H. Matsutani, "Prediction Router: Yet Another Low Latency On-Chip Router Architecture", Proc. HPCA 2009.
[23] J.Schaffer et al., "A study of control parameters affecting online performance of genetic algorithms for function optimization," Proc. of International Conference on Genetic Algorithms, pp.51-60, 1989.
[24] K. Puttaswamy, G.H.Loh, "Thermal Herding: Microarchitecture Techniques for Controlling Hotspots in High-Performance 3D-Integrated Processors", Proc. HPCA 2007, pp. 193-204.
[25] Y. Liu, et al., "Fine Grain 3D Integration for Microarchitecture Design Through Cube Packing Exploration", Proc. ICCD, 2007.
[26] D. Park et al. "MIRA: A Multi-layered On-Chip Interconnect Router Architecture", Proc. ISCA 2008, pp. 251-261.
[27] K. Puttaswamy, G. H. Loh, "Implementing caches in a 3D technology for high performance processors" Proc. ICCD 2005.
[28] Y. Xu, et al., "A Low-Radix and Low-Diameter 3D Interconnection Network Design", Proc. HPCA 2009.
[29] S. Murali, C. Seiculescu, L. Benini, G. De Micheli, "Synthesis of Networks on Chips for 3D Systems on Chips", Proc. ASPDAC 2009.
[30] C. Seiculescu, et al., "SunFloor 3D: A Tool for Networks on Chip Topology Synthesis for 3D Systems on Chips", Proc. DATE 2009.
[31] S. Yan, B. Lin, "Design of Application-Specific 3D Networks-on-Chip Architectures", Proc. ICCD, 2008.
[32] P. Zhou, P.-H. Yuh, S. Sapatnekar, "Application-Specific 3D Network-on-Chip Design Using Simulated Allocation", Proc. ASPDAC 2010
[33] S. Pasricha, "Exploring Serial Vertical Interconnects for 3D ICs", Proc. DAC, 2009.
[34] A. Eiben, et al, "Introduction to Evolutionary Computing", Springer 2003.
[35] S. Nakaya et al., "An Adaptive Genetic Algorithm For Vlsi Floorplanning Based On Sequence-Pair", Proc. ISCAS 2000.
[36] W. Huang, et al. "Differentiating the Roles of IR Measurement and Simulation for Power and Temperature-Aware Design." Proc. ISPASS 2009.
[37] S.C. Woo et al."The SPLASH-2 programs: Characterization and methodological considerations", Proc. ISCA, 1995.
[38] A. Kahng, et al., "ORION 2.0: A Fast and Accurate NoC Power and Area Model for Early-Stage Design Space Exploration", Proc. DATE, 2009.
[39] CACTI 6.5, http://www.hpl.hp.com/research/cacti/

An Ultra-low Power Symbol Detection Methodology and Its Circuit Implementation for a Wake-up Receiver in Wireless Sensor Nodes

Deepak Kumar Meher
deepak.meher@sandisk.com

Arunkumar Salimath
arunk22.10@gmail.com

Achintya Halder
achintya@ece.iikgp.ernet.in

Department of E&ECE, IIT Kharagpur, INDIA.

Abstract— An RF envelope detector (ED) and an asynchronous latching circuit have been designed for a wake-up receiver in 400 MHz MICS and 433 MHz / 915 MHz ISM band. The architecture is designed to tolerate significant process, supply-voltage and temperature variations. An alternative bit encoding technique has been used, which eliminates the need for symbol synchronization and the associated circuitry. The power consumption of the entire circuit, which is designed using 1.8 V supply voltage and 180 nm CMOS process, is limited to 43 uW during symbol detection and is limited to 34 uW when no input signal activity is present in the receiver RF front-end. For an input current swing of ±3 μA from the RF front-end, the circuit successfully detects up to a 2.5 Mbps input data rate.

Keywords- wake-up receiver, envelope detector, symbol synchronation, process and temperature invariant, wireless sensor network

I. INTRODUCTION

In any wireless sensor network, individual wireless sensor nodes (WSNs) spend most of the time in monitoring the wireless medium to start communicating with the gateway WSN. To minimize the power consumption of the WSN, a wake-up receiver is used, which is always active, whereas the rest of the functional blocks of the WSN may stay in idle/sleep/powered-down mode. Upon detecting the wake-up signal from the wireless channel and decoding the received command, the rest of the blocks (e.g. data transmitter, data receiver, data processing units, etc.) are powered-on/activated. Various architectures for wake-up receiver have already been proposed [1,2]. The input sensitivity at RF and the power consumption are the primary specifications of any wake-up receiver.

Simple modulation techniques, e.g. On-Off Keying (OOK), are used in the wake-up receivers [1, 2] in order to reduce the hardware complexity and the subsequent power consumption. These OOK receivers make use of envelope detection for bit demodulation after amplifying the input RF signal using a low noise amplifier (LNA) in the receiver RF front-end. To increase the input sensitivity of the wake-up receiver, the gain of the envelope detector may be increased instead of increasing the gain of the RF LNA alone, a power consuming approach..

In this work, a new symbol detection methodology and the corresponding architecture have been proposed, which consist of a novel high gain, high bandwidth and low power

envelope detector and low-power asynchronous latching circuitry. In essence, the proposed architecture improves the input sensitivity and uses a much simpler approach for bit detection, which eliminates the use of external clock and clock recovery circuitry, therefore, lowering the overall power consumption of the wake-up receiver.

The paper is organized as follows. A review of various existing wake-up receiver architectures is provided in Section II. In Section III, the proposed architecture, internal circuit blocks and the bit-decoding and synchronization technique are explained in detail. Simulation results are presented in Section IV, followed by the conclusion in Section V.

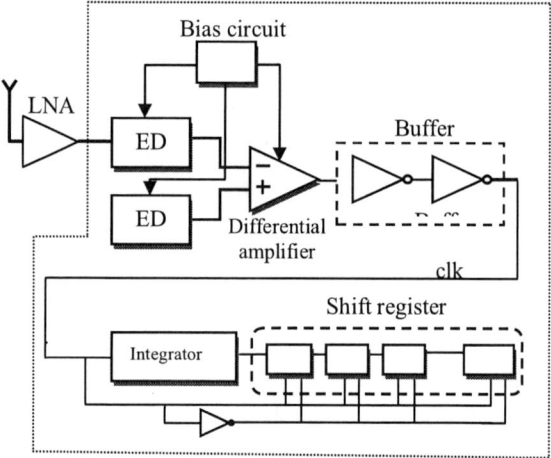

Fig 1. Block diagram of complete receiver: the blocks enclosed by the dotted line is the designed circuit

II. A REVIEW OF WAKE-UP RECEIVER ARCHITECTURE AND SYMBOL SYNCHRONIZAION TECHNIQUES

Conventional super-heterodyne receivers which use a precise local oscillator (LO), is not preferred for wake-up receiver, design. The power consumption of local oscillator itself exceeds the power budget of an entire wake-up receiver.

The work presented in [1] does use a local oscillator realized in the form of a simple ring oscillator, which consumes little power at the cost of a high drift in LO frequency. The resultant IF signal varies accordingly and, therefore, is extracted using a wide-band filter and is converted to baseband signal using another envelope detector. The complex baseband signal processing, consumes extra

978-1-4673-0438-2/12 $31.00 © 2012 IEEE   274

power and the circuit noise is integrated over a large bandwidth.

A tuned RF architecture for wake-up receiver has been proposed in [2]. It is similar to architecture shown in Fig-1 and it eliminates the power hungry local oscillator. The RF signal is amplified and then down-converted to baseband using an envelope detector. However, it adopts a complex bit detection algorithm. In effect, the lowering of the power consumption in the RF front-end is largely offset by the increased power consumption in the baseband.

The envelope detectors using non-linear diode input-output characteristics are simple and consume low power but signal strength should be sufficiently high to overcome the forward cut-in voltage. A BJT based low power, high frequency envelope detector has been proposed in [3], which is not compatible with CMOS process. MOSFET in sub-threshold region may be used to replace the BJT at the cost of a significantly poor gain and input bandwidth.

To achieve a high gain with high bandwidth, current mode envelope detectors may be used. The current mode full wave rectifier proposed in [4] uses the square law of MOSFET

to rectify its input. The primary problem in the circuit in [4] is the unequal gain in the positive and the negative half-cycles, resulting in high ripple in the baseband signal, which affects the symbol detection process. Another improved version of the structure in [4] uses Wilson current mirror [5]. This uses a negative feedback to reduce the input impedance. It requires a high input current and has a limited bandwidth of 100 MHz. In contrast, in the proposed work, the envelope detector has a high gain, high bandwidth with little static power consumption (see Section IV).

Various bit encoding and symbol synchronization techniques are described in [6]. Open-loop synchronizers extract the clock from the incoming data. First, the bit frequency component is extracted and, then, by using a high gain amplifier the clock is extracted. In contrast, a closed-loop synchronizer has a free running VCO, whose phase and frequency are matched with the incoming signal (shown in Method-1, Fig-2). This type of synchronizer is power hungry, but produces nearly accurate synchronization of clock and incoming data.

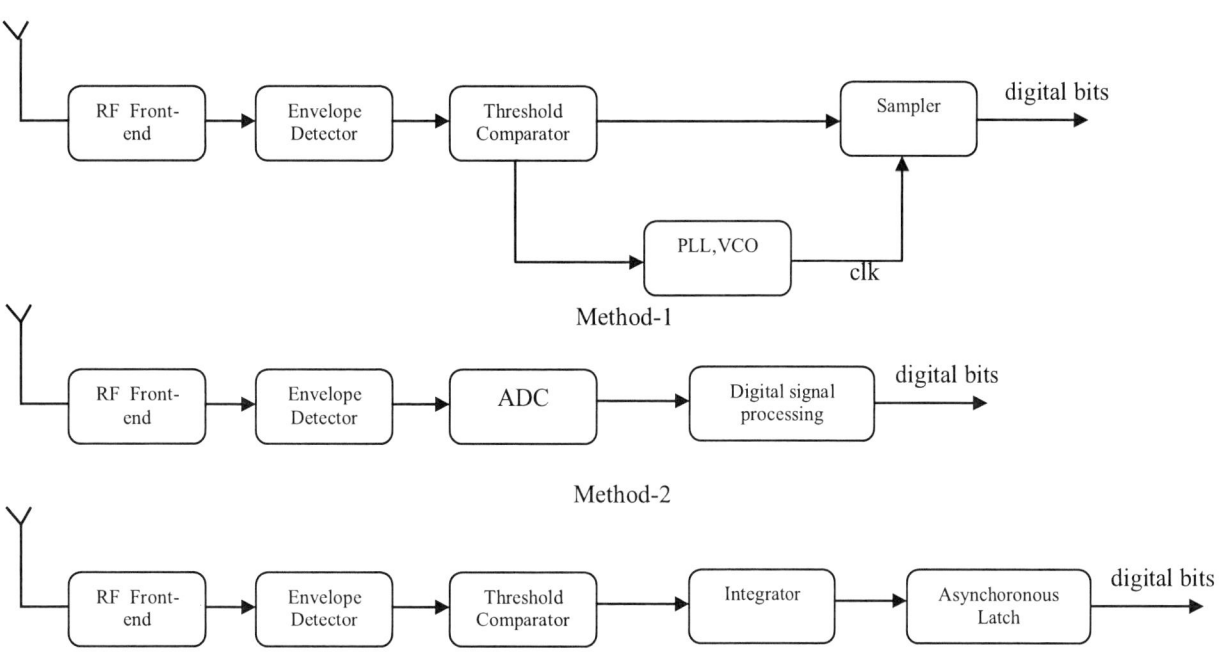

Fig 2. Methods for synchronization in receiver

Bit detection can also be done in the digital domain (as shown by Method-2 in Fig-2) at the cost of high power consumption. In this method, the envelope detector output is first converted into digital values by an ADC and then signal processing is done to extract the incoming data bits. Such a receiver is presented in [9] and the synchronization scheme accounts for 70% of the total power to achieve maximum sensitivity.

The synchronization technique can further be simplified if clock is embedded with the data itself. Manchester coding is

an example of this, which is obtained by taking the XOR of the data bit and the clock signal. Synchronization technique based on Manchester coding for wireless sensor network has been proposed in [7]. In [7], an edge detector and a counter have been used to detect the bits. However, an external clock is needed to run the counter. In the proposed bit encoding scheme the use of external clock is completely avoided.

III. PROPOSED RECEIVER ARCHITECTURE AND INDIVIDUAL BLOCKS

A. Proposed Receiver Architecture

A tuned RF architecture, which is able to demodulate On-Off keying (OOK) signal is chosen to simplify the receiver architecture. The LNA should ideally have a high gain and high quality factor to remove out of band noise. The envelope detector, which takes the input from LNA, down-converts the OOK signal to baseband. A dummy envelope detector is used to reduce the effects of temperature and bias variations. The pseudo differential signal is then amplified by a differential amplifier. The rail-to-rail level restoration is achieved by the buffering action of the inverter stages.

To simplify the bit detection and synchronization, a different bit encoding is used. Signal having a carrier presence for 20% of the bit-period represents symbol '0' and that for 80% of the bit-period represents symbol '1'. Therefore, unlike the conventional OOK signal, every bit-period of the rail-to-rail level restored signal has a falling-edge, which is used for clocking the shift register. The data is recovered from the level restored signal by integrating and comparing it with a particular threshold. The wake-up signal bits are thus stored in the shift register and depending on the bit value comparison (for the address-bit portion of the shift-register) the WSN wakes up and takes necessary actions.

B. The Circuit Blocks

Various blocks in the proposed architecture are: envelope detector, differential amplifier, inverter buffer and bias circuit.

The proposed envelope detector uses an improved version of current mode rectifier described in [4]. The envelope detector circuit (see Fig-3) works with a differential input current and has a single-ended voltage output. When a current is injected at the input node Vx, the node voltage Vx rises whereas the node voltage Vy falls; the output current flows in M3. Similarly, when a current is injected at node Vy, the output current passes through M0. In both the cases the direction of current are same and the output current through the load resistor, R, is the rectified version of input. The output current of the rectifier can be expressed as:

$$I_{OUT} = 2K_n \left[\left(\frac{V_b - V_{tn1} - V_{tn0}}{2} \right)^2 + \left(\frac{I_{in}}{2K_n(V_b - V_{tn0} - V_{tn1})} \right)^2 \right] \quad (1)$$

where, V_{tn1} and V_{tn0} are the threshold voltage of M1 and M0. The output rectified current has a dc component and a component dependent on the input current strength multiplied by the circuit gain. The input resistance, R_{in}, is constant and independent of input current and is given by:

$$R_{in} = \frac{1}{g_{m1} + g_{m0}} = \frac{1}{2K_n(V_b - V_{tn1} - V_{tn0})} \quad (2)$$

where, $K_n = \mu_n C_{ox} \frac{W}{2L}$

The bandwidth is adjusted by selecting a proper W/L ratio and the bias voltage Vb. The envelope detector has a high bandwidth due to low parasitic capacitance associated to the input. Since the envelope detector has a symmetric structure, the gain for positive input half-cycle and the gain for negative input half-cycle are equal. The rectified output current passes through the filter (realized by using a parallel combination of R and C) and produces an output voltage. A high value of R is chosen to get a high output swing and the C is realized by the input capacitance of the next stage (differential amplifier). The filter removes the fundamental and higher harmonics of the input RF carrier.

The DC component of the envelope detector output is highly sensitive to temperature, bias and process variation since it is biased using a very low current (2.5 μA). To cancel the PVT variation effects, a dummy envelope detector is used, which produces an equal DC voltage at the other input of the differential amplifier under PVT variations. The difference-signal is further amplified by this differential amplifier.

To obtain an approximately rail-to-rail voltage swing, a buffer consisting of three CMOS inverter, is used following the differential amplifier. The threshold voltages of the inverters are set by proper sizing of the transistors. The output of the differential amplifier as well as the threshold voltage of the inverters varies with temperature and bias variation. The threshold voltage can be written as,

$$V_{TH} = \frac{\sqrt{\frac{K_p}{K_n}} (V_{DD} - |V_{tp}|) + V_{tn}}{1 + \sqrt{\frac{K_p}{K_n}}} \quad (3)$$

Let $K_p = a^2 K_n$,

$$V_{TH} = \frac{a(V_{DD} - |V_{tp}| + V_{tn})}{1 + a} \quad (4)$$

$$\frac{\partial V_{TH}}{\partial V_{DD}} = \frac{a}{1 + a} \quad (5a)$$

$$\frac{\partial V_{TH}}{\partial T} = \frac{(1 - a)}{(1 + a)} \frac{\partial V_{tn}}{\partial T} \quad (5b)$$

Assuming $V_{tn} = |V_{tp}|$, it can be seen that for a=1, $V_{TH} = V_{DD}/2$.

From Equation (5a) and (5b), V_{TH} has a bias sensitivity of 50% and no temperature dependency. That means 180 mV (10% of the nominal supply voltage) change in V_{DD} will cause 90 mV change in V_{TH}. The inverter threshold voltage, V_{TH}, is affected more by the bias variation compared to the temperature variation. Therefore, V_{TH} is set lower a lower value than $V_{DD} / 2$.

The bias circuit (in Fig-3) consisting of Mb1, Mb2, Mb3 provides bias voltage to the two envelope detectors and the differential amplifier. As the temperature varies the threshold voltage of M1, M0 also varies. The Vb of bias circuit has also a dependency upon the threshold voltage of Mb2 and Mb3.

Therefore, the output is little sensitive to temperature variation.

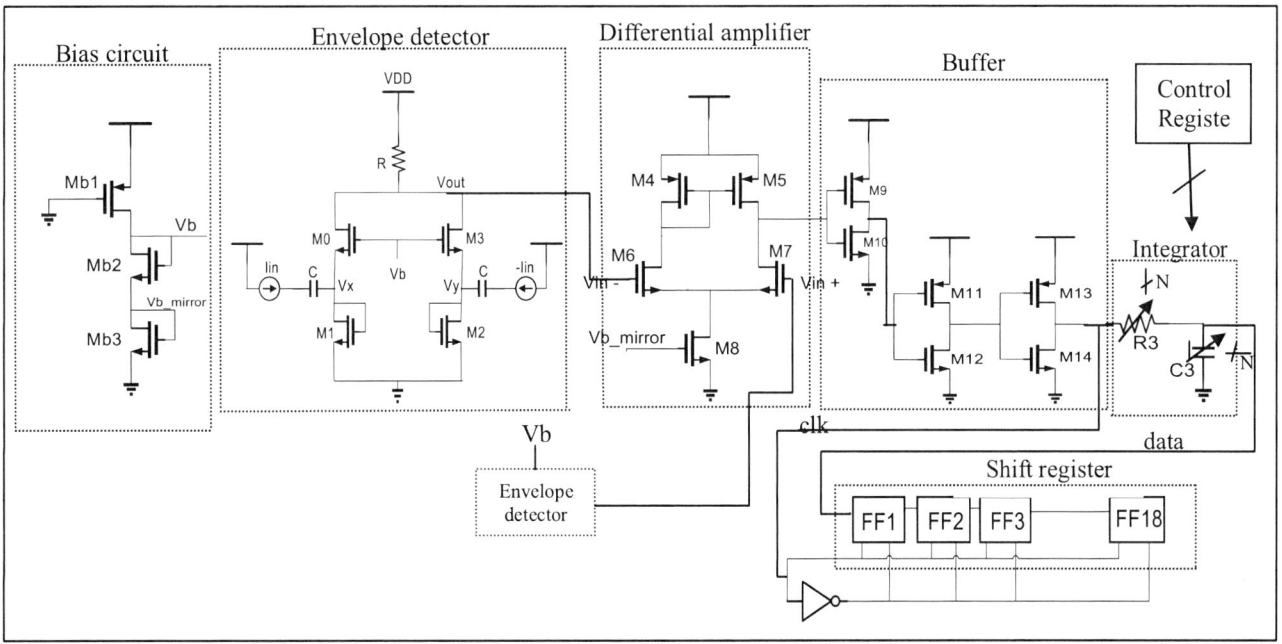

Fig 3. Schematic diagram of complete receiver

C. Bit Encoding and Symbol Detection Technique

In the proposed technique, the bit '0' is represented by the signal having the carrier-presence of 20% of the entire bit-period, whereas the bit '1' is represented by the signal having carrier presence for 80% of the entire bit-period (see Fig-4).

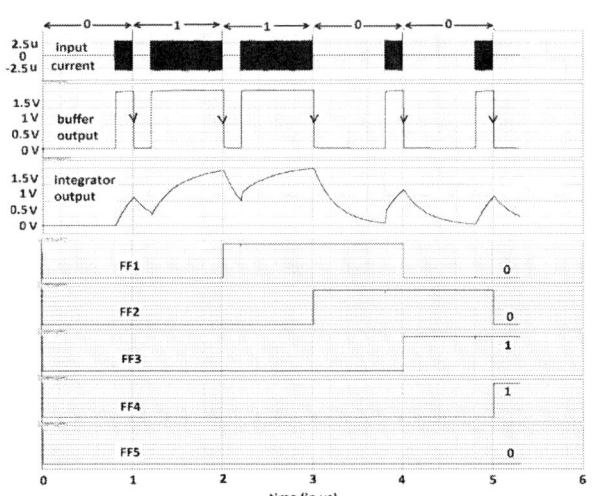

Fig 4. Bit encoding and bit detection technique

After demodulation, every bit-period has a negative edge at the bit-boundary, whereas the location of the rising edge depends on the bit value. The output of the buffer can directly be used as a clock to the negative-edge triggered shift

register. The buffered signal is further integrated to recover the data bits which are sampled by the shift register.

The time constant of integrator is selected such that the bit '1' voltage level and the bit '0' voltage level are separated out across the switching threshold of the inverter located within the very first flip-flop (FF1) of the shift register. The shift register is realized by cascading several D-flip flops consisting of static transmission-gate latches connected in master-slave configuration. The R and C values in the inductor are digitally selected by the control unit and thus account for the spread in RC values due to process variations.

IV. SIMULATION RESULTS

The envelope detector pairs, bias circuit, buffer, integrator and shift register are designed and laid-out in 180 nm National Semiconductor CMOS process. The envelope detector needs an OOK signal having ± 3 µA current amplitude and it produces an 80 mV swing at the output having a ripple amplitude of 7 mV at the typical PVT corner (see Fig-8). The rectifier used in the envelope detector has a bandwidth of 900 MHz at the extreme corner of PVT variation, requiring an OOK input signal having a minimum of 2.5 uA amplitude (see Fig-7). The bandwidth has a direct dependency on Vb. The differential amplifier is biased using a 5 µA tail current and it has an input common mode range of 0.8V- 1.5V. The differential amplifier output voltage swing is 320 mV. The threshold voltage of the first inverter is designed to be ~700 mV, while that of the second inverter is designed to be ~900 mV.

The power dissipation at 1 Mbps with 80% - 20% duty cycle, when bit '0's are detected is 37 μW, when bit '1's are detected is 43 μW, and when no symbol present is 34 μW. The power dissipation when no symbol is detected is important, because the receiver is going to remain in that state for a long time when the wireless medium is idle. The power dissipation of individual blocks is given in Table I. A series of '1's and '0's are passed into the envelope detector and the average power drawn from the battery is calculated, which includes both the static power and the dynamic power. The marginal difference in power consumption for bit '1' and bit '0' occurs due to the difference in their duty cycle values. The receiver is not in power-down mode during no symbol presence. The shift-register consumes more power for bit '1' because all the outputs of the shift-register are reset to '0' initially.

TABLE I
POWER CONSUMPTION OF INDIVIDUAL BLOCKS
@1Mbps WITH 80% - 20% DUTY CYCLE

Circuit block	Power consumption No symbol	Bit '0' detected	Bit '1' detected
Envelope detector in the signal path	4.7 μW	4.8 μW	5.2 μW
Additional envelope detector	4.7 μW	4.7 μW	4.7 μW
Differential amplifier	13.9 μW	14.4 μW	16 μW
Buffer	1.6 μW	4.4 μW	6 μW
Bias circuit	9 μW	9 μW	9 μW
Shift register	0 μW	0.2 μW	1.8 μW
Total	34 μW	36.8 μW	42.7 μW

The sensitivity of the circuit is given by the minimum amount of current required at the input of the envelope detector so that bits are detected without error and are stored in the shift register. Fig-5 and Fig-6 show the input-output characteristics of the rectifier.

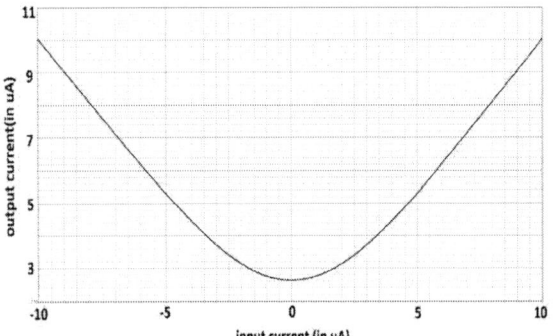

Fig 5. Input output characteristics of the rectifier in envelope detector

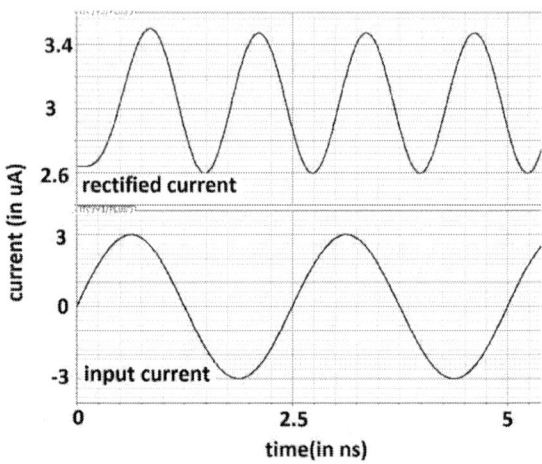

Fig 6. Transient response of current mode rectifier in envelope detector

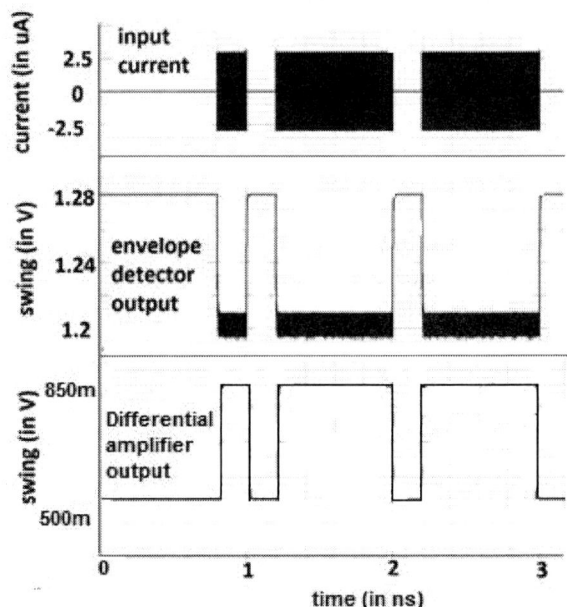

Fig 7.Time domain input-output characteristics of envelope detector

Fig-7 corresponds to the time domain input-output characteristics of the envelope detector and differential amplifier at the typical corner. The envelope detector signal swing of 80 mV at the typical corner varies by +/-15 mV over a temperature range of 100 C and 10% bias variation. Similarly the differential amplifier voltage swing of 323 mV at the typical corner varies by +/- 30mV over a temperature range of 100 C and 10% bias variation. Fig-8 illustrates the input sensitivity of the envelop detector vs. the input signal frequency at the extreme corner of the process, voltage and temperature variations. Physical layout of the entire block is shown in Fig-9. The layout occupies an area of 0.0105 mm².

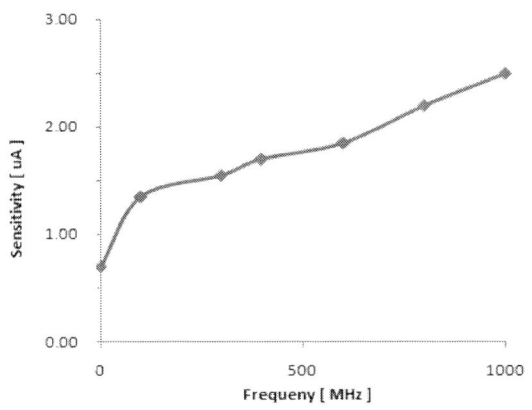

Fig 8. Input sensitivity of envelope detector vs. input signal frequency

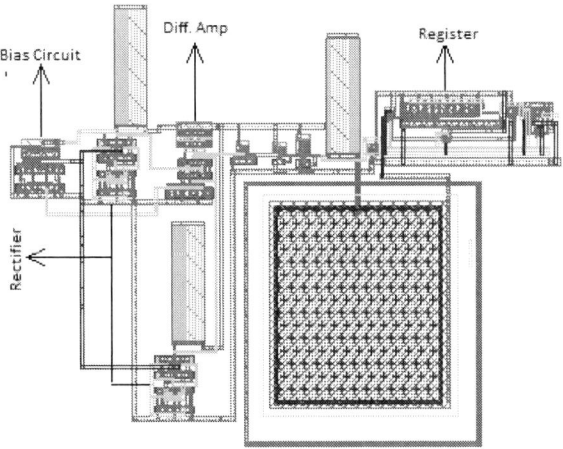

Fig 9. Layout of the envelope detector in 180 nm process

TABLE II
PERFORMANCE COMPARISON OF DIFFERENT WAKE-UP RECEIVERS

Work	Process	Supply	Power	Remark
[1]	90 nm CMOS	0.5 V	52 μW	Power consumption in baseband signal processing not included
[2]	90 nm CMOS	0.5 V	65 μW	Power consumption for ADC and DAC not included Complex bit detection algorithm
Presen-ted Work	180 nm CMOS	1.8 V	34μW (+ RF front-end power)	Low-power wake receiver with PVT variation tolerance

V. CONCLUSION

A novel technique consisting of envelope detection and low-power synchronization has been presented for wake-up receivers in WSNs. In the proposed technique, the simple bit detection mechanism helps in lowering the power consumption in the baseband and the high gain, high bandwidth and low-power envelope detector helps in improving the input sensitivity of the wake-up receiver. The design is targeted to operate in 402-405 MHz MICS band and 433 MHz / 915 MHz ISM band. In the MICS the sensitivity of the envelope detector is 2.5uA. The design is made robust to operate over a large range of PVT variations. Without the RF front-end, the proposed wake-up receiver consumes 43 uW. In conjunction with a front-end similar to that of [1,2], the entire wake-up receiver can operate within sub-100 uW power budget.

The technique along with the existing RF front ends [1,2], is expected to constitute a robust WuRx with sub-100uW power budget. The design is compliant to operate in MICS (402 MHz-405 MHz) and ISM (433 MHz and 915 MHz) bands.

ACKNOWLEDGEMENTS

The authors thank National Semiconductor Inc, USA for providing access to the 180nm CMOS Process Design Kit used.

REFERENCES

[1] N. M. Pletcher, S. Gambini, J. M. Rabaey, "A 2GHz 52μW Wake-Up Receiver with -72dBm Sensitivity Using Uncertain-IF Architecture," *ISSCC Dig. Tech. Papers*, pp. 524-525, Feb. 2008

[2] N. Pletcher, S. Gambini, J. Rabaey, "A 65 μW, 1.9GHz RF to Digital Baseband Wakeup Receiver for Wireless Sensor Nodes, " Proc. *IEEE Custom Integrated Circuits Conf. (CICC)*, pp. 539-542, Sep. 2007

[3] R. G. Meyer, "Low-Power Monolithic RF Peak Detector Analysis," *IEEE J.Solid-State Circuits*, vol. 30, no. 1, Jan. 1995

[4] C. C. Chang, S. I. Liu, "Current-Mode Full-Wave Rectifier and Vector Summation Circuit," *Electronic Letters*, vol. 36, no. 19, Sep. 2000

[5] S. Khucharoensin, V. Kasemsuwwan, "A High Performance CMOS Current-Mode Precision Full-Wave Rectifier," IEEE Conference on Circuits and Systems, 2003

[6] B. Sklar, "Digital Communications Fundamentals and Applications," 2nd Ed., Prentice Hall, 2001

[7] N. C. MacEwen, L. H. Crockett, E. Pfann, R.W. Stewart, "Symbol Synchronisation Implementation for Low-Power RF Communication in Wireless Sensor Networks," IEEE Conference on Signals, Systems and Computers, Nov 2005

[8] X.Huang et al., "A 2.4GHz/915MHz Wake-up Receiver with Offset and Noise Suppression," *ISSCC Dig. Tech. Papers*, pp. 222-223, Feb. 2010

[9] D.Daly et al., "Energy Efficient OOK Transcievers for Wireless Sensor Networks, " *IEEE J.Solid-State Circuits*, vol. 42, no. 5, May. 2007

2012 25th International Conference on VLSI Design

Low-Power Self Reconfigurable Multiplexer Based Decoder for Adaptive Resolution Flash ADCs

Chetan Vudadha, Goutham Makkena, M Venkata Swamy Nayudu, Sai Phaneendra P, Syed Ershad Ahmed,
Sreehari Veeramachaneni, N Moorthy Muthukrishnan, M.B. Srinivas
Department of Electrical Engineering,
Birla Institute of Technology and Science-Pilani, Hyderabad Campus, Hyderabad, India.
{chetan, h2009002, h2009007, h2009009, syed, srihari, moorthy, mbs}@bits-hyderabad.ac.in

Abstract—**This paper presents a new improved multiplexer based decoder for flash analog-to-digital converters. The proposed decoder is based on 2:1 multiplexers. It calculates the binary code for low operand length thermometer code at initial stages and groups the output of initial stages to generate the final result. The proposed decoder can be configured to operate on thermometer code with reduced length without any extra overhead. This 'self-reconfigurable' property is particularly useful in adaptive resolution analog-to-digital converters. Simulation results indicate that the proposed decoder results in reduced delay, power and power delay product when compared to existing digital decoders for flash analog-digital converters.**

Keywords-Flash converter; Multi-precision; Reconfigurable; Low power; Thermometer-to-binary ;

I. INTRODUCTION

Analog –to-digital converter (ADC) is a key functional block in the design of mixed signal, system on chip and signal processing applications. Many types of ADCs have been developed for different applications [1]. High speed ADCs are often based on flash structure [2-4] shown in Figure 1. Recently flash ADCs with adaptive resolution have also been proposed [5-6]. In the flash ADC implementation the input signal is applied to inputs of 2^N-1 comparators, where N indicates the resolution of the ADC. Each comparator is connected to a reference voltage commonly generated by a resistive ladder. The output of comparator is high if the input voltage is larger than the reference voltage at the input of comparator otherwise the output is low. Hence the output pattern corresponds to thermometer code. The thermometer code is converted in to binary code by $(2^N$-1)-to-N decoder generally called as Thermometer-to-Binary decoder. Many implementations have been proposed for thermometer to binary conversion [7-13]. A comparison of different decoders was presented in [7]. Folding techniques for the decoders have been presented in [13] [15]. These techniques result in reduced hardware complexity.

For low resolution and low speed ADCs, the inputs to the decoder will be perfect thermometer code. As the resolution and speed of operation of the ADC increase, bubble errors are introduced in the thermometer code. The bubble errors are unwanted digital zeros introduced in the thermometer code and are result of many sources, for example: clock

jitter, device mismatch, meta-stability and error probability of the comparators etc.

This paper presents an improved Multiplexer (MUX) based decoder for flash ADCs. The proposed decoder can be configured to operate on thermometer code with reduced operand length without any extra overhead and is suitable for adaptive resolution ADC designs. The decoder can also be configured to operate on multiple thermometer codes of reduced length. The proposed decoder results in reduced delay and power than the existing digital decoders.

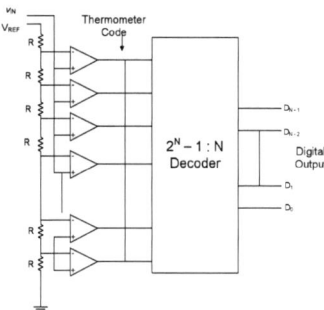

Figure 1.   Illustration of Flash ADC

The rest of the paper is organized as follows: Section II describes the related work. Section III describes the proposed improved MUX based decoder. This section also explains the self reconfigurable property of the proposed decoder. Simulation results are presented in section IV and conclusions are drawn in section V.

II. RELATED WORK

A common approach to decode the thermometer code is to use a gray or binary ROM based decoder [8-9]. The basic structure of the ROM based decoder is shown in Figure 2. The ROM based approach has 2 stages. In the first stage the thermometer code is converted in to 1-out of-2^N-1 code. This can be done by using array of NAND gates as shown in the Figure 2. The second stage is the ROM structure which takes the 1-out-of 2^N-1 code as input and selects appropriate row in the ROM. Although ROM decoder approach is simple and straight forward to design, it is however slow and consumes large power due to a constant static current used to preset the ROM encoder [9]. Another problem of binary ROM decoder

978-1-4673-0438-2/12 $31.00 © 2012 IEEE

280

is the bubble error. When bubble error occurs there are more than one '1's in 1-out of-2^N-1 code. Hence more than one row in the ROM will be enabled resulting in erroneous binary code.

Many digital decoders for converting thermometer code to binary code have been presented in the literature. The straight forward approach is the Wallace tree based decoder [10] [7], which count the number of number of '1's. This approach has the benefit of bubble suppression. Another advantage of using ones counter as a decoder is that depending on the speed of ADC a suitable ones counter topology may be used by a speed power trade off. The disadvantage of this approach is that it results in large delay and power.

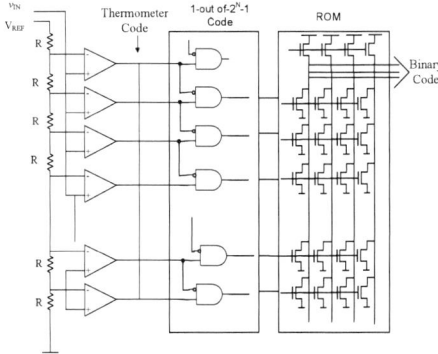

Figure 2.   Flash ADC implmentation with ROM Decoder

A more power and delay efficient approach of converting thermometer to binary code is to use Fat tree based decoder [11]. Fat tree structure has two stages. The first stage converts the thermometer code to 1-out of-2^N-1 code. The second stage converts the 1-out of-2^N-1 code to binary code using multiple trees of OR gates. This results in reduced area and delay when compared to Wallace tree based decoder. Figure 3 shows the implementation of fat tree based decoder for 15-bit thermometer code input.

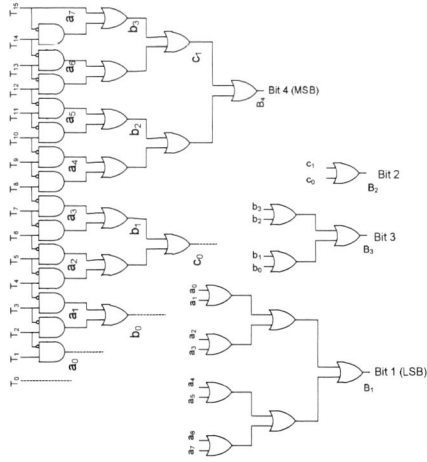

Figure 3.   Fat Tree Based Decoder [11]

A more optimized implementation of the fat tree based encoder is presented in [12]. This approach reduces the array of OR gates into NAND-NOR pairs. The NAND-NOR gates were implemented using a pseudo-dynamic CMOS logic.

A MUX based thermometer to binary decoder is proposed in [13]. This decoder results in short critical path and small area. At each level, the input thermometer code is divided in two and one of the bits in the binary output is calculated. Figure 4 shows the implementation of MUX based decoder for 15-bit thermometer code input. The disadvantage of this approach is that it results in huge fan-out in the critical path. The MUX based implementation of 15-bit thermometer code to binary decoder results in fan-out of 7 in level 1 and 4 in level 2. The increased fan-out results in increased power consumption and delay.

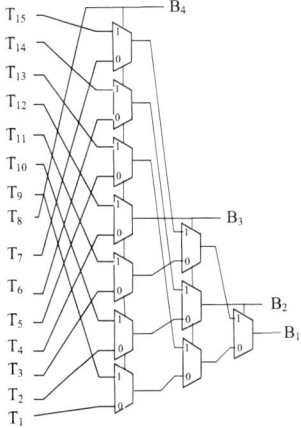

Figure 4.   Existing MUX Based Decoder [13]

III.   PROPOSED IMPROVED MUX-BASED DECODER

A. Basic Idea

The main idea behind the design of the proposed MUX based decoder is to group the results of smaller length MUX-based decoders to form a larger decoder for thermometer to binary conversion. This idea is explained by designing a 7-bit thermometer code to binary decoder using a 3-bit thermometer to binary code decoder.

A simple circuit to convert 3-bit thermometer code to binary code along with the truth table for the same is shown in the Figure 5.   Here T_3-T_1 represents the input Thermometer code and B_2-B_1 represents the binary code.

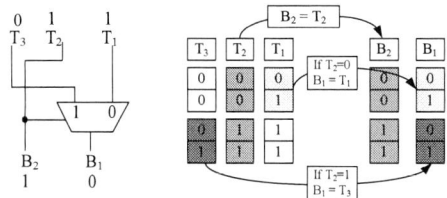

Figure 5.   3-bit Thermometer to binary decoder for 2-bit Flash ADC

Now the 3-bit thermometer to binary decoder can be used to design a 7-bit thermometer to binary decoder. The truth table and pictorial representation of the design of 7-bit thermometer to binary decoder using the 3-bit thermometer to binary decoder is shown in Figure 6. In the truth table, T_7-T_1 represents the input thermometer code and B_3-B_1 represents the binary code.

As seen from truth table

$$B_3 = T_4, \text{ and}$$

When $T_4 = 0$ the B_2-B_1 are equivalent to the outputs of 3-bit thermometer to binary decoder with $T_3T_2T_1$ as inputs.

When $T_4 = 1$ the B_2-B_1 are equivalent to the outputs of 3-bit thermometer to binary decoder with $T_7T_6T_5$ as inputs.

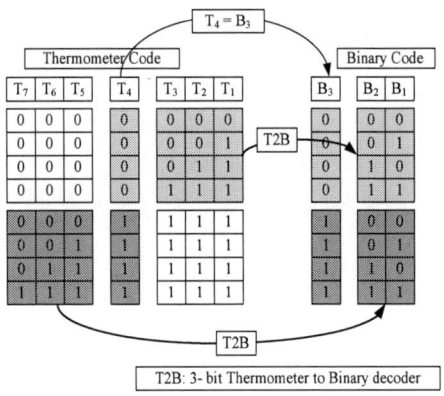

Figure 6.   Truth table for 7-bit thermometer to binary decoder

Hence using a T_4 as selection signal, 7-bit thermometer to binary decoder can be constructed using 3-bit thermometer to binary decoders and array of MUXs. Such an implementation of 7-bit Thermometer to binary decoder is shown in the figure 7, the dotted block indicates 3-bit thermometer to binary decoder.

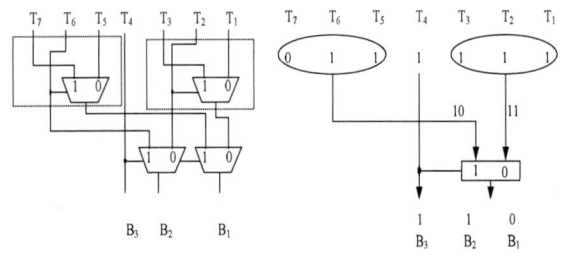

Figure 7.   Proposed 7-bit Thermometer to binary decoder for 3-bit Flash ADC

The 7-bit thermometer to binary converter uses two 3-bit thermometer to binary converters. Two MUXs are used for implementing the selection. The outputs of 3-bit decoders are fed to these MUXs and the selection signal T_4 selects outputs from either one of the dotted 3-bit decoders.

Consider example where the 7-bit thermometer code T_7-T_1 is given by 0111111, since the T_4 signal being '1' indicates that all the bits to right hand side of T_4 i.e. T_3-T_1 are '1' making the binary code B_3-B_1 greater than "100". The LSB bits B_2-B_1 are now defined by the number of '1's

present in the left hand side of T_4 signal, which is equal to two i.e. "10". This now represents the LSB bits B_2-B_1 of the final binary code.

This design methodology can be extended to implement a 2^N-1 bit decoder that can be used for an N-bit flash ADC. A generalized implementation of a 2^N-1 bit decoder is shown in the figure 8.

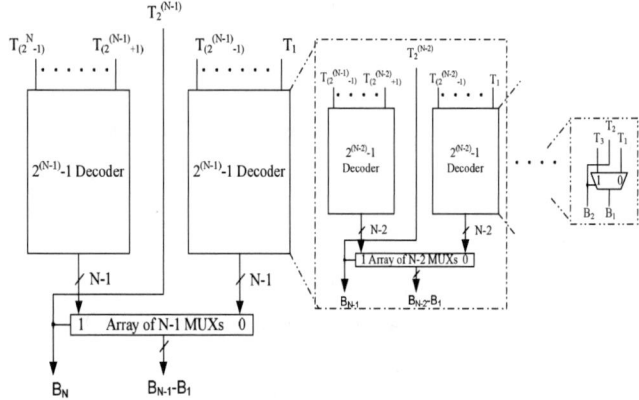

Figure 8.   Generalized Implmentation of 2^N-1 thermometer to binary decoder for N-bit Flash ADC

A 2^N-1 bit decoder can be designed by using two 2^{N-1}-1 bit decoders and an array of MUXs to generate the binary code B_N-B_1 from the thermometer code $T_{(2^N-1)}$ - T_1. Each of the 2^{N-1}-1 bit decoders can further be designed using two 2^{N-2}-1 bit decoders and an array of MUXs. This iterative implementation can be done until the basic element, i.e. a 3-bit thermometer to binary decoder, is reached.   This implementation results in a more regular structure.

The proposed 15-bit thermometer to binary decoder is shown in figure 9(d).This decoder results in a more regular structure, same number of gates and less maximum fan-out than the existing MUX based decoder [13]. The proposed decoder results in reduced number of gates when compared to Wallace tree decoder and fat tree decoders.

B. Critical pathdelay comparision

The critical path delay for existing and proposed decoders is shown in figure 9. The proposed MUX based decoder has three gates delay in its critical path for a 15-bit thermometer to binary decoder shown in figure 9(d). Although the existing 15-bit thermometer to binary MUX based decoder has a 3 gate delay in the critical path, a maximum fan-out of 7 for the input signal at the first stage results in increased delay. The proposed decoder has a maximum fan-out on input signal in the last stage of the circuit and hence does not occur in the critical path.

The 15-bit Wallace tree based decoder and Fat tree based decoder has a critical path delay of 7 gates and 4 gates respectively. The critical path delay in terms of gates for different 63-bit decoders, which can be used in 6-bit flash ADC are presented in Table I.

978-1-4673-0438-2/12 $31.00 © 2012 IEEE   282

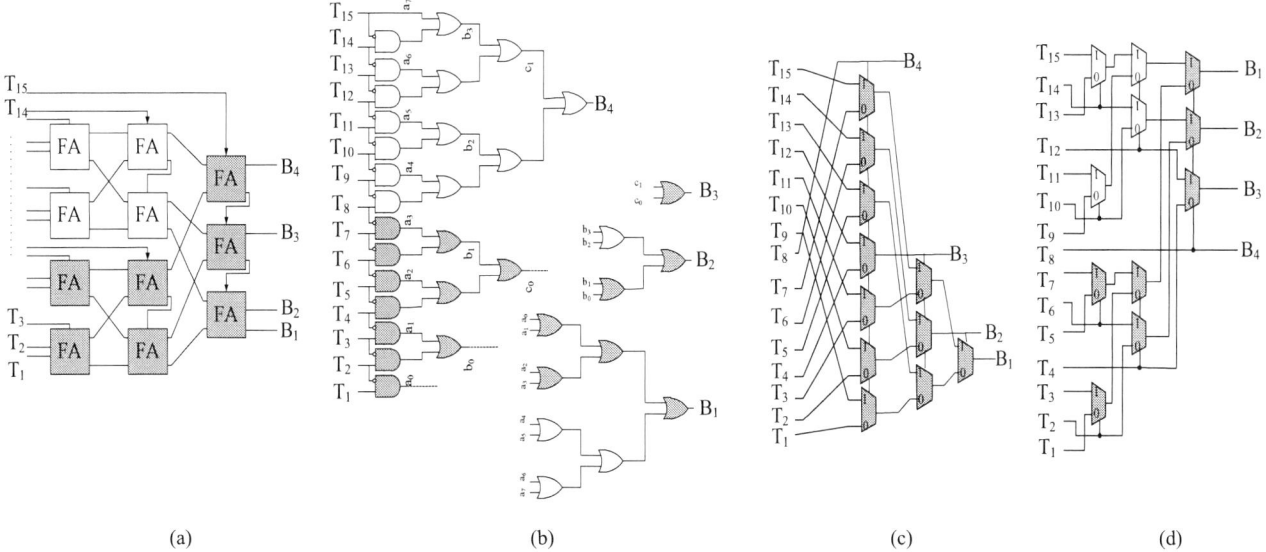

Figure 9. Different 15-bit Thermometer to Binary Decoders (a) Wallace Tree Based (b) Fat-tree Based (c) Existing MUX Based (d) Proposed

TABLE I. CRITICAL PATH DELAY COMPARISON FOR DIFFERENT DECODERS

63-bit Decoder (for 6-bit Flash ADC)	No. of gates in Critical Path Delay	Maximum fan-out in the Critical path
Wallace Tree	18	2
Fat Tree Decoder	6	2
Existing MUX-based	5	31
Proposed MUX-based	5	2

C. Self-Reconfigurable Property

The proposed thermometer to binary decoder is designed by grouping the signals generated from the smaller thermometer to binary converters. Hence the proposed decoder has a unique self reconfigurable property.

Consider a 15-bit thermometer to binary decoders shown in the Figure 9(d). These decoders can be configured to operate as a 7-bit thermometer to binary decoders by making the MSB bits T_8-T_{15} as logic zero. Since the MSB bits are tied to logic zero the gates to which these signals are fed do not have switching activity. The gates with switching activity for the existing and proposed decoders are shown in grey in the Figure 9. The Table II shows the number of gates with switching activity for different decoders, when 15-bit decoders are used to operate on Thermometer codes of 7-bit i.e. when 4-bit flash ADC is used for 3-bit resolution.

TABLE II. GATES WITH SWITCHING ACTIVITY FOR LOWER RESOLUTION OPERATION

15-bit decoders used as 7-bit decoder by zero padding (Figure 6)	Decoder	No. gates with switching activity
	Wallace Tree	21
	Fat Tree Decoder	18
	Existing MUX-based	11
	Proposed MUX-based	7

As seen from the Table II the gates with switching activity are less in the proposed decoder when compared to the existing decoder designs. The switching activity is directly related to dynamic power, which forms a major component of the total power consumption. Hence the proposed decoder results in low power consumption when operated for thermometer codes with smaller length and is ideally suited for adaptive resolution ADCs.

Further the proposed decoder can be configured to operate on multiple thermometer codes. The 15- bit thermometer to binary decoder can be operated as two 7-bit thermometer to binary decoders by making the T_8 signal as logic zero and latching the intermediate outputs of the 7-bit thermometer to binary decoder which has T_9-T_{15} as inputs. This can also be achieved by making the T_8 signal as logic one and latching the intermediate outputs of the 7-bit thermometer to binary decoder which has T_1-T_7 as inputs. This property is unique to the proposed MUX based decoder and is not present in any of the existing decoder designs.

Since the proposed decoder can be configured to operate on single or multiple thermometer codes of shorter length without any extra circuitry, it is said to be self-reconfigurable.

D. Bubble Error Correction

Wallace tree based decoder has inherent bubble suppression property, where as for other decoders, including the proposed decoder, need a bubble correction circuit for correcting the bubble errors. The different bubble correction circuits that can be used for different decoders are presented in [14]. Since the basic operation of the proposed MUX based decoder is similar to the existing MUX based decoder, the same bubble correction circuit that was used for MUX based decoder in [14] can be used for the proposed improved MUX based decoder. Figure 10 shows the bubble error correction circuits for single bubble and double bubble error respectively. In the Figure 10 T1-T8 signals represents thermometer code with bubble errors.

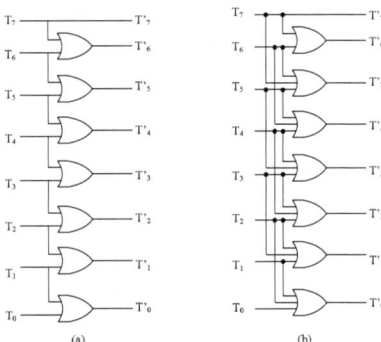

Figure 10. Circuit for (a) Single Bubble Error Correction. (b) Double Bubble Error Correction.

E. Heterogenous Decoders

In the proposed methodology of designing thermometer to binary decoders for flash ADC, smaller decoders were used to implement larger decoders as shown in the Figure 8. Any of the existing decoders like Wallace tree decoder or Fat tree decoder can be used for implementing the smaller decoders. This results in a family of heterogeneous decoders.

Consider as an example a 7-bit decoder shown in Figure 5. The dotted block in the Figure 5 indicates a 3-bit MUX based decoder. This 3-bit decoder can be a fat tree based decoder of a Wallace tree based decoder.

One such implementation which uses a combination of 3-bit counter and proposed methodology is shown in the Figure 11. Since the selection signal used for MUXs is critical they can be made bubble tolerant. The remaining signals are used as inputs to a 3-bit counter.

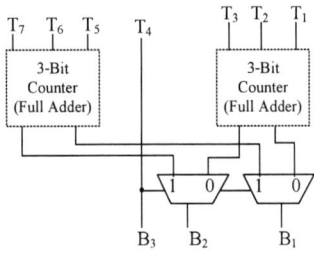

Figure 11. Heterogeneous Decoder

IV. SIMULATION RESULTS AND COMPARISIONS

All the architectures were structurally described using Verilog HDL and simulated using Cadence Incisive Unified Simulator (IUS) v6.1 covering all functional combinations. The designs were mapped on the TSMC 180nm Technology with slow_normal library (operating conditions 1.8 V, 25°C), and slow_highVt(operating conditions 0.9 V, 125°C) using Cadence RTL Compiler v7.1.The power analysis is done on all designs with 50% toggle rate at 500MHz frequency.

Table III and IV show simulation results for all the architectures for 63-bit thermometer to binary decoders i.e., for a 6-bit resolution, for slow_normal and slow_highvt libraries respectively.

TABLE III. SIMULATION RESULTS FOR 63-BIT (FOR 6-BIT FLASH ADC) THERMOMETER TO BINARY DECODERS FOR SLOW_NORMAL LIBRARY

	Delay (nS)	Area (µm²)	Power (µW)	Power-Delay Product (fJ)
Wallace Tree	1.926	1276.430	351.428	676.850
Fat Tree	0.701	752.170	62.412	43.751
Existing MUX Based	1.021	361.973	44.745	45.685
Proposed	0.546	361.973	44.732	24.424

TABLE IV. SIMULATION RESULTS FOR 63-BIT (6-BIT RESOLUTION) THERMOMETER TO BINARY DECODERS FOR SLOW_HIGHVT LIBRARY

	Delay (nS)	Area (µm²)	Power (µW)	Power-Delay Product (fJ)
Wallace Tree	2.582	1276.430	340.879	880.15
Fat Tree	0.931	752.170	60.054	55.910
Existing MUX Based	1.334	361.973	42.953	57.299
Proposed	0.752	361.973	42.565	32.009

Wallace tree decoder has more power, area and delay compared to the other architectures. Delay of the fat tree based decoder falls in between the proposed architecture and the existing MUX based decoder. Because of its inherent tree structure, the area occupied by the fat tree based decoder is more compared to existing MUX based and proposed decoder.

The design presented in [12] optimizes the fat tree based decoder at transistor level using pseudo dynamic CMOS logic. In this paper, we have concentrated on gate level implementation of the design and hence comparison of [12] with the proposed decoder has not been done.

Although the proposed and existing MUX based decoders have same area, the proposed decoder results in lower delay, as it removes high fan-out in the critical path which the existing MUX based decoder suffers from.

The proposed architecture has the same number of gates as the existing MUX based design and hence the power dissipated by both the decoders is almost the same.

A. Power Results for re-configurability

The table V and VI show the power consumption of all the decoders of 63-bit length (for 6-bit flash ADC), when operated for lesser bit lengths i.e., for different resolutions, for slow_normal and slow_highvt libraries respectively.

The existing MUX based decoder has no self-reconfigurable property. This results in higher power consumption for lower resolution inputs when compared to fat tree and proposed decoders. The power consumption of the existing MUX based decoder for lower resolution is lower than its higher resolution counterpart because the activity is less in lower resolution inputs.

Wallace tree, fat tree and proposed decoders have self-reconfigurable property as discussed in section III. But the number of gates the Wallace tree and fat tree based decoders require is high, when compared to the proposed decoder and hence consume more power.

TABLE V.   Power Consumption for different operand lengths in slow_normal corner library. All units are in μW

	7-bit (for 3-bit flash ADC)	15-bit (for 4-bit flash ADC)	31-bit (for 5-bit flash ADC)
Wallace Tree	56.516	104.859	184.889
Fat Tree	9.737	17.048	31.812
Existing MUX Based	11.523	19.995	32.202
Proposed	8.413	13.571	23.538

TABLE VI.   Power Consumption for different operand lengths in slow_highVt corner library. All units are in μW

	7-bit (for 3-bit flash ADC)	15-bit (for 4-bit flash ADC)	31-bit (for 5-bit flash ADC)
Wallace Tree	50.702	98.317	177.047
Fat Tree	7.229	14.558	29.367
Existing MUX Based	10.055	18.363	30.107
Proposed	7.013	12.116	21.979

B. Bubble Error Correction Results

The tables VII and VIII show the results of the proposed decoder with single and double error bubble correction circuits shown in Figure 7.

TABLE VII.   Simulation results for Proposed Decoder with Single and Double bubble Error Correction in slow_normal library

Bubble Error Correction	Delay (nS)	Area (μm^2)	Power (μW)	Power-Delay Product (fJ)
Single	0.698	580.709	63.028	43.994
Double	0.739	624.456	52.278	38.633

TABLE VIII.   Simulation results for Proposed Decoder with Single and Double bubble Error Correction in slow_highVt library

Bubble Error Correction	Delay (nS)	Area (μm^2)	Power (μW)	Power-Delay Product (fJ)
Single	0.950	580.709	54.672	51.9384
Double	1.006	624.456	49.443	49.740

The proposed decoder with single bubble correction circuit has 21.7%, 37.9% and 29% delay, area and power overheads respectively under slow_normal library conditions.

The proposed decoder with double bubble correction circuit has 26.1%, 42.15% and 14.4% delay, area and power overheads respectively under slow_normal library conditions.

V.   CONCLUSION

A new improved multiplexer based decoder for flash analog-to-digital converters is proposed, which converts thermometer code to binary code. The proposed decoder is designed by grouping the signals generated from the smaller thermometer to binary decoders. It can be configured to operate on thermometer code with reduced length without any extra overhead which is suitable for adaptive resolution analog to digital converters. Simulation results indicate that the proposed decoder results in better performance when compared to the existing decoders in terms of power, delay and area.

REFERENCE

[1] R. J. Van de Plassche, *CMOS Integrated Analog-to-Digital and Digital-to-Analog Converters* , Kluwer Academics Publishers,2nd Edition, 2005.

[2] Uyttenhove, K.; Marques, A.; Steyaert, M.; , "A 6-bit 1 GHz acquisition speed CMOS flash ADC with digital error correction," *Custom Integrated Circuits Conference, 2000. CICC. Proceedings of the IEEE 2000* , vol., no., pp.249-252, 2000.

[3] Kaess, F.; Kanan, R.; Hochet, B.; Declercq, M.; , "New encoding scheme for high-speed flash ADC's," *Circuits and Systems, 1997. ISCAS '97., Proceedings of 1997 IEEE International Symposium on* , vol.1, no., pp.5-8 vol.1, 9-12 Jun 1997.

[4] Abed, K.H.; Nerurkar, S.B.; , "High speed flash analog-to-digital converter," *Circuits and Systems, 2005. 48th Midwest Symposium on* , vol., no., pp.275-278 Vol. 1, 7-10 Aug. 2005.

[5] Jincheol Yoo; Daegyu Lee; Kyusun Choi; Jongsoo Kim; , "A power and resolution adaptive flash analog-to-digital converter," *Low Power Electronics and Design, 2002. ISLPED '02. Proceedings of the 2002 International Symposium on* , vol., no., pp. 233- 236, 2002.

[6] Veeramachanen, S.; Kumar, A.M.; Tummala, V.; Srinivas, M.B.; , "Design of a Low Power, Variable-Resolution Flash ADC," *VLSI Design, 2009 22nd International Conference on* , vol., no., pp.117-122, 5-9 Jan. 2009.

[7] Sall, E.; Vesterbacka, M.; Andersson, K.O.; , "A study of digital decoders in flash analog-to-digital converters," *Circuits and Systems, 2004. ISCAS '04. Proceedings of the 2004 International Symposium on* , vol.1, no., pp. I-129- I-132 Vol.1, 23-26 May 2004.

[8] Agrawal, N.; Paily, R.; , "An improved ROM architecture for bubble error suppression in high speed flash ADCs," *Student Paper, 2008 Annual IEEE Conference* , vol., no., pp.1-5, 15-26 Feb. 2008.

[9] Yao-Jen Chuang; Hsin-Hung Ou; Bin-Da Liu; , "A novel bubble tolerant thermometer-to-binary encoder for flash A/D converter," *VLSI Design, Automation and Test, 2005. (VLSI-TSA-DAT). 2005 IEEE VLSI-TSA International Symposium on* , vol., no., pp. 315- 318, 27-29 April 2005.

[10] Wallace, C. S.; , "A Suggestion for a Fast Multiplier," *Electronic Computers, IEEE Transactions on* , vol.EC-13, no.1, pp.14-17, Feb. 1964.

[11] Daegyu Lee; Jincheol Yoo; Kyusun Choi; Ghaznavi, J.; , "Fat tree encoder design for ultra-high speed flash A/D converters," *Circuits and Systems, 2002. MWSCAS-2002. The 2002 45th Midwest Symposium on* , vol.2, no., pp. II-87- II-90 vol.2, 4-7 Aug. 2002.

[12] Hiremath, V.; Saiyu Ren; , "An ultra high speed encoder for 5GSPS Flash ADC," *Instrumentation and Measurement Technology Conference (I2MTC), 2010 IEEE* , vol., no., pp.136-141, 3-6 May 2010.

[13] Sail, E.; Vesterbacka, M.; , "A multiplexer based decoder for flash analog-to-digital converters," *TENCON 2004. 2004 IEEE Region 10 Conference* , vol.D, no., pp. 250- 253 Vol. 4, 21-24 Nov. 2004.

[14] Bui Van Hieu; Seunghyun Beak; Seunghwan Choi; Jongkook Seon; Jeong, T.T.; , "Thermometer-to-binary encoder with bubble error correction (BEC) circuit for Flash Analog-to-Digital Converter (FADC)," *Communications and Electronics (ICCE), 2010 Third International Conference on* , vol., no., pp.102-106, 11-13 Aug. 2010.

[15] Yen-Tai Lai and Chia-Nan Yeh. 2010, "A folding technique for reducing circuit complexity of flash ADC decoders," *Analog Integr. Circuits Signal Process.* 63, 2 (May 2010), 339-348.

978-1-4673-0438-2/12 $31.00 © 2012 IEEE

A 1.25GHz 0.8W C66x DSP Core in 40nm CMOS

Raguram Damodaran, Timothy Anderson, Sanjive Agarwala, Rama Venkatasubramanian, Michael Gill,
Dhileep Gopalakrishnan, Anthony Hill, Abhijeet Chachad, Dheera Balasubramanian, Naveen Bhoria,
Jonathan Tran, Duc Bui, Mujibur Rahman, Shriram Moharil, Matthew Pierson, Steve Mullinnix,
Hung Ong, David Thompson, Krishna Gurram, Oluleye Olorode, Nuruddin Mahmood, Jose Flores,
Arjun Rajagopal, Soujanya Narnur, Daniel Wu, Alan Hales, Kyle Peavy, Robert Sussman
Texas Instruments Inc, 12500 TI Boulevard, Dallas TX 75243
Email: ramav@ti.com

Abstract—The next-generation C66x DSP integrated fixed and floating-point DSP implemented in TSMC 40nm process is presented in this paper. The DSP core runs at 1.25GHz at 0.9V and has a standby power consumption of 800mW. The core transistor count is 21.5 million. The DSP core features 8-way VLIW floating point Datapath and a two level memory system and delivers 40 GMACS or 10 GFLOPS floating point MAC performance at 1.25GHz.

Keywords: VLIW, Multicore, DSP processor

I. INTRODUCTION

Digital Signal Processors (DSPs) are widely used in wireless communications, imaging, audio and video applications. Due to the energy efficiency considerations and increased power density constraints, both general purpose processors and DSP cores have started following the trend of adding multiple cores in a System-on-chip (SoC) rather than increasing the core frequency of operation. The latest trends in multi-core DSP platforms is summarized in [1].

Table I
C66X DSP SPECIFICATIONS

Process	40nm (TSMC 40G) 9LM
Frequency	1.25GHz
Voltage Range	$0.72V - 1.0V$ (Min VDD_SRAM_Array = $0.81V$)
Standby Power	DSP Core - 0.8W 1MB L2 - 0.27W
Dynamic Power	DSP Core - 0.28W for average power scenario
Transistors	21.5 Million
Memory	32KB L1I 32KB L1D 1MB L2
Clock Skew	60ps
DSP Core size	Core area = $3.24mm^2$ 1MB L2 area = $3.24mm^2$

DSP cores are extensively used in wireless communication infrastructure, wireless handheld devices, multimedia gateway applications and other high compute DSP intensive applications [2][3]. As wireless communications and video applications are becoming dominant driving forces, it is clear that latest Digital Signal Processors need to not only perform fixed point arithmetic but also floating point. Further, support for high memory bandwidth and very good energy efficiency are features that are sought after in the DSPs.

This paper presents the C66x DSP core which is a very high performance DSP core with integrated fixed and floating point datapath implemented in 40nm CMOS process. The DSP core specifications are shown in Table.I. The chip micrograph is shown in Fig. 1(a) and the micrograph with bump is shown in Fig. 1(b). The DSP core features 8-way VLIW floating point datapath and a two level memory system. The DSP core delivers 40 GMACS 16-bit fixed point performance (or) 10 GFLOPS 32-bit floating point MAC performance at 1.25GHz and has a standby power consumption of 800mW. The DSP core architecture is explained in detail in Section II. The physical design considerations and implementation details are explained in detail in Section III. The energy efficiency improvements in the core and

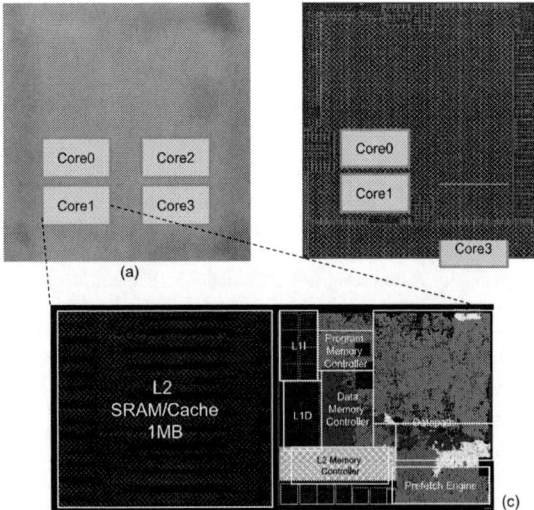

Figure 1. C66x DSP Core Floorplan and micrograph

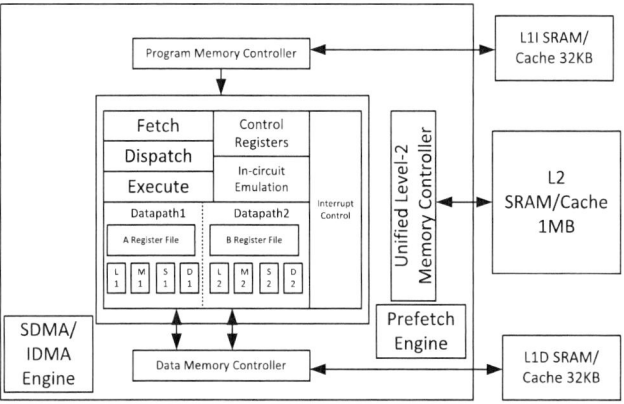

Figure 2. C66x DSP Block Diagram

comparison with state of the art DSP cores is shown in Section IV and the paper is concluded in Section V.

II. DSP CORE ARCHITECTURE

The C66x DSP Block diagram is shown in Fig. 2. It shows the CPU, memory subsystem along with the 32KB L1I, 32KB L1D and 1MB L2 memories.

A. CPU architecture details

The DSP core features 8-way VLIW integrated fixed and floating point datapath consisting of 64 general-purpose registers and eight functional units. The enhanced CPU architecture [4] offers 4X the peak 16-bit fixed-point MAC performance of the previous generation C64x+ DSP core reported in [2]. The core delivers 40 GMACS 16-bit fixed point performance (or) 10 GFLOPS 32-bit floating point MAC performance at 1.25GHz. One GMAC is the ability to compute one billion Multiply-accumulate operations in one second and one GFLOP is the ability to compute one billion floating point operations in one second. The Instructions per cycle (IPC) has been improved significantly with an enhanced instruction set that is backwards-compatible with TMS320C64x+[TM]DSP instruction set [2]. This new CPU has improved vector processing capabilities with the addition of SIMD instructions that operate on 128-bit vectors of 16-bit or 32-bit quantities, achieving 32 multiply operations per cycle. General processing power is also improved by shortening the latencies of key floating point operations like multiply and addition. Also, more specific instructions for complex linear algebra and video processing have been added. The new CPU provides up to 5 times improvement over the previous generation CPU on key wireless benchmarks such as FFT (Fast Fourier Transform) and MMSE (Minimum Mean Square Error).

B. Memory system architecture

The memory system features two levels of on-die caches - 32KB direct mapped L1 instruction cache, 32KB 2-way set associative writeback L1 data cache and 1MB 4-way private unified L2 cache. Both L1I and L2 feature ECC and parity protection. Memory protection is offered in hardware on all the levels of caches. The L1 and L2 memories can be configured as SRAM or cache or a mix of both [5]. All the controllers operate at the highest clock frequency in order to achieve the lowest CPU read latency. Emulator and debugger modules are embedded in the core. The core implements:

1) A slave DMA (SDMA) engine to move data from an external master to any of the internal memories.
2) An internal DMA (IDMA) engine to move data between internal memories.

A stream based prefetch engine has been implemented to improve the memory access latency. Full hardware coherency is supported between the two levels of on-die caches. A programmable pipelined SRAM is implemented to achieve the full throughput and low latency in the L2 memory interface. The burst mode operation of a 4 virtual bank (VB) pipelined memory is shown in Fig. 3. Overall, the memory system provides about 40% speedup on memory access latency as compared to [2].

C. Integrated powerdown controller

An integrated power-down controller has been implemented in the core which allows software-driven power management of the C66x modules. The DSP can power-down all or part of itself through the power-down controller based on its own execution thread or in response to an external stimulus from a host or global controller. Clocks to specific modules can be shut off when CPU is in IDLE mode.

III. PHYSICAL DESIGN

The DSP Core is implemented in high performance 40nm process with 9 metal interconnect layers. A 12 track cell

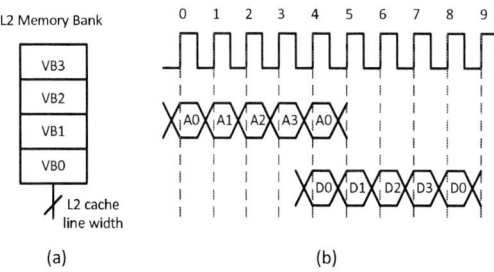

Figure 3. L2 pipelined sram operation

978-1-4673-0438-2/12 $31.00 © 2012 IEEE

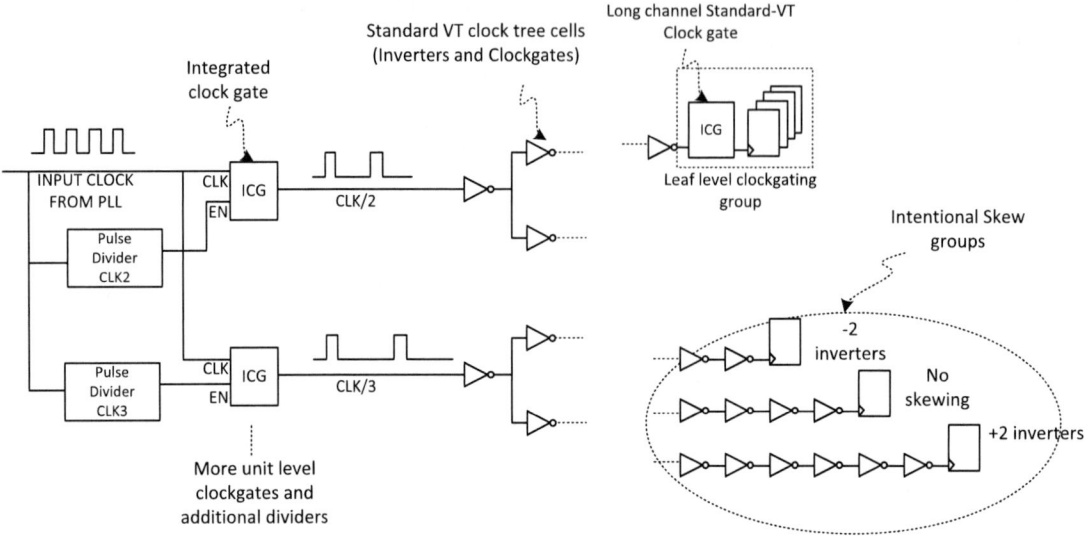

Figure 4. Clock Tree Implementation

library architecture has been used to implement the standard cells. The cell library has about 4900 cells - which includes a mix of 2 Vt transistors, each with two channel length variants for performance vs. leakage trade-off. Custom pass-gate input cells were developed and have been used in critical paths. Area optimized high performance flops have been used to reduce the total flip flop area by about 20%. Pass-gate based one-hot multiplexers have been used in the critical paths. Auto place and route (APR) was used for implementation of the core. The floorplan with major blocks highlighted is shown in Fig. 1(c).

A. Placement and routing considerations

The DSP core being a VLIW machine has wide instruction and data buses. Channelizing the buses is very important to ensure placeability and routability of the design. Based on the dataflow in the design, bounds and structured placement were used to achieve an optimal placement for the datapath and the memory system. Each and every unit of the synthesized datapath module was bounded for optimal placement. Nets in critical paths were custom routed in upper metal layers. The most critical parts of the datapath such as the fixed point adder and register files are coded with direct cell instantiations and the tools are only allowed to size the cells up or down. This way, we can strictly control the synthesis result and ensure that the tools yield expected results. The rest of the design is coded with mostly structural RTL to help guide the tools achieve desired timing closure. Timing analysis was performed using static timing analysis with built-in signal integrity support to handle extra delay due to noise and capacitive coupling.

B. Multicore SoC considerations

System-on-chip (SoC) level considerations to facilitate multiple instantiations of the core in a multi-core SoC added additional constraints on the placement of the design. Further, the core has been designed to support multiple L2 sizes as shown in Fig.5. Eventhough only two L2 size configurations are shown, the concept can be extended for any L2 size as long as the memory meets the timing requirements. This requirement added additional placement constraints and structured placement requirements to the design.

C. Clock tree design

The core has multiple synchronous clock domains and few asynchronous clock domains. The 1.25GHz high speed clock from the PLL is divided locally to generate the necessary lower frequency synchronous clock domains. Clock division is achieved by suppressing the enables to the clockgates that control the clock to each module in the design. Local pulse clock division reduces on-chip variation significantly. This in-effect results in lesser hold buffer area overhead in the core. The clock tree implementation in the core is shown in Fig. 4.

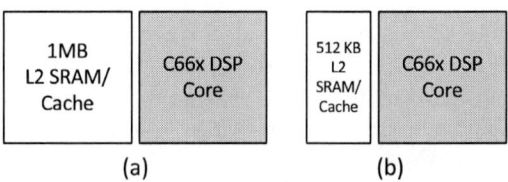

Figure 5. DSP core support for multiple L2 sizes

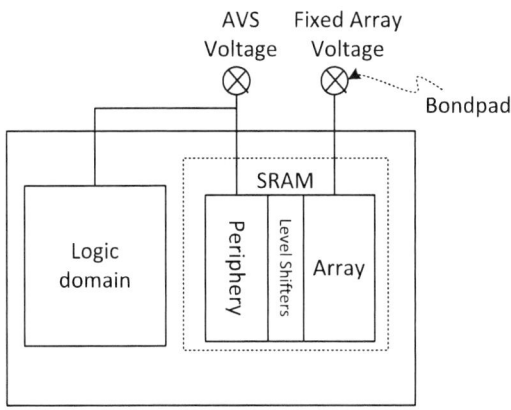

Figure 6. Power Management hookup

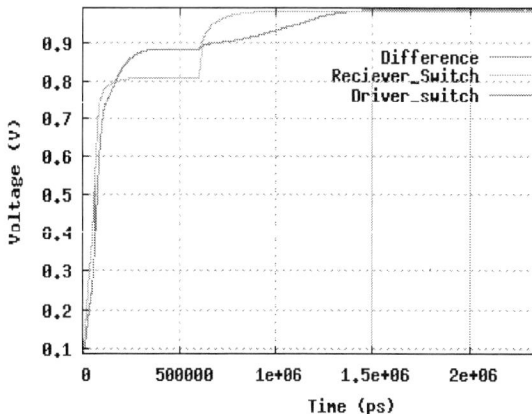

Figure 7. Differential voltage between any two power switches during wakeup

The DSP clock architecture deliberately avoids half cycle paths in the design. Clock pulse width shrinkage due to P/N skew, differential aging and clock jitter has become a major problem in sub-65nm regime. In order to guarantee minimum pulse width at the flip flops and the SRAM clock pins, lots of additional design considerations were enforced. Clock tree metrics defined in [6] were strictly enforced in the design implementation. The clock tree in the core is implemented using inverters. About 98% of the flops in the core are clock gated. This reduces the dynamic power consumption of the clock tree significantly. Even though it is desired to implement the entire clock tree using Standard-Vt clock tree cells for better tracking across all PTV corners, a significant portion of the clock tree leakage power is attributed to the leaf level clock tree cell. So, the leaf level clocktree cells were implemented using Long channel Standard-Vt cells. This is illustrated in Fig. 4. This reduces the clock tree leakage power consumption by about 40%.

Controlled intentional skewing was applied during design synthesis phase and the intentional skews were preserved all the way through physical design flow. The skews were determined based on architectural analysis and physical design considerations to help achieve the 1.25GHz timing closure goal. Mismatch effects are more pronounced with increase in number of useful skew levels in the clock tree. Hence intentional skewing in the clock tree was limited to +/- 6 inverter levels as shown in Fig. 4.

D. Power domains and power management of the core

All SRAM macros implement split-rail power supply. The core logic and SRAM periphery are controlled by the scalable Automatic voltage scaling (AVS) power grid. The SRAM Array V_{DD} was clamped at fixed array voltage from the bond pad. This is illustrated in Fig. 6.

1) Power grid design: The DSP core implements integrated power switches. The power grid IR drop goal being

$30mV$, experiments were run to determine the optimal power switch spacing in order to achieve the desired IR drop goal. Array of power switches were placed every $20\mu m$ and this added an area overhead of 5% to the overall core area. SRAM's have integrated power switches and the outputs are isolated when the SRAM is powered down. The integrated powerdown controller follows specific sequencing for shut down and wakeup between logic and SRAM domains in order to guarantee proper electrical isolation between the domains.

The differential voltage between any two power switches during wakeup was used to determine the hookup order of the power grid chain. A goal of 180mV was set for the differential voltage between two power switches during wakeup and is illustrated in Fig. 7.

The ability to independently shut down core and SRAM power domains enables significant reduction in active power in a multicore SoC usage scenario.

2) Array source biasing: The use of array source biasing by raising the ground level of the SRAM has been proven to be effective in reducing the static power consumption [7][8]. Recollect that full throughput and low latency in the L2 memory interface has been achieved using a programmable pipelined SRAM. The L2 SRAM memory array is very large and implements array source biasing to reduce the leakage power of the memory. The SRAM has dynamically controlled sector biasing scheme where during functional mode only one sector of the whole array is in un-biased state while all other sectors of the array are in source-biased state, thereby reducing the array leakage significantly. The leakage current is reduced by 2X as compared to the normal mode by using the sector biasing scheme. Each local bank array is split into multiple sectors. Each sector comprises of a certain number of physical rows. Whenever a block is accessed, the array bias is removed only for the sector that is accessed and the memory readout is carried out. Once the read is

978-1-4673-0438-2/12 $31.00 © 2012 IEEE

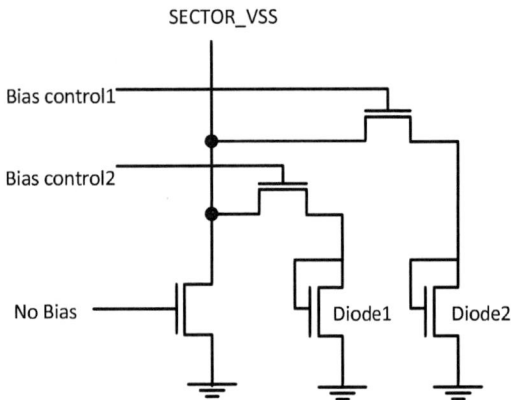

Figure 8.   L2 SRAM Array - sector source biasing

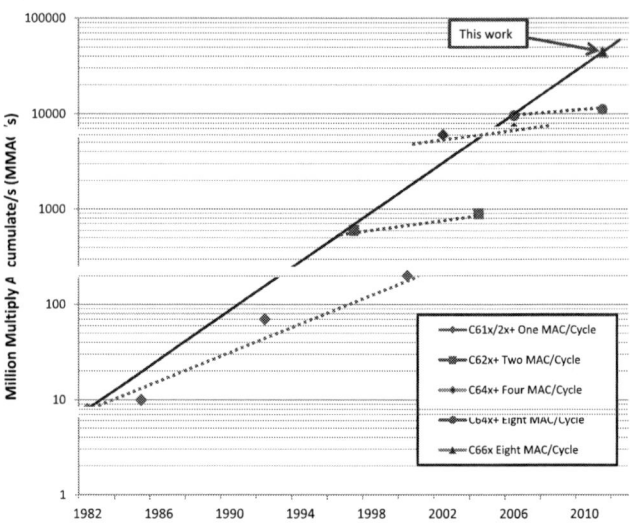

Figure 9.   Comparison of five generations of DSPs - extended from [1]. The trend in each generation correspond to the increase in performance due to clock increases within an architecture. The solid line shows the increase due to both the clock increase and the parallel processing.

complete, the array block bias is restored to put the sector back in the low-leakage state. This is illustrated in Fig.8. Two different diodes optimized for different operating points are controlled by static inputs biascontrol1 and biascontrol2 to generate different sector bias voltages.

E. Design for testability

With the size of DSP cores and SoC's increasing multi-fold, numerous test structures were integrated in the core to ensure testability of the design. Transition fault testing (TFT) is becoming increasingly important to debug at-speed failures in silicon. The clock dividers were architected to enable full at-speed transition fault testing. Test logic was integrated into the pass-gate based one-hot multiplexers to achieve good DFT coverage.

IV. DSP CORE PERFORMANCE RESULTS

Fig. 9 shows the MMAC/s (Million multiply accumulate operations per cycle) trend over five generations of DSP architectures. The MMAC improvement trend attributed to both the clock increase and parallel processing across generations is consistent for this next generation C66x DSP as well.

Typically all the architectural and physical design innovations implemented on the DSP core are targeted towards improving the overall power and energy efficiency of the core. Power efficiency metric of a DSP core is defined as the number of MMAC's supported by one milli-watt of power. The power efficiency of the C66x core (this work) and the previous generation core [2] is shown in Table. II. It can be seen that there is a 4.5X improvement in power efficiency that has been achieved with respect to the previous generation DSP core.

The BDTI benchmarks are a set of typical algorithms that are run on processor cores. They are used to assess the architectural improvements over multiple generations of a core as well as to determine the competitiveness of the core [9]. The BDTI score of the C66x DSP (this work) and the previous generation fixed point DSP (C64x+ [2]) and floating point DSP (C67x) is shown in Fig. 10. It is shown that there is an 8X improvement in the floating point score and 1.5X in the fixed point score over previous generation DSP cores. Further more, the floating point performance of the C66x DSP core is 3X more than the floating-point DSPs available in the industry [9] and 1.5X more than the fixed-point DSPs available in the industry [9].

V. CONCLUSION

The C66x DSP core features many innovations that advance the state of the art in DSP computing. The floating point performance of this core is 3X more than the floating

Table II
POWER EFFICIENCY OF THE DSP CORE FAMILY

Parameter	C64x+ [2]	C66x	Comment
Process and Voltage	65nm, 1.1V	40nm, 1V	
Operating speed	1GHz	1.25GHz	
Total power	960mW	1180 mW	
16-bit Fixed point MMAC	8000	40000	
Power Efficiency			4.5X
(MMAC/mW)	8.3	37.0	improvement

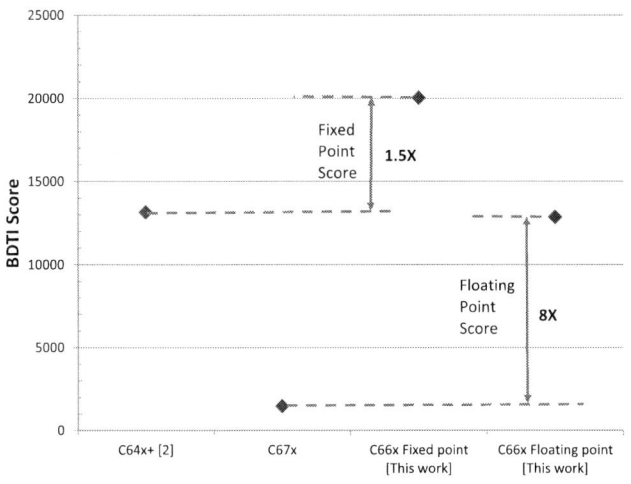

Figure 10.   BDTI Benchmark comparison

point DSPs available in the industry. Early silicon results show that C66x core silicon was fully functional, operating at 1.25GHz at 0.9V with a standby power of 800mW. There is a 4.5X improvement in power efficiency of the DSP core as compared to the previous generation DSP core.

REFERENCES

[1] Karam, L.J.; AlKamal, I.; Gatherer, A.; Frantz, G.A.; Anderson, D.V.; Evans, B.L.; , "Trends in multicore DSP platforms," Signal Processing Magazine, IEEE , vol.26, no.6, pp.38-49, November 2009

[2] Agarwala, S.; et.al , "A 65nm C64x+ Multi-Core DSP Platform for Communications Infrastructure," Solid-State Circuits Conference, 2007. ISSCC 2007. Digest of Technical Papers. IEEE International , vol., no., pp.262-601, 11-15 Feb. 2007

[3] MSC8256   High   Performance   Multi-core   DSP http://cache.freescale.com/files/dsp/doc/data_sheet/MSC8256.pdf

[4] Anderson, T.; Bui, D.; Moharil, S.; Narnur, S.; Rahman, M.; Lell, A.; Biscondi, E.; Shrivastava, A.; Dent, P.; Yan, M.; Mahmood, H.; , "A 1.5 GHz VLIW DSP CPU with Integrated Floating Point and Fixed Point Instructions in 40nm CMOS ," 20th IEEE Symposium on Computer Arithmetic, 2011. ARITH-20 2011.

[5] S. Agarwala. et. al., "Unified memory system architecture including cache and directly addressable static random access memory," U.S. Patent 6606686, Aug. 2003.

[6] Rajaram, A.; Damodaran, R.; Rajagopal, A.; , "Practical Clock Tree Robustness Signoff Metrics," Quality Electronic Design, 2008. ISQED 2008. 9th International Symposium on , vol., no., pp.676-679, 17-19 March 2008

[7] Wang, Y.; Ahn, H.; Bhattacharya, U.; Coan, T.; Hamzaoglu, F.; Hafez, W.; Jan, C.-H.; Kolar, R.; Kulkarni, S.; Lin, J.; Ng, Y.; Post, I.; Wel, L.; Zhang, Y.; Zhang, K.; Bohr, M.; , "A

1.1GHz 12?A/Mb-Leakage SRAM Design in 65nm Ultra-Low-Power CMOS with Integrated Leakage Reduction for Mobile Applications," Solid-State Circuits Conference, 2007. ISSCC 2007. Digest of Technical Papers. IEEE International , vol., no., pp.324-606, 11-15 Feb. 2007

[8] Zhang, K.; Bhattacharya, U.; Zhanping Chen; Hamzaoglu, F.; Murray, D.; Vallepalli, N.; Yih Wang; Zheng, B.; Bohr, M.; , "SRAM design on 65-nm CMOS technology with dynamic sleep transistor for leakage reduction," Solid-State Circuits, IEEE Journal of , vol.40, no.4, pp. 895- 901, April 2005

[9] BDTI Benchmark
http://www.bdti.com/Resources/BenchmarkResults/
BDTIMark2000

A Reconfigurable On-die Traffic Generator in 45nm CMOS for a 48 iA-32 Core Network-on-Chip

Praveen Salihundam[1], Mohammed Asadullah Khan[1], Shailendra Jain[1], Yatin Hoskote[2], Satish Yada[1],
Shasi Kumar[1], Vasantha Erraguntla[1], Sriram Vangal[2] and Nitin Borkar[2]
Microprocessor Research, Intel Labs, Intel Corporation
[1]Bangalore, India, [2]Hillsboro, OR, USA

Abstract - **A reconfigurable on-die Traffic Generator (TG) is proposed to test the packet switched 2D-mesh network of a 48 iA-32 core Single-chip Cloud Computer. The Single-chip Cloud Computer (SCC) is an experimental processor created by Intel Labs. The 24-tile Network-on-Chip (NoC) consists of a Traffic Generator per tile which can be programmed to generate deterministic and random traffic patterns. It also consists of reconfigurable activity control, (non)-cacheable reads and writes, message class and route control bits to feed synthetic traffic to the network to investigate NoC functional, protocol issues and to measure the key power-performance metrics. In this paper, we present the architecture and design details of the Traffic Generator, operating modes, reconfigurability and the testing procedures. This semi-custom design has a transistor count of 54K, which is 0.1% of tile transistor count, and occupies $0.3mm^2$ area which is 0.9% of tile area. The estimated power consumption is only 23mW at 1.1V and at 50^0C, 0.02% of the total chip power in 45nm high-K nine metal CMOS process.**

Keywords-2D mesh; Single-chip Cloud Computer; on-die Traffic Generator; Packet switching; NoC; on-die testing

I. INTRODUCTION

Many-core architectures are inevitable to accommodate maximum computing power within a given power envelope. The performance of many-core NoCs is greatly influenced by the interconnect bandwidth [1]. To accommodate the ever increasing demand for performance, NoCs have become increasingly complex and thereby difficult to test[2][3][4]. Some of the challenges in NoC validation are:

1. Isolation and diagnosis of faults in NoCs, which utilize packet based inter-core communication is difficult since it is nearly impossible to identify whether the fault is originating in a core or in a network component without an appropriate testing methodology.

2. NoC post-silicon testing poses a unique challenge since the network components (viz. routers and wires) might be slower or faster than the cores. Further it is not essential that the interconnect fabric should be operated at the same frequency or voltage as the rest of the processing cores.

Generating synthetic traffic with reconfigurable flow control *at-speed* is not possible with JTAG based testing methods. Testing network topologies using a conventional JTAG based method makes it difficult to benchmark the performance of the interconnect fabric in isolation, since the performance of many-core chips is closely tied to the core performance in such a test scenario. Hence the use of traffic generators has been proposed to test NoCs [2][3][4].The usage of Traffic Generators (TG) as a test mechanism has been proposed in literature. This technique has the potential to improve test coverage and reduce the test time [5][6][7]. However in the absence of preliminary and rigorous network tests, which rules out faults in the network, it becomes difficult to use the aforementioned technique. With stand alone JTAG based testing and characterization, it is also difficult to measure the key metrics like the network throughput and energy efficiency of the network. There is no silicon-proven standard procedure available in literature for testing these network protocols. Hence a hybrid testing and characterization technique which involves both JTAG and a traffic generator [2][3][4] is proposed.

An on-die traffic generator was designed to test the network of the Single-chip Cloud Computer (SCC) [8]. To improve network throughput, flow control protocols such as virtual channels (VC), speculation and route pre-computation techniques were developed. The seed data, activity control, flow control and route control bits can be programmed through JTAG to generate the traffic across the chip in order to test some of the link layer protocols such as dimension-ordered XY routing, Virtual Channel (VC) allocation strategy for message classes, valid packets and bubbles, debit based flow control, speculative VC [12] allocation, VC reservation and the route pre-computation. In addition, testing these features and checking the resiliency of these protocols in case of any errors requires a dedicated reconfigurable packet generator. Network throughput and power numbers are also be obtained for various synthetic work-loads. These advantages are the motivation to design a separate hardware to test the network of the SCC.

The remainder of the paper is organized as follows. Section 2 describes the SCC architecture, the network features and protocols. TG architecture and functional features are explained in section 3. Various operating modes of the TG and the interface between the TG and the Network Interface (NI) are also described in this section. Section 4 deals with the physical implementation details of

978-1-4673-0438-2/12 $31.00 © 2012 IEEE

the TG. Configuration methods and testing procedures of the key NoC features are detailed in section 5. TG layout and chip micrograph, Silicon measurements are given in Section 6. Section 7 concludes by summarizing the advantages of TG based network testing and by highlighting the important silicon results.

II. SINGLE-CHIP CLOUD COMPUTER ARCHITECTURE

The Single-chip Cloud Computer (SCC) experimental processor [8] is a 48-core 'concept vehicle' created by Intel Labs as a platform for many-core software research. It consists of 24 tiles with each tile comprising of two Pentium[TM] cores, two 256KB L2 caches and two cache controllers. The cores, 4 DDR3 memory controllers and a Voltage Regulator Controller (VRC) are interconnected with a 6x4 packet-switched 2D mesh using a 5-port router [8]. Cores connect to the network through a Network Interface (NI) unit which packetizes and de-packetizes the instructions from the cache controllers and is responsible for maintaining the network related protocols. Each tile also consists of a Traffic Generator (TG) to characterize the mesh with different traffic profiles. The core and TG modes are mutually exclusive and these modes can be programmed on a per-tile basis.

The interconnect fabric consists of 24 five port packet-switched routers. The interconnect fabric is a 6x4 2D mesh capable of providing 256GB/s of bisection bandwidth (each router providing 64 GB/s of link bandwidth) at 2GHz. This mesh uses a packet based protocol [9] to route packets across different tiles. The packets are classified into two types, viz., requests and responses. Each router has 5 ports, 4 of which connect it to adjacent tiles and a local port which is connected to the core with a 16B bidirectional point to point link via a Network Interface. The router uses virtual cut-through switching with eight virtual channels. Out of these 6 VCs are in a free pool and 2 VCs are reserved for requests and responses, which ensures deadlock-free routing and maximum bandwidth utilization [10]. Each VC can accommodate a full packet which can hold a single 32B cache line, and enables fast cut-through routing.

III. TRAFFIC GENERATOR DESIGN

This section describes the design and implementation of an on-die traffic generator that was used to characterize the packet-switched 2D mesh under different traffic profiles. The *in situ* traffic patterns generated are similar to that of a cache controller to mimic the actual core-to-core data transfers. The message transfers are in packets with each packet split into 128-bit FLow control unITs (FLIT) and can hold up to 3 FLITs [10]. The Network Interface (NI) packetizes the message coming from the core and the TG and sends them onto the network as shown in the Fig. 1.

The traffic generator can be programmed to bypass the core traffic and inject the synthetic traffic with a

tgPathEnable (*tgpathen*) bit setting as illustrated in the Fig. 1. The TG always operates in "*request-only*" mode and doesn't acknowledge the receipt of a packet.

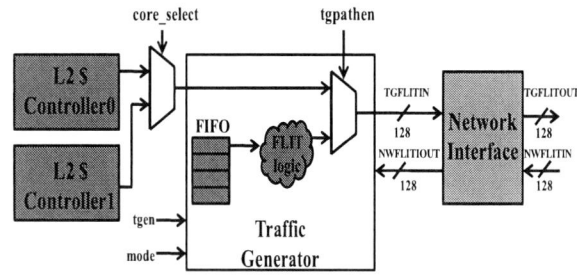

Fig. 1 Core-TG bypass path and TG mode selection

A. *Architecture and functional features*

The functional block diagram of the traffic generator is shown in the Fig. 2. The 32-deep FIFO in the TG is loaded through JTAG with the 32-bit seed data payload and the 48-bit control sideband. FIFO can be read either sequentially or randomly by the 5-bit address generator. It consists of a Linear Feedback Shift Register (LFSR) and a binary counter. LFSR generates the random addresses while binary counter generates the sequential FIFO addresses. Time stamp generators generate the local time stamps once the TGs are enabled and are attached to the outgoing FLIT. On the reception of the header FLIT, the receiving TGs compare the incoming time stamp with the local time stamp to compute the latency of the received FLIT. The received FLIT is then stored in the FIFO with transaction, source and destination IDs of the FLIT. The FIFOs are scanned out to check the functionality, latency, and the parity errors.

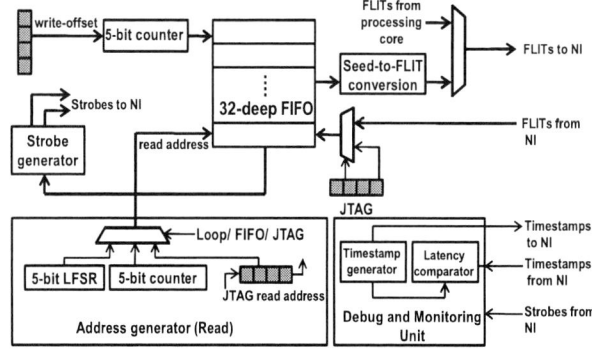

Fig. 2 Traffic generator block diagram

Fig. 3 details the entry definition in the FIFO which consists of 32-bits used as a seed to generate 128-bit data payload. The 48-bit sideband consists of valid (*valid*), activity factor enable (*ActFactEn*), activity factor (*ActFact*) and repeat (*Rpt*) to control the activity of the data of the outgoing FLIT. The packet type fields control the message class (*MC*), read or write (*Rd/Wr*) and cacheable or non-cacheable (*C/NC*) fields of the FLIT and these are used to test the behavior of the network for cacheable requests and responses.

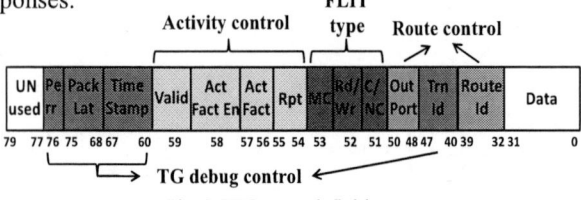

Fig. 3 FIFO entry definition

The route of the packet can be controlled with the outport (*OutPort*) and the destination tile ID (*RouteID*). The router directs the incoming packet to the pre-computed output port coming from the downstream node which is a router, when on network, and TG for local tiles. It also holds the TG debug control fields such as parity error (*Perr*), packet latency (*PackLat*), timestamp (*TimeStamp*) and the transaction ID (*TrnID*) to trace packet traversal on the network. A packet can be traced using the corresponding transaction ID, source and destination tile IDs on network.

B. *Operating modes*

TG supports 4 operating modes to provide a deterministic or random traffic and to generate infinite or finite traffic patterns across the chip. These modes are set through JTAG with mode and loop bit settings as defined in the Table 1. Mode bit selects addressing mode (random or sequential) while the loop bit chooses whether to loop around the FIFO once it has read all the 32 entries. The FIFOs in the sending and receiving TGs are read and written in a sequential order in mode0 and the test stops automatically once all the 32 entries are sent. FIFO writes are controlled by both TG FIFO write control logic and the NI write strobe. In mode2, the FIFO entries are read randomly (generated by LFSR) and NI allows write strobes to the receiving TG.

TABLE I OPERATING MODES

TG mode	Mode	Loop	Mode of Operation
Mode0	0	0	FIFO mode (Data can be written to and read from the FIFO by NI)
Mode1	0	1	Loop mode with sequential addressing(No NI write to FIFO)
Mode2	1	0	FIFO mode(NI read/write to FIFO allowed. Write pointer offset can be introduced via scan)
Mode 3	1	1	Loop mode with random addressing (No NI write to FIFO addresses generated by LFSR)

The FIFO local write enables are overridden by the NI global write strobes (decoded from the TG mode bit settings) as the NI enables the write strobes based on the MCs, C/NC and Rd/ Wr and valid bits of a FLIT before sending to the TG. In addition, a write pointer offset can be introduced in mode2 through JTAG in order not to overwrite the existing FIFO entries below the offset. As the FIFO writes are active in both of these modes, mode0 and

mode2 are typically used to study the network behavior for an emulated inter-core traffic.

The loop modes (mode1 and 3) disable the NI write strobes and are used to generate indefinite traffic with deterministic and random FLITs respectively. A stable and sustained traffic for longer duration is required on the network to be able to measure the power consumption in the network. Hence, these modes are used to measure power under various traffic profiles.

C. *Interfacing with the network*

The Traffic Generator communicates with the network through a Network Interface unit [8]. The interfacing and handshaking between the TG and the NI are described in this section. NI decodes the message class for each incoming FLIT and keeps track of the number of empty slots and the available Virtual Channels (VC) in the local router.

The FIFO entries and TG mode bits are loaded through JTAG and tgPathEnable (*tgpathen*), TGenable (*tgen*) are deactivated with reset asserted. After the initial scan load, *tgpathen* and *tgen* are activated to enable the TG-NI path and to trigger the TG respectively. The reset is then deactivated to allow the TGs to communicate. Fig. 4(a) shows a case with a cache read followed by a write. The current FLIT is read from the TG once it gets a *grant* from the NI. The network interface generates a grant (*grant*) if at least one VC is available in NI-Router link. TG decodes the *Rd/Wr* and *C/NC* bits in the current entry and generates read (*tg2niRdStrobe)* and write *(tg2niWrStrobe)* strobes. In the current scenario with (*C/NC* = 0), tg2niRdStrobe and tg2niWrStrobes are asserted to indicate a cache read and write respectively. Given a *grant,* the TG places the remote cache read address (single cycle) and then a write address followed by the data to be written.

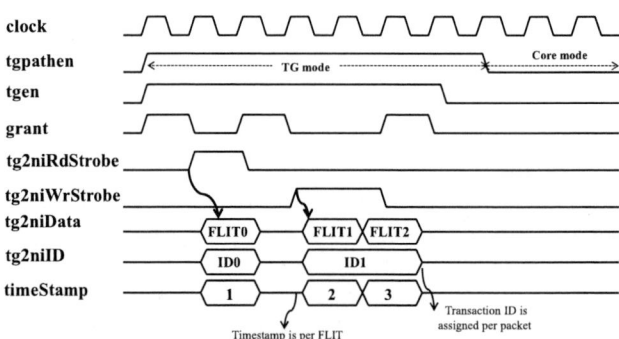

Fig. 4(a) TG-NI timing diagram (read followed by a write)

The 32B cache write takes 2 cycles in which the first FLIT carries the cache write address and the next FLIT carries a 16B data along with transaction ID (tg*2niID*) for each packet and timestamp (timeStamp) per FLIT. On the

978-1-4673-0438-2/12 $31.00 © 2012 IEEE   294

receiving side, upon reception of the header FLIT of a packet the NI sends a strobe (ni2tgIDStrobe) along with a transactionID (tg2niID) from the incoming packet and places it on the bus along with the 128-bit data payload (NI2TGData) and the time stamp as shown in Fig.4(b).

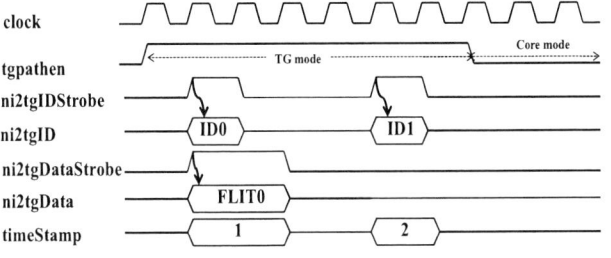

Fig. 4(b) NI-TG timing diagram (Cacheable write (C/NC=0))

The data from the NI is valid only when the dataStrobe (ni2tgDataStrobe) is high and both of these strobes are enabled per packet.

IV. PHYSICAL IMPLEMENTATION DETAILS

Traffic generator was designed to operate at 2GHz at 1.1V with a single pipe-stage. The semicustom-designed TG was partitioned into FIFO, control logic, data multiplexers and the repeater channels. The FIFO was implemented using 8T Register-File (RF) cells. The control logic was implemented using industry-standard synthesis and place & route tools. FIFO reads are domino-based to enable high performance operation as shown in Fig. 5(a). The local and global bit lines are routed in metal2 and are fully shielded.

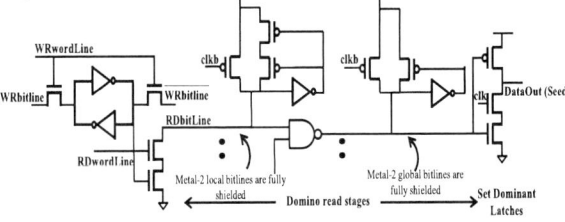

Fig. 5(a) FIFO RF cell and read circuitry

The register file uses the 2-level folded bitline topology to save area and power with two levels of domino nodes followed by an SDL (Set Dominant Latch) for dynamic to static logic conversion to optimize the read access time.

The control logic has an overwrite and overflow circuit to disable the local writes. Both reads and writes are active in modes normal (mode0) and write offset (mode2) modes. The write pointer is frozen for overwrite (OV) and overflow (OF) conditions as shown in the Fig. 5(b). Incoming FLITs get dropped at the receiving TG once the FIFO control unit detects the OV and OF flags. Read, write pointers always track each other so that the 32-deep FIFO can be shared

between transmit and receive data to reduce the power consumption.

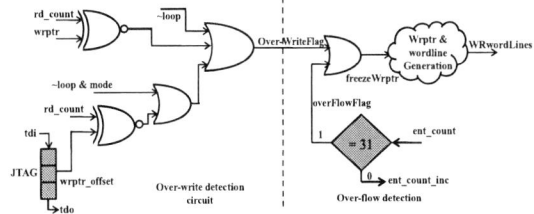

Fig. 5(b) Overwrite and overflow detection

The data multiplexers and the repeater stations are custom designed and the area of the data multiplexers and the repeater station is reduced by sharing the metal-6 tracks between the CC and the TG signals due to mutually exclusive operating modes.

V. CONFIGURING THE TG TO TEST THE KEY NoC FEATURES

This section describes the methods to configure the TG to measure and test some of the key metrics and functional features respectively. To measure the maximum frequency of the network, all TGs are enabled with tgpathen = 1; and setting the TG to operate in mode0 with valid single-FLIT packets and without any bubbles with maximum data activity. The FIFOs are scanned out to check the packet transmission and Fmax can thus be obtained. Testing can be done for all 24 tiles or between 2 tiles to obtain the Fmax variation to understand the effect of with-in-die variations on Fmax. For power measurements, FIFO contents are scanned in without changing the outport (outPortID) and destination tile (routeID), data payload is toggled in subsequent entries to have maximum activity. The activity factor enable (actFctEn), activity factor (actFct) and the repeat (rpt) bits are set to 1, 11 and 11 respectively to generate maximum activity on the data bus. Seed-to-FLIT conversion is illustrated in Fig. 6 with actFctEn, actFct and rpt bits set to 1, 11 and 10 respectively. The corner tiles are programmed to have the traffic changed from horizontal to vertical routeID after every 16 cycles. The mode bits are set to 11 (mode3) to run the TGs indefinitely. The data link power is estimated with the FIFO entries having maximum and zero activity. The difference in power consumption can come from multiple sources such as buffer writes, reads, crossbar traversal and the link traversal.

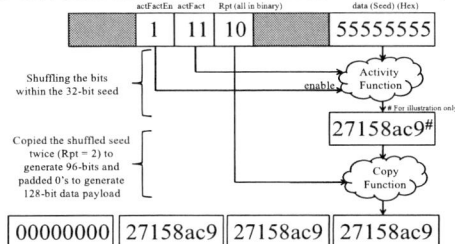

Fig. 6 Activity control and seed-to-FLIT conversion

Similarly, the control path power is estimated with the entries having no activity on the data seed and having the *outPortID* and *routeID* changed in each cycle which stresses the control paths of the router more with minimum activity on the datapath.

As the network follows dimension ordered-XY routing protocol [10] to avoid the routing conflicts, *RouteID*s of the individual TGs are selected one at a time and the *outport ID*s, which are directly injected into the local router, are selected to send the packets to adjacent tiles. Local router directly routes these packets to pre-computed *outport*s without computing the *outport*. The route path still follows the XY routing irrespective of the *outportID*s loaded in the TG. This is to test XY-routing for erroneous *outportID*s and to check the resiliency of the routing protocols.

VC allocation strategy for valid FLITs and bubbles can be tested as follows. Each router in the mesh network uses 8 VCs for all requests and responses to increase network throughput and the point of network saturation. After reset, all VCs are free and each outgoing packet is assigned an available VC. Each VC of a router is defined between an *outportID* of the current router and the corresponding in-port of the adjacent router and not per router. Payload FIFOs in the router stores a packet and the corresponding incoming VC and these VCs can be scanned-out. TG FIFOs are filled with valid FLITs and bubbles and the TG-NI-Router links are activated. The data FLITs and the VCs are scanned out from Payload FIFOs to check if the network is capturing any bubbles and to detect erroneous VC allocation which reduce the network throughput very significantly. The network is designed to suppress the bubbles and does not allocate any VCs for these bubble packets. Hence TG-NI-Router links may have some bubbles but Router-Router-NI-TG links do not see any bubbles.

Two VCs are kept in reserved pool to service the requests and responses to avoid network deadlock and starvation issues in case of flooding of requests or responses. TG FIFO contents are pre-loaded with packets with same *outPortID* and with continuous request and responses message classes and one response and request message classes respectively. The VCs in the FIFO, corresponding to the outport and inport links, are scanned out to check the pattern of VCs at each router along the route. The TG contents in the receiver side are also scanned out and compared against the sent packets and any deadlock and starvation issues are tested based on the difference between the sender and the receiver.

Mesh behavior for intermediate bubbles affects the throughput of the network though it is not a functional issue. Intermediate bubbles are not allowed to consume VCs when on the network which otherwise take a significant amount of the network bandwidth. Bubbles are allowed in the TG-NI link at the sending end and these bubbles get over-written by the valid packets.

The scanned-out VC pattern from the router connected to the local router is used to check if any VCs are allocated to the bubbles. Each router in the network operates on a request-acknowledge mode. The upstream router maintains a free list of VCs available in the current router and assigns an available VC to an outgoing packet in a FIFO order. Once it assigns a VC, it locks the VC and does not release until the current router sends the current packet tagged with the locked VC. The current router sends an acknowledgement packet in reverse direction to the upstream router with a *VCID* tag associated with a packet that has just exited the current router forming a credit loop to make sure successful transmission of packet at each hop. The VC pattern scanned-out from the current router is used to understand the credit loop cycle.

VI. SILICON MEASUREMENTS

The chip was fabricated in a 45nm nine metal high-K [13] CMOS process. The die occupies 567 mm^2 of area with 1.3 B transistor count and enclosed in a 1567 pin land grid array (LGA) package. Silicon testing of the TG has been done and is fully functional. Post-silicon validation of 2D mesh typically takes 5 weeks using standard JTAG testing whereas the on-die TG reduces test time to 1 week. A 5X reduction in silicon testing time comes with minimum TG overhead. TG occupies only 0.9% of total tile area with each tile occupying 18.7 mm^2. The total transistor count of the TG is only 54K which is 0.1% of tile transistor count and each TG in the chip consumes only 23mW which is 0.02% of total chip power at 1.1V, 2GHz, 50^{0}C. The TG layout is shown in Fig 7 with key building blocks highlighted.

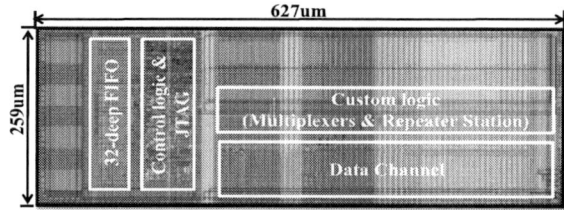

Fig. 7 TG layout

The network traffic is generated as shown in the Fig. 8 (a) with a bit in each tile showing the *tgpathen* setting. The middle TGs are not activated as these routers are used as carriers.

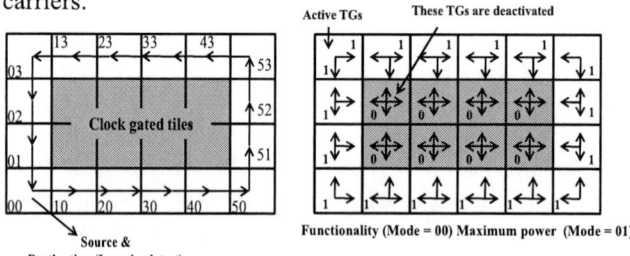

Fig. 8(a) Many-to-Many (b) Loop-back traffic patterns

Table 2 shows the router power for a many-to-many traffic pattern (Fig. 8(a)) and for a loop-back pattern (Fig. 8(b)) with 10% seed data activity. The router consumes 550mW and 525mW respectively for a many-to-many pattern with 10% and 0% seed activity respectively. The total router power consumption numbers for a loop-back test with and without clock gating the middle tiles (Fig. 8(b)) are 397mW and 457mW respectively which shows a 13% savings in total power with clock gating.

TABLE II MEASURED ROUTER POWER FOR VARIOUS TRAFFIC PATTERNS

Traffic pattern	Seed data activity	Tile clock gating	Router power (mW)
Many-to-Many	10%	No	550
Many-to-Many	0%	No	525
Loop-back	10%	No	457
Loop-back	10%	Yes	397

The network throughput measurements at 50^0C show 5.2X energy savings for 3.2X bandwidth reduction [10]. The router Fmax and power measurements at different voltages are shown the Fig. 9. The Fmax and power numbers of the router at the nominal voltage (1.1V) are 2GHz and 550mW respectively. Fmax scales from 2.35GHz to 725MHz whereas the power consumption reduces from 965mW to 64mW for 1.25V to 0.7V voltage change. The TG consumes 15.9mW and 26.5mW at 1.1V and at 1.25V respectively based on post-lay power simulations for a 10% data activity which shows TG consumes only 3% router power.

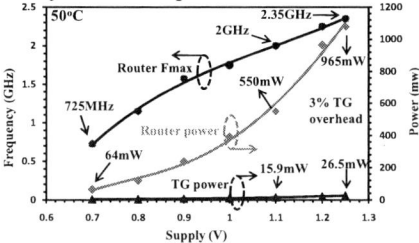

Fig. 9 Router (Measured), TG (Post-layout) Fmax, power

Injection rate of the traffic generator scales from 600Gbps to 180Gbps (3.3X) for 1.25V to 0.7V voltage change. This can be traded off for 4.1X energy savings at 0.7V as shown in the Fig 10.

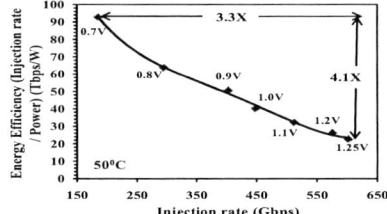

Fig. 10 TG injection rate Vs. Energy efficiency

VII. CONCLUSION

We have demonstrated an energy efficient independent testing mechanism for a 2D mesh network of SCC using an on-die reconfigurable traffic generator. This can also be used to test various types of networks by programming the seed FIFOs based on the packet structure. Post-silicon validation time of the network is reduced by 5X using TG over standard JTAG procedure with minimum overhead. TG occupies only 0.9% of tile area, 0.1% of tile transistor count and 0.02% of total chip power. NoC protocol features are tested and the network power-performance metrics are measured for various workloads using the TG. As the NoCs become more widely adopted and complex to meet the demands of future computing needs, on-die traffic generators play a significant role in testing and characterization of on-chip networks.

ACKNOWLEDGEMENTS

The authors thank T. Jacob, J. Howard, S. Dighe, N. Kothari, V. Gupta, M. Acksen and the entire design team for execution, V. De, S. Borkar, J. Schutz and J. Rattner for help, encouragement and support, the entire mask design team for the TG layout.

REFERENCES

[1] Luca Benini, Giovanni De Micheli, "Networks on Chips: A New SoC Paradigm," Computer , vol. 35, no. 1, pp. 70-78, January, 2002

[2] P.P. Pande et al, "Design, synthesis, and test of Networks on Chips", *IEEE Design & Test of Computers*, vol. 22, no. 5, pp. 404-413, September–October 2005.

[3] S. Mahadevan et al, "A Network traffic generator model for fast network-on-chip simulation", Proc of the Conference on Design, Automation and Test in Europe, pp. 780-785, 2005.

[4] L. Tedesco et al, "Traffic Generation and Performance Evaluation for Mesh-based NoCs", Integrated Circuits and Systems Design, pp. 184-189, 2005.

[5] E.Cota et al,"The impact of NoC reuse on the testing of core-based systems,"Proc. 21st VLSI Test Symposium, pp. 128-133, Apr. 2003.

[6] E. Cota and C. Liu, "Constraint-Driven Test Scheduling for NoC-Based Systems," IEEE Trans. On Computer-Aided Design of Integrated Circuits and Systems, vol.25, no. 11, pp.2465-2478, Nov. 2006.

[7] Xiao Zhang, H. G. Kerkhoff, B. Vermeulen, "On-chip Scan-Based Test Strategy for a Dependable Many-Core Processor Using a NoC as a Test Access Mechanism," Digital System Design: Architectures, Methods and Tools (DSD), 2010 13th Euromicro Conference on , pp.531-537, Sept. 2010

[8] J. Howard et al, "A 48-Core IA-32 Processor in 45nm CMOS using On-die Message-Passing and DVFS for Performance and Power Scaling" ISSCC, 2010.

[9] Dally,Towles, "Principles and Practices of Interconnection Networks", Morgan Kaufmann, 2004.

[10] P.Salihundam et al, "A 2Tb/s 6X4 mesh network for a Single-chip Cloud Computer with DVFS in 45nm CMOS", *Journal of solid state circuits*, vol. 46, no. 4, pp 757-766, Apr 2011.

[11] William J. Dally and B. Towles, "Route Packets, Not Wires: On-chip Interconnection Networks," DAC, 2001

[12] Li-Shiuan Peh, William J Dally, "A delay model and speculative architecture for pipelined routers" Proc. Seventh International Symposium on High Performance Computer Architecture (HPCA-7), January 2001.

[13] K. Mistry et al., "A 45 nm logic technology with high k+metal gate transistors, strained silicon, 9Cu interconnect layers, 193 nm dry patterning, and 100% pb-free packaging," in IEDM Dig. Tech. Papers, Dec. 2007, pp. 247–250.

Efficient Online RTL Debugging Methodology for Logic Emulation Systems

Somnath Banerjee, Tushar Gupta
Mentor Graphics India Pvt. Ltd.
NOIDA, INDIA
{somnath_banerjee, tushar_gupta}@mentor.com

Abstract - **The offline debugging model provided by logic emulation systems has some specific disadvantages. Since analysis of signal traces and bug fixing is decoupled from emulation run, validation of a potential fix requires a costly iteration through design recompilation and mapping process, followed by fresh emulation run. This slows down overall verification process. This paper presents an online debugging methodology to achieve rapid verification closure with capability to execute the design back and forward for debug. On encountering an error, the design under test (DUT) can be reverse executed step-by-step to locate source of the error. A two pass emulation technique is used to generate checkpoints and traces needed to support reverse execution. Easy and efficient reverse execution based debug is supported using an innovative technique called optimized design slicing, which allows debug along a meaningful design portion likely to cause the error being investigated. Once the source of error is located, potential bug fixes can be evaluated online by forcing a set of signals to desired values, without going through the design recompilation process and restarting emulation from time 0. Benchmarks on several customer designs have shown that the methodology enhances verification performance significantly.**

I. INTRODUCTION

For today's high-end multi-million gates designs, and with increasing need to run more and more verification cycles, it is becoming difficult to achieve verification closure based on simulator based online debug in acceptable time-to-market schedules. Field Programmable Gate Array (FPGA) based logic emulators [1, 2] are capable of verifying complex logic designs at clock speeds four to six orders of magnitude higher than a software simulator, however with emulators, designs are generally debugged "post-mortem" or "off-line" [3]. During emulation, a large trace file is written, containing value changes for signals in design in standard formats like Value

Change Dump (VCD) or a vendor specific optimized format. Upon completion, the designer analyzes the traces using a commercial waveform viewer to debug functional errors. Once the cause of error is understood, the designer modifies the design netlist, typically described in some Hardware Description Language (HDL) files, recompile the design for target emulator and rerun emulation (figure 1).

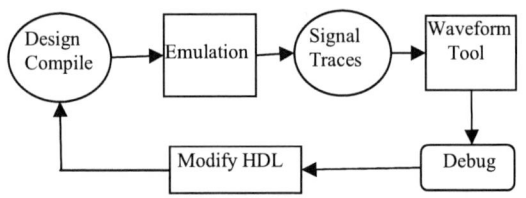

Figure 1: Offline Debug Flow

This kind of debugging methodology has specific limitations mentioned below.

• Signal traces provide efficient and fast 100% visibility with limited depth (i.e number of cycles). This depth is determined by the depth of storage memory (DRAM) that stores the signal traces. Providing visibility for infinite depth requires periodic flushing of DRAM contents to a hard-disk connected over the network, pausing emulation run whenever flushing is needed. This impacts emulation speed. Moreover waveform tools may take long time in loading very large trace files.

• After determining the source of a bug by analyzing the waveforms, designer needs to correct his HDL description and recompile in order to emulate the modified design. The recompilation process comprises front-end compile, back-end synthesis and FPGA vendor specific place and route (P&R). This recompilation process can take significant time for big designs. Such iterations potentially for each bug in the DUT impact overall verification efficiency and increase time-to-market.

This paper presents an alternative online debugging system which addresses the specific concerns with above debug model. Design execution can be stopped on detecting functional error either manually or via user defined breakpoints. The functional error is generally manifested in terms of unexpected value on one or more signals. From this point, the system can be reverse executed step by step (one cycle at a time) to locate the source of bug. Reverse execution happens only along a design slice, or a portion of design likely to impact the output of the misbehaving signal and variables/signals belonging to the slice being debugged are displayed in a visual browser for easy navigation and error tracing. After tracing the cause of bug, designer can explore a potential fix by forcing one or more signals to desired values and subsequently advancing the design clock to check functional correctness. This process can be repeated to evaluate multiple possible fixes without going through HDL modification, design synthesis and P&R process. The sequence of reverse/forward execution can be repeated as many times as needed to detect multiple bugs in the design. Fresh emulation can be run after correcting all such bugs in HDL code. The fast reverse execution capability is achieved through an innovative optimized slice-based checkpointing methodology, along with a mechanism to automatically determine interesting design slices.

The rest of the paper is organized as follows. Section 2 describes previous works and highlights their differences with current work. Section 3 gives an overview of the online debugging strategies. Overview of the debugging system and infrastructure is presented in section 4. Section 5 gives a description of overall verification flow. Experimental results and conclusions are presented in section 6 and 7 respectively.

II. PREVIOUS WORKS

Few past researches have focused on building online RTL debugging capabilities on logic emulator. An integrated online debugging environment for FPGA based logic emulation systems has been presented in [4]. It supports setting breakpoints on signals, automatic halting of design execution on maturing of breakpoints and viewing of variables for debug. A way to "rewind" system state is also provided, however it restores values of only a set of signals selected by designer. This is a difficult use model since designer may not have prior knowledge about the set of signals that should be observed for error tracing. In our approach a true reverse execution mechanism is supported, which restores a full design slice instead of a set of signals. [4] does not offer any solution to explore potential fixes online, unlike our approach. Slightly different debug methodologies combining emulation and simulation, with superior debug capabilities have also been studied [5, 6]. These

systems typically run long test sequences in emulator, and on error detection, switch over to simulation based debugging with the help of saved checkpoints/traces. Such systems are known to have scalability issues since simulators cannot load large designs with satisfactory performance, because of large memory requirements. These systems also do not provide any capability to evaluate potential fixes without going through the design synthesis and P&R process.

III. ONLINE DEBUGGING STRATEGY

The proposed online debugging strategy comprises two techniques – step-by-step reverse execution for error tracing and online fix exploration using forced signal analysis. These two techniques are demonstrated with an example below.

A. Reverse Execution

Design execution is usually stopped by the emulation OS when a user defined breakpoint matures, due to an unexpected value on a signal. From that point, the system is taken back one clock cycle at a time and the values of the signals that can cause the error are examined. This step is performed interactively again and again moving backwards in time, leading all the way to the source of error. To illustrate this concept better, let us take the example of the data buffering circuit shown in figure 2.

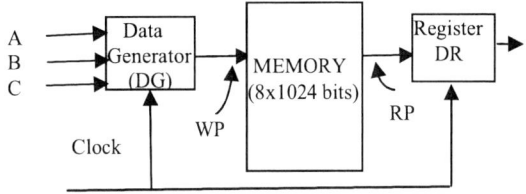

Figure 2: Data buffering circuit

The data generator (DG) module takes 3 inputs A, B, C and generates an 8 bit data, which gets stored in an 8x1024 bits memory via a write port. The data register (DR) module is connected to a read port of the memory. The data read in DR is fed to its fan-out circuit. A write or read port comprises 3 different buses – data, address, write/read enable. DG and DR are synchronous circuits driven by same design clock. At any clock cycle, read and write port of the memory may have different values, hence DG usually writes to a location different from the one read by DR.

Let us take the case when after 100 cycles, DR contains an erroneous value of 44. Design execution stops at this point. Say at that point, DR reads from memory address 511. Now the design is reverse executed step by step to determine the source of error, i.e when this erroneous value was written to address location 511. At each step, the address and data at the

978-1-4673-0438-2/12 $31.00 © 2012 IEEE   299

write port are examined. After stepping back 10 cycles, it is found that at the end of 90 cycles, the address bus reads 511 and the data being written in 44. So it is asserted that DG is producing a wrong output for the set of inputs A, B and C at cycle 90. The values of signals A, B and C are examined and found to be all 1. Now the RTL code for DG is reviewed and it is determined that for this particular set of inputs, DG produces a wrong output.

B. Forced Signal Analysis

Forced signal analysis provides user ability to quickly evaluate a potential bug fix by forcing the value of one or more signals to desired values. Multiple fixes can be evaluated in the same session, without restarting emulation from beginning. Forcing a value on a signal means that even when clock(s) are run, the signal cannot get a different value regardless of set of values in its fan-in cone. The system also allows for forcing a signal to different values at different clock cycles, via a force input file. For example, it is possible to force a signal to value 1 at clock cycle 100, to value 0 from clock cycle 101 to 201, and again to 1 from cycle 202 to 255. Finally a forced signal can be released, so that it gets driven normally by its fan-ins. Generally a bug in design manifests itself in the form of wrong values on one or more signals. Potential fixes can be evaluated by forcing those misbehaving signals to appropriate values and advancing clock(s). If the fix does not give correct results, the system can be taken back to the state before forcing happened and a different fix can be explored by forcing different values/signals. This also allows emulation to move on in spite of existing errors, allowing the opportunity to discover more bugs in a single session.

In the example of figure 2, say the correct output of DG for all 1 inputs is 22. But at cycle 90, this set of inputs led to erroneous DG output of 44. By forced signal analysis, output of DG can be forced to a value of 22 at cycle 90 and design clock(s) can be advanced to verify whether it leads to generation of correct output at DR. After execution of some cycles, other issues may be seen and fixes for them can be evaluated in the same run, reducing number of iterations.

IV. SYSTEM OVERVIEW

In this section, an overview of the system and the processes to support debugging are discussed.

A. Platform and Runtime Interface

The logic emulator consists of an array of FPGAs, connected by special interconnects and cross-bars. The array of FPGAs is connected to a host workstation. The host workstation runs the OS which controls every low

level operation and offers a wealth of services for deploying higher level applications. It also runs the software compiler, which takes design netlist as input and produces configuration bitstreams that can be directly downloaded onto the emulator via the OS.

The runtime interface can run independently on even a remote system. It interacts with the OS for controlling emulation. The interface comprises a general-purpose shell with a variety of commands to load, stop and debug a design on emulator plus a visual browser to display design connectivity, signal values. Some of the shell commands are shown in table 1.

Table 1: Runtime shell commands

Load	Load a design onto emulator
Run	Run design for one or more cycles
Stop	Stop design clock
Runrev	Reverse execute design for 1 cycle
Force	Force a signal
Release	Release an already forced signal
Breakpt	Set, unset breakpoints
View	View a variables current value

B. Optimized Design Slices

Program slicing is an analysis technique that extracts a set of statements from a program that may affect the values of a set of variables at some point during program execution. It was first introduced by Weiser [7] and is used in fields such as software debugging, maintenance and testing. In RTL verification space, it has mainly been applied to formal verification space [8, 9]. For a specified variable in a given HDL description, a slice is a subset of the original design that produces same behavioral for the variable (known as slice target variable or STV). The slice is usually much smaller and less complex than the original design. During reverse execution, it is sufficient to restore the slice corresponding to the variable on which unexpected value was seen. We introduce the concept of an optimized slice (figure 3) specifically suited for reverse execution. During reverse execution based debug, designer may not need visibility into some blocks inside a slice. These blocks can be third party vendor IPs, blocks already verified by standalone simulation or ROMs that have constant data during runtime and hence do not need to be examined for debugging. Designer marks these blocks as blackboxes in a slice. Elimination of resource intensive blocks makes a slice even smaller and makes it more efficient to checkpoint/restore on emulator. The software compiler is presented with a list of STVs and blackboxes and it computes optimized slices for each of them. It generates a HDL description of the slices and a list of state elements in each of them. The input ports of the slices are also specially instrumented for checkpoint/restore purpose as they act as primary inputs when restoring a corresponding slice. The state elements falling in the blackboxes are excluded from the list since they do not have to be checkpointed and restored.

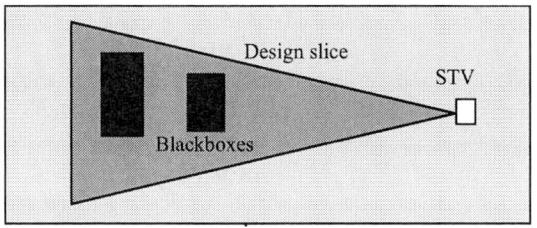

Figure 3: An optimized design slice

C. Checkpoint and Traces

A checkpoint approach is developed for periodic capturing of system state from the emulator and constructing a database necessary for piecewise reverse execution from a point of error. Two kinds of checkpoints are a typically saved – full design checkpoint (D-Checkpoint) and optimized slice checkpoint (S-Checkpoint). A D-Checkpoint represents a snapshot of the whole design at a given clock cycle and has all necessary state information to initialize the DUT to jumpstart emulation from checkpoint cycle. The state information should include the following items –

- State outputs (flip-flips, latches and memories)
- Primary inputs

When all the states are initialized with appropriate values and the DUT is stimulated with the primary input data, combinational signals automatically get restored to proper values [3]. The state information is stored as compressed buffers in host workstation and optionally can be offloaded to a hard drive. The host maintains a data structure called checkpoint queue to store the D-Checkpoints in chronological order. To minimize the impact of saving state information during emulation, D-Checkpoints are saved periodically after a certain number of clock cycles, known as checkpoint period.

S-Checkpoint is a smaller checkpoint that can be saved and restored relatively quickly. Its contents are similar to that of a D-Checkpoint, but only for a design slice instead of full DUT. Even visibility into all parts of a slice may not be needed during error tracing via reverse execution, because of presence of 3rd party IPs or blocks already verified by simulation. So a slice is optimized by all such blocks from an S-Checkpoint.

D. Variable Tagging

In order to facilitate the later stages of debugging process, it is necessary for the designer to provide certain hints to the system of which parts of the system may need to be explored at runtime. Debugging logic

is inserted by the software compiler underneath user's design. A way to tag variables is provided via a set of library components which abstracts away details of underlying synthesized logic. In software programs, a variable is solely represented by a data type and its location in memory, which can be easily monitored and modified by a debugger. However, in hardware designs, a variable can either be a stored value (register) or a simple wire connecting two logic gates. In the later case, the wires can end up being lost in hardware synthesis and mapping process due to logic optimization. Software compiler inserts special synthesized logic around the tagged signals in hardware to be able to sample and/or force them at runtime. Three different types of variable tagging are supported – breakpoints, forcible variables and blackboxes.

1) Breakpoints: Just like in a software debugger, designer can set breakpoints on a set of signals. Design execution is automatically stopped by OS whenever a breakpoint matures, i.e a signal on which breakpoint has been set assumes specified value. In case the signal is a flip-flop, its value is sampled by OS at the end of each clock edge. If the signal is a combinatorial one, a flip-flop is inserted in parallel for sampling.

2) Forcible variables: Designers can tag variables that may need to be forced during forced signal analysis. Synthesizing a variable as forcible carries overhead in terms of extra logic, hence the software compiler does not make every variable to be forcible by default. A forcible variable is replaced by logic shown in figure 4. The original signal and output of a D flip-flop (Force Value FF) are connected to a 2-input multiplexor (mux) and the select pin of the mux is driven by another D flip-flop (SelectFF). D and clock inputs of both the FFs are tied to ground so that their outputs (Q) can only be modified by external commands. By default the SelectFF is set to 0 and hence the mux output is same as the original variable. For forcing the original variable to a certain value, the Force Value FF is set to desired value (0/1) and SelectFF is set to 1, so that the mux output becomes equal to forced value irrespective of original signal value. For releasing the forced variable, SelectFF is set to 0.

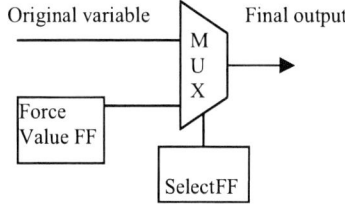

Figure 4: Forcible Variable

3) Blackbox: Blocks that need not be looked into during error tracing are tagged as blackboxes, so that their contents are not saved during checkpoint and not restored during reverse execution. To ensure that the fan-out logic of a blackbox are driven with correct value during reverse execution, even when the blackbox itself does not get restored to appropriate contents, its output ports are instrumented with logic shown in figure 5. Similar logic is synthesized around slice input ports, to ensure correct input stimuli during reverse execution of a slice. A D flip-flop (Trace FF) is added in parallel to the port. The port and output of corresponding Trace FF is connected to inputs of a 2 input mux. During forward execution, the mux selects the port as its output but during reverse execution, its select pin (driven by another D flip-flop called SelectFF) is set to appropriate value to select output of Trace FF as the final output, regardless of the value at the port.

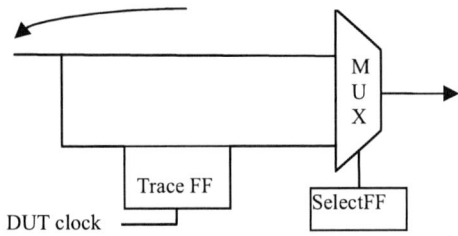

Figure 5: Instrumentation for blackbox output port and slice input ports

V. VERIFICATION FLOW

Instead of continuous tracing, a two-pass emulation strategy is used to generate information needed for debug. During forward execution (i.e first pass), D-Checkpoints are periodically saved after a predetermined number of clock cycles [10]. This interval is known as checkpoint period. This period is chosen to be long enough to cause minimal impact on execution speed. When design execution is halted due to a matured breakpoint, second emulation pass is initiated to construct debug information. The DUT is restored to last saved D-Checkpoint and run 1 cycle at a time saving appropriate S-Checkpoint at each cycle. This S-Checkpoint is for the slice corresponding to the signal on which breakpoint matured. Once S-Checkpoints are saved for each cycle till the point of error, rerevse execution can start. The system is taken back one cycle at a time, effectively loading the S-Checkpoint corresponding to the appropriate clock cycle. After restoring an S-Checkpoint, SelectFFs corresponding to all the blackbox output ports and slice inputs ports are set to 1. At any cycle, user has full visibility into all variables in the slice (except the blackboxes) to assist him in debugging. If the last saved D-Checkpoint is reached while backtracking without the source of error found, the same process of preparing the design for reverse execution (second emulation pass) has to be repeated with previous D-Checkpoint. This iterative process is repeated until the source of the error is located. It is necessary to choose a checkpoint period larger than the latency of most errors in the system, so that only one or two D-Checkpoints are needed to debug most of the errors.

After the source of error is detected, designer can start a forced signal analysis. First the DUT is restored to the last saved D-Checkpoint before the cycle when breakpoint matures. For example, in the case shown in section 2, the cycle when breakpoint matures is 100 and let the previous D-Checkpoint was saved at cycle 50. Hence the DUT must be restored to the D-Checkpoint corresponding to cycle 50. Now normal forward execution is run up to the cycle of error (in the example case up to cycle 90). Now one or more forcible signals can be forced to desired values and clocks can be executed forward to see their effect on design behavior. In case a particular set of forced signals/values does not yield a solution, the DUT can again be taken back to the D-Checkpoint before error injection (at cycle 50 in case of example) and a different set of forced signals/values can be tested.

VI. EXPERIMENTAL RESULTS

The system has been prototyped on an industry standard FPGA based emulator capable of emulating designs comprising multi-million gates. Experiments have been performed on following three large commercial designs –

1) A graphics processing unit (GPU)
2) A multimedia SoC (MSoC)
3) A multi core central processing unit (CPU)

A number of signals are tagged for debugging. Table 2 shows the total number of user gates required to map the designs on emulator and variables tagged as breakpoints, forcible and blackbox variables. Tagged variables are quite lesser in number compared to total design size and hence resource overhead due to extra synthesized logic is negligible.

Table 2: Tagged variables

Design	Total Gates (M)	Var. tagged (breakpoint)	Var. tagged (forcible)	Var. tagged (blackbox)
GPU	256	1634	4870	1800
MSoC	168	1423	3900	2190
CPU	182	1480	3745	1760

Table 3 shows the forward execution performance of the 3 designs when ran with periodic capturing of D-Checkpoints. The checkpoint period is kept at 10M cycles for each of the designs. 2nd column shows normal forward execution performance without checkpoint saving, 3rd and 4th column show time and memory requirements of a single D-Checkpoint and 5th column shows percentage impact of periodic checkpoint saving on forward execution speed.

Table 3: Forward execution performance

Design	Emu. Freq. (without checkpoint saving) (MHz)	D-Chkp Time (sec)	D-Chkp memory (MB)	Emu. Freq. (with checkpoint saving) (MHz)
GPU	1.3	3	550	1.0
MSoC	1.4	1.2	420	1.2
CPU	1.4	1.4	450	1.15

The performance of reverse execution based debug has been shown in table 4. Column 2 shows the average size of a design slice in terms of number of emulation gates, column 3 shows average time to compute slices and column 4 shows average time to load an S-Checkpoint for debugging. It can be noticed that the average size of a slice is only 10-15% of the total design, indicating a significant advantage in debugging a slice instead of the whole unsliced design.

Table 4: Reverse execution performance

Design	Average Slice Size (M gates)	Average Slice compute time (s)	Average time to load S-Checkpt (s)
GPU	1.3	3	2
MSoC	1.4	1.2	1.7
CPU	1.4	1.4	1.5

Table 5 shows overall verification statistics and specifically display the gain in performance as a result of using online debugging techniques. 2nd column shows the approximate design synthesis and P&R time, i.e the time required to prepare the design for emulation after fixing a bug in HDL. 3rd column lists the total number of bugs discovered and fixed. Maximum number of D-Checkpoints needed during reverse execution to trace source of bugs is shown in column 4. Most bugs are located using either 1 or 2 D-Checkpoints, indicating the checkpoint period is longer than latency of most errors. Only one bug in the CPU design had a longer latency and required 3 D-Checkpoints to be loaded. Column 5 shows total number of iterations (design mapping + emulation) needed to complete verification of the system (bug detection + fix). Forced signal analysis allowed multiple bugs to be detected in same emulation session. It can be argued that without forced signal analysis, fixing each bug and validating the fix would take at least one design iteration. This estimate is still optimistic, since it does not consider wasted iterations due to incorrect fixes. Based on this estimate, gain in overall verification performance has been calculated and presented in column 6. For example, in case of GPU, 32 bugs are caught and fixed/verified in 5 iterations, without our method it would have taken at

least 32 iterations. So the overall gain in verification time is approximately 6x.

Table 5: Overall verification performance

Design	Design compile time (hour)	No of Bugs	No of D-Chkp for rev. exec	No of design map+ emu iteration	Reduction in verif. time
GPU	7	32	2	5	6x
MSoC	6	42	2	8	5x
CPU	4	26	3	5	5x

VII. CONCLUSION

In this paper, we presented an online debugging methodology for logic emulation systems. The methodology addresses various shortcomings of traditional offline debug techniques, which result in overall reduction in functional verification speed Experiments on three real customer designs show that the proposed methodology significantly enhances verification performance.

REFERENCES

[1] Varghese, J.; Butts, M.; Batcheller, J.;, "An efficient logic emulation system," *Very Large Scale Integration (VLSI) Systems, IEEE Transactions on*, vol.1, no.2, pp.171-174, June 1993

[2] Chan, T.; Yeh, B.; Hu, E.;, "A first pass ASIC development methodology using logic emulation," *ASIC Conference and Exhibit, 1994. Proceedings., Seventh Annual IEEE International*, vol., no., pp.214-218, 19-23 Sep 1994

[3] Marantz, J.;, "Enhanced visibility and performance in functional verification by reconstruction," *Design Automation Conference, 1998. Proceedings*, pp. 164- 169, 15-19 Jun 1998

[4] Camera, K.; Brodersen, R.W.; , "An integrated debugging environment for FPGA computing platforms," *Field Programmable Logic and Applications, 2008. FPL 2008. International Conference on*, pp.311-316, 8-10 Sept. 2008

[5] Zan Yang; Byeong Min; Gwan Choi;, "Si-emulation: system verification using simulation and emulation," *Test Conference. Proceedings. International*, pp.160-169, 2000

[6] Chin-Lung Chuang; Wei-Hsiang Cheng; Liu, C.-N.J.; Dong-Jung Lu; , "Hybrid Approach to Faster Functional Verification with Full Visibility," *Design & Test of Computers, IEEE*, vol.24, no.2, pp.154-162, 2007

[7] Mark Weiser; "Program slicing". In *Proceedings of the 5th international conference on Software engineering* (ICSE '81). IEEE Press, Piscataway, NJ, USA, 439-449.

[8] Jen-Chieh Ou; Daniel G. Saab; Jacob A. Abraham; , "HDL Program Slicing to Reduce Bounded Model Checking Search Overhead," *Test Conference, 2006. ITC '06. IEEE International*, pp.1-7, Oct. 2006

[9] Vasudevan, S.; Viswanath, V.; Abraham, J.A.;, "Efficient Microprocessor Verification using Antecedent Conditioned Slicing," *VLSI Design, 2007., 20th International Conference on*, pp.43-49, 6-10 Jan. 2007

[10] Banerjee, S.; Gupta, T.;, "Automatic error recovery in targetless logic emulation," *Quality Electronic Design, 2009. ASQED 2009. 1st Asia Symposium on*, pp.380-384, 15-16 July 2009

2012 25th International Conference on VLSI Design

SCARE: Side-Channel Analysis based Reverse Engineering for Post-Silicon Validation

Xinmu Wang, Seetharam Narasimhan, Aswin Krishna, and Swarup Bhunia
Department of Electrical Engineering and Computer Science
Case Western Reserve University, Cleveland, OH 44106, USA
Email: xxw58@case.edu

Abstract—Reverse Engineering (RE) has been historically considered as a powerful approach to understand electronic hardware in order to gain competitive intelligence or accomplish piracy. In recent years, it has also been looked at as a way to authenticate hardware intellectual properties in the court of law. In this paper, we propose a beneficial role of RE in post-silicon validation of integrated circuits (IC) with respect to IC functionality, reliability and integrity. Unlike traditional destructive RE approaches, we propose a fast non-destructive side-channel analysis approach that can hierarchically extract structural information from an IC through its transient current signature. Such a top-down side-channel analysis approach is capable of reliably identifying pipeline stages and functional blocks. It is also suitable to distinguish sequential elements from combinational gates. For extraction of random logic structures (e.g. control blocks and finite state machines) we combine side-channel analysis with logic testing based Boolean function extraction. The proposed approach is amenable to automation, scalable, and can be applied as part of post-silicon validation process to verify that each IC implements exclusively the functionality described in the specification and is free from malicious modification or Trojan attacks. Simulation results on a pipelined DLX processor demonstrate the effectiveness of the proposed approach.

Index Terms—Reverse engineering, side-channel analysis, logic testing, self-referencing.

I. INTRODUCTION

Since the Cold War era, reverse engineering (RE) has been considered as a powerful tool to analyze electronic hardware for gaining competitive intelligence or for commercial piracy. Although regarded illegal in common belief, in most countries around the globe, RE is allowed for analysis, evaluation or teaching purposes [1]. In military and many mission-critical applications, RE can provide enabling technology for post-silicon validation of integrity and reliability of complex chips, which are designed and fabricated in untrusted environments [2]. For semiconductor industry, RE has become an attractive (and often, the only) option for claiming hardware Intellectual Property (IP) rights in the court of law. This requirement has led to the formation of a number of industrial entities, e.g. ChipWorks [3], dedicated to reverse engineering and the analysis of microchips and electronic systems.

In recent years, IC trust has emerged as a critical concern in semiconductor industry. Dictated by economic reasons, modern semiconductor design and fabrication flow involves third party IP cores, outsourced design and test services, as well as CAD tools supplied by third-party vendors. Lack of control on the design and fabrication steps greatly increases the vulnerability to malicious design modifications, called *hardware Trojan* attacks [4]. An attacker can mount these Trojan attacks to cause malfunction during field operation or leak secret information from inside a chip. Both side-channel and logic-testing based non-invasive approaches have been proposed earlier in the context of Trojan detection [5] when golden chip instances are available. However, due to untrusted fabrication facility in most cases, golden chips, which are needed to benchmark and detect compromised chips, are hard to achieve and demand reverse-engineering.

Image recognition based structural extraction involving de-packaging and de-layering an IC has been conventionally used as a reverse engineering approach [3]. Such a method is highly expensive, time-consuming, and destroys the chip. Since the chip "validated" in this way cannot function properly anyway, it can no longer be used as the benchmark for detecting other potentially compromised chips. On the other hand, some functional RE approaches have been investigated in recent years, e.g. [6] and [7]. Yet the complexity of logic testing approaches increases dramatically with the circuit size, especially in absence of full-scan testability in the design. More importantly, logic testing based approaches rely on random test vector generation, which can fail to detect extraneous undesired functions reliably if the functions are activated and observed only under rare conditions [4]. This implies that logic testing approaches aim to identify only the Boolean functions while considering the actual structural connectivity information transparent, which itself implies potential ignorance of design-parameter-violation-Trojans.

In this paper, we propose a top-down, hierarchical unified side-channel and logic testing approach that can extract both structural and functional information from a manufactured

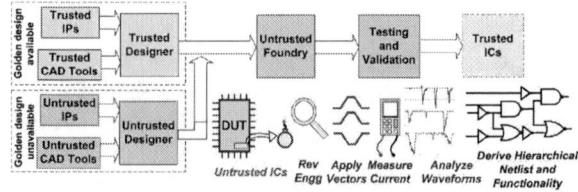

Fig. 1. Untrusted stages of the IC manufacturing flow. Steps of the proposed methodology to perform non-invasive RE and trust validation.

978-1-4673-0438-2/12 $31.00 © 2012 IEEE

304

IC. The method assumes the availability of a golden design (not golden chip instance), and can be extended to scenarios without the golden design. Fig. 1 illustrates the proposed top-down approach. This approach is valuable in two contexts:

(1) For validating a golden chip instance as Trojan detection benchmark, it is a significantly more low-cost, time-efficient and reliable choice compared to image recognition and logic testing based reverse-engineering approaches.

(2) When the method is considered directly as a Trojan detection approach, it is applicable to detecting comprehensive types of Trojans with no need of a golden chip by providing circuit structural information with the resolution of a single gate. Also, by using temporal and spatial self-referencing, this approach is invulnerable to significant process noise. Comparatively, conventional logic testing and side-channel based approaches are limited by their effective Trojan ranges, lower resolution, vulnerability to environmental noise, and need of a golden chip. When extended to no-golden-design scenarios, the proposed approach can depend only on the datasheet specifications to detect malicious hardware inserted in any stage of the design and fabrication flow. The hierarchical approach is scalable to large designs.

II. BACKGROUND

Malicious insertions (or *Hardware Trojans*, Fig. 2), are usually cleverly designed so as to be rarely triggered during normal operation. The reasons for the failure of logic testing based approach to detect hardware Trojans are as follows:

(1) Exhaustive enumeration is impractical for large designs, especially for sequential designs with/without scan-chains, creating chances of omitting rare events which trigger hardware Trojans. Fig. 2(a) shows an example of a combinational rarely-triggered Trojan, which can evade non-exhaustive testing.

(2) Trigger of sequential malicious insertions requires a sequence of unknown rare events, which can hardly be achieved even with exhaustive testing. An example of such sequential Trojan is given in Fig. 2(b), which cannot be triggered during one-time exhaustive enumeration. Moreover, state-elements in such sequential Trojans could use rare switching activity of internal circuit nodes as their clock signal, as illustrated in Fig. 2(c), which again lowers Trojan trigger possibility rendering them almost transparent in logic-based circuit extraction.

On the other hand, side-channel analysis based approaches [5] using transient current (IDDT), quiescent current

or path delay fingerprint have been proposed for Trojan detection in untrusted ICs. The main deterrent to such approaches is the large amount of process-induced parameter variations [8] which can mask the effect of malicious circuitry on the measured side-channel parameter. To overcome this drawback, various statistical techniques have been proposed to make process-invariant self-similarities in the design get reflected in the measured side-channel parameter such as transient supply current [9]. While logic values at the primary output reflect only the Boolean function with respect to the present state and primary inputs, the current waveform contains information about relative timing of different paths in the form of glitches, which can reveal significant information about internal structure, such as number of switching gates for particular vector pairs and their connectivity. Similarly, quiescent leakage current [10] contains information about all the gates in an IC, but it is difficult to observe the effect of small Trojans on the total leakage current, hence such methods have decreasing sensitivity for large designs. Therefore we choose an IDDT based side-channel approach.

III. METHODOLOGY

Transient current (IDDT) signature of an IC in response to input transitions contains structural information of an IC including connectivity and dependency among blocks. However, to identify structural blocks of an IC from its current signature, two major challenges have to be addressed: (1) avoid the aliasing effect due to simultaneous switching of multiple blocks; (2) eliminate the effect of process variations and measurement noise. We adopt a novel side-channel analysis approach, referred as self-referencing, which compares an IC with itself - either spatially between two or more regions or temporally between two time instances. The idea of spatial self-referencing can be explained using Fig. 3, which shows that the self-similarity of circuit blocks can be exploited hierarchically to identify constituent logic sub-blocks in structured logic. Similarly, temporal self-similarities in current signature are used to build a transient current *signature library* containing process and technology independent current signatures for each datapath block. The overall flow of the automated reverse engineering approach is illustrated in Fig. 4. Next, we describe key steps in detail with specific examples.

From the golden structural block diagram, functional blocks are defined along with their input/output dependencies. Next, functional vector sets are generated targeting activation of specific blocks [9]. In circuits with pipeline stages, temporal self-referencing can be used to restrict the switching activity to one stage by appropriate choice of vectors. Spatial self-

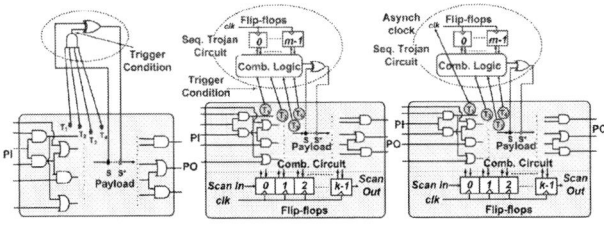

Fig. 2. Examples of different forms of malicious insertions: (a) Combinational Trojan. (b) Sequential Trojan using global clock. (c) Sequential Trojan using internal node activity as clock signal.

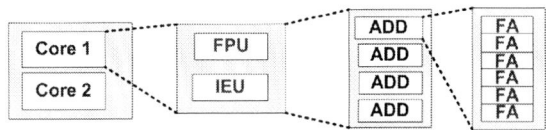

Fig. 3. Spatial self-referencing for identifying hierarchical functional blocks.

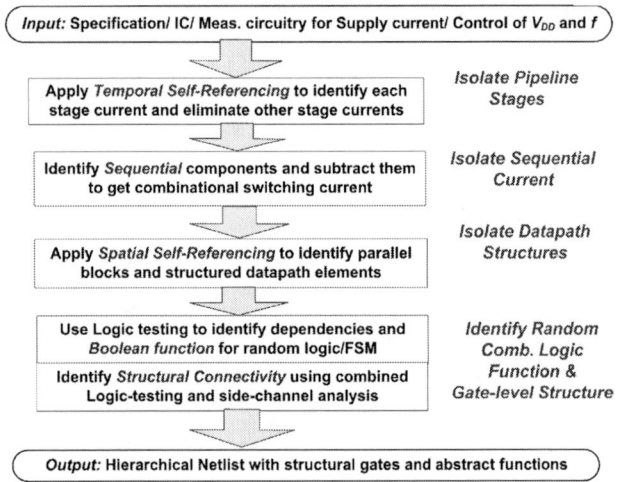

Fig. 4. Main steps of the proposed approach for IC reverse engineering.

referencing can also be used to identify parallel structural blocks and homogeneous array structures such as memory.

The next step is to isolate the sequential and combinational parts of switching current by using the correlations between the switching at the positive and negative edges of the clock. By using a slow clock, all the combinational switching can be confined to the positive half-cycle and the switching in the negative half-cycle corresponds only to the master stages of the flip-flops and clock coupling current. After the sequential current and the switching current during memory access has been subtracted from the stage current, the switching current caused by combinational circuit activity can be isolated out.

Due to their regular structures, standard datapath elements exhibit technology and process independent transient current features in response to specific input test patterns, which can be exploited to identify their specific types and implementation. A *signature library* based on relative features of transient current shapes, e.g. waveform correlation and number of observable ripples, is built after comprehensive characterization of different datapath elements and their standard

implementations. One can match the measured signature with macro-elements from the library to confirm the implementation specified in the golden netlist. Signature characterization is performed with the following perspectives:

(1) *Architecture-specific signature information*: One can sensitize different paths in a circuit which relate to some particular functional behaviors, and manifest information of structural features. For example, overall topology (e.g. flattened structure or blocked structure) of an adder can be revealed by transitions involving carry propagation.

(2) *Temporal self-referencing*: Transient current signatures can be obtained by comparing switching current for different transitions that trigger the same part of the circuit.

(3) *Spatial self-referencing*: Structural symmetry causes similar transient current for different transitions, helping in detection of repeated structures at high level (e.g. parallel structures) and low level (e.g. repeated full-adders in multi-bit adder).

Adder: Fig. 5 provides an instance demonstrating all the above three perspectives. Current waveforms for two test sets containing 3 vector pairs each are obtained.

(1) **Set S1** contains vectors to perform *single bit addition without carry propagation*. In particular, three vector pairs i, ii, and iii are used to perform single-bit addition at *bit0*, *bit1* and *bit8* and the current waveforms for two types of adders are shown in Fig. 5(a) and (b) for two technology nodes. Test vectors used on the Ripple Carry Adder (RCA) give closely matching current waveforms for all three vectors, implying that RCA contains a repeated bit-wise structure. In the case of Carry Save Adder (CSA), the shape of switching current for different operations depends on the relative bit position inside its block (4 bits are grouped as a block). This can be observed in Fig. 5(a), where current waveforms match for addition in the same relative positions inside each block. Besides, from Fig. 5(a) and (b) we can see the invariance in shape in terms of relative features across different technology nodes.

(2) **Set S2** consists of vectors to *activate carry propagation paths of different lengths* to explore self-similarity inside the adder architectures, by propagating the carry from the *carry-in bit*, *bit3* and *bit7*. In the top sub-figure of Fig. 5(c) we

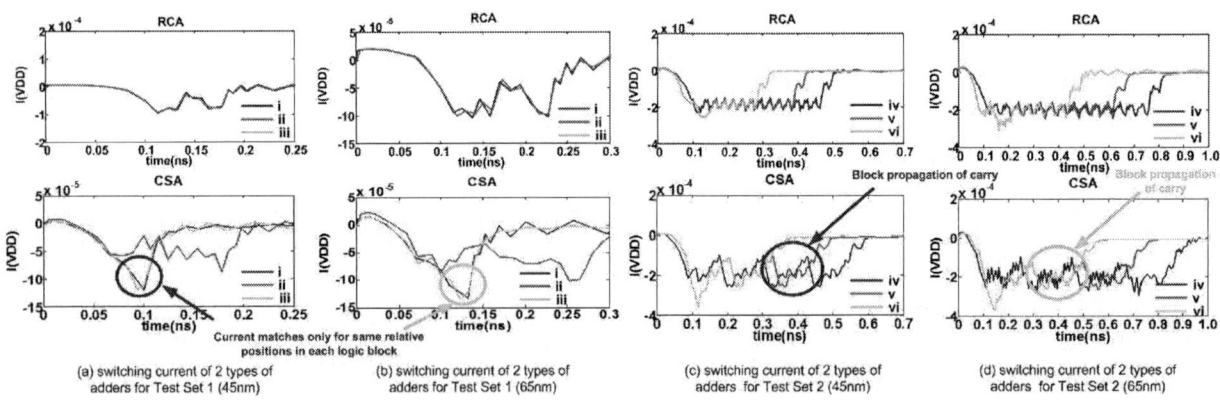

Fig. 5. Current signatures of RCA and CSA adders for 45nm and 65nm technology nodes used for self-referencing based reverse engineering.

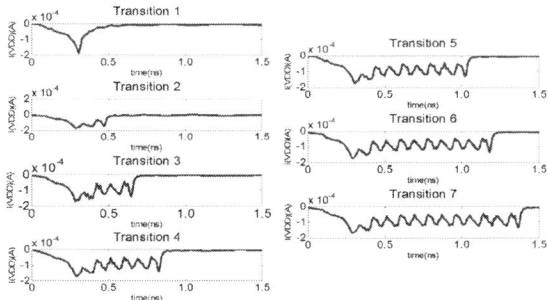

Fig. 6. Self-referencing current signatures of 8-bit Array Multiplier.

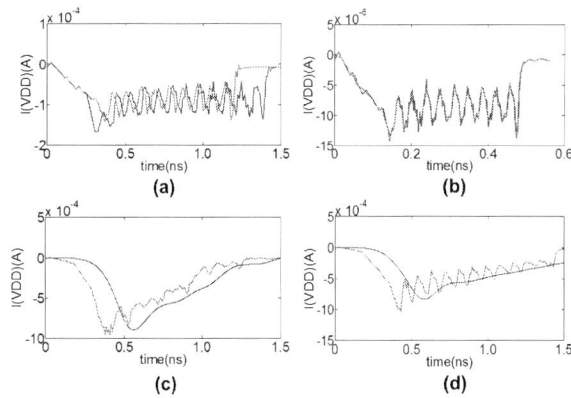

Fig. 7. Self-referencing current signatures of an 8-bit Wallace Tree Multiplier and the corresponding current of Array Multiplier for comparison.

can clearly see the rippling effect in supply current which indicates the carry propagating to the Most Significant Bit (MSB) for RCA. The overlapping current for the 3 vectors confirms the ripple propagation of the most significant 8 bits. For the blocked CSA, if the carry is at the input of a block, the triggered blocks have the same switching activity, forming the block propagation signature(red and blue traces in the center sub-figure of Fig. 5(c)). However, if generated inside a block, the block propagation waveform will only appear when the carry propagates to the next block (the green curve). Similar signatures can be derived at another technology node (65nm), as shown in Fig. 5(d), again confirming the technology and process independent nature of the signatures. Quantitatively, cross-correlation is performed between pairs of shapes to measure the similarity. Then the correlation values in response to different vector pairs are digitized and multiplied together to obtain the overall correlation with respect to one test set. Finally, signatures from all test sets collaboratively define the actual signature of a datapath element implementation.

Multiplier: Current signatures exploring structural self-similarity of multipliers can be obtained in a similar way. For 8-bit Array Multiplier, the following transitions are applied: **T1:** (0x02, 0x00)→(0x02, 0x01); **T2:** (0x04, 0x00)→(0x04, 0x01); **T3:** (0x08, 0x00)→(0x08, 0x01); **T4:** (0x10, 0x00) ↘(0x10, 0x01); **T5:** (0x20, 0x00)→(0x20, 0x01); **T6:** (0x40, 0x00)→(0x40, 0x01); **T7:** (0x80, 0x00)→(0x80, 0x01). The corresponding switching current is shown in Fig. 6. In each transition, only one partial product is made to be 1 and propagate to one primary output through a series of full adders. Regularly increasing number of ripples in the switching current indicates an array structure.

On the other hand, the structure of a Wallace Tree Multiplier (WTM) is relatively irregular. Test vector pairs T1, T2 and T3 are applied for triggering current signatures. In particular, T1 sensitizes the longest path with no carry propagation (Fig. 7(a) red curve), indicating a shorter path than that of an 8-bit array multiplier (Fig. 7(a) blue curve), thus implying WTM. T1 and T2 sensitize two different paths with exactly the same structure, which is specific to WTM. The identical waveforms form a signature verifying this self-similarity.
T1: (0x20, 0x00)→(0x20, 0x08); **T2:** (0x80, 0x00)→(0x80, 0x01); **T3:** (0x00, 0xff)→(0x80, 0xff). Another feature of

WTM is that the switching activities are more focused on the former levels compared to other types of multipliers to reduce the critical path delay, which is explored by T3. We first pre-process the current waveform by filtering out the high frequency components, then use a "normalized slope" of the rising part to represent the signature metric.
Metric 1: The ratio of the peak current value (I_{peak}) over that of the middle time point of the rising part of the switching current (I_{mid}). (WTM (Fig. 7(c)) > 2.3, Array multiplier (Fig. 7(d)) < 2)
Metric 2: The ratio of a normalized switching current amplitude over a normalized switching current duration. The former one is defined as (I_{peak} -I_{end})/I_0, whereas the latter one is T_{tran}/T_0. I_{end} is the current value of the last time point in the post-filtering waveform, T_{tran} is the switching duration of the real switching waveform, while I_0 and T_0 are the peak current and switching duration of a 1-bit full adder, which can be obtained from both multipliers by applying certain test vectors. (WTM > 3.2, Array multiplier < 2.3)

After obtaining datapath element structures, the remaining combinational logic is grouped as random logic with no pre-determined current signature. By applying test vectors to trigger each small group of gates, different gate-level transient current signatures can be obtained and compared with a pre-characterized signature library, e.g. trigger certain paths while setting other inputs to non-controlling values.

Scenarios Without A Golden Design: Unavailability of a golden design makes reverse engineering gate-level implementation of random logic to be a remarkably difficult task. Because there is no golden netlist to verify, test vector generation is not oriented. In this case, we adopt the approach described in the flow chart in Fig. 8. First the logic expression obtained from logic testing [7] is synthesized to a gate-level netlist. Then iterative side-channel based verification is performed based on this *initial guess*, during which the predicted netlist is updated with the confirmation or modification of each predicted gate. For each logic level, test vectors are intelligently generated to *focus the switching activity* on a small number of gates. Considering F=A&B′, the dual manners to implement

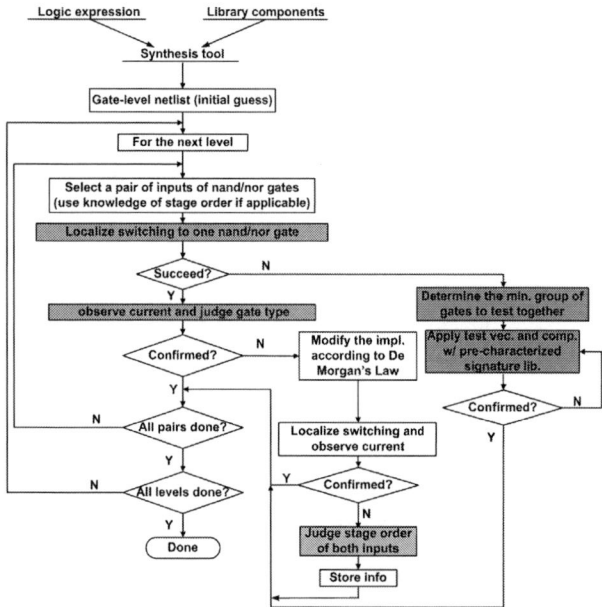

Fig. 8. Steps of random logic structure identification.

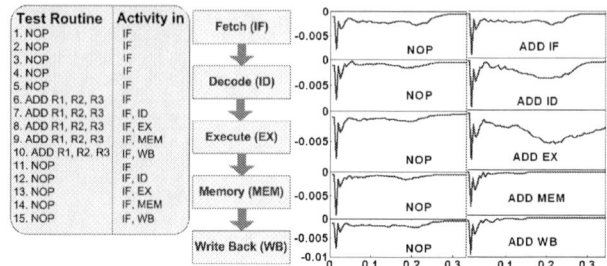

Fig. 10. Temporal self-referencing helps to identify the pipeline stage currents of a DLX processor.

this function using a reduced library is shown in Fig. 9(a) and (b). Considering Fig. 9(a) as the initial guess, to verify this gate, input B is set to the controlling value '1' while switching A. If the prediction is correct, switching of a single inverter should appear; while for the case of (b), no switching activity is expected. Repeating the test with A kept constant helps confirm that A and B are direct inputs of the gate. A case where B indirectly limits the switching caused by A for function F is given in Fig. 9(c). Here, both $0\rightarrow1$ and $1\rightarrow0$ at B would cause significant switching even if A is set to its controlling value.

However, in this step two exceptions might be encountered. First, if neither of the dual implementations can be confirmed, it implies mis-prediction of the existence of a gate; hence a different set of nodes have to be tried as the inputs. The other exception occurs when the switching activity cannot be limited to one NAND/NOR gate according to the predicted netlist, which could happen because of *shared input logic cone* that leads to loss of independent controllability of different gates. In this case test vectors are generated targeting multiple gates as a group, followed by a current signature comparison step.

The hierarchical top-down reverse engineering process, as described above, is very amenable to automation. The side-channel analysis steps at different levels of hierarchy can also be fully automated. However, the only step that requires

manual intervention (and hence can only be partially automated) is the high-level test generation based on functional specifications. This needs to be based on the functional block-diagram for a chip and can vary widely from design-to-design. The final result of the RE process is the complete gate-level implementation, along with hierarchical functional and structural description. Any undesired gate or function is easily identified as a malicious modification or Trojan circuit.

IV. CASE STUDY: DLX PROCESSOR

We perform the automated reverse engineering procedures on a 32-bit DLX processor to prove its effectiveness. All simulation results are obtained by performing HSPICE simulations in Predictive Technology Model (PTM) 70nm [11] technology.

1. Partitioning sequential space using Temporal Self-Referencing: By filling the processor pipeline with the same instruction, we can ensure that only one pipeline stage has switching activity in each clock cycle. Special instructions such as NOPs and JUMPs are used to characterize the background switching current of program counters and state transition of the pipeline stage control FFs. Once all the background current information is obtained, it is subtracted from the total current to focus on the individual pipeline stages such as Instruction Decode (ID), Execute (EX), Memory (MEM) and Write Back (WB). As shown in Fig. 10, the current signature for each stage corresponding to an ADD instruction is different from that for NOP, and the current for each stage has a unique signature in terms of peak current, delay and other transient current shape information.

2. Identifying and isolating sequential current component: As shown in Fig. **??**, for structured sequential cir-

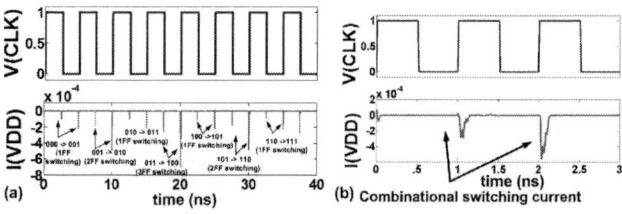

Fig. 9. An example of the verification unit: (a),(b) Dual implementations of function F=A&B'. (c) Here, B indirectly limits the switching caused by A.

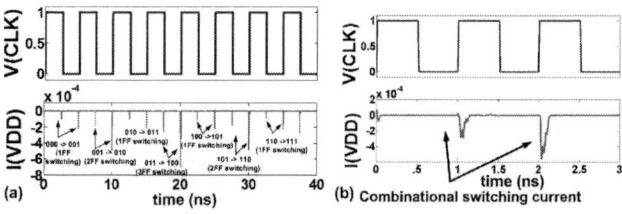

Fig. 11. Extraction of combinational logic current by subtracting sequential current component: (a) 3-bit binary counter shows the FF switching pattern of 1-2-1-3 which can be easily identified from the current at the positive or negative edge of CLK. (b) Extracting combinational current.

978-1-4673-0438-2/12 $31.00 © 2012 IEEE   308

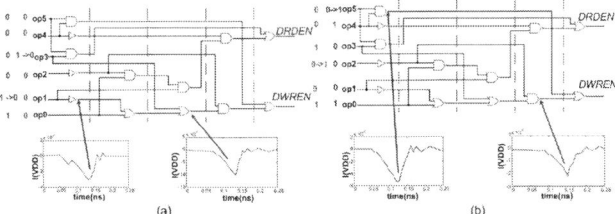

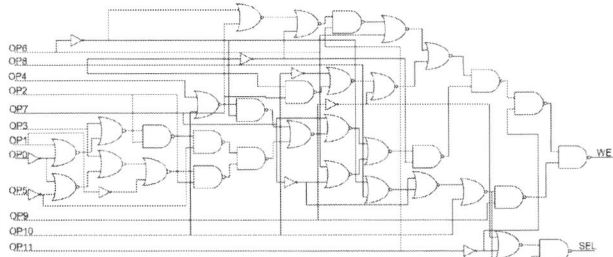

Fig. 12. Transient current signatures corresponding to specific vectors used to identify random logic structure isolated from the MEM stage of the DLX processor, with dependence on (a) a0, a1 and a3; and (b) a2, a4 and a5.

Fig. 13. Random logic structure of WB stage of the DLX processor.

cuits such as shift registers and counters, there are process-independent current signatures which are clearly identifiable and can be detected and eliminated. Similarly, memory access instructions such as LOAD/STORE can be used to find current specific to memory access circuitry. By careful selection of instructions, we can estimate width of memory, structure of address decoders and other peripheral logic, and timing of memory access relative to other operations.

3. Identifying datapath elements by Spatial/Temporal Self-Referencing: By exploring self-similarity of datapath elements using temporal/spatial self-referencing, we reverse engineer the implementation of the structured datapath elements. For example, we identified a CSA and a WTM in EX stage by applying vectors as described in Section III.

4. Identifying random logic and datapath sub-structure by combining side-channel analysis with logic testing: In this step, we successfully reverse engineer random logic in MEM and WB stages of the DLX processor after subtracting out background current due to other stages, the sequential current, and memory current. In MEM stage, two output logic cones structures *DRDEN* and *DWREN* with function are derived, where *DRDEN* is data read enable signal and *DWREN* is data write enable signal for memory access, respectively. The Boolean functions obtained from logic testing approach:
$DRDEN = op5 \; \& \; op4' \; \& \; op3 \; \& \; (op1' \,|\, op2' \; \& \; op0)$
$DWREN = op5 \; \& \; op4 \; \& \; op3' \; \& \; op2' \; \& \; (op1' \,|\, op0)$
Particularly, by switching different input bits of the MEM stage, we first figure that they are functionally dependent on input bits *op5*, *op4*, *op3*, *op2*, *op1*, *op0*. Then the Boolean equations are derived by applying exhaustive test vectors at these six inputs. Based on this, we obtained the predicted netlists using synthesis tool. After applying the verification

procedure, the actual circuits are found as illustrated in Fig. 12, in which some transient current waveforms are also shown to demonstrate the netlist verification process. Similarly, in WB stage, we reverse engineer the structure of logic for MUX select signal *SEL* and write-back enable signal *WE*. The schematics for the actual logic are illustrated in Fig. 13.

V. CONCLUSION

We have presented a novel reverse engineering based IC trust validation process which combined transient current based side-channel analysis with logic testing based function extraction. We have shown that RE can be used for trust validation in two scenarios: 1) when golden design is available; 2) without golden design (i.e. with functional specification only). Although we focus on using RE for trust validation, the process can also be adapted to improve the effectiveness of conventional manufacturing test. The validation steps can be easily automated to minimize the cost and time of trust validation. Since the technique works at multiple levels of hierarchy, it is easily scalable to large designs. The approach can work without scan, although presence of scan can be leveraged to improve the logic function extraction process. The proposed RE based trust validation can be used in conjunction with other existing protection approaches. For example, low-cost hardware Trojan detection approaches using static/transient current signature can be used for fast security screening of manufactured ICs, while the proposed approach can be used to increase the level of trust significantly. Future investigation would focus on developing an automation framework and validation with measurement results from commercial ICs.

ACKNOWLEDGMENT

We acknowledge the support from NSF grant CNS-1054744.

REFERENCES

[1] D. James. "Reverse engineering delivers product knowledge, aids technology spread". *Electronic Design*, 2006. [Online]. Available: http://electronicdesign.com/Articles/Index.cfm?AD=1&ArticleID=11966
[2] DARPA, "Integrity and Reliability of Integrated Circuits (IRIS)", 2010. [Online]. Available: https://www.fbo.gov/index?id=342ac5ed191ae7b8b03357fead590c4e
[3] Chipworks, Inc., "Semiconductor manufacturing - reverse engineering of semiconductor components, parts and process". [Online]. Available: http://www.chipworks.com
[4] R.S. Chakraborty, F. Wolff, S. Paul, C. Papachristou and S. Bhunia, "MERO: A statistical approach for hardware Trojan detection", *CHES Workshop*, 2009.
[5] M. Tehranipoor and F. Koushanfar. "A survey of hardware Trojan taxonomy and detection". *IEEE Design and Test of Computers*, 2010.
[6] M.C. Hansen, H. Yalcin, and J.P. Hayes. "Unveiling the ISCAS-85 Benchmarks: A case study in reverse engineering". *IEEE Design and Test of Computers*, vol. 16, no. 3, pp. 72-80, 1999.
[7] D.G. Saab, V. Nagabudi, F. Kocan, and J. Abraham. "Extraction based verification method for off the shelf Integrated Circuits". *ASQED*, 2009.
[8] S. Borkar *et al*, "Parameter variations and impact on circuits and micro-architecture", *DAC*, 2003.
[9] D. Du, S. Narasimhan, R.S. Chakraborty and S. Bhunia, "Self-referencing: A scalable side-channel approach for hardware Trojan detection", *CHES*, 2010.
[10] R. Rad, J. Plusquellic and M. Tehranipoor, "A sensitivity analysis of power signal methods for detecting hardware Trojans under real process and environmental conditions", *IEEE TVLSI*, 2010.
[11] Predictive Technology Model, [Online] http://www.eas.asu.edu/~ptm/

Kriging-Assisted Ultra-Fast Simulated-Annealing Optimization of a Clamped Bitline Sense Amplifier

Oghenekarho Okobiah[1], Saraju P. Mohanty[2], Elias Kougianos[3], and Oleg Garitselov[4]

NanoSystem Design Laboratory (NSDL, http://nsdl.cse.unt.edu)

University of North Texas, Denton, TX 76207, USA.

E-mail ID: oo0032@unt.edu[1], saraju.mohanty@unt.edu[2], eliask@unt.edu[3], and omg0006@unt.edu[4]

Abstract—Simulations using SPICE provide accurate design exploration but consume a considerable amount of time and can be infeasible for large circuits. The continued technology scaling requires that more circuit parameters are accounted for along with the process variation effects. Regression models have been widely researched and while they present an acceptable accuracy for simulation purposes, they fail to account for the strong correlation effect between parameters on the design. This paper presents an ultra-fast design-optimization flow that combines correlation-aware Kriging metamodels and a simulated annealing algorithm that operates on them. The Kriging-based method generates metamodels of a clamped bitline sense amplifier circuit which take into account the effects of correlation among the design and process parameters. A simulated annealing based optimization algorithm is used to optimize the circuit through the Kriging metamodel. The results show that the Kriging metamodels are very accurate with very low error. The optimization algorithm finds an optimized precharge time while keeping power consumption as constraint in an average execution time of 2.78 ms, as compared to a 45 minutes for an exhaustive search of the design space; i.e. close to $10^6 \times$ faster. To the best of the authors' knowledge this is the first paper that uses Kriging and simulated annealing for nano-CMOS design.

Keywords-Kriging Methods, Metamodeling, DRAM, Sense Amplifier, Fast Design Optimization, Simulated Annealing

I. INTRODUCTION AND CONTRIBUTIONS

Computer simulations for the design and optimization of analog/mixed-signal (AMS) circuits often consumes a considerable amount of time. The continued scaling and increasing complexity of nanoscale technology increases the number of design factors and process parameters that affect the performance of AMS circuits. In addition, the effects of process variation now has to be taken into consideration during the design process. These effects increase the already enormous time for an exhaustive simulation search of the design space and makes design optimization a very time consuming task. To increase the speed of design space exploration, designers resort to other alternatives such as interpolating functions, fast algorithms and metamodeling.

Metamodels are approximations of the behavior, output, or figure-of-merit (FoM), of a simulated design model in response to inputs or design parameters [1]. In essence, a metamodel is an abstraction of the design model itself. The use of metamodels in circuit design allows the designer to efficiently explore the design space. With metamodels, the time for design optimization is significantly reduced while providing a reasonably close output when compared to an exhaustive search for an optimal design. Commonly used metamodeling techniques include linear and low-order polynomial regressions [1], [2], [3], [4], and neural networks [5], [6], [7], [8], [9]. Interpolating functions, which include linear and low-order polynomial regression techniques, are one of the most popular methods used by designers. They provide an accurate description of the local design space but are not effective when applied globally [5], [10]. Regression based techniques assume that errors due to process variation across the design space are random, and they approximate this error equally over the surface points on the metamodel. For designs and processes in which the error due to variation is significantly correlated between the design parameters across the local and global space, regression based metamodels do not provide an accurate fit. The technology scale into deep nanometer regions significantly increases the correlation effects between parameters, hence there is a need for design methods which accurately capture and model these effects in the design process.

This paper presents a design methodology that uses a metamodeling technique based on Kriging prediction methods and uses a simulated annealing based optimization algorithm for design optimization. The Kriging based metamodel takes into consideration the error correlation between design inputs. Kriging prediction techniques were originally used in the geostatistics field and have now been used in other fields [2], [11], [12] and only recently in VLSI [13]. In generating the metamodel, the Kriging technique predicts responses based on regression with observed data from surrounding data points. This differs from conventional regression techniques because for each predicted point, a new set of weights is calculated based on the correlations and variance of the design points in the local space. As a case study, a Kriging based metamodel is generated for a clamped bitline sense amplifier. The generated metamodel is then optimized using a simulated annealing based optimization algorithm. This methodology improves process aware design optimization reducing computational expense while providing an optimized result.

The **novel contribution of this paper** is a fast Kriging based metamodel design flow which is optimized with a simulated annealing based algorithm.

The rest of this paper is organized as follows. A brief discus-

[0]This research is supported in part by NSF awards CNS-0854182 and DUE-0942629 and SRC award P10883.

978-1-4673-0438-2/12 $31.00 © 2012 IEEE

sion of selected related research is presented in Section II. The proposed design methodology is introduced in Section III. A brief background and fundamentals of Kriging metamodels are presented in Section IV. Section V briefly describes the design and characterization of the clamped bitline sense amplifier. Section VI describes the Gaussian Kriging based metamodels used in this work. Section VIII presents the conclusion and future research directions.

II. RELATED PRIOR RESEARCH

The use of metamodels for design simulation has been well researched. The most popular metamodeling technique has been the low order polynomial regression technique [12], [4], [1], [3]. In [4], a comparison of different sampling techniques used for metamodel creation is presented. While low order polynomial regression techniques are capable of generating accurate models for local optimizations, they are not very accurate in a global design space [5], [9], [14]. The weighting systems used in regression techniques are independent and are averaged over the design space. This fails to account for the spatial autocorrelation effects between input design variables. In [14] circuit designs are expressed as equations in polynomial forms. These circuit equations are reduced to form convex problems which are solved by geometric programming. This method ensures global optimization but does not result in accurate surfaces due to approximations for the circuit equations.

Neural networks (NN) have also been used to generate metamodels which have been shown to outperform regression techniques [8], [9], [7], [6]. Neural networks use a learning approach to train and adjust the weights in developing metamodels for the underlying design system. In [8], [7], well known simulation problems are used to test the accuracy of NN metamodels. Optimal metamodel generation based on neural networks is still researched actively particularly for determining the optimal network structures and the application of neural network metamodeling for point targets.

III. THE PROPOSED KRIGING-ASSISTED ACCURATE AND ULTRA-FAST DESIGN OPTIMIZATION METHODOLOGY

Computationally intensive simulations are very expensive. To reduce this cost, metamodels are generated to aid the design process and its optimization. Commonly used metamodel functions do not take into account the error correlation between design parameters, which is increasingly becoming significant in the deep nanometer technology range. A new methodology, Kriging Assisted Ultra-Fast Simulated Annealing Optimization design flow is proposed. Kriging techniques take into consideration the error correlation effects between design parameters. The generated metamodel is then optimized using a simulated annealing based algorithm. The methodology is incorporated in the design flow shown in Fig. 1. The design flow can be broken into 4 steps as described below.

A. Design and Netlist Optimization

The first step in the design flow is to create a model of the circuit design that meets the design specifications. The circuit

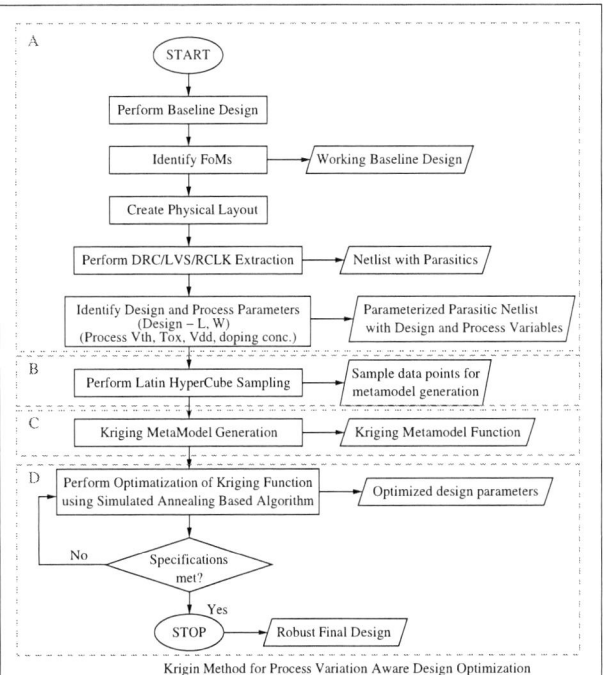

Fig. 1.   The proposed Kriging assisted ultra-fast design optimization flow.

schematic is drawn and simulated using a CAD tool. After the design is verified for key performance characteristics, the physical layout design is created using Design Rule Checks (DRC) as a guide. Once the DRC is complete, a layout vs. schematic (LVS) verification is also completed to ensure that the physical design matches the circuit schematic. A parasitic netlist, including resistance, capacitance and self and mutual inductance (RCLK) is then extracted from the physical design and used for further simulations to give a more accurate description of the design. The design and process parameters are identified in the netlist which is then parameterized and used for sample point generations in the next step. In this flow, the design parameters chosen are the transistor gate length L and width W. For process parameters, threshold voltage (V_{th}), oxide thickness (T_{ox}), supply voltage (V_{DD}), and doping concentration are considered.

B. Latin HyperCube Sampling (LHS)

Latin Hypercube Sampling (LHS) techniques are one of the commonly used methods for generating sample data points for Kriging based metamodels. LHS generates n random sample points based on a range of specified inputs. The LHS technique divides the input range into n intervals of equal length, from which it randomly selects points from each interval such that the interval appears once in each row and column of a design matrix. Data points may be selected uniformly, randomly, from midpoints or in any distribution form in each interval. When the distribution used to sample points from each interval is the midpoint, the technique is called Middle Latin HyperCube Sampling (MLHS). The design points L and W are used as the sampling corners while the process parameters are varied to model the effects of process variation.

C. Kriging Based MetaModel Generation

The sample design points generated by LHS are used with the Kriging based algorithm to generate the metamodel surface. The Kriging technique generates predicted output response points of design inputs based on observations from the sampled data. The generated metamodel is a function of the design parameters L and W, and process parameters. Two Kriging methods, ordinary Kriging and simple Kriging, are used to generate metamodels for each of the FoMs (precharge time T_{PC}, sense delay T_{SD}, and sense margin V_{SM}), of the clamped bitline sense amplifier. A total of 8 metamodels are generated and are compared to an accurate model generated by exhaustive simulation.

D. Optimization using Simulated Annealing Based Algorithm

A simulated annealing based algorithm is used to optimize the Kriging metamodels. The metamodels can be optimized for each of the identified FoMs. In this paper, the precharge time (T_{PC}) is used as the objective while the average power consumption (P_{SA}) is used as a design constraint.

IV. FUNDAMENTALS OF KRIGING METAMODELS

Kriging methods were originally proposed in the early 1950's by Daniel Krige (hence the term "Kriging") for use in geostatistical methods. Its application has since spread into many other fields. The fundamental idea behind Kriging is that the predicted outputs are weighted averages of sampled data. The weights are unique to each predicted point and are a function of the the distance between the point to be predicted and observed points. The weights are chosen so that the prediction variance is minimized [15], [2].

The general expression of a Kriging model is as follows:

$$ y(\mathbf{x_0}) = \sum_{j=1}^{L} \lambda_j B_j(\mathbf{x}) + z(\mathbf{x}), \qquad (1) $$

where $y(\mathbf{x_0})$, is the predicted response at design point $(\mathbf{x_0})$ $\{B_j(\mathbf{x}), j = 1, \cdots, L\}$ is a specific set of basic functions over the design domain D_N, λ_j are fitting coefficients (also known as weights) to be determined and $z(\mathbf{x})$ is the random error. Kriging differs from common least squares based approaches in that $z(\mathbf{x})$ is assumed to be a random process and not independent, unique to each weight and not distributed identically. It is assumed that the process has a known mean, variance σ^2, and correlation function. The correlation function, called the *variogram* in geophysics, is expressed as follows:

$$ r(\mathbf{s}, \mathbf{t}) = \text{Corr}(z(\mathbf{s}), z(\mathbf{t})). \qquad (2) $$

The variogram is used to derive the Kriging weights, λ_j. The autocorrelation of the design points is characterized by the covariance function [16]. The weights are chosen so that the Kriging variance is minimized. There are different variations of Kriging models. Two methods explored in this paper are the ordinary and simple Kriging techniques. Ordinary Kriging

assumes a mean that is constant in the local domain of a predicted point, while simple Kriging assumes a constant and known mean over the global domain.

For ordinary Kriging techniques, the weights are chosen to minimize the Kriging variance under the unbiasedness constraint that $E(\widehat{Z}(x) - Z(x)) = 0$. Hence the weights are chosen so that the following expression is satisfied:

$$ \sum_{j=1}^{n} \lambda_j = 1. \qquad (3) $$

This condition is not required for simple Kriging. The weights then for ordinary Kriging are given by the following:

$$ \begin{pmatrix} \lambda_1 \\ \vdots \\ \lambda_n \\ \mu \end{pmatrix} = \Gamma^{-1} \begin{pmatrix} \gamma(e_1, e_0) \\ \vdots \\ \gamma(e_n, e_0) \\ 1 \end{pmatrix}, \qquad (4) $$

where μ is a Lagrange multiplier used to ensure equation (3). Γ is the covariance matrix of the observed points and for ordinary Kriging is given by:

$$ \Gamma = \begin{pmatrix} \gamma(e_1, e_1) & \cdots & \gamma(e_1, e_n) & 1 \\ \vdots & \ddots & \vdots & 1 \\ \gamma(e_n, e_1) & \cdots & \gamma(e_n, e_n) & 1 \\ 1 & 1 & 1 & 0 \end{pmatrix}, \qquad (5) $$

where

$$ \gamma(e_1, e_2) = E\left(|z(e_1) - z(e_2)|^2\right). \qquad (6) $$

The last row and column are absent in simple Kriging method.

V. THE 45 NM CLAMPED BITLINE SENSE AMPLIFIER: A CASE STUDY CIRCUIT

A. The Clamped Bitline Sense Amplifier Circuit Design

The clamped bitline sense amplifier is a variation of the conventional sense amplifier used in DRAMs. The advantage of the clamped bitline is that it is clamped to a stable voltage after a sensing operation. This reduces the capacitive effect of the bitlines during the sensing operation, hence resulting in a decreased dynamic power consumption and sense delay time [17], [18]. Fig. 2(a) shows the circuit schematic design of the clamped bitline sense amplifier. Transistors MP1, MP2, MN1 and MN2 form the cross-coupled inverters, while transistors MN3 and MN4 provide a low impedance between the bitlines through V_{CLAMP}.

The initial design parameters for the transistors are length L_n, L_p = 45 nm, width W_n = 120 nm, and W_p = 240 nm. These dimensions are based on the nominal 45 nm technology node values and similar designs in [19]. The clamped bitline sense amplifier needs matched transistors for optimal performance, making it a good test circuit to model the effects of process variation. The physical layout design is shown in Fig. 2(b). The extracted SPICE netlist from the layout includes the parasitics of the design which impact the its performance as seen in Table I.

978-1-4673-0438-2/12 $31.00 © 2012 IEEE

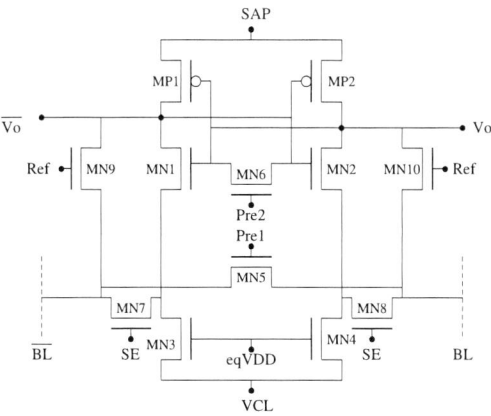

(a) Schematic design.

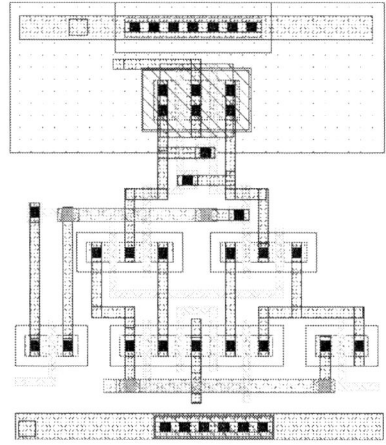

(b) Physical design.

Fig. 2. Circuit and layout for the clamped bitline sense amplifier.

B. Characterization of 45 nm Clamped Bitline Sense Amplifier

In characterizing the performance of the sense amplifier design, the following figures of merit (FoM) were selected based on previous publication [19].

Precharge and Voltage Equalization Time is the time required to equally precharge both bitlines BL and $\overline{BL}$. This reduces power consumption during the sense operation by reducing the voltage swing. The capacitance of the bitline significantly affects the precharge time.

Power Consumption is the average power consumed by the clamped bitline sense amplifier. The average power measured includes dynamic power, subthreshold leakage and gate oxide leakages. With technology scaling now in the deep nanometer regions, the leakage power components now contribute significantly to power consumption.

Sense Delay is the minimum amount of time required for sufficient voltage to appear on the bitlines that can be correctly

detected by the sense amplifiers. The cell data value affect the sense delay. The impact of the bitline capacitance on the sense delay is reduced by the design of the clamped bitline.

Sense Margin is the minimum voltage that can be correctly detected by the clamped bitline.

The circuit schematic and the physical design were both simulated for verification and characterization. The performance was characterized based on the selected FoMs. Table I shows a summary of these values. The last column also shows the area of the physical design.

VI. Kriging Metamodeling of the Clamped Bitline Sense Amplifier

A. Kriging Model Generation for the FoMs

The extracted netlist from the physical layout is parameterized and used to generate sample data points using the LHS technique. Two Kriging methods are used to generate the metamodels: (1) Simple Kriging and (2) Ordinary Kriging. As discussed in section III-C, each Kriging predicted point is calculated with a different weight. The weights are based on the empirical semivariogram. Hence, the covariance functions were determined to obtain the spatial autocorrelation of the design parameters. For this paper, to simplify the analysis, only W_n has been used as a design parameter. A parametric analysis varying W_n and W_p shows that the FoMs are dominated by W_n. The topology of the circuit supports this trend: there are 10 NMOS transistors compared to 2 PMOS transistors. The use of only W_n has been used to illustrate the proposed methodology and in future work, the approach will be extended to designs with multiple design parameters.

The empirical variogram is estimated from the created variogram. It is then fitted with the theoretical spherical model, which was the best fit for the sampled data points. Each FoM can be expressed based on the general form of the Kriging function. For example, the predicted precharge time $\widehat{Y}_{pr}$ at an unknown design point W_n^* is expressed as:

$$\widehat{Y}_{pr}\left(W_n^*\right) = \sum_{i=1}^{N} \lambda\left(W_n^*\right)_i Y_{pr}\left(W_{n_i}\right), \qquad (7)$$

where $Y_{pr}(W_{n_i})$ are the observed precharge values for the given N W_{n_i} $(i = 1, 2, \ldots, N)$ sample points. The weights $\lambda(W_n^*)$ are unique for each predicted point W_n^* and are calculated from Eqn. (4). Using similar equations, the values for the other FoMs of the sense amplifier are predicted.

B. Kriging Metamodels and Accuracy Analysis

The generated metamodels for the FoMs are presented in this section. An exhaustive baseline simulation was also done to compare the accuracy of the Kriging predicted models. A total of 1000 design points were simulated to densely capture the design space compared to the 20 and 100 LHS points used to generate the Kriging surfaces.

The predicted curves for the ordinary Kriging based metamodels are shown in Fig. 3 with W_n as the design input. The results for simple Kriging are very similar and are omitted due

978-1-4673-0438-2/12 $31.00 © 2012 IEEE

TABLE I
FIGURES OF MERIT OF THE OPTIMAL CLAMPED BITLINE SENSE AMPLFIER.

Design	Precharge time, T_{PC} (ns)	Sense delay, T_{SD} (ns)	Power, P_{SA} (μW)	Sense Margin, V_{SM} (mV)	Area μm^2
Schematic	10.31	1.79	1.84	26.91	-
Layout	10.40	1.91	1.88	26.86	6.045
Optimized	8.16	1.68	1.98	28.03	6.356
Change	21.54 %	12.04 %	-5.32 %	-4.36 %	5.15 %

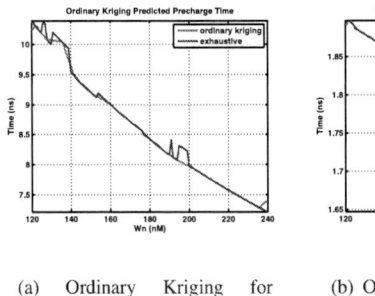

(a) Ordinary Kriging for precharge time

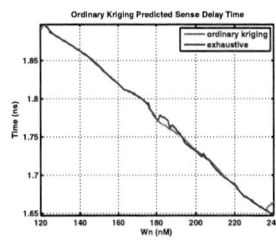

(b) Ordinary Kriging for sense delay

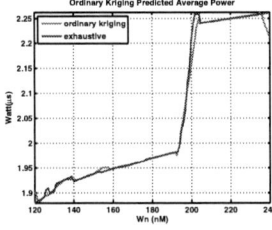

(c) Ordinary Kriging for average power

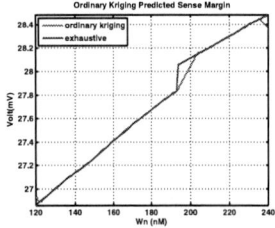

(d) Ordinary Kriging for sense margin

Fig. 3. Ordinary Kriging responses using W_n as the design parameter.

TABLE II
STATISTICAL ANALYSIS OF THE KRIGING PREDICTED VALUES.

FoMs	Ordinary Kriging		Simple Kriging	
Samples	20	100	20	100
Precharge				
MSE	6.02×10^{-21}	3.85×10^{-19}	5.32×10^{-21}	3.63×10^{-19}
$RMSE$	7.76×10^{-11}	6.20×10^{-10}	7.29×10^{-11}	6.02×10^{-10}
R^2	0.9931	0.5560	0.9939	0.5810
STD	6.95×10^{-11}	6.09×10^{-10}	6.60×10^{-11}	5.91×10^{-10}
Sense Delay				
MSE	1.12×10^{-23}	8.27×10^{-24}	7.49×10^{-24}	4.02×10^{-24}
$RMSE$	1.02×10^{-10}	2.88×10^{-12}	2.73×10^{-12}	2.00×10^{-12}
R^2	0.9984	0.9985	0.9987	0.9993
STD	8.62×10^{-11}	2.64×10^{-12}	2.29×10^{-12}	1.79×10^{-12}
Power				
MSE	3.64×10^{-15}	4.35×10^{-15}	3.56×10^{-15}	4.69×10^{-15}
$RMSE$	6.24×10^{-11}	6.60×10^{-08}	5.96×10^{-08}	6.85×10^{-08}
R^2	0.9957	0.8145	0.8486	0.8003
STD	5.75×10^{-11}	6.40×10^{-08}	5.69×10^{-08}	6.66×10^{-08}
Sense Margin				
MSE	2.79×10^{-09}	6.31×10^{-09}	2.56×10^{-09}	4.32×10^{-09}
$RMSE$	5.28×10^{-05}	7.94×10^{-05}	5.06×10^{-05}	6.57×10^{-05}
R^2	0.9987	0.9753	0.9900	0.9831
STD	2.58×10^{-05}	7.73×10^{-05}	4.79×10^{-05}	6.41×10^{-05}

to space constraints. The plots also show the exhaustive design points simulations. From the plots it is seen that the predicted Kriging metamodels for both the ordinary and simple Kriging techniques closely match the exhaustive simulation.

A statistical analysis on both responses shows that the accuracy of the Kriging method is very high. A summary of the statistical analysis is shown in Table II for both ordinary and simple Kriging metamodels compared to the exhaustive design surface. The metrics used for comparison are the Mean Square Error (MSE), the Root Mean Square Error (RMSE) and the correlation coefficient R^2:

$$RMSE = \sqrt{\frac{1}{N} \sum_{i=1}^{N} \left(Y_i - \widehat{Y}_i \right)^2}, \qquad (8)$$

where N is the number of design points predicted.

From an analysis of the results in Table II the predicted points have an average R^2 of 0.99. The simulation time for the generation of the metamodels was 3 mins compared to 72 hrs used for exhaustive simulation.

C. Experimental Setup

The Cadence virtuoso platform was used for the initial circuit schematic design and the physical layout. The extracted and parameterized netlists were used to write Ocean Scripts that were used to run the exhaustive simulation and gather LHS sample data points. The Spectre analog simulator was used to perform the simulations. The algorithm used to generate the Kriging metamodels was written using MATLAB with the help of the toolboxes mGstat [20] and SUMO [21].

VII. SIMULATED ANNEALING BASED OPTIMIZATION

Simulated annealing optimization is based on the Monte Carlo algorithm and was originally used to simulate the annealing process used in metallurgy. This gives the simulated annealing algorithm random characteristics. Successive runs of the algorithm will produce different results. The optimization steps are presented in Algorithm 1.

The algorithm takes random walks through the design space starting from the middle point of each design parameter, looking for points with low energies. In each step, the probability of taking a step is determined by the Boltzmann distribution, $p = \left(e^{\frac{\Delta_{T_{PC}}}{T}} \right)$ if $\Delta_{T_{PC}}$ is high, and p = 1 when $\Delta_{T_{PC}}$ is low. Therefore a step will occur if a new value is better than the previous one. If the new value is worse, the transition can still

---

Algorithm 1 Simulated-Annealing Based Optimization of the Clamped-Bitline Sense Amplifier.

1: Initialize iteration counter: $counter \leftarrow 0$.
2: Initialize temperature Θ.
3: Initialize $Cooling\_Rate$.
4: Start with an initial solution $\widehat{CBSA_i}$.
5: Calculate the FoMs for $\widehat{CBSA_i}$ using the Kriging models.
6: Consider the objective of interest T_{PC_i}.
7: $result \leftarrow \Delta_{T_{PC}} \leftarrow T_{PC_i}$.
8: **while** ($\Delta_{T_{PC}}! = 0$) **do**
9: $\quad counter \leftarrow max\_Iteration$.
10: $\quad$ **while** ($counter > 0$) **do**
11: $\quad\quad$ Generate random transition from solution $\widehat{CBSA_i}$ to $\widehat{CBSA_j}$.
12: $\quad\quad$ Calculate the FoMs for $\widehat{CBSA_j}$ using the Kriging models.
13: $\quad\quad$ **if** ($T_{PC_j} < result$) **then**
14: $\quad\quad\quad result \leftarrow T_{PC_j}$.
15: $\quad\quad\quad \widehat{CBSA_i} \leftarrow \widehat{CBSA_j}$.
16: $\quad\quad$ **else**
17: $\quad\quad\quad \Delta_{T_{PC}} \leftarrow T_{PC_i} - T_{PC_j}$.
18: $\quad\quad\quad$ **if** ($\Delta_{T_{PC}} < 0$, random$(0,1) < e^{\frac{\Delta_{T_{PC}}}{T}}$) **then**
19: $\quad\quad\quad\quad T_{PC_i} \leftarrow T_{PC_j}$.
20: $\quad\quad\quad\quad \widehat{CBSA_i} \leftarrow \widehat{CBSA_j}$.
21: $\quad\quad\quad$ **end if**
22: $\quad\quad$ **end if**
23: $\quad\quad counter \leftarrow counter - 1$.
24: $\quad$ **end while**
25: $\quad \Theta \leftarrow \Theta \times Cooling\_Rate$.
26: **end while**
27: **return** $result$ and $\widehat{CBSA_i}$.

---

occur, and its likelihood is proportional to the temperature T and inversely proportional to $\Delta_{T_{PC_i}}$.

The finalized values for the design are shown in Table I. T_{PC} has been reduced by 21.54 % while P_{SA} was increased by 5.32 %. T_{SD} and V_{SM} was also improved by 12.04 % and 4. 36%, respectively. The area for the final layout design was also increased by 5.15%. The simulated annealing based algorithm finds optimized values in 2.78 ms compared to a run of 45 minutes for an exhaustive search optimization. In other words, the proposed design flow could speedup the optimization process by a factor approximately $10^6 \times$.

VIII. CONCLUSIONS AND FUTURE RESEARCH

This paper presented a new methodology that uses Kriging metamodels and the simulated annealing algorithm for sense amplifier optimization. Kriging methods generate metamodel functions that accurately capture the global design space while taking into account the spatial autocorrelation of the input design parameters. Comparisons with exhaustive simulations show that Kriging predicted models are very accurate with very low RMSE and high R^2. The simulated annealing based algorithm optimized the generated metamodel function for

the precharge T_{PC} FoM, improving it by 21.54%. In future research, the methodology will be extended to multiple design parameters and multi-objective optimization algorithms.

REFERENCES

[1] W. E. Biles, J. P. C. Kleijnen, W. C. M. van Beers, and I. van Nieuwenhuyse, "Kriging Metamodeling in Constrained Simulation Optimization: An Explorative Study," in *Proceedings of the 39th Winter Simulation Conference*, 2007, pp. 355–362.
[2] W. Van Beers, "Kriging Metamodeling in Discrete-Event Simulation: An Overview," in *Proceedings of the Winter Simulation Conference*, 2005, pp. 202–208.
[3] G. Dellino, J. Kleijnen, and C. Meloni, "Robust Simulation-Optimization using Metamodels," in *Proceedings of the Winter Simulation Conference (WSC)*, Dec. 2009, pp. 540–550.
[4] O. Garitselov, S. Mohanty, E. Kougianos, and P. Patra, "Nano-CMOS Mixed-Signal Circuit Metamodeling Techniques: A Comparative Study," in *Proceedings of the International Symposium on Electronic System Design (ISED)*, 2010, pp. 191–196.
[5] B. Ankenman, B. Nelson, and J. Staum, "Stochastic Kriging for Simulation Metamodeling," in *Proceedings of the Winter Simulation Conference*, Dec. 2008, pp. 362–370.
[6] R. A. Kilmer, A. E. Smith, and L. J. Shuman, "Computing Confidence Intervals for Stochastic Simulation Using Neural Network Metamodels," *Computers and Industrial Engineering*, vol. 36, no. 2, pp. 391–407, 1999.
[7] I. Sabuncuoglu and S. Touhami, "Simulation Metamodelling With Neural Networks: An Experimental Investigation," *International Journal of Production Research,*, vol. 40, no. 11, pp. 2483–2505, 2002.
[8] C. W. Zobel and K. B. Keeling, "Neural Network-based Simulation Metamodels for Predicting Probability Distributions," *Computers and Industrial Engineering*, vol. 54, pp. 879–888, May 2008.
[9] A. Khosravi, S. Nahavandi, and D. Creighton, "Developing Optimal Neural Network Metamodels Based on Prediction Intervals," in *International Joint Conference on Neural Networks*, June 2009, pp. 1583–1589.
[10] J. Staum, "Better Simulation Metamodeling: The Why, What, and How of Stochastic Kriging," in *Proceedings of the Winter Simulation Conference (WSC)*, Dec. 2009, pp. 119–133.
[11] B. Harrington, Y. Huang, J. Yang, and X. Li, "Energy-Efficient Map Interpolation for Sensor Fields Using Kriging," *IEEE Transactions on Mobile Computing*, vol. 8, no. 5, pp. 622–635, May 2009.
[12] M. Zakerifar, W. Biles, and G. Evans, "Kriging Metamodeling in Multi-objective Simulation Optimization," in *Proceedings of the Winter Simulation Conference (WSC)*, 2009, pp. 2115–2122.
[13] H. You, M. Yang, D. Wang, and X. Jia, "Kriging Model Combined with Latin Hypercube Sampling for Surrogate Modeling of Analog Integrated Circuit Performance," in *Proceedings of the International Symposium on Quality of Electronic Design*, 2009, pp. 554–558.
[14] V. Aggarwal, "Analog Circuit Optimization using Evolutionary Algorithms and Convex Optimization," Master's thesis, Massachusetts Institute of Technology, May 2007.
[15] N. A. C. Cressie, *Statistics for Spatial Data*. New York: Wiley, 1993.
[16] G. Bohling, "Kriging," Kansas Geological Survey, Tech. Rep., 2005.
[17] I. Arsovski, "High-Speed Low-Power Sense Amplifier Design," University of Toronto, Tech. Rep., 2001.
[18] T. Blalock and R. Jaeger, "A Subnanosecond Clamped-Bit-Line Sense amplifier for 1T Dynamic RAMs," in *Proceedings of the International Symposium on VLSI Technology, Systems, and Applications*, 1991, pp. 82–86.
[19] O. Okobiah, S. P. Mohanty, E. Kougianos, and M. Poolakkaparambil, "Towards Robust Nano-CMOS Sense amplifier Design: A Dual-Threshold versus Dual-Oxide Perspective," in *Proceedings of the 21st ACM Great Lakes Symposium on VLSI*, 2011, pp. 145–150.
[20] *mGstat: A Geostatistical Matlab Toolbox*. [Online]. Available: mgstat. sourcefourge.net
[21] D. Gorissen, I. Couckuyt, P. Demeester, T. Dhaene, and K. Crombecq, "A Surrogate Modeling and Adaptive Sampling Toolbox for Computer Based Design," *J. Mach. Learn. Res.*, vol. 11, pp. 2051–2055, August 2010.

978-1-4673-0438-2/12 $31.00 © 2012 IEEE

Fast-Accurate Non-Polynomial Metamodeling for nano-CMOS PLL Design Optimization

Oleg Garitselov[1], Saraju P. Mohanty[2], and Elias Kougianos[3]
Nano-Systems Design Laboratory (http://nsdl.cse.unt.edu)[1,2,3]
Department of Computer Science and Engineering [1,2]
Department of Engineering Technology [3]
University of North Texas, Denton, USA.[1,2,3]
Email-ID: omg0006@unt.edu[1], saraju.mohanty@unt.edu[2], and eliask@unt.edu[3]

Abstract—At the nanoscale domain, the simulation, design, and optimization time of the circuits have increased significantly due to high-integration density, increasing technology constraints, and complex device models. This necessitates fast design space exploration techniques to meet the shorter time to market driven by consumer electronics. This paper presents non-polynomial metamodels (surrogate models) using neural networks to reduce the design optimization time of complex nano-CMOS circuit with no sacrifice on accuracy. The physical design aware neural networks are trained and used as metamodels to predict frequency, locking time, and power of a PLL circuit. Different architectures for neural networks are compared with traditional polynomial functions that have been generated for the same circuit characteristics. Thorough experimental results show that only 100 sample points are sufficient for neural networks to predict the output of circuits with 21 design parameters within 3% accuracy, which improves the accuracy by 56% over polynomial metamodels. The generated metamodels are used to perform optimization of the PLL using a bee colony algorithm. It is observed that the non-polynomial (using neural networks) metamodels achieve more accurate results than polynomial metamodels in shorter optimization time.

Index Terms—**Metamodeling, Neural Networks, Nano-CMOS, PLL, Polynomial, Modeling, Circuit Optimization**

I. INTRODUCTION

The design constraints on the designer with competitive time to market discourages the use of slow exhaustive design space exploration to reach fully optimal performance. At the same time accurate circuit level, full-blown parasitic netlist based design optimization may be intractable for current nanoscale high-density complex circuits. In many cases optimization is very sub-optimal and the design is within certain design margins. For nanoscale circuits, the simulation, design, and optimization time has increased significantly due to high-integration density, increasing technology constraints, and complex device models. It is especially true for analog/mixed-signal (AMS) circuits. Following Amdahl's law, considering the slowest part in the optimization process, circuit simulations have the highest priority. Simulation is needed for AMS circuits since numerical methods cannot predict the parasitics that influence the circuit performance greatly after the physical

[0]This research is supported in part by SRC award P10883 and NSF awards CNS-0854182 and DUE-0942629.

design has been derived, hence it is hard to predict the outcome of the actual circuit. Different approaches have been proposed to either simplify the circuit (i.e. macromodeling) or predict the output values for the circuit using surrogate approaches like metamodeling, multiple regression techniques, and Design of Experiments [1], [2].

Metamodeling based design flows are investigated as approaches to reduce design cycle time. The proposed non-polynomial metamodeling design flow speeds up design process by creating accurate metamodels and uses them for optimization. Metamodeling is used in variety of different fields to predict the values of time consuming or expensive processes [3]. Creation of the metamodel starts with sampling the design space and then using mathematical approaches to create formula(s) for output prediction. For circuits, the sampling is performed using circuit simulations. There are different approaches to create the predictive formula(s). Polynomial least square regression is the most common and very widely used [1]. Its simplicity is very attractive, but it is not efficient for very high dimensional circuits (many parameters) due to the number of coefficients which is limited by memory space. To improve polynomial regression models, splines can be used. But they also have the same limitations as regular polynomials for high dimensional datasets.

Neural networks may be an answer for creating very high dimensional models [4]. For a limited amount of simulations, the trained neural network preforms almost equally well for any number of parameters. Multi-layer networks are trained in parallel for every input by adjusting corresponding weights for non-linear and linear functions. Once the network learns and conducts final adjustments for weights of the internal functions, it is able to predict the values with only the number of parameters times the number of layers functions in the model. This makes neural networks very robust. Finding the right architecture for a neural network usually requires some experimentation. A few techniques are considered in this work: different non-linear functions and varying the amount of neurons in the hidden layer. The data to create the neural networks are directly generated by SPICE. Once the neural network is generated it can predict the output value very fast, due to its small complexity. Hence, the prediction is much faster than SPICE. A trained neural network can also be used by constrained single and multi objective optimization algorithms, just like regular metamodels.

The rest of the paper is organized as follows: Section II discusses related previous research. A brief overview of the PLL circuit that was used in this paper is presented in Section III. Section IV describes the creation and use of neural networks for the PLL circuit. Experimental results are presented in Section V. The paper is concluded with directions for future research in Section VI.

II. RELATED RESEARCH AND CONTRIBUTIONS

The current literature is rich in research trying to speedup the design process of complex AMS circuits. Design space exploration approaches from high level descriptions of analog circuits are given in [5]. Posynomial modeling for gate sizing is presented in [6]. A layout-aware modeling approach for analog synthesis is given in [7]. A single manual design iteration design flow is proposed in [8] for fast design optimization of VCOs.

The following are selected research works that have applied neural networks in VLSI design. In [9], the author shows that neural networks can be used for circuit analysis. In [10], the authors introduce the creation of neural networks for estimating the output of operational amplifiers from a high level perspective which does not account for parasitics. In [11], optimal and Hopfield binary neural networks are used for testing stuck at fault and delay faults in digital circuits. In [12], neural networks are trained on multi-dimensional mapping between geometrical variables and the values of independent circuit elements to predict of electromagnetic behavior of vias. In [13], the authors propose to speed up simulations by replacing repeated simulation data such as polynomial and look up models with well trained neural networks. In [14], a Hopfield neural network model is used to represent digital circuit behavior. In [15], a feed-forward dynamic neural network model is developed for amplifier and mixer circuits directly from input-output large-signal measurements, without having to rely on internal details of the circuit. In [16], neural networks are used for electromagnetic susceptibility analysis and optimization of electronic devices.

The **novel contributions of this paper** are as follows. A non-polynomial metamodel based design optimization flow for analog/mixed-signal circuits is presented. For non-polynomial metamodeling different architectures of neural networks are considered to perform tradeoff analysis between speed and accuracy. As a practical demonstration of the use of the non-polynomial neural network metamodels, a physical design optimization of a 180 nm CMOS PLL is undertaken. A biologically inspired tool, the bee colony algorithm is used for optimization of the PLL physical design that uses the metamodels instead of the actual circuit (i.e. the parasitic aware netlist). It is demonstrated that the non-polynomial neural network metamodel assisted optimization is faster and more accurate compared to the polynomial metamodel.

III. THE CASE STUDY CIRCUIT: A 180NM CMOS PLL

The phase locked loop (PLL) is a closed feedback loop system which is implemented as shown in Fig. 1. The detailed baseline design of this circuit is discussed in [17].

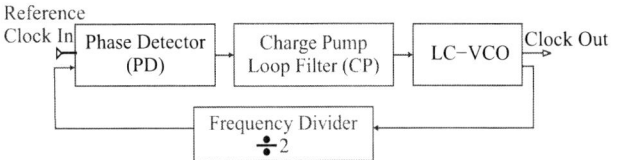

Fig. 1. Block diagram of a phase locked loop (PLL).

The physical design is shown in Fig. 2. A parasitic-aware netlist, including resistance (R), capacitance (C) and inductance (both self and mutual) (LK) is extracted from the layout. The netlist is then parameterized and used for simulations for the input data sets for each metamodel. Once the data are received from SPICE simulations, they are processed by an external tool (Matlab).

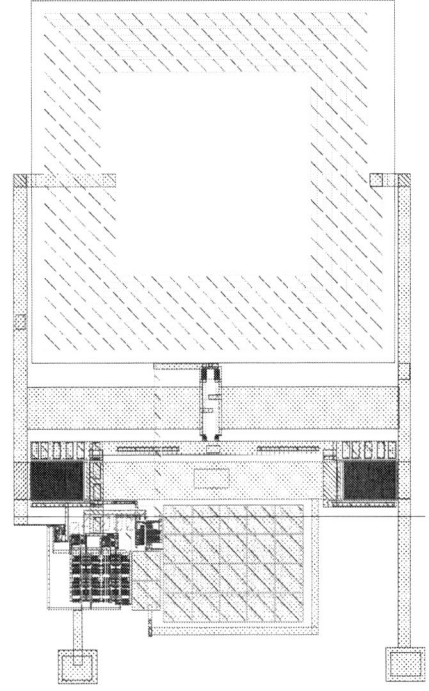

Fig. 2. Physical design of the PLL with area 525×326 μm.

SPICE simulation results of the circuit are shown in Fig. 3, which shows the frequency over time plot, with the PLL locking at 24.58 μs. The baseline phase noise diagram in Fig. 4 shows that the circuit has acceptable phase noise, (-163 dBc/Hz at 10 Hz offset). The average power consumption of the PLL is 9.28 mW.

978-1-4673-0438-2/12 $31.00 © 2012 IEEE   317

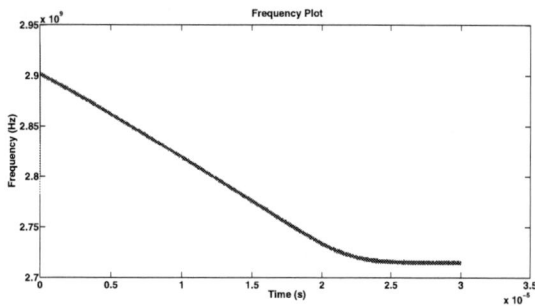

Fig. 3.   Frequency plot for the PLL circuit.

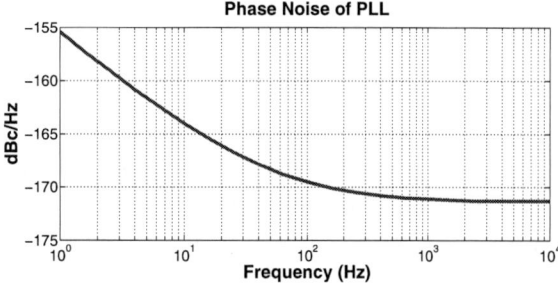

Fig. 4.   Phase noise of the baseline PLL circuit.

IV. NON-POLYNOMIAL METAMODELING OF THE PLL

A. Metamodeling Design Flow

The proposed design flow using non-polynomial metamodels is shown in Fig. 5. The physical design is parameterized and used twice, once for training samples and once more for verification. The non-polynomial metamodels are created for each output set of the design. Computationally expensive optimization algorithms can be applied using the fast non-polynomial metamodels as they are ultra fast compared to the actual RCLK netlist. The optimized values are then used to adjust the initial physical layout to create the near optimal design. This design flow only uses 2 iterations for physical design, at the beginning and the end. Overall, the design flow is as accurate as the parasitic-aware netlist of the circuit but ultra-fast due to the metamodel abstraction, which in turn minimizes the amount of time the designer needs to spend on the design optimization of the circuit.

B. Neural Network Exploration for Non-Polynomial Metamodeling

Neural network models are composed of a mass of fairly simple computational elements and rich interconnections between the elements. Neural networks operate in a parallel and distributed fashion which may resemble biological neural networks. Most neural networks have some sort of "training" rule by which the weights of connections are adjusted on the basis of presented patterns. They normally have great potential for parallelism, since the computations of the components are independent of each other. It has been proven in the universal approximation theorem that a neural network with one hidden layer can estimate any continuous function that maps to real numbers.

Over-fitting is the phenomenon where the network becomes worse instead of improving after a certain point during training when it is trained to as low errors as possible. This is because excessive training or a large amount of neurons in the hidden layer may make the network memorize the training patterns and stop adjusting the weights. There are several methods to avoid over-fitting. One method is regularization which tries to limit the complexity of the network such that it is unable to learn peculiarities. Another method is early stopping which aims to stop training at the point of optimal generalization.

1) Multilayer Neural Networks: A multiple layer neural network consists of an input, a with nonlinear activation function in hidden layer, and linear activation function in the output layer. Multilayer networks are very flexible and powerful due to their ability to represent nonlinear as well as linear functions. The multilayer network needs to have at least one non-linear function, otherwise a composition of linear functions becomes just another linear function.

The two common nonlinear activation functions that are usually used for the hidden layer are [18]:

$$b_j\left(v_j\right) = \left(\frac{1}{1 + e^{-\lambda v_j}}\right), \text{ or} \tag{1}$$

$$b_j\left(v_j\right) = \tanh(\lambda v_j), \tag{2}$$

where j denotes a neuron in the hidden layer, b_j and v_j are its input and output, respectively, and λ is the neuron transfer function steepness. The predicted output is given by:

$$\hat{y} = \sum_{j=1}^{d} \beta_j b_j(v_j) + \beta_0, \tag{3}$$

where β_j is the weight in the output due to neuron j, d is the number of neurons in the hidden layer and β_0 a constant. On the other hand, a linear layer function has the following format:

$$v_i = \sum_{i=1}^{s} w_{ji} x_i + w_{j0}, \tag{4}$$

Where w_{ji} is the weight connection between the jth component in the hidden layer and the ith component of the input.

The network training is performed to minimize the least square criterion:

$$E = \sum_{k=1}^{n} (y_k - \hat{y_k})^2, \tag{5}$$

where y_k and $\hat{y_k}$ are the actual and predicted responses, respectively, at the k-th training point (of n).

2) Radial Neural Networks: Radial neural networks are also two-layer networks. The first layer has radial base neurons, and calculates its weighted inputs with distance and its net input with a radial function. The second layer has linear neurons, and calculates its weighted input with a dot product function and its net inputs by combining its weighted

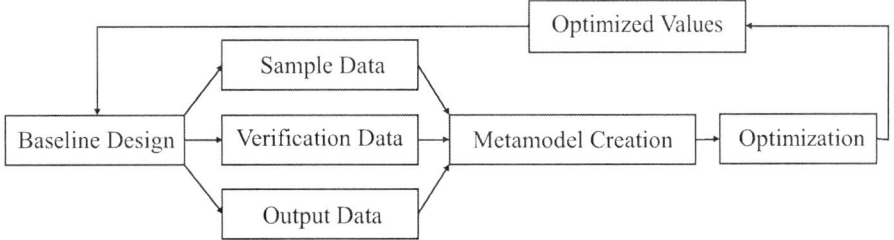

Fig. 5. The non-polynomial metamodeling based design flow.

inputs and biases. Both layers have biases. The radial network mathematical model is as follows:

$$y = \sum_{i=1}^{N} a_i \rho \left(\| x - c_i \| \right), \qquad (6)$$

where c_i is the center vector of neuron i, x is the prediction point, ρ is the neuron's transfer function and a_i are the weighs of the linear neuron.

Initially the radial basis layer has no neurons. The following steps are repeated until the network's mean squared error falls below the desired goal. The network is simulated. The input vector with the greatest error is found. A radial basis neuron is added with weights equal to that vector. The linear layer weights are then recalculated to minimize the error.

Each neuron in the radial basis layer will output a value according to how close the input vector is to each neuron's weight vector. Thus, radial basis neurons with weight vectors quite different from the input vector have will have outputs near to zero. These small outputs have only a negligible effect on the linear output neurons.

In contrast, a radial basis neuron with a weight vector close to the input vector produces a value near 1. If a neuron has an output of 1 its output weights in the second layer pass their values to the linear neurons in the second layer.

C. Standardizing or Normalizing Data

The data set is generated from the RLCK parasitic aware netlist simulations. The input data set is the same for every metamodel and is generated using Latin Hyper Cube Sampling (LHS). LHS supports any amount of planes and is proven to work better than Monte Carlo due to the more even distribution of points with still the random factor that helps to detect nonlinearity. LHS divides each plane (parameter) into Latin squares and randomly picks a point from each square. Output is generated for each run from SPICE simulations saving each needed value to its own data set. Hence, each metamodel has its own target data set. This paper targets neural networks that have a single output with multiple inputs.

The validation and test data must be standardized or normalized using the statistics computed from the training data. It is desirable to either normalize or standardize the input data as the input dataset has large dynamic range. If not, the training of higher values can outweigh the lower and the neural network will not train properly. In this paper, 2 commonly used methods are applied to standardize the data:

1) Normalizing to mean (μ) 0 and standard deviation (σ) 1.
2) Standardizing to midrange 0 and range 2 (from -1 to 1).

D. Metamodel Generation from Trained Neural Network Data

For comparison purposes, the data was fitted into partial polynomial equations. Since the full polynomial function would result in a very large amount of coefficients for 21 variables, partial polynomial functions of order of 1 through 6 are considered. Further, the stepwise regression method is used to filter out the coefficients that do not contribute to the function's outcome.

E. Metamodel Selection Criteria

There may be numerous metamodels created from the same sampled set. RMSE and R^2 are common metrics used for goodness of fit. The Root Mean Square Error (RMSE) is derived from sum of square errors (SSE):

$$RMSE = \sqrt{\frac{1}{N} SSE} \qquad (7)$$

$$= \sqrt{\frac{1}{N} \sum_{k=1}^{N} (y(x_k) - \hat{y}(x_k))^2}. \qquad (8)$$

Where y are the actual simulation result values and $\hat{y}$ are the results of the metamodel at the same location as the simulation point. R^2 is the coefficient of determination, which predicts the probability of a future result to be predicted by the model and is also used to verify the model accuracy.

The created model may fit perfectly to the training data set but may not qualify as a good model to represent the output for the given process at other points. For this reason, the verification data set is created so that the points are at different locations than the original sample. It is a good idea not to use the verification set for training, since it will defeat the purpose of testing the metamodel on totally unbiased points. If the verification RMSE and R^2 values do not differ very much from the training values, then the model has been trained correctly, otherwise it has not.

F. Design Optimization of the PLL Circuit

The best (may be the most accurate or the fastest, depending on the requirement) non-polynomial metamodels from the previous section need to be selected for optimization. The optimization algorithm that is being used is the Bee Colony

Algorithm (BCA). BCA is the artificial representation of a bee colony behavior as bees try to find the best food source [19], [20]. More information about the algorithm in the context of AMS circuit optimization can be found in previous research [17]. This algorithm was found to be effective for use on AMS circuits with metamodels. As a specific objective and constraint optimization, the PLL circuit is characterized for output frequency, power, and locking time. A separate metamodel is created for each Figure-of-Merit (FoM) from the same sample set. Each single simulation calculates all FoMs so the number of simulations that are needed does not depend on the number of the metamodels that need to be created.

V. EXPERIMENTAL RESULTS

Given that each SPICE simulation for the PLL circuit takes approximately 10 minutes to converge, the amount of simulation runs are limited. In this work 100 simulations for training and 30 simulations for verification have been chosen. Different architectures of neural networks are evaluated. For feed-forward networks two differentiable transfer functions (tanh - tansig, and logarithmic - logsig) are used for the hidden layer. In addition, the experimental results also consider the difference between raw regular input data in comparison to normalized and standardized input sets.

The verification data set is also chosen using LHS, but it is ensured that none of the points match the training set. After the neural network training is completed, the input values for verification set are fed into the network and the RMSE value is calculated for the verification set. The R^2 values are calculated for training and verification sets for each combination of the above neural networks. Selected results are summarized in Tables I and II for brevity. The statistics of the best created polynomial functions that were created from [1] are listed in the last rows of the tables for comparison purposes.

From the data it is observed that neural networks with no standardization of the input data perform the worst. Even though polynomials show best results without standardization or normalization, this is not the case for neural networks. Also, it can be concluded that neural networks perform better fitting for this circuit, mostly because of the non-linear and linear flexibility of the neural networks. The data also demonstrates which architecture and normalization or standardization should be used, i.e. which has the best performance.

Fig. 6 shows the progression of the FoM as the BCA optimization progresses. The frequency metamodel is used to filter (constraint) the results within 0.05% of the required 2.7 GHz operational locking frequency of the PLL, which can be used in Multichannel Multipoint Distribution Systems (MMDS). The constraint narrows down the search criteria for the algorithm. If the model is within range for frequency, the power and locking time metamodels are used to calculate a composite FoM, which is defined by:

$$FoM = \left(\frac{1}{power \times lockingTime} \right). \quad (9)$$

Table III shows the optimized values for both polynomial and neural network metamodels.

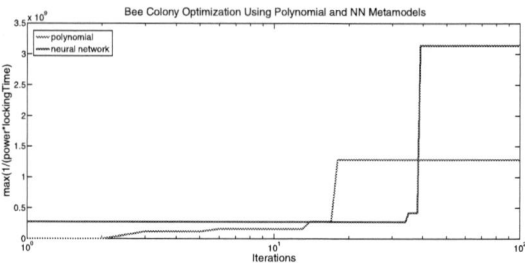

Fig. 6. Bee Colony optimization conducted on polynomial and neural network metamodels for optimizing power and locking time within 0.5% of 2.7 GHz frequency constraint.

TABLE III
PLL CIRCUIT PARAMETERS AFTER OPTIMIZATION.

FoM	Polynomial	Neural Network
Power	3.9 mW	3.9 mW
Locking Time	8.476 μs	3.3147 μs
Frequency	2.6909 GHz	2.7026 GHz

VI. CONCLUSION AND FUTURE RESEARCH

This paper explored the generation and usage of neural networks for metamodels of a PLL circuit. The bee colony algorithm with both non-polynomial and polynomial metamodels has been used for optimization. Neural networks are reusable and can be used as a system of equations to accurately represent the needed output. Neural networks show on average 56% increase in accuracy of prediction over the polynomial metamodels that have been generated from the same input data samples. In addition, neural network prediction, which is on average within 3.2% of SPICE output, is enormously faster than SPICE simulation and is shown to find better solution during the optimization phase of design. Even though the circuit that this paper uses as an example is parameterized with 21 parameters, in future work higher and more complex circuits that can have hundreds of parameters will be investigated.

REFERENCES

[1] O. Garitselov, S. P. Mohanty, and E. Kougianos, "Nano-CMOS Mixed-Signal Circuit Metamodeling Techniques: A Comparative Study," in *Proc. Int. Symp. Elec. Des.*, 2010, pp. 191–196.

[2] S. Basu, B. Kommineni, and R. Vemuri, "Variation-Aware Macromodeling and Synthesis of Analog Circuits Using Spline Center and Range Method and Dynamically Reduced Design Space," in *Proc. Int. Conf. VLSI Des., VLSID*, 2009, pp. 433–438.

[3] R. Barton, "Simulation Optimization Using Metamodels," in *Proceedings of the 2009 Winter Simulation Conference (WSC).*, December 2009, pp. 230–238.

[4] L. Wang, "A Hybrid Genetic Algorithm- Neural Network Strategy for Simulation Optimization," *Applied Mathematics and Computation*, vol. 170, no. 2, pp. 1329–1343, 2005.

[5] A. Doboli and R. Vemuri, "Exploration-based High-level Synthesis of Linear Analog Systems Operating at low/medium Frequencies," *IEEE Transactions on Computer-Aided Design of Integrated Circuits and Systems*, vol. 22, no. 11, pp. 1556–1568, November 2003.

[6] S. Roy, W. Chen, C. Chung-Ping Chen, and Y. H. Hu, "Numerically Convex Forms and Their Application in Gate Sizing," *IEEE Transactions on Computer-Aided Design of Integrated Circuits and Systems*, vol. 26, no. 9, pp. 1637–1647, September 2007.

TABLE I
FREQUENCY NON-POLYNOMIAL METAMODEL COMPARISON OF THE PLL.

Function	Data Filtering	R^2-test	R^2-verification	RMSE	Neurons
logsig→purelin	none	0.802	0.723	52.74 MHz	4
tansig→purelin	none	0.839	0.713	51.24 MHz	3
radial→purelin	none	0.020	0.490	81.51 MHz	
logsig→purelin	minmax	0.917	0.664	48.89 MHz	9
tansig→purelin	minmax	0.855	0.699	53.65 MHz	1
radial→purelin	minmax	0.844	0.712	50.88 MHz	
logsig→purelin	meanstd	0.843	0.733	53.60 MHz	1
tansig→purelin	meanstd	0.793	0.762	51.64 MHz	5
radial→purelin	meanstd	0.848	0.749	48.97 MHz	
	Data Filtering	R^2	Order	RMSE	Number of Coefficients
polynomial	none	0.930	4	77.96 MHz	48

TABLE II
LOCKING TIME NON-POLYNOMIAL METAMODEL COMPARISON OF THE PLL.

Function	Data Filtering	R^2-Test	R^2-Verification	RMSE	Neurons
logsig→purelin	none	0.828	0.873	1.30 μs	1
tansig→purelin	none	0.850	0.723	1.44 μs	9
radial→purelin	none	0.078	0.830	2.26 μs	
logsig→purelin	minmax	0.826	0.870	1.29 μs	1
tansig→purelin	minmax	0.839	0.942	1.12 μs	10
radial→purelin	minmax	0.931	0.508	1.65 μs	
logsig→purelin	meanstd	0.826	0.906	1.22 μs	2
tansig→purelin	meanstd	0.737	0.939	1.12 μs	3
radial→purelin	meanstd	0.963	0.691	1.23 μs	
	Data Filtering	R^2	Order	RMSE	Number of Coefficients
polynomial	none	0.877	4	1.91 μs	56

[7] A. Pradhan and R. Vemuri, "A Layout-aware Analog Synthesis Procedure Inclusive of Dynamic Module Geometry Selection," in *Proceedings of the 18th ACM Great Lakes symposium on VLSI*, ser. GLSVLSI '08. New York, NY, USA: ACM, 2008, pp. 159–162.

[8] D. Ghai, S. P. Mohanty, and E. Kougianos, "Design of Parasitic and Process-Variation Aware Nano-CMOS RF Circuits: A VCO Case Study," *IEEE Trans. VLSI Syst.*, vol. 17, no. 9, pp. 1339–1342, 2009.

[9] G. Wolfe and R. Vemuri, "Extraction and Use of Neural Network Models in Automated Synthesis of Operational Amplifiers," *IEEE Transactions on Computer-Aided Design of Integrated Circuits and Systems.*, vol. 22, no. 2, pp. 198–212, February 2003.

[10] L. Xia, I. Bell, and A. Wilkinson, "A Robust Approach for Automated Model Generation," in *DTIS '09. 4th International Conference on Design Technology of Integrated Systems in Nanoscal Era, 2009.*, April 2009, pp. 281–286.

[11] P. Zhongliang, "Neural Network Model for Testing Stuck-at and Delay Faults in Digital Circuit," in *Proceedings 17th International Conference on VLSI Design. 2004.*, 2004, pp. 499–504.

[12] Y. Cao and Q.-J. Zhang, "Neural Network Techniques for Fast Parametric Modeling of Vias on Multilayered Circuit Packages," in *2010 IEEE Electrical Design of Advanced Packaging Systems Symposium (EDAPS)*, December 2010, pp. 1–4.

[13] A. Zaabab, Q.-J. Zhang, and M. Nakhla, "A Neural Network Modeling Approach to Circuit Optimization and Statistical Design," *IEEE Transactions on Microwave Theory and Techniques.*, vol. 43, no. 6, pp. 1349–1358, June 1995.

[14] E. Macii and M. Poncino, "Estimating Power Consumption of CMOS Circuits Modelled as Symbolic Neural Networks," *IEE Proceedings Computers and Digital Techniques.*, vol. 143, no. 5, pp. 331–336, September 1996.

[15] J. Xu, M. Yagoub, R. Ding, and Q. Zhang, "Feedforward Dynamic Neural Network Technique for Modeling and Design of Nonlinear Telecommunication Circuits and Systems," in *Proceedings of the International Joint Conference on Neural Networks, 2003.*, vol. 2, July

2003, pp. 930–935.

[16] Z. Aimin, Z. Hang, L. Hong, and C. Degui, "A Recurrent Neural Networks Based Modeling Approach for Internal Circuits of Electronic Devices," in *2009 20th International Zurich Symposium on Electromagnetic Compatibility*, January 2009, pp. 293–296.

[17] O. Garitselov, S. P. Mohanty, E. Kougianos, and P. Patra, "Bee Colony Inspired Metamodeling Based Fast Optimization of a Nano-CMOS PLL," in *Proceedings of the 2nd IEEE International Symposium on Electronic System Design (ISED)*, 2011.

[18] K.-T. Fan, R. Li, and A. Sudjianto. *Design and Modeling for Computer Experiments.* 23-25 Blades Court, London SW15 2NU, UK: Chapman and Hall/CRC, 2006.

[19] D. Karabora and B. Akay, "A Comparative Study of Artificial Bee Colony Algorithm," *Applied Mathematics and Computation*, no. 214, pp. 108–132, 2009.

[20] R. Hedayatzadeh, B. Hasanizadeh, R. Akbari, and K. Ziarati, "A Multi-Objective Artificial Bee Colony for Optimizing Multi-Objective Problems," in *2010 3rd International Conference on Advanced Computer Theory and Engineering (ICACTE)*, vol. 5, August 2010, pp. 277–281.

Circuit Optimization at 22nm Technology Node

Angada B. Sachid
Dept. of Electrical Engineering and Computer Sciences
University of California
Berkeley, USA
Email: angada@eecs.berkeley.edu

P. Paliwal, S. Joshi, M. Shojaei, D. Sharma, V. Rao
Department of Electrical Engineering
Indian Institute of Technology
Bombay, India
Email: pallavip@ee.iitb.ac.in

Abstract—**With every new technology node, scaling down of Device-to-Interconnect Capacitance ratio causes Interconnect delay to become bottleneck for circuit perfomance. To mitigate this effect, interconnect routing area on-chip should be minimized for improved power-delay product. In this aspect, FinFET with multiple fins per lithographic pitch gains more advantage, in comparison to Planar Device; since, such FinFET devices allow increase of electrical width without increasing device layout area and thus, interconnect capacitance is comparatively lower. Therefore, minimum delay could be achieved for lesser device width, and thus, with lower power. This paper proves the perfomance enhancement with such FinFET Device for Mux Circuit; and aims to find out Optimum Design Space for Mux Circuit, at 22nm technology node, with practical value of Interconnect Capacitive load (extrapolated from circuit layout in current technology node).**

Keywords-**22nm design; Interconnect parasitics; FinFET; Scaling trend.**

I. INTRODUCTION

For current generation planar device, minimum-sized transistors usually give required perfomance for digital circiuts. But, the same assumption may not hold true on moving to lower technology nodes, wherein Interconnect Capacitance becomes comparable to Device Capacitance. In this case, Device electrical width should be increased, without increasing Interconnect Capacitance, to improve circuit perfomance. Hence, FinFET gain edge over Planar Device, since, its fin height could be increased to increase electrical width, while keeping device footprint and thus, Interconnect Capacitive load constant. Also, if multiple fins could be fabricated on same Device Footprint, better perfomance could be achieved, in comparison to FinFET having only single fin per device footprint, as would be shown in Section VI for Mux Circut at 22nm Technology Node.

Optimum Design Point for digital circuit can be obtained by increasing transistor's electrical width to reduce delay and decreasing Supply Voltage to reduce power, whose intersection space results in minimum Power-Delay Product.

At 22nm Technology Node, Device Width/Power Scaling Analysis for SRAM Cell has been derived in [1]. In this work, we have derived Optimum design Space for 32-bit Mux Circuit, with estimated interconnect capacitive load at 22nm technology node. Interconnect Capacitance Load for 22nm Mux circuit is derived by extracting wire parasitics from Mux Circuit Layout drawn in available technology nodes[180nm-65nm], and thereafter extrapolating wire parasitics value to 22nm, with method shown in Section IV.

The application of this work lies in the fact that Large Fan-In Mux circuits are fundamental blocks in several applications [2], with one of the usage as Column Decoders in memory circuits. With the results derived in this paper for 32-bit Mux Circuit, its overlapping point with Design Space of SRAM Cell (while considering its practical Interconnect Load) can be found. Intersection of Design Space of Mux-Circuit and SRAM Cell would result in memory circuit design with minimum overall delay through SRAM block.

Section II justifies analytically, that for digital circuits designed with FinFET (in sub-22nm node), optimum design point has to be found with simultaneous reduction in supply voltage and increase in electrical width. Section III emphasizes on selection of correct High Fan-in Mux architecture and circuit layout for optimum perfomance [referred from [2]]. Section IV explains method for extrapolating wire capacitance to 22nm Mux circuit, from circuit layout in existing technology nodes. Section V describes method for finding optimum design space for digital circuit, with example of Mux 4:1 circuit (simulated with its interconnect load) in circuit simulator-"Sequel". Section VI evaluates Power-Delay reduction for 32:1 Mux circuit, when circuit is designed with FinFET instead of planar device. In Section VII, optimum design point for 32:1 Mux is being found.

II. POWER-DELAY TRADE-OFF IN SUB-22NM CIRCUIT

FinFET is a 3D device in which most of the electrical width is perpendicular to the wafer surface (Fig. 1). The electrical width (W_{ELEC}) of a FinFET is given by:

$$W_{ELEC} = N(2H_{FIN} + T_{FIN}); H_{FIN} = A_R T_{FIN} \quad (1)$$

H_{FIN} is the fin height, T_{FIN} is the fin thickness and N is the number of fins. H_{FIN} is related to T_{FIN} by the fin aspect ratio, A_R, which is a technology parameter. To explain our technology aware design approach, let us consider that W_{ELEC} can be increased in the same device footprint and it proportionately increases I_{DRIVE} and C_{GATE}. The

978-1-4673-0438-2/12 $31.00 © 2012 IEEE

Figure 1. 3D model of the FinFET showing fin height (H_{FIN}) perpendicular to the wafer surface

delay (τ) and dynamic power (P_{dyn}) is represented by the first-order expressions :

$$\tau = \frac{C_{LOAD}V_{DD}}{I_{DRIVE}} = \frac{(C_{GATE}+C_{INT})V_{DD}}{I_{DRIVE}} \quad (2)$$

$$P_{dyn} = \alpha C_{LOAD}V_{DD}^2 f = \alpha(C_{GATE}+C_{INT})V_{DD}^2 f$$

where, C_{INT} represents Interconnect Capacitance.

When W_{ELEC} is increased by k times,

$$\tau^{new} = \frac{(kC_{GATE}+C_{INT})V_{DD}}{kI_{DRIVE}} < \tau \ (for \ k>1) \quad (3)$$

$$P_{dyn}^{new} = \alpha(kC_{GATE}+C_{INT})V_{DD}^2 f > P_{dyn}$$

For τ^{new},

$$V_{DD}^{new} = \frac{(kC_{GATE}+C_{INT})V_{DD}}{k(C_{GATE}+C_{INT})} < V_{DD} \quad (4)$$

$$P_{dyn}^{new} = \alpha(kC_{GATE}+C_{INT})V_{DD}^2 f$$

$$\therefore P_{dyn}^{new} = \alpha \frac{(kC_{GATE}+C_{INT})^3 V_{DD}^2 f}{k^2(C_{GATE}+C_{INT})}$$

From Eqns. [3], if $C_{INT} << C_{GATE}$, then τ remains unchanged and P_{dyn} increases by k times. If $C_{INT} >> C_{GATE}$, then τ decreases by k times and P_{dyn} remains unchanged. If $C_{INT} = C_{GATE}$, then at the same VDD, $(\tau_{new}/\tau) = $ (k+1/2k), which is less than 1, for $k > 1$ and $(P_{dyn}^{new}/P_{dyn}) = $ (k+1) which is greater than 1. In this case, the circuit is faster and consumes more power. From Eqns. [2], it is also evident that the sensitivity of VDD to delay and power are different: $\tau \alpha VDD$ and $P_{dyn} \alpha VDD^2$. Since, $(\tau^{new}/\tau) < 1$, the circuits are faster than required (Eqn. [3]), VDD can be reduced so that τ^{new} can be made equal to τ (Eqn. [4]). In this case, P_{dyn} is reduced. Our technology aware design methodology uses non-minimal electrical width and reduced supply voltage to simultaneously reduce delay and power. This technique is applicable to circuits in which interconnect capacitance can not be neglected, which is usually the case with a large number of circuits at sub-22nm nodes.

The limitation of using Eqn. [2] is that it overestimates the delay. A more accurate equation uses $I_{EFF} = \frac{(I_{ON}+I_{LIN})}{2}$ [3] or other forms of I_{EFF} [4]. By using any of these equations, we get similar insights about the approach used in this paper.

To simultaneously reduce delay and power in sub-22nm circuits, we propose to simultaneously reduce VDD and increase the electrical width of a device in the same device footprint using FinFET and FinFET-like devices. Our methodology is unique to FinFET and FinFET-like devices at sub-22nm nodes where C_{INT} is non-negligible. In planar devices like bulk MOSFET or UTB-SOI MOSFET, increasing W_{ELEC} increases the device/logic gate footprint which increases C_{INT}. Since C_{INT} is proportional to the perimeter of the logic gate/cell footprint, increasing W_{ELEC} by k times increases C_{INT} by k times. Hence, in planar devices, using Eqn. [2] and Eqn.[4], P_{dyn} increases by k times and τ remains unchanged.

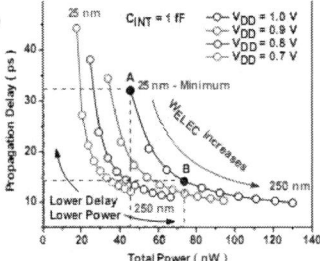

Figure 2. Impact of increasing electrical width per device footprint (W_{ELEC}) on delay and power of inverter (C_{INT} = 1 fF) [5]

III. MINIMUM DELAY MUX DESIGN

In heterogeneous tree structure [Fig.3(a)], Mux Delay has to be minimized by choosing correct number of stages and correct number of switches per group, while considering effect of Interconnect Capacitive Load, right from the start of the design.

From post-layout simulation at 180nm/90nm, it was verified that minimum delay is achieved for high Fan-in circuits (Mux-16/Mux-32), if Mux-circuit is broken into 2 stages, with 1st stage having group of 8 switches (as suggested in [2]).The reason being that if more number of stages is kept (or) 1st Stage is broken into less than 8 Switches per Group, then Device Capacitance dominates and increases delay. On the other hand, if more number of switches is included per group in Stage-1 (instead of 8 switches/group), Interconnect Wire joining 1st stage and 2nd stage becomes too long, and thus, Interconnect Capacitance dominates.

Following considerations are made while laying out Mux Circuit (as suggested in [2]) :- (i) Transmission Gates for Stage-2 are placed at centre of Mux-8 group of Stage-1, to make Mux-Delay independent of switching input. (ii) In Fig.3, interconnect capacitance of incoming Horizontal Wire is considered negligible and Interconnect Capacitance is extracted only for Vertical Long Wire.

Accordingly, Total Device Capacitance is obtained as, addition of C_{gd} of Stage-1 switches, with $C_{gs} + C_{gd}$ of Stage-2 switches.

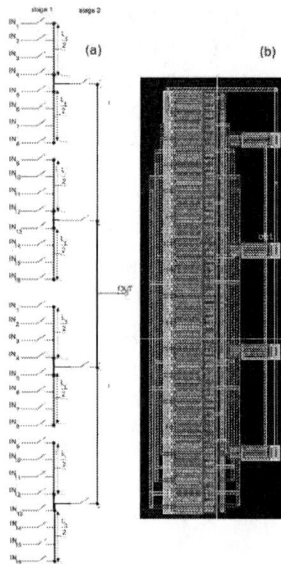

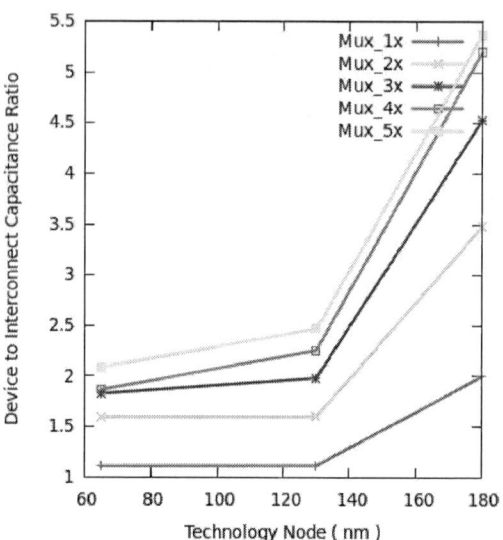

Figure 3. (a) 32:1 Mux Floorplan divided into 2 stages, with Stage-1 divided into 8 switches per group[referred from [2]]. (b) 32:1 Mux Layout in 180nm, with 2 stages of switches from input to output

Figure 4. Scaling Trend of Device-Interconnect Capacitance Ratio (from 180nm to 65nm Technology Node) for 32:1 Mux Circuit. In figure, 2x,3x,4x,5x indicates, with what multiple of minimum width transistor is Mux-32 circuit designed.

Layout of 32:1 Mux in 180nm Technology Node is shown in Fig. 3(b), wherein 8 selection lines are activating 1 input switch in each group of Stage-1, and 4 selection lines are activating 1 out of 4 Transmission Gate switch in Stage-2. In Mux Circuit simulation in Section. V, it is assumed that control signal on selection lines, for transmission gates is already available. So, during simulation, only data input to Mux circuit is changing.

IV. INTERCONNECT CAPACITANCE ESTIMATION AT 22NM

With wire parasitics extracted from Mux Circuit Layout in available technology nodes[180nm to 65nm], Device-Interconnect Capacitance Ratio and Wire Capacitance scaling trend is observed below. Interconnect Capacitance at 22nm is extrapolated based on Wire Capacitance scaling trend information from ITRS taken as reference.

A. Scaling trend of Device-Interconnect Capacitance Ratio for existing Technology Nodes

For a fixed circuit, for every new technology node, Interconnect Capacitance value keeps on coming closer to Device Capacitance. As shown in Fig.4 for 32:1 Mux, at 180nm Device-Interconnect Ratio is comparatively high (ratio varies from 5.5 to 2) and towards 65nm, Device-Interconnect ratio has decreased to 1.11 (for Mux-32 circuit with minimum sized transistors). At 22nm also, Planar Device is expected to follow the same trend, with Interconnect Capacitance becoming comparable to Device Capacitance.

B. Extrapolation methods for 22nm-Mux Wire Capacitance

1) Interconnect Capacitance Reference Trend Line (from ITRS): For extrapolating Interconnect capacitance for Mux Circuits, from 65nm to 22nm, a Reference Curve (Fig. 6) is needed, whose slope can be followed while extending line from 65nm to 22nm. Capacitance per unit length for Intermediate Wires is taken from ITRS (for 90nm-22nm), and from PTM Interconnect Calculator (for 180nm-130nm), since according to the model in Fig. 5, Capacitance stated by ITRS and PTM considers Coupling Capacitance with adjacent wires.

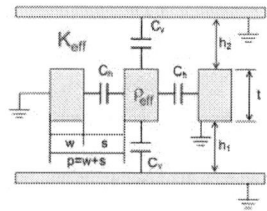

Figure 5. Interconnect Model used by ITRS

2) Interconnect Capacitance Extrapolated using Graph Imposition: Though, according to ITRS Interconnect Model shown in Fig.5 , coupling capacitance is included within the stated Capacitance value; but, still, capacitance/unit length extracted from Mux Circuit Layout (using UMC Model Library for 180nm - 65nm technology node) is higher than that stated by ITRS. This results in difference in slope between ITRS Curve and extracted Mux Capacitance curve

978-1-4673-0438-2/12 $31.00 © 2012 IEEE

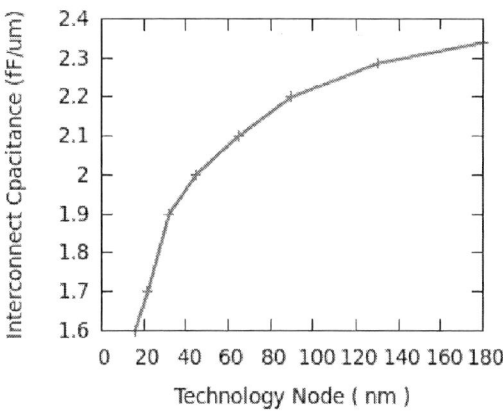

Figure 6. Reference Trend Line from ITRS (Linear Scale), for Interconnect Capacitance per unit length in different technology nodes

for existing Technology Nodes, due to which Reference Trend Line from ITRS cannot be used on Linear Scale.

As shown in Fig. 7 for 16:1 Mux circuit, Reference Curve and Extracted Capacitance Curves for available Technology Nodes, coincide to some extent on Log Scale. Thus, slope from ITRS curve (log scale) could be used for predicting 22nm Interconnect Capacitive Load.

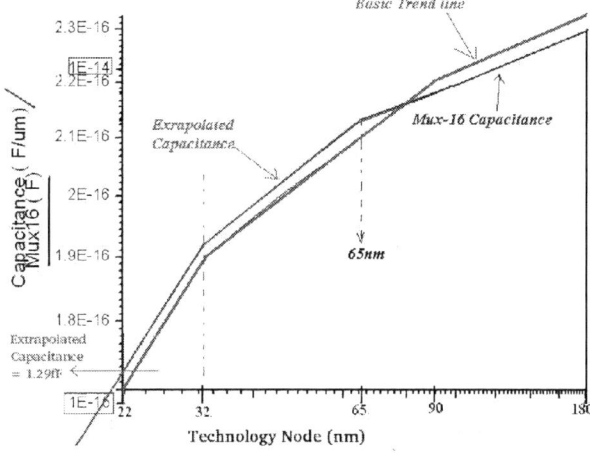

Figure 7. Extrapolation of Interconnect Capacitance at 22nm, for Mux-16 circuit, by superimposing Reference Interconnect Capacitance (fF/um) Curve from ITRS, on extracted Wire Capacitance(fF) curve for Mux-16 layout in 180nm-65nm technology nodes. (with both axis on Log Scale)

C. Extrapolation results for Interconnect Capacitance

Interconnect Capacitance extrapolated from Graph (Fig. 7) is considered correct, since, it is based on slope of ITRS Reference Curve (Fig. 6) which would be considering all the Interconnect related scaling factors more closely.

Table I
EXTRAPOLATED WIRE CAPACITANCE FOR 22NM MUX CIRCUIT

22nm Mux Circuit	Extrapolated Interconnect Capacitance
Mux 4	0.2318fF
Mux 8	0.4809fF
Mux 16	1.29fF
Mux 32	2.578fF
Mux32-2x	4.11fF
Mux32-3x	7.503fF
Mux32-4x	9.97fF
Mux32-5x	10.3fF

V. CIRCUIT SIMULATION IN SEQUEL

Dual-Gate FinFET Model (based on look-up table format) is used with Sequel *(Circuit Simulator Tool)* to predict optimum Design Space for Mux circuits at 22nm Node. Model of FinFET with non-conducting top is being used, because it has lesser parasitic than tri-gate FinFET.

A. 22nm FinFET Model used in Mux Circuit

FinFET Model used for simulation has parameters :-
Gate Length (L_G) = 22nm
Equivalent Oxide Thickness (EOT) = 1nm
Fin Thickness (t_{fin}) = 8nm.
Source-Drain Contact Pitch = 100nm

All resistive and capacitive parasitics related to device is included in FinFET LUT Model. Therefore, only parasitics corresponding to interconnect is estimated (from Mux Layout in 180nm - 65nm Technology node) and added as load in circuit netlist at 22nm node.

For keeping Device Aspect Ratio in range 3 to 5, height is varied between 25nm to 40nm for single fin. For FinFETs that can be fabricated with 4 fins within same device footprint, its electrical width can be varied from 50nm to 320nm. As Device Layout Area is not changing, same Interconnect Capacitive Load is considered for all simulations, while increasing Device Width to find optimum design point.

B. Device/Wire Capacitance Ratio for 22nm Mux-4 Circuit

- *Interconnect Capacitance*
  For 4:1 Mux at 22nm node, Interconnect Capacitive load of **0.2318fF** is added in circuit, as per extrapolated wire capacitance for this circuit derived in Table. I.

- *Device Capacitance*
  In 22nm FinFET device model, Gate Capacitance (C_{gg}) = 1fF/um. Therefore, Output Capacitance of Transmission Gate ($C_{gs,n} + C_{gd,p}$) = 1fF/um
  For 4:1 Mux designed with minimum sized transistors, Electrical Width = 50nm. So, Output Device Capacitance of Mux-4 circuit = 4 x 1fF/um x 0.05um = **0.2fF**

Above values of Device/Interconnect Capacitance for 4:1 Mux shows that Device to Interconnect Capacitance Ratio has scaled down to unity, for Digital Circuits designed with minimum sized transistors. Since, Interconnect Capacitance have become comparable to Device Capacitance of minimum-sized transistors, FinFETs Width should be increased while keeping Interconnect Capacitance constant to give better performance.

C. Finding Optimum Design Point for 22nm Mux-4 Circuit

For designing Mux circuit with minimum power-delay product, device width is increased from 50nm to 320nm (considering FinFET with 1,2,4 fins could be fabricated within same device footrint) to reduce delay. Thereafter, for each device width, power-delay values are found for different supply voltages (from 1.0V - 0.6V), to check for which Supply Voltage and Device Width, minimum Power and Delay is obtained. Capacitance of value 0.2318fF is added as load to Mux-4 circuit, to consider Interconnect loading effect in circuit simulation. Here, Dynamic power dissipation is calculated for Mux circuit having input, which is changing with frequency 5GHz.

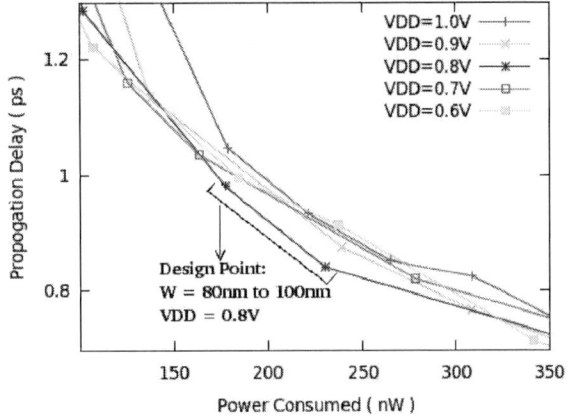

Figure 8. Variation of Width and Supply Voltage to find Optimal Design Space for 4:1 Mux

D. Simulation Result for Mux-4 circuit

As shown in Fig. 8, (i) For Constant Power, Maximum Improvement in Speed can be 22.4%. (ii) For Constant Speed, Maximum Improvement in Power can be 15%. (iii) For obtaining optimal Power-Delay Product, 4:1 Mux ciruit should be operated at 0.8V, with Device Width in range 80nm-100nm.

VI. POWER-DELAY TREND COMPARISON BETWEEN FINFET AND PLANAR DEVICE

For high fan-in Mux circuits, before finding its optimum design space with FinFET device, we are assessing below, as to how much reduction in Delay and Power could be brought about by FinFET device in comparison to Planar Device at 22nm Technology Node.

To imitate the perfomance of planar device (whose electrical width cannot be increased, without increasing device layout area), FinFET model with only single fin per device

is used for 32:1 Mux circuit simulation. With such FinFET device, after maximum possible height is reached for single fin, another transistor has to be added in parallel to increase width, and thus, device layout area increases. In this case, to model the impact of increased interconnect routing (caused by increased device area), load capacitance at 32:1 Mux output is increased, as per values extrapolated from Mux-32 layout (designed with increased transistor sizes) in available technology nodes (as shown in Table. I).

For FinFETs that can be fabricated with multiple fins on same device footprint, interconnect load remains constant while device width is increased, to improve circuit perfomance. Perfomance improvement of such multi-fin FET device over single-fin FET device is being derived below.

A. Delay Trend comparison for Mux-32

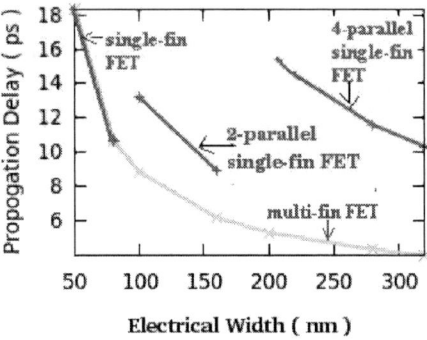

Figure 9. Delay Comparison for Mux32 circuit designed with (a) Single-Fin FET device (b) Multi-Fin FET device

As shown in Fig. 9, delay decreases for single-fin FET, when fin's height is increased upto allowed value; since, interconnect capacitance remains constant during this variation in transistor width. After maximum height is reached for single fin, when parallel transistor is added to increase width for single-fin FET, Interconnect Length increases with increase in routing area. This increase in wire length causes Interconnect Capacitance Load to increase proportionally, and therefore delay increases with addition of parallel transistor (shown in Fig. 9, as rise in delay for electrical width between 80nm to 100nm and for width between 160nm to 200nm).

For multi-fin FET, delay keeps on decreasing with increased device width, since, addition of fins on same device area keeps interconnect load constant.

In Mux-32 design, delay reduction in multi-fin FET (with No. of Fins = 4) over single-fin FET device is about 64%.

B. Power Trend comparison for Mux-32

As shown in Fig. 10, power consumed by single-fin FET increases rapidly, on addition of parallel transistor to increase width. The reason being that interconnect capacitance (which is comparable to device capacitance for

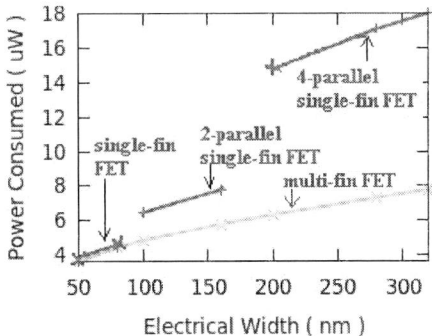

Figure 10. Power Comparison for Mux32 circuit designed with (a) Single-Fin FET device (b) Multi-Fin FET device

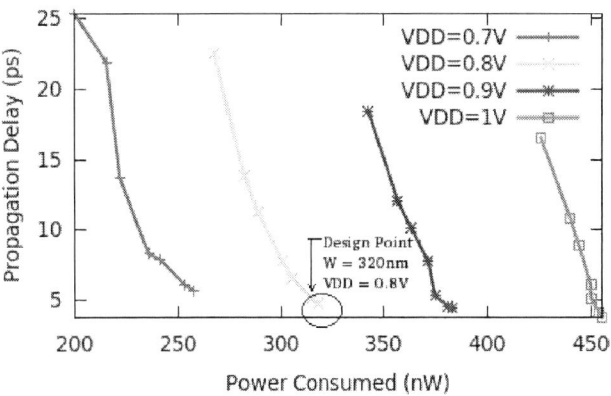

Figure 11. Variation of Width and Supply Voltage to find Optimal Design Space for 32:1 Mux

lower technology nodes) increases and takes more time to charge/discharge. Power consumed by single-fin FET increases only slightly, when only height of fin is changed to increase width, since, its wire parasitics remain constant.

For multi-fin FET, power increases at gradual rate with increase in transistor width, because interconnect load remains constant upto 320nm device width (with fins added in same device footprint).

VII. OPTIMUM DESIGN POINT FOR MUX-32 CIRCUIT WITH FINFET

A. Device/Wire Capacitance Ratio for 22nm Mux-32 Circuit

- *Interconnect Capacitance* According to Table. I, Interconnect capacitive load for 32:1 Mux circuit is 2.578fF. While increasing device width by increasing number of fins, this interconnect load is kept same, since, device area remains constant with upto 4 fins fabricated on same device footprint.
- *Device Capacitance* Output Capacitance of Transmission Gate $(C_{gs,n} + C_{gd,p})$ = 1fF/um.
  For 32:1 Mux designed with minimum sized transistors, Electrical Width = 50nm. Hence, Output Device Capacitance of Mux-32 circuit = 32 x 1fF/um x 0.05um + 4 x 2 x 1fF/um x 0.05 = **0.2fF**

B. Simulation Result for Mux-4 circuit

Delay reduction with increase in transistor's electrical width and Power reduction with decrease in supply voltage gives design point for Mux-32 circuit, wherein power-delay product is minimum. (Here, Dynamic Power dissipation corresponds to input data changing with 1GHz frequency.)

As shown in Fig. 11, optimum design point for Mux32 exists for Device Width = 320nm and Supply Voltage = 0.8V. Design Point obtained for Mux Circuit falls in accordance with analysis described in section. II.

VIII. CONCLUSION

In this work, with example of Mux Circuit, we have illustrated that for any digital block with regular structure,

optimum design point for that block could be obtained by varying Device Width/Supply Voltage, while simulating FinFET Model with realistic value of Interconnect Capacitance for that circuit. We have shown how interconnect capacitance could be predicted for lower technology nodes, by extrapolating wire parasitics extracted from circuit layout in available technology nodes. As future work, optimum design space of SRAM cell with realistic value of Interconnect capacitive load can be found, whose intersection with design point of Mux circuit, will give optimized perfomance in designed Memory Circuit.

REFERENCES

[1] A. B. Sachid et al., "Sub-20 nm Gate Length FinFET Design: Can High-K spacers Make a Difference?," *Electron Devices Meeting, 2008. IEDM 2008.*, IEEE, 2008, pp.1-4.

[2] M. Alioto and G. Palumbo, "Interconnect-Aware Design of Fast Large Fan-In CMOS Multiplexers," *Circuits and Systems II: Express Briefs, IEEE Transactions on* , vol.54, no.6, pp.484-488, June 2007.

[3] E. Yoshida et al., "Performance Boost using a New Device Design Methodology Based on Characteristic Current for Low-Power CMOS," *Electron Devices Meeting, 2006. IEDM '06. International*, Dec. 2006, pp.1-4.

[4] K. von Arnim et al., "An Effective Switching Current Methodology to Predict the Performance of Complex Digital Circuits," *Electron Devices Meeting, 2007. IEDM '07. IEEE International*, Dec. 2007, pp.483-486.

[5] A. Sachid, M. Baghini, D. Sharma, and V. Rao, "Alternate scaling strategies for Multi-Gate FETs for high-performance and low-power applications," *SoC Design Conference (ISOCC), 2010 International*, Nov. 2010, pp.256-259.

Synthesis of Reversible Circuits using Heuristic Search Method

Kamalika Datta[*], Gaurav Rathi[†], Indranil Sengupta[†] and Hafizur Rahaman[*]

[*]Department of Information Technology, Bengal Engineering & Science University, Shibpur, Howrah 711103, India
Email: kdatta.iitkgp@gmail.com, rahaman_h@it.becs.ac.in
[†]Department of Computer Science & Engineering, Indian Institute of Technology, Kharagpur 721302, India
Email: isg@iitkgp.ac.in, gaurav.rathi01@gmail.com

Abstract—Reversible circuits are of vital importance in many applications involving low power design. One of the principle areas where reversible circuits play great role is quantum computing. One of the foremost requirements of quantum computation is that it requires all the circuits that are used should be reversible in nature. Reversible circuit is one which maps an individual input vector to a singular output vector. Because of its application in many areas including quantum computing, many synthesis approaches have been developed. In this paper we focus on a synthesis approach which is based on permutation theory and heuristic search. An artificial intelligence based search technique A^* is used to find near optimal solutions. Experimental results demonstrate that the proposed approach provides solutions within a very reasonable span of time.

I. INTRODUCTION

During the past few years research in the field of reversible computing is gaining momentum. Reversible logic [3][5] is one which maps an n-bit input to an n-bit output. Even though reversible logic is very different from irreversible or traditional logic, it is one of the profoundly investigated research topics.

In modern circuit design, power dissipation is one of the most important factors which must be addressed. Landauer [2] showed that loss of information during computation is equivalent to loss of power. It is also proved that the heat generated during computation is not because of processing of information but because of information loss in the computation process. Irreversible logic gates which are the vital building blocks in the design of modern-day circuits always dissipate energy. But circuits which theoretically dissipate zero power is required to be information preservative. And in this regard reversible logic is a viable choice since here inputs can be recovered from the outputs, i.e. information are bijectively transformed.

Even though the energy loss in present-day circuits is due to some imperfect properties of transistors and other materials, reversible logic has become an important alternative because of the theoretical zero power dissipation concept. Implementations of reversible circuits have already been reported in [23] which conforms to the fact that no additional power supply is required and the outputs are computed from the input signals only. Also, one of the significance of reversible logic lies in its applications in quantum computing. A quantum circuit works on unitary functions and hence it is reversible in nature.

Precisely due to these reasons, synthesis of reversible circuits is now an active area of research.

The main objective of circuit synthesis is that, given a gate library how to design a circuit which realizes the desired specification. In the scenario of quantum computing, the gates used in the synthesis process accord to physical operations on quantum states known as $q$ubits. And in the process of synthesis reducing number of gates gives rise to effective implementation. Various synthesis processes [8], [12], [14] for reversible circuits have been explored over the past few years. Broadly the different synthesis approaches are based on function representations such as permutation [9], BDD [20], and truth tables [21]. Different approaches have distinct advantages and disadvantages.

In this paper we have used a permutation based tree search approach for synthesis. A tree is dynamically created using permutations for different inputs. Thereafter without performing exhaustive search, a heuristic based search technique known as A^* is used to efficiently explore the search space. Rest of the paper is organized as follows. Section II describes the review works related to reversible logic, section III gives the details of the proposed technique, section IV provides the experimental observations and finally section V concludes the work.

II. LITERATURE SURVEY AND PRELIMINARIES

In this section we will briefly discuss about reversible circuits and reversible gates, and some of the important works that have been reported for the synthesis of Boolean functions using reversible gates.

In classical circuit synthesis, given a desired function and a gate library, the main aim is to find a circuit which realizes the function. One of the main objectives is to reduce the number of gates used to implement the function. We can consider reversible circuit synthesis as a special case of classical synthesis problem with certain restrictions.

A reversible circuit is defined as one with the following features:

- number of inputs must be equal to number of outputs,
- no fanouts are allowed, and
- the gates used must be reversible in nature.

As reversible functions permute the input vectors to generate the output vectors which has a bijective property, the input vectors can also be uniquely determined from the output vectors. Figure 1 depicts a reversible circuit with cascade of three reversible gates. The first gate in the cascade is the Toffoli gate, followed by a NOT gate and a CNOT gate. There exists various reversible gates in literature but here we discuss about Controlled-NOT, NOT and TOFFOLI gates only which constitute the so-called CNT library.

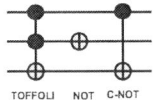

Fig. 1.   Example reversible circuit

Controlled-NOT gate, also known as Feynman gate [13], is one of the main components of quantum circuits. It is a two input gate which works as follow; the first input is passed to the output directly and the second input changes only if the first input is 1. NOT gate is simple and works on single line of input. It just inverts the value at the input. Toffoli gate [12] is very similar to C-NOT gate with the only difference in the number of inputs. For a three input gate, it passes the first two inputs unchanged and changes the third if the first two inputs are both 1. This definition of Toffoli gates can be generalized to more than 3 inputs.

In [9] Shende et al. investigated the synthesis of reversible circuits with minimum number of gates which contains no redundant input-output line pairs. They proposed new constructions for reversible circuits composed of CNT gate library based on permutation theory. Herein the structure of reversible circuits were studied and an optimal circuit synthesis method was proposed which performed better than its counterpart but suffered from the drawback of exponential runtime. The main focus of their algorithm was on the metric of reducing the gate count.

In [15] Maslov et al. proposed an iterative network synthesis algorithm based on Reed-Muller (RM) spectra which selects a Toffoli gate which conforms to a functional application specification based on metric of lower RM cost. However this algorithm fails to converge for certain large benchmark specifications, thereby necessitating an alternative algorithm based on RM spectrum which focuses on the characterization of templates for certain functional specifications. The advantage of the new approach is in its adaptability with more number of input functions.

In [18] Kole et al. proposed a permutation based optimal synthesis approach which uses a hybrid DFS-BFS synthesis technique to search the permutation tree. The main idea is to find an optimal solution and the cost metric used is the gate count. The main disadvantage of this technique is that, as the number of inputs increases the size of the tree will also increase exponentially, correspondingly increasing the search complexity. If the desired permutation is present at a deep leaf node, then the entire subtree above the leaf node has to be traversed.

In [7] Younis et al. proved that some reversible circuits exist which can be made asymptotically energy-lossless if the delay in the circuits can be changed to be arbitrarily large.

In [16] Gupta et al. proposed a tool for synthesis of reversible circuits. As the search space is large, they have used some heuristics to reduce the search space. This algorithm performs well and gives good result and are able to synthesize circuits with four and five inputs.

In [17] Saeedi et al. proposed a non-search based moving forward synthesis algorithm for CNOT-based quantum circuits. This algorithm guarantees to provide results with much less steps as compared to other search based methods.

As many works have been done, but still there is a huge scope of improvement as far as synthesis of reversible circuits are concerned, due to its importance in upcoming quantum circuits. Reducing the search space is one of the major issue in the synthesis process using permutation. So here in this paper we suggest a heuristic search algorithm which decreases the search space and finds near optimal solution for the desired specification.

III. THE PROPOSED APPROACH

In the synthesis of reversible circuits, given a specification of a circuit and a gate library, we derive a sequence of gates from the library to realize the given specification. In this work we perform a similar process, where the specification is the desired permutation and gate library used is the CNT gate library. There can be many measures to evaluate a synthesis process, like number of gates, quantum and transistor costs [22]. In this paper we have considered number of gates as the metric and reported quantum and transistor costs as well. We have used a heuristic search based technique A^* to find the desired permutation. In this section we will first discuss about the synthesis approach, then we will explain the A^* algorithm, and finally how the A^* algorithm is applied to the synthesis process of reversible circuits.

A. Synthesis approach

The synthesis approach used in this paper uses the concept of permutation. A circuit which is reversible in nature can be comprehended as a network of generalized Toffoli gates. So provided with a specification, our aim is to derive a network of Toffoli gates, that recognizes the desired specification. In [18] Kole et al. proposed a synthesis method that searches the permutation tree level by level to produce optimal results. They have used DFS with iterative deepening approach which they call as DFS-BFS method to arrive at the desired permutation. In our proposed work, we have also used the notion of permutation to define the search space, and used a heuristic search algorithm A^* to find the desired permutation.

A generalized Toffoli gate with n inputs define a permutation on the set $\{0, 1, 2, 3, ..., (2^n - 1)\}$. As mentioned in [18], the number of possible generalized Toffoli gates with n inputs is $n*2^{n-1}$. These possible gates are numbered from 0 to $n*2^{n-1} - 1$. Each of these gates impose a unique permutation on the input vectors. Figure 2 shows all possible 2-input Toffoli gates along with the encoded values and the corresponding permutations.

T - Gate	Encoded value	Permutation
	0	(1, 0, 3, 2)
	1	(0, 1, 3, 2)
	2	(2, 3, 0, 1)
	3	(0, 3, 2, 1)

Fig. 2.   Encoding for 2-input Toffoli gates

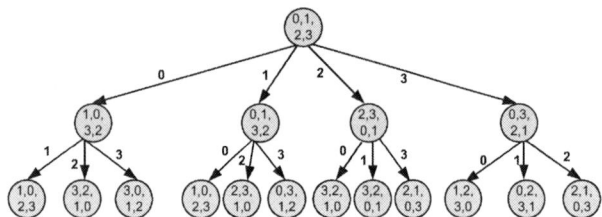

Fig. 3.   A complete synthesis tree for number of inputs=2 till level=2

Figure 3 shows the synthesis tree for number of input $n = 2$ and maximum level $l = 2$. Here the root of the synthesis tree represents the permutation $\{0, 1, 2, ..., (2^n - 1)\}$. The number of children from root node will be all possible Toffoli gate of size n, which is $n * 2^{(n-1)}$. And for the remaining levels, number of child nodes will be one less than the number of child nodes from root to next level, i.e. $n * 2^{(n-1)} - 1$. This is because of the fact that it is redundant to use the same Toffoli gate two times consecutively in cascade.

In general, there is no limit to the number of levels, and hence the tree can grow indefinitely. And in each subsequent levels the number of nodes are also increasing. At level $l = 1$ number of nodes are m, where m is the number of possible Toffoli gates. At level $l = 2$ number of nodes will be $m(m - 1)$; and at level $l = 4$ it will be $m(m - 1)^2$, and so on. Hence as we move deeper into the levels, number of nodes increases exponentially. Because of the exponential sized search space, polynomial time deterministic algorithms and exhaustive search algorithms are not feasible. However in the implementation due to memory constraint we have limited the search to a finite number of levels.

Here in this paper to reduce the complexity of searching the huge search space, we use the heuristic search algorithm A^* to arrive at the desired permutation.

B. A^* algorithm

Exhaustive search methods for finding the desired solution are sometime infeasible and uses lot of memory space and

time. So there is always a limit to the use of such methods, and more effective search algorithms should be in place as an alternative to these exhaustive search techniques. In this paper we use some heuristics, i.e. some task dependent information to advance to the goal node. There exists some heuristics which exceptionally reduces search efforts but do not ensure in finding minimal cost paths. But in many cases heuristic search helps in finding optimal solution. And in practical scenarios we are more concerned in reducing combination of the cost of the path and the cost which is incurred in searching for the path. In our algorithm we have used number of gates used for synthesis as the measure which we are trying to reduce.

The A^* algorithm is a heuristic search algorithm which uses the best-first search and a modified evaluation function to estimate the length of the path from start node to the solution node. In turn A^* reduces the total path cost. Sometime it is also observed that A^* gives cheapest solution under proper conditions (when the heuristic function is admissible). As the problem of reversible logic synthesis can be viewed as a search problem we can apply a A^* in an intelligent manner.

In the proposed technique the node of the tree is defined as a structure with:

- An array containing the permutation.
- A gate which links to its parent node.
- A heuristic function $h(n)$ which is an estimate of the number of gates required from the current node n to the goal node.
- The level of the node which is the number of gates taken to reach the node from the root. This is denoted as $g(n)$.
- The evaluation function f at any specified node n, $f(n)$, which is the sum of the costs of the path from start node to node n, i.e. $g(n)$, and cost of minimal cost path from the node n to the goal node, i.e. $h(n)$:

$$f(n) = g(n) + h(n) \qquad (1)$$

Algorithm 1 *Reversible Logic Synthesis*

Input: The desired permutation vector P and number of input m

Output: Sequence of Tofolli gates to implement P

begin

    $found = 0$;
    $S = (0, 1, 2, ..., 2^m - 1)$;   /* Initial Permutation */
    $add\_to\_OPEN(S)$;
    do
        $n = remove\_from\_OPEN()$;
        if ($n == NULL$)
            exit with failure;
        $add\_to\_CLOSED(n)$;
        if ($n == P$)   /* goal node found */;
            begin
                $found = 1$;
                Obtain the solution by tracing a path along
                the pointers from n to S;
            end

978-1-4673-0438-2/12 $31.00 © 2012 IEEE

        else
            begin
                Q = set of child nodes of n corresponding
                to all Toffoli gates of size m (excluding the
                gate from parent of n);
                for all $q \epsilon Q$
                    $add\_to\_OPEN(q)$;
                end for
            end
    while $(found == 0)$;
end

Function $add\_to\_OPEN(x)$

begin
    if $x \epsilon CLOSED$ return;
    Calculate $f(x)$;   /* $f(x) = g(x) + h(x)$ */
    Insert x into the priority queue OPEN based on the
        value of $f(x)$;
end

The function $remove\_from\_OPEN$ returns the next node from the priority queue $OPEN$. The function add_to_CLOSED adds a node to a list $CLOSED$, which contains nodes that have been already explored.

In the proposed algorithm we start with the root node with default permutation $(0, 1, \ldots 2^m - 1)$ and enqueue into the priority list OPEN. As this is not the desired permutation, we remove that from OPEN and place that on CLOSED list. Then we create all child nodes of the initial permutation with all the possible Toffoli gates. For each such node n, we calculate the net expected cost $f(n)$ which is the combination of current cost $g(n)$ and heuristic cost $h(n)$, after which we enqueue the nodes according to the expected cost into the priority queue OPEN. Thereafter we dequeue the node which is the first element in the list. Now the current node is checked for desired permutation; if it does not match then again we expand it, i.e. create $m * 2^{m-1} - 1$ child nodes. The same process is repeated till we get the desired permutation or the time budget expires. Depending on the heuristic function there is no guarantee that the algorithm will converge to an optimal solutions. Moreover since we are limiting the depth of the search, for some circuits we are not able to get the solution. So, by relaxing this constraint the above problem can be avoided.

C. Choice of the heuristic function $h(x)$

A very simple way to estimate the value of the heuristic function $h(x)$ is to count the number of mismatches in the permutation positions between the current node x and the goal node. However, one problem with this measure is that the number of mismatches in the permutation positions is often not correlated to the number of gates required for implementation. A single NOT gate, for example, will cause a mismatch in all the permutation positions.

For this reason, we have come up with an alternative measure which directly takes into account the fact that one gate can change only one bit of an input vector, and that

too possibly not for all input combinations. In the calculation of h, we look at all the bit positions of the permutations, and estimate the *distance* by counting how many bit positions differ. If the same bit changes for all the permutation positions, we can say that the goal permutation is ONE gate away. If the bit changes for half the permutation positions, then it is estimated that the goal is TWO gate away, and so on. This can be generalized as follows.

For m inputs, for a particular bit position,
if the bits change for $\frac{p}{2^i}$ positions, then
we add $(i + 1)$ to the heuristic estimate,   (2)
where $p = 2^m$.

The algorithm for calculating $h(x)$ is stated below.

Algorithm 2 *Heuristic Function Calculation*

Input: Current permutation $C = (C_0, C_1, ..., C_{p-1})$
            Goal permutation $G = (G_0, G_1, ..., G_{p-1})$
Output: Heuristic estimate h of C
begin
    $h = 0$;
    for $i = 0$ to $(m - 1)$ **do**
        k = number of positions in which i^{th} bit of
            C_j and G_j differ, $0 \le j \le p - 1$;
        if $(k > 0)$
            $h = h + log_2(\frac{p}{k}) + 1$; /* Follows from (2) */
    end for
end

The following examples illustrate how the function $h(n)$ is calculated.

Example 1
Let $n = 2$, $C = (1, 2, 0, 3)$, $G = (3, 1, 2, 0)$.
            $C : 01\ 10\ 00\ 11$
            $G : 11\ 01\ 10\ 00$
For the least significant bit, there is change in 2 positions.
For the most significant bit, there is change in 4 positions.
Therefore: h $= \left[log_2 \left(\frac{4}{2} \right) + 1 \right] + \left[log_2 \left(\frac{4}{4} \right) + 1 \right] = 2 + 1 = 3$

Example 2
Let $n = 2$, $C = (2, 3, 0, 1)$, $G = (3, 2, 1, 0)$
            $C : 10\ 11\ 00\ 01$
            $G : 11\ 10\ 01\ 00$
For the least significant bit, there is change in 4 positions.
For the most significant bit, there is no change.
Therefore: h $= \left[log_2 \left(\frac{4}{4} \right) + 1 \right] + 0 = 1$

The overall synthesis algorithm is illustrated below with the help of an example. Consider
            $n = 2$, $G = (3, 2, 1, 0)$
The steps of execution are shown in the Table I.

The partial tree that is searched in the process is shown in Figure 4. The solution is gate 2 followed by gate 0.

978-1-4673-0438-2/12 $31.00 © 2012 IEEE   331

TABLE I
STEPWISE ILLUSTRATION OF THE ALGORITHM

Node No.	Gate No.	Permutation	g(x)	h(x)	f(x)	OPEN	Action
0	-	0,1,2,3	0	2	2	-	Expand 0
1	0	1,0,3,2	1	1	2	1	
2	1	0,1,3,2	1	3	4	1,2	
3	2	2,3,0,1	1	1	2	3,1,2	
4	3	0,3,2,1	1	3	4	3,1,4,2	Expand 3
5	0	3,2,1,0	2	0	2	5,1,4,2	
6	1	3,2,0,1	2	2	4	5,1,6,4,2	
7	3	2,1,0,3	2	3	5	5,1,6,4,2,7	Match 5

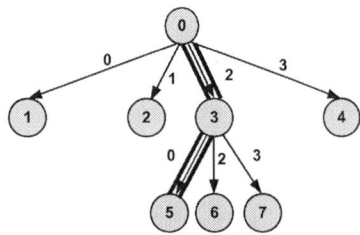

Fig. 4. Solution of synthesis

TABLE II
RESULTS OF THE EXPERIMENTATION

Given Permutation	Gate Count (GC)				Proposed method		
	[16]	[17]	[18]	[19]	GC	QC	TC
(1 0 3 2 5 7 4 6)	4	4	4	3	4	8	32
(7 0 1 2 3 4 5 6)	3	3	3	3	3	7	24
(0 1 2 3 4 6 5 7)	3	3	3	3	3	7	32
(0 1 2 4 3 5 6 7)	5	7	5	4	5	9	48
(1 2 3 4 5 6 7 0)	3	3	3	3	3	7	24
(3 6 2 5 7 1 0 4)	7	8	7	6	7	19	72
(1 2 7 5 6 3 0 4)	7	8	6	6	6	14	56
(4 3 0 2 7 5 6 1)	6	6	6	5	6	10	48
(7 5 2 4 6 1 0 3)	7	6	7	5	7	19	64
(1 2 3 4 5 6 7 8 9 10 11 12 13 14 15 0)	4	4	4	3	4	22	48
(0 7 6 9 4 11 10 13 8 15 14 1 12 3 2 5)	4	3	4	4	5	9	48
(7,1,4,3,0,2,6,5) 3_17	-	-	-	-	6	14	42
(0,7,4,3,2,5,1,6) ham3	-	-	-	-	5	9	48
(24,9,2,3,12,29,6,7,0,1,26,11,4,5,14,31,8,25,18,19,28,13,22,23,16,17,10,27,20,21,30,15) 4mod5(M)	-	-	-	-	5	13	48

IV. EXPERIMENTAL RESULTS

The proposed algorithm has been tested on a dual-core Pentium based PC with 2.4 GHz clock and 2 GB main memory running Ubuntu. We have used C language for the implementation. The inputs to the program are the number of gates and the desired permutation, and the output is the gate sequence. The results are reported in Table II. The assessment metric used in this method is gate count(GC). We have also reported quantum cost (QC) and transistor cost (TC) for the circuits [22].

The heuristic search algorithm A^* performs well in reaching optimal solutions for many of the example benchmarks within reasonable time. We have used the benchmark circuits used in [16] [17] [18] [19] to test our algorithm. We have also compared our results with some of the recent reported works. It has been observed that our proposed algorithm performs better than [16] and [17] in terms of number of gate count. Comparing the result with [18], our algorithm works better in terms of time taken. In [18] Kole et al. used a hybrid DFS-BFS method which exhaustively explores the search space and hence always generates optimal solutions. However, in [19] Li et al., the number of gates required for some of the benchmarks, for instance $(1, 0, 3, 2, 5, 7, 4, 6)$ and $(0, 1, 2, 4, 3, 5, 6, 7)$, have been shown to be even less than the optimal as reported by [18], which seems contradictory. In both the methods they have used generalized Tofolli gates.

V. CONCLUSION

A heuristic search algorithm A^* is proposed in this paper to synthesize a reversible circuit. The evaluation metric used here is the number of gate count. The heuristic function that we have used gives optimal solutions for most of the benchmark circuits within reasonable time, whereas time taken to search the first solution takes negligible amount of time which gives sub-optimal solution. Some improvements can be incorporated by tuning the heuristic function with some probabilistic value of the estimates.

REFERENCES

[1] J. P. McGregor and R. B. Lee, *Architectural enhancements for fast subword permutations with repetitions in cryptographic applications* ICCD, pp. 453-461, 2001.

[2] R. Landauer, *Irreversibility and heat generation in computing process* IBM Research and Development, vol. 5, pp. 183-191, 1961.

[3] C. H. Bennett, *Logical reversibility of computation* IBMJ. Research and Development, vol. 17, pp. 525-532, 1973.

[4] G. E. Moore, *Cramming more components onto integrated circuits* Electronics, vol. 38, no. 8, 1965.

[5] M. Nielsen and I. Chuang, *Quantum Computation and Quantum Information* Cambridge University. Press, Sept. 2000.

[6] P. Picton, *A universal architecture for multiple valued reversible logic* MVL Journal, vol. 5, pp. 27-37, 2000.

[7] S. Younis and T. Knight, *Asymtotically zero energy split-level charge recovery logic* Workshop on Low Power Design, 1994.

[8] A. Mischenko and M. Perkowski, *Reversible Maitra cascades for single output functions* IEEE/ACM 11th International Workshop on Logic Synthesis (IWLS), pp. 197-202, 2002.

[9] V. V. Shende and A. K. Shende and I. L. Markov and J. P. Hayes , *Synthesis of reversible logic circuits* IEEE Trans. on CAD of Integrated Circuits and Systems,vol. 22, no. 6, pp. 710-722, 2003.

[10] V. V. Shende and A. K. Shende and I. L. Markov and J. P. Hayes, *Reversible logic circuit synthesis* IEEE/ACM 11th International Workshop on Logic and Synthesis (IWLS) ,pp. 125-130, 2002.

[11] A. De Vos and B. Raa and L. Storme, *Generating the group of reversible logic gates* Journal of Physics A: Mathematical and General,vol. 35, pp. 7063-7078, 2002.

[12] T. Tofolli, *Reversible computing* Tech. Memo-MIT/LCS/TM-151, MIT Lab for Comp. Sci., 1980.

[13] R. Feynman, *Quantum mechanical computers* Optic News, vol. 11, pp. 11-20, 1985.

[14] E. Fredkin and T. Tofolli, *Conservative logic* Int. J. of Theoretical Physics, vol. 21, pp. 219-253, 1982.

[15] D. Maslov and G. W. Dueck and D. M. Miller, *Techniques for the synthesis of reversible Tofolli networks* ACM Trans. Des. Autom. Electr. Systems, vol. 12, no. 4, Article 42, pp. 42.1-42.28, 2007.

[16] P. Gupta and A. Agrawal and N. K. Jha, *An algorithm for synthesis of reversible logic circuits* IEEE Trans. on CAD, vol. 25, no. 11, pp. 2317-2329, 2006.

[17] M. Saeedi and M. Saheb Zamani and M. Sedighi, *Moving forward: A nonsearch based synthesis method toward efficient CNOT-based quantum circuit synthesis algorithms* In Proc. ASPDAC, pp. 83-88, 2008.

[18] D. Kole and H. Rahaman and D.K. Das and B. Bhattacharya, *Optimal Reversible Logic Circuits Synthesis based on a Hybrid DFS-BFS Technique* International Symposium on Electronic System Design (ISED2010), pp. 208-212, 2010.

[19] M. Li and Y. Zheng and M.S. Hsiao and C. Huang, *Reversible Logic Synthesis Through And Colony Optimization* Design, Automation, Test in Europe (DATE2010), pp. 208-212, 2010.

[20] R. Wille and R. Drechsler *BDD-based synthesis of reversible logic for large functions* In proceedings of Design Automation Conference (DAC), pp. 270-275, 2009.

[21] D. M. Miller and D. Maslov and G. W. Dueck *A transformation based algorithm for reversible logic synthesis* In proceedings of Design Automation Conference (DAC), pp. 318-323, 2003.

[22] R. Drechsler and A. Finder and R. Wille *Improving ESOP-Based Synthesis of Reversible Logic Using Evolutionary Algorithms* EvoApplications (2), Springer, pp. 151-161, 2011.

[23] B. Desoete and A. D. Vos *A reversible carry-look-ahead adder using control gates* INTEGRATION, the VLSI journal, vol. 33, Issue. 1, pp. 89-104, 2002.

[24] D. Grosse and X. Chen and G. W. Dueck and R. Drechsler *Exact SAT-based Toffoli Network Synthesis* 17th Great Lakes Symposia on VLSI, pp. 96-101, Lago Maggiore, Italy, 2007.

978-1-4673-0438-2/12 $31.00 © 2012 IEEE

Minimum Cost Fault Tolerant Adder Circuits in Reversible Logic Synthesis

Sajib Kumar Mitra
Department of Computer Science and Engineering
University of Dhaka
Dhaka-1000, Bangladesh
Email: sajibmitra.csedu@yahoo.com

Ahsan Raja Chowdhury
Faculty of Engineering and Technology
University of Dhaka
Dhaka-1000, Bangladesh
Email: farhan717@univdhaka.edu

Abstract—Conventional circuit dissipates energy to reload missing information because of overlapped mapping between input and output vectors. Reversibility recovers energy loss and prevents bit error by including Fault Tolerant mechanism. Reversible Computing is gaining the popularity of various fields such as Quantum Computing, DNA Informatics and CMOS Technology etc. In this paper, we have proposed the fault tolerant design of Reversible Full Adder (RFT-FA) with minimum quantum cost. Also we have proposed the cost effective design of Carry Skip Adder (CSA) and Carry Look-Ahead Adder (CLA) circuits by using proposed fault tolerant full adder circuit. The regular structures of *n*-bit Reversible Fault Tolerant Carry Skip Adder (RFT-CSA) and Carry Look-ahead Adder (RFT-CLA) by composing several theorems. Proposed designs have been populated by merging the minimization of total gates, garbage outputs, quantum cost and critical path delay criterion and comparing with exiting designs.

Index Terms—Reversible Logic, Fault Tolerant, Carry Skip Adder, Full Adder, Quantum Cost

I. INTRODUCTION

Higher level of integration and use of fabrication processes have dramatically reduced the heat loss over the last decades. Landauer [1] proved that logic computation that are not reversible, necessarily generate $kT*\log 2$ joules energy per bit information loss, where k means Boltzman's constant and T is the absolute root temperature where computation is performed. Reversible circuit doesn't loss information by considering a unique mapping between input and output. By using reversible computation zero power dissipation circuits is possible [2]. Reversible circuits are fundamentally different from traditional irreversible and are used to emphasis future technology. However reversible computation admits to generate multiple functions simultaneously. Quantum Computation is also gaining popularity as some exponentially hard problem can be solved in polynomial time and reversibility can be used to construct Quantum circuits [3].

Different arithmetic operations were realized by using reversible primitives since few decades earlier. Existing designs of Full Adder circuit were proposed in [4], [5], [6], [7] and finally generalized by [3] but any of these designs has no fault detection capability. Fault Tolerant full adder circuit was proposed in [8], [9], [10] without any generalization or cost effective structure. We have achieved more compact design of *n*-bit adder circuit which shows better performance than

existing designs [6], [7], [8], [9], [12], [11]. Here we have pictured the regular structure of Fault Tolerant adder circuits by using Fault Tolerant gates (Fig. 1 shows the design of Fredkin (FRG) [13] and New Fault Tolerant (NFT) [14] gates).

Rest of the paper is organized as follows: Section II discusses the construction of Reversible Logic, Fault Tolerant method, Quantum realization and Arithmetic Full Adder circuit. Section III illustrates the proposed cost effective design of Reversible Fault Tolerant Full Adder (RFT-FA), Carry Skip Adder (RFT-CSA) and Carry Look-Ahead Adder (RFT-CLA) by attaching the comparison with existing designs. Section IV describes a brief overview of the performance of proposed designs. Section V ends the paper with concluding remarks.

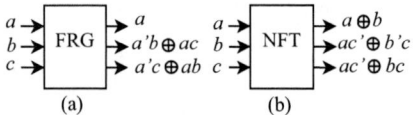

Fig. 1. (a) Fredkin Gate and (b) New Fault Tolerant Gate

II. BACKGROUND STUDY

In this section, we have discussed about the basic definitions and properties of Reversible Logic, Fault Tolerant mechanism and Quantum realization of reversible circuit.

A. Reversible Logic

Reversible Logic always retrains an unique mapping between input and corresponding output vectors.

Definition 1. The unit logic entity of reversible circuit is called **Reversible Gate** where the number of inputs is equal to the number of outputs and there is an one to one mapping between input and output vectors [3].

Let, the input vector, $I_v = \{I_1, I_2, ..., I_n\}$ and output vector, $O_v = \{O_1, O_2, ..., O_n\}$ then according to the above definition the relationship is $I_v \leftrightarrow O_v$.

Definition 2. The input vector, I_v and output vector, O_v for 2×2 **Feynmen Gate (FG)** [15] is defined as follows:

$$I_v = \{a, b\} \text{ and}$$
$$O_v = \{a, a \oplus b\}$$

978-1-4673-0438-2/12 $31.00 © 2012 IEEE

Fig. 2.   2×2 Feynman/CNOT Gate

TABLE I
TRUTH TABLE OF FEYNMAN GATE

Input		Output	
a	b	a	$a{\oplus}b$
0	0	0	0
0	1	0	1
1	0	1	1
1	1	1	0

Fig. 2 shows the block diagram of Feynman Gate and Table I shows the unique mapping between input and output vectors of Feynman gate.

Definition 3. The **Garbage Output** of any reversible gate or circuit is unwanted or unused output which will not be used in future rather than for checking reversibility [3].

For example, the Exclusive-OR operation can be realized by using only one Feynman Gate which produces an extra dummy output (a) along with its principle output signal ($a \oplus b$) to preserve reversibility (shown in Fig. 2).

Definition 4. Delay of any circuit is the number of maximum gate(s) from any input to any output where both ends preserve a continuous communication line. Total delay to generate EX-OR function is 1 (shown in Fig. 2) [3].

Definition 5. The input vector, I_v and output vector, O_v of 3×3 **Fredkin Gate (FRG)** [13] is defined as follows:

$$I_v = \{a,\ b,\ c\} \text{ and}$$
$$O_v = \{a,\ \bar{a}b \oplus ac,\ \bar{a}c \oplus ab\}$$

The block representation of FRG is shown in Fig. 1(a).

Definition 6. The 3×3 dimensional **Feynman Double Gate (F2G)** [16] is another reversible gate where the input vector, I_v and the output vector, O_v are defined as follows:

$$I_v = \{a,\ b,\ c\} \text{ and}$$
$$O_v = \{a,\ a \oplus b,\ a \oplus c\}$$

The Block Diagram and the unique mapping between input and output vectors of Feynman Double (F2G) gate can be shown as Fig. 3 and Fig. 4 respectively.

Definition 7. The input vector, I_v and output vector, O_v of 3×3 **New Fault Tolerant (NFT) gate** [14] as follows:

$$I_v = \{a,\ b,\ c\} \text{ and}$$
$$O_v = \{a \ominus b,\ a\bar{c} \oplus \bar{b}c,\ a\bar{c} \oplus bc\}$$

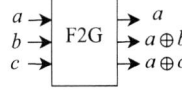

Fig. 3.   3×3 Feynman Double Gate

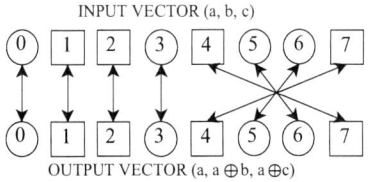

Fig. 4.   Feynman Double Gate (F2G) preserves Fault Tolerance over input-output unique mapping

The block diagram of NFT gate is shown in Fig. 1(b).

Definition 8. The input and output vectors of 4×4 **Modified TSG (MTSG) gate** [3] are I_v and O_v respectively and can be defined as follows:

$$I_v = \{a,\ b,\ c,\ d\} \text{ and}$$
$$O_v = \{a,\ a \oplus b,\ ab \oplus c,\ (a \oplus b)c \oplus ab \oplus d\}$$

Another popular reversible gates are Peres Gate (PG) [18], Toffoli Gate (TG) [17] and New Gate (NG) [19] etc.

B. Fault Tolerant Method

Reversibility recovers bit loss but is not able to detect bit error in circuit. Fault Tolerant reversible circuit is capable to prevent error at outputs.

Definition 9. Fault Tolerant (FT) gate, also called Conservative Reversible Gate [9] which means the Hamming weight of its input and output are equal.

Let, the input and output vectors of any Fault Tolerant gate are $I_v=\{I_0, I_1, ..., I_{n-1}\}$ and $O_v=\{O_0, O_1, ..., O_{n-1}\}$ where the following equations (1) and (2) must be preserved:

$$I_v \leftrightarrow O_v \tag{1}$$
$$I_0 \oplus I_1 \ominus ... \oplus I_{n-1} = O_0 \oplus O_1 \oplus ... \oplus O_{n-1} \tag{2}$$

For example, the Fault Tolerance property of Feynman Double gate (shown in Fig. 3) can be verified from Fig. 4 where square and circle represents ODD and EVEN parities respectively and the equivalent decimal values of input-output vectors are represented as corresponding decimal number (0-7).

F2G, FRG and NFT are 3×3 dimensional and MIG is 4×4 dimensional fault tolerant gate having unique mapping between Input and Output vectors. The input and corresponding output parities of Fault Tolerant gates are same [16]. In early, fault tolerant Gate is also called Parity Preserving Gate.

C. Quantum Realization

Quantum realization is another fact to judge the efficiency of reversible circuit which uses matrix multiplication rather than conventional Boolean operations. In Quantum Mechanics, the states of a particle is represented by qubits instead of bits. The operations over on qubits are matrix multiplication specified by using quantum gates (shown in Fig. 5) [20].

Definition 10. Quantum Cost (QC) of any reversible circuit is the total number of 2×2 quantum primitives that are used to realize equivalent quantum circuit [3], [21].

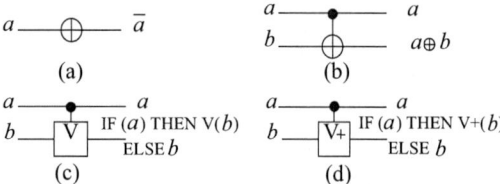

(a)   (b)

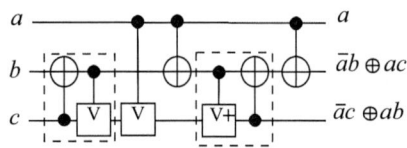

(c)   (d)

Fig. 5. Elementary Quantum Logic gates: (a) NOT, (b) Exclusive-OR, (c) Square Root of NOT (SRN) and (d) Hermitian matrix of SRN

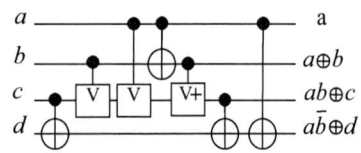

Fig. 6. Quantum circuit realization of Fredkin gate [13]

For example, the quantum cost of Feynman gate (shown in Fig. 2) is one because single 2×2 Quantum EX-OR gate is enough to realize its operations. The quantum circuit of several reversible gates is presented in [3]. Fig. 6 shows the quantum representation of Fredkin gate and the cost is 5.

Here we have proposed the quantum realization of reversible fault tolerant Modified IG (MIG) [8] and New Fault Tolerant (NFT) [14] gates by using quantum EX-OR, Square Root of NOT (SRN or V) and Hermitian of SRN (V^+) gates as shown in Fig. 7. According to design, the quantum cost of Modified IG (MIG) (New Fault Tolerant (NFT)) gate is 7(5).

D. Arithmetic Adder Circuit

Adder is an essential part of digital circuits to implement most of the mathematical operations. Single bit adder or Full Adder is the unit entity of any kind of computing devices.

Definition 11. A **Full Adder** is a digital circuit which takes two bits from two operands and carry bit from prior stage as input and generates the summation of three bits and corresponding carry as output [3].

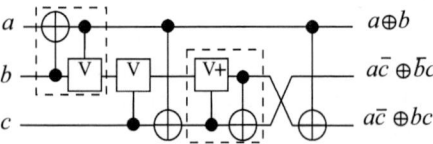

(a) Modified IG gate, Quantum Cost= 7

(b) New Fault Tolerant Gate, Quantum Cost = 5

Fig. 7. Quantum equivalent circuit realization of MIG [8] and NFT [14] by using quantum primitives

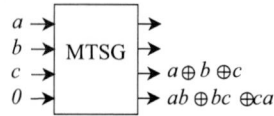

Fig. 8. Full adder Realization by using MTSG [3]

Let, a and b both are the single bit operand and c_{in} is the carry of previous stage then outputs of full adder circuit, Sum (s) and Carry (c_{out}) can be defined as follows:

$$s = a \oplus b \oplus c_{in}$$
$$c_{out} = ac_{in} \oplus bc_{in} \oplus ab$$

For example, single MTSG gate able to realize reversible full adder circuit as shown in Fig. 8.

Following section has described the proposed designs of Fault Tolerant Full Adder (RFT-FA) circuit followed by Carry Skip Adder (RFT-CSA) and Carry Look-ahead Adder (RFT-CLA) by including their performances over existing designs.

III. PROPOSED DESIGN

In this section, first we have described the proposed design of cost effective Reversible Fault Tolerant Full Adder by using New Fault Tolerant (NFT) and Feynman Double (F2G) gates. Then we have described the design of Fault Tolerant Carry Skip (RFT-CSA) and Carry Look-ahead (RFT-CLA) adders by using proposed design of Fault Tolerant Full Adder.

A. Fault Tolerant Full Adder Design

The quantum cost of New Fault Tolerant (NFT)(shown in Fig. 7(b)) and Fredkin (FRG) gates are same i.e. 5. But the quantum cost of Feynman Double (F2G) gate is 2. We have used New Fault Tolerant (NFT) and Feynman Double (F2G) gates because of reusability of proposed adder circuit for Carry Skip Adder and Carry Look-ahead Adder.

Definition 12. Single NFT Full Adder (SNFA) is a Fault Tolerant full adder circuit which consists of one New Fault Tolerant (NFT) gate and three Feynman Double (F2G) gates where the quantum cost is 11 and the total number of garbage output is 3 (shown in Fig. 9).

Theorem 1: The minimum number of garbage bit to realize Reversible Fault Tolerant Full Adder circuit is 3.

Proof: Let, a, b and c_{in} are the inputs of a full adder circuit where s and c_{out} are the corresponding outputs. There are three different states at the inputs (a, b and c_{in}) where the outputs (s and c_{out}) produce same patterns as shown in Table II. For any parity preserving reversible circuit, total number of EVEN or ODD parity at input or output is equal. Table II shows that the all input patterns are EVEN but the corresponding output patterns are ODD. Turning three ODD patterns at output into EVEN by adding two extra bits is not possible. Because two bits can represent 2^2 different states where 00 and 11 (01 and 10) are EVEN (ODD) only. So, Reversible Full Adder circuit requires at least 3 garbage bits to make itself Reversible Fault Tolerant Full Adder.

978-1-4673-0438-2/12 $31.00 © 2012 IEEE

TABLE II
INPUT-OUTPUT PATTERNS OF FULL ADDER

Input			Output	
a	b	c_{in}	s	c_{out}
0	1	1	0	1
1	1	0	0	1
1	0	1	0	1

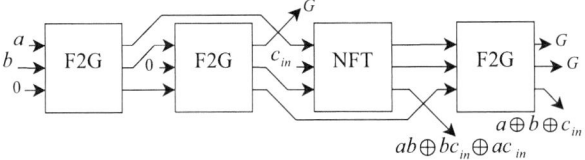

Fig. 9.   Proposed design of Fault Tolerant Single NFT Full Adder (SNFA)

TABLE III
COMPARISON BETWEEN PROPOSED AND EXISTING DESIGNS OF
REVERSIBLE FAULT TOLERANT FULL ADDER

Fault Tolerant Full Adder	Total Gates		GB	QC
	3×3	4×4		
Proposed	4	0	3	11
Existing [8]	0	2	3	14
Existing [9]	5	0	4	25
Existing [10]	6	0	6	18

GB= Garbage bits, QC= Quantum Cost

The performance analysis between the proposed design and all other existing designs is shown in Table III.

Table III shows that the Ref. [8] uses two gates and the corresponding quantum cost is 14 where proposed design uses low dimensional four gates and the quantum cost is only 11.

B. Fault Tolerant Carry Skip Adder

Fast carry emission is the main concern of Carry Skip Adder and it depends on: firstly, if any operand is equal to logical 1 then the full adder propagates c_{in} to c_{out} and secondly, it also generates carry itself (c_{out} independent on c_{in}).

Definition 13. Propagate is a simple XOR operation between two operands which is responsible for only bypassing the carry of previous stage to next stage [6].

Let, $X = (x_0, x_1, x_2, \ldots, x_{n-1})$ and $Y = (y_0, y_1, y_2, \ldots, y_{n-1})$ are two n-bit operands where Propagate p_i of ith stage can be defined from x_i and y_i as follows:

$$p_i = x_i \oplus y_i$$

Definition 14. Generate is an AND operation which enables current stage of adder to generate carry for next stage [6]. So the Generate of ith stage as follows:

$$g_i = x_i y_i$$

Definition 15. Reversible Fault Tolerant Carry Skip Adder (RFT-CSA) consists of SNFAs and FRGs to perform summation and propagate carry respectively which reduce the delay or bypassing carry due to the recalculation of carry for the next stage. If any input is equal to a logical 1, then it propagates the carry input to the carry output [9].

Proposition 1. n-bit RFT-CSA can be realized by using $(3n+1)$ F2Gs, n FRGs and n NFTs.

Proof: Let, n SNFAs are needed to realize n-bit RFT-CSA (each SNFA consists of three F2Gs and one NFT) to generate sum (s_i) and Propagate (p_i) where $i = 0, 1, 2, \ldots, (n-1)$. And n FRGs are needed for performing AND operation among n Propagates with c_{in}. So the calculation of the number of NFT (NFT_{CSA}), the number of FRG (FRG_{CSA}) and the number of F2G ($F2G_{CSA}$) to implement n-bit RFT-CSA is as follows:

$$NFT_{CSA} = n,$$
$$FRG_{CSA} = n \text{ and}$$
$$F2G_{CSA} = 3n$$

But RFT-CSA needs another extra F2G to generate final carry, c_{out} by performing EXOR operation between c_{n-1} and $(p_{n-1}p_{n-2} \cdots p_0 c_{in})$.

$$F2G_{CSA} = 3n + 1$$

So, n-bit RFT-CSA can be realized by using $(3n+1)$ F2Gs, n FRGs and n NFTs.

Proposition 2. n-bit RFT-CSA can be realized with $(n+5)$ Critical Path Delay.

The proposed design of 4-bit RFT-CSA is shown in Fig. 10 which uses proposed full adder (SNFA) circuit.

Finally, the total garbage (GB_{CSA}) and Quantum Cost (QC_{CSA}) of n-bit RFT-CSA can be written as follows:

$$GB_{CSA} = 4n$$
$$QC_{CSA} = 5 * 2n + 2 * (3n + 1)$$
$$= 16n + 2$$

The comparison between proposed RFT-CSA and existing designs is shown in Table IV. Carry Skip Adder is more reliable in case of hardware implementation where circuit cost is another factor of design with respect to Delay. In Table IV, the QC of proposed design is 66 which is minimum than all existing designs. Although the number of 3×3 gates of [7] is about equal to proposed design but QC of proposed design has been improved 20% because of using cost effective Feynman Double gate (QC of Feynman Double gate is only 2). The proposed cost-effective design of Fault Tolerant CSA (RFT-CSA) has improved cost factor having fault detection as well.

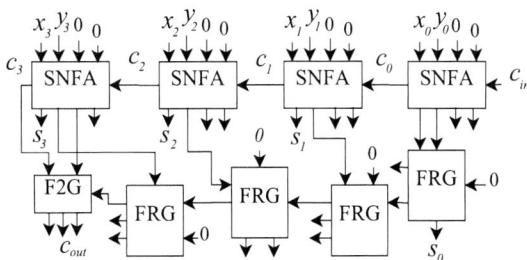

Fig. 10.   Proposed design of Reversible Fault Tolerant Carry Skip Adder

978-1-4673-0438-2/12 $31.00 © 2012 IEEE

TABLE IV
COMPARISON BETWEEN PROPOSED AND EXISTING DESIGNS OF CARRY SKIP ADDER

Carry Skip Adder	Total Gates		GB	DL	QC
	3×3	4×4			
Proposed†	21	0	16	9	66
Existing [7]§	22	0	26	16	88
Existing [8]†	6	8	19	13	80
Existing [9]§	24	0	23	14	120

† Fault Tolerant, § with fan-out
GB= Garbage bits, DL= Delay, QC= Quantum cost

TABLE V
COMPARISON BETWEEN PROPOSED AND EXISTING DESIGN [8] OF FAULT TOLERANT CARRY LOOK-AHEAD ADDER

Fault Tolerant Carry Look-ahead Adder	Total Gates		GB	DL	QC
	3×3	4×4			
Proposed (4-bit)	16	0	12	7	44
Existing [8] (2-bit)	15	4	28	12	73

GB= Garbage bits, DL= Delay, QC= Quantum cost

C. Fault Tolerant Carry Look-ahead Adder Circuit

This section introduces the design of Reversible Fault Tolerant Carry Look-ahead Adder (RFT-CLA) circuit overlaps the performance of all existing designs. Proposed design of RFT-CLA is based on New Fault Tolerant (NFT) and Feynman Double (F2G) gates where the carry is generated before sum.

Definition 16. Reversible Fault Tolerant Carry Look-ahead Adder (RFT-CLA) consists of serial attachment of n SNFAs but the work as a carry generator itself where the carry output of ith stage (c_i) is produced before sum s_i where $i=$ 0, 1, 2, ..., $(n-1)$.

Proposition 3. n-bit RFT-CLA can be realized by using the combination of n NFTs and n F2Gs.

Proposition 4. The Delay of n-bit Reversible Fault Tolerant CLA ($D_{RFT-CLA}$) can be minimized to $(n+3)$.

Proof: According to Definition 4, the Delay of any circuit is the number of maximum gates laying on contiguous path of any input to output. The Delay of SNFA, $D_{SNFA}=$ 4 to generate sum not carry. Delay of parallel adder circuit depends on carry propagation (from c_{in} to c_{out}) of every stage. Any n-bits RFT-CLA needs n SNFAs where Delay of RFT-CLA, $D_{RFT-FA}\neq 4n$. Because carry input (c_i) of i_{th} stage is generated by spending 1 units Delay where $i=$ 0, 1, 2, ..., $(n-1)$. In first stage, extra two units Delay is added because

of first carry output (c_0) generation is related to operands at first stage. On the other hand, last stage has extra single unit Delay because the final sum is generated after one stage of generation of final carry (c_{out}). So the Delay calculation for n-bits RFT-CLA is as follows:

$$D_{RFT-FA} = n+3$$

Therefore an n-bit RFT-CLA can be realized by using $(n+3)$ unit Delay.

Proposition 5. An n-bit Reversible Fault Tolerant Carry Look-ahead Adder (RFT-CLA) can be realized with minimum Quantum Cost $11n$.

Proposition 6. An n-bit Reversible Fault Tolerant Carry Look-ahead Adder (RFT-CLA) can be realized with minimum Garbage $3n$.

Fig. 11 shows the proposed design of 4-bit Reversible Fault Tolerant Fast Adder and Table V shows the performance of proposed design by comparing with existing [8] design of 2-bit reversible Fault Tolerant Carry Look-ahead Adder.

IV. PERFORMANCE ANALYSIS

Previous two sections have discussed about the design of Reversible Fault Tolerant Full Adder, Carry Skip Adder and Carry Look-ahead Adder and the comparison of corresponding existing designs. This section have presented an abstract overview of proposed designs and the complexity analysis for n-bit Reversible Fault Tolerant Adder circuits. Fig. 12 represents that the performance of proposed 4-bit Reversible Fault Tolerant Carry Skip Adder is better compared to existing [8]. The number of gates in proposed design is greater than existing design [8] because of dimensional impact (lower dimension is preferable) which can be treated as negligible because of other factors (delay, garbage and quantum cost). Table VI describes the another evolutionary observation between proposed n-bit Reversible Fault Tolerant Carry Skip Adder and Carry Look-ahead Adder designs. We have already given the comparative study between proposed and existing designs of reversible fault tolerant Carry Skip Adder (Carry Look-ahead Adder) in Table IV (Table V) individually. Proposed designs of CSA and CLA demand better performance than existing designs in terms of number of gates, garbage outputs, delay and quantum cost. Along with the lower dimensional (3×3) fault tolerant gates our proposed designs have got more flexibility in reversible CMOS [22] realization. Pictorial representation of performance evaluation of Reversible Fault Tolerant Carry Look-ahead adder circuit over Carry Skip Adder circuit is shown in Fig. 13.

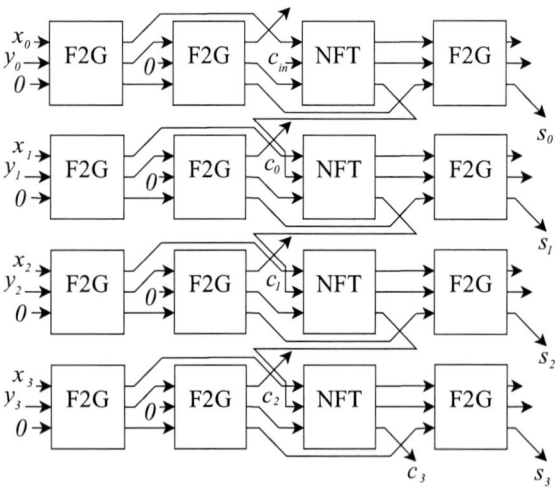

Fig. 11. Efficient Design of 4-bit Reversible Fault Tolerant Carry Look-ahead Adder

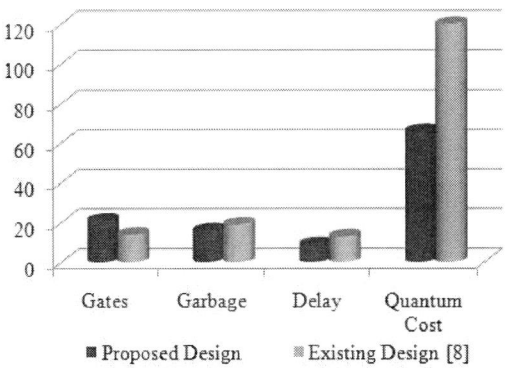

Fig. 12. Comparison between Proposed design and Existing design [8] of 4-bit Carry Skip Adder

TABLE VI
COMPARISON AMONG PROPOSED REVERSIBLE FAULT TOLERANT FAST, CARRY SKIP AND CARRY LOOK AHEAD ADDER

Proposed Adder Cirucits	Total Gates		GB	DL	QC
	FRG	F2G			
RFT-CLA †	n	$3n$	$3n$	$n+3$	$11n$
RFT-CSA	$2n$	$3n+1$	$4n$	$n+5$	$16n+2$

† Minimum Delay and Minimum cost
GB= Garbage bits, DL= Delay, QC= Quantum cost

V. CONCLUSION

In our proposed designs, we have combined all marginal cost factors (Gate cost, Delay, Garbage and Quantum cost) to generate optimized architecture of Reversible Fault Tolerant adder circuits which gather better performance than existing all fault tolerant designs. This paper has covered the designs of minimum cost fault tolerant Carry Skip Adder (RFT-CSA) and Carry Look-ahead Adder Circuits. Both designs have used the proposed structure of fault tolerant Full Adder (RFT-FA or SNFA) circuit has minimum quantum cost 11. Several number of theorems have been proposed to make the designs of RFT-CSA and RFT-CLA more generalized for n-bit fault tolerant adder circuitry. Finally, we have attached the evolutionary report of performance of proposed designs.

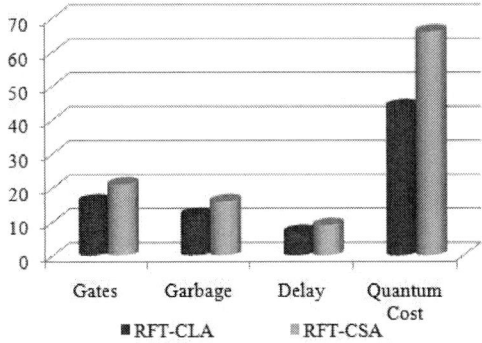

Fig. 13. Comparison between Proposed design of 4-bit Reversible Fault Tolerant Carry Skip Adder and Carry Look-ahead Adder

REFERENCES

[1] R. Landauer, "Irreversibility and heat generation in the computing process," *IBM Journal Of Research and Development*, vol. 5, pp. 183–191, 1961.

[2] C. H. Bennett, "Logical reversibility of computation," *IBM Journal Of Research and Development*, vol. 17, pp. 525–532, 1973.

[3] A. K. Biswas, M. M. Hasan, A. R. Chowdhury, and H. M. H. Babu, "Efficient approaches for designing reversible binary coded decimal adders," *Microelectronics Jounrnal*, vol. 39, no. 12, pp. 1693–1703, December 2008.

[4] H. M. H. Babu and A. R. Chowdhury, "Design of a compact reversible binary coded decimal adder circuit," *Elsevier Journal Syst. Archit.*, vol. 52, no. 5, pp. 272–282, January 2006.

[5] H. M. H. Babu, M. R. Islam, A. R. Chowdhury, and S. M. A. Chowdhury, "Reversible logic synthesis for minimization of full-adder circuit," in *17th International Conference on VLSI design'04*, Mumbai, India, 2004, pp. 757–760.

[6] H. Thapliyal and M. B. Srinivas, "A novel reversible tsg gate and its application for designing reversible carry look-ahead and other adder architectures," in *Asia-Pacific Computer Systems Architecture Conference*, 2005, pp. 805–817.

[7] P. K. Lala, J. P. Parkerson, and P. Charaborty, "Adder designs using reversible logic gates," *WSEAS TRANSACTIONS on CIRCUITS and SYSTEMS*, June 2010.

[8] M. S. Islam, M. M. Rahman, Z. begum, and M. Z. Hafiz, "Efficient approaches for designing fault tolerant reversible carry look-ahead and carry-skip adders," *MASAUM Journal of Basic and Applied Sciences*, vol. 1, no. 3, 2009.

[9] J. W. Bruce *et al.*, "Efficient adder circuits based on a conservative reversible logic gates," in *ISVLSI '02: Proceedings of the IEEE Computer Society Annual Symposium on VLSI*. Washington, DC, USA, 2005, pp. 83–88.

[10] M. Haghparast and K. Navi, "Design of a novel fault tolerant reversible full adder for nanotechnology based systems," *World Applied Science Journal*, vol. 3, no. 1, pp. 114–118, 2008.

[11] B. Desoete and A. D. Vos, "A reversible carry-look-ahead adder using control gates," *Intregation the VLSI Journal*, vol. 33, no. 1, pp. 89–104, 2002.

[12] J. Lim, D. Kim, and S. Chae, "A 16-bit carry-lookahead adder using reversible energy recovery logic for ultra-low-energy systems," *IEEE J. Solid-State Circuits*, vol. 34, pp. 898–903, 1999.

[13] E. Fredkin and T. Toffoli, "Conservative logic," *International Journal Of Theoretical Physics*, vol. 21, pp. 219–253, 1982.

[14] M. Haghparast and K. Navi, "Novel fault tolerant gate for nanotechnology based systems," *American Journal of Applied Sciences*, vol. 5, no. 5, pp. 519–523, 2008.

[15] R. P. Feynman, "Quantum mechanical computers," *Opt. News*, vol. 11, no. 2, pp. 11–20, 1985.

[16] B. Parhami, "Fault tolerant reversible circuits," in *In Proc. of 40th Asimolar Conference Signals, Systems and Computers*. Pacific Grove, CA, 2006, pp. 1726–1729.

[17] T. Toffoli, "Reversible computing," *MIT Lab for Computer Science*, vol. 85, pp. 632–644, 1980.

[18] A. Peres, "Reversible logic and quantum computers," *Physics Review A.*, vol. 32, no. 6, pp. 3266–3276, December 1985.

[19] M. H. A. Khan and M. A. Perkowski, "Logic synthesis with cascades of new reversible gate families," in *6th International Symposiumon Representation and Methodology of Future Computing Technology (Reed-Muller)*, March 2003, pp. 43–55.

[20] W. N. N. Hung, X. Song, G. Yang, J. Yang, and M. Perkowski, "Quantum logic synthesis by symbolic reachability analysis," in *41st Conference on (DAC'04), Design Automation Conference*, May 2004, pp. 838–841.

[21] M. Perkowski and et al, "A hierarchical approach to computer-aided design of quantum circuits," in *6th International Symposium on Representations and Methodology of Future Computing Technology*, 2003, pp. 201–209.

[22] Y. V. Rentergem and A. D. Vos, "Optimal design of a reversible full adder," *International Journal of Unconventional Computing*, vol. 1, pp. 339–355, 2005.

2012 25th International Conference on VLSI Design

Width-Aware Fine-Grained Dynamic Supply Gating: A Design Methodology for Low-Power Datapath and Memory

Lei Wang, Somnath Paul and Swarup Bhunia
Department of Electrical Engg. & Computer Science
Case Western Reserve University
Cleveland, USA
Email: {lxw185, sxp190, skb21}@case.edu

Abstract—With increasing contribution of leakage in total active power, run-time leakage control techniques are becoming extremely important. Supply gating provides an effective, low-overhead and technology scalable approach for active leakage reduction through the well-known "stacking effect". However, conventional supply gating approaches are typically coarse-grained in both space and time - i.e. are applied to large datapath or memory blocks when an entire logic/memory block is idle for sufficiently long period. They suffer from limited applicability at run time. On the other hand, fine-grained supply gating is constrained primarily by the large wake-up delay and wake-up power overhead. In this paper, we propose a novel fine-grained width-aware dynamic supply gating (WADSG) approach to reduce both active leakage and redundant switching power in datapath and embedded memory (e.g. L1/L2 cache). The approach exploits the abundance of narrow-width (NW) operands in general-purpose and embedded applications to "supply-gate" unused parts of integer execution units and memory blocks while they are in use. We introduce a novel levelized gating strategy to virtually eliminate the wake-up delay overhead. We employ the proposed WADSG approach to a superscalar processor. To reduce the wake-up power we use a width-aware instruction issue policy. In case of L1 and L2 cache, we store the width information per "ways" of associative cache and supply-gate the most significant bits of the NW ways. We also propose a width-aware block allocation and replacement policy to maximize the number of NW ways. Simulation results for 45nm technology with Spec2k benchmarks show major savings (*34.5%*) in total processor power (considering both switching and active leakage power) with *no* performance impact. As a by-product, the proposed scheme also improves the thermal profile of both datapath and memory.

Keywords-Narrow-width operands, Dynamic Supply Gating, Processor Datapath, Cache, Width-Aware Issue

I. INTRODUCTION

Power consumption has emerged as a primary design constraint for integrated circuits (ICs) in both general-purpose and embedded applications [1-2]. The active power in an IC comprises of switching power and active leakage in logic and memory circuits. In the nanometer technology regime, leakage power has become a major component of total power [1] despite breakthrough advances in device technologies, such high-K metal gate devices. Due to exponential dependence of subthreshold leakage (considered to be the largest leakage component) to temperature, leakage current increases significantly at run time in the high-activity regions of a chip e.g. datapath. Active leakage in datapath has introduced both power and power-density induced reliability concerns. Moreover, active leakage in large on-chip memory (e.g. processor caches) can be very high due to large number of memory cells present in an embedded memory. Fig. 1(a) and 1(b) show the relative magnitude of leakage and dynamic power across different technology nodes, for a processor datapath (64-bit logarithmic adder) and cache (2MB), respectively, with predictive

We acknowledge Intel Corporation for funding part of the research presented in this work.

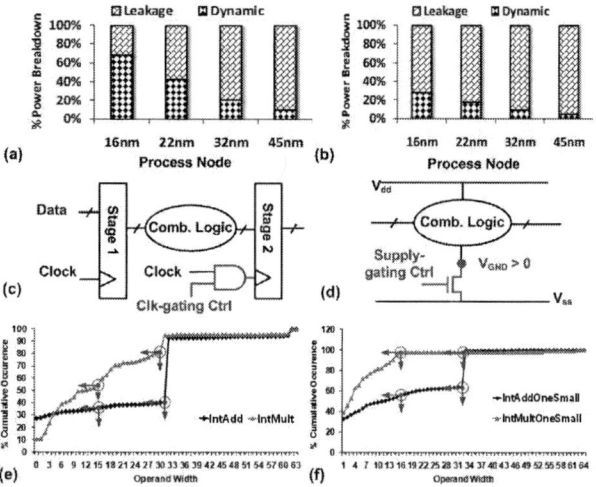

Fig. 1. Percentages of switching and leakage power for: a) functional units and b) on-chip cache in a processor across technology nodes; c) clock gating of sequential elements; d) supply-gating of combinational logic; e) cumulative value of narrow width (NW) operations with both operands NW; and f) only one operand NW for a set of benchmark applications.

technology models (PTM) [15]. We observe that for both datapath (Fig. 1(a)) and memory (Fig. 1(b)) active leakage plays a major role in total power. Hence, it is extremely important to employ run-time leakage control techniques in datapath and embedded memory.

Power gating has been widely explored earlier [7, 9, 14] and used in practice as an effective run-time power management technique. Power gating comes in two forms: 1) clock gating and 2) supply gating, as illustrated in Fig. 1(c) and (d). A majority of exiting power gating approach focus on clock gating to reduce active power in datapath modules. They work by "gating" the clock to select registers, thereby preventing the dataflow across pipe stages. While clock gating has been shown to be generally effective to reduce switching power as a coarse-grained power-gating approach, it cannot be effectively used for leakage control in datapath. Besides, they cannot be applied to memory. On the other hand, supply gating (Fig. 1(d)) has been shown to be highly effective to reduce standby leakage in both datapath and memory due to the well-known "stacking effect". Use of supply gating for active leakage reduction faces two major challenges: *1) identifying the idle periods for a functional unit (FU); and 2) minimizing the wake-up delay/power overhead for the gated FU*. While the clock gating circuit typically imposes minimal delay and power overhead for switching between gated and ungated modes,

978-1-4673-0438-2/12 $31.00 © 2012 IEEE

supply gating can introduce large overhead in delay and power due to gating logic. Two classes of dynamic supply gating solutions have been explored to address the above challenges: 1) a coarse (block-level) gating strategy that chooses to "gate" an entire FU [4] [13]; and 2) leakage-aware synthesis strategy [3] that requires re-design of the datapath. The first class of approaches miss the opportunity of spatial as well as temporal fine-grained gating of a logic block, while the second class suffers from scalability to large design and incurs large area overhead. Supply gating and its variants (e.g. source biasing) has also been explored for leakage reduction in embedded memory [12]. However, they typically suffer from considerable power and performance overhead due to the gating logic. Besides, they cannot take advantage of fine-grained gating (e.g. within a subarray and a words of a subarray).

In this paper, we propose a fine-grained (in both spatial and temporal sense) dynamic supply gating approach that can be employed to both datapath and embedded memory . It can lead to drastic reduction in active leakage and redundant switching power. The proposed approach is based on the observation that most of the operands in general purpose applications are narrow-width (NW) [14] i.e. the higher bits of the operand (64 to 32 or 32 to 16) are 0, while the lower bits are non-zero. Fig. 1(e) shows the occurrence of NW operands in SPEC2KInt benchmarks compiled for 64-b Alpha and simulated using Simplescalar toolset [6] with reference data. From Fig. 1(e), we note that for integer addition, *36%* and *93%* of the operations have both operands less than and equal to 16 and 32 bits, respectively. For integer multiplication, the percentages are *58%* and *81%*. Fig. 1(f) shows the cumulative occurrence of *minimum operand width* which further establishes that the scenario of one NW operand is more frequent than when both operands are NW. We exploit this abundance of NW operands to develop a width-aware dynamic supply gating (*WADSG*) methodology which supply gates unused logic gates(cells) in datapath (memory). In particular, the key technical contributions of this work are as follows:

1. We propose a methodology for fine-grained (in space and time) active power reduction for both datapath and embedded memory by exploiting the abundance of NW operations in general-purpose and embedded applications. The proposed approach reduces both active leakage as well as redundant switching power. We provide extensive analysis on the effectiveness of the proposed width-aware dynamic supply gating approach in a 64-bit superscalar processor by employing WADSG to both processor execution units and L1/L2 cache.

2. The key challenges with DSG are identified as a) hiding wake-up delay; b) minimizing the wake-up power; and c) minimizing the width computation and storage overhead. To address these challenges, we follow a circuit-architecture codesign approach. First, we propose a novel levelized gating strategy to hide the wake-up delay of the gating transistors. Instead of using a single shared gating transistor, it uses multiple ones across logic levels which masks the wakeup latency by the logic propagation delay. Second, we introduce width-aware allocation policies, where operations are allocated to the FUs in a manner that minimizes the wakeup overhead while maximizing the leakage saving. Third, we propose low-overhead circuit/microarchitectural modifications to minimize the width computation overhead.

3. We extend the concept of *WADSG* to achieve both dynamic and leakage power savings for on-chip caches. We exploit the associativity of processor caches to store the width information per "way" of a cache, instead of each block/word. This minimizes the overhead associated with wakeup as well as storing width information. Furthermore, we propose a width-aware way allocation scheme which

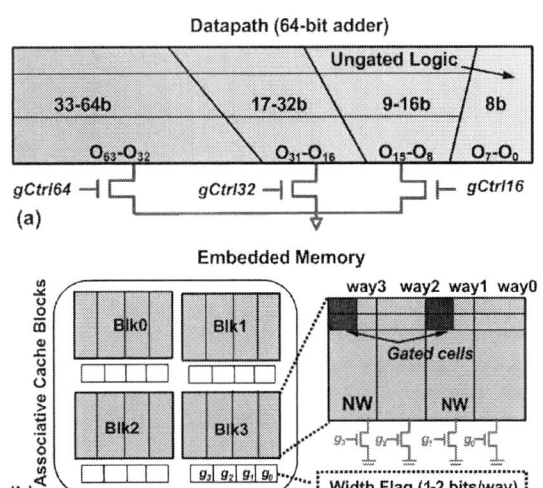

Fig. 2. a) Application of WADSG to processor execution units; b) application of WADSG to cache blocks.

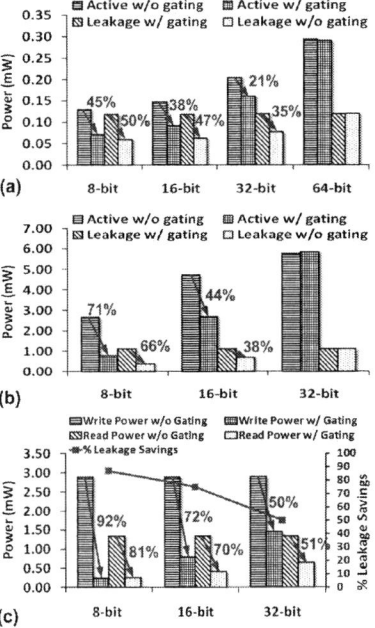

Fig. 3. Power consumption in a) 64-b integer adder and b) 32-b integer multiplier with and without width-aware supply gating; c) Read, write and leakage power saving in a 2Kb SRAM array with width-aware gating.

tries to increase the number of NW ways.

4. We analyze the impact of the proposed scheme on system power saving. We also show that, as a by-product, the proposed approach can improve the die thermal profile by reducing the datapath power density. Finally, we show that WADSG can be applied to floating point datapath and can be extended to embedded domains, which also show abundance of NW operands.

II. WIDTH-AWARE SUPPLY GATING

A. Overall Approach

Fig. 2 illustrates the proposed gating approach. The entire logic for a datapath is partitioned into different blocks corresponding to different output bits. A levelized gating approach is then applied to each block as shown in Fig. 2(a). Based on 1-2 bits which denote the width of the input operands to the datapath, one or more blocks are "waked-up" in a levelized manner to minimize the wake-up overhead. The approach is applied to cache blocks by assigning 1-2 status bit(s) to each cache way to indicate if they are narrow width (NW) or full width (FW) (Fig. 2(b)). In the following sections, we describe the circuit/architecture co-design approaches to realize the *WADSG* scheme for a conventional processor pipeline.

B. Impact on Power Saving

1) Functional Units: Fig. 3(a) and (b) present the power saving results for a 64-b integer logarithmic adder and a 32-b Wallace tree multiplier through *WADSG* scheme with 45nm PTM [15]. Active power was measured by simulating for 1K input random vectors for a clock frequency of 1GHz with 10% input activity. The delay was measured by sensitizing the circuit critical timing path. From Fig. 3(a) we note that for the integer adder, WADSG can achieve average savings of *35%* and *44%* in dynamic and leakage power, while for the multiplier, savings are higher *57%* and *52%*, respectively.

2) L1 and L2 Data Caches: Power consumed in reading and writing NW words to the data array of on-chip caches can be substantially minimized through *WADSG*. Unused bit locations in a word can be also supply-gated to reduce leakage power. Fig. 3(c) shows the read, write and leakage power savings for a 2Kb SRAM array with peripheral read/write circuitry. With WADSG, we note that average read, write and leakage power savings are *67%*, *71%* and *70%*, respectively.

C. Challenges: Wake-up and Width Computation Overhead

Waking up the entire 32-b or 64-b FU from the gated to active state incurs significant delay and power overhead. For our 64-b adder, this delay overhead was calculated to be *25%*. Compared to the power consumed during normal 64-b operation, the wake-up power overhead is *58%*. For the multiplier, the wake-up delay and power overheads are *22%* and *45%* of the normal 32-b operation. Considering that the width computation hardware is realized using simple 2-input cascaded OR logic, for 64-b operands it will require 6 logic stages. When synthesized at 45nm node, it incurs a delay and power overhead of *272ps* and *22μW*, respectively. It is important to hide the width-computation delay and minimize the power overhead due to it.

III. ARCHITECTURE-LEVEL MODIFICATIONS

A. Issue and Wake-up/Select Logic

1) FU Allocation Strategies: Challenges for the *WADSG* scheme as outlined in Section II.C can be addressed through novel width-aware FU allocation policies. These policies were implemented for the baseline processor configuration (Table I).

A. Transition-Aware FU Allocation: Due to large wake-up power overhead, an execution unit which is gated to a NW state should remain in the same state and make minimum number of transitions to FW state. The transition-aware FU allocation aims to minimize $|W_{t,i} - W_{t-1,i}|$, where $W_{t,i}$ and and $W_{t-1,i}$ are the maxm. operand widths to execution unit i in cycle t and $t-1$ respectively. In this policy, $W_{t-1,i} \forall i = 1\ to\ N$, is checked and the operand is assigned to unit i, such that i will undergo minm transition from W_i. For the proposed scheme, width information W_i for each FU is encoded using 1 or 2 bits of state elements.

B. Leakage-Aware FU Allocation: This policy targets at keeping some of the execution units in NW state for major portion of the execution time. For example, in the *Leakage Aware (3:3)* and *Leakage*

TABLE I
PROCESSOR CONFIGURATION

Processor	8-way issue, 128 RUU, 64 LSQ, 6 integer ALUs, 2 integer mul/div units, 4 FP ALUs, 4 FP mul/div units, 2 Wr/Rd ports
Branch Prediction	Combined, 32-entry RAS, 2K 4-way BTB, 8 cycle mis-prediction penalty
Caches	64KB 2-way 2 cycle I/D L1, 2MB, 4-way 12 cycle L2
Main Memory	100 cycle latency, 32-byte wide bus

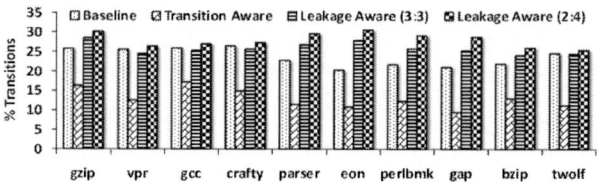

Fig. 4. Total number of transitions from NW to FW operands expressed as a percentage of the total number of operations for baseline and different FU allocation policies.

Aware (2:4) policies, the last 3 and 4 integer ALUs are targeted for NW operation. Wide operand operation is allowed on these units, only if the number of wide units are not enough to support all the ready instructions in a given window.

Fig. 4 shows the transitions, expressed as a percentage of the total number of operations for the baseline configuration as well as for the proposed allocation policies. For the baseline case, integer ALU and the multiplier undergoes transitions for *24%* of the operations on an average. Rate of transition is higher for leakage-aware allocation policies (~26% for 3:3 and ~28% for 2:4). However, the transition-aware allocation strategy reduces the average number of transitions to *12%* of the total operations.

2) Issue Logic Modification: Fig. 5(a) shows the scope for *WADSG* scheme in a superscalar based pipeline model [5]. Fig. 5(a) shows that the *WADSG* scheme can be applied to i) the reorder buffer and the register file; ii) functional units and iii) on-chip cache (both L1 and L2). In order to realize this, the width is first calculated when an operand is written into the on-chip memory from the main memory and is encoded into 1-2 bits. The overhead for this width computation logic is reported in Section II.C. Note that this width computation can be performed in parallel to the error correction code (ECC) calculation which is performed in case of an L2 miss, and therefore does not incur any additional delay overhead. In order to propagate this width information along the pipeline, each ROB and register file entry needs to be augmented with additional bits ($W_1 W_0$) for storing the width information for each operand. In the FU allocation policy implemented, the operand width information for the source operands coming from ROB or register file is utilized by a modified issue logic to select either a NW or FW FU. The logic for selecting the ready instructions from the issue window is implemented as a tree of priority encoders (Fig. 5(b)). In the modified FU selection logic illustrated in Fig. 5(b), D_n indicates the previous state of FU_n. For the leakage-aware FU allocation scheme, once a request (Req_k) is selected by the root of the priority encoder tree (Fig. 5(b)), the modified logic checks whether one of the source operands is FW or both are NW. In case of FW operand(s), the request is first forwarded to wide FUs (FU which is not currently gated). If no wide FUs are left for allocation, a FU which is targeted for NW operation is *waked-up* and the wide operation is assigned to this FU. For NW operations, the request is first forwarded to FUs targeted for NW operation. Fig.

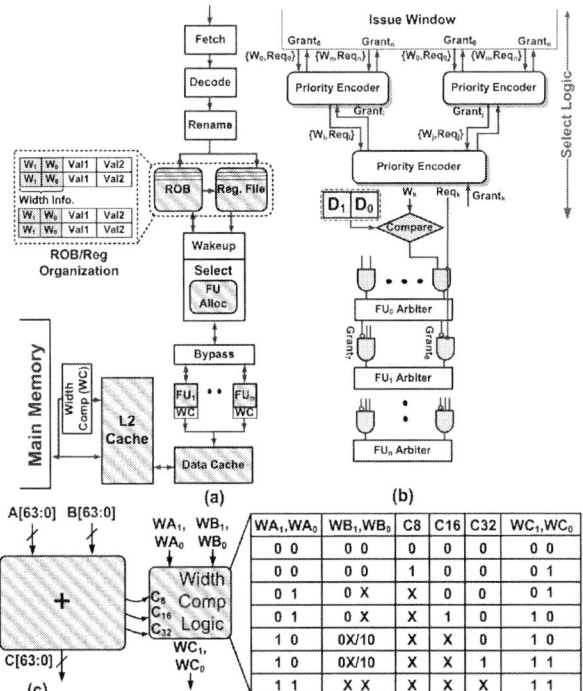

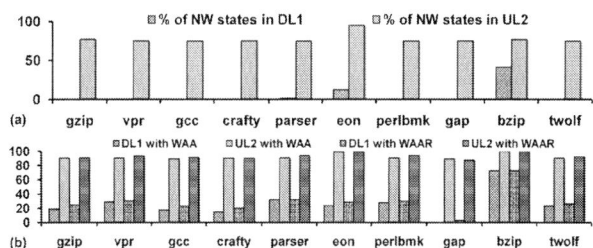

Fig. 7. a) % of NW states in DL1 and UL2 for SPECInt2K benchmarks simulated on the baseline processor; b)% improvement in NW states in DL1 and UL2 with width-aware way allocation and replacement policies.

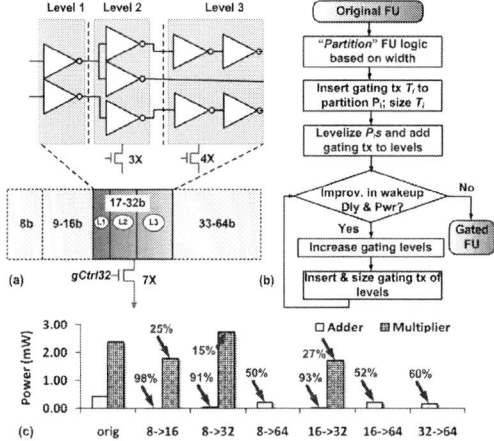

Fig. 5. a) Scope of *WADSG* in the superscalar processor pipeline and modifications to the reorder buffer (ROB) design; b) Detailed implementation of the modified select logic; c) Width calculation logic at the output of the ALU incurs minimal design overhead.

Fig. 8. a) Optimized FU design with levelized gating scheme; b) Methodology to realize a FU design with levelized gating; c) Reduction in wake-up power overhead with proposed FU design; d) Optimized FU design considering one of the two operands is NW.

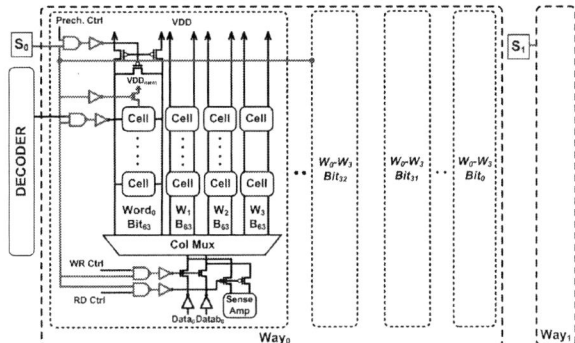

Fig. 6. Width-aware gating of cache line. Two state elements (S_1, S_0) are required to hold the width information for a 2-way set associative cache with each block of size 32B.

5(c) shows the logic for width computation logic for the adder which is based on the width of the input operands and the output sum bits. It requires only 12 gates, incurs minimal power overhead ($2.52\mu W$) and the delay overhead is completely masked by the delay for computing the most significant bits. Note that width for immediate operands are calculated in the *rename* stage and therefore does not affect the performance.

B. Width-Aware Gating in Cache

The idea of width-aware gating can be extended to save leakage power in caches at all levels. To minimize the storage ($S_0, S_1, ...S_n$) required to store the width information, we restrict the number to only a few per cache subarray dictated by the associativity of the cache and the size of the cache block (Fig. 6). For exam-

ple, for a 2-way set associative cache with block size of 32B which is organized in a 4-way bit-interleaved fashion, number of state elements per cache block holding the width information is $(\# \ of \ ways) * (Cache \ line \ size)/(Cache \ word \ size) * (Degree \ of \ interleaving) = 2 * 32 * 8/(64 * 4) = 2$. Here we assume, that the width information is encoded into 2 levels, i.e. only bits 32 to 63 are considered for gating. Assuming the cache is divided into subarrays of size 4KB, number of state elements required for a 32KB L1 data cache (DL1) and a 2MB unified L2 (UL2) are 16 bits and 1Kb, respectively. Fig. 7(a) shows the percentage of state elements in L1 data cache and L2 unified cache which remain in NW state during simulation of 1B instructions for different SPECInt2K benchmarks. This was obtained by noting the number of state elements holding NW information at the end of each simulation cycle and updating them in case of a store or replacement. We noted that for the SPECInt2K suite, $\sim6\%$ of the L1 data cache and 77.5% of UL2 are on an average occupied by NW operands.

To further improve the NW states, we explored a novel width-aware allocation (WAA) and replacement (WAAR) policies. For WAA policy, during write operation to a set-associative cache, if invalid cache blocks are present in multiple ways and the operand to be written is a NW operand, it is allocated to the way which is already storing NW operands. Fig. 7(b) shows that with the proposed WAA policy, percentage of NW states improves by *21%* for DL1 and by *18.5%* for UL2 over the baseline configuration with no allocation. Percentage of NW states can be further improved by modifying

978-1-4673-0438-2/12 $31.00 © 2012 IEEE

the the LRU replacement policy for the baseline configuration such that cache blocks with FW operands is given higher priority during replacement. This increases the percentage of NW state elements and improves leakage saving. A *3%* increase in NW states for both DL1 and UL2 is observed with the *WAAR* compared to *WAA* approach. Penalty is however incurred due to higher DL1 and UL2 miss rates (avg. *1.4%*).

IV. CIRCUIT-LEVEL OPTIMIZATIONS

A. Minimizing Wake-up Overhead

To minimize the wake-up overhead of gated FU designs, we propose a novel FU design methodology based on the concept of *levelized gating*. The proposed design is illustrated in Fig. 8(a). The basic idea is to exploit the logic propagation delay inside a FU (adder or multiplier) to *mask* the FU wake-up delay. We observe that this can be achieved if the FU logic corresponding to the first few logic levels is left ungated while the logic for the following levels is gated. This allows the logic for the higher levels to be already waked up by the time the inputs from the first few logic levels arrive. To achieve a FU design with levelized gating, we first partition the FU logic based on output width (Fig. 8(b)). Each partition is then subdivided into levels, a gating transistor is then added to this level and is properly sized. The process of levelization is continued until no more improvement in wake-up delay and power is achieved. For width aware gating, the adder logic was partitioned into 4 portions corresponding to the output range of $\leq 8, 8-16, 16-32$ and $32-64$. Logic for the lower 8 output bits was left ungated. The multiplier was also divided into 4 portions and similarly gated. Fig. 8(c) shows that on an average *74%* and *12%* improvement in wake-up power can be achieved for the optimized adder and multiplier designs. Our simulations show that with *levelized gating*, neither adder nor the multiplier incurs any wake-up delay penalty.

V. SIMULATION RESULTS

To validate the effectiveness of the *WADSG* scheme, detailed circuit level simulations considering the effect of gating transistor were performed to estimate the impact on dynamic and leakage power. These savings were incorporated into the Wattch [8] architecture level simulator. Power savings for SPECInt2K due to *WADSG* was then accurately estimated by fast-forwarding these applications for 500M instructions and simulating them for 1B instructions on the baseline processor configuration. For estimating the total power saving in the FU, we considered ADD, MULT and logic operations. With power traces obtained during these simulations, Hotspot 5.0 [16] was used to estimate the maximum die temperature for the Alpha processor floorplan.

A. Power Saving in FUs

Fig. 9 shows the active power savings for the integer ALU and multiplier with different FU allocation policies. For the integer applications, we see that when the 64-b output range is divided into 4 bins, the best average savings (*25%*) is achieved with a leakage-aware FU allocation scheme. For the integer multiplier, leakage-aware allocation achieves better savings *45%*.

B. Power Saving in L1 and L2 Data Caches

Fig. 10(a) shows the read and write power savings achieved through width-aware way allocation for both L1 and L2 caches. As we note from Fig. 10, for 4 bins of operand width, average savings in read and write power for DL1 are *10%* and *62%* respectively. Savings for L2 are much higher (*45%* and *71%*). Fig. 10(b) shows the percentage

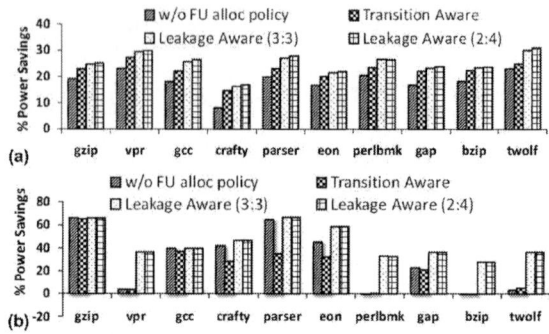

Fig. 9. % power savings in a) Integer ALU and b) integer multiplier for different issue policies, considering 2 bits to encode the operand width.

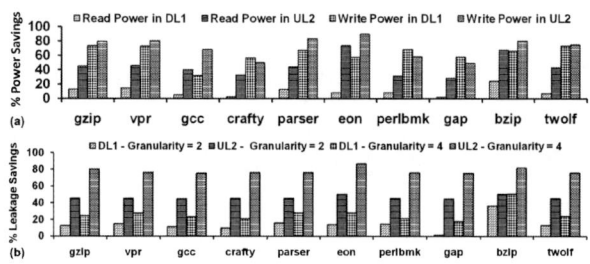

Fig. 10. a) Read and write power savings with *WAAR* in DL1 and UL2 caches; b) Leakage power savings in DL1 and UL2 caches considering 1 and 2 bits to encode operand width.

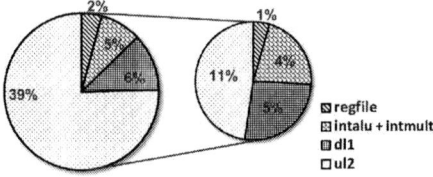

Fig. 11. Improvement in power consumption of individual processor components after application of *WADSG*.

savings for DL1 and L2 compared to the baseline configuration. Average leakage savings for DL1 and L2 considering 1-bit encoding are *15%* and *46%*, respectively. The savings improve to *27%* and *78%* for 2-bit encoding.

C. Overall Power Reduction

We have evaluated the total power savings for the processor considering the leakage contribution to the total power to be 40% for high-performance systems [10]. We assume that the *WADSG* scheme is applied to the integer ALUs, multipliers, register files [11] and the on-chip caches. With these considerations, we achieve *11.5%* saving in active power and *69.3%* saving in leakage power. Total system-level power saving considering the power overhead due to width computation on a L2 miss is calculated to be *34.5%*. Fig. 11 shows the contributions of the processor components to which *WADSG* has been applied to the total power consumption before and application of *WADSG* scheme. Note that FPUs and result buses, which are also amenable to width-aware gating is likely to improve the savings further.

978-1-4673-0438-2/12 $31.00 © 2012 IEEE

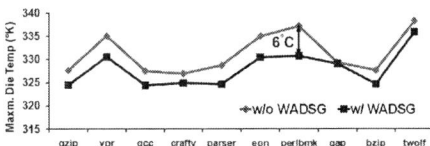

Fig. 12. Reduction in maxm. die temperature with *WADSG* scheme

D. Improvement in Thermal Profile

Improvement in active power (dynamic and leakage) translates directly to reduction in power density and hence temperature for critical hotspots, e.g. integer units. Fig. 12(a) shows that *WADSG* can be reduce the maximum die temperature by $6°C$.

VI. DISCUSSION

A. Addressing the Ldi/dt effect

During wake-up from a NW state to a FW state, both IntALU and IntMult draws significant current, almost 2X the normal FW operation. If all the IntALUs and the IntMults in a processor are switching from NW to FW in the same clock cycle, the di/dt would be large enough to cause a voltage drop in the power line. From our simulations, we note that such a scenario is rare and can be easily addressed by modifying the FU allocation policy to prevent all the IntALUs or Mults from switching together without incurring any significant performance overhead.

B. Extension to Register File, Reorder Buffer and FPUs

A width-aware gating to reduce dynamic power during read and write has already been investigated for register files [11]. Our investigation shows that *WADSG* can be applied to register files to achieve drastic reduction in leakage power (*21%* for a 32-entry register file using 45nm PTM HP model). The same applies to the reorder buffer. Finally, WADSG can be applied to FPU of a processor. We observed that mantissa addition and multiplication, especially for double precision FP operands, contribute heavily to the total power dissipated in FP units. As these are integer operations, they are amenable to power reduction through *WADSG*. In this case, *WADSG* benefits from the large percentage of FP operands with NW significands in SPECFP2K benchmarks compiled for 64-b Alpha (Fig. 13(a)). We note that for *31%* and *35%* of FP additions, mantissa for both the operands can be effectively expressed into 8 and 16 significant bits, respectively.

C. Effectiveness for Embedded Applications

The proposed approach can also be highly effective for embedded domains e.g. for multimedia and DSP applications. We have noted the frequency of NW operands for the Mibench benchmark suite using the large reference input data set. Fig. 13(b) shows the percentage cumulative occurrence of operand width for Mibench benchmarks compiled for 64-b Alpha. From these distributions, we note that *38%* and *56%* of the integer addition and multiplication operations are NW. These observations establish that *WADSG* employed to embedded processors would reap similar power savings as general-purpose applications.

VII. CONCLUSION

We have presented a fine-grained dynamic supply gating approach for datapath and memory. An architecture-circuit co-design approach is followed to maximize the power saving advantage while minimizing the impact on performance and design complexity. Simulation

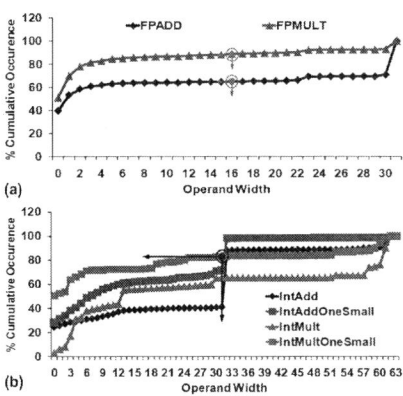

Fig. 13. % Cumulative occurrence of *Maximum Operand Width* for: a) floating point operations in SPECFP2K benchmarks; and b) integer operations for Mibench benchmarks, both compiled for 64-b Alpha.

results for general-purpose and embedded applications with varying data formats show significant power saving in processor execution units and L1/L2 caches. We have shown that the effectiveness of the approach can be enhanced using a width-aware instruction issue and memory block allocation strategy. Besides integer FUs and embedded memory, the approach can be applied to register file, reorder buffer and FPUs which experience width-dependent change in activity. As a by-product, the proposed approach can improve the thermal profile of the FUs. The proposed approach is scalable across technology nodes due to increased effectiveness of transistor stacking. Future work would include gating of sequential elements and other datapaths; application to graphics and digital signal processors; and exploring additional implications of WADSG e.g. reduction of ECC overhead.

REFERENCES

[1] T. Mudge and U. Hölzle, "Challenges and Opportunities for Extremely Energy-Efficient Processors", *IEEE Micro*, 2010.
[2] J. Balfour et al., "An Energy-Efficient Processor Architecture for Embedded Systems", *IEEE Comp. Arch. Letters*, 2008.
[3] S. Bhunia et al.. "A novel synthesis approach for active leakage power reduction using dynamic supply gating", *DAC*, 2005.
[4] J.W. Tschanz et al. "Dynamic Sleep Transistor and Body Bias for Active Leakage Power Control of Microprocessors", *JSSC*, 2003.
[5] S. Palacharla et al., "Complexity-Efective Superscalar Processors", *ISCA*, 1997.
[6] Simplescalar Toolset V3.0: [Online] http://www.simplescalar.com/
[7] H. Li et al, "DCG: Deterministic Clock-Gating for Low-Power Microprocessor Design", *IEEE TVLSI*, 2004.
[8] Wattch v1.02d: [Online] http://www.eecs.harvard.edu/~dbrooks/wattch-form.html.
[9] R. Canal, A. Gonzalez and J.E. Smith, "Very Low Power Pipelines using Significance Compression", *ISCA*, 2000.
[10] J. Lee and N.S. Kim, "Optimizing Total Power of Many-Core Processors Considering Voltage Scaling Limit and Process Variations", *ISLPED*, 2009.
[11] S. Wang et al., "Exploiting Narrow-Width Values for Thermal-Aware Register File Designs", *DATE*, 2009.
[12] K. Flautner et al., "Drowsy Caches: Simple Techniques for Reducing Leakage Power", *ISCA*, 2002.
[13] Z. Hu et al., "Microarchitectural Techniques for Power Gating of Exccution Units", *ISLPED*, 2004.
[14] D. Brooks and M. Martonosi, "Dynamically Exploiting Narrow Width Operands to Improve Processor Power and Performance", *HPCA*, 1999.
[15] Predictive Technology Model [Online] http://ptm.asu.edu/
[16] Hotspot 5.0: Temperature Modeling Tool [Online] http://lava.cs.virginia.edu/HotSpot/

Eliminating Performance Penalty of Scan

Ozgur Sinanoglu
Computer Engineering Department
New York University - Abu Dhabi

Abstract

Stringent performance requirements magnify the performance degradation impact of Design-for-Testability (DfT) techniques. As more aggressive performance optimizations are being employed, resulting in high-performance designs with reduced logic depth, the impact of scan multiplexers is becoming even more magnified. In this work, we propose a scan cell transformation technique that transfers the scan multiplexer delay from the input of the flip-flop to its output, enabling the removal of the scan multiplexer delay off the critical paths. By inserting a few shadow flip-flops properly, the proposed transformation technique retains test development (test data, quality, etc.) and application (test time, power dissipation, etc.) intact, fully complying with the conventional design and test flow. Experimental results justify the efficacy of the proposed techniques in eliminating the performance penalty of scan quickly and cost-effectively, and thus in enhancing functional speed of integrated circuits.

1 Introduction

Stringent performance requirements magnify the performance degradation impact of Design-for-Testability (DfT) techniques. Controllability and observability of each flip-flop have been ensured via the insertion of a scan multiplexer yet at the expense of functional path prolongation by a multiplexer delay. While such a transformation on every flip-flop eliminates the sequentiality of the test generation problem, critical path prolongation and thus functional speed degradation is the end-result, undermining the expected fulfillment of the stringent performance requirements. As more aggressive performance optimizations are being employed, resulting in high-performance designs with reduced logic depth, the impact of scan multiplexers is thus becoming even more magnified.

Traditionally, partial scan has been the approach for eliminating/alleviating the performance penalty of scan. An extensive amount of research has been conducted in partial scan design, targeting the removal of scan multiplexers on a set of selected flip-flops. The consequent benefit is potential alleviation of the performance penalty of scan, in addition to other benefits such as test time, data volume and power reduction. The previously proposed techniques in partial scan can be classified mainly into three categories: structure-based techniques that typically involves breaking the cycles and/or reducing scan depth [1, 2, 3, 4, 5, 6, 7, 8, 9, 10, 11], testability-based techniques that select scan flip-flops based on testability improvements [5, 6, 12, 13, 14, 15, 16, 17, 18, 19, 20, 21], and

test generation-based techniques which intertwine test generation and scan flip-flop selection [22, 23, 24, 25, 26, 27]. Other partial scan techniques include those driven by layout constraints [5], timing constraints [28], re-timing [2, 29], and toggling rate of flip-flops and entropy measures [30]. *These techniques typically necessitate the utilization of sequential ATPG (or combinational ATPG with time frame expansion), however, to generate test patterns on the partially scanned design, not only failing to comply with the existing design/test flow that industry utilizes today but also incapable of ensuring the quality of full scan.*

In this paper, we propose a transformation technique that is based on transferring the scan multiplexer delay from the input of the flip-flop to its output for a target set of flip-flops, in order to remove the multiplexers off from (near) critical paths. For each transformed flip-flop, we insert a shadow flip-flop in such a way that any type of test can still be applied intact with the cost-quality metrics fully preserved; *this, to the best of our knowledge, is a unique aspect of the proposed technique.* Thus, by only inserting a few additional flip-flops, penalty of scan can be eliminated while retaining test patterns and test application process intact. This way, the functional performance of integrated circuits can be further improved. To summarize, the proposed technique:

- Can **eliminate the performance penalty of scan**.

- **Retains** test development process (**test data, test application time, quality, pattern count**) **intact**.

- Retains test application (static, at-speed tests) intact.

- Requires the transformation of a very few flip-flops at a very low area cost (**less than 0.1%** for larger circuits).

- Is orthogonal to and can be utilized in conjunction with any other DfT approach (test compression, test power reduction, etc.).

The remainder of the paper is organized as follows. In Section 2, we present the proposed scan cell transformation technique that eliminates the performance penalty of scan. Sections 3 and 4 elaborate on the flow of application and the retainment of test quality, respectively, when the proposed transformations are utilized. Sections 5 and 6 present the experimental results and conclusions, respectively.

2 Proposed Scan Cell Transformation

In this section, we present the proposed transformation technique that transfers the scan multiplexer delay from the

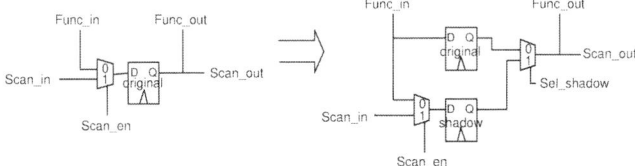

Figure 1. Proposed scan cell transformation

input of the flip-flop to its output, which, with the support of a *shadow flip-flop*, restores the controllability and observability loss, fully retaining the test capabilities of conventional scan.

Figure 1 illustrates the application of this technique on a flip-flop (referred to as the original flip-flop), transferring the multiplexer delay from the input of the flip-flop to its output[1]. Such a transformation necessitates the insertion of a test-only shadow flip-flop and a multiplexer that is driven by a Sel_shadow signal.

The transformed scan cell operates as follows. The original flip-flop always captures the functional input fed by the combinational logic. The shadow flip-flop, which is used only during the test mode[2], latches the output of the preceding scan cell when the scan enable signal $Scan\_en$ is high; during the shift mode, the shadow flip-flop is connected to the preceding scan cell. The shadow flip-flop latches output of the combinational logic when $Scan\_en$ is low; during capture, both the original and the shadow flip-flops of a transformed scan cell capture the same response bit of the combinational logic.

The succeeding scan cell and the combinational logic are both driven by the transferred multiplexer, which selects between the original and the shadow flip-flops based on the Sel_shadow signal. In the test mode, this signal is always high except for the very first shift cycle following the capture window. As a result, only during the first shift cycle subsequent to the capture window, the original flip-flop drives the scan-out signal; in this first shift cycle, the succeeding scan cell receives its input from the original flip-flop. During all the other shift cycles and capture window, the shadow flip-

---
[1]Because of the delay due to the tap-off in front of the original flip-flop, replacing a multiplexer with the tap-off point delivers a gain that may be slightly less than a multiplexer delay.

[2]A typical TAP controller generates two signals $Test$ and $Scan\_en$. During the test mode, the $Test$ signal is always high, while it is low during the normal mode; the $Scan\_en$ signal is high during the shift mode, and low during the capture mode.

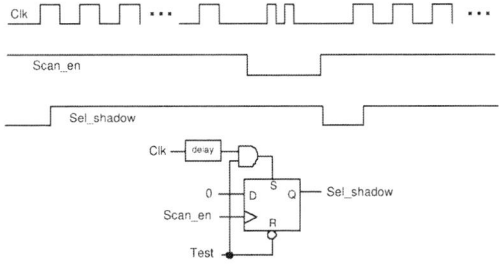

Figure 2. Timing & generation of Sel_shadow **signal**

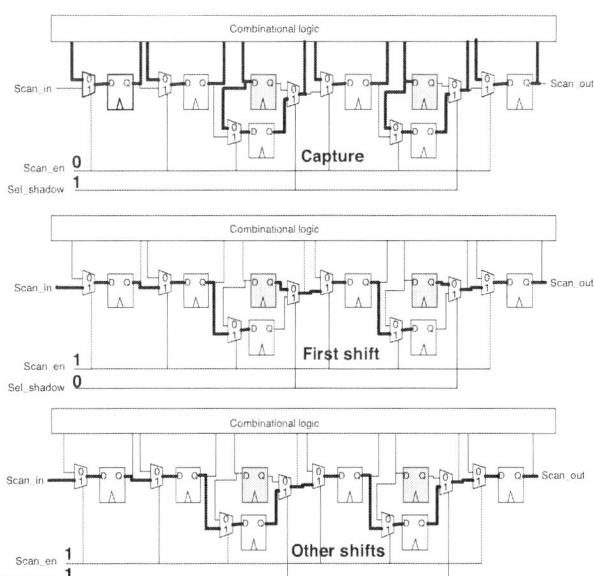

Figure 3. Scan chain operations with two scan cells transformed (third and fifth from the left)

flop drives the combinational logic (during capture) and the succeeding scan cell in the scan chain (during all shift cycles but the first one). Also, the Sel_shadow signal is low during the normal mode, to ensure that the functional operations are carried out by selecting the original (and faster) flip-flop.

The Sel_shadow signal can be easily generated out of the conventional test signals as shown in Figure 2. In this simple circuitry, the $test$ signal is used to ensure that the original flip-flop is selected ($Sel\_shadow = 0$) during the normal mode. In the test mode, an active clock sets the Sel_shadow signal. When the clock signal is inactive, a rising edge on the $Scan\_en$ signal (indicating the beginning of shift cycles) resets Sel_shadow to 0. The first clock pulse during shift operations sets this signal back to 1, which is preserved until the end of the next capture window. Delayed version of the clock is utilized in order to ensure that the signal remains low until *after* the first shift operation has been completed; the magnitude of this delay should be adjusted to overcome the clock skew. It should be noted that the Sel_shadow signal is **not** timing-sensitive; this signal transitions either when design switches from capture mode to shift mode (where dead-time is typically inserted) or between the first and second shift cycles (where the shift operations are typically conducted at a low speed). Therefore, this signal can be easily routed to the few transformed scan cells; if the routing of this signal is a concern for any reason, it can be locally generated out of the $Scan\_en$ signal by utilizing the simple circuitry in Figure 2.

The shift and capture mode operations are illustrated on a scan chain example with two transformed scan cells in Figure 3; as two scan cells (third and fifth from the left) are transformed in this example, two shadow flip-flops are inserted through which scan operations are carried out. The topmost part of the figure illustrates the capture mode operation,

978-1-4673-0438-2/12 $31.00 © 2012 IEEE   347

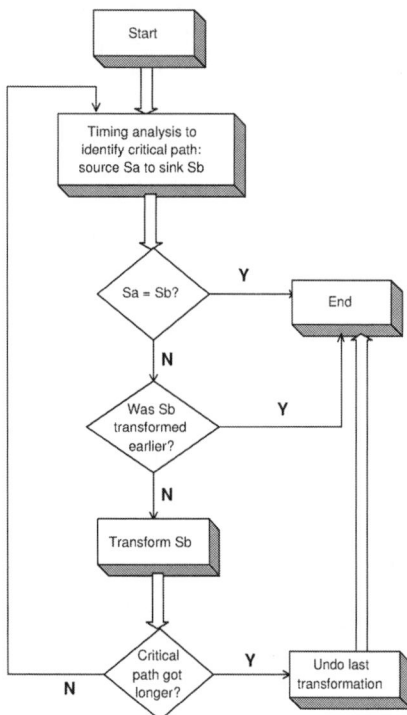

Figure 4. Iterative application flow

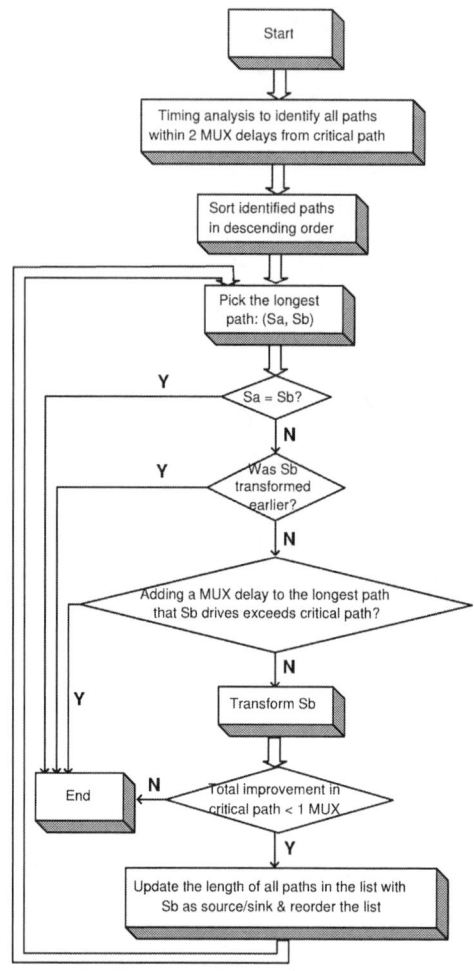

Figure 5. Cumulative application flow

wherein both the original and the shadow flip-flops capture the output of the combinational logic, while the shadow flip-flop drives the combinational logic. The middle part of the figure illustrates the first shift cycle subsequent to the capture window, wherein shadow flip-flop is connected to the preceding scan cell and the original flip-flop drives the succeeding scan cell. The bottommost part of the figure illustrates all the subsequent shift cycles, wherein the scan chain logically goes through the shadow flip-flops. The only remaining combination, which is $Scan\_en = 0$ and $Sel\_shadow = 0$, corresponds to the functional mode operation wherein the original flip-flop is logically connected to the combinational logic.

3 Application Flow

The proposed technique targets the flip-flops that are at the sink of the critical paths. The transformation is applied on such a flip-flop, effectively transferring the multiplexer delay from the input of the flip-flop to its output, and thus off that critical path. Such a transformation shortens the targeted critical path, while making other paths that stem from the transformed flip-flop longer; in the worst case, the critical path both originates and terminates at the same flip-flop, rendering this technique ineffective. Thus, upon every transformation, the new critical paths should be identified, and the transformations should be applied on other flip-flops where the new critical paths terminate. This technique terminates when no further performance improvement can be achieved.

Figure 4 illustrates an iterative application flow for the pro-

posed technique. In every iteration, a static timing analysis tool identifies the new critical path in the design. Unless the critical path originates and terminates at the same flip-flop, or the sink scan cell has already been transformed in an earlier iteration, in which case the technique terminates, the scan cell at the sink of the critical path is transformed by applying the changes in Figure 1. It is of course possible that the transformation can render a near-critical path (stemming from the sink scan cell) longer than the current critical path; in this case, this last transformation is undone, and the technique terminates.

While the iterative flow is accurate timing-wise, repetitive application of the timing analysis tool may be computationally costly. An alternative flow, which we refer to as the cumulative flow, is provided in Figure 5. In this alternative flow, the timing analysis is executed once to identify all the near-critical paths that should be included in the analysis; as paths can be shortened or prolonged by a multiplexer delay, the paths under analysis are those within two multiplexer delays from the longest path. Subsequently, the proposed technique is dynamically applied to shorten the paths, one at a time, starting from the longest path; every time a scan cell is transformed, the length of all paths affected by this transformation are updated

978-1-4673-0438-2/12 $31.00 © 2012 IEEE   348

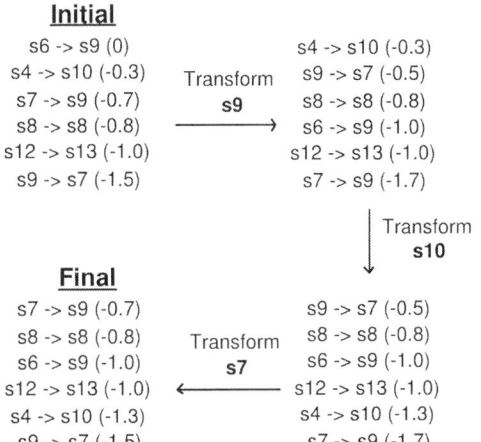

Initial

s6 -> s9 (0)		s4 -> s10 (-0.3)
s4 -> s10 (-0.3)	Transform	s9 -> s7 (-0.5)
s7 -> s9 (-0.7)	**s9**	s8 -> s8 (-0.8)
s8 -> s8 (-0.8)	⟶	s6 -> s9 (-1.0)
s12 -> s13 (-1.0)		s12 -> s13 (-1.0)
s9 -> s7 (-1.5)		s7 -> s9 (-1.7)

Transform **s10**

Final

s7 -> s9 (-0.7)		s9 -> s7 (-0.5)
s8 -> s8 (-0.8)	Transform	s8 -> s8 (-0.8)
s6 -> s9 (-1.0)	**s7**	s6 -> s9 (-1.0)
s12 -> s13 (-1.0)	⟵	s12 -> s13 (-1.0)
s4 -> s10 (-1.3)		s4 -> s10 (-1.3)
s9 -> s7 (-1.5)		s7 -> s9 (-1.7)

Figure 6. Example application of cumulative flow

and the list is sorted again. The termination condition is the same as in the iterative flow, except that the critical path improvement of one complete multiplexer delay also terminates the proposed technique, successfully in this case as this is the best case scenario.

The application of the cumulative flow on an example is illustrated in Figure 6. The top-left part of the figure provides the output of the timing analysis tool that lists all the six near-critical paths that can potentially become the final critical path upon transformations. For each path, the source and the sink scan cells are provided in addition to the difference of the near-critical path length from the critical path length in parenthesis; this length difference is given as a multiple of one multiplexer delay. At a first glance, it can be seen that the best possible gain in this example is 0.8 multiplexer delay, as the proposed technique cannot improve the path that originates and terminates at $s8$.

In the first iteration, $s9$ is successfully transformed, as the longest path stemming from $s9$ ($s9$ to $s7$) remains below the critical path despite the prolongation by a multiplexer delay. This transformation is reflected to the list of paths by updating the length of the three paths affected, two shortened and one prolonged. In the second iteration, $s10$ is transformed, changing the length of only the currently longest path in the list. In the third iteration, the proposed technique transforms $s7$, affecting the length of two paths in the list; this transformation increases the length of $s7$ to $s9$ path yet to a value under the length of the longest path. The fourth iteration is when the proposed technique terminates, as $s9$ was already transformed in the first iteration. The critical path of the design is thus improved by 0.7 multiplexer delay.

The timing accuracy of the iterative flow and the computational efficiency of the cumulative flow can be combined in a hybrid flow for the application of the proposed transformation techniques, wherein the timing analysis tool can be executed only once upon every n cumulative transformations. This way, the timing information is *refreshed* with every n transformations, rather than performing all transformations based

on a single timing analysis run. The value of n can be chosen properly depending on the desired levels of timing accuracy and computational efficiency.

4 Test Quality Considerations

In this section, we elaborate on test capabilities in the presence of transformed scan cells.

In conventional scan-based testing, **scan flush test** is applied by running a few patterns through the scan chain with no capture operation in between. In the proposed scheme, all these patterns are shifted through scan chain that traverses the untransformed scan cells and the shadow flip-flops[3] of transformed scan cells as illustrated in bottommost part of Figure 3. This way, all the faults on the scan path are covered as in the conventional scan flush testing. As the scan path does not traverse the transformed (original) flip-flops, the faults on their input and output are not covered during scan flush testing, however, which will be discussed later in this section.

Scan load/unload operations are performed slightly differently (yet transparently through the use of on-chip generated Sel_shadow signal) compared to conventional scan-based testing, with the only difference being the first shift cycle subsequent to capture window, as illustrated in Figure 3, wherein the succeeding scan cell in the chain receives the value captured in the original flip-flop. Upon the completion of all the shift operations, the chain is completely loaded with the new stimulus (through the untransformed scan cells and the shadow flip-flops of transformed scan cells) while the content of all original flip-flops will have been unloaded and observed in an identical manner as in conventional testing.

In **static (stuck-at fault) testing**, the stimulus loaded into the untransformed scan cells and the shadow flip-flops of transformed scan cells are applied to the circuit under test, while all flip-flops capture; the original and shadow flip-flop of a transformed scan cell capture the same fault-free bit. The faults at the input of the original flip-flop, which remained undetected upon scan flush testing, manifest in the original flip-flop upon capture, and are detected by the end of the shift cycles, as the content of the original flip-flop is shifted into the succeeding scan cell in the first shift cycle. Also, as the same patterns are loaded and applied, and the same responses are unloaded and observed, the same set of faults in the combinational logic is covered as in conventional testing.

In **launch-off-capture testing**, upon the loading of the stimulus pattern into the untransformed scan cells and the shadow flip-flops of transformed scan cells, transitions are launched from untransformed scan cells and the shadow flip-flops, and captured in all flip-flops. Upon the first capture operation (transition launch) the content of the original and the shadow flip-flops of a transformed scan cell become identical (as they both capture the same value), and transitions are fired from the untransformed scan cells and the shadow flip-flops; the second capture operation helps test the at-speed operation

---

[3] A single dummy clock pulse may be required prior to all the clock pulses in order to set the Sel_shadow signal to 1; both $Scan\_en$ and Sel_shadow signals are high throughout scan flush testing.

978-1-4673-0438-2/12 $31.00 © 2012 IEEE   349

of the paths terminating at the original flip-flops as their content is shifted out and observed. Therefore, the same set of transition faults in the combinational logic is covered as in conventional testing.

In **launch-off-shift testing** also, transitions are launched from untransformed scan cells and the shadow flip-flops. The $Scan\_en$ signal switches from high to low at-speed, and these transitions propagate into both the original and the shadow flip-flops, while the content of the original flip-flops is shifted out and observed, covering the same set of transition faults in the combinational logic as in conventional testing. Furthermore, the only timing-sensitive signal is still the $Scan\_en$ signal, as there is no at-speed timing requirement imposed on the Sel_shadow signal. Thus, the proposed transformations retain both the test quality and the timing requirements intact.

The only faults covered differently compared to conventional testing are those at the output of the original flip-flops, which remained undetected upon scan flush testing. In conventional scan chains, scan flush testing covers these faults, while due to proposed transformations, they are detected *indirectly*. A fault on the output of the original flip-flop, if activated, will be captured in the succeeding scan cell upon the first shift cycle, and will be shifted out and detected in the observed responses. Its activation is guaranteed as throughout the course of testing, every original flip-flop will have captured either binary value at some point, activating either stuck-at fault on the output of the original flip-flop. Therefore, these faults can be assumed to be covered, necessitating no additional test generation effort expended for them. It should be noted that in conventional testing, there are other faults that are detected *indirectly*, similar to the output faults of the original flip-flops: stuck-at-1 faults on the $Scan\_en$ wire attached to a scan cell. Whenever a scan cell with such a fault captures a value (from combinational logic) that differs from the value shifted into its preceding scan cell, this fault is *indirectly* detected. None of these faults needs to be considered in test generation, as their indirect detection is guaranteed.

Slight test generation and design effort may have to be expended into the detection of the transition faults on the output of the original flip-flops, however. With the design support capable of feeding a clock pulse to *only* the original flip-flop of the transformed scan cells upon the completion of shift cycles[4], a single pulse may justify the original flip-flop to the desired value prior to the double capture that would launch the transition from the original flip-flops and capture them in all flip-flops. Of course, these special patterns require $Shadow\_sel$ to be low during capture mode to enable transition launch from the original flip-flop.

As both the test data/quality and the number of shift cycles remain intact upon the proposed transformations, the performance enhancement benefits are reaped transparently.

---

[4]The pattern to be loaded into the untransformed scan cells and shadow flip-flops is the two patterns merged together: the pattern for stuck-at-v fault on the original flip-flop input and the pattern for the transition fault (from v' to v) at the output of the original flip-flop. This way, the pre-capture pulse justifies the original flip-flop to v' prior to double capture.

Circuit	# scan cells	# trans. cells	% MUX-delay saving	% Overall Perf. Imp.	% Area Cost	Runtime (s)
s713	19	6	**76.2**	1.7	10.0	<1
s953	29	10	**78.6**	2.1	10.7	<1
s1423	74	3	**64.3**	0.9	2.1	<1
s5378	179	1	**75.0**	1.4	0.4	<1
s9234	228	2	**80.0**	1.4	0.3	<1
s13207	669	2	**85.7**	0.9	0.2	2
s15850	597	2	**94.1**	2.4	0.2	34
s35932	1728	0	**0**	0.0	0.0	7
s38417	1636	1	**93.3**	1.6	<0.1	12
s38584	1452	1	**93.7**	2.2	<0.1	11
b20	490	2	**93.3**	1.1	0.1	26
b21	490	3	**93.3**	1.1	0.1	25
b22	735	3	**93.3**	1.1	0.1	41

Table 1. Results of the proposed technique.

5 Experimental Results

We have implemented and applied the proposed transformation technique on various ISCAS89 and ITC99 benchmark circuits to gauge the performance savings that can be delivered. In our experiments, we have utilized the netlist tracing tool[5] that we have implemented to identify the critical paths, and to measure their length. Subsequently, we transformed a few scan cells by making the proper changes in the netlist. Fault coverage and test pattern count are identical before and after transformations, and are thus not reported.

Table 1 provides the results of the proposed technique. Columns 2 and 3 provide the number of total and transformed flip-flops, respectively. Column 4 denotes what portion of a multiplexer delay has been saved by the proposed transformation technique (100% in this column denotes that scan penalty is completely eliminated), while the corresponding percentage improvement in overall functional performance is given in column 5. Finally, columns 6 and 7 provide the percentage area cost[6] and computational run-time in seconds.

It can be seen that the proposed technique is capable of delivering almost a multiplexer delay saving (with a slight deviation of a fan-out delay), and thus alleviating the penalty of scan in all circuits except for s35932. For the two smallest circuits, this technique ends up transforming 6 and 10 flip-flops, respectively, necessitating the insertion of this many flip-flops and multiplexers. For the other designs, however, at most 3 flip-flops and multiplexers are inserted to reap the benefit of saving a multiplexer delay and improving performance. For s35932, no improvement can be delivered, as the critical path originates and terminates at the same flip-flop.

The proposed technique runs within a minute for all the benchmark circuits and achieves its goal fully for almost all the cases. The transformation of a few scan cells incurs an area cost of less than 0.1% for the larger designs, which is quite negligible. The overall improvement in functional performance of the design varies between 1% and 2.5%; for high performance designs where one multiplexer delay constitutes

---

[5]In our netlist tracing tool, we have utilized static delays for gates obtained from an sdf file, implementing the cumulative application flow discussed earlier. The iterative execution of the static timing analysis can be integrated into our framework, delivering more accurate performance saving results.

[6]Area cost, which is measured by the weighted gate count (i.e., gate count multiplied by the average fanin), of the base case includes the scan overhead.

a significant portion of the functional cycle time, the proposed scheme can be very useful in eliminating this penalty quickly and cost-effectively.

6 Conclusions

In this work, we propose a scan cell transformation technique in order to eliminate the performance penalty of scan. The proposed technique is based on transferring the multiplexer delay from the input of the flip-flop to its output, shortening the critical path. This technique restores controllability and observability by inserting a very few shadow flip-flops through which scan operations are conducted, fully retaining the test patterns and test application process intact; these transformations incur an area cost of less than 0.1% for larger designs. In almost all the designs we experimented with, the performance penalty of scan is eliminated quickly and cost-effectively. Thus, and as the proposed technique fully complies with the conventional design and test flow, it can be utilized to attain significant functional performance enhancements especially for high-performance designs.

References

[1] P. Ashar and S. Malik, "Implicit computation of minimum-cost feedback-vertex sets for partial scan and other applications," *Design Automation Conference*, pp. 77–80, Jun. 1994.

[2] S.T. Chakradhar, A. Balakrishnan, and V.D. Agrawal, "An exact algorithm for selecting partial scan flip-flops," *Design Automation Conference*, pp. 81–86, Jun. 1994.

[3] K.-T. Cheng and V.D. Agrawal, "A partial scan method for sequential circuits with feedback," *IEEE Transactions on Computers*, vol. 39, no. 4, pp. 544–548, Apr. 1990.

[4] K.-T. Cheng, "Single clock partial scan," *IEEE Design Test of Computers*, vol. 12, no. 2, pp. 24–31, 1995.

[5] V. Chickermane and J.H. Patel, "An optimization based approach to the partial scan design problem," *International Test Conference*, pp. 377–386, Sep. 1990.

[6] V. Chickermane and J.H. Patel, "A fault oriented partial scan design approach," *International Conference on Computer-Aided Design*, pp. 400–403, Nov. 1991.

[7] R. Gupta and M.A. Breuer, "The ballast methodology for structured partial scan design," *IEEE Transactions on Computers*, vol. 39, no. 4, pp. 538–544, Apr. 1990.

[8] A. Kunzmann and H. J. Wunderlich, "An analytical approach to the partial scan design problem," *Journal of Electronic Testing: Theory and Applications*, vol. 1, pp. 163–174, 1990.

[9] D.H. Lee and S.M. Reddy, "On determining scan flip-flops in partial-scan designs," *International Conference on Computer-Aided Design*, pp. 322–325, Nov. 1990.

[10] J. Park, S. Shin, and S. Park, "A partial scan design by unifying structural analysis and testabilities," *International Symposium on Circuits and Systems*, vol. 1, pp. 88–91, 2000.

[11] S.-E. Tai and D. Bhattacharya, "A three-stage partial scan design method using the sequential circuit flow graph," *International Conference on VLSI Design*, pp. 101–106, Jan. 1994.

[12] M. Abramovici, J.J. Kulikowski, and R.K. Roy, "The best flip-flops to scan," *International Test Conference*, p. 166, Oct. 1991.

[13] V. Boppana and W.K. Fuchs, "Partial scan design based on state transition modeling," *International Test Conference*, pp. 538–547, Oct. 1996.

[14] P. Kalla and M. Ciesielski, "A comprehensive approach to the partial scan problem using implicit state enumeration," *IEEE Transactions on Computer-Aided Design of Integrated Circuits and Systems*, vol. 21, no. 7, pp. 810–826, Jul. 2002.

[15] K.S. Kim and C.R. Kime, "Partial scan by use of empirical testability," *International Conference on Computer-Aided Design*, pp. 314–317, Nov. 1990.

[16] P. S. Parihk and M. Abramovici, "Testability-based partial scan analysis," *Journal of Electronic Testing: Theory and Applications*, vol. 7, pp. 47–60, Aug. 1995.

[17] G.S. Saund, M.S. Hsiao, and J.H. Patel, "Partial scan beyond cycle cutting," *International Symposium on Fault-Tolerant Computing*, pp. 320–328, Jun. 1997.

[18] E. Trischler, "Incomplete scan path with an automatic test generation methodology," *International Test Conference*, pp. 153–162, 1980.

[19] D. Xiang, S. Venkataraman, W.K. Fuchs, and J.H. Patel, "Partial scan design based on circuit state information," *Design Automation Conference*, pp. 807–812, Jun. 1996.

[20] D. Xiang and J.H. Patel, "A global algorithm for the partial scan design problem using circuit state information," *International Test Conference*, pp. 548–557, oct. 1996.

[21] D. Xiang and J.H. Patel, "Partial scan design based on circuit state information and functional analysis," *IEEE Transactions on Computers*, vol. 53, no. 3, pp. 276– 287, Mar. 2004.

[22] V.D. Agrawal, K.-T. Cheng, D.D. Johnson, and T.S. Lin, "Designing circuits with partial scan," *IEEE Design Test of Computers*, vol. 5, no. 2, pp. 8–15, Apr. 1988.

[23] M.S. Hsiao, G.S. Saund, E.M. Rudnick, and J.H. Patel, "Partial scan selection based on dynamic reachability and observability information," *International Conference on VLSI Design*, pp. 174–180, Jan. 1998.

[24] H.-C. Liang and C. L. Lee, "An effective methodology for mixed scan and reset design based on test generation and structure of sequential circuits," *Asian Test Symposium*, pp. 173–178, 1999.

[25] X. Lin, I. Pomeranz, and S.M. Reddy, "Full scan fault coverage with partial scan," *Design, Automation and Test in Europe*, pp. 468–472, 1999.

[26] I. Park, D. S. Ha, and G. Sim, "A new method for partial scan design based on propagation and justification requirements of faults," *International Test Conference*, pp. 413–422, Oct. 1995.

[27] S. Sharma and M.S. Hsiao, "Combination of structural and state analysis for partial scan," *International Conference on VLSI Design*, pp. 134–139, 2001.

[28] J.-Y. Jou and K.-T. Cheng, "Timing-driven partial scan," *International Conference on Computer-Aided Design*, pp. 404–407, Nov. 1991.

[29] D. Kagaris and S. Tragoudas, "Retiming-based partial scan," *IEEE Transactions on Computers*, vol. 45, no. 1, pp. 74–87, Jan. 1996.

[30] O. Khan, M. L. Bushnell, S. K. Devanathan, and V. D. Agrawal, "Spartan: A spectral and information theoretic approach to partial scan," *International Test Conference*, p. Paper 21.1, 2007.

A Silicon Testing Strategy for Pulse-Width Failures

Srinivas Vooka, Khushboo Agarwal, Abhijeet Shrivastava , Pranav Murthy, Venkatraman Ramakrishnan
Texas Instruments (India) Pvt. Ltd.

Abstract—With the increasing clock frequencies to multiple Gigahertz and increasing need to achieve it at lower voltages for keeping operating power lower, frequency of operation is not only limited by the datapath delay scaling but also by the behavior of clock signals. Failures induced by violation of minimum clock pulse width required, at higher frequencies is more commonly seen due to increased operating frequencies and usage of voltage scaling techniques for achieving higher frequencies [1]. Silicon failures caused by minimum pulse width violation due to clock shrinkage can be due to variety of reasons right from duty cycle distortion at clock generator (PLL) to clock pulse distortion due to slew degradation along the clock path. In this paper authors discuss different techniques that enable us to minimize pulse width failures as the mechanism that limit the frequency of operation of the device. In this paper we propose a simple and novel technique to detect and diagnose pulse width failures. Silicon results from 40nm multi-million gate industrial SoC is presented to indicate the effects of pulse width degradation on performance of the device and to evaluate the effectiveness of the proposed pattern generation technique.

Keywords- Pulse shrinkage, pulse width violation, clock failures, corner lots, duty cycle

I.   INTRODUCTION

Traditionally the maximum achievable frequency of operation across different voltage/process points has mostly been limited by the scaling of logic delays between the registers commonly known as the datapath logic. But with the advent of voltage scaling/DVFS techniques[2,3], the operating voltages have been scaled to achieve required performance. Maximum frequency of operation of device can either be limited by violation of setup timing of the flop or by violation of minimum pulse width required for proper operation at higher frequencies. As the scaling of minimum pulse width requirement across different voltages may not be similar to datapath delay scaling, minimum pulse width violation related failures due to clock pulse shrinkage at higher frequencies can be one of the primary mechanisms that limits maximum frequency of operation. Inability of present day structural test methods to clearly differentiate between failure caused by setup timing violation and pulse width violation is obscuring the details on magnitude of the failures caused by clock pulse width violation at higher frequencies.

Flip-flops form the basic building blocks of SoC (System on Chip) designs. Among various kinds of flip-flops, D-flip flops are the most commonly used in present-day designs. For proper operation of flip-flops, minimum pulse width constraints also need be satisfied, in addition to meeting setup and hold times. Clock pulse width is commonly measured by measuring the time between rising and falling edges of the clock signal at the 50% transition point for various clock input slopes (transition times) as shown in Figure 1. A D-flip-flop is normally built using two latches commonly known as master and slave latches as shown in Figure 2. For a positive-edge triggered D flip-flop, the data is allowed to enter the master latch when CLK is low and the slave latch is closed (not allowing the data to enter the slave latch). When the clock goes from low to high, the master latch is closed, while the slave latch is opened to allow data to enter the latch from master latch. At other moments, the output of D flip-flop remains the same. Minimum low clock pulse width constraint is necessary to ensure that the data transfer to master latch from D input happens successfully and minimum high clock pulse width is required to ensure internal data transfer between master and slave latch has adequate time to complete successfully.

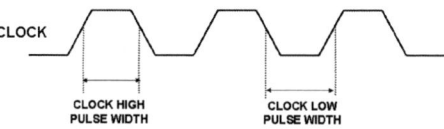

Figure 1. Clock pulse width measurement

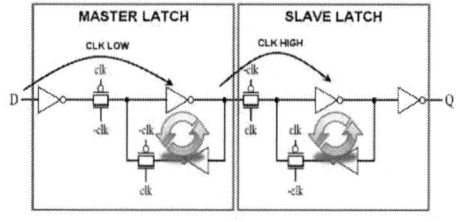

Figure 2. Positive edge triggered D-flip flop

Flops have three distinct regions of operation in reference to clock pulse width as shown in Figure 3. Region1 where the minimum clock pulse requirement of flop is met and clk-q(Clock to Q) delay of the flop is within the specified limit as provided in the library model. A narrow region called as Region2 where the minimum clock pulse width requirement for a flop is violated but the violation results in increase of clk-q delay more than specified amount but it still does not result in logical failure. Region3 in which the pulse width is less than threshold pulse width required for operation of flop, in this region violation results in logical failure.

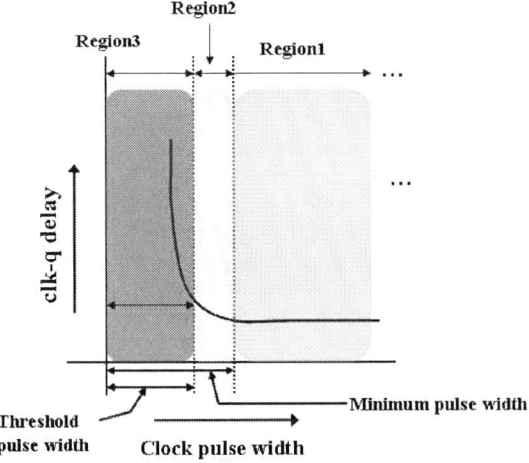

Figure 3. Output delay to pulse width relationship

With the increasing need for devices working at higher frequencies, devices working at higher than 2GHz clock frequency are common in 40nm process node. A 2GHz signal translates to 250ps pulse width assuming 50% duty cycle. Even a modest design has transition time around 50ps leaving a relatively small stability period. The minimum pulse widths of flip-flops are in the same order as the clock pulse width at these higher clock frequencies. Increasing insertion delay of clocks due to increase in size of the designs make pulse width a very sensitive parameter given that the clock signal propagates through the distribution network, and if the signal quality is not controlled, this can limit the maximum operating frequency of the device.

The rest of the paper is organized as follows. Section II discusses on various design and test approaches that can be used to avoid early and false pulse width failures. Section III discusses on the shortcomings of present day testing processes in isolating and differentiating the pulse width failures from setup timing related failures. In section IV, we propose a novel and simple technique that can be used for isolating the pulse width related failures. Silicon comparative results are presented on a 40nm industrial design in section V followed by conclusion in section VI.

II. CAUSES & CARE ABOUTS FOR AVOIDING PULSE WIDTH FAILURES

In this section we investigate various reasons for clock width failures in the design and methods for minimizing them.

A. PLL source induced clock pulse distortion

On-chip PLLs (phase locked loops) as shown in Figure 4 are employed in most of the present day SoCs for generation of high speed clock pulses on-chip. It should also be noted that the fluctuation of clock pulse commonly known as clock jitter increases with clock frequency [4]. The clock pulse width fluctuation of conventional PLLs using differential VCO mainly originates from differential to single-ended signal conversion process. Lower pulse width at the source for higher frequencies can cause a failure due to violation of minimum clock pulse width. A simple solution for this problem would be generating 50% duty-cycle by running VCO at twice as high as the desired clock frequency, and then dividing the frequency by 2 using post divider. Figure 5 illustrates the performance improvement for a 40nm industrial design by using divided clock instead of raw clock. This method, however, wastes power and loses design flexibility especially if VCO is not capable of generating higher frequencies. In such cases duty cycle correction circuitry should be used in design for performance enhancement [5].

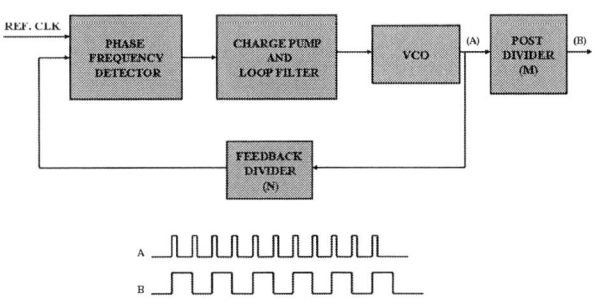

Figure 4. Block diagram of a typical on-chip PLL (Waveform shown at point A and B assuming M=2)

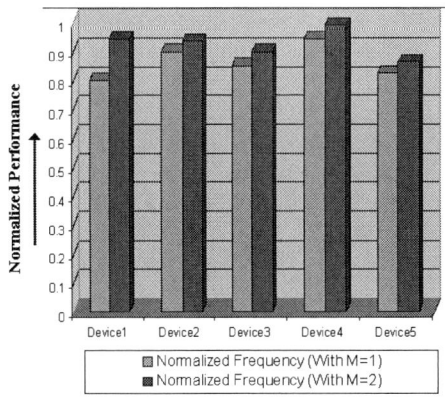

Figure 5. Data illustrating performance improvement due to pulse width correction (using clock division)

B. Clock distortion due to rail to rail swing issues at lower voltages

It is very important to verify the effective rail-to-rail voltage swing on clock nets of the design [6]. At clock frequencies close to 1 GHz, inability to swing from rail-to-rail voltage can cause potential pulse-swallowing resulting in functional failures as shown in Figure 6.

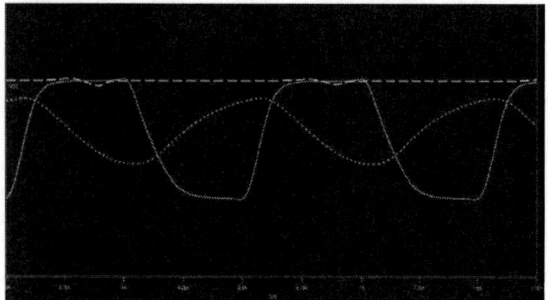

Figure 6. SPICE simulation illustrating pulse gobbling at higher frequencies: Solid waveform represents a full rail-to-rail swing at a driver located up in clock tree. The dotted line shows a non-rail-to-rail swing at the input of a cell having its input connected to a high load in the tree. The dashed line waveform shows no toggle on a downstream buffer output indicating a gobbled pulse down the tree.

A non rail-rail swing clock like in the simulation snapshot show in Figure 7 can also cause a frequency dependent skew due to residual charge. The exponential tail is largely dependent on the resistance of the wire.

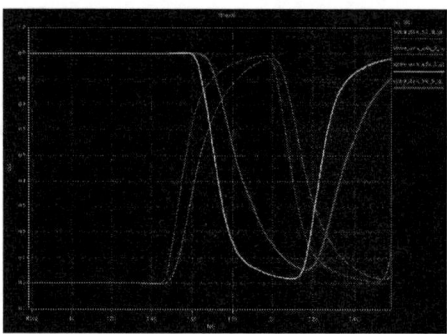

Figure 7. SPICE simulation illustrating a non rail-to-rail swing

One way of knowing the existence of issues related to incomplete rail swing during static timing analysis is to compute an extrapolated transition time by multiplying the default (transition measured between voltage trip points) slew by a suitable factor. This suitable factor can be heuristically derived from SPICE simulations [6].

C. Clock distortion due to local mismatches

Clock tree buffers are used in the clock network for routing the clock to destination flops from the clock source [14]. As the rise and fall time of clock pulse is affected by the N-MOS transistor and P-MOS transistors strength respectively, the transistor strengths are matched during the design of clock tree buffers in order to minimize clock pulse width degradation. The wave-forms in Figure-7 are indicative of this scenario as

well – in any given wave-form, it is evident that the rise transition-time is very different from the fall transition-time. An analysis of propagation delays (not indicated in the figure) would indicate a similar difference between rise delays and fall delays. Such imbalances impact the pulse-width of the signal being propagated, and can be accumulated over multiple repeater stages.

An added consideration would be intra-die variation. Due to increased variability during manufacturing process of transistors in deep submicron technologies mismatch between PMOS and NMOS is unavoidable [4], leading to increased pulse-width impact.

Corner lots analysis[15] provide a way to understand behavior of the device under different possible process variations. Drain current of the transistor commonly known as Idsat is used as the measure of strength of the transistor. Corner lots [13] are provided by the foundry and are typically determined by Idsat characterization data for N and P channel transistors. There are therefore five possible corners: typical-typical (TT), fast-fast (FF), slow-slow (SS), fast-slow (FS), and slow-fast (SF). Figure 8 is the plot with device maximum operating frequency on Y-Axis and ratio of Idsat of N-MOS and P-MOS transistors on X-axis, across corner lot devices. Figure 8 clearly indicates the trend of reduction in operating frequency of the device with increased ratio or mismatch between N and P transistor strength. The reduction in operating frequency is caused by reduction in high pulse width due to higher NMOS strength in comparison to PMOS strength.

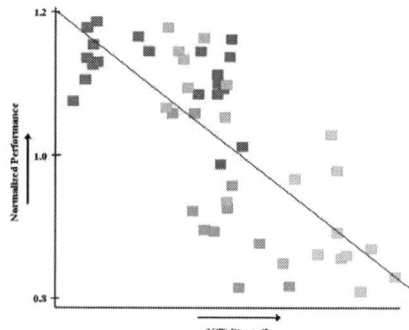

Figure 8. Plot Fmax and ratio of N & P-MOS strength

Burnin is a common technique used for screening early failures by accelerating the failures by using elevated temperature and voltage during testing [7]. This scenario accelerates a process of transistor performance degradation (called ageing). Extensive usage of clock gates in the design for power reduction [8,9] causes portions of the clock tree to be inactive at different points in time, resulting in differential ageing scenarios on the clock tree. If some of the clocks are kept gated at all times during the burnin process, there can result an asymmetric stress of N & P transistors causing the additional N/P strength mismatches that can give rise to duty cycle distortion.

D. Mismatch in frequency division methodology during test and functional mode of operation

With increase in size of the device and integration of multiple peripherals in to single SoC, peripherals operating of different clock frequencies are becoming very common. For synchronous clock domains, lower frequency clocks are usually derived from source clock (commonly a PLL) using frequency division. Flop based division and clock gate based division are two common methods used for frequency division. Flop based frequency division provides 50% duty cycle clock at the output. Clock gate based division commonly known as pulse division that swallows the clock pulses divides the clock frequency as required but doesn't provide 50% duty cycle clock at the output. Figure 9 illustrates the clock waveforms for divide by 2 using flop and clock-gate method. While flop based methods creates a clock waveform with 50% duty cycle (Equal high and low pulse width), clock gate based method of frequency division creates a clock of 25% duty cycle. (Low pulse width is equal to 3 times high pulse width).Clock gate based frequency division methodology is commonly employed during transition fault testing as it eases the clock pulse generation process from high speed clock in multi-clock domain design during capture phase of transition fault testing [10].

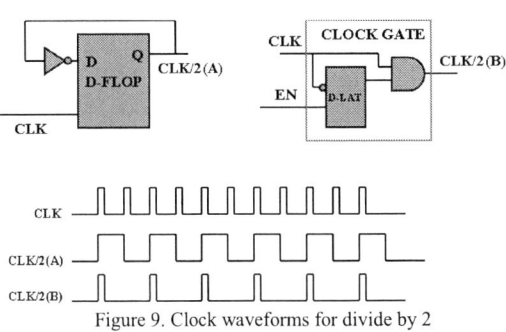

Figure 9. Clock waveforms for divide by 2

Differences in the frequency division technique between functional and test mode can either give rise to additional yield fallout or test escapes. In case flop based division is used in functional mode and pulse based division technique is used in test mode, it can result in additional yield fallout as minimum clock pulse width limit is reached earlier in test mode than functional mode of operation. Alternatively if functional mode uses pulse division and test mode uses flop based division, screening tests can result in optimistic frequency of operation that might not be met in functional mode of operation, resulting in customer returns. Therefore there is a need to ensure that structural delay tests not only use the functional mode frequency during test but also match clock pulse width to the one that is used during functional mode of operation.

III. PULSE WIDTH FAILURE DETECTION & DIAGNOSTICS USING EXISTING TESTS

Stuck-at and transition fault patterns are most commonly used structural patterns for screening the defective devices. Stuck-at patterns use single slow speed clock pulse during capture phase as shown in Figure 10a and transition fault patterns that screen delay defect uses two back-to-back at-speed pulses during capture phase as shown in Figure 10b.

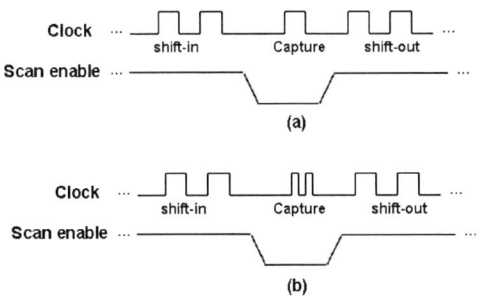

Figure 10. Clock and scan enable waveforms during stuck-at and transition fault testing

As stuck-patterns use single low speed clock pulse during capture phase as shown in Figure 10a, they are not very effective in helping to detect failures induced by pulse width violation. If the transition fault patterns use similar clock as functional with careabouts mentioned in section III.D taken care of, then they would be effective in detection of failures induced by pulse width violation. Even though transition fault patterns are effective in detecting the failures induced by pulse width violation, they fall short in isolating the failure location. If the failure is caused in Region2(of Figure 3) of operation of the flop, using transition fault patterns we would not be able to differentiate whether the failure is result of datapath delay induced defects or minimum clock pulse width failure. In case failure is result of logical failure as illustrated in region3 (of Figure 3), due to usage of two clock pulses the failure caused by the first clock pulse propagates into second level of logic making it difficult to diagnose the reason for failure.

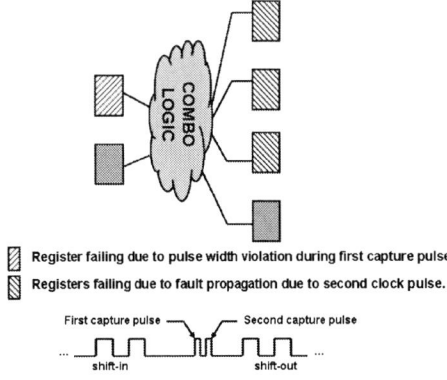

Figure 11. Failure propagation to second logic level

978-1-4673-0438-2/12 $31.00 © 2012 IEEE

IV. PROPOSED APPROACH FOR ENABLING DIAGNOSIS

Given these limitations of traditional stuck-at patterns and transition-fault patterns, in order to enable to detection and diagnosis of the failure we propose a method in which only single clock pulse from rated frequency clock is leaked during capture phase. Pulse width used for testing can be modulated by increasing or decreasing the frequency of at-speed clock. The higher the clock frequency, the lower the pulse width of single clock pulse that is leaked for testing during capture.

A. S-A fault pattern generation with TFT clocking mechanism

In this approach stuck-at 0 and stuck-at 1 faults are added at the D-input of the flops (Functional data input) and patterns are generated using ATPG tool. Instead of using external slow speed clock during capture as is normally done during stuck-at fault testing, a single pulse is leaked from high speed internal clock using clock shaper as shown in the clock waveform of Figure 12. Both type of faults are added at D-input as shown in Figure 12 , as pulse width requirement can be different based on whether flop is transitioning from low to high or high to low. Table I shows the pattern count for 3 different industrial 40nm designs using the above approach. This accounts approximately to 10% of stuck-at pattern count.

TABLE I. PATTERN COUNT DETAILS

Design	Flop Count	Pattern Count
DesignA	45K	1460
DesignB	60K	2035
DesignC	110K	6210

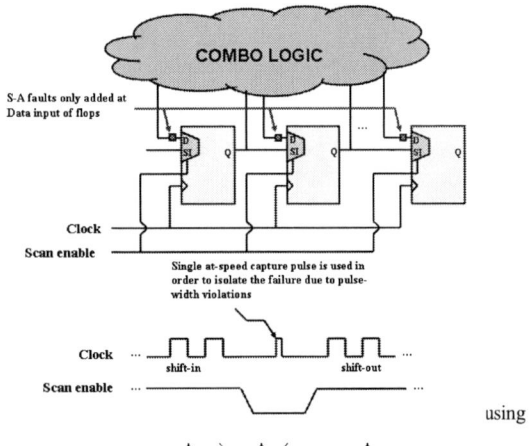

F   using

B. Compact Scan chain based approach for enabling searches

In this approach instead of using functional data input, scan input is used for testing pulse width violation. This helps in reduction of pattern count as the combinational logic is completely bypassed while testing for pulse width violations of the flop. Figure 13 illustrates the concept explained above.

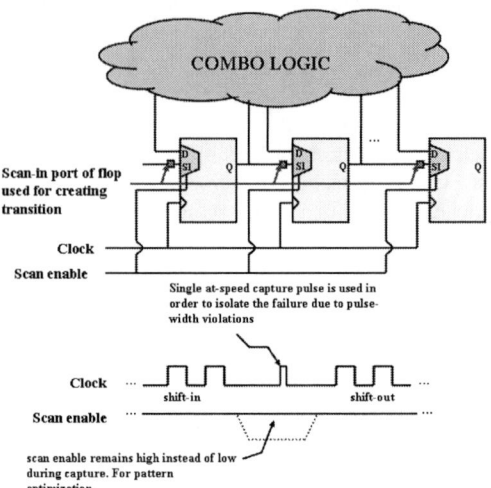

Figure 13. Directed pulse width fault detection pattern generation using scan input (SI-input) of the flop

A compact two patterns that scans-in repetitive 0101 and 1010 pattern to enable low to high and high to low transitions on all the flops in the design can be used. For some designs scanning in 0101/1010 pattern that transition every alternate cycle is an issue due to power concerns [12], in such cases alternate 4 pattern set(Option 2) or 6 pattern set(option 3) can be used as shown in Table II.

TABLE II. SCAN-IN PATTERN OPTIONS

Option	Scan-in Patterns (Repetative)	Total Number of Patterns
Option1	10 , 01	2
Option2	1100 , 0110 , 0011 , 1001	4
Option3	111000 , 011100 , 001110 , 000111 , 100011 , 110001	6

If the pulse width requirement for a flop type is dependent upon the scan enable value then this methodology of pattern generation can't be used for those flops. This can happen for some flop types due to the internal circuit architecture of the flops. Earlier approach can be used in such cases for testing the flop.

Regular stuck-at pattern passing and directed pulse width pattern failing clearly indicates pulse width violation at failing flop(s). If multiple flops are failing then source of the problem could be understood by using common failure point analysis.

V. SILICON RESULTS

Figure 12 shows the plot of maximum frequency of operation (Normalized Fmax) data collected using Fmax searches on 50 devices across process corner lot for a 500M gate 40nm industrial SoC for a domain operating of highest frequency(> 1GHz). Region1 consists of set of devices where the datapath logic limits the maximum frequency of operation clear from the fact that maximum frequency of operation (Fmax) of directed pulse width patterns is higher than Fmax of transition fault patterns, Region2 has the collection of devices where the maximum frequency of operation is limited by pulse width failure, clear by the fact that directed pulse width fault patterns and transition fault patterns having same Fmax.

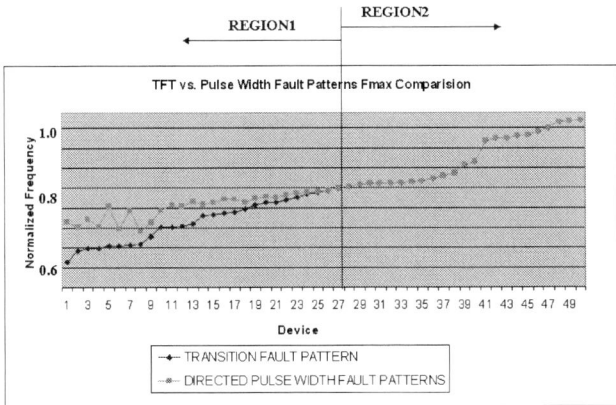

Figure 14. Fmax Comparison

VI. CONCLUSION

In this paper we have proposed multiple test and design techniques that can help in minimizing the probability of pulse width failure mechanism limiting maximum frequency of operation of the device. A novel and simple pattern generation technique is proposed to differentiate and diagnose the pulse width related failure from rest of the failure mechanisms. Effectiveness of the proposed technique is demonstrated based on silicon results across corner lots on 40nm industrial SoC. Proposed technique is very effective in diagnosing the logical failure(Region3 in Figure 3) and this requires pulse width to be less than or equal to threshold pulse width (Threshold pulse width < Minimum pulse width as shown in Figure 3). Authors are presently investigating on extending the work on techniques to diagnose the pulse width failures that results in increased delay but not logical failures.

REFERENCES

[1] T .Chawla , S.Marchal , A. Amara, A. Vladimirescu, "Pulse width variation tolerant clock tree using unbalanced cells for low power design," IEEE International Midwest Symposium on Circuits and Systems, 2009. pp. 443 – 446.

[2] Yu-Wei Yang, K. Shu-Min Li, "Temperature-aware dynamic frequency and voltage scaling for reliability and yield enhancement," Asia and South Pacific Design Automation Conference, 2009 ,pp. 49 – 54.

[3] Bo Zhai, D. Blaauw, D. Sylvester,K. Flautner, "The limit of dynamic voltage scaling and insomniac dynamic voltage scaling," IEEE Transactions on Very Large Scale Integration (VLSI) Systems, 2005 , pp. 1239 – 1252.

[4] Liang-Teck Pang,B. Nikolic, "Measurement and analysis of variability in 45nm strained-Si CMOS technology," IEEE Custom Integrated Circuits Conference, 2008., pp. 129 – 132.

[5] T. Gawa,K. Taniguchi, "A 50% duty-cycle correction circuit for PLL output," IEEE International Symposium on Circuits and Systems, 2002. pp.IV-21 - IV-24.

[6] Pranav Murthy, Sanju Nair Attoor, Rajagopal K. A., "Timing Analysis on a Large High-performance 40nm Video SoC", SNUG India 2010 proceedings.

[7] T. Barrette, V. Bhide, K. De,M. Stover, E. Sugasawara, "Evaluation of early failure screening methods, " IEEE International Workshop on IDDQ Testing, 1996., Page(s): 14 – 17.

[8] Sanghyeon Baeg, "Delay Fault Coverage Enhancement by Partial Clocking for Low-Power Designs With Heavily Gated Clocks," IEEE Transactions on Computer-Aided Design of Integrated Circuits and Systems, 2007 , pp. 2215 – 2221.

[9] A. Chakraborty,G. Ganesan,A. Rajaram, D.Z Pan, "Analysis and optimization of NBTI induced clock skew in gated clock trees," Design, Automation & Test in Europe Conference & Exhibition, 2009. pp. 296 – 299.

[10] Xiao-Xin Fan, Yu Hu, Laung-Terng Wang,"An On-Chip Test Clock Control Scheme for Multi-Clock At-Speed Testing," Asian Test Symposium, 2007., pp. 341 – 348.

[11] J. Saxena, K. M. Butler, J. Gatt, Raghuraman R., S. P. Kumar, S. Basu, D. J. Campbell, J. Berech, "Scan-Based Transition Fault Testing - Implementation and Low Cost Test Challenges," IEEE International Test Conferance, 2002, pp. 1120-1129.

[12] K. M. Butler, J. Saxena, T. Fryars, T. Hetherington, A. Jain, J. Lewis, "Minimizing Power Consumption in Scan Testing: Pattern Generation and DFT Techniques," IEEE International Test Conferance,2004, pp. 355-364.

[13] Process corners: http://en.wikipedia.org/wiki/Process_corners

[14] G. E. Tellez and M. Sarrafzadeh, "Minimal buffer insertion in clock trees with skew and slew rate constraints," IEEE Transactions on CAD, 1997, pp. 333-342.

[15] http://dnenni.wordpress.com/2009/11/23/moores-law-and-40nm-yield : Moore's Law and 40nm Yield

978-1-4673-0438-2/12 $31.00 © 2012 IEEE

At-speed Testing of Asynchronous Reset De-assertion Faults

Arvind Jain, Maheedhar Jalasutram, Srinivas Vooka, Prasun Nair
Texas Instruments (India) Pvt. Ltd,
({a-jain, Maheedhar, vsrinivas}@ti.com)
Neeraj Pradhan (BITS-Pilani, Goa)

Abstract

In sub-threshold technology nodes, device failure due to timing related defects (setup & hold timing) is on rise due to extreme process variability and increasing use of voltage scaling techniques for achieving required performance. High coverage using stuck-at fault patterns, which can effectively screen static defects is no longer sufficient to control DPPM (Defective parts per million). High test coverage of timing defects that is induced by process variation is required for controlling DPPM. Lot of work has been done to find the ways to increase the delay test coverage of industrial circuits including the various methods to cover inter-domain clock faults but very little or no work is done on the ways to effectively cover the asynchronous reset paths to the memory registers for timing defects. In this paper we propose a novel methodology that allows us to effectively detect the failures induced by timing defects on asynchronous reset path of the registers. This problem is further complicated by the fact that commercially available ATPG tools are not capable of generating test patterns due to modeling limitations. Results from 45nm industrial multi-million gates design is presented to illustrate the effectiveness of the proposed methodology.

Keywords: reset synchronizer, delay fault, de-assertion, ATPG.

1 Introduction

Reset is one of the most important applications for any system and present day SOC (System on chip) is not an exception for it. As power-up state of a SoC is unknown, asserting reset helps to bring SoC into a fixed initial state (Commonly known as "reset" state) and makes the behavior of a chip predictable [1], in addition to ensuring a predictable initial state, the reset pin is also used to force the chip into a known "reset" state whenever it enters into some kind of a brownout condition.

Most test generation procedures for synchronous sequential circuits assume the existence of hardware reset [2-5]. These procedures assume that the reset circuitry is fault free. Faults in the reset can cause high DPPM since they may not be detected by a test sequence for other faults in the circuit, yet they may affect the operation of the circuit. In addition, they may invalidate test sequences generated under the assumption that the reset is fault free.

Lot of work has been done to find the ways to increase the test coverage of industrial circuits including the various methods to cover inter-domain clock faults [6] but there is not much prior work in addressing this problem of at-speed reset fault detection. Single reset line control in ATPG mode to detect stuck-at faults on the reset pins of all the flops was proposed in [7]. This approach has been used in most industrial designs. In this

approach, all the functional resets are muxed with a single ATE pin which can be asserted or de-asserted to achieve stuck-at fault coverage on reset pins of all the pins. This is a standard approach used for stuck-at fault testing of reset faults but approach cannot be extended to delay fault testing of reset faults due to tester limitations and clock-reset synchronization. [8] Provides further insight into easier stuck-at fault testing of reset faults without using ATE pin. In this approach, reset is controlled by a scan flip-flop during capture and hence tool can target stuck-at fault coverage for reset. But this approach is restricted to only stuck-at fault testing. But in sub-threshold technology nodes, speed defects are prominent due to process variability [9] and increased use of voltage scaling techniques [10]. Hence most industrial designs rely on functional tests for at-speed reset fault coverage which is expensive and incomplete. None of the ATPG tools supports structural at-speed coverage of reset de-assertion faults. At-speed reset faults in most industrial ATPG tools are considered untestable by default and hence they are not even targeted during delay ATPG.

In this paper, we propose a novel technique to detect at-speed reset de-assertion faults. Reset assertion in flops is an asynchronous operation; however, reset de-assertion happens synchronously so as to ensure that there are no race conditions in the circuit, once reset is de-asserted. Hence this paper addresses only reset de-assertion delay fault testing as it is functionally at-speed operation unlike reset de-assertion.

This paper is organized into eight sections. Section 2 provides background of functional reset operation and reset synchronization. Section 3 discusses problem statement. Basic approach and all possible solutions are discussed in section 4. Section 5 provides detail about proposed approach and flow. Experimental setup and results are presented in Section 6 and 7. Section 8 concludes the paper.

2 Functional reset operation

2.1 Synchronizing reset de-assertion

In most SoCs, the reset signal coming into the chip is an asynchronous input. Although both the active and the inactive edge of such an input is asynchronous, it is not necessary to synchronize the active edge of reset if it can be guaranteed that the reset signal will be held active for a long enough period to allow all flip-flops to enter into the "reset" state". However after reset de-assertion, program execution begins and the chip continually changes its state. Therefore, irrespective of whether a design uses reset in a synchronous or an asynchronous manner, it is extremely important that reset de-assertion is always synchronized in the receive-clock domain. This is necessary to prevent two potential problems [11]: -

- Failures because of flops entering into a metastable state due to violation of the reset recovery time and
- Functional failures because of reset removal occurring in different cycles for different flip-flops.

The other thing that needs to be ensured with regard to reset removal is not to synchronize the asynchronous reset signal more than once (parallel synchronizers) in the same clock domain. Since signal delay through a synchronizer is between one and two clock periods in the receive clock domain, there exists a possibility that reset removal through one synchronizer might occur earlier than the other. This can result in reset removal happening across different cycles even in the same clock domain.

2.2 Reset Synchronizer

If a design uses synchronous reset to initialize all flops, it is necessary that the clocks are toggling when reset is active. This can be a problem in chips which are pin limited and more than one peripheral share the same set of pins in different device configurations. Under such circumstances, external clocks are gated off from reaching the module until a particular mix of peripherals is selected after reset. Another issue for synchronous resets is that the reset pulse must be wide enough to be recognized by an active edge of the clock. This can become a serious issue for peripherals that run of slow external clocks.

To circumvent both these problems and still use synchronous reset, the reset synchronizer shown in Figure 1 can be used. At reset time, even though clocks are not running, both stages of the synchronizer will get cleared asynchronously (assuming active low reset) on reset assertion and the output will become low. The output will still remain low even after reset is de-asserted if the clocks are not toggling [12].

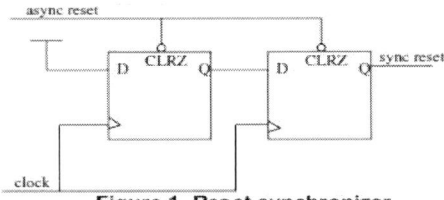

Figure 1. Reset synchronizer

De-assertion of reset through the reset-synchronizer will only take place once the clocks start running and when reset de-assertion has propagated through both the stages of the synchronizer. This means that the logic fed by the synchronizer output will see reset for two cycles once the clocks start toggling. Hence reset de-assertion becomes at-speed operation from reset synchronizer to reset pin of the flops whereas reset assertion still remain asynchronous.

2.3 Test operation of reset synchronizer

If functionally reset are controlled directly from reset synchronizer, reset reaching to the flops is synchronous (driven from a flop) and cannot be controlled from the primary inputs of the chip, care should be taken to gate off reset during scan shifting. There are two ways to do this: (Figure 2)

    a) Use the scan-enable signal to mux synchronous (functional) and ATE reset pin.

    b) Use the scan-mode (data-register) signal to mux synchronous (functional) and ATE reset pin.

The advantage of using method (a) over method (b) is that extra stuck-at fault coverage can be obtained on synchronizer logic and it also allows at-speed testing of reset de-assertion path if synchronizer flops are on scan. This mode of operation (where dft_rst_bypass follow scan_enable behavior in figure 2) will be assumed in rest of the paper.

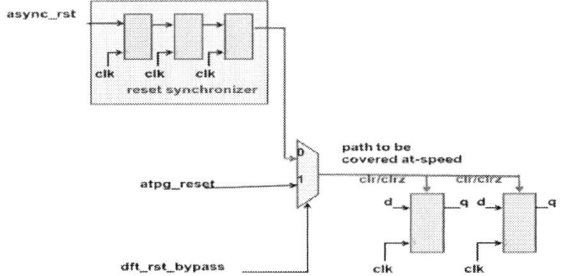

Figure 2: Test operation of Reset synchronizer (rstgen)

3 Problem statement

Reset assertion in flops is an asynchronous operation; however, reset de-assertion happens synchronously so as to ensure that there are no race conditions in the circuit, once reset is de-asserted. A reset synchronizer module synchronizes the reset de-assertion with the domain clock. (Figure 2). The path from the reset synchronizer which synchronizes the reset de-assertion of flops (Figure 2), to the reset pins of the flops is not covered by usual ATPG methods for at-speed operations, due to below reasons:

1. ATPG tools treat the reset pin on flops as clock signal. Two clock signals on a flop toggling together are considered illegal operation hence ATPG tool sets the output of such flops to 'X'. (Figure 3)

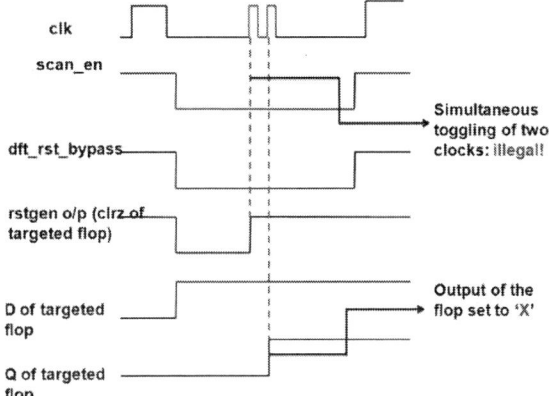

Figure 3: Simultaneous toggling of reset and clock

2. Reset de-assertion does not guarantee a state change. There is no simple way to verify whether reset de-assertion has happened successfully to meet the timing requirement, because there is no definite state of the flop that can be checked.

Therefore, until now, only stuck-at faults have been considered along this path. However, since the path is an at-speed path during de-assertion functionally, it needs to be covered for at-speed testing to decrease defect level.

4 Other Possible Solutions

4.1 Re-modelling the flip-flops

This approach involves changing the UDP (User defined primitives) of flops so that the reset pin is modelled as a data pin which may solve the issue 1 discussed in section 3. The disadvantage to this approach is that it involves changes at the very basic UDP level, which may not be possible after the design process is over and lead to incompatibilities. Issue 2 may not be addressed with this solution. Hence this approach will not be addressed further in the paper.

4.2 Separate clock for reset synchronizer and design flops

This approach involves feeding different clock inputs to the synchronizer flops and the target flops. Please note from the timing diagrams (Figure 3) that the second clock pulse is redundant for the synchronizer flops (as both the first and second stage flops in the synchronizer module store the same value) and the first clock pulse is redundant for the target flops, as we would like to check the output only on the second pulse (the capture pulse). Hence by having separate clocks for synchronizer and design flops, we can avoid race condition on design flops. Again this solution target issue 1 and requires lot of design changes which may not be feasible. Having separate clocks for design flops and synchronizer flops is not acceptable for functional use case hence it will require separate clocking for test and functional use which will be huge overhead. This approach will not be discussed further in the paper.

5 Proposed approach

5.1 Basic Approach

The basic approach which will be followed for extending coverage for at-speed tests to the paths mentioned above, involves the following measures:

- To constrain the synchronizer flops to certain values (Figure 4) which would result in reset de-assertion signal on the output of the synchronizer on successive clock pulses during the capture phase.
- To ensure that dft_rst_bypass follows Scan Enable (SE), so that during the scan phase the control for reset rests with ATPG Reset Bypass, and during the capture phase, the control for reset rests with the synchronizer modules.

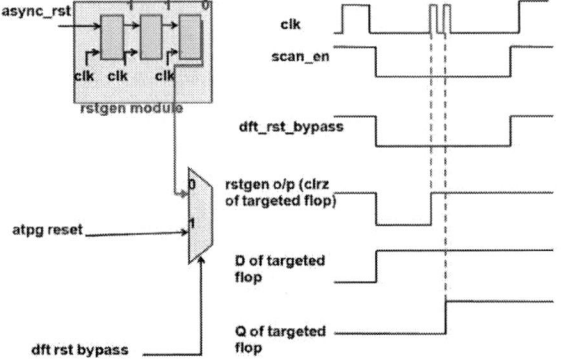

Figure 4: Creating async reset de-assertion (Input pattern generation)

Please note from Figure 4, that two at-speed pulses are provided during the capture phase to the design. In the first clock pulse, this reset signal is de-asserted as the next value on the synchronizer flop is fed out. If reset de-assertion has happened successfully, it can be checked on the next clock pulse (the capture pulse) only if the value on the 'D' pin of the target flop is opposite to the reset state of the flop. But as discussed above, current ATPG tools do not handle this situation hence we need to follow different

approach to generate pattern to cover reset de-assertion. This approach requires pattern creation in two phases

- Generating the input pattern from the ATPG Tool.
- Constructing the output pattern outside the ATPG tool using good simulation values

5.2 Implementation details
Input pattern generation from ATPG tool

In Atpg tool, Reset synchronizer flops should be constraint to certain values which would result in reset de-assertion signal on the output of the synchronizer on successive clock pulses during the capture phase (As shown in Figure 4).
For example in Figure 4:

Rstgen flop stage 1- constrained to value 1
Rstgen flop stage 2- constrained to value 1
Rstgen flop stage 3- constrained to value 0

We also need to ensure that dft_rst_bypass follows Scan Enable (SE). Following above constraints, ATPG tool generated pattern will be used as an input pattern to cover reset de-assertion faults.

Constructing the Output part of the pattern

Please note from the timing diagram (Figure 5), that the 'D' to 'Q' capture will only matter if the 'D' value is opposite to the reset state of the flop because only then would we have a state change to indicate valid capture.

- We need to find all the flops whose D values are opposite to the reset state of the flops, prior to the second capture pulse. These are our 'flops of interest'.
- We can then construct the output outside the tool by strobing for only these 'flops of interest' and masking all the others.
- The number of distinct rstgen module-flop pairs gives us the coverage obtained for reset de-assertion paths.

There is a finer point to be noted here. The flops, which do not qualify under 'flops of interest' (if during the normal functioning of the device the 'D' pin is maintained at the same state as the reset state of the flop), should not be considered during the at-speed testing of reset de-assertion paths. **Because an at-speed fault, even if it exists on the given path, will not lead to any functional failure.**

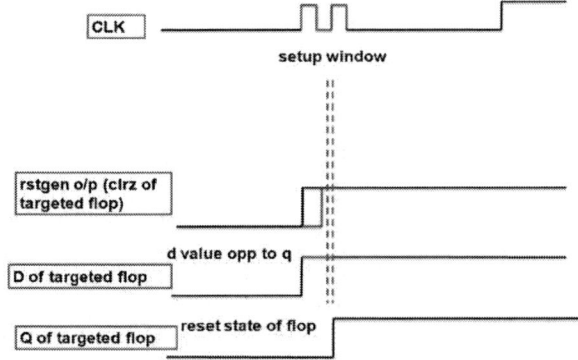

Figure 5: Timing diagram for 'flops of interest'

5.3 Flow/Algorithm details

The primary flow description is given in figure 6. The following steps describe the algorithm for reset de-assertion testing flow

978-1-4673-0438-2/12 $31.00 © 2012 IEEE   360

- Set up the transition fault testing setup, by including the constraints on the synchronizer flops. dft_rst_bypass should follow Scan Enable (SE).
- With these constraints in place, generate patterns in ATPG tool.
- Perform a good circuit simulation of the pattern generated after removing all the faults.
- Get the entire flop list value dump (i.e. the values on the pins of all the flops in the design during good simulation of the pattern) from the ATPG tool.
- From the flop list value dump, identify the 'flops of interest'. These are the flops which satisfy all the following three criteria:
    1. Received a valid clock pulse (i.e. two at-speed clock pulses, the launch and the capture pulse)
    2. Received a valid reset signal. For instance, for clrz, the valid reset signal will be 0-1-1, where the first value indicates the reset pin value prior to receiving the first pulse, and the next two indicate the reset pin value on receiving the first and the second clock pulse.
    3. Have a 'D' value opposite to the reset state of the flop, prior to the second clock pulse (i.e. the capture pulse).
- Reconstruct the output part of the TDL by strobing for only the flops of interest' and masking all the other flops in the design

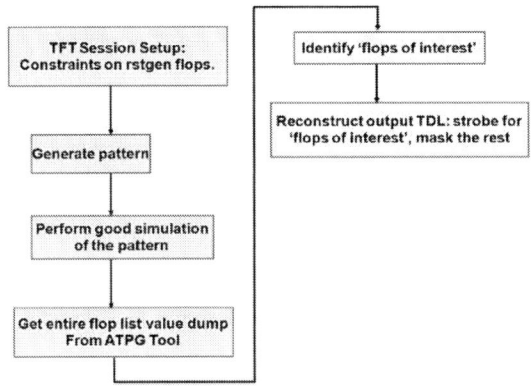

Figure 6: Proposed flow/algorithm

5.4 Advantages and limitation of proposed approach

Advantages of Proposed Approach: This approach has following advantages over other discussed scheme discussed in previous section.

- No design changes required.
- No flop remodeling required.
- Uses the current features present in ATPG tools, without any modification.

Limitations of the Proposed Methodology: Proposed approach suffers from some demerits which have been enumerated below:

- This method covers only a fraction of paths from the reset synchronizer modules to the target flops, as only a fixed number of patterns can be generated at a given time.
- Since only 'flops of interest' are being taken into account, a number of flops are neglected from at-speed testing of reset de-assertion signal. As remarked earlier, while we may be justified in neglecting some of these flops.
- Since during the capture phase, the entire design is reset, there is great uncertainty whether crucial components on the chip such as the clock/reset generators etc. themselves are not resetted during the capture cycle. This would adversely impact coverage.

5.5 Enhancements to primary flow

To further increase the coverage for reset de-assertion paths, some additional enhancements were done to the primary flow described in figure 6. These enhancements help in optimizing the fault coverage obtained

- **Generating multiple patterns:** Instead of generating only one pattern, we generate multiple patterns and select the most optimal pattern. This is defined as the pattern which gives the highest path coverage. Improvement results are shown in next section.
- **Activating one synchronizer module at a time:** Instead of activating all synchronizer modules together, we activate them one at a time. Total coverage is the sum of coverage obtained for each module. For each module, multiple patterns are generated and the most optimal is chosen. Improvement results are shown in next section.
- **Using multiple patterns per rstgen:** Instead of selecting single pattern for every rstgen, we can select multiple input patterns which provide maximum coverage (after considering overlap between two patterns). Improvement results are shown in next section.

6 Experimental setup

Proposed flow was implemented on a production wireless IP design. This design is a 45nm hardware accelerator IP for cellular camera phone market. It has about 222K scan flip-flops, 8 SIs and 8 SOs, and 2 clock domains. Design has ~40 different rstgen (reset synchronizer) controlling reset of design flops. The experimental set up below has been used

1. Synopsys Tetramax[13] was used to generate input patterns for proposed.
2. Several perl scriptware were developed for below steps
    o For creating ATPG setup with constraints for rstgen modules.
    o For parsing, value dump from ATPG.
    o To identify, flops of interest based on criteria discussed in section 5.3.
    o Generating simulatable/Tester ready patterns.
3. Synopsys VCS-MX [14] was used as simulation tool to verify script ware generated patterns.

7 Results

This section enumerates some of the important results obtained. The results clearly indicate how including the additional enhancements to the flow (discussed in Section 5.5) results in better coverage.

7.1 Generating multiple patterns for a rstgen module

Table 1 shows the results obtained for one rstgen module (module number 4). Total Targeted flops refer to the number of flops which get a valid clock pulse and valid reset/reset de-assert signals. Out of these, the number of flops which have a 'D' value opposite to the reset state of flops gives the flop coverage. These are simply our 'flops of interest'.

Table 1:Generating multiple patterns*

Pattern Number	No of Flops Covered	Total Targeted Flops
0	3	9
1	5	7
2	97	130
3	92	125
4	71	129
5	97	127
6	96	126
7	459	4306
8	101	144
9	5	17
10	1	8

* Results shown above are only for one rstgen module. Only the first 11 patterns out of 99 simulated are shown.

It clearly shows how generating different patterns is beneficial as the coverage obtained by various patterns differs vastly. The most optimal pattern in terms of the highest number of flops covered is finally selected by the flow. In this case, pattern number 7, which gives coverage of 459 flops. Also note the coverage of patterns 4 and 5 (in bold). While pattern 4 has higher number of flops that receive a valid clock pulse and reset signal, it is pattern 5 that has higher flop coverage. This shows that some patterns are better suited than others for the testing scheme proposed.

7.2 Activating one synchronizer module at a time

A similar procedure outlined above is followed, wherein each rstgen module is activated individually, and the pattern giving the highest coverage is selected for that module. Table 2 (shows the results for each rstgen module. When rstgen modules are activated individually, the highest coverage of 17473(219K flops) paths is achieved. This is the sum of the coverage obtained for all synchronizer modules. When rstgen modules are activated simultaneously, the highest coverage of 3967 paths is achieved. It clearly shows that individual activation of rstgen module gives much higher coverage compare to activation of all the rstgen together.

7.3 Using multiple patterns per rstgen

Figure 7 shows the effects of the third enhancement in place, i.e. the case when more than one pattern is finally selected for each rstgen module. In this case, the results are again for rstgen module 4, but this time the best 4 patterns are selected, which are 11, 61, 33 and 73. Note that these are not simply the best 4 patterns

giving the highest coverage from the table. For instance, pattern 42 has a higher coverage of 120 flops as compared to pattern 73 which has coverage of 104 flops, but pattern 73 is selected. This is because here is some overlap in the coverage for paths by different patterns. As mentioned earlier, the algorithm tries to find the best 4 patterns which give the highest coverage for unique paths. Therefore, the coverage given for the 4 patterns is the coverage obtained for unique flops in the design and is different from the coverage reported in the tool.

8 Conclusion

In synchronous sequential circuits, reset de-assertion is an at-speed operation. It is important to have at-speed defect coverage on reset de-assertion path as functional test are expensive and incomplete for this coverage. Several possible solutions were analyzed to cover reset de-assertion faults. We proposed a novel technique/flow to cover reset de-assertion path using ATPG tool based approach. Implementation details and detailed ` flow was discussed in the paper. Several enhancements to the primary flow were also evaluated on Industrial circuit. Proposed flow can be extended to any SoC/IP with any ATPG tool.

References

1. I. Pomeranz and S. Reddy, "Applications of homing sequences for synchronous sequential circuits," Technical Report 4-7, A52242, 1992, Electrical and Computer Engineering Department, University of Iowa, Iowa City

2. H-K.T. Ma, S. Devadas, A.R. Newton, and A. S-incentelli, "Test Generation for Sequential Circuits", IEEE Trans. on Computer-Aided Design, Oct. 1988, pp. 1081-1093.

3. A. Ghosh, S. Devadas and A. R. Newton, "Test Generation and Verification for Highly Sequential Circuits", IEEE Trans. On Computer-Aided Design, May 1991, pp. 952-667.

4. H. Cho, G.D. Hachtel and F. Somenzi, "Fast Sequential ATPG Based on Implicit State Enumeration", in Proc.1991 Intl. Test Conf. pp. 67-74.

5. G. *Cabodi*, P. Camurati and S. Quer, "Full Symbolic ATGP for Large Circuits", in Proc. 1994 Intl. Test Conf., Oct. 1994, pp. 980-988.

6. Xiao-Xin Fan, Yu Hu, Laung-Terng Wang,"An On-Chip Test Clock Control Scheme for Multi-Clock At-Speed Testing," Asian Test Symposium, 2007, pp. 341 – 348.

7. I. Pomeranz, S.M. Reddy, "On the Detection of Reset Faults in Synchronous Sequential Circuits," VLSID, pp.470, Tenth International Conference on VLSI Design: VLSI in Multimedia Applications, 1997.

8. T.L. McLaurin, R. Slobodnik, Kun-Han Tsai, A. Keim, "Enhanced testing of clock faults," Int. Test Conf., Oct 2007.

9. Yu-Wei Yang, K. Shu-Min Li, "Temperature-aware dynamic frequency and voltage scaling for reliability and yield enhancement," Asia and South Pacific Design Automation Conference, 2009, pp. 49 – 54.

10. Liang-Teck Pang,B. Nikolic, "Measurement and analysis of variability in 45nm strained-Si CMOS technology," IEEE Custom Integrated Circuits Conference, 2008., pp. 129 – 132.

11. "A Metastability Primer," Philips Semiconductors Application Note AN219,

http://www.semiconductors.philips.com/acrobat/applicationnotes/AN219_1.pdf, Nov 15, 1989.

12. Subrangshu Das, Subash Chandar, Ashutosh Tiwari, "Reset Careabouts in a SoC Design," VLSID, pp.788, 17th International Conference on VLSI Design, 2004.

13. Synopsys Tetramax user guide, 2010.
14. Synopsys VCS-MX user guide, 2010.

Table 2:Individually Activating Rstgens*

sync module number	Pattern number	Flops Covered	Total Flops	sync module number	Pattern number	Flops Covered	Total Flops	sync module number	Pattern number	Flops Covered	Total Flops
1	74	126	837	14	1	140	649	27	32	86	360
2	12	414	4298	15	15	381	4302	28	26	448	4320
3	81	313	4319	16	89	414	4301	29	53	105	166
4	7	459	4306	17	84	394	4330	30	15	434	4828
5	11	388	4379	18	80	77	128	31	30	337	7085
6	37	410	4815	19	91	334	4301	32	55	462	1140
7	84	402	4315	20	37	435	4447	33	26	347	4302
8	32	86	125	21	72	388	4672	34	26	505	4317
9	21	390	4313	22	13	494	4309	35	63	530	5728
10	66	312	4645	23	4	2849	16805	36	79	361	4438
11	74	630	1144	24	20	402	4527	37	52	193	13902
12	97	740	1479	25	48	542	5822	38	63	459	4500
13	16	423	4308	26	95	362	4317	39	7	901	1355
								All enable	58	3967	36366

- When rstgen modules are activated simultaneously, the highest coverage of 3967 paths is achieved.
- When rstgen modules are activated individually, the highest coverage of 17473(219K flops) paths is achieved. This is the sum of the coverage obtained for all synchronizer modules.

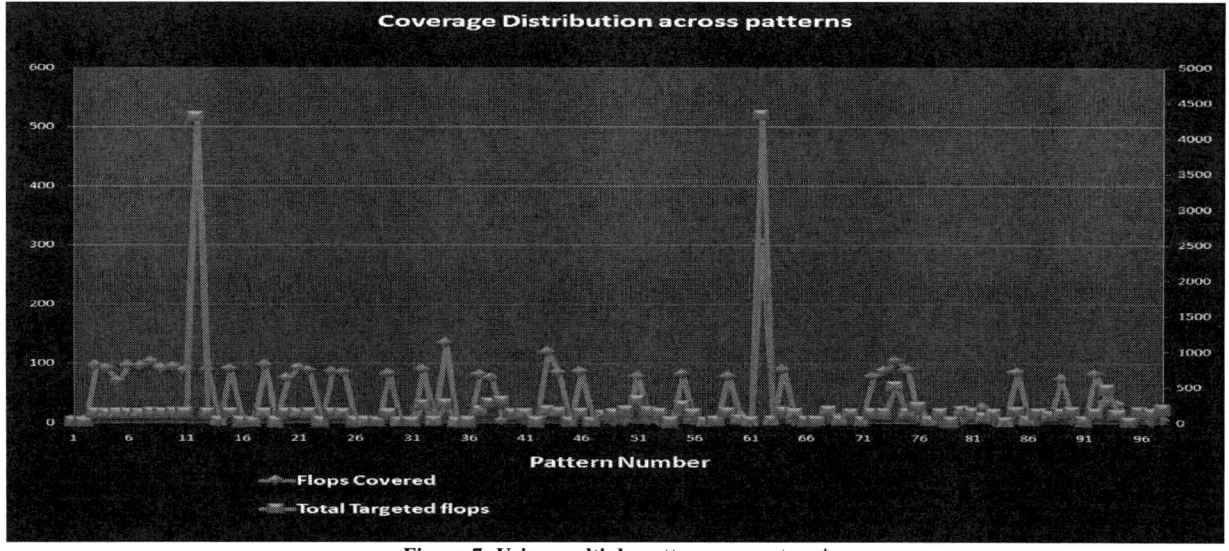

Figure 7: Using multiple patterns per rstgen*

* The highest coverage of 511 flops is obtained with pattern 11. The 4 patterns selected are: 11:61:33:73 based on highest unique coverage.

2012 25th International Conference on VLSI Design

A Library for Passive Online Verification of Analog and Mixed-Signal Circuits

Debjit Pal   Pallab Dasgupta
Dept. of Computer Science and Engg.
Indian Institute of Technology Kharagpur, India
Email: {debjit,pallab}@cse.iitkgp.ernet.in

Siddhartha Mukhopadhyay
Dept. of Electrical Engg.
Indian Institute of Technology Kharagpur, India
Email: smukh@ee.iitkgp.ernet.in

Abstract—The development and use of assertions in the Analog and Mixed-signal (AMS) domain is a subject which has attracted significant attention lately from the verification community. Recent studies have suggested that natural extensions of assertion languages (like PSL and SVA) into the AMS domain are not expressive enough to capture many AMS behaviors, and that a library of auxiliary AMS functions are needed along with the assertion language. The integration of auxiliary functions with the core fabric of a temporal logic is non-trivial and can be challenging for a verification engineer. In this paper we propose a purely library-based verification approach, where libraries for checking elementary properties can be naturally connected with libraries for auxiliary functions to monitor complex AMS behaviors. We study the modeling of behaviors with the proposed library, and outline the main challenges and their solutions towards implementing the verification library over commercial AMS simulators.

I. INTRODUCTION

Assertions are widely used in simulation based verification for monitoring complex temporal behaviors in digital integrated circuits. Assertions languages like *Property Specification Language* (PSL) [3] and *SystemVerilog Assertions* (SVA) [2] derive their syntactic fabric from temporal logics, like *Linear Temporal Logic* (LTL) [15]. The task of extending assertion languages towards capturing Analog and Mixed-Signal (AMS) behaviors is being seriously pursued by the research community [11], [12], [14], [13] as well as industry consortia [1].

Most of the recent efforts towards developing AMS assertion languages consider natural extensions of existing assertion languages. These extensions broadly consider two aspects. Firstly in the AMS domain, we must consider *dense real time* behaviors as opposed to *synchronous discrete time* or *clocked* behavior as in digital circuits. Real time extensions of Linear Temporal Logic (LTL), such as Metric Interval Temporal Logic (MITL) [9], [10], provide the desired semantic capability. Secondly, AMS behaviors are described over real valued variables as opposed to pure Boolean variables used in the digital domain. Since temporal logics are defined over Boolean atomic propositions, the recent AMS extensions of these logics restrict the use of real variables to within *analog predicates*. For example, if $V(y)$ denotes the real valued voltage at a net y, then $V(y) < 5V$ is an analog predicate.

This work was supported by SRC/GRC Research Contract 1835.001.

Encapsulating real valued variables within analog predicates essentially allows us to retain the semantics of the underlying temporal logic, as explained through the following example:

$$\mathcal{G}(x \Rightarrow \mathcal{F}_{[3,5]} \ y)$$

The above property says that whenever the Boolean signal, x, is high at time t, the Boolean signal, y, must be high sometime between $t + 3$ to $t + 5$. Now consider the following property which has a very similar syntactic structure:

$$\mathcal{G}((V(in) > 5V) \Rightarrow \mathcal{F}_{[3,5]} \ (V(out) > 3V))$$

This property says that whenever the real valued signal $V(in)$ is above $5V$ at time t, the real valued signal, $V(out)$, must be above 3V sometime between $t + 3$ to $t + 5$.

Though these recent AMS extensions of temporal logics have a neat semantics which easily follows from the semantics of the underlying temporal logic, the encapsulation of real variables within analog predicates severely constrains the expressibility of AMS behaviors. Researchers have studied the use of auxiliary AMS functions with the assertion language [13] to partially address this limitation. For example, consider the following property:

Example 1.1: *After the signal* enable *becomes high, the output voltage,* $V(out)$, *will rise following the curve* $f(t) = (1 - e^{ct})$ *with a tolerance of* ϵ *for the next* $20\mu s$.

In order to monitor this requirement, the function, $f(t)$, has to be encoded (say, in Verilog-AMS) as an auxiliary function in the simulation framework. This auxiliary model will have to be triggered by the *enable* signal. Let z denote the output of this auxiliary model (that is, z represents $f(t)$). We can now express the required property using z as follows:

$$\mathcal{G}(enable \Rightarrow \mathcal{G}_{[0,20\mu s]} \ (|V(out) - z| < \epsilon)$$

This example shows that for capturing quite simple AMS behaviors, the verification engineer must use a combination of auxiliary functions and the assertion language. We believe that this can potentially be quite confusing at times for the verification engineer.

In this paper we propose an alternative to using AMS assertions, namely the use of a verification library. It is interesting to note that the Open Verification Library (OVL) [4] was a reasonable popular alternative to the use of assertions

978-1-4673-0438-2/12 $31.00 © 2012 IEEE   364

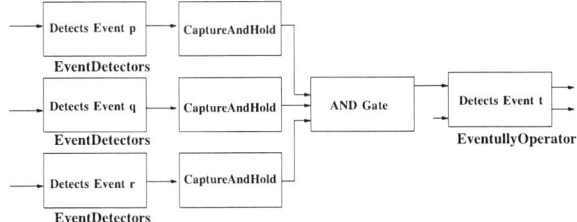

Fig. 1.   Multiple Event Detection

in the digital domain. OVL was accepted into industrial practice primarily because mixed-mode simulation was in its infancy and therefore AMS simulators did not support assertion languages like Open Vera Assertions (OVA) [6] and SVA. Another reason for the acceptance of OVL was that verification engineers found it more convenient to graphically compose OVL modules to develop complex monitors as compared to developing assertions which capture the same behavior.

We believe that verification libraries have some definite advantages in the AMS domain. There are distinct advantages in unifying auxiliary function libraries with verification libraries and creating a common framework for capturing AMS behaviors. Moreover, some behaviors which are easily expressed using the libraries cannot be succinctly expressed using some of the existing AMS extensions of temporal logics. For example, consider the property which says:

Example 1.2: *If events p, q and r occur in any order, then event t will occur subsequently.*

It is easy to express that p, q and r occurs in any sequence as $\mathcal{F}p \wedge \mathcal{F}q \wedge \mathcal{F}r$ in LTL, and also easy to express that t will follow any of these events (say p) as:

$$\mathcal{G}(p \;\Rightarrow\; \mathcal{F}t)$$

but it is not easy to express that all three events will be followed by t. For example, the property:

$$\mathcal{G}(\; \mathcal{F}p \wedge \mathcal{F}q \wedge \mathcal{F}r \;\Rightarrow\; \mathcal{F}t)$$

does not capture the desired intent, since it does not force event t to occur *after* events p, q and r. In order to express the desired intent in LTL, we have to enumerate all six possible sequences in which p, q and r occur and then for each sequence expression that it will be followed by t. Figure 1 shows a monitor for the desired property using verification libraries. The *Eventdetectors* detects occurrence of events *p, q and r* in any order and they get latched in the *CaptureAndHold* modules. The output of the *CaptureAndHold* modules are logically ANDed, which triggers checking of event *t*. The option of using state elements such as latches (i.e. *CaptureAndHold*) and combinational elements such as gates along with the monitors provided in the proposed verification library is a significant advantage over pure formal properties.

In our approach, we have used existing standardized resources (like Verilog-AMS) [5] to develop a library of checkers to be used in a passive online monitoring methodology. We

call it online since the checkers check the simulation trace as soon as it is generated. We call it passive as the outputs of checkers do not modify the testbenches on-the-fly. We demonstrate the use of these libraries in the verification of large real world AMS designs from the power management domain, like BUCK regulators, Linear Drop-Out regulators (LDO), and their components. There exists several fundamental differences between our work and the work reported in the only other paper we could locate on AMS verification libraries, namely [14]. These are as follows:

1) The libraries reported in [14] work on clock boundaries only, which is not suitable for AMS properties. AMS simulators use their own sampling algorithms which typically produce irregular sampling intervals. Our libraries handshake with the AMS simulator and instruct the simulator to improve accuracy near the events by declaring events of interest through *cross events* in Verilog-AMS. Hence in our approach the risk of missing an event is marginalized.

2) The libraries reported in [14] miss a very important point, namely that two independent matches of a property may overlap in time. In order to handle this aspect through a verification library, the library modules must be *reentrant*. In the proposed work, this is achieved by implementing the modules in such a way that they can spawn concurrent threads to handle overlapping matches.

For the second case above, it is important to note that due to the dense time semantics of AMS behaviors, it is possible to have a continuum of matches for a property unless it is triggered by a discrete event like a cross event. This leads to a potential explosion in the number of states for an online monitor, though it is possible to develop offline monitors which work by analyzing real time intervals. Since our goal is to develop a library of *online* monitors, we chose to impose the restriction that all libraries are triggered by events. Note that this was also the case in [14] with the clock event being the trigger.

The paper is organized as follows. Section 2 introduces the AMS Verification Library (AMSVL) and its modules. Section 3 presents the tool flow and implementation issues. Section 4 presents experimental results using industrial test cases. Section 5 presents our conclusions.

II. THE STRUCTURE OF AMSVL

The proposed AMS verification library, called AMSVL, extends the OVL approach to the AMS domain. Like the OVL modules, AMSVL modules can be interconnected to create composite properties from the atomic properties represented by individual modules. The AMSVL library consists of broadly three types of modules, as explained below:

• **Latch modules** are of two types namely (i) *CaptureAndHold* and (ii) *GenerateDelay*. The first type captures and holds any input digital signal forever until simulation is over. The second type holds any input digital signal for a specified delay time. This module is typically used as a delay operator.

- **Simple Arithmetic and Boolean Operation Modules** consist of six types of modules namely:

  1) *ArithmeticOperator* : This is used to generate sum or difference of voltage of analog input signals of the modules. With the addition of unit resistors at the input / output ports, sum or difference of currents of the analog input nets can also be calculated.

  2) *EventDetector* : This is required to monitor analog cross events on an analog input signal with respect to a specified threshold parameter. An extended version of this module, called *EventDetector_Extended* is used to monitor analog events based on the relative values of their input analog signals.

  3) *PredicateEvaluator* : This is required to compare any analog signal with user specified threshold value. It is important to note the difference of this operator with the EventDetector. For an analog predicate like $V(in) > 5V$, the PredicateEvaluator module will simply assert a match if the predicate is true at the time it is triggered, where as the EventDetector module will wait for $V(in)$ to cross $5V$ and then assert a match signal. An extended version of this module, called *PredicateEvaluator_Extended* is used to evaluate predicates which compare two analog signals based on their relative values.

  4) *BoolOperator* : This is standard Boolean logical operators.

- **Interval Operation Modules** These modules represent the standard temporal operators:

  1) *GlobalOperator* : This is used to check the truth of an expression over a specified period of time.

  2) *EventuallyOperator* : This is used to check whether an expression ever becomes true within a specified time frame.

  3) *UntilOperator* : This is required to check whether an expression remains true over a time window until another event occurs.

  4) *PredicateAssert* : This is required to check the truth of an expression when a particular condition is satisfied over a specified period of time.

Due to lack of space, it is not possible to demonstrate the use of all of these operators. The following three examples highlight the use of some of these modules.

Example 2.1: Figure 2 shows the AMSVL realization of the property:

After v_{in} crosses 2.0 volts, v_{out} should cross 2.5 volts sometime between 10µs and 20µs.

In Figure 2, the *EventDetector* module monitors the event of v_{in} crossing 2.0 volts. The *match* pulse of this module triggers the *EventuallyOperator* module. The *PredicateEvaluator* compares the voltage at v_{out} and keeps its *assertE* pin high as long as v_{out} remains above 2.5 volts. The *EventuallyOperator* module has parameters *MinimumDelay* and *MaximumDelay*

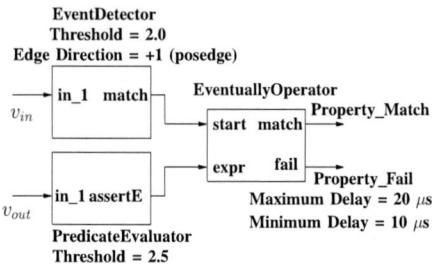

Fig. 2.   AMSVL Realization of Example 2.1

which, in this case, are set to 10µs and 20µs respectively. If *assertE* goes high at some time between 10µs and 20µs of receiving its *start* pulse, the *EventuallyOperator* will assert its *Property_Match* pin, thereby signaling a *match*. On the other hand, if *assertE* does not rise within the specified time bound, then *Property_Fail* will be asserted at the end of 20 µs (that is, after *MaximumDelay*).

We now present a more complex example in which the network of AMSVL modules uses a feedback loop to express a recurring property expression.

Example 2.2: Figure 3 shows the AMSVL realization of the property:

After 190 µs of entering startup mode, the buck regulator will enter its steady state mode, where its steady state voltage will remain within 0.5 V with a tolerance of 0.05 V for the next 10 µs. The steady state voltage should remain in this range at a sampling granularity of 20 µs.

In Figure 3, *BUCK_EN* is the enable pin of the buck regulator, and *BUCK_FB* is a feedback pin of the BUCK regulator used to implement our property.

In Part A of Figure 3, the network detects whether the buck regulator has entered its startup mode. In Part B of the same figure, the network detects whether the buck regulator has entered its steady state mode. In Part C of the figure, the *GenerateDelay* operator activates the *GlobalOperator* module 190µs after start up is detected (by Part A). Within the next 10µs the *GlobalOperator* module will either assert its *match* signal or its *fail* signal, following which the feedback path will be activated to reactivate the *GlobalOperator* module after another 10µs (using the second *GenerateDelay* module). Since the loop delay consisting of the *GlobalOperator* and the second *GenerateDelay* module is upperbounded by 20µs, the constraint on sampling granularity is satisfied.

Example 2.3: Figure 4 shows the AMSVL realization of the property:

When the PLL gets locked, the frequency of oscillation of pll_refclk and pll_fdbkclk should be equal. The frequencies should remain equal till PLL remains locked.

In Figure 4, *pll_en* is the enable pin, *pll_refclk* is the reference clock pin and *pll_fdbkclk* is the feedback clock pin of the PLL used to implement our property. Stability in the voltage of the pin *vctrl* of PLL is used to determine the locking condition.

978-1-4673-0438-2/12 $31.00 © 2012 IEEE   366

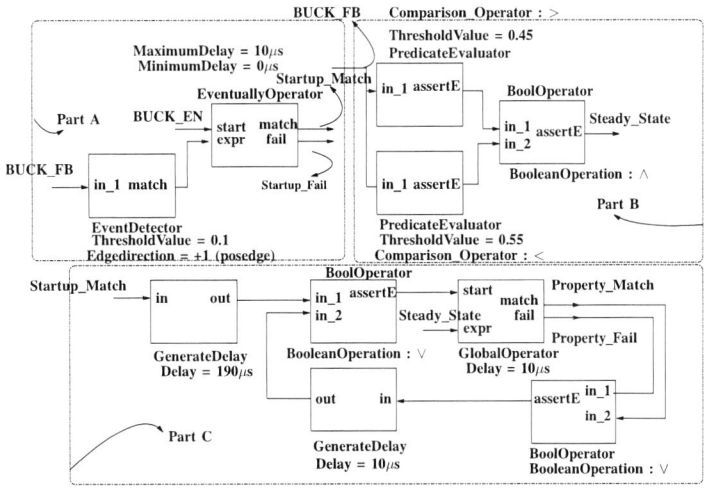

Fig. 3.   AMSVL Realization of Example 2.2

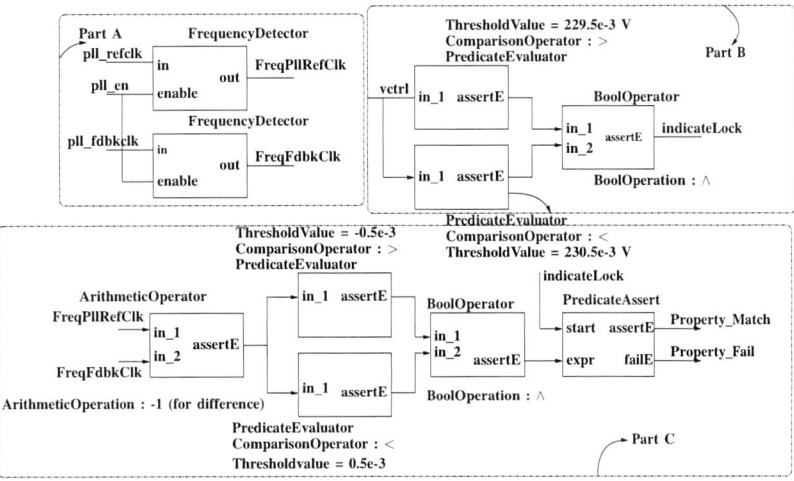

Fig. 4.   AMSVL Realization of Example 2.3

In Part A of Figure 4, the network uses and auxiliary function namely FrequencyDetector to detect the frequency of oscillation of *pll_refclk* and *pll_fdbkclk*. In Part B of the same figure, the network detects whether the PLL has entered into the lock state by checking the voltage of the *vctrl* pin. In Part C of the figure, *ArithmeticOperator* calculates the difference of the frequency of two pins. The two *PredicateEvaluators* check whether the difference is within certain tolerance value as shown in the figure. The *indicateLock* (as detected in Part B) activates the *PredicateAssert* module and keeps it activated as long as the PLL is locked. Depending upon the frequency difference value, the *PredicateAssert* will accordingly either assert *assertE* or *failE*.

III. TOOL FLOW AND IMPLEMENTATION ISSUES

Figure 5 shows the tool flow for the verification of AMS behaviors using AMSVL along with auxiliary functions. In the present version of the tool, auxiliary functions are encoded in Verilog-AMS.

The library is implemented as a package consisting of source code of the modules and the symbols in Cadence CDBA format. The library can be installed in Cadence AMS Virtuoso Environment by simply adding the library through Library Manager. Default values of parameters specified at the cell level can be overridden by specifying parameter values at the instance level which can be done through symbols in Cadence Virtuoso. For complete reference of the library, please see [8]. These parameters are called CDF (Component Description Format). Detailed description can be found in Cadence reference manuals [7]. We show in Figure 6 how symbols can be used to develop checker networks.

Some of the main implementation issues are described in the following subsections.

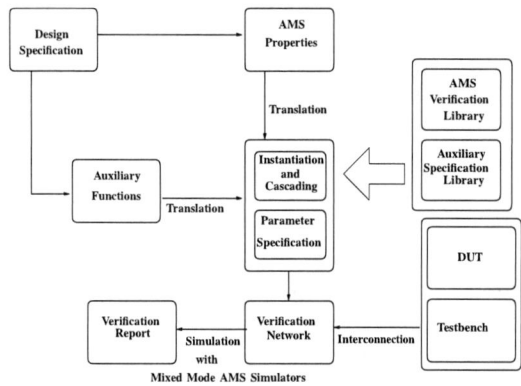

Fig. 5.   Tool Flow of AMS Verification Library

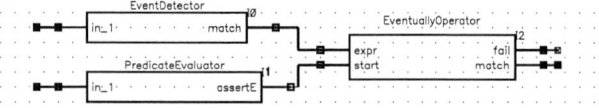

Fig. 6.   Schematic of Example 2.1

A. Synchronization with the AMS simulator

We use *cross_events* to check properties. The accuracy with which the predicates are evaluated with the help of the *cross_events* is controlled by *TimeTolearnce* and *Value-Tolerance* parameters of the *cross_events*. The *TimeTolerance* parameter specifies the maximum allowable error on the real time scale between the estimated crossing point and the true crossing point and the *ValueTolerance* parameter specifies the maximum allowable error on real value scale between estimated crossing point and the true crossing point. For example, the change in the truth value of the predicate ($V(out) > 2.0$) can be monitored by the *PredicateEvaluator* module with the help of the following cross event. Here match is a logic signal which is asserted as soon as $V(out)$ crosses 2.0 volts in the positive edge direction (denoted by +1 in the *cross* statement below).

```
initial begin
    match = 1'b0;
end
always @(cross(V(out) - 2.0, +1, TimeTolerance,
                    ValueTolerance))
        begin
            match = 1'b1;
        end
```

The use of cross events creates simulation overhead because the AMS simulator has to insert additional simulation points to report the cross within the specified value and time tolerance. The time and value tolerances in the *PredicateEvaluator* and *EventDetector* modules should be carefully chosen by the user keeping in mind that overconstraining may lead to degradation in simulation performance.

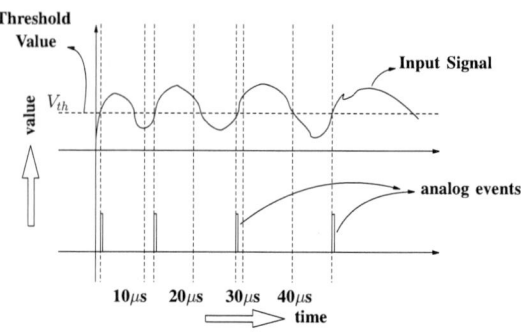

Fig. 7.   Scenario for Property Checking in Parallel Threads

B. Spawning Threads for Overlapping Matches

It is often the case that multiple matches of a property overlap in time along a simulation trace. In order to ensure that matches and violations are not missed, the AMSVL modules need to be able to handle such overlaps. The following example illustrates one such scenario.

Example 3.1: Suppose we are interested in monitoring the following property:

After v_{in} crosses V_{th}, v_{out} should cross V'_{th} sometime between 15μs and 25μs.

Figure 7 shows the v_{in} waveform, highlighting the places where it crosses V_{th}. Since successive crossings happen earlier than the time interval of the property, that is, $[15\mu s, 25\mu s]$, this is a candidate situation where overlapping matches/failures are possible.

In AMSVL, overlapping matches are handled by spawning a new thread whenever a module is triggered. This is necessary only in the *Interval Operation Modules*, that is, the modules representing the temporal operators. To implement this feature, we have used the *task* construct and the *fork-join* construct of Verilog-AMS. We present a code fragment of the *EventuallyOperator* module to explain the implementation.

```
event trig_match;
event trig_fail;
task my_task();
begin
    fork
        begin : test_maxdelay_time
            #MaximumDelay;
            -> trig_fail;
            disable test_expr;
        end
        begin : test_expr
            #MinimumDelay;
            if (expr)
                ->trig_match;
            else begin
                wait(expr);
                if((\$abstime - TimeOfStart[NumberOfStart])
                    < MaximumDelay)

                    ->trig_match;
                else
                    ->trig_fail;
            end
            disable test_maxdelay_time;
        end
        begin : start_again
            @(posedge start);
```

978-1-4673-0438-2/12 \$31.00 © 2012 IEEE

```
        if(NumberOfStart == MaxNumberOfStart)
            NumberOfStart = 0;
        else
            NumberOfStart = NumberOfStart + 1;
        TimeOfStart[NumberOfStart] = $abstime;
        my_task;
    end
  join
end
endtask
```

IV. EXPERIMENTAL RESULTS

We studied the verification of three industrial test cases, all of which are from the power management domain. The first circuit contained six Low Drop-Out (LDO) regulators. For this circuit, monitors for 28 properties were developed using AMSVL. The second circuit contained two buck regulators, and we developed monitors for 10 properties using AMSVL. The third circuit was a Integrated Circuit having four LDOs and one buck regulator. We developed monitors for 33 properties in AMSVL for this circuit. Each property required a carefully selected network of AMSVL modules. Moreover, in all of these cases, auxiliary functions were developed and used seamlessly with the AMSVL modules. The approximate number of cross_events encountered are 248 for the LDO circuit, 100 for the BUCK regulator circuit and 600 for the Integrated circuit.

Table I shows the overhead incurred by the simulator towards handling the auxiliary functions and the AMSVL modules. The simulations were carried out on a 2.33 GHz, Intel-Xeon server with 32GB RAM. Table II reports some details about the circuits that are used as test cases in Table I.

TABLE I
CPU TIME FOR SIMULATIONS OF CIRCUITS

Cross Event Precision		Sim Time	CPU Time (secs)			Over-head (%)
time (sec)	value (V)	(μsec)	Design	Design + Aux Func	Design + Aux Func + Prop Check	
Test Case I: LDO Circuit						
1e-9	1e-6	700	83.89	84.51	97.12	15.77
1e-6	1e-4	700	83.89	84.51	96.01	14.45
1e-4	1e-3	700	83.89	83.51	95.42	13.75
Test Circuit II: BUCK Circuit						
1e-9	1e-6	500	90.25K	92.31K	98.05K	8.65
1e-6	1e-4	500	90.25K	92.31K	97.78K	8.35
1e-4	1e-3	500	90.25K	92.31K	97.66K	8.22
Test Case III: IC Netlist						
1e-9	1e-9	600	75.44K	75.82K	81.24K	7.69
1e-6	1e-4	600	75.44K	75.82K	81.09K	7.49
1e-4	1e-3	600	75.44K	75.82K	80.89K	7.23

The following observations may be made from the experimental results:

1) The overhead of property checking is non-trivial, but the overhead becomes marginal with increase in the size of the circuits. For example, the LDOs are lightweight circuits as compared to buck regulators, and hence the overhead is more visible for LDOs as compared to buck regulators and PMUs.

TABLE II
DESCRIPTION OF THE TESTCASES

Test Cases	No. of Nodes	No. of Transistors	No. of Capacitors	No. of Resistors
LDO circuit	8604	2016	5166	7608
BUCK circuit	3586	4910	700	990
IC Netlist	7529	3799	3794	5567

2) It is interesting to see that the auxiliary functions have an insignificant contribution in the overhead. The AMSVL modules are directly responsible for the overhead. This is largely due to the cross events that the AMSVL modules introduce, thereby increasing the number of simulation points near the occurrences of those events.

V. CONCLUSION

We believe that the library based verification approach will find acceptance in industrial practice. In the AMS domain, auxiliary functions appear to be significant value, and AMSVL modules can be used seamlessly with auxiliary functions. Our results show that AMSVL modules do have simulation overhead, but we believe that the online debugging capability that AMSVL monitors provide will outweigh the simulation overhead.

The current version of AMSVL is compatible with all mixed mode simulation platforms which support Verilog-AMS. It will be interesting to migrate AMSVL to Matlab since Matlab has a very rich library of functions which can serve as built-in auxiliary functions.

ACKNOWLEDGEMENT

The authors would like to thank Dr. Scott Little of FreeScale Inc. and Subhankar Mukherjee for their kind discussions.

REFERENCES

[1] Accellera: http://www.accellera.org/activities/verilog-ams/
[2] IEEE Std 1800-2009, "IEEE Standard for System Verilog: Unified Hardware Design, Specification and Verification Language, IEEE, 2010."
[3] IEEE Std 1850-2010, "IEEE Standard for Property Specification Languages (PSL), IEEE, 2010"
[4] Open Verification Library: http://www.accellera.org/activities/ovl
[5] Accellera Verilog-AMS Language Reference Manual Analog and Mixed Signal Extensions to Verilog-HDL, version 2.4 edn (November 2006).
[6] Open Vera: http://www.open-vera.com/
[7] Cadence AMS Simulator: http://www.vtvt.ece.vt.edu/vlsidesign/ tutorialmixedsignal_intro.php
[8] http://www.facweb.iitkgp.ernet.in/~pallab/manual.pdf
[9] R. Alur and T.A. Henzinger, 1990. Real-Time Logics: Complexity and Expressiveness, Information and Computation 104, 390 - 401.
[10] R. Alur and T.A. Henzinger, 1994. A Really Temporal Logic. Journal of ACM vol. 41, no. 1, pp. 181 - 203, 1994.
[11] O. Maler, D. Nickovic : Monitoring Temporal Properties of Continuous Signals in FORMATS / FTRTFT. Springer pp. 152-166 (2004)
[12] O. Maler, D. Nickovic, A. Pnueli : Checking Temporal Properties of Discrete, Timed and Continuous Behaviors. In: Pillars of Computer Science. pp. 475-505 (2008)
[13] S. Mukherjee, P. Dasgupta : Auxiliary State Machines and Auxiliary Functions : Constructs for Extending AMS Assertions. in Proceedings of the IEEE International Conference on VLSI Design (2011)
[14] R. Mukhopadhyay, S.K. Panda, P. Dasgupta, J. Gough : Instrumenting AMS Assertion Verification on Commercial Platforms. in ACM Transactions Design Automation Electronic Systems (TODAES) vol. 14(2) (2009)
[15] A. Pnueli. 1977. the Temporal Logic of Programs. In Proceedings of Foundations of Computer Science (FOCS). 46 - 57.

A Fast Equation Free Iterative Approach to Analog Circuit Sizing

Supriyo Maji and Pradip Mandal

Department of Electronics and Electrical Communication Engineering,
Indian Institute of Technology, Kharagpur, India-721302
Email: supriyomj@gmail.com, pradip@ece.iitkgp.ernet.in

Abstract—A fast equation free iterative approach for sizing of analog circuit is proposed. Equation based sizing approach has been popular as it removes time consuming simulation effort. If equations are cast in posynomial inequality format, a special optimization technique called geometric programming (GP) can be deployed. The advantage of formulating the problem in GP form is that, it ensures global optimality and can return the final design point instantly even in the presence of hundreds of equation and thousands of variable. But main limitation comes in deriving performance equations in posynomial inequality format. In this context, we develop one novel methodology for fast sizing of analog circuit. This method does not require any such performance expressions. It is based on the meaningful presentation of only device constraints. Infeasibility is handled iteratively making suitable changes on those constraints. Due to the simplicity of formulation, fully automated flow is achieved.

Keywords-analog; circuit; sizing; equation; device; macro-model; design-centering;

I. INTRODUCTION

Though in an integrated system, analog circuitry occupies a small physical area compared to its digital counterpart, due to complex trade offs among numerous performance metrics, designing of analog blocks come out to be the main bottleneck in the design time reduction. Advent of many CAD techniques has helped to achieve the design without much intervention from the designer, but delivering the design within stipulated time limit is the primary concern and still remains unresolved. In the present context, where technology is changing fast in comparison to the architectures, it calls for a tool which not only can yield solution independent of expert designer but should also live up to fast moving market requirements.

Existing approaches of automatic circuit sizing are broadly classified into two main categories, namely simulation based and equations based. Simulation based technique as stated in [1][2], suffers from several drawbacks. As it requires time consuming simulation effort at transistor level, a topology with many transistors cannot be handled with comfort. Moreover, final design point as obtained is only a suboptimal design. On the other hand, equation based approaches not only can guarantee fast convergence but based on the smoothness of feasible region, sometimes global optimal solution can also be ensured. It therefore appears that an equation based constrained optimization

method is the most promising approach for automatic circuit sizing.

The objective of [3] has been to propose an equation-based constrained optimization method that is fast, robust and ensures global optimal solution. But the main limitation comes in terms of posing performances in posynomial inequality format. In fact, it again asks for rigorous simulation effort for circuits where existing non-posynomial performance equations cannot be manually approximated. Here, we should mention that in analog circuits there are mainly two types of performance constraints i.e. dc and ac. [3][4] show that dc performances (input common mode range (ICMR), output swing (OS)) for analog circuits can be expressed in terms of over drive, threshold voltage and other biasing conditions. So, automatic generation of dc constraints only asks for graph based analysis. But main hurdle comes in terms of developing ac performance expressions. Symbolic analysis based ac performance expression generation [5] cannot guarantee posynomiality conditions. In addition, symbolic analyzer involves high computational effort even for a simple circuit. Automatic generation of ac performances in posynomial format has been reported in [6] but addressing a complex circuit will require enormous simulation effort. In [7], macromodel based one efficient methodology has been proposed for ac & dc performance expression generation in posynomial format. The proposed method is used for topology synthesis [8] & selection [9].

On the contrary, we propose one novel methodology where ac performance expressions are not required. The methodology is based on relating ac performances to device parameters and then formulating the problem as design centering [10][11]. Thus only input to the optimizer are constraints which are efficiently formed in terms of device parameters. Device constraints are tweaked appropriately to capture infeasibility over the iteration. Here, it must be noted that in all the equation based approaches so far, main difficulty to automate the entire flow has been due to the development effort needed for complex circuit equations and then reshaping the problem for suitability into optimizer. By removing the requirement of ac performance expressions, we ease complexity of formulation leading to the realization of fully automated design flow.

The rest of the paper has been organized as follows. We have stated the methodology in detail in section [II]. Device

978-1-4673-0438-2/12 $31.00 © 2012 IEEE

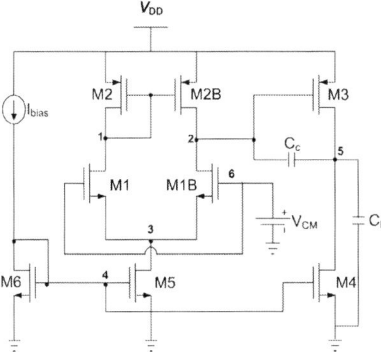

Figure 1. Schematic of two stage op-amp

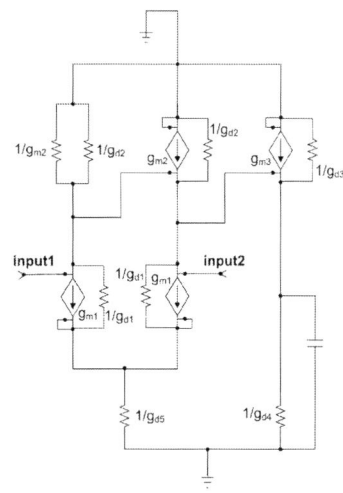

Figure 2. AC macromodel of two stage op-amp

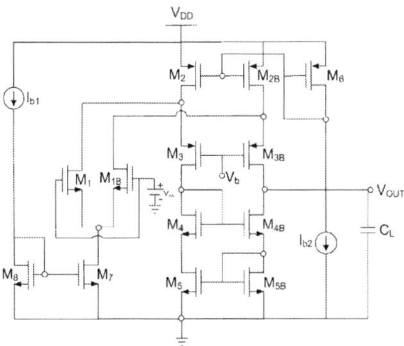

Figure 3. Schematic of folded cascode op-amp

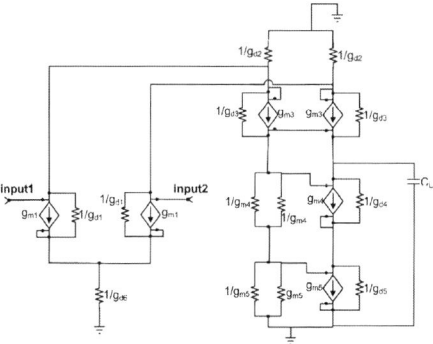

Figure 4. AC macromodel of folded cascode op-amp

models as used in our program is illustrated in section [III]. Experimental results comes next in section [IV]. The conclusion has been drawn in section [V].

II. METHODOLOGY

A. Typical geometric programming (GP) formulation for circuit sizing

Circuit performances are cast in posynomial or monomial format in terms of device models.

$$P_i = f(g_m, g_d, v_{ov}) \qquad (1)$$

Here, P_i denotes performances of circuit like gain (A_v), phase margin (PM), unity gain frequency (UGF), common mode rejection ratio (CMRR) which are posynomial or monomial function of device parameters (g_m, g_d, v_{ov}).

$$g_m, g_d, v_{ov} = f(W_1, L_1, I_1..) \qquad (2)$$

Here, W_1, L_1, I_1 are basic design variables.

Now, the problem can be formulated as GP in the following form:

$$minimize \ \ P_0(\hat{x}) \ over \ \hat{x} = [w_1, l_1, i_1...]^T$$

$$subject \ to \ \ P_i(x) \leq 1 \ \ i = 1, 2, ...p$$

$$and \ \ P_j(x) = 1 \ \ j = 1, 2, ...q$$

$$and \ \ x_k > 0 \ \ k = 1, 2, ...r$$

As clearly evident from the formulation, it requires two sets of model information. One is to take care of technology effects through device models. Other is performances to be modeled in terms of device parameters. In all the sizing approaches so far they have required these information. They depend on manual formulation of the whole sizing problem. But, manual approximation cannot guarantee posynomiality condition as desired by GP. In [6], it is reported that models can be generated automatically in suitable format but the method is crippled with extensive simulation effort and run time. With number of transistors increasing in a circuit, simulation is a costly affair. In addition, modeling inaccuracy is a serious issue.

B. Explaining our perspective with example

In this present context, we develop a methodology where explicit performance expressions are not required. It must be noted that, most analog circuits behave in the linear region. So, device models or small signal model parameters become crucial in analyzing circuit performances. Our main trick is to handle device parameters in a meaningful way

Performance parameters	g_{m1}	g_{d1}	g_{m2}	g_{d2}	g_{m3}	g_{d3}	g_{d4}	g_{d5}	C_c	C_L
Gain	↑	↓	–	↓	↑	↓	↓	–	–	–
CMRR	↑	↓	↑	↓	–	–	–	↓	–	–
UGF	↑	–	–	–	–	–	–	–	↓	–
PM	↓	–	–	–	↑	–	–	–	↑	↓

Table I

DEPENDENCIES OF DIFFERENT CIRCUIT PERFORMANCES ON DEVICE PARAMETERS FOR TWO STAGE OP-AMP

Performance parameters	g_{m1}	g_{d1}	g_{d2}	g_{m3}	g_{d3}	g_{m4}	g_{d4}	g_{m5}	g_{d5}	g_{d6}	C_L
Gain	↑	↓	↓	↑	↓	↑	↓	↑	↓	–	–
CMRR	↑	↓	↓	↑	↓	↑	–	↑	–	↓	–
UGF	↑	–	–	–	–	–	–	–	–	–	↓
PM	↓	–	–	–	–	–	–	↑	–	–	↑

Table II

DEPENDENCIES OF DIFFERENT CIRCUIT PERFORMANCES ON DEVICE PARAMETERS FOR FOLDED CASCODE OP-AMP

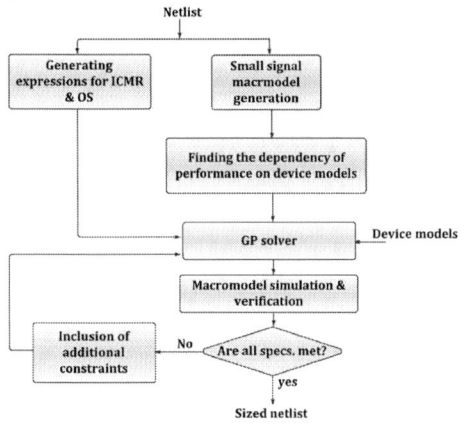

Figure 5.   Flow diagram

so that, they effectively represent circuit performances. The complete flow diagram is shown in Fig. 5. From the circuit netlist, we automatically develop dc performance constraints from dc macromodel following graph based expression generation methodology for different node potentials [7]. A small signal macromodel is also developed for the netlist. This step is more about reading transistors connections properly for the netlist. A small signal macromodel of two stage op-amp (Fig. 1) is shown in Fig. 2. Next, it is to vary device parameters. We capture how different performances are related to device parameters. If we increase g_{m1}, gain improves. While decreasing g_{d5} enhances CMRR. These dependencies are shown in Table I for two stage op-amp. Small signal macromodel for folded cascode op-amp (Fig. 3) is shown in Fig. 4. Dependencies for folded cascode are shown in Table II. We also calculate how strongly or weakly parameters are related to performances. This dependency is reflected in an weightage (w_{i_i}) which is later used in sizing as stated in subsection [II-E]. This dependency calculation is based on finding coefficient strength of a higher order

function as shown below.

$$f(P) = a_0 + a_1.x + a_2.x^2 + ... + a_n.x^n \qquad (3)$$

Here, x is device model parameter and *f(P)* is performance. We vary x keeping other model parameter values at constant and performances are extracted by macromodel simulation. The extracted data is used to fit model template as shown in (3). Based on observation of template coefficient strengths, weightage is assigned. Suppose coefficient w.r.t. n^{th} order term (a_n) is dominant then model parameter is assigned an weightage of n. For simulation, fast behavioral simulator can be used instead of transistor level spice simulation. One important aspect is that, relation of parameters to performances are technology independent so can be treated as knowledge database. Now, the trick is to maximize or minimize these device model parameters according to their requirements (i.e. after finding whether dependency is direct or inverse). This helps to the push design point away from the boundary of constraints. In the next section, we briefly discuss design centering concept and later illustrate the technique in the context of present problem.

C. Design centering

In general, a sizing formulation involves one objective function which is to be minimized along with other inequality and equality constraints. If model inaccuracy is large, unless left sufficient margin, final solution may reach boundary and can even violate specifications. Inaccuracies of model parameters can also be incorporated in sizing formulation [12] by considering maximum model fitting error to make performance prediction more robust. But this causes over design [12]. On the other hand, design centering based approach [10][11] attempts to the center design within performance space and maximize inscribed ellipsoid lying inside performance constraints. Thus with process variation or mismatch, even if there is deviation in the final design point, chances of violating specification is

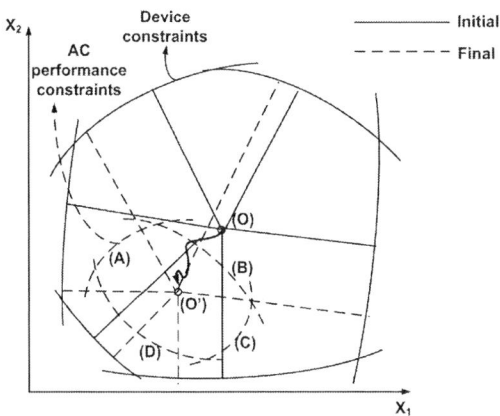

Figure 6. Illustrating translation of design point over the iteration

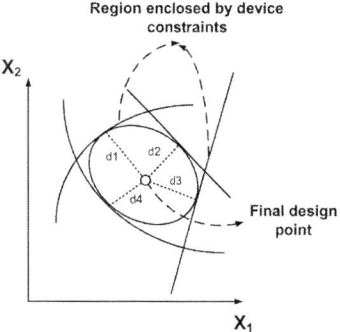

Figure 7. Illustration of performance centering where feasible space is enclosed by device constraints

reduced. Another important aspect is that, design centering formulation follows GP rules.

D. Problem statement

We define problem statement for our formulation as follows.

$$minimize \prod_{i=1}^{n}(1/d_i) \qquad (4)$$

$$d_i.(g_m^k) \leq 1 \quad or \quad d_i.(g_d^k) \leq 1 \qquad (5)$$

Input common mode range (ICMR) and output swing (OS) constraints of any circuits can be written in following format.

$$d_i.(\sum_{i=1}^{n} k_1.V_{ovi} + \sum_{i=1}^{m} k_2 V_{bi}) \leq 1 \qquad (6)$$

$$d_i \geq 1 \qquad (7)$$

d is the set of variables representing length of ellipsoidal axes as shown in Fig. 7, while k is 1 or -1 depending on the relation with performances. n and m denote the number of transistors and biasing conditions respectively in the topology and k_1, k_2 are either 0 or 1. DC constraints like ICMR and OS ensure input and output transistors are in saturation.

Now to keep other transistors in saturation we make sure that intrinsic gain for individual transistor is high [9]. We know for amplifier topologies, transistors need to be kept in saturation which alternately means high intrinsic gain. We formulate the problem as following.

$$d_i g_{di}/g_{mi} \leq 1 \qquad (8)$$

$$d_i \geq 1 \qquad (9)$$

E. Handling infeasibility through iteration

Now, there are parameters like g_{m1} and C_c (for two stage op-amp) which are having conflicting relationships with performances. So, they do not fall in the rule of maximization or minimization. For this reason, in the first iteration we cannot guarantee that all specifications will be met. But at least we can say, values assigned to those parameters will not be arbitrary because transistors are ensured to behave in saturation by ICMR, OS and intrinsic gain maximization constraints. We let these variables stay in the formulation without any explicit bound and let the optimizer decide its value based on other constraint profiles. Now, with the design point obtained in first iteration, we simulate macromodel and see the performances which have failed the specifications. Based on individual performance deviation (P_{diff}) from specification ($P_{specification}$), weightage is assigned to parameters. Suppose if gain is more strongly met than UGF then in next iteration, ellipsoidal axes length corresponding to parameters (here, g_{d1}, g_{d2}, g_{m3}, g_{d3}, g_{d4} for two stage op-amp) which improve gain are given less weightage while parameter (here, g_{m1}) corresponding to UGF is given more weightage. So our idea is to calculate distance from boundary and then assigning an weightage accordingly on axes length so that over the iteration all specifications are successfully met.

F. More insights on the methodology

Let us discuss issues on finding a design point which meets ac specifications without explicitly putting those constraints. To start with, we can say boundaries formed by device models are more relaxed. It covers more region compared to the region formed by only ac performance constraints. This is illustrated in Fig. 6. In consequence, it leads us to a design point (O) which may fail some specifications (specs. A, D, C are met but B fails). It happens because in first iteration constraints on some device parameters (for example g_{m1} and C_c for two stage op-amp) cannot be judged. Now, let us define what is the difference between a design point bounded by ac performance constraints (O')

Performance(Unit)	Spec.	Our pred.	Spice pred.	Err.(%)	Pred. with equation based approach	Spice pred.	Err.(%) in equation based approach
Gain(dB)	≥ 60	67	67.79	1.18	66.05	67.33	1.93
CMRR(dB)	≥ 65	70	69.05	1.36	75.34	76.66	1.75
PM(deg)	≥ 60	61	62.3	2.13	63	59.73	5.19
UGF(MHz)	max	32.64	33.66	3.03	27.64	29.61	7.13
ICMR(V)	0.8-1.4	0.68-1.52	0.63-1.53	4.41	0.74-1.58	0.7-1.54	4.05
Swing(V)	0.4-1.5	0.27-1.67	0.25-1.68	6.4	0.22-1.7	0.23-1.65	4.54

Table III

OPTIMIZED PERFORMANCE DATA FOR TWO STAGE OP-AMP USING DEVELOPED METHODOLOGY AND COMPARISON W.R.T. EQUATION BASED GENERAL APPROACH TO SIZING

Performance(Unit)	Spec.	Our pred.	Spice pred.	Err.(%)	Pred. with equation based approach	Spice pred.	Err.(%) in equation based approach
Gain(dB)	≥ 60	63	63.67	1.06	63.51	64.29	1.23
CMRR(dB)	≥ 70	77	77.39	0.51	72.64	70.86	2.45
PM(deg)	≥ 60	81	82.3	1.6	85	88.7	4.35
UGF(MHz)	max.	25.25	24.55	1.67	21.76	20.61	5.28
ICMR(V)	0.7-1.3	0.61-1.54	0.63-1.52	3.27	0.64-1.53	0.66-1.51	3.12
Swing(V)	0.5-1.5	0.34-1.72	0.32-1.74	5.88	0.42-1.72	0.45-1.76	7.14

Table IV

OPTIMIZED PERFORMANCE DATA FOR FOLDED CASCODE OP-AMP USING DEVELOPED METHODOLOGY AND COMPARISON W.R.T. EQUATION BASED GENERAL APPROACH TO SIZING

1. Solve Eq. (3)-(7) and run macromodel simulation with the d_i assigned a weightage ($d_i \leftarrow d_i/w_{1\_i}$) as decided in [II-B].

2. Calculate difference of individual performance specification and performance achieved in each iteration
$$P_{i\_diff} = P_{i\_macro\_simulation} - P_{i\_specification}$$

3. If, $P_{i\ diff} > 0$; all specifications are met; go to step 6, otherwise calculate the weightage according to the following definition
$$w_{2\ i} = \frac{P_{i\_diff}}{\sum_{i=1}^{n} P_{i\_diff}} \text{ for all } P_{i\ diff} > 0.$$

4. Find parameters of performance (P) corresponding to max. and min. of $P_{i\_diff} > 0$.

5. $d_i \leftarrow d_i/w_{2\_i}$; Here d_i is the ellipsoidal axes length of the first iteration corresponding to performances (P) with only max or min of P_{i_diff} as found in step 4. Go to step 1.

6. End.

Figure 8. Algorithm to capture infeasibility through iteration

and device constraints (O). It is only the length of the axes that has changed. So we anticipate, starting from say (O), by changing the axes length, over the iteration, we can converge to the final design point (O') which meets all the specifications. Now, this is possible by iteratively varying weightages of axes length or to say other way, by controlling how strongly or weakly specifications are met. We have explained the algorithm in Fig. 8. It should be noted that convergence of O to O' depends on how the algorithm is formed and number of iterations to be run. We have experimented with varied specifications. In all the conditions, specifications have been met successfully except doubling or maximum tripling convergence time.

Our methodology has been codified in matlab and for the optimization purpose we have used cvx [13]. Ocean script is invoked from matlab to run the simulation. The methodology has been tested on several op-amps.

G. A comparison to simulation based approach

It must be noted that in all the simulation based sizing approaches [1][2], transistor level spice simulation has been used which is time consuming. On the contrary, we run macromodel based simulation in iteration, which is significantly faster. While technology dependency is taken care by device models. So, we use GP to quickly obtain initial design point and after that, using device model based constraints, macromodel simulation is run. Attained design point is verified and accordingly direction is set for the next iteration.

III. DEVICE MODELING

Simplified square-law based device model as used in [4] does not work satisfactorily at deep sub-micron resulting in mismatch between GP and SPICE prediction. In [3][4], it is shown that device models should be in monomial format to make ac performance metrics posynomial. Thus posynomial device modeling approach as described in [14] may prove to be highly accurate but cannot be used in GP based application. In [12], device parameters which are modeled as convex piecewise linear function for different ranges of overdrive voltage result in improved accuracy compared to simple square-law based model. We have developed codes using matlab and ocean scripts to extract small signal model parameters which are valid in saturation region of operation [15]. Developed script is independent of technology. In any new technology, models can be extracted in one full day.

978-1-4673-0438-2/12 $31.00 © 2012 IEEE

To capture first order effect, monomial model template is used while detailed second order effect is captured through empirical equations. As second order equations have non-convexity issues, they are kept outside GP formulation and updated over the iteration. Numerical values as obtained through updation is reflected in the change in monomial model coefficients. Iteration stops if change from one to next is small.

IV. EXPERIMENTAL RESULTS

We show experimental results obtained for two stage op-amp in Table III. In the same Table, we have also provided results if equation based general methodology is followed. General approach produces more error in predicting performances w.r.t. spice. This error is contributed by monomial device models and posynomial performance expressions. While our method involves only one set of modeling error (i.e. device models). It should be noted that attained Gain or CMRR meet specifications while UGF barely. This happens as in the first iteration there is conflicting requirements of some parameters, so explicit constraints could not be imposed on them as discussed before. Over the iteration performances improve. In fact, by increasing number of iteration, requirements can be strongly met. Here, we have obtained final result with only six iterations. Folded cascode design example is shown in Table IV. It must be noted that ICMR and OS errors are almost comparable for both these approaches. It happens because dc saturation constraints are represented directly by device models involving only one set of error. In the present design example, we have taken supply voltage of 1.8V and op-amps driving a load of 3pF.

In all the equation based approaches so far, run time of the tool does not indicate exact measurement of sizing time because they have assumed that performance expressions are available to them in suitable format. In the given example of two stage op-amp, final design point is obtained under seven minutes. This run time indicates time taken by the optimizer and macromodel simulation for verification. For folded cascode circuit, experimentation has been successfully completed under ten minutes for all type of specifications.

V. CONCLUSION

A fast approach to analog circuit sizing is presented. Equation based approach proves superior to simulation based approach in providing rapid convergence and in some occasion optimal design. But the approach is crippled with complex development effort needed to derive complex circuit equations in suitable format which again asks for rigorous simulation. In view of this, we have developed methodology which does not use any ac performance expressions. This approach is based on relating device parameters to different performances and then enclosing feasible region meaningfully by device constraints. Infeasibility is handled iteratively by varying ellipsoidal axes length. This methodology asks

for minimum macromodel simulation effort so, it is very fast. Due to the simplicity of formulation, method is fully automated.

REFERENCES

[1] J. Yuan, N. Farhat, and J. V. D. Spiegel, *"GBOPCAD: a synthesis tool for high-performance gain-boosted opamp design"*, IEEE Trans. on CAS-I, Vol. 52, pp. 1535-1544, Aug, 2005.

[2] R. Phelps, M. Krasnicki, R. Rutenbar, L. R. Carley, and J. R. Hellums, *"Anaconda: simulation-based synthesis of analog circuits via stochastic pattern search"*, IEEE Trans. CAD, vol. 19, pp. 703-717, June, 2000.

[3] P. Mandal, and V. Visvanathan, *"CMOS Op-AMP Sizing Using a Geometric Programming Formulation"*, IEEE Trans. CAD, vol. 20, pp. 22-38 Jan, 2001.

[4] M. Hershenson, S. Boyd, and T. H. Lee, *"Optimal Design of a CMOS Op-amp via Geometric Programming"*, IEEE Trans. CAD, vol. 20, pp. 1-21, Jan, 2001.

[5] F. V. Fernandez, P. Wambacq, G. Gielen, A. Rodriguez-Vazquez, and W. Sansen, *"Symbolic Analysis of large analog integrated circuits by approximation during expression generation"*, ISCAS, 1994.

[6] W. Daems, G. Gielen, and W. Sansen, *"Simulation-Based Generation of Posynomial Performance Models for the Sizing of Analog Integrated Circuits"*, IEEE Trans. CAD, vol. 22, pp. 517-534, May, 2003.

[7] S. Maji, S. Dam, and P. Mandal, *"Automatic Generation of Saturation Constraints and Performance Expressions for Geometric Programming based Analog Circuit Sizing"*, ISQED, 2011.

[8] S. Maji, and P. Mandal, *"A Geometric Programming Aided Knowledge based Approach for Analog Circuit Synthesis and Sizing"*, GLSVLSI, 2011.

[9] S. Maji, and P. Mandal, *"A CAD Methdology for Automatic Topology Selection & Sizing"*, IEEE International SOCC, 2011.

[10] H. L. Abdel-Malek, and A. -K. S. O. Hassan, *"The ellipsoidal technique for design centering and region approximation"*, IEEE Trans. CAD, vol. 10, pp. 1006-1014, Aug, 1991.

[11] X. Li, J. Wang, L. T. Pileggi, T. -S. Chen, and W. Chiang, *"Performance-Centering Optimization for System-Level Analog Design Exploration"*, ICCAD, 2005.

[12] J. Kim, J. Lee, L. Vandenberghe, and C. -K. K. Yang, *"Techniques for improving the accuracy of geometric-programming based analog circuit design optimization"*, ICCAD, 2004.

[13] M. Grant, S.P. Boyd and Y. Ye, cvx users guide, Available at http://www.stanford.edu/ boyd/cvx.

[14] V. Aggarwal, and U. -M. O'Reilly, *"Simulation-based Reusable posynomial models for MOS transistor parameters"*, DATE, 2007.

[15] S. DasGupta, and P. Mandal, *"An Improvised MOS Transistor Model Suitable for Geometric Program Based Analog Circuit Sizing in Sub-micron Technology"*, VLSID, 2010.

Iterative Performance Model Upgradation in Geometric Programming based Analog Circuit Sizing for Improved Design Accuracy

Samiran Dam
Department of Electrical &
Electronics Communication Engineering
Indian Institute of Technology
Kharagpur, West Bengal 721302
Email: samiran.dam@gmail.com

Pradip Mandal
Department of Electrical &
Electronics Communication Engineering
Indian Institute of Technology
Kharagpur, West Bengal 721302
Email: pradip@ece.iitkgp.ernet.in

Abstract—**In this paper, we propose a technique to improve the accuracy of the final design predicted by Geometric Programming based CMOS analog circuit sizing methodology. Here we use a multi-level AC performance modeling paradigm to develop the empirical models of circuit performance metrics. Performance models are then upgraded over iterations of design cycle. This iterative model upgradation in a sequence of geometric programming guides the final design to converge with better accuracy. The methodology is validated by designing a two-stage amplifier cascaded with a Class-A (source-follower) output buffer stage in UMC 0.18 μm technology.**

I. INTRODUCTION

Geometric programming (GP) based sizing of analog integrated circuits is well established [1], [2]. GP has the advantage of fast convergence to the global optimal solution with high accuracy and without any dependence on initial guess. In GP, circuit performance constraints are formulated as posynomial or monomial expressions of primary design parameters. Deriving the performance constraints for large and complex circuits in posynomial or monomial forms may not be possible using traditional circuit analysis techniques. This has been a hindrance for GP-based sizing methods proposed so far. This is true for any equation-based circuit optimization technique. There are several attempts to generate the circuit performances automatically. Existing approaches can be classified into two major categories - *symbolic analysis based approaches* and *empirical model based approaches*.

Fernández, F. V. et al. [3], proposed an algorithm that generates approximated symbolic expressions for small-signal characteristics of large analog integrated circuits. Although this algorithm is reasonably fast, it has some drawbacks. First, if accurate expressions are needed, the order of the problem proliferates exponentially. Second, some circuit performances such as sensitivities, symbolic poles, and zeros cannot be extracted in general. Verhaegen, W. et al. [4] presented a technique for generating approximated symbolic expressions for network functions in linearized analog circuits. It is based on the compact *Determinant Decision Diagram* (DDD) representation of the circuit. DDD is used as the core computation

engine in some modern symbolic circuit simulators. Although DDD is a much more efficient algorithm, it still encounters bottleneck when applied to larger analog circuits containing more than 20 transistors.

To overcome the difficulties encountered by the symbolic-analysis based methods, researchers focused to generate simulation-based empirical models of analog circuit performances. The advantage is that any circuit performance that can be simulated in SPICE can be empirically modeled. Daems, W. et al. [5] and Eeckelaert, T. et al. [6] presented a direct-fitting method to generate posynomial performance models. Performance metrics are directly modeled in terms of basic design parameters i.e. width, length, bias current and voltage. Disadvantage of these works is that due to a large number of design parameters in a complex circuit, it demands a large number of simulation data and hence large number of terms in the model template. In addition, for meaningful data collection through simulation, it has to ensure that all transistors are in valid region of operation. With the change of process technology, performances need to be modeled again which is not desirable.

Here, in this work, we utilize an efficient multi-level performance modeling methodology [7] to generate GP-compatible, *technology-independent* performance models. Key difference of this modeling paradigm with respect to the previous attempts is that performances are modeled as posynomial or monomial functions of device parameters (i.e. g_m, g_d and *capacitances* of transistors). This helps to generate technology-independent models. However, though for some circuit performances it is possible to develop global posynomial or monomial models accurate uniformly over a large design parameter space, it may not be feasible for some other performances to generate models which would be accurate globally. For example, in case of a multi-pole system, if the locations of non-dominant poles vary significantly from one design point to another, then it is extremely difficult to develop a unified model of first non-dominant pole. Along with this, in high-frequency applications, effect of parasitic poles can

978-1-4673-0438-2/12 $31.00 © 2012 IEEE

no longer be neglected. Beside this, due to inaccuracy in the device models [8], actual circuit-level SPICE simulation result deviates from that predicted by the optimizer. This paper shows how traditional GP-based circuit sizing approach can be adapted to compensate various seeds of inaccuracies to yield a more accurate and reliable design. After each iteration performance models are updated based on design point obtained in previous iteration and aids the final design to converge over few iteration with more accuracy. The technique has been validated by citing a design example of a two-stage opamp cascaded with a Class-A (source follower) output buffer stage.

The paper has been organized as follows. Section II revisits the basics of geometric programming. Section III states the AC performance modeling paradigm. Section IV explains the proposed technique. The design example is demonstrated in section V. Finally the conclusion is drawn in section VI.

II. GEOMETRIC PROGRAMMING

Following equation describes a canonical geometric programming formulation:

$$
\begin{aligned}
\underset{[\mathbf{x}]}{\text{minimize}} \quad & F_0(\mathbf{x}) \\
\text{subject to} \quad & G_i(\mathbf{x}) \le 1, \ i = 1, 2, ...m \\
& H_j(\mathbf{x}) = 1, \ j = 1, 2, ...n \\
& x_k > 0, \ k \in k = 1, 2, ...N
\end{aligned} \tag{1}
$$

Where, m is the number of inequality constraints and n is the number of equality constraints. $G_i([x])$ can be *posynomial* or *monomial* function but $H_j([x])$ should be strictly *monomial* functions of design parameters, $[x]$. A *posynomial* function of $[x]$ is expressed as:

$$
f([x]) = \sum_{i=1}^{T} C_i \prod_{j=1}^{R} x_j^{\alpha_{ij}}, C_i \ge 0 \tag{2}
$$

Where, R is the number of design parameters. Each product term of the above *posynomial* expression is called a *monomial*. In the context of CMOS analog circuit sizing, $[x]$ are transistor widths (W), lengths (L) and bias current (I_b). AC and DC performance equations are the constraint functions. Objective function $[f_0([x])]$ of the problem can be any of the performance metrics which the designer wants to minimize or maximize.

However, above said nominal design approach tends to minimize the objective function with highest priority and as a result other performance metric values are pushed towards the margin of their respective specified boundaries. Severe consequence of this situation is that due to process variation and variation of other operating conditions if the design point shifts from the nominal value then some of the performances may violate respective specifications. To encounter this adversity an adapted version of GP formulation is proposed in [9] to force the design point to be at the center of the feasible performance space so that even if the design point shifts from the nominal value it does not violate the specifications. In our design

problem formulation we have considered *yield* of the circuit as the objective function instead of considering any particular performance metric as the objective. The *yield*-maximization problem is formulated as follows:

$$
\begin{aligned}
\underset{[\epsilon, \mathbf{x}]}{\text{minimize}} \quad & \prod_{i=1}^{m} \epsilon_i \\
\text{subject to} \quad & \epsilon_i G_i(\mathbf{x}) \le 1, \ i = 1, 2, ...m \\
& H_j(\mathbf{x}) = 1, \ j = 1, 2, ...n \\
& \epsilon_i > 1, \ i = 1, 2, ...m \\
& x_k > 0, \ k = 1, 2, ...N
\end{aligned} \tag{3}
$$

Here instead of setting any particular performance metric to be the objective function, volume of an ellipsoid inscribed within the feasible performance space is considered to be the objective. ϵ_i determines that how far the value of ith performance metric would be from its specification boundary. More the ϵ_i value more the performance will maintain a safe margin from its specification boundary.

III. AC PERFORMANCE MODEL GENERATION

Objective of AC performance model generation is to generate efficient technology-independent and GP-compatible expressions of different AC performance metrics. AC performance model generation scheme is based on the concept of *levels of abstraction* as shown in Fig. 1.

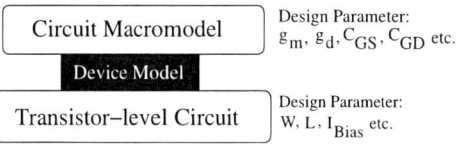

Fig. 1. Levels of Abstraction

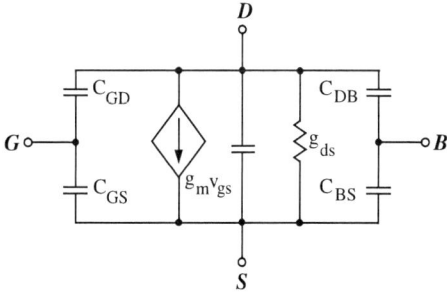

Fig. 2. Macromodel of a NMOS device

Performances metrics are empirically modeled, based on SPICE simulation, as functions of the device model parameters such as g_m, g_d, *parasitic* and *load capacitances*. These intermediate design parameters are replaced later by their monomial expression (device models [8]) which are functions of primary design parameters such as transistor *widths*, *lengths* and *bias currents*. At the higher level of the abstraction, actual circuit is replaced by its macromodel. Circuit macromodel is generated using device macromodels. Fig. 2 shows the macromodel of a NMOS device. *Device model* bridges two levels of abstraction. Fig. 3 outlines the flow of the AC *performance model generator* (PMG).

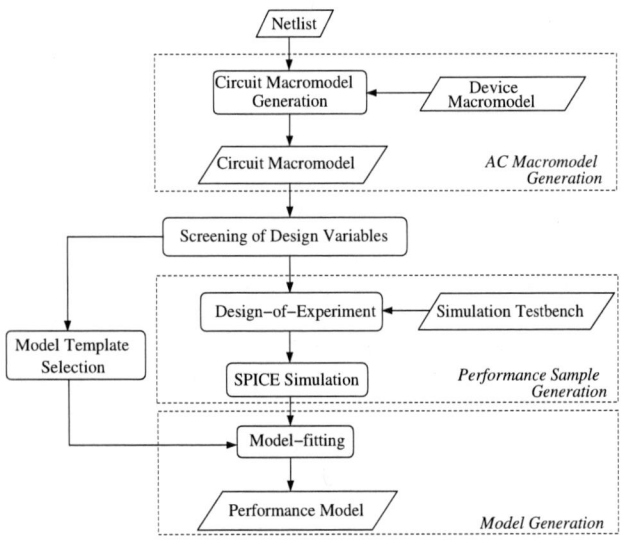

Fig. 3.   Performance Model Generator

The method implements an efficient screening technique to prune a large design parameter space by reducing its dimension. The screening process is followed by a formulation of a design for the experiment (SPICE simulations). *Non-negativity Least Square* regression technique is followed for posynomial approximation of SPICE-simulated training points. Significant advantages of this method are -

1) Macromodel simulation requires *significantly less time* compared to transistor-level simulation.
2) Since small signal parameters are the design parameters, *maintaining transistors in saturation is not a criterion* at this level of abstraction. Therefore, standard design of experiment can be *easily* explored.
3) *Technology-independent* performance models can be generated. It facilitates an *easy technology-porting*.
4) Last but not the least is *automatic generation of GP-compatible* performance models.

IV. PROPOSED TECHNIQUE

Fig. 4 describes the general flow of the proposed cell-level design methodology. Traditional GP-based sizing method is adapted to take care of the performance metrics for which, global posynomial/monomial models, uniformly accurate over a large design parameter space, are not feasible to develop. Initially a set of coarse empirical models are generated by the PMG. These initial models may be inaccurate, for some performance metrics, if the design point moves significantly away from the modeling region in which the models were developed. An obvious consequence of this is that the optimizer may yield an inaccurate or, in worst-case, an infeasible design. This problem can be dealt by including the sequential upgradation

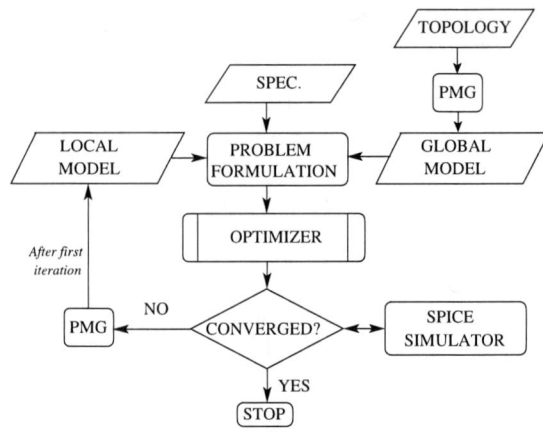

Fig. 4.   Design Flow

of the models by invoking PMG. In a given iteration, models are upgraded based on the new design point obtained from previous iteration. The decision, whether a model need to be updated or not, is made via checking the amount of inaccuracy present in the corresponding performance metric value obtained in SPICE simulation with respect to the predication made by the optimizer.

As the models are accurate within a small neighborhood of design parameter space, it is important to start the next iteration from the previous state after the models are upgraded locally. Otherwise, optimization may not converge if it starts from the initial state. This process continues till all the performances meet their specifications within a given tolerance of accuracy. The concept is illustrated in Fig. 5. As the iteration increases, size of the neighborhood defining the local design parameter space is tightened to increase the accuracy of the model. This method proves extremely effective in case of modeling the frequency-domain behavior of a multi-pole system where the location of non-dominant poles and zeros vary significantly from one design point to another. Ability of the proposed methodology to produce a feasible design is demonstrated in the next section.

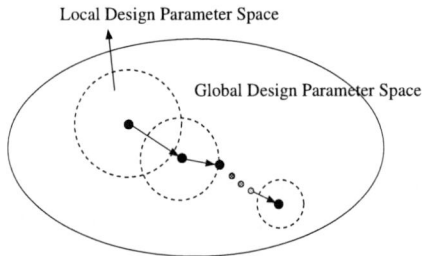

Fig. 5.   Displacement of design points from iteration to iteration

V. DESIGN EXAMPLE

Accuracy of the developed models and proposed design methodology in GP-based environment is corroborated by sizing a two-stage opamp cascaded with a class-A buffer stage [Fig. 6]. Buffer stage is required to drive large load capacitance. Load capacitance (C_L) considered in this example is $10\ pF$.

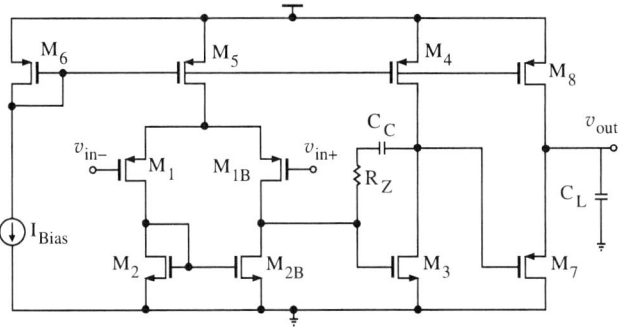

Fig. 6.   Buffered Opamp

Empirical models of *Open-loop gain* and *CMRR* for the topology shown in Fig. 6 are same as that of a normal two-stage opamp as derived in [7]. These models are shown in Eq. (4) and Eq. (5) respectively.

$$\frac{1}{GAIN} = 1.012\left[\frac{g_{d1}g_{d3}}{g_{m1}g_{m3}}\right] + 1.013\left[\frac{g_{d1}g_{d4}}{g_{m1}g_{m3}}\right]$$
$$+ 0.993\left[\frac{g_{d2}g_{d3}}{g_{m1}g_{m3}}\right] + 0.993\left[\frac{g_{d2}g_{d4}}{g_{m1}g_{m3}}\right] \quad (4)$$

$$\frac{1}{CMRR} = 0.493\left[\frac{g_{d1}g_{d5}}{g_{m1}g_{m2}}\right] + 0.487\left[\frac{g_{d2}g_{d5}}{g_{m1}g_{m2}}\right] \quad (5)$$

Now let us look at the two most critical performance metrics responsible for ensuring the stability of the overall circuit performance - *unity gain bandwidth* and *phase margin*. The buffer introduces additional poles apart from the poles arising from first two stages. Large sizes of output transistors (M_7 and M_8) cause large parasitic capacitances C_{GS7}, C_{GD7} and C_{GD8} that can no longer be neglected. Thus, obtaining an optimal design by manually adjusting these critical parameters proves to be difficult.

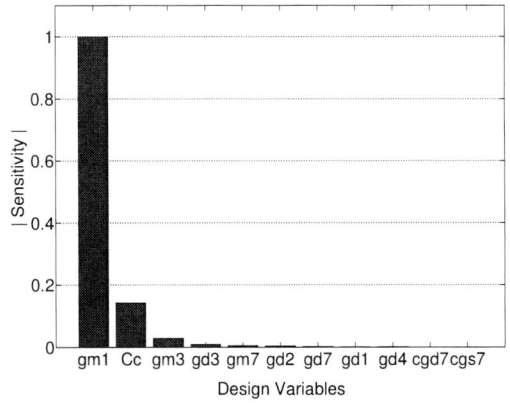

Fig. 7.   Sensitivity of UGB on design parameters

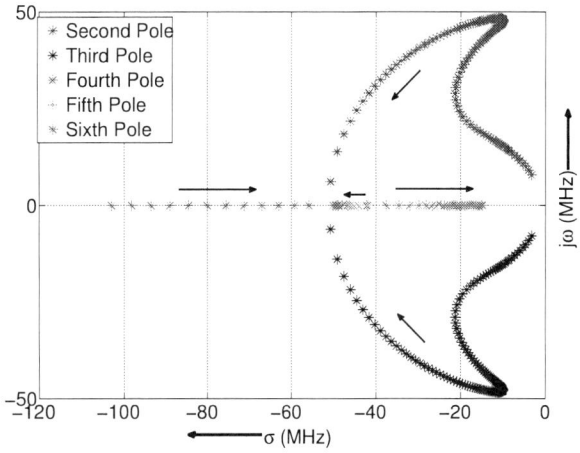

(a) with increasing size of M_3

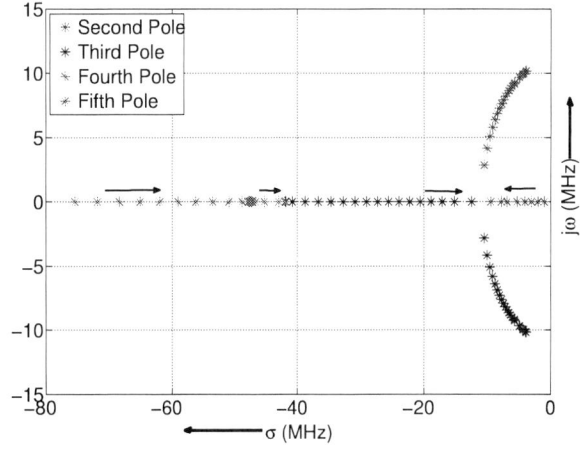

(b) with increasing size of M_7

Fig. 8.   Root Locus Plots

Introduction of the compensation capacitance, C_C ensures the that UGB is primarily defined by C_C and transconductance of M_1 (g_{m1}). Screening process identifies that transconductance of M_3 (g_{m3}) has a non-trivial effect on UGB too as shown in the Fig. 7. Hence, in the screening process, template chosen for *Unity Gain Bandwidth* model is:

$$UGB(MHz) = c_0 g_{m1}^{c_1} g_{m3}^{c_2} C_C^{c_3} \quad (6)$$

The non-dominant poles are highly dependent on the sizes of transistors M_3 and M_7. So, it is difficult to develop an accurate global model for first non-dominant pole that can be used for formulating phase margin constraint. Thus an iterative upgradation of model coefficients is required. Complexity of this situation can be interpreted from the root locus plots [Fig. 8] generated from the macromodel simulation in the screening phase. Given a fixed size of transistor M_3, if transconductance of M_7 (g_{m7}) increases, the complex conjugate pole pair (second and third pole) tends to move close to the imaginary

TABLE I
OPTIMIZATION RESULTS

Design Points		Performance Values				
Transistor	W/L ($\mu m/\mu m$)	Performance Metric	User Specification	Optimizer	SPICE	Error (%)
Iteration: 1						
M_1	17.48/1.5	$Area(\mu m^2)$	1000	996.57	-	-
M_2	1/0.5	$Power(\mu Watt)$	800	800	800.65	0.08
M_3	6.78/0.5	$R_{OUT}(\Omega)$	200	188.27	187	-0.68
M_4	61/0.83	$Gain(dB)$	60	69.13	68.05	-0.4
M_5	20.36/0.83	$CMRR(dB)$	66	73.58	72.16	-1.97
M_6	39.78/0.83	$UGB(MHz)$	10	11.63	10.46	-11.18
M_7	437.76/0.5	$PhaseMargin(^o)$	60	81.1	76.82	-5.56
M_8	747.36/0.83	$f_{non-dom}(MHz)$	-	-24.85±j56.64	-32.81±j51.82	-
Iteration: 2						
M_1	17.48/1.5	$Area(\mu m^2)$	1000	998.86	-	-
M_2	1/0.5	$Power(\mu Watt)$	800	800	800.65	0.08
M_3	**11.5/0.5**	$R_{OUT}(\Omega)$	200	200	198.77	-0.62
M_4	**104.16/0.84**	$Gain(dB)$	60	69.05	68.08	-1.43
M_5	20.52/0.84	$CMRR(dB)$	66	73.61	72.44	-1.62
M_6	40.08/0.84	$UGB(MHz)$	10	**10.51**	**10.46**	**-0.51**
M_7	**411.6/0.5**	$PhaseMargin(^o)$	60	79.36	78.18	**-1.51**
M_8	**708.9/0.84**	$f_{non-dom}(MHz)$	-	**-44.96±j60.45**	**-44.49±j61.96**	-

axis. Whereas, increasing size of M_3, i.e. higher value of g_{m3} tends to push the complex conjugate pair to higher frequency while placing the fourth pole (a real pole) at a lower frequency. While designing, it is very difficult to anticipate the locations of the poles as both the transistors, M_3 and M_7, and other design parameters affect simultaneously. In case of an aggressive specification, even a slim change of the design point may force the non-dominant poles to cross each other. Thus it is very important to identify all the design parameters affecting the non-dominant poles. Screening process for the non-dominant pole identifies the dominant design parameters as shown by Eq. (7).

$$f_{non-dom}(MHz) = f(g_{m3}, g_{m7}, g_{d3}, g_{d4}, C_{GS7}, C_{GD7}, C_C) \quad (7)$$

Here, it is important to mention the way phase margin constraint is formulated in case where a complex-conjugate pair is first non-dominant pole. The formulation is shown in Eq. (8).

$$\left[\frac{\lambda_1.UGB}{tan(90^o - PM_{SPEC})} + \lambda_2^2 UGB^2\right] \leq 1 \quad (8)$$

Where λ_1 and λ_2 are the coefficients of the second order transfer function $\left[\frac{1}{1+\lambda_1 s+(\lambda_2 s)^2}\right]$ resulted by the complex-conjugate pole-pair; PM_{SPEC} is phase margin specification. It is assumed that the dominant pole has already contributed 90° phase at unity gain bandwidth (UGB) frequency.

Final design along with the optimized values of the design

parameters after two iterations is shown in Table. I. As the reader can see that the error in UGB and PM between optimizer prediction and SPICE simulation at final design point is reduced in the second iteration in comparison to the first where these errors were significantly large. In this case, it is observed that further iterations does not render any significant improvement of the accuracy. But, for more stringent specification, the cycle may require more iteration. Fig. 9 shows the frequency response of the designed circuit.

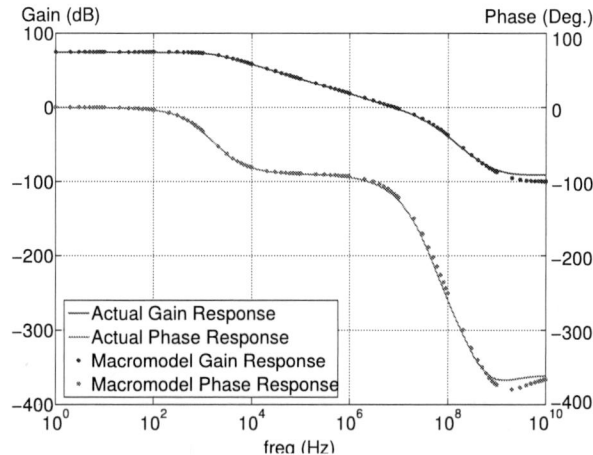

Fig. 9. Frequency Response of the designed circuit

VI. CONCLUSION

Main feature of the proposed technique is the formulation of the analog circuit design problem as a sequence of geometric programs based on automatically-generated reusable performance models. Because of this iterative formulation, as the iteration proceeds and the coefficient of empirical performance models are updated via circuit macromodel simulation, the design converges to an optimal/near optimal solution with more accuracy. A good agreement between the responses of actual circuit and its macromodel invariably confirms the correctness of the empirical models. This technique can be employed to design most of the linear CMOS analog circuits having complex design tradeoffs.

ACKNOWLEDGMENT

The authors would like to thank the Department of Information technology, Government of India for sponsoring the project and Michael Grant, Stephen Boyd and Yinyu Ye for making free optimization toolbox cvx [10] available online.

REFERENCES

[1] P. Mandal and V. Visvanathan, *CMOS Op-amp Sizing Using a Geometric Programming Formulation*, Computer-Aided Design of Integrated Circuits and Systems, IEEE Transactions on, vol. 20, no. 1, pp. 22–38, Jan 2001.

[2] M. Hershenson, S. Boyd, and T. Lee, *Optimal Design of a CMOS Op-amp via Geometric Programming*, Computer-Aided Design of Integrated Circuits and Systems, IEEE Transactions on, vol. 20, no. 1, pp. 1–21, Jan 2001.

[3] F. V. Fernández and P. Wambacq and G. G. E. Gielen and Á. Rodríguez-Vázquez and W. M. C. Sansen, *Symbolic Analysis of Large Analog Integrated Circuits by Approximation During Expression Generation*, in ISCAS, 1994, pp. 25–28.

[4] W. Verhaegen and G. G. E. Gielen, *Efficient DDD-Based Symbolic Analysis of Large Linear Analog Circuits*, in Proceedings of the 38th annual Design Automation Conference, ser. DAC 01. New York, NY, USA: ACM, 2001, pp. 139–144.

[5] W. Daems, G. G. E. Gielen, and W. M. C. Sansen, *A Fitting Approach to Generate Symbolic Expressions for Linear and Nonlinear Analog Circuit Performance Characteristics*, in Proceedings of the conference on Design, automation and test in Europe, ser. DATE 02. Washington, DC, USA: IEEE Computer Society, 2002, pp. 268–273.

[6] T. Eeckelaert, W. Daems, G. G. E. Gielen, and W. M. C. Sansen, *Generalized Posynomial Performance Modeling*, in Proceedings of the conference on Design, Automation and Test in Europe - Volume 1, ser. DATE 03. Washington, DC, USA: IEEE Computer Society, 2003.

[7] S. Maji, S. Dam, and P. Mandal, *Automatic Generation of Saturation Constraints and Performance Expression for Geometric Programming Based Analog Circuit Sizing*, in Proceedings of the 12th International Symposium on Quality Electronic Design, ISQED 2011, 14-16 March 2011, pp. 761–768.

[8] S. DasGupta and P. Mandal, *An Improvised MOS Transistor Model Suitable for Geometric Program Based Analog Circuit Sizing in Submicron Technology*, in VLSI Design, 2010. VLSID 10. 23rd International Conference on, Jan. 2010, pp. 294–299.

[9] S. Deyati and P. Mandal, *An Automated Design Methodology for Yield Aware Analog Circuit Synthesis in Submicron Technology*, in Quality Electronic Design (ISQED), 2011 12th International Symposium on, March 2011, pp. 1–7.

[10] M. Grant, S. P. Boyd, and Y. Ye, *CVX Users Guide*, http://www.stanford.edu/ boyd/cvx.

2012 25th International Conference on VLSI Design

Analysis of Reachable Sensitisable Paths in Sequential Circuits with SAT and Craig Interpolation

Matthias Sauer*   Stefan Kupferschmid*   Alexander Czutro*   Sudhakar Reddy[†]   Bernd Becker*

* Albert-Ludwigs-University Freiburg
Georges-Köhler-Allee 051
79110 Freiburg, Germany
{ sauerm | skupfers | aczutro | becker }
@informatik.uni-freiburg.de

[†] University of Iowa
5324 Seamans Center
Iowa City, IA 52242, United States
reddy@engineering.uiowa.edu

Abstract—Test pattern generation for sequential circuits benefits from scanning strategies as these allow the justification of arbitrary circuit states. However, some of these states may be unreachable during normal operation. This results in non-functional operation which may lead to abnormal circuit behaviour and result in over-testing.

In this work, we present a versatile approach that combines a highly adaptable SAT-based path-enumeration algorithm with a model-checking solver for invariant properties that relies on the theory of Craig interpolants to prove the unreachability of circuit states. The method enumerates a set of longest sensitisable paths and yields test sequences of minimal length able to sensitise the found paths starting from a given circuit state. We present detailed experimental results on the reachability of sensitisable paths in ITC 99 circuits.

I. INTRODUCTION

Scan-based manufacturing tests are universally used to screen out defective VLSI devices. Scan allows to justify arbitrary circuit states that may be unreachable and thus do not occur during functional operation of the circuits under test. Non-functional operation during test is known to not only cause abnormal switching activity, which causes abnormal power dissipation and supply-voltage droops, but also to sensitise non-functional paths [1]. These effects may lead to yield loss as good devices could fail test [2].

Several works have proposed methods for the generation and application of tests that introduce lower switching activity during test [3]. One way to avoid non-functional operation during scan-based test is to scan in only states that are reachable from the reset state or from a state reached after circuit synchronisation [4]. A simulation-based test generation procedure to compute such tests for detection of transition-delay faults was investigated in [5] and a sequential ATPG-based procedure was proposed in [6]. Other previous work on reachability in sequential circuits is often based on probabilistic methods. [7] and [8] combine random simulation with a BDD-based approach to compute the complete set of reachable states. However, these methods scale poorly as the possible search space grows exponentially with the number of flip-flops.

This work has been supported by the German Research Council (DFG) as part of the Transregional Collaborative Research Center "Automatic Verification and Analysis of Complex Systems" (SFB/TR 14 AVACS, *www.avacs.org*), and under grants GRK 1103, BE 1176-14/2 and BE 1176-15/2.

Furthermore, it is not possible to directly obtain the necessary assignments required to get into a reachable state. In [9] a genetic algorithm is used for sequential ATPG. However, due to the randomness of evolutionary approaches, such methods are heuristic and hence neither complete nor optimal.

Tests for delay faults are used to address several goals of manufacturing test of VLSI devices. These include determining the maximum frequency of operation using tests for several selected critical paths, and detection of small-delay defects by testing for gate-delay faults through longest paths [10].

In this work, we consider the generation of tests for delay faults using a versatile approach based on recent advances in SAT-based test generation. Thanks to sophisticated learning strategies that have been incorporated into modern SAT-solvers in the last few years, the classification of hard-to-detect and identification of redundant faults has recently been proved to be managed by SAT-based tools in a more efficient way than classical structural ATPG [11], [12], [13]. Also, the straightforward formalism of SAT-based problem formulation permits an easy combination of SAT-based approaches with other optimisation approaches [14], [15].

However, previous related SAT-based approaches for the test of sequential delay faults [16], [17] can not provide an unreachability proof as the search is executed until an upper bound is reached.

We present an approach that takes all these factors into consideration. Given a small delay fault in a sequential circuit with scanning capabilities, we derive a two-pattern test using the SAT-based path-enumeration tool PHAETON [18]. Then, the approach attempts to determine whether the computed assignment to the flip-flops represents a reachable state.

This is done by formulating the problem as a model-checking problem (MC) which is passed to the CIP-solver (Craig Interpolation Prover) [19]. CIP has been originally designed for formal verification problems and matches the performance of the best pure Bounded-Model-Checking(BMC)-solvers, while providing competitive performance when proving safety properties, i.e. proving that a certain (unsafe) state is not reachable. In this work, CIP is used for the first time in the context of test.

Furthermore, given a set of circuit paths, the approach can determine the functional sensitisability of every path starting at reachable states. It can also determine if the paths are robustly

978-1-4673-0438-2/12 $31.00 © 2012 IEEE

or non-robustly testable starting at reachable states. In addition, it is complete and therefore generates tests when they exist. For the results we report, we consider functional sensitisation. If a path is not functionally sensitisable, then it is known as a false path independently of circuit delays [20]. Also, the method is able to provide optimal test sequences. Here, by test sequence we mean a sequence of assignments for the primary circuit inputs such that the sequence constitutes a test for a target path with respect to a given initial state. An optimal test sequence is a test sequence of minimal length.

The remainder of the paper is structured as follows. A brief overview of SAT-based path enumeration and Craig-interpolant-based model checking is provided in Section II. The combination of both tools is introduced in Section III. Experimental results are reported in Section IV. Section V concludes the paper.

II. PRELIMINARIES

In this section, a brief overview of SAT-based path enumeration (PHAETON) and Craig-interpolant-based model checking (CIP) is provided.

A. PHAETON

The inputs of PHAETON are a circuit (gate-level net list) and a list of gates to be analysed. For every gate in the list, the method searches for the longest sensitisable paths that pass through that gate.

Only complete paths are considered, i.e. paths that start at an input and end at an output. A path is defined as *functionally sensitisable* if there is a test pattern pair that produces a rising or a falling transition at all gate outputs along the path. The *length* of a path is defined as the sum of the delays assigned to its gates according to the employed delay model. The longest sensitisable path is therefore the path with the maximum delay.

The supported delay model assigns every gate a fixed integer delay that depends on which gate input is the on-path input (pin-to-pin), and on whether the transition at the gate's output is a rising or a falling one. Since the tool works with integer delays only, the real-valued delays used in the experimental set-up are mapped to integer delays between 1 and a user-defined constant α called the *delay resolution*. In order to minimise the global path error, a sophisticated rounding strategy is employed.

A given target gate G is processed in two stages. In the first stage, the length L_G of the longest sensitisable path through the gate is measured by means of iterative SAT-solving using a binary search over a search space that starts with the length of the shortest structural path and ends with the length of the longest structural path. Several learning strategies are employed to narrow down the search space.

After the application of the first stage to all target gates, the second stage can be applied in two different modes either to all gates or to a selection of gates depending on the application. This stage can either enumerate all sensitisable paths through the target gate G that have a length between $L_G - r$ and L_G, for a user-specified parameter r known as the *length range*; or it can enumerate the K longest sensitisable paths through G for a user-specified target number K of paths.

The enumeration of all paths of a fixed length l is done by formulating a SAT-instance $S_G[= l]$ that is satisfiable if and only if a sensitisable path through G of length l exists. If such a path is identified, i.e. if a Boolean solution is found, clauses are added to the SAT-instance such that it remains satisfiable only if a found sensitisable path of length l differs from the first one. This is repeated until the SAT-instance becomes unsatisfiable, which means that there are no more sensitisable paths of that length. Please refer to [18] for details.

Furthermore, the tool is able to consider additional conditions imposed on the path-enumeration algorithm, such as whether the sensitisation has to be robust or non-robust, and also conditions regarding the scanning technique (broadside, skewed-load, full enhanced scan, etc).

B. BMC and Craig interpolation

This section briefly introduces Bounded-Model-Checking (BMC) [21], Craig interpolants [22] and McMillan's work on SAT-based model checking [14].

Among other applications, BMC is employed to derive error traces in sequential circuits that are required to satisfy a certain property P but in fact do not satisfy this property. The circuit's structure and the problem conditions are encoded as a propositional formula of the form

$$BMC_k = I_0 \wedge T_{0,1} \wedge \ldots \wedge T_{k-1,k} \wedge P_k \qquad (1)$$

I_0 encodes the initial state of the circuit, i.e. the initial contents of the flip-flops. The terms of the form $T_{i,i+1}$ represent the so-called transfer function or transition relation that defines one system step from time point i to time point $i + 1$. The last predicate P_k stands for a desired property whose satisfiability after k steps is to be tested. If the property never holds independently of the value of k, BMC_k is unsatisfiable, whereas BMC_k is satisfiable if there exists a path in the transition system that starts at I_0 and, after k transition steps, reaches a state in which P_k holds.

In order to prove that a certain state of a transition system cannot be reached independently of k, i.e. that the desired property never holds, the circuit is unfolded until reaching its diameter. However, very large k-values may be reached. Hence, there are several approaches that attempt to find a fixed point sooner. Among others including k-induction [23] and BDD-based approaches [24], there are methods based on the theory of Craig interpolation [14]. This approach is used by the CIP-solver that we employ in this work.

A classical BMC approach searches for the smallest k-value for which the desired property holds by attempting to solve a series of problem instances. The first one is $BMC_0 = I_0 \wedge P_0$ (cf. Equation 1). It is satisfiable if the property holds in the initial state. If the instance is not satisfiable, BMC tests whether taking one more step into consideration will satisfy the property, i.e. whether the formula $BMC_1 = I_0 \wedge T_{0,1} \wedge P_1$ is satisfiable. This is repeated until P_k holds for some k, or until a user-defined maximal bound is reached [21].

The drawback of this classical BMC approach is that it is not applicable for larger circuits, since it is not viable to attempt to solve as many instances as the circuit's diameter, which can become very large even for medium-sized circuits.

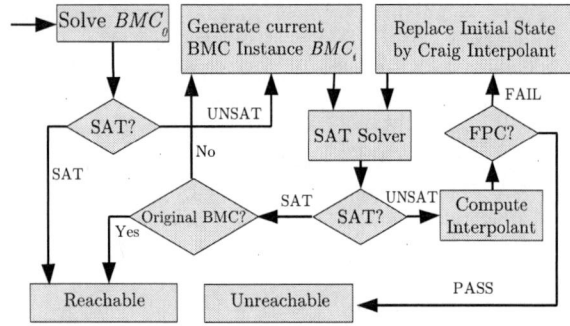

Fig. 1.   BMC and Craig interpolation

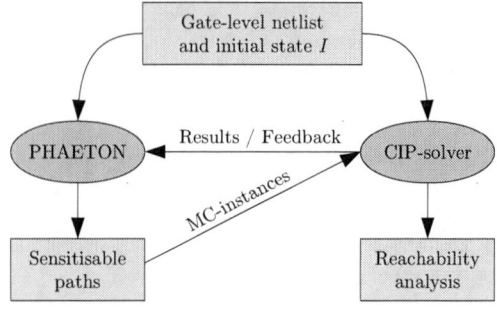

Fig. 2.   Overall flow

In order to cope with this issue, McMillan introduced in [14] a modified MC-approach that extends a classical BMC-procedure by using Craig interpolants. The Craig interpolant represents an over-approximation of all reachable states after a certain number of transition steps. This over-approximation is recomputed by an iterative algorithm until a fixed point is reached.

Formally, Craig interpolants [22] are defined as follows:

Theorem 1 (Craig): Let A and B be two propositional formulas such that their conjunction is unsatisfiable. Then, a formula C with the following properties exists:

1) C contains only variables which occur in both A and B.
2) $A \rightarrow C$
3) $C \rightarrow \neg B$

C is called a *Craig interpolant* of A and B.

Figure 1 illustrates the MC-algorithm which combines BMC and Craig interpolation. As in classical BMC, the first step consists in testing whether the property holds in the initial state I_0, i.e. whether BMC_0 is satisfiable. The next problem instance to be tested is $BMC_1 = I_0 \wedge T_{0,1} \wedge P_1$. If this instance is unsatisfiable, by Craig's theorem, there is a formula C_1 such that $I_0 \wedge T_{0,1} \rightarrow C_1$ and $C_1 \rightarrow \neg P_1$. Moreover, C_1 contains only variables that occur in both sub-formulas, i.e. it contains only variables that represent the flip-flop contents. Given that $I_0 \wedge T_{0,1} \rightarrow C_1$, all flip-flop assignments (states in the sequential circuit) reachable after one transition step starting at the initial state, are over-approximated by C_1. Then, a so-called fixed-point check (FPC) is performed. It consists in checking whether the Craig interpolant implies the initial state, i.e. whether the set of states that are reachable after one transition step are initial states themselves. If so, the algorithm terminates as no new states can be reached starting at the current step. If FPC fails, the next problem to solve is not BMC_2, but a variation of BMC_1 in which the initial state is replaced by the found Craig interpolant. If this instance is also not satisfiable, a new Craig interpolant is computed and used for a new iteration of the algorithm. In contrast, if the formula is satisfiable, verification is performed by solving the original BMC_i formula corresponding to the current unfolding depth, since the Craig interpolant is an over-approximation and may therefore contain non-reachable states. In case that the verification formula BMC_i is unsatisfiable, the overall algorithm restarts by computing a new Craig interpolant parting from BMC_i. See [14] for details.

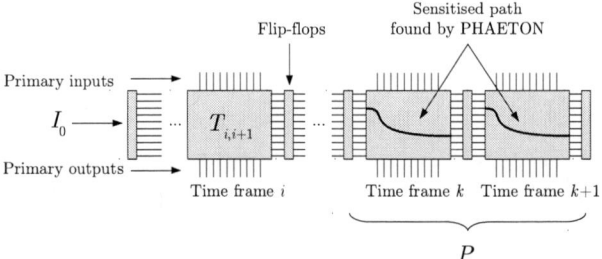

Fig. 3.   MC formulation

The construction of a Craig interpolant can be done on-the-fly by the underlying SAT-solver's conflict analysis routine.

III. REACHABILITY OF SENSITISABLE PATHS

In this section, we introduce the overall flow that combines the SAT-based path-enumeration and the model-checking tool in order to calculate sets of reachable sensitisable paths. The overall data flow is illustrated in Figure 2.

The inputs of the problem are a sequential circuit (gate-level net list along with timing data needed by PHAETON) and an assignment of Boolean variables to the flip-flops in the circuit, representing an initial circuit state I (flip-flop contents after a system reset or after the application of a synchronising sequence). The aim of the algorithm is to find a set of longest reachable sensitisable paths, as well as test sequences that sensitise those paths starting at the given initial state. A sensitisable path is defined as *reachable* if a test sequence of length k exists for some $k \in \mathbb{N}$ such that the application of that test sequence to the circuit in state I sensitises the path after k time frames. If a test sequence of length k is found for a path p, we call k p's *depth*.

First, PHAETON is used to extract a set of longest sensitisable paths. Then, for each found path p, a MC-instance $MC(I, T_{i,i+1}, P)$ is generated and passed to the CIP-solver. If the MC-instance is determined to be unsatisfiable, p is proved as unreachable. On the other hand, if the MC-instance is satisfiable, p is reachable and there exists an appropriate test sequence that can be extracted from the Boolean solution computed by the CIP-solver.

An MC-instance consists of three predicates namely the initial state I, the transfer function $T_{i,i+1}$, and a certain property P. Figure 3 illustrates how these predicates are connected in order to check whether a functional sequence starting from the initial state I_0 justifying the target property P performing k transition steps exits or not. We call $MC(I, T_{i,i+1}, P)$ satisfiable if there exists a functional sequence for some value k.

One arbitrary step of the sequential circuit from time point i to time point $i+1$ corresponds to one application of the transfer function $T_{i,i+1}$. The transfer function is given by the SAT-representation of the circuit's combinational logic that can be obtained by performing a so-called Tseitin-transformation [25]. A Boolean formula encoding multiple time frames of a sequential circuit is obtained by connecting the transfer function multiple times, i.e. $T_{0,1} \wedge T_{1,2} \wedge \ldots T_{k,k+1}$.

The initial state I_0 defines the starting state of the sequence and corresponds to the initial logical values of the flip-flops. Analogously, the target property P imposes the justification requirements that need to hold at the end of the sequence. Note that our method supports justification conditions for internal circuit lines over multiple time frames.

The complete MC-instance is passed to the CIP-solver that returns a Boolean solution, or classifies the instance as unsatisfiable which means that no solution exists.

Let p be one of the sensitisable paths found by PHAETON. Since PHAETON is a SAT-based tool, the values assigned to the side inputs of the gates along p can be extracted from the Boolean solution at no additional expense. These values comprise the set of necessary assignments needed to guarantee the sensitisation of the path. Note that this set of necessary assignments automatically respects additional conditions imposed on the path-enumeration algorithm.

The algorithm can be applied in three different modes that differ in the definition of the target property P. In the first mode, the desired property P_k passed to the CIP-solver is composed of the necessary assignments extracted from the found path's side inputs as explained above. This mode is called *Path-Targeting Mode*. In the second mode, the desired property P_k is composed only of the values that have to be assigned to the flip-flops in the last-but-one time frame. That means, in this mode, the CIP-solver attempts to justify the first test pattern of the test pair found by PHAETON. This mode is called *Pattern-Targeting Mode*. The last mode, called *Second-Pattern-Targeting Mode*, is a combination of the other modes. While Path-Targeting is used for the first time frame, we use pattern targeting for the second one. Therefore, the solver is free to change the first pattern to a functional one while keeping the second one unchanged.

If the MC-instance is determined to be unsatisfiable, p is proved to be unreachable. On the other hand, if the MC-instance is satisfiable, p is reachable and there exists an appropriate test sequence that can be extracted from the Boolean solution computed by the CIP-solver.

Note that, due to the iterative nature of the CIP-solver's algorithm, the found sequence is guaranteed to have the shortest possible length. Also, the method is guaranteed to find a solution if one exists, provided that the CIP-solver is allocated unlimited solving time and memory.

We also developed a technique to speed up the algorithm by shortening the number of iterations required by the CIP-solver. At the beginning of the overall flow, only one initial circuit state I is given. However, whenever the CIP-solver identifies

a reachable path of depth d greater than or equal to a user defined limit d_{min}, the last reached state can be regarded as an additional pseudo-initial state. Subsequent applications of the CIP-solver can target one of the learnt pseudo-initial states in addition to the original initial state I. Since the triangle inequality holds, an upper bound for the optimal depth of new reachable paths can be computed. As the distance between I and all learnt pseudo-initial states is known (it equals the depth of the paths for which they were identified), the upper bound is equal to the sum of the distance between I and the targeted pseudo-initial state I' and the length of the test sequence that starts at I'.

IV. EXPERIMENTAL RESULTS

The flow described in the previous section was applied to sequential ITC 99 benchmark circuits. All measurements were performed on an AMD Opteron computer using one 2.6 GHz-core and up to 4 GB RAM. In all experiments, the CIP-solver was set to classify an MC-instance as an abort after a time out of 10 seconds. All run-times listed in this section are given in seconds. As input data for all experiments, we extracted 500 globally longest sensitisable paths for each circuit using PHAETON [18]. Additionally, each path is sensitisable using broadside tests [26]. For circuits with less than 500 paths, we applied the flow to all sensitisable paths in the circuit.

We assume all circuits to be resetable and therefore use the all-zero-state (all flip-flops set to 0) as the initial state in all experiments. Please note that our flow itself can handle any set of initial states.

The results for the application of the path-targeting mode are shown in Table I. The table distinguishes between four sets of paths: sensitisable paths found by PHAETON (columns labelled "Total"), sensitisable paths found to be reachable according to the CIP-solver, sensitisable paths found to be unreachable, and sensitisable paths not classified by the CIP-solver within the timeout of 10 seconds. For each of the possible classifications, the table contains four data columns. Column "Count" shows the number of paths that fall into this classification. The run-time needed is given in the next column. The last two columns contain the maximal and the average iteration depth, i.e. the number of time frames needed to reach the desired state starting at the all-zero-state.

The results show that many MC-instances that are proven to be unsatisfiable, i.e. the target property is unreachable, have very low iteration depths. This is explained by the integration of Craig interpolants that accelerate the process of finding a fixed-point. Consequently, in contrast to previous purely fixed-point-based approaches, our flow is applicable for the proof of unreachability as well as for the proof of reachability.

Furthermore, the results are optimal in terms of the generated sequence lengths. Hence, the tool flow finds not only the correct classifications without the use of heuristics, but even computes the shortest possible assignments to the primary inputs.

Figure 4 provides a visualisation of the percentage of reachable, unreachable and aborted instances in the form of stacked-bar charts. The thick black line (secondary y-axis) shows the average iteration depth that was reached.

For many circuits, a high proportion or even all of the sensitisable paths are also sensitisable starting at reachable

978-1-4673-0438-2/12 $31.00 © 2012 IEEE

TABLE I
APPLICATION TO ITC 99 BENCHMARK CIRCUITS

	Total				Reachable				Unreachable				Aborts			
			Depth				Depth				Depth				Depth	
Circuit	Count	Time	Max	Avg	Count	Time	Max	Avg	Count	Time	Max	Avg	Count	Time	Max	Avg
b01	46	0.22	4	1.46	46	0.21	4	1.46	0	-	-	-	0	-	-	-
b02	20	0.09	4	1.90	20	0.09	4	1.90	0	-	-	-	0	-	-	-
b03	348	3.40	14	5.76	88	1.00	5	1.91	260	2.29	14	7.06	0	-	-	-
b04	500	28.30	6	4.68	500	27.08	6	4.68	0	-	-	-	0	-	-	-
b05	500	67.21	131	7.55	3	0.63	44	44.00	497	65.28	131	7.33	0	-	-	-
b06	39	0.22	5	2.05	38	0.21	4	1.97	1	-	5	5.00	0	-	-	-
b07	500	165.07	159	19.85	6	1.08	27	23.17	493	152.13	159	19.59	1	11.10	108	108.00
b08	130	3.00	34	11.19	118	2.85	34	11.62	12	0.10	7	7.00	0	-	-	-
b09	424	12.72	54	21.19	294	7.49	20	16.59	130	5.09	54	31.61	0	-	-	-
b10	162	2.10	11	3.83	132	1.85	11	3.53	30	0.18	10	5.13	0	-	-	-
b11	500	105.67	78	16.23	358	100.88	78	20.34	142	3.62	48	5.88	0	-	-	-
b12	500	4482.18	497	233.74	34	185.38	77	46.18	120	42.08	181	17.96	346	4253.72	497	93.27
b13	203	537.32	379	83.37	80	27.91	222	24.43	82	4.23	116	17.39	41	505.06	379	246.98
b14	500	337.27	4	3.02	319	246.84	4	3.04	181	72.66	3	3.00	0	-	-	-
b15	500	815.80	42	6.76	8	47.58	18	14.50	450	268.62	42	5.27	42	481.19	20	14.52
b17	500	3452.91	97	17.78	0	-	-	-	393	1668.09	41	12.31	107	1709.08	97	20.09
b18	500	4131.32	5	3.26	0	-	-	-	500	3917.37	5	3.26	0	-	-	-
b20	500	1614.53	6	5.81	500	1570.51	6	5.81	0	-	-	-	0	-	-	-
b21	500	1669.89	6	5.52	500	1625.34	6	5.52	0	-	-	-	0	-	-	-
b22	500	2580.25	7	4.25	500	2507.55	7	4.25	0	-	-	-	0	-	-	-

states, while for some circuits none or a very small number of paths is sensitisable starting at reachable states.

In nearly all cases the needed iteration depth is very low. Therefore, we did not employ the addition of multiple start states, as explained at the end of Section III. However, in a different experiment we used that optimisation and achieved a reduction of the iteration depths of more than 40% for b12. In addition the number of aborts was significantly reduced.

In industry, broadside tests are often preferred over skewed-load tests [27] since the state of the second pattern of two-pattern tests is obtained through the functional logic of the circuit. Hence, broadside tests can be regarded as more functional. In order to verify this intuitive observation on the functionality of broadside tests, we performed the following experiment which also demonstrates the versatility of the tool presented in this work. We generated broadside tests for the same functionally sensitisable paths selected for Table I and then we used the tool to determine if the states of the tests are reachable. Results of this experiment are given in Table II.

The table compares the reachability numbers achieved by the tool using the three different targeting modes introduced in Section III. For each targeting mode, the run-time, the average iteration depth and the number of reachable, unreachable and

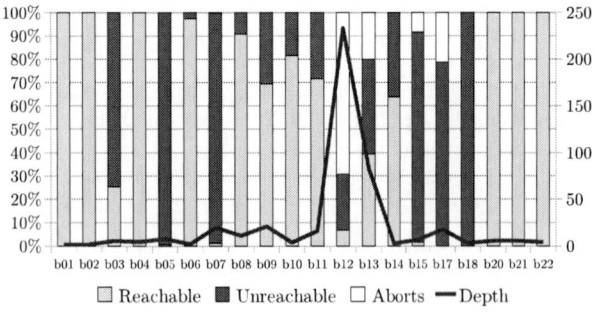

Fig. 4. Quantitative reachability results

aborted instances are given.

As can be seen, targeting the second pattern leads to higher numbers of reachable paths as compared to targeting the first test pattern. This supports the claim that the second test pattern of broadside testing is more functional than the first.

The second observation is that targeting the path's necessary assignments requires higher run-times than targeting the first or targeting the second test pattern. Targeting a test pattern restricts the search space, since it does not allow the solver to choose other assignments to the primary inputs. Therefore, this mode provides a check whether a given test pattern is reachable or not. However, targeting the path's side inputs is equivalent to performing a sequential test pattern generation. The CIP-solver performs a search for paths that are known to be sensitisable without restriction to the given test pattern. In many cases, this mode corroborates the reachability of paths that the pattern-targeting mode would otherwise classify as unreachable, e.g. because the given test pair uses illegal states. Hence, the reachability numbers of the path-targeting mode are higher. Extreme cases are e.g. circuits b21 and b22, where the pattern-targeting modes classify all 500 paths as unreachable although test sequences exist. These test sequences are found by the path-targeting mode.

The higher percentage of reachable instances in path targeting mode demonstrates that explicitly targeting functional justification requirements during test generation is more effective than generating the test pair first and then performing the functional justification.

V. CONCLUSION AND FUTURE WORK

In this paper, we presented a versatile technique to verify the reachability of sensitisable paths. The technique is complete and can yield the shortest existing test sequence. Furthermore, the method can provide a formal proof that no justification is possible regardless of the number of allowed time frames. Consequently, in contrast to previous purely fixed-point-based approaches, our flow is applicable for the proof of unreachability as well as for the proof of reachability.

TABLE II
CLASSIFICATION COMPARISON USING DIFFERENT TARGETING MODES

Circuit	Pattern Targeting					Second-Pattern Targeting					Path Targeting				
	Time	Depth	Reach.	Unreach.	Abort	Time	Depth	Reach.	Unreach.	Abort	Time	Depth	Reach.	Unreach.	Abort
b01	0.19	1.93	46	0	0	0.22	2.33	46	0	0	0.21	1.46	46	0	0
b02	0.08	2.00	20	0	0	0.08	2.00	20	0	0	0.09	1.90	20	0	0
b03	1.88	3.59	10	338	0	2.60	7.34	38	310	0	3.29	5.76	88	260	0
b04	7.77	3.67	18	482	0	16.21	7.02	161	339	0	27.08	4.68	500	0	0
b05	5.21	4.56	0	500	0	74.30	34.42	0	500	0	65.91	7.55	3	497	0
b06	0.19	2.31	36	3	0	0.18	2.36	37	2	0	0.21	2.05	38	1	0
b07	5.22	5.16	0	500	0	5.76	6.25	0	500	0	164.31	19.85	6	493	1
b08	9.02	19.39	17	113	0	13.01	26.17	39	91	0	2.95	11.19	118	12	0
b09	4.93	13.01	50	374	0	9.27	21.00	103	321	0	12.58	21.19	294	130	0
b10	1.57	8.25	22	140	0	2.38	10.03	36	126	0	2.03	3.83	132	30	0
b11	41.48	15.37	68	432	0	72.61	19.06	83	417	0	104.50	16.23	358	142	0
b12	65.10	5.43	0	496	4	1589.50	51.75	0	392	108	4481.18	233.74	34	120	346
b13	1.49	3.16	0	203	0	1.47	3.70	0	203	0	537.20	83.37	80	82	41
b14	57.26	3.21	3	497	0	404.33	8.58	304	196	0	319.50	3.02	319	181	0
b15	82.73	3.06	0	500	0	97.76	4.05	0	500	0	797.39	6.76	8	450	42
b17	329.50	3.00	0	500	0	283.62	3.10	0	500	0	3377.17	17.78	0	393	107
b18	917.19	3.00	0	500	0	1069.91	3.00	0	500	0	3917.37	3.26	0	500	0
b20	128.78	3.00	0	500	0	159.30	3.96	0	499	1	1570.51	5.81	500	0	0
b21	129.36	3.00	0	500	0	143.81	3.56	0	500	0	1625.34	5.52	500	0	0
b22	196.34	3.00	0	500	0	203.14	3.06	0	500	0	2507.55	4.25	500	0	0
ØITC99	99.26	5.46	14.50	353.90	0.20	207.47	11.14	43.35	319.80	5.45	975.82	22.96	177.20	164.55	26.85

Our experimental results demonstrate the versatility of the approach by showing its application for the computation of a path's reachability. Future application may include the computation of initialisation and syncronisation sequences.

One further extension of the method is the extraction of conflict information learned by the CIP-solver to increase the efficiency of the method.

REFERENCES

[1] J. Rearick, "Too Much Delay Fault Coverage is a Bad Thing," in *Int'l Test Conference*, 2001, pp. 624 –633.

[2] J. Saxena, K. Butler, V. Jayaram, S. Kundu, N. Arvind, P. Sreeprakash, and M. Hachinger, "A Case Study of IR-Drop in Structured At-Speed Testing," in *Int'l Test Conference*, vol. 1, 30-oct. 2, 2003, pp. 1098 – 1104.

[3] P. Girard, "Low Power Testing of VLSI Circuits: Problems and Solutions," in *Int'l Symp. on Quality Electronic Design*, 2000, pp. 173 –179.

[4] I. Pomeranz, "On the Generation of Scan-Based Test Sets with Reachable States for Testing under Functional Operation Conditions," in *Design Automation Conference*, july 2004, pp. 928 –933.

[5] I. Pomeranz and S. Reddy, "Generation of Functional Broadside Tests for Transition Faults," *Trans. on Computer-Aided Design of Integrated Circuits and Systems*, vol. 25, no. 10, pp. 2207 –2218, oct. 2006.

[6] H. Lee, I. Pomeranz, and S. Reddy, "On Complete Functional Broadside Tests for Transition Faults," *Trans. on Computer-Aided Design of Integrated Circuits and Systems*, vol. 27, no. 3, pp. 583 –587, march 2008.

[7] J.-T. Tsai, C.-Y. Wang, and K.-J. Chang, "Reachability Analysis of Sequential Circuits," in *VLSI Design, Automation and Test*, april 2010, pp. 181 –184.

[8] Y.-M. Kuo, C.-H. Lin, C.-Y. Wang, S.-C. Chang, and P.-H. Ho, "Intelligent Random Vector Generator Based on Probability Analysis of Circuit Structure," in *Int'l Symp. on Quality Electronic Design*, march 2007, pp. 344 –349.

[9] L. Shen, "Genetic Algorithm Based Test Generation for Sequential Circuits," in *Asian Test Symposium*, 1999, pp. 179 –184.

[10] A. Majhi, J. Jacob, L. Patnaik, and V. Agrawal, "On Test Coverage of Path Delay Faults," in *Int'l Conf. on VLSI Design*, jan 1996, pp. 418 –421.

[11] S. Bommu, K. Chandrasekar, R. Kundu, and S. Sengupta, "CONCAT: CONflict Driven Learning in ATPG for Industrial designs," in *Int'l Test Conference*, oct. 2008, pp. 1 –10.

[12] A. Czutro, I. Polian, M. Lewis, P. Engelke, S. M. Reddy, and B. Becker, "Thread-Parallel Integrated Test Pattern Generator Utilizing Satisfiability Analysis," *International Journal of Parallel Programming*, vol. 38, no. 3-4, pp. 185–202, June 2010.

[13] R. Drechsler, S. Eggersglüß, G. Fey, A. Glowatz, F. Hapke, J. Schlöffel, and D. Tille, "On Acceleration of SAT-based ATPG for Industrial Designs," *IEEE Trans. on CAD*, vol. 27, no. 7, pp. 1329–1333, 2008.

[14] K. L. McMillan, "Interpolation and SAT-Based Model Checking," in *Int'l Conference Computer Aided Verification*, 2003, pp. 1–13.

[15] M. Sauer, A. Czutro, I. Polian, and B. Becker, "Estimation of Component Criticality in Early Design Steps," in *IEEE Int'l Online Testing Symp.*, 2011, pp. 104–110.

[16] M. Prasad, M. Hsiao, and J. Jain, "Can SAT be Used to Improve Sequential Atpg Methods?" in *Int'l Conf. on VLSI Design*, 2004, pp. 585 – 590.

[17] G. Parthasarathy, M. Iyer, K. Cheng, and L. Wang, "Efficient Reachability Checking Using Sequential SAT," in *Asia and South Pacific Design Automation Conference*, jan. 2004, pp. 418 – 423.

[18] M. Sauer, A. Czutro, T. Schubert, S. Hillebrecht, I. Polian, and B. Becker, "SAT-based Analysis of Sensitisable Paths," in *IEEE Design and Diagnostics of Electronic Circuits and Systems*, 2011, pp. 93–98.

[19] S. Kupferschmid, M. Lewis, T. Schubert, and B. Becker, "Incremental Preprocessing Methods for Use in BMC," *Formal Methods in System Design*, pp. 1–20, 2011, 10.1007/s10703-011-0122-4. [Online]. Available: http://dx.doi.org/10.1007/s10703-011-0122-4

[20] K.-T. Cheng and H.-C. Chen, "Delay Testing for Non-Robust Untestable Circuits," in *Int'l Test Conference*, oct 1993, pp. 954 –961.

[21] E. Clarke, A. Biere, R. Raimi, and Y. Zhu, "Bounded Model Checking Using Satisfiability Solving," *Journal of Formal Methods in System Design*, pp. 93–98, 2001.

[22] W. Craig, "Linear Reasoning: A New Form of the Herbrand-Gentzen Theorem," *Journal of Symbolic Logic*, pp. 250–268, 1957.

[23] M. Sheeran, S. Singh, and G. Stålmarck, "Checking Safety Properties Using Induction and a SAT-Solver," in *Int'l Conference on Formal Methods in Computer-Aided Design*, 2000, pp. 108–125.

[24] J. R. Burch, E. M. Clarke, D. E. Long, K. L. Mcmillan, and D. L. Dill, "Symbolic Model Checking for Sequential Circuit Verification," *IEEE Trans. on CAD*, vol. 13, pp. 401–424, 1994.

[25] G. Tseitin, "On the Complexity of Derivation in Propositional Calculus," *Studies in Constructive Mathematics and Mathematical Logic*, 1968.

[26] J. Savir and S. Patil, "Broad-side delay test," *Computer-Aided Design of Integrated Circuits and Systems, IEEE Transactions on*, vol. 13, no. 8, pp. 1057 –1064, aug 1994.

[27] J. Savir, "Skewed-Load Transition Test: Part I, calculus," in *Test Conference, 1992. Proceedings., International*, sep 1992, p. 705.

2012 25th International Conference on VLSI Design

FORMAL VERIFICATION OF GALOIS FIELD MULTIPLIERS USING COMPUTER ALGEBRA TECHNIQUES

Jinpeng Lv, Priyank Kalla
Department of Electrical and Computer Eng.
University of Utah, Salt Lake City, UT-84112
{lv, kalla}@eng.utah.edu

Abstract: *Finite (Galois) field arithmetic finds applications in cryptography, error correction codes, signal processing, etc. Multiplication usually lies at the core of all Galois field computations and is a high-complexity operation. This paper addresses the problem of formal verification of hardware implementations of modulo-multipliers over Galois fields of the type $\mathbb{F}_{2^k}$, using a computer-algebra/algebraic-geometry based approach. The multiplier circuit is modeled as a polynomial system in $\mathbb{F}_{2^k}[x_1, x_2, \cdots, x_d]$ and the verification test is formulated as a Nullstellensatz proof over the finite field. A Gröbner basis engine is used as the underlying computational framework. The efficiency of Gröbner basis computations depends heavily upon the variable (and term) ordering used to represent and manipulate the polynomials. We present a variable (and term) ordering heuristic that significantly improves the efficiency of Gröbner basis engines. Using our approach, we can verify the correctness of up to 96-bit multipliers, whereas contemporary BDDs/SAT/SMT-solver based methods are infeasible.*

I. INTRODUCTION

Finite (Galois) field theory is extensively applied in Elliptic Curve Cryptography, error correction codes, digital signal processing, etc. Therefore, dedicated hardware implementations of Galois field arithmetic abound. Multiplication lies at the core of most Galois field computations – where two k-bit inputs A, B are multiplied modulo an irreducible polynomial $P(x)$ over the field $\mathbb{F}_{2^k}$. Incorrect (buggy) multiplication can lead to full leakage of the secret key [1] in cryptosystems. Therefore, it is of utmost importance to verify the correctness of hardware implementations of finite field multipliers residing at the core of such systems. This paper addresses formal verification of multiplier circuits over (binary) Galois fields of the type $\mathbb{F}_{2^k}$ using computer algebra techniques.

Problem Statement: Let F represent a word-level (multiplier) specification such that $F = A \cdot B \pmod{P(x)}$, where A and B are elements in $\mathbb{F}_{2^k}$ (k-bit inputs), and $P(x)$ is an irreducible polynomial of $\mathbb{F}_{2^k}$. Let G be a gate-level circuit implementation of F. Our objective is to formally prove that the circuit G correctly computes the multiplication for all possible values of the inputs A, B. Otherwise, we have to produce a counter-example that excites the bug in the design.

Approach and Contributions: We model the multiplier circuit as a polynomial system in $\mathbb{F}_{2^k}[x_1, x_2, \cdots, x_d]$ and then for-

This work is sponsored in part by a grant from NSF #CCF-546859.

mulate the equivalence test using the theory of Hilbert's weak Nullstellensatz over Galois Fields [2] (a proof-by-refutation approach). A Gröbner basis engine [3] is used as the computational framework to solve the underlying polynomial decision problems.

A straight-forward application of our approach allows to verify only upto 48-bit multipliers. The source of this limitation is in the efficiency of Gröbner basis engines – which is highly susceptible to the effects of variable and term orderings used to represent and manipulate the polynomials. Therefore, to improve the efficiency and scalability of our approach, we present *a new heuristic to derive a variable and term order for polynomial representation*. Using this new term ordering, we are able to significantly enhance the efficiency of the Gröbner basis engine, and consequently scale our technique to *verify up to 96-bit multipliers*. Our experiments also demonstrate that contemporary approaches (BDDs, SAT- and SMT-solver based) are *unable to verify* the correctness of multipliers that are larger than 16-bits wide.

Our approach is generic enough to verify the implementation of any Galois field arithmetic circuit against its given polynomial specification. However, for this paper, we concentrate only on modulo multipliers over $\mathbb{F}_{2^k}$ fields. Our approach can prove correctness or detect the *presence of bugs*. It cannot, however, find input vectors that excite the bugs in flawed designs (counter-examples). In such cases, we show that SMT-solvers are able to generate the counter-examples, complementing our technique.

Motivation for this work: The main motivation behind this work is to formally prove the correctness of Galois field multipliers for applications in Elliptic Curve Cryptography. In many organizations, cryptography implementations have to be *certified correct* before they can be deployed for secure communications. While SMT-solvers are able to catch bugs quickly, none of the contemporary decision procedures are able to formally prove the correctness of modulo-multiplier implementations (as demonstrated in our experiments). Our approach provides an automatic formal verification engine for correctness proof.

II. PRELIMINARIES: GALOIS FIELDS & MULTIPLICATION

We briefly describe the relevant finite field concepts (details can be found in the textbook [4]) and modular multiplier design over such fields [5] [6] [7] [8].

A finite field, also called a Galois field, is a field with a finite number of elements. The number of elements q of the

978-1-4673-0438-2/12 $31.00 © 2012 IEEE   388

finite field is a power of a prime integer – i.e. $q = p^k$, where p is a prime integer, and $k \geq 1$. Galois fields are denoted as $\mathbb{F}_q$ and also $GF(q = p^k)$. We are interested in fields where $p = 2$ and $k > 1$; i.e. *binary Galois extension fields* $\mathbb{F}_{2^k}$, as employed in Elliptic Curve Cryptography implementations.

To construct $\mathbb{F}_{2^k}$, we take the polynomial ring $\mathbb{F}_2[x]$ and an irreducible polynomial $P(x) \in \mathbb{F}_2[x]$ of degree k, and construct $\mathbb{F}_{2^k}$ as $\mathbb{F}_2[x] \pmod{P(x)}$. For example, $\mathbb{F}_8 = \mathbb{F}_2[x]$ $\pmod{x^3 + x + 1}$. All the field operations are performed modulo the irreducible polynomial $P(x)$ and the coefficients are reduced modulo $p = 2$. Any element $A \in \mathbb{F}_{2^k}$ can be represented in polynomial form as $A = a_0 + a_1 x + \cdots + a_{k-1} x^{k-1}$, where $a_i \in \mathbb{F}_2 = \{0, 1\}$ and x is the root of the irreducible polynomial.

We will employ an important property of Galois fields [4]: For all elements $A \in \mathbb{F}_q, A^q = A$, and hence $A^q - A = 0, \forall A \in \mathbb{F}_q$.

We now illustrate multiplication over $\mathbb{F}_{2^k}$ through the following example.

Example II.1: Let us consider the field $\mathbb{F}_{2^4}$. We take as inputs: $A = a_0 + a_1 \cdot x + a_2 \cdot x^2 + a_3 \cdot x^3$ and $B = b_0 + b_1 \cdot x + b_2 \cdot x^2 + b_3 \cdot x^3$, along with the primitive polynomial $P(x) = x^4 + x^3 + 1$. We have to perform the multiplication $F = A \times B \pmod{P(x)}$. The coefficients of $A = \{a_0, \ldots, a_3\}, B = \{b_0, \ldots, b_3\}$ are in $\mathbb{F}_2 = \{0, 1\}$. So we can perform this multiplication as shown below:

		a_3	a_2	a_1	a_0
$\times$		b_3	b_2	b_1	b_0
		$a_3 \cdot b_0$	$a_2 \cdot b_0$	$a_1 \cdot b_0$	$a_0 \cdot b_0$
	$a_3 \cdot b_1$	$a_2 \cdot b_1$	$a_1 \cdot b_1$	$a_0 \cdot b_1$	
	$a_3 \cdot b_2$	$a_2 \cdot b_2$	$a_1 \cdot b_2$	$a_0 \cdot b_2$		
$a_3 \cdot b_3$	$a_2 \cdot b_3$	$a_1 \cdot b_3$	$a_0 \cdot b_3$			
s_6	s_5	s_4	s_3	s_2	s_1	s_0

In polynomial expression, we have the result as: $S = s_0 + s_1 \cdot x + s_2 \cdot x^2 + s_3 \cdot x^3 + s_4 \cdot x^4 + s_5 \cdot x^5 + s_6 \cdot x^6$, where, $s_0 = a_0 \cdot b_0$, $s_1 = a_0 \cdot b_1 + a_1 \cdot b_0$, $s_2 = a_0 \cdot b_2 + a_1 \cdot b_1 + a_2 \cdot b_0$, and so on. Here the multiply "$\cdot$" and add "$+$" operations are performed modulo 2, so they can be implemented in a circuit using AND and XOR gates. Note that unlike integer multipliers, there are no carry-chains in the design, as the coefficients are always reduced modulo $p = 2$. However, the result is yet to be reduced modulo the primitive polynomial $P(x) = x^4 + x^3 + 1$. This is shown below:

s_3	s_2	s_1	s_0	
s_4	0	0	s_4	$\Leftarrow s_4 \cdot x^4 \pmod{P(x)} = s_4 \cdot (x^3 + 1)$
s_5	0	s_5	s_5	$\Leftarrow s_5 \cdot x^5 \pmod{P(x)} = s_5 \cdot (x^3 + x + 1)$
s_6	s_6	s_6	s_6	$\Leftarrow s_6 \cdot x^6 \pmod{P(x)} = s_6 \cdot (x^3 + x^2 + x + 1)$
g_3	g_2	g_1	g_0	

The final result (output) of the circuit is: $G = g_0 + g_1 x + g_2 x^2 + g_3 x^3$; where $g_0 = s_0 + s_4 + s_5 + s_6$; $g_1 = s_1 + s_5 + s_6$; $g_2 = s_2 + s_6$; $g_3 = s_3 + s_4 + s_5 + s_6$.

The above multiplier design is called the *Mastrovito multiplier* [6] [5]. In cryptosystems, multiplication is often performed repeatedly – e.g., for exponentiation. For such applications, Montgomery multiplier architectures [7] [8] over Galois fields are employed for faster computation (at the expense of area).

In our experiments, we verify custom implementations of both Mastrovito and Montgomery multipliers. Because of their obvious complexity, Montgomery multipliers are harder to verify than the ones based on Mastrovito design style.

III. LIMITATIONS OF PREVIOUS WORK

Contemporary graph-based canonical DAG representations of Boolean functions such as BDDs [9], OKFDDs [10], BMDs [11], etc. are ill-suited for verification of such modulo-multiplication applications, particularly over large finite fields (as also demonstrated in [12]). This verification problem is also very hard for SAT solvers, due to the large size and the presence of AND-XOR terms in the design. Contemporary Satisfiability Modulo Theory (SMT) solvers employ a mixture of theories for reasoning – however, none of them employ polynomial equation solving over finite fields (which is a very hard problem in itself). Therefore, for our applications, we use the SMT solvers to model the circuit constraints using the theory of fixed-size bit-vectors (QF-BV). Our experiments, described in later sections, reveal that BDDs/SAT/SMT solvers cannot prove equivalence beyond 16-bit multipliers.

The theorem-proving approach of [13] comes closest to ours, as they also verify a finite field $\mathbb{F}_{2^k}$ implementation against a given specification. They devise a decision procedure based on polynomial division, variable elimination, term re-writing, etc., to verify Galois field arithmetic. However, their correctness criterion is 2^k-field-size independent. If this condition is not satisfied, then their approach requires decision over $\mathbb{F}_2$, and that really limits the scalability of their approach.

Gröbner basis based algebraic techniques have been utilized for SAT solving and formal verification: [14] [15] [16] [17]. However, these methods are employed for verification by modeling constraints over $\mathbb{F}_2 = \mathbb{Z}_2$ which leads to the inefficiency of these approaches. The recent work of [18] employs computer algebra over finite integer rings $\mathbb{Z}_{2^k}$ for verification of finite word-length *integer datapaths*. None of the above works have successfully solved the modulo-multiplier verification problem over $\mathbb{F}_{2^k}$. The work of [19] verifies a given composite field $\mathbb{F}_{(2^m)^n}$ multiplier against its specification using a Gröbner basis approach which is similar to the setup in this paper. However, that approach has the limitation that it requires that the circuit decomposition hierarchy be made available to the verification engine.

IV. VERIFICATION SETUP AND POLYNOMIAL MODELING

Our verification setup is formulated as an *equivalence check* between the *polynomial specification* and the *circuit implementation*. The setup (miter) is depicted in Fig. 1. Given the specification polynomial $F = A \cdot B \pmod{P(x)}$, and the circuit implementation G (as shown in Example II.1), we want to prove that for all possible inputs, F is always equal to G over $\mathbb{F}_{2^k}$. This can, conversely, also be solved as proving that $F \neq G$ has no solutions. The existence of no solutions to $F \neq G$ proves their equivalence. Whereas, a solution to $F \neq G$ implies a bug.

To test whether $F \neq G$ has no solutions: (i) we represent all

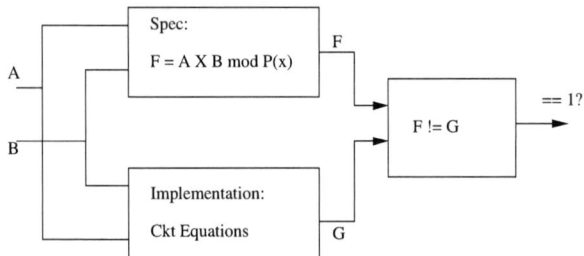

Fig. 1: The Equivalence Verification Setup: Miter

the constraints corresponding to our setup of Fig. 1 as a polynomial system over the Galois field $\mathbb{F}_{2^k}$; and subsequently (ii) we use techniques from symbolic computer algebra, specifically Hilbert's Nullstellensatz and Gröbner bases [2], to deduce whether or not the polynomial system has any solutions over the given field.

The *specification* is provided in word-level polynomial form as follows: $A = a_0 + a_1 \cdot x + \cdots + a_{k-1} \cdot x^{k-1}$, $B = b_0 + b_1 \cdot x + \cdots + b_{k-1} \cdot x^{k-1}$ and $F = A \cdot B \pmod{P(x)}$. where $A, B \in \mathbb{F}_{2^k}$ ($a_i, b_i \in \mathbb{F}_2$) symbolically represent the inputs, and $F \in \mathbb{F}_{2^k}$ represents the result of the computation.

The *implementation* is given as a circuit netlist. We map the gate-level Boolean operators (AND, OR, NOT, XOR) to polynomials over $\mathbb{F}_2 (\subset \mathbb{F}_{2^k})$ (as shown in Example II.1) using the following equalities:
$f : B \to F_2$

$$\neg a \iff a+1 \pmod 2; \quad a \vee b \iff a+b+a \cdot b \pmod 2$$

$$a \wedge b \iff a \cdot b \pmod 2; \qquad a \oplus b \iff a+b \pmod 2$$

Therefore, we can represent the implementation also in polynomial form over $\mathbb{F}_{2^k}$. Let G symbolically denote word-level result of the implementation (output of the circuit).

The $F \neq G$ constraint is also modeled as a polynomial over $\mathbb{F}_{2^k}$ as follows: $t(F - G) = 1$, where $t \in \mathbb{F}_{2^k}$. The correctness of the above constraint modeling can be shown as follows: (i) When $F = G, F - G = 0$, so $t \cdot 0 = 1$ has no solutions. (ii) When $F \neq G, (F - G) \neq 0$. Over any field, every non-zero element has a multiplicative inverse. Let $t^{-1} = (F - G)$. Then $t \cdot t^{-1} = 1$ will always have a solution.

To summarize, the constraints corresponding to the entire verification instance (miter) – the specification, implementation, and the $F \neq G$ constraint – can be modeled as a polynomial equation system in $\mathbb{F}_{2^k}$. Subsequently, we can reason whether or not solutions exist to this polynomial system using a computer algebra approach – which is described next.

V. VERIFICATION USING HILBERT'S NULLSTELLENSATZ AND GRÖBNER BASES

Let $\mathbb{F}$ be any field and $\mathbb{F}[x_1, \ldots, x_d]$ the polynomial ring over $\mathbb{F}$ with indeterminates $x_1, \ldots, x_d$. An *ideal* generated by $f_1, \ldots, f_s \in \mathbb{F}[x_1, \ldots, x_d]$ is $\langle f_1, \ldots, f_s \rangle = \{h = \sum_{i=1}^{s} g_i \cdot f_i : g_i \in \mathbb{F}[x_1, \ldots, x_d]\}$. Suppose that we are given a set of polynomials $f_1, f_2, \ldots, f_s \in \mathbb{F}[x_1, \ldots, x_d]$ and that we wish to find solutions to the polynomial system $f_1 = f_2 = \cdots = f_s = 0$. The set of all

solutions to a given system of polynomial equations is called the **variety**, and is denoted by $V(f_1, \ldots, f_s)$.

Theorem V.1: **Weak Nullstellensatz over** $\mathbb{F}_q$ [20]: Let $f_1, \ldots, f_s \in \mathbb{F}_q[x_1, \ldots, x_d]$ generate an ideal $I = \langle f_1, \ldots, f_s \rangle \subseteq \mathbb{F}_q[x_1, \cdots, x_d]$. Let ideal $I_0 = \langle x_1^q - x_1, x_2^q - x_2, \ldots, x_d^q - x_d \rangle$ and denote $\langle I, I_0 \rangle = \langle f_1, \ldots, f_s, x_1^q - x_1, \ldots, x_d^q - x_d \rangle$. Let $V_{F_q}(I)$ denote the variety of I over the Galois field $\mathbb{F}_q$. Then $V_{F_q}(I) = \emptyset$ if and only if $1 \in \langle I, I_0 \rangle$.

A detailed proof of this result is given in [20]. We briefly describe the significance of this result and how it applies to our work.

• We represent the polynomial constraints corresponding to our verification setup of Fig. 1 as the ideal $I = \langle f_1, \ldots, f_s \rangle \subset \mathbb{F}_q[x_1, \ldots, x_d]$, where $q = 2^k$.

• Then we generate another set of polynomials representing ideal $I_0 = \langle x_1^q - x_1, \ldots, x_d^q - x_d \rangle$, for all the variables $\{x_1, \ldots, x_d\}$ in our system.

• Subsequently, we append the polynomials of I_0 to those of I and generate the ideal $\langle I, I_0 \rangle = \langle f_1, \ldots f_s, x_1^q - x_1, \ldots, x_d^q - x_d \rangle$.

• Now we have to test if the constant polynomial 1 is a *member of the ideal* $\langle I, I_0 \rangle$.

• If indeed $1 \in \langle I, I_0 \rangle$, then $V_{F_q}(I) = \emptyset$; i.e. the polynomial system $f_1 = \cdots = f_s = 0$ has no solution over the given field $\mathbb{F}_q$. This implies that our miter constraints are infeasible, and the implementation (G) is equal to the specification (F).

• On the other hand, if $1 \notin \langle I, I_0 \rangle$, then it means that there exist a *finite number of solutions within the field* $\mathbb{F}_q$, $(V_{F_q}(I) \neq \emptyset)$, and that there are definitely bug(s) in the implementation [20].

The significance of ideal I_0: Recall from Section II, for all elements A of a Galois field $\mathbb{F}_q$, $A^q = A$, or $A^q - A = 0$. In other words, $A^q - A$ *vanishes* $\forall A \in \mathbb{F}_q$. The ideal $I_0 = \langle x_i^q - x_i \rangle$ is, therefore, also called the ideal of all vanishing polynomials of the field $\mathbb{F}_q$. In general, the Weak Nullstellensatz deduces the variety over an algebraically closed field. Galois fields are, however, not algebraically closed. These vanishing polynomials $\langle x_i^q - x_i \rangle$ *restrict the variety to* $\mathbb{F}_q$, discarding the solutions over the algebraic closure of $\mathbb{F}_q$.

To solve our verification problem, we have to test if $1 \in \langle I, I_0 \rangle$. For this test, it is required to compute a *Gröbner basis* of $\langle I, I_0 \rangle$, for which, Buchberger's algorithm is used, as shown in Algorithm 1.

Input: : $F = \{f_1, \ldots, f_s\}$
Output: : $G = \{g_1, \ldots, g_t\}$
  $G := F$;
  REPEAT
    $G' := G$
    For each pair $\{f, g\}, f \neq g$ in G' DO
      $S(f, g) \xrightarrow{G'}_+ r$
      IF $r \neq 0$ THEN $G := G \cup \{r\}$
  UNTIL $G = G'$

Algorithm 1: Buchberger's Algorithm

For Gröbner basis computation, a monomial (term) ordering is fixed to ensure that polynomials are manipulated

978-1-4673-0438-2/12 $31.00 © 2012 IEEE

in a consistent manner. Buchberger's algorithm then takes pairs of polynomials in the basis and combines them into "S-polynomials" to cancel leading terms. An S-polynomial is defined as:

$$S(f,g) = \frac{L}{lt(f)} \cdot f - \frac{L}{lt(g)} \cdot g \qquad (1)$$

where $L = \text{LCM}(lm(f), lm(g))$, where $lm(f)$ is the leading monomial of f, and $lt(f)$ denotes the leading term of f. The S-polynomial is then reduced (divided) by all elements of G' to a remainder r, denoted as $S(f,g) \xrightarrow{G'}_+ r$. Multivariate polynomial division is used for this reduction step. This process is repeated for all unique pairs of polynomials, including those created by newly added elements, until no new polynomials are generated; ultimately constructing the Gröbner basis.

A Gröbner basis can be "reduced" further (details given in [3]). A reduced Gröbner basis is a canonical representation of the ideal w.r.t. the imposed monomial order, and it provides a *decision procedure* to test if $1 \in \langle I, I_0 \rangle$.

Theorem V.2: (From [3]) Let $I = \langle f_1, \ldots, f_s \rangle \subset \mathbb{F}[x_1, \ldots, x_d]$ be an ideal. Let G denote the *reduced Gröbner basis* of I. Then $V(I) = \emptyset$ if and only if $G = \{1\}$.

Thus, to test if $1 \in \langle I, I_0 \rangle$, we need to compute the *reduced Gröbner basis G* of the ideal $\langle I, I_0 \rangle$ and check if $G = \{1\}$.

Using this formulation, we have performed verification for correct and buggy implementations of both Mastrovito and Montgomery designs. We use the SINGULAR computer algebra tool [21] to compute the Gröbner basis. The experiments are described in Section VII, and we discuss them later. However, we make the following observations here: To compute a Gröbner basis, an ordering needs to be imposed on the monomial terms of the polynomials. Conventionally, lexicographic (lex), degree-lexicographic (deg-lex) and degree-reverse-lexicographic (degrevlex) orderings are used [2]. Term orderings have a significant impact on the efficiency of Gröbner basis computations. When we use these conventional lex, deglex and degrevlex orderings, we are able to verify only upto 48-bit circuits in $\mathbb{F}_{2^{48}}$; beyond which SINGULAR runs out of memory. To improve the scalability of our approach, we now present a variable (and term) ordering heuristic that allows to verify upto 96-bit circuits ($\mathbb{F}_{2^{96}}$).

VI. Term Ordering to Improve Gröbner Basis Computation

The critical computation in Gröbner bases algorithm is that of S-polynomials and their subsequent reduction: $S(f,g) \xrightarrow{I}_+ r$. A rudimentary implementation of Gröbner bases algorithm has the limitation that too many S-polynomials might be created, many of which may actually reduce to 0, thus wasting computation time. Leading monomials play a major role in computing $S(f,g)$. Different monomial orderings may lead to different S-polynomials and hence a different Gröbner basis. This is demonstrated below:

Example VI.1: Consider the polynomials: $f_1 = x_0 x_1 + x_2$, $f_2 = x_1 x_2 + x_3$. The above representation is obtained if the *lex* term order, with variable order $x_0 > x_1 > x_2 > x_3$, is chosen.

Here, the leading monomials $lm(f_1) = x_0 x_1$, $lm(f_2) = x_1 x_2$, and $f_3 = \text{S}(f_1, f_2) = -x_0 x_3 + x_2^2$. The Gröbner basis computed with this order contains these three polynomials $\{f_1, f_2, f_3\}$. If, however, a *lex* order with $x_3 > x_2 > x_1 > x_0$ is used, then $f_1 = x_2 + x_1 \cdot x_0$, with $lm(f_1) = x_2$ and $f_2 = x_3 + x_2 \cdot x_1$ with $lm(f_2) = x_3$ and $\text{S}(f_1, f_2) = x_3 x_1 x_0 - x_2^2 x_1$. We see the benefit of this order as this S-poly reduces to 0 modulo $\{f_1, f_2\}$, i.e. $S(f_1, f_2) \xrightarrow{f_1, f_2}_+ 0$, implying that $\{f_1, f_2\}$ is itself a Gröbner basis with this term order.

We wish to heuristically derive a variable and term order such that fewer S-polynomials are computed in the Gröbner basis algorithm. Our variable ordering heuristic is inspired by Buchberger's *Product Criterion* [22]:

Lemma VI.1 (Product Criterion) Let $f, g \in \mathbb{F}[x_1, \cdots, x_d]$ be polynomials. If the equality $lm(f) \cdot lm(g) = LCM(lm(f), lm(g))$ holds, then $S(f,g) \xrightarrow{G'}_+ 0$.

The above result states that when the leading monomials of f, g are relatively prime, then $S(f,g)$ always reduces to 0 modulo G'. Thus $S(f,g)$ need not be considered. Modern computer algebra engines always perform this check before computing $S(f,g)$. Note that, in Example VI.1, the lex term order with $x_3 > x_2 > x_1 > x_0$ makes $lm(f_1)$ and $lm(f_2)$ relatively prime, which obviates the need to compute $\text{S}(f_1, f_2)$ and thus $G = \{f_1, f_2\}$ itself constitutes a Gröbner basis. *Our variable ordering heuristic exploits the above result to derive a variable and term order such that leading monomials of most polynomials become relatively prime, and that speeds-up the computation.*

Algorithm 2 presents our WVAO (*Weighted Variable Activity Order*) ordering heuristic. For every variable, we introduce a measure (or count), called *Weighted Variable Activity (WVA)*. Intuitively, WVA corresponds to the frequency of a variable's occurrence among polynomials, weighted according to the size of the term in which the variable appears. Moreover, if a variable appears in more than one polynomial, its WVA measure is further enhanced by a multiplicative factor. The algorithm inputs the polynomials and iteratively updates the WVA measure for each variable. Finally, the variables are sorted according to *ascending order of their WVA count. Using this variable order, a lexicographic (lex) term ordering is imposed on the polynomials.*

The motivation behind this approach is that if a variable has lower WVA count, then it appears infrequently in the terms. If terms with variables that have lower WVA count are made leading terms, then they are more likely to be relatively prime; and thus avoid S-polynomials computations. We illustrate our algorithm using the example below:

Example VI.2: Consider the following polynomials extracted from a 2-bit multiplier over $\mathbb{F}_{2^2}$: $f_1 = a_0 \cdot b_0 + r_0$; $f_2 = a_0 \cdot b_1 + a_1 \cdot b_0 + r_1$; $f_3 = a_1 \cdot b_1 + r_2$; $f_4 = r_0 + r_2 + y_0$; $f_5 = r_1 + r_2 + y_1$. Polynomial f_1 is input first; it contains two terms $a_0 \cdot b_0$ and r_0. Since variables a_0, b_0, r_0 are encountered for the first time, (line 5 in the algorithm), their WVAs are assigned according to the size of their terms (Line $6-9$): WVA(a_0)=2, WVA(b_0)=2, WVA(r_0)=1. Next, f_2 is input;

978-1-4673-0438-2/12 $31.00 © 2012 IEEE   391

```
Input:  polys: polynomials extracted from given circuit
Output: Static Variable Order
1:   WVA= Initialize to zero for input variables;
2:   for (i=0; i < number of polynomials; i++) do
3:       term_list[i]=Extract_Term(polys[i]);
4:       for all var,term ∈ term_list[i] do
5:           occurrence=Search(var,var_list[0,...,j − 1]);
6:           if occurrence==j then
7:               /*First occurrence of var*/
8:               WVA[var]=Var_Num_In_Term(term);
9:           end if
10:          if occurrence < j && Same_Poly(var)==1 then
11:              /*var has occurred in same polynomial*/
12:              WVA[var]+=2·Var_Num_In_Term(term);
13:          end if
14:          if occurrence<j && Same_Poly(var)==0 then
15:              /*var already appeared in a different polynomial*/
16:              WVA[var]+=4·Var_Num_In_Term(term);
17:          end if
18:      end for
19:  end for
20:  /*Sort variables in terms of WVA in ascending order*/
21:  Sort(WVA[0,...,j]);
22:  return WVA;
```

Algorithm 2: WVA Ordering Algorithm

it contains three terms $a_0 \cdot b_1$, $a_1 \cdot b_0$ and r_1. Since a_0 has already appeared in another polynomial f_1, its WVA is updated according to lines $(14 − 17)$: WVA(a_0)=10. Similarly, WVA(b_0)=10. Since a_1 and b_1 appear for the first time, their WVAs are both 2. Similarly, WVA(r_1)=1. In this fashion, the WVA counts for all variables are updated until all polynomials are analyzed. This is illustrated in Table I. The variables are sorted in ascending order of their WVA counts, which results in $y_0 > y_1 > r_0 > r_1 > r_2 > a_0 > a_1 > b_0 > b_1$. Finally, a *lex* term order, with the above variable order, is used to represent the polynomials. When the polynomials are represented with the generated term order, we can see that $lm(f_1) = r_0$, $lm(f_2) = r_1$, $lm(f_3) = r_2$, $lm(f_4) = y_0$, and $lm(f_5) = y_1$. The leading terms of all polynomials become relatively prime, and no S-polynomial needs to be computed – and $\{f_1, f_2, f_3, f_4, f_5\}$ is itself a Gröbner basis using this order.

TABLE I: WVAO Algorithm execution for Example VI.2.

	a_0	a_1	b_0	b_1	r_0	r_1	r_2	y_0	y_1
f_1	2	0	2	0	1	0	0	0	0
f_2	10	2	10	2	1	1	0	0	0
f_3	10	10	10	10	1	1	1	0	0
f_4	10	10	10	10	5	1	5	1	0
f_5	10	10	10	10	5	5	9	1	1
	Obtained WVAO in Ascending Order								
WVAO	y_0	y_1	r_0	r_1	r_2	a_0	a_1	b_0	b_1

VII. EXPERIMENTAL RESULTS

We have conducted experiments to verify custom implementations of both Mastrovito and Montgomery multipliers, which were derived from [6] [5] [7] [8]. We use the computer algebra tool SINGULAR [v. 3-1-2] [21] to compute the Gröbner basis using the *slimgb* command. Our experiments are conducted on a desktop with 2.40GHz Intel Core(TM)2 Quad CPU and 2GB memory running 64-bit Linux.

Experiments with Singular: When our circuits are correctly designed, we do observe that the reduced Gröbner basis $GB(I, I_0) = \{1\}$, thus proving the equivalence. Results of the verification of Mastrovito multipliers using SINGULAR are shown in Table II. The results are shown for various variable orderings. For lp (lex), dp (degrevlex) and Dp (deglex), the best variable order found was "primary inputs > intermediate variables > primary outputs". Of these, the lex order performs best and we can verify upto 48-bit multipliers. Beyond that, the Gröbner basis creates too many polynomials and SINGULAR runs out of memory. However, using our proposed order (WVAO), we are able to extend the verification for upto 96-bit multipliers. Moreover, our term order also leads to faster reduced Gröbner basis computation. For example, for 32-bit multipliers, our term order completes verification within 52.76s, as compared to 2509.57s required for the Dp order.

The experiments for Montgomery multiplier verification are shown in Table III. The implementation of Montgomery multiplier was verified against the specification. Using the same lp, Dp and dp orders, we can verify only upto 44-bit multipliers, whereas our WVA order extends the verification upto 64-bit fields. Beyond the 64-bit multipliers, our WVA order also results in memory overflow in the Gröbner basis computation.

Table IV shows the verification results using BDDs, SAT and SMT solvers. To verify that the multiplier implementation is bug-free, we use BDDs/SAT/SMT-solvers to prove that the miter of Fig. 1 is infeasible. None of BDDs, SAT or SMT solvers can verify the correctness of circuits that are larger than 16-bits wide.

TABLE II: Verification of correct Mastrovito Multipliers using SINGULAR. Run-time given is seconds. MO= Memory Out.

Stats	Word size of the operands k-bits					
	8	16	32	48	64	96
#variables	180	665	2523	5593	9849	21945
#terms	507	2005	7947	17813	31605	70965
lex	0.03	3.94	537.17	7598	*MO*	*MO*
dp	0.02	13.12	2024.35	*MO*	*MO*	*MO*
Dp	0.03	14.32	2509.57	*MO*	*MO*	*MO*
WVAO	**0.01**	**0.74**	**52.76**	**595**	**2934**	**59428**

Experiments with Bugs: Our method can only detect the presence or absence of bugs (depending upon whether $1 \in \langle I, I_0 \rangle$). It cannot generate counter-examples that excite the bug. However, SMT solvers can identify bugs and gener-

TABLE III: Verification of correct Montgomery Multipliers using SINGU-LAR. Run-time given is seconds. MO= Memory Out.

Solver	Word size of the operands k-bits				
	8	16	32	44	48
#variables	428	1492	5540	10255	12147
#terms	1713	3669	24489	45811	54391
lex	0.28	17.4	1523.45	9001.23	*MO*
dp	0.37	36.88	6479.40	*MO*	*MO*
Dp	0.37	35.79	6169.90	*MO*	*MO*
WVAO	**0.07**	**4.30**	**454.90**	**2159.96**	**4024.21**

TABLE IV: Verification Results for bug-free modular multipliers over F_{2^k} for BDDs, SAT, SMT-solver based methods. TO = timeout of 10hrs.

Solver	Word size of the operands k-bits		
	8	12	16
MiniSAT	22.55	*TO*	*TO*
CryptoMiniSAT	7.17	16082.40	*TO*
PicoSAT	14.85	*TO*	*TO*
Yices	10.48	*TO*	*TO*
Beaver	6.31	*TO*	*TO*
CVC	*TO*	*TO*	*TO*
Z3	85.46	*TO*	*TO*
Boolector	5.03	*TO*	*TO*
SimplifyingSTP	14.66	*TO*	*TO*
ABC	242.78	*TO*	*TO*
BDD	0.10	14.14	1899.69

ate counter-examples quickly. We created buggy implementations by incorrectly connecting some signals in the design. Bug-catching results are shown in Table V. The SMT-solver YICES outperforms all others for bug catching.

TABLE V: Verification of Designs with Bugs. SAT, SMT, BDD. TO = Time-out limit of 1 hour.

Solver	Word-size of the operands: k-bits					
	8	12	16	32	64	96
MiniSAT	0.03	92.40	770.18	*TO*	*TO*	*TO*
CryptoMiniSAT	0.08	2.62	33.39	*TO*	*TO*	*TO*
PicoSAT	0.01	559.01	*TO*	*TO*	*TO*	*TO*
Yices	**0.00**	**0.00**	**0.00**	**0.02**	**0.03**	**0.12**
Beaver	0.04	0.07	0.13	0.42	2.15	87.08
CVC	50.20	*TO*	*TO*	*TO*	*TO*	*TO*
Z3	0.05	0.09	0.07	0.84	19.85	48.15
Boolector	0.01	0.03	0.06	0.60	11.19	156.51
SimplifyingSTP	0.03	0.05	0.09	0.36	2.47	12.69
ABC	234.00	*TO*	*TO*	*TO*	*TO*	*TO*
BDD	0.11	13.68	1823.17	*TO*	*TO*	*TO*

VIII. CONCLUSIONS

This paper has presented a formal approach to model and verify multiplier circuits over Galois fields $\mathbb{F}_{2^k}$ using a computer-algebra based approach. We model the verification test as a Nullstellensatz proof over $\mathbb{F}_{2^k}$ using a Gröbner basis engine. We analyze the verification constraints and derive a term order for efficient Gröbner basis computation. Using our approach, we are able to verify the correctness of upto 96-bit multipliers over $\mathbb{F}_{2^{96}}$, whereas conventional techniques based on SAT/SMT/BDD solvers are infeasible.

REFERENCES

[1] E. Biham, Y. Carmeli, and A. Shamir, "Bug Attacks," in *Proc. Annual Conf. on Cryptology: Advances in Cryptology*, ser. CRYPTO 2008, 2008, pp. 221–240.

[2] D. Cox, J. Little, and D. O'Shea, *Ideals, Varieties, and Algorithms: An Introduction to Computational Algebraic Geometry and Commutative Algebra.* Springer, 2007.

[3] W. W. Adams and P. Loustaunau, *An Introduction to Grobner Bases.* American Mathematical Society, 1994.

[4] R. J. McEliece, *Finite Fields for Computer Scientists and Engineers.* Kluwer Academic Publishers, 1987.

[5] E. Mastrovito, "VLSI architectures for computation in Galois fields," Ph.D. dissertation, Linköping University, Sweden, 1991.

[6] E. D. Mastrovito, "VLSI Designs for Multiplication Over Finite Fields gf(2^m)," *Lecture Notes in Computer Science*, vol. 357, pp. 297–309, 1989.

[7] C. Koc and T. Acar, "Montgomery Multiplication in GF(2^k)," *Designs, Codes and Cryptography*, vol. 14, no. 1, pp. 57–69, Apr. 1998.

[8] H. Wu, "Montgomery Multiplier and Squarer for a Class of Finite Fields," *IEEE Transactions On Computers*, vol. 51, no. 5, May 2002.

[9] R. E. Bryant, "Graph Based Algorithms for Boolean Function Manipulation," *IEEE Trans. on Computers*, vol. C-35, pp. 677–691, August 1986.

[10] R. Drechsler, A. Sarabi, M. Theobald, B. Becker, and M. Perkowski, "Efficient Representation and Manipulation of Switching Functions based on Ordered Kronecker Functional Decision Diagrams," in *DAC*, 1994, pp. 415–419.

[11] R. E. Bryant and Y.-A. Chen, "Verification of Arithmetic Functions with Binary Moment Diagrams," in *DAC*, 95.

[12] D. Mukhopadhyaya, G. Sengar, and D. Chowdhury, "Hierarchical Verification of Galois Field Circuits," *IEEE Trans. on CAD*, 2007.

[13] S. Morioka, Y. Katayama, and T. Yamane, "Towards Efficient Verification of Arithmetic Algorithms over Galois Fields $gf(2^m)$," *Proc. Computer-Aided Verification*, vol. 2102, pp. 465–477, 2001.

[14] M. Clegg, J. Edmonds, and R. Impagliazzo, "Using the Gröbner Basis Algorithm to Find Proofs of Unsatisfiability," in *ACM Symp. on Theory of Computation*, 1996, pp. 174–183.

[15] G. Avrunin, "Symbolic Model Checking using Algebraic Geometry," in *Proc. Computer-Aided Verification (CAV) Conf.*, 1996, pp. 26–37.

[16] C. Condrat and P. Kalla, "A Gröbner Basis Approach to CNF formulae Preprocessing," in *TACAS*, 2007.

[17] Y. Watanabe, N. Homma, T. Aoki, and T. Higuchi, "Application of Symbolic Computer Algebra to Arithmetic Circuit Verification," in *Proc. ICCD*, October 2007, pp. 25–32.

[18] O. Wienand, M. Wedler, D. Stoffel, W. Kunz, and G. Gruel, "An Algebraic Approach to Proving Data Correctness in Arithmetic Datapaths," in *Intl. Conf. on Computer-Aided Verification*, 2008.

[19] J. Lv, P. Kalla, and F. Enescu, "Verification of Composite Galois Field Multipliers over GF($(2^m)^n$) using Computer Algebra Techniques," in *Proc. High-Level Design Validation and Test Workshop*, 2011.

[20] S. Gao, "Counting Zeros over Finite Fields with Gröbner Bases," Master's thesis, Carnegie Mellon University, 2009.

[21] W. Decker, G.-M. Greuel, G. Pfister, and H. Schönemann, "SINGULAR 3-1-3 — A computer algebra system for polynomial computations," 2011, http://www.singular.uni-kl.de.

[22] B. Buchberger, "A Criterion for Detecting Unnecessary Reductions in the Construction of a Groebner Bases," in *EUROSAM*, 1979.

978-1-4673-0438-2/12 $31.00 © 2012 IEEE

A Novel SMT-Based Technique for LFSR Reseeding

Sarvesh Prabhu*, Michael S. Hsiao*, Loganathan Lingappan[†] and Vijay Gangaram[†]

*Department of Electrical and Computer Engineering, Virginia Tech, Blacksburg, Virginia, 24060
Email: {sarvesh, mhsiao}@vt.edu

[†]Intel Corporation, USA
Email: {loganathan.lingappan, vijay.gangaram}@intel.com

Abstract—**In order for logic built-in-self-test (LBIST) to achieve coverages comparable with deterministic tests, multiple (and frequently many) seeds are often needed. Unlike previous methods that attempt to chain/compact the number of seeds, we present a novel Satisfiability Modulo Theory (SMT) based technique that can reduce the number of seeds significantly while simultaneously achieving high coverage for LBIST. In this technique we integrate the process of deterministic test generation and seed generation in one SMT process to eliminate the problems of chaining the separately generated deterministic patterns. Experimental results show the promise of the approach.**

Index Terms—**LBIST; LFSR-reseeding; SMT;**

I. INTRODUCTION

Logic Built-In Self-Test (LBIST) is a popular technique for on-chip at-speed testing. The basic block diagram of a typical LBIST is shown in Figure 1. It consists of an on-chip test generator that generates the test vectors which are then applied to the Device Under Test (DUT). An output analyzer is usually a multiple input shift register (MISR) that compresses the output responses into a signature. At the end of the test session, the LBIST controller compares the signature of the MISR with the expected fault-free signature stored to determine if the underlying device is defective.

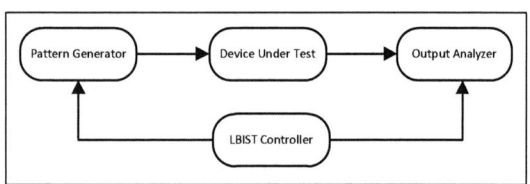

Fig. 1. Block Diagram of LBIST

The on-chip test generator typically consists of a pseudo random test generator in the form of Linear-Feedback-Shift-Register (LFSR). Starting from a random seed (initial state of the LFSR), a number of patterns can be derived and applied to the DUT, with which a number of the faults may be detected. For the remaining faults, one may apply additional sequences from other random seeds. However, this usually does not help to detect them, which most likely are either random-resistant or hard to test. Instead, an automatic test pattern generator

supported in part by SRC grant 2011-TJ-2134.

(ATPG) is often used to generate an incompletely-specified test suite for these remaining faults. A test is incompletely specified if there exists don't-care bits in the test vector. Storing this entire test set directly on chip would be too expensive if the number of tests is large. So, one approach is to compress these ATPG vectors in such a way that these deterministic tests can somehow be derived from one (or few) seeds. This idea of LFSR reseeding was introduced in [1], whereby the derived ATPG vectors are chained together and a seed is computed that could reproduce the same ATPG vectors. Though promising, chaining all the ATPG vectors so that they can be produced from a single seed can be very challenging, especially if the number of ATPG vectors is large.

A number of other approaches for LFSR reseeding have been proposed in the past [2]–[8], most of which focus on finding and concatenating several incompletely specified test cubes. The linear equations for the specified bits in concatenated test cubes are then solved to find a suitable LFSR seed. In the following, we will illustrate how existing methods work as well as point out the challenges involved.

A. Motivation

An example of conventional seed generation by chaining the test vectors is first given below. Let the characteristic polynomial of the LFSR be $X^4 + X^3 + 1$, and let the deterministic test set consist of 6 incompletely-specified vectors: 11XXX, X0XX1, X1X0X, XX1X1, 1XXX0 and 0X1XX. For simplicity of discussion and illustration, we will concatenate 2 vectors for computing one seed. From the 6 ATPG vectors, the first concatenated 2-vector string would be 11XXXX0XX1. To find a seed such that the LFSR will generate the concatenated vector, the following set of linear equations needs to be solved:

$$b0 = A0$$
$$b1 = A1$$
$$b6 = A0 + A1 + A2 + A3$$
$$b9 = A0 + A2$$

Here, $b0 - b9$ are the bits of the concatenated vector and $A0 - A3$ are the bits of the LFSR seed. Essentially, these linear equations specify that the initial state of the LFSR contains the first vector 11XXX, and five clocks later, the second vector X0XX1 would be derived. The linear equations can be modified to suit other constraints, such as one vector is produced with every new clock, etc. After solving the linear equations, the seed found is $A0 = 1$, $A1 = 1$, $A2 = 0$, $A3 = 1$. The same process can be repeated for computing the other seeds for the remaining four vectors. At the end, one would end up with 3 seeds for these six test vectors.

From the above example, we see that in such a method the deterministic test set generation and seed computation are two separate and independent processes. In other words, the vectors are first generated for a set of undetected faults by an ATPG, followed by the linear equation solving to find the seed(s). The results obtained using this approach depends critically on the number of vectors concatenated for each seed and the LFSR polynomial. Longer concatenations will allow for fewer seeds, but it makes the set of linear equations harder to solve (or could be unsolvable). To tackle this problem a multiple polynomial LFSR reseeding method was presented in [9].

TABLE I
LARGE NUMBER OF SEEDS NEEDED FOR CONVENTIONAL ATPG VECTORS

ckt	seeds	polynomials	ckt	seeds	polynomials
c432	16	2	s953	46	4
c499	23	5	s1196	82	6
c880	19	5	s1238	84	7
c1355	41	1	s1423	26	8
c1908	67	5	s1488	56	3
c2670	59	6	s1494	55	3
c3540	48	5	s5378	129	3

Table I shows the number of seeds and the number of distinct polynomials needed if the number of concatenated vectors for each seed is set to 8 [2]. Note that as the number of concatenated vectors increases, the complexity of solving this set of linear equations also becomes harder. In almost all cases, multiple polynomials were needed to achieve full coverage. In total, we see that the number of seeds is always greater than 10, and more seeds were often needed for larger circuits. In addition to generating many seeds, the previous method of seed generation has the following disadvantages:

- It is unclear how many m (out of n) ATPG vectors should be chained for one seed.
- The set of m vectors to be chained has a direct effect on the result.
- Determining the order in which the m vectors are chained can be challenging.
- If the linear equations for a concatenated vector has no solution, then
  - We need to change the order of concatenation, or
  - We need to change the set of m vectors to be concatenated, or
  - Find a different ATPG test cubes for some of the faults in concatenated vector.

These aforementioned challenges render such an approach for seed generation inefficient. We seek to propose and offer a completely different seed generation that would overcome the aforementioned challenges, and simultaneously reduce the number of seeds using only a single polynomial. To the best

of our knowledge, this is the first SMT-based formulation for LFSR reseeding, for which a single process of test generation and seed computation is proposed. Experimental results show that very few seeds are needed to achieve full coverage.

The rest of the paper is organized as follows. Section II describes the proposed SMT-based reseeding technique for LFSR reseeding. Section III shows the experimental results. Section IV concludes the paper.

II. THE PROPOSED TECHNIQUE

We propose a new SMT-based technique for LFSR reseeding in which we integrate the process of LFSR seed computation together with the test generation process, such that the test vectors generated would already be chained by the derived LFSR seed. Similar to the conventional LBIST technique, we first apply vectors starting from a random seed and drop all the detected faults. To detect the remaining faults, the method finds one or few LFSR seeds that can generate vectors to detect the remaining faults.

In the proposed SMT formulation, we represent every signal of the k-identical copies of the circuit as a bit vector of size k, as shown in Figure 2. Note that this circuit is not an unrolled circuit, since full-scan (could be with the STUMPS architecture [10]) is assumed for the DUT. The k copies of the signal $g1$ would be combined into a single bit-vector. This is done for every signal in the circuit. Hence, in a sense, we only have one circuit, rather than k copies, since we are using bit-vectors to represent each signal.

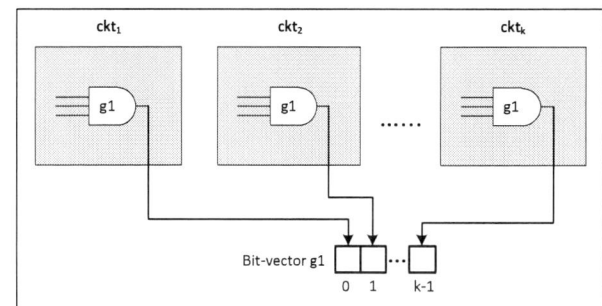

Fig. 2. Bit vector representation of k-identical copies of the circuit

We then add the LFSR constraints on the bit-vectors of the primary inputs of this k-copy circuit. The complete formula is then equivalent to a k-copy circuit with the inputs constrained by LFSR (Figure 2). The complete illustration is shown in Figure 3.

Next, the constraints for m target faults are added to the SMT formula. Note that a fault can be detected in any of the k copies, thereby implicitly solving the problem of vector-ordering faced by the previous equation-solving approach. The details for these constraints are described in the following subsections. The conventional way of detection of these m target faults would be to construct one fault-free circuit followed by m faulty circuits with one distinct fault in each of the faulty circuits. However, this would result in a huge circuit structure and become too cumbersome and expensive to solve in both

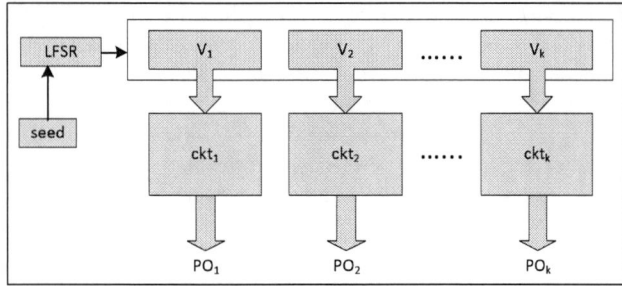

Fig. 3. Addition of LFSR constraints to the setup

time and space. Hence, we excite all m faults in one circuit and detect all m faults in another circuit. Consequently, there are only two (k-copy) circuits. Note that in both circuits, each signal is a bit-vector of size k as explained earlier. Also, there is no constraint on which fault should be detected by which of the k vectors/cycles.

Note that the constraints for detecting the faults (described subsequently) can be automatically derived for each target fault. Further, SMT is ideal for solving such an instance since we have to target m multiple faults in the k-copy circuit simultaneously. Conventional ATPGs target single faults and would not be able to target multiple faults.

A. Excitation Constraints

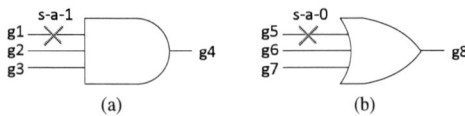

Fig. 4. Example of stuck-at fault at input of a gate

To excite a fault we add constraints to ensure that each of the m faults is excited by at least one of the k vectors. In addition, if a fault is at the input of a gate, we also ensure that it is propagated across the corresponding gate. For example, consider an input stuck-at-1 fault at an AND gate $g4$ as shown in Figure 4a. To excite this fault and propagate it across $g4$, we need the following constraint:

$$((not\ g1)\ and\ g2\ and\ g3)) \neq 00...0$$

Note that $g1$, $g2$ and $g3$ are all bit-vectors. Thus, bit-wise AND of (not $g1$), $g2$, and $g3$ is also a bit-vector. Let $exc\_vec1 = ((not\ g1)\ and\ g2\ and\ g3))$, then whenever exc_vec1 contains a logical 1 bit, it indicates that the corresponding bits in $g1$, $g2$, and $g3$ have to be 0, 1, 1, respectively. In other words, the stuck-at 1 fault at $g1$ would have been excited and propagated across this AND gate. Therefore, we need to constrain this bit-vector to not equal to the all-0 vector to ensure that the stuck fault is excited at least once. In this example, if exc_vec1 is 10101 (in a 5-copy circuit), then the fault is excited by vectors $v1$, $v3$ and $v5$. Similarly, for an input stuck-at-0 fault at an OR gate as shown in Figure 4b, the constraint is simply

$$((not\ g5)\ or\ g6\ or\ g7)) \neq 11...1$$

One can deduce from the preceding discussion that if any bit in the above bit-vector is 0, the fault is excited by the

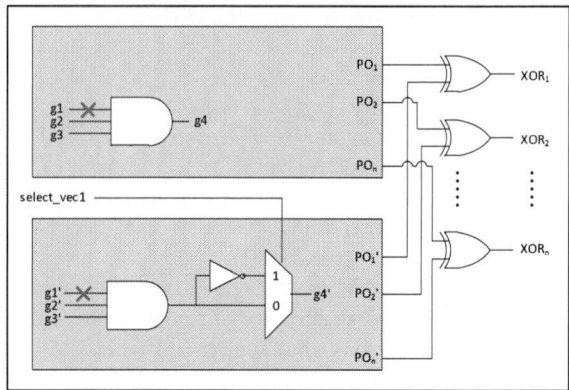

Fig. 5. Addition of fault detection constraints

corresponding vector. This is because a 0-bit indicates that the corresponding bits in $g5$, $g6$, and $g7$ would have been 1, 0, 0, respectively.

For stuck-at faults directly on the gate output, it is sufficient to make sure that the gate output is set to the correct value to excite the fault. For example, consider a stuck-at-0 fault at output of the AND gate $g4$ in Figure 4a; then the constraint added for excitation is simply

$$(g4 \neq 00...0)$$

Likewise, for the stuck-at-1 fault at output of OR gate $g8$, the constraint for fault excitation is

$$(g8 \neq 11...1)$$

B. Detection Constraints

As explained earlier, we excite each fault in the fault-free k-copy circuit and detect the corresponding fault in the other k-copy faulty circuit. Consider the fault at an input $g1$ of the AND gate shown earlier. The fault is excited whenever any bit of exc_vec1 is set to 1. (Recall that $exc\_vec1 = ((not\ g1)\ and\ g2\ and\ g3)$.) To inject this fault in the faulty circuit, we add a multiplexer at the output of the faulty gate, and set the select line of the multiplexer to be $select\_vec1$, also a bit vector of size k. If any bit of $select\_vec1$ is 0 we allow the fault-free value to pass to the output of the MUX, i.e., no fault is injected. But if any bit of $select\_vec1$ is 1, we reverse the value at the gate output, i.e., fault is injected.

From the previous section we know that the fault is excited when any bit of exc_vec1 is set to 1. Hence, we only need to inject the fault whenever the bit of exc_vec1 is 1 and do not inject when exc_vec1 is 0. Thus, the constraint added to ensure this is simply $select\_vec1 == exc\_vec1$.

However, since all m injected faults share the same circuit, it may be possible that some fault is hyperactive, i.e., excited frequently but may not be detected. In such a case, we may not want to inject the fault whenever the fault is excited since they can cause excessive masking (discussed later in the paper). To handle such cases, we change the constraint to be such that the fault is not injected whenever exc_vec1 is 0:

$(not\ exc\_vec1) \rightarrow (not\ select\_vec1)$, which can be re-written as

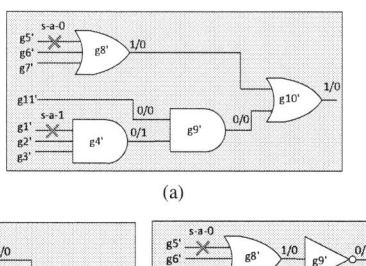

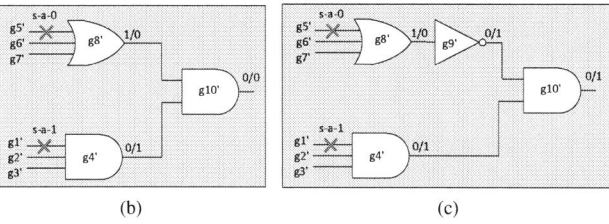

Fig. 6. Different scenarios of fault masking

$(select\_vec1 \text{ and } (not\ exc\_vec1)) = 00...0$

To ensure detection of the fault, we create a miter circuit. In a miter, the bit-vectors of the corresponding POs are XORed as shown in Figure 5. If any bit within $select\_vec1$ is 1 and any of the XOR gate in the fanout-cone of that fault is also 1, then the fault is considered to be detected. So, if the AND gate had p POs in its fanout-cone, the detection constraint added for detection of this fault is

$((select\_vec1 \text{ and } XOR_1) \text{ or } (select\_vec1 \text{ and } XOR_2) ... (select\_vec1 \text{ and } XOR_p)) \neq 00...0$

The excitation and detection constraints for every fault are ANDed. After adding all of the constraints, the final formula is given to the SMT solver, YICES [11]. If the formula is satisfiable, then the SMT solver returns a seed that can generate a k-vector sequence that detect all the m target faults. However, if the constraints are unsatisfiable then the SMT solver returns a seed that detects maximum number of faults amongst the m faults.

C. Use of Polarized z-sets and Dominators to prevent masking

In the SMT formulation described thus far, there may exist some fault masking, i.e., some faults may not be detected even if the constraints for that fault are satisfied. This is due to the following reasons.

- The approach assumes that whenever $((select\_vec1 \text{ and } XOR_1) \text{ or } ... (select\_vec1 \text{ and } XOR_p)) \neq 00...0$ is satisfied the target fault is detected. This is true if the output cones of all m faults are non intersecting. If the same vector excites many faults (that have the same output cones) but detects only one of them, then according to the current formulation all faults having the PO at which the fault is detected in its output cone will be considered detected.
  For example, consider the circuit fragment shown in Figure 6a. Let the s-a-1 fault at input of gate $g4'$ be $f1$ and s-a-0 at input of gate $g8'$ be $f2$. Note that for a given vector that sets $g11' = 0$, $f1$ should have been blocked. However, if both faults are injected in the same circuit copy and $f2$ has been detected, $f1$ is also considered

detected since $g10'$ lies in the fanout of both the faults. Hence, if $f2$ is detected by some vector, then based on the current formulation, $f1$ would also be considered detected (as we only use one faulty circuit for all m faults) even if the vector does not detect $f1$.

- If the presence of one fault is masking the effect of another fault in the faulty circuit, then the fault is considered undetected even if the vector actually detects the fault.
  Consider the circuit fragment shown in Figure 6b. It can be seen that because of simultaneous injection of both faults the propagation of fault $f1$ (input of $g4'$) is blocked at gate $g10'$ due to the 1/0 value at $g8'$ (since we use one faulty circuit for all m faults). Hence, $f1$ is considered undetected due to a fault effect from another fault even if this vector actually detects it.

- Similarly, the presence of one fault may falsely claim detection of another fault. In this case, the fault is considered detected even if the vector actually does not detect the fault.
  Consider a circuit fragment shown in Figure 6c. If both faults are injected, then the fault effects of both faults propagate through gate $g10'$. But if only one fault is injected then the propagation of either fault effect would have been blocked at gate $g10$. Hence both the faults may be considered detected even if this vector does not detect any of the two faults.

z-set: Consider a circuit with n outputs denoted as z_0, $z_1,...,z_{n-1}$. For a gate g in the circuit, the z-set(g) is the set that contains every output z_i such that there is a directed path in the circuit from g to z_i [12].

Polarized z-set: Polarized z-set of a fault gate g is a set of tuples $\{(z_0, p_0), (z_1, p_1)...(z_n, p_n)\}$ such that $\{z_0, z_1....z_n\}$ is the z-set of gate g and $\{p_0, p_1...p_n\}$ are the polarities of faulty values that can be propagated at the respective POs in the z-set. The polarity p_i of PO z_i is 0(1) if $D(\bar{D})$ can be propagated to the PO z_i through *all* paths from gate g to z_i. The polarity is set to x if both D and $\bar{D}$ can be propagated to the PO z_i through different paths.

For a target fault to be considered as detected, we ensure that the correct value is propagated to the PO where the fault is detected. For example, if the polarized z-set of a fault $f1$ is $\{(z_1, 1), (z_3, 0)\}$, then the fault is considered to be detected only if either $\bar{D}$ is propagated to the PO z_1 or D is propagated to PO z_3. If a D is propagated to PO z_1 (and no effect propagated to z_3), then it is considered to be the fault effect of some other fault and fault $f1$ is considered to be undetected. For the PO having polarity x, the fault is considered to be detected if either D or $\bar{D}$ is propagated to that PO. Thus, an injected fault is considered detected only if a fault effect of the correct polarity is propagated to at least one of the gates in its polarized z-set.

If a PO $z1$ has a polarity 1, the constraint added for detection is

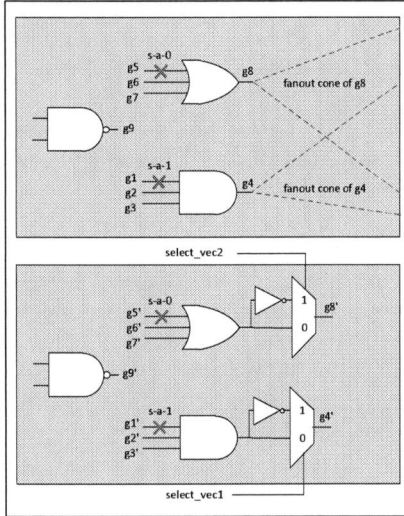

Fig. 7. Example of equivalent gates

$$(select\_vec1 \ and \ ((not \ z1) \ and \ z\_1)) \ \neq \ 00...0$$

Here $z1$ and z_1 are the corresponding gates in the fault free and faulty copy of the circuit. This constraint ensures that in at least one copy when $select\_vec1$ was 1 (i.e. the fault was excited and injected), the fault free value of $z1$ is 0 and faulty value is 1.

Similarly the constraint added if polarity of $z1$ is 0 is

$$(select\_vec1 \ and \ (z1 \ and \ (not \ z\_1))) \ \neq \ 00...0$$

If the polarity of $z1$ is x, then we simply add the XOR constraint explained in the earlier section:

$$(select\_vec1 \ and \ (z1 \ xor \ (not \ z\_1))) \ \neq \ 00...0$$

This constraint ensures that either D or $\bar{D}$ is propagated to $z1$.

To further mitigate the effect of the fault masking, in addition to the polarized z-sets, we add additional constraints to ensure that fault effect of correct polarity propagates to all the dominators of the target fault. The dominators for a signal g is a set of gates that any path from g to any PO must go through. We first find the dominators and their polarities for each fault. Then we add constraints similar to the constraints added for polarized z-set for the fault-free and faulty gate variables of each dominator of the faulty gate. This ensures that a fault is considered detected only if the correct fault effect is present at every dominator gate.

We note that the use of polarized z-sets and gate dominators does not completely eliminate the possibility of fault masking, but it helps to reduce masking considerably.

D. Equivalent Gates

Since we are using one fault-free and one faulty copy of the circuits, the number of variables needed to represent the gates in the SMT formula is equal to twice the number of gates in the circuit. By finding equivalent gates in the two copies of the circuit, we can enforce that the values of the corresponding gates are the same. This can reduce the time required by the SMT solver to find a satisfying model.

In order to find the equivalent gates between the fault-free and faulty circuits, we first find the fanout cones of all the m faults that are injected. We then find the union of all these m fanout cones. The gates which are not in union of these fanout cones are considered outside of the propagation cones. Therefore, they should always be at the same logic values. We replace the variables for these gates in the faulty copy of the circuit with the corresponding values in the fault-free circuit.

For example, Figure 7 shows two faults injected in the faulty circuit. These are faults at the inputs of gate $g4$ and $g8$. After finding the union of the fanout cones, we observe that gate $g9$ does not lie in the union. Hence gate $g9$ is unaffected by the injection of the two faults, and the value of $g9$ in the faulty copy of the circuit, $g9'$, can be safely replaced by the fault-free value $g9$. This can reduce the number of variables in the formula significantly and help to reduce the time required by the SMT solver to find a solution.

E. Benefits of the proposed technique

The following summarizes the benefits of our technique:

- Since all faults are taken care of in a single instance of the SMT formulation, there is no question of selection of which faults to target.
- The formulation has no constraint on which fault is detected by which of the k vectors.
- Our technique implicitly takes care of the vector-ordering problem in the previous linear-equation-based approach.
- The formulation is compact as it combines bit-level and word-level reasoning, eliminating the need to explicitly duplicate the bit-level circuit.

III. EXPERIMENTAL RESULTS

Experiments were performed on Red Hat Linux workstations with Intel® Pentium® D 3.0GHz CPU, 8GB memory and 4 cores. A number of ISCAS85 and full scan versions of ISCAS89 circuits were used to demonstrate the effectiveness of the proposed approach. SMT solver YICES [11] was used as the underlying SMT solver. A single LFSR polynomial was used for every circuit, and the LFSR had the same size as the number of inputs (including flops) of the circuit. The polynomials were referred from the library by Xilinx [13]. We note that the LFSR size can easily be modified and the overall formulation would be similar.

In Yices, the number of iterations during $max-sat$ can be controlled by setting $max\_num\_iterations\_in\_maxsat$ (referred to as n in Table II) parameter. If this bound is reached then Yices returns $unknown$ and prints the current best solution. By default, the value of this parameter is 10000. If value of this parameter is reduced then Yices terminates early but with suboptimal results.

We first fault simulate 1000 vectors starting from a random seed and drop all the detected faults. For the remaining set of faults we add the excitation and detection constraints. Table II shows the results for this technique. Column 1 reports

978-1-4673-0438-2/12 $31.00 © 2012 IEEE   398

TABLE II
EXPERIMENTAL RESULTS

circuit	total faults	faults det by init. seed	faults inj.	k=0				k=10				k=20			
				$n=10000$		$n=100$		$n=10000$		$n=100$		$n=10000$		$n=100$	
				seed	time	seed	time	seed	time	seed	time	seed	time	seed	time
c880	942	921	21	3	419	5	110	4	1110	3	346	3	6071	3	431
c1355	1574	1536	38	6	10346	6	3532	4	14794	5	6144	4	30917	5	14416
c1908	1879	1823	76	14	50129	16	12978	12	46546	10	18656	12	73002	14	18245
s953f	1079	969	110	17	98432	17	16238	17	126523	17	20615	15	183809	16	23451
s1423f	1515	1467	48	7	9663	11	3105	9	24414	14	6523	9	92346	13	5040
s1488f	1486	1444	42	15	142436	13	33215	10	127539	13	37533	13	159219	13	43804

the name of the circuit. Column 2 reports the total number of collapsed faults. Column 3 reports the number of faults detected by the initial random vectors. Column 4 reports the number of faults injected, which is m in our formulation. Column 5 reports the number of seeds needed to detect all the m faults, including the initial random seed when the number of vectors (i.e., faults injected + 'k' in our formulation) is set to be equal to the number of faults injected and n is at default value (i.e.,10000). Column 6 reports the cumulative execution time (in seconds) needed to generate all seeds. Columns 7 and 8 report the corresponding results when n is set to 100. Similarly, columns 9 to 16 report the number of seeds and time needed to generate these seeds for k equal to 10 and 20.

According to the results, it can be seen that very few seeds are needed to detect all the faults. For example, for circuit c1355 with a total of 1574 faults, 1536 faults were detected with an initial random seed, leaving 38 faults. From column 13 of Table II ($k = 20$) we see that to detect all remaining faults, 58 (38+20) copies of the circuit were used. The SMT solver (with $n=10000$) could not detect all 38 faults with a single seed, but 3 seeds were needed and found by the SMT approach. Together with the initial random seed, there is a total of 4 seeds. Note that in our SMT formulation, we need not detect a fault in each and every vector produced by the seed. Hence, it would be more flexible than the previous vector-chaining approaches. In a sense, it would be trying to chain a number of vectors such that we allow for any arbitrary number of don't-care vectors inserted between every pair of vectors. Finally, the time needed to generate a seed depends directly on the number of faults injected. Hence, the most amount of time is needed to generate the first seed and the time needed reduces after every seed since there would be fewer faults left after each seed. For c1355, 30917 seconds were taken by the SMT solver to compute the 3 seeds. If n is reduced to 100 then as expected the time needed for seed generation is reduced to 14416 seconds but now one additional seed is needed to detect all remaining faults.

From the table we can also see that the time needed for seed generation increases with larger values of k. For example, with $n = 10000$, the time needed for seed generation for c1355 is 10346, 14794 and 30917 seconds for k set to 0, 10 and 20, respectively. Also the number of seeds needed decreases with increase in k for most of the circuits. We would also like to point out the state of the SMT solvers is still in its infancy, and we believe the performance of SMT solvers will steadily increase in the coming years.

IV. CONCLUSION

We have proposed a new SMT-based formulation for the LFSR reseeding problem for LBIST. Instead of separate engines to compute the vectors and chaining them, our method unifies the two steps into one, eliminating the need to chain ATPG-generated vectors. Excitation and detection constraints are encoded as SMT constraints, and Polarized z-sets are proposed as well to enhance the distinguishability between detected faults. Preliminary results show potential and promise of the approach, whereby very few seeds are needed to achieve full coverages. The proposed approach also shows that a single LFSR polynomial with few seeds is sufficient to achieve complete coverage. Future work includes improved treatments of hyperactive but hard-to-detect faults.

REFERENCES

[1] B. Koenemann, "LFSR-coded test patterns for scan designs," in *European Test Symposium*, 1991.

[2] S. Hellebrand, B. Reeb, S. Tarnick, and H. Wunderlich, "Pattern generation for a deterministic BIST scheme," in *IEEE/ACM International Conference on Computer-Aided Design. ICCAD-95. Digest of Technical Papers.*, Nov. 1995, pp. 88 –94.

[3] P. Trouborst, "LFSR reseeding as a component of board level BIST," in *Proceedings, International Test Conference*, Oct. 1996, pp. 58 –67.

[4] S. Neophytou, M. Michael, and S. Tragoudas, "Efficient Deterministic Test Generation for BIST schemes with LFSR reseeding," in *12th IEEE International On-Line Testing Symposium, 2006. IOLTS 2006.*, p. 6 pp.

[5] Z. Wang, K. Chakrabarty, and M. Bienek, "A Seed-Selection Method to Increase Defect Coverage for LFSR-Reseeding-Based Test Compression," in *12th IEEE European Test Symposium. ETS '07.*, May 2007, pp. 125 –130.

[6] Y.-H. Fu and S.-J. Wang, "Test Data Compression with Partial LFSR-Reseeding," in *Proceedings. 14th Asian Test Symposium*, Dec. 2005, pp. 343 – 347.

[7] Y.-Z. Yan, H. Wang, Z.-J. Yang, and S. Yang, "A New LFSR Reseeding Method for BIST," in *8th International Conference on Solid-State and Integrated Circuit Technology. ICSICT '06.*, Oct. 2006, pp. 2145 –2147.

[8] C. Krishna and N. Touba, "Reducing Test Data Volume using LFSR Reseeding with Seed Compression," in *Proceedings. International Test Conference, 2002.*, pp. 321 – 330.

[9] S. Hellebrand, S. Tarnick, J. Rajski, and B. Courtois, "Generation Of Vector Patterns Through Reseeding Of Multiple-Polynomial Linear Feedback Shift Registers," in *Proc. of International Test Conference*, 1992, pp. 120–129.

[10] P. H. Bardell and W. H. McAnney, "Self-testing of multichip logic modules," in *Proceedings., International Test Conference.*, 1982, pp. 200–204.

[11] B. Dutertre and L. D. Moura, "The Yices SMT solver," Tech. Rep., 2006.

[12] I. Pomeranz, S. Venkataraman, S. Reddy, and B. Seshadri, "Z-Sets and Z-Detections: Circuit Characteristics that Simplify Fault Diagnosis," in *Proceedings. Design, Automation and Test in Europe Conference and Exhibition.*, vol. 1, Feb. 2004, pp. 68 – 73.

[13] (1996) Efficient Shift Registers, LFSR Counters, and Long Pseudo-Random Sequence Generators. [Online]. Available: http://www.xilinx.com/support/documentation/application_notes/xapp052.pdf

Two Graph based Circuit Simulator for PDE-Electrical Analogy

Yogesh Dilip Save *, H. Narayanan [†] and Sachin B. Patkar[‡]
Indian Institute of Technology, Bombay, Mumbai 400076.
*Email: *syogesh,[†]hn,[‡]patkar@ee.iitb.ac.in*

Abstract—The aim of the paper is to develop an efficient circuit simulator to solve circuits arising out of an electrical analogy for Partial Differential Equations (PDEs). This electrical analogy arises when we solve PDE through finite element method (FEM). The paper also proposes an optimal method for simulation of such circuits. We have built simulators based on Modified Nodal Analysis and Two Graph method for solution of PDEs through electrical analogy and compared their timing performance with commercial simulators. The timing performance of circuit simulators is improved for special PDE problems (such as Convection-diffusion) by an efficient implementation of iterative Cholesky with Two Graph method. The method is based on a graph representation of linear systems of equations. Such iterative methods would not be feasible with MNA. Using this method, we have been able to simulate circuits arising from the Convection-Diffusion problem with approximately 1.6 million nodes and 47 million edges in less than 8 minutes.

Keywords-Circuit Simulator, Partial Differential Equations, Electrical Analogy, Iterative Cholesky

I. INTRODUCTION

The electrical equivalent approach is useful in the analysis of coupled problems where electrical circuit and devices governed by Partial Differential Equations (PDEs) closely interact. For example, in VLSI design, a device cum circuit simulator helps circuit designers in predicting the effect of variation in technology on circuit design. It is natural therefore, to conceive of the solution of such coupled systems by converting them to a single network using electrical analogy [1]. Then the resultant circuit can be solved with the help of a circuit simulation tool. In this paper we describe a circuit simulator for efficient solution of circuits arising out of such electrical analogy for Partial Differential Equations.

The electrical analogy arises when we solve PDE through the finite element method (FEM) [1]. The resulting equations can be made to appear as those corresponding to an electrical network by transforming the 'elements' (triangles) of FEM into equivalent electrical subcircuits. The subcircuits combine together to form an equivalent circuit corresponding to PDE. The equivalent circuits contain static as well as dynamic devices. Generally for dynamic devices, a static model is constructed that is valid for a particular time point. Further the static electrical network is decomposed into linear and nonlinear devices. Generally for nonlinear devices, a linear model is constructed that is valid only locally around a point [2]. The overall performance of circuit

simulator depends on how circuit equation formulation is done for linear devices i.e., resistors, independent and dependent voltage and current sources. In this paper, we describe circuit formulation methods and identify an efficient method for electrical equivalent circuit arising from PDEs.

The circuit formulation method plays an important role in the performance of a circuit simulator. The formulation method needs to be flexible, computationally efficient, and economical with storage. The nodal approach for formulating circuit equations is a classical method which satisfies these requirements. But in its basic form it is incapable of handling circuits with voltage sources (which is used to represent boundary condition in the equivalent circuit of the PDE). The Modified Nodal Analysis (MNA) method [3] is an extended version of the nodal method which preserves its sparsity advantages. The coefficient matrix obtained with this method is sparse, but not symmetric and positive definite for resistive circuits. Another method called Two Graph method [4] exploits the properties of devices and derives two smaller graphs. The coefficient matrix obtained with this method is symmetric positive definite for resistive circuits. The Two Graph method is useful when the network contains a large number of voltage sources [5]. Thus, the method is very useful while handling electrical equivalent circuits arising from PDE problems in which a large number of boundary nodes are required to accurately represent boundary conditions. This method was originally designed for networks with resistors, current sources and voltage sources (RVJ). But the equivalent circuits arising from some special PDE problems e.g., convection-diffusion problem, also contain voltage controlled current sources. Therefore, modifications in the Two Graph Method are required to solve such kinds of PDE problems.

The main contribution of this paper is an efficient circuit formulation method for circuits arising from PDE-Electrical analogy. The paper also discusses an efficient solution technique to solve such circuits. The solution methodology of the present paper is based on graph theoretic techniques and different from those available in the literature. For example, in [6], permutation techniques for linear equations generated from FEM are presented.

In section II, we extend Two Graph Method to handle RVJ circuits (circuit containing resistor, independent current and voltage sources) with voltage controlled current sources. The method is explained through an example in section III.

978-1-4673-0438-2/12 $31.00 © 2012 IEEE

In section IV, some modifications in implementation and solution technique in Two Graph method are presented in order to solve circuits arising out of PDE problem efficiently.

II. TWO GRAPH METHOD FOR VOLTAGE CONTROLLED CURRENT SOURCE (VCCS)

The Two Graph method [4] is an elementary technique, which is very effective when the circuit contains a significant number of voltage sources. The Two Graph method has a number of advantages in terms of size of matrix and structure of matrix but it is capable of handling networks with resistors, voltage sources and current sources only. But the circuits arsing from PDE problem also contains voltage controlled current source. In this section, the Two Graph method is extended in order to handle circuits with voltage controlled current source (VCCS).

Two Graph method for VCCS is based on the idea of writing Kirchhoff's Current Equations (KCE) of underlying graph of DC network in terms of KCE of more than one of its contractions. First, we short circuit voltage source branches in the input circuit and write KCE for the network so obtained. Then we short circuit a tree of the underlying graph of this network in the original network and write KVE for it. Let $\mathcal{G}$ be the graph of the given network $\mathcal{N}$, containing no cutsets of current sources and no loops of voltage sources (to ensure unique solution). In voltage controlled devices, we have added a $0A$ current source as controlling branch without disturbing the incidence relationship of existing edges (i.e., the addition is 'soldering type') and its voltage is used for calculating the value of the devices.

Let $\mathcal{G}_1$ denote the graph obtained by short circuiting the independent voltage sources E. It follows that $\mathcal{G}_1$ contains no current source cutsets. Let $\mathcal{G}_t$ be a tree of $\mathcal{G}_1$. Let $\mathcal{G}_2$ be the graph obtained from $\mathcal{G}$ by short circuiting the branches in $\mathcal{G}_t$ and replacing them by self loops. Clearly, in $\mathcal{G}_2$, the voltage sources contain no loops. The computation of solution for linear circuits can be explained formally as follows:-

Let $\left[\mathbf{A}'_{rG}, \mathbf{A}'_{rJ}, \mathbf{A}'_{rT_1}\right]$ be a reduced incidence matrix of $\mathcal{G}_1$ (obtained by deleting one row per component) and let $\left[\mathbf{A}''_{rG}, \mathbf{A}''_{rJ}, \mathbf{A}''_{rT_1}, \mathbf{A}''_{rE}\right]$ be a reduced incidence matrix of $\mathcal{G}_2$ (T denotes the controlled branch).

The KCE of G are equivalent to

$$
\begin{bmatrix} \mathbf{A}'_{rG} & \mathbf{A}'_{rJ} & \mathbf{A}'_{rT_1} & \mathbf{0} \\ \mathbf{A}''_{rG} & \mathbf{A}''_{rJ} & \mathbf{A}''_{rT_1} & \mathbf{A}''_{rE} \end{bmatrix} \begin{bmatrix} \mathbf{i}_G \\ \mathbf{i}_J \\ \mathbf{i}_{T_1} \\ \mathbf{i}_E \end{bmatrix} = \begin{bmatrix} \mathbf{0} \\ \mathbf{0} \end{bmatrix} \quad (1)
$$

The KVE of G are equivalent to

$$
\begin{bmatrix} \mathbf{A}'^T_{rG} & \mathbf{A}''^T_{rG} \\ \mathbf{A}'^T_{rJ} & \mathbf{A}''^T_{rJ} \\ \mathbf{A}'^T_{rT_1} & \mathbf{A}''^T_{rT_1} \\ \mathbf{0} & \mathbf{A}''^T_{rE} \end{bmatrix} \begin{bmatrix} \mathbf{v}'_n \\ \mathbf{v}''_n \end{bmatrix} = \begin{bmatrix} \mathbf{v}_G \\ \mathbf{v}_J \\ \mathbf{v}_{T_1} \\ \mathbf{v}_E \end{bmatrix} \quad (2)
$$

The characterstic of resistor is given by,

$$
\mathbf{i}_G = \mathbf{G}\mathbf{v}_G \quad (3)
$$

From network constraints equation (1) and equation (2), and device characteristic (3), get

$$
\left(\mathbf{A}'_{rG}\mathbf{G}\mathbf{A}'^T_{rG}\right)\mathbf{v}'_n + \mathbf{A}'_{rT}\mathbf{i}_{T_1} = -\mathbf{A}'_{rJ}\mathbf{i}_J - \left(\mathbf{A}'_{rG}\mathbf{G}\mathbf{A}''^T_{rG}\right)\mathbf{v}''_n \quad (4)
$$

The characterstic of VCCS is given by,

$$
\mathbf{i}_{T_1} = \mathbf{T}_1\mathbf{v}_{C_1} = \mathbf{T}_1\left(\mathbf{A}'^T_{rC_1}\mathbf{v}'_n + \mathbf{A}''^T_{rC_1}\mathbf{v}''_n\right) \quad (5)
$$

Using value of $\mathbf{i}_{T_1}$, the equation (4) becomes

$$
\begin{aligned} (\mathbf{A}'_{rG}\mathbf{G}\mathbf{A}'^T_{rG} + \mathbf{A}'_{rT_1}\mathbf{T}_1\mathbf{A}'^T_{rC_1})\mathbf{v}'_n &= -\mathbf{A}'_{rJ}\mathbf{i}_J \\ -(\mathbf{A}'_{rG}\mathbf{G}\mathbf{A}''^T_{rG})\mathbf{v}''_n &- (\mathbf{A}'_{rT_1}\mathbf{T}_1\mathbf{A}''^T_{rC_1})\mathbf{v}''_n \end{aligned} \quad (6)
$$

$$
\mathbf{G}'_m\mathbf{v}'_n = -\mathbf{A}'_{rJ}\mathbf{i}_J - \mathbf{G}''_m\mathbf{v}''_n \quad (7)
$$

where,

$$
\mathbf{G}'_m = \mathbf{A}'_{rG}\mathbf{G}\mathbf{A}'^T_{rG} + \mathbf{A}'_{rT_1}\mathbf{T}_1\mathbf{A}'^T_{rC_1}
$$
$$
\mathbf{G}''_m = \mathbf{A}'_{rG}\mathbf{G}\mathbf{A}''^T_{rG} + \mathbf{A}'_{rT_1}\mathbf{T}_1\mathbf{A}''^T_{rC_1}
$$

In the equation (7), $\mathbf{v}'_n$ is the unknowns. The value of $\mathbf{v}''_n$ is computed from equation (2). In practice, $\mathbf{v}''_n$ is easily computed by graph operation on $\mathcal{G}_2$ (explained in section III).

III. ILLUSTRATIVE EXAMPLE

We illustrate the Two Graph method for VCCS with an example. In Figure 1 a linear circuit with VCCS is given.

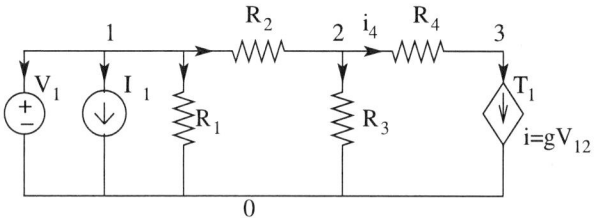

Figure 1.   linear circuit with VCCS ($\mathcal{N}$)

Shorting all the voltage sources in the network $\mathcal{N}$ results in the network as shown Figure 2. Note that node 0 of the network $\mathcal{N}_1$ is composed of (0,1) of $\mathcal{N}$. $\mathcal{N}_t$ is a tree of $\mathcal{N}_1$ composed of resistors and voltage sources.

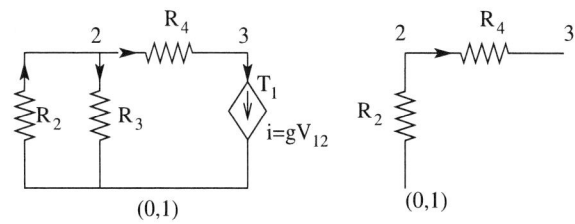

Figure 2. $\mathcal{N}_1$ network and its tree $\mathcal{N}_t$

978-1-4673-0438-2/12 $31.00 © 2012 IEEE

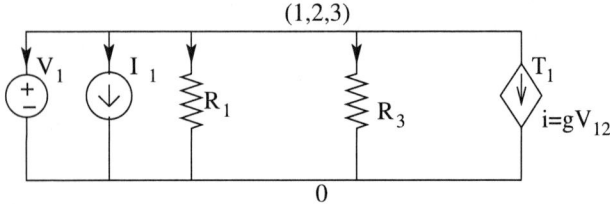

Figure 3. $\mathcal{N}_2$ Network

Shorting $\mathcal{N}_t$ in the network $\mathcal{N}$ results in the network $\mathcal{N}_2$ as shown in Figure 3.

First we compute the node voltages ($\mathbf{v}_n''$) of the network $\mathcal{N}_2$. Note that $\mathcal{N}_2$ has a tree composed entirely of voltage sources. Hence its node voltages can be computed in linear time using BFS [7]. Once the node voltages are known for $\mathcal{N}_2$, all its branch currents are computed. Then, the second term (involved $\mathbf{v}_n''$) in the RHS of equation (7) is computed. It is equal to the net current coming out of nodes in the network $\mathcal{N}_1$, where the currents through its branches are equal to the current flowing through them in the network $\mathcal{N}_2$.

In the next step, the first term in the RHS is computed. It is the contribution of the current sources present in the network $\mathcal{N}_1$.

We form the matrix on the LHS side explicitly. The matrix is stored in sparse storage format (column compressed format) [8]. Then its LU factors are computed using standard LU routine. Using Backward and Forward substitution, unknowns are computed. Having computed value of the node voltages, branch voltages are computed. Then the resistor branch current is obtained from relationship $\mathbf{i} = \mathbf{G}\mathbf{v}_G$. Using device characteristic of VCCS current through it is obtained. Finally voltage source branch currents are computed from equation (1). Then KCL check is performed on each node to verify the solution of the network.

IV. TWO GRAPH METHOD WITH ITERATIVE CHOLESKY SOLVER

The circuit arising from PDE contains resistors, independent current and voltage sources, and voltage controlled current sources (VCCS). Due to presence of VCCS the system matrix resulting from Two Graph method becomes asymmetric. This increases the storage requirement. Also, the system matrix may not be positive definite. If we consider the value of controlling voltage as known then VCCS become simple current sources and the system matrix corresponding to such circuit becomes symmetric and positive definite. Then one can solve such circuit iteratively with updated value of VCCS to obtain final solution. In this section, we explain modification in implementation of Two Graph method and solution techniques to solve the circuit resulting from PDE problem while preserving the advantages of Two Graph method.

Consider a circuit with resistors, independent current and voltage sources, and VCCS. Then equation (4) becomes

$$\left(\mathbf{A}_{rG}'\mathbf{G}\mathbf{A}_{rG}'^T\right)\mathbf{v}_n' + \mathbf{A}_{rT_1}'\mathbf{i}_{T_1} = -\mathbf{A}_{rJ}'\mathbf{i}_J - \left(\mathbf{A}_{rG}'\mathbf{G}\mathbf{A}_{rG}''^T\right)\mathbf{v}_n'' \tag{8}$$

Using device characteristics (5) and equation (2), the equation (8) becomes

$$(\mathbf{A}_{rG}'\mathbf{G}\mathbf{A}_{rG}'^T + \mathbf{A}_{rT_1}'\mathbf{T}_1\mathbf{A}_{rC_1}'^T)\mathbf{v}_n'$$
$$= -\mathbf{A}_{rJ}'\mathbf{i}_J - (\mathbf{A}_{rG}'\mathbf{G}\mathbf{A}_{rG}''^T)\mathbf{v}_n'' - (\mathbf{A}_{rT_1}'\mathbf{T}_1\mathbf{A}_{rC_1}''^T)\mathbf{v}_n'' \tag{9}$$

$$(\mathbf{M} - \mathbf{N})\mathbf{v}_n' = \mathbf{b} \tag{10}$$

where,

$$\mathbf{M} = \mathbf{A}_{rG}'\mathbf{G}\mathbf{A}_{rG}'^T \qquad \mathbf{N} = -\mathbf{A}_{rT_1}'\mathbf{T}_1\mathbf{A}_{rC_1}'^T$$
$$\mathbf{b} = -\mathbf{A}_{rJ}'\mathbf{i}_J - (\mathbf{A}_{rG}'\mathbf{G}\mathbf{A}_{rG}''^T)\mathbf{v}_n'' - (\mathbf{A}_{rT_1}'\mathbf{T}_1\mathbf{A}_{rC_1}''^T)\mathbf{v}_n''$$

$$\mathbf{M}\mathbf{v}_n' = \mathbf{b} + \mathbf{N}\mathbf{v}_n' \tag{11}$$

We can solve equation (11) iteratively by treating right side as known. Thus, at $(k+1)^{th}$ iteration,

$$\mathbf{M}\mathbf{v}_n'^{(k+1)} = \mathbf{b} + \mathbf{N}\mathbf{v}_n'^{(k)} \tag{12}$$

Here, $\mathbf{M} = \mathbf{A}_{rG}'\mathbf{G}\mathbf{A}_{rG}'^T$ is symmetric positive definite matrix and easy to form. Thus, we can find Cholesky factor of $\mathbf{M}$. By putting $\mathbf{M} = \mathbf{L}\mathbf{L}^T$ in equation (12),

$$\mathbf{L}\mathbf{L}^T\mathbf{v}_n'^{(k+1)} = \mathbf{b} + \mathbf{N}\mathbf{v}_n'^{(k)} \tag{13}$$

Once Cholesky factorization of $\mathbf{M}$ is found, we can reuse the same factor in every iteration.

Let, $\mathbf{L}\mathbf{L}^T\mathbf{v}_n' = \mathbf{z}$ and $\mathbf{L}^T\mathbf{v}_n' = \mathbf{y}$.

$$\mathbf{L}\mathbf{y} = \mathbf{z} \tag{14}$$

By forward substitution, $\mathbf{y}$ can be easily found.

$$\mathbf{L}^T\mathbf{v}_n' = \mathbf{y} \tag{15}$$

By backward substitution, $\mathbf{v}_n'$ can be easily found.

Formation of right side

The right side of equation (13) contains fixed term ($\mathbf{b}$) and variable term ($\mathbf{N}\mathbf{v}_n'^{(k)}$). Computation of $\mathbf{b} = -\mathbf{A}_{rJ}'\mathbf{i}_J - (\mathbf{A}_{rG}'\mathbf{G}\mathbf{A}_{rG}''^T)\mathbf{v}_n'' - (\mathbf{A}_{rT_1}'\mathbf{T}_1\mathbf{A}_{rC_1}''^T)\mathbf{v}_n''$ is already discussed in section III. Here, we discuss how to compute $\mathbf{N}\mathbf{v}_n'^{(k)}$ efficiently, where $\mathbf{N} = -\mathbf{A}_{rT_1}'\mathbf{T}_1\mathbf{A}_{rC_1}'^T$. Thus, we need to compute $-\mathbf{A}_{rT_1}'\mathbf{T}_1\mathbf{A}_{rC_1}'^T\mathbf{v}_n'^{(k)}$. Here, $\mathbf{A}_{rC_1}'^T\mathbf{v}_n'^{(k)}$ is branch voltages of controlling branch and easily found using node voltages. $\mathbf{T}_1$ is a diagonal matrix. Thus, $\mathbf{T}_1(\mathbf{A}_{rC_1}'^T\mathbf{v}_n'^{(k)})$ is just a scaling of the branch voltages. We can treat them as branch currents. Then, $-\mathbf{A}_{rT_1}'(\mathbf{T}_1\mathbf{A}_{rC_1}'^T\mathbf{v}_n'^{(k)})$ are the currents entering into the nodes.

978-1-4673-0438-2/12 $31.00 © 2012 IEEE

V. EXPERIMENTAL RESULTS

In this section, we present the results of experiments performed on general purpose circuits using in-house simulators based on MNA and Two Graph method and a comparison of its performance with standard circuit simulators (C1 and C2). We also present the results of experiments performed on circuits arising from PDE electrical analogy to determine best circuit equation formulation method and linear solution technique for a circuit simulator to solve PDE problems.

Experiments were performed on a 3 GHz Pentium-IV processor with 4 GB RAM. In our case we use KCL error tolerance of 10^{-6}. We have used "Triangle" software [9] for mesh generation. In the table, the entity $Triangle$ represents the number of triangles after meshing. The entity $Edges$ gives the number edges and $Nodes$ gives the number nodes in the planar electrical circuit. t_{MNA}, t_{TG}, t_{c1} and t_{c2} are the time taken in seconds by our simulators (MNA and Two Graph method) and commercial simulators (C1 and C2) respectively to solve the circuit.

Table I contains results on the circuit with resistors and current sources comparing the timing performance of our simulators with that of commercial simulators. Our simulators are much faster than the commercial simulators. There are no difference in the timing performance of our circuit simulators (MNA and TG) for RJ circuits. The performance of our simulators is linear. This is because we have used sparse LU method to solve a linear system of equation and the method works well on a planar network [5].

Table I
RJ CIRCUIT WITH 20% CURRENT SOURCES

$Nodes$	$Edges$	t_{c1}	t_{c2}	t_{MNA}	t_{TG}
1e3	2.50e3	0.2	0.17	0.03	0.02
5e3	1.25e4	2.17	0.74	0.16	0.16
1e4	2.50e4	7.94	1.61	0.35	0.35
5e4	1.25e5	328	13.6	2.02	1.94
1e5	2.50e5	1676	50.8	4.25	4.24

Table II contains results on the circuit with resistors, voltage sources (20%) and current sources (20%) comparing the timing performance with that of commercial simulators. It clearly shows that in the presence of voltage sources the performance of Two Graph based simulator improves in contrast to MNA based simulators. Figure 4 shows the comparison of circuit simulators for RJ and RVJ circuits.

Table II
RVJ CIRCUIT WITH 20% CURRENT AND 20% VOLTAGE SOURCES

$Nodes$	$Edges$	t_{c1}	t_{c2}	t_{MNA}	t_{TG}
1e3	2.50e3	0.25	0.22	0.03	0.02
5e3	1.25e4	3.34	1.84	0.17	0.13
1e4	2.50e4	14.7	8.51	0.38	0.28
5e4	1.25e5	716	281	2.2	1.58
1e5	2.50e5	3576	1220	4.7	3.66

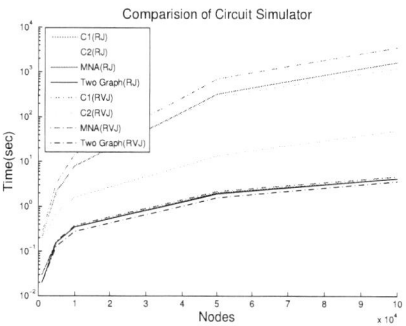

Figure 4.   Comparison of Simulators for RJ and RVJ circuits

Figure 5 shows the approximation of geometry (hole in a square box) with different number of boundary points for different regions. To approximate outer part i.e., square (regular shape) a small number of boundary points are sufficient. On other hand to model inner part i.e., circle, a large number boundary points are required. The approximation of complex geometry improves with more boundary points. Figure 5 which gives reasonable approximation of contains boundary points equal to 10% of total number of nodes. Also, generally a variation of solution is more near curved surface. To get an accurate solution finer meshing is required. This eventually results into a large number of boundary points. In electrical analogy presented in [1], voltage sources are used to represent the boundary condition. So Two Graph method based simulator gives good timing performance in case of PDE problem with complex geometry.

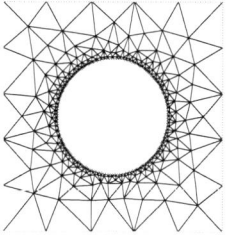

Figure 5.   Approximation of circle with 200 points on boundary

To evaluate performance of Two Graph based simulator for solution of PDE problems we have generated equivalent circuits for Poisson problem. Poisson equation has wide application in electrostatic, electromagnetics, mechanical system and so on. The problem in the given domain (Figure 5) is defined by the equation,

$$-\nabla(c\nabla u) = f \qquad (16)$$

The equivalent electrical circuit corresponding to the PDE is obtained through the FEM. Firstly an electrical subcircuit corresponding to each element (triangle) is derived from the equations arising from FEM and then combine them

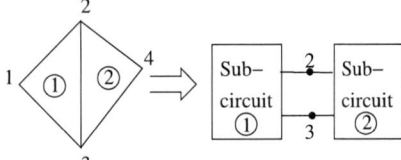

Figure 6. Equivalent circuit for PDE

appropriately as shown in Figure 6. Boundary conditions are imposed with voltage sources. The equivalent circuit corresponding to an element is as shown in Figure 7. The

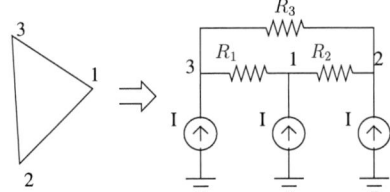

Figure 7. Circuit corresponding to Poisson Equation for a single element

value of components is given in [1]. In order to vary size of circuit we have varied area of element. Table III contains results on equivalent circuit generated from the problem. The entity Bnd represents the number of boundary nodes. From the table, it is clear that the Two Graph method based simulator performs better at large node sizes.

Table III
SIMULATION RESULTS ON CIRCUIT GENERATED FROM POISSON PDE

Nodes	Triangle	Edges	Bnd	t_{MNA}	t_{TG}
34601	67550	238893	1650	2.81	2.68
100380	197658	696447	3100	10.07	8.99
256603	507404	1784606	5800	31.05	23.8
638954	1266906	4450650	11000	76.54	63.2

To evaluate performance of Two Graph based simulator for solution of various kind of PDE problems we have generated equivalent circuits for an elliptical PDE. The problem in a square domain is defined by the equation,

$$-\nabla(c\nabla u) + ku = f \qquad (17)$$

The equivalent circuit corresponding to an element (triangle) is as shown in Figure 8. The value of components is given in [1]. Table IV contains results on equivalent circuit generated from the problem. The Two Graph method based simulator performs better due to presence of voltage sources.

Table V contains results on the linear circuit with voltage controlled current sources comparing the timing performance with that of commercial simulators. Here, the number of controlled source branches is kept at 5% of total number of circuit edges. From these results it is clear that the performance of our simulators is better than that of

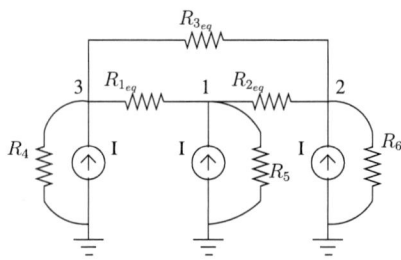

Figure 8. Circuit corresponding to Elliptical PDE for a single element

Table IV
SIMULATION RESULTS ON CIRCUIT GENERATED FROM ELLIPTICAL PDE

Nodes	Triangle	Edges	t_{MNA}	t_{TG}
8.80e1	1.50e2	2.35e2	0.01	0.01
8.09e2	1.54e3	2.35e3	0.12	0.12
7.92e3	1.55e4	2.35e4	1.5	1.31
7.79e4	1.55e5	2.35e5	16.1	14.2

commercial simulators. Also timing performance of the our simulators is linear.

Table V
RVJ CIRCUIT WITH VOLTAGE CONTROLLED CURRENT SOURCES

Nodes	Edges	t_{c1}	t_{c2}	t_{MNA}	t_{TG}
1e3	2.50e3	0.2	0.02	0.03	0.02
1e4	2.50e4	20.5	0.78	0.35	0.34
5e4	1.25e5	1616	298	2.11	1.95
1e5	2.50e5	-	1265	4.48	3.98
1e6	2.50e6	-	-	65.9	50.3

To evaluate performance of Two Graph based simulator for solution of PDE problems we have generated equivalent circuits for Convection-Diffusion problem. The problem in the given domain is defined by the equation,

$$-\nabla(c\nabla u) + k\nabla u = f \qquad (18)$$

The equivalent circuit corresponding to an element (trian-

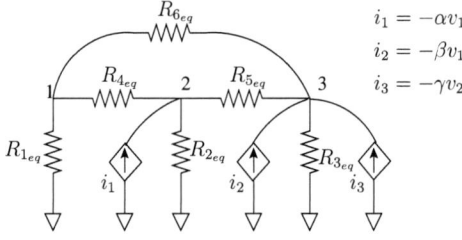

Figure 9. Circuit corresponding to Convection-Diffusion for an element

gle) is as shown in Figure 9. The value of components is given in [1]. In order to vary size of circuit we have varied area of mesh. Table VI contains results on equivalent circuit generated from the problem. The equivalent circuit contains voltage sources, current sources, resistors and voltage controlled current sources [1]. The timing performance of MNA and Two Graph based simulator is not linear at large size. This is because the equivalent circuit contains a large number of voltage controlled current source (nearly 20%).

We have integrated iterative Cholesky method with Two Graph based simulator as discussed in section IV. However, iterative Cholesky will not work efficiently with MNA because presence of voltage sources leads asymmetry and non-positive definiteness. In the table, t_{Chol} is the time taken in seconds by Two Graph based simulator with iterative Cholesky to solve the circuit. The entity Itr represents the number iterations required to converge the solution. The timing performance of the our simulator is linear as shown in Figure 10. The number of iteration required to converge the solution is less than 10 for all the cases.

Table VI
SIMULATION RESULTS ON CIRCUIT GENERATED FROM CONVECTION-DIFFUSION PROBLEM

Nodes	Edges	VCCS	t_{MNA}	t_{TG}	Itr	t_{Chol}
8.80e1	2.30e3	4.50e2	0.01	0.01	6	0.01
8.09e2	2.35e4	4.63e3	0.12	0.11	7	0.15
7.79e4	2.35e6	4.65e5	16.1	15.6	9	17.2
7.77e5	2.35e7	4.65e6	378	340	10	218
1.55e6	4.70e7	9.31e6	984	946	10	457

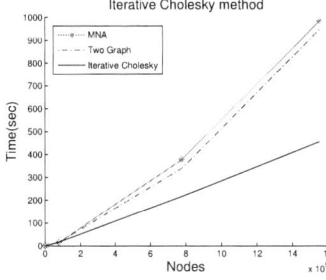

Figure 10.   Comparison of Simulators for Convection-Diffusion problem

VI. CONCLUSION

In this paper we have described a circuit simulator to solve circuits arising out of an electrical analogy for Partial Differential Equations (PDEs). We have presented circuit equation formulation methods for simulation of linear circuits. We have built two in-house simulators based on Modified Nodal Analysis and Two Graph method and compared timing performance of our simulators with commercial simulators. It has been found that for larger sizes, our simulators are much faster than commercial simulators. For RJ circuit (circuit containing Resistor and current source only) the timing performance of both of our simulators is same. But when circuit contains voltage sources(V) in addition to R, J, the simulator based on Two Graph method outperformed MNA based simulator. This is because Two Graph method takes advantage of the presence of voltage sources and produces a symmetric positive definite system matrix (representing

circuit equations) with less number of unknowns. Thus, the simulator based on Two Graph method is useful to handle circuits arising out of PDE problems with complex geometry as to represent such geometry a large number boundary points are required which translates to large number of voltage sources.

Two Graph method in its basic form is capable of handling only RVJ circuits. We have extended the method to solve circuits containing voltage controlled current sources. Timing performance of the simulator on RVJ circuit with VCCS (nearly 5% of number of edges) is linear. But with circuits arising from PDE problems, timing performance is not linear at higher node sizes. This is because such circuits contain large number of VCCS (nearly 20 %). We used a special method (Iterative Cholesky) to solve for circuit containing RVJ and VCCS. Implementation of this method is based on an efficient application of graph theoretical techniques. Though this method does not converge for general purpose problems, it produces excellent results on circuits arising out of PDE problems.

At present, we have used sparse LU and Cholesky method for direct solution. We need to explore the use of iterative methods such as the Preconditioned Conjugate Gradient method, BiCG, GMRES etc. for PDE problems.

REFERENCES

[1] Y. D. Save, H. Narayanan, and S. B. Patkar, "Solution of Partial Differential Equations by electrical analogy," *Journal of Computational Science*, vol. 2, no. 1, pp. 18 – 30, 2011.

[2] E. J. Mastascusa, *Computer-Assisted Network and System Analysis.*   John Wiley and Sons, 1987.

[3] C. W. Ho, A. E. Ruehli, and P. A. Brennan, "The modified nodal approach to network analysis," *IEEE Trans. on Circuits and Systems*, vol. 22, pp. 504–509, Jun 1975.

[4] S. Batterywala and H. Narayanan, "Efficient DC analysis of RVJ circuits for moment and derivative computations of interconnect networks," in *12th International Conference on VLSI Design, 1999*, Jan. 1999, pp. 169 –174.

[5] G. Trivedi, M. P. Desai, and H. Narayanan, "Fast DC analysis and its application to combinatorial optimization problems," in *Proc. of the 19th Intl. Conf. on VLSI Design*, 2006, p. 695.

[6] O. Schenk, S. Rollin, and A. Gupta, "The effects of unsymmetric matrix permutations and scalings in semiconductor device and circuit simulation," *IEEE Trans. on Comput.-Aided Design Integr. Circuits Syst.*, vol. 23, no. 3, pp. 400 – 411, March 2004.

[7] H. Narayanan, *Submodular functions and electrical networks*. Amsterdam: North-Holland, 1997.

[8] W. J. McCalla, *Fundamentals of computer-aided circuit simulation.*   Deventer, The Netherlands: Kluwer, Boston, 1988.

[9] J. R. Shewchuk, "Triangle: Engineering a 2D quality mesh generator and delaunay triangulator," in *Applied Computational Geometry: Towards Geometric Engg.*, May 1996, pp. 203–222.

978-1-4673-0438-2/12 $31.00 © 2012 IEEE

2012 25th International Conference on VLSI Design

Modeling of Partially Depleted SOI DEMOSFETs With a Sub-circuit Utilizing the HiSIM-HV Compact Model

Tarun Kumar Agarwal and M. Jagadesh Kumar, *Senior Member*, IEEE

Department of Electrical Engineering

Indian Institute of Technology, Delhi

New Delhi, India 110016

Email: tarun.agrawal90@gmail.com; mamidala@ieee.org

Abstract

This paper presents a sub-circuit model for partially depleted SOI drain extended MOSFETs (DEMOS) based on the HiSIM-HV model suitable for circuit simulator implementation. Our model accounts both for the high voltage and the floating body effects such as the quasi saturation effect, the impact ionization in the drift region and the famous kink effect. The model is validated for a set of channel and drift lengths to demonstrate the scalability of the model. The accuracy of the proposed sub-circuit model is verified using 2-D numerical simulations.

1. Introduction

Recently, high-voltage (HV) devices such as the drain-extended (DE) MOS device has been commonly used in many smart power IC's. Example circuits are the input/ouput interface circuits and power management switches that regulate power from battery or system supplies [1]. To optimally design the power circuits, an accurate model for PD SOI DEMOSFETs is required including the high voltage and the floating body effects.

To model bulk HV MOS and SOI HV MOS with body contact, various modeling approaches are reported in literature such as HV-EKV, MM20, MOOSE and HiSIM-HV. To incorporate the new effects in the existing transistor models, a frequently followed approach is to use a sub-circuit. Some compact models are examples of this approach such as BSIM3SOI, PSP-SOI and HiSIM-SOI. These models are formulated within the framework of the bulk MOSFET compact models such as BSIM3, PSP and HiSIM and account for the floating body effects observed in the partially depleted SOI devices. Moreover, the HiSIM-HV model [2], an industry standard surface potential

based compact model, accounts for the high voltage effects. To the best of our knowledge, there is no surface potential based compact model in the literature for PD SOI DEMOSFETs taking into account the floating body and the high voltage effects simultaneously.

The aim of this paper is, therefore, to develop a sub-circuit model for PD SOI DEMOSFETs simultaneously considering all the above special dc effects by utilizing the HiSIM-HV compact model. The accuracy of the proposed sub circuit model is verified by 2-D device simulations.

2. Device structure and simulation results

Fig. 1(a) shows the cross section of a floating body PD SOI DEMOS device, which with a body contact shows the same dc behavior as a bulk high-voltage MOS device shown in Fig. 1(b) except the self-heating. On the other hand, without a body contact, the dc behavior of floating body PD SOI DEMOS devices can be categorized into the high voltage and the floating body effects. And, to analyze the special dc behavior of floating body PD SOI DEMOSFETs, the transistor is divided into the surface MOS and the parasitic BJT region as shown in Fig. 1(a). The device parameters for simulation are given in Table. 1 [3]. The device simulations are done using ATLAS, a 2-D numerical simulator. The drift region doping for the reference device is chosen to have a breakdown voltage higher than 15 V for the floating body device and 20 V for the body contacted device.

The surface MOS region of a PD SOI DEMOS device can be further divided into the channel and the drift region. This classification and the concept of the intrinsic drain potential (V_K) have been a powerful tool to understand the high voltage effects such as the quasi saturation effect and the impact ionization in the drift region [4]. As Fig. 1(a) shows, V_K denotes the surface

978-1-4673-0438-2/12 $31.00 © 2012 IEEE   406

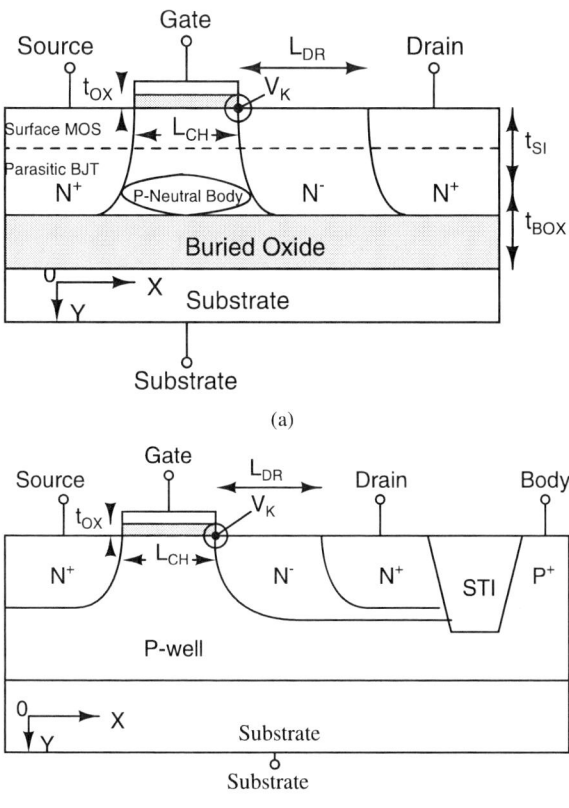

Figure 1. Cross sectional schematics of the n-channel devices. (a) A floating body partially depleted drain extended MOSFET (DEMOS). (b) A drain extended bulk MOSFET.

Table 1. Device parameters used in the simulation, derived from a reference device in [3].

Symbol	Description	Value
t_{OX}	Gate oxide thickness	30 nm
t_{SI}	Silicon film thickness	180 nm
t_{BOX}	Buried oxide thickness	1 μm
N_{ch}	Channel doping concentration	$1.5 \times 10^{17}\ cm^{-3}$
N_{dr}	Drift region doping concentration	$4 \times 10^{16}\ cm^{-3}$
L_{CH}	Channel length	0.64 μm
L_{DR}	Drift region length	1 μm

potential at the junction of the channel region and the drift region and Fig. 2 shows the variation of V_K with the gate and the drain voltage. Based on the variation of V_K with the gate voltage, the two modes of operation for HV MOSFETs are defined. First being, a low voltage FET (the channel region) dominant mode of operation in which V_K increases with the gate voltage due to an increase in the surface potential and reaches to a maximum value (V_{KMAX}). The second mode of operation is when the drift region starts dominating and

V_K starts decreasing from V_{KMAX} with an increase in the gate voltage. Based on the variation of V_K, the quasi saturation and the impact ionization in the drift region are physically explained using Kirk effect [5].

It is observed that in the low-voltage FET mode of a floating body PD SOI DEMOS transistor operation, with an increase in the drain voltage, impact ionization at the body-drain junction increases with increase in number of holes collected by the floating body, which in effect increases the body potential and lowers the threshold voltage (V_T) of the device, resulting in a kink in the output characteristics of floating body PD SOI DEMOSFETs. As it is shown in Fig. 3(a), in the drift dominant mode of operation, the floating body potential rises at lower drain voltages due to the impact ionization in the drift region and the kink in the output characteristics occurs at lower drain voltages as shown in Fig. 3(b). In conclusion, to accurately model the kink behavior in the output characteristics, all the body current components, mainly the impact ionization current, are to be accurately modeled.

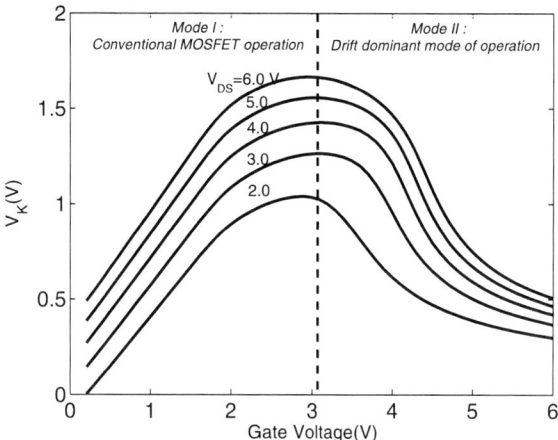

Figure 2. Variation of V_K with gate bias the gate and drain biases obtained from ATLAS.

3. Modeling strategy

Based on the physical insights gained from the simulations, a model for partially depleted SOI DE-MOSFETs can be derived from a HiSIM-HV model and a sub-circuit. Here, the quasi saturation effect and the impact ionization in the drift region are taken into account in HiSIM-HV model by using a bias dependent resistance and a current source for the impact ionization in the drift region.

Fig. 4 shows the SPICE sub circuit implementation, containing a bulk high voltage MOSFET (accounting

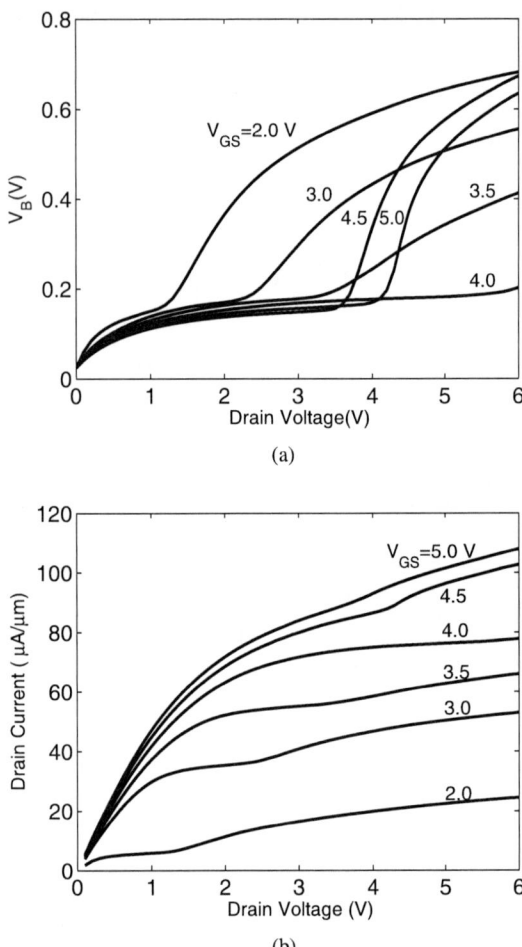

(a)

(b)

Figure 3. Numerically simulated characteristics of a floating body PD SOI DEMOSFET (a) Body potential variation with drain bias. (b) Output characteristics showing kink effect.

for the surface MOS region), a pair of diodes, a parasitic BJT and the external current-controlled current source. As modeling of the breakdown phenomenon is not taken in consideration, the drain to source current (I_{DS}) and the body current (I_B) are expressed explicitly in terms of the external node voltages at the nodes D (Drain), G (Gate), S (Source), B (Body) and X (Substrate). Here, X is connected to ground.

3.1. Bulk high-voltage MOSFET or surface MOS region modeling

A body-contacted PD SOI DEMOS device shows the same dc behavior as a bulk high-voltage MOSFET shown in Fig. 1(a) except the self-heating. The surface MOS region (the channel and the drift region), which

results in high voltage effects, is modeled using a HiSIM-HV compact model. The HiSIM-HV model shows good capability in modeling the high voltage effects such as quasi saturation with channel and drift length scaling. In the HiSIM-HV model, the channel

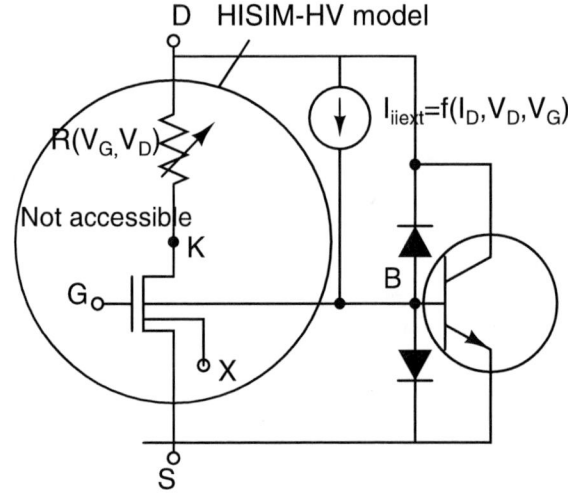

Figure 4. Sub-circuit implementation in SPICE with the HiSIM-HV compact model.

region is modeled using a surface potential based model, HiSIM2, and the drift region is modeled using a bias-dependent resistance. The HiSIM-HV model also accounts for the impact ionization in the drift region by adding an extra current source which is a function of the drain to source current (I_{DS}) and the surface potentials at the nodes D, K and S.

3.2. Floating body effect and parasitic BJT modeling

Based on the physical insights gained from the simulations, it is concluded that to model the floating body effects observed in a PD SOI DEMOS device, the floating body potential needs to be accurately calculated. And, to accurately calculate the inferred body potential in the floating body PD SOI DEMOS transistor, all the body current components such as the impact ionization current, diode current and the parasitic BJT currents need to be accurately modeled. Therefore, to accurately calculate the diode and the parasitic BJT currents, BSIM3SOI modeling strategy is used [6]. The accurate modeling of the impact ionization current is given in the next sub-section.

3.3. Accurate modeling of the impact ionization in the drift region

To accurately model the impact ionization in the drift region, an external current-controlled current source is added in the sub-circuit as shown in Fig. 4. The equation of the external current controlled current source used in this sub-circuit is developed from a well known equation [7]: A general body current equation due to impact ionization in the drift region can be approximated as:

$$I_{ii2} = alpha \cdot (V_D - (EC*l)) \cdot I_{DS} \cdot exp\left(-\frac{B \cdot L_{DR}}{(V_D - (EC*l))}\right) \quad (1)$$

Here, $EC * l$ is the potential at y=0 (at body-drift junction) and V_D is the potential at y=L_{DR} where, l is the channel length of the device. Since I_B increases with gate voltage continuously until the device breaks down after the transistor enters the triode region. To model this behavior, eq. 1 is modified as:

$$I_{ii2} = alpha \cdot (V_D - (EC*l)) \cdot I_{DS} \cdot exp\left(-\frac{B}{E_{eff}}\right) \quad (2)$$

where,

$$E_{eff} = \left(\frac{(V_D - (EC*l))^{coeff2} \cdot (V_{GS} - V_{OFF})^{coeff1}}{L_{DR}}\right) \quad (3)$$

Model parameters in the current-controlled current source model (eq. 2) are *alpha, B, V_{OFF}, EC, coeff1* and *coeff2*.

4. Results and discussion

The model parameters of the developed model are extracted using a commercial extraction software package, ICCAP, after implementing the SPICE sub-circuit shown in Fig. 4 in Spectre. The potential of the floating node "B" is calculated within the simulator using the in-built model, HiSIM-HV version 1.11, and the sub-circuit. A body-contacted PD SOI DEMOS device minimizes the floating body and the parasitic BJT effects. Therefore, the model parameters of HiSIM-HV compact model are extracted using the scalable data for the body-contacted PD SOI DEMOS device. The HiSIM-HV compact model is capable of modeling the quasi saturation. And, to model the floating body effects and the impact ionization in the drift region, a sub-circuit consisting of a pair of diodes, a parasitic BJT and an external current-controlled current source is added to the HiSIM-HV compact model, as shown in Fig. 4. The calculated results from the proposed model

are calibrated with the 2-D numerically simulated characteristics of both the body contacted and the floating body PD SOI DEMOSFETs for different dimensions. Fig. 5 shows that the calculated output character-

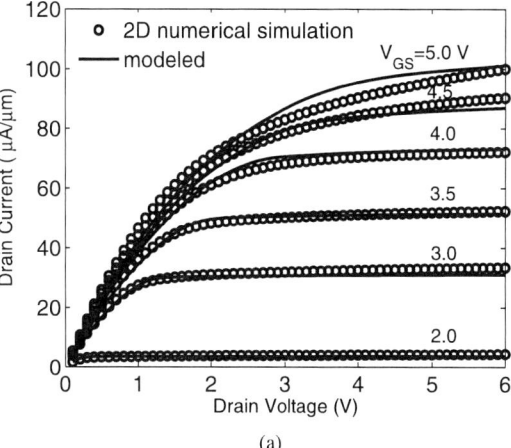

(a)

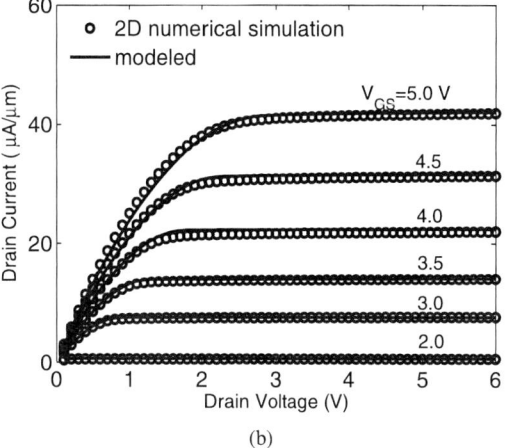

(b)

Figure 5. (symbols) 2-D numerically simulated and (solid lines) calculated output characteristics of a body contacted PD SOI DEMOS using HiSIM-HV approach for (a) L_{CH}=0.64 μm and L_{DR}=1 μm. (b) L_{CH}=3 μm and L_{DR}=1 μm.

istics of the body contacted transistor have a good consistency with the 2-D numerically simulated data for different channel lengths. The HiSIM-HV model captures the quasi saturation effect and the impact ionization in the drift region with the internal drift resistance model and an extra impact ionization current, respectively. Moreover, including an external current-controlled current source in the proposed sub-circuit, the kink behavior in the output characteristics is accurately modeled as shown in Fig. 6(b).

Thus, it is clear from Fig. 6 that accurate modeling

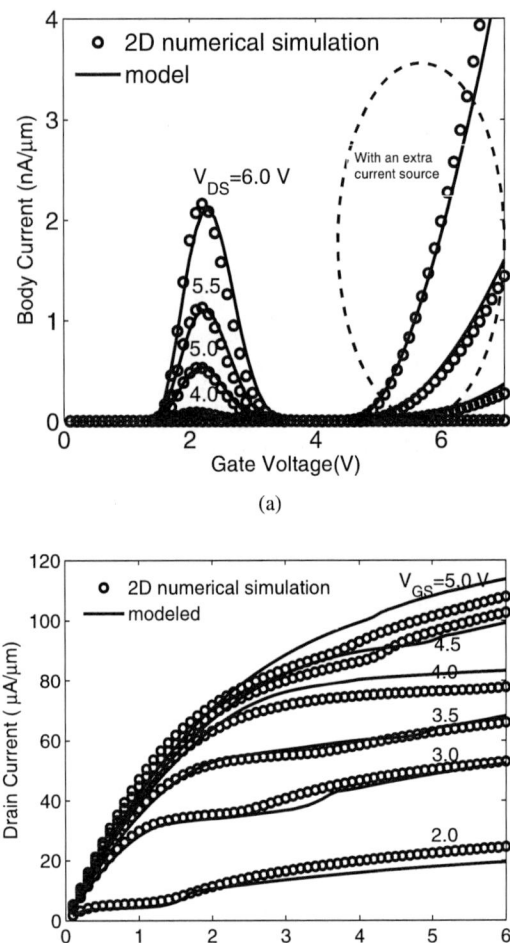

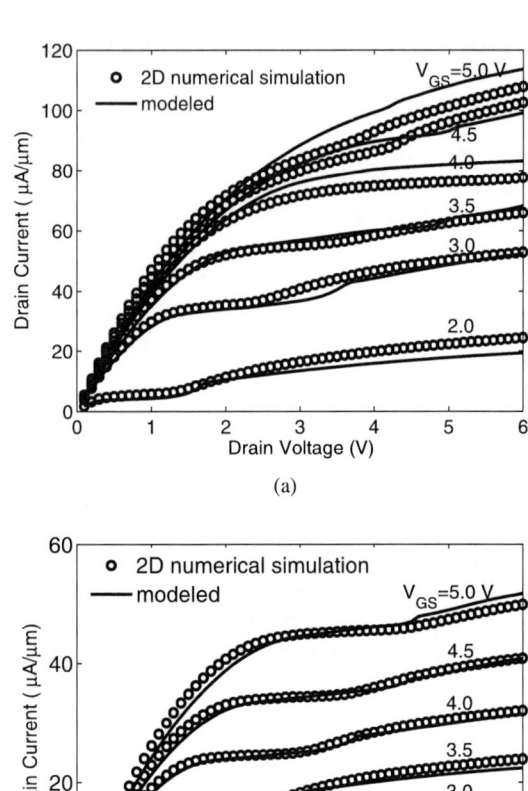

Figure 6. (symbols) 2-D numerically simulated and (solid lines) calculated characteristics of a PD SOI DEMOS transistor using HiSIM-HV approach for L_{CH}=0.64 μm and L_{DR}=1 μm. (a) the body current characteristics showing the impact ionization in the drift region. (b) the output characteristics showing the kink effect.

Figure 7. (symbols) 2-D numerically simulated and (solid lines) calculated output characteristics of a floating body PD SOI DEMOS using HiSIM-HV approach for (a) L_{CH}=0.64 μm and L_{DR}=1 μm. (b) L_{CH}=3 μm and L_{DR}=1 μm.

of the impact ionization in the drift region gives better fitting of the kink behavior in the output characteristics of the floating body device. Fig. 7 shows that the calculated output characteristics of the floating body PD SOI DEMOS transistor have a good consistency with the 2-D numerically simulated data for different dimensions.

5. Conclusion

A scalable model for high voltage floating body PD SOI devices is developed using a proposed sub-circuit approach by utilizing the HiSIM-HV compact model. The developed model includes the high voltage and the floating body effects observed in floating body PD SOI DEMOS devices. To study and model the floating body and high voltage effects, the device is divided into the surface MOS and the parasitic BJT region. Using ATLAS, a 2-D numerical simulator, the physical description of the specific dc behavior of PD SOI DEMOSFETs is presented. In the proposed modeling approach, the surface MOS region is modeled using the HiSIM-HV compact model while to capture the floating body effects, the BSIM3SOI model approach is used with the HiSIM-HV compact model. The model performance is demonstrated for the 20-V DEMOS device by implementing the SPICE sub-

978-1-4673-0438-2/12 $31.00 © 2012 IEEE

circuit in Spectre (Cadence). Generally, the proposed compact models can be run on any SPICE simulator, since the model equations of the standard HiSIM-HV and BJT (Level=1) compact models are not changed. Therefore, the developed compact models here will enable the accurate design of complex circuits, using PD SOI DEMOS devices based on SPICE simulation.

Acknowledgment

The authors would like to thank V. Subramanian, A. R. Trivedi, Y. S. Chauhan, A. Bandyopadhyay and folks at IBM SRDC, Bangalore for their interesting comments.

References

[1] J. Mitros, C.-Y. Tsai, H. Shichijo, M. Kunz, A. Morton, D. Goodpaster, D. Mosher, and T. Efland, "High-voltage drain extended MOS transistors for 0.18 μm logic CMOS process," *IEEE Transactions on Electron Devices*, vol. 48, no. 8, pp. 1751-1755, Aug 2001.

[2] HiSIM HV 1.0.1 Users Manual, Copyright 2008, Hiroshima University and STARC, http://home.hiroshima-u.ac.jp/usdl/HiSIM_HV/C-Code/HiSIM_HV_1.0.2_UsersManual.pdf.

[3] J.G. Fiorenza, D.A. Antoniadis, J.A. del Alamo, "RF power LDMOSFET on SOI," *IEEE Electron Device Letters*, vol.22, no.3, pp.139-141, Mar 2001.

[4] C. Anghel, N. Hefyene, A. Ionescu, M. Vermandel, B. Bakeroot, J. Doutreloigne, R. Gillon, S. Frere, C. Maier, and Y. Mourier, "Investigations and Physical Modelling of Saturation Effects in Lateral DMOS Transistor Architectures Based on the Concept of Intrinsic Drain Voltage," in *IEEE European Solid-State Device Research Conference (ESSDERC)*, Sept. 2001, pp. 399-402.

[5] L. Wang, J. Wang, C. Gao, J. Hu, P. Li, W. Li and S.H.Y.Yang, "Physical Description of Quasi-Saturation and Impact-Ionization Effects in High-Voltage Drain-Extended MOSFETs," *IEEE Transactions on Electron Devices*, vol.56, no.3, pp.492-498, Mar. 2009.

[6] M. Chan, P. Su, H. Wan, C.-H. Lin, S. K. H. Fung, A. M. Niknejad, C. Hu, and P. K. Ko, "Modeling the floating-body effects of fully depleted, partially depleted, and body-grounded SOI MOSFETs," *Solid State Electron.*, vol. 48, no. 6, pp. 969-978, Jun. 2004.

[7] N. Arora, *MOSFET Models for VLSI Circuit Simulation: Theory and Practice.* Secaucus, NJ, USA: Springer-Verlag New York, Inc., 1993.

2012 25th International Conference on VLSI Design

Implications of Halo Implant Shadowing and Backscattering from Mask Layer Edges on Device Leakage Current in 65nm SRAM

H. C. Srinivasaiah
Department of Telecommunication Engineering,
Dayananda Sagar College of Engineering (DSCE), Bangalore, India
E-mail: hcsrinivas@dayanandasagar.edu or hcsrinivas@ieee.org

Abstract—Effect of shadowing and/or backscattering of halo implant species from halo implant mask layer on leakage current of NMOS driver transistor of a 65nm SRAM cell is studied. The halo implant mask layer thickness has been varied from 100nm to 3000nm in steps; in response to this variation, leakage behavior of this NMOS transistor is observed. The leakage current of this transistor is shown to be a strong function of halo implant mask layer thickness with implant window width W=0.27µm. The poly gate is located approximately in the middle of this implant window. The leakage current is seen to increase monotonically by more than an order of magnitude with the increase in thickness till 500nm. When this thickness is increased beyond 500nm, the leakage current variation fits approximately into a damped oscillatory curve whose period is seen to be proportional to the width W of the halo implant window. The leakage current observed is 22nA for this NMOS device (with gate width W_n=120nm) at 500nm of halo implant mask layer thickness. Further when the halo implant window width W is increased beyond 0.27µm with the halo mask layer thickness fixed at 500nm, the leakage attained a minimum value of 0.54nA. All the leakage currents that are observed are in saturation region at cell V_{dd}=1.2V.

Keywords: *Shadow effect, Process technology, Halo implantation, 65nm process technology, Low power SRAM, Embedded SRAM, and 6T-SRAM*

I.   INTRODUCTION

Ever since invention of semiconductor ICs, design metrics of major importance are power, performance, size, cost, etc. Power and performance are traded-off with each other [1]. The speed of the memory sub-system is the determinant of overall performance of digital systems. Low power hierarchical sub-array memory design technique is the key in implementation of powerful processors with on-chip cache [1, 2].

Complex ICs have multi-level cache memory to achieve increased throughput [2]. Low power memories are designed using low power design techniques based on controlling charging capacitances, operating voltages, and dc static currents. In order to reduce charging capacitances traditionally, multi-divided data and word line memory architecture is followed [1, 3].

Embedded SRAM (eSRAM) memories achieve lower latency when compared to embedded DRAM (eDRAM)

memories but they are power hungry elements [2]. Integration of on chip SRAM consuming low operating power and low standby power is very challenging due to the issues that are arising from lithography and etching process steps with scaling [4]. In deeply scaled technologies total chip current due to subthreshold leakage is shown to dominate in total active current [1, 4].

This paper presents techniques to address subthreshold leakage issues from process technology perspective. We highlight the importance of developing and optimizing manufacturing process, and the manufacturing process integration requirements. These two aspects are observed to be key factors in minimizing leakage power in Deep-Sub-Micron (DSM) ICs [4, 5]. Traditionally, halo implant process is one of the dominant factors in controlling the device's leakage [5, 6]. The excessive device leakage in 65nm SRAM memory resulting from issues associated with halo implant process integration steps is systematically analyzed in this work; likely remedies for the leakage arising due to issues associated with halo implant process integration steps are proposed.

Section II of this paper discusses expected halo implant mask layer shadowing and backscattering effect from the edges of halo implant mask layer. In section III, impact of halo implant mask layer shadow and/or backscattering effect on driver NMOS transistor of a 6T, 65nm SRAM cell is discussed. This section also deals with various physical models that are used specific to short channel devices, during three dimensional (3D) process/device simulations using Technology CAD (TCAD) tool. Section IV, deals with the quantitative assessment of leakage as a function of halo implant mask layer thickness. This section also discusses halo implant related design rule implications on leakage, and the solution explored in this work to mitigate leakage issues. In section V, we conclude with the discussion on the results, highlighting the proposed solutions to mitigate SRAM leakage current.

II.   MASK   LAYER   SHADOWING   AND   DOPANTS
BACKSCATTERING EFFECT

Modern DSM CMOS devices have deep source/drain (s/d) regions, and shallow s/d or lightly-doped-drain (LDD) region (Fig. 1(a)). Deep s/d and LDD regions are necessary to mitigate short channel effect (SCE) in DSM devices [5].

978-1-4673-0438-2/12 $31.00 © 2012 IEEE

The LDD region also alleviates hot carrier effect by reducing electric fields near channel [7]. The LDD s/d formation is complemented by pocket halo implant to further control SCE, improving the leakage current [5, 6]. Device leakage currents have exponential relation with threshold voltage (V_t) role off ΔV_t [8]. The halo implant is the key process step in controlling ΔV_t [5, 6, 8] in DSM devices.

The placement of pocket halo under gate region, close to LDD junction is very crucial in sub-100nm devices, to effectively control the leakage current [6]. If the average concentration of dopants during the pocket halo implant goes high in the channel under the gate region, we expect undesirable reverse SCE (RSCE) [8, 9]. The placement of the pocket halo in the region under the gate is done using tilted quad halo implantation [5, 6, 10]. Halo implant dose, energy, and tilt are very significant in controlling the leakage behavior of the DSM CMOS devices [5].

The shadow effect for tilted implants arises during patterning of implant window using photolithographic processes [6, 7, 11]. This shadow effect is very critical for tilted halo implant in DSM devices, as this implant decides their leakage behavior [5, 6]. The halo implant mask layer blocks the halo implant from reaching the area under the gate [6] (see Fig. 1(a)). The halo implant for NMOS and PMOS devices is done using separate masks. The halo implant is tilted at an angle θ between $25°$ to $30°$ from the normal to the wafer [5, 6]. This tilted implant serves the purpose of placing the halo implant concentration (called pocket halo), laterally under the gate, close to LDD region as shown in Fig. 1(a). As it is evident from this figure, the halo implant mask layer thickness Δy, the tilt angle θ and, the s/d diffusion area length Δx plays an important role in positioning the pocket halo under the gate stack. For a given Δy and θ, the Δx is given by, $\Delta x = \Delta y \times Tan\theta$, [9]. Thus, for a given angle θ, Δx and Δy decides the implant dose received by the part of the wafer which is exposed to halo implantation.

Fig. 1(b) and 1(c) shows implant beam incident at an angle $\theta°$, on a wafer with wafer rotation angle 0° and 180° respectively. The triangular grey regions indicate implant shadow areas where the ion beam is shadowed. Although the ion beam is shadowed in the triangular grey area in the current 0° wafer position (Fig. 1(b)), it receives halo implant when the wafer is rotated by 180°, as shown Fig. 1(c). Similarly the triangular grey area shadowed in 180° wafer position will be exposed to halo implant beam at 0° wafer position. Thus these grey regions receive only 50% of the total expected halo implant dose. Superimposing Fig. 1(b) and Fig. 1(c), we get Fig. 1(d). Further, if the implant mask layer thickness increases, we get multiple shadowing along the direction of the depth with spacing S between successive shadows given as $S=W \times Tan\ (90-\theta)=0.468\mu m$, (i.e. $S/2=0.234\mu m$) with the width of implant window $W =$

$0.27\mu m$, (see Fig. 1(a)) and $\theta=30°$ in this study, a simple trigonometric calculation.

In Fig. 1(d) with mask thickness labeled 1, the gate stack receives direct halo; at mask thickness labeled 2, the gate stack is in shadow area, and at mask thickness labeled 3 the wafer gets exposed to backscattered (or reflected) implant beam from mask wall. As the mask thickness increased further, we come across shadow areas with alternate white exposed areas. Those white exposed areas are receiving indirect backscattered halo dose. Depending on the mask layer material characteristics the back scattered implant dose expected to reach wafer surface may get successively reduced. Finally we may have a minimum average backscattered halo concentration reaching the wafer surface underneath the gate stack. At this mask thickness, it ends up having maximum device leakage.

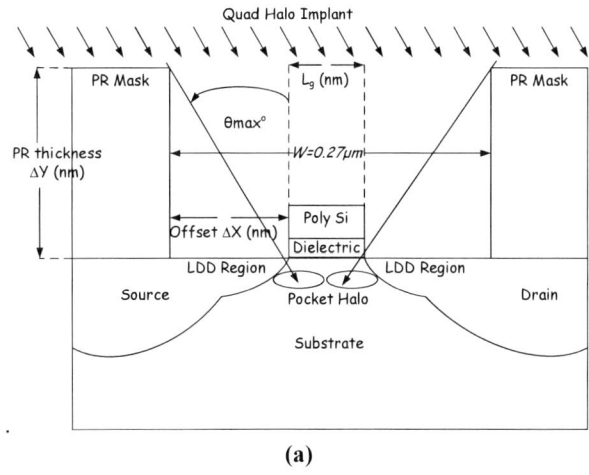

(a)

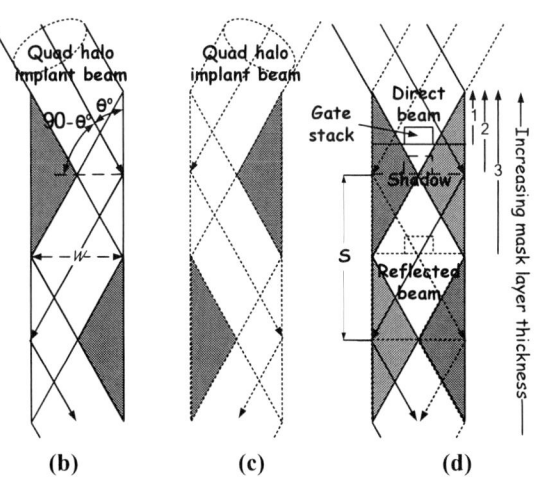

(b)　(c)　(d)

Figure 1: Halo implant shadow effect, (a) thicker halo implant mask layer blocking direct halo implant beam, (b) halo implant window for tilted ion beam at 0° rotated wafer (c) halo implant window for tilted ion beam at 180° rotated wafer (d) superimposed Fig. 1(b) and Fig. 1(c)

As the mask layer thickness is increased in the direction of arrow shown in Fig. 1(d), the poly gate stack passes through shadow grey areas and exposed white areas. When gate lies in grey area, corresponding device leakage increases and when it lies in exposed white area, the leakage of the same device decreases, relatively. The severity of halo implant mask layer shadowing depends on circuit topology. A 6T SRAM cell of Fig. 2 with 65nm transistor gate length is used to study the impact of halo implant mask layer shadow effect. If two devices are sharing the s/d as in the case of transistors M1 and M5, and transistors M2 and M6 in Fig. 2(a) and (b) (whose layout is in Fig. 3(a)), the side of the poly gate close to the mask layer edge, is severely affected by shadowing. It is understood from Fig. 1(a), that if Δx is increased, the halo implant shadowing will be minimum for a given Δy, and θ thus minimizing leakage and vice versa.

Fig. 2(a) is the circuit schematic of 6T SRAM cell with each transistor and important nodes labeled for ease of referencing. Fig. 2(b) is the simplified view of 3D structure of 65nm 6T SRAM cell process simulated using Sentaurus TCAD tool. The hypothetical interconnection among all the 6 transistors is in accordance with circuit schematic of Fig. 2(a). In Fig. 2(b) interlayer dielectric (ILD) and trench oxide is stripped for better view of the internal details of the SRAM cell

III.   65nm SRAM CELL AND DRIVER NMOS DEVICE STRUCTURE

Dependence of leakage current of M1 transistor in 65nm SRAM cell (Fig. 2) on halo implant mask layer thickness Δy and the halo implant window width W (Fig. 1(a)) is explored using Sentaurus TCAD tool [10]. Fig. 3(a) shows the snapshot of 65nm SRAM cell layout. A rectangular boundary enclosing this layout is the domain of process simulation. This rectangle contains all the layers that are required to simulate/emulate the 3D structure of Fig. 2(b). The layers in this layout define diffusion area, poly gate, all metals, contacts, etc., superimposed. Fig. 3(b) is a further simplified view of Fig. 2(b) showing junction profiles of all the 6 transistors, M1 to M6 in the SRAM cell under study.

The simulation domain of the NMOS driver transistor M1 is shown by a rectangle enclosing it, juxtaposed with simulation domain of transistor M3 (with M1 and M3 forming an inverter). Widths of driver transistor M1 and M2 is $W_1=W_2=120nm$ and that of transistor M3, M4, M5, and M6, respectively are $W_3=W_4=W_5=W_6=90nm$, the gate lengths $L_g=65nm$ for all the transistors. The area of the SRAM cell under study is $0.594\mu m^2$ with γ=width of M1÷width of M5 (with gate length of M1 to M6 equal to 65nm) = 1.33. In Fig. 3(b), NMOS and PMOS transistors are isolated by 120nm wide and 200nm deep trench.

The simplified structure of M1 with various electrodes is shown in Fig. 4(a). Poly overhanging on active area is 1λ, $\Delta x=3\lambda$ and the halo implant tilt angle $\theta=30°$. Drain metal contact is extended over bulk region providing the necessary

bulk contacts for device simulation. Gate stack consists of 15Å of SiO_2 over which is 65nm thick polysilicon. On top of polysilicon copper is deposited for electrical connectivity. In the current view the boundaries of 15Å $SiO2$ gate dielectric is not noticeable, as it is extremely thin compared to other thicknesses. Removal of trench oxide and ILD saves mesh point for device simulation.

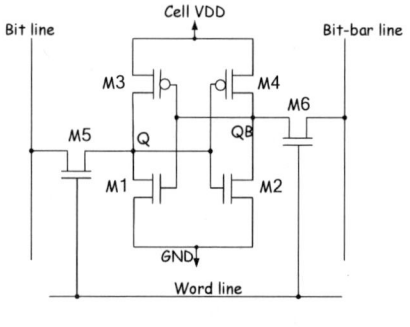

(a)

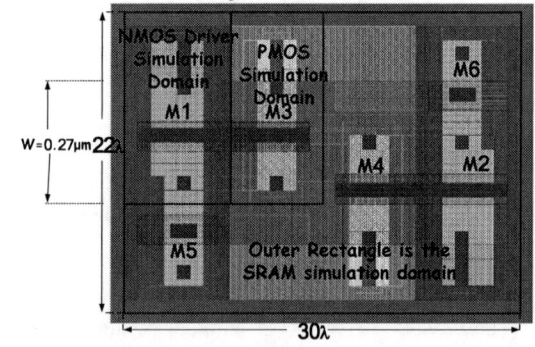

(b)

Figure 2: A 6T SRAM cell (a) circuit schematic, (b) process simulated/emulated 3D structure of 65nm SRAM cell with hypothetical metal connections among cell transistors.

(a)

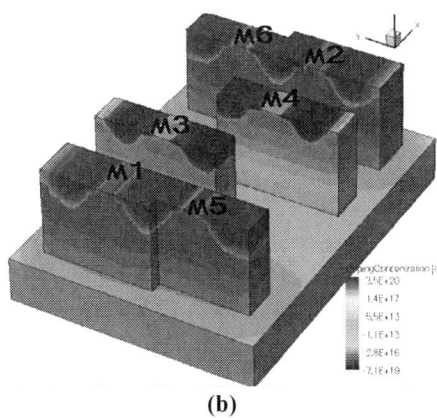

(b)

Figure 3: Topography of 6T SRAM cell under study (a) Cell layout (λ=30nm≈Lg/2=32.5nm), (b) Junction profiles of all the 6 process simulated/emulated transistors.

The deep s/d and LDD junction profile of this transistor is shown in simplified view of Fig. 4(b). The depth of deep s/d junction is 65nm whereas the depth of shallow or LDD region is 25nm to 30nm. We can see asymmetry in LDD region as manifestation of halo implant being shadowed and/or back scattered by halo implant mask layer (Δy=500nm). The term halo implant mask layer refers to a specified thickness of (patterned) photoresist (PR), in this study.

A process recipe targeting ITRS specifications for 65nm technology node is developed, and optimized to simulate/emulate NMOS and PMOS transistors of the 65nm SRAM cell. Fig. 5 shows simplified process flow used in process simulation/emulation of the device of Fig. 4, and SRAM cell of Fig. 2. In Fig. 5, we begin with a wafer with initial Boron concentration 1×10^{16}/cm^3. This is followed by anti latchup implantation. Following this implantation, anti punch through implantation is performed for both N/PMOS devices. Indium and antimony implant species are used in this step for N/PMOS devices respectively [5]. Subsequently, gate patterning is performed to define gate electrodes. The N/PMOS gate stack consists of *15Å* of *SiO$_2$*, over which a 65nm thick polysilicon is deposited and patterned using respective mask (Fig. 3(a)).

A 65nm nitride spacer is integrated by isotropic deposition and anisotropic etching, which helps to achieve self aligned deep s/d implant. About 65nm deep s/d junctions are achieved for both N/PMOS devices, using As/B implantations respectively. Separate masks define implant windows for N/PMOS devices during this implant step. During the deep s/d implant the polysilicon gates of N/PMOS transistor also receive As/B implantation. Following the self aligned deep s/d implantation, a 1050°C, 20sec annealing is simulated to activate the deep s/d and poly implant species. Next, the 65nm nitride spacer is stripped.

Subsequently, LDD and halo implantation is performed for both N/PMOS transistors. During LDD implantation,

As/B species are used for N/PMOS transistors, respectively. Immediately after LDD implant step, 30° tilted B/P halo is implanted for N/PMOS transistors respectively.

After LDD/halo implantation, a 1050°C, 4sec spike annealing is simulated to activate LDD/halo implantation species. Finally s/d, well/bulk, and gate contacts are added for device simulation. Resulting SRAM cell structures (after metallization simulation), and transistor M1 (after contacts added) are shown in Fig. 2(b) and Fig. 4(a) respectively.

Word "emulation" is used, as some of the structural parameters are process emulated by Sentaurus TCAD tool's 3D geometric operation capability [10], which is computationlly economical. The process steps such as implantation, annealing, etc., are simulated, and the process steps such as etch, and deposition are emulated.

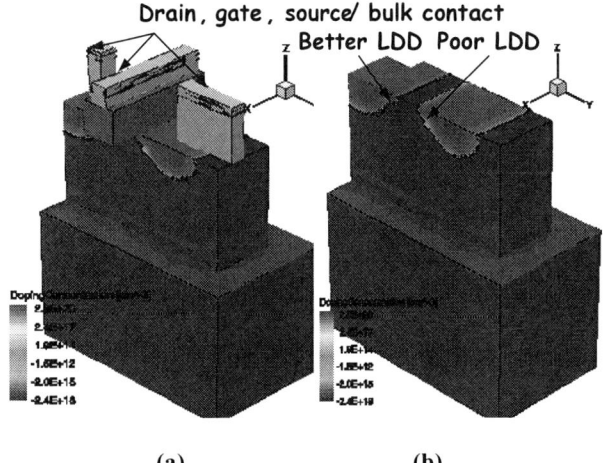

(a)   **(b)**

Figure 4: Simplified view of process simulated/emulated transistor M1 for Δy=500nm, (a) device structure with contacts, (b) junction profile of M1.

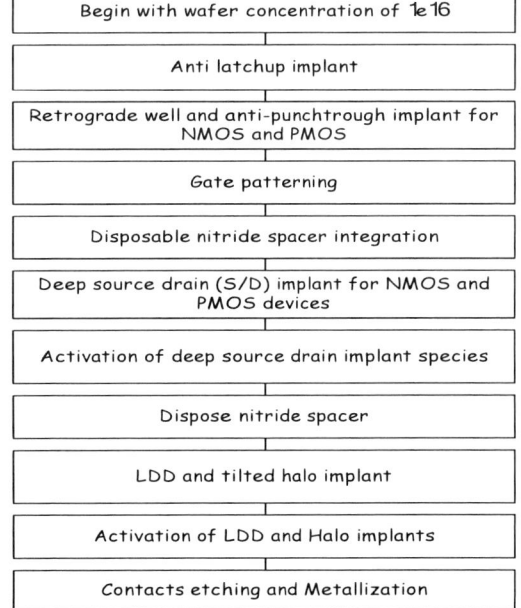

Figure 5: Process flow for simulation of transistors/SRAM cell.

The computational complexity of 3D process/device simulation is memory and processor intensive. Mesh points used for structure of Fig. 4(a) were in the range of 25,000 to 30,000. In order to save computation resources, unwanted grid points were avoided by stripping off trench oxide and other ILDs (Fig. 4), before starting device simulation. Doping criteria based adaptive meshing is used by specifying minimum and maximum mesh size in the direction along 3 coordinates of the 3D structure of Fig. 4(a). Relatively dense mesh was built in junctions and channel areas of M1 transistor.

Sentaurus "sprocess" and "sdevice" tools are used for process simulation and device simulation respectively. Visualization tool called "inspect" is used to plot and view all the I_d-V_{gs} curves. A script was written to facilitate automatic extraction of current values. Hydrodynamic carrier transport and Vandort's [10] models were used to account for carrier velocity overshoot and channel carrier quantization effects, respectively, during "sdevice" simulations.

IV. RESULTS AND DISCUSSION

Fig. 6 shows I_d-V_{gs} characteristics of transistor M1, process simulated/emulated without halo (i.e. non-halo); with halo and halo implant mask layer thicknesses $\Delta y=100nm$, $500nm$, and $2000nm$ respectively; and Δx relaxed, in the saturation region of device operation. Both saturation region leakage currents $I_{dsatleak}$ and saturation region ON state currents $I_{dsatdrive}$ at $V_{dd}=1.2V$ are extracted and tabulated in Table 1 for all 5 splits of M1 transistor. The non-halo device had a very high leakage and drive current, as expected. Device with $\Delta y=100nm$ had low leakage of 0.87nA; but PMOS transistors are degraded in performance due to thin Δy which causes B halo penetration into PMOS area. The device with Δx relaxed had 0.54nA, leakage which is lowest among all the 5 splits. These current values are when the gate width of M1, $W_n=120nm$.

Fig. 7 is the plot of $I_{dsatleak}$ as a function of halo implant mask layer thickness Δy for driver transistors M1. Halo implant mask layer thickness Δy is varied from 100nm to 3000nm in 100nm steps during process simulation/emulation of driver transistor M1. The $I_{dsatleak}$ is extracted for all the 30 splits of M1 transistors resulting in the curve of Fig. 7.

In Fig. 7, the leakage current increases from 0.87nA at $\Delta y=100nm$ monotonically till $\Delta y\approx700nm$. After 700nm the leakage currents fits into an oscillatory curve till $\Delta y\approx1700nm$. When calculated, the distance between a minima and subsequent maxima from the graph of Fig. 7 is approximately $0.234\mu m$, which is S/2, where S is given by an equation discussed in section II. Referring back to the discussion of Fig. 1(d), minimum corresponds to gate stack exposure to halo, and maxima correspond to shadowing. When Δy is further increased from 1700nm, till 2100nm the

leakage current is approximately constant at a value of 24nA. Subsequently there are 2 peaks both $\approx28nA$, around $\Delta y=2200nm$ and $\Delta y=2900nm$. As Δy is varied from 100nm to 3000nm the leakage current varies by more than an order of magnitude. The dependence of leakage on the size of the implant window is one of the important observations made in this work.

The dependence of leakage current on halo implant mask layer thickness variations is a source of device characteristics fluctuation [12]. In real-time manufacturing process it is impossible to achieve constant PR thickness across entire wafer when PR is spin coated. Obviously the leakage currents of all transistors characterized by halo implant process will vary from site to site on a given wafer in response to PR mask layer thickness variations. The leakage current is varied by more than an order of magnitude, when PR mask layer thickness varies from 100nm to beyond 500nm monotonically. Discrete stochastic nature of dopants in the channel region results in significant fluctuation in device threshold voltage V_t [4, 12], and hence fluctuations in leakage current [8]. The combined effect of discrete dopants, and shadowing/back-scattering effect will increase the overall fluctuation in the leakage currents.

TABLE 1: SATURATION REGION LEAKAGE AND DRIVE CURRENTS FOR 5 SPLITS OF DEVICES OF FIGURE 6; WIDTH OF M1 TRANSISTOR=120nm

Condition of halo process simulation	$I_{dsatleak}$ (nA)	$I_{dsatdrive}$ (mA)
Non-halo (without halo)	365	0.21
With halo and $\Delta y=100nm$ and $\Delta x=3\lambda$	0.87	0.116
With halo and $\Delta y=500nm$ and $\Delta x=3\lambda$	22.04	0.165
With halo and $\Delta y=2000nm$ and $\Delta x=3\lambda$	24.13	0.1610
With halo, $\Delta y=500nm$, and $\Delta x=\infty$ (relaxed, i.e. $W>>0.27\mu m$)	0.54	0.11

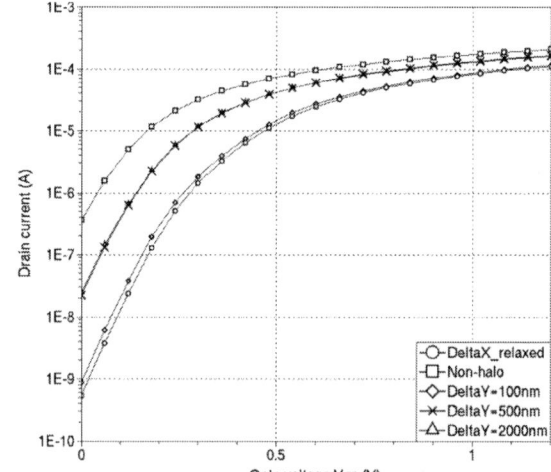

Figure 6: I_d-V_{gs} curves (log scale) for 5 splits of M1 transistors of Table 1

978-1-4673-0438-2/12 $31.00 © 2012 IEEE

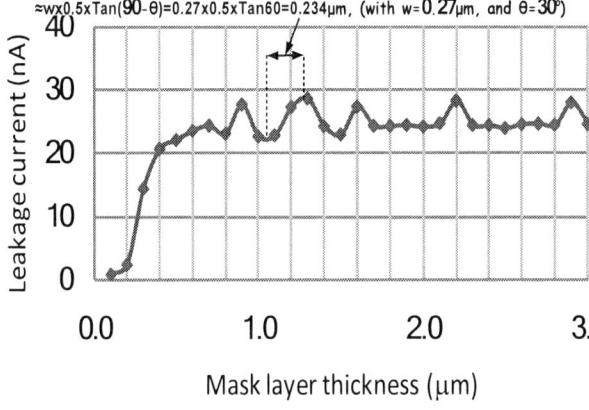

$\approx w \times 0.5 \times Tan(90-\theta) = 0.27 \times 0.5 \times Tan60 = 0.234 \mu m$, (with $w = 0.27 \mu m$, and $\theta = 30°$)

Figure 7: Leakage current of M1 transistor plotted as a function of halo implant mask layer thickness (Δy).

The backscattering become even worst below 65nm technologies because of the PR mask layer edge roughness becomes comparable with the implant window width. This edge roughness is due to the PR mask layer edges being modulated by the standing wave patterns of the light in the exposed areas of the physical masks [13]. The spatially modulated intensity in the standing wave spatially creates unevenness or roughness in PR mask layer.

As a solution proposed in this work to mitigate shadowing/back scattering by halo implant mask layer thickness effect, multiple NMOS or PMOS transistors need to be laid out in a common implant window. This will relax the value of Δx for most of the transistors, so that the shadowing effect is minimized, and hence leakage can be mitigated. The fact of minimum leakage for relaxed Δx is corroborated by one of the I_d-V_{gs} curve (with circle markers) in Fig. 6 for driver transistor M1.

V. CONCLUSIONS

The impact of halo implant photolithography process step on the NMOS driver transistor leakage current behavior in 65nm bulk SRAM cell has been investigated systematically. The dependence of SRAM cell transistor leakage current on halo implant mask layer thickness, Δy is explored, when Δy is varied from 100nm to 3000nm at constant halo implant window width $W=0.27\mu m$. More than an order of magnitude variation in leakage is noticed as the halo implant mask layer thickness is increased in 100nm steps from its minimum value of 100nm. During this study a minimum leakage of 0.87nA is achieved for $\Delta y=100$nm; with this value of Δy PMOS transistors are expected to degrade. In order to achieve good leakage performance of N/PMOS devices, we need at least $\Delta y=500$nm and relatively large value of Δx ($\geq 3\lambda$); this combination resulted in 0.54nA of leakage which is the least value among 5 splits of M1 transistor (in Fig. 6) as tabulated in Table 1. Keeping this observation in view, transistors of type (N/PMOS) maybe placed and oriented such that multiple transistors are halo implanted through common halo implant window of relatively large dimension. This approach facilitates for thicker (than 500nm) halo implant mask layer and larger offset (Δx) values, mitigating leakage currents, effectively.

ACKNOWLEDGMENT

Author would like to acknowledge Visvesvaraya Technological University (VTU), Belgaum, Karnataka, for funding this research project. Authors also convey special thanks to Dr. D. Premachandra Sagar, Vice-chairman, Dayananda Sagar group of institutions (DSI) for all his support and constant encouragement for this research work.

REFERENCES

[1] Kiyoo Itoh, Katsuro Sasaki and Yoshinobu Nakagome, *"Trends in Low-Power RAM circuit Technologies"*, in proc. of IEEE, Vol. 83, No.4, pp. 524-543, Apr. 1995.

[2] Barth, *et al*, *"A 45nm SOI Embedded DRAM Macro for the POWER Processor 32 MByte On-Chip L3 Cache"*, IEEE Journal of Solid State Circuits (JSSC), Vol. 46, No. 1, pp. 64-75, Jan. 2011.

[3] Tadahiko Sugibayashi, *et al* *"A 30-ns 256-Mb DRAM with a Multidivided Array Structure"*, IEEE JSSC, Vol. 28, No.11, pp. 1092-98, Nov. 1993.

[4] ITRS, *"2009_PIDS.pdf"*, 2009 edition, available at: http://www.itrs.net/Links/2009ITRS/2009Chapters_2009Tabl es/2009_PIDS.pdf

[5] H. C. Srinivasaiah, and Navakanta Bhat, *"Characterization of Sub-100nm CMOS Process Using Screening Experiment Technique"*, Solid State Electronics, Vol. 49, pp. 431-436, Mar. 2005.

[6] Mark Brandon. Fuselier, Jon D. Check, Frederick N. Hause, and Marilyn I. Wright, *"A Method of Analyzing the Effects of Shadowing of Halo Implant"*, US Patent No. US 6,426,262 B1, Jul. 30, 2002.

[7] Frank K. Baker, and James R. Pfiester, *"The Influence of Tilted Source-Drain Implants On High-Field Effects in Sub-micrometer MOSFETs"*, IEEE TED, Vol. 35, No. 12, pp. 2119-2124 Dec. 1988.

[8] Yuan Taur, and Tak H. Ning, *"Fundamentals of Modern VLSI Devices"*, First edition, Cambridge University Press, 1998.

[9] Terence B. Hook, Jeffery S. Brown, Matthew Breitwisch, Dennis Hoyniak, and Randy Mann, *"High-Performance Logic and High-Gain Analog CMOS Transistors Formed by a Shadow-Mask Technique With a Single Implant Step"*, IEEE TED, Vol. 49, No. 9, pp. 1623-27, Sept. 2002.

[10] *Sentaurus TCAD release-10 Manual*, 2010.

[11] Gadi Krieger, Gianpaolo Spadini, Peter P. Cuevas, and John Schuur, *"Shadowing Effects Due to Tilted Arsenic Source/Drain Implant"*, IEEE Transactions on Electron Devices (TED), Vol. 36. No. 11, pp. 2458-2461, Nov. 1989.

[12] Keith A. Bowman, Steven G. Duvall, and James D. Meindl, *"Impact of Die-to-Die and Within-Die Parameter Fluctuations on the Maximum Clock Frequency Distribution for Gigascale Integration"*, IEEE Journal of Solid-State Circuits, Vol. 37, No. 2, pp. 183-190, Feb. 2002.

[13] Eugene D. Fabricius, *"Introduction to VLSI design"*, McGraw Hill edition 1990.

Customizing Instruction Set Extensible Reconfigurable Processors using GPUs

Unmesh D. Bordoloi[1], Bharath Suri[1], Swaroop Nunna[2], Samarjit Chakraborty[2], Petru Eles[1], Zebo Peng[1]

[1]Linköpings Universitet, Sweden   [2]TU Munich, Germany

[1]E-mail:{ bhasu733@student.liu.se }, {unmbo, petel, zebpe}@ida.liu.se
[2]E-mail: {swaroop.nunna, samarjit.chakraborty}@rcs.ei.tum.de

Abstract—**Many reconfigurable processors allow their instruction sets to be tailored according to the performance requirements of target applications. They have gained immense popularity in recent years because of this flexibility of adding custom instructions. However, most design automation algorithms for instruction set customization (like enumerating and selecting the optimal set of custom instructions) are computationally intractable. As such, existing tools to customize instruction sets of extensible processors rely on approximation methods or heuristics. In contrast to such traditional approaches, we propose to use GPUs (Graphics Processing Units) to efficiently solve computationally expensive algorithms in the design automation tools for extensible processors. To demonstrate our idea, we choose a custom instruction *selection* problem and accelerate it using CUDA (CUDA is a GPU computing engine). Our CUDA implementation is devised to maximize the achievable speedups by various optimizations like exploiting on-chip shared memory and register usage. Experiments conducted on well known benchmarks show significant speedups over sequential CPU implementations as well as over multi-core implementations.**

I. INTRODUCTION

Instruction set extensible reconfigurable processors have become increasingly popular over the last decade. Their popularity is driven by the fact that they strike the right balance between the flexibility of general purpose processors and the performance of ASICs. The existing instruction cores of extensible processors may be extended with *custom instructions* to meet the performance requirements of the target application.

Our contributions: Design automation problems for customizing instruction sets are computationally intractable (NP-hard) [17]. In this paper, we propose the use of GPUs to accelerate the running times of design automation tools for customizable processors. To show the applicability of GPUs in instruction set customization algorithms, we choose a custom instruction *selection* problem and accelerate it using CUDA (Compute Unified Device Architecture), NVIDIA's parallel computing architecture based on GPUs [12]. Our contribution is interesting because we show how the custom instruction selection problem can be engineered to exploit on-chip memory on GPUs and other CUDA features. We choose custom instruction *selection* problem because (i) of its intractability (see Section II) and (ii) it has received lot of attention in recent years (see Section I-A). We would like to note that in contrast to traditional approaches (like approximation schemes [2]), our GPU-based technique provides *optimal* solutions.

Our contribution is also practically relevant because instruction set customization techniques are incorporated into compilers [18] and such compilers are invoked repeatedly by designers. Typically, a designer would choose the values of certain system parameters (e.g., processor frequency, deadlines) once an implementation version of the application has been fixed and then invoke the compiler to determine whether the constraints (like performance and area) are satisfied. If the compiler returns a negative answer, then some of the parameters are modified (e.g., an optimized version of the implementation is chosen or processor frequency is scaled) and the compiler is invoked once again. Thus, the designer iteratively interacts with the tools to adjust the parameters and functionalities till the performance constraints are satisfied. If each invocation of the tool takes long time to run to completion, the interactive design sessions become tedious affecting the design productivity. Hence, by bringing down the running times of the tools by significant margins using GPUs, the usability of such tools may be improved. Further, this comes at no additional cost because most desktop/notebook computers today are already equipped with a commodity GPU.

Finally, given that the combinatorial optimization problem mapped to GPU in this paper is a variant of the knapsack problem, our results might be meaningful to a wider range of problems in the design automation domain.

Overview of the problem: Given a library of custom instruction candidates the goal is to select a subset of instructions such that the performance is enhanced while keeping the area costs at minimum. In such a scenario, conflicting tradeoffs are inherent because while the performance of a system may be improved by the use of custom instructions, the benefits come at the cost of silicon area. Hence, a designer is not interested in identifying one solution which meets the performance requirements, but would rather like to identify all the conflicting tradeoffs between performance and area. The designer can then inspect all solutions and pick one which suits his/her design.

Note that in the above setup, part of the application is implemented as software on a programmable processor and the rest in hardware as custom instructions on a sea of FPGA. In this setting, a good metric for performance is the *processor utilization* because it is a measure of the load on the processor. Moreover, in this paper, we assume that the processor is running hard real-time tasks and processor utilization is a well known metric that is used to capture the feasibility of such systems (for details, see Section II). Formally, let (c, u) denote the hardware cost c, arising from the use of custom instructions and the corresponding utilization u, of the processor. We are then interested in generating the Pareto-optimal curve [4] $\{(c_1, u_1), \ldots, (c_n, u_n)\}$ in a multi-objective design space. Each (c_i, u_i) in this set has the property that there does not exist any implementation choice with a performance vector (c, u) such that $c \leq c_i$ and $u \leq u_i$, with at least one of the inequalities being strict. Further, let $\mathcal{S}$ be the set of performance vectors corresponding to all implementations choices. Let $\mathcal{P}$ be the set of performance vectors $\{(c_1, u_1), \ldots, (c_n, u_n)\}$ corresponding to all the Pareto-optimal solutions. Then for any $(c, u) \in \mathcal{S} - \mathcal{P}$ there exists a $(c_i, u_i) \in \mathcal{P}$ such that $c_i \leq c$ and $u_i \leq u$, with at least one of these inequalities being strict (i.e., the set $\mathcal{P}$ contains *all* performance tradeoffs). The vectors $(c, u) \in \mathcal{S} - \mathcal{P}$ are referred to as *dominated solutions*, since they are "dominated" by one or more Pareto-optimal solutions.

A. Related Work

Note that in this work we focus on the custom instruction *selection* phase. In recent years, lot of research has been devoted to custom instruction selection techniques so as to optimize either performance or hardware area [2], [9]. In this paper, we have considered a more general problem formulation by focusing on multi-objective optimization instead of optimizing for a single objective.

Custom instruction selection techniques assume that a library of custom instruction candidates is given. Such a library of custom instructions may be *enumerated* by extracting frequently occurring computation patterns from the data flow graph of the application [14], [17]. We believe our paper would motivate researchers to explore the possibility of deploying GPUs in this phase as well.

Motivation for using GPU: It should be mentioned in this section that our paper has been motivated by the recent trend of applying GPUs to accelerate non-graphics applications. Applications that have harnessed the computational power of GPUs span across numerical algorithms, computational geometry, database processing, image processing, astrophysics and bioinformatics [13]. Of late, there has also been lot of interest in accelerating computationally expensive algorithms in the computer-aided design of electronic systems [7], [3], [5]. There are many compelling reasons behind exploiting GPUs for such non-graphics related applications. First, modern GPUs are extremely powerful. For example, high-end GPUs, such as the NVIDIA GeForce GTX 480 and ATI Radeon 5870, have 1.35 TFlops and 2.72 TFlops of peak single precision performance, whereas a high-end general-purpose processor such as the Intel Core i7-960, has a peak performance of 102 Gflops. Additionally, the memory bandwidth of these GPUs is more than $5\times$ greater than what is available to a CPU, which allows them to excel even in low compute intensity but high bandwidth usage scenarios. Finally, GPUs are now commodity items as their costs have dramatically reduced over the last few years. The attractive price-performance ratios of GPUs gives us an enormous opportunity to change the way design automation tools like compilers for instruction set customization perform, with almost no additional cost.

However, implementing general purpose applications on a GPU is not trivial. The GPU follows a highly parallel computational paradigm. Since many threads run in parallel, it must be ensured that they do not have arbitrary data dependency on each other. Hence, the challenge is to correctly identify the data parallel segments so that dependency constraints of the application mapped to the GPU are not violated. Secondly, in order to exploit the high bandwidth on-chip shared memory, it is important to identify the frequently accessed data structures so that they can be pre-fetched in the shared memory.

II. PROBLEM DESCRIPTION

In this section, we discuss our system model and formally present the multi-objective optimization problem.

System Model: We assume a multi-tasking hard real-time system. Formally, we use the sporadic task model [1] in a preemptive uniprocessor environment. Thus, we are interested in selecting custom instructions for a task set $\tau = \{T_1, T_2, \ldots, T_m\}$ consisting of m hard real-time tasks with the constraint that the task set is schedulable. Any task T_i can get triggered independently of other tasks in τ. Each task T_i generates a sequence of jobs; each job is characterized by the following parameters:

- *Release Time*: the release time of two successive jobs of the task T_i is separated by a minimum time interval of P_i time units.
- *Deadline*: each job generated by T_i must complete by D_i time units since its release time.
- *Workload*: the worst case execution requirement of any job generated by T_i is denoted by E_i.

Throughout this paper, we assume the underlying scheduling policy to be the earliest deadline first (EDF). Assuming that for all tasks T_i, $D_i \geq P_i$, the schedulability of the task set τ can be given by the following condition $(U = \sum_{i=1}^{m} \frac{E_i}{P_i}) \leq 1$, where U is the processor utilization due to τ [1].

Problem Statement: For a given processor P, let each of the tasks T_i have n_i number of custom instruction choices which can be implemented in hardware. For simplicity of exposition, assume that the processor P's clock frequency is constant and all the execution times of the tasks are specified with respect to this clock frequency. The objective is to minimize P's utilization (by mapping certain custom instructions onto hardware) and at the same time also minimize the total hardware cost. In other words, our goal is to compute the *cost-utilization* Pareto curve $\{(c_1, u_1), \ldots, (c_n, u_n)\}$ for a prespecified clock frequency of P. Note that it is possible that the given task set (without utilizing custom instructions) is already schedulable on the processor, i.e., $U < 1$. In these cases, the Pareto curve reveals how the utilization can be further reduced at the cost of hardware area. This is interesting because the designer can then use the processor for soft real-time tasks or clock the processor at a lower frequency to save power. On other hand, if the original task set is not schedulable, the Pareto curve reveals the hardware costs at which the task set becomes schedulable. Note that the designer can also choose to clock the processor

Algorithm 1 Custom Instruction Selection

Require: The task set τ, and a set S_i for each task T_i.
1: $U_{0,0} \leftarrow \sum_{i=1}^{m} E_i / P_i$
2: **for** $j \leftarrow 1$ to mC **do**
3: $\quad U_{0,j} \leftarrow \infty$
4: **end for**
5: **for** $i \leftarrow 1$ to m **do**
6: $\quad$ **for** $j \leftarrow 0$ to mC **do**
7: $\quad\quad$ For each pair $(\delta_{i,k}, c_{i,k})$ that belongs to the set S_i
8: $\quad\quad U_{i,j} \leftarrow \min\{U_{i-1,j}, U_{i-1,j-c_{i,k}} - \delta_{i,k}/P_i\}$
9: $\quad$ **end for**
10: **end for**

at a higher frequency to make the task set schedulable. By revealing the utilization points for $U > 1.0$, our results will expose the higher frequencies at which the processor may be clocked for the system to be schedulable.

Each of the n_i choices of the task T_i is associated with a certain hardware cost. Choosing the jth implementation choice for the task T_i lowers its execution requirement on P from E_i to $e_{i,j}$. Equivalently, the amount by which the execution requirement of T_i gets lowered on P is $\delta_{i,j} = E_i - e_{i,j}$. Hence, for each task T_i we have a set of choices $S_i = \{(\delta_{i,1}, c_{i,1}), \ldots, (\delta_{i,n_i}, c_{i,n_i})\}$, where $c_{i,j}$ is the hardware cost associated with the jth implementation choice. Let $x_{i,j}$ be a Boolean variable that is assigned 1 if the jth implementation choice for the task T_i is chosen and is assigned 0, otherwise. In this setup, the objective is to minimize the utilization $U(S) = \sum_{i=1}^{m} \frac{E_i - x_{i,j}\delta_{i,j}}{P_i}$ and the cost $C(S) = \sum_{i=1}^{m} c_{i,j} x_{i,j}$, where S is the chosen implementation among the various available options.

NP-hardness: The NP-hardness of the problem can be shown by transforming the knapsack problem [10] into a special instance of this problem. Towards this, corresponding to each item in the knapsack problem, we have a task with performance gain equal to the profit and the hardware cost equal to the weight of the item. A complete proof is omitted due to space constraints.

Algorithm: An algorithm to compute optimally the Pareto curve described above consists of two parts. First, a dynamic programming algorithm (Algorithm 1) computes the minimum utilization that might be achieved for each possible cost. The second part finds all undominated solutions (*cost-utilization* Pareto curve) from the entire solution set found by the dynamic programming algorithm. We denote this part as 'Retain Undominated' — a straightforward sequential implementation on CPU. Below, we discuss Algorithm 1.

Let $U_{i,j}$ be the minimum utilization that might be achieved by considering only a subset of tasks from $\{1, 2, \ldots, i\}$ when the cost is exactly j. If no such subset exists we set $U_{i,j} = \infty$. Let the maximum cost be represented by C i.e. $C = max_{(i=1,2,\ldots,n;j=1,2,\ldots,n_i)} c_{i,j}$. Clearly, mC is an upper bound on the total cost that might be incurred. Lines 1 to 4 of Algorithm 1 initialize $U_{0,0}$ to $\sum_{i=1}^{m} E_i/P_i$, and $U_{0,j}$ to ∞ for $j = \{1, 2, \ldots, mC\}$. The values $U_{i,j}$ for $i = 1$ to $i = m$ are computed using the iterative procedure in lines 5 to 10. Thus, any non-infinity value $U_{n,j}$ for $j = \{1, 2, \ldots, mC\}$ implies that there exists a design choice of the task set with utilization

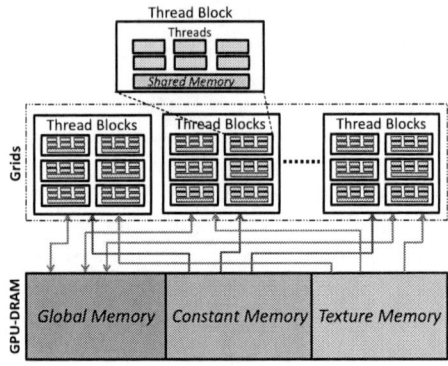

Fig. 1. CUDA programming model

$U_{n,j}$ and cost j. It can be easily verified that the running time of Algorithm 1 is $O(nmC)$, where $n = \sum_{i=1}^{m} n_i$, and its space complexity is $O(m^2 C)$. The algorithm runs in pseudo-polynomial time, and hence, turns out to be a computationally expensive kernel. In this paper, we accelerate the running times of this algorithm by mapping it to the GPU and obtain optimal and exact solutions as described in Section IV.

III. CUDA

In this section, we provide a brief description of CUDA. For a complete description, we refer the reader to NVIDIA's guide [12]. CUDA abstracts the GPU as a powerful multi-threaded coprocessor capable of accelerating data-parallel, computationally intense operations. The data parallel operations, which are similar computations performed on *streams* of data, are referred to as *kernels*. Essentially, with its programming model and hardware model, CUDA makes the GPU an efficient streaming platform.

In CUDA, *threads* execute data parallel computations of the kernel and are clustered into blocks of threads referred to as thread blocks. These thread blocks are further clustered into grids. During implementation, the designer can configure the number of threads that constitute a block. Each thread inside a block has its own registers and local memory. The threads in the same block can communicate with each other through a memory space shared among all the threads in the block and referred to as *Shared Memory*. The *Shared Memory* space of the thread block and is typically in the order of KB. However, an explicit communication and synchronization between threads belonging to different blocks is only possible through GPU-DRAM. GPU-DRAM is the dedicated DRAM for the GPU in addition to DRAM of the CPU. It is divided into *Global Memory*, *Constant Memory* and *Texture Memory*. We note that the *Constant* and *Texture Memory* spaces are read-only regions whereas *Global Memory* is a read-write region. Figure 1 illustrates the above described CUDA programming model. In case a memory location being accessed, by a CUDA memory instruction, resides in GPU-DRAM, i.e., either in *Global*, *Texture* or *Constant Memory* spaces, the memory instruction consumes an additional 400 to 600 cycles. On the other hand, if the memory location resides on-chip in the registers or *Shared Memory*, there will be almost no additional latencies in the absence of memory access conflicts. Note that in contrast to the GPU-DRAM, the *Shared Memory*

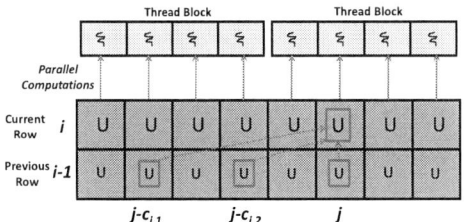

Fig. 2. Data dependency graph for Algorithm 1

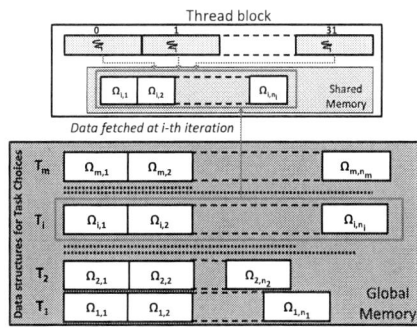

Fig. 3. Data fetched into shared memory at i-th iteration of the algorithm

region is a on-chip memory space. The additional latencies on GPU-DRAM might obscure the speedups that can be achieved due to parallelization and hence the on-chip shared memory must be judiciously exploited.

IV. PROPOSED FRAMEWORK ON CUDA

As described in Section II, the computation of the *cost-utilization* Pareto curve to expose the design tradeoffs at custom instruction selection phase involves a pseudopolynomial algorithm (Algorithm 1). In this section, we present our CUDA based framework to implement Algorithm 1 to accelerate its running times. This involves the following major challenges. First, we need to identify and isolate the data parallel computation of the algorithm so that they may be compiled as the *kernels*. Recall that kernels are executed by data parallel threads on CUDA. Secondly, we must devise the algorithm such that it can exploit the on-chip *Shared Memory* and registers to enhance the achievable speedups. Finally, thread block size must be appropriately configured. In light of these challenges, we now provide a systematic implementation of Algorithm 1 in the following.

Identifying data parallelism: As mentioned above, our first goal is to identify the data-parallel portions (*kernels*) in Algorithm 1 which can be computed by CUDA threads in a SIMD fashion. The kernels must not have any data dependencies (on each other) because they will be executed by threads running in parallel. Towards this we first identify the data dependencies in Algorithm 1. Algorithm 1 (lines 5 - 10) builds a dynamic programming (DP) matrix. The i-th row of the matrix corresponds to the i-th task T_i in the task set described in Section II. Each cell in the i-th row represents the value $U_{i,j}$ where $j = \{0, 1, 2, \ldots, mC\}$. According to Algorithm 1 (line 8), the computation of these values depends only on the values present in the previously computed rows. Figure 2 illustrates this for the cell $U_{i,j}$. This implies that the values of the cells of the same row in the DP-based matrix can be computed independently of each other by using different CUDA threads in SIMD fashion. Therefore, we isolate (line 8 of Algorithm 1) as the *kernel* of our CUDA implementation. In the following, we explain the effective usage of the on-chip share memory.

Memory usage: We store the DP-matrix in the *Global Memory* space (GPU-DRAM). Note that we use *Global Memory* space instead of *Constant* or *Texture Memory* because *Constant* and *Texture Memory* are read-only regions. During the computation of our DP-matrix we need to perform both read (to fetch values from the previous rows computed earlier)

and write (to update the DP-matrix with the values of the row computed in the current iteration) operations which can only be done explicitly with *Global Memory*. Also, note that we have so far not used the on-chip *Shared Memory* because the size of the *Shared Memory* is typically quite small (see Section III) and the DP-matrix cannot fit into it.

However, the on-chip *Shared Memory* can be exploited to store other frequently accessed data structures. To identify such data structures, we once again focus on the kernel operations of our algorithm (line 8 of Algorithm 1). We note that the computation of each of the $U_{i,j}$ values corresponding to the task T_i (i.e., the i-th row of our DP-based matrix) needs the values of all the n_i hardware implementation choices of T_i. Now let us denote the choice tuple $(\delta_{i,k}, c_{i,k})$ by $\Omega_{i,k}$ for $k = \{1, 2, \ldots, n_i\}$. Thus, from line 8 of Algorithm 1, the computation of the i-th row in the DP-matrix requires the values $\Omega_{i,1}, \Omega_{i,2}, \ldots, \Omega_{i,n_i}$.

This set, $\{\Omega_{i,1}, \Omega_{i,2}, \ldots, \Omega_{i,n_i}\}$, is essentially a subset of the overall specification of the task set. Also, in iteration i of computing the DP-matrix this set of required data structure remains constant, i.e., information about the other parts of the task set is not required. This set changes only at the next iteration (iteration $i + 1$) because this iteration corresponds to a different task in the task set which might have a different set of hardware implementation choices. This observation provides an opportunity to significantly reduce the GPU based execution times by loading these values $\{\Omega_{i,1}, \Omega_{i,2}, \ldots, \Omega_{i,k}\}$ to the on-chip *Shared Memory* at the beginning of each iteration. Compared to the DP-matrix, this set of values is much smaller and can fit into the on-chip *Shared Memory*. Figure 3 illustrates our scheme of prefetching the required data structure from *Global Memory* to *Shared Memory* at the start of each iteration. The figure shows a thread block (which consists of 32 threads) fetching the required data from the *Global Memory* at the i-th iteration.

Register usage: The threads of CUDA access registers (used to store the local variables) which have very low access latencies like the *Shared Memory*. If the total number of required registers is greater than that available in the processor for the current set of thread blocks, then CUDA will schedule less thread blocks simultaneously to cope with the situation. This will decrease the degree of parallelism offered by CUDA. In Algorithm 1 there is a division operation (line 8) that is known to contribute to high

978-1-4673-0438-2/12 $31.00 © 2012 IEEE   421

register usage. Hence, in our implementation, we convert it into a multiplication operation to optimize the register usage.

Thread block: We recall from Section III that the on-chip memory is shared only between the threads within a single block. Hence, configuring the thread blocks to an appropriate size is also important to effectively exploit the GPU on-chip memory. For example, if we choose a very small thread block size, then the computation of each row in our DP-based matrix will involve lot of thread blocks. However, only the threads within a thread block share the same chunk of on-chip memory. This implies that data from the *Global Memory* to *Shared Memory* will have to be transferred for a large number of thread blocks, inspite of the fact that all the threads in a single iteration need the same data structures - $\{\Omega_{i,2}, \ldots, \Omega_{i,k}\}$, as described above.

We note, however, that thread block size cannot be increased arbitrarily to increase performance. As an example, consider the Tesla GPU from NVIDIA that allows a maximum of 1024 threads in a thread block. Interestingly, the total number of threads that can be active simultaneously is 1536, as limited by the hardware. If we set thread block size to 1024, only one thread block (i.e., 1024 threads) will run in parallel. This is because it is not possible for the GPU to run only some threads of a thread block. On the other hand, if we set thread block size as 768, two thread blocks (i.e., 1536 threads in total) can run in parallel because all the threads can be activated simultaneously. Hence, for Tesla, we choose 768 as the thread block size. Under certain conditions (like register spillage, *Shared Memory* capacity overrun), it is possible that a thread block size of 1024 delivers a better performance than with 768. Our optimizations on *Shared-Memory* and register usage, as described above, ensure that such scenarios do not occur for the problem addressed in this paper.

V. EXPERIMENTAL RESULTS

In this section, we report the experimental results that were obtained by running our CUDA-based implementation on 5 different task sets that were constructed using well known benchmarks. We compared these results with those obtained by running sequential CPU-based implementation as well as multi-core implementations.

Experimental Setup: We created 5 task sets with number of tasks between 8 and 12. These task sets comprise of 5 benchmarks (*compress, jfdctint, ndes,edn, adpcm*) from WCET [15], 3 benchmarks (*aes, sha, rijndael*) from MiBench [8], 3 benchmarks (*g721encoder, djpeg, cjpeg*) from MediaBench [11] and one benchmark (*ispell*) from Trimaran [16]. Table I shows the combination of benchmarks incorporated in each of the task sets and the sizes of the task sets.

We chose the Xtensa [6] processor platform from Tensilica for our experiments. Xtensa is a configurable processor core allowing application-specific instruction-set extensions. The custom instruction configurations from the benchmarks were obtained by using the XPRES compiler from Tensilica. First, the workload E_i is computed for each task T_i which refers to the workload without any custom instruction enhancement. Assuming Tensilica identifies n_i custom instructions for each

Task Set	Benchmarks	Size
1	aes, djpeg, g721decode, rijndael, adpcm jfdctint, cjpeg, edn, ispell, sha, ndes, compress	12
2	djpeg, g721decode, rijndael, adpcm jfdctint, cjpeg, edn, ispell, sha, ndes, compress	11
3	aes, djpeg, g721decode, rijndael jfdctint, cjpeg, edn, ispell, sha, ndes	10
4	adpcm, rijndael, cjpeg, ispell sha, ndes, djpeg, compress, edn	9
5	cjpeg, ispell, edn, sha g721decode, djpeg, compress, ndes	8

TABLE I
TASK SETS

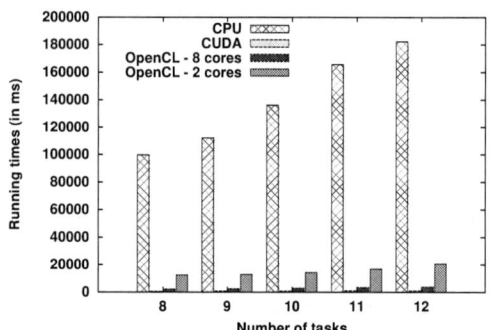

Fig. 4. Comparison of all implementations.

task T_i, we compute — (i) the performance improvement $\delta_{i,j}$ and (ii) the area cost $c_{i,j}$ — for each of the jth custom instruction configurations. The workload is in terms of Multiply-Accumulate (MAC) operation's cycles and the hardware area is in terms of number of adders.

We set P_i for the tasks, such that the $U = \sum_{i=1}^{N} \frac{E_i}{P_i}$ is 0.80, 1.00, 1.05, 1.08 and 1.10 for the 5 different tasks set. The GPU used for evaluating our experiments was a NVIDIA Tesla M2050 GPU. This GPU was connected via on-board PCI express slot to the host machine with 2 Xeon E5520 CPUs, each with 4 cores, i.e., 8 cores overall and each core ran at 2.27GHz. We compared the performance of our CUDA implementation against an OpenCL implementation on the multi-core host. We also compared the results with an OpenCL implementation on a dual core laptop, each core running at 2.1 GHz. We also implemented (in C) a sequential version of the algorithm that was run on a single core of the Xeon host machine. Thus, we have four implementations overall — CUDA, OpenCL 8-core, OpenCL 2-core and a sequential CPU.

Results: To illustrate the benefits of our CUDA implementation, we compared the running times for computing the Pareto

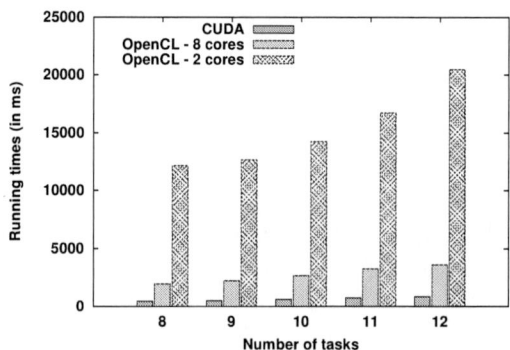

Fig. 5. Comparison of OpenCL multi-core and CUDA implementations.

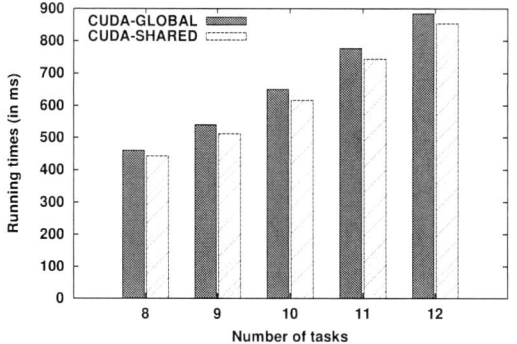

Fig. 6. Running times of CUDA-Shared and CUDA-Global implementations.

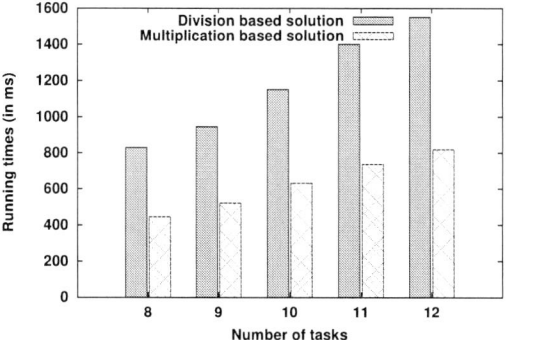

Fig. 7. The speedup obtained after minimizing register spillage using multiplication operation instead of division.

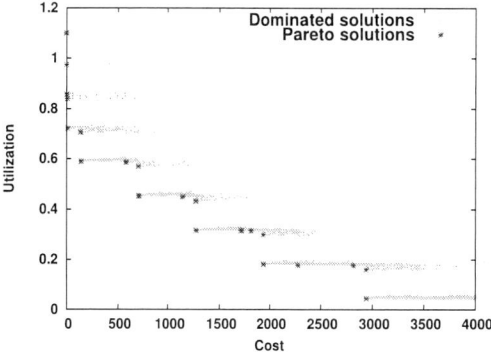

Fig. 8. The Pareto curve for task set 1. The points on the Pareto curve are highlighted with an asterisk.

to retain the undominated solutions after Algorithm 1 (see Section II). This part is implemented in the CPU because it is not amenable to parallelization and its running time is significantly less that Algorithm 1 (always less than 10 milliseconds). However, for accuracy its running times has also been included in the running times reported here.

VI. CONCLUDING REMARKS

We presented a technique to implement a custom instruction selection algorithm on GPUs. To the best of our knowledge, this is the first paper on instruction set customization using GPUs. Our technique exploits not just the parallelism but also the shared memory features offered by GPU architectures in order to achieve significant speed ups.

REFERENCES

[1] S. Baruah, A.K. Mok, and L.E. Rosier. Preemptively scheduling hard-real-time sporadic tasks on one processor. In *IEEE RTSS*, 1990.
[2] U. D. Bordoloi, H. P. Huynh, S. Chakraborty, and T. Mitra. Evaluating design trade-offs in customizable processors. In *DAC*, 2009.
[3] D. Chatterjee, A. De Orio, and V. Bertacco. GCS: High-performance gate-level simulation with GP-GPUs. In *DATE*, 2009.
[4] K. Deb. *Multi-Objective Optimization Using Evolutionary Algorithms.* John Wiley & Sons, 2001.
[5] J. Feng, S. Chakraborty, B. Schmidt, W. Liu, and U. D. Bordoloi. Fast schedulability analysis using commodity graphics hardware. In *RTCSA*, 2007.
[6] R. E. Gonzalez. Xtensa: A configurable and extensible processor. *IEEE Micro*, 20(2):60–70, 2000.
[7] Kanupriya Gulati and Sunil P. Khatri. Towards acceleration of fault simulation using graphics processing units. In *DAC*, 2008.
[8] M. R. Guthaus et al. Mibench: A free, commercially representative embedded benchmark suite. In *IEEE Annual Workshop on Workload Characterization*, 2001.
[9] H. P. Huynh and T. Mitra. Instruction-set customization for real-time embedded systems. In *DATE*, 2007.
[10] H. Kellerer, U. Pferschy, and D. Pisinger. *Knapsack problems.* Springer, 2004.
[11] C. Lee, M. Potkonjak, and W. H. Mangione-Smith. Mediabench: a tool for evaluating and synthesizing multimedia and communicatons systems. In *MICRO*, 1997.
[12] NVIDIA. CUDA Programming Guide version 1.0, 2007.
[13] J. D. Owens, D. Luebke, N. Govindaraju, M. Harris, J. Krüger, A. E. Lefohn, and T. J. Purcell. A survey of general-purpose computation on graphics hardware. *Computer Graphics Forum*, 26(1):80–113, 2007.
[14] N. Pothineni, A. Kumar, and K. Paul. Application specific datapath extension with distributed i/o functional units. In *VLSI Design*, 2007.
[15] F. Stappert. WCET benchmarks. http://www.c-lab.de/home/en/download.html.
[16] Trimaran:. An infrastructure for research in backend compilation and architecture exploration. http://www.trimaran.org.
[17] A. K. Verma, P. Brisk, and P. Ienne. Rethinking custom ISE identification: a new processor-agnostic method. In *CASES*, 2007.
[18] P. Yu. Design methodologies for instruction-set extensible processors. *PhD Thesis, C.Sc. Dept., National University of Singapore*, Jan. 2009.

curve for the different task sets for all four implementations discussed above. Figure 4 plots the running times for these implementations. Our CUDA implementation is 220× faster than the CPU implementation (on single processor). Due to such tremendous speedups, the bar graph showing the CUDA implementation almost co-incides with the x-axis. To better illustrate the speedups when compared to the multi-cores, Figure 5 shows only the CUDA implementation along with the multi-core implementations. As seen in this figure, even compared to a dual-core (on a laptop) and 8-core implementations (on Intel Xeon), our GPU-based implementation is 24× and 8× faster, respectively.

To illustrate the benefits of *Shared-Memory* usage and register size optimization, we conducted further experiments. Figure 6 shows the running times of CUDA-Global (where we do not utilize *Shared-Memory*) and CUDA-Shared (where we exploit the on-chip *Shared-Memory* as discussed in Section IV). Using the on-chip shared memory leads to an improvement of 6% on an average. Similarly, our optimization to manage the register spillage (based on the conversion of the division operation as a multiplication) also yields significant speedups (on average around 85%) as shown in Figure 7. Finally, in Figure 8 we illustrate the Pareto curve that was obtained for task set 1. Note that that the solution space is significantly huge, but for clarity of illustration, we have plotted only a part of the graph and the x-axis is truncated when cost is 4000.

We would like to mention that all the running times reported here include the time taken for transfer of data from the host machine to the GPU and vice-versa. Also, recall that computing the Pareto curve involves a straightforward algorithm

2012 25th International Conference on VLSI Design

Energy-Efficient Application Mapping in FPGA through Computation in Embedded Memory Blocks

Anandaroop Ghosh, Somnath Paul and Swarup Bhunia
Department of Electrical Engineering and Computer Science
Case Western Reserve University, Cleveland, OH, USA
Email: *axg468@case.edu*

Abstract—**FPGAs have emerged as the preferred prototyping and accelerator platform for diverse application domains such as digital signal processing (DSP), security and multimedia, which often impose real-time performance requirements. Most applications in these domains demand efficient implementation of complex datapaths or functions e.g. transcendental functions, which are spatially mapped in the configurable logic or embedded DSP blocks of a FPGA device. Requirement of elaborate computational resources to realize these operations impose a major barrier to energy efficiency. In this paper, we propose to use embedded memory blocks in FPGA for computing to significantly improve energy efficiency of the applications which are dominated by complex datapaths and/or functions. Complex operations are decomposed or fused into large multiple input/output lookup tables (LUTs); mapped to embedded memory blocks and evaluated through memory access over single or multiple cycles. Different parts of an application are selectively mapped into memory or logic/DSP blocks in a heterogeneous mapping framework to maximize energy efficiency. We explore optimal energy configuration of embedded memory for mapping operations of varying input size and develop a complete mapping flow including decomposition, fusion and packing. Effectiveness of the proposed flow is evaluated for a set of applications using a commercial state-of-the-art FPGA system (Altera Stratix IV). Finally, the proposed framework is extended to drastically trade-off energy versus accuracy at run time for common signal processing applications.**

Index Terms—**FPGA, Embedded RAM, Memory Based Computing, Energy-Efficiency, Energy-Accuracy Trade-off.**

I. INTRODUCTION

Reconfigurable computing platforms, such as Field Programmable Gate Array (FPGA), are increasingly used in embedded applications (such as DSP, multimedia, security and graphics) due to the flexibility in application mapping, reduced design cost and largely improved time-to-volume. However, these platforms are well-known to suffer from poor energy efficiency, primarily due to large overhead of their elaborate programmable interconnect (PI) fabric. Currently PI accounts for 80% of power and 60% of delay in FPGAs at nanoscale process technologies [3]. There is a growing need to address the power/energy issues in reconfigurable platforms while retaining their performance and flexibility advantages.

FPGA vendors such as Xilinx and Altera, as well as researchers from academia have investigated various device engineering options (such as low-k dielectric, multiple device thresholds) as well as architecture-level techniques (e.g. clustered architecture) to improve the energy consumption of the FPGA devices [6]. These optimization approaches however, cannot provide adequate solution to reduce the energy requirement for many compute-intensive applications. Besides, as the interconnect delay does not scale as significantly as logic delay, fine-grained architectures of FPGAs suffer from poor technological scalability of performance and energy.

On the other hand, an efficient application mapping methodology that can drastically reduce the need of PIs for an application while using conventional FPGA architecture can be extremely effective in reducing energy consumption. Such an approach can be attractive to both FPGA vendors and users alike, since it does not require modifications in FPGA hardware and hence saves device design/fabrication

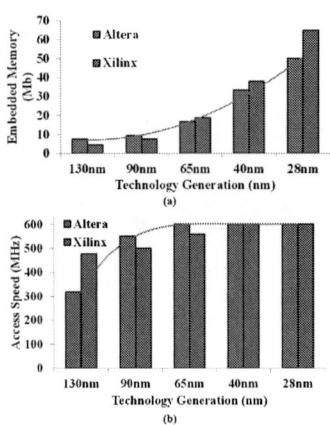

Fig. 1. The trend in embedded memory in FPGA: a) size (Mb), and b) access speed (MHz) for Altera Stratix [9] and Xilinx Virtex [10] series of FPGA devices across technology generations.

cost. Moreover, it is also able to work on legacy hardware. In this paper, we propose a novel application mapping methodology that uses the embedded memory arrays in FPGA for mapping complex compute-intensive parts of an application. We note that modern FPGAs come with large number of embedded memory blocks (EMBs) - e.g. Altera Stratix-IV devices are equipped with 17-33 Megabits (Mb) of block random access memory (RAM) in addition to the one-dimensional lookup tables (LUT) for configurable logic blocks (CLBs). As shown in Fig. 1, driven by aggressive technology scaling, FPGA devices from different vendors are integrating larger amount of RAM with faster access speed in each technology generation. In many applications, large part of embedded memory resources remain unused. Hence, through opportunistic use of these RAM blocks for computation - in particular, to realize complex datapaths or functions, we can significantly improve both energy efficiency and resource utilization compared to conventional logic based implementation.

Several previous investigations have earlier considered fine-grained application mapping in EMBs inside a FPGA [15], [16]. Although these investigations lay the foundation of using EMBs for computation, they are limited to mapping small Boolean functions in fine-grained applications (e.g. control logic) and do not consider coarse-grained mapping of complex operations (e.g. datapaths or transcendental functions). Moreover, they primarily focus on improving performance and do not analyze the impact on energy dissipation. Finally, they do not study the complete application mapping process and the effectiveness of computation with memory at nanoscale technology nodes using commercial FPGAs.

Compared to existing mapping approaches, the proposed mapping approach achieves significant improvement in energy-efficiency since

978-1-4673-0438-2/12 $31.00 © 2012 IEEE 424

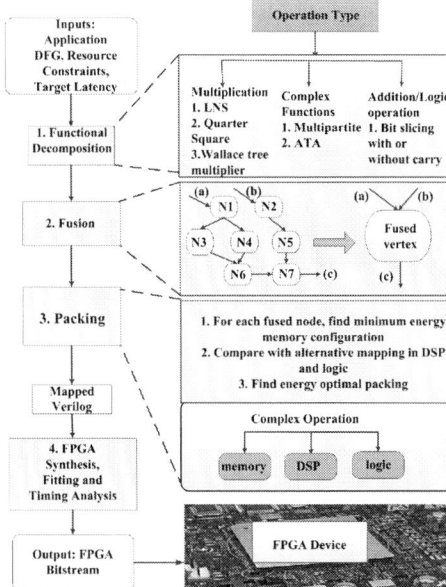

Fig. 2. Proposed application mapping steps using EMBs for computation.

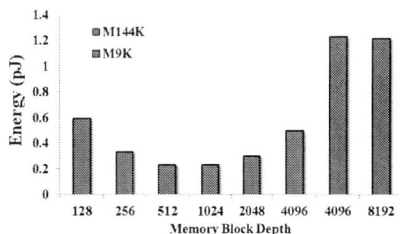

Fig. 3. Variation in energy consumption of a 12-bit input 12-bit output memory with varying memory type and block depth for a commercial FPGA.

the entire function is computed typically in one or two LUT accesses leading to significant improvement in operational latency and energy due to large reduction in PI overhead. In particular, the key contributions of the paper are:

1) It analyzes the effectiveness of mapping coarse-grained compute-intensive operations in an application to embedded memory blocks in FPGA. It shows that opportunistic mapping of complex datapaths and functions in memory in the form of large multi-input multi-output LUTs can significantly improve energy efficiency. It explores the most energy-efficient memory configuration in a FPGA for mapping operations with varying memory requirements.

2) It then develops a heterogeneous application mapping framework, which combines conventional application mapping in logic and DSP blocks (for DSP-enhanced FPGA devices) with judicious mapping of specific computations in memory. It determines the complete mapping methodology including functional decomposition, which partitions complex functions into multi-input/multi-output LUTs of manageable size, fusion of small operations into a large one, and finally optimal packing of operations into a combination of EMBs and logic array.

3) It validates the effectiveness of the proposed methodology by mapping a number of common scientific and graphics applications on a commercial state-of-the-art FPGA platform (Altera Stratix IV, 40nm process). It compares the performance and energy efficiency for the proposed heterogeneous mapping approach with conventional mapping strategies in FPGA.

4) It shows that computation in memory can be an effective vehicle for energy-accuracy tradeoff at run time. In this case, operand bitwidth for complex operation can be dynamically truncated to achieve exponential reduction in memory space leading to large savings in computation energy. The impact on the quality of service (due to operand truncation) is minimized through choice of optimal values for truncated bits and preferential truncation of the operations. We demonstrate the effectiveness of such dynamic trade-off for a common filtering application.

II. USE OF EMBEDDED MEMORY BLOCKS FOR COMPUTATION

It is well-known that a majority of scientific and graphics applications include a set of common compute-intensive kernels, which essentially constitute of basic mathematical operations such as addition, multiplication as well as complex functions such as sine, cosine, reciprocal, arctan, square root, exponentiation, and logarithm [1]. Traditionally, in a FPGA framework the transcendental functions are mapped using CORDIC approach [4] or Taylor series expansion [5], which either require large number of computing resources or suffers from large latency. Evaluation of these functions by holding the output response of a function as LUT in the embedded memory array is an attractive solution in terms of performance and energy consumption. For larger bitwidth, efficient decomposition techniques (as proposed in [7] and [8]) have also been employed. The focus of our work is to achieve energy efficient mapping of applications dominated by complex functions using EMBs. The following computations have been identified to be amenable for mapping to memory.

1) Any function satisfying the maximum LUT input size in a FPGA device. The function can be a regular one (such as $Sine(x)$ where x has a maximum resolution of 16-bits) or any arbitrary function obtained by fusing many simple ones.

2) Complex datapath having constant coefficients like two constant coefficient multiplications followed by an addition.

3) Any function that is easily bit-sliceable like logic and Galois field operations and some datapath operations like addition, magnitude comparison, multiplication, etc.

4) Functions where operands can be truncated with minimal degradation of output quality, e.g. DSP or multimedia applications.

5) Functions which are amenable to decomposition using a different number system. For example, a multiplication of two operands can be converted to a set of single-operand logarithmic operations followed by an arithmetic addition.

III. APPLICATION MAPPING METHODOLOGY

In this section, we describe the proposed application mapping methodology which uses a combination of EMBs and other FPGA resources to maximize energy efficiency for a given input application.

A. Mapping Flow

Figure 2 shows the proposed application mapping methodology. A comprehensive software flow has been developed in C language which implements the proposed methodology. The input to our application mapping flow is a Verilog netlist containing a data flow graph (DFG) representation of the target circuit. Major steps of the mapping process are described below.

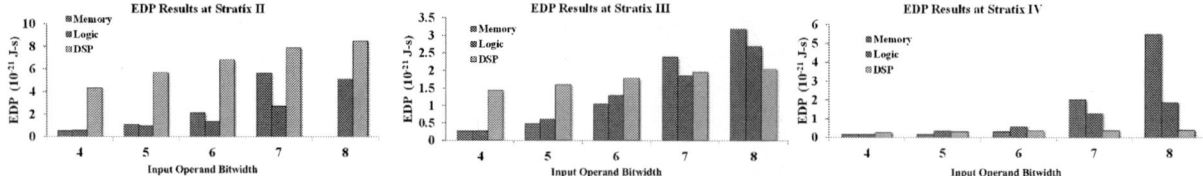

Fig. 4. Comparison of EDP with varying input resolution for constmult-add in Altera Stratix II, Stratix III and Stratix IV series of devices.

TABLE I

EDP IMPROVEMENTS WITH THE PROPOSED HETEROGENEOUS MAPPING APPROACH COMPARED TO MAPPING IN LOGIC AND DSP BLOCKS FOR THREE
GENERATIONS OF ALTERA FPGA FAMILIES

Input bitwidth	Stratix II		Stratix III		Stratix IV	
	% Impr. over Logic	% Impr. over DSP	% Impr. over Logic	% Impr. over DSP	% Impr. over Logic	% Impr. over DSP
4	10.26	87.90	1.16	81.18	1.12	33.80
5	-11.58	80.96	22.90	70.14	52.92	47.20
6	-54.62	68.64	18.74	41.34	44.37	14.04
7	-108.30	28.24	-28.60	-21.88	-56.83	-431.72
8	*	*	-18.51	-56.30	-195.79	-1299.69

*cannot place due to lack of memory resources

1) Functional Decomposition: In this step, nodes of the input DFG which violate the maximum LUT input constraint are decomposed into multiple LUTs with varying input sizes to limit the memory resource requirement. The maximum LUT input size is determined by the available memory sizes of the underlying device, for e.g. 16 in the case of Stratix IV devices. The decomposition step employs function-specific decomposition process, such as bit-slicing for large addition and logarithmic number system (LNS) for multiplication. For mapping complex transcendental functions, we employ multi-partite decomposition [8] whenever feasible or Addition Table Addition (ATA) approaches [7] for other functions. Nodes which satisfy the maximum LUT input constraint are not decomposed.

2) Fusion: Fusion involves opportunistic reduction in the total number of operations after decomposition (represented as vertices in the DFG) by combining multiple operations into a single operation which can be suitably mapped in a 2D LUT. Nodes in the modified DFG obtained after decomposition are clustered to achieve maximum input cone sharing [11] without violating the maximum LUT input (M) and output constraint (N).

3) Packing: The fused DFG is packed or allocated into different functional units like CLB, DSP or memory. We follow a greedy heuristic-based approach by packing each operation to a resource which provides optimal energy. In case of memory, the LUT contents of the functional blocks are also generated along with the fused DFG. Finally after packing, the allocated resources and their interconnections are converted to appropriate verilog netlist which consists of the required resources and the EMB contents which is readily mappable into the actual FPGA device with resource and I/O constraints.

B. Energy-Efficient Configuration of RAM Blocks

In the packing step we explore the optimal energy configuration of the EMBs for all the memory blocks instantiated in any application. For energy-efficient configuration of RAM blocks in a FPGA, we explore the design space for a given input and output size by varying memory block type and block depth [14].

Figure 3 shows the variation in energy consumption by varying the block type and depth of a 12-bit input 12-bit output memory. We can observe that the optimal configuration is obtained for a block type of M9K and block depth of 1024.

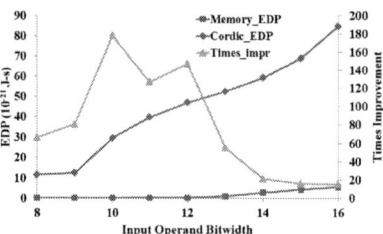

Fig. 5. Variation in EDP with varying input resolution for CORDIC based operations.

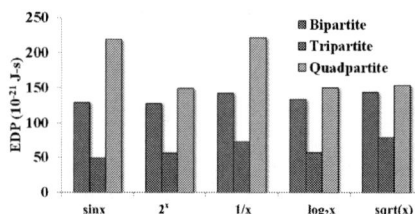

Fig. 6. EDP trends with bipartite, tripartite and quadpartite decompositions.

C. Mapping Complex Datapath in Memory

For smaller bitwidths, datapaths with large number of combinational levels are expected to have significantly better energy efficiency while mapping in memory compared to normal logic or DSP based mapping. To analyze this effect, we mapped a common datapath for DSP applications, namely two constant coefficient multiplications followed by an addition (constmult-add). We found optimal memory configurations for mapping it to memory and obtained energy and delay results across different technology generations, namely Stratix II, Stratix III and Stratix IV. We compared these results with the corresponding mapping results in logic and DSP blocks. Figure 4 compares the energy delay product (EDP) for varying input bitwidths in case of three alternative mapping schemes. Table I shows the corresponding EDP improvement compared to mapping only in logic and DSP. For Stratix IV, we observe EDP improvement for up to 6-bit input resolution, as shown in Table I.

978-1-4673-0438-2/12 $31.00 © 2012 IEEE

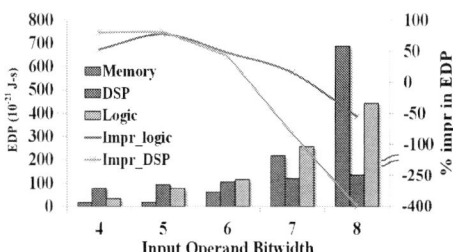

Fig. 7. Variation in EDP with varying input resolution for 8-tap FIR filter.

D. Mapping Complex Functions in Memory

Conventionally complex transcendental functions are computed in FPGA using CORDIC based approaches [12] where the latency of the computation increases linearly with the increase in input resolution. On the other hand, a LUT storing the output response of a function is capable of evaluating the function in one or few lookup operations, providing tremendous improvement in terms of energy consumption and latency. In a Stratix IV FPGA, for up to 16-bit input resolution, any function can be mapped in a single lookup table due to adequate embedded memory resources. Figure 5 shows the improvement in energy efficiency or EDP of LUT based approach with respect to normal CORDIC based uni-variable function realization in Stratix IV devices. The EDP improvement is minimum (15X) in case of 16-bit operands and maximum (178X) in case of 10-bit input operands. If the size of the input bitwidth does not match with the size of the lookup table that can be supported, then we need to employ efficient decomposition techniques, such as, multipartite and ATA method [7] in order to minimize the memory requirement at the cost of graceful increase in latency. These decomposition techniques have been investigated in order to achieve energy-efficient mapping of complex transcendental functions. Figure 6 shows the energy efficiency results in case of Stratix IV using bipartite, tripartite and quadpartite decompositions [8] of transcendental functions with 24-bit fixed point inputs. In general, tripartite decompositions give the most energy optimal mapping for 24-bit input uni-variable functions as shown in Fig. 6.

IV. APPLICATION MAPPING RESULTS

In this section, we evaluate the effectiveness of the proposed approach in improving energy efficiency for several common applications, which require complex datapaths and/or functions.

A. 8-tap FIR Filter

Implementation of a 8-tap FIR includes eight constant coefficient multiplications followed by the addition of the product terms. The EDP trends in Stratix IV for 8-tap FIR filter is shown in Fig. 7 with variation in input bitwidth. In case of the proposed heterogeneous mapping of memory and logic, significant EDP improvement is achieved for up to 7-bit inputs (16.7% improvement for 7-bit) when compared to logic only implementation and 6-bit inputs (41.2% improvement for 6-bit input) when compared to a heterogeneous implementation of logic and DSP blocks.

B. Calculation of Approximation Coefficient in DWT

The datapath for the approximation coefficient computation in a discrete wavelet transform (DWT) consists of two constant coefficient multiplications of the odd samples followed by one addition. After a single level of truncation, the intermediate output is added to the even sample of the current level. The EDP results with the variation

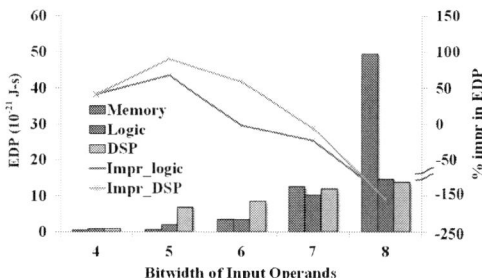

Fig. 8. Variation in EDP with varying input resolution for DWT.

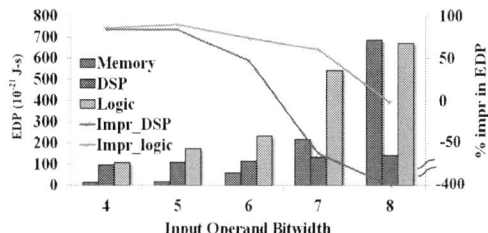

Fig. 9. Variation in EDP with varying input resolution for coherence calculation of an 8-point cluster.

in input bitwidth is shown in Fig. 8 and compared with that of logic and DSP based implementation in Stratix IV. EDP improvements are observed for up to 6-bit input width compared to both logic only (nearly 1% improvement for 6-bit input) and a heterogeneous implementation of logic and DSP (58.7% improvement in EDP for 6-bit input).

C. Coherence Calculation in a Cluster

For calculation of coherence in a cluster, first, the absolute distance between the different points allocated to a particular cluster from its centroid is computed and then the distances for all the points from their corresponding cluster centroid are squared and added which define the coherence of the cluster. Memory based computation in coherence calculation of an 8-point cluster is extremely helpful and gives significant advantage for up to 6-bit input over a DSP based implementation (47.4% improvement for 6-bit input) and up to 7-bit input over a logic only implementation (58.2% improvement for 7-bit input) The advantages in EDP over different bitwidth of inputs are shown in Fig. 9 for a Stratix IV platform.

D. Newton Raphson Method of Root Finding

Newton Raphson's (NR) method is widely used for the evaluation of roots for polynomial functions. Consider a 3rd order polynomial given by

$$f(x) = a_3 x^3 + a_2 x^2 + a_1 x + a_0 \quad (1)$$

NR method employs the following iteration formula in order to move closer to the actual root:

$$X_{n+1} = X_n - \frac{f(x)}{f'(x)} = X_n - \frac{a_3 X_n^3 + a_2 X_n^2 + a_1 X_n + a_0}{3a_3 X_n^2 + 2a_2 X_n + a_1} \quad (2)$$

For a 24-bit fixed point input, the proposed framework takes 106 KB of memory through the ATA based decomposition method [7]. The energy and EDP improvements achieved with respect to conventional mapping in FPGA is shown in Table II. The results of conventional mapping for all applications are obtained by choosing the better EDP results between logic and DSP-based implementation.

E. Solution to Schrodinger Equation 1-D

Finding solution of a time-independent Schrodinger wave equation for arbitrary periodic potentials is a common scientific application. Similar to many other scientific applications, it is dominated by transcendental functions. For single dimension, the function is given by:

$$\psi_n(x) = \sqrt{\frac{2}{L}} sin(\frac{n\pi x}{L}) \qquad (3)$$

A FPGA based evaluation of this function using the proposed mapping approach for 24-bit fixed-point input operand requires 193.96 KB of memory. Due the increased latency of CORDIC based computation, the proposed framework has significant improvement in energy and EDP over conventional mapping in FPGA, as shown in Table II.

F. Lighting in Computer Graphics

Lighting is a very well-known compute-intensive kernel in computer graphics, which involves computing several transcendental functions in parallel. The algorithm for lighting is as shown below:

$$I = K_a + c((K_d + K_s sin(v,t))sin(l,t) - \\ K_s exp[sln(cos(v,t))]cos(l,t)) \qquad (4)$$

where $c = not(msb(l.n))$ and K_a, K_d and K_s are the ambient, diffuse and specular coefficients, s is the surface shininess, l, n, v, t are the light, normal, view and tangent unit vectors on the surface and $sin(v,t)$ denotes the sine of the angle between the vectors v and t. The actual inputs to the computing framework are the vectors v, t and l whereas the coefficients or the constants are assumed to be known beforehand. We find that for a 24-bit fixed point input, proposed computing framework requires 335.84 KB of memory. The energy and EDP improvements achieved with respect to conventional mapping in FPGA is shown in Table II.

G. Gaussian Random Noise Generator

Generation of random numbers with Gaussian probability distribution is used in wide range of applications, such as simulation of economic systems, scientific applications like molecular dynamics, and cryptography. Popular Box-Muller Method of generating the Gaussian noise involves evaluation of three complex transcendental functions, namely $f1$, $f2$, $f3$ where $f1 = \sqrt{-ln(u_1)}$, $g1 = \sqrt{2} sin(2\pi u_2)$ and $g2 = \sqrt{2} cos(2\pi u_2)$ where $u1$ is 24-bit and $u2$ is 16-bit wide. The input bitwidth determines the maximum possible value of the Gaussian sample. Here it is assumed that that the input bitwidth is 40 bits and the output bitwidth is 24 bits. The total memory requirement is 270.76 KB. The energy and EDP improvements achieved with respect to conventional mapping in FPGA is shown in Table II.

V. DYNAMIC ENERGY ACCURACY TRADE-OFF

As demonstrated in the previous sections, heterogeneous mapping can provide significant improvement in energy consumption for specific datapath bitwidth in case of applications dominated by complex datapaths or functions. As a result, such a heterogeneous mapping methodology can be utilized to trade-off energy consumption and accuracy requirement dynamically for DSP and multimedia applications. In this case, operand bitwidth can be reduced at run time through truncation such that the memory requirement in mapping the complex datapaths/functions can be exponentially reduced leading to large savings in energy. Clearly, operand truncation would have impact on the output quality. However, judicious choice of bit values assigned at truncated bit positions and a non-uniform truncation

approach that aggressively truncates inputs of only the less critical components can lead to modest impact on output. Next, we describe the truncation approaches in details.

1) Uniform Truncation: First, we study the effect of truncating inputs uniformly, i.e. changing input resolution uniformly across the board to see the effect on primary output values of a FIR filter. We truncate inputs at all taps equally and generate the LUTs by optimal value allocation at the truncated bits. The designed filter is essentially a low-pass equiripple 32-tap FIR filter with a passband frequency of 9.6 kHz and a stopband frequency of 12 kHz. The original filter input operand resolution is 8 bits. Three bits of the inputs, starting from the LSB have been allocated different values and the impact in the output mean square error (MSE) for different value allocation at the inputs is shown in Fig. 10. The inputs applied to the filter is an impulse input of magnitude $8'b11111111$. It can be inferred that for 3-bit truncation, LUT representation assuming $3'b011$ or $3'b100$ values at the truncated positions provides minimum error at the output.

2) Preferential Truncation: To minimize the effect on output quality due to truncation while maintaining the benefit of energy reduction, we propose a preferential truncation approach, which truncates specific inputs more aggressively than others. This leverages on the observation that most DSP applications have some components which are more critical than others in terms of output quality [17]. Hence, unlike the uniform truncation approach discussed earlier, we propose truncating the inputs of a less significant component more aggressively than the significant ones.

The variation in the stopband ripple due to zeroing different tap coefficients is shown in Fig. 11. It shows the sensitivity of different taps to output quality. We observe that the middle taps are more critical than the side ones. Based on this observation, less truncation is applied to the more significant taps while more aggressive truncation is applied at the non-critical taps in order to achieve minimum impact on output quality at iso-energy consumption. In particular, the middle tap inputs are truncated by one bit, the inputs corresponding to the 1st and 2nd order coefficients are truncated by 2 and 3 bits, respectively, and so on. We refer this preferential truncation scheme as $Config.1$. We also explore two other variations of preferential truncations. In the second ($Config.2$) and third schemes ($Config.3$), 2 and 3 bits are truncated respectively at the middle with increasing truncation at the sides.

Figure 12 presents the output MSE and energy consumption trade-off for different truncation scenarios in case of a 32-tap FIR filter. Figure 12(a) compares between two uniform truncation approaches. Zero value allocation denotes blind truncation with $1'b0$ at the truncated bits. The three design points in case of zero value allocation denote $1'b0$, $2'b0$ and $3'b0$ truncations at primary inputs. Similarly, optimal value allocation denote that optimal bit values are assigned at the truncated bits ($1'b0$, $2'b01$, $3'b011$, respectively). We observe that optimal value allocation gives lesser MSE compared to zero value

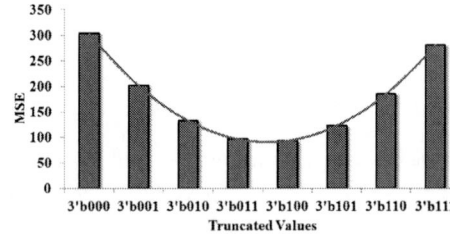

Fig. 10. Variation in mean square error (MSE) with different value assignments at the truncated bits for a 32-tap FIR filter.

978-1-4673-0438-2/12 $31.00 © 2012 IEEE

TABLE II
COMPARISON OF ENERGY, LATENCY AND EDP FOR SEVERAL COMMON SCIENTIFIC AND GRAPHICS APPLICATIONS FOR 24-BIT FIXED POINT INPUTS

Applications	Conventional Mapping in FPGA			Proposed Mapping Framework				%Impr. in Energy
	Latency (ns)	Energy (pJ)	EDP $(10^{-18}J-s)$	Latency (ns)	Energy (pJ)	EDP $(10^{-18}J-s)$	Memory (KB)	
Polynomial evaluation (3rd order)	49.0	379.1	18.9	7.6	53.7	0.4	106.0	85.8
Schrodinger equation (1D)	274.4	1258.1	345.2	25.0	368.7	9.2	193.9	70.7
Lighting (computer graphics)	380.0	3684.0	1400.2	90.0	1540.4	138.6	335.8	58.2
Gaussian noise generator	180.0	232.3	41.8	27.6	157.2	4.3	270.8	32.3

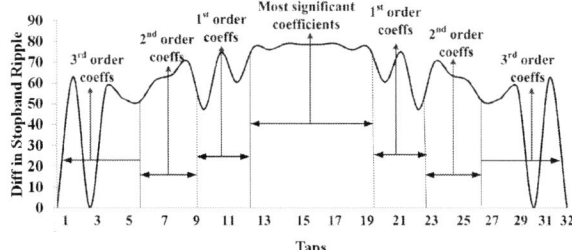

Fig. 11. Variation in stopband ripple magnitude by zeroing different coefficients of a 32-tap FIR filter.

allocation but consumes the same amount of energy.

Figure 12(b) compares between uniform (with optimal value assignment) and preferential truncation approaches. The three design points for preferential truncation correspond to $Config.1$, $Config.2$ and $Config.3$ as described above. Preferential truncation can provide significant improvement in output quality at similar energy levels compared to optimal value allocation. In particular, for $Config.2$, the proposed preferential truncation methodology results in 87% and 96.9% improvement in output MSE over optimal value and zero value allocation respectively at iso-energy consumption.

VI. CONCLUSION

We have presented a novel application mapping process for FPGA that exploits the embedded memory blocks in FPGA for computation to minimize the overhead of programmable interconnects. We show that the proposed mapping process can significantly improve energy efficiency for many applications which require complex datapath, complex functions or both. Scalability of the proposed mapping to different bitwidth is addressed with detailed analysis. Software architecture to perform functional decomposition, fusion of operations and packing of individual coarse-grained operations is presented. The effectiveness of the proposed mapping has been validated for a commercial FPGA platform, namely Altera Stratix IV FPGA. Finally, computation in embedded memory in FPGA is extended to

dynamically trade-off between energy and accuracy through use of judicious bitwidth truncation at run time. With emergence of novel high-density memory technologies- both volatile and non-volatile- FPGA devices are expected to integrate more memory with improved performance, which can significantly benefit the proposed approach. Future work will include extending the framework to other domain of applications, support for different data format and precision, and more efficient functional decomposition.

REFERENCES

[1] "The landscape of parallel computing research: A view from Berkeley" [Online] http://www.eecs.berkeley.edu/Pubs/TechRpts/2006/EECS-2006-183.html.

[2] T. Good and M. Benaissa, "AES on FPGA from thr Fastest to the Smallest," em Cryptographic Hardware and Embedded Systems, 2005.

[3] A. Rahman, S. Das, A. P. Chandrakasan and R. Reif, "Wiring Requirement and Three-Dimensional Integration Technology for Field Programmable Gate Arrays," IEEE Transactions on Very Large Scale Integration Systems, Vol. 11, No. 1, 2003.

[4] W.B. Ligon III, G. Monn, D. Stanzione, F. Stivers and K.D. Underwood, "Implementation and Analysis of Numerical Components for Reconfigurable Computing," Aerospace Applications Conference,1999.

[5] C. Brunelli, H. Berg and D. Guevorkian, "Approximating sine functions Using Variable Precision Taylor Polynomials," IEEE Workshop on Signal Processing Systems, 2009.

[6] J. Lamoureux and W. Luk, "An Overview of Low Power Techniques for Field Programmable Gate Arrays," NASA/ESA Conference on Adaptive Hardware and Systems, 2008.

[7] W.F. Wong and E. Goto, "Fast Evaluation of Elementary Functions in Single Precision," IEEE Transactions on Computers, Vol. 44, No. 3, 1995.

[8] F. D. Dinechin and A. Tisserand, "Multipartite Table Methods," IEEE Transactions on Computers, Vol. 54, No. 3, 2005.

[9] "Stratix FPGA: Low Power, High Performance" [Online] http://www.altera.com/devices/fpga/stratix-fpgas/stratix/stratix/stx-index.jsp

[10] "Xilinx Virtex series of FPGAs" [Online] http://www.xilinx.com/products/index.htm

[11] S. Paul, S. Chatterjee, S. Mukhopadhyay and S. Bhunia, "A Circuit-Software Co-Design Approach for Improving EDP in Reconfigurable Frameworks," International Conference on Computer-Aided Design, 2009.

[12] T. Lang and E. Antelo, "High Throughput CORDIC Based Geometry Operations for 3D Computer Graphics," IEEE Transactions on Computers, Vol. 54, No. 3, 1997.

[13] D.U. Lee, W. Luk, J.D. Villasenor and P.Y.K. Cheung, "A Gaussian Noise Generator for Hardware Based Simulations," IEEE Transactions on Computers,Vol. 53, No. 12, 2004.

[14] R. Tessier, V. Betz, D. Neto, A. Egier and T. Gopalsamy, "Power-Efficient RAM Mapping Algorithms for FPGA Embedded Memory Blocks," IEEE Transactions on Computer-Aided Design of Circuits and Systems, Vol. 26, No. 2, 2007.

[15] J. Cong and S. Xu, "Technology Mapping for FPGAs with Embedded Memory Blocks," International Symposium on Field Programmable Gate Arrays, 1998.

[16] S. Wilton, "SMAP: Heterogeneous Technology Mapping for Area Reduction in FPGAs with Embedded Memory Arrays," International Symposium on Field Programmable Gate Arrays, 1998.

[17] N. Banerjee, J.H. Choi and K. Roy, "A Process Variation Aware Low Power Synthesis Methodology for fixed point FIR filters," International Symposium on Low Power Electronics and Design, 2007.

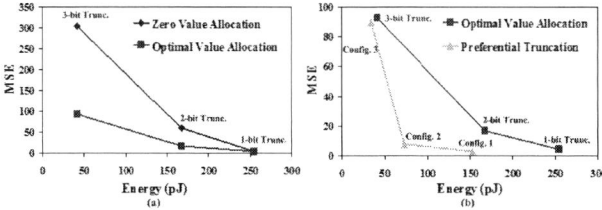

Fig. 12. MSE and energy consumption trade-off for a 32-tap FIR filter using three variations of heterogeneous mapping approach - (a) zero vs. optimal value allocation, (b) optimal value allocation vs. preferential truncation.

978-1-4673-0438-2/12 $31.00 © 2012 IEEE

Intra-Task Dynamic Cache Reconfiguration*

Hadi Hajimiri, Prabhat Mishra
Department of Computer & Information Science & Engineering
University of Florida, Gainesville, Florida, USA
{hadi, prabhat}@cise.ufl.edu

ABSTRACT

Optimization techniques are widely used in embedded systems design to improve overall area, performance and energy requirements. Dynamic cache reconfiguration (DCR) is very effective to reduce energy consumption of cache subsystems. Finding the right reconfiguration points in a task and selecting appropriate cache configurations for each phase are the primary challenges in phase-based DCR. In this paper, we present a novel intra-task dynamic cache reconfiguration technique using a detailed cache model, and tune a highly-configurable cache on a per-phase basis compared to tuning once per application. Experimental results demonstrate that our intra-task DCR can achieve up to 27% (12% on average) and 19% (7% on average) energy savings for instruction and data caches, respectively, without introducing any performance penalty.

1. INTRODUCTION

Energy conservation has been a primary optimization objective in designing embedded systems. Several studies have shown that memory hierarchy accounts for as much as 50% of the total energy consumption in many embedded systems [1]. Unlike desktop-based systems, embedded systems are designed to run a specific set of well-defined applications (tasks). Moreover, different applications require highly diverse cache configurations for optimal energy consumption in the memory hierarchy. Thus it is possible to have a cache architecture that is tuned for those applications to have both increased performance as well as lower energy consumption. Traditional dynamic cache reconfiguration (**DCR**) techniques reduce cache energy consumption by tuning the cache to applications need during runtime on task-by-task basis. For each task only one cache configuration is assigned to the task, and it is not changed during the task execution. These techniques are referred as inter-task DCR. Studies have shown that *inter-task* DCR can achieve significant energy savings [2].

Due to task-level granularity, inter-task DCR loses the energy savings opportunity that can be achieved by increasing the reconfiguration granularity. A modern processor executes billions of instructions per second and a program's behavior can change many times during that period. The behavior of some programs changes drastically, switching between periods of high and low performance, yet system design and optimization typically focus on average system behavior. Instead of assuming average behavior, it is highly beneficial to model and optimize phase-based program behavior. *Intra-task* tuning techniques tweak system parameters for each application phase of execution. Parameters are varied during execution of an application, as opposed to keeping fixed as in an application-based (inter-task) tuning methodology. Furthermore, inter-task DCR is not beneficial in a single-task environment (or in a multi-task environment where execution time of one task is dominant) because the cache configuration is determined on a per task basis. Since many small-size embedded-mobile applications are based on a single-task model, inter-task DCR cannot provide the best possible energy savings for such systems.

These limitations lead to the idea of intra-task DCR where a given task is partitioned into several phases, and different cache configurations

are assigned for each phase. There have been limited attempts [3] [4] for developing an intra-task DCR but they provide no systematic methodology of selecting the best program locations where DCR can be applied (phase detection). Furthermore, these approaches either perform exhaustive exploration (can be infeasible in many scenarios) or select suboptimal cache configurations. In this paper, we propose an intra-task DCR approach based on static analysis of a target application achieving significant improvements in energy consumption. It also can be applied to a single-task environment since it reconfigures the cache within each task. We propose a phase detection technique that fully exploits drastic changes in program behavior and finds boundaries between phases of high and low performance. In addition, we propose a dynamic programming based cache assignment algorithm that finds the optimal cache solution and reduces the time complexity of design space exploration.

The rest of the paper is organized as follows. Section 2 provides an overview of related research activities. Basic notations, cache and energy model are described in Section 3. Our proposed intra-task DCR methodology is presented in Section 4. Experimental results are discussed in Section 5. Finally, Section 06 concludes the paper.

2. RELATED WORK

DCR has been extensively studied in several works [5] [6] [7]. The problem is to determine the best cache configuration for a particular application. Most such methods configure cache size, line size, and associativity for only a single level of cache. Existing techniques can be classified into dynamic and static analysis. By dynamic analysis different cache configurations are evaluated on-line (i.e., during runtime) to find the best configuration. However, it introduces significant performance/energy overhead which may not be feasible in many embedded systems with real-time constraints. During static analysis, variety of cache options can be explored thoroughly and the best cache configuration is chosen for each application [5]. Regardless of the tuning method, the predetermined best cache configuration can be stored in a look-up table or encoded into specialized instructions [5]. The reconfigurable cache architecture proposed by Zhang et al. [8] determines the best cache parameters by using Pareto-optimal points trading off energy consumption and performance. Chen and Zou [9] introduced a novel reconfiguration management algorithm to efficiently search the large space of possible cache configurations for the optimal one.

Peng and Sun [3] introduced a phase-based self-tuning algorithm, which can automatically manage the reconfigurable cache on a per-phase basis. Their method used dynamic profiling of applications and limited to only four choices of cache configurations for L1 cache. Gordon-Ross et al. [4] proposed an intra-task DCR where each task is partitioned into fixed-length timeslots. It [4] shows limited improvements in the energy reduction (only 3% on average). Moreover, it provides no systematic methodology for selecting the best program locations where DCR can be profitable (programs divided into equal phases). These techniques solved the cache assignment either by performing exhaustive exploration (can be infeasible in many scenarios) or selecting suboptimal cache configurations. Our methodology outperforms existing approaches using novel phase detection and cache selection algorithms.

---

* This work was partially supported by NSF grant CCF-0903430 and SRC grant 2009-HJ-1979.

3. BACKGROUND AND MOTIVATION

3.1 Inter-task versus Intra-task DCR

Fig. 1 illustrates how energy consumption can be reduced by using inter-task (application-based) cache reconfiguration in a simple system supporting three tasks. In application-based cache tuning, dynamic cache reconfiguration happens when a task starts its execution or it resumes from an interrupt (either by preemption or when execution of another task completes). Fig. 1 (a) depicts a traditional system and Fig. 1 (b) depicts a system with a reconfigurable cache. For the ease of illustration let's assume cache size is the only reconfigurable parameter of cache (associativity and line size are ignored). In this example, Task1 starts its execution at time P1. Task2 and Task3 start at P2 and P3, respectively. In a traditional approach, the system always executes using a 4096-byte cache. We call this cache as the **base cache** throughout the paper. This cache is the best possible cache configuration (in terms of energy consumption) for this set of tasks. In Fig. 1(b), Task1, Task2, and Task3 execute using 1024-byte cache starting at P1, 8192-byte cache starting at P2, and 4096-byte cache starting at P3, respectively.

Although inter-task DCR provides significant energy savings compared to using only the *base cache*, it has several practical limitations as discussed in Section 1. Hence it may be more efficient in terms of energy consumption to utilize different cache configurations in different phases of a task. Fig. 1 (c) depicts intra-task DCR where reconfiguration can be done per phase basis. A task may need larger cache size for only a small phase of execution. Increasing the cache size for this phase would boost performance and decrease both cache misses and energy consumption. However, in some of the program phases the application may need a lower cache size thus the cache size can be reduced without loss of performance to produce savings in energy consumption. In these cases, intra-task DCR is able to fulfill cache needs of application perfectly while minimizing the energy consumption.

3.2 Energy Model

In this subsection, we describe the energy model for the reconfigurable cache. We assume that DCR is available in the target system. Specifically, we have a highly configurable cache architecture, with re-

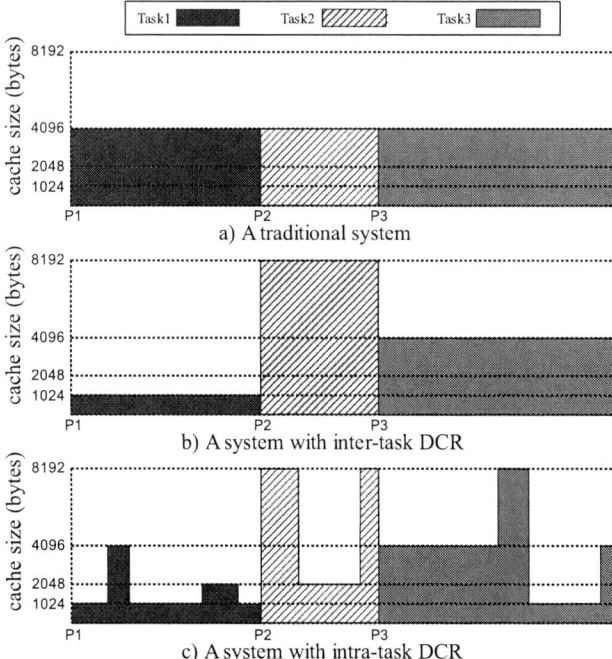

a) A traditional system

b) A system with inter-task DCR

c) A system with intra-task DCR

Fig. 1: DCR for a system with three tasks

configurable parameters including cache size, line size and associativity, which can be tuned to m different configurations $C = \{c_1, c_2, c_3, \dots , c_m\}$. Cache energy consumption consists of dynamic energy E_{cache}^{dyn} and static energy E_{cache}^{stat} [10]: $E_{cache} = E_{cache}^{dyn} + E_{cache}^{stat}$. The number of cache accesses *num_accesses*, cache misses *num_misses* and clock cycles CC are obtained from simulation using SimpleScalar [11] for any given task and cache configuration. Let E_{access} and E_{miss} denote the energy consumed per cache access and miss, respectively. Therefore, we have:

$$E_{cache}^{dyn} = num\_accesses . E_{access} + num\_misses . E_{miss}$$

$$E_{cache}^{stat} = P_{cache}^{stat} . CC . t_{cycle}$$

Where P_{cache}^{stat} is the static power consumption of cache. We collect E_{access} and P_{cache}^{stat} from CACTI [12] for all cache configurations and adopt E_{miss} and other numbers for other parameters from [8].

4. INTRA-TASK DCR

We define a phase as a set of intervals (or time slices) within a program's execution that has similar behavior. The key observation for discovering phases is that the cache behavior of a program changes greatly during execution. We can find this phase behavior and classify it by examining the number of cache misses in each interval. We collect this information through static profiling of the program. We begin the analysis of phases with an illustrative example of the time-varying behavior of *epic-encode* from MediaBench [13]. To characterize the behavior of this program, we have simulated its execution using a 1024-byte cache with one-way associativity and 32-byte line size. Fig. 2 shows the cache behavior of the program, measured in terms of cache miss statistics using two cache configurations (C_1 and C_2).

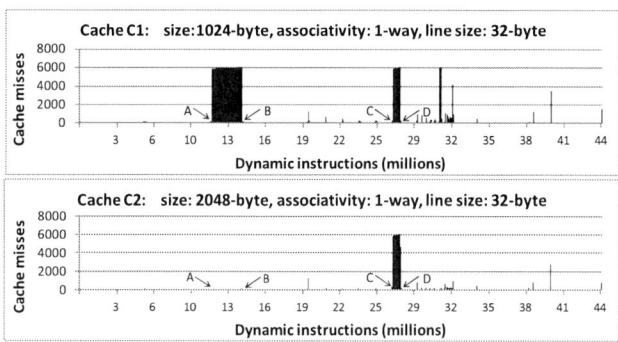

Fig. 2: Instruction cache miss for *epic-encode* benchmark

Each point on the graph represents the frequency of instruction cache misses taken over 100,000 instructions of execution (an *interval*). Two important aspects can be observed from this graph. First, average behavior does not sufficiently characterize a program's behavior in all phases of execution. For example, in *epic-encode* the number of instruction cache misses varies by several orders of magnitude. Second, the program can exhibit stable behavior for millions of instructions and then suddenly change. As a result, *epic-encode*'s behavior alternates greatly between phases. These two aspects, imply that significant energy savings can be achieved by accurately reconfiguring the cache to satisfy long-term execution behavior.

For *epic-encode* benchmark, we first need to find cache miss statistics in order to find potential reconfiguration points. Note that the least energy cache configuration for *epic-encode* benchmark is a 2048-byte cache with associativity of 1 and line size of 32 (cache C_2 in Fig. 2) chosen by inter-task cache configuration method. From Fig. 2, it can be observed that up to point A (around the dynamic instruction 12 million) miss rates are nearly the same for both caches C_1 and C_2. Starting from A to point B the miss rates are greatly different. We find

978-1-4673-0438-2/12 $31.00 © 2012 IEEE

A and B as potential reconfiguration points for this example. For the ease of illustration, let's assume only configurations C_1 and C_2 are available. Since C_2 is larger than C_1, C_2 is beneficial for performance and dynamic energy but detrimental for leakage energy compared to C_1. To reduce energy consumption we can run the program using configuration C_1 up to A then reconfigure the cache and use configuration C_2 from A to B and then again reconfigure the cache back to C_1.

In this paper, with the aim of energy optimization, we present a method to enable automatic partitioning of a program's execution into a set of phases that will quantify the changing behavior over time. The goal is that after finding phases, each phase would use a specific cache configuration suitable for that phase to reduce energy consumption without performance loss. We define the following terms that we will use in the rest of the paper:

- An *interval* is a section of continuous execution, a time slice, within a program. We chose intervals of equal length, as measured by the number of instructions executed during program execution. In this paper we choose 100,000 instructions as the length of intervals[2].
- A *phase* is a set of consecutive intervals within a program's execution that have similar and stable behavior. Boundaries of each phase are determined by reconfiguration points. For example, Fig. 2 has three phases; *start of execution* to A, A to B, and B to *end of execution*.
- A *potential reconfiguration point* is a point in the execution of a program at which a noticeable and sudden change in program behavior happens going from one phase to another phase. For example, A and B are potential reconfiguration points.
- The *profitability* of a reconfiguration point is a metric that shows how well a reconfiguration point can distinguish two different phases of a program. We use this metric for building a spectrum of energy savings while the number of reconfigurations is limited. We describe this metric in Section 4.1.

Fig. 3 shows an overview of our intra-task DCR approach. Our approach has two major steps (represented by ovals): phase detection and cache assignment. During a program's lifetime it can execute millions or billions of instructions each of which can be a reconfiguration point. The challenge is to choose a small number of profitable points from these millions of points. Moreover, the reconfiguration overhead is not constant and is different based on the point where the reconfiguration happens and can be found by actually reconfiguring and flushing the cache at that point during simulation. Thus finding the best set of reconfiguration points that is capable of separating program phases and guarantees energy savings is a difficult problem. We instead, find a set of potential reconfiguration points. Next, we choose if reconfiguring the cache is feasible at each point and if yes to what cache configuration. In order to find the potential reconfiguration points we compare frequency of misses in each interval. In addition, the energy consumption of a phase using a particular cache can vary depending on whether the previous phase has executed using the same cache (reconfiguration is needed if the cache is different). These are the main challenges we address in our approach. In the remainder of this section we explain each of these steps in detail.

4.1 Phase Detection
A phase is a set of consecutive intervals determined by two reconfiguration points (starting interval and ending interval). Finding best possible set of potential reconfiguration points is the objective of this step. First, we generate cache miss statistics (using simulation) for all possible cache configurations and find frequency of misses in each interval. Next, we compute the difference of frequency of misses (for all

---

[2] We chose the interval length to be small (100,000 instructions) to increase granularity of cache miss information.

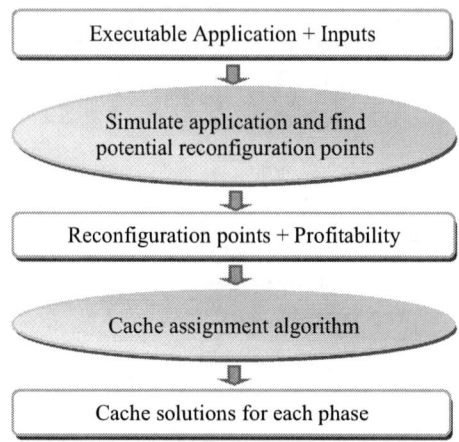

Fig. 3: Overview of our intra-task DCR

possible pairs of cache configurations) to discover the potential reconfiguration points. The statistics for data and instruction caches are gathered separately. Miss data is then used to calculate the frequency of cache misses in each interval of 100,000 dynamic instructions.

We use the example in Fig. 2 to explain our phase detection algorithm. Fig. 2 shows the miss frequency of the application *epic-encode* using cache configurations C_1 and C_2. Every point in the chart represents the frequency of misses (in thousands) in an interval. For example, the frequency of misses at A, for cache configuration C_1 is 6000 while it is nearly zero using C_2. We compare frequency of misses for cache configurations C_1 and C_2 to discover potential reconfiguration points. We include the edge of the regions in which the magnitude of the difference is greater than the threshold (we choose threshold to be 1000 in this example) into the set of reconfiguration points. For example, the magnitude of the difference in intervals A to B is greater than 1000 so we take the edge points of this region (the first instruction in A and the last instruction in B) as potential reconfiguration points. Considering the edge points of A to B as our potential reconfiguration points will create the phases, P_{start} (*start of execution* to A), P_2 (A to B), and so on.

Analyzing a miss frequency by itself may not necessarily lead to finding reasonable reconfiguration points since changes in cache miss frequencies may happen for all caches due to the cache behavior of a program. For instance, in Fig. 2, at the interval C to D we observe a significant change in cache misses. However this change is nearly the same in both cases. Since both cache configurations have the same behavior these points are not good candidates for a reconfiguration point. We find reconfiguration points as phase boundaries so that we would reconfigure the cache and use a different cache configuration. If all of the caches exhibit the same behavior this means program can continue with the same cache it was executing before. For this reason we compare miss frequencies of different cache configurations instead of scrutinizing frequencies solely.

Algorithm 1 outlines our heuristic to find a set of reconfiguration points. We compare frequency of misses ($f1$ and $f2$) for all pairs of cache configurations using a dynamic threshold to find potential reconfiguration points with their profitability. Every element in arrays $f1$ and $f2$ keeps the frequency of misses in an interval (for example $f1_k$ represents the number of misses in the kth interval). We treat the frequency of misses as a pattern (a time-varying quantity). So basically we use (compare) the intersection of two patterns and exploit their differences to discover potential reconfiguration points. The array *Profitability* (in Algorithm1) is determined by the magnitude of the differences between two frequencies of misses and is used as a metric that represents effectiveness of a point in discovering boundaries of

Algorithm 1: Finding potential reconfiguration points
Input: Cache miss statistics for each cache configuration
Output: List of potential cache configuration points
Begin
    th = the starting threshold;
    n = number of intervals;
    li = an empty list to store potential reconfiguration points;
    for i=*0 to 17* **do**
        for j=i *to 17* **do**
            f1 = array of frequency of misses for cache C_i for all intervals
            f2 = array of frequency of misses for cache C_j for all intervals
            Profitability = differences of *f1* and *f2*;
            for k=0 *to n* **do**
                if ($Profitability_k > th$) then
                    add the pair (po_k , $Profitability_k$) to li;
            end for
        end for
    end for
    return li;
end

phases (for instance *A* and *B* in Fig. 2). We include a point, po_k, (starting point of the kth interval) into the set of reconfiguration points only if the profitability of this interval, $Profitability_k$, is greater than the threshold. Note that these points are potential reconfiguration points and reconfiguration may not actually happen at these points.

Finding potential reconfiguration points may seem simple; however, there are several challenges in finding points that are beneficial in practice. First, the absolute number of misses can be significantly different for each cache configuration not because of changes in interactions between cache and program but due to the difference in cache/line size. For example total number of misses for a 1024-byte cache may be several orders of magnitude greater than number of misses for an 8192-byte cache. This makes comparison of frequencies unfair and biased towards the smaller cache. Second, since the length of intervals is relatively small, reconfiguration points may be chosen very close to each other (with a distance of one interval). Reconfiguring cache after such a short period of time does not seem reasonable due to the reconfiguration overhead. Third, when a program is not in a stable phase it may have a chaotic cache behavior; many ups and downs will be present in the miss frequencies. Thus comparison of miss patterns is prone to fluctuation from glitches in frequency of misses that will result in numerous unnecessary reconfiguration points that are ineffective in separating program phases.

To cope with these challenges we carried out several improvements to the algorithm. First, in order to perform an unbiased comparison between frequencies we normalize them before comparing so that the sum of number of misses over all program execution is a constant value for all caches. This way we ignore the absolute number of misses while we keep the information about the behavior of cache. Second, we limit the minimum distance between two reconfiguration points to be at least 500,000 instructions. This will be the minimum length of a phase that is reasonable to reconfigure the cache. This length is mainly determined by the reconfiguration overhead. Therefore, if there are multiple reconfiguration points that are close to each other (in the minimum distance range) we choose the one with the highest profitability. Third, we ignore the short-term fluctuations in frequency of misses (cache behavior) when comparing cache miss patterns.

4.2 Cache Assignment Algorithm
Finding the best cache configuration for each program phase using potential reconfiguration points from the previous step is the goal of this step. We call this problem as cache assignment since we are assigning a cache for each of the program phases. In this step we employ

a dynamic programming based algorithm for optimal cache assignment which significantly reduces the time complexity of cache selection. Solution for instruction cache can affect the energy of data cache by increasing/decreasing the execution time of a phase. This will change the static energy of the chosen cache. However, according to simulation results reconfiguration overhead for data/instruction caches mostly consist of dynamic energy hence we can solve the cache selection problem for data and instruction cache independently.

After finding the potential reconfiguration points we need to find the exact energy consumption and the execution time for each phase for all possible cache configurations. We use simulation to obtain cache statistics (time and energy) for the possible 18 cache configurations[3]. We modified SimpleScalar [11] to reconfigure and flush the cache at the reconfiguration points. Reconfiguration overhead for all phases and for instruction and data caches are computed separately. For phase P_i we find energy/time for each of the two cases. Case1 is when the chosen cache for the phase is the same as the selected cache for phase P_{i-1}. In this case the cache is not flushed and will keep the data (no reconfiguration). Case2 is when the chosen cache for this phase is different from the selected cache for phase P_{i-1}, where reconfiguration takes place and the cache should be flushed. Therefore simulation starts this phase with an empty cache (accounts for reconfiguration overhead).

Table 1: Notations

Symbol	Representing
E_{P_i,C_j}	Energy consumption of phase P_i using cache C_j while P_{i-1} also used C_j (no reconfiguration)
$E^f_{P_i,C_j}$	Energy consumption of phase Pi using cache C_j starting with a flushed cache (includes reconfiguration overhead), i.e., P_{i-1} does not use C_j.
$S_{i,C_j} =$ $\{C_{P_{start}}, \dots, C_{P_i} = C_j\}$	The most profitable solution for the set of consecutive phases P_{start} to P_i assigning $C_{P_{start}}$ to $P_{start},\dots,$ $C_{P_{i-1}}$ to P_{i-1} and C_j to P_i, i.e., the last phase uses C_j
$E_{S_{i,C_j}}$	Energy consumption of solution S_{i,C_j} (for phases P_{start} to P_i)

We present a recursive approach to find the optimal solution for each of the phases. Table 1 includes a set of notations we use in the rest of this section. In our recursive approach, in the general case, when there are m phases and n available cache configurations, we can find the best cache configuration for the phases P_{start} to P_i using the following formula:

$$\begin{cases} E_{S_{i,C_1}} = min(E_{S_{i-1,C_1}} + E_{P_i,C_1}, \dots , E_{S_{i-1,C_n}} + E^f_{P_i,C_1}) \\ \vdots \\ E_{S_{i,C_n}} = min(E_{S_{i-1,C_1}} + E^f_{P_i,C_n}, \dots , E_{S_{i-1,C_n}} + E_{P_i,C_n}) \end{cases} \quad \textbf{(Eq. 1)}$$

$$\text{with the initial state:} \begin{cases} E_{S_{start,C_1}} = E^f_{P_{start},C_1} \\ \vdots \\ E_{S_{start,C_n}} = E^f_{P_{start},C_n} \end{cases}$$

We observe that storing all possible cache combinations is not needed (for finding the optimal solution) in each iteration. We only need to keep the one with the lowest energy consumption from all possible solutions ending with a particular cache. All other combinations ending with the same cache can be discarded.

---

[3] In our work we use a 4KB L1 cache architecture proposed in [16]. Since the reconfiguration of associativity is achieved by way concatenation, 1KB L1 cache can only be direct-mapped as three of the banks are shut down. For the same reason, 2KB cache can only be configured to direct-mapped or 2-way associativity. Therefore, there are 18 (=3+6+9) configuration candidates for L1.

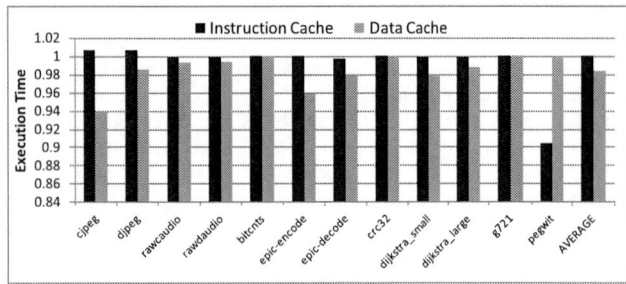

Fig. 5: Execution time normalized to the least-energy cache configuration found by inter-task DCR

Algorithm 2 shows an iterative implementation of our cache assignment approach. In each iteration, we evaluate C_l for the phase P_i considering all of the solutions found from P_{start} to P_{i-1} in the last iteration. By comparing these solutions we find the best solution for phases P_{start} to P_i ending with cache C_l. For each of the possible cache configurations we find the minimal energy option ending with that cache (chosen for P_i) and keep it for next iteration discarding the other solutions. Similar computation is done for caches C_2 to C_n. For our final least energy cache solution we use:

$$E_{S_{final}} = min(E_{S_{end,C_1}}, ..., E_{S_{end,C_n}}) \tag{Eq. 2}$$

In the general case, suppose m is the number of phases (number of potential reconfiguration points) and n is the number of possible cache configurations (18 in our case). Having n different cache options for each phase we can count the total number of possible solutions for cache assignment:

$$n^m = \underbrace{n \times n \times ... \times n}_{m\ times}$$

Therefore, finding the optimal cache assignment in a brute force manner (trying all possible solutions), takes the time complexity of $O(n^m)$. In our approach, $E_{S_{end,C_1}}, ..., E_{S_{end,C_n}}$ are computed in m iterations starting with the initial state. Computing (Eq. 1) in each iteration needs $n \times n$ comparisons. Therefore, in the recursive approach we reduce the time complexity of finding the optimal solution to $O(mn^2)$.

It should be noted that since the length of each phase is relatively long we can assume that reconfiguration (flushing) at the beginning of phase P_i only has an impact on the energy/time of phase P_i and will fade out for the next phase. In other words, if no reconfiguration occurs at the beginning of phase P_{i+1} we can assume no reconfiguration has happened prior to phase P_{i+1} in estimating time/energy of this phase. Therefore, energy/time of a cache for phase P_{i+1} is only dependent on the cache selected for the previous phase, P_i. Reconfiguration should be done when the selected caches for two consecutive phases are different and reconfiguration overhead should be accounted.

5. EXPERIMENS

5.1 Experimental Setup
In order to quantify effectiveness of our approach, we examined *cjpeg*, *djpeg*, *epic* (*encode* and *decode*), *adpcm* (*rawcaudio* and *rawdaudio*), *pegwit*, *g.721* (*encode*) benchmarks from the MediaBench [13] and *dijkstra*, *crc32*, *bitcnt* from MiBench [14] compiled for the PISA [11] target architecture. All applications were executed with the default input sets provided with the benchmarks suites.

We utilized the configurable cache architecture developed by Zhang et al [8] with a four-bank cache of base size 4 KB, which offers sizes of 1 KB, 2 KB, and 4 KB, line sizes ranging from 16 bytes to 64 bytes, and associativity of 1-way, 2-way, and 4-way. The reconfigurable cache was reported to have negligible performance and energy over-

Algorithm 2: Finding Cache Assignment
Input: Cache energy and time for all caches for each phase
/* m = total number of phases */
Output: Cache configuration for each phase
Begin
    S = a 2-dimentional list to store the best caches found;
    for phase j=0 to m **do**
        for cache i=0 to 17 **do**
            Find the best cache assignment from phase P_{start} up to P_j for cache configuration i using (Eq. 1);
            Update S_i;
        end for
    end for
    return the minimal energy solution in S using (Eq. 2);
end

head compared to non-configurable cache [8]. For comparison purposes, we used the *base cache* configuration set to be a 4 KB, 2-way set associative cache with a 32-byte line size, a common configuration that meets the average needs of the studied benchmarks [8].

To obtain cache hit and miss statistics, we modified the SimpleScalar toolset [11]. The modified version was able to dump dynamic instructions of cache misses as well as energy and time statistics for each program phase for both cases of starting with a flushed cache or a cache keeping previous data. The reconfiguration overhead (energy/time) is computed by flushing the cache at the reconfiguration points. Note that flushing the data cache requires all dirty blocks to be written back to main memory whereas flushing the instruction cache will just reset the valid bits for all cache blocks. The reconfiguration overhead also includes memory access latency/energy of bringing the data/instructions (that were previously in the cache) back to the cache. We applied the same energy model used in [8], which calculates both dynamic and static energy consumption, memory latency, CPU stall energy, and main memory fetch energy.

5.2 Energy versus Performance
Fig. 4 shows energy consumption using our intra-task DCR for all benchmarks normalized to the energy consumption for the least-energy cache found by inter-task DCR. Note that inter-task DCR is shown to achieve up to 53% cache subsystem energy savings in studies [10]. Our intra-task DCR approach achieves up to 27% (12% on average) energy savings compared to inter-task DCR for instruction cache. Energy savings of up to 19% (7% on average) is gained for data cache subsystem using our approach. It should be noticed that only nominal modifications are needed to make a working system using inter-task DCR to benefit from our intra-task DCR.

Fig. 5 demonstrates the execution time for all benchmarks normalized to the execution time for the least-energy cache configuration found by inter-task DCR. It is important to note that intra-task DCR introduces nearly no performance loss compared to the conventional inter-

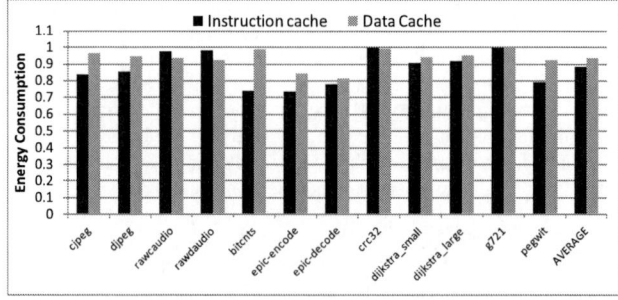

Fig. 4: Energy consumption normalized to the best cache configuration found by inter-task DCR

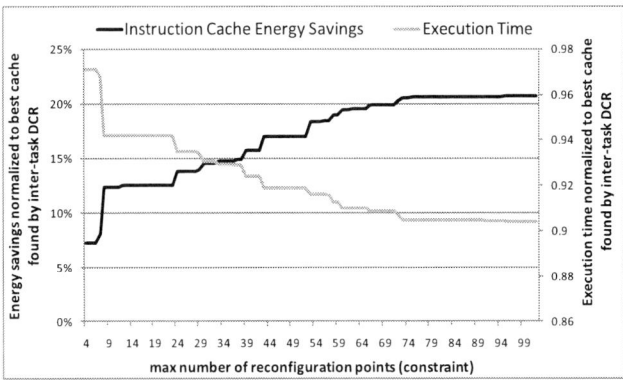

Fig. 6: Energy savings and execution time spectrum for instruction cache using *pegwit* benchmark

task DCR. Less than 1% performance loss observed using intra-task DCR for instruction cache. However, in some cases it actually achieves better performance (10% in the case of *pegwit* benchmark). Interestingly, incorporating intra-task DCR for data cache gains performance by 2% on average (up to 6% using *cjpeg* benchmark). We observed that intra-task DCR does not achieve energy savings compared to inter-task DCR in some applications. By further analysis it turned out that these applications have nearly the same cache behavior (either stable or chaotic) throughout their entire execution. This means that these applications cannot be separated into phases based on their cache behavior. In other words, they consist of only one phase. As expected, our intra-task DCR chooses only one cache configuration in this case which is same as inter-task DCR.

5.3 Overhead versus Energy Savings

The reconfiguration overhead is higher for data cache compared to instruction cache since by flushing the data cache all dirty blocks are needed to be written back to main memory whereas flushing the instruction cache only resets the valid bits for all cache blocks. Depending on the behavior of application in a particular phase, number of dirty blocks can be very high. When number of dirty blocks is relatively high, flushing the cache would result in high energy/performance overhead. Our cache assignment algorithm does not reconfigure cache at the reconfiguration point for a phase starting with high number of dirty blocks. Nonetheless, applications happen to show significant changes in cache requirements that can be exploited using intra-task DCR to compensate for the reconfiguration overhead.

In a given system, if it allows only a limited number of cache reconfigurations, we study the effect of our approach by constructing a spectrum of energy savings. We limit the number of potential reconfiguration points to *l* (using profitability) so that at most *l* number of reconfigurations can happen. Note that this does not mean that reconfiguration will happen at exactly *l* points. Fig. 6 illustrates the number of reconfiguration points – energy savings tradeoffs using *pegwit* benchmark (we observed similar pattern for other benchmarks as well). As we increase *l*, higher energy savings can be achieved. Interestingly, execution time has an inverse relation to energy savings. In other words, increasing *l* increases energy savings and decreases exe-

cution time (improves performance). This is expected since one of the best ways to reduce energy consumption is to shrink execution time. Intra-task DCR can reduce execution time by reducing cache misses through assigning larger caches in the cache intensive phases of programs. There are number of saturation points in the graph where increasing the number of potential reconfiguration points (*l*) does not change the result of our intra-task DCR. This is due to the fact that the newly added reconfiguration point is not a good choice (effective phase delimiter) to do the reconfiguration and is ignored by our cache assignment algorithm.

6. CONCLUSION

Optimization techniques are widely used in embedded systems design to improve overall area, energy and performance requirements. Dynamic cache reconfiguration is very effective to reduce energy consumption of cache subsystem. In this paper we presented a intra-task dynamic cache reconfiguration approach using novel phase detection and cache selection algorithms. Our experimental results demonstrated up to 27% reduction (12% on average) and 19% (7% on average) in overall energy consumption of instruction and data cache, respectively, compared to existing inter-task DCR techniques.

REFERENCES

1 A. Malik, B. Moyer, and D. Cermak. A low power unified cache architecture providing power and performance flexibility. *ISLPED* (2000).

2 H. Hajimiri, K. Rahmani, P. Mishra. Synergistic integration of dynamic cache reconfiguration and code compression in embedded systems. *International Green Computing Conference (IGCC)* (2011).

3 M. Peng, J. Sun, Y. Wang. A Phase-Based Self-Tuning Algorithm for Reconfigurable Cache. (), ICDS 07.

4 A. Gordon-Ross, J. Lau, B. Calder. Phase-based Cache Reconfiguration For a Highly-Configurable Two-Level Cache Hierarchy. *GLSVLSI 08.*

5 Gordon-Ross et al. Fast configurable-cache tuning with a unified second level cache. ISLPED (2005).

6 P. Vita. Configurable Cache Subsetting for Fast Cache Tuning. DAC (2006).

7 D. H. Albonesi. Selective Cache Ways: On-Demand Cache Resource *Allocation (2000).*

8 C. Zhang, F. Vahid, W. Najjar. A highly-configurable cache architecture for embedded systems. *ISCA* (03).

9 L. Chen, X. Zou, J. Lei, Z. Liu. Dynamically Reconfigurable Cache for Low-Power Embedded System. *ICNC (2007).*

10 W.Wang and P. Mishra. Dynamic Reconfiguration of Two-Level Caches in Soft Real-Time Embedded Systems. ISVLSI (2009).

11 Burger et al. Evaluating future microprocessors: the simplescalar toolset. University of Wisconsin-Madison, Technical Report CS-TR-1308 (2000).

12 CACTI. HP Labs, CACTI 4.2, http://www.hpl.hp.com/.

13 Lee et al. MediaBench: A Tool for Evaluating and Synthesizing Multimedia and Communications Systems. *Micro.* (1997).

14 Guthaus et al. MiBench: A free, commercially representative embedded benchmark suite. WWC (2001).

15 A. Gordon-Ross, F. Vahid, N. Dutt. Automatic tuning of two-level caches to embedded *applications. DATE* (2004).

A Diagnosability Metric for Test Set Selection targeting better Fault Detection*

Subhadip Kundu[1], Santanu Chattopadhyay[2], Indranil Sengupta[1], Rohit Kapur[3]

[1]Dept. of CSE, Indian Institute of Technology Kharagpur, India. Email: {subhadip, isg}@iitkgp.ac.in
[2]Dept. of E & ECE, Indian Institute of Technology Kharagpur, India. Email: santanu@ece.iitkgp.ernet.in
[3]Synopsys Inc., Mountain View, California, USA. Email: Rohit.Kapur@synopsys.com

Abstract- Diagnosis is the methodology to identify the reason behind the failure of manufactured chips. This is particularly important from the yield enhancement viewpoint. The primary focus of a diagnosis algorithm is to accurately narrow down the list of suspected candidates. But for any diagnosis algorithm, the effectiveness will depend on the test set in use. If the test set used is not good enough to distinguish between fault pairs, the diagnosis algorithm can never be able to distinguish between a good number of faults. This problem leads us to find a metric which can characterize test sets in terms of their diagnostic power. In literature, several methods have been proposed for assessment of the diagnostic power of a test set. Though the methods are accurate in nature, the bottleneck is the space and time complexity. Thus, given a number of test sets (with same fault coverage) for a circuit, it is very difficult to select one of them for better diagnosis. In this paper, we have proposed a probability based approach to find out a metric to describe diagnostic power of a test set. We call this metric, the diagnosibility of the test set for a given circuit. Our method uses almost 99% less space compared to the proposed methods and is well accurate.

Keywords: fault diagnosis, fault dictionary, clustering of faults.

I. INTRODUCTION

Fault diagnosis is an important operation for the production of good chips. It has a major role in fast yield ramp up. Defects can be modeled at the logic level by faults that affect single or multiple circuit locations and produce erroneous output responses for one or more input test vectors. Fault diagnosis uses the observed failing responses and the structure of the circuit under diagnosis (CUD) to search for potential fault locations. Information provided by the logic diagnosis process is, therefore, used to guide the circuits physical observation during failure analysis. The quality of diagnosis impacts directly the time-to-market and the total product cost.

Classical algorithms targeting logic diagnosis are based on two paradigms namely Cause-Effect and Effect-Cause [1], [2]. The Cause-Effect paradigm, which is usually based on fault simulation, builds a fault dictionary containing circuit responses for a given test set in presence of a given set of faults [3], [4]. The main drawback of this approach is the dictionary size. The second paradigm, called Effect-Cause, resorts to an error backtracing process, such as the Critical Path Tracing [5]. It starts from the failing primary outputs of the circuit under test (CUT) to reach the primary inputs. Each CUT line visited by the backtrace process is considered as a possible source of faulty behavior. Compared to the Cause–Effect methods, Effect–Cause techniques are more memory

efficient and can cope better with larger designs. In recent years, several papers on fault diagnosis [6-14] have been published. In [13], authors proposed a particle swam optimization (PSO) based diagnosis algorithm which can successfully diagnose multiple stuck at faults.

The primary importance of any diagnosis algorithm is to accurately narrow down the list of suspected candidates. For that, all the diagnosis algorithms depend on the failure information produced by the tester. Some diagnosis algorithms [13][14] also use the pass patterns to narrow down the list further. Overall, the backbone of any diagnosis algorithm is the test set in use.

Consider two faults f_1 and f_2 which have similar responses for all the patterns in the test set in use. If any of these two faults occur, all diagnosis algorithms (which use the same test set) will report both the faults with same rank. So, if a test set cannot distinguish between a pair (or more) of faults, no diagnosis algorithm using that test set can.

This leads us to the problem of assessing test sets in terms of their diagnosing capability. Surprisingly, not much works have been done on this problem. The exact brute force method is to record the output of the circuit in presence of each of the faults for each of the test vectors in the test set. This can be carried out via fault simulation. The record thus obtained is known as the fault dictionary. Once the fault dictionary has been obtained, it is easy to check for each of the pairs of faults, whether they are distinguishable or not. So we can get the exact count of the number of pairs of faults which are distinguishable. The detail of the algorithm is given in Section II. In [15], authors have proposed a single structure to store all the information related to all F faults in the circuit with respect to a pattern k. They have used a $F*F$ distinguishability matrix D_k. The generic element $d_k (i, j)$ of the matrix is 1 iff the pattern k can distinguish f_i and f_j, 0 otherwise. This method creates distinguishability matrix for all the test patterns in the test set. From this, the global distinguishability matrix, D, can be generated by bitwise *OR*-ing of all the matrices. The number of undistinguishable fault pairs is equal to the number of 0s present in D. But, though the method is accurate and produces exact result, the space taken by this method is $O(F^2*T)$ where T is the total number of test patterns in the test set. With increasing size of VLSI circuits, the dictionary based approaches are no longer a feasible solution because of their growing size.

With this background, we propose a simple, probability based metric for assessing the diagnosibility of a test set. The metric is based on the count of detection of a fault. First, we

---

*This work is partially supported by the consultancy project sponsored by Synopsys Inc., USA.

distribute the faults into different clusters (pockets) depending on their detection count. The faults which fall in the same pocket are identified by the same number of patterns at same number of outputs. It may be noted that we don't need to remember either the patterns or the outputs themselves, only two counts are required. Clearly, the faults which lie in different pockets are automatically distinguishable. For the faults in the same pocket, we find out the probability of them being distinguishable. The proposed method is fairly accurate and superior to the dictionary based approaches in terms of time and space complexities. In [17], authors have proposed a similar metric to find out the diagnostic coverage of a test set. For a set of test patterns, they also have distributed the faults in different groups. Faults in different groups are distinguishable whereas faults in the same group are not. But, while distributing the faults in different groups, they need to use the exact method given in Algorithm 1 (described in Section II), whereas in our case, we group the faults depending only on their detection count.

The rest of the paper is organized as follows. In Section II, we discuss the problem statement and the exact brute force method to solve the problem. In Section III, we describe the proposed method to distribute faults in different pockets according to their detection counts. In Section IV, we analyze each cluster (pockets) of faults in detail. The calculation of the metric is given in Section V. Experimental results are noted in Section VI. We also analyze the space and time complexity of the proposed method and compare it with the exact method in the same section. Finally, Section VII summarizes our observation.

II. PROBLEM STATEMENT

The following is the problem we consider in this work:

Problem: *Given two (or more) test sets with equal fault coverage for a given circuit, decide which one is better for fault diagnosis in a time and space efficient manner.*

We have done the analysis based on single stuck at fault assumption. We also assume that the test sets have approximately same fault coverage for the given circuit (otherwise, the set with higher fault coverage is always preferable). Thus, the pattern set which can distinguish more pairs of faults f_i and f_j (for every pair of i and j and $i \neq j$), will be considered as a better test set for diagnosis. This parameter is similar to the Diagnostic Resolution (DR) described in [16].

If we have to solve the problem using exact brute force method, we need to store the response of each fault for every test pattern in the test set. The detail of the algorithm is given in algorithm 1.

Once the number of distinguishable pairs have been calculated for a given test set, it can be compared with other test sets using this number. The higher the number of distinguishable pairs, the better is the test set for fault diagnosis. But for a single test set, memory required for storing the dictionary is $O(F*T*O)$, where F is the total number of faults, T is the number of patterns in the test set and O is the number of outputs of the circuit. For every test set we want to compare, we need to store this dictionary. If we use

the method proposed in [15], we need $O(F^2*T)$ space for each test pattern set. So, for large circuits, neither of them is affordable.

Algorithm 1: Exact Method

Input: *A Circuit C, A test set TS, A set of faults F.*
Output: *Number of distinguishable fault pairs.*
num_dp = 0; {num_dp is the number of distinguishable pairs.}
for all f_i in F **do**
    for all t_j in TS **do**
        Simulate C in presence of f_i for input t_j and store the outputs;
    end for
end for
for all f_i and f_j in F such that $i \neq j$ **do**
    for all t_k in TS **do**
        if the output of f_i and f_j are different for test vector t_k,
            then num_dp++ and **break** the inner loop;
        end if
    end for
end for
output num_dp;

III. PROPOSED METHOD

Since the exact method is not practical, we develop a metric based on some probabilistic analysis which is much more time and space efficient. We define the ***test detection set*** of a fault as the subset of the test set which contains the test patterns that detect the fault, and the ***output detection set*** as the subset of output lines which contains only those outputs where the fault can be propagated by the test set. The idea is inspired by the fact that if the test detection set of a fault contains at least one pattern which is not there in test detection set of another fault, the two faults can be distinguished. Similar argument can be given for output detection set of a fault. Clearly, if two faults are identified by different number of test patterns or at different number of outputs, the faults are automatically distinguished. So, we only need to consider the faults which are detected by same number of test patterns and at same number of outputs. Based on the above discussion, we propose the following method:

We divide faults into several pockets. The faults which fall in the same pocket are identified by the same number of patterns at same number of outputs. It may be noted that for two faults to share a pocket, it is not necessary that they be identified by the same subset of patterns, only the number of patterns in two sets be same. Similar is the case about outputs. A pocket has the following attributes associated with it:

- **Number of Test Detection (x):** It indicates the number of patterns by which each fault in the pocket is detected by the test set.

- **Number of Output Detection (y):** It indicates the number of outputs at which each fault in the pocket is detected by the test set.

Above two act as identifier of a pocket. The other attributes are:

- **Number of faults (m):** It indicates the number of faults in the pocket. Each of the m faults is detected by x number of patterns at y number of outputs.
- **Total Test pattern Count (T):** It counts the total number patterns associated with the pocket. We take the union of each fault's (associated with the pocket) test detection set and count the number.
- **Total Output pattern Count (O):** Similarly for output also we count the total number of outputs associated with the pocket.

Though it may appear that the total number of pockets will be close to the product of number of patterns in the test set and the number of outputs for the circuit, actually it is very less. Because it is very unlikely that there exists even a single fault which is detected by all the patterns at all the outputs. Clearly, the faults which lie in different pockets are automatically distinguishable. For the faults in the same pocket, we find out the probability of them being distinguishable.

IV. ANALYSIS OF THE POCKETS

Each pocket can be of one of the following five types:

- **Case 1: $x = 0$ and $y = 0$**
  Clearly, the faults in the pocket are undetectable faults. We do not consider these faults because both the test sets have same fault coverage.
- **Case 2: $m = 1$, $x \neq 0$ and $y \neq 0$**
  Since there is only one fault in the pocket, clearly it is distinguishable.
- **Case 3: $((x * (m-1) < T) \,\|\, (y * (m-1) < O))$**
  These pockets contain more than one fault. But each of the faults in a pocket is distinguishable from other faults in the same pocket. Consider a pocket with the following values: $x = 4$, $y = 3$, $m = 2$, $T = 5$, and $O = 3$.
  Let f_1 and f_2 be the two faults in the pocket. Now, each of the faults is detected by 4 test patterns. But, the total number of test patterns associated with the pocket is 5. Clearly, there should be at least one test pattern which detects only one of the faults, not the other. So the two faults are distinguishable. Similarly, this analysis is also valid for outputs.
  In a generalized format, the condition is:

  if $((x * (m-1) < T) \,\|\, (y * (m-1) < O))$, all faults in the pocket are distinguishable.
- **Case 4: none of the above types**
  These are the normal pockets. For these pockets, we have done a probabilistic analysis to find the measure. The analysis is as follows:
  Total no. of ways to choose x patterns from $T = C_1 = T_{C_x}$
  Total no. of ways to choose y outputs from $O = C_2 = O_{C_y}$
  Thus, total number of combinations possible $C = C_1 * C_2 = T_{C_x} * O_{C_y}$
  A fault can be detected on any one of these combinations. If another fault is detected on any combination other than the previous combination, then two faults are distinguishable. Let P be the probability that these two faults are distinguished from each other. P is calculated as

follows:

$$ P = \frac{C*(C-1)}{C^2} = \frac{C-1}{C} $$

Clearly, if we extend this formula for m faults, the probability P_m (probability that all the faults in the pocket are distinguishable) is calculated as:

$$ P_m = \frac{(C-1)*(C-2)*....*(C-(m-1))}{C^{m-1}} \quad\text{............. (1)} $$

where m is the number of faults in the pocket.
It may be noted that eqn. 1 leads to a non-zero value only if $m \leq C$. This gives cese to another case detailed next.

- **Case 5: $m > C$**
  If the number of faults m in the pocket is greater than C, clearly all the faults are not distinguishable. So, P becomes 0.

Once we have distributed the faults among pockets, we measure the probability of distinction of faults within each pocket using case 1 to case 5. After calculation of these, we move to the calculation of the metric based on these probabilities.

V. CALCULATION OF THE PROPOSED METRIC

Since, we assume that the test set which can distinguish more pairs of faults is better, our metric is also based on the probability that any pair of faults are distinguished.

Let F be the number of faults which can be detected by a test set.

F = Total Faults – Undetected Faults

Therefore, total number of pairs of faults: F_{C_2}.

For a pocket, we find the metric by multiplying the total pair of faults (m_{C_2}) with the probability of all the faults being distinguished (P_m). For case 2 and 3, $P_m = 1$ and for case 5, $P_m = 0$. For case 4, P_m is the value obtained from eqn 1. Faults belonging to different pockets are automatically distinguishable. Let the number of faults in the pockets be m_1, m_2, m_3, ... m_n. Therefore, there are $(F_{C_2} - (m_{1_{C_2}} + m_{2_{C_2}} + ... + m_{n_{C_2}}))$ pairs of faults which are automatically distinguishable. Thus the metric can be calculated as:

$$ M = (F_{C_2} - (m_{1_{C_2}} + m_{2_{C_2}} + ... + m_{n_{C_2}})) * 1.0 + m_{1_{C_2}} * P_1 + m_{2_{C_2}} * P_2 + ... + m_{n_{C_2}} * P_n \text{............ (2)} $$

where, P_1, P_2, ..., P_n are the probabilities associated with each pocket.

We also find the Normalized Metric (NM) by dividing M with total number of pairs of faults.

$$ NM = M / F_{C_2} \text{............... (3)} $$

The test set having higher value of M or NM will be considered as a better one for diagnosis.

Algorithm 2 calculates the proposed metric, given a circuit and a test set. We have used two tables, *Fault-Test Table* and *Fault-Output Table*, to store the necessary information. The generic element for *Fault-Test Table* (also for *Fault-Output table*), x_{ij}, is *1* if i^{th} fault is detected by j^{th} test pattern (or i^{th} fault is detected on j^{th} output), 0 otherwise (same for *Fault-Output table*). After performing the simulation, we used eqn.

1,2, and 3 to find out the diagnosability metric. Once the metric has been calculated for a given test set for a circuit, it can be compared with other test sets using this metric.

Algorithm 2 Metric Calculation

Input: *A Circuit C, A test set TS, A set of faults F.*

Output: *Diagnosibility metric*

diagnosibility = 0 {diagnosibility is the proposed metric}
for all f_i in F **do**
    for all t_j in TS **do**
        Simulate C in presence of f_i for input t_j;
        if   the output is different from fault free circuit
            then
                Store t_j in test detection set of f_i;
                Store the mismatched outputs in output detection set of f_i;
        end if
    end for
end for
put the faults in appropriate pockets according to the size of their test detection set and output detection set (a set of pockets P is created);
for all pockets p_i in P **do**
    Take union of test detection set of all faults f_j in pocket p_i to get Total Test pattern Count;
    Take union of output detection set of all faults f_j in pocket p_i to get Total Output pattern Count;
    if the pocket is of type 2 or 3, **then**
        Set Probability $P_i = 1.0$;
    else
        Calculate P_i according to equation 1;
    end if-else
end for
Calculate diagnosibility according to equation 2 and modify it according to equation 3;
output diagnosibility;

VI. EXPERIMENTAL RESULTS

In order to judge the figure of merit of the proposed method, we have experimented with two well-known test pattern generation algorithms. One of the tools, TACAD in our discussion is a free software available in the academia, while the other one TINDUS is a standard industrial test pattern generator. The tools have been utilized to generate test patterns for full-scan version of ISCAS89 benchmarks. The diagnosability metric (using Algorithm 2) are calculated for the test pattern sets and compared with exhaustive approach (using Algorithm 1). The results are reported in Table 1. The number of patterns generated by each test set is reported in columns 2 and 3. Generally, TINDUS produces lesser number of patterns compared to TACAD.

First, the exact method (Algorithm 1) has been applied on the test sets to find the number of pairs of faults which are indistinguishable by the test sets. We have compared our result with exact method only because method proposed in [15] will produce the same results as the exact one. The results

are given under the column heading "*Actual results*". The entries in TACAD and TINDUS column show the number of pairs of faults which are indistinguishable for the given circuit. For example, in case of s208, TACAD pattern set could not distinguish between 25 pairs of faults whereas for TINDUS, it is 29. In the next column, under the heading "*By Metric based Approach*", we have reported the normalized metric value for each test set computed using Algorithm 2.

We have only reported a rounded value up to 3 digits after decimal point. In the next column, under the heading "*Better Test set*", we indicate which test set is better for diagnosis. It has two sub-columns indicating the actual result and the decision based on the proposed method. The rows highlighted and shaded give the false results. The rows in italics and underlined with broken lines, though indicate that the two test set have similar results, if we extend the NM value to four digits after decimal point, the decision produced by the proposed method is same as the actual result. In most of the cases the proposed method successfully identifies the better test set for diagnosis.

For case 5 faults, we have tried to solve the issue using an MISR based approach. For all faults in the pocket, we simulate them once again, and compact the responses using an MISR as described in [1]. The output of the MISR is known as the signature of the fault. Then, we compare the signature between any two faults. If the signatures match, we consider the faults to be indistinguishable, otherwise they are distinguishable. So, for every pocket of type 5, we have found the count and added them with our metric. But, surprisingly, it did not change the metric much. The reasons are: first of all, there are not many faults in these pockets, and secondly, according to equation 2, the metric is dominated by the first term of the equation, which is $(F_{C_2} - (m_{1_{C_2}} + m_{2_{C_2}} + ... + m_{n_{C_2}}))$ * 1.0. Let us consider a small circuit, $s208$, to understand the effect. $s208$ has 217 collapsed stuck at faults. So the total number of fault pairs is 23436. For TACAD pattern set, the total number of fault pairs combining all the pockets is 1895. So, the value of 1st term in equation 2 becomes 21541. So, almost 92% of the fault pairs are distinguishable by just dividing the faults into the pockets. For larger circuits, the percentage is even higher. So, even if we take the correct decision using MISR for case 5, it does not alter the metric much. For this reason, we have not reported the MISR based approach in the experimental results.

There are some sources of error in the analysis. Consider a pocket with $x = 2$, $y = 2$, $m = 2$, $T = 2$, $O = 2$. Therefore, the faults are detected by the same two patterns at same two outputs. Let the two faults are f_1 and f_2 and test patterns are t_1 and t_2 which detect them at o_1 and o_2. By our analysis, f_1 and f_2 are indistinguishable. But consider a case when t_1 detects f_1 at o_1 and f_2 at o_2 and t_2 detects f_1 at o_2, and f_2 at o_1. Clearly, f_1 and f_2 are distinguishable, however our method, will not be able to resolve this aliasing.

Another interesting point we found during our experiment is that for every circuit, special pockets (case 2, 3, and 5) are more than the normal pockets (case 4). The normal pockets are only 10-20% of the total pockets. This implies that we can take certain decision about whether the faults are distinguishable or not in most of the cases.

A. Space Complexity Analysis

The method proposed in [15] needs $O(F^2*T)$ area which is higher than the brute force method described in Algorithm 1. The space required for exact method is clearly F*T*O bits. We do not consider any compaction algorithm to compact the size of the dictionary because the same can be applied to compact Fault-Test Table and Fault-Output Table, used in Algorithm 2, in a similar fashion.

For Proposed Method: We can perform our analysis by any of the following two methods:

Method 1: We can perform fault simulation only once and store a fault's test detection set and output detection set (as described in Algorithm 2). In that case, the maximum size required by the proposed method is:

S = F*T + F*O + Number of Pockets * Size of a Pocket

Note that F*T and F*O tables are highly sparse. But since we have not counted that, for the time being we are taking the entire space.

A pocket contains 5 integers, one float and a pointer as noted below in the pocket data structure (detail of each variable is given in Section III).

```
struct pocket{
    int      x;
    int      y;
    int      m;
    int      T;
    int      O;
    float    P_m;
    struct   pocket   *next;
};
```

Hence, Size of the Pocket = (5*4 + 4 + 4) * 8 = 224 bits.
Based on method 1, we have compared our result with exact method and reported in Table 2. We achieved almost 99% reduction in space for every bigger circuit.

Method 2: If we perform fault simulation twice, we can reduce the size even further. In the first pass, we will only count the size of the detection sets of each fault instead of storing the detection set. Once, we have the count, we distribute the faults among different pockets. In the next pass, we can simulate the faults in a pocket and then find out the total number of test patterns and total number of outputs associated with each pocket. So the size required in this case:

S = O(F) + O(O) + O(T) + Number of Pockets * Size of a Pocket

So, the size required in this case is even lesser. We did not compare the results with actual method because we already have a reduction of 99% for method 1. But, we did compare method 2 results with method 1 and have reported the results in Table 2. We have achieved, on an average, more than 96% reduction with respect to method 1.

B. Time Complexity Analysis

We also have performed the time complicity analysis.
For Exact method: Clearly, the time complexity of the exact method id is: $O(F^2*T*O)$.

For Proposed Method: The time complexity of the proposed method will be:

O(F*T) + O(F*O) + O(Number of Pocket). Though the maximum number of pockets is of the order O(T*O), actually it is much less than that. Thus, time complexity of the proposed method will be: MAX(O(F*T) , O(F*O)).

So, we could achieve an order improvement compare to the exact method.

VI. CONCLUSION

In this paper, we have proposed a novel technique for assessing the diagnosis power of test set. Using this technique, we are able to select test sets based on their diagnostic power. The proposed method is quite accurate and reduces space and time requirements by order of magnitude. This method can also be used while generating diagnostic patterns (*Automatic Diagnostic Test Generation (ADTG)*). We are now working on to extend this metric for multiple fault diagnosabilty.

REFERENCES

[1] L.T. Wang, C.W. Wu, X. Wen,"VLSI Test Principles and Architectures: Design for Testability," 1st Edition, Elsevier, 2006.

[2] J.A. Waicukauski and E. Lindbloom, "Failure Diagnosis of Structured VLSI," IEEE Design & Test of Computers, vol. 6, no. 4, pp. 49-60, 1989.

[3] I. Pomeranz, "On Pass/Fail Dictionaries for Scan Circuits," Proc. IEEE Asian Test Symp., pp. 51-56, 2001.

[4] I. Pomeranz and S.M. Reddy, "On Dictionary-Based Fault Location in Digital Logic Circuits," IEEE Trans. Computers, vol. 46, no. 1, pp. 48-59, 1997.

[5] M. Abramovici, P.R. Menon, and D.T. Miller, "Critical Path Tracing-an Alternative to Fault Simulation," IEEE Design & Test of Computers, vol. 1, no. 1, pp. 83-92, 1984.

[6] T. Bartenstein, D. Heaberlin, L. Huisman, and D. Sliwinski, "Diagnosing combinational logic designs using the single location at-a-time (SLAT) paradigm," in Proc. Int. Test Conf., pp. 287–296, 2001.

[7] B. Boppana, R. Mukherjee, J. Jain, and M. Fujita, "Multiple error diagnosis based on Xlists," in Proc. 36th Design Automation Conf., pp. 660–665, 1999.

[8] S P. Y. Chung and I. N. Hajj, "Diagnosis and correction of multiple logic design errors in digital circuits," IEEE Trans. Very Large Scale (VLSI) Integr. Syst., vol. 5, no. 2, pp. 233–237, 1997.

[9] S.-Y. Huang, "On improving the accuracy of multiple defect diagnosis," in Proc. 19th IEEE VLSI Test Symp., pp. 34–39, 2001.

[10] J. C. M. Li and E. J. McCluskey, "Diagnosis of sequence dependent chips," in Proc. 20th IEEE VLSI Test Symp., pp. 187–192, 2002.

[11] H. Takahashi, K. O. Boateng, K. K. Saluja, and Y. Takamatsu, "On diagnosing multiple stuck-at faults using multiple and single fault simulation in combinational circuits," IEEE Trans. Comput.-Aided Des. Integr. Circuits Syst., vol. 21, no. 3, pp. 362–368, 2002.

[12] H. Takahashi, N. Yanagida, and Y. Takamatsu, "Enhancing multiple fault diagnosis in combinational circuits based on sensitized paths and EB testing," in Proc. 4th Asian Test Symp., pp. 58–64, 1995.

[13] S. Kundu, S. Chattopadhyay, I. Sengupta, and R. Kapur, "Multiple fault diagnosis based on multiple fault simulation using Particle Swarm Optimization," in Proc. 24th IEEE Conference on VLSI Design (VLSID), 2011.

978-1-4673-0438-2/12 $31.00 © 2012 IEEE

[14] T. Z.Wang, M. Marek-Sadowska, K-H Tsai, J. Rajski, "Analysis and Methodology for Multiple-Fault Diagnosis," IEEE Trans. Comput.-Aided Des. Integr. Circuits Syst., vol. 25, no. 3, March 2006.

[15] P. Camurati, D. Medina, P. Prinetto, M. Sonza Reorda, "Assessing the diagnostic power of test pattern sets," in

Microprocessing and Microprogramming, vol 30, pp. 413-420, 1990.

[16] P. Camurati, D. Medina, P. Prinetto, M. Sonza Reorda, "Diagnostic test pattern generation algorithm," in Proc European Design Automation Conf., pp. 470-474, 1990.

[17] Y. Zang and V.D.Agrawal, "A Diagnostic Test Generation System," in Proc. Int. Test Conf., pp. 1–9, 2010.

TABLE 1: Test set comparison report

Circuit	#Patterns in the test set		Actual Results (in number of indistinguishable pairs of faults)		By Metric based Approach (NM)		Better Test set	
	TACAD	TINDUS	TACAD	TINDUS	TACAD	TINDUS	Actual	By Metric
s208	36	32	25	29	0.944	0.928	TACAD	TACAD
s298	33	27	27	34	0.965	0.949	TACAD	TACAD
s349	22	17	19	24	0.984	0.966	TACAD	TACAD
s382	34	28	26	27	0.972	0.971	TACAD	TACAD
s386	76	74	4	6	0.890	0.916	TACAD	TINDUS
s400	33	33	48	46	0.965	0.978	TINDUS	TINDUS
s420	76	76	45	46	0.952	0.946	TACAD	TACAD
s444	35	34	103	102	0.968	0.976	TINDUS	TINDUS
s510	62	60	11	23	0.969	0.970	TACAD	TINDUS
s526	62	60	42	38	0.970	0.969	TINDUS	TACAD
s641	62	29	12	18	0.983	0.983	TACAD	Both are Same
s713	54	36	178	186	0.988	0.988	TACAD	Both are Same
s820	110	109	86	80	0.869	0.896	TINDUS	TINDUS
s832	113	103	87	100	0.865	0.854	TACAD	TACAD
s838	147	152	90	98	0.946	0.949	TACAD	TINDUS
s953	95	92	7	9	0.987	0.987	TACAD	Both are Same
s1196	140	148	60	40	0.988	0.991	TINDUS	TINDUS
s1238	149	151	73	66	0.987	0.990	TINDUS	TINDUS
s1423	70	41	187	197	0.995	0.995	TACAD	Both are Same
s1488	127	119	70	74	0.950	0.954	TACAD	TINDUS
s1494	124	119	67	68	0.955	0.955	TACAD	Both are Same
s5378	257	124	574	578	0.997	0.998	TACAD	TINDUS
s9234	377	147	1789	1679	0.986	0.987	TINDUS	TINDUS
s13207	470	273	2464	2256	0.982	0.996	TINDUS	TINDUS
s15850	436	131	3178	2948	0.998	0.999	TINDUS	TINDUS
s38417	895	100	4746	4173	0.997	0.999	TINDUS	TINDUS
s38584	653	146	2892	3307	0.999	0.997	TACAD	TACAD

TABLE 2: Space comparison with bigger circuits

Circuit	Space taken by actual method in bits		Space taken by proposed approach in bits				% Reduction by Method 1 w.r.t Actual method		% Reduction by Method 2 w.r.t Method 1	
			Method 1		Method2					
	TACAD (x1)	TINDUS (y1)	TACAD (x2)	TINDUS (y2)	TACAD (x3)	TINDUS (y3)	TACAD ((x1-x2) / x1)*100%	TINDUS ((y1-y2) / y1)*100%	TACAD ((x2-x3) / x2)*100%	TINDUS ((y2-y3) / y2)*100%
s9234	6.53E+08	2.55E+08	4656381	2978499	320706	235804	99.29	98.83	93.11	92.08
s13207	3.38E+09	1.96E+09	12141550	10168559	397325	337992	99.64	99.48	96.73	96.68
s15850	3.43E+09	1.03E+09	13502115	9770524	536300	373819	99.61	99.05	96.03	96.17
s38417	4.43E+10	4.95E+09	78570964	53154936	1447061	595738	99.82	98.92	98.16	98.88
s38584	4.1E+10	9.16E+09	87418671	68510630	1029432	524477	99.79	99.25	98.82	99.23
Avg. Reduction							99.63	99.11	96.60	96.61

978-1-4673-0438-2/12 $31.00 © 2012 IEEE

Test Planning for Core-based 3D Stacked ICs with Through-Silicon Vias

Breeta SenGupta   Urban Ingelsson   Erik Larsson

Department of Computer and Information Science
Linköping University, SE-581 83 Linköping, Sweden
Email: {breeta.sengupta, urban.ingelsson, erik.larsson} (at) liu.se

Abstract—Test planning for core-based 3D stacked ICs with trough-silicon vias (3D TSV-SIC) is different from test planning for non-stacked ICs as the same test schedule cannot be applied both at wafer sort and package test. In this paper, we assume a test flow where each chip is tested individually at wafer sort and jointly at package test. We define cost functions and test planning optimization algorithms for non-stacked ICs, 3D TSV-SICs with two chips and 3D TSV-SICs with an arbitrary number of chips. We have implemented our techniques and experiments show significant reduction of test cost.

Index Terms—Test Scheduling, 3D stacked IC, JTAG, Test Architecture, Through Silicon Via.

I. INTRODUCTION

3D stacked ICs with trough-silicon vias (3D TSV-SICs) are emerging and have attracted a fair amount of research [1]–[6]. As the cost of test, which is highly related to test time and the additional design-for-test (DfT) hardware, accounts for a considerable part of the total manufacturing cost, it is important to develop a test plan minimizing the overall test cost. The testing of non-stacked ICs is well-defined; each IC is tested twice during the manufacturing process: during wafer sort, the bare chip (die) is tested, and during package test, the packaged IC is tested. For non-stacked ICs, the same tests are applied to the chip both during wafer sort and package test; hence, the same test schedule is used twice. However, for testing 3D TSV-SICs it is different. First, the test-flow is not well-defined. For 3D TSV-SICs, there are more test alternatives; testing can be performed on each individual IC, partial stacks, and/or the final stack [7]. Second, as the number of tests are different in each of these steps, test schedules are to be developed for each step (each individual IC, partial stacks, and the final stack), which is the focus of this paper.

Much work on test scheduling for non-stacked ICs have been performed [8]–[11]. For example, Chou *et al.* proposed a test scheduling technique that organize the tests in sessions such that the test time is minimized while power constraints are met [9]. Muresan *et al.* [8] proposed a test scheduling technique with the same optimization goal as Chou *et al.* While, the test architecture is unclear in the approach by Muresan *et al.* [8], Iyengar *et al.* [12]–[14] and Marinissen *et al.* [15] proposed test scheduling techniques and test architecture optimization for IEEE 1500. However, no work has addressed test scheduling in an IEEE 1149.1 environment. An increasing amount of work address testing of 3D TSV-SICs [1]–[4], [7], [16], [17].

In our previous work [7], we have defined a cost efficient test flow, while maximizing the yield. The scheme proposes that each individual IC is tested and then the complete stack is tested [7]. Marinissen *et al.* accounted for the variations in hardware required for various test schedules, although the overall test cost has not been optimized [16]. DfT hardware optimization has been addressed in [15], [18]–[20]. However, no work has addressed test scheduling for scan tested core based ICs. And, no work has defined test cost models and test planning algorithms that optimizes the overall test cost for 3D TSV- SICs in an IEEE 1149.1 environment.

In this paper, we assume the test flow that we introduced in our previous work [7], an IEEE 1149.1 environment, and we define test cost functions and test planning optimization algorithms for non-stacked ICs, 3D TSV- SICs with two chips and 3D TSV-SICs with an arbitrary number of chips.

The rest of the paper is organized as follows. In Section II, the JTAG test architecture assumed in our work is detailed. The problem definition is in Section III. In Section IV, we show a motivational example on the test scheduling problem for 3D TSV-SICs. The proposed test scheduling techniques are in Section V. The paper is concluded with experimental results in Section VI and conclusions in Section VII.

II. TEST ARCHITECTURE

The test architecture of a non-stacked IC, that has been assumed in this paper, is shown in Fig. 1. Here a chip is considered to consist of a number of cores that are accessed by an on-chip JTAG infrastructure [7]. The JTAG test access port (TAP) may have up to five terminals, namely Test Data Input (TDI), Test Data Output (TDO), Test Mode Select (TMS), Test Clock (TCK) and an optional Test Reset (TRST). In Fig. 1 only the TDI and TDO pins are shown, as the test interface terminals. Each core on a chip is accessed by the JTAG TAP via test data registers (TDRs). One TDR may be used to connect multiple cores on a single chip. In Fig. 1, the IC contains three cores: Core1, Core2 and Core3. Core1 and Core2 share a common TDR, while Core3 has an exclusive TDR. Only one TDR can be accessed at a time. Thus, if tests for more than one core of a chip are to be executed concurrently, in a session, as shown in Fig. 2, these cores are to be connected in series on the JTAG interface in one TDR. Since, Core1 and Core2 are tested in the same session as in Fig. 2, denoted by $(1, 2)$, the two cores are connected

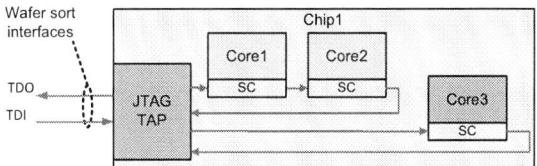

Fig. 1. Test architecture of a non-stacked chip with JTAG

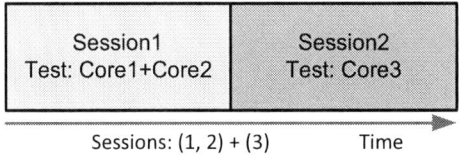

Fig. 2. Sessions formed by core tests

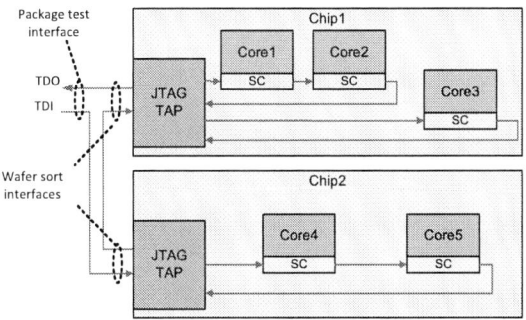

Fig. 3. Test architecture of 3D TSV-SIC with JTAG

to the JTAG TAP by the same TDR, as seen in Fig. 1. Correspondingly, in Session2, only Core3 is tested, denoted by (3) in Fig. 2, which is connected to the JTAG TAP by a single TDR.

During the package test of the 3D TSV-SIC, the TDO of the lower JTAG TAP in the stack serves as the TDI of the corresponding JTAG TAP of the chip on top of it. The TDO of the topmost chip is directed out via TSVs. The TDI of the lowermost chip and the TDO of the topmost chip serve as the package test interfaces as shown in Fig. 3. A session of tests from one chip can be performed concurrently with a session of tests from another chip by selecting the corresponding TDRs by the respective on-chip JTAG TAPs of to the two chips.

III. PROBLEM DEFINITION

In this section the test cost for non-stacked IC, 3D TSV-SIC with two chips in the stack and 3D TSV-SIC with N chips in the stack, are defined. The overall objective is a test plan with a minimal cost in terms of test application time (TAT) and hardware (number of TDRs), defined as:

$$Cost(TAT, TDR) = \alpha \cdot TAT + \beta \cdot TDR \qquad (1)$$

where, α and β are constants set by the designer depending on the particular system.

A. Non-stacked IC

For a non-stacked IC with C cores, having a test schedule with S sessions, we assume for a core c_{ij}, $1 \leq i \leq C, 1 \leq j \leq S$, having a scan chain of length l_{ij} and requiring p_{ij} test patterns. The test time for a core c_{ij} is given by:

$$Time(c_{ij}) = (l_{ij} + \delta) \cdot p_{ij} + l_{ij} \qquad (2)$$

where, δ accounts for the number of clock cycles required by the JTAG for apply and capture, which is equal to 5.

A test schedule for the C cores consists of S sessions, where each core c_{ij} belongs to an unique session s_j, $1 \leq j \leq S$. The number of cores that are tested in a session s_j is given by m_j. The test time t_j for a session s_j is denoted by:

$$t_j = \left(\delta + \sum_{\forall i \in m_j} l_{ij} \right) \cdot max_{\forall i \in m_j}(p_{ij}) + \sum_{\forall i \in m_j} l_{ij} \qquad (3)$$

The overall test time for a test schedule is given as:

$$Time = \sum_{j=1}^{S} t_j \qquad (4)$$

The hardware cost is directly related to the number of sessions, since each session corresponds to a TDR; hence, $TDR = S$.

In the case of non-stacked ICs, the same schedule is applied at wafer sort and at package test; hence, $TAT = 2 \cdot Time$.

The cost function in Eq.1 is in the case of non-stacked ICs given as:

$$\begin{aligned} Cost(TAT, TDR) &= \alpha \cdot TAT + \beta \cdot TDR \\ &= \alpha \cdot 2 \cdot Time + \beta \cdot S \end{aligned} \qquad (5)$$

The problem is to find a test schedule such that the TAT and the number of TDRs required result in a minimized cost.

B. 3D TSV-SIC with two chips in the stack

For a 3D TSV-SIC design having a stack of two chips, Chip1 and Chip2, we assume that Chip1 and Chip2 have C_1 and C_2 cores, respectively. During wafer sort, Chip1 and Chip2 have test schedules with S_1 and S_2 sessions respectively. For each core c_{1im}, $1 \leq i \leq C_1$, $1 \leq m \leq S_1$, in Chip1, the length of the scan chain is l_{1im} and the number of patterns required is p_{1im}, while for each core c_{2jn}, $1 \leq j \leq C_2$, $1 \leq n \leq S_2$, in Chip2, the length of the scan chain is l_{2jn} and the number of patterns required is p_{2jn}. For wafer sort, Chip1 and Chip2 have test schedules with S_1 and S_2 sessions respectively. Each core c_{1im} belongs to an unique session s_{1m}, and each core in Chip2 c_{2jn} belongs to an unique session s_{2n}. The number of cores that are tested in a session s_{1m} (s_{2n}) is given by m_{1m} (m_{2n}). The test time t_{1m} for a session s_{1m} session is denoted by:

978-1-4673-0438-2/12 $31.00 © 2012 IEEE   443

$$t_{1m} = \left(\delta + \sum_{\forall i \in m_{1m}} l_{1im}\right) \cdot max_{\forall i \in m_{1m}}(p_{1im}) + \sum_{\forall i \in m_{1m}} l_{1im} \tag{6}$$

and the test time T_{2n} for a session s_{2n} session is denoted by:

$$t_{2n} = \left(\delta + \sum_{\forall j \in m_{2n}} l_{2jn}\right) \cdot max_{\forall j \in m_{2n}}(p_{2jn}) + \sum_{\forall j \in m_{2n}} l_{2jn} \tag{7}$$

Given Eq.6, the test time for wafer sort for Chip1 is given as:

$$t_{1WS} = \sum_{m=1}^{S_1} t_{1m} \tag{8}$$

and given Eq.7, the test time for wafer sort for Chip2 is given as:

$$t_{2WS} = \sum_{n=1}^{S_2} t_{2n} \tag{9}$$

The total time taken for wafer sort is:

$$t_{WS} = t_{1WS} + t_{2WS} \tag{10}$$

For package test of Chip1 and Chip2 a test schedule with S_3 sessions is formed. Each core c_{1im} (c_{2jn}) belongs to a unique session s_{3o}, $1 \leq o \leq S_3$. The number of cores that are tested in a session s_{3o} is given by the set m_{3o}. The test time t_{3o} for a session s_{3o} is denoted by:

$$t_{3o} = \left(\delta + \sum_{\forall i,j \in m_{3o}} (l_{1im} + l_{2jn})\right) \cdot max_{\forall i,j \in m_{3o}}(p_{1im}, p_{2jn})$$
$$+ \sum_{\forall i,j \in m_{3o}} (l_{1im} + l_{2jn}) \tag{11}$$

Given Eq.11, the test time for package test for Chip1 and Chip2 is given as:

$$t_{PT} = \sum_{t=1}^{S_3} t_{3o} \tag{12}$$

The TAT is given by

$$TAT_{2chip} = t_{1WS} + t_{2WS} + t_{PT} \tag{13}$$

The hardware required is the sum of the number of TDRs required during wafer sort of Chip1 and Chip2:

$$TDR = S_1 + S_2 \tag{14}$$

The overall test cost can be expressed by the following equation:

$$Cost_{2chip}(TAT, TDR) = \alpha \cdot TAT + \beta \cdot TDR$$
$$= \alpha \cdot TAT_{2chip} + \beta \cdot (S_1 + S_2) \tag{15}$$

The problem is to find the test schedules for wafer sort of Chip1 and Chip2 individually, and package test for jointly testing Chip1 and Chip2 such that the TAT and the total number of TDRs required by Chip1 and Chip2 during wafer sort result in a minimized cost.

C. 3D TSV-SIC with N chips in the stack

The cost minimization problem for a 3D TSV-SIC with N chips forming the stack can be generalized from the two problems stated above. Any chip in the stack n_i, $1 \leq i \leq N$, has C_i cores, each denoted by c_{ijk}, $1 \leq j \leq C_i$, $1 \leq k \leq S_i$, each having a scan chain of length l_{ijk}, and requiring p_{ijk} patterns. During wafer sort, the test schedule of a chip n_i has S_i sessions, each denoted by s_{ik}, with m_{ik} tests in each session. Then, the test time t_{ik} for a session s_{ik} is given by

$$t_{ik} = \left(\delta + \sum_{\forall j \in m_{ik}} l_{ijk}\right) \cdot max_{\forall j \in m_{ik}}(p_{ijk}) + \sum_{\forall j \in m_{ik}} l_{ijk} \tag{16}$$

The time taken by each chip n_i during wafer sort is

$$t_{iWS} = \sum_{k=1}^{S_i} t_{ik} \tag{17}$$

Thus, the total time taken for wafer sort of the 3D TSV-SIC is

$$t_{N.WS} = \sum_{i=1}^{N} T_{iWS} = \sum_{i=1}^{N} \left(\sum_{k=1}^{S_i} t_{ik}\right) \tag{18}$$

For package test of the 3D TSV-SIC, a test schedule is formed with S_N sessions. Each core c_{ij} belongs to a unique session s_o, $1 \leq o \leq S_N$. The number of cores that are tested in a session s_o is given by m_o. The test time t_o is denoted by:

$$t_o = \left(\delta + \sum_{\forall j \in m_o} \sum_{i=1}^{N} l_{ijo}\right) \cdot max_{\forall j \in m_o}(p_{ijo}) + \sum_{\forall j \in m_o} \sum_{i=1}^{N} l_{ijo} \tag{19}$$

Given Eq.19, the test time for package test is given as:

$$t_{N.PT} = \sum_{o=1}^{S_N} t_o \tag{20}$$

Hence, the overall cost is

$$Cost_N(TAT, TDR) = \alpha \cdot TAT + \beta \cdot TDR$$
$$= \alpha \cdot t_{N.PT} + \beta \cdot \left(\sum_{\forall i \in N} S_i\right) \tag{21}$$

The problem is to find the test schedules with S_1 sessions for wafer sort of Chip1, S_2 sessions for wafer sort of Chip2, and S_3 sessions for package test for jointly testing of Chip1 and Chip2 such that the TAT and the total number of TDRs required by all the N chips during wafer sort result in a minimized cost.

TABLE I
GIVEN L, P VALUES FOR EACH CORE OF THE 3D TSV-SIC

	Chip 1			Chip 2	
	Core1	Core2	Core3	Core4	Core5
Scan chain length (l_{ijk})	50	40	30	20	10
Patterns required (p_{ijk})	50	40	30	20	10

TABLE II
TEST SESSION ALTERNATIVES

Cases	Wafer Sort (T_{ws})		Package Test (T_{pt})	Total Time	Cost	No. of TDRs
	Chip 1	Chip 2				
1	(1, 2, 3)	(4, 5)	(1, 2, 3)+(4, 5)	14200	15000	2
2	(1, 2, 3)	(4)+(5)	(1, 2, 3)+(4) + (5)	14100	15300	3
3	(1, 2)+(3)	(4, 5)	(1, 2)+(3)+(4, 5)	13300	14500	3
4	(1)+(2)+(3)	(4, 5)	(1)+(2)+(3)+(4, 5)	12900	14500	4
5	(1, 2)+(3)	(4)+(5)	(1, 2)+(3)+(4)+(5)	13200	14800	4
5	(1)+(2)+(3)	(4)+(5)	(1)+(2)+(3)+(4)+(5)	12800	14800	5

IV. MOTIVATIONAL EXAMPLE

Here we present an example to demonstrate the variation of cost incurred due to the trade-off between test time and hardware required. Given is a 3D SIC with two chips in the stack, illustrated in Fig. 3. The lengths of the scan chains and the number of patterns required for each core is listed in Table I. We assume that the cost of a single TDR is equivalent to 400 time units.

The time taken for wafer sort, t_{WS}, for the configuration shown, as in case 3 in Table II, *i.e.*, Core1 and Core2 with a common TDR, forming session s_{11}, Core3 forming session s_{12}, Core4 and Core5: session s_{21} is:

$$
\begin{aligned}
t_{WS} &= t_{11} + t_{12} + t_{21} \\
&= max(p_{111}, p_{121}) \cdot (l_{111} + l_{121} + 5) + (l_{111} + l_{121}) \\
&\quad + (l_{132} + 5) \cdot p_{132} + l_{132} \\
&\quad + max(p_{241}, p_{251}) \cdot (l_{241} + l_{251} + 5) + (l_{241} + l_{251}) \\
&= 50 \cdot 95 + 90 + 30 \cdot 35 + 30 + 20 \cdot 35 + 30 \\
&= 6650 \ time \ units \ (t.u.)
\end{aligned}
$$

Performing the tests in the same order on package test as in wafer sort would result in this case

$$
T_{ws} = T_{pt} \tag{22}
$$

Therefore the total test time becomes,

$$
T = T_{ws} + T_{pt} = 6650 + 6650 = 13300 \ t.u. \tag{23}
$$

In this case we require three TDRs for testing the chip. Hence, we can calculate the total test cost from Eq.1:

$$
\begin{aligned}
Cost_{case3} &= \alpha \cdot TAT + \beta \cdot TDR \\
&= 13300 + 400 \cdot 3 \\
&= 14500 \ units
\end{aligned}
$$

Similarly, considering separate TDRs for all five cores would give, $T = 12800 \ t.u$, as shown in case6 in Table II. But, the schedule results in more sessions, thus an increased

hardware cost. The total cost incurred in case6 is $Cost_{case6} = 14800$ units.

The minimum number of sessions is obtained when during wafer sort Core1, Core2 and Core3 are in s_{11} and Core4 and Core5 are in s_{21}, while during package test all five cores are in the same session. The total time leads to $T = 14200 \ t.u.$, which is significantly higher than the alternative distribution of sessions discussed above. Although, in this case, the hardware requirement is minimum. The overall cost incurred in case1 is $Cost_{case1} = 15000$, which is higher than case3 and case6 discussed above.

In case2, where Core1, Core2 and Core3 are tested in session s_{11}, while Core4 is tested in session s_{21} and Core5 in session s_{22}, the cost incurred is $Cost_{case2} = 15300$ units.

In case4, where Core1, Core2 and Core3 are tested in three different sessions, while Core4 and Core5 are tested in the same session, the total test cost is $Cost_{case4} = 14500$ units. We can see that the cost incurred in performing case4 is minimum compared to the rest of the five cases in Table II.

Therefore, from the above studies on the distribution of TDRs in a 3D SIC it was seen that the test time can be reduced by increasing the number of TDRs, thereby increasing the number of sessions. Although, an increased number of sessions implies increased hardware cost. Hence, in this paper, we try to obtain a trade-off between the hardware cost and the test time, in order to give the minimum total effective cost.

V. PROPOSED APPROACHES

In this section we propose three algorithms, for non-stacked IC, 3D TSV-SIC with two chips in the stack and 3D TSV-SICs with any number of chips in the stack, to arrive at a test plan which requires minimal overall test cost, in terms of TAT and the number of TDR, as defined in Eq.1.

A. Non-stacked IC

By the following steps of the algorithm we arrive at the reduced cost for non-stacked ICs.

- Given is the list of C cores c_{ij}, $1 \leq i \leq C$, $1 \leq j \leq S$, in a chip, sorted by the number of patterns required p_{ij}. The length of the scan chains are denoted by l_{ij}.
- The constants of the cost function defined by Eq.1, α and β are also provided.
- Initially, TAT is set equal to the test time of core c_{11}.
- The number of sessions, S is initially set equal to one. The first session, s_1, in the test schedule contains the test of core c_{11}. Core c_{11} is then removed from the sorted list.
- Each core c_{ij}, remaining in the sorted list, is descended in the following way:
  The increase in TAT for each core c_{ij} is calculated by including it in all existing sessions. If the cost of a single TDR is less than the cost incurred by including the core test in any of the existing sessions due to the increased test time, the core test forms a new session.
  Once the core is assigned a session, it is excluded from the sorted list.

978-1-4673-0438-2/12 $31.00 © 2012 IEEE

TABLE III
TAT AND TDR FOR NON-STACKED IC

No.	Design	Minimal Test Cost			Cost with Maximum TDR (= No. of cores)				Cost with Minimum TDR (= 1)			
		Cost	TAT	TDR	Cost	TAT	TDR	Cost	Inc.(%)	TAT	TDR	Cost	Inc.(%)
1	p22810	501490	7	534250	474489	22	577449	8.1	2022377	1	2027057	279.4
2	p93791	614233	4	633701	589394	13	652665	3.0	1990806	1	1995673	214.9
3	g1023	46885	4	51813	42429	12	57213	10.4	137727	1	138959	168.2
4	d695	35757	4	40689	34331	8	44195	8.6	80369	1	81602	100.6
5	h953	271381	2	305483	230771	7	350128	14.6	418607	1	435658	42.6
6	d281	117946	2	144992	97310	5	164925	13.8	186458	1	199981	37.9

- The test plan is achieved when test of each core c_{ij}, has been assigned its respective session s_j.

B. 3D TSV-SIC with two chips in the stack

The wafer sort test schedules for the two chips forming the 3D TSV-SIC, Chip1 and Chip2 are obtained by applying the algorithm for test scheduling of non-stacked ICs. The test planning algorithm for package test is discussed below:

- Given is the list of sessions S_1 of Chip1 and sessions S_2 of Chip2, denoted by s_{1m} and s_{2n} respectively.
  The lists of sessions of Chip1 and Chip2, s_{1m} and s_{2n}, are sorted in descending order of their test times, t_{1m} and t_{2n}.
- The test schedule for the package test is obtained by simultaneously initiating the sessions s_{1m} and s_{2n} for all $m = n$. The total number of sessions during package test is S_1 if $S_1 > S_2$, and S_2 otherwise.
- The reduction in test time for each new session formed during package test of the two chip 3D TSV-SIC is the test time of the session s_{1m}, if $s_{1m} < s_{2n}$ and s_{2n} otherwise.
  The sum of the reduction in test time over all the sessions formed during package test gives the overall reduction in the TAT.

C. 3D TSV-SIC with N chips in the stack

The algorithm used for scheduling tests for 3D TSV-SICs with two chips in the stack can be extended for 3D TSV-SICs with N chips in the stack.

- Given is the list of sessions S_i of each chip n_i, each denoted by s_{ik}, $1 \leq k \leq S_i$.
  All the sessions of each chip n_i are sorted in descending order of their test times.
- The test schedule for the package test is obtained by simultaneously initiating the kth session, S_{ik} of each chip n_i, $\forall i \in (1 to N)$.
  The total number of sessions during the package test of the 3D TSV-SIC with N chips in the stack is $max(S_i)$ and the time taken by each session is $max(t_{ik})$, $1 \leq k \leq max(S_i)$.

VI. EXPERIMENTAL RESULTS

In this section we illustrate the benefits of the proposed approach on the three configurations described earlier, namely,

TABLE IV
REDUCTION IN TAT FOR 3D TSV-SIC FOR 2, 3 AND 4 CHIPS

No. of chips	Design nos.	Cost Naive Approach	Cost Reduced	Percentage Reduction
2	1,2	4029631	1185943	70.57
	2,3	2143565	730389	65.93
	3,4	222748	94954	57.37
	4,5	503976	284600	43.53
	5,6	649111	475849	26.69
3	1,2,3	3298487	1951685	40.83
	2,3,4	2605603	1292321	50.40
	3,4,5	1759284	737453	58.08
	4,5,6	2925324	727092	75.15
4	1,2,3,4	3897924	1951685	49.93
	2,3,4,5	3051605	1292321	57.65
	3,4,5,6	3639169	737453	79.73

non-stacked IC, 3D TSV-SIC with two chips in the stack and 3D TSV-SIC with N chips in the stack.

Experiments have been performed on the six ITC'02 benchmark system on chip (SOC) designs mentioned below: p22810, p93791, g1023, d695, h953 and d281.

The following assumptions were made when constructing 3D TSV-SICs from the non-stacked SOC benchmarks :

- The modules in the benchmark SOC designs are projected as cores in a non-stacked IC
- All scan elements (inputs, outputs, and scan cells) at a core are connected to a single scan-chain
- 3D TSV-SICs are constructed by vertically stacking any number of the non-stacked designs
- The constant α in Eq.1 for all designs is considered to be one
- The constant β in Eq.1 for all designs is calculated by dividing the test time of the first core in the sorted list, $Time(c_1)$, by the number of cores, C.

A. Non-stacked IC

Table III compares the minimized overall cost for non-stacked ICs to the overall cost when the test time cost is minimal and to the overall cost when the cost of hardware is minimum, i.e., there is only one TDR. In Table III, each row corresponds to a SOC benchmark design, which is shown in the second column. The costs of three different test schedules are compared in the following three groups of columns. The first group of three columns shows the minimal test cost of the respective designs as obtained by the algorithm proposed in Section V. Next is the cost incurred when the TAT is minimum; in other words the hardware cost is maximum,

978-1-4673-0438-2/12 $31.00 © 2012 IEEE

with the number of TDRs equal to the number of cores in the IC. The last column in the group of four columns evaluates the increase in the test cost *wrt* the minimal test cost. The rightmost group of four columns shows the test cost when all cores share a common TDR, thereby maximizing TAT. In Table III, it can be seen that the maximum reduction in cost *wrt* minimized TAT is up to 15% for h953 and *wrt* minimized number of TDRs is up to 280% for p22810.

In Table IV, the package test cost for various 3D TSV-SIC designs made by stacking the six benchmark designs in Table III are shown. The number of chips that have been stacked to make the 3D TSV-SIC is shown in the leftmost column. The group of five rows have 3D TSV-SICs with two chips in the stack, followed by a group of four rows having three chips in the stack and the group of three rows at the bottom are designs made by stacking four chips. The second column from left shows the benchmark designs that have been used to make the stack, which correspond to the serial number used in Table III. For instance, the first 3D TSV-SIC design contains two chips in the stack, 1 and 2, which refers to p22810 and p93791 respectively. The third column lists the test times obtained by summing up the test times, of each design forming the stack, corresponding to the minimal cost, as obtained in Table III. The next column shows the reduced test time by applying the algorithm proposed in Section V. In the rightmost column, the relative reduction in the test time is evaluated. We can see that the test time can reduce up to 75%, when chips g1023, d695, h953 and d281 are stacked.

VII. CONCLUSION

In this paper, we define test cost as a function of TAT and the number of TDRs for non-stacked ICs, 3D TSV-SIC with two chips in the stack and 3D TSV-SIC with N chips in the stack. The test cost is minimized by co-optimizing TAT and the number of TDRs. We propose an algorithm for scheduling tests, which addresses the following three problems:

1) For a non-stacked IC, in an IEEE 1149.1 environment, where the same test schedule is applied during wafer sort and package tests, the tests of all the cores are grouped in sessions such that the cost is minimized by co-optimizing the TAT and the number of TDRs required. We find that the cost can increase by 280%, when either one of the variables are minimized.

2) For a 3D TSV-SIC, having two chips, each chip is tested individually during wafer sort and jointly during package test. The cost is minimized by forming sessions from different chips concurrently during the package test. Results show that by applying the algorithm, the test time can be reduced by up to 70%.

3) The algorithm for test scheduling of 3D TSV-SICs with two chips is extended to 3D TSV-SICs with any number of chips forming the stack. Experimental results show significant reductions in the overall test cost. The reduction in test time is up to 75%.

REFERENCES

[1] E. J. Marinissen and Y. Zorian, "Testing 3D Chips Containing Through-Silicon Vias," in *IEEE International Test Conference (ITC)*, 2009, pp. 1–11.

[2] H.-H. S. Lee and K. Chakrabarty, "Test Challenges for 3D Integrated Circuits," in *IEEE Design and Test of Computers, Special Issue on 3D IC Design and Test*, Oct. 2009, pp. 26–35.

[3] D. L. Lewis and H.-H. S. Lee, "A Scan-Island Based Design Enabling Pre-bond Testability in Die-Stacked Microprocessors," in *IEEE International Test Conference (ITC)*, 2007, pp. 1–8.

[4] X. Wu, P. Falkenstern, and Y. Xie, "Scan Chain Design for Three-Dimensional Integrated Circuits (3D ICs)," in *International Conference on Computer Design (ICCD)*, 2007, pp. 208–214.

[5] Y.-J. Lee and S. K. Lim, "Co-Optimization of Signal, Power, and Thermal Distribution Networks for 3D ICs," in *Electrical Design of Advanced Packaging and Systems Symposium*, 2008, pp. 163–166.

[6] R. Weerasekera. "System Interconnection Design Trade-offs in Three-Dimensional (3-D) Integrated Circuits," in *KTH Information and Communication Technology*, 2008.

[7] B. SenGupta, U. Ingelsson, and E. Larsson, "Scheduling Tests for 3D Stacked Chips under Power Constraints," in *Accepted to be published in Journal of Electronic Testing: Theory and Applications (JETTA)*, 2011.

[8] V. Muresan, X. Wang, V. Muresan, and M. Vladutiu, "Greedy Tree Growing Heuristics on Block-Test Scheduling Under Power Constraints," in *Journal of Electronic Testing: Theory and Applications*, 2004, pp. 61–78.

[9] R. M. Chou, K. K. Saluja, and V. D. Agrawal, "Scheduling tests for VLSI systems under power constraints," in *IEEE Transactions on VLSI Systems*, vol. 5, no. 2, Jun. 1997, pp. 175–185.

[10] Y. Zorian, "A Distributed BIST Control Scheme for Complex VLSI devices," in *IEEE VLSI Test Symposium (VTS)*, Apr. 1993, pp. 6–11.

[11] E. Larsson and Z. Peng, "An Integrated Framework for the Design and Optimization of SOC Test Solutions," in *Journal of Electronic Testing: Theory and Applications, Special Issue on Plug-and-Play Test Automation for System-on-a-Chip*, vol. 18, no. 4, Aug. 2002, pp. 385–400.

[12] V. Iyengar, K. Chakrabarty, and E. J. Marinissen, "Wrapper/TAM Co-Optimization, Constraint-Driven Test Scheduling, and Tester Data Volume Reduction for SOCs," in *IEEE VLSI Test Symposium (VTS)*, no. 44.3, Jun. 2002, pp. 685–690.

[13] ——, "Test Access Mechanism Optimization, Test Scheduling, and Tester Data Volume Reduction for System-on-Chip," in *IEEE Transactions on Computers*, vol. 52, no. 12, Dec. 2003, pp. 1619–1632.

[14] ——, "Test Wrapper and Test Access Mechanism Co-Optimization for System-on-Chip," in *Journal of Electronic Testing: Theory and Applications*, vol. 18, 2002, pp. 213–230.

[15] E. J. Marinissen, R. Kapur, M. Lousberg, T. McLaurin, M. Richetti, and Y. Zorian, "On IEEE P1500s Standard for Embedded Core Test," in *Journal of Electronic Testing: Theory and Applications (JETTA)*, vol. 18, 2002, pp. 365–383.

[16] E. J. Marinissen, J. Verbree, and M. Konijnenburg, "A Structured and Scalable Test Access Architecture for TSV-Based 3D Stacked ICs," in *IEEE VLSI Test Symposium (VTS)*, Apr. 2010, pp. 1–6.

[17] B. Noia, S. K. Goel, K. Chakrabarty, E. J. Marinissen, and J. Verbree, "Test-Architecture Optimization for TSV-Based 3D Stacked ICs," in *IEEE European Test Symposium (ETS)*, May 2010, pp. 24–29.

[18] E. J. Marinissen, K. Chakrabarty, and V. Iyengar, "A Set of Benchmarks for Modular Testing of SOCs," in *International Test Conference (ITC)*, no. 19.1, 2002, pp. 519–528.

[19] L.-T. Wang, C.-W. Wu, C.-W. Wu, and X. Wen, "VLSI test principles and architectures:design for testability," in *Academic Press*, 2006.

[20] S. Goel, "Test-Access Planning and Test Scheduling for Embedded Core-Based System Chips," in *University Press, Eindhoven, The Netherlands*, 2005.

Externally Tested Scan Circuit With Built-In Activity Monitor and Adaptive Test Clock

Priyadharshini Shanmugasundaram
NVIDIA
Santa Clara, CA 95050, USA
priyas@nvidia.com

Vishwani D. Agrawal
Auburn University
Auburn, AL 36849, USA
Email: vagrawal@auburn.edu

Abstract—We reduce the test time of external test applied from an automatic test equipment (ATE) by speeding up low activity cycles without exceeding the specified peak power budget. An activity monitor is implemented as hardware or as pre-simulated and stored test data for this purpose. The achieved test time reduction depends upon the input and output activity factors, α_{in} and α_{out}, of the scan chain. When on-circuit built-in hardware control is used, test time reductions of about 50% and 25% are possible for vectors with low input activity ($\alpha_{in} \approx 0$) and moderate input activity ($\alpha_{in} = 0.5$), respectively, in ITC02 benchmark circuits. When stored pre-simulated test data is used, test time reduction of up to 99% is shown for vectors with low input and output activities.

I. INTRODUCTION

Reducing the time of scan testing while keeping the power consumption low is a challenging problem. A recent proposal [6], [7] suggests dynamic control of scan clock frequency in both self test and externally tested scan circuits. It is assumed that the circuit activity is proportional to the activity in the scan register. The slowest scan clock is determined under the assumption of maximum activity, i,e., every flip-flop toggling at every clock. In this paper, we relax that assumption. We propose techniques to reduce test time in externally tested scan circuits through dynamic control of scan clock frequency for which peak activity factor may or may not be pre-computed. The test time reduction achieved is better than those of prior proposals [6], [7].

Section II discusses implementations of the proposed technique. Section III gives a mathematical analysis of the scheme. Section IV explains the experimental results obtained. Section V discusses the conclusions drawn from this work.

II. IMPLEMENTATION

During external test [4], an automatic test equipment (ATE) applies test patterns through scan-in pin(s) of the circuit and receives the response through scan-out pin(s). The expected response of the good circuit under test (CUT) are stored in the ATE and the CUT passes if every response matches the expected response. The clock frequency at which test patterns are scanned in can be varied based on the activity (signal transitions) the patterns produce to control the power consumption of the CUT.

P. Shanmugasundaram was formerly with Auburn University, Dept. of ECE, Auburn, AL 36849, USA.

This research was supported in part by the National Science Foundation Grant CNS-0708962.

A. Using hardware control

In this technique, hardware is used to dynamically control the scan clock frequency. This hardware can be either added on-chip or kept off-chip on external test fixture mounted on the ATE. We define the *activity factor* α for a signal as the average number of transitions per clock that signal makes. Thus, for a clock signal, $\alpha = 2$. For a glitch-free non-clock signal, $\alpha \leq 1.0$, attaining a peak value (α_{peak}) of 1.0.

1) Circuits with single scan chain and $\alpha_{peak} = 1$: We consider a design with a single scan chain and assume that every pattern captured in the scan chain potentially generates the worst-case activity upon scan-out. So, the scan begins with the slowest clock. Scan-in bits are monitored and allow speed up of clock. Implementations of such a scheme for external test [6] and for self-test [7] may be found in recent papers.

2) Circuits with multiple scan chains and $\alpha_{peak} = 1$: Here the scan-in transitions for all chains are monitored and their combined count is used to control the scan clock frequency. Details of such a scheme are available in recent documents [5], [6].

3) Circuits with single scan chain and $\alpha_{peak} < 1$: The two cases mentioned above work well when the captured data causes close to peak activity, $\alpha_{peak} = 1$, in the scan register. However, it is unreasonable to expect that all captures will correspond to this worst case. In general, scan clock frequency can be computed based on a peak activity factor (α_{peak}) lower than 1, which may be determined from simulation or analysis. That leads to a new type of design [5] as proposed in this paper.

Given a value of the peak scan chain activity α_{peak} due to captured values, the slowest scan clock frequency is given by [5]–[7]:

$$f_{test} = \frac{2P_{budget}}{\alpha_{peak}CV^2} \qquad (1)$$

where P_{budget} is the maximum power that a good circuit can consume during test without malfunctioning, C is the total node capacitance for all gates, and V is the supply voltage. For the conventional, non-adaptive, scan test the entire testing will be done with a clock of constant frequency, f_{test}.

Figure 1 shows an adaptive clock implementation for single scan chain and peak activity factors less than 1. The activity monitor comprises of an XNOR gate connected between the input and output of the first flip-flop, and an XNOR gate

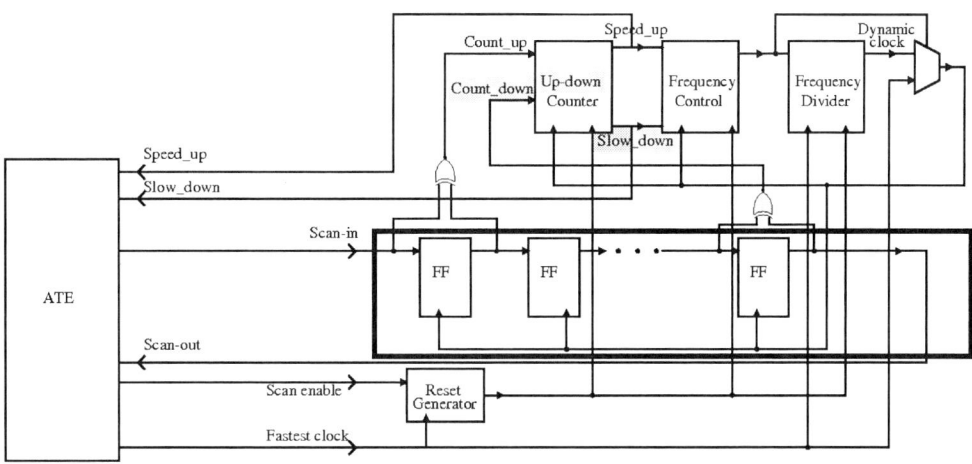

Fig. 1. Implementation in single scan chain circuits, $\alpha_{peak} < 1$.

connected between the input and output of the last flip-flop. The former monitors the number of non-transitions entering the scan chain and the latter monitors the number of non-transitions leaving the scan chain. An up-down counter keeps track of the number of non-transitions in the scan chain. Thus, the input XNOR drives the count_up signal and the output XNOR drives the count_down signal of the up-down counter. The increase in the number of non-transitions in the scan chain during scan-in is the difference between those entering the scan chain and those leaving the scan chain.

The up-down counter is reset to 0 at the start of each scan-in. When a non-transition enters the scan chain, the counter counts up and when a non-transition leaves the scan chain, the counter counts down. When the counter counts up to a certain threshold value, the speed_up signal is set to 1 for one clock cycle, the frequency control block increases the frequency of scan clock and the counter is reset to 0. Similarly, when the counter counts down to 0, the slow_down signal is set to 1 for one clock cycle, the frequency control block lowers the frequency of scan clock and the counter is reset to the threshold value. Thus, whenever the number of non-transitions in the scan chain increases, the frequency is increased and when the number reduces, the frequency is decreased. The rest of the circuitry functions the same as in the scheme where the peak activity factor is 1 [5]–[7].

As shown in Figure 1, the ATE supplies a *fastest clock* whose frequency is determined from the circuit characteristics such as the functional or structural critical path delay and the fastest possible operation of the scan register. The circuit can run at this clock frequency if there were no power constraints. The adaptive scan would use this frequency when the activity drops almost to zero.

A *slowest clock* frequency is determined from power considerations according to Eq. 1. The largest frequency division ratio in the frequency divider of Figure 1 generates the slowest clock from the ATE-supplied fastest clock.

At the start of scan-in of a vector, the slowest clock is employed since the activity factor of the vector captured in the scan chain before the start of scan-in is assumed to be α_{peak}. In this design, scan clock frequency is never increased beyond the fastest clock or decreased below the slowest clock regardless of the signal from the counter.

The implementation of Figure 1 is general and can be customized for $\alpha_{peak} = 1$ by removing the XNOR gate from the flip-flop at the end of the scan chain and tying the count_down signal of the up-down counter to 0. That leads to the previous designs [6], [7].

During testing a two-way data transfer occurs between the ATE and CUT. Test inputs flow from the ATE to CUT and CUT responses flow to ATE. In synchronization with the fastest clock provided by the ATE, the CUT generates its own adaptive clock using speed_up and slow_down signals. These two signals are sent from CUT to ATE so that it can send and receive data at the adaptive clock rate.

4) Circuits with multiple scan chains and $\alpha_{peak} < 1$: When CUT has multiple scan chains, the activity of all chains must be monitored. XNOR gates are added across the input and output of the first flip-flop and across the input and output of the last flip-flop in every scan chain. The outputs of the XNOR gates at the inputs of the scan chains are fed to the count_up inputs of a parallel counter [9] which counts up by the number of 1s at its count_up inputs. Similarly, the outputs of the XNOR gates at the ends of scan chains are fed to the count_down inputs of the same parallel counter that counts down by the number of 1s at its count_down inputs. The rest of the circuitry remains unaltered and still resembles Figure 1. When the count reaches a certain threshold value, the frequency is stepped up and the counter is reset to 0. When the count reaches 0, the frequency is stepped down and the counter is reset to the threshold value. Except for the use of the parallel counter this control scheme is similar to that of Figure 1.

The case of multiple chains with $\alpha_{peak} = 1$ assumption has been discussed before [5], [7]. It requires only one XNOR per scan chain and a parallel up counter.

B. Using pre-simulated and stored test data

An alternative to the hardware control is a dynamically controlled scan clock through the use of pre-simulated and stored test data. The clock frequencies at which test patterns should be applied can be found based on activity factors of the input bits and that of the response bits. This information is then stored in the test program. Since the expected response and the input scan bits are given in the test program, the activity in the scan chain is know accurately in every clock cycle. Thus, the test time can be optimized much more effectively compared to the hardware control technique described earlier.

C. Using a combination of hardware control, and pre-simulated and stored test data techniques

Two techniques to reduce test time in scan circuits, one using hardware control and the other using pre-simulated and stored test data, have been described above. Though both effectively reduce test time, each has its own drawbacks. The hardware control technique has lower reduction in test time while the pre-simulated and stored test data technique increases test data volume, which now includes additional control bits, and requires regeneration of control bits every time a test pattern is modified.

To utilize the benefits of both techniques, we propose a combination. The activity factor of the captured response bits is used to determine the clock frequency with which the next test pattern scan-in should begin. This information is pre-stored with test data and is sent by ATE to the counter and frequency control block in CUT so that they are reset appropriately at the start of each scan-in. The hardware then monitors the activity in the subsequent scan-in cycles. Thus, the activity in the scan chain for every clock cycle is precisely known without tremendous increase in test data volume. If a test pattern is modified, the initial scan frequency for that pattern can be set to the lowest scan clock frequency without compromising heavily on test time reduction.

III. ANALYSIS

Let α_{peak} be the peak activity factor of the test vectors, α_{in} be the activity factor of the scan-in vector, α_{out} be the activity factor of the vector captured in the scan chain prior to scan-in, and v be the number of frequencies. The period T of the fastest scan clock is v times shorter than the slowest clock. Thus, the period of the slowest clock is vT. If the vectors were scanned in with the slowest clock, the total scan-in time per vector would be NvT, where N is the number of flip-flops in the scan chain. Some details are omitted in the following analysis due to space limitation. The reader may refer to a recent report [5].

We estimate the reduction in test time assuming that the majority of time is spent in scanning. The non-adaptive (conventional) test runs with the slowest clock giving a reference scan time per test as NvT. Thus,

$$\text{Test time reduction} = \frac{NvT - \text{Adaptive clock scanin time}}{NvT} \tag{2}$$

TABLE I
SCAN-IN TIME REDUCTION VS. NUMBER OF SCAN CLOCK SPEEDS FOR ACTIVITY FACTOR $\alpha_{in} = 0.5$.

Number of scan clock speeds, v	Test time reduction (%)	
	Simulation	Eq. 4
1	0.00	0.00
2	0.34	0.00
4	12.64	12.50
8	18.78	18.75
16	22.03	21.88
32	23.56	23.44
64	25.17	24.22
128	27.41	24.61

A. Using hardware control

In this technique, scan-in of vectors initially begins with the slowest clock frequency, which is gradually increased in steps based on the activity factor of the scan-in vector.

1) $\alpha_{peak} < 1$: This analysis considers uniform α_{in} and α_{out}. Thus, if $\alpha_{in} > \alpha_{out}$ the number of non-transitions in the scan chain never decreases and hence there will be no change in scan clock frequency. However, if $\alpha_{in} < \alpha_{out}$, the number of non-transitions in the scan chain increases and the scan clock frequency is continuously increased. The scan-in speed of test vectors, starting at the slowest clock rate, is continuously increased. The test time reduction for this model is given by,

$$\text{Test time reduction} = \frac{\alpha_{out} - \alpha_{in}}{2\alpha_{peak}} - \frac{1}{2v} \tag{3}$$

2) $\alpha_{peak} = 1$: In this model, the scan chain is assumed to be filled with most transitions prior to scan-in and hence, the scan-in vector is assumed to be the sole contributor of non-transitions in the scan chain. These non-transitions enter the scan chain at a rate of $(1 - \alpha_{in})$ per cycle. The reduction in test time for this model is given by,

$$\text{Test time reduction} = \frac{1 - \alpha_{in}}{2} - \frac{1}{2v} \tag{4}$$

A C program was written to generate random vectors for a circuit with 1,000 flip-flops. The test time reduction for these vectors was estimated, and compared with the values obtained from the formula. Table I shows the test time reduction versus number of frequencies for scan-in bit activity factor $\alpha_{in} = 0.5$. Table II shows the variation of test time reduction with activity factor when the number of frequencies $v = 8$. Both tables compare the test times estimated for random vectors (column 2), with those obtained from Eq. 4 (column 3).

It can be observed from Tables I and II that for a chosen number of frequencies, vectors with lower transition densities achieve higher reduction in scan-in time. Also, the test time reduction increases as more frequencies are used. The scan-in time reduces rapidly until 8 frequencies after which the reduction becomes gradual.

B. Using pre-simulated and stored test data

If the pre-simulated and stored test data technique is used, the activity in the scan chain is precisely known in every clock

TABLE II
SCAN-IN TIME REDUCTION VS. ACTIVITY FACTOR α_{in} FOR $v = 8$ SCAN-IN CLOCK SPEEDS.

Activity factor, α_{in}	Test time reduction (%)	
	Simulation	Eq. 4
0	43.75	43.75
0.1	38.63	38.75
0.2	34.00	33.75
0.3	28.97	28.75
0.4	23.51	23.75
0.5	18.78	18.75
0.6	14.92	13.75
0.7	9.60	8.75
0.8	4.79	3.75
0.9	0.00	0.00
1	0.00	0.00

cycle. Thus, the scan-in of every test pattern need not be started with the slowest clock frequency. Assuming that the scan-in vector has a uniform activity factor (α_{in}) that is higher than the activity (α_{out}) produced by the captured vector, the number of non-transitions entering the scan chain will be lower than that leaving it and hence the scan clock frequency is continuously decreased. However, if α_{in} lower than α_{out}, then the scan clock frequency will be continuously increased.

1) Case 1: $\alpha_{in} < \alpha_{out}$: The test time reduction is [5],

$$\frac{((\alpha_{peak} - \alpha_{in}) + (\alpha_{peak} - \alpha_{out}))}{2\alpha_{peak}} - \frac{1}{2v} \qquad (5)$$

This formula can be extended to the implementation discussed in the previous subsection as well. If the peak activity factor is assumed to be 1, then the activity factor of the captured vector is also 1, i.e., $\alpha_{out} = \alpha_{peak} = 1$. The test time reduction is,

$$\frac{(1 - \alpha_{in})}{2} - \frac{1}{2v} \qquad (6)$$

which is the same as that obtained earlier.

2) Case 2: $\alpha_{in} > \alpha_{out}$: The test time reduction is,

$$\frac{((\alpha_{peak} - \alpha_{in}) + (\alpha_{peak} - \alpha_{out}))}{2\alpha_{peak}} + \frac{1}{2v} \qquad (7)$$

IV. ATE EXPERIMENTS

A. Verification of BIST

The test times for ISCAS89 benchmark circuits were examined using the Advantest T2000GS ATE. In order to dynamically control the scan clock frequency using hardware control technique, a modification in the ATE software is essential. Without this modification, test patterns cannot be applied to the CUT at a dynamically changing clock rate. To overcome this obstacle, we generated the patterns on-chip using a test-per-scan BIST scheme.

In verilog netlists of the ISCAS89 benchmark circuits flip-flops were added at all primary inputs and primary outputs. All flip-flops were converted to scan types and chained together. Thus, the number of flip-flops in the circuit is the sum of the number of primary inputs, number of primary outputs and number of D-type flip-flops. A 23-bit linear feedback shift

TABLE III
REDUCTION IN TEST TIME FOR ISCAS89 CIRCUITS - SINGLE SCAN CHAIN, $\alpha_{peak} = 1$.

Circuit	Number of scan flip-flops	Number of frequencies, v	Test time reduction (%)	Increase in area (%)
s27	8	2	7.49	14.72
s298	23	4	14.57	16.25
s386	20	4	15.25	15.29
s9234	286	4	14.01	5.82
s13207	852	8	19.00	3.98
s38584	1768	8	18.91	2.13

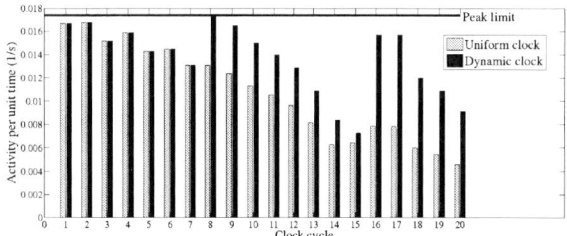

Fig. 2. Activity vs. number of clock cycle for s386 circuit.

register (LFSR), a 23-bit signature analysis register (SAR), and a test-per-scan BIST controller were implemented [1], [8]. A single bit output of the LFSR supplied the scan input and the scan output was fed into the SAR. The counter, frequency control circuitry, and frequency divider circuitry for dynamic frequency control were implemented as shown in the unshaded portion of Figure 1. The number of frequencies for each circuit was chosen according to the size of the circuit or the number of scan flip-flops. A suitable number for random patterns to achieve sufficient fault coverage for each circuit as mentioned in [3] was incorporated in the BIST controller.

The circuits were implemented with and without the dynamic frequency control circuitry on CycloneII FPGAs. The clock signal to the FPGA was supplied from the Advantest T2000GS ATE. The time required for test application with and without dynamic scan clock frequency control was recorded in each case. DesignCompiler, a synthesis tool from Synopsys, was used to analyze the area of the circuits with and without the dynamic frequency control circuitry.

Table III shows the results. The number of frequencies chosen for each circuit is shown in column 3. The percentage reduction in test time with respect to the test time for the core circuit is shown in column 4 and the percentage increase in area with respect to the area of the core circuit is shown in column 5.

Since an LFSR generates pseudo-random patterns, it is safe to assume that the input activity factor of the test patterns is around 0.5. It can be seen that the results conform to the theoretical values shown in Table I.

At any node, the capacitance and the voltage are constant. Therefore, power dissipated at the node is proportional to the product of activity and frequency. Hence, the activity per unit time is a direct measure of power dissipated in the circuit. Therefore, an analysis to find activity per unit time was performed on the s386 benchmark circuit. The Synopsys power analysis tool, PrimeTime PX, was used. The activity per unit time in every cycle was found for the circuit for a scan vector with an activity factor of 1. The peak among these values was set as the limit for activity per unit time. The values of activity per unit time for the circuit in every cycle were found for a vector with an activity factor of 0.25 using uniform clock and dynamic clock, respectively. The results are shown in Figure 2. Notably, the activity per unit time in every cycle is closer to the peak limit when dynamic clock method is used. Also, the peak limit is never exceeded in either case. A reduction of 11.25% was observed with the dynamic clock method.

The results for multiple scan chain implementation would be very similar to that obtained for single scan chain. The test time will not vary much since the activity of the circuit will be very similar in both single and multiple chain implementations. However, there would be a marginal increase in area due to the additional XNOR gates at the first flip-flop of every scan chain and also due to the use of a parallel counter as opposed to the simple counter used for the single scan chain.

These results for reduction in test time conform to the theoretical results given in Tables I and II. Two trends are clearly observed in Table III. As circuit size increases, the area overhead drops and test time reduction improves. These circuits are not very large from today's standard and we can expect better results as predicted by the analysis.

B. External ATE Test

1) Using hardware control:

$\alpha_{peak} = 1$: In order to estimate the test time reduction for larger circuits, an accurate mathematical analysis was applied to ITC02 circuits. Test vectors with different activity factors ($\alpha_{in} \approx 0$, $\alpha_{in} = 0.5$ and $\alpha_{in} \approx 1$) were generated. Test vectors were generated randomly to achieve $\alpha_{in} = 0.5$. In order to generate test vectors with low activity factors ($\alpha_{in} \approx 0$), one transition was randomly placed per test vector. Test vectors with high activity factors ($\alpha_{in} \approx 1$) were generated to resemble clock signals. The analysis was performed assuming that the peak activity factor (α_{peak}) was 1.

The test time reduction with the proposed implementation was computed for the generated test vectors. Table IV shows the results. The number of scan flip-flops in column 2 is the sum of number of inputs, number of outputs and number of flip-flops. The number of frequencies for circuits are shown in column 3. The test time reductions achieved for best, moderate and worst case activity factors are shown in columns 4, 5 and 6, respectively. A simulation tool was not used for these circuits due to the large sizes of the circuits. However, it is important to note that any simulation tool would produce the same results since the input activity at the scan chain was closely monitored during estimation of test time. Evidently, more test time reduction can be achieved in larger circuits. The reduction in test time varies from 0% for patterns causing very high activity to 50% for patterns with almost no activity.

TABLE IV
TEST TIME REDUCTION FOR ITC02 CIRCUITS, $\alpha_{peak} = 1$.

Circuit	Scan flip-flops	Number of clock freq.	Test time reduction (%)		
			$\alpha_{in} \approx 0$	$\alpha_{in} = 0.5$	$\alpha_{in} \approx 1$
u226	1416	8	46.68	18.75	0
d281	3813	16	46.74	21.81	0
h953	5586	32	48.32	23.38	0
f2126	15593	64	49.15	24.18	0
p34392	23005	128	49.53	24.57	0
t512505	76714	512	49.85	24.87	0
a586710	41411	256	49.73	24.77	0

TABLE V
TEST TIME REDUCTION FOR t512505 CIRCUIT, $\alpha_{peak} < 1$.

α_{in}	α_{out}							
	0	0.1	0.2	0.3	0.4	0.5	0.6	0.65
0	0	7.59	15.29	22.98	30.67	38.36	46.06	49.9
0.1	0	0	7.59	15.29	22.98	30.67	38.36	42.21
0.2	0	0	0	7.59	15.29	22.98	30.67	34.52
0.3	0	0	0	0	7.59	15.29	22.98	26.83
0.4	0	0	0	0	0	7.59	15.29	19.13
0.5	0	0	0	0	0	0	7.59	11.44
0.6	0	0	0	0	0	0	0	3.75
0.65	0	0	0	0	0	0	0	0

$\alpha_{peak} < 1$: In order to estimate the reduction in scan-in time achieved with the model proposed for dynamic scan clock frequency control in circuits with peak activity factors lower than 1, the t512505 ITC02 benchmark circuit was chosen. This circuit is large enough to employ 512 different scan clock frequencies because it has 76714 scan flip-flops.

The pattern sets of various large benchmark circuits were studied to analyze trends in peak activity factors. The mean value of peak activity factor (α_{peak}) in these pattern sets was found to be around 0.57 and the standard deviation (σ) was around 0.025. The value of mean + 3σ was found to be around 0.65. This indicates that the probability that the peak activity factor of the test patterns of a circuit would lie below 0.65 is 99.7%. Therefore, the peak activity factor for the t512505 circuit was set at 0.65. The pattern sets generated by TetraMAX ATPG for large benchmark circuits were analyzed and it was found that the peak activity factor in these test vectors never exceeded 0.65. The value of 0.65 for peak activity factor can be used only for large circuits having flip-flop numbers in the range of a several hundred. For smaller circuits with flip-flop numbers on an order of a few tens, the peak activity factor was found to be 1.

Accurate mathematical analysis was used to estimate the reduction in scan-in time achieved for the t512505 circuit when $\alpha_{peak} = 0.65$ and 512 frequency steps are used. The activity factor of the captured vector was assumed to be 0.65 and the activity was monitored at the input and output of the scan chain. Test vectors with different activity factors ranging from 0 to 0.65 were generated and the test time reduction obtained using the proposed implementation was determined for these vectors. The results are listed in Table V. It shows the variation of scan-in time reduction as a function of α_{in} and α_{out}. Table V shows that when the activity factor of the scan-out vector (α_{in}) is greater or equal to the activity factor of the captured vector (α_{out}), there is no reduction in scan-in time. The frequency is increased only when the number of non-transitions in the scan chain increases. However, when $\alpha_{in} > \alpha_{out}$ the number of non-transitions (as counted by the counter) never increases and hence the scan-in is carried out at the starting frequency which is the frequency employed when dynamic scan clock frequency control is not implemented. Thus, the reduction in scan-in time is 0% in such cases.

Table V indicates that scan-in time reduction is greater for smaller values of α_{in} and for higher values of α_{out}. This can be explained from the perspective of number of non-transitions in the scan chain. If α_{in} is low, the number of non-transitions entering the scan chain is high and if α_{out} is high, the number of non-transitions leaving the scan chain is low. Thus, the net number of non-transitions in the scan chain is high giving a higher reduction in scan-in time.

2) Using pre-simulated and stored test data: The reduction in scan-in time achieved in the t512505 circuit when pre-simulated and stored test data technique or a combination of hardware control, and pre-simulated and stored test data techniques is used was estimated through mathematical analysis. The results are listed in Table VI.

In Table VI, scan-in time reduction is more for smaller α_{in} and α_{out}. When α_{in} is low, more non-transitions enter the scan chain and the counter counts faster. Thus, the reduction in scan-in time is higher. When α_{out} is low, the frequency at which scan-in begins is low since the frequency is predetermined based on the activity factor of the captured vector. Thus, there is a larger reduction in scan-in time.

This technique performs well due to the availability of information about the number of transitions or non-transitions present in the scan chain in every cycle. This increases the efficiency of the dynamic scan clock frequency control technique and results in greater reduction of scan-in time.

C. Justification

When an ATPG tool is used to generate test vectors for external tests, the vectors may have very few care bits. The don't care bits can be filled in using heuristics [2] to minimize scan transitions. Then, a dynamic control of scan clock will provide a large reduction in test time. This is illustrated using the ISCAS89 benchmark s38584. The Synopsys ATPG tool TetraMAX was used to generate two sets of vectors, a set of 961 vectors with no don't care bits and another set of 14,196 vectors with don't care bits. The vector set without don't cares was found to have an activity factor around 0.5 and the vector set with don't care bits had a low activity factor around 0.01 when care bits were filled using a minimum transition heuristic [2]. Test time reductions of 18.8% and 43.14% were achieved in the circuits with and without don't care bits, respectively.

In a different scenario, a test set may initially contain few (say, 10%) high activity ($\alpha_{in} = 0.5$) vectors. These resemble fully-specified random vectors and achieve about 70-75% fault coverage. The latter 90% vectors then detect about 20-25% hard-to-detect faults and contain many don't cares, which may

TABLE VI
TEST TIME REDUCTION FOR T512505 WITH PRE-SIMULATED AND STORED TEST DATA.

α_{in}	α_{out}							
	0	0.1	0.2	0.3	0.4	0.5	0.6	0.65
0	99.8	92.2	84.5	76.8	69.1	61.4	53.8	49.9
0.1	92.4	84.7	76.8	69.1	61.4	53.8	46.1	42.2
0.2	84.7	77.0	69.3	61.4	53.8	46.1	38.4	34.5
0.3	77.0	69.3	61.6	53.9	46.1	38.4	30.7	26.8
0.4	69.3	61.6	53.9	46.3	38.6	30.7	23.0	19.1
0.5	61.6	53.9	46.3	38.6	30.9	23.2	15.3	11.4
0.6	53.9	46.3	38.6	30.9	23.2	15.5	7.8	3.8
0.65	50.1	42.4	34.7	27.0	19.3	11.6	3.9	0

be filled in for reduced ($\alpha_{in} \leq 0.05$) activity. The dynamic scan clock control technique will be potentially beneficial.

V. CONCLUSION

The adaptive test in externally tested scan circuits dynamically varies the scan clock frequency. This achieved reduction in test times on all ISCAS89 benchmark circuits. Notably, the area overhead of the built-in circuitry is low. A test time reduction of about 19% was achieved using test-per-scan BIST system tested with Advantest T2000GS tester. An analysis on ITC02 benchmark circuits with on-chip hardware showed a test time reduction of 50% when scan vectors with very low activity ($\alpha \approx 0$) were used. For scan vectors with moderate activity ($\alpha = 0.5$), a test time reduction of 25% was observed. Higher reduction in test time was observed when pre-simulated and stored test data was used.

This work points to a deficiency in the present ATE test programming systems that do not allow interactive testing. A capability allowing the ATE to dynamically change the test startegy based on feedback from CUT will be beneficial.

REFERENCES

[1] V. D. Agrawal, C. R. Kime, and K. K. Saluja, "A Tutorial on Built-In Self-Test, Part 1: Principles," *IEEE Design & Test of Computers*, vol. 10, pp. 73–82, Mar. 1993.

[2] N. Badereddine, P. Girard, S. Pravossoudovitch, C. Landrault, and A. Virazel, "Minimizing Peak Power Consumption during Scan Testing: Test Pattern Modification with X Filling Heuristics," in *Proc. Int. Conf. on Design and Test of Integrated Systems in Nanoscale Technology*, Sept. 2006, pp. 359–364.

[3] F. Brglez, D. Bryan, and K. Kozminski, "Combinational Profiles of Sequential Benchmark Circuits," in *Proc. Int. Symp. Circuits and Systems*, May 1989, pp. 1929–1934.

[4] M. L. Bushnell and V. D. Agrawal, *Essentials of Electronic Testing for Digital, Memory and Mixed-Signal VLSI Circuits.* Springer, 2000.

[5] P. Shanmugasundaram, "Test Time Optimization in Scan Circuits," Master's thesis, Auburn University, Dec. 2010.

[6] P. Shanmugasundaram and V. D. Agrawal, "Dynamic Scan Clock Control for Test Time Reduction Maintaining Peak Power Limit," *Proc. 29th IEEE VLSI Test Symp.*, pp. 248–253, May 2011.

[7] P. Shanmugasundaram and V. D. Agrawal, "Dynamic Scan Clock Control in BIST Circuits," *Proc. Joint IEEE Int. Conf. on Industrial Electronics and 43rd Southeastern Symp. on System Theory*, pp. 237–242, Mar. 2011.

[8] C. Stroud, *A Designer's Guide to Built-In Self-Test.* Springer, 2002.

[9] E. E. Swartzlander, Jr., "A Review of Large Parallel Counter Designs," in *Proc. IEEE Computer Society Annual Symposium on VLSI*, Feb. 2004, pp. 89–98.

978-1-4673-0438-2/12 $31.00 © 2012 IEEE

Author Index

Abraham, Jacob A. 155	Chary, Veerabadra 80
Adhikari, Sumit 11	Chattaraj, Nilanjan 39
Agarwal, Khushboo 352	Chatterjee, Abhijit 143, 245
Agarwal, Tarun Kumar 406	Chatterjee, Shouri 51
Agarwala, Sanjive 286	Chattopadhyay, Santanu 436
Aggarwal, Supriya 57	Chaudhuri, Ritesh Ray 185
Agrawal, Vishwani D. 1, 448	Chaudhuri, Sourindra 238
Ahmed, Syed Ershad 280	Chavali, K.V.R. Suryakiran 257
Ahuja, Jaswinder S. 3	Chowdhury, Ahsan Raja 334
Akademi, Seer 38	Czutro, Alexander 382
Amrutur, Bharadwaj 173	Dam, Samiran 376
Anderson, Timothy 286	Damm, Markus 11
Ansari, Allmin 101	Damodaran, Raguram 286
Arrawatia, Mahima 209	Das, Isha 85
Atienza, David 25	Dasgupta, Pallab 364
B, Kameswara Rao 29	Dasgupta, Parthasarathi 227
B, Muralidhar Reddy 29	Datta, Kamalika 328
B, Ravi Kishore 29	Devarakond, Shyam Kumar 143
Babitch, Dan 179	Dhar, Anindya Sundar 39
Baghini, Maryam Shojaei 209	Diddi, Varish 209
Bala, Phalguni 92	Dubash, Noshir 179
Balachandran, Shankar 197	Dutt, Nikil 22
Balasubramanian, Dheera 286	Eles, Petru 418
Balsara, Poras T. 125, 221	Erraguntla, Vasantha 292
Banerjee, Debashis 143	Evans, Donald 80
Banerjee, Gaurab 173	Fiez, Terri 20
Banerjee, Indrajit 227	Flores, Jose 286
Banerjee, Somnath 298	Furth, Paul M. 31, 131
Barlas, Irtaza 245	Gangaram, Vijay 394
Becker, Bernd 382	Garimella, Annajirao 31, 131
Bhat, K.N. 173	Garitselov, Oleg 310, 316
Bhat, Navakanta 173	Gautham, Vikas 38
Bhattacharya, Anirban 185	Ghosh, Amitava 85
Bhattacharyya, Tarun K. 45	Ghosh, Anandaroop 424
Bhattacharyya, Tarun Kanti 185	Gill, Michael 286
Bhoria, Naveen 286	Goel, Ankur 80
Bhunia, Swarup 18, 304, 340, 424	Gopalakrishnan, Dhileep 286
Bordoloi, Unmesh D. 418	Grimm, Christoph 11
Borkar, Nitin 292	Gupta, Breeta Sen 442
Bui, Duc 286	Gupta, Gauri 62
Chabloz, Jean-Michel 191	Gupta, Nitin 92
Chachad, Abhijeet 286	Gupta, Rajesh 7, 22
Chakrabarti, P.P. 38	Gupta, Shalabh 96, 101
Chakrabarti, Pinaki 233	Gupta, Tushar 298
Chakraborty, Samarjit 9, 418	Gurram, Krishna 286
Chandrachoodan, Nitin 149	Gyselinckx, Bert 5

Author Index

Hajimiri, Hadi.. 430
Halder, Achintya...................................... 85, 274
Hales, Alan.. 286
Hanumolu, Pavan.. 20
Hati, Manas Kumar... 45
Hemani, Ahmed.. 191
Hemmady, Shankar... 27
Hill, Anthony.. 286
Honkote, Vinayak.. 137
Hoskote, Yatin.. 292
Hsiao, Michael S.. 394
Ingelsson, Urban.. 442
Jadcherla, Srikanth.. 38
Jain, Arvind.. 358
Jain, Shailendra.. 292
Jalasutram, Maheedhar....................................... 358
Janraj, C.J... 251
Jha, Niraj K.. 238
Joshi, Ajay.. 28
Joshi, S.. 322
Kagliwal, Ankit... 197
Kalla, Priyank.. 388
Kalyan, T. Venkata.. 251
Kapadia, Nishit... 262
Kapur, Rohit.. 436
Khan, Mohammed Asadullah.................................... 292
Khare, Kavita.. 57
Kochar, Harsha.. 209
Kothamasu, Siva.. 35
Kougianos, Elias...................................... 310, 316
Krishna, Aswin.. 304
Kudithipudi, Dhireesha...................................... 167
Kumar, Girish... 209
Kumar, M. Jagadesh.. 406
Kumar, Saravana.. 51
Kumar, Shasi.. 292
Kundu, Subhadip... 436
Kupferschmid, Stefan.. 382
Lakshminarayanan, G... 257
Larsson, Erik... 442
Lingappan, Loganathan....................................... 394
Lu, Chao.. 215
Lv, Jinpeng... 388
Mahendale, Mahesh.. 38
Mahmood, Nuruddin... 286
Maji, Supriyo... 370

Makkena, Goutham.. 280
Mandal, A.S.. 62
Mandal, Pradip... 370, 376
Manohar, Sujan K...................................... 125, 221
Mazumdar, Bodhisatwa.. 113
Meher, Deepak Kumar... 274
Merkel, Cory E.. 167
Mishra, Prabhat....................................... 161, 430
Mishra, Prateek... 238
Mitra, Sajib Kumar.. 334
Mitra, Subhashish.. 22
Mohanty, Saraju P..................................... 310, 316
Moharil, Shriram.. 286
Moon, Un-Ku.. 20
More, Ankit... 137
Moyade, Pawan Kumar... 101
Mukherjee, Ritwik... 227
Mukhopadhyay, Debdeep....................................... 113
Mukhopadhyay, Siddhartha.................................... 364
Mullinnix, Steve.. 286
Murthy, Pranav.. 352
Muthukrishnan, N. Moorthy................................... 280
Mutyam, Madhu... 251
Nair, Prasun.. 358
Nambath, Nandakumar... 101
Nandy, Tapas... 92
Narasimhan, Seetharam....................................... 304
Narayanan, H.. 400
Narnur, Soujanya.. 286
Natarajan, Jayaram.. 245
Nayudu, M. Venkata Swamy.................................... 280
Nunna, Swaroop.. 418
Okobiah, Oghenekarho.. 310
Olorode, Oluleye.. 286
Ong, Hung... 286
Pal, Debjit... 364
Paliwal, P.. 322
Parameswaran, Sri... 203
Park, Junyoung.. 155
Park, Sang Phill.. 215
Pasricha, Sudeep...................................... 262, 268
Patil, Rajesh A.. 62
Patkar, Sachin B.. 400
Pattanam, Sathyam K.. 38
Paul, Somnath... 340, 424
Peavy, Kyle... 286

Author Index

Pecheux, François...11	Shah, Jimit...107
Peddersen, Jorgen..203	Shanmugasundaram, Priyadharshini.............448
Peng, Zebo..418	Sharma, D..322
Phaneendra, P. Sai.......................................280	Shojaei, M...322
Pierson, Matthew...286	Shridhar, Arvind..25
Prabhu, Sarvesh...394	Shrivastava, Abhijeet....................................352
Pradhan, Neeraj...358	Sinanoglu, Ozgur..346
Pramod, M...173	Singh, Mohit...96
Pudota, Raju Bala Showry...............................38	Somasundar, Vinod K.....................................125
PV, Shantha Kumari..33	Sootkaneung, Warin..74
Raghunandan, K.S...107	Srinivas, M.B...280
Raghunathan, Vijay..................................36, 215	Srinivasaiah, H.C..412
Rahaman, Hafizur....................................227, 328	Srivastava, Mani..22
Rahman, Mujibur...286	Stephani, Richard..80
Rai, Dharmendra...80	Suri, Bharath...418
Rajagopal, Arjun..286	Surkanti, Punith...31
Ramakrishnan, Venkatraman............................352	Surkanti, Punith R..131
Ramamurthy, Praveen C..................................173	Sur-Kolay, Susmita...18
Ranganathan, Nagarajan....................................13	Sussman, Robert..286
Ranka, Sanjay..161	Tamarapalli, Nagesh...16
Rao, M Kalyana Kumar......................................33	Tang, Liang..203
Rao, Preeti..119	Taskin, Baris...137
Rao, V..322	Thapliyal, Himanshu...13
Rathi, Gaurav..328	Thompson, David..286
Reddy, K. Srinivasa..68	Tran, Jonathan...286
Reddy, Sudhakar..382	Ustun, H. Mert...155
Reddy, Venkateswara..80	Vangal, Sriram...292
Roy, Anindya Lal..185	Varghese, Kuruvilla..107
Roy, Kaushik...215	Veeramachaneni, Sreehari...............................280
Sachid, Angada B..322	Venkataraman, Srikanth.....................................16
Sahoo, S.K...68	Venkatasubramanian, Rama.............................286
Sahula, Vineet...62	Venkatasubramanian, Ramakrishnan...........125, 221
Salihundam, Praveen.......................................292	Verma, Prateek..119
Salimath, Arunkumar.......................................274	Vooka, Srinivas..352, 358
Saluja, Kewal K..74	Vudadha, Chetan..280
Samanta, Tuhina...227	Wang, Lei...340
Sarma, Deepa N...257	Wang, Xinmu...304
Sathisha, N...80	Wang, Zhe...161
Sauer, Matthias..382	Warrier, Tripti..251
Save, Yogesh Dilip..400	Wells, Joshua W...245
Sellappan, Boopalan...33	Wu, Daniel..286
Sen, Shreyas...143	Yada, Satish..292
Sen, Subhajit...179	Yoge, Dhiraj Reddy Nallapa..............................149
Sengupta, Indranil.............................113, 328, 436	

IEEE Computer Society Conference Publications Operations Committee

CPOC Chair
Roy Sterritt
University of Ulster

Board Members
Mike Hinchey, *Co-Director, Lero-the Irish Software Engineering Research Centre*
Larry A. Bergman, *Manager, Mission Computing and Autonomy Systems Research Program Office (982), JPL*
Wenping Wang, *Associate Professor, University of Hong Kong*
Silvia Ceballos, *Supervisor, Conference Publishing Services*
Andrea Thibault-Sanchez, *CPS Quotes and Acquisitions Specialist*

IEEE Computer Society Executive Staff
Evan Butterfield, *Director of Products and Services*
Alicia Stickley, *Senior Manager, Publishing Services*
Thomas Baldwin, *Senior Manager, Meetings & Conferences*

IEEE Computer Society Publications
The world-renowned IEEE Computer Society publishes, promotes, and distributes a wide variety of authoritative computer science and engineering texts. These books are available from most retail outlets. Visit the CS Store at *http://www.computer.org/portal/site/store/index.jsp* for a list of products.

IEEE Computer Society *Conference Publishing Services* (CPS)
The IEEE Computer Society produces conference publications for more than 250 acclaimed international conferences each year in a variety of formats, including books, CD-ROMs, USB Drives, and on-line publications. For information about the IEEE Computer Society's *Conference Publishing Services* (CPS), please e-mail: cps@computer.org or telephone +1-714-821-8380. Fax +1-714-761-1784. Additional information about *Conference Publishing Services* (CPS) can be accessed from our web site at: *http://www.computer.org/cps*

Revised: 1 March 2009

CPS Online is our innovative online collaborative conference publishing system designed to speed the delivery of price quotations and provide conferences with real-time access to all of a project's publication materials during production, including the final papers. The ***CPS Online*** workspace gives a conference the opportunity to upload files through any Web browser, check status and scheduling on their project, make changes to the Table of Contents and Front Matter, approve editorial changes and proofs, and communicate with their CPS editor through discussion forums, chat tools, commenting tools and e-mail.

The following is the URL link to the ***CPS Online*** Publishing Inquiry Form:
http://www.ieeeconfpublishing.org/cpir/inquiry/cps_inquiry.html